Mammals of North America - Volume 2

Sergio Ticul Álvarez-Castañeda

# Mammals of North America - Volume 2

## Systematics and Taxonomy

Springer

Sergio Ticul Álvarez-Castañeda
Centro de Investigaciones Biológicas del Noroeste
La Paz, Baja California Sur, México

ISBN 978-3-031-50827-1 ISBN 978-3-031-50825-7 (eBook)
https://doi.org/10.1007/978-3-031-50825-7



This Springer imprint is published by the registered company Springer Nature Switzerland AG
The registered company address is: Gewerbestrasse 11, 6330 Cham, Switzerland

# Preface

The keys are presented in alphabetical rather than evolutionary order. A summary of the orders, families, subfamilies, genera, subgenera, and species information is included. This organization was selected to facilitate species identification, aiming to support non-specialists in the group. Presenting them in phylogenetic order implies a general knowledge of the group of mammals. The keys are arranged hierarchically, first into orders, then by family within each order and by subfamilies within each family, as needed. The arrangement of genera and species is based on the characteristics that support the most precise and didactic separation.

The book considers the species distributed in México, the United States (mainland), and Canada as of April 2021 and includes specific keys for each order. In total, 12 orders, 49 families, 48 subfamilies, 223 genera, and 782 species are presented. Preparing this version involved a compilation work and countless previous versions spanning more than 38 years, which have been reviewed and tested under different conditions.

***How to Use the Keys*** It is convenient to have a measurement method suitable for the species to be identified. A vernier or caliper and a magnifying glass of 10× minimum are recommended to observe various characteristics of the skull of small mammals that are described; in the laboratory, a stereo microscope is recommended.

The keys have been written using a simple language deemed accessible for non-specialists. However, the use of technical words could not be avoided in some cases, and support material is then provided as illustrations based on photographs from specimens of the species of interest. The intention is to use morphological traits of organisms. However, sometimes these are difficult to appreciate by a non-specialist or in morphologically cryptic species whose differentiation is based on genetic analyses; in these cases, the geographic distribution was selected, as it is easier to determine and distinguish.

Each key contains external and cranial characteristics. For most species, only one of them can be used. However, users are encouraged to review both types of characteristics together to confirm the identification of species; therefore, it is convenient to combine the two options as many times as possible. The elaboration of keys by combining external and skull characters was complex since, on occasions, they are not fully compatible to build a logical identification key. Both types of characters are equally important in the keys; it was impossible to have both options in some cases, so the one considered most appropriate was used for differentiating between similar species.

When only the external or cranial characters are mentioned, it is due to the lack of diagnostic information for the other set of characteristics. The identification should proceed using both options found in the dichotomous key and analyzing which of the two yields the closest fit in the description and distribution. These issues will arise mainly in some subfamilies of bats, rodents, and shrews, where the differences between species are very subtle in many cases.

The identification guides are composed of dichotomous keys, with two options for identification, in which one of them is the best suits of the characteristics of the specimen to be identified. After one of the keys, the known distribution of the species is given, based on its general patterns. Subsequently, the valid scientific name of the species is given as of December 2022. For each species, a paragraph provides the description of the species, with its diagnostic characteristics highlighted in bold. The external characteristics are described in the first place, including the coloration pattern and morphological characters.

The coloration was described based on literature descriptions and a review of voucher specimens, highlighting the characteristics that allow for distinguishing the species. The coloration can vary greatly between regions, subspecies, and even seasons of the year, so the most common one is described. Secondly, the cranial and two length measurements define the specimen's size, one that can be recorded in the field when the specimen is alive, and the cranial one during laboratory work. In both cases, it should be considered that these measurements were compiled from intervals reported in the literature and others recorded directly from voucher specimens, which include the representativeness of the species. The aim has always been to consider the variation of specimens or populations. However, for several species, data are available from a single population only, so specimens from different populations may be slightly larger or smaller in size than the interval presented; the coloration may also show slight variations. In addition, as living specimens are somehow subject to selection processes, there may be some lying outside the range that do not fully match the characteristics described. Concerning dimensions, two measurements were used at the same time, expressed in millimeters in all cases, which may vary depending on their frequency of use in each group. The first is a somatic metric: total length or forearm length in the case of bats; the second is cranial: skull length or condylobasal length. The same set of measurements is always maintained between species that are comparable. A few measurements are used because the aim is to use them as a reference for size; there are several species that are morphologically similar, but with significant differences in meristic aspects. It is important to consider that the intervals given for the measurements are relative since there may be variations depending on the geographic region and subspecies.

***Literature Used*** The literature used was varied and addressed aspects of taxonomy, nomenclature, and distribution, which are the axes of this work. In the keys of species, the section of the description compiles information from different sources cited in the file of each species. All the literature information describes the physical characteristics of the species and includes a comparison with other species sharing a potentially sympatric distribution. In many cases, aspects of the descriptions of genera, subgenera, and species are strongly influenced by the syntheses written in the early twentieth century, when the major taxonomic reviews of the species were carried out. In virtually all cases, a physical examination of specimens was conducted to verify the characteristics mentioned in the literature; additional information was recorded and mentioned since information on the description of several species is scarce and limited to the characteristics that allow distinguishing between species of the same genus that are found in the same or a close range.

***Translation*** The work is presented in English, and the proper names of the locations are mentioned in the original language. In many cases, the difference in language is restricted to using a stress mark; for example, México (with a stress mark as it is the name of the country in Spanish) and New Mexico (no stress mark as it is the name of a state in the USA, in English). As a rule, the names of states are not translated as some lack a translation, such as Nuevo León. The distribution of subspecies is not translated either since subspecies are essentially named after the states where they are found.

***Data Used for Maps*** The data used to elaborate the maps were gathered from the Vertnet database of different museums. (I accept that data from misidentifications may be included in the maps) For each species, the unique localities of specimens were obtained and sorted according to 25-year intervals. The localities were reviewed to confirm that they were within the theoretical distribution area of the species; localities outside the theoretical distribution range of the species were reviewed or sought to confirm that they were included in a publication that would allow for validating their identification. If this validation was not possible following either of these two approaches, these outlier localities were eliminated from the maps. This is significant for species from recent periods, especially those that have undergone taxonomic changes or considered cryptic species in some cases. An example is the division of *Peromyscus maniculatus* into different species, so the localities considered were adjusted to the theoretical distribution areas proposed by the authors for each of the new species. The distribution range may change over time and as detailed studies of these groups are conducted. For some species, some literature records have not been included in the databases of the museums consulted. In the map, these cases are marked with a box instead of a point, meaning that this location comes from the literature reviewed rather than the museum database.

In general, for old records, i.e., prior to 1980, the georeferences from the geographic location are calculated, so they are approximate. Thus, there may be minor variations in their actual location, considering the continental scale used for elaborating the maps. With the information gathered on the localities, ecology, types of associated vegetation, and physiography, a theoretical distribution area was drawn. In some cases, the range shown is far from being an accurate reflection of the current range, especially for large mammals. The new localities are shown in a darker color and the old ones in a lighter one, so an approximate variation of the distribution over time can be visualized, showing whether it has shrunk or expanded. Occasionally, tropical species—particularly those from southern México—have very few locations compared to those living in colder environments. In addition, these species are associated with tropical forests undergoing deforestation processes. Thus, a large part of their distribution could be considered as theoretical.

The delimitation of the geographic range of subspecies is an approximation because I did not have the opportunity of review all the material. The boundaries of the distribution between subspecies were based on previous publications, including publications of some species or genera, mammal species, and the publication of Hall (1981). For some species on which I have experience, the ecology or physiography across the distribution area can modify the distribution boundaries. The distribution of subspecies is an approximation and should be taken as such because only a detailed examination of specimens can define the assigned subspecies.

***Photos*** All the photos were taken and edited by the author, some were used and modificated from "Key for the identification of the Mexican Mammals" (Álvarez-Castañeda et al. 2017a). Some of the photos edited to further emphasize the characteristic mentioned in the key. The photos were taken from museum specimens of different institutions which are mentioned in the acknowledgments.

***Protection*** At present, it is hugely important to know the protection and risk categories of the species according to the laws and regulations in the country where you work; this is why it was decided to include this information. When carrying out fieldwork and *in situ* identification, it is essential to determine the protection and risk category of the species before proceeding further, because doing so may have legal implications for the collector.

The information obtained on the species listed as endangered is limited due to the scarce studies conducted on many of these species. The causes affecting a given species may include direct uses, habitat modification, or the introduction of exotic species, among others. This is why the early identification of the species allows us to obtain valuable information for their understanding and the implementation of conservation measures.

When working with species included in protection lists, it is highly recommended to collect the greatest amount of data possible, such as reproductive status, somatic measurements, sex, activity, behavior, ecological and niche aspects, associations with other animal and plant species, surrounding vegetation, degree of habitat alteration, and others. Taken together, this information can provide key points for understanding the species and subsequently support conservation actions and policies. Working with mammal species should always be supported by specific permits/licenses from the respective authority, particularly for species listed under a protection and risk category. On the other hand, the information generated can be used by the personnel in charge of wildlife protection for decision-making about any relevant actions for immediate implementation.

La Paz, Baja California Sur, México Sergio Ticul Álvarez-Castañeda

## References

Álvarez-Castañeda, S. T., T. Álvarez, and N. González-Ruiz. 2017a. Keys for identifying Mexican Mammals. The Johns Hopkins University Press. Baltimore, U.S.A.

Hall, E. R. 1981. The mammals of North America. Second edition. John Wiley and Sons. New York, U.S.A.

# Acknowledgments

A work of this type and scope needed a great work team who contributed throughout the various process stages. In most cases, the persons who supported me did so voluntarily and as a practice to gain knowledge about particular mammal taxa. The group of collaborators includes researchers from different institutions who provided advisory or reviewed specific parts of the text. Staff associated with my work group and students of different academic levels actively participated in the review of the keys and tested them with actual collection specimens. I received feedback from all of them, on parts that needed clarification or that should be illustrated for a better understanding of the work.

The idea to carry out this project was first proposed by Ticul Álvarez some 40 years ago as a "tropicalization" of the North American keys, aiming to improve the analysis and better differentiate the species in the same genus or similar genera with sympatric or close distribution. In this large-scale project, we first worked on the keys of Mexican bats, *Claves para los murciélagos de México* (Álvarez et al. 1994). Subsequently, given the significant changes in taxonomy and nomenclature, we decided to address all the mammals in México with the Keys for identifying Mexican Mammals (Álvarez-Castañeda et al. 2017a). When this work was completed, I realized that one of the key reference books for North America—Mammals of North America (Hall 1981)—was already obsolete for many species. Thus, a more updated version was necessary to include new species and the recent taxonomic and distribution changes, but preserving the essence of being useful for the differentiation of species, especially in a world in which fieldwork and taxonomists have been reduced. That is why I continued with the initiative that emerged at the dawn of my scientific career in Mammalogy.

The second cornerstone in the development of this work was James Patton. I had the opportunity to learn from his extensive knowledge when I did a sabbatical stay at the Museum of Vertebrate Zoology at the University of California, Berkeley. He was then the driving force in one way or another taking on the task of preparing a synthetic work on the mammals of North America. I must confess that I was completely unfamiliar with several of the groups and species of the subcontinent at this point since my field of experience was centered in México. However, this challenge served as a personal learning project to broaden my knowledge of the group in at least this region of the world. The MVZ collection allowed me to start learning and understanding the different groups of mammals, in addition to other collections visited. Having access to Jim's huge knowledge was also pivotal in this respect. I am deeply grateful for his help in the revision of the language in some parts of the draft.

My very special thanks to Leticia Cab Sulub, who supported me with the database of localities for the elaboration of maps of the species distribution. This activity was a very arduous task because, besides the projection of localities and distribution areas, several versions had to be made for different scales, drawing types, and projections, in addition to working on the optimal design of the possible distribution areas. For most species, several versions of maps were drafted until I was convinced that the final map was the best theoretical distribution based on the sources of information consulted.

Alina Gabriela Monroy Gamboa and Patricia Cortés Calva provided continued and valuable support in the repeated reading of the different draft versions, always making highly pertinent observations that helped to clarify the writing and spotting errors in different parts. I am hugely grateful for the time spent by them in the detailed reviews. I should also highlight their great personal support. Thanks also to Consuelo Lorenzo, who reviewed the final version of the whole manuscript, contributing a great number of excellent recommendations and highlighting multiple areas that needed to be improved. To Carmen Izmene Gutierrez Rojas for its revision in detail of many specific task in the different drafts.

Noé González was key in many aspects of this process. Together, we discussed several approaches for the keys, the characteristics of the different taxa, and the reviews of several groups, particularly Chiroptera and Cricetidae. In this second work, he decided not to participate actively; however, I frequently consulted with him to resolve the keys, address specific characteristics, or discuss systematic changes, and occasionally talked about recently published literature.

Maria Elena Sánchez Salazar supported me in different aspects of the writing and editing of the manuscript, so it can be as clear as possible. I deeply thanks for all the support you have provided throughout my professional development, even when when I was a student of bachelor. Her recommendations improved the overall editorial quality of the text and have made the writing easier to read and understand.

This work involved the review of several scientific collections, which was possible thanks to the support of the personnel in charge of them and, in some cases, to the donation of material to the Collection of the Centro de Investigaciones Biológicas del Noroeste (CIB), which was used to observe details of the species and for testing the keys. The collections reviewed and the persons I wish to thank are the following. Museum of Vertebrate Zoology, University of California, Berkeley: James L. Patton, Michael Nachman, Cristopher Conroy, and Eileen Lacey. Smithsonian Institute: Alfred Gardner, Neal Woodman, Michael Carleton, Jesús Maldonado, Darrin Lunde, and Robert Fisher. University of New Mexico: Joseph Cook, Terry Yates, and Jonathan Dunnum. Brigham Young University: Duke Rogers. Field Museum: Bruce Patterson and Adam Ferguson. University of Kansas: Robert Timm and Maria Eifler. Museum of Texas Tech University; Robert Baker, Robert Bradley, and Heath Garner. American Museum of Natural History: Robert Voss. San Diego Natural History Museum: Philip Unitt and Scott Tremor. Angelo State University: Robert Dowler. Humboldt State University: Thorvald "Thor" Holmes. Escuela Nacional de Ciencias Biológicas del Instituto Politécnico Nacional: Ticul Álvarez, Aurelio Ocaña, and Juan Carlos López Vidal. Instituto de Biología del Universidad Nacional Autónoma de México: Oscar Sánchez, Fernando Cervantes, Julieta Vargas, and Yolanda Hortelano. El Colegio de la Frontera Sur, San Cristóbal de las Casas: Consuelo Lorenzo and Jorge Bolaños. Centro Interdisciplinario de Investigación para el Desarrollo Integral Regional, unidad Oaxaca del Instituto Politécnico Nacional: Miguel Briones and Natalia Martín Regalado. Instituto Nacional de Antropología e Historia: Joaquín Arroyo and Aurelio Ocaña. Universidad Autónoma Metropolitana, unidad Iztapalapa: José Ramírez Pulido and Noé González. Thanks also to the personnel in charge of the collection of Centro de Investigaciones Biológicas del Noroeste: Patricia Cortés Calva, Mayra de la Paz Cuevas, and Carmen Izmene Gutiérrez Rojas.

Parts, sections, or taxonomic groups in the manuscript were reviewed by different specialists in the groups that gave me their highly qualified opinion, which was extremely useful and was taken into account. In this respect, I am very grateful to James L. Patton, Robert Bradley, Noé González, Patricia Cortés, Lázaro Guevara, Consuelo Lorenzo, Gabriela Monroy, Sonia Gallina, Francisco X. González, and Elizabeth Arellano. Thanks also to Connor Burgin, Robert Timm, Ronald Pine, and Luis Ruedas for their multiple comments and points of view on the list of species included in this manuscript. The keys were also tested and evaluated by my students to identify the critical points that should be reviewed and addressed: Alina Gabriela Monroy Gamboa, Evelyn Patricia Ríos Mendoza, Ana Lilia Trujano Álvarez, Eduardo Aguilera Miller, Cintya A. Segura Trujillo, Leticia Cab Sulub, Horacio Cabrera Santiago, Issac Camargo Pérez, Gabriela Suárez Gracida, Luis Ernesto Pérez Montes, Lilia Isabel López Pérez, Erika Patricia González Quintero, Francisco Javier Navarro Frías, Hortensia Santillán Ortiz, Nansy Sánchez, Jorge Villalpando, Jacqueline Tun Balam, and Arturo Hernández Gutiérrez. Special thanks to Lia C. Méndez Rodríguez and Lia Montserrat Álvarez Méndez for the reviews to the different versions and their huge unconditional support at all times.

***Taxonomic Criteria*** This work is a compilation of the existing information as of the date of publication. The nomenclature is based on the available literature. In cases when two different points of view are expressed, one is adopted but the other one is also mentioned; in these cases, the rationale for selecting a given author is explained. The nomenclatural review above the family level was not addressed in this work. The following two current classifications are taken for reference.

## References

Álvarez, T., S. T. Álvarez-Castañeda, and J. C. López-Vidal. 1994. Claves para los murciélagos de México. Publicación Especial, Centro de Investigaciones Biológicas de Baja California Sur y Escuela Nacional de Ciencias Biológicas, Instituto Politécnico Nacional. La Paz, México.

Álvarez-Castañeda, S. T., T. Álvarez, and N. González-Ruiz. 2017a. Keys for identifying Mexican Mammals. The Johns Hopkins University Press. Baltimore, U.S.A.

Hall, E. R. 1981. The mammals of North America. Second edition. John Wiley and Sons. New York, U.S.A.

# Contents

# Order Rodentia

## **Order Rodentia** Bowdich, 1821

Rodentia includes all rodents, such as agoutis, squirrels, jutias, beavers, hamsters, dormice, marmots, prairie dogs, rats, mice, tuco-tuco, and gophers. It is the most diverse order of mammals, containing approximately 50% of all species. This order has, as a unique characteristic, **a pair of open-rooted (ever-growing) upper and lower incisors, each elongated, with thick layers of enamel on the front and a softer dentine on the back**. The differential wear from gnawing creates perpetually sharp chisel edges; the absence of other incisors and canine teeth results in a diastema on the upper and lower jaws; the **articulation of the lower jaw branches ensures that the incisors do not meet when food is chewed and that the upper and lower molariforms do not come into contact while the animal gnaws;** the molariform teeth may be either hypsodont or brachydont; the **masseter muscle is highly specialized, and its arrangement serves as a basis for differentiating the suborders;** feet clawed; fibula not articulating with the calcaneus. There are three suborders in North America: Ctenohystrica, Eusciurida, and Supramyomorpha. They are currently of cosmopolitan distribution. The keys were elaborated based on the review of specimens and the following sources: Lawlor (1979), Hall (1981), and Álvarez-Castañeda et al. (2017a).

1. No zygomatic plate; infraorbital foramen greatly enlarged, much larger than the foramen magnum; origin of the angular process deflected laterally (Fig. 1) ........................................ suborder Ctenohystrica (p. 2)

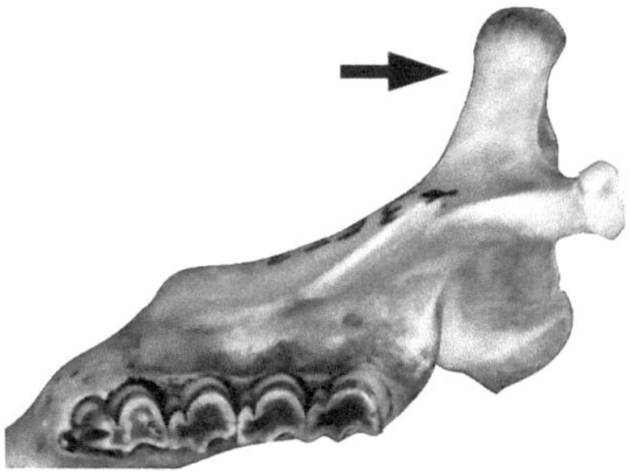

**Fig. 1** Origin of the angular process deflected laterally

1a. Well-defined zygomatic plate (Fig. 2); infraorbital foramen smaller than the foramen magnum and either piercing the zygomatic plate or positioned on the flanks of the rostrum anterior to the zygomatic plate; origin of the angular process directly ventral to the sheath of the lower incisors (Fig. 3) ........................................ 2



S. T. Álvarez-Castañeda, *Mammals of North America - Volume 2*, https://doi.org/10.1007/978-3-031-50825-7_1

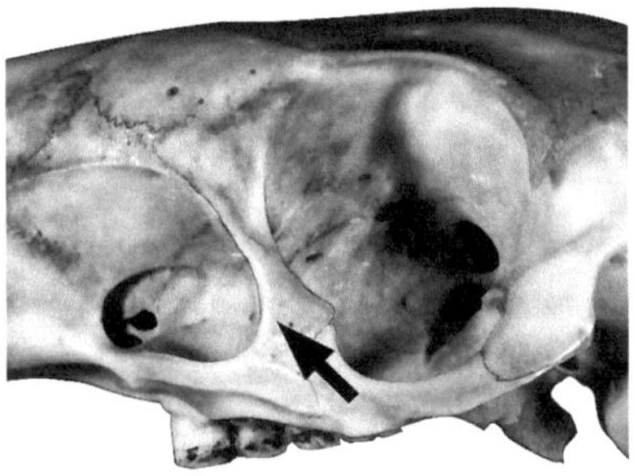

**Fig. 2** Zygomatic plate well defined

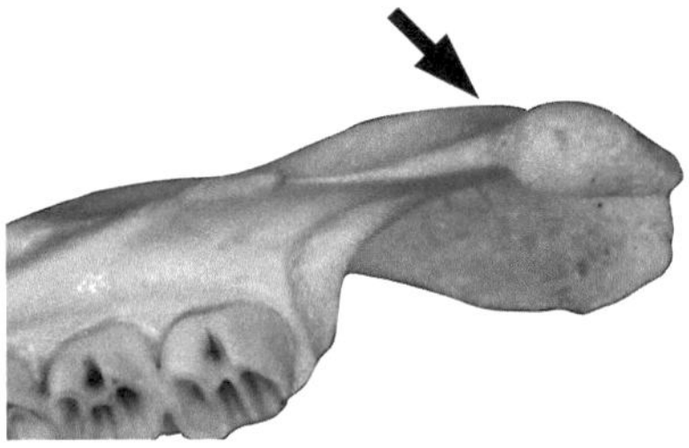

**Fig. 3** Origin of the angular process directly ventral to the sheath of the lower incisors

2. Infraorbital foramen positioned on the flanks of the rostrum anterior to the zygomatic plate; four maxillary teeth ........ suborder Supramyomorpha (p. 134)

2a. Infraorbital foramen piercing the zygomatic plate; four or five maxillary teeth ........ suborder Eusciurida (p. 11)

## **Suborder Ctenohystrica** Huchon, Catzeflis, and Douzery, 2000

Auditory bullae with weakly developed septae; vagina with a closure membrane; penis with a sacculus urethralis; presence of a scapuloclavicularis muscle; malleus and incus fused (Luckett and Hartenberger 1985; Huchon et al. 2000). The Suborder Ctenohystrica is new and includes all extant Ctenodactylidae and Hystricognathi. The keys were elaborated based on the review of specimens and the following sources: Lawlor (1979), Hall (1981), and Álvarez-Castañeda et al. (2017a).

1. Body covered by hair modified like spines or very rough; edge of the angular mandible process strongly curved inward ........ 2

1a. Body covered with silky or rough pelage but not modified like spines; edge of the angular mandible process not strongly curved inward ........ 3

2. Body covered with very rough pelage not modified like spines; hind feet webbed; paroccipital process distinctly long and curved under the auditory bullae ........ family Echimyidae (p. 7)

2a. Body covered by hair modified like spines; hind feet not webbed; paroccipital process distinctly short ........ family Erethizontidae (p. 8)

3. Dorsal pelage brown with white spots; hindfoot length greater than 160.0 mm; zygomatic arches large and shaped as a corrugated plate (Fig. 4); skull length greater than 120.0 mm (from northern Veracruz and southern Oaxaca south through South America, including the Yucatán peninsula, México) ........ family Agoutidae; *Cuniculus paca* (p. 3)

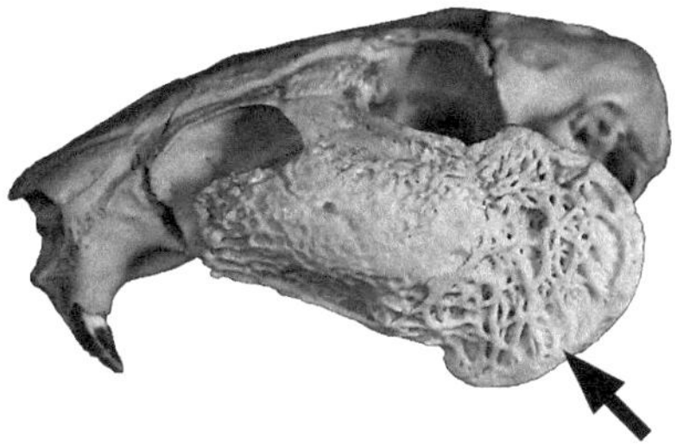

**Fig. 4** Zygomatic arches are large and shaped as corrugated plate

3a. Dorsal pelage speckled without spots; hindfoot length less than 160.0 mm; zygomatic arches thin and not forming a wide and corrugated plate (Fig. 5); skull length less than 120.0 mm .................................. family Dascyproctidae (p. 4)

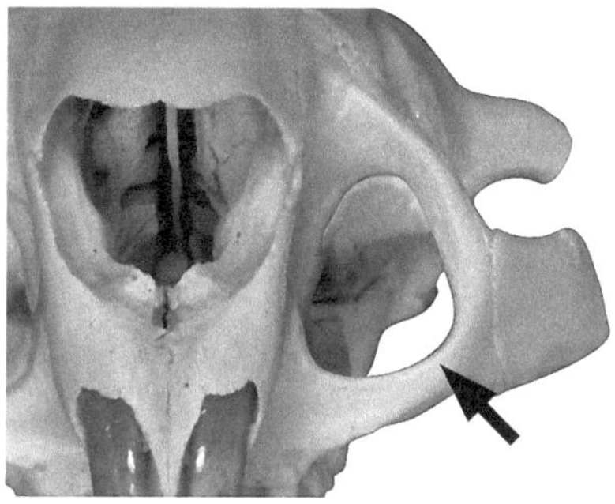

**Fig. 5** Zygomatic arches are thin and do not form a wide and corrugated plate

## **Family Agoutidae** Gray, 1821

Family that includes pacas, whose specimens are characterized by **a robust external form; head large; ears and eyes large; tail vestigial; jugal and maxillary expanded into a conspicuous cheek-plate, its surface becoming excessively rugose in adults; cheek-teeth strongly hypsodont, but enamel structures not completely multilaminar;** cheek-teeth hypsodont, semirooted, having deep reentrant enamel folds that form isolated narrow lakes as they wear out. Distribution restricted to tropical America. No subfamilies are recognized.

### Genus ***Cuniculus*** Brisson, 1762

**Pacas have** the **pelage stiff and appressed light brown to orange brown; external form robust; head large; ears and eyes large; tail vestigial; four toes on each forefoot; five toes on each hindfoot; jugal and maxillary arches expanded into a conspicuous cheek-plate, its surface becoming excessively rugose in adults; cheek-teeth strongly hypsodont, but enamel structures not completely multilaminar.**

#### ***Cuniculus paca*** (Linnaeus, 1766)

Lowland Paca, tepezcuintle

*C. p. nelsoni* (Goldman, 1913). From southern Tamaulipas and southern Chiapas to Central America including the Yucatán peninsula.

**Conservation:** *Cuniculus paca* is listed as Least Concern by the IUCN (Emmons 2016a).

**Characteristics:** Lowland Paca is large sized; **total length 600.0–00.0 mm and skull length 145.0–167.0 mm. Dorsal pelage reddish brown with a darker back, four lines of white creamy spots along the body, and other lines only found on the hips and rump; top line consisting of small isolated spots; second and third lines very distinctive, almost as continuous lines on the flanks but discontinuous on the hips and shoulders; fourth line discontinuous along its length;** ears big and almost naked; throat and muzzle underparts white; region around the nose pinkish; fingers strong, thick, closed, with nails almost like hooves; tail vestigial; teeth hypsodont with deep reentrant enamel forming a number of islands; **buccal and maxillary arches expanded as a plaque on the cheeks, with the surface and frontal bones rugged in adults**.

**Comments:** *Cuniculus paca* was recognized in the family Cuniculidae (McKenna and Bell 1997; Rowe and Honeycutt 2002); it was previously in the subfamily Agoutinae of the family Dasyproctidae (Hall 1981). *C. paca* ranges from lowlands of northern Veracruz and southern Oaxaca south through South America, including the Yucatán Peninsula (Map 1). The most similar species to the genus *Cuniculus* is *Dasyprocta mexicana* and *D. punctata*, from which it differs by having the dorsal pelage with spots; larger average size, hindfoot length greater than 160.0 mm and skull length greater than 120.0 mm; jugal and maxillary arches expanded into conspicuous cheek-plate, surface excessively rugose in adults.

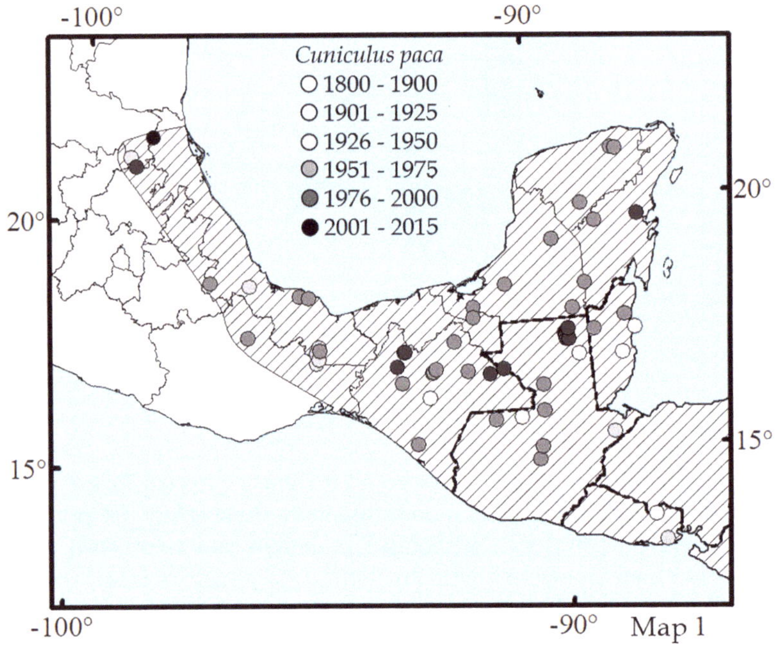


**Map 1** Distribution map of *Cuniculus paca*

**Additional Literature:** Ellerman (1940), Husson (1978), Nelson and Shump (1978), Hall (1981), Pérez (1992), Lacher (2016a).

## **Family Dascyproctidae** Miller and Gidley, 1918

The family Dasyproctidae includes agouti rats characterized by **hind feet long, narrow, tridactyl, bearing hooflike claws; hind feet with three digits and forefeet with four; tail obsolete;** cheek-teeth hypsodont, semirooted, having deep reentrant enamel folds that form isolated narrow lakes as they wear out. Distribution restricted to tropical America. No subfamilies are recognized.

## Genus ***Dasyprocta*** Illiger, 1811

Dorsal pelage with grizzly hair in different colors; **body form slender, cursorial;** ears rounded and almost naked; **hind feet long, narrow, tridactyl, bearing hooflike claws; hind feet with three digits and forefeet with four; tail obsolete; hair on the posterior part of the body long and thick; skull slender, somewhat rectangular dorsally; nasal bones shorter than the frontal bones; frontal bones broad, relatively flat; sagittal crest small and only present in adults; auditory bullae relatively large; paroccipital processes prominent; palatine foramina short, placed well anteriorly; angular process of the mandible reflected strongly outward.** The keys were elaborated based on the review of specimens and the following sources: Hall (1981) and Álvarez-Castañeda et al. (2017a).

1. Ventral hair with color contrasting strongly between the base and the tip; dorsum and rump dark; molariform teeth with enamel islets due to wear (restricted to Veracruz, Tabasco, and Oaxaca, México)................ *Dasyprocta mexicana* (p. 5)

1a. Ventral hair homogeneous in color; head and nape blackish; molariform teeth without enamel islets due to wear (from eastern Tabasco, central Chiapas, and the Yucatán Peninsula, México, south through Panama)........................................ ..........................................................................................................................................*Dasyprocta punctata* (p. 6)

### ***Dasyprocta mexicana*** Saussure, 1860

Mexican agouti, guaqueque negro

Monotypic.

**Conservation:** *Dasyprocta mexicana* is listed as Critically Endangered (A2c: population reduction in the last 10 years 80% or less and may not be reversible by decline in the area of occupancy) by the IUCN (Vázquez et al. 2008).

**Characteristics:** Mexican Agouti is large sized; total length 490.0–620.0 mm and skull length 102.0–152.0 mm. **Dorsal pelage with gray grizzly hair with other colors, middle back and hips blackish; ventral region tawny whitish, with hair paler at the tip**; ears long, rounded, and almost naked; limbs blackish and long; hind feet with three digits and forefeet with four; tail small or absent; digits strong, thick, and closed; claws almost like hooves; teeth hypsodont with deep reentrant enamel forming a number of islands, dorsally square; nasal bones shorter than the frontal bones.

**Comments:** *Dasyprocta mexicana* is restricted to southern Veracruz, central Tabasco, and lowlands of northeastern Oaxaca, and can be found in sympatry, or nearly so, with *D. punctata* in eastern Tabasco (Map 2). It can be distinguished by its contrasting bicolored ventral hair, back blackish and enamel islets in the molar teeth due to wear. From *Cuniculus paca*, by its smaller average size;, hindfoot length less than 160.0 mm; skull length less than 120.0 mm; dorsal pelage speckled without spots; zygomatic arches thin and not forming a wide and corrugated plate.

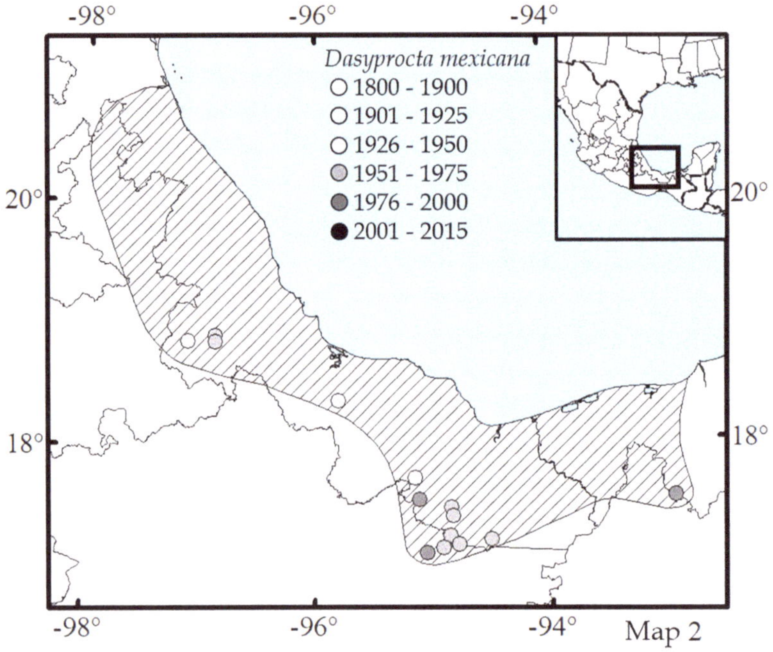


**Map 2** Distribution map of *Dasyprocta mexicana*

**Additional Literature:** Woods and Kilpatrick (2005), Lira-Torres and Briones-Salas (2011, 2012), Pérez-Irineo and Santos-Moreno (2012), Samudio (2016).

## *Dasyprocta punctata* Gray, 1842

Central American agouti, guaqueque centroamericano

1. *D. p. chiapensis* Goldman, 1913. Coast of Chiapas.
2. *D. p. yucatanica* Goldman, 1913. Throughout the Yucatán Peninsula and eastern Campeche.

**Conservation:** *Dasyprocta punctata* is listed as Least Concern by the IUCN (Emmons 2016b).

**Characteristics:** Central American Agouti is large sized; total length 572.0–586.0 mm and skull length 105.6–118.8 mm. **Dorsal pelage orange brown with different yellowish to orange shades interspersed with blackish and longer bristles on the hips, head, and nape; fur of underparts with hair paler at the tip**; ears rounded and almost naked; hind limbs blackish and long, with three digits, forefeet with four; tail small or absent; claws strong, thick and closed almost like hooves; teeth hypsodont with deep reentrant enamel forming a number of islands, square dorsally, and nasal bones shorter than the frontal bones.

**Comments:** *Dasyprocta punctata* ranges from lowlands of eastern Tabasco and central Chiapas south through Panama, including the Yucatán Peninsula (Map 3). It can be found in sympatry, or nearly so, with *D. mexicana* in eastern Tabasco and northwestern Chiapas. It can be distinguished by its contrasting bicolored ventral hair, back blackish and enamel islets in the molar teeth due to wear. From *Cuniculus paca*, by its smaller average size, hindfoot length less than 160.0 mm and skull length less than 120.0 mm; dorsal pelage speckled without spots; zygomatic arches thin and not forming a wide and corrugated plate.

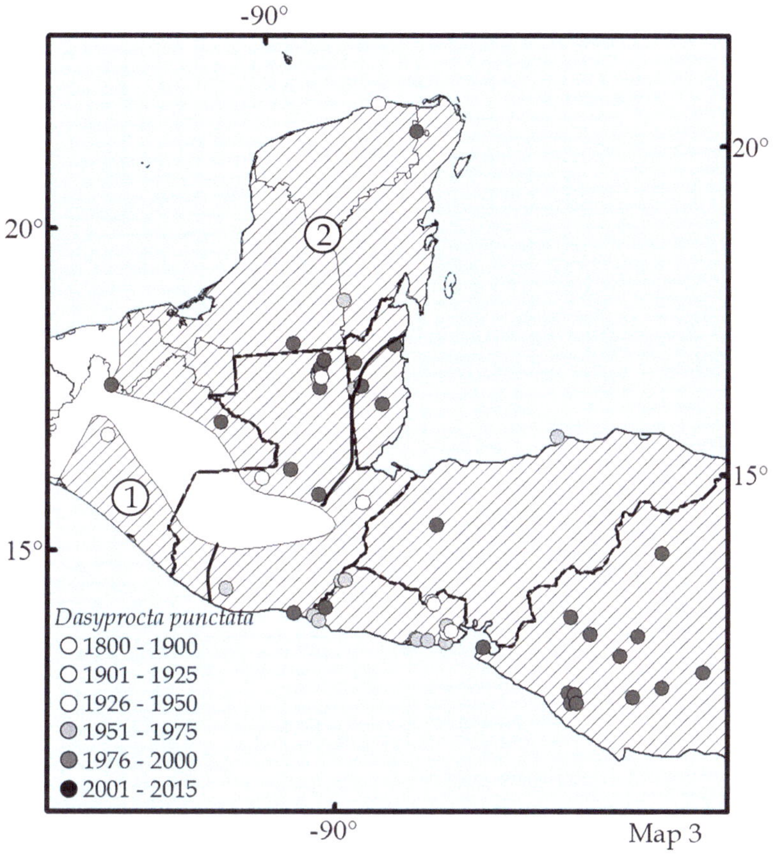


**Map 3** Distribution map of *Dasyprocta punctata*

**Additional Literature:** Handley (1976), Hall (1981); Emmons and Feer (1990), Patton and Emmons (2015), Samudio (2016).

## **Family Echimyidae** Gray, 1825

Spiny rats have stiff, pointed hairs, or spines in a large number of species, which presumably provide protection from predators.

### Myocastorini

### ***Myocastor*** Kerr, 1792

Individuals are large sized; robust; compact; adapted for aquatic life; feet short; pentadactyl and webbed posteriorly; skull heavy, well ridged; the paraoccipital process greatly elongated and curved slightly anteriorly at the distal end; frontal bones broad and flattened; small postorbital process present; palate constricted anteriorly; no accessory canal in the infraorbital foramen.

### ***Myocastor coypus*** (Molina, 1782)

Nutria, coipú

**Conservation:** *Myocastor coypus* is listed as an introduced species and Least Concern by the IUCN (Ojeda et al. 2016).

**Characteristics:** Nutria is large sized; total length 800.5–1050.0 mm. Dorsal pelage rich glossy brown or chestnut, with a soft dense gray under fur; fur of underparts paler; tail nude and scaly. *Myocastor coypus* resembles a very large rat with the rostrum and size of a castor; hind feet webbed; incisors large and bright orange-yellow; tail long and rounded; ears proportionally shorter.

**Comments:** *Myocastor coypus* was introduced in the United States and now can be found in British Columbia, California, Florida, Louisiana, Oklahoma, Ohio, Ontario, Oregon, Nebraska, Kansas, Minnesota, Montana, New Mexico, Texas, and Washington (not mapped). *M. coypus* can be distinguished from *Ondatra zibethicus* by its larger size; tail rounded, not laterally flattened. From *Castor canadensis*, by not having the tail flattened as paddle-like.

**Additional Literature:** Glass and Halloran (1961), Davis (1966b), Hall (1981), Fabre et al. (2016), Ojeda et al. (2016), D'Elia et al. (2019).

## **Family Erethizontidae** Bonaparte, 1845

The family Erethizontidae includes New World porcupines or coendu characterized by heavy-looking body (500.0–1150.0 mm); **hair stiff as spines, each with hooks on the distal part;** tail short in *Echinoprocta* and *Erethizon* or long and prehensile in *Chaetomys* and *Coendou*; limbs relatively short; **pollex reduced and hallux with small, long, and curved claws;** distributed in America. No subfamilies are considered, following Emmons (2005). The keys were elaborated based on the review of specimens and the following sources: Hall (1981) and Álvarez-Castañeda et al. (2017a).

1. No hallux; four digits on the hind foot; tail long and prehensile; back of the skull triangular with the interorbital region higher than the rostrum and the occipital bone (from southeastern San Luis Potosí, Veracruz, and Oaxaca south through Central America, including the Yucatán peninsula, México) .................. subgenus *Coendou*; *Coendou mexicanus* (p. 8)

1a. Hallux well developed; five digits on the hind foot; tail short and not prehensile: back of the skull straight (throughout Canada and the United States, except for the Great Plains and southeastern United States, and northern México).......... ...........................................................................................................................................*Erethizon dorsatum* (p. 10)

## Genus ***Coendou*** Lacépède, 1799

**Body clothed above with short and thick spines or quills; tail prehensile, modified for contact on the distal dorsal surface, with quills dorsally and proximally ventrally; forefeet and hind feet with four digits; spines not mixed with hair, without woody hair; skull broad, robust; frontal region markedly broadened and strongly arched in dorsal view; nasals short; palate relatively broad; tympanic well inflated; molariform teeth with one persistent internal and one persistent external fold, two additional outer folds tend to become isolated with wear.**

### Subgenus ***Coendou*** Lacépède, 1799

**Prehensile-tailed porcupines with the nasofrontal sinuses inflated; spines not mixed with hairy covering.**

## *Coendou mexicanus* (Kerr, 1792)

Mexican hairy dwarf porcupine, puercoespín tropical

1. *C. m. mexicanus* (Kerr, 1792). From southeastern San Luis Potosí, northern Veracruz, and Oaxaca south through Central America.
2. *C. m. yucataniae* (Thomas, 1902). Restricted to the Yucatán Peninsula.

**Conservation:** *Coendou mexicanus* is listed as Threatened in the Norma Oficial Mexicana (DOF 2019) and as Least Concern by the IUCN (Vázquez et al. 2016e).

**Characteristics:** Mexican hairy dwarf porcupine is large sized; **total length 625.0–900.0 mm and skull length 85.7–102.9** mm. Dorsal pelage blackish to dark brown with **spines yellowish and black at the tip**; head with a larger number of yellowish spines; fur of underparts dark; **tail long, thin, and prehensile**; nose large, bulbous, naked, and pinkish; four digits in each forefeet; body covered with rigid hair as short spines; toothrow with flat crowns; lower edge of the jaw angular process hook shaped; zygomatic arches with the anterior border in front of the maxillary first molars; skull very wide and robust, **frontal region almost 30% wider than the rostrum width**; skull strongly arched in lateral view.

**Comments:** *Coendou mexicanus* was considered under the genus *Sphiggurus* (Husson 1978; Bonvicino et al. 2000). Molecular analyses showed that *Coendou* and *Sphiggurus* are different genera, and the species of North America is within the genus *Coendou* (Voss et al. 2013); for this reason, *Coendou mexicanus* is used here. *C. mexicanus* ranges in the lowlands from southeastern San Luis Potosí, northern Veracruz, and Oaxaca south through Central America, including the Yucatán Peninsula (Map 4). *C. mexicanus* and *Erethizon dorsatum* have allopatric distribution.

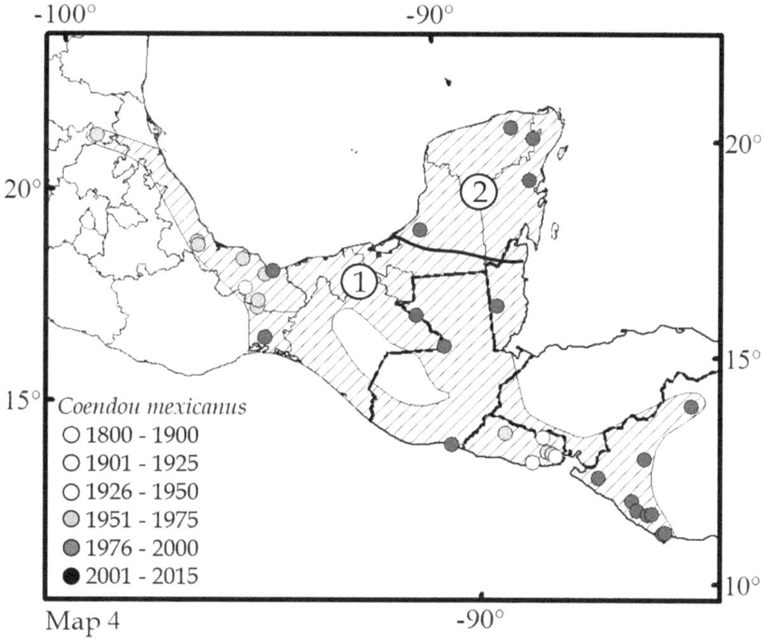


**Map 4** Distribution map of *Coendou mexicanus*

**Additional Literature:** Ellerman (1940), Wood (1993), Nowak (1991), Voss (2011), Barthelmes (2016).

## Genus *Erethizon* Cuvier, 1823

**Individuals are large sized; body stout; tail short, thick, non-prehensile; four digits on the forefeet and five on the hind feet; skull compact, broad, heavily constructed; nasals broad; frontal bones broad, heavily ridged, with the ridges converging posteriorly and forming a sagittal crest; zygoma uncomplicated, jugal deeper anteriorly; mandible with low coronoid and angular processes; upper cheek-teeth with two persistent folds, one internal and one external fold, two additional outer folds tending to become isolated with wear.**

### *Erethizon dorsatum* (Linnaeus, 1758)

North American porcupine, puercoespín norteamericano

1. *E. d. bruneri* Swenk, 1916. From western North Dakota and eastern Montana south through northwestern Texas and central Oklahoma.
2. *E. d. couesi* Mearns, 1897. From western California east through northern Texas, including southeastern Utah and southwestern Colorado, and south through Sonora to Coahuila.
3. *E. d. dorsatum* (Linnaeus, 1758). From the Northwestern Territories, Alberta, eastern North and South Dakota, and from Indiana east through the Atlantic coast.
4. *E. d. epixanthum* Brandt, 1835. From southern Alberta and Saskatchewan south through California and New Mexico, west through western California and Oregon, and east through Colorado and Montana.
5. *E. d. myops* Merriam, 1900. From Alaska east through northern British Columbia and Alberta, including the Northern Territories.
6. *E. d. nigrescens* Allen, 1903. British Columbia and Washington.

**Conservation:** *Erethizon dorsatum* is listed as Endangered in the Norma Oficial Mexicana (DOF 2019) and as Least Concern by the IUCN (Emmons 2016c).

**Characteristics:** North American porcupine is large sized; **total length 648.0–1030.0 mm and skull length 97.0–107.8 mm**. Dorsal pelage blackish to dark brown with hair between the spines; **tail short, thin, and not prehensile**; four digits in each forefoot and five in each hind foot; body heavy, covered with rigid hair as long **dark spines**; toothrow with flat crowns; lower edge of the jaw angular process hook shaped; zygomatic arches with the anterior border in front of the maxillary first molars; **frontal bones wide with a prominent crest converging posteriorly** and forming the sagittal crest.

**Comments:** *Erethizon dorsatum* ranges from Alaska east through the Labrador Peninsula, Canada, and from northern Canada south through Sonora, Chihuahua, and Coahuila, México. In the United States, it is absent from Minnesota south through Texas and east through Pennsylvania to Florida, with isolated populations in the Appalachian Mountains (Map 5). *E. dorsatum* and *Coendou mexicanus* have allopatric distribution.

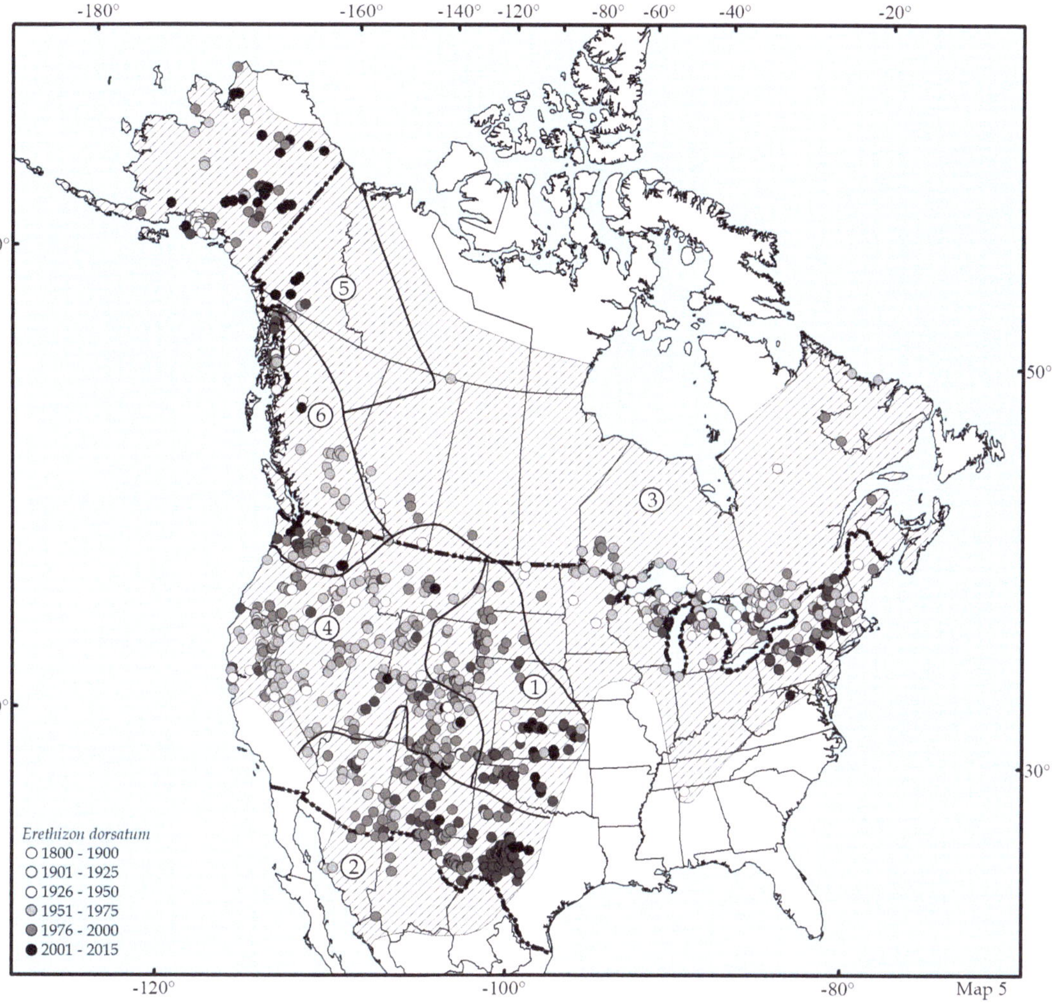


**Map 5** Distribution map of *Erethizon dorsatum*

**Additional Literature:** Woods (1973), Hall (1981), Gatica-Colima et al. (2014), Barthelmes (2016).

## **Suborder Eusciurida** Flynn et al. 2019

Jaw structure primitive; no hypocones on the upper molars and primitively low crowned cheek-teeth; incisor enamel derived, as highly organized and advanced as multiserial enamel; with "uniserial enamel," which is a decussating prism pattern organized into thin and thick bands (Flynn et al. 2019). Two families occur in North America: Sciuridae and Aplodontiidae. The keys were elaborated based on the review of specimens and the following sources: Howell (1918), Lawlor (1979), and Hall (1981).

1. Skull broad, flat, and triangular in dorsal view; auditory bullae flask shaped (Fig. 6); mandible angular process strongly inflected (ranges in disjunct groups: from southwestern British Columbia, Canada, south through northwestern California; Mendocino County and northern San Francisco Bay, California; eastern California and western Nevada, United States)........................................................................................................family Aplodontiidae; *Aplodontia rufa* (p. 12)

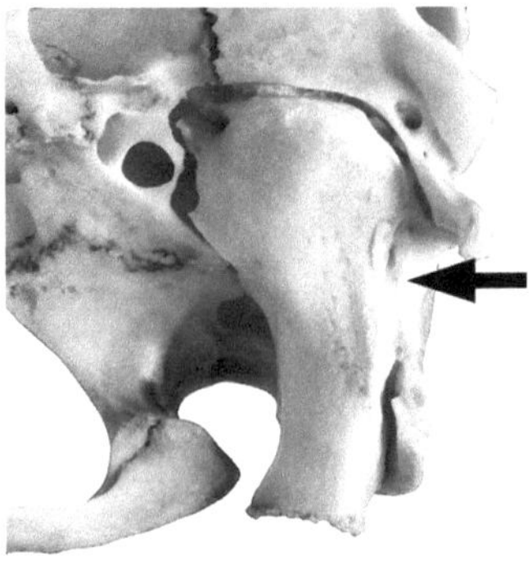

**Fig. 6** The auditory bullae flask shaped

1a. Skull not broad and flat; auditory bullae not flask shaped; mandible angular process not strongly inflected ........................................................................ family Sciuridae (p. 13)

## **Family Aplodontiidae** Brandt, 1855

Zygomatic plate narrow and horizontal; masseter muscle wholly beneath the infraorbital foramen and no part of the muscle reaches the rostrum; tibia and fibula always separate; **angular process of the mandible markedly inflected; skull widened and flattened posteriorly**.

### Genus *Aplodontia* Richardson, 1829

**Skull unusually broad and flat, especially posteriorly**; five upper and four lower cheek-teeth; first upper cheek-teeth as a small simple peg; **auditory bullae flask shaped;** palate extending posteriorly beyond the tooth-rows; the **angular process of the mandible greatly inflected;** coronoid process high.

#### *Aplodontia rufa* (Rafinesque, 1817)

Mountain Beaver, castor de montaña

1. *A. r. californica* (Peters, 1864). Restricted to the western California highlands, from Sierra Nevada northward.
2. *A. r. humholdtiana* Taylor, 1916. Restricted to the northwestern coast of California.
3. *A. r. nigra* Taylor, 1914. Restricted to Point Arena, California.
4. *A. r. pacifica* Merriam, 1899. Western coast of Oregon and extreme northwestern California.
5. *A. r. phaea* Merriam, 1899. Restricted to Lagunitas, northern San Francisco Bay, California.
6. *A. r. rainieri* Merriam, 1899. Central-western Washington and southern British Columbia.
7. *A. r. rufa* (Rafinesque, 1817). Coastal areas of Washington and southwestern British Columbia.

**Conservation:** *Aplodontia rufa* is listed as Least Concern by the IUCN (Fellers et al. 2016).

**Characteristics:** Mountain beaver is medium sized; total length 310.0–470.0 mm and skull length 51.0–65.5 mm. Dorsal pelage relatively uniformly colored gray to brown, with geographic variation to reddish or blackish shades; body compact; limbs short and stout; eyes small; **tail well furred and exceedingly short; pale patch under each small ear.** See the description of the genus for characteristics.

**Comments:** *Aplodontia rufa* ranges in four disjunct groups: from southwestern British Columbia south through northwestern California; restricted to Point Arena, Mendocino County, California; restricted to Lagunitas, northern San Francisco Bay, California; and from Mountain Shasta south through Sierra Nevada Range in California and western Nevada (Map 6). *A. rufa* is the only living species of the family.

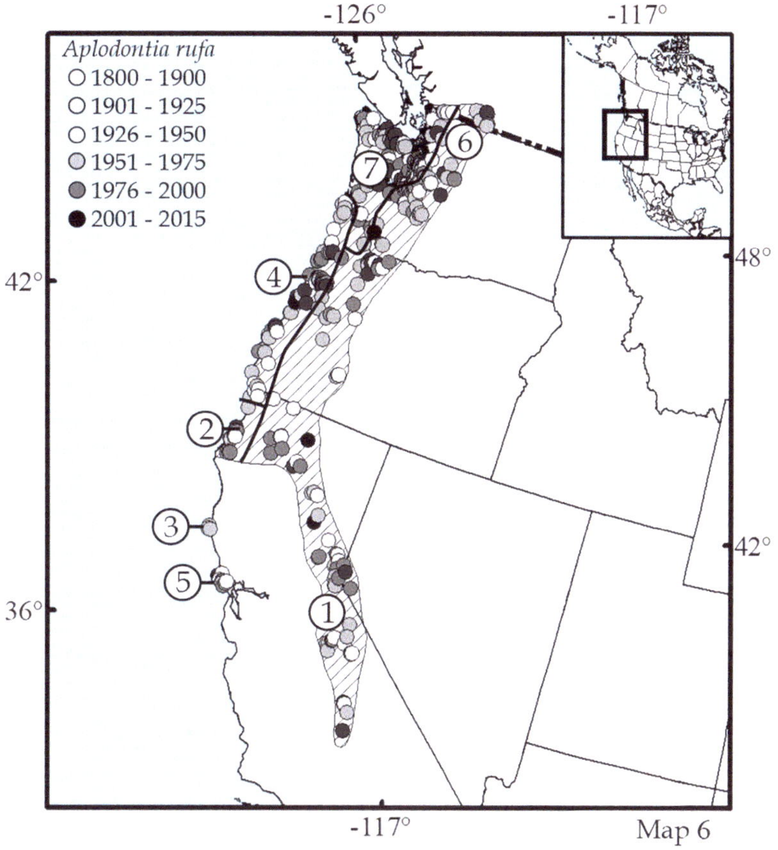


**Map 6** Distribution map of *Aplodontia rufa*

**Additional Literature:** Taylor (1918), Hall (1981), Carraway and Verts (1993), Fellers et al. (2016).

## **Family Sciuridae** Fischer de Waldheim, 1817

The family Sciuridae includes squirrels, prairie dogs, and marmots characterized by a thin body shape (70.0–1000.0 mm); tail short or long, often with abundant hair; ears either very small or relatively large, without a tragus; postorbital process large. Cosmopolitan distribution, except for Australia. One subfamily ranges in North America with two tribes: Pteromyini and Sciurini. The keys were elaborated based on the review of specimens and the following sources: Howell (1918), Hall (1981), and Thorington et al. (2012).

1. Tail long and fluffy; one or two upper premolars on each side, when two, the first one reduced to a spicule; outer face of the zygomatic arches usually not twisted into a horizontal plane .................................... subfamily Sciurinae (p. 14)

1a. Tail usually short, when long not fluffy; two upper premolars on each side; outer face of the zygomatic arches usually twisted into a horizontal plane ........................................................................................subfamily Xerinae (p. 42)

## **Subfamily Sciurinae** Hemprich, 1820

1. Skinfold present between anterior and hind limb; tail with very long lateral hair, giving appearance of being flattened; the zygomatic plate low, slightly tilted upwards; braincase breadth greater than 44% of the skull length ........................................................................ tribe Pteromyini (p. 14)

1a. No skinfold between anterior and hind limbs; tail with short or long hair, but not giving appearance of being flattened; the zygomatic plate tilted strong upwards; braincase breadth less than 44% of the skull length ........................................................................ tribe Sciurini (p. 18)

## Tribe Pteromyini

### Genus *Glaucomys* Thomas, 1908

Dorsal pelage grayish brown with membrane edge blackish; fur of underparts white; area below the eyes to the cheeks white; **tail with very long lateral hair giving the appearance of being flattened**; ears naked; **lateral membrane attached frontward to the arms from the wrist to the ankle in hind limbs**; hair very silky; in relation to the other sciurids skull very flat; braincase unusually deep and inflated; nasals depressed anterior and end up very straight; nose tip straight from the interorbital to occipital region; narrow interorbital breadth; interorbital foramen oval. The keys were elaborated based on the review of specimens and the following sources: Howell (1918), Hall (1981), and Arbogast et al. (2017).

1. Smaller size, total length less than 260.0 mm; fur of underparts white and not gray at the base; skull less than 36.0 mm (from Nova Scotia south through Florida and west through Minnesota and Texas, United States. In México at highlands of Sonora, Chihuahua, east through Tamaulipas and south through Guerrero, Oaxaca, and Chiapas) ........................................................................ *Glaucomys volans* (p. 17)

1a. Larger size, total length greater than 260.0 mm; fur of underparts white and gray at the base; skull greater than 36.0 mm ........................................................................ 2

2. Darker pelage; in association with more humid habitats, milder Pacific Coast (from southwestern British Columbia, Canada, south through southern California, United States) ........................................ *Glaucomys oregonesis* (p. 14)

2a. Paler pelage; in association with more dryer habitats, inland (from Alaska east through Newfoundland and Labrador, Canada, and south through the Appalachian Mountains in Tennessee, and south through Idaho and Utah, with a population western South Dakota, United States) ........................................ *Glaucomys sabrinus* (p. 16)

### *Glaucomys oregonensis* Bachman, 1839

Humboldt's flying squirrel, ardilla voladora de Oregon

1. *G. o. californicus* (Rhoads, 1897). San Bernardino and San Jacinto Mountains, southern California.
2. *G. o. flaviventris* Howell, 1915. Northern California.
3. *G. s. fuliginosus* (Rhoads, 1897). From Cascade Range south through Siskiyou Mountains, northwestern California.
4. *G. o. klamathensis* (Merriam, 1897). Central Oregon, east of Cascade Mountains.
5. *G. o. lascivus* (Bangs, 1899). Sierra Nevada, California.
6. *G. o. oregonensis* (Bachman, 1839). Coastal region, from British Columbia southward to Oregon.
7. *G. o. stephensi* (Merriam, 1900). Coastal region of northern California.

**Conservation:** The population *Glaucomys oregonensis* is listed by the United States Endangered Species Act (FWS 2022) and Canada, and (as *Glaucomys sabrinus*) as Least Concern by the IUCN (Cassola 2016ah).

**Characteristics:** Humboldt's flying squirrel is medium sized**; total length 264.0–328.0 mm and** skull length 38.2–42.6 mm. Dorsal pelage grayish brown with a blackish membrane edge; **fur of underparts lead colored at the base**, middle and tip white; area below the eyes to the cheeks white; tail dark gray dorsally and cream or yellow ventrally; skull very similar between species, the main difference versus other species is its smaller size.

**Comments:** *Glaucomys oregonensis* was recognized as a distinct species (Arbogast et al. 2017). *G. oregonensis californicus*, *G. o. flaviventris*, *G. o. klamathensis*, *G. o. lascivus*, *G. o. oregonensis*, and *G. o. stephensi* previously were subspecies of *G. sabrinus* (Howell 1918; Arbogast et al. 2017; Koprowski et al. 2017). *G. s. columbiensis* and *G. s. fuliginosus* can contain individuals of both species *G. sabrinus* and *G. oregonensis*. *G. oregonensis* ranges from southwestern British Columbia south through southern California (Map 7). It can be found in sympatry, or nearly so, with *G. sabrinus* in northwestern Washington and southwestern British Columbia. *G. oregonensis* can be distinguished from *G. sabrinus* by its darker pelage. From all other squirrel species, by its skinfold between the fore and hind limbs; tail with very long lateral hair giving the appearance of being flattened; the zygomatic plate nearly horizontal and braincase breadth greater than 44% of the skull length.

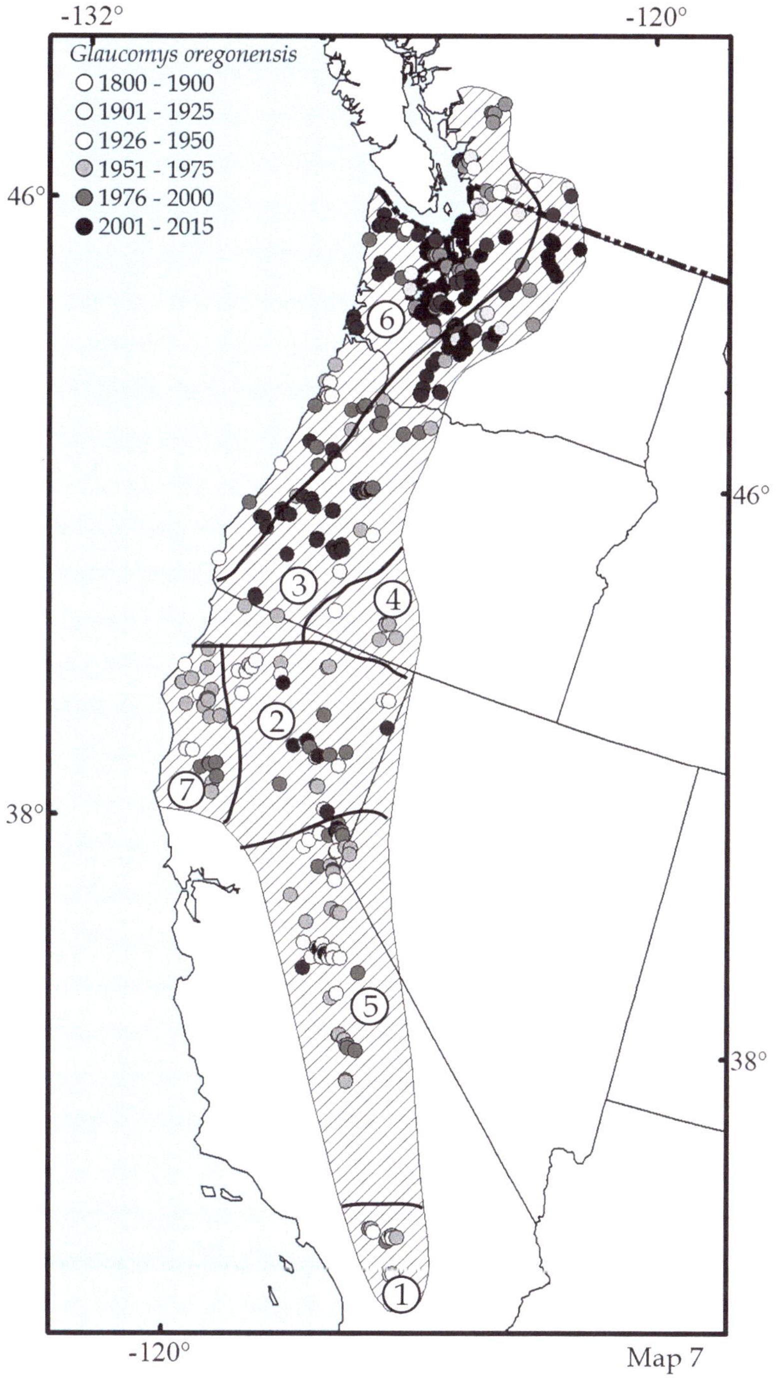


**Map 7** Distribution map of *Glaucomys oregonensis*

**Additional Literature:** Wells-Gosling and Heaney (1984), Arbogast (1999).

## *Glaucomys sabrinus* (Shaw, 1788)

Northern flying squirrel, ardilla voladora del norte

1. *G. s. alpinus* (Richardson, 1828). From British Columbia and southwestern Alberta north through Yukon.
2. *G. s. bangsi* (Rhoads, 1897). From eastern Oregon to northwestern Wyoming.
3. *G. s. canescens* Howell, 1915. Southern Manitoba, eastern North Dakota, and western Minnesota.
4. *G. s. coloratus* Handley, 1953. Area between North Carolina and Tennessee.
5. *G. s. columbiensis* Howell, 1915. Interior valleys of southern British Columbia and northern Washington.
6. *G. s. fuscus* Miller, 1936. West Virginia.
7. *G. s. goodwini* Anderson, 1943. Northern New Brunswick.
8. *G. s. gouldi* Anderson, 1943. Prince Edward Island and Nova Scotia.
9. *G. s. griseifrons* Howell, 1934. Lake Bay, Prince of Wales Island, Alaska.
10. *G. s. latipes* Howell, 1915. From British Columbia and Washington east through Montana and Idaho.
11. *G. s. lucifugus* Hall, 1934. Utah.
12. *G. s. macrotis* (Mearns, 1898). From New Brunswick west through Minnesota.
13. *G. s. makkovikensis* (Sornborger, 1900). From Labrador west through Québec.
14. *G. s. murinauralis* (Musser, 1961). Southwestern Utah.
15. *G. s. reductus* Cowan, 1937. Midcoastal areas of British Columbia.
16. *G. s. sabrinus* (Shaw, 1788). From the Mackenzie River east through southern Québec and south through Minnesota and Wisconsin.
17. *G. s. yukonensis* (Osgood, 1900). From Yukon west through Alaska.
18. *G. s. zaphaeus* (Osgood, 1905). From northern British Columbia north through Alaska.

**Conservation:** The population of *Glaucomys sabrinus coloratus* is considered Endangered by the United States Endangered the Species Act and *G. sabrinus* as Least Concern by the IUCN (Cassola 2016ah).

**Characteristics:** Northern Flying squirrel is the largest flying squirrel, **usually weighing more than 150 g; total length 275.0– 342.0 mm** and skull length 38.0–44.2 mm. Dorsal pelage grayish brown with the membrane edge blackish; **fur of underparts lead colored at the base**, middle and tip white; area below the eyes to the cheeks white; tail dark gray dorsally and cream or yellow ventrally; skull very similar between species, the main differences is in size.

**Comments:** The subspecies *Glaucomys sabrinus californicus*, *G. s. flaviventris*, *G. s. klamathensis*, *G. s. lascivus*, *G. s. oregonensis*, and *G. s. stephensi* were proposed as a distinct species *Glaucomys oregonensis* (Arbogast et al. 2017). *G. s. columbiensis* and *G. s. fuliginosus* can contain individuals of both species *G. sabrinus* and *G. oregonensis*. Those specimens from inland to the former and those of the coast to the latter (Arbogast et al. 2017). *G. sabrinus* ranges through much of Alaska east through Newfoundland and Labrador and south through Appalachian Mountains in Tennessee, and south through Idaho and Utah, with a population in the Black Hills western South Dakota and different populations in the Rocky Mountains (Map 8). *G. sabrinus* can be found in sympatry with *G. volans* in southeastern Canada and northeastern United States from Minnesota to Nova Scotia, including Ontario and Québec. With *G. oregonensis* in northwestern Washington and southwestern British Columbia. *G. sabrinus* can be distinguished from *G. oregonensis*, by having larger body sizes; lighter pelage and be in association with dryer habitats, absent from the coastal wet areas of western Washington, southwestern British Columbia, and all California. From *G. volans* mainly by its greater size, with a total length from greater than 260.0 mm and skull length than 36.0 mm; fur of the underparts white, but gray at the base; usually weighing more than 100 g and the skull is very similar. From all other squirrels species by its skinfold between the fore and hind limbs; tail with very long lateral hair giving the appearance of being flattened; the zygomatic plate nearly horizontal and the braincase breadth greater than 44% of the skull length.

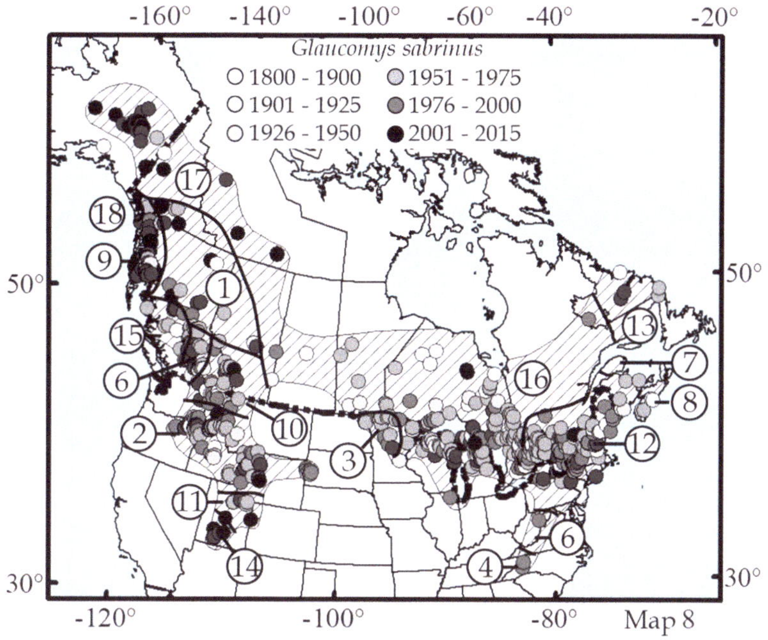


**Map 8** Distribution map of *Sciurus sabrinus*

**Additional Literature:** Howell (1918), Jackson (1961), Hall (1981), Wells-Gosling and Heaney (1984), Thorington et al. (2012).

## *Glaucomys volans* (Linnaeus, 1758)

Southern flying squirrel, ardilla voladora del sur

1. *G. v. chontali* Goodwin, 1961. Sierra Madre del Sur, Oaxaca.
2. *G. v. goldmani* (Nelson, 1904). Highlands of central Chiapas.
3. *G. v. guerreroensis* Diersing, 1980. Sierra Madre del Sur, Guerrero.
4. *G. v. herreranus* Goldman, 1936. Tamaulipas, Nuevo León, San Luis, Jalisco, and Estado de México.
5. *G. v. madrensis* Goldman, 1936. Chihuahua and Durango.
6. *G. v. oaxacensis* Goodwin, 1961. Highlands of Oaxaca.
7. *G. v. querceti* (Bangs, 1896). Florida and Georgia.
8. *G. v. saturatus* Howell, 1915. From Carolinas west through Oklahoma and Kentucky.
9. *G. v. texensis* Howell, 1915. Restricted to Texas.
10. *G. v. volans* (Linnaeus, 1758). From Kansas to Virginia north through to the border with Canada.

**Conservation:** *Glaucomys volans* is listed as Threatened by the Norma Oficial Mexicana (DOF 2019) and as Least Concern by the IUCN (Cassola 2016ai).

**Characteristics:** Southern Flying squirrel is the smallest flying squirrel**, usually weighing approximately 60 g; total length 210.0– 253.0 mm and skull length 32.0–36.0 mm. Dorsal pelage grayish brown with membrane edge blackish**; **fur of the underparts whitish at the base** and white, cream, or yellowish in the middle or at the tip; patagium dorsal and ventral surfaces dark brownish at edge; area below the eyes to the cheeks white; head gray; tail brown dorsally and cream or yellow ventrally; skull very similar between species, the main difference is the size.

**Comments:** *Glaucomys volans* ranges from Nova Scotia south through Florida and west through Minnesota and Texas; in México, from the highlands of Sonora, Chihuahua, and Jalisco east through Tamaulipas and south through Oaxaca, Chiapas, and Guerrero. *G. sabrinus* can be found in sympatry with *G. volans* in southeastern Canada and northeastern United States from Minnesota east through Nova Scotia, including Ontario and Québec (Map 9). *G. volans* can be distinguished from *G. sabrinus* mainly by its smaller size, with a total length less than 260.0 mm and skull length less than 36.0 mm; fur of underparts white, not gray at the base; skull very similar. From all other squirrel species, by its skinfold between the fore and hind limbs; tail with very long lateral hair giving the appearance of being flattened; the zygomatic plate nearly horizontal and braincase breadth greater than 44% of the skull length.

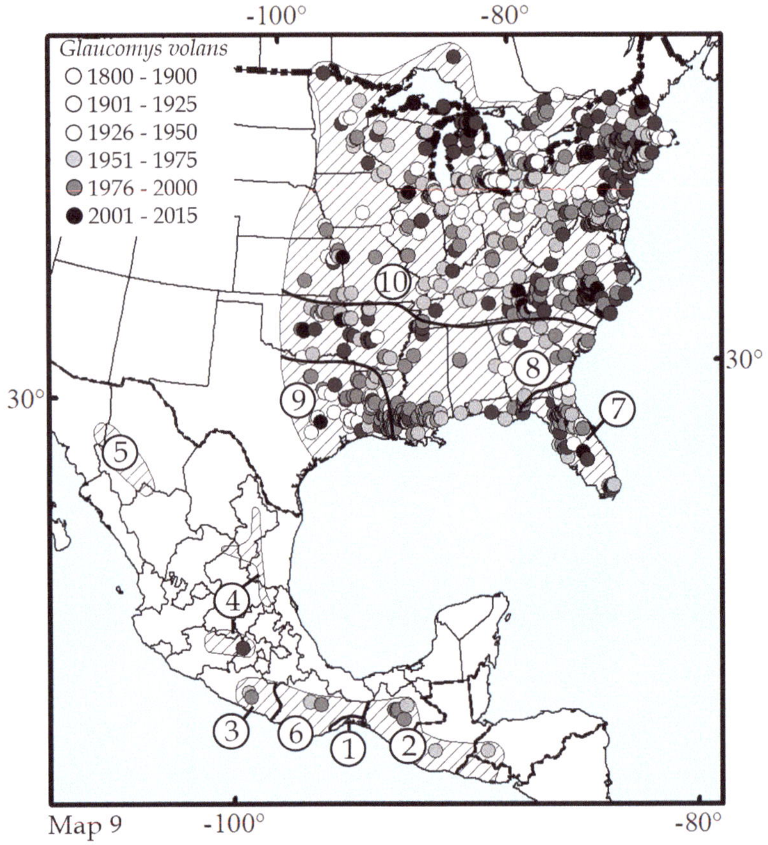


**Map 9** Distribution map of *Glaucomys volans*

**Additional Literature:** Bangs (1896), Howell (1915a, 1918), Nadler and Sutton (1967), Arbogast (1999), Dolan and Carter (1977), Thorington et al. (2012), Arbogast et al. (2017).

## Tribe Sciurini

1. Lateral stripe dark gray or charcoal; auditory bullae with three septa well distinguished externally (Fig. 7); anterior margin of the zygomatic arches at the level of the fourth upper molariforms, counted from the back to the front .................................................................................................................................................... *Tamiasciurus* (p. 38)

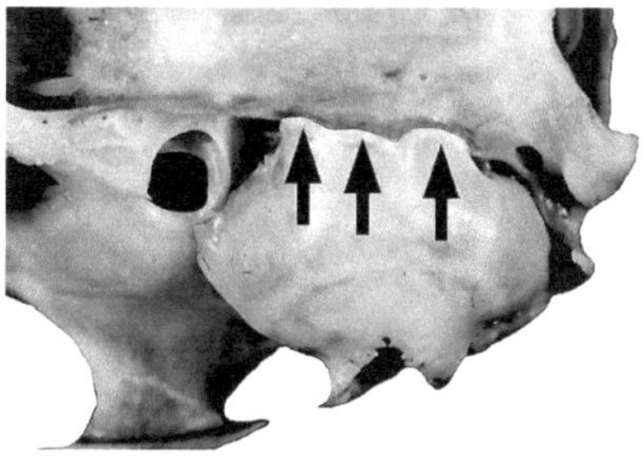

**Fig. 7** The auditory bullae with three septa well distinguished externally

1a. Darker lateral stripe absent; auditory bullae with two septa well distinguished externally, never three (Fig. 8); anterior margin of the zygomatic arches at the level of the second or third upper molariforms, counted from the back to the front .......... *Sciurus* (p. 19)

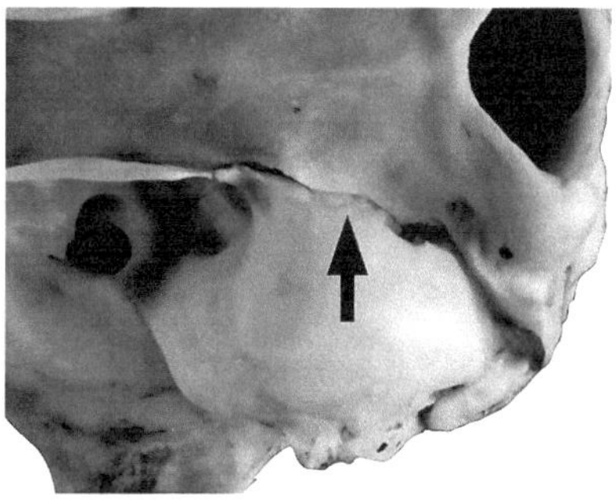

**Fig. 8** The auditory bullae with two septa well distinguished externally, never three

(NOTE: Several species of this genus may exhibit strong variations among individuals, populations, or seasons of the year, so the overall pattern is mentioned, but it can hardly be a criterion for species identification unless specified in the key).

## Genus *Sciurus* Linnaeus, 1758

Squirrels have great variations in coloration in general, from pale gray to black dorsally, and can have spots of different colors on the hips, back, shoulders, and flanks, frequently including melanistic individuals; fur of underparts showing great variations in color, but tending to be reddish in most specimens; ears of different size; **tail always large and fluffy;** infraorbital foramen always forming a channel; first and second upper molars with four transverse ridges; palate broad with two upper premolars; back of the jugal crooked. *Sciurus* has four subgenera: *Echinosciurus* [*S. aureogaster*, *S. colliaei*, *S. deppei*, *S. variegatoides*, and *S. yucatanensis*], *Hesperosciurus* [*S. aberti* and *S. griseus*], *Neosciurus* [*S. carolinensis*], and *Parasciurus* [*S. alleni*, *S. arizonensis*, *S. nayaritensis*, *S. niger*, and *S. oculatus*]. The keys were elaborated based on the review of specimens and the following sources: Howell (1918), Hall (1981), Thorington et al. (2012), and Álvarez-Castañeda et al. (2017a).

1. Dorsal pelage gray grizzled with silver; jugal straight in dorsal view in such a way that its middle part cannot be seen (from Washington, United States, south through northern Baja California, México) .......... subgenus *Hesperosciurus; Sciurus griseus* (p. 28)

1a. Dorsal pelage in different colors and patterns, but not gray grizzled with silver; jugal twisted in dorsal view in such a way that its middle part can be seen .......... 2

2. Ears long and broad with a large hair tuft at the tip; tail comparatively short; frontal region of the skull flattened, postorbital breadth smaller than the interorbital breadth; interorbital breadth equal to the nasal length (from Colorado and Utah, United States, south through Sonora, Chihuahua, and Durango, México) .................................................................... subgenus *Hesperosciurus; Sciurus aberti* (p. 26)

2a. Ears without a terminal hair tuft at the tip; tail comparatively long; frontal region of the skull not flattened, postorbital breadth approximately equal to the interorbital breadth; interorbital breadth generally smaller than the nasal length.................................................................... 3

3. General coloration salt and pepper, but not in a solid color; one upper molar on each side; ranges only in Mexico, except for one that has a terminal brush at the ears.................................................subgenus *Parasciurus* 4

3a. All types of coloration; two upper molars on each side; ranges throughout North America.................................... 8

4. Ventral pelage pale tawny; basal length less than 49.0 mm (from Saskatchewan and Manitoba, Canada, east through the Atlantic coast in New York and Pennsylvania, and south through Florida and west through Texas, Coahuila, Nuevo León, and Tamaulipas, México). Introduced populates can be found in California, United States) ....................................................................*Sciurus niger* (p. 35)

4a. Ventral pelage ocher or white; basal length greater than 49.0 mm.................................................... 5

5. Fur of underparts ocher to yellowish .................................................................................... 6

5a. Fur of underparts whitish................................................................................................ 7

6. Dorsal pelage grizzly with yellow, black, and white shades; total length usually greater than 550.0 mm; tail length generally greater than 270.0 mm; hindfoot length greater than 75.0 mm, zygomatic breadth greater than 37.1 mm (from Arizona, United States, south through Jalisco, México)...........................................*Sciurus nayaritensis* (p. 34)

6a. Dorsal pelage gray with the middle area darker; total length usually less than 550.0 mm; tail length generally less than 270.0 mm; hindfoot length less than 75.0 mm; zygomatic breadth less than 37.1 mm (from San Luis Potosí, Hidalgo, Veracruz, Puebla, Querétaro, Michoacán, Ciudad de México, Estado de México, and Guanajuato, México)..............................................................................*Sciurus oculatus* (p. 37)

7. Dorsal pelage gray or yellowish brown; total length usually less than 500.0 mm; hindfoot length generally less than 68.0 mm; basal length less than 50.0 mm; interorbital breadth less than 18.5 mm (from Coahuila and central Nuevo León west through Tamaulipas and south through San Luis Potosí, México)................................*Sciurus alleni* (p. 32)

7a. Dorsal pelage grayish mixed with yellow at the shoulders; total length generally greater than 500.0 mm; hindfoot length generally greater than 68.0 mm; basal length greater than 50.0 mm; interorbital breadth greater than 18.5 mm (from Arizona, western New Mexico, United States, and Sonora, México)....................................*Sciurus arizonensis* (p. 33)

8. Third upper premolars small, not reaching the crown level of the four upper premolars; nasals narrow, moderately V shaped on the posterior border (from Saskatchewan, Canada, east through the Atlantic coast and southward to Texas, including the Florida Peninsula, United States)................................. subgenus *Neociurus; Sciurus carolinensis* (p. 30)

8a. Third and four upper premolars of the same size and crowns at the same level; nasals wide, not moderately V shaped on the posterior border ....................................................................... subgenus *Echinosciurus* 9

9. Total length less than 447.0 mm; tail length less than 195.0 mm; dorsal pelage dark ocher with reddish or brown shades; fur of underparts white or yellowish with reddish shades; basal length less than 43.6 mm; zygomatic breadth less than 30.1 mm (from Nuevo León and Tamaulipas, Mexico, south through Central America, including the Yucatán Peninsula) ..................................................................... subgenus *Echinosciurus; Sciurus deppei* (p. 23)

9a. Total length greater than 447.0 mm; tail length greater than 195.0 mm; dorsal pelage not as described above; basal length greater than 43.5 mm; zygomatic breadth greater than 30.0 mm ............................................. 10

10. Total length generally less than 500.0 mm; tail length less than 225.0 mm; hindfoot length less than 65.0 mm; fur of underparts paler or dark gray; interorbital breadth less than 17.0 mm; skull length less than 57.3 mm (from Yucatán Peninsula and western Chiapas, Mexico, south through Central America) ..................................................................... subgenus *Echinosciurus; Sciurus yucatanensis* (p. 25)

10a. Total length generally greater than 500.0 mm; tail length greater than 225.0 mm; hindfoot length greater than 65.0 mm; fur of underparts varying in color; interorbital breadth greater than 17.0 mm; skull length greater than 57.3 mm .................................................................................................... 11

11. Fur of underparts not white, it can be practically any shade of yellow, brown, and gray, even melanistic (from Nayarit, Guanajuato, and Nuevo León, México, south through Guatemala, and was introduced to Elliott Key, Florida, United States)........................................................................................ subgenus *Echinosciurus; Sciurus aureogaster* (p. 21)

11a. Fur of underparts white, if not, at least with white patches on the groin, armpits, and neck ........................................ 12

12. Upper edge of the ears black; back of the ears with a large spot; total length usually greater than 525.0 mm; tail length generally greater than 275.0 mm; skull length greater than 59.0 mm (from southern Chiapas, México, south through Central America)..........................................................................subgenus *Echinosciurus; Sciurus variegatoides* (p. 24)

12a. Upper edge of the ears not black; back of the ears with a small or undefined spot; total length usually less than 525.0 mm; tail length generally less than 275.0 mm; skull length less than 59.0 mm (from Sonora and Chihuahua south through Colima, México) ..........................................................................................................subgenus *Echinosciurus; Sciurus colliaei* (p. 22)

## Subgenus ***Echinosciurus*** Trouessart, 1880

Large-sized squirrel; tail long, about 50% of the total length; coloration variable; pelage texture soft or hispid, according to the environment; skull broad, dorsal outline flattened or slightly swollen at the frontal region; rostrum short, nasal length about 95% or more of the interorbital breadth. One lower and two upper premolarson each side; third upper premolars usually small and slender; molar dentition not distinctive (Allen 1915b). All species currently treated under the subgenus *Echinosciurus* (Allen 1915b; de Abreu-Jr et al. 2020a, b) were previously under the subgenus *Sciurus*.

### ***Sciurus aureogaster*** Cuvier, 1829

Red-bellied squirrel, ardilla de vientre rojo

1. *S. a. aureogaster* Cuvier, 1829. Eastern coast of México.
2. *S. a. nigrescens* Bennett, 1833. Central and western coast of México south through Guatemala.

**Conservation:** *Sciurus aureogaster* is listed as Least Concern by the IUCN (Koprowski et al. 2017).

**Characteristics:** Red-bellied squirrel is medium sized; total length 518.0–573.0 mm and skull length 57.9–61.9 mm. This species of squirrel shows the greatest variation in general coloration, pale gray grizzled white to oxide grayish red dorsally, **sometimes with spots of different colors on the nape, hips, shoulders, and flanks**, frequently including melanistic individuals; fur of underparts with great variation in color, from white through orange to chestnut, but mostly reddish; **ears small;** tail color variegated grayish buff with a pale to orange red ventral region, or chestnut with deep orange shades.

**Comments:** *Sciurus aureogaster* ranges from Nayarit, Guanajuato, and Nuevo León south through Guatemala (Map 10), and was introduced to Elliott Key, Florida. *S. aureogaster* can be found in sympatry, or nearly so, with *S. alleni*, *S. colliaei*, *S. deppei*, *S. nayaritensis*, *S. ocultus*, *S. yucatanicus*, and *S. variegatoides*. *S. aureogaster* can be differentiated from all the other species of Sciuridae by its spots of different colors on the nape, hips, and shoulders. From *S. alleni*, *S. nayaritensis*, and *S. ocultus*, by having two upper molars. From *S. deppei*, by its larger size, greater than 447.0 mm. From *S. colliaei*, by not having the fur of the underparts white. From *S. yucatanicus*, by having prominent dorsal patches and sphenopalatine vacuities closed or very small. From *S. variegatoides*, by having the venter orange or chestnut, frequently spotted with black.

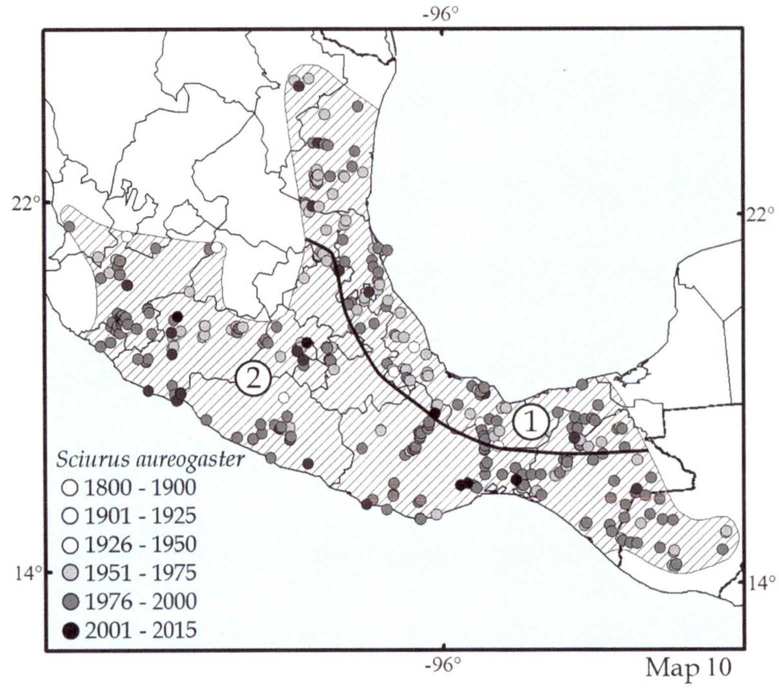


**Map 10** Distribution map of *Sciurus aureogaster*

**Additional Literature:** Musser (1968, 1970), Hall (1981), Coates-Estrada and Estrada (1986), Hoffmeister and Hoffmeister (1991), Jiménez-Guzmán et al. (1999), Koprowski et al. (2017).

## *Sciurus colliaei* Richardson, 1839

Collie's squirrel, ardilla de la Sierra Madre Occidental

1. *S. c. colliaei* Richardson, 1839. Restricted to Nayarit and most southern Sinaloa.
2. *S. c. nuchalis* Nelson, 1899. Jalisco and Colima.
3. *S. c. sinaloensis* Nelson, 1899. Restricted to southern Sinaloa.
4. *S. c. truei* Nelson, 1899. From central Sinaloa to southern Sonora, including western Chihuahua and Durango.

**Conservation:** *Sciurus colliaei* is listed as Least Concern by the IUCN (de Grammont et al. 2016).

**Characteristics:** Collie's squirrel is medium sized; total length 440.0–578.0 mm and skull length 56.0–64.5 mm. Dorsal pelage grizzled gray with a yellowish wash that continues to the tail base, flanks paler gray than in the dorsum; fur of underparts usually white, but it can be pale orange; hips, shoulders, limbs and ears can vary in color, usually dark gray to reddish; fur of underparts whitish; tail base blackish grizzled gray to dark gray with yellow shades dorsally, edges white, the rest blackish with grizzly white in different proportions; ears with the top edge not black; postauricular spot small or undefined; population on the northern parts of the range with only one upper premolar on each side.

**Comments:** *Sciurus colliaei* ranges from Sonora and Chihuahua south through Colima (Map 11). It can be found in sympatry, or nearly so, with *S. aberti*, *S. aureogaster*, and *S. nayaritensis*. *S. colliaei* can be distinguished from *S. aureogaster* by not having spots of different colors on the nape, hips, and shoulders, and by having one upper molar on each side and fur of the underparts in different colors, but not white. From *S. aberti*, by not having a hair tuft in the ears extending 2.0–3.0 cm above each ear. From *S. nayaritensis*, by having the underparts whitish and with one upper molar on each side.

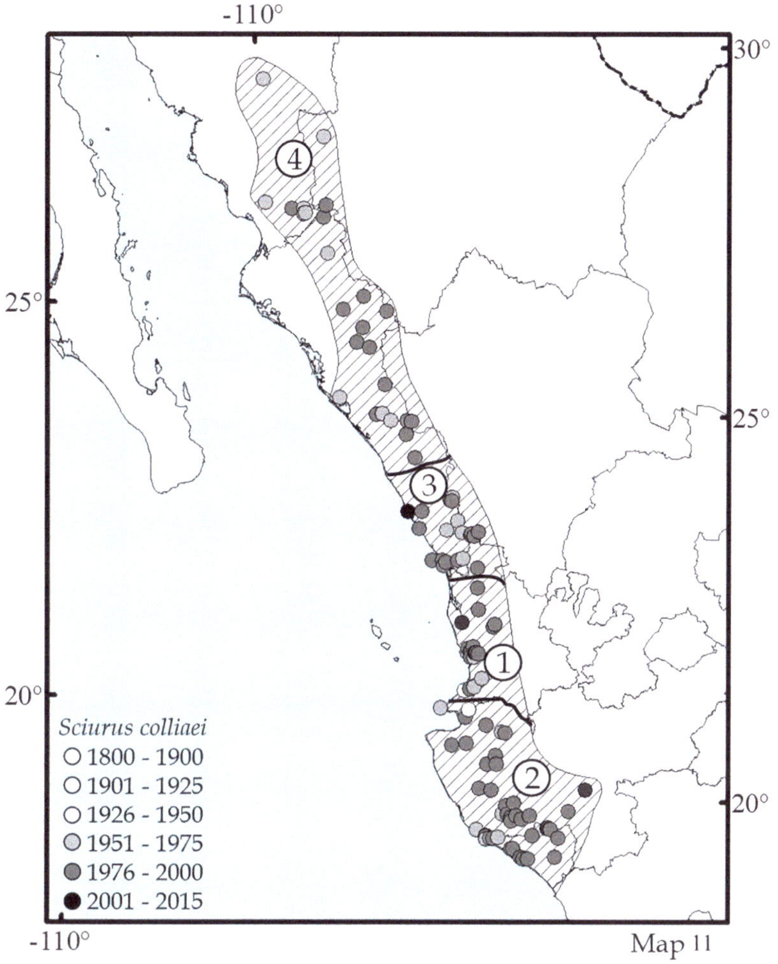


**Map 11** Distribution map of *Sciurus colliaei*

**Additional Literature:** Nelson (1899), Anderson (1962), Musser (1968), Caire (1978), Hall (1981), Hoffmeister and Hoffmeister (1991), Best (1995f).

## *Sciurus deppei* Peters, 1863

Deppe's squirrel, ardilla tropical

1. *S. d. deppei* Peters, 1863. From northern Veracruz south through Chiapas.
2. *S. d. negligens* Nelson, 1898. Tamaulipas, San Luis Potosí, Hidalgo, and northern Veracruz.
3. *S. d. vivax* Nelson, 1901. From the Yucatán peninsula through Costa Rica.

**Conservation:** *Sciurus deppei* is listed as Least Concern by the IUCN (Koprowski et al. 2016).

**Characteristics:** Deppei's squirrel is medium sized; total length 343.0–387.0 mm and skull length 49.5–53.0 mm. **Dorsal pelage gray to brown ocher to oxide red, external limbs and feet dark gray**; individuals show considerable variations in color; fur of underparts yellowish white to yellowish red, **but always with yellow spots on the groin, armpits, and neck**; face gray; tail blackish grizzled dorsally, with white shades and reddish oxide to ocher ventrally, white at the tip.

**Comments:** *Sciurus deppei* ranges from Nuevo León and Tamaulipas south through Central America, including the Yucatán Peninsula (Map 12). It can be found in sympatry, or nearly so, with *S. alleni*, *S. aureogaster*, *S. oculatus*, and *S. yucatanensis*. *S. deppei* can be distinguished from all the species of *Sciurus* by its smallest size, less than 390.0 mm.

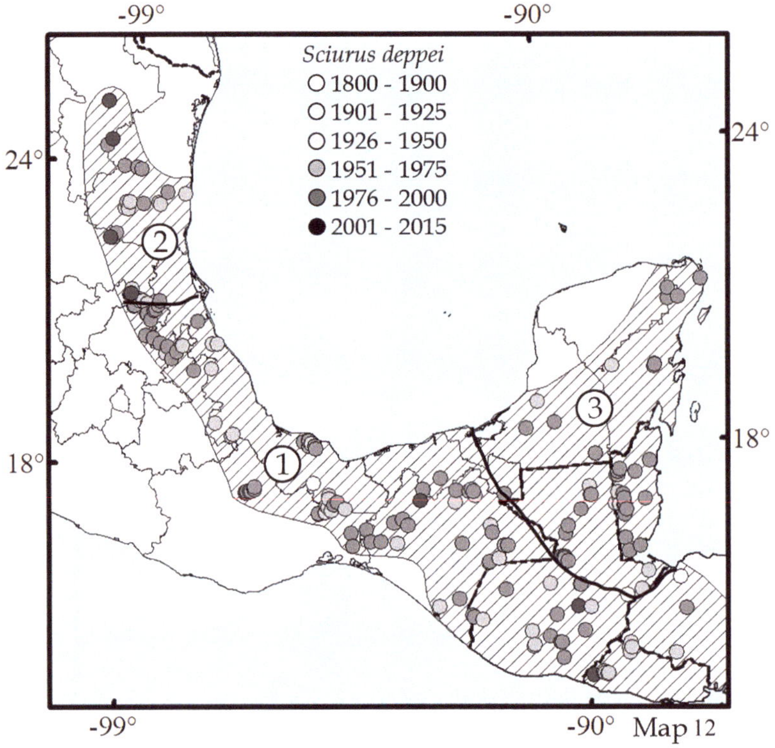


**Map 12** Distribution map of *Sciurus deppei*

**Additional Literature:** Nelson (1898), Allen (1915b), Goodwin (1946), Dalquest (1953), Hall (1981), Heaney (1983), Jones et al. (1983a), Hoffmeister and Hoffmeister (1991), Best (1995e).

## *Sciurus variegatoides* Ogilby, 1839

Variegated squirrel, ardilla centroamericana

*S. v. goldmani* Nelson, 1898. Southeastern Chiapas.

**Conservation:** *Sciurus variegatoides* is listed as Special Protection in the Norma Oficial Mexicana (DOF 2019) and as Least Concern in the IUCN (Reid 2016a).

**Characteristics:** Variegated squirrel is medium sized; total length 510.0–560.0 mm and skull length 54.0–58.5 mm. Dorsal pelage varying greatly in coloration from almost completely steel gray to grizzled gray with a yellow wash, with a distinct brown dorsal patch, or completely reddish, or with a rust-colored neck, flanks, and limbs; fur of underparts whitish; limbs white; tail blackish dorsally with a large amount of grizzly white and fawn or reddish hairs ventrally; usually no spots on the shoulders and hips; ears with the top edge black; postauricular spot large.

**Comments:** *Sciurus variegatoides* ranges from southern Chiapas south through Central America (Map 13), and none of the species of *Sciurus* has sympatric distribution. *S. variegatoides* can be distinguished from *S. aureogaster* by having the venter white.

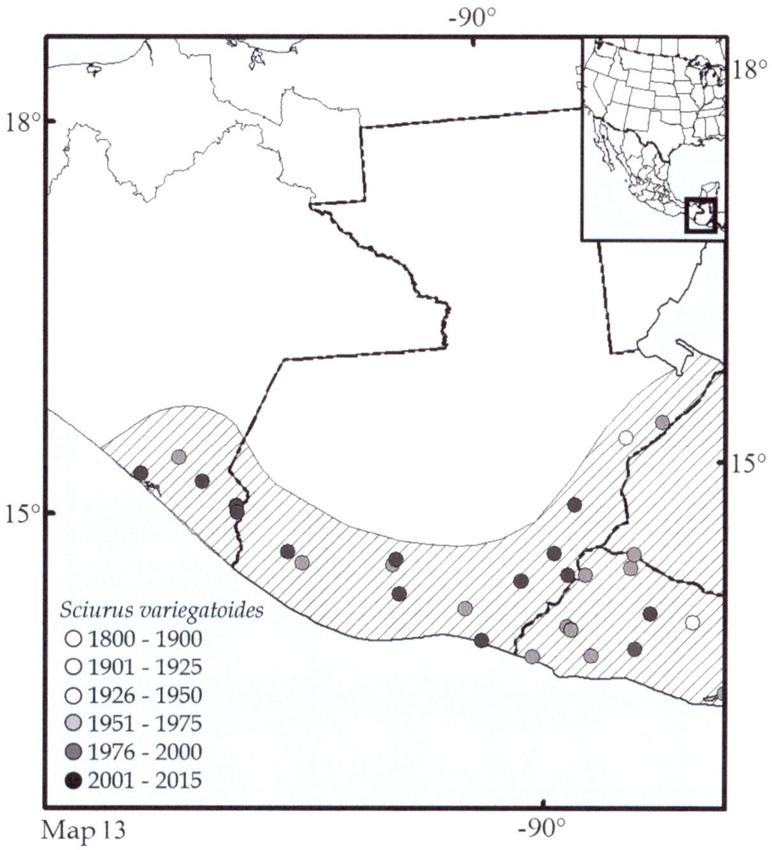


**Map 13** Distribution map of *Sciurus variegatoides*

**Additional Literature:** Goldman (1920), Harris (1937), Musser (1968), Hoffmeister and Hoffmeister (1991), Best (1995a).

## ***Sciurus yucatanensis*** Allen, 1877

Yucatan squirrel, ardilla yucateca

1. *S. y. baliolus* Nelson, 1901. From the central Yucatán peninsula, southern Campeche, and Tabasco south through Guatemala and Belize.
2. *S. y. phaeopus* Goodwin, 1932. Restricted to eastern Chiapas and Guatemala.
3. *S. y. yucatanensis* Allen, 1877. Northern Yucatán peninsula, including Campeche, Quintana Roo, and Yucatán.

**Conservation:** *Sciurus yucatanensis* is listed as Least Concern by the IUCN (Vázquez et al. 2016d).

**Characteristics:** Yucatán squirrel is medium sized; total length 450.0–500.0 mm and skull length 55.1–60.2 mm. **Dorsal pelage blackish gray grizzled with yellowish ocher shades,** often with an olive brown or tawny wash that appears rough in texture; fur of underparts varying from whitish gray grizzled with yellowish gray to black; tail black at the core and frosted with white to buff.

**Comments:** *Sciurus yucatanensis* is restricted to the Yucatán Peninsula, Campeche, Yucatán, Quintana Roo, and from western Chiapas south through Guatemala and Belize (Map 14). It can be found in sympatry, or nearly so, with *S. deppei*, from which it can be differentiated by its larger size, greater than 390.0 mm. From *S. aureogaster*, by not having a dorsal coloration pattern; however, *S. aureogaster* has melanistic specimens, and sphenopalatine vacuities open.

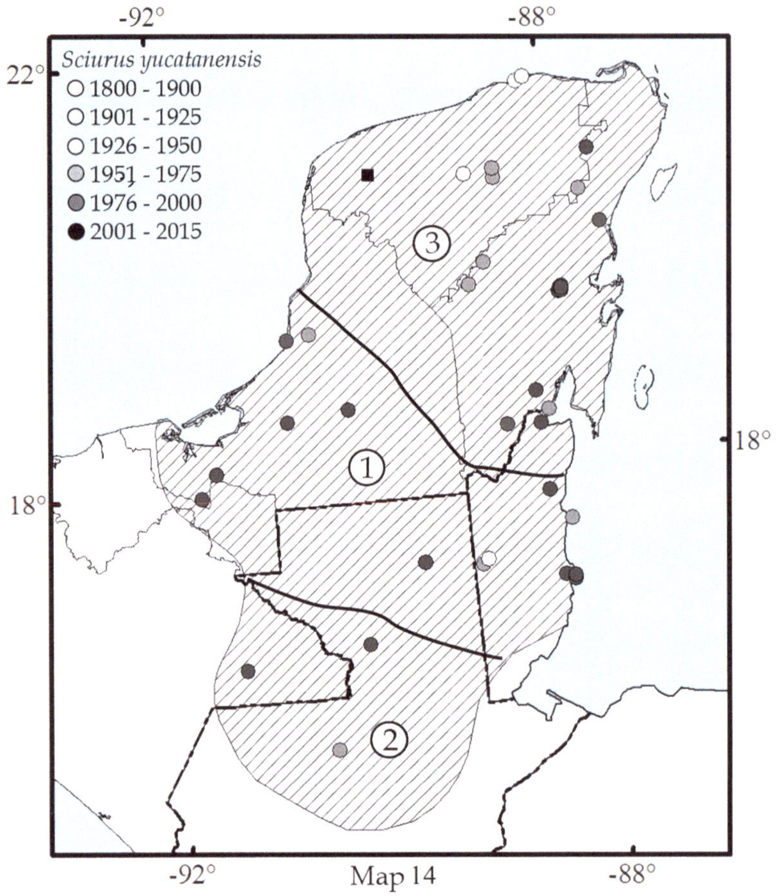


**Map 14** Distribution map of *Sciurus yucatanensis*

**Additional Literature:** Nelson (1899), Musser (1968), Hall (1981), Hoffmeister and Hoffmeister (1991), Best (1995a).

## Subgenus *Hesperosciurus* Nelson, 1899

**Very large-sized squirrel;** tail very long and full (about 50% of the total length); **dorsal pelage gray and fur of the underparts white, without special markings;** skull massive, heavily built; **heavier malar, process of the superior border better developed;** one lower and two upper premolars on each side; third upper premolars usually **heavily developed; third upper molars with a single strongly developed conical cusp, with the accessory cusplets nearly suppressed** (Allen 1915b)**.**

### *Sciurus aberti* Woodhouse, 1853

Abert's squirrel, ardilla orejas peludas

1. *S. a. aberti* Woodhouse, 1853. Northern Arizona and New Mexico.
2. *S. a. barberi* Allen, 1904. Northwestern Chihuahua.
3. *S. a. chuscensis* Goldman, 1931. Northwestern Arizona and extreme northeastern New Mexico.
4. *S. a. durangi* Thomas, 1893. Restricted to western Durango highlands.
5. *S. a. ferreus* True, 1894. Rocky Mountains, central Colorado.
6. *S. a. kaibabensis* Merriam, 1904. Kaibab Plateau, northern Arizona.
7. *S. a. mimus* Merriam, 1904. Northern New Mexico and southern Colorado.
8. *S. a. navajo* Durrant and Kelson, 1947. Restricted to southeastern Utah.
9. *S. a. phaeurus* Allen, 1904. Northwestern Durango and southwestern Chihuahua.

**Conservation:** The populations of *Sciurus aberti barberi*, *S. a. durangi*, and *S. a. phaeurus* in México are listed as Special Protection in the Norma Oficial Mexicana (DOF 2019) and as Least Concern by the IUCN (Cassola 2017a).

**Characteristics:** Abert's squirrel is large sized; total length 463.0–584.0 mm and skull length 58.1–62.9 mm. **Dorsal pelage grizzled gray to charcoal, often with reddish shades and a darker band in the middle of the back;** fur of underparts white; flanks with one blackish stripe; **ear tips with a large hair tuft, especially prominent in winter**; tail gray frosted with white dorsally and white ventrally, some individuals blackish; tail short and unusually broad; skull proportionally shorter and wider; frontal region of the skull flattened; braincase wide and low; postorbital breadth less than the interorbital breadth; interorbital breadth equal to the nasal length.

**Comments:** *Sciurus aberti* was included under the subgenus *Sciurus;* the current treatment includes it under the subgenus *Echinosciurus* (Allen 1915b; de Abreu-Jr et al. 2020a, b). *S. kaibabensis* was reinstated as a subspecies of *S. aberti* (Cockrum 1961); it was previously regarded as a full species (Merriam 1904a). *S. aberti* ranges from Colorado and Utah south through Sonora, Chihuahua, and Durango (Map 15). It can be found in sympatry, or nearly so, with *S. arizonensis*, *S. colliaei*, and *S. nayaritensis*. *S. aberti* can be differentiated from the species of *Sciurus* by having the ears with a hair tuft, which extends 2.0–3.0 cm above each ear; the general coloration is a gray coat with a rusty colored stripe on the back; fur of underparts white.

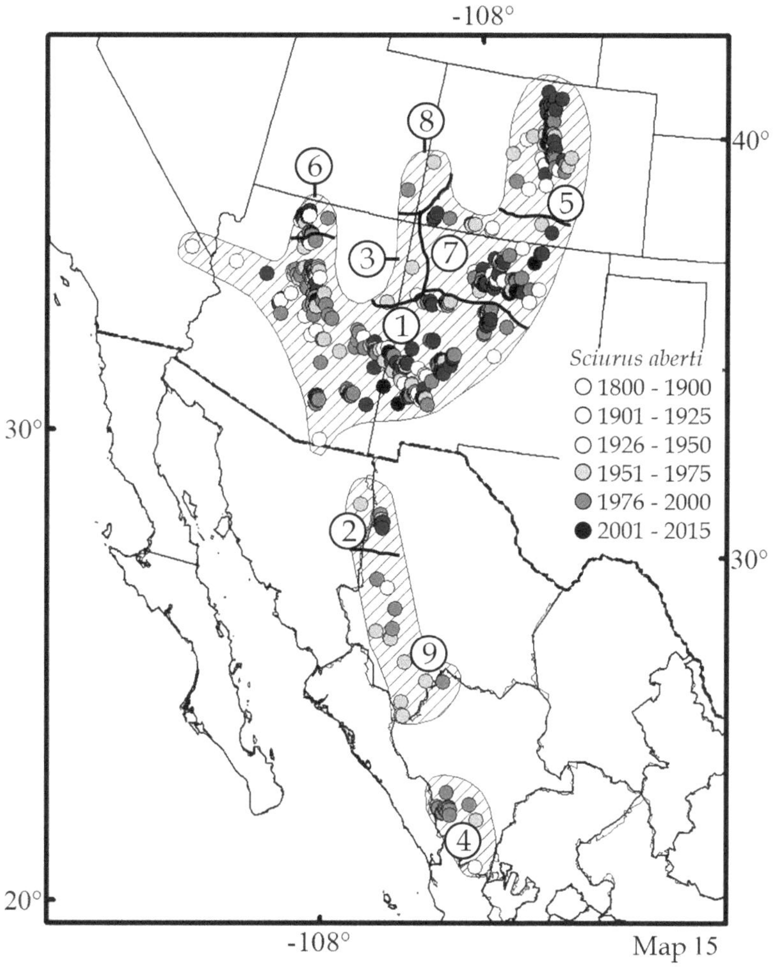


**Map 15** Distribution map of *Sciurus aberti*

**Additional Literature:** Allen (1915b), Warren (1942), Hall and Kelson (1959), Keith (1965), Armstrong (1972), Findley et al. (1975), Ramey and Nash (1976), Nash and Seaman (1977), Hoffmeister and Diersing (1978).

## ***Sciurus griseus*** Ord, 1818

Western gray squirrel, ardilla gris del oeste

1. *S. g. anthonyi* Mearns, 1897. From Sequoia National Park and San Luis Obispo, California, south through northern Baja California.
2. *S. g. griseus* Ord, 1818. From the Sequoia National Park and San Francisco Bay inland northward through Washington.
3. *S. g. nigripes* Bryant, 1889. Coast Range, from San Francisco Bay south through San Luis Obispo.

**Conservation:** *Sciurus griseus* is listed as Threatened in the Norma Oficial Mexicana (DOF 2019) and as Least Concern by the IUCN (Lacher et al. 2016m).

**Characteristics:** Western gray squirrel is medium sized; total length 510.0–770.0 mm and skull length 57.0–60.5 mm. **Dorsal pelage gray grizzled with silver shades;** in summer, it may have grizzled reddish cinnamon shades on the back; fur of underparts white, sometimes with a faint wash of buff; a white spot around the eyes; upperparts of the forefeet gray and of the hind feet dark gray; **ears with a steel or silver gray coloration and quite prominent, nearly double the ear length in many other squirrels**; **tail gray grizzled with silver shades and lateral hair white without yellow hair**; **jugals relatively thin and crooked in vertical plane**; braincase broad at the parietal level; suture of both nasal bones with the frontal bones not at the same level; molars proportionally large.

**Comments:** *Sciurus griseus* ranges from Washington south through northern Baja California (Map 16), and can be found in sympatry, or nearly so, with *S. carolinensis* that has been introduced in the western coast of the United States, from which it can be differentiated by having the larger ears in a steel or silver gray coloration and quite prominent, nearly double the ear length in many other squirrels; venter pure white; tail long and plumose and the largest size, greater than 510.0 mm.

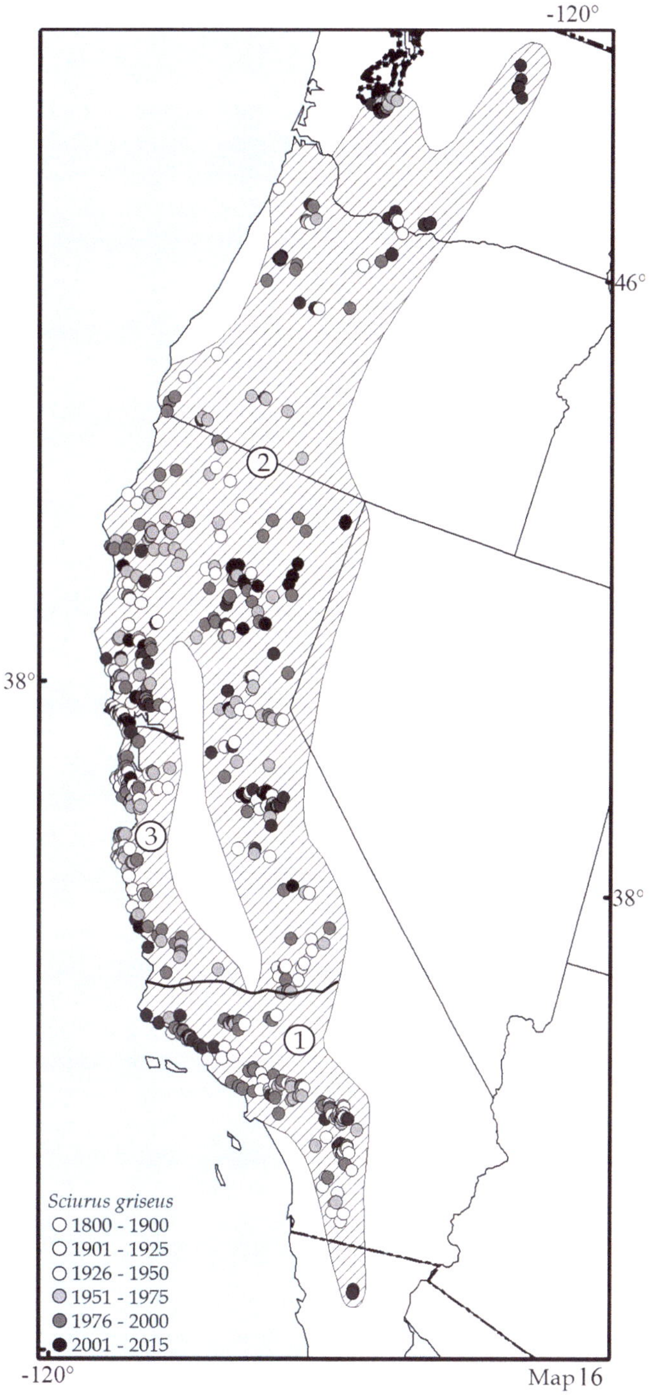


**Map 16** Distribution map of *Sciurus griseus*

**Additional Literature:** Nelson (1899), Allen (1915b), Bailey (1936), Bryant (1945), Ingles (1965), Levin and Flyger (1971), Crase (1973), McLaughlin (1984), Carraway and Verts (1994), de Abreu-Jr et al. (2020a, b).

## Subgenus *Neosciurus* Trouessart, 1880

Medium-sized squirrels; **tail medium in length (about 46% of the total length), broad and full**; dorsal pelage gray and fur of underparts white; skull long and narrow, dorsal outline only slightly convex anterior to the frontoparietal suture; rostrum long and narrow, **nasals narrow, moderately V shaped on the posterior border, about 33% of the skull length, 93% of the interorbital breadth;** zygomatic breadth 55% of the total length. One lower and two upper premolars on each side; **third upper premolars usually small, not reaching the crown level of the four upper premolars**; dentition not distinctive (Allen 1915b).

### *Sciurus carolinensis* Gmelin, 1788

Eastern gray squirrel, ardilla gris del este

1. *S. c. carolinensis* Gmelin, 1788. From southwestern Kansas and Texas east through Atlantic coast.
2. *S. c. extimus* Bangs, 1896. Central and southern Florida.
3. *S. c. fuliginosus* Bachman, 1839. Central and southern Louisiana and Mississippi.
4. *S. c. hypophaeus* Merriam, 1886. From Wisconsin and Minnesota north through Ontario, Manitoba, and Saskatchewan.
5. *S. c. pennsylvanicus* Ord, 1815. From Kansas to South Dakota east through the Atlantic coast, including southern Ontario and Québec.

**Conservation:** *Sciurus carolinensis* is listed as Least Concern by the IUCN (Cassola 2016db).

**Characteristics:** Eastern gray squirrel is medium sized; total length 430.0–500.0 mm and skull length 57.2 –63.3 mm. Dorsal pelage grizzled pale to slate gray, sometimes washed with cinnamon shades, mainly on the hips; fur of underparts white to buff to cinnamon; postauricular patches white to buff, more conspicuous in winter; white eye ring; tail of similar coloration to the back, with a pale frosting of white.

**Comments:** *Sciurus carolinensis* was included under the subgenus *Sciurus*; the current treatment includes it under the subgenus *Neosciurus* (Allen 1915b; de Abreu-Jr et al. 2020a, b). *S. carolinensis* ranges from Saskatchewan east through the Atlantic coast and southward to Texas, including the Florida Peninsula (Map 17). *S. carolinensis* is sympatric, or nearly so, with *S. niger*, but it has been introduced in some cities near San Francisco Bay and Sacramento in California, so in these areas it is sympatric, or nearly so, with *S. griseus*. *S. carolinensis* can be differentiated from other genera by its long and fluffy tail; one or two upper premolars on each side; when two, the first premolar is reduced to a spicule; zygomatic arches with the outer face usually not twisted into a horizontal plane and auditory bullae with two septa. From the other species of *Sciurus*, by its larger body; guard hair white at the tip; lack of a distinctive hair tuft at the tip of the ears; lack of peg-like third upper premolars; sphenopalatine foramen larger than the optic foramen. From *S. niger*, by its smaller size and fur of underparts white to buff to cinnamon. From *S. griseus*, by the dorsal pelage grizzled pale to slate gray; fur of underparts white to buff to cinnamon and one upper molar on each side.

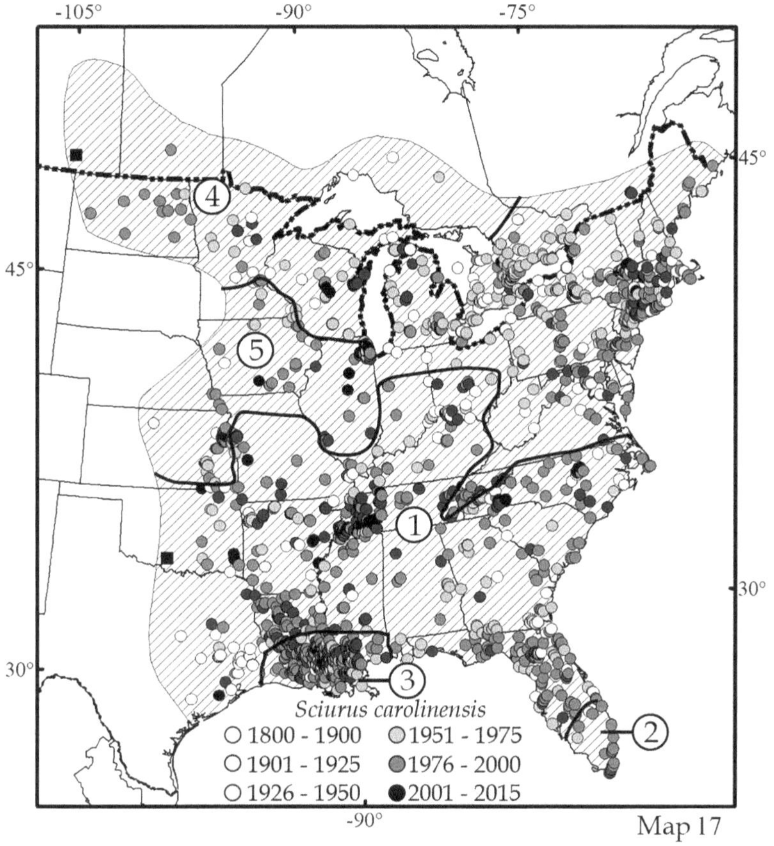


**Map 17** Distribution map of *Sciurus carolinensis*

**Additional Literature:** Barkalow and Shorten (1973), Flyger and Gates (1982), McGrath (1987), Koprowski (1994a).

## Subgenus ***Parasciurus*** Trouessart, 1880

**Large-sized squirrels, among the largest of American tree squirrels;** tail long and broad, about 50% of the total length; pelage thick and soft; gray dorsally (generally dark gray with fulvous suffusion); fur of underparts white or buffy, sometimes ferruginous; skull broad and heavy, dorsal outline flattened over the frontal region, and occipital region relatively slightly depressed; rostrum and nasals broad, the latter well produced posteriorly, forming about 33% of the skull length, their length about equal to the interorbital breadth; zygomatic breadth about 58% of the skull length; **one upper** and one lower premolar on each side; **third upper premolars usually with a strong cusp on the fronto-lateral border of the crown**; molars not distinctive (Allen 1915b). All species currently treated under the subgenus *Parasciurus* (Allen 1915b; de Abreu-Jr et al. 2020a, b) were previously under the subgenus *Sciurus*.

## *Sciurus alleni* Nelson, 1898

Allen's squirrel, ardilla de la Sierra Madre Oriental

Monotypic.

**Conservation:** *Sciurus alleni* is listed as Least Concern by the IUCN (de Grammont and Cuarón 2016d).

**Characteristics:** Allen's squirrel is medium sized; total length 145.0–493.0 mm and skull length 57.5–60.8 mm. Dorsal pelage yellowish brown grizzled with black and gray shades, darker on the back than on the flanks; fur of underparts whitish; postauricular spots weakly gray; tail grizzly blackish dorsally with yellowish hair at the base, sometimes with the ventral region yellowish brown to grayish yellow; melanism has been reported; basal length less than 50.0 mm; interorbital breadth less than 18.5 mm; **one upper premolar on each side**.

**Comments:** *Sciurus alleni* has been considered a subspecies of *S. oculatus* (Moore 1960). *S. alleni* ranges from southeastern Coahuila and central Nuevo León west through Tamaulipas and south through San Luis Potosí (Map 18). It can be found in sympatry, or nearly so, with *S. aureogaster*. *S. alleni* can be differentiated from *S. aureogaster* and *S. deppei* by having only one upper premolar on each side. From *S. oculatus*, by its postauricular patches usually absent and, when present, usually consisting of a tiny area with pelage slightly shorter and grayer than the surrounding area of the head. From *S. aureogaster*, by not having spots of different colors on the nape, hips, shoulders, and flanks, and one upper molar on each side.

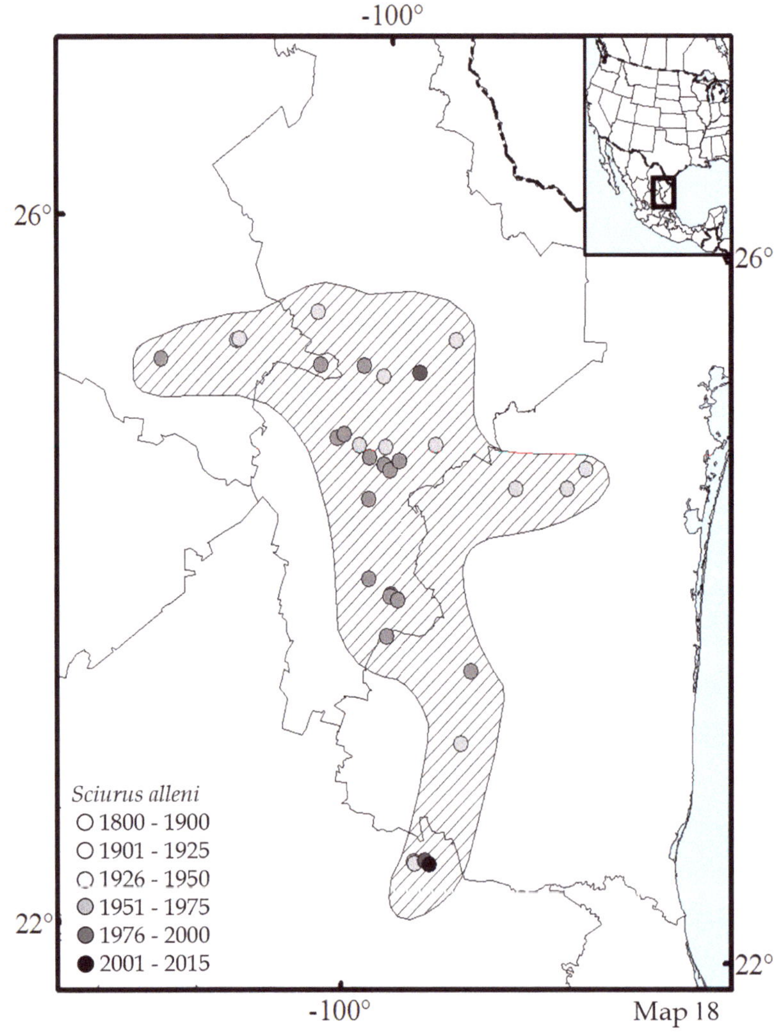


**Map 18** Distribution map of *Sciurus alleni*

**Additional Literature:** Allen (1915b), Dice (1937), Dalquest (1950, 1953), Baker (1956), Musser (1968), Best (1995g).

## *Sciurus arizonensis* Coues, 1867

Arizona gray squirrel, ardilla gris de Arizona

Monotypic.

**Conservation:** *Sciurus arizonensis* is listed as Threatened in the Norma Oficial Mexicana (DOF 2019) and as Data Deficient by the IUCN (Linzey et al. 2019).

**Characteristics:** Arizona gray squirrel is medium sized; total length 506.0–568.0 mm and skull length 60.1–66.1 mm. **Dorsal pelage uniform gray grizzled with silver and in less proportion brownish yellow on the nape and the back**; fur of underparts white to cream; **tail with white hair laterally and yellow hair ventrally;** eye ring cream to white, highly contrasting; basal length greater than 50.0 mm; interorbital breadth greater than 18.5 mm.

**Comments:** *Sciurus arizonensis* has been considered a possible subspecies of *S. niger* (Lee and Hoffmeister 1963; Brown 1984). Three subspecies was recognized within *Sciurus arizonensis*, *S. a. arizonensis*, *S. a. catalinae*, and *S. a. huachuca* (Hall 1981); however, Hoffmeister (1986) considered monotypic. *S. arizonensis* ranges in Arizona, western New Mexico, and Sonora (Map 19). It can be found in sympatry, or nearly so, with *S. aberti*. *S. arizonensis* can be differentiated from *S. aberti* by having smaller ears, without a hair tuft and tail with a reddish black stripe; skull longer and flatter; auditory bullae less inflated; broader rostrum and with one molar on each side instead of two. From *S. nayaritensis*, by having the fur of the underparts whitish; limbs and feet brownish interspersed with white; rostrum narrow; distance across the maxillary-premaxillary suture less than 15.0 mm, and distance between the infraorbital canals usually equal to o less than 13.4 mm.

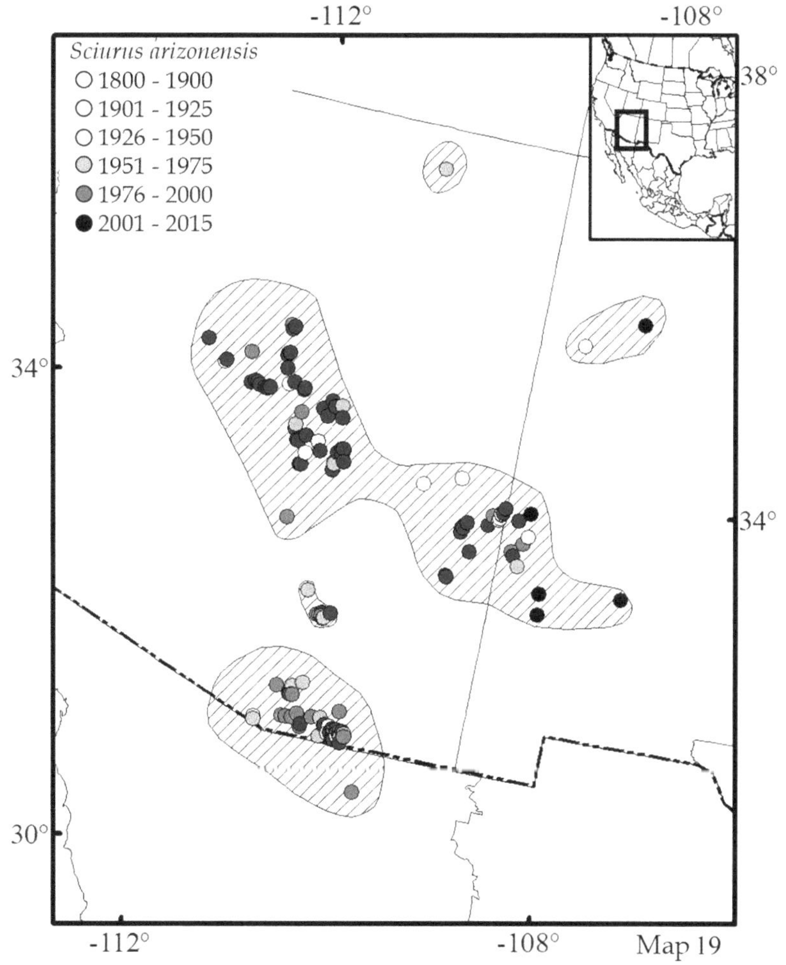


**Map 19** Distribution map of *Sciurus arizonensis*

**Additional Literature:** Allen (1915b), Lee and Hoffmeister (1963), Brown (1984), Hoffmeister and Hoffmeister (1991), Best and Riedel (1995).

## *Sciurus nayaritensis* Allen, 1890

Mexican fox squirrel, ardilla de Nayarit

1. *S. n. apache* Allen, 1893. From southern Durango north through Sonora and Chihuahua.
2. *S. n. chiricahuae* Goldman, 1933. Restricted to the Chiricahua Mountains, southeastern Arizona.
3. *S. n. nayaritensis* Allen, 1890. From southern Durango south through Jalisco.

**Conservation:** *Sciurus nayaritensis* is listed as Least Concern by the IUCN (Cassola 2016dc).

**Characteristics:** Mexican fox squirrel is medium sized; total length 530.0–575.0 mm and skull length 48.0–53.2 mm. Dorsal pelage brown grizzly with yellowish shades; **fur of underparts and limbs yellow bright to orange**; spot around the eyes and limbs ocher to yellowish white; **tail much longer than the head-and-body**; zygomatic breadth greater than 37.1 mm.

**Comments:** *Sciurus nayaritensis* ranges from the Chiricahua Mountains in southeastern Arizona south through Sierra Madre Occidental into Jalisco (Map 20). It can be found in sympatry, or nearly so, with *S. aberti*, *S. aureogaster*, and *S. colliaei*. *S. nayaritensis* can be differentiated from *S. arizonensis* by having the fur of the underparts orangish; limbs and feet brownish interspersed with reddish or ochraceous shades; rostrum wide; distance across the maxillary-premaxillary suture greater than 15.0 mm and distance between the infraorbital canals usually equal to or greater than 13.4 mm. From *S. aureogaster*, by not having spots of different colors on the nape, hips, and shoulders, and by having one upper molar on each side. From *S. aberti*, by not having a hair tuft at the tip of the ears that extends 2.0–3.0 cm beyond each ear. From *S. colliaei*, by having the fur of the underparts yellow to bright orange and two upper molars on each side.

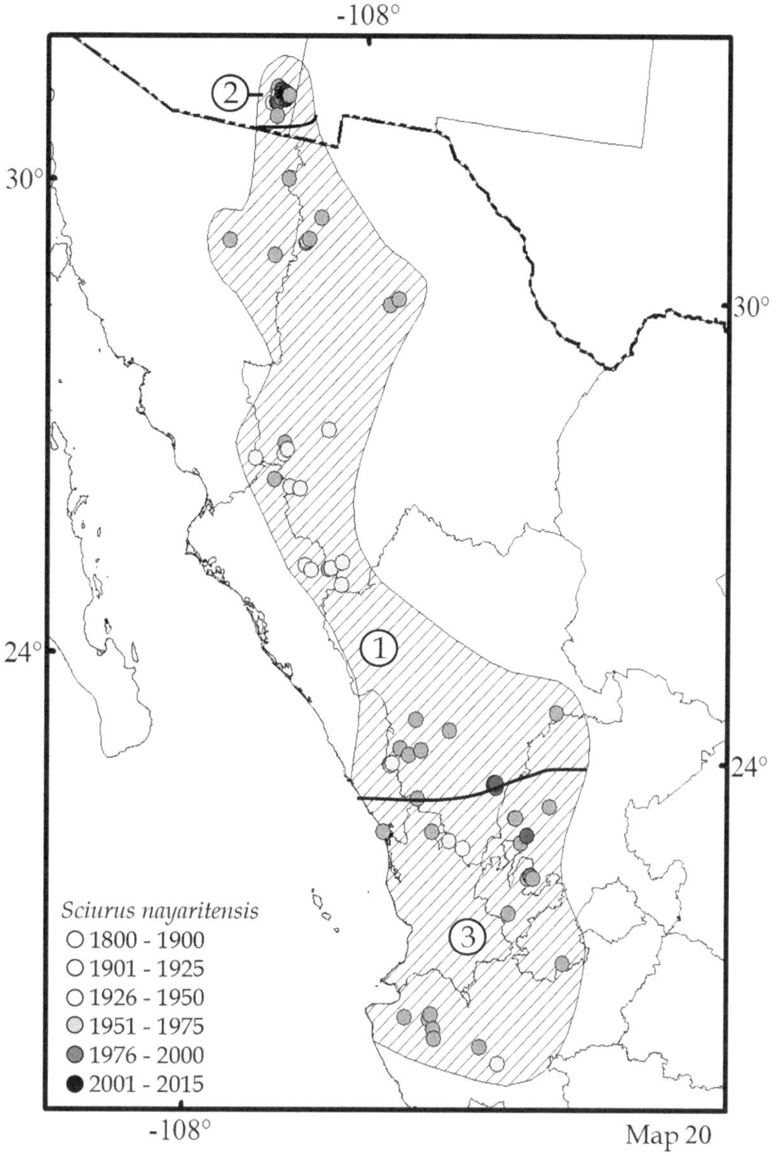


**Map 20** Distribution map of *Sciurus nayaritensis*

**Additional Literature:** Baker and Greer (1962), Lee and Hoffmeister (1963), Anderson (1972), Caire (1978); Hall (1981), Hoffmeister (1986), Matson and Baker (1986), Hoffmeister and Hoffmeister (1991), Best (1995d), Best and Riedel (1995).

## *Sciurus niger* Linnaeus, 1758

Eastern fox squirrel, ardilla zorro del este

1. *S. n. avicennia* Howell, 1919. Southern Florida.
2. *S. n. bachmani* Lowery and Davis, 1942. Mississippi and Alabama.
3. *S. n. cinereus* Linnaeus, 1758. Delmarva Peninsula, Delaware, Maryland, and Virginia.
4. *S. n. limitis* Baird, 1855. Central Texas and northern Coahuila.
5. *S. n. ludovicianus* Custis, 1806. Eastern Texas and western Louisiana.
6. *S. n. niger* Linnaeus, 1758. From North and South Carolina south through northern Florida.
7. *S. n. rufiventer* Geoffroy Saint-Hilaire, 1803. From the western Appalachian Mountains east through Oklahoma to Manitoba.

8. *S. n. shermani* Moore, 1956. Northern and central Florida.
9. *S. n. subauratus* Bachman, 1839. From central Arkansas south through Louisiana.
10. *S. n. vulpinus* Gmelin, 1788. Eastern and southern Pennsylvania, West Virginia, Maryland, and northern Virginia.

**Conservation:** *Sciurus niger* is listed as Least Concern by the IUCN (Linzey et al. 2016m).

**Characteristics:** Eastern fox squirrel is medium sized; total length 454.0–698.0 mm and skull length 58.8–67.0 mm. Dorsal pelage with three color patterns: in the north range, grizzled with a mix of buff to orange; fur of the underparts white to cinnamon, usually rufous. In the southern range, grizzled buff to gray to blackish; nose, ears, and feet white or cream; crown and nape black. In the northeastern coastal range, silvery gray washed with buff to a reddish shade on the hips, feet, and head; tail pale gray; fur of the underparts white to pale gray, sometimes cinnamon. Melanism is common in the south.

**Comments:** *Sciurus niger* ranges from Saskatchewan and Manitoba east through the Atlantic coast in New York and Pennsylvania, south through Florida, and west through Texas, Coahuila, Nuevo León, and Tamaulipas (Map 21). Introduced populations can be found in California and in sympatry, or nearly so, with *S. carolinensis;* it has been introduced into some cities in California, like *S. griseus. S. niger* can be differentiated from *S. carolinensis* by its larger size and fur of the underparts light tawny. From *S. griseus*, by having the dorsal pelage buff to orange and with one upper molar on each side.

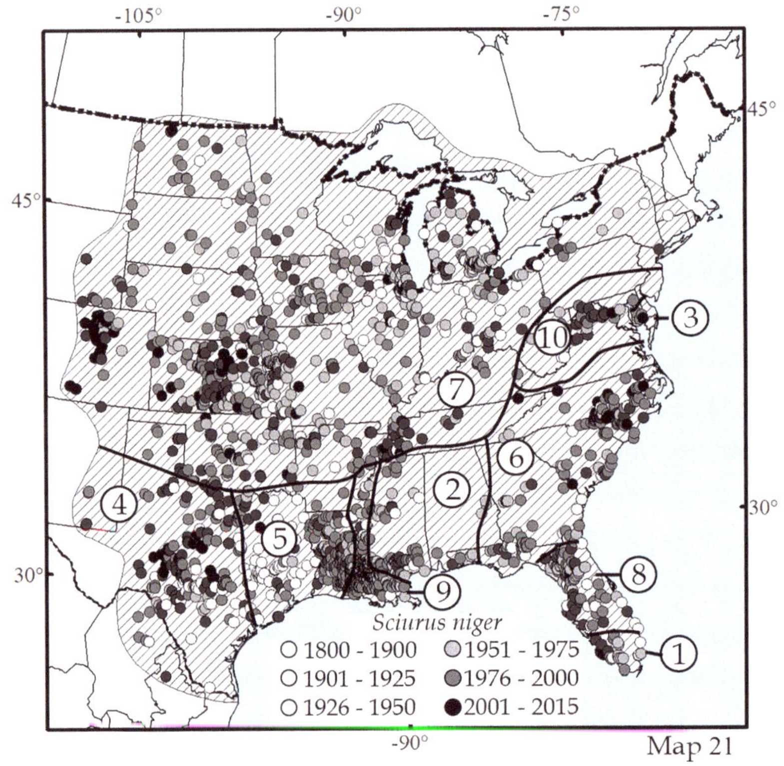


**Map 21** Distribution map of *Sciurus niger*

**Additional Literature:** Baumgartner (1943), Musser (1968), Hall (1981), Flyger and Gates (1982), Hoffmann et al. (1993), Koprowski (1994b).

## *Sciurus oculatus* Peters, 1863
Peters's squirrel, ardilla mexicana

1. *S. o. oculatus* Peters, 1863. Puebla, Hidalgo, Tlaxcala, and western Veracruz.
2. *S. o. shawi* Dalquest, 1950. Restricted to San Luis Potosí.
3. S. *o. tolucae* Nelson, 1898. Ciudad de México, Estado de México, Guanajuato, Querétaro, and Michoacán.

**Conservation:** *Sciurus oculatus* is listed as Special Protection in the Norma Oficial Mexicana (DOF 2019) and as Least Concern in the IUCN (Álvarez-Castañeda et al. 2016l).

**Characteristics:** Peters's squirrel is large sized relative to other squirrels, total length 530.0–560.0 mm and skull length 54.5–56.6 mm. Dorsal pelage grizzled in gray or with a medium dorsal stripe, or heavily interspersed with black shades; fur of underparts varying from whitish yellow to ochraceous shades; ears and spot around the eyes white to yellowish or buff; tail with grizzly black and white shades; zygomatic breadth less than 37.1 mm.

**Comments:** *Sciurus alleni* has been considered to be a subspecies of *S. oculatus* (Moore, 1960). *S. oculatus* ranges in San Luis Potosí, Hidalgo, Veracruz, Puebla, Querétaro, Michoacán, Estado de México, Ciudad de México, and Guanajuato (Map 22). It can be found in sympatry, or nearly so, with *S. aureogaster* and *S. deppei*. *S. oculatus* can be differentiated from *S. aureogaster* by not having spots of different colors on the nape, hips, and shoulders, and by having one upper molar on each side. From *S. deppei*, by its larger total size.

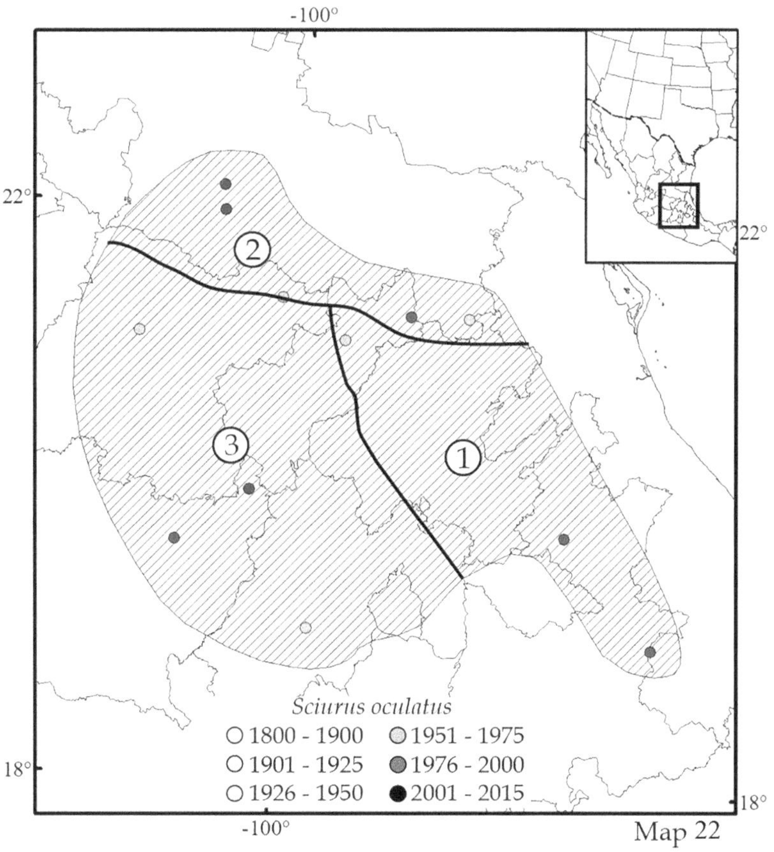


**Map 22** Distribution map of *Sciurus oculatus*

**Additional Literature:** Nelson (1899), Dalquest (1950, 1953), Baker (1956), Musser (1968), Hoffmeister and Hoffmeister (1991), Best (1995c, g).

## Genus ***Tamiasciurus*** Trouessart, 1880

Dorsal pelage varying from grayish to reddish brown, frosted with brown; limbs and head have the same coloration as the back; fur of underparts pale buff to reddish orange or white, depending on the species; **lateral stripe dark gray or charcoal generally present**; tail of the same color as the back, fringed with yellow or white-tipped hairs; ear hair tuft small, most evident in winter; anterior inner margin of the zygomatic arches at the middle of the fourth upper molariforms; **auditory bullae with three septa well distinguished externally**. The keys were elaborated based on the review of specimens and the following sources: Howell (1918), Hall (1981), Arbogast et al. (2001), Thorington et al. (2012), and Hope et al. (2016).

1. Fur of underparts with some level of rust or yellowish coloration (from British Columbia, Canada, south through northern Baja California, México) ............................................................................................*Tamiasciurus douglasii* (p. 38)

1a. Fur of underparts white or in paler coloration ..........................................................................................................2

2. Occurs from Alaska east through Newfoundland, and south through Virginia, Kentucky, and the Appalachian Mountains; in the west, south through Arizona and New Mexico, United States, including Vancouver Island, Canada ...................... .............................................................................................................................*Tamiasciurus hudsonicus* (p. 41)

2a. Only occurs in southern Utah, Colorado, Arizona, and New Mexico, United States............*Tamiasciurus fremonti* (p. 39)

### ***Tamiasciurus douglasii*** (Bachman, 1839)

Douglas' squirrel, ardilla de la costa oeste

1. *T. d. douglasii* (Bachman, 1839). Pacific coast in Washington and Oregon.
2. *T. d. mearnsi* (Townsend, 1897). Restricted to the Sierra de San Pedro Mártir, Baja California.
3. *T. d. mollipilosus* (Audubon and Bachman, 1841). From southern British Columbia south through Sierra Nevada, California.

**Conservation:** *Tamiasciurus douglasii mearnsi* (as *Tamiasciurus mearnsi*) is listed as Threatened by the Norma Oficial Mexicana (DOF 2019), and as Endangered (B1ab (iii, v): extent of occurrence less than 5000 km$^2$ in five or less populations and under continuing decline in area, extent and quality of habitat, and number of mature individuals) by the IUCN (de Grammont and Cuarón 2018a). *Tamiasciurus douglasii* is listed as Least Concern by the IUCN (Cassola 2016du).

**Characteristics:** Douglas' squirrel is small sized; total length 270.0–348.0 mm and skull length 46.4–50.5 mm. Dorsal pelage grayish to gray brown, often with a dark or chestnut midline, frosted with brown shades; in summer, it has a lateral black stripe; limbs of the same coloration as the back; **fur of underparts pale buff to reddish orange**; **lateral stripe dark gray when present**; head and tail of the same color as the back, but the tail has more black shades and is fringed with yellow or white-tipped hairs; eye rings paler cinnamon; ears with a small hair tuft, most evident in winter.

**Comments:** *Tamiasciurus douglasii mearnsi* was previously considered a distinct species (Lindsay 1981; Thorington and Hoffmann 2005; Koprowski et al. 2016a, b), but the current treatment considers it a subspecies of *T. douglasii* (Hope et al. 2016). *T. d. albolimbatus* is considered a junior subspecies of *T. d. mollipilosus. T. douglasii* ranges from British Columbia south through northern Baja California, México (Map 23), and its distribution is allopatric with respect to the other species of *Tamiasciurus*. However, *T. douglasii* can be differentiated from *T. hudsonicus* by having the fur of underparts pale buff to reddish orange. From *T. mearnsi* when it was considered a different species, it can be differentiated by having a long and fluffy tail; lateral stripe dark gray or charcoal, strongly contrasting.

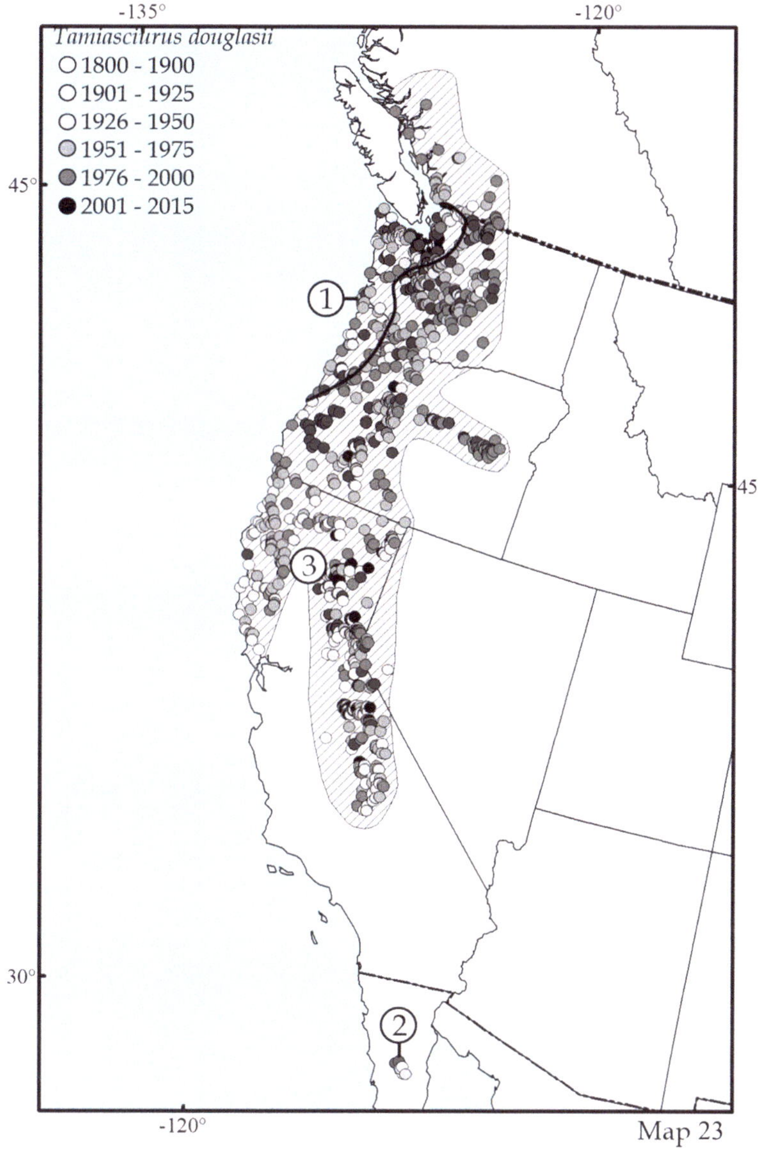


**Map 23** Distribution map of *Tamiasciurus douglasii*

**Additional Literature:** Dalquest (1948), Cowan and Guiguet (1956), Ingles (1965), Lindsay (1981, 1982), Flyger and Gates (1982), Gurnell (1987), Steele (1999), Yensen and Valdés-Alarcón (1999), Arbogast et al. (2001), Thorington et al. (2012).

## *Tamiasciurus fremonti* (Audubon and Bachman, 1853)

Fremont's squirrel, Ardilla roja de Nuevo Mexico

1. *T. f. fremonti* (Audubon and Bachman, 1853). Colorado.
2. *T. f. grahamensis* (Allen, 1894). Mount Graham, southeastern Arizona.
3. *T. f. lychnuchus* (Stone and Rehn, 1903). Central New Mexico.
4. *T. f. mogollonensis* (Mearns, 1890). Arizona and New Mexico.

**Conservation:** *Tamiasciurus hudsonicus fremonti* is listed as Endangered by the Species Act in the United States. *Tamiasciurus fremonti* (as *Tamiasciurus hudsonicus*) is not recognized as a full species by the IUCN (Cassola 2016dv).

**Characteristics:** Fremont's squirrel is small sized; total length 285.0–350.0 mm. Dorsal pelage reddish to ferruginous brown in the southern populations and grayish in the northern populations, with the dorsal reddish line faint; fur of underparts white to reddish; flanks with a line charcoal to black; eye ring usually white to buffy, similar in color to the snout and chin; tail of similar color to the back.

**Comments:** *Tamiasciurus fremonti* was reinstated as a distinct species by Hope et al. (2016); it was previously considered a subspecies of *T. hudsonicus* (Hardy 1950). In the phylogenetic study of *T. fremonti* (Arbogast et al. 2001; Hope et al. 2016), no subspecies were assigned, but using the theorical distribution of the clade, the previous subspecies *T. hudsonicus*, *T. h. fremonti*, *T. h. grahamensis*, *T. h. lychnuchus*, and *T. h. mogollonensis* should be considered subspecies of *T. fremonti* (Koprowski et al. 2016a). *T. fremonti* ranges in the southern Rockies, Sacramento Mountains, southwestern sky-islands and central Rockies in Colorado, Utah, New Mexico, and Arizona (Map 24), and has allopatric distribution with respect to the other species of *Tamiasciurus*.

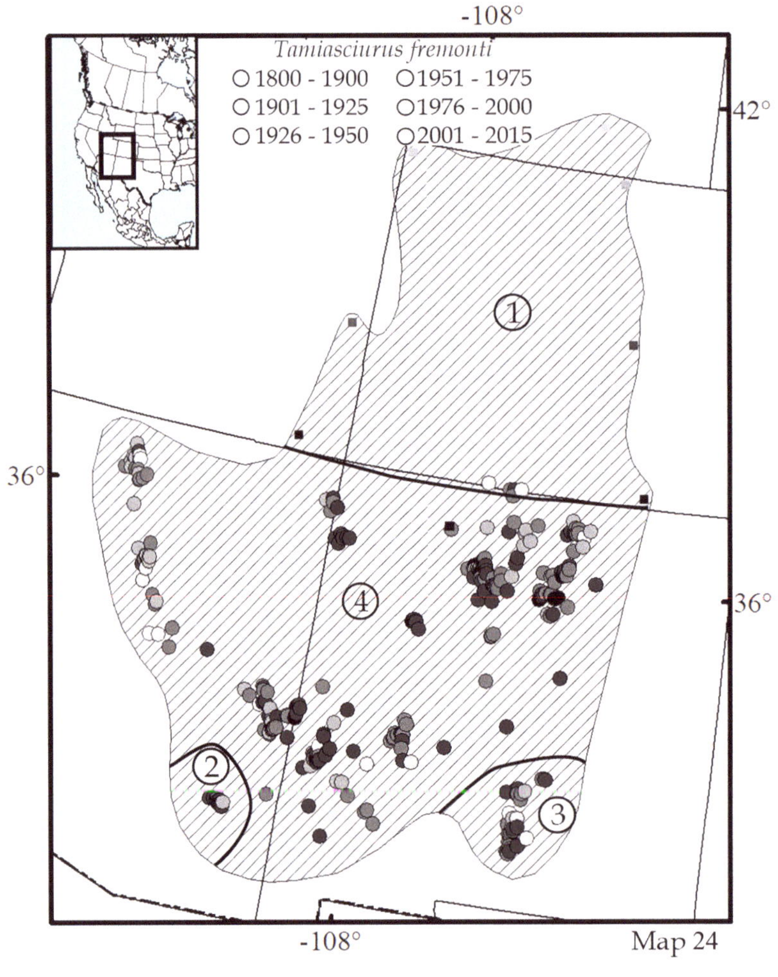


**Map 24** Distribution map of *Tamiasciurus fremonti*

**Additional Literature:** Stone and Rehn (1903).

## *Tamiasciurus hudsonicus* (Erxleben, 1777)

Red squirrel, ardilla roja

1. *T. h. abieticola* (Howell, 1929). Southern Appalachian Mountains.
2. *T. h. baileyi* (Allen, 1898). From northern-central Montana east through southern-central Wyoming.
3. *T. h. columbiensis* Howell, 1936. From southern Yukon to central British Columbia and southwestern Alberta.
4. *T. h. dakotensis* (Allen, 1894). Southeastern Montana, Wyoming, and western South Dakota.
5. *T. h. dixiensis* Hardy, 1942. Southwestern Utah.
6. *T. h. gymnicus* (Bangs, 1899). From Vermont, New Hampshire, and Maine north through Saint Lawrence River and Nova Scotia.
7. *T. h. hudsonicus* (Erxleben, 1777). Manitoba, Ontario, northern Minnesota, and northern Michigan.
8. *T. h. kenaiensis* Howell, 1936. Southeastern Alaska.
9. *T. h. lanuginosus* (Bachman, 1839). Vancouver Island and southwestern British Columbia.
10. *T. h. laurentianus* Anderson, 1942. Southeastern Québec, mostly north of the mouth of the Saint Lawrence River.
11. *T. h. loquax* (Bangs, 1896). From Michigan east through New York, southern New England, Ontario, and from Québec south through Virginia.
12. *T. h. minnesota* (Allen, 1899). From eastern North Dakota east through northern Iowa.
13. *T. h. pallescens* Howell, 1942. From North Dakota and Saskatchewan east through Manitoba.
14. *T. h. petulans* (Osgood, 1900). Coastal area of southeastern Alaska, Yukon, and northwestern British Columbia.
15. *T. h. picatus* (Swarth, 1921). Coastal area and islands of central British Columbia.
16. *T. h. preblei* Howell, 1936. South of the tundra in Alaska, northern Canada, Alberta, and Saskatchewan.
17. *T. h. regalis* Howell, 1936. Isle Royale in Lake Superior, northern Michigan.
18. *T. h. richardsoni* (Bachman, 1839). Southeastern British Columbia, western Oregon, central Idaho, and western Montana.
19. *T. h. streatori* (Allen, 1898). Southern British Columbia, eastern Washington, and northern Idaho.
20. *T. h. ungavensis* Anderson, 1942. Northern Québec and the Labrador Peninsula.
21. *T. h. ventorum* (Allen, 1898). From southern-central Montana south through Utah.

**Conservation:** *Tamiasciurus hudsonicus grahamensis* (as *T. fremonti*) is listed as Endangered by the Species Act in the United States. *Tamiasciurus hudsonicus* is listed as Least Concern by the IUCN (Cassola 2016dv).

**Characteristics:** Red squirrel is small sized relative to the other species of the genus; total length 270.0–385.0 mm and skull length 43.5–52.0 mm. Dorsal pelage reddish, ferruginous brown, or grayish, frosted with pale reddish shades; limbs and feet reddish brown; **fur of underparts white**; **lateral stripe with charcoal borders;** head and tail of the same color as the back, but tail blackish with blackish to yellowish or rusty shades at the tip; eye ring white.

**Comments:** *Tamiasciurus hudsonicus* ranges from Alaska east through Newfoundland and south through Virginia, Kentucky, and the Appalachian Mountains; and in the west, south through Arizona and New Mexico, including Vancouver Island (Map 25). Its distribution is allopatric with respect to the other species of *Tamiasciurus*. However, *T. hudsonicus* can be differentiated from *T. douglasii* by the fur of underparts white or in paler coloration.

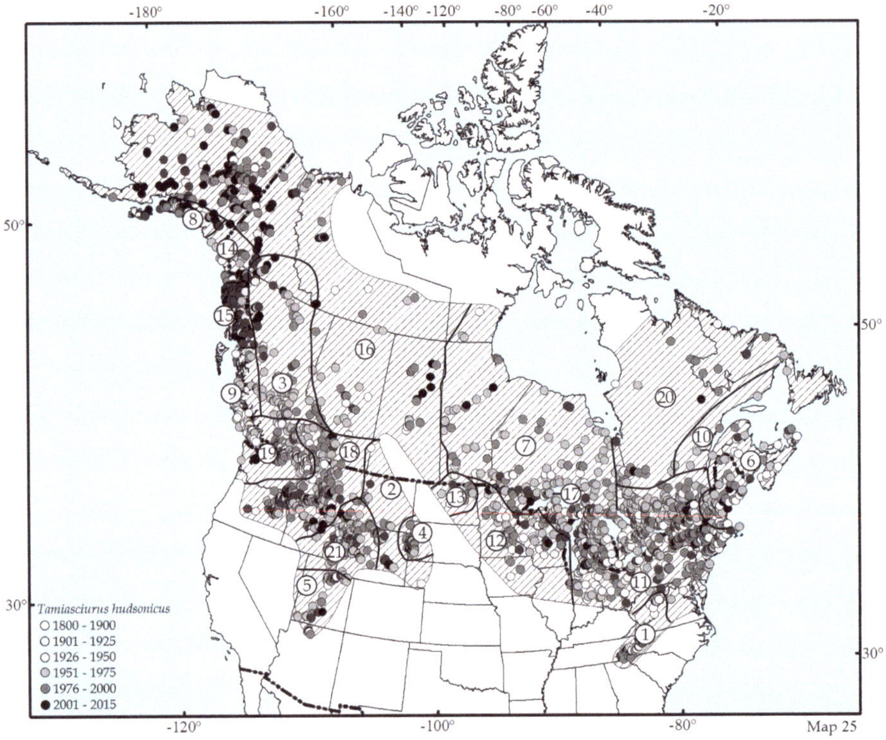


**Map 25** Distribution map of *Tamiasciurus hudsonicus*

**Additional Literature:** Cowan and Guiguet (1956), Ingles (1965), Hall (1981), Lindsay (1981), Gurnell (1987), Steele (1998, 1999), Arbogast et al. (2001), Hope et al. (2016), Koprowski et al. (2016a).

## **Subfamily Xerinae** Osborn, 1910

Xerinae is a subfamily that comprises many terrestrial squirrels. It includes three tribes: Marmotini, Xerini, and Protoxerini. The keys were elaborated based on the review of specimens and the following sources: Howell (1918), Hall (1981), Helgen et al. (2009), and Thorington et al. (2012).

1. Dorsal pelage at most with two pale bands or dots, but not as previously described; with an antero-infraorbital channel (Fig. 9)........................................tribe Marmotini (p. 43)

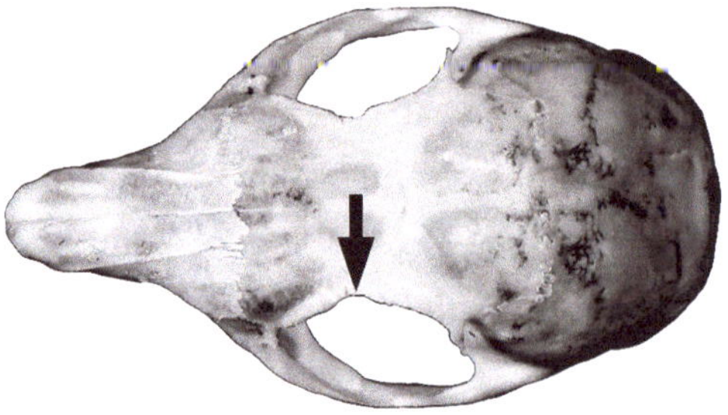

**Fig. 9** The antero-infraorbital channel is present

1a. Dorsal pelage with five dark bands equally spaced and subequal in breadth; two lateral bands shorter and diffuse; no antero-infraorbital channel (Fig. 10)........................................tribe Tamiini (p. 100)

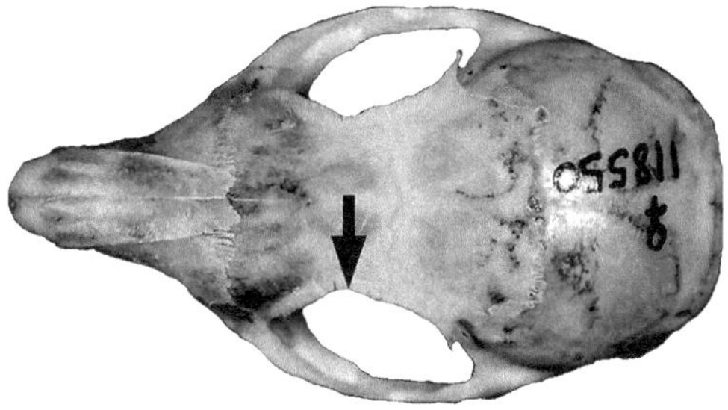

**Fig. 10** The antero-infraorbital channel is absent

## Tribe Marmotini

1. Tail length less than 30% of the total length; body robust; first upper molariforms wider than the upper incisors; upper toothrow converging on the back, considering the lingual edge............................................................*Cynomys* (p. 55)

1a. Tail length greater than 30% of the total length; body elongated; first upper molariforms thinner than the upper incisors; upper toothrow not converging on the back, considering the lingual edge......................................................................2

2. Two well marked white stripes at the flanks, one on each side, not defined by a blackish line; auditory bullae large, its length one-half longer than the upper toothrow length; first upper premolars very small................................................ ......................................................................................................................................... *Ammospermophilus* (p. 44)

2a. Different types of dorsal marks, but never two well marked white stripes on the flanks; if present, they are defined by a blackish line; auditory bullae of normal size, its length less or slightly greater than the upper toothrow length; first upper premolars well developed and peg shaped..........................................................................................................................3

3. Tail relatively long, greater than 70% of the head-and-body length..................................................................................4

3a. Tail relatively short, less than 70% of the head-and-body length.......................................................................................5

4. Six nipples; ear length less than 18.0 mm; supraorbital foramen closed (Fig. 11)............................*Notocitellus* (p. 74)

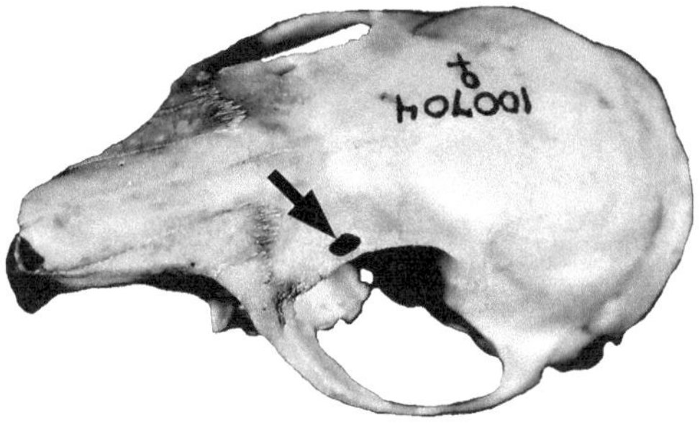

**Fig. 11** Supraorbital foramen closed as in *Notocitellus*

4a. More than six nipples; ear length greater than 18.0 mm; supraorbital foramens open..............*Otospermophilus* (part, p. 76)

5. Total length greater than 370.0 mm; tail length greater than 150.0 mm............................ *Otospermophilus* (part, p. 76)

5a. Total length less than 370.0 mm; tail length less than 150.0 mm.......................................................................................6

6. Ear length greater than 15.0 mm; with pale and dark dorsal stripes greater than 8.0 mm in width .................................. .........................................................................................................................................*Callospermophilus* (p. 49)

6a. Ear length less than 15.0 mm; no stripes, but can have them spotted or mottled ...........................................................7

7. Dorsal pelage with pale spots in a series of longitudinal lines ............................................................ *Ictidomys* (p. 61)

7a. Dorsal pelage with or without pale spots; when present, they are not in series of longitudinal lines ............................ 8

8. Dorsal pelage salt and pepper; first upper molars bearing two cups and a functional cutting edge (from Alberta, Saskatchewan, Manitoba, and Ontario eastward to Illinois, Indiana, Wisconsin and south through Kansas, Missouri, Illinois and Indiana, United States)........................................................................... *Poliocitellus franklinii* (p. 81)

8a. Dorsal pelage not salt and pepper; first upper molars simple ........................................................................... 9

9. Limbs proportional to the body; ears and tail not relatively short, either spotted or mottled; upper first and second molar joining protocones with an abrupt change of direction.................................................................. *Urocitellus* (p. 82)

9a. Limbs proportionally shortrelative to the body; ears and tail relatively short, not spotted or mottled; upper first and second molar joining protocones without an abrupt change of direction ............................................ *Xerospermophilus* (p. 95)

## Genus *Ammospermophilus* Merriam, 1892

Dorsal color variations brownish to grayish with grizzly appearance; fur of underparts pale cream to gray; **flanks with a white stripe from the shoulders to the hips**; no facial lines on the head; hair bushy with a combination of white, gray, and black hairs; white spot around the eyes; interorbital region narrower than the postorbital constriction; skull arch shaped in lateral view; molariforms with a short crown; **auditory bullae large, its length 1.5-fold the upper toothrow length; first upper premolars very small**. The keys were elaborated based on the review of specimens and the following sources: Hall (1981), Thorington et al. (2012), and Álvarez-Castañeda et al. (2017a).

1. Tail grizzly with white gray ventrally and blackish or in brown shades dorsally; upper toothrow length greater than 7.0 mm ............................................................................................................................. 2

1a. Tail ventrally white, at least in its proximal half, always in gray shades; upper toothrow length less than 7.0 mm ........ 3

2. Tail grizzly with white, gray, and blackish hair ventrally, body in gray shades (restricted to Arizona, New Mexico, United States, and Sonora, México) ................................................................... *Ammospermophilus harrisii* (p. 44).

2a. Tail grizzly with white, brown, and blackish hair ventrally, body in brown shades (restricted to San Joaquin Valley, California, United States)...................................................................................*Ammospermophilus nelsoni* (p. 48)

3. Tail hair white with two black rings at the tip; rostrum wide; braincase profile straight (from New Mexico, Texas, Chihuahua, Coahuila, and Durango)..............................................................*Ammospermophilus interpres* (p. 45)

3a. Tail hair white with one black ring at the tip; rostrum thin; braincase profile arched (from southeastern Oregon southward to the Baja California Peninsula and westward to Colorado and New Mexico, United States) ............................... ..........................................................................................................*Ammospermophilus leucurus* (p. 46)

### *Ammospermophilus harrisii* (Audubon and Bachman, 1854)

Harris's antelope squirrel, ardilla listada sonorense

1. *A. h. harrisii* (Audubon and Bachman, 1854). Along the east bank of the Colorado River in Arizona southward to the coastal area of Sonora.
2. *A. h. saxicola* (Mearns, 1896). Central and eastern Arizona, Sonora, and New Mexico.

**Conservation:** *Ammospermophilus harrisii* is listed as Least Concern by the IUCN (Timm et al. 2016a).

**Characteristics:** Harris's antelope squirrel is small sized; total length 216.0–267.0 mm and skull length 37.6–41.9 mm. Dorsal pelage gray, suffused with reddish to yellowish brown shades near the head and along the fore and hind limbs; fur of underparts white; **tail ventral half with a combination of white, gray, and black hairs, without rings**; auditory bullae not corrugated, without the presence of two transverse constrictions, smaller, and less inflated.

**Comments:** *Ammospermophilus harrisii* is restricted to Arizona, New Mexico, and Sonora (Map 26), and with allopatric distribution with respect to the other species of *Ammospermophilus*. However, *A. harrisii* can be differentiated from *A. interpres* and *A. leucurus* by its tail ventral half grizzly with white gray and blackish hair, without rings; upper toothrow length greater than 7.0 mm. From *A. nelsoni*, by its smaller size; zygomatic arches narrower and not widely spreading; auditory bullae smaller, less inflated, and not corrugated, without two transverse constrictions.

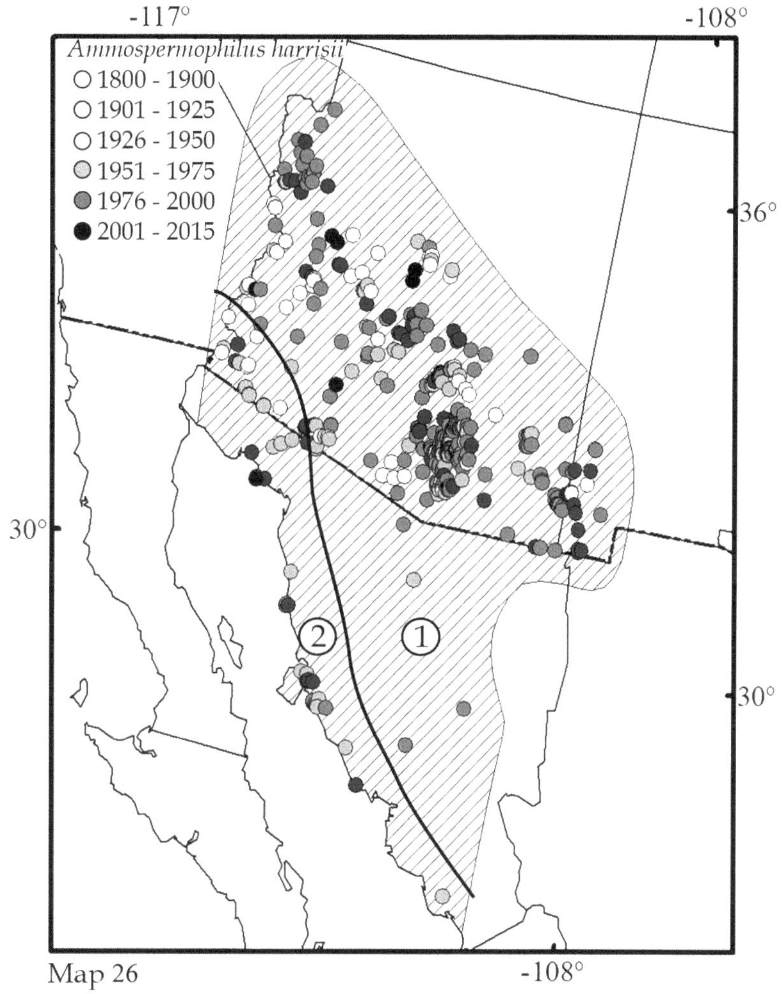


**Map 26** Distribution map of *Ammospermophilus harrisii*

**Additional Literature:** Merriam (1893), Howell (1938), Findley et al. (1975), Caire (1978), Hall (1981), Hafner (1984), Hoffmeister (1986), Best et al. (1990a, b), Harrison et al. (2003), Thorington et al. (2012), Koprowski et al. (2017).

## *Ammospermophilus interpres* (Merriam, 1890)

Texas Antelope squirrel, ardilla listada texana

Monotypic.

**Conservation:** *Ammospermophilus interpres* is listed as Least Concern by the IUCN (Timm et al. 2016b).

**Characteristics:** Texas antelope squirrel is small sized; total length 220.0–235.0 mm and skull length 37.7–40.6 mm. Dorsal pelage gray suffused with brown near the head and along the fore and hind limbs, grading to gray through the tail; fur of underparts white; **tail with two black rings at the tip;** skull relatively broad and short, with a robust rostrum and shallow braincase.

**Comments:** *Ammospermophilus interpres* is found in the Chihuahuan desert at New Mexico, Texas, Chihuahua, Coahuila, and Durango (Map 27), and has allopatric distribution with respect to the other species of *Ammospermophilus*. However, *A. interpres* can be differentiated from other species of *Ammospermophilus* by having the tail hair white with two black rings at the tip. From *A. harrisii*, by the upper toothrow length less than 7.0 mm, and from *A. leucurus*, by its slightly smaller size with a coloration uniformly darker; heavier rostrum, and a more flattened braincase.

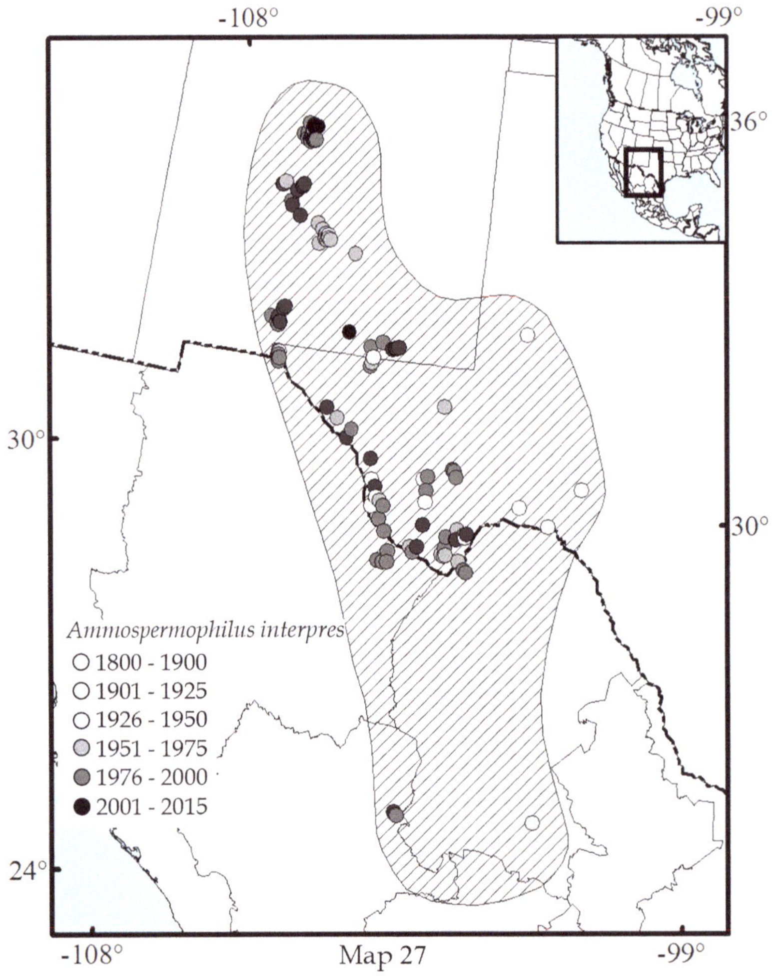


**Map 27** Distribution map of *Ammospermophilus interpres*

**Additional Literature:** Merriam (1893), Howell (1938), Baker (1956), Anderson (1972), Schmidly (1977), Hafner (1984), Best et al. (1990a, b), Harrison et al. (2003), Thorington et al. (2012), Koprowski et al. (2017).

## *Ammospermophilus leucurus* (Merriam, 1889)

White-tailed antelope squirrel, juancito o ardilla listada de cola blanca

1. *A. l. canfieldiae* Huey, 1929. Central areas of the Baja California Peninsula.
2. *A. l. cinnamomeus* (Merriam, 1890). Northern Arizona, southeastern Utah, and western Colorado.
3. *A. l. escalante* (Hansen, 1955). Southwestern Utah.
4. *A. l. extimus* Nelson and Goldman, 1929. Southern Baja California Peninsula.
5. *A. l. insularis* Nelson and Goldman, 1909. Espíritu Santo Island, Baja California Sur.
6. *A. l. leucurus* (Merriam, 1889). From the northern Baja California Peninsula north through Oregon.
7. *A. l. notom* (Hansen, 1955). Northeastern Utah.
8. *A. l. peninsulae* (Allen, 1893). Northwestern Baja California Peninsula.

9. *A. l. pennipes* Howell, 1931. Western Colorado and New Mexico.
10. *A. l. tersus* Goldman, 1929. Northwestern Arizona.

**Conservation:** *Ammospermophilus leucurus insularis* is listed (as *A. insularis*) as Threatened in the Norma Oficial Mexicana (DOF 2019). *A. leucurus* is unlisted by the United States Endangered Species Act (FWS 2022) and as Least Concern by the IUCN (Linzey et al. 2016b).

**Characteristics:** White-tailed antelope squirrel is small sized; total length 194.0–239.0 mm and skull length 36.7–41.5 mm. Dorsal pelage gray to slate gray on the back and head; fur of underparts whitish; **tail with one black ring at the tip**; auditory bullae not corrugated, without the presence of two transverse constrictions, smaller, and less inflated.

**Comments:** *Ammospermophilus leucurus insularis* is considered a junior synonym of *A. l. peninsulae* (Álvarez-Castañeda et al. 2017a); however, Mantooth et al. (2013) considered it a full species, including all the southern Baja California Peninsula, from the Vizcaíno Desert to the south, but was not accepted (Koprowski et al. 2017). *A. leucurus* ranges from southeastern Oregon southward to the Baja California Peninsula and west through Colorado and New Mexico, including the San Marcos and Espíritu Santo islands, Gulf of California, Baja California Sur (Map 28), and has allopatric distribution with respect to the other species of *Ammospermophilus*; however, *A. leucurus* can be differentiated from other species of *Ammospermophilus* by having one black ring on the tip of the tail; rostrum thin and braincase profile arched. From *A. leucurus*, by its slightly larger size with a paler reddish coloration; skull with a thin rostrum and braincase less flattened. From *A. nelsoni*, by its smaller size, narrower zygomatic arches and not widely spreading; auditory bullae smaller, less inflated, and not corrugated, without two transverse constrictions.

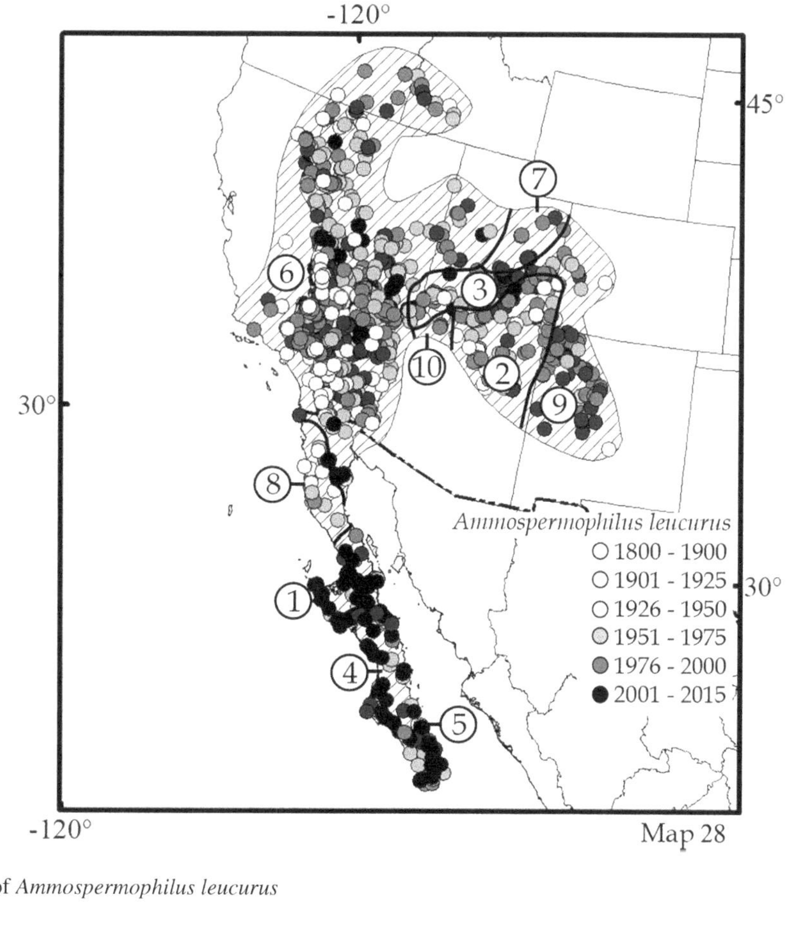


**Map 28** Distribution map of *Ammospermophilus leucurus*

**Additional Literature:** Merriam (1893), Nelson and Goldman (1909), Howell (1938), Bryant (1945), Hall (1946), Best et al. (1990a, b), Belk and Smith (1991), Harrison et al. (2003), Whorley et al. (2004), Thorington et al. (2012).

## ***Ammospermophilus nelsoni*** (Merriam, 1893)

Nelson's Antelope squirrel, ardilla listada del Valle de San Joaquín

Monotypic.

**Conservation:** *Ammospermophilus nelsoni* is listed as under protection by the United States Endangered Species Act (FWS 2022) and as Endangered (B2ab (ii, iii): area of occupancy less than 500 km$^2$ in five or less populations and under continuing decline in area of occupancy and extent and/or quality of habitat) by the IUCN (Koprowski 2017).

**Characteristics:** Nelson's antelope squirrel is small sized; total length 234.0–267.0 mm and skull length 41.5–43.1 mm. Dorsal pelage buff to tan; back suffuse with yellow shades; fur of underparts white to cream; **tail buff gray dorsally and creamy white ventrally**; **auditory bullae distinctly corrugated, with two transverse constrictions**, conspicuously larger, and more inflated.

**Comments:** *Ammospermophilus nelsoni* is restricted to the San Joaquin Valley of southern California (Map 29), and have allopatric distribution with respect to the other species of *Ammospermophilus*. However, *A. nelsoni* can be differentiated from other species of *Ammospermophilus* by having the hair in brown coloration. From *A. leucurus* and *A. interpres*, by the upper toothrow length greater than 7.0 mm. From *A. leucurus* and *A. harrisii*, by its larger size; zygomatic arches widely spread; auditory bullae conspicuously larger, more inflated, and distinctly corrugated, with two transverse constrictions.

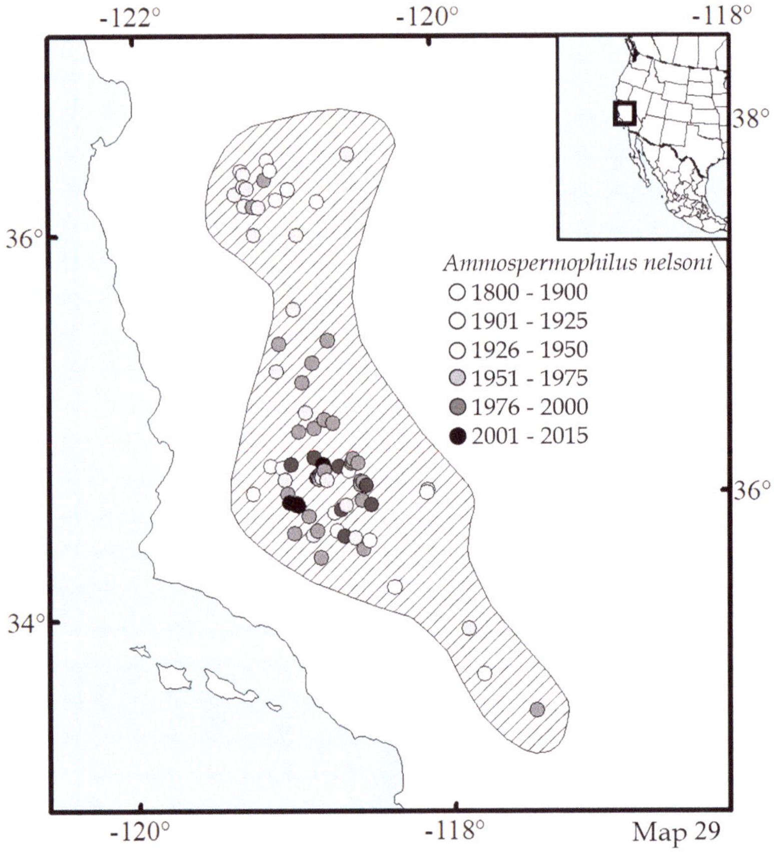


**Map 29** Distribution map of *Ammospermophilus nelsoni*

**Additional Literature:** Merriam (1893), Grinnell and Dixon (1918), Grinnell (1933), Howell (1938), Hall (1981), Hafner (1984), Best et al. (1990c), Harrison et al. (2003), Thorington et al. (2012), Koprowski et al. (2017).

# Genus *Callospermophilus* Merriam, 1897

**Dorsal pelage has six stripes; two dark stripes on the each flank with a paler one in the middle**; the darker stripes blackish with intraspecific variation; the paler stripes buff cream to pale reddish brown; central portion of the back, rumps, hips, and upperparts of the legs brown and frosted with blackish brown shades, and buff because the hair has four brown-blackish-buff-blackish bands, so the hair length gives the dorsal pattern; fur of underparts yellowish buff, including the neck and tail, whitish in the groin and neck; neck flanks and back of the face reddish cinnamon; **eyes with two paler stripes, one above and the other below**; snout reddish brown; tail blackish a mixture of buff hairs above and mainly reddish buff below. *Callospermophilus* was a subgenus of *Spermophilus*, but genetic and morphometrical analyses highlight that it should be considered at the genus level (Helgen et al. 2009). The keys were elaborated based on the review of specimens and the following sources: Hall (1981), Helgen et al. (2009), Thorington et al. (2012), and Álvarez-Castañeda et al. (2017a).

1. Dorsal stripes with less contrast relative to the general coloration; black stripes more a shadow than a deep-colored stripe; skull narrow (restricted to the Chihuahua and Durango highlands, México) ........................................................................ *Callospermophilus madrensis* (p. 51)

1a. Dorsal stripes contrasting with the general coloration; black stripes present; skull wide ........................................ 2

2. Gray stripes paler and with a black stripe at each flank, normally well developed; nasal bones not reduced in width at the symphysis with the frontal bones (Fig. 12; from central British Columbia south through California, Nevada, Arizona, and New Mexico, and west through Wyoming and Colorado, United States)........... *Callospermophilus lateralis* (p. 50)

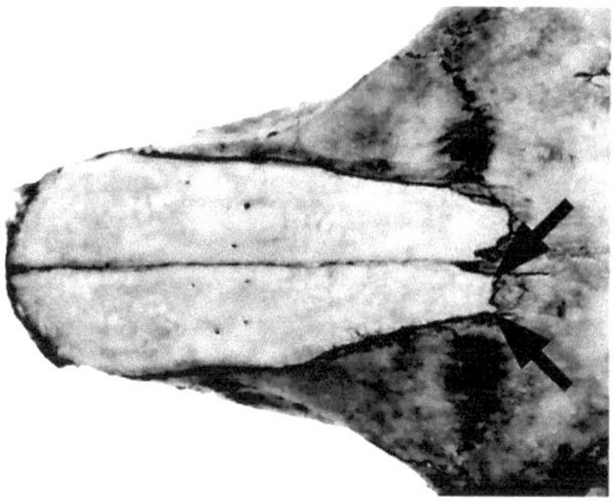

**Fig. 12** Nasal bones not reduced in width at the symphysis with the frontal bones like in *Callospermophilus lateralis*

2a. Gray stripes shadowed and only the outer stripe black; the inner stripe absent or poorly developed; nasal bones reduced in width at the symphysis with the frontal bones (Fig. 13; restricted to the Cascade Mountain in British Columbia and Washington, United States)........................................................................ *Callospermophilus saturatus* (p. 52)

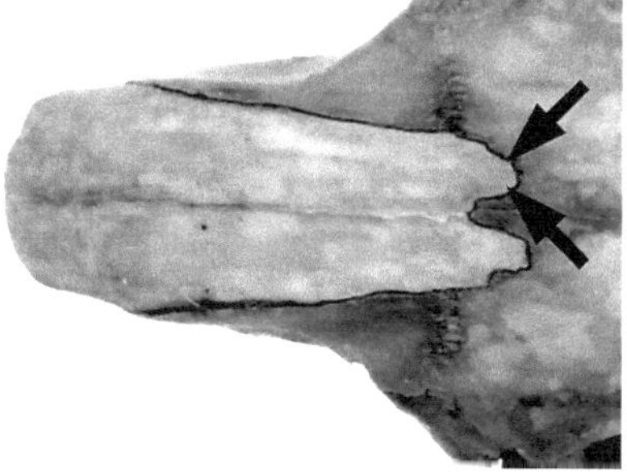

**Fig. 13** Nasal bones reduced in width at the symphysis with the frontal bones like in *Callospermophilus saturatus*

## ***Callospermophilus lateralis*** (Say, 1923)

Golden-mantled ground squirrel, ardilla de manto dorada

1. *C. l. arizonensis* Bailey, 1913. San Francisco Mountains, Arizona, and New Mexico.
2. *C. l. bernardinus* (Merriam, 1898). San Bernardino Mountains, central-southern California.
3. *C. l. castanurus* (Merriam, 1890). From the Wasatch Mountains, central Utah north through Wyoming.
4. *C. l. certus* Goldman, 1921. Charleston Mountains, southern Nevada.
5. *C. l. chrysodeirus* (Merriam, 1890). From the Columbia River south through Yosemite, central-eastern California.
6. *C. l. cinerascens* (Merriam, 1890). From Montana to the Yellowstone National Park, central-eastern California.
7. *C. l. connectens* Howell, 1931. Restricted to Homestead, northeastern Oregon.
8. *C. l. lateralis* (Say, 1923). Mountain slopes from Wyoming south through Arizona and New Mexico.
9. *C. l. mitratus* Howell, 1931. South Yolla Bolly Mountain, northern California.
10. *C. l. tescorum* Hollister, 1911. Idaho, Montana, Alberta, and British Columbia.
11. *C. l. trepidus* Taylor, 1910. Idaho, Nevada, Oregon, and Utah.
12. *C. l. trinitatus* Merriam, 1901. Trinity Mountains, east of Hoopa Valley, northern California.
13. *C. l. wortmani* (Allen, 1895). From Sweetwater County, Wyoming, south through the northwestern corner of Colorado.

**Conservation:** *Callospermophilus lateralis* is listed as Least Concern by the IUCN (Cassola 2016e).

**Characteristics:** Golden-mantled ground squirrel is medium sized; total length 235.0–295.0 mm and skull length 39.6–45.6 mm. Dorsal pelage grizzled gray brown to charcoal on the back, suffused with buff to ocher shades; **only the outer stripe black; the inner stripe absent or poorly developed; white stripes shorter**; **top, head, flanks, and shoulders with a less contrasting golden mantle; head larger**; **nasal bones not reduced in width at the symphysis with the frontal bones**.

**Comments:** *Callospermophilus lateralis* ranges from central British Columbia south through California, Nevada, Arizona, and New Mexico, and west through Wyoming and Colorado (Map 30). It is found in the Rocky Mountains with fragmented distribution in its southern ranges, and have allopatric distribution with respect to the other species of *Callospermophilus;* however, *C. lateralis* and *C. saturatus* ranges in British Columbia and Washington. *C. lateralis* can be differentiated from the species of *Neotamias* by the absence of the medial black stripe, and a smaller and less bushy tail. From *C. madrensis*, by having the mantle more pronounced and the dorsal stripes more prominent, with shorter white stripes, the black stripe longer, and the tail longer; and from *C. saturatus*, by its distinctive golden mantle with only the outer stripe black, the inner stripe absent or poorly developed, and a larger head; nasal bones reduced in width at the symphysis with the frontal bone.

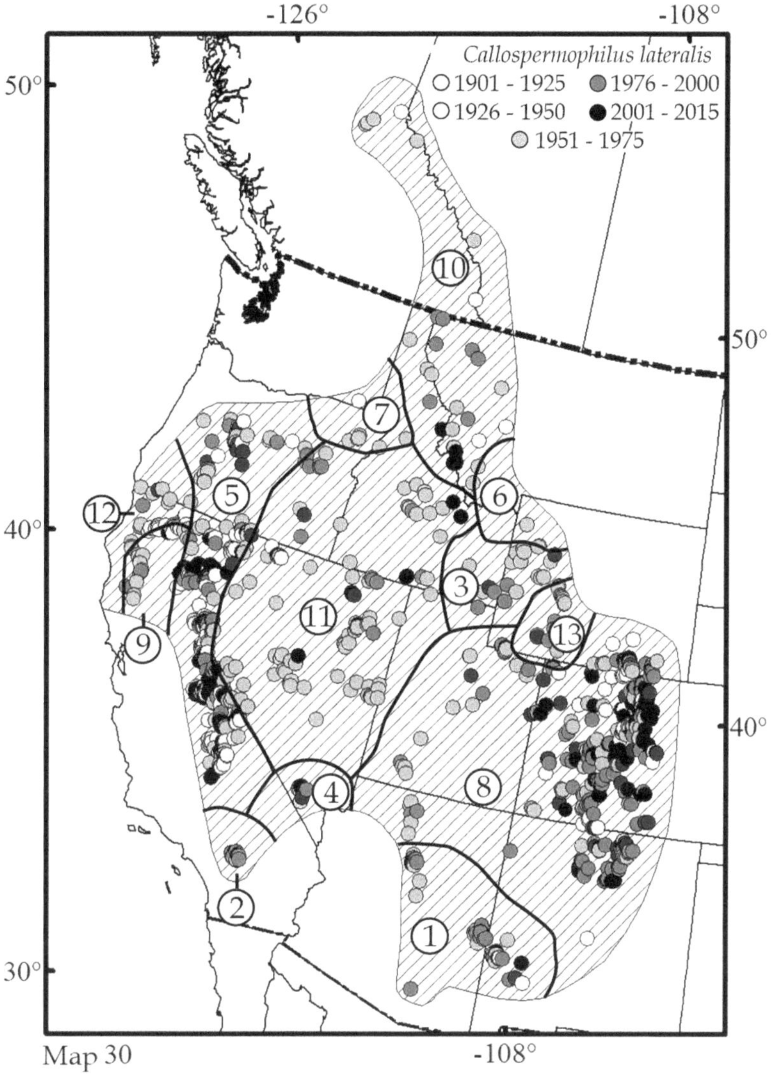


**Map 30** Distribution map of *Callospermophilus lateralis*

**Additional Literature:** Howell (1938), Gordon (1943), Cowan and Guiguet (1965), MacClintock (1970), Hall (1981), Hoffmeister (1986), Rickart and Yensen (1991), Bartels and Thompson (1993), Thorington et al. (2012), Koprowski et al. (2017).

## *Callospermophilus madrensis* Merriam, 1901

Sierra Madre ground squirrel, ardilla de manto de la Sierra Madre

Monotypic.

**Conservation:** *Callospermophilus madrensis* is listed as subject to Special Protection in the Norma Oficial Mexicana (DOF 2019) and as Near Threatened by the IUCN (Álvarez-Castañeda et al. 2016ad).

**Characteristics:** Sierra Madre ground squirrel is medium sized, the smaller member within the genus; **total length 215.0–243.0 mm** and skull length 44.1–44.4 mm. Dorsal pelage brown interspersed with gray h; **dorsal stripes less contrasting relative to the general coloration, black stripes more a shadow than a deep-colored stripe, and white stripes longer**; head crown with brownish orange hair; fur of underparts of the same shade; head brown without a fringe, except for two narrow creamy stripes that outline the eyes; **tail shorter,** with long hair and interspersed with blackish hair.

**Comments:** *Callospermophilus madrensis* has a restricted ranges to the Sierra Madre of Chihuahua and Durango (Map 31), and has allopatric distribution with respect to the other species of *Callospermophilus*. *C. madrensis* can be differentiated from the species of *Neotamias* by the absence of the median black stripe and tail smaller and less bushy. From the other species of *Callospermophilus* by its dorsal stripes less contrasting in relation to the general coloration; black stripes more as a shadow than a deep color stripe and longer white stripes; smaller body size and tail shorter; skull narrower, and braincase more arched than in *C. lateralis*.

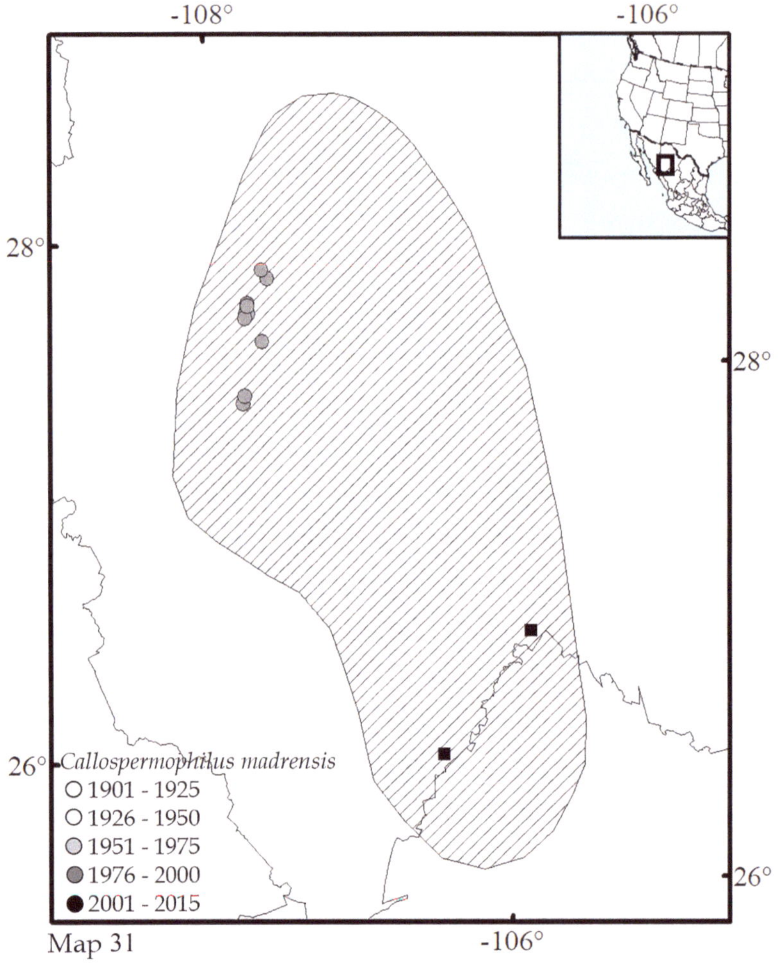


**Map 31** Distribution map of *Callospermophilus madrensis*

**Additional Literature:** Howell (1938), Baker and Greer (1962), Anderson (1972), Best and Thomas (1991b), Rickart and Yensen (1991), Bartels and Thompson (1993), Thorington et al. (2012), Koprowski et al. (2017).

## ***Callospermophilus saturatus*** (Rhoads, 1895)

Cascade golden-mantled ground squirrel, ardilla de manto de las montañas Cascadas

Monotypic.

**Conservation:** *Callospermophilus saturatus* is listed as Least Concern by the IUCN (Cassola 2016f).

**Characteristics:** Cascade golden-mantled ground squirrel is medium sized; total length 286.0–315.0 mm and skull length 43.5–48.4 mm. Dorsal pelage grizzled gray brown to charcoal on the back, suffused from buff to ocher; **with an outer and an inner well marked black stripes;** fur of underparts, tail and feet buff; a **white stripe bordered dorsally and poorly defined black stripes ventrally;** shoulders, top and flanks of the head **less contrastingly golden mantle; head smaller; ears tawny;** skull narrower; braincase more arched than in *C. lateralis*.

**Comments:** *Callospermophilus saturatus* was placed as a subspecies of *C. lateralis* (Elliot 1901, 1905; Miller 1912a). *C. saturatus* is restricted to the Cascade Mountains of British Columbia and Washington (Map 32), and has allopatric distribution with respect to the other species of *Callospermophilus*; however, *C. lateralis* and *C. saturatus* range in British Columbia and Washington. *C. saturatus* can be differentiated from the species of *Neotamias* by the absence of the medial black stripe, and with a smaller and less bushy tail. From *C. madrensis*, by its mantle more pronounced and dorsal stripes more prominent, with shorter white stripes. From *C. lateralis*, by a less contrasting golden mantle, with an outer and an inner well marked black stripes; head smaller; nasal bones not reduced in width at the symphysis with the frontal bones.

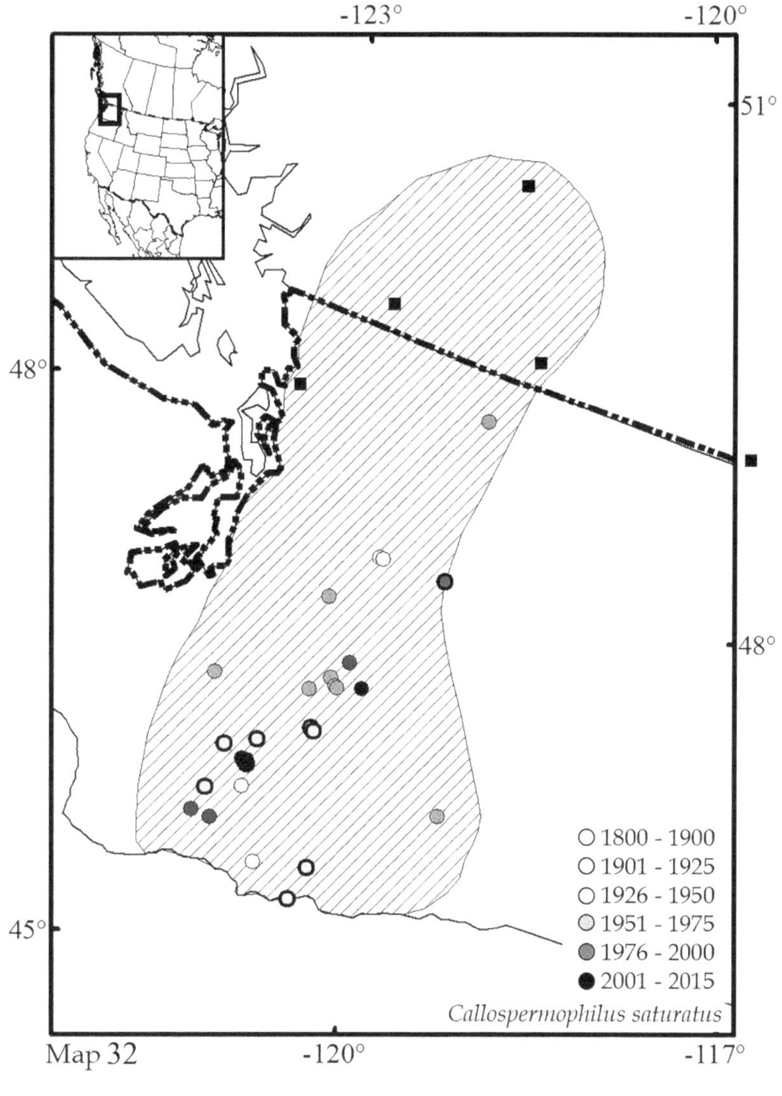


**Map 32** Distribution map of *Callospermophilus saturatus*

**Additional Literature:** Howell (1938), Hall (1981), Stephen (1988), Trombulak (1988), Rickart and Yensen (1991), Thorington et al. (2012), Koprowski et al. (2017).

## Genus *Cynomys* Rafinesque, 1817

Dorsal pelage pale brown to grayish with grizzly blackish shades; fur of underparts of paler coloration; **tail length less than one-half of the head-and-body length; skull massive; sagittal crest well developed in the posterior area of the brain-case and slightly narrower than in anteriorly; zygomatic arches robust;** third upper molars with one additional series of ridges than the other upper molars; molariforms elliptical; nasals wide and rounded at the anterior edge; jugal plate triangular underdeveloped. *Cynomys* has two subgenera: *Cynomys* [*C. mexicanus* and *C. ludovicianus*] and *Leucocrossuromys* [*C. gunnisoni*, *C. parvidens*, and *C. leucurus*]. The keys were elaborated based on the review of specimens and the following sources: Howell (1918), Hall (1981), and Thorington et al. (2012).

1. Tail black at the tip, on average greater than one-fifth of the total length, larger than the hind limbs; third molariforms wide, greater than one-third of the toothrow length........................................................................ subgenus *Cynomys* 2

1a. Tail not black at the tip, in paler colors, on average less than one-fifth of the total length, of similar length to the hind limbs; third molariforms narrow, less than one-third of the toothrow length.................... subgenus *Leucocrossuromys* 3

2. Tail distal half black; inflection of the back edge of the jaw almost at right angle in relation to its main axis (Fig. 14; restricted to Coahuila, Nuevo León, San Luis Potosí, and Zacatecas, México) .....................*Cynomys mexicanus* (p. 56)

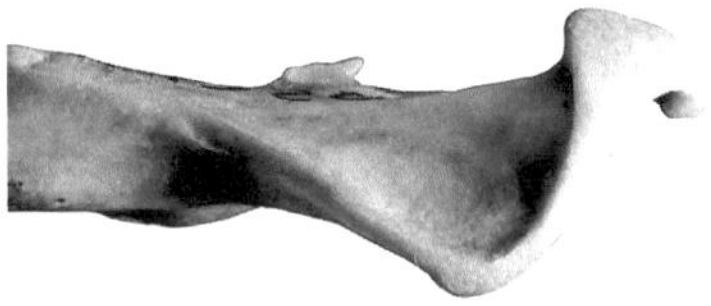

**Fig. 14** Inflection of the back edge of the jaw almost at right angle in relation to its main axis like in *Cynomys mexicanus*

2a. Tail distal third black; inflection of the back edge of the jaw almost at 45° relative to its main axis (Fig. 15; from Saskatchewan south through northern Sonora and Chihuahua, east through Nebraska near the border with Iowa and west through northern Montana)........................................................................................................................ *Cynomys ludovicianus* (p. 55)

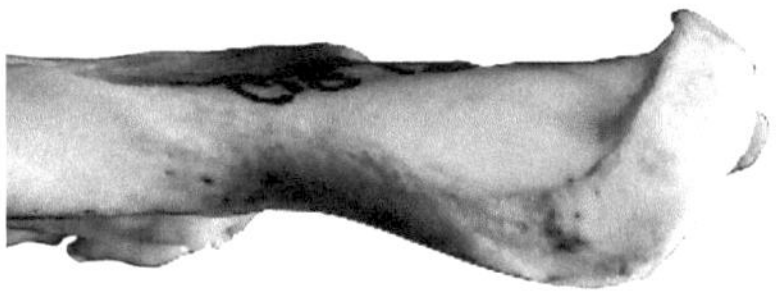

**Fig. 15** Inflection of the back edge of the jaw almost at 45° relative to its main axis like in *Cynomys ludovicianus*

3. Tail distal third bordered with white at the tip, center gray (from Arizona, Colorado, New Mexico, Wyoming, and Utah, United States).........................................................................................................*Cynomys gunnisoni* (p. 58)

3a. Tail distal third whitish, without a dark center..........................................................................................................4

4. Tail mainly of the same color throughout; upper and lower spots around the eyes blackish (seen as a perpendicular stripe); interorbital breath greater than 13.5 mm; supraorbital notch very deep (restricted to Utah, United States).......... ..........................................................................................................................*Cynomys parvidens* (p. 60)

4a. Tail with the distal half paler; upper and lower spots around the eyes darker (seen as a two dots); interorbital breath less than 13.5 mm; supraorbital notch absent (from Idaho, Montana, Wyoming, Nebraska, Colorado, and Utah, United States).........................................................................................................................*Cynomys leucurus* (p. 59)

## Subgenus ***Cynomys*** Rafinesque, 1817

**Tail black at the tip**; larger specimens; do not hibernate each year; **large molars; third molariforms wide, greater than one-third of the toothrow length; broad jugal bones.**

### ***Cynomys ludovicianus*** (Ord, 1815)

Black-tailed prairie dog, perrito de las praderas colinegra

1. *C. l. arizonensis* Mearns, 1890. Southeastern Arizona, western New Mexico, and Texas, and northern Sonora and Chihuahua.
2. *C. l. ludovicianus* (Ord, 1815). From western New Mexico and Texas north through North Dakota, Montana, and southern Saskatchewan.

**Conservation:** *Cynomys ludovicianus* is listed as Threatened by the Norma Oficial Mexicana (DOF 2019) and the Species Act of Canada, and as Least Concern by the IUCN (Cassola 2016n).

**Characteristics:** Black-tailed prairie dog is large sized; total length 317.0–415.0 mm and skull length 57.0–64.0 mm. Dorsal pelage buff to brown to cinnamon, frosted with a pale buff similar to the coloration of the limbs; fur of underparts pale buff to white on the venter; head of similar color to the back, without grizzle in black; neck and cheeks paler, similar to the underpart coloration; **tail longer than one-fifth of the total length, and with one-third of the tip blackish;** claws blackish; region above the eyes and around the snout blackish, without grizzle in black; **inflection of the back edge of the jaw almost at 45° relative to its main axis;** broad and angular; zygomatic arches wide and with conspicuous processes; upper surface of the maxillary root of the zygoma bordering the premaxillary and frontal bones narrow; auditory bullae comparatively small.

**Comments:** *Cynomys ludovicianus* ranges from southern Saskatchewan and Montana west through North Dakota and south through Arizona, Chihuahua, New Mexico, Sonora, and Texas (Map 33). *C. ludovicianus* can be differentiated from *C. leucurus* by having the distal third of the tail black. From *C. mexicanus*, it is remarkably similar in all aspects. The differences are the non-overlapping geographic range and the tail color and length: *C. ludovicianus* has the proximal third black and in *C. mexicanus* the proximal half is black.

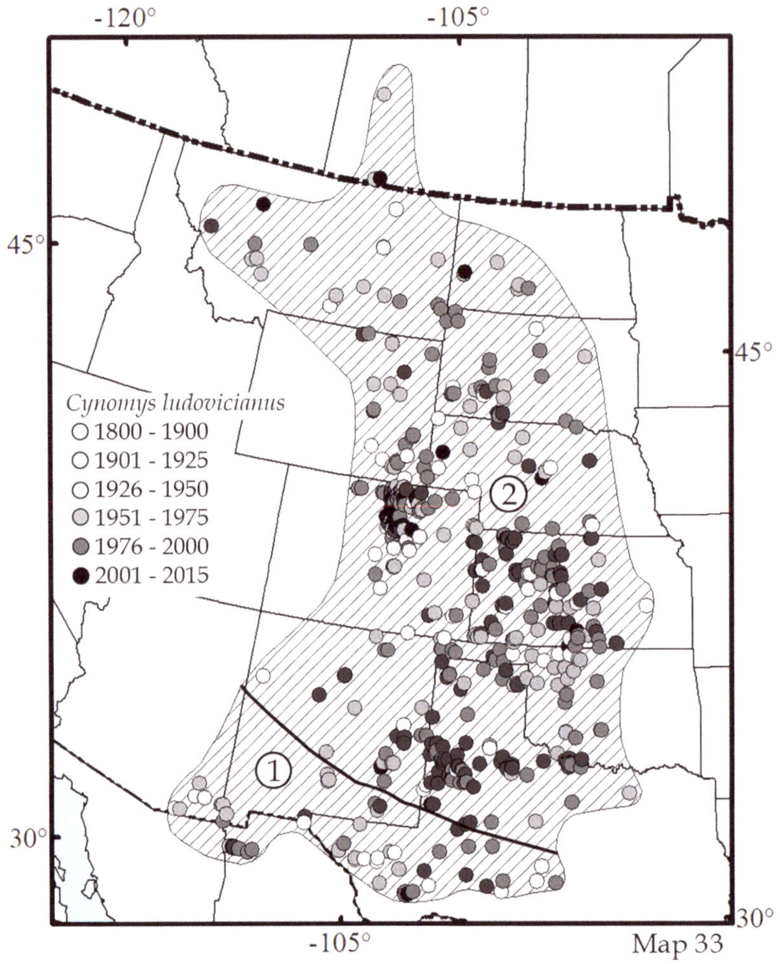


**Map 33** Distribution map of *Cynomys ludovicianus*

**Additional Literature:** Merriam (1892a), Chesser (1983), Hollister (1916), Tate (1947), Clark et al. (1971), Hoffmann et al. (1993), Hoogland (1995, 1996), Thorington et al. (2012), Koprowski et al. (2017).

## *Cynomys mexicanus* Merriam, 1892

Mexican prairie dog, perrito de las praderas mexicano

Monotypic.

**Conservation:** *Cynomys mexicanus* is listed as Endangered by the Norma Oficial Mexicana (DOF 2019) and as Endangered (B1ab (i, ii, iii, iv, v): extent of occurrence less than 5000 km$^2$ in five or less populations and under continuing decline in extent of occurrence, area of occupancy, quality of habitat, number of subpopulations, and number of mature individuals) by the IUCN (Álvarez-Castañeda et al. 2018a).

**Characteristics:** Mexican prairie dog is large sized; total length 390.0–430.0 mm and skull length 59.0–61.0 mm. Dorsal pelage grizzled buff, hair with three color bands, buff at the base, middle pale buff and blackish buff at the tip. The coloration effect is related to hair length; fur of underparts yellowish but skin blackish; head often slightly darker; tail long relative to other prairie dogs and distal half tipped in black; muzzle lower; neck with patches of buff; ears small and pressed against the head; claws blackish; **inflection of the back edge of the jaw almost at a right angle relative to its main axis,** broad and angular; zygomatic arches wide and with conspicuous processes; upper surface of the maxillary root of the zygoma bordering the premaxillary and frontal bones narrow; auditory bullae comparatively small.

**Comments:** *Cynomys mexicanus* is endemic to northern-central México, ranges in Coahuila, Nuevo León, San Luis Potosí and Zacatecas (Map 34), and has allopatric distribution with respect to the other species of *Cynomys*. *C. mexicanus* and *C. ludovicianus* are remarkably similar in all aspects. The differences are the non-overlapping geographic ranges and the tail color and length, in *C. mexicanus* approximately the proximal half is black and in *C. ludovicianus* the proximal third is black.

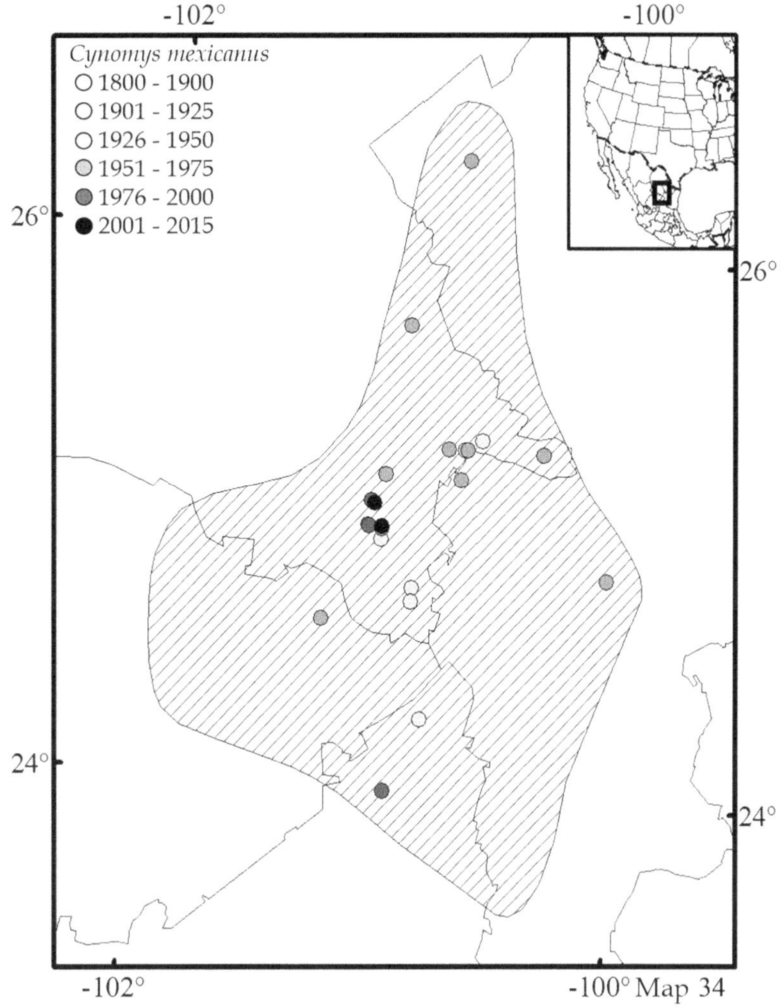


**Map 34** Distribution map of *Cynomys mexicanus*

**Additional Literature:** Merriam (1892a), Hollister (1916), Clark et al. (1971), Pizzimenti (1975), Ceballos-G and Wilson (1985), Treviño-Villareal (1991), Hoffmann et al. (1993), Thorington et al. (2012), Koprowski et al. (2017).

## Subgenus *Leucocrossuromys* Hollister, 1916

**Tail white or gray at the tip; smaller size than the species of the other subgenera;** hibernate each year; smaller molars; **third molariforms narrow, less than one-third of the toothrow length; thin jugal bones.**

## *Cynomys gunnisoni* (Baird, 1855)

Gunnison's prairie dog, perrito de las praderas de la planicie de Colorado

1. *C. g. gunnisoni* (Baird, 1855). Arizona, Colorado, Nevada, and New Mexico.
2. *C. g. zuniensis* Hollister, 1916. Colorado and New Mexico.

**Conservation:** *Cynomys gunnisoni* is listed as Least Concern by the IUCN (Cassola 2016l).

**Characteristics:** Gunnison's prairie dog **is the smallest of the prairie dogs**; total length 308.0–370.0 mm and skull length 53.0–61.5 mm. Dorsal pelage buff to pale yellow, grizzled with small amounts of black; fur of underparts white to cream, with a gradual transition to a darker back; **head top, cheeks, and "eyebrows" noticeably darker than other parts of the pelage**; neck paler without black grizzle; **tail fading to a pale buff or white at the tip, but hair in the middle of the back with a blackish band**; often a faint black patch between the eyes and the snout flanks; maxillary arm of the zygoma spreading broadly; mastoids small and more oblique; auditory bullae small; occiput view from behind high and not markedly broadened.

**Comments:** *Cynomys gunnisoni* has been suggested as conspecific with *C. parvidens* and *C. leucurus* (Burt and Grossenheider 1964). *C. gunnisoni* ranges in Arizona, Colorado, New Mexico, Utah, and Wyoming (Map 35), and has allopatric distribution with respect to the other species of *Cynomys*. However, *C. gunnisoni* and *C. ludovicianus* range in Arizona and New Mexico, and *C. parvidens* in Utah. *C. gunnisoni* can be differentiated from *C. ludovicianus* by having the tail hairs paler at the tip and on average less than one-fifth of the total length, proximally similarly large than the hind legs, and third molariforms narrow, less than one-third of the toothrow length. From *C. parvidens*, by its smaller size and the tail fades to a pale buff or white at the tip, but the middle blackish.

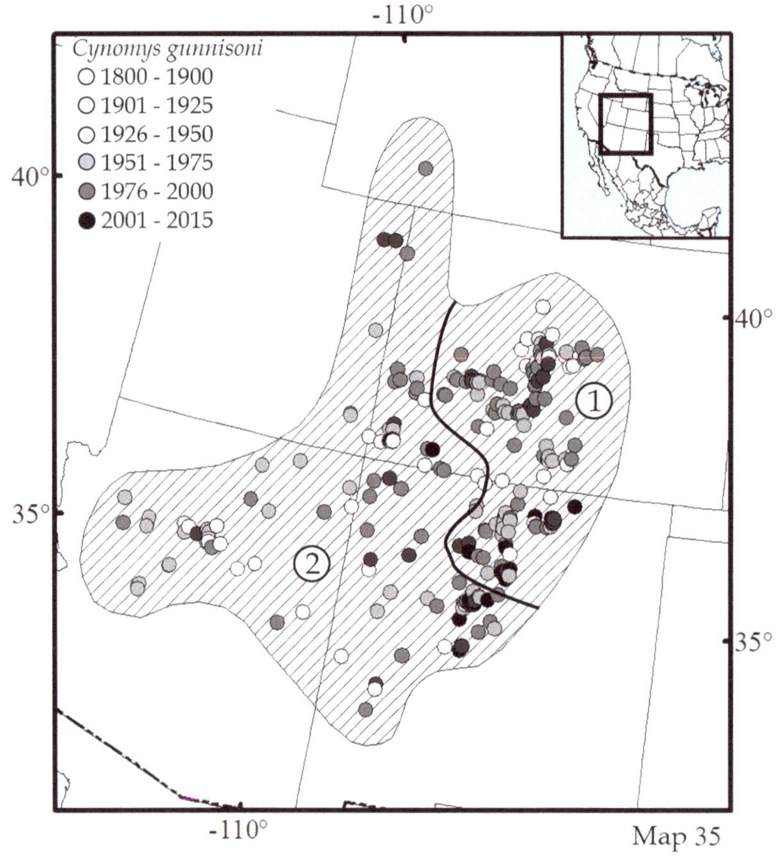


**Map 35** Distribution map of *Cynomys gunnisoni*

**Additional Literature:** Merriam (1890b), Hollister (1916), Durant (1952), Clark et al. (1971), Pizzimenti and Hoffmann (1973), Pizzimenti (1976a, b), Hoffmann et al. (1993), Thorington et al. (2012), Koprowski et al. (2017).

# *Cynomys leucurus* Merriam, 1890

White-tailed prairie dog, perrito de las praderas coliblanca

Monotypic.

**Conservation:** *Cynomys leucurus* is listed as Least Concern by the IUCN (Cassola 2016m).

**Characteristics:** White-tailed prairie dog is large sized; total length 340.0–370.0 mm and skull length 56.0–61.3 mm. Dorsal pelage yellowish buff frosted with black; fur of the underparts paler; head is colored similarly to the dorsum, but grizzled with black; neck and cheeks paler with no black grizzle; **tail buff to white at the base, with a suffusion of cinnamon and white at the tip**; **with a dark brown or black stripe perpendicular to the rostrum from above to below each eye**; supraorbital notch absent.

**Comments:** *Cynomys leucurus* has been suggested as conspecific with *C. gunnisoni* and *C. parvidens* (Burt and Grossenheider 1964). *C. leucurus* is restricted to a small area in Idaho, Montana, Wyoming, Nebraska, Colorado, and Utah (Map 36), and has allopatric distribution with respect to the other the species of *Cynomys*. However, *C. leucurus* can be found near the distribution of *C. ludovicianus* in Montana and Wyoming. *C. leucurus* can be differentiated from all other species of *Cynomys* by having the tail buff to white at the base, with a suffusion of cinnamon and white at the tip; with a dark brown or black stripe perpendicular to the rostrum from above and below each eye.

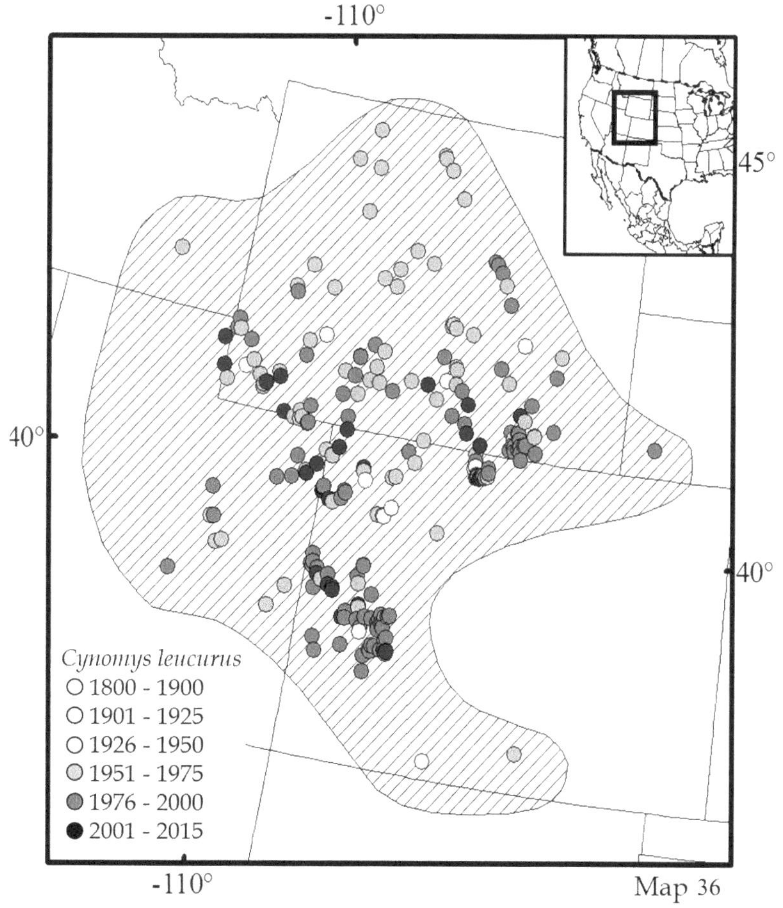


**Map 36** Distribution map of *Cynomys leucurus*

**Additional Literature:** Hollister (1916), Durant (1952), Hall and Kelson (1959), Long (1965), Hoffmann et al. (1969, 1993), Clark et al. (1971), Pizzimenti (1976a, b), Hoffmann et al. (1969, 1993), Thorington et al. (2012), Koprowski et al. (2017).

## *Cynomys parvidens* Allen, 1905

Utah prairie dog, perrito de las praderas de Utah

Monotypic.

**Conservation:** *Cynomys parvidens* is listed as Threatened by the United States Species Act and as Endangered (B1ab (iii): extent of occurrence less than 5000 km$^2$ in five or less populations and under continuing decline in area, extent, and/or quality of habitat) by the IUCN (Roach 2018b).

**Characteristics:** Utah prairie dog is large sized; total length 305.0–360.0 mm and skull length 55.0–60.0 mm. Dorsal pelage buff to cinnamon to clay; fur of underparts pale cinnamon to pale buff; tail grading from buff proximally to white distally; upper lip and chin pale buff; **with a blackish spot above and below each eye**; supraorbital notch very deep; **interorbital breadth wide; zygomatic arches relatively weak; nasals longer and narrow posteriorly, with the posterior end rounded; teeth small**.

**Comments:** *Cynomys parvidens* has been suggested as conspecific with *C. gunnisoni* and *C. leucurus* (Burt and Grossenheider 1964). *C. parvidens* is restricted only to southwestern and southern-central Utah (Map 37), and has allopatric distribution with respect to the other species of *Cynomys*. However, *C. gunnisoni* is present in Utah. *C. parvidens* can be differentiated from *C. gunnisoni* by its larger size and the tail without a mid blackish band above.

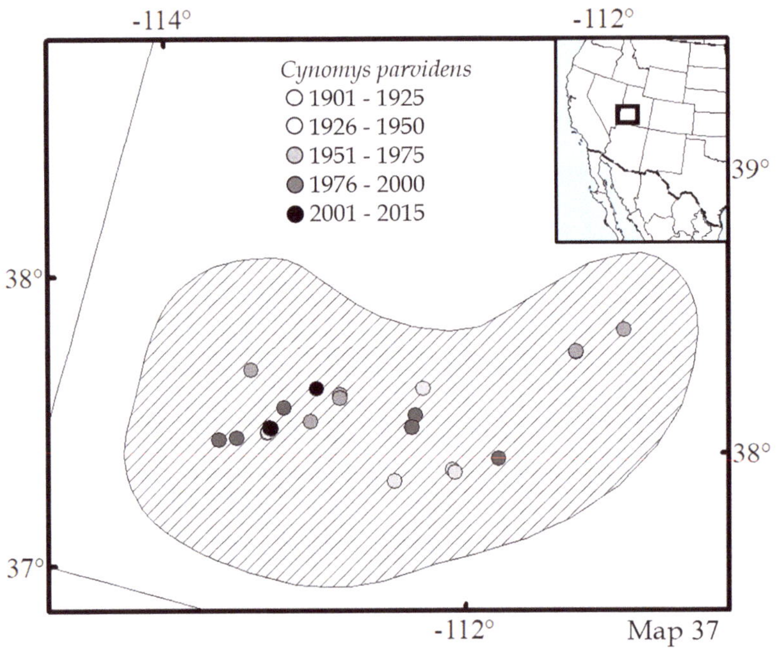


**Map 37** Distribution map of *Cynomys parvidens*

**Additional Literature:** Hollister (1916), Durant (1952), Burt and Grossenheider (1964), Clark et al. (1971), Pizzimenti and Nadler (1972), Pizzimenti and Collier (1975), Hoffmann et al. (1993), Thorington et al. (2012), Koprowski et al. (2017).

## Genus ***Ictidomys*** Allen, 1877

Dorsal pelage with a pattern of general pale buff over a cinnamon to dark sepia; **9 or 13 dorsal stripes, which can be either lines or series of spots**; fur of underparts buff to cinnamon buff; head darker than the back, with many small spots similar to those in the back; head flanks and neck sometimes of the same color as the underparts or darker, depending on the species; spots, limbs, and feet of similar color; tail with different patterns among the species, but in all cases it is a combinations of fuscous black with blackish and buff; **postorbital constriction narrow; parietal ridges prominent, meeting at an acute angle well in front of the superior nuchal line; external margins of the infraorbital foramina usually slanted ventrolaterally**. *Ictidomys* was a subgenus of *Spermophilus*, but genetic and morphometrical analyses highlight that it should be considered at the genus level (Helgen et al. 2009). The keys were elaborated based on the review of specimens and the following sources: Hall (1981), Helgen et al. (2009), Thorington et al. (2012), and Álvarez-Castañeda et al. (2017a).

1. Back with stripes and spots in linear series (from southern-central Canada, Alberta, Saskatchewan, and Manitoba south through Utah, Arizona, New Mexico, and Texas, and from Montana and Utah east through Michigan and Ohio, United States)........................................ *Ictidomys tridecemlineatus* (p. 63)

1a. Back area with spots in linear series and without stripes........................................2

2. Total length less than 320.0 mm; hindfoot length less than 46.0 mm; tail length less than 130.0 mm; skull length less than 45.0 mm (from New Mexico and Texas south through Coahuila, Nuevo León, and Tamaulipas)........................................ *Ictidomys parvidens* (p. 62)

2a. Total length greater than 320.0 mm; hindfoot length greater than 46.0 mm; tail length greater than 120.0 mm; skull length greater than 45.0 mm (from eastern Jalisco and southern Zacatecas east through Puebla)........................................ *Ictidomys mexicanus* (p. 61)

### ***Ictidomys mexicanus*** (Erxleben, 1777)

Mexican ground squirrel, ardilla manchada mexicana

Monotypic.

**Conservation:** *Ictidomys mexicanus* is listed as Least Concern by the IUCN (Linzey et al. 2016c).

**Characteristics:** Mexican ground squirrel is medium sized; **total length 322.0–380.0 mm** and skull length 45.3–52.5 mm. **Dorsal pelage with nine evident dorsal and lateral stripes of pale spots, no lines present**; fur of underparts buff to cinnamon buff; head darker than the back, with many very small spots which can be seen as frosted with pale buff; head flanks darker than the underparts, spots, limbs, and feet; chin, cheeks, eye rings, and nose flanks buff to cinnamon buff; tail grizzled black, suffused and with a frosted coloration below, without a pattern; skull arch shaped in lateral view; and auditory bullae proportionally large.

**Comments:** *Ictidomys parvides* was recognized as a distinct species (Harrison et al. 2003; Herron et al. 2004), previously regarded as a subspecies of *I. mexicanus* when it was first described (Mearns 1896). *I. mexicanus* ranges from eastern Jalisco and southern Zacatecas east through Puebla (Map 38). It can be found in sympatry, or nearly so, with *I. tridecemlineatus* in Texas and New Mexico, from which it can be differentiated by not having 13 lines on the back.

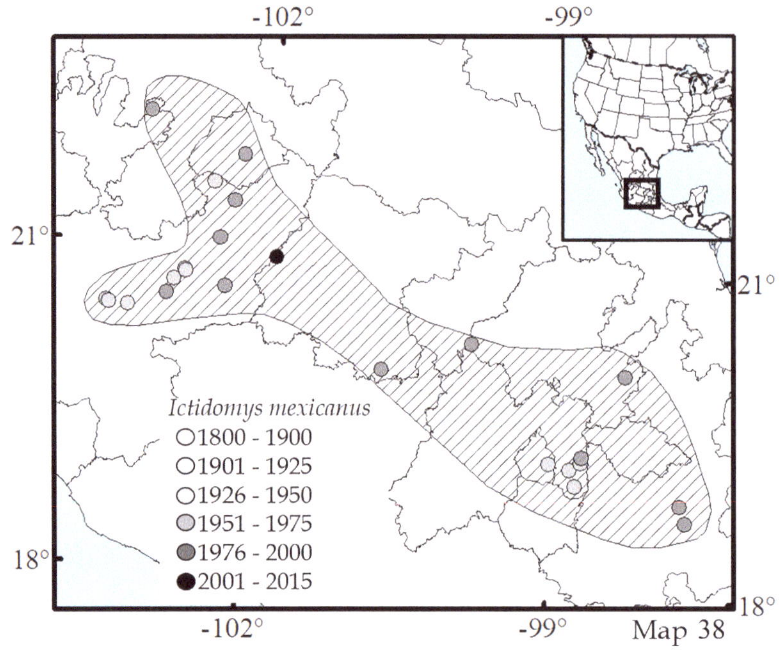


**Map 38** Distribution map of *Ictidomys mexicanus*

**Additional Literature:** Bryant (1945), Young and Jones (1982), Rickart and Yensen (1991), Thorington and Hoffmann (2005), Thorington et al. (2012), Koprowski et al. (2017).

## *Ictidomys parvidens* (Mearns, 1896)

Rio Grande ground squirrel, ardilla manchada texana

Monotypic.

**Conservation:** *Ictidomys parvidens* (as *Ictidomys mexicanus*) is listed as Least Concern by the IUCN (Linzey et al. 2016c).

**Characteristics:** Rio Grande ground squirrel is medium sized; **total length 276.0–312.0 mm and skull length 41.1–45.3 mm. Dorsal pelage with nine evident dorsal and lateral stripes of paler spots, no lines present**; fur of underparts buff to cinnamon buff; head darker than the back with many very small spots that can be seen as frosted with pale buff, head flanks darker than the underparts, spots, limbs and feet; chin, cheeks, eye rings, and nose flanks buff to cinnamon buff; tail grizzled black, suffused and with a frosted coloration without a pattern ventrally; skull arch shaped in lateral view; auditory bullae proportionally large.

**Comments:** *Ictidomys parvidens* was recognized as a distinct species (Harrison et al. 2003; Herron et al. 2004), previously regarded as a subspecies of *I. mexicanus* (Mearns 1896). The hybridization between *Ictidomys parvidens* and *I. tridecemlineatus* has been recorded. *I. parvidens* ranges from New Mexico and Texas south through Coahuila, Nuevo León, and Tamaulipas (Map 39). It can be found in sympatry, or nearly so, with *I. tridecemlineatus* in Texas and New Mexico. *I. mexicanus* can be differentiated from *I. tridecemlineatus* by its nine stripes of series of spots.

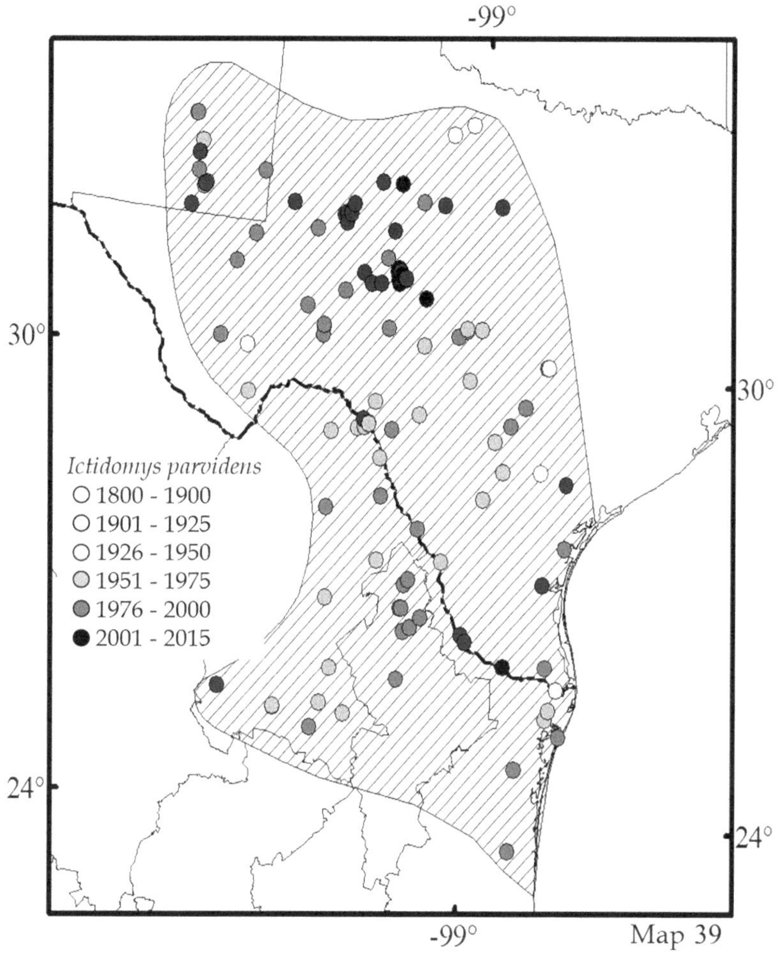


**Map 39** Distribution map of *Ictidomys parvidens*

**Additional Literature:** Mearns (1907), Howell (1938), Bryant (1945), Baker (1956), Davis (1974), Schmidly (1977), Thorington and Hoffmann (2005), Thorington et al. (2012).

## *Ictidomys tridecemlineatus* (Mitchill, 1821)

Thirteen-lined ground squirrel, ardilla manchada de trece líneas

1. *I. t. alleni* (Merriami, 1898). Northwestern Wyoming.
2. *I. t. arenicola* (Howell, 1928). Eastern New Mexico, southwestern Colorado, western Kansas, western Oklahoma, and northwestern Texas.
3. *I. t. blanca* (Armstrong, 1971). Southern-central Colorado.
4. *I. t. hollisteri* (Bailey, 1913). Central New Mexico.
5. *I. t. monticola* (Howell, 1928). Eastern-central Arizona and western-central New Mexico.
6. *I. t. olivaceous* (Allen, 1895). Northeastern Wyoming and western-central South Dakota.
7. *I. t. pallidus* (Allen, 1874). Northern Great Plains, from Colorado and Nebraska north through south Saskatchewan.
8. *I. t. parvus* (Allen, 1895). Southern-central Wyoming, northeastern Utah, and northwestern Colorado.
9. *I. t. texensis* (Merriami, 1898). Central Texas, Oklahoma, southeastern Kansas, and southwestern Missouri.
10. *I. t. tridecemlineatus* (Mitchill, 1821). From central Kansas to North Dakota eastward to Ohio, Alberta, Saskatchewan, Manitoba, and northwestern Montana.

**Conservation:** *Ictidomys tridecemlineatus* is listed as Least Concern by the IUCN (Cassola 2016ak).

**Characteristics:** Thirteen-lined ground squirrel is medium sized; total length 170.0–297.0 mm and skull length 34.1–46.0 mm. **Dorsal pelage with 13 dorsal and lateral stripes, which can be a continuous line or a line broken in a series of spots in line;** additional spot lines on the flanks, paler stripes separated by cinnamon to dark sepia; fur of underparts buff to cinnamon buff; head darker than the back with many small spots similar to those on the back; flanks of the head and neck of the same color as the paler stripes, spots, limbs and feet; tail fuscous black dorsally with three colors, buff at the center and peripherally with an intermedial blackish ring; hair with three color bands, which give the tail a buff-blackish-buff coloration; tail ventrally russet brown to cinnamon buff and buff at the tip; chin, cheeks, eye rings, and nose flanks buff to cinnamon buff; skull long, narrow, and lightly built; molariform toothrows only slightly convergent posteriorly.

**Comments:** The hybridization between *Ictidomys mexicanus* and *I. tridecemlineatus* has been recorded. *Ictidomys tridecemlineatus* ranges in the Great Plains from southcentral Canada, Alberta, Saskatchewan, and Manitoba south through Utah, Arizona, New Mexico, and Texas, and from Montana and Utah east through Michigan and Ohio (Map 40). *I. tridecemlineatus* is sympatric, or nearly so, with *I. parvidens* in Texas and New Mexico. It can be differentiated from all the other squirrels, including *I. parvidens*, by its 13 stripes of line and spot series on the back, and head darker than the back, with many small spots similar to those on the back.

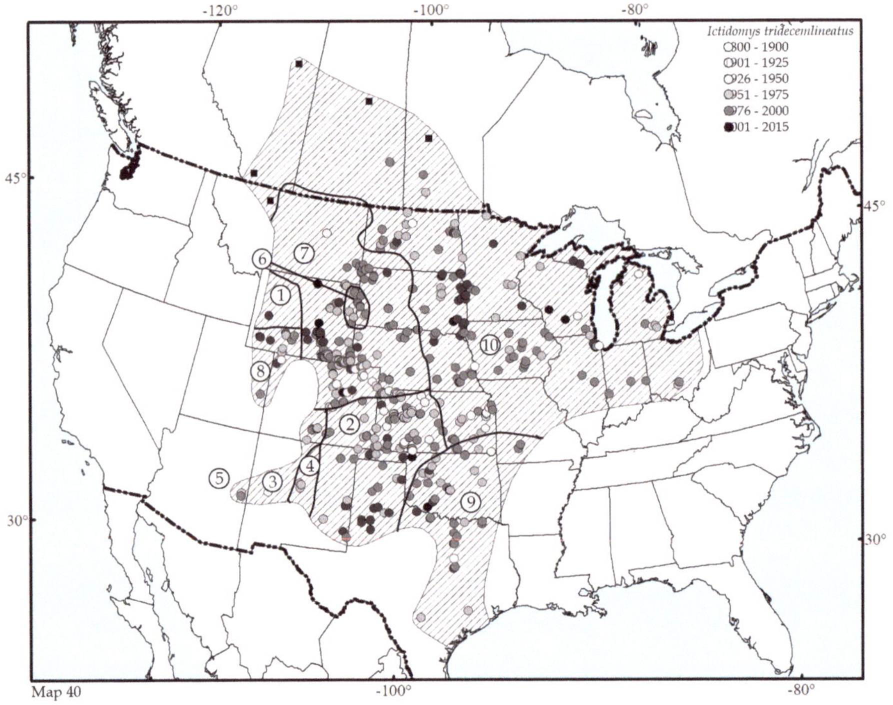


**Map 40** Distribution map of *Ictidomys tridecemlineatus*

**Additional Literature:** Howell (1938), Hall and Kelson (1959), Armstrong (1971, 1972), Streubel and Fitzgerald (1978b), Rickart and Yensen (1991), Thorington et al. (2012), Koprowski et al. (2017).

## Genus *Marmota* Blumenbach, 1779

Dorsal pelage tawny, some individuals grizzly with white shades on the head between the eyes; feet pale buff to dark brown; neck flanks with conspicuous buffy patches; fur of underparts ochraceous to deep cinnamon; **tail length less than 30% of the total length; skull massive; sagittal crest well developed in the posterior part of the braincase; zygomatic arches robust; third upper molars with one more series of ridges than other upper molars; molariforms elliptical; nasal bones wide and rounded at the anterior edge** and posterior edge of varying width, depending on the species**; postorbital process well developed; palatal generally wide;** incisive foramina varying depending on the species. *Marmota* has two subgenera: *Marmota* [*M. broweri* and *M. monax*] and *Petromarmota* [*M. caligata*, *M. flaviventris*, *M. olympus*, and *M. vancouverensis*]. The keys were elaborated based on the review of specimens and the following sources: Howell (1918), Hall (1981), and Thorington et al. (2012).

1. Dorsal pelage soft, guard hairs black-tipped, with a subterminal pale band; head crown black from the nose to the nape, without spot of the different colors (restricted to northern Alaska, United States) ........................................................ *Marmota* (*Marmota*) *broweri* (p. 66)

1a. Dorsal pelage coarse, many guard hairs paler at the tip, or not, or else uniform dark brown or black; head crown black or brown, from the nose to the nape, with the presence of white, brown, or grizzle spots ........................ 2

2. Pelage uniformly dark brown, guard hairs not paler at the tip; posterior border of the nasals deeply V shaped (restricted to Vancouver Island) ........................ *Marmota* (*Petromarmota*) *vancouverensis* (p. 73)

2a. Pelage of varying coloration and patterns, but not uniformly dark brown, guard hairs paler at the tip; posterior border of the nasals deeply U shaped ........................ 3

3. No white hairs on the head, except around the nose and no solid colors in the upperparts; eight pair of nipples, one abdominal; upper toothrows parallel (from central Alaska east through Labrador Peninsula and south through Washington, and Oklahoma, Arkansas, Mississippi, and Alabama, and east through North Carolina, United States) ........................ *Marmota* (*Marmota*) *monax* (p. 67)

3a. White hairs on the head, sometimes with solid colors in the upperparts; ten pair of nipples, two abdominal; upper toothrows diverging anteriorly ........................ 4

4. Fur of underparts ochraceous to deep cinnamon; neck flanks with conspicuous buffy patches; feet buffy, hazel, or tawny; posterior palatal wide, width of the palatal, at the root of the last molar equal to or greater than the length of the three first molariforms (from southern British Columbia and Alberta south through California, Nevada, Utah, Colorado, and northern New Mexico, United States) ........................ *Marmota* (*Petromarmota*) *flaviventris* (p. 70)

4a. Fur of underparts with different shades but not deep cinnamon; neck flanks lacking buffy patches; feet blackish or blackish brown; posterior palatal narrow, width of the palatal at the root of the last molars is equal to or greater than the length of the two first molariforms ........................ 5

5. Anterior part of the body predominantly black and white with grizzle of tawny hair on the posterior part (from southern Alaska south through Washington, Idaho, and Montana, United States) ........................ *Marmota* (*Petromarmota*) *caligata* (p. 68)

5a. All the body of similar coloration, not black and white in the anterior part; dorsal parts pale brown mixed with whitish shades (restricted to the Olympic peninsula, Washington, United States) ........................ *Marmota* (*Petromarmota*) *olympus* (p. 72)

## Subgenus *Marmota* Blumenbach, 1779

**Eighth mammae (nipples; only one abdominal pair);** dorsal pelage with pale-tipped guard hairs and a pale-colored nose, but lacking the contrasting pale-and-dark pattern on the face as seen in *Petromarmota*; **upper toothrows parallel.**

## ***Marmota broweri*** (Hall and Gilmore, 1934)

Alaska marmot, marmota de Alaska

Monotypic.

**Conservation:** *Marmota broweri* is listed as Least Concern by the IUCN (Cassola 2016an).

**Characteristics:** Alaska Marmot is large sized; total length 539.0–652.0 mm and skull length 96.0–100.0 mm. **Dorsal pelage dark, from gray to charcoal, mixed with cinnamon, mainly near the hips**; **fur of underparts gray to charcoal**; neck flanks paler; head dark brown to charcoal and relatively uniform in color; cheeks and eyes paler; tail uniformly gray to brown; postorbital process projecting back of the line drawn across their bases, back of the process in approximately 80° or less than the skull longitudinal axes; posteriorpart of the palatal narrow, palatal width at the root of the last molars longer than the two first molariforms; posterior area of the nasals similar in width to the maxillary branch; incisive foramina of equal width or constricted posteriorly. Species group *Monax*.

**Comments:** *Marmota broweri* was recognized as a distinct species (Rausch and Rausch 1965, 1971; Hoffmann et al. 1979), previously regarded as a subspecies of *M. caligata* (Hall and Gilmore 1934). *M. broweri* ranges from northern Alaska north through near the Artic Sea (Map 41), and has allopatric distribution with respect to the other species of *Marmota*. *M. broweri* can be differentiated from *M. caligata* by having a black cap only in the head top, with a white patch from the forehead to the nose; shorter mandibular angular process, less than 52.8 mm, rostrum less than 27.8 mm and postorbital width greater than 16.3 mm, less well developed sagittal and lambdoidal crests.

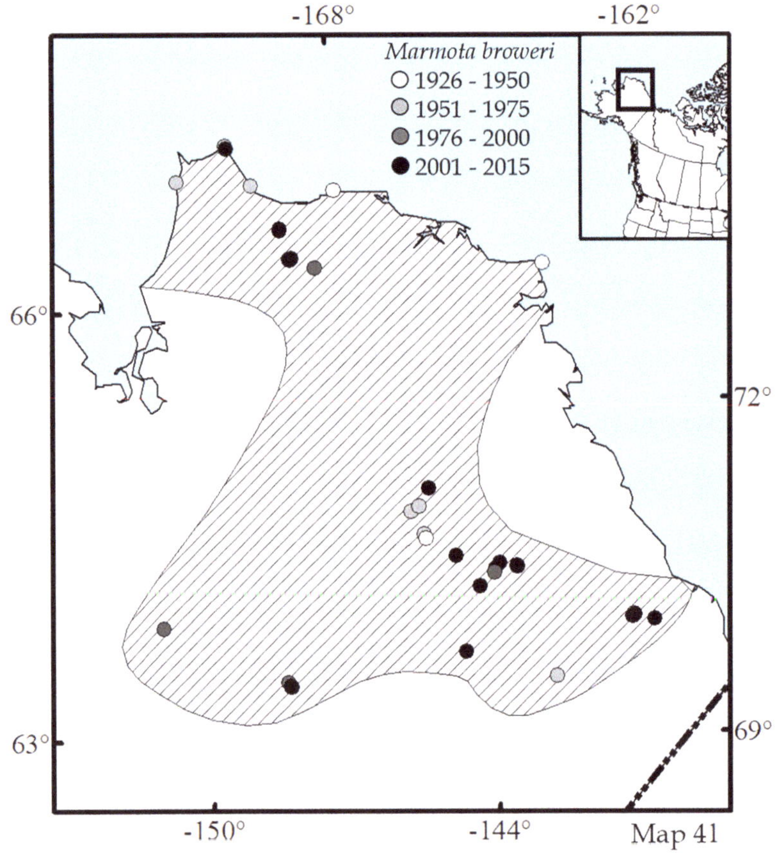


**Map 41** Distribution map of *Marmota broweri*

**Additional Literature:** Frase and Hoffmann (1980), Steppan et al. (1999), Braun et al. (2011b), Thorington et al. (2012), Koprowski et al. (2017).

## ***Marmota monax*** (Linnaeus, 1758)

Woodchuck, marmota americana

1. *M. m. bunkeri* Black, 1935. Eastern Oklahoma, Kansas, and Nebraska.
2. *M. m. canadensis* (Erxleben, 1777). From the British Columbia and northwestern Territories eastward to Labrador.
3. *M. m. ignava* (Bangs, 1899). Northern Québec and Labrador.
4. *M. m. johnsoni* Anderson, 1943. Endemic to the extreme southeastern Québec on the south bank of the Saint Lawrence Seaway.
5. *M. m. monax* (Linnaeus, 1758). From Arkansas, Missouri, and Iowa east through the Atlantic coast.
6. *M. m. ochracea* Swarth, 1911. Alaska, Yukon, and northern British Columbia.
7. *M. m. pretrensis* Howell, 1915. British Columbia, northern Idaho, and northwestern Washington.
8. *M. m. preblorum* Howell, 1914. Throughout the New England region.
9. *M. m. rufescens* Howell, 1914. From the Dakotas east through New York, including Ontario.

**Conservation:** *Marmota monax* is listed as Least Concern by the IUCN (Cassola 2016ap).

**Characteristics:** Woodchuck is large sized; total length 420.0–660.0 mm and skull length 99.8–104.8 mm. Dorsal pelage grizzled gray to cinnamon to dark brown, often appearing frosted, pale at the tip; forelimbs sometimes blackish mixed with rufous to cinnamon; fur of underparts gray to reddish to dark brown; **head darker than the back, with white to tan patches surrounding the nose,** above the eyes, and the lower jaw apex; neck flanks lacking buffy patches; if present, feet from pale buff to blackish brown; tail relatively short, reddish brown to blackish; postorbital process projecting to the back of the line drawn across their bases, back of the process at approximately 90° relative to the longitudinal skull axis; posterior palatal narrow, palatal width at the last molar root equal to or greater than the length of the two first molariforms and **extending 2.0–4.0 mm backward**; **upper toothrows parallel; sagittal crest well developed; interorbital region broad**; posterior nasals broader than the maxillary branch; incisive foramina not of the same width or constricted posteriorly; posterior border of the nasals deeply U shaped. Species group *Monax*.

**Comments:** *Marmota monax* ranges widely from central Alaska east through the Labrador Peninsula and south through Washington and Oklahoma, Arkansas, Mississippi, and Alabama, and east through North Carolina (Map 42), and has allopatric distribution with respect to the other species of *Marmota*. *M. monax* can be differentiated from *M. caligata* and *M. flaviventris* by being smaller on average; tail black, dark brown, bushy, almost flattened, and short, less than the 25% of the total length; head without white markings (except around the nose); neck flanks similar in color to the dorsum; forelegs with overlaid deep reddish brown hairs, and upper toothrows parallel.

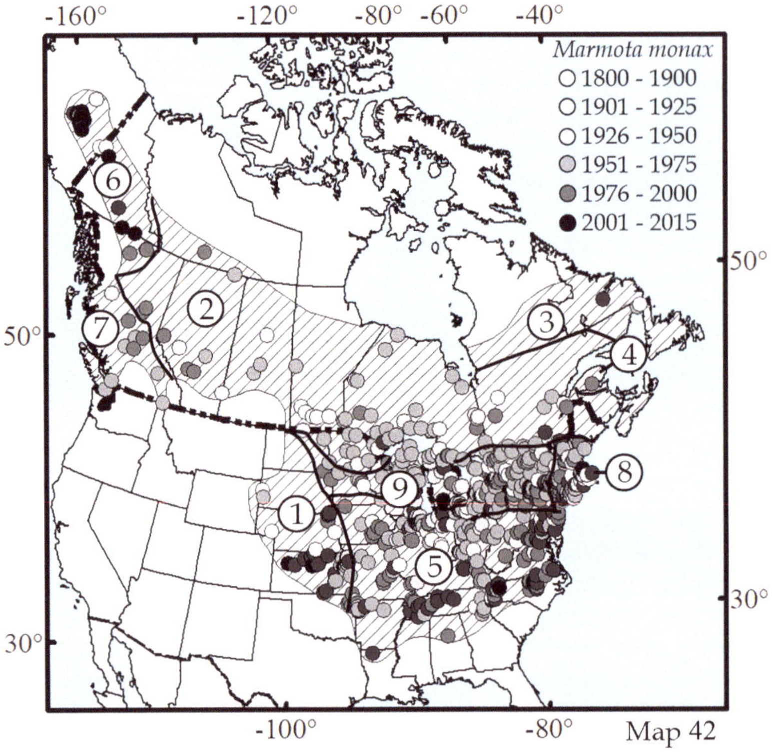


**Map 42** Distribution map of *Marmota monax*

**Additional Literature:** Howell (1914b, 1915b), Frase and Hoffmann (1980), Hall (1981), Kwiecinski (1998), Steppan et al. (1999), Thorington et al. (2012), Koprowski et al. (2017).

## Subgenus ***Petromarmota*** Steppan et al., 1999

**Ten mammae (nipples; two pairs in the abdomen);** dorsal pelage with pale-tipped guard hairs; nose and chin white; pale patch between or just in front of the eyes, separated from the white nose by a dark brown to black transverse stripe; **upper toothrows diverging anteriorly.**

### ***Marmota caligata*** (Eschscholtz, 1829)

Hoary marmot, marmota canosa

1. *M. c. caligata* (Eschscholtz, 1829). Western Yukon and most of Alaska south through the Yukon River, including Montague Island and the Glacier Bay.
2. *M. c. cascadensis* Howell, 1914. Southern and western British Columbia and Washington.
3. *M. c. nivaria* Howell, 1914. Alberta, Idaho, and Montana.
4. *M. c. okanagana* (King, 1836). Southern Yukon and northwestern Territories into eastern British Columbia and western Alberta south through Idaho and Montana.
5. *M. c. oxytona* Holister, 1912. Southeastern Yukon and central British Columbia.
6. *M. c. raceyi* Anderson, 1932. Western British Columbia.
7. *M. c. sheldoni* Howell, 1914. Restricted to Montague Island, Alaska.
8. *M. c. vigilis* Heller, 1909. Restricted to the Glacier Bay, Alaska.

**Conservation:** *Marmota caligata* is listed as Least Concern by the IUCN (Cassola 2018).

**Characteristics:** Hoary marmot is large sized; total length 620.0–820.0 mm and skull length 90.8–108.8 mm. Dorsal coloration with the anterior part of the dorsum pale cream to white and yellow to tan posteriorly; fur of underparts pale cream to white; rump and tail blackish at the tip; head cream to white, with dark brown to black patches around the snout, crown, and chin; tail bushy with hairs with three bands, at the base and at the tip blackish, middle buff; **tail length less than 30% of the total length**; postorbital process projecting to the back of the line drawn across their bases; back process at approximately 80° or less relative to the skull longitudinal axis; posterior palatal narrow; palatal width at the last molar root longer than the two first molariforms; posterior nasals similarly broader than the maxillary branch; incisive foramina different in shape among populations; posterior border of the nasals deeply U shaped. Species group *Caligata.*

**Comments:** *Marmota caligata* ranges from the Yukon River eastward to southern Alaska south through Washington, Idaho, and Montana, including southwestern Yukon (Map 43), and has allopatric distribution with respect to the other species of *Marmota. M. caligata* can be differentiated from all the other species of *Marmota* by its larger size, greater than 630.0 mm; mixed black-and-white or solid brownish pelage; plantar pad of the hind feet with a circular or subcircular shape; angular process of the mandible deeper and more well-developed sagittal and lambdoidal crests. From *M. broweri*, by having a black cap extending from the nose to the neck, with a white patch between the eyes and nose, and black lips; longer angular process of the mandible, greater than 52.8 mm; rostrum greater than 27.8 mm, smaller postorbital width, less than 16.3 mm, and more well developed sagittal and lambdoidal crests. From *M. olympus*, by having a black-and-white pelage; black feet; postorbital region and postorbital width less than 24.2 mm and 16.2 mm, respectively, and zygomatic width greater than 64.2 mm. From *M. vancouverensis*, by having a black-and-white pelage; deeply V shaped posterior border of the nasal and lacrimal bones rectangular. From *M. monax*, by its larger average size; tail great than the 25% of the total length; head with white markings; forelegs without deep reddish brown hairs; upper toothrows not parallel.

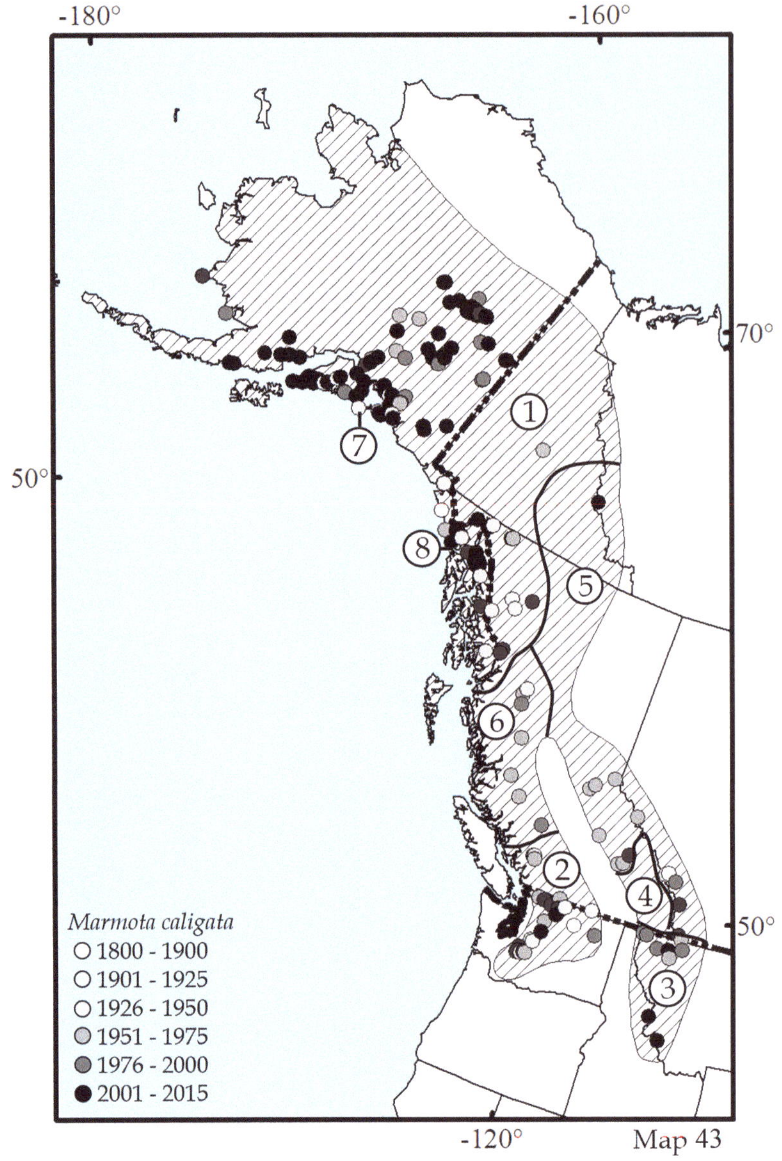


**Map 43** Distribution map of *Marmota caligata*

**Additional Literature:** Howell (1915b), Rausch (1953), Hoffmann et al. (1979), Frase and Hoffmann (1980), Steppan et al. (1999), Braun et al. (2011b), Thorington et al. (2012).

## *Marmota flaviventris* (Audubon and Bachman, 1841)

Yellow-bellied Marmot, marmota de vientre amarillo

1. *M. f. avara* (Bangs, 1899). Nevada, Oregon, Washington, and British Columbia.
2. *M. f. dacota* (Merriam, 1889). Restricted to the Black Hills, South Dakota, and Wyoming.
3. *M. f. engelhardti* Allen, 1905. Restricted to central to southwestern Utah.
4. *M. f. flaviventris* (Audubon and Bachman, 1841). Cascade and northern and central Sierra Nevada of Oregon and California.
5. *M. f. fortirostris* Grinnell, 1921. Restricted to the White Mountains on the California and Nevada border.
6. *M. f. luteola* Howell, 1914. Southern-central Rocky Mountains, from southern Wyoming south through northern New Mexico.
7. *M. f. nosophora* Howell, 1914. Idaho, Montana, Wyoming, and Alberta.

8. *M. f. notioros* Warren, 1934. Restricted to southern-central Colorado.
9. *M. f. obscura* Howell, 1914. Northern-central New Mexico and southern Colorado.
10. *M. f. parvula* Howell, 1915. Restricted to central Nevada.
11. *M. f. sierrae* Howell, 1915. Restricted to southern Sierra Nevada, California.

**Conservation:** *Marmota flaviventris* is listed as Least Concern by the IUCN (Cassola 2016ao).

**Characteristics:** Yellow-bellied Marmot is large sized; total length 470.0–700.0 mm and skull length 70.0–101.5 mm. Dorsal pelage generally reddish brown frosted with buff, some individuals grizzly with white shades; fur of underparts buff to yellow, more reddish in the venter and neck; head usually reddish dark brown with paler patches between the eyes and the snout; snout flanks and lower jaw apex with a white spots; **neck flanks with conspicuous buffy patches; feet pale buff to dark brown**; postorbital process projecting to the back of the line drawn across their bases, and the back part is approximately 90° relative to the skull longitudinal axis; **postorbital processes long and slender, projecting slightly back; posterior palatal wide; palatal width at the last molar root longer than the three first molariforms**; incisive foramina of the same width or constricted posteriorly; posterior nasals not broader than the maxillary branch and deeply U shaped. Species group *Caligata*.

**Comments:** *Marmota flaviventris* ranges from southern British Columbia and Alberta south through California, Nevada, Utah, Colorado, and northern New Mexico. There is also an isolated population from South Dakota (Map 44). It has allopatric distribution with respect to the other species of *Marmota*. *M. flaviventris* can be differentiated from *M. monax* by its larger average size; tail greater than 25% of the total length; head with white markings, neck flanks with conspicuous buffy patches; forelegs without deep reddish brown hairs; upper toothrows not parallel.

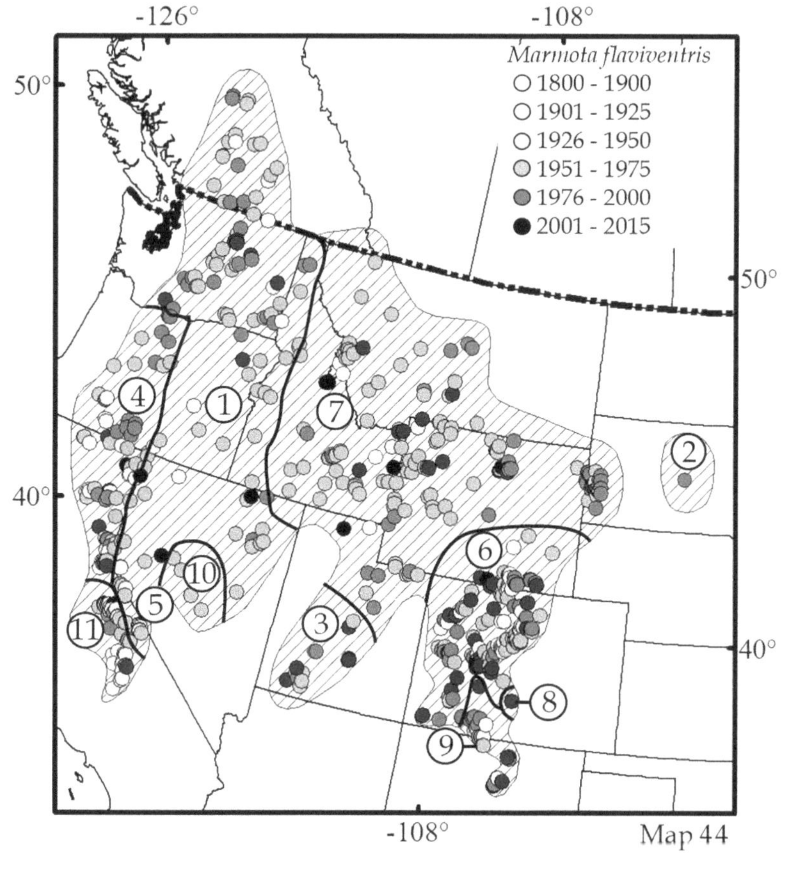


**Map 44** Distribution map of *Marmota flaviventris*

**Additional Literature:** Howell (1915b), Warren (1936), Hall and Kelson (1959), Armstrong (1972), Armitage et al. (1979), Frase and Hoffmann (1980), Steppan et al. (1999), Thorington et al. (2012).

## ***Marmota olympus*** (Merriam, 1898)

Olympic Marmot, marmota de la península de Olimpia

Monotypic.

**Conservation:** *Marmota olympus* is listed as Least Concern by the IUCN (Cassola 2016ap).

**Characteristics:** Olympic Marmot is large sized; total length 680.0–785.0 mm and skull length 98.8–108.2 mm. Dorsal pelage yellow brown to pale brown, in some individuals frosted with a paler coloration; fur of underparts of similar coloration to the back, grayish to deep cinnamon; head of similar color to the back, snout and stripe between eyes buff paler; feet pale buff to dark brown; neck flanks with conspicuous buffy patches; tail long and densely haired, of similar color to the back; postorbital process projecting back of the line drawn across their bases, the back process approximately at 80° or less relative to the skull longitudinal axis; posterior palatal narrow; palatal width at the last molar root longer than the two first molariforms; posterior nasals similarly broader than the maxillary branch; posterior border of the nasals deeply U shaped. Species group *Caligata*.

**Comments:** *Marmota olympus* is restricted to the Olympic Mountains of western Washington (Map 45), and has allopatric distribution with respect to the other species of *Marmota*. *M. olympus* can be differentiated from *M. caligata* by having light brownish pelage and feet; postorbital region and postorbital width greater than 24.2 mm and 16.2 mm, respectively, and zygomatic width less than 64.2 mm; from *M. vancouverensis*, by having a uniform pale brown pelage, including the limbs and rump; posterior border of the nasal bones U shaped.

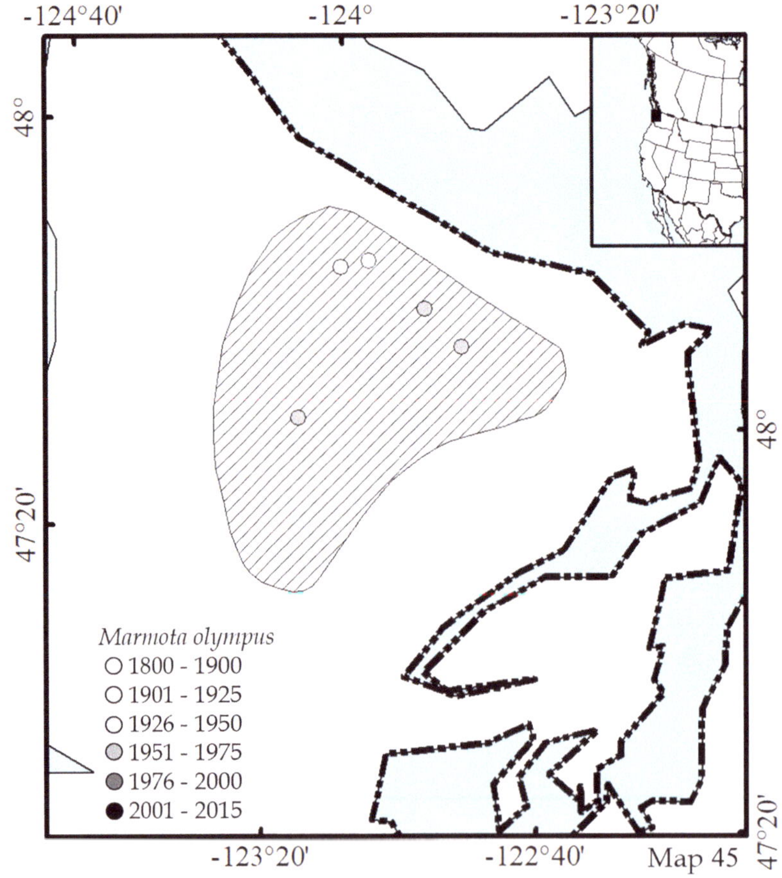


**Map 45** Distribution map of *Marmota olympus*

**Additional Literature:** Howell (1915b), Hoffmann et al. (1979), Frase and Hoffmann (1980), Nagorsen (1987), Steppan et al. (1999), Edelman (2003), Braun et al. (2011b), Thorington et al. (2012).

## *Marmota vancouverensis* Swarth, 1911

Vancouver Island marmot, marmota de isla Vancouver

Monotypic.

**Conservation:** *Marmota vancouverensis* is listed as Endangered by the Species Act in Canada, and as Critically Endangered (C1 + 2a (i): small population size projected continuing decline in 25% within three years + under continuing decline in numbers of mature individuals and less than 50 mature individuals in each subpopulation) by the IUCN (Roach 2017a).

**Characteristics:** Vancouver Island marmot is large sized relative to other squirrels, total length 670.0–750.0 mm and skull length 100.5–104.8 mm. Dorsal pelage dark brown to sepia, some individuals grizzly with white shades; fur of underparts dark brown to sepia or small patches of white on the venter, sometimes with white patches aggregated as a line along the venter midline; head of similar coloration to the back, with a whitish patch surrounding the snout and lower jaw, and sometimes small patches also on the forehead; neck flanks with conspicuous buffy patches; feet pale buff to dark brown; the postorbital process projecting back of a line drawn across their bases, the back process at approximately 80° or less to the skull longitudinal axis; posterior palatal narrow; palatal width at the last molar root longer than the two first molariforms; posterior nasals similarly broader than the maxillary branch; **posterior border of the nasals deeply V shaped.** Species group *Caligata*.

**Comments:** *Marmota vancouverensis* is restricted to Vancouver Island (Map 46), and has allopatric distribution with respect to the other species of *Marmota*. *M. vancouverensis* can be differentiated from *M. caligata* and *M. olympus* by having a uniform brown olive pelage, with the anterior part of the snout white and limbs of the same color as the back; posterior border of the nasal bones not deeply V shaped, and lacrimal bones not rectangular.

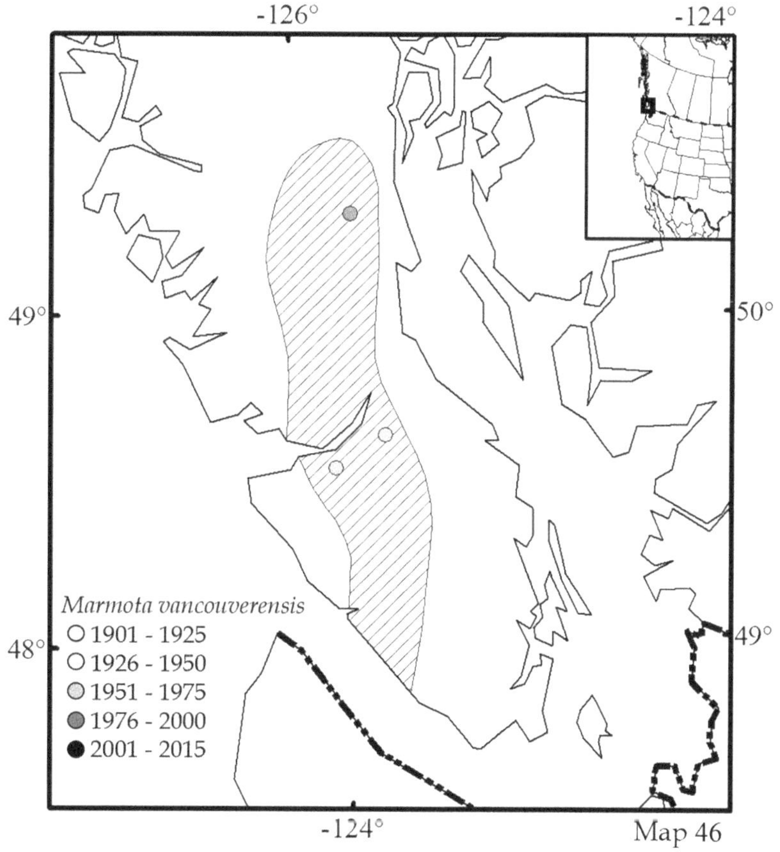


**Map 46** Distribution map of *Marmota vancouverensis*

**Additional Literature:** Howell (1915b), Hoffmann et al. (1979), Nagorsen (1987), Steppan et al. (1999), Braun et al. (2011b), Thorington et al. (2012).

## Genus *Notocitellus* Howell, 1938

**Dorsal pelage rough**; fur of underparts yellowish; chin, throat, and nose flanks yellowish ocher; neck, shoulders, forefeet, ears, and hind feet hazel, although the last two can be tawny; **tail almost same in size as the body;** six nipples; skull with a rounded profile; supraorbital processes well developed; **rostrum long and narrow**. *Notocitellus* was a subgenus of *Spermophilus*, but genetic and morphometrical analyses highlight that it should be considered at the genus level (Helgen et al. 2009). The keys were elaborated based on the review of specimens and the following sources: Howell (1918), Hall (1981), Thorington et al. (2012), and Álvarez-Castañeda et al. (2017a).

1. Tail with approximately 15 black rings and tail length greater than 175.0 mm; skull length greater than 52.0 mm; interorbital breadth 45% of the zygomatic breadth (from Nayarit south through northern Guerrero, México) ........................................ *Notocitellus annulatus* (p. 75)

1a. Tail without black rings and tail length less than 175.0 mm; skull length less than 52.0 mm; interorbital breadth 49% of the zygomatic breadth (Jalisco, Michoacán, and Guerrero, México) ........................................ *Notocitellus adocetus* (p. 74)

### *Notocitellus adocetus* (Merriam, 1903)

Tropical ground squirrel, cuinique o juancito del Balsas

1. *N. a. adocetus* (Merriam, 1903). Michoacán, northwestern Guerrero, and eastern Jalisco.
2. *N. a. infernatus* (Álvarez and Ramírez-Pulido, 1968). Southeastern Michoacán.

**Conservation:** *Notocitellus adocetus* is listed as Least Concern by the IUCN (de Grammont and Cuarón 2016b).

**Characteristics:** Tropical ground squirrel is small sized; total length 315.0–353.0 mm and skull length 41.6–46.2 mm. Dorsal pelage hispid grayish with hairs grizzly with pale brown cinnamon shades, without markings; fur of underparts pale yellowish; head, back, and bushy tail darker; with a faint pale stripe above and below the eyes; ears shorts; cheeks grizzled and washed with fulvous shades; **tail long without black rings along its length**; **interorbital region 49% of the zygomatic breadth**.

**Comments:** *Notocitellus adocetus* is endemic to the Eje Volcánico Trasnversal, occurring in Jalisco, Michoacán, and Guerrero (Map 47). It can be differentiated from *N. annulatus* by its smaller size; paler and not reddish; tail without black rings; jugal narrow; in the jaw, the coronoid and angular processes shorter.

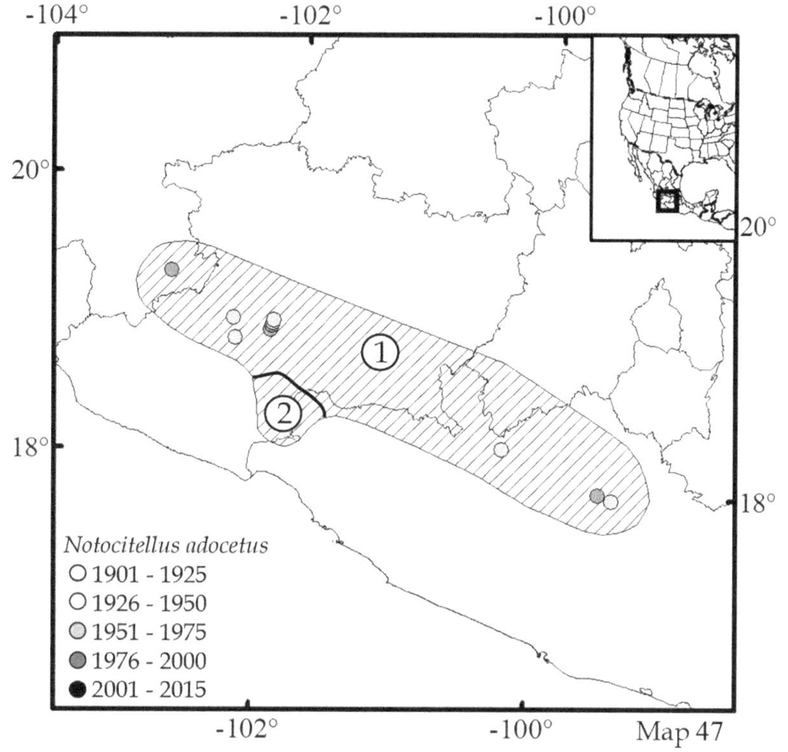


**Map 47** Distribution map of *Notocitellus adocetus*

**Additional Literature:** Merriam (1903), Howell (1938), Álvarez and Ramírez-Pulido (1968), Birney and Genoways (1973), Alvarez et al. (1987), Rickart and Yensen (1991), Best (1995i).

## *Notocitellus annulatus* (Audubon and Bachman, 1842)

Ring-tailed ground squirrel, ardilla de tierra cola anillada

1. *N. a. annulatus* (Audubon and Bachman, 1842). Northern Jalisco and Nayarit.
2. *N. a. goldmani* (Merriam, 1902). From Jalisco southward to western Guerrero.

**Conservation:** *Notocitellus annulatus* is listed as Least Concern by the IUCN (Álvarez-Castañeda et al. 2016m).

**Characteristics:** Ring-tailed ground squirrel is small sized; total length 383.0–470.0 mm and skull length 52.0–57.0 mm. Dorsal pelage black hairs grizzly with brown cinnamon; fur of underparts yellowish; head, back, and bushy tail darker; with faint pale stripe above and below the eyes; **tail with approximately 15 black rings in its length**; **interorbital region 45% of the zygomatic breadth**.

**Comments:** *Notocitellus annulatus* ranges from Nayarit south through northern Guerrero (Map 48) and can be differentiated from *N. adocetus* by its larger size; tail darker and reddish with approximately 15 black rings; jugal broader; in the jaw, the coronoid and angular processes are longer.

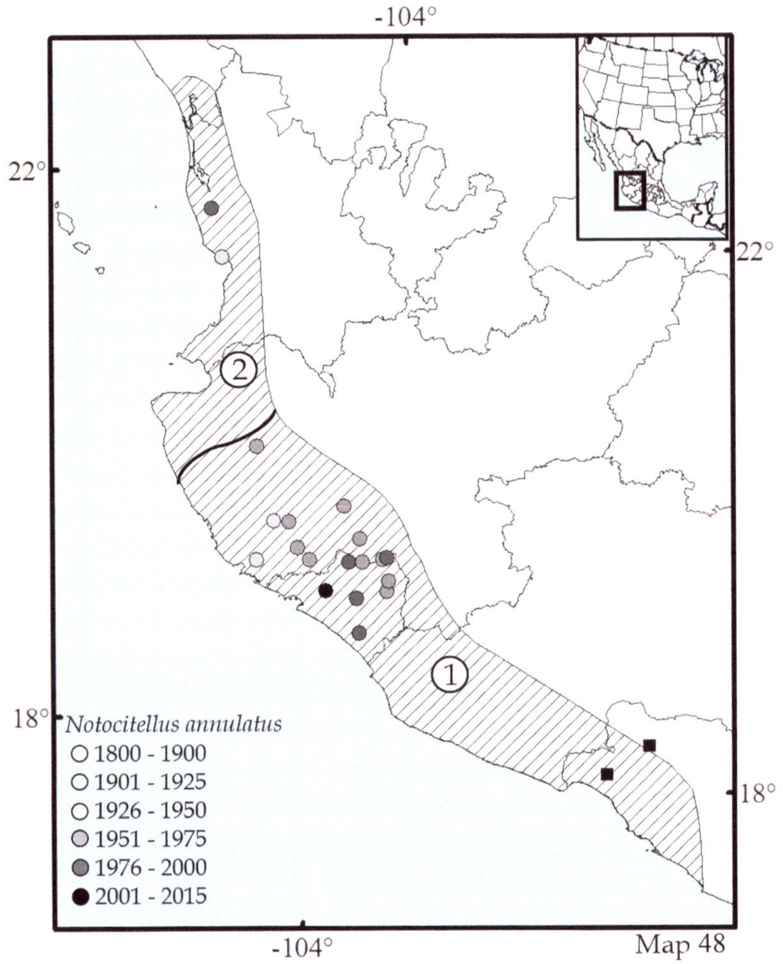


**Map 48** Distribution map of *Notocitellus annulatus*

**Additional Literature:** Merriam (1903), Howell (1938), Hall (1981), Rickart and Yensen (1991), Best (1995h), Thorington and Hoffmann (2005).

## Genus *Otospermophilus* Brandt, 1844

Dorsal pelage grizzled in white and black, with the overall shade depending on the proportion of each color; fur of underparts pale gray; **tail long, greater than 70% of the total length, bushy, gray dorsally and paler ventrally, looking as ringed;** postorbital constriction wider than the interorbital width; occlusal surface of the first and second upper molars square shaped; third upper molars slightly larger than the second upper molars; **supraorbital foramen open**. *Otospermophilus* was a subgenus of *Spermophilus*, but genetic and morphometrical analyses highlight that it should be considered at the genus level (Helgen et al. 2009). The keys were elaborated based on the review of specimens and the following sources: Howell (1918), Hall (1981), Helgen et al. (2009), Thorington et al. (2012), and Álvarez-Castañeda et al. (2017a).

1. Shoulders and nape of the same color as the rest of the back (from eastern Nevada, Utah, Colorado, and Oklahoma, United States, south through Michoacán, Estado de México, Ciudad de México, and Puebla, México).......................... ........................................................................................................................ *Otospermophilus variegatus* (p. 80)

1a. Shoulders and nape of different color than the rest of the back, generally blackish .......................................................2

2. From north of the Sacramento–San Joaquin River Delta north through southern Washington ........................................................................................................................................................... *Otospermophilus douglasii* (p. 79)

2a. From south of the Sacramento–San Joaquin River Delta, United States south through the Baja California Peninsula, México ........................................................................................................... *Otospermophilus beecheyi* (p. 77)

## ***Otospermophilus beecheyi*** (Richardson, 1829)

California rock squirrel, ardillón de California

1. *O. b. atricapillus* (Bryant, 1889). Restricted to Baja California Sur.
2. *O. b. beecheyi* (Richardson, 1829). Coastal California from southern Los Angeles Valley north through San Francisco Bay.
3. *O. b. fisheri* (Merriam, 1893). Central Valley, California, and Northwestern Nevada.
4. *O. b. nesioticus* (Elliot, 1904). Restricted to Santa Catarina Island, California.
5. *O. b. nudipes* (Huey, 1931). Southern California and northern Baja California.
6. *O. b. parvulus* (Howell, 1931). From the San Bernardino Mountains to Owen Valley, California.
7. *O. b. rupinarum* (Huey, 1931). Central Baja California Peninsula.
8. *O. b. sierrae* (Howell, 1938). Restricted to the Sierras of eastern California.

**Conservation:** *Otospermophilus beecheyi* is listed as Least Concern by the IUCN (Timm et al. 2016g).

**Characteristics:** California rock squirrel is medium sized; total length 357.0–500.0 mm and skull length 51.6–62.4 mm. Dorsal pelage grizzled in white and black, with the overall shade depending on the proportion of each color; **paler large spots from the nape to the shoulders**; fur of underparts yellowish white to pale brown; tail bushy, gray dorsally and paler ventrally, looking ringed; white rings around the eyes; skull without differences from the other species of the genus.

**Comments:** *Spermophilus atricapillus* is considered a junior synonym of *Otospermophilus beecheyi* (Álvarez-Castañeda and Cortés-Calva 2011). *O. beecheyi* ranges from south of the Sacramento–San Joaquin River delta south through the Baja California Peninsula (Map 49), and is not found in sympatry with *O. variegatus*. However, in central-northern California, Lake Almanor in Plumas County, it can be found in sympatry with *O. douglasii*. *O. beecheyi* can be differentiated from *O. douglasii* only by its distribution and genetic data, but no differences in morphological characteristics have been recorded; and from *O. variegatus*, by having paler large spots from the nape to the shoulders; head and shoulders not black.

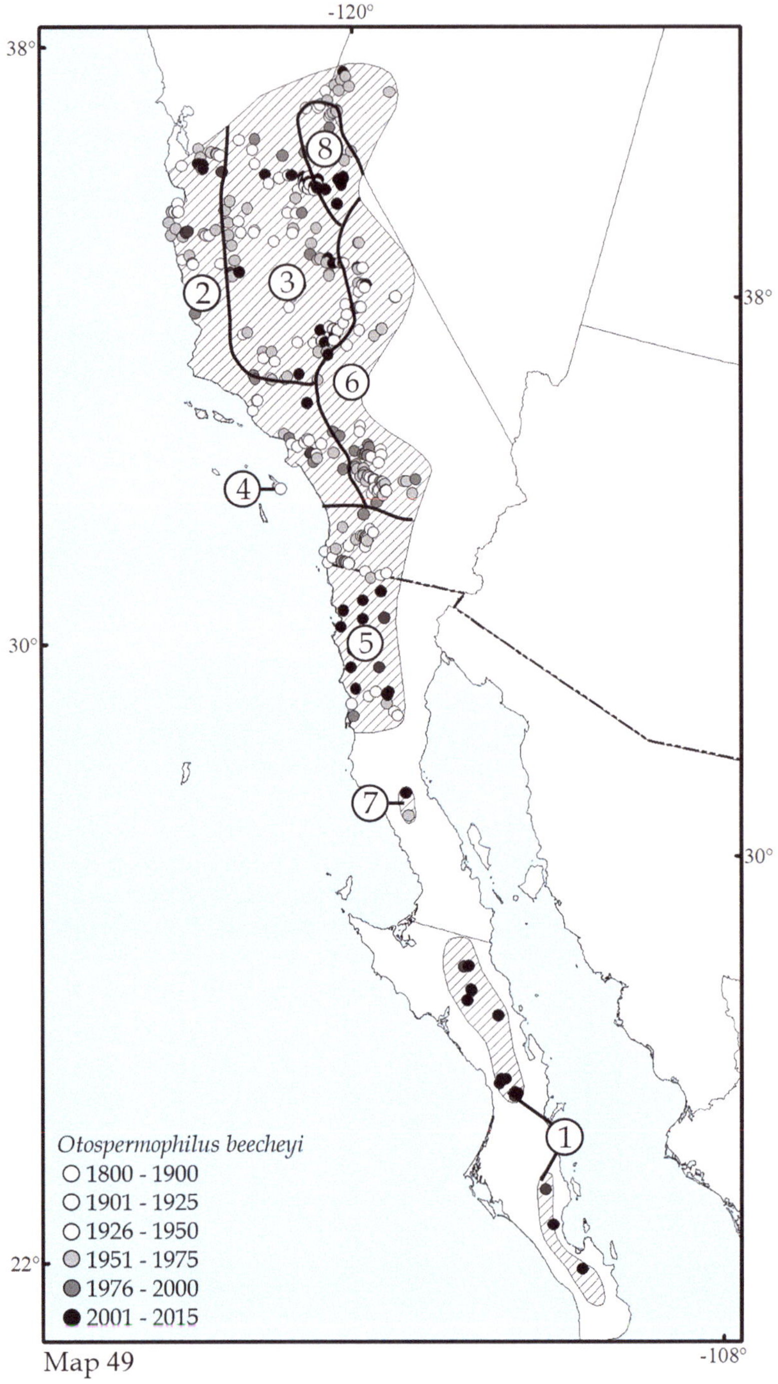


**Map 49** Distribution map of *Otospermophilus beecheyi*

**Additional Literature:** Álvarez-Castañeda et al. (1996), Rickart and Yensen (1991), Thorington and Hoffmann (2005), Smith et al. (2016), Thorington et al. (2012).

## *Otospermophilus douglasii* (Richardson, 1829)

Douglas's rock squirrel, ardillón del norte de California

Monotypic.

**Conservation:** *Otospermophilus douglasii* (as *O. beecheyi*) is listed as Least Concern by the IUCN (Timm et al. 2016g).

**Characteristics:** Douglas's rock squirrel is medium sized; total length 429.0–525.0 mm and skull length 56.0–67.7 mm. Dorsal pelage grizzled in white and black, with the overall shade depending on the proportion of each color; **paler large spots from the nape to the shoulders**; fur of underparts pale gray; tail bushy, gray dorsally and paler ventrally, looking ringed; skull with no differences from the other species of the genus.

**Comments:** *Otospermophilus douglasii* was elevated to full species (Phuong et al. 2014); it was previously regarded as a subspecies of *O. beecheyi* (Grinnell 1913). *O. douglasii* ranges from southern Washington south through the northern Sacramento–San Joaquin River Delta in California (Map 50), and is not found in sympatry with *O. variegatus*. However, in central-northern California, Lake Almanor in Plumas County, it can be found in sympatry with *O. beecheyi*. *O. douglasii* can be differentiated from *O. beecheyi* only by its distribution and genetic data, but no differences in morphological characteristics have been recorded; from *O. variegatus*, by having large paler spots from the nape to the shoulders.

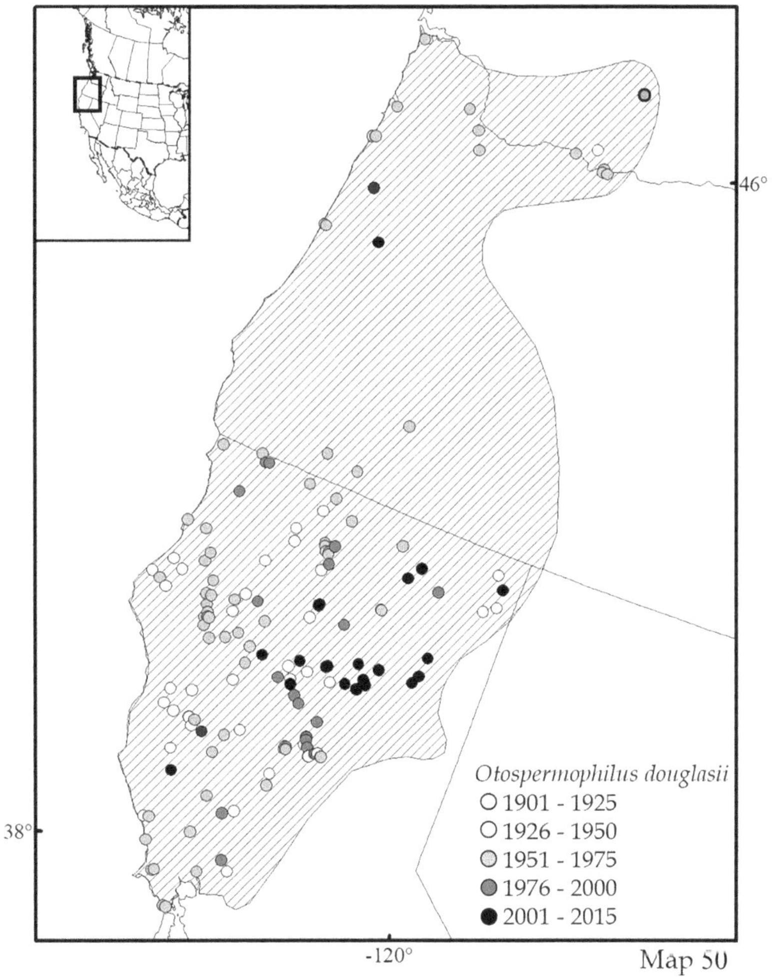


**Map 50** Distribution map of *Otospermophilus douglasii*

**Additional Literature:** Howell (1938).

## *Otospermophilus variegatus* (Erxleben, 1777)
Rock squirrel, ardillón común

1. *O. v. buckleyi* (Slack, 1861). Restricted to central Texas.
2. *O. v. couchii* (Baird, 1855). From Coahuila and eastern Chihuahua south through eastern Zacatecas, northern San Luis Potosí, and Tamaulipas.
3. *O. v. grammurus* (Say, 1823). From southeastern Utah and Colorado south through Sonora, Chihuahua, and Texas.
4. *O. v. robustus* (Durrant and Hansen, 1954). Restricted to Nevada and westernmost Utah.
5. *O. v. rupestris* (Allen, 1903). From southern Sonora and Chihuahua south through Nayarit and Zacatecas.
6. *O. v. tularosae* (Benson, 1932). Restricted to Lincoln County, central New Mexico.
7. *O. v. utah* (Merriam, 1903). Utah, southern Nevada, and northwestern Arizona.
8. *O. v. variegatus* (Erxleben, 1777). Central México, from Nayarit east through southern Tamaulipas south through Puebla and Michoacán.

**Conservation:** *Otospermophilus variegatus* is listed as Least Concern by the IUCN (Lacher et al. 2016d).

**Characteristics:** Rock squirrel is medium sized; total length 430.0–525.0 mm and skull length 56.0–67.7 mm. Dorsal pelage grizzled in white and black, with the overall shade depending on the proportion of each color; the central back occasionally with larger amount of black hair; **without large paler spots from the nape to the shoulders**; fur of underparts pale gray; tail gray dorsally and paler ventrally; it has the bushiest tail of all land squirrels; skull with no differences from the other species of the genus.

**Comments:** *Otospermophilus variegatus* ranges from eastern Nevada, Utah, Colorado, and Oklahoma south through Michoacán, Estado de México, Ciudad de México, and Puebla (Map 51), and is not found in sympatry with *O. beecheyi* or *O. douglasii*. *O. variegatus* differs from *O. douglasii* by not having large paler spots from the nape to the shoulders.

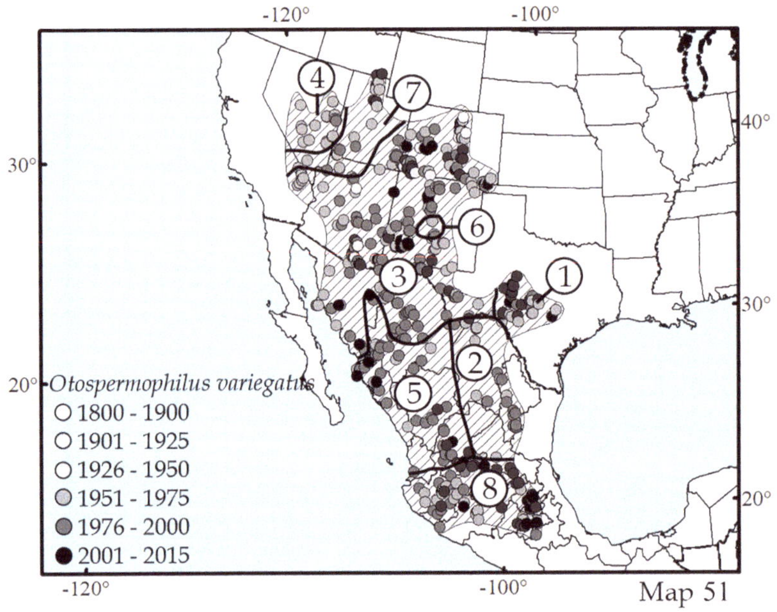


**Map 51** Distribution map of *Otospermophilus variegatus*

**Additional Literature:** Bryant (1945), Howell (1938), Dalquest (1953), Baker (1956), Hoffmeister (1956), Hall (1981), Oaks et al. (1987), Rickart and Yensen (1991).

# Genus ***Poliocitellus*** Howell, 1938

Dorsal pelage tawny olive to clay; head grayish; fur of underparts pinkish buff or buffy white; **tail dorsally and ventrally blackish mixed with buff, overlaid and bordered with creamy white**; sk**ull long and narrow; postorbital and interorbital constrictions of approximately equal width; first and second upper molars subquadrate in occlusal outline;** first and second upper molars broadly V shaped; anterior cingulum usually joining the protocone with an abrupt change of direction; **metalophs complete and mesostyles present on the first upper premolars,** and first and second upper molars; third upper molars slightly larger than the second upper molars; occlusal outline of the first and second lower molars rhomboidal. *Poliocitellus* was a subgenus of *Spermophilus*, but genetic and morphometrical analyses highlight that it should be considered at the genus level (Helgen et al. 2009).

## ***Poliocitellus franklinii*** (Sabine, 1822)

Franklin's ground squirrel, ardilla de tierra gris

Monotypic.

**Conservation:** *Poliocitellus franklinii* is listed as Least Concern by the IUCN (Cassola 2016cv).

**Characteristics:** Franklin's ground squirrel is medium sized; total length 380.0–400.0 mm and skull length 52.0–55.0 mm. Dorsal pelage salt and pepper, tawny olive to clay; fur of underparts pinkish buff to buffy white; head gray; **tail dorsally and ventrally blackish mixed with buff, bordered with creamy white**; **skull long and narrow; postorbital and interorbital constrictions similar in width**; **first and second upper molars with a subquadrate occlusal outline; trigon on the upper premolars**, first and second upper molars broadly V shaped; anterior cingulum usually joining the protocone with an abrupt change of direction; **metalophs complete and mesostyles present in the upper premolars and the first and second molars.**

**Comments:** *Poliocitellus franklinii* is the only species of the genus. It ranges from Alberta, Saskatchewan, Manitoba, and Ontario eastward to Illinois, Indiana, Wisconsin, and south through Kansas, Missouri, Illinois, and Indiana (Map 52). *P. franklinii* can be found in sympatry with species of *Ictidomys*, *Cynomys*, *Sciurus*, *Neotamias*, and *Urocitellus*, from which it can be differentiated by having the dorsal pelage salt and pepper and the head gray; tail dorsally and ventrally blackish mixed with buff; skull long and narrow with the postorbital and interorbital constrictions similar in width; occlusal outline of the first and second upper molars subquadrate; anterior cingulum usually joining the protocone with an abrupt change of direction; metalophs complete and mesostyles present in the upper premolars and the first and second molars.

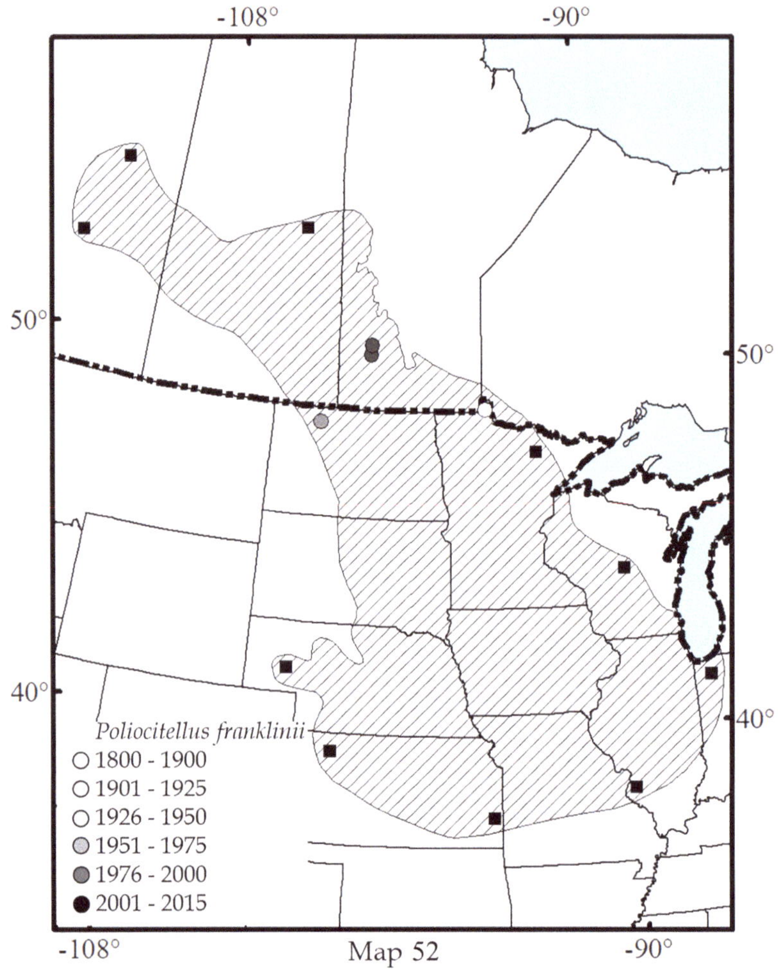


**Map 52** Distribution map of *Poliocitellus franklinii*

**Additional Literature:** Murie (1973), Jones et al. (1983b), Choromanski-Norris et al. (1989), Rickart and Yensen (1991), Smith (1993), Kurta (1995), Ostroff and Finck (2003).

## Genus *Urocitellus* Obolenskij, 1927

**Dorsal pelage varying broadly among species from individuals in solid colorations to those spotted or mottled**; fur of underparts paler than in the dorsum; eyes with two creamy white stripes above and below; tail small, in some species very small, less than 20% of the head-and-body length. *Urocitellus* was a subgenus of *Spermophilus*, but genetic and morphometrical analyses highlight that it should be considered at the genus level (Helgen et al. 2009). The keys were elaborated based on the review of specimens and the following sources: Howell (1918), Hall (1981), Helgen et al. (2009), Thorington et al. (2012), and Álvarez-Castañeda et al. (2017a).

1. Upper parts unspotted and unmottled ........ 2

1a. Upper parts spotted or mottled ........ 8

2. Tail length greater than 20% of the total length and bushy; hindfoot length greater than 39.0 mm ........ 3

2a. Tail length less than 20% of the total length and not bushy; hindfoot length less than 39.0 mm ........ 6

3. Tail ventrally grayish; wing of pterygoids nearly not parallel and skull size greater than 27.0 mm (from southeast Idaho and southern Montana south through Wyoming and Utah, United States) ........ *Urocitellus armatus* (p. 84)

3a. Tail ventrally buffy or reddish; wing of pterygoid wing nearly parallel or not and of different skull size ........ 4

4. Tail shorter and reddish ventrally; without cinnamon color on the nose and underparts (from center-south Oregon and northeastern California east through Idaho, Nevada, and Utah, United States) ..................... *Urocitellus beldingi* (p. 85)

4a. Tail longer and buffy ventrally; with cinnamon shades on the nose and underparts ........................................................ 5

5. Hindfoot length less than 43.0 mm; ranges south of 45.5° N, except for Montana (from Oregon, Nevada, Idaho, Montana, Utah, Wyoming, Colorado, and Nebraska, United States) ........................................................... *Urocitellus elegans* (p. 89)

5a. Hindfoot length greater than 43.0 mm; ranges north of 45.0° N, except for South Dakota and Minnesota (from Alberta, Saskatchewan, Canada, and Montana east through Minnesota, United States) .......................... *Urocitellus richardsonii* (p. 92)

6. Dorsal pelage pale smoke gray (restricted to southern Washington, United States)........... *Urocitellus townsendii* (p. 93)

6a. Dorsal pelage brownish, pale, or dark gray or salt and pepper ......................................................................................... 7

7. Total length less than 220.0 mm; tail with short hair at the tip; dorsal pelage salt and pepper or pale smoke gray; skull wide, braincase wide greater than 45% of the total length (restricted to southern Oregon, western Idaho and northwestern Nevada, United States) ................................................................................................................................. *Urocitellus canus* (p. 87)

7a. Total length greater than 220.0 mm; tail with large hair at the tip; dorsal pelage brownish or smoke gray; skull narrow, braincase wide. Less than 45% of the total length (from California, Oregon, Nevada, Utah, and Idaho, United States) ................................................................................................................................. *Urocitellus mollis* (p. 90)

8. Total length greater than 300.0 mm; hindfoot length greater than 43.0 mm; skull length greater than 41.0 mm ............ 9

8a. Total length less than 300.0 mm; hindfoot length less than 43.0 mm; skull length less than 41.0 mm......................... 10

9. Dorsal stripes whitish; tail long, equal to or greater than 30% of the total length; palatal posterior margin without a spine; auditive conduct of the auditory bullae well defined (Fig. 16; from Alaska, United States and British Columbia east through Hudson Bay, Canada) ................................................................................................................. *Urocitellus parryii* (p. 91)

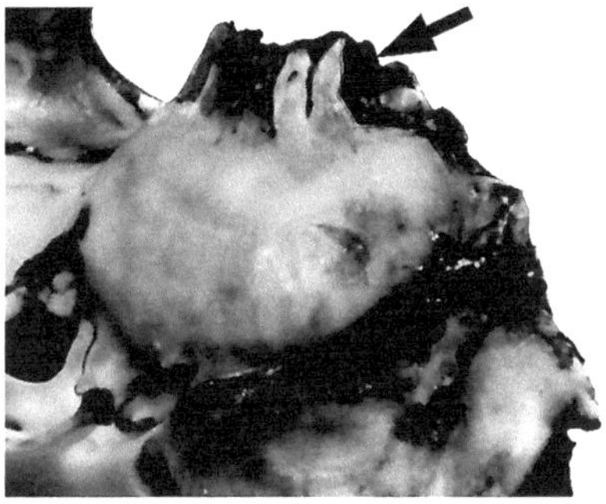

**Fig. 16** Auditive conduct of the auditory bullae as well defined as in *Urocitellus parryii*

9a. Dorsal stripes buffy; tail short, less than 30% of the total length; palatal posterior margin with a thin spine; auditory conduct of the auditory bullae not well defined (Fig. 17; restricted to Alberta, British Columbia, Idaho, Montana, Oregon, and Washington, United States)........................................................................................... *Urocitellus columbianus* (p. 88)

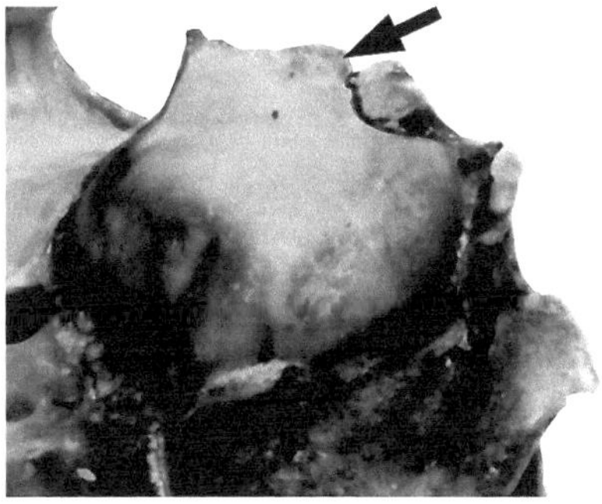

**Fig. 17** Auditory conduct of the auditory bullae as not well defined as in *Urocitellus columbianus*

10. Dorsal pelage grayish; tail short, less than 22% of the total length; ears inconspicuous (restricted to Washington and Oregon, United States)........ *Urocitellus washingtoni* (p. 94)

10a. Dorsal pelage brownish; tail long, greater than 22% of the total length; ears conspicuous (restricted to western Idaho, United States)........ *Urocitellus brunneus* (p. 86)

## *Urocitellus armatus* (Kennicott, 1863)

Uinta ground squirrel, ardilla de tierra de las montañas Uinta

Monotypic.

**Conservation:** *Urocitellus armatus* is listed as Least Concern by the IUCN (Cassola 2016eb).

**Characteristics:** Uinta ground squirrel is small sized; total length 205.0–230.0 mm and skull length 46.3–48.5 mm. **Dorsal pelage gray to gray brown with reddish brown shades**, without spots or mottled on the back, frosted with blackish brown and buff, but the buff is dominant, same color throughout the back, except for grayish buff postauricular spots; flanks with reduced blackish frosted areas; feet and limbs pale buff; fur of underparts pale buff, same as in the neck**; head coloration similar to the back, more reddish above the nose**; eyes with two creamy white stripes above and below, with low contrast; tail ventrally of the same color as the back, but more grizzled with black, more conspicuous, paler at the tip, and paler below.

**Comments:** *Urocitellus armatus* ranges from southeast Idaho and southern Montana south through Wyoming and south-central Utah (Map 53). It can be found in sympatry, or nearly so, with *U. beldingi*, *U. columbianus*, *U. elegans*, and *U. mollis*. *U. armatus* can be differentiated from *U. brunneus*, *U. columbianus*, *U. parryii*, and *U. washingtoni* by not having spotted or mottled upperparts. From *U. beldingi*, by its tail grayish above and a larger skull, greater than 27.0 mm. From *U. elegans*, by not having the fur of the underparts cinnamon and the tail buff above; from *U. mollis*, by a bushy and larger tail, greater than 20% of the head-and-body length.

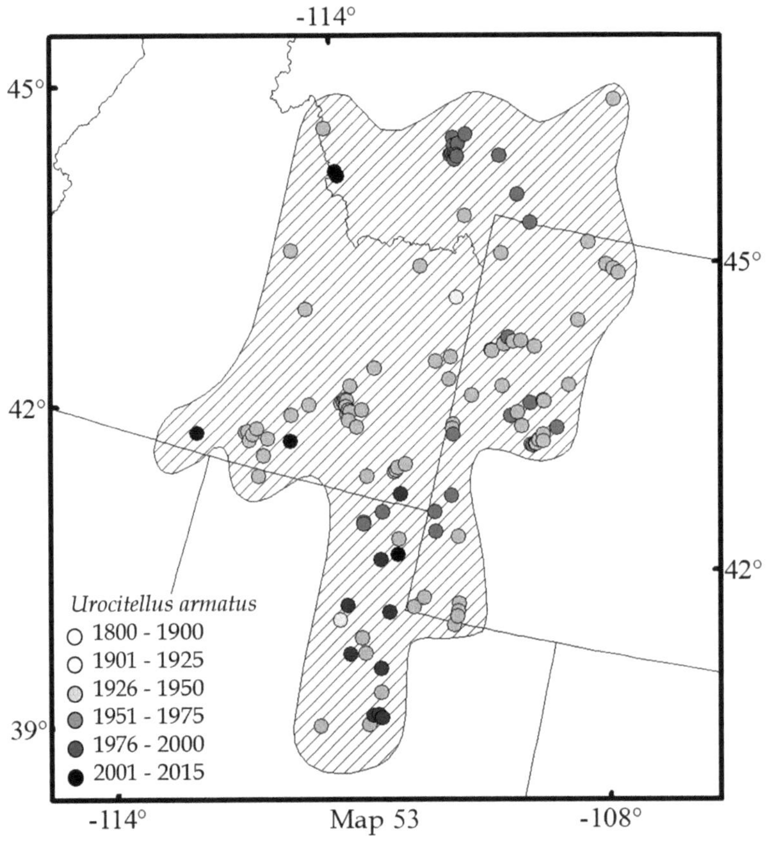


**Map 53** Distribution map of *Urocitellus armatus*

**Additional Literature:** Howell (1938), Robinson and Hoffman (1975), Hall (1981), Nadler et al. (1982), Eshelman and Sonnemann (2000), Rickart and Yensen (1991).

## *Urocitellus beldingi* (Merriam, 1888)

Belding's ground squirrel, ardilla de tierra

1. *U. b. beldingi* (Merriam, 1888). Northern-central California and extreme western Nevada.
2. *U. b. creber* (Hall, 1940). Eastern Oregon, Nevada, Idaho, and Utah.
3. *U. b. oregonus* (Merriam, 1898). Oregon and northeastern California.

**Conservation:** *Urocitellus beldingi* is listed as Least Concern by the IUCN (Cassola 2016ec).

**Characteristics:** Belding's ground squirrel is small sized; total length 180.0–215.0 mm and skull length 41.2–46.5 mm. **Dorsal pelage reddish brown in central back** and shoulders paler, without spots or mottled on the back, frosted with blackish brown and buff; flanks, feet, and legs pale buff with less frosted shades; fur of underparts buff to straw yellow, same in the neck; **ears and limbs small relative to other ground squirrels**; head more grayish brown with the flanks paler; eyes with two creamy white stripes above and below; **tail short, flat, bushy, and lacking stripes or spots, tip with banded tricolor hairs, red, black, and white; fur of underparts reddish**.

**Comments:** *Urocitellus beldingi* ranges from central-southern Oregon and northeastern California east through eastern Idaho and northeastern Utah, including northern Nevada (Map 54). It can be found in sympatry, or nearly so, with *U. armatus*, *U. brunneus*, *U. canus*, *U. columbianus*, *U. elegans*, *U. mollis*, and *U. washingtoni*. *U. beldingi* can be differentiated from *U. armatus* by having the tail buff or reddish below and a smaller skull, less than 27.0 mm. From *U. columbianus* and *U. washingtoni*, by not having the dorsal parts mottled. From *U. elegans*, by its shorter tail; without cinnamon color on the nose and underparts; tail not buffy ventrally; and from *U. canus* and *U. mollis*, by its tail bushy and larger, great than 20% of the head-and-body length.

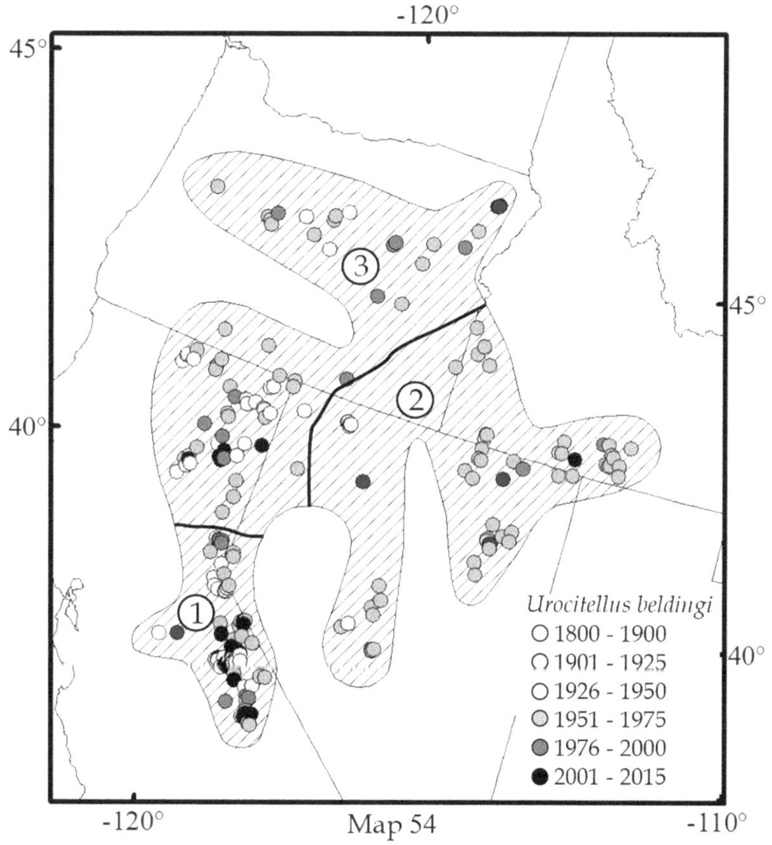


**Map 54** Distribution map of *Urocitellus beldingi*

**Additional Literature:** Grinnell and Dixon (1918), Davis (1939), Hall (1946), Nadler (1966), Nadler et al. (1982), Jenkins and Eshelman (1984), Rickart and Yensen (1991).

## *Urocitellus brunneus* (Howell, 1928)

Idaho ground squirrel, ardilla de tierra de Idaho

Monotypic.

**Conservation:** *Urocitellus brunneus* is considered Threatened by the Species Act in the United States, and is listed as Critically Endangered (B1ab (iii) + 2ab (iii): extent of occurrence less than 100 km$^2$ in only one population and under continuing decline in area, extent and quality of habitat + B2ab (iv, v): area of occupancy less than 10 km$^2$ with only one population and under continuing decline in area, extent and quality of habitat, number of locations or subpopulations, or number of mature individuals) by the IUCN (Yensen 2000).

**Characteristics:** Idaho ground squirrel is small sized; total length 209.0–258.0 mm and skull length 36.1–42.5 mm. Dorsal pelage reddish brown with white spots; fur of underparts washed with buff shades; **it is recognized by its small head and body**; eyes with a distinct ring; nose and limbs yellow pink to orange; **tail short, yellow pink to orange ventrally**.

**Comments:** *Urocitellus brunneus endemicus* was considered a full species (Rickart and Yensen 1991; Hoisington-Lopez et al. 2012), but is a subspecies of *U. brunneus* (McLean et al. 2016). *U. brunneus* ranges in western-central Idaho. *U. brunneus* is found in only three isolated areas: between the Seven Devils and Cuddy mountains and east of West Mountains (Map 55). It can be found in sympatry, or nearly so, with *U. beldingi*, *U. canus*, *U. columbianus*, and *U. mollis*. *U. brunneus* can be differentiated from *U. washingtoni* by having larger ears, 13.0–18.0 mm, a darker brown coloration; dorsal spots smaller and less distinct; lateral line less contrasting, with more pigmented rufous patches above the nose and on the thighs; tail longer, more than 22% of the total length. From *U. columbianus* and *U. parryii*, by its smaller size, less than 300.0 mm of the total length; tail not bushy and skull length less than 41.0 mm. From *U. washingtoni*, by its larger pinnae and darker pelage. From the other species of *Urocitellus*, by being spotted on the upperparts.

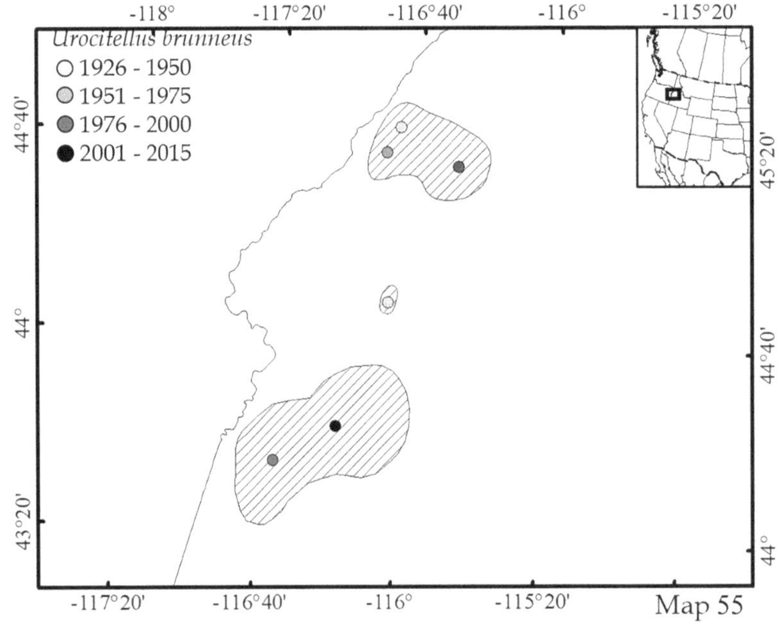


**Map 55** Distribution map of *Urocitellus brunneus*

**Additional Literature:** Howell (1938), Rickart and Yensen (1991), Gill and Yensen (1992), Yensen and Sherman (1997).

## *Urocitellus canus* (Merriam, 1898)

Merriam's ground squirrel, ardilla de tierra canosa

1. *U. c. canus* (Merriam, 1898). Oregon, northwestern Nevada and northeastern California.
2. *U. c. vigilis* (Merriam, 1913). Restricted to eastern Oregon, the Snake River western bank, western Idaho.

**Conservation:** *Urocitellus canus* is listed as Least Concern by the IUCN (Yensen and NatureServe 2017a).

**Characteristics:** Merriam's ground squirrel is **the smallest member of the genus**; total length 165.0–265.0 mm and skull length 31.8–43.2 mm. Dorsal pelage dark gray with no visible stripes or spots; cheeks and hind legs of similar color to the back, but with a pinkish or buffy wash; **ears short**; fur of underparts whitish; tail short and narrow, gray dorsally and cinnamon ventrally.

**Comments:** *Urocitellus canus* was recognized as a distinct species (Nadler 1968; Hoffmann et al. 1993); it was previously regarded as a subspecies of *U. townsendii* (Howell 1938). *U. canus* ranges from eastern Oregon, northwestern Nevada, and northeastern California to the western bank of Snake River in Idaho and northwestern Nevada (Map 56). It can be found in sympatry, or nearly so, with *U. armatus*, *U. beldingi*, *U. columbianus*, *U. elegans*, and *U. washingtoni*. *U. canus* can be differentiated from *U. brunneus*, *U. columbianus*, *U. parryii*, and *U. washingtoni* by not having spotted or mottled upperparts. From *U. armatus*, *U. beldingi*, and *U. elegans*, by its tail not bushy and shorter than 20% of the head-and-body length, and from *U. mollis*, by its smaller size, less than 270.0 mm; skull length broad and braincase greater than 15% of the skull length.

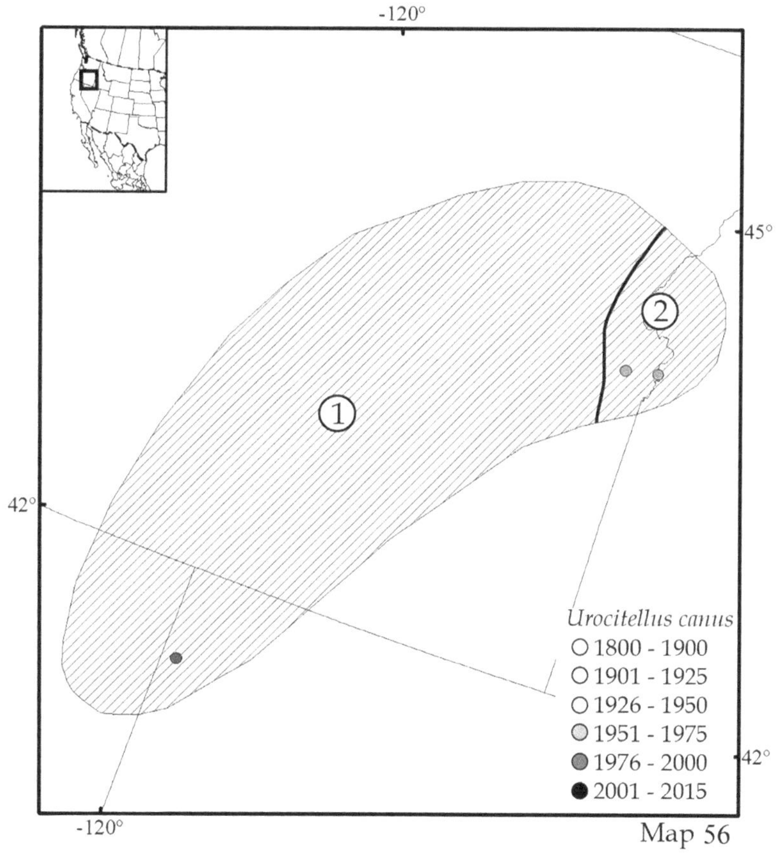


**Map 56** Distribution map of *Urocitellus canus*

**Additional Literature:** Verts and Carraway (1998), Cole and Wilson (2009), Helgen et al. (2009).

## *Urocitellus columbianus* (Ord, 1815)

Columbian ground squirrel, ardilla de tierra de la Columbia Británica

1. *U. c. columbianus* (Ord, 1815). British Columbia, Alberta, Washington, and Montana, including Idaho and Montana.
2. *U. c. ruficaudus* (Howell, 1928). Northeastern Oregon.

**Conservation:** *Urocitellus columbianus* is listed as Least Concern by the IUCN (Cassola 2016ed).

**Characteristics:** Columbian ground squirrel is small sized; total length 327.0–410.0 mm and skull length 49.5–57.5 mm. The coloration among individuals of these species is highly variable. Dorsal pelage reddish brown, without spots or mottled on the back, strongly frosted with blackish brown and buff, more grayish at nape and the back neck; neck flanks buff; flanks of the same color as the back; limbs with a strong cinnamon coloration that continues to the feet; underparts, throat and neck hairs reddish, paler in the pectoral area; head cinnamon in the forehead, cheeks less intensely colored; eyes with two narrow creamy white stripes above and below; tail short, cinnamon with a mixture of paler hairs.

**Comments:** *Urocitellus columbianus* ranges from southeastern British Columbia and southwestern Alberta south through eastern Oregon, central Idaho, and western Montana (Map 57). It can be found in sympatry, or nearly so, with *U. armatus*, *U. beldingi*, *U. brunneus*, *U. canus*, *U. elegans*, *U. mollis*, *U. towsendii*, and *U. washingtoni*. *U. columbianus* can be differentiated from *U. parryii* by having whitish dorsal stripes; tail longer, 30% or more of the total length, a vestigial posterior palatal spine, and the bulla additive conduct well defined. From *U. brunneus* and *U. washingtoni*, by its larger size, greater than 300.0 mm of the total length; tail bushy and skull length greater than 41.0 mm. From *U. parryii*, by its less buffy and more reddish general coloration; smaller average size, total length 233.0 mm and tail length 50.0 mm; and from the other species of *Urocitellus*, by being spotted on the upperparts; larger size and robust body, and dense pelage.

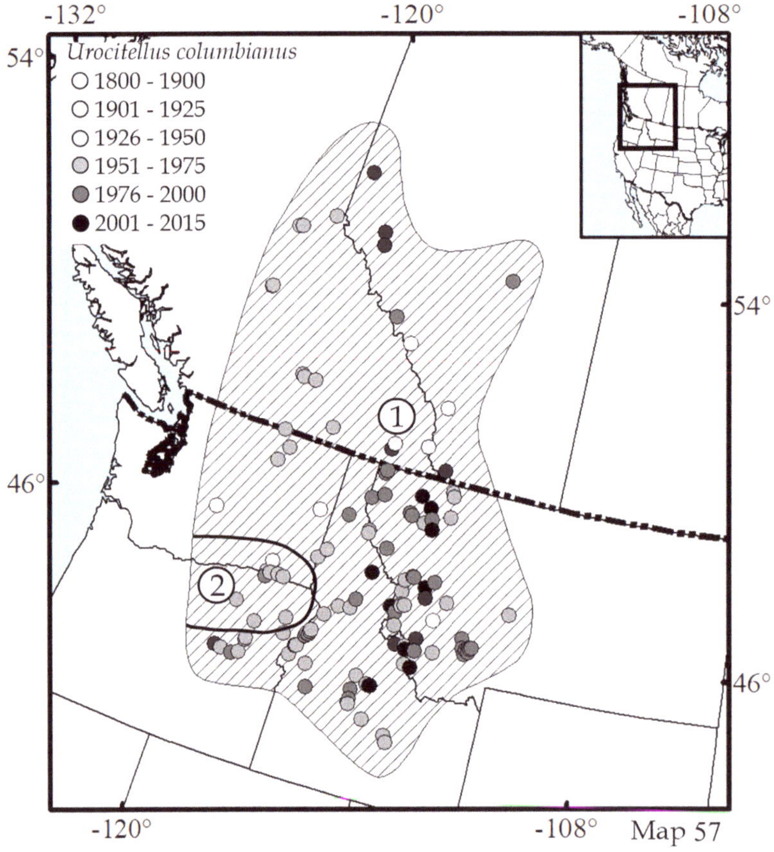


**Map 57** Distribution map of *Urocitellus columbianus*

**Additional Literature:** Howell (1938), Hansen (1954), Kivett et al. (1976), Hall (1981), Elliott and Flinders (1991), Rickart and Yensen (1991).

## *Urocitellus elegans* (Kennicott, 1863)

Wyoming ground squirrel, ardilla de tierra de Wyoming

1. *U. e. aureus* (Davis, 1939). From central Idaho to southwestern Montana.
2. *U. e. elegans* (Kennicott, 1863). From northwestern Utah to southwestern Nebraska (not in the map), south through central Colorado.
3. *U. e. nevadensis* (Howell, 1928). Northeastern Nevada, southeastern Oregon, and southwestern Idaho.

**Conservation:** *Urocitellus elegans* is listed as Least Concern by the IUCN (Yensen and NatureServe 2017b).

**Characteristics:** Wyoming ground squirrel is small sized; total length 253.0–272.0 mm and skull length 42.0–44.8 mm. **Dorsal pelage mixed gray, buff and dusky, without spots or mottled on the back, frosted with grayish brown and buff, shoulders paler and nape of different color** than the **throat**; flanks with reduced dark frosted shades; feet and limbs buff; fur of underparts buff to cinnamon, of the same color as the neck; head more reddish brown with the flanks showing a similar pattern to the back; eyes with two creamy white stripes above and below; **ears large**; **tail relatively long**, flat and of similar color to the back, tip with three bands, buff, black, and buff.

**Comments:** *Urocitellus elegans* was considered a subspecies of *U. richardsonii* (Bailey 1893). *U. elegans* ranges from southeastern Oregon and northeastern Nevada east through southwestern Montana, Wyoming, northeastern Utah, and northern-central Colorado. *U. elegans* is distributed in three disjoint geographic ranges, one for each of the three subspecies (Map 58). It can be found in sympatry, or nearly so, with *U. armatus*, *U. beldingi*, *U. canus*, *U. columbianus*, *U. mollis*, and *U. richardsoni*. *U. elegans* can be differentiated from *U. beldingi* by having a longer tail, with cinnamon color on the nose and underparts; tail underside buffy. From *U. armatus* and *U. townsendii*, by having the fur of the underparts cinnamon and the tail buff ventrally. From *U. brunneus*, *U. columbianus*, *U. parryii*, and *U. washingtoni*, by not having spotted or mottled upperparts. From *U. canus* and *U. mollis*, by having the tail bushy and larger, greater than 20% of the head-and-body length, and from *U. richardsoni*, by having the tail reddish ventrally and smaller in total and skull length, usually less than 280.0 mm and 43.0 mm, respectively.

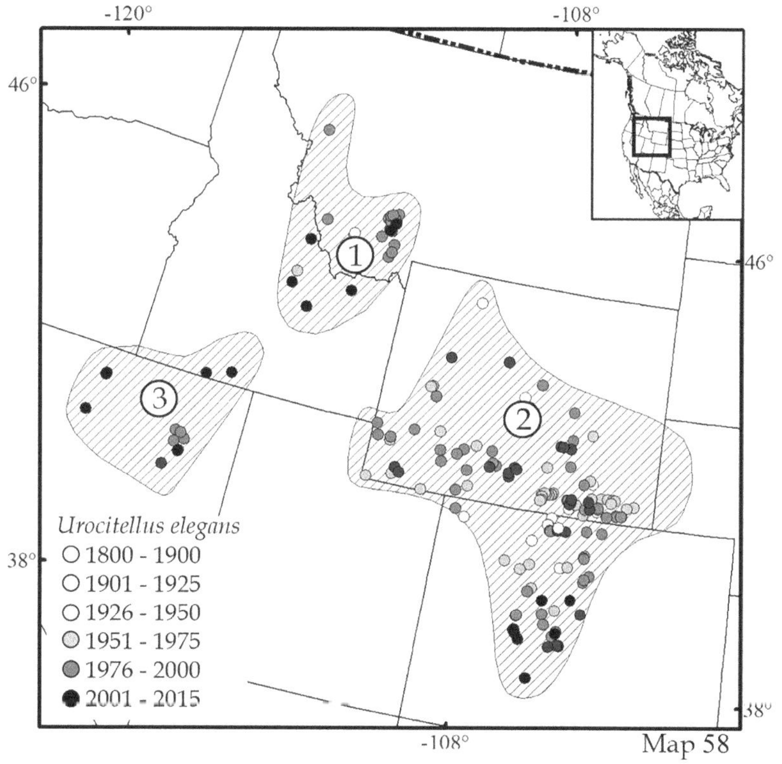


**Map 58** Distribution map of *Urocitellus elegans*

**Additional Literature:** Howell (1938), Long (1965), Armstrong (1972), Zegers (1984), Hall (1981), Rickart and Yensen (1991).

## ***Urocitellus mollis*** (Kennicott, 1863)

Piute ground squirrel, ardilla de tierra del pueblo Piute

1. *U. m. artemesiae* (Merriam, 1913). Snake River plain, central Idaho.
2. *U. m. mollis* (Kennicott, 1863). Western California, southeastern Oregon, southern Idaho, Utah, and Nevada.
3. *U. m. idahoensis* (Merriam, 1913). Western-central Idaho, northeast of the Snake River.

**Conservation:** *Urocitellus mollis* is listed as Least Concern by the IUCN (Yensen 2019).

**Characteristics:** Piute ground squirrel is small sized; total length 167.0–270.0 mm and skull length 32.0–43.5 mm. **Dorsal pelage evenly colored, pale smoke gray suffused with pinkish buff**; fur of underparts white to cream, washed with pinkish buff; cheeks and hind limbs washed with red to rust shades; tail grizzled smoke gray dorsally and cinnamon cast ventrally.

**Comments:** *Urocitellus mollis* was formerly considered part of *U. townsendii*, but it differs by having a chromosome count of $2n = 36$. *U. mollis* ranges in Idaho, Nevada, and western Utah, plus an isolated population in Oregon and California (Map 59). It can be found in sympatry, or nearly so, with *U. armatus*, *U. beldingi*, *U. brunneus*, *U. canus*, and *U. elegans*. *U. mollis* can be differentiated from *U. armatus*, *U. beldingi* and *U. elegans* by its tail not bushy and shorter than the 20% of the head-and-body length. From *U. canus*, by its larger size, greater than 220.0 mm, skull narrow and braincase less than 15% of the skull length, and from *U. brunneus*, *U. columbianus*, *U. parryii*, and *U. washingtoni*, by not having spotted or mottled upperparts.

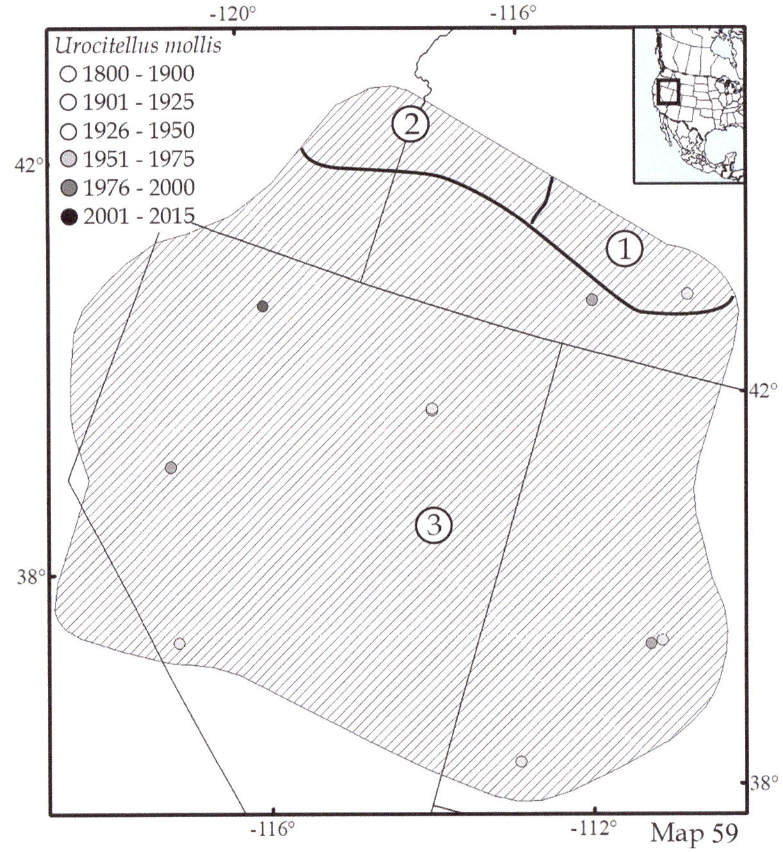


**Map 59** Distribution map of *Urocitellus mollis*

**Additional Literature:** Howell (1938).

## *Urocitellus parryii* (Richardson, 1825)

Arctic ground squirrel, ardilla de tierra del ártico

1. *U. p. ablusus* (Osgood, 1903). Western and southwestern Alaska.
2. *U. p. kennicottii* (Ross, 1861). Northern Alaska, Yukon, and northwestern Mackenzie.
3. *U. p. kodiacensis* (Allen, 1874). Restricted to Kodiak Island, southwestern Alaska.
4. *U. p. lyratus* (Hall and Gilmore, 1932). Restricted to Saint Lawrence Island, western Alaska.
5. *U. p. nebulicola* (Osgood, 1903). Restricted to the Shumagin Islands, southwestern Alaska.
6. *U. p. osgoodi* (Merriam, 1900). Restricted to south of Brooks Range, central Alaska.
7. *U. p. parryii* (Richardson, 1825). Northern coast of Mackenzie and Keewatin.
8. *U. p. plesius* (Osgood, 1900). Southeastern Alaska, Yukon, Northwest Territories, Nunavut, and northern British Columbia.

**Conservation:** *Urocitellus parryii* is listed as Least Concern by the IUCN (Cassola 2016ee).

**Characteristics:** Arctic ground squirrel is small sized, but is the largest member of the genus; total length 330.0–495.0 mm and skull length 50.5–66.0 mm. Coloration among individuals of this species is highly variable. Dorsal pelage pale buff or grizzled buff to ocher or rich chestnut to cinnamon to fuscous, **with white to buff spots on the back, and frosted with blackish brown shades**; flanks, feet, and limbs paler buff to tawny without frosted shades, gray in winter; fur of underparts white to straw yellow to cinnamon buff; head more reddish brown with the flanks paler; eyes with a creamy white ring spot; tail dorsally of the same color as the back, but grizzled with black, which increases toward thetip, ventrally tawny to russet and blackish at the tip.

**Comments:** *Urocitellus parryii* ranges in the northern of the American continent, from Alaska and northern British Columbia east through Hudson Bay (Map 60). No species of *Urocitellus* are sympatric with *U. parryii*. *U. parryii* can be differentiated from *U. brunneus* and *U. washingtoni* by its larger size, greater than 300.0 mm in total length; tail bushy and skull length greater than 41.0 mm. From *U. columbianus*, by having a more buffy and less reddish general coloration; larger average size, total length 245.0 mm; tail length 55.0 mm; from the other species of *Urocitelus*, by having spotted upperparts.

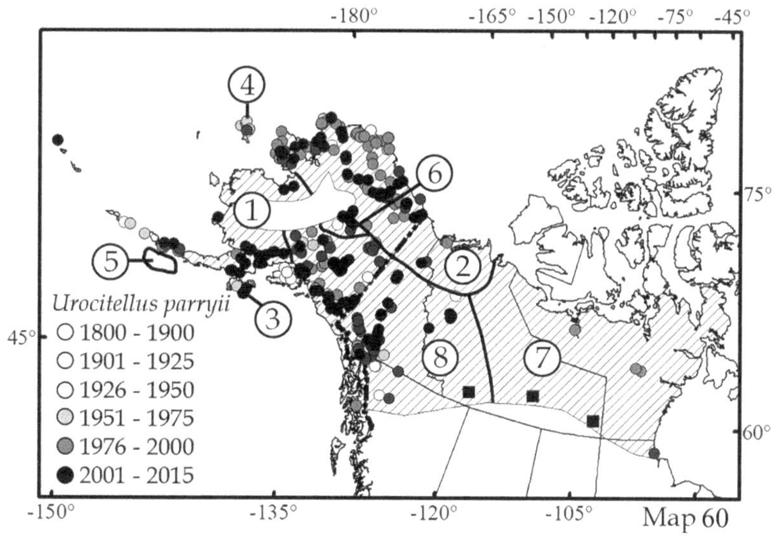


**Map 60** Distribution map of *Urocitellus parryii*

**Additional Literature:** Howell (1938), Batzli and Sobaski (1980), Rickart and Yensen (1991), Buck and Barnes (1999), Kryštufek and Vohralik (2013), McLean (2016).

## *Urocitellus richardsonii* (Sabine, 1822)

Richardson's ground squirrel, ardilla de tierra

Monotypic.

**Conservation:** *Urocitellus richardsonii* is listed as Least Concern by the IUCN (Cassola 2016ef).

**Characteristics:** Richardson's ground squirrel is small sized; total length 277.0–315.0 mm and skull length 45.1–48.4 mm. **Dorsal pelage fuscous cinnamon to gray buff, without spots or mottled on the back, frosted with grayish brown and buff with a dappled effect; shoulders paler with less dark frosted; nape of t**he **same color** as that of the **throat**; flanks, nape, and dorsal limbs paler, of similar color to the back; fur of underparts clay to buff to cinnamon; neck whitish; head more reddish than the back; eyes with two creamy white stripes above and below, with no contrast with the paler head flanks; tail short, flat dorsally and cinnamon buff to clay ventrally, tip with banded tricolor hairs, cinnamon, black, and buff.

**Comments:** *Urocitellus richardsonii* was considered a species with four subspecies; however, three of them were considered different species and are under *U. elegans: U. e. aureus*, *U. e. elegans*, and *U. e. nevadensis*. *U. richardsonii* ranges in the northern Great Plains from Alberta, Saskatchewan, Manitoba, Montana, Idaho, North Dakota, South Dakota, and Minnesota (Map 61). It can be found in sympatry, or nearly so, with *U. columbianus* and *U. elegans*. *U. richardsonii* can be differentiated from *U. brunneus*, *U. columbianus*, *U. parryii*, and *U. washingtoni* by not having spotted or mottled upperparts, and from *U. elegans*, by its tail buffy ventrally and of larger size in total and skull length, usually greater than 280.0 mm and 43.0 mm, respectively.

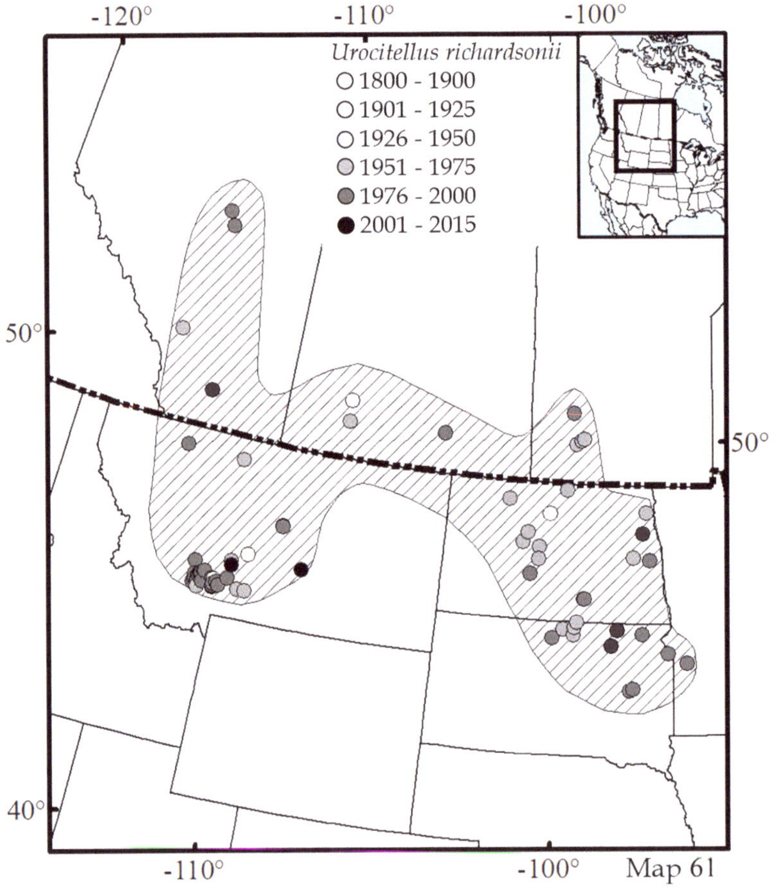


**Map 61** Distribution map of *Urocitellus richardsonii*

**Additional Literature:** Howell (1938), Nadler et al. (1971, 1982), Michener and Koeppl (1985), Rickart and Yensen (1991), McLean (2018).

## *Urocitellus townsendii* (Bachman, 1839)

Townsend's ground squirrel, ardilla de tierra del Río Columbia

1. *U. t. nancyae* (Nadler, 1968). Washington, north of Yakima River, and west of Columbia River.
2. *U. t. townsendii* (Bachman, 1839). Washington, south of Yakima River, and north of Columbia River.

**Conservation:** *Urocitellus townsendii* is listed as under protection by the United States Endangered Species Act (FWS 2022) and as Vulnerable (B1ab (iii, v): Extent of occurrence less than 20,000 km$^2$ with ten or less populations and under continuing decline in area, extent and quality of habitat, and number of mature individuals) by the IUCN (NatureServe 2016a).

**Characteristics:** Townsend's ground squirrel is small sized; total length 170.0–271.0 mm and skull length 32.4–43.0 mm. Dorsal pelage uniform pale smoke gray mixed with buff; fur of underparts white to cream, washed with pinkish buff; **general coloration paler but contrasting between the upper and underparts**; head of similar color to the back; tail grizzled smoke gray dorsally, cinnamon ventrally; cheeks and hind limbs washed with red to rust shades; skull short and broad; zygomata heavy and widely expander with a broad braincase; rostrum with nearly parallel flanks; supraorbital borders slightly elevated, and postorbital processes long, slender, and decurved.

**Comments:** *Urocitellus townsendii* was considered a species with seven subspecies. However, two were considered different species and are under *U. canus*, *U. elegans canus*, and *U. e. vigilis*, and three under *U. mollis*, *U. m. mollis*, *U. m. artemesiae*, and *U. m. idahoensis*. *U. townsendii* ranges in southern-central Washington, north of the Columbia River (Map 62). It can be found in sympatry, or nearly so, with *U. columbianus* and *U. washingtoni*. *U. townsendii* can be differentiated from *U. brunneus*, *U. columbianus*, *U. parryii*, and *U. washingtoni* by not having spotted or mottled upperparts, and from *Poliocitellus*, by its small size, less than 250.0 mm.

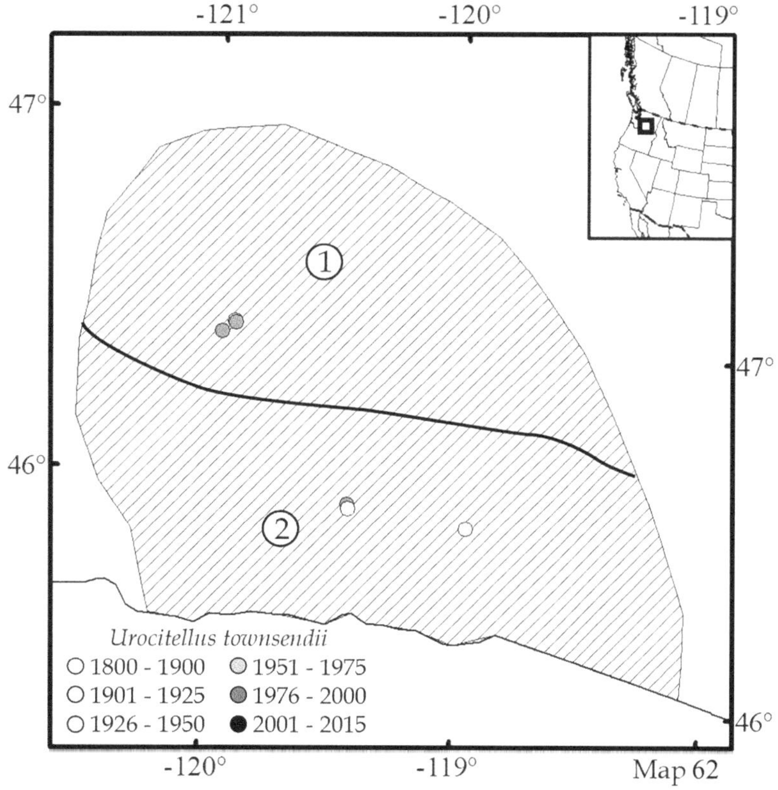


**Map 62** Distribution map of *Urocitellus townsendii*

**Additional Literature:** Howell (1938), Davis (1939), Scheffer (1941), Hansen (1954), Nadler (1968), Rickart (1982, 1986, 1987), Rickart and Yensen (1991).

## ***Urocitellus washingtoni*** (Howell, 1938)

Washington ground squirrel, ardilla de tierra de Washington

Monotypic.

**Conservation:** *Urocitellus washingtoni* is listed as under protection by the United States Endangered Species Act (FWS 2022) and Near Threatened by the IUCN (NatureServe 2016b).

**Characteristics:** Washington ground squirrel is small sized; total length 185.0–246.0 mm and skull length 35.0–42.0 mm. **Dorsal pelage pale smoke gray frosted with blackish brown and cream to buff flecks, no stripes or linear arrangements of large spots**; fur of underparts grayish white washed with buff; limbs paler and grayish buff; feet white to buff; head of similar coloration to the back, with pale frosted shades; eyes with creamy white stripes below and above; cheeks creamy white; tail very short, grizzled gray, of similarcolor as the back, but the frosted shade not in flecks; underside with less color mixture and reddish; skull short and broad; zygomata heavy and widely expanded with a broad braincase; rostrum with nearly parallel flanks; supraorbital borders slightly elevated, and postorbital processes long, slender, and decurved.

**Comments:** *Urocitellus washingtoni* ranges in eastern Washington and northeastern Oregon (Map 63). It can be found in sympatry, or nearly so, with *U. beldingi*, *U. columbianus*, and *U. towsendii*. *U. washingtoni* can be differentiated from *U. beldingi* and *U. towsendii* by having mottled dorsal parts. From *U. brunneus*, by its larger dorsal spots, lateral line more distinct, smaller ears, and paler pelage. From *U. columbianus* and *U. parryii*, by its smaller size, less than 300.0 mm of the total length; tail not bushy and skull length less than 41.0 mm. From the other species of *Urocitellus*, by its spotted upperparts. From *Poliocitellus*, by its small size, less than 250.0 mm.

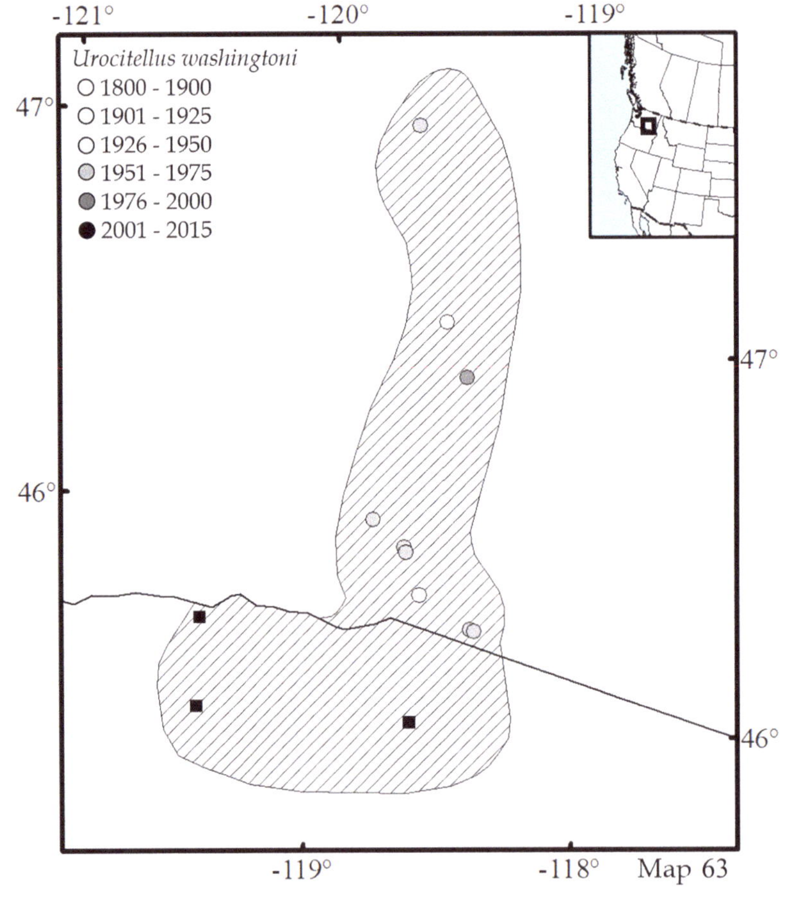


**Map 63** Distribution map of *Urocitellus washingtoni*

**Additional Literature:** Howell (1938), Scheffer (1941), Hall (1981), Rickart and Yensen (1991).

## Genus *Xerospermophilus* Merriam, 1892

Dorsal pelage varying in different combinations according to the substrate of its ranges, from cinnamon to dark smoke gray with different brown shades; individuals with or without spots on the back, depending on the species; fur of underparts in pale shades; **limbs relatively short relative to the head-and-body length; ears and tail relatively short;** skull arch shaped in lateral view; rostrum and interorbital region tend to be wide; auditory bullae proportionally large; the **postorbital process relatively short and broad**. *Xerospermophilus* was a subgenus of *Spermophilus*, but genetic and morphometrical analyses highlight that it should be considered at the genus level (Helgen et al. 2009). The keys were elaborated based on the review of specimens and the following sources: Howell (1918), Hall (1981), Helgen et al. (2009), Thorington et al. (2012), and Álvarez-Castañeda et al. (2017a).

1. Tail white ventrally; skull length on average greater than 38.5 mm (restricted to southern California, United States)........................................ *Xerospermophilus mohavensis* (p. 95)

1a. Tail normally cinnamon ventrally, but not white; skull length on average less than 38.5 mm........................................2

2. Dorsal pelage without spots, but if present, not numerous and not concentrated on the hips (From California, Arizona, and Nevada, United States, south through Baja California and Sonora, México)..............*Xerospermophilus tereticaudus* (p. 98)

2a. Dorsal pelage with many pale spots, mainly on the hips........................................3

3. Dorsal spots large, conspicuous, whitish; fur of underparts whitish; head without blackish hairs; braincase lower; auditory bullae narrow and inflated (from Wyoming, South Dakota, and Nebraska, United States, south through Jalisco, Guanajuato, San Luis Potosí, and Tamaulipas)........................................*Xerospermophilus spilosoma* (p. 97)

3a. Dorsal spots small, not particularly conspicuous, buffy; fur of underparts buffy; head with blackish hairs; braincase higher; auditory bullae broad and flat (restricted to eastern Puebla and western Veracruz, México)................................ ........................................*Xerospermophilus perotensis* (p. 96)

### *Xerospermophilus mohavensis* (Merriam, 1889)

Mohave ground squirrel, ardilla de tierra de desierto del Mohave

Monotypic.

**Conservation:** *Xerospermophilus mohavensis* is listed as Near Threatened by the IUCN (Roach and Naylor 2016b).

**Characteristics:** Mohave ground squirrel is small sized; total length 210.0–230.0 mm and skull length 38.1–40.0 mm. **Dorsal pelage nearly uniform pale brown with no stripes or spots**; fur of underparts white to cream; head with cinnamon shades; feet pale buff to cinnamon; tail short, broad, and frosted with white to cream shades, fuscous dorsally and white to cream ventrally; skull smooth, with the palate shelf backward as a long and slender spine; braincase short and broad; rostrum short; nasals end nearly in line with the premaxillae; zygomatic arches heavy and widely expanded; postorbital processes broad.

**Comments:** *Xerospermophilus mohavensis* is endemic to the northwestern Mohave Desert and Owens Valley of southern California (Map 64), and has allopatric distribution with respect to the other species of *Xerospermophilus;* however, *X. tereticaudus* has been recorded in south California. The hybridization between these two species in the Mojave River may have occurred, so these two species may represent semispecies with incomplete reproductive isolation. *X. mohavensis* can be differentiated from *X. spilosoma* and *X. tereticaudus* by its tail white dorsally, shorter, broader, and haired; cheeks are stouter and brownish.

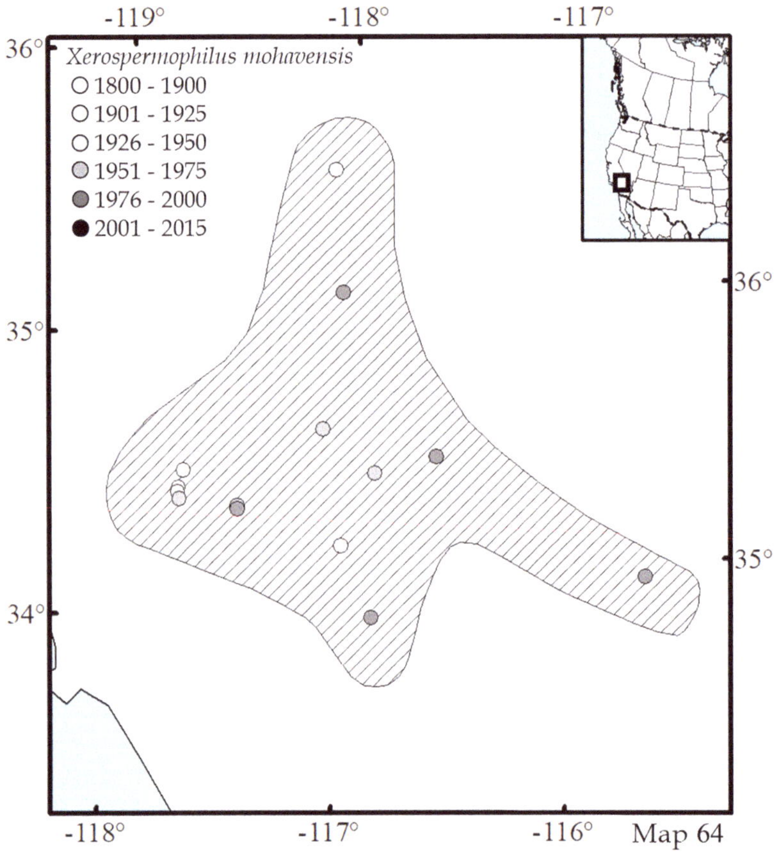


**Map 64** Distribution map of *Xerospermophilus mohavensis*

**Additional Literature:** Merriam (1889c), Grinnell and Dixon (1918), Howell (1938), Hafner and Yates (1983), Hafner (1992), Best (1995b).

## *Xerospermophilus perotensis* (Merriam, 1893)

Perote ground squirrel, ardilla de tierra de Perote

Monotypic.

**Conservation:** *Xerospermophilus perotensis* is listed as Threatened in the Norma Oficial Mexicana (DOF 2019) and as Endangered (B1ab (iii): extent of occurrence less than 5000 km$^2$ in five or less populations and under continuing decline in area, extent and/or quality of habitat) by the IUCN (Álvarez-Castañeda et al. 2016aa).

**Characteristics:** Perote ground squirrel is small sized; total length 243.0–261.0 mm and skull length 42.2–44.5 mm. **Dorsal pelage grizzled yellowish-brown, with small buffy spots**; fur of underparts buffy; white spot outlining the eyes; feet buffy; tail grizzled yellowish-brown dorsally, ochraceous buff ventrally, and black at the tip; braincase high; auditory bullae broad and flat.

**Comments:** *Xerospermophilus perotensis* was recognized as a distinct species (Piaggio and Spicer 2001; Harrison et al. 2003; Herron et al. 2004). *X. perotensis* is restricted to the Cuenca Oriental, western Puebla and eastern-central Veracruz (Map 65), and has allopatric distribution with respect to the other species of *Xerospermophilus*. *X. perotensis* can be differentiated from *X. spilosoma* by its larger size; tail shorter; coloration more yellowish; dorsal spots buffy, less conspicuous, and smaller; fur of the underparts buffy and head marked with white shades; skull larger and relatively narrow; higher braincase; auditory bullae broader and flatter.

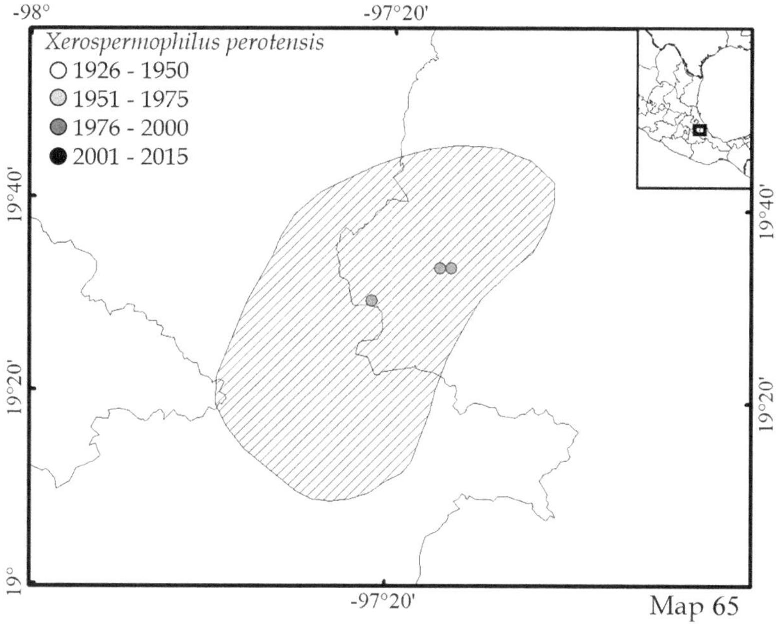


**Map 65** Distribution map of *Xerospermophilus perotensis*

**Additional Literature:** Howell (1938), Best and Ceballos (1995), Hoffmann and Thorington (2005), Fernández (2012), Thorington et al. (2012).

## *Xerospermophilus spilosoma* (Bennett, 1833)

Spotted ground squirrel, ardilla de tierra de desierto manchada

1. *X. s. altiplanensis* (Anderson, 1972). Western-central Chihuahua.
2. *X. s. ammophilus* (Hoffmeister, 1959). Extreme northern Chihuahua.
3. *X. s. annectens* (Merriam, 1893). Southern Texas and the Rio Grande Valley.
4. *X. s. bavicorensis* (Anderson, 1972). Restricted to the Laguna de Babícora Basin, western-central Chihuahua.
5. *X. s. cabrerai* (Dalquest, 1951). San Luis Potosí and southern Nuevo León.
6. *X. s. canescens* (Merriam, 1890). Southern-central New Mexico, southeast Texas, and Chihuahua.
7. *X. s. cryptospilotus* (Merriam, 1890). Four Corners area in Arizona, Utah, Colorado, and New Mexico.
8. *X. s. marginatus* (Bailey, 1890). Southeastern Colorado, southwestern Kansas, New Mexico, western Texas, and fringe of Oklahoma.
9. *X. s. obsoletus* (Kennicott, 1863). Southeastern Wyoming, southern South Dakota, northeastern Colorado, Nebraska, and Kansas.
10. *X. s. oricolus* (Álvarez, 1962). Coast of Tamaulipas.
11. *X. s. pallescens* (Howell, 1928). Chihuahua, Coahuila, Durango, and Zacatecas.
12. *X. s. pratensis* (Merriam, 1890). Northern Arizona.
13. *X. s. spilosoma* (Bennett, 1833). Durango, Zacatecas, San Luis Potosí, Aguascalientes, and northern Michoacán.

**Conservation:** *Xerospermophilus spilosoma* is listed as Least Concern by the IUCN (Lacher et al. 2016q).

**Characteristics:** Spotted ground squirrel is small sized; total length 185.0–253.0 mm and skull length 34.1 to 42.7 mm. Dorsal pelage with different coloration according with the substrate of its range, cinnamon to dark smoke gray through various brown shades, **in all cases with a series of spots on the back and flanks, which may vary geographically**, head darker than the back, and spotless; fur of underparts light gray to white; white spot outlining the eyes; tail blackish at the tip.

**Comments:** *Xerospermophilus spilosoma perotensis* was elevated to full species (Piaggio and Spicer 2001; Harrison et al. 2003; Herron et al. 2004). *X. spilosoma* ranges from Wyoming, south Dakota, and Nebraska south through Jalisco, Guanajuato, and San Luis Potosí, and a second branch through southern Texas and Tamaulipas, and from Idaho and Arizona east through Kansas, Oklahoma, and Texas (Map 66), and has allopatric distribution with respect to the other species of *Xerospermophilus*. *X. spilosoma* can be differentiated from *X. mohavensis*, by having its tail not white ventrally. From *X. perotensis*, by its smaller size; tail longer; coloration more whitish to gray; dorsal spots whitish, more conspicuous, and larger; fur of the underparts white and head without white marks; smaller skull, relatively wide and lower braincase; auditory bullae narrow and globose. From *X. tereticaudus*, by having many paler spots on the back, mainly on the hips.

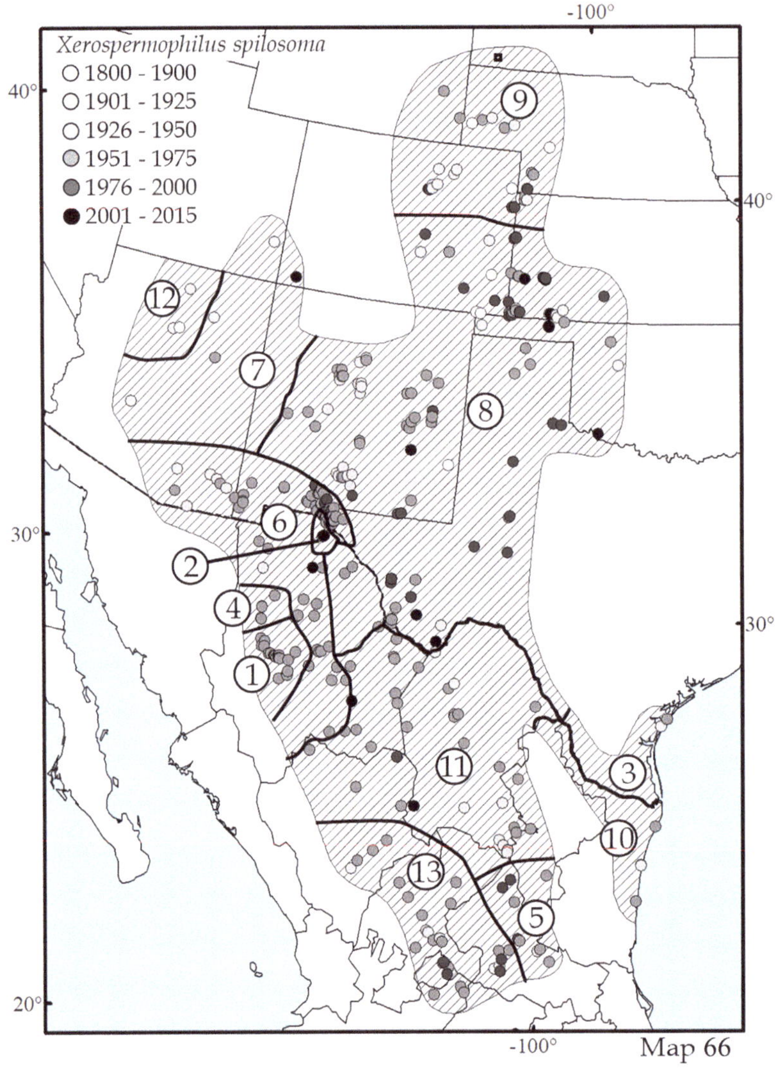


**Map 66** Distribution map of *Xerospermophilus spilosoma*

**Additional Literature:** Howell (1938), Streubel and Fitzgerald (1978a), Best and Ceballos (1995), Harrison et al. (2003); Herron et al. (2004), Hoffmann and Thorington (2005), Thorington et al. (2012).

## *Xerospermophilus tereticaudus* (Baird, 1858)

Round-tailed ground squirrel, ardilla de tierra de desierto de cola redonda

1. *X. t. apricus* (Huey, 1927). Restricted to Valle de la Trinidad, northern Baja California.
2. *X. t. chlorus* (Elliot, 1904). Restricted to the Coachella Valley, southern California.
3. *X. t. neglectus* (Merriam, 1889). Eastern of the Colorado River, southern and western Arizona, Sonora, and Nayarit.
4. *X. t. tereticaudus* (Baird, 1858). Western of the Colorado River, southern Nevada, California, and Baja California.

**Conservation:** *Xerospermophilus tereticaudus* is listed as Least Concern by the IUCN (Lacher et al. 2016r).

**Characteristics:** Round-tailed ground squirrel is small sized; total length 204.0– 278.0 mm and skull length 34.9–39.3 mm. **Dorsal pelage uniform, without spots;** three main shades: pale gray, pale cinnamon, and pale brown; flanks whitish yellow; fur of underparts whitish; tail of the same color as the back, blackish at the tip and yellowish to tan ventrally; cheeks white to pale clay, head crown dark; skull smooth, with the palate shelf backward in the form of a long and slender spine; braincase short and broad; rostrum short; nasals end nearly in line with the premaxillae; zygomatic arches heavy and widely expanded; postorbital processes broad.

**Comments:** *Xerospermophilus tereticaudus* ranges from southern California, Arizona, and Nevada southward to northeastern Baja California, western Sonora, and Nayarit (Map 67), and has allopatric distribution with respect to the other species of *Xerospermophilus*; however, *X. mohavensis* ranges in south California and hybridization between these two species has been detected in the Mojave River, so these two species may represent semi-species with incomplete reproductive isolation. *X. tereticaudus* can be differentiated from *X. mohavensis* by its tail not white ventrally, longer, narrower, and haired; cheeks stouter and whitish; and from *X. spilosoma*, by not having many paler spots on the back, mainly on the hips.

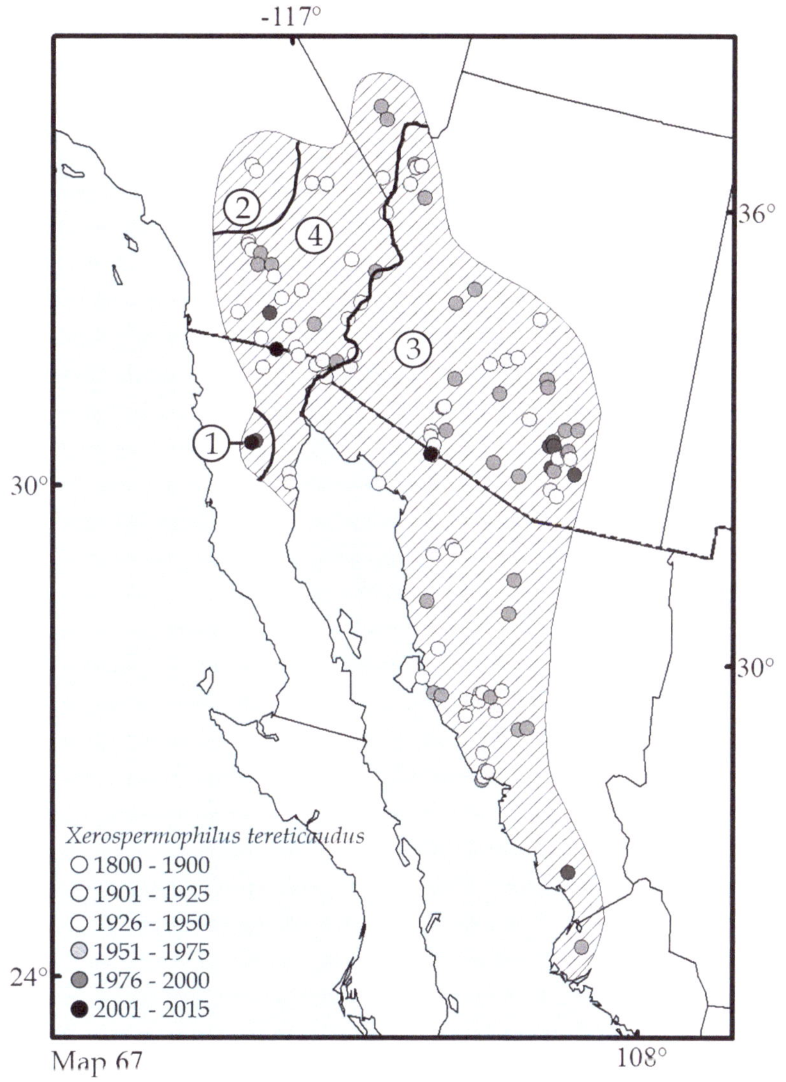


**Map 67** Distribution map of *Xerospermophilus tereticaudus*

**Additional Literature:** Hafner and Yates (1983), Ernest and Mares (1987), Hafner (1992), Thorington et al. (2012).

## Tribe Tamiini

1. Seven dorsal stripes; with no internal white stripes, but with a wide space, equivalent to less than the width of two of the existing stripes, of the same coloration as the neck and hips; third upper molars absent (from southern Manitoba east through Nova Scotia, and south through Louisiana to Georgia, United States; it was introduced to Newfoundland)....... ............ *Tamias striatus* (p. 133)

1a. Nine dorsal stripes present, which can be poorly defined, but all of approximately the same width; third upper molars present............ *Neotamias* (p. 100)

## Genus *Neotamias* Howell, 1929

**The dorsal pattern of *Neotamias* includes nine longitudinal stripes, five darker and four paler, with the paler stripes in between the darker ones. Variation in the presence, absence, coloration, shade, length, and intensity of nine particular stripes** that can be used for the identification of the species. The nomenclature used in this key for the stripes is as follows. For darker stripes, dark central stripe on the back middlemost part, usually from the nape to the rump, dark middle stripes on the flanks of the back, and outer darker stripes on the flanks. Inner paler stripes between the dark central stripe and the dark middle stripe. Outer paler stripes between the dark middle stripe and the dark outer stripe; **small vestigial upper premolars present**. The genus *Neotamias* is used (Patterson and Norris 2016), with five species groups: *alpinus* [*N. alpinus*], *amoenus* [*N. amoenus* and *N. panamintinus*], *quadrivittatus* [*N. bulleri*, *N. cinereicollis*, *N. durangae*, *N. palmeri*, *N. quadrivittatus*, *N. ruficaudus*, *N. rufus*, *N. solivagus*, *N. speciosus*, and *N. umbrinus*], *minimus* [*N. canipes*, *N. minimus*], and *townsendii* [*N. dorsalis*, *N. merriami*, *N. obscurus*, *N. ochrogenys*, *N. quadrimaculatus*, *N. senex*, *N. siskiyou*, *N. sonomae*, and *N. townsendii*]). The genus *Neotamias* is used following Patterson and Norris (2016); however, Koprowski et al. (2017) still use *Tamias.*

**Note:** For many *Neotamias* species, the genital bones and vocalizations are used for differentiating between species. In this key, we attempted to use only morphological characteristics, which are not very clear for identification. The distribution patterns can be valuable to determine particular species. The keys were elaborated based on the review of specimens and the following sources: Howell (1918), Hall (1981), and Thorington et al. (2012).

1. Only the dorsal midline of contrasting coloration; the other eight stripes very poorly defined or absent (from southern Idaho and Wyoming, United States, south through Sonora and Durango, México)............ *Neotamias dorsalis* (p. 109)

1a. At least five stripes can be clearly seen on the back............ 2

2. Dorsal pelage in dark shades, none of the paler stripes contrast strongly............ 3

2a. Dorsal pelage in brown shades or paler gray, at least the outer paler stripes contrast............ 11

3. Paler dorsal stripes practically absent, if present, there is a scarce interpellated clear line in the place where stripes should be............ 4

3a. Paler dorsal stripes present, at least the outer ones............ 6

4. Postauricular spot practically absent or markedly reduced, dirty gray (from southern British Columbia, Canada and Washington, United States)............ *Neotamias townsendii* (p. 130)

4a. Postauricular spot present although reduced, in gray............ 5

5. Restricted to southwestern California and the Baja California Peninsula; nasal bones separated by a small notch; tip of the baculum very small in relation to the shaft length (from San Francisco Bay and Sierra Nevada, United States, south through the northernmost Baja California peninsula, México)............ *Neotamias merriami* (p. 112)

5a. Restricted to the California Van Duzen River northern coast to Sonoma; nasal bones not separated by a small notch; baculum L shaped (along the northern coast of California, United States)............ *Neotamias ochrogenys* (p. 116)

6. Dark dorsal middle stripes well defined in blackish shades, region posterior to the neck not white grayish, in dark shades............ 8

6a. Dark dorsal middle stripes not very well defined: if well defined in brownish shades, the region posterior to the neck white grayish, in gray shades ........ 8

7. Hind legs of a paler color than the hips; ears with with a posterior edge, without a contrasting white stripe; nasal bones separated by a small notch; baculum tip less than one-half of the shaft length (from northwestern California, Siskiyou County south through San Francisco Bay, California, United States) ........ *Neotamias sonomae* (p. 128)

7a. Hind legs of the same coloration as the hips; ears with with a posterior edge with a contrasting white stripe; nasal bones not separated by a small notch; baculum tip longer than the shaft (from Oregon south through coast of northern California, United States) ........ *Neotamias siskiyou* (p. 126)

8. Dorsal pelage with no reddish stripes ........ 9

8a. Dorsal pelage with reddish shades and with paler and darker stripes ........ 22

9. From Sierra Nevada, California, southward; nasal bones separated by a small notch; tip of the baculum less than one-fifth of the shaft length (restricted to San Bernardino, San Jacinto, and San Francisco in California, United States, Baja California, and northern Baja California Sur, México) ........ *Neotamias obscurus* (p. 115)

9a. From Sierra Nevada, California, northward; nasal bones not separated by a small notch; tip of the baculum greater than one-fifth of the shaft length (from central Oregon south through Sierra Nevada, United States) ........ *Neotamias senex* (p. 124)

10. Coloration of the flanks and shoulders intensely reddish cinnamon ........ 11

10a. Coloration of the flanks and shoulders not intensely reddish cinnamon, although they may have cinnamon shades ........ 12

11. Postauricular spots white and large, larger than the ear size (restricted to Sierra Nevada and western-central Nevada, United States) ........ *Neotamias quadrimaculatus* (p. 120)

11a. Postauricular spot grayish and small, smaller than the ear size (from British Columbia, Alberta, and Washington east through Montana, United States) ........ *Neotamias ruficaudus* (p. 122)

12. Postauricular spot large and whitish, occupying part of the shoulders; general coloration in reddish brown shades; head in paler grayish shades; dark stripes pale brown or dark, but not blackish ........ 13

12a. Postauricular spot small and grayish, not occupying part of the shoulders; general coloration in dark brown shades, even grayish; head in grayish shades, not paler; dark stripes dark brown or blackish ........ 18

13. Middle dark stripe well defined; outer paler stripes well marked, stand out from the general coloratio ........ 14

13a. Middle dark stripe in brown shades and not contrasting; outer paler stripes present, but not standing out from the general coloration ........ 15

14. Middle faces stripe strongly contrasting with flanks; total length less than 215.0 mm (from northern California to San Bernardino Mountains and western-central Nevada, United States) ........ *Neotamias speciosus* (p. 129)

14a. Middle faces stripe does not strongly contrast with flanks; total length greater than 215.0 mm (from California east trough Wyoming and Colorado, United States) ........ *Neotamias umbrinus* (p. 131)

15. Restricted to eastern Utah, western Colorado, and northeastern Arizona, United States ........ *Neotamias rufus* (p. 123)

15a. Occurs in California or Nevada ........ 16

16. Restricted to Spring Mountain, Nevada ........ *Neotamias palmeri* (p. 118)

16a. Not present in Spring Mountain, Nevada ........ 17

17. Total and skull length less 196.0 mm and 32.5 mm, respectively. Restricted to the Sierra Nevada highlands, California, United States ........ *Neotamias alpinus* (p. 102)

17a. Total and skull length greater than 192.0 mm and 32.5 mm, respectively; restricted to the mountains of eastern California and southwestern Nevada, United States ........ *Neotamias panamintinus* (p. 119)

18. Dark outer stripes well defined and blackish brown; consequently, pale outer lines strongly contrasting; skull length less than 34.5 mm ........ 19

18a. Dark outer stripes not well defined and brownish, clear outer stripes not contrasting strongly; skull length greater than 34.5 mm ........ 20

19. Inner paler stripes well defined; fur of underparts yellowish; bent tip of the baculum length greater than 25% of the shaft length........ 20

19a. Inner paler stripes mainly absent such, instead, same coloration pattern on all the body, similar to the rump coloration; fur of underparts whitish; bent tip of the baculum length less than 25% of the shaft length ........ 21

20. General coloration dark gray, with no marked contrast between the stripes; baculum length greater than 4.5 mm and middle part broad, greater than 10% of the total length (restricted to the Craters of the Moon area, Butte County, central-south Idaho)........ *Neotamias cratericus* (p. 108)

20a. General coloration medium gray with the flanks cinnamon, clear contrast between stripes; baculum length less than 4.5 mm and middle part narrow less than 10% of the total length (from central British Columbia and Alberta, Canada, south through central California and east through central Montana and northwestern Wyoming, United States) ........ *Neotamias amoenus* (p. 103)

21. (Restricted to central Washington)........ *Neotamias grisescens* (p. 111)

21a. From Yukon east through western Québec, in the west south through California, Arizona, and New Mexico, and from Washington and Oregon, in the east south through the Dakotas, Colorado, and New Mexico ........ *Neotamias minimus* (p. 114)

22. Hind legs Gray (restricted to New Mexico and Texas, United States)........ *Neotamias canipes* (p. 106)

22a. Hind legs brown ........ 23

23. Cheeks, neck, shoulders, and rump pale gray (restricted to the mountains of central south Arizona and New Mexico, United States)........ *Neotamias cinereicollis* (p. 107)

23a. Cheeks, neck, shoulders, and rump not pale gray (from western Utah and northeast Arizona east through western Oklahoma, United States) ........ *Neotamias quadrivittatus* (p. 121)

24. Endemic to Coahuila, México........ *Neotamias solivagus* (p. 127)

24a. Endemic to the Sierra Madre Occidental, México ........ 25

25. Restricted to Chihuahua and Durango, México ........ *Neotamias durangae* (p. 110)

25a. Restricted to southern Durango, Zacatecas, and Jalisco, México ........ *Neotamias bulleri* (p. 105)

## ***Neotamias alpinus*** (Merriam, 1893)

Alpine chipmunk, chichimoco alpino

Monotypic.

**Conservation:** *Neotamias alpinus* is listed as Least Concern by the IUCN (Cassola 2016bh).

**Characteristics:** Alpine chipmunk is small sized relative to other squirrels; total length 166.0–196.0 mm and skull length 29.0–32.0 mm. Dorsal pelage with a dark central stripe, darker than the others, and blackish; middle stripes with low contrast, brownish and similar in color to the general pattern; flank stripes very inconspicuous, in cinnamon coloration when present; inner stripes paler in some individuals but very narrow and not contrasting; outer stripes wide and white, but pale in some individuals; fur of underparts orange. General coloration grayish with cinnamon shades; ears of the same color as the head crown, with white postauricular patches; hind feet paler gray; rump grayish; facial stripes well developed but not contrasting; back of the head grayish; tail blackish at the tip. Species group *alpinus*.

**Comments:** *Neotamias alpinus* is endemic to the highlands of Sierra Nevada, California (Map 68) and can be found in sympatry, or nearly so, with *N. amoenus*, *N. quadrimaculatus*, *N. minimus*, *N. senex*, *N. speciosus*, and *N. umbrinus*. It can be differentiated from *N. amoenus*, *N. quadrimaculatus*, *N. senex*, *N. speciosus*, and *N. umbrinus* by its smaller size. From *N. minimus*, by its shorter tail and larger ears.

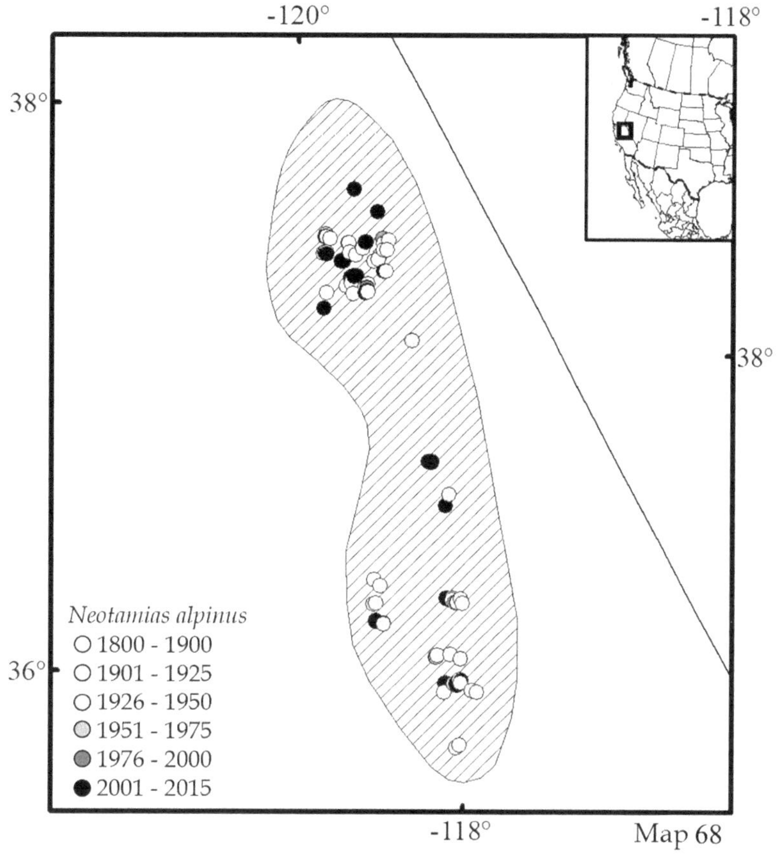


**Map 68** Distribution map of *Neotamias alpinus*

**Additional Literature:** Merriam (1893), Grinnell and Storer (1924), Howell (1929), Callahan (1976), Sutton (1992), Clawson et al. (1994a), Thorington et al. (2012).

## *Neotamias amoenus* Allen, 1890

Yellow-pine chipmunk, chichimoco del pino ponderosa

1. *N. a. affinis* (Allen, 1890). Central British Columbia and Washington.
2. *N. a. albiventris* (Booth, 1947). Northwestern Oregon and southeastern Washington.
3. *N. a. amoenus* (Allen, 1890). Northern California, central eastern Oregon, and central Washington.
4. *N. a. canicaudus* (Merriam, 1903). From eastern Oregon to most of western Montana.
5. *N. a. caurinus* (Merriam, 1898). Olympic Mountains, western Washington.
6. *N. a. celeris* (Hall and Johnson, 1940). Humboldt County, northwestern Nevada.
7. *N. a. felix* (Rhoads, 1895). Mount Baker Range, coastal British Columbia, and extreme northwestern Washington.
8. *N. a. ludibundus* (Hollister, 1911). Central British Columbia and central-western Alberta.
9. *N. a. luteiventris* (Allen, 1890). From central British Columbia and Alberta southward to Wyoming, including northeastern Washington, northern Idaho, and Montana.
10. *N. a. monoensis* (Grinnell and Storer, 1916). Sierra Nevada, central-eastern California, and western Nevada.
11. *N. a. ochraceus* (Howell, 1925). Siskiyou Mountain, northern California, and Oregon.
12. *N. a. septentrionalis* (Cowan, 1946). Central British Columbia.
13. *N. a. vallicola* (Howell, 1922). Restricted to the Bitterroot Valley, western Montana.

**Conservation:** *Neotamias amoenus* is listed as Least Concern by the IUCN (Cassola 2016bi).

**Characteristics:** Yellow-pine chipmunk is small sized; total length 180.0–245.0 mm and skull length 31.0–35.5 mm. Dorsal pelage with a central dark stripe and middle blackish stripe from the nape to the tail base, but not well defined and less contrasting than the central and middle stripes. Inner paler stripes well defined but hoary; outer stripes paler and well defined; fur of underparts yellowish. General coloration medium gray with the flanks cinnamon; ears large with a black spot in the outer upper front part; hind feet grayish; rump medium gray; facial stripes well developed but not strongly contrasting; head medium gray dorsally; tail blackish at the base with reddish brown grizzle; **bent tip of the baculum length greater than 25% of the shaft length**. Species group *amoenus*.

**Comments:** *Neotamias amoenus cratericus* was recognized as a distinct species (Herrera et al. 2022). *N. amoenus* ranges from central British Columbia and Alberta south through central California, northern Nevada, and Utah, and east through western Montana and Wyoming (Map 69). It can be found in sympatry, or nearly so, with *N. alpinus*, *N. merriami*, *N. minimus*, *N. quadrimaculatus*, *N. quadrivittatus*, *N. rufus*, *N. senex*, *N. siskiyou*, *N. sonomae*, *N. speciosus N. townsendii*, and *N. umbrinus*. *N. amoenus* can be differentiated from *N. alpinus* by its larger size. From *N. minimus*, by having the inner paler stripes well defined and fur of the underparts yellowish. From *N. speciosus*, by its smaller size; shorter and wider ears; less sharply contrasting pale and dark stripes; facial stripes more washed with ochraceous shades and the dark facial stripe usually not darker; skull less massive, with a more pointed rostrum and smaller incisive foramina. From *N. panamintinus*, it is similar in size, but with larger feet and ears; darker pelage; darker dorsal stripes and a narrow skull; and from *N. ruficaudus*, by its smaller body and skull, and its tail not reddish.

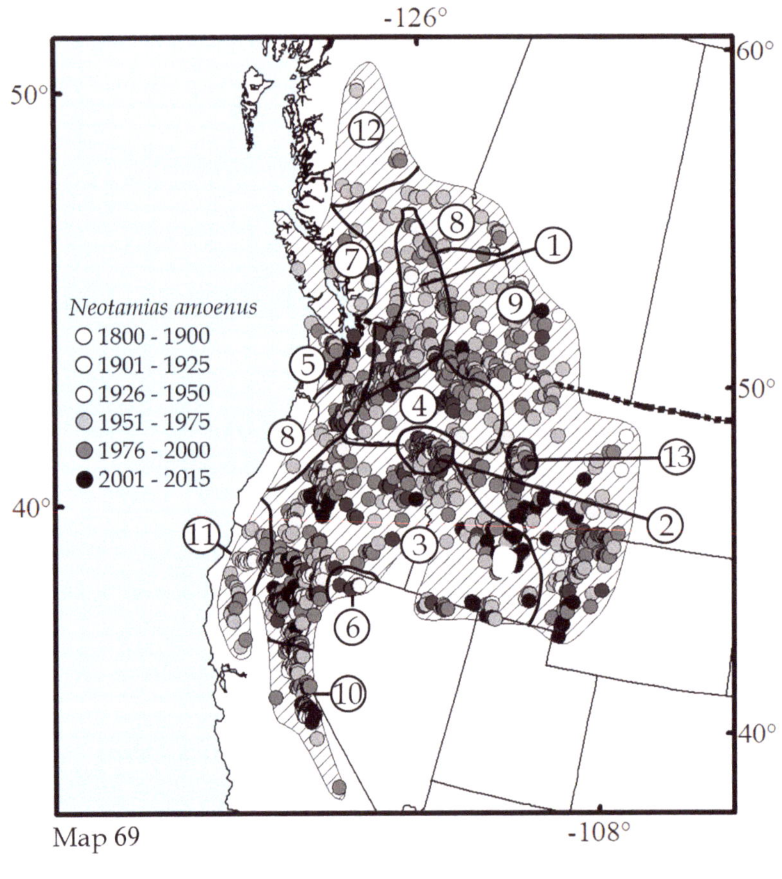


**Map 69** Distribution map of *Neotamias amoenus*

**Additional Literature:** Howell (1929), White (1953), Bryant (1945), Ellerman (1940), Patterson (1984), Patterson and Heany (1987), Best et al. (1994a, b, c), Sutton (1992), Thorington et al. (2012).

## *Neotamias bulleri* (Allen, 1889)

Buller's chipmunk, chichimoco de Zacatecas

Monotypic.

**Conservation:** *Neotamias bulleri* is listed as under protection by the Norma Oficial Mexicana (DOF 2019) and as Vulnerable (B1ab (iii): Extent of occurrence less than 20,000 km$^2$ with ten or less populations and under continuing decline in area, extent and/or quality of habitat) by the IUCN (Álvarez-Castañeda et al. 2016ab).

**Characteristics:** Buller's chipmunk is small sized; total length 222.0–248.0 mm and skull length 35.7–39.6 mm. Dorsal pelage with a central dark blackish stripe from the nape to the base of the tail, wide, contrasting with the general back coloration; middle stripes wide, brownish to blackish brown. Flank stripes very narrow, brown yellow, paler than the middle stripes. Inner paler stripes forming a narrow stripe, dark gray to reddish gray, not contrasting with the central and middle stripes; outer stripes similar to the inner paler stripes, but some individuals with slightly paler stripes; fur of underparts whitish. General coloration brownish reddish gray; ears whitish on the back outer edge and anterior side blackish, **hair inside the ears of rust coloration**; postauricular patches gray; hind feet pale to medium gray; rump brownish reddish gray; facial stripes contrasting with the general coloration of the head; tail brownish gray, paler at the tip. Species group *quadrivittatus*.

**Comments:** *Neotamias bulleri* is restricted to the Sierra Madre Occidental in southern Durango, western Zacatecas, and northern Jalisco (Map 70), and has allopatric distribution with respect to the other species of *Neotamias*.

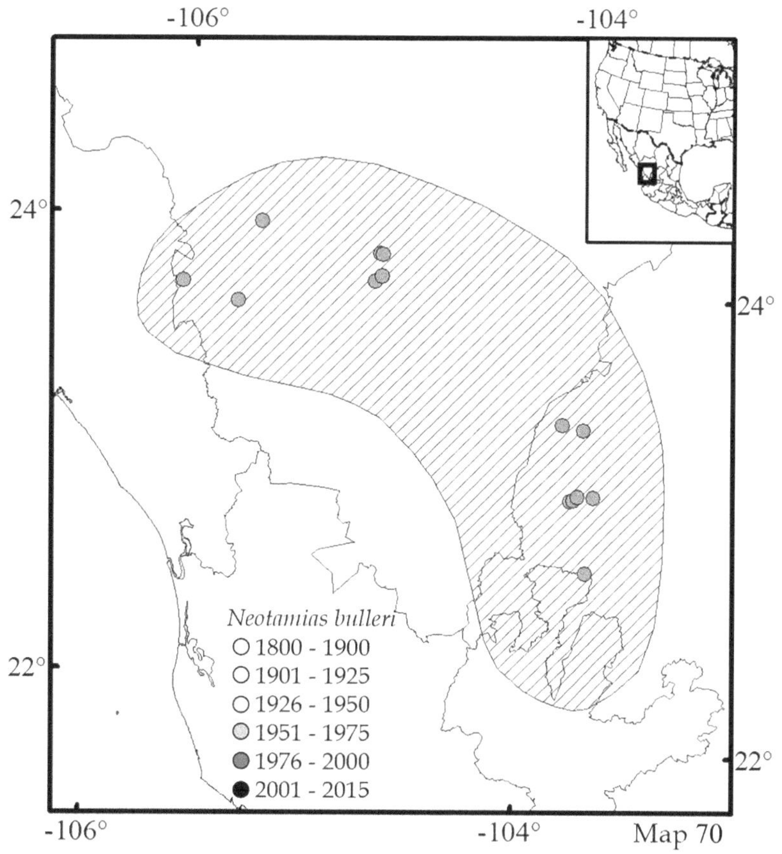


**Map 70** Distribution map of *Neotamias bulleri*

**Additional Literature:** Allen (1890, 1903), Howell (1929), White (1953), Callahan (1980), Sutton (1992), Bartig et al. (1993), Thorington et al. (2012).

## *Neotamias canipes* (Bailey, 1902)

Gray-footed chipmunk, chichimoco de patas grises

1. *N. c. canipes* (Bailey, 1902). Restricted to Capitan, Jicarilla, Gallinas, central New Mexico, Sierra Diablo, and Guadalupe Mountains, western Texas.
2. *N. c. sacramentoensis* (Fleharty, 1960). Restricted to the Sacramento Mountains, southern New Mexico.

**Conservation:** *Neotamias canipes* is listed as Least Concern by the IUCN (Cassola 2016bj).

**Characteristics:** Gray-footed chipmunk is small sized; total length 227.0–264.0 mm and skull length 36.0–39.0 mm. Dorsal pelage with the dark central stripe blackish contrasting with the inner paler stripes, wide from the nape to near the tail base; middle stripes blackish to brownish at the front, and back ends more brownish, at the middle of similar coloration to the central stripe; flank stripes mainly absent or similar in coloration to the flanks, which are reddish brown. Inner paler stripes very narrow and grizzle with brown, not strongly contrasting; outer stripes very narrow, contrasting, and without grizzle; fur of the underparts pale grayish. General coloration brownish gray; ears with similar coloration to the head dark stripe; **hind feet darker gray**; rump darker gray with brownish shade in the middle; facial stripes contrasting with the general face coloration; head top darker gray with a mixture of paler grizzle hairs; tail dark gray with some mix of cinnamon hairs, ventrally cinnamon. Species group *minimus*.

**Comments:** *Neotamias canipes* is restricted to mountains Capitan, Jicarilla, Gallinas, and Sacramento, southeastern New Mexico, and Sierra Diablo and Guadalupe western Texas (Map 71), and has allopatric distribution with respect to the other species of *Neotamias*.

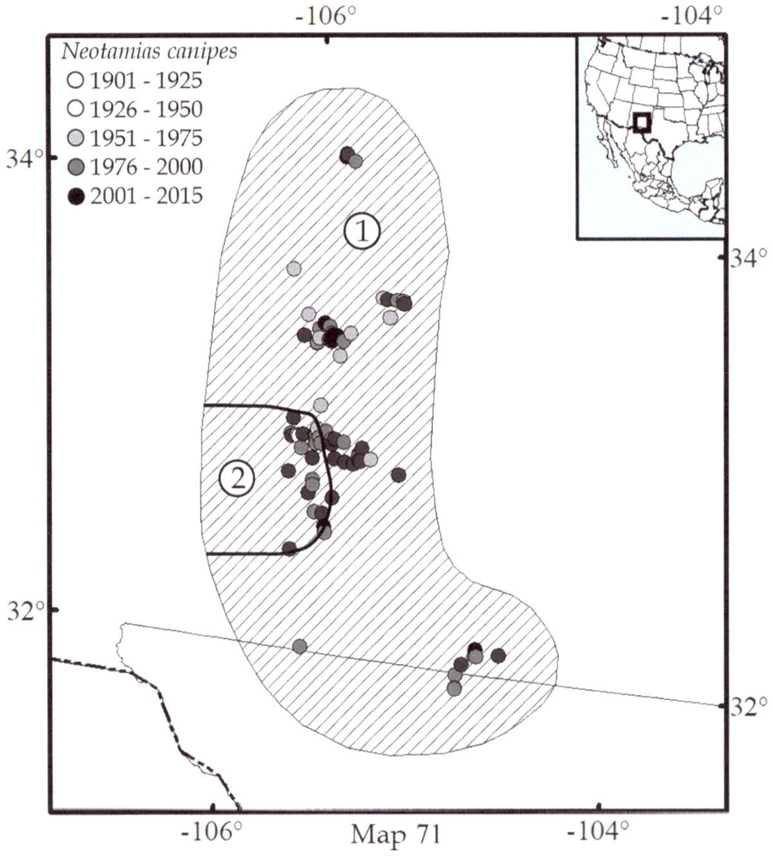


**Map 71** Distribution map of *Neotamias canipes*

**Additional Literature:** Bailey (1902a), Howell (1929), Fleharty (1960), Findley et al. (1975), Patterson (1980b, 1984), Best et al. (1992), Sutton (1992).

## *Neotamias cinereicollis* (Allen, 1890)

Gray-collared chipmunk, chichimoco de collar gris

1. *N. c. cinereicollis* (Allen, 1890). Arizona and western New Mexico.
2. *N. c. cinereus* (Bailey, 1913). Western-central New Mexico.

**Conservation:** *Neotamias cinereicollis* is listed as Least Concern by the IUCN (Cassola 2016bk).

**Characteristics:** Gray-collared chipmunk is small sized; total length 208.0–250.0 mm and skull length 35.0–38.4 mm. Dorsal pelage with a central dark stripe blackish contrasting with the inner paler stripes that are wide and well defined from the nape to the tail base; middle stripes with a similar pattern to the central stripe and wider in the middle part; flank stripes only a small narrow line in the middle of the outer paler stripes. Inner stripes paler with a mix of brownish hairs; outer stripes similar to the inner paler stripes, but some individuals with less mixture of brownish hairs in the middle and posterior areas; fur of underparts pale grayish. General coloration grayish with brownish shades; ears of a similar coloration to the dark stripe of the head; hind feet grayish; stripes continue to the rump, and area without stripes brownish gray; facial stripes well marked; head top dark grayish brown, brownish at the tip; **cheeks, neck, shoulders, and rump pale gray**; tail darker gray dorsally, with brownish hairs at the tip; fur of underparts cinnamon. Species group *quadrivittatus*.

**Comments:** *Neotamias cinereicollis* is restricted to Arizona and New Mexico (Map 72), and has allopatric distribution with respect to the other species of *Neotamias*.

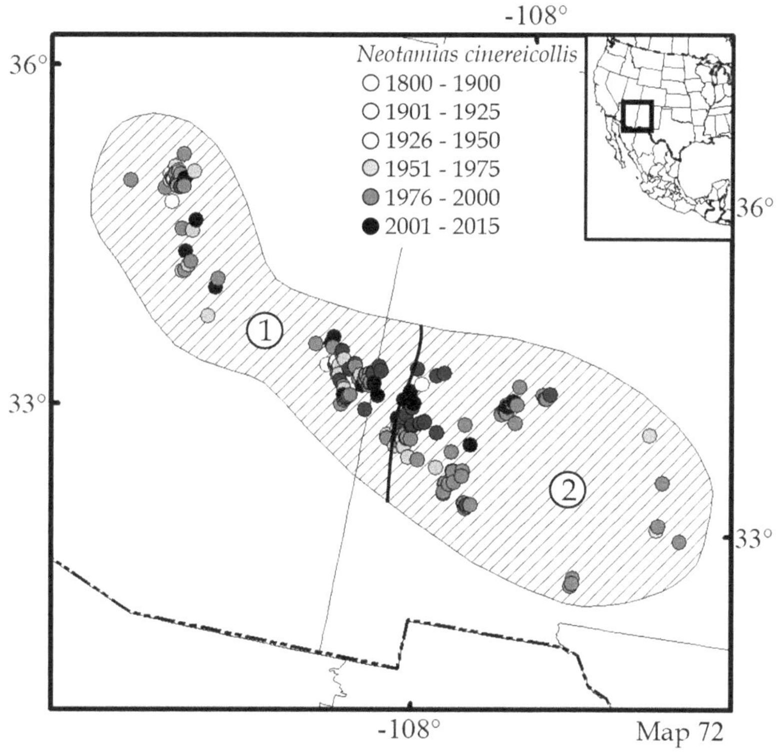


**Map 72** Distribution map of *Neotamias cinereicollis*

**Additional Literature:** Bailey (1902a), Mearns (1907), Howell (1929), White (1953), Hall (1981), Hoffmeister 1986), Sutton (1992), Hilton and Best (1993), Thorington et al. (2012).

## *Neotamias cratericus* (Blossom, 1937)

Craters of the moon chipmunk, chichimoco de los crateres de la luna

Monotypic.

**Conservation:** *Neotamias cratericus* (as *Neotamias amoenus*) is listed as Least Concern by the IUCN (Cassola 2016bi).

**Characteristics:** Craters of the moon chipmunk is small sized; total length 178.0–240.0 mm and skull length 31.0–35.0 mm. Dorsal pelage with a central dark stripe and middle blackish from the nape to the tail base, but not well defined and less contrasting than the central and middle stripes. Inner paler stripes well defined but hoary; outer stripes paler and well defined; fur of underparts yellowish. General coloration medium gray with the flanks cinnamon; ears large with a black spot in the outer upper frontal part; hind feet grayish; rump medium gray; facial stripes well developed but not strongly contrasting; dorsal part of the head medium gray; tail blackish at the base with reddish brown grizzle shades; bent tip of the baculum length greater than 24% of the shaft length, **and its length greater than 4.5 mm**. Species group *amoenus*.

**Comments:** *Neotamias cratericus* was recognized as a distinct species (Herrera et al. 2022); it was previously regarded as a subspecies of *N. amoenus* (Blossom 1937). It is endemic to Butte County, central-southern Idaho (Map 73), where it can be found in sympatry, or nearly so, with *N. amoenus* and *N. minimus*. *N. cratericus* can be differentiated from *N. minimus* by having the inner paler stripes well defined and fur of the underparts yellowish; and from *N. amoenus*, using genetic data and its distribution range.

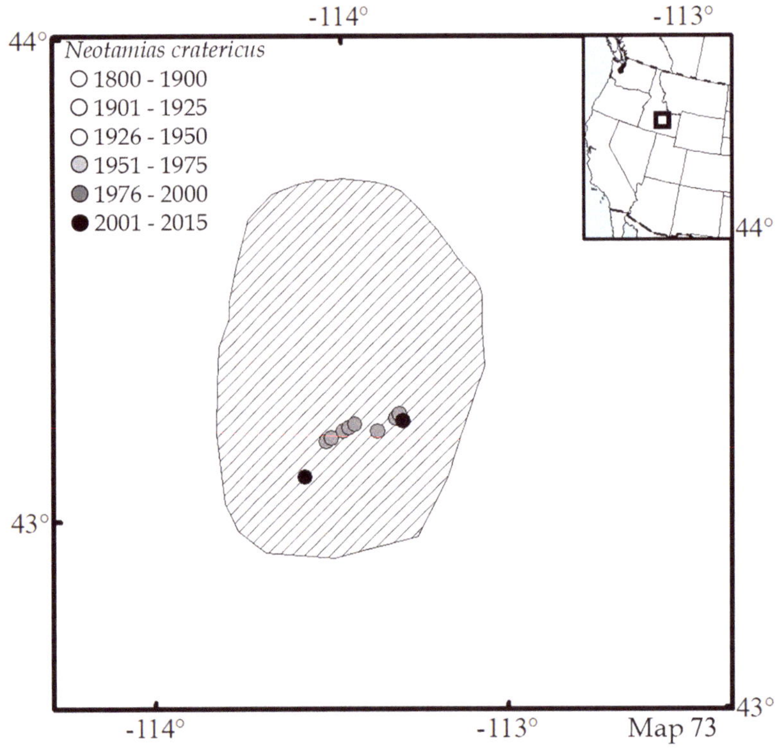


**Map 73** Distribution map of *Neotamias cratericus*

**Additional Literature:** Howell (1929), White (1953), Bryant (1945), Sutton (1992), Thorington et al. (2012).

## *Neotamias dorsalis* (Baird, 1855)

Cliff chipmunk, chichimoco de los acantilados

1. *N. d. carminis* (Goldman, 1938). Two isolated and disjunct localities in Coahuila.
2. *N. d. dorsalis* (Baird, 1855). From Arizona and New Mexico south through Durango.
3. *N. d. grinnelli* (Burt, 1931). Nevada.
4. *N. d. nidoensis* (Lidicker, 1960). Central Chihuahua.
5. *N. d. sonoriensis* (Callahan and Davis, 1977). Coastal Sonora.
6. *N. d. utahensis* (Merriam, 1897). Northwestern Arizona, eastern Nevada, Utah, Idaho, Wyoming, and Colorado.

**Conservation:** *Neotamias dorsalis* is listed as Least Concern by the IUCN (Lacher et al. 2016s).

**Characteristics:** Cliff chipmunk is small sized; total length 208.0–278.0 mm and skull length 35.5–40.1 mm. Dorsal pelage with a central dark stripe, blackish from the nape to the tail base, narrow and contrasting with the back general coloration; **middle and flank stripes absent, similar coloration to the rump. Inner paler stripes gray, not contrasting with the other stripes and the back general coloration; outer stripes gray, not contrasting with the other stripes and the back general coloration**; fur of underparts whitish. General coloration gray pale, not contrasting with the stripes, except for the central dark one, and others with few hoary; ears of similar color to the top of the head; postauricular patches whitish contrasting with the back coloration; hind feet paler brownish; rump brownish medium gray; facial stripes strong and contrasting; head top brownish gray, paler at the tip; tail blackish with a large amount of the hair paler, giving a hoary effect; cinnamon ventrally. Species group *townsendii*.

**Comments:** *Neotamias dorsalis* ranges from southern Idaho, Nevada, Utah, southern Wyoming, and northwestern Colorado south through Sonora, Chihuahua, Sonora, and Durango. There are two disjunction population segments (Map 74), and can be found in sympatry, or nearly so, with *N. merriami*, *N. rufus*, and *N. speciosus*. *N. dorsalis* can be differentiated from these species by not having the middle and flank dark stripes, of similar coloration to the rump, and the inner paler stripes gray; general dorsal coloration and outer stripes gray. None of the stripes contrasts with the general dorsal coloration.

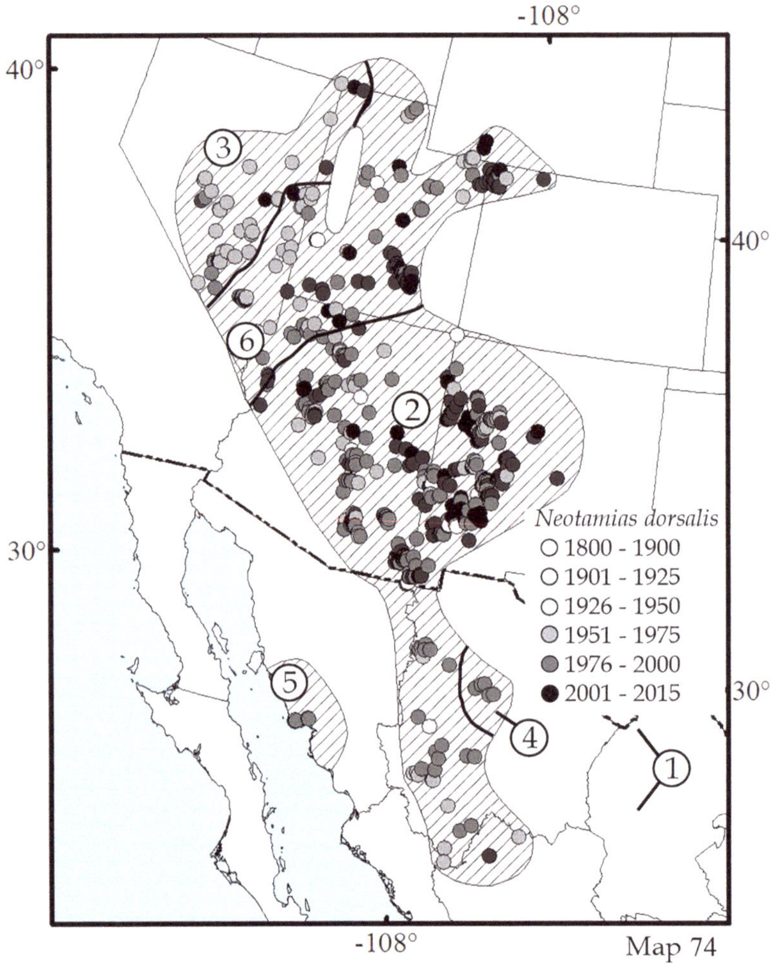


**Map 74** Distribution map of *Neotamias dorsalis*

**Additional Literature:** Mearns (1907), Howell (1929), Hall (1946), White (1953), Lidicker (1960c), Long (1965), Callahan and Davis (1976), Hart (1992), Sutton (1992), Thorington et al. (2012).

## *Neotamias durangae* (Allen, 1903)

Durango chipmunk, chichimoco de Durango

Monotypic.

**Conservation:** *Neotamias durangae* is listed as Least Concern by the IUCN (Álvarez-Castañeda et al. 2016z).

**Characteristics:** Durango chipmunk is small sized; total length 220.0–249.0 mm and skull length 35.5–39.7 mm. Dorsal pelage with a central dark stripe blackish from the nape to the tail base, wide, contrasting with the back general coloration; middle stripes brownish to blackish brown and wide; flank stripes very small, paler than the middle stripes. Inner paler stripes forming a narrow stripe, dark gray to reddish gray, not contrasting with the central and middle stripes; outer stripes similar to the inner paler stripes, but slightly paler in some individuals; fur of underparts whitish. General coloration brownish reddish gray; ears whitish on the back outer and anterior blackish; postauricular patches gray; hind feet light to medium gray; rump brownish reddish gray; facial stripes contrasting with the head general coloration; tail brownish gray, paler at the tip. Species group *quadrivittatus*.

**Comments:** *Neotamias durangae* ranges in the Sierra Madre Occidental, from Chihuahua south through Durango (Map 75), and has allopatric distribution with respect to the other species of *Neotamias*.

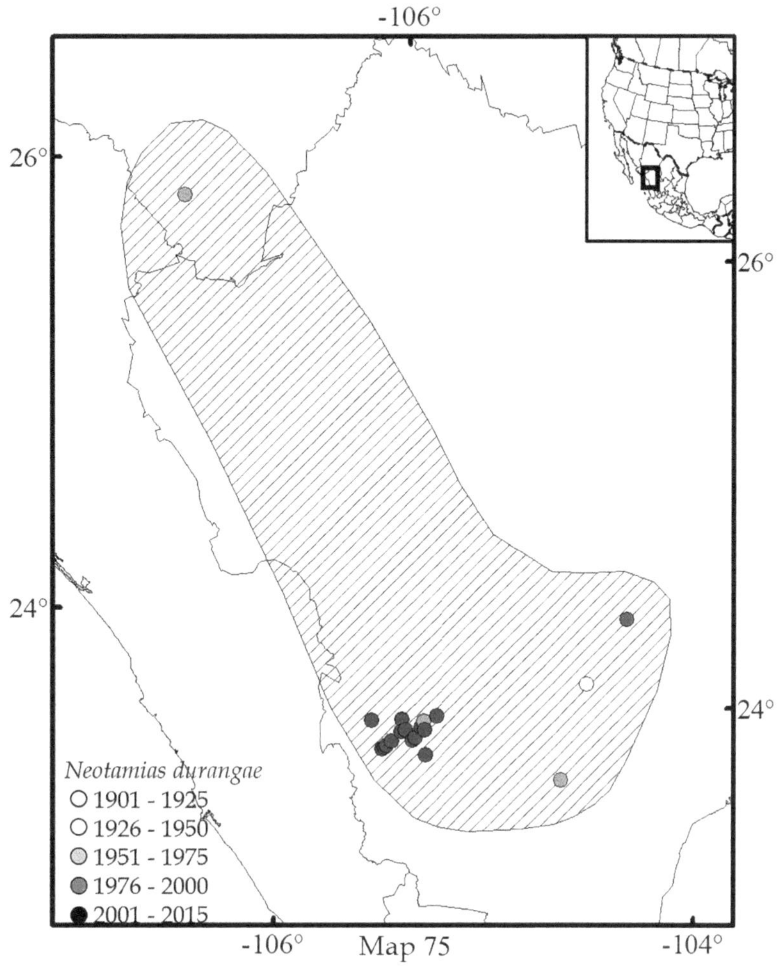


**Map 75** Distribution map of *Neotamias durangae*

**Additional Literature:** Allen (1903), Howell (1929), Baker (1956), Fleharty (1960), Anderson (1972), Callahan (1980), Best et al. (1993), Thorington et al. (2012).

## *Neotamias grisescens* (Howell, 1925)

Least gray chipmunk, chichimoco pigmeo gris

Monotypic.

**Conservation:** *Neotamias grisescens* (as *Neotamias minimus*) is listed as Least Concern by the IUCN (Cassola 2016bl).

**Characteristics:** Least gray chipmunk is the **smallest in size**; total length 165.0–220.0 mm and, skull length 28.2–34.0 mm. Dorsal pelage with a central dark stripe blackish from the nape to the tail base, wider at the middle, contrasting with the back general coloration; middle stripes less contrasting with the central stripe and more reddish; flank stripes paler but not strongly contrasting with the others. Inner paler stripes mainly absent, in their place, same coloration as the general pattern of the individual, similar to the rump coloration; outer stripes are the only paler stripes present but not markedly contrasting; fur of underparts whitish. General coloration gray frosted, of similar color to the rump, part of the flanks, and head top; ears similar to the general coloration; postauricular patches whitish but smaller; hind feet pale gray; rump similar to the general coloration; facial stripes present but not strongly contrasting; crown head and tail base similar to the general coloration, hair with dark coloration increasing at the tip, cinnamon ventrally. Baculum bent tip length less than 25% of the shaft length. Species group *minimus*.

**Comments:** *Neotamias grisescens* was recognized as a distinct species (Herrera et al. 2022); it was previously regarded as a subspecies of *N. minimus* (Howell 1925) and is endemic to Central Washington (Map 76). It can be found in sympatry, or nearly so, with *N. amoenus*, *N. minimus*, and *N. ruficaudusinus*. *N. minimus* can be differentiated from *N. amoenus* by having the inner paler stripes not well defined and the fur of the underparts whitish. From *N. ruficaudus*, by its smaller head-and-body length and its tail not reddish; and from *N. minimus*, using genetic data and its distribution range.

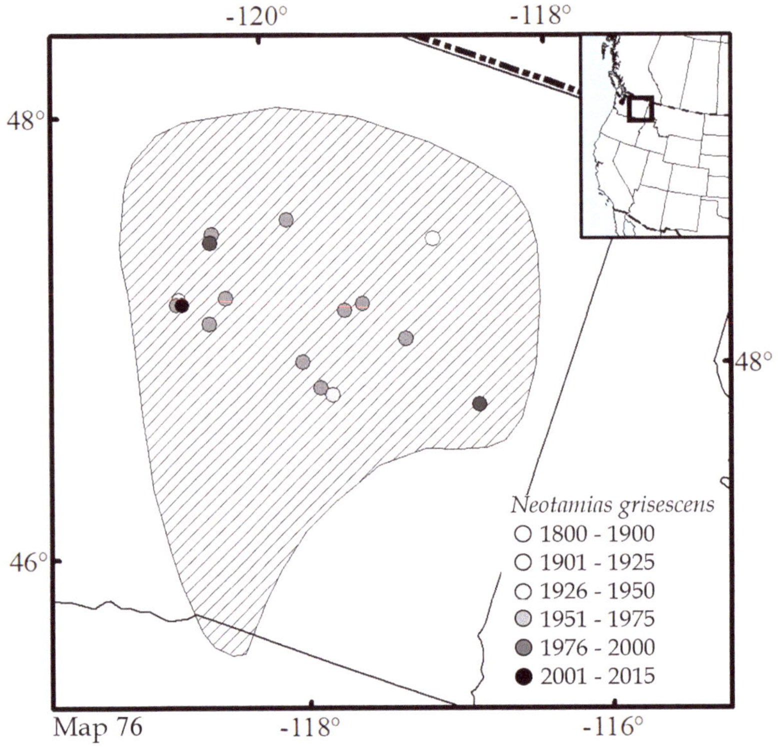


**Map 76** Distribution map of *Neotamias grisescens*

**Additional Literature:** Howell (1929), Hall (1946, 1981), White (1953), Verts and Carraway (2001b), Thorington et al. (2012).

## *Neotamias merriami* (Allen, 1889)

Merriam's chipmunk, chichimoco del sur de California

1. *N. m. kernensis* (Grinnell and Stoner, 1916). Eastern slopes of southernmost portion of the Sierra Nevada, California.
2. *N. m. merriami* (Allen, 1889). From the northernmost Baja California Peninsula north through the mountain range to Sierra Nevada in the east and San Luis Obispo in the west, California.
3. *N. m. pricei* (Allen, 1895). Coastal areas of California, from San Luis Obispo north through San Francisco Bay.

**Conservation:** *Neotamias merriami* is listed as Special Protection by the Norma Oficial Mexicana (DOF 2019) and as Least Concern by the IUCN (Álvarez-Castañeda et al. 2016a).

**Characteristics:** Merriam's chipmunk is small sized; total length 233.0–277.0 mm and skull length 35.5–40.7 mm. Dorsal pelage with the central dark stripe blackish in many individuals only in the middle and the back not markedly contrasting, the anterior part very narrow in some individuals or similar to the general coloration. In many individuals, hair tips paler; middle stripes similar to the central stripe, but paler; flank stripes mainly absent. Inner paler stripes medium gray, with hairs paler at the tip giving a hoary effect, not markedly contrasting with the dark stripes; outer stripes similar to the inner paler ones, but having more contrast; fur of underparts yellowish white. General coloration gray, contrasting with the stripes, but with the hairs paler, including those in the stripes; ears paler; postauricular patches gray, not markedly contrasting; hind feet pale cinnamon; rump brownish gray, with hairs paler at the tip; facial stripes paler, but not markedly contrasting; back of the head medium gray, similar to the nape; tail blackish with a small area with paler hair. **Tip of the baculum very small relative to the shaft length.** Species group *townsendii*.

**Comments:** *Neotamias obscurus* was recognized as a distinct species (Callahan 1977); it was previously regarded as a subspecies of *N. merriami* (Howell 1929). *N. merriami* ranges from San Francisco Bay and Sierra Nevada south through the northernmost Baja California peninsula (Map 77). It can be found in sympatry, or nearly so, with *N. amoenus*, *N. dorsalis*, *N. obscurus*, *N. quadrimaculatus*, *N. paramintinus*, *N. senex*, *N. speciosus*, and *N. umbrinus*. *N. merriami* can be differentiated from these species by its larger size and its long bushy tail, nearly of the same size as the head-and-body length; long narrow ears and dark dorsal stripes usually equal in width. From *N. ochrogenys*, by its longer tail.

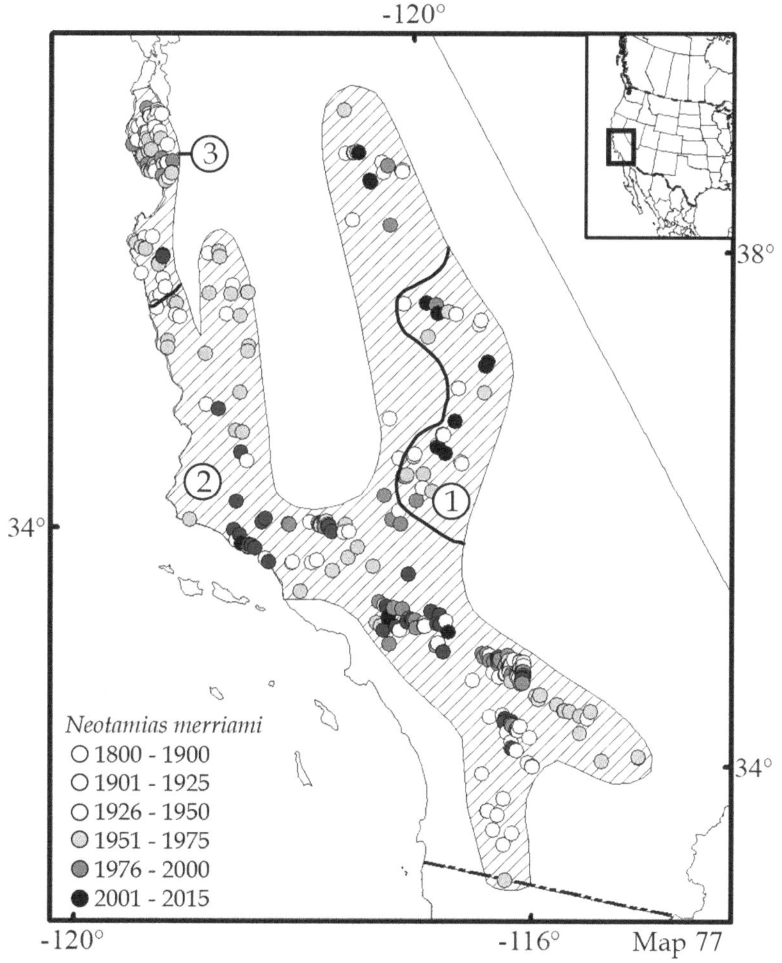


**Map 77** Distribution map of *Neotamias merriami*

**Additional Literature:** Johnson (1943), Callahan (1976, 1977), Sutton (1992), Best and Granai (1994b), Thorington et al. (2012).

## *Neotamias minimus* (Bachman, 1839)

Least chipmunk, chichimoco pigmeo

1. *N. m. arizonensis* (Howell, 1922). Restricted to the Prieto Plateau, Green Lee County, eastern Arizona.
2. *N. m. atristriatus* (Bailey, 1913). Restricted to Peñasco, northern New Mexico.
3. *N. m. borealis* (Allen, 1877). From British Columbia and northern Montana east through Ontario.
4. *N. m. cacodemus* (Cary, 1906). Restricted to the Bad Lands National Monument, southwestern South Dakota.
5. *N. m. caniceps* (Osgood, 1900). From Yukon south through northern British Columbia.
6. *N. m. caryi* (Merriam, 1908). San Luis Valley, southern-central Colorado.
7. *N. m. chuskaensis* Sullivan and Petersen, 1988. Chuska Mountains in Arizona and New Mexico.
8. *N. m. confinis* (Howell, 1925). Bighorn Mountains, northern Wyoming.
9. *N. m. consobrinus* (Allen, 1890). From Utah and Colorado north through the boundary between Idaho and Wyoming.
10. *N. m. hudsonius* (Anderson and Rand, 1944). Northern Manitoba.
11. *N. m. jacksoni* (Howell, 1925). Oneida County, northern Wisconsin.
12. *N. m. minimus* (Bachman, 1839). Restricted to Wyoming.
13. *N. m. neglectus* (Allen, 1890). From southern Manitoba and northern Minnesota east through western Québec.
14. *N. m. operarius* (Merriam, 1905). Eastern edge of Utah, Wyoming, Colorado, and northern New Mexico.
15. *N. m. oreocetes* (Merriam, 1897). Southernmost boundary between Alberta, British Columbia, and Montana.
16. *N. m. pallidus* (Allen, 1874). Montana, eastern North Dakota, northeastern Wyoming, South Dakota, and Nebraska.
17. *N. m. pictus* (Allen, 1890). Southern Idaho and northern Utah.
18. *N. m. scrutator* (Hall and Hatfield, 1934). From southern-central Washington to Sierra Nevada, California, and west Nevada, including southwestern Idaho.
19. *N. m. selkirki* (Cowan, 1946). Restricted to the Toby Creek, southeastern British Columbia.
20. *N. m. silvaticus* (White, 1952). Restricted to the boundary between Wyoming and South Dakota.

**Conservation:** *Neotamias minimus* is listed as Least Concern by the IUCN (Cassola 2016bl).

**Characteristics:** Least chipmunk is the **smallest in size**; total length 167.0–225.0 mm and skull length 28.6–34.0 mm. Dorsal pelage with a central dark stripe blackish from the nape to the tail base, wider at the middle, contrasting with the back general coloration; middle stripes less contrasting with the central stripe and more reddish; flank stripes paler, but not strongly contrasting with the others. Inner paler stripes mainly absent, in their place, same coloration as the general pattern, similar to the rump coloration; outer stripes are the only paler stripes present but not markedly contrasting; fur of underparts whitish. General coloration gray frosted, similar in color to the rump, part of the flanks, and head top; ears similar to the general coloration; postauricular patches whitish but smaller; hind feet pale gray; rump similar to the general coloration; facial stripes present, but not strongly contrasting; crown head and tail base similar to the general coloration, with hairs with dark coloration increasing at the tip, cinnamon ventrally. **Baculum bent tip length less than 25% of the shaft length**. Species group *minimus*.

**Comments:** *Neotamias minimus grisescens* was recognized as a distinct species (Herrera et al. 2022). *N. minimus* ranges from Yukon east through western Québec; from the western range of its distribution south through California, Arizona, and New Mexico; and from Washington, Oregon, and California east through North Dakota, South Dakota, Colorado, and New Mexico. From its eastern range, south through Wisconsin and Minnesota (Map 78). It can be found in sympatry, or nearly so, with *N. alpinus*, *N. amoenus*, *N. panamintinus*, *N. quadrimaculatus*, *N. quadrivittatus*, *N. ruficaudus*, *N. rufus*, *N. senex*, *N. siskiyou N. speciosus*, and *N. umbrinus*. *N. minimus* can be differentiated from *N. alpinus* by its larger tail and shorter ears. From *N. amoenus*, by having the inner paler stripes not well defined and the fur of the underparts whitish. From *N. panamintinus*, by being less reddish; and from *N. ruficaudus*, by its smaller head-and-body length and the tail not reddish.

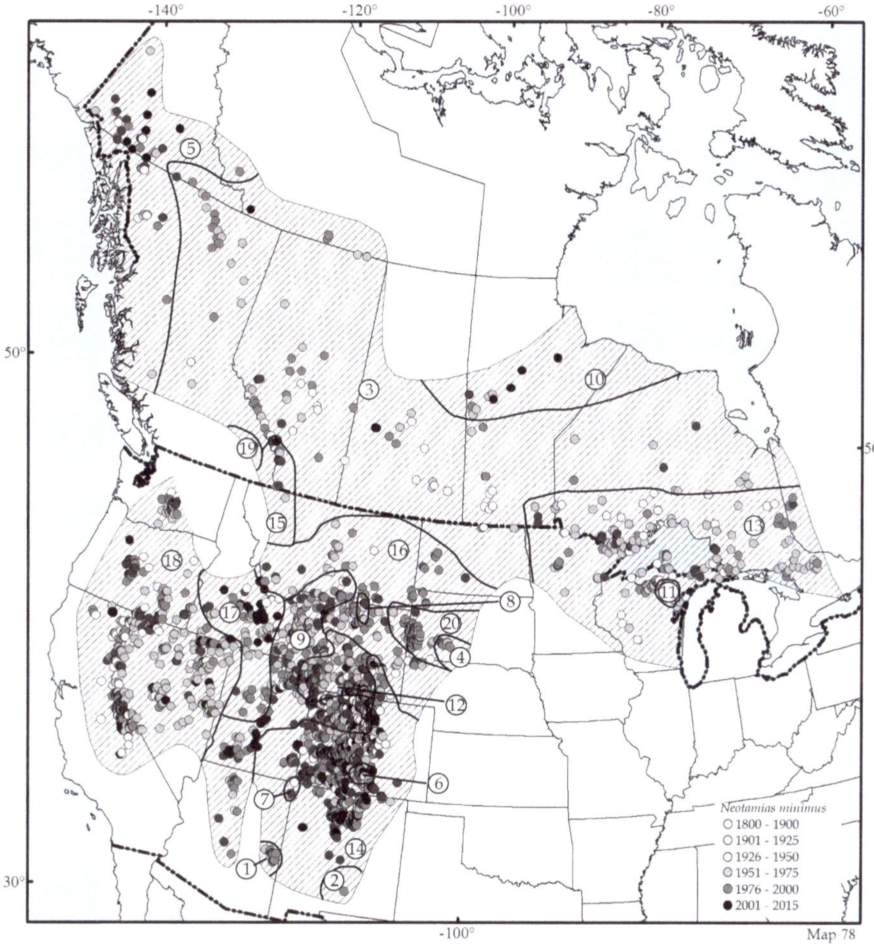


**Map 78** Distribution map of *Neotamias minimus*

**Additional Literature:** Howell (1929), Hall (1946, 1981), White (1953), Jones and Manning (1992), Sutton (1992), Verts and Carraway (2001b), Thorington et al. (2012).

## ***Neotamias obscurus*** (Allen, 1890)

California chipmunk, chichimoco de California

1. *N. o. davisi* (Callahan, 1977). San Bernardino and Palm Spring mountains, southern California.
2. *N. o. meridionalis* (Nelson and Goldman, 1909). Isolated in Sierra de San Francisco and San Borja, central Baja California Peninsula.
3. *N. o. obscurus* (Allen, 1890). Extreme southern California southward to San Pedro Mártir, Baja California Peninsula.

**Conservation:** *Neotamias obscurus* is listed as Least Concern by the IUCN (Álvarez-Castañeda et al. 2016b).

**Characteristics:** California chipmunk is small sized; total length 208.0–240.0 mm and skull length 33.2–39.5 mm. Dorsal pelage with the dark central stripe blackish in many individuals, only very contrasting in the middle and posterior parts of the back; in some individuals, anterior part very narrow and similar to the general color, or absent; middle stripes similar to the central stripe, but paler brownish and more as a shadow in some individuals; flank stripes mainly absent. Inner paler stripes medium gray or absent, in its place the general coloration of the individuals similar or less pale than the rump; outer stripes similar to the inner paler stripes; fur of the underparts whitish. General coloration gray, paler at the tip, and low stripes contrasting; ears similar in coloration to the head crown; postauricular patches gray, not stronger contrast; hind feet pale cinnamon; rump similar to the general coloration; facial stripes clear, but not stronger contrast; back of the head medium gray similar to the nape; tail blackish with hair paler at the tip. **Tip of the baculum less than one-fifth of the shaft length**. Species group *townsendii*.

**Comments:** *Neotamias obscurus* was recognized as a distinct species (Callahan 1977); it was previously regarded as a subspecies of *N. merriami* (Howell 1929). *N. obscurus* ranges in San Bernardino, San Jacinto mountains of California, Sierra de Juárez and San Pedro Mártir, Baja California, and four disjunction populations in northern Baja California Sur (Map 79). It can be found in sympatry, or nearly so, with *N. merriami*. *N. obscurus* can be differentiated from *N. merriami* by being slightly smaller in body size and cranial dimensions, and by having a grayer venter.

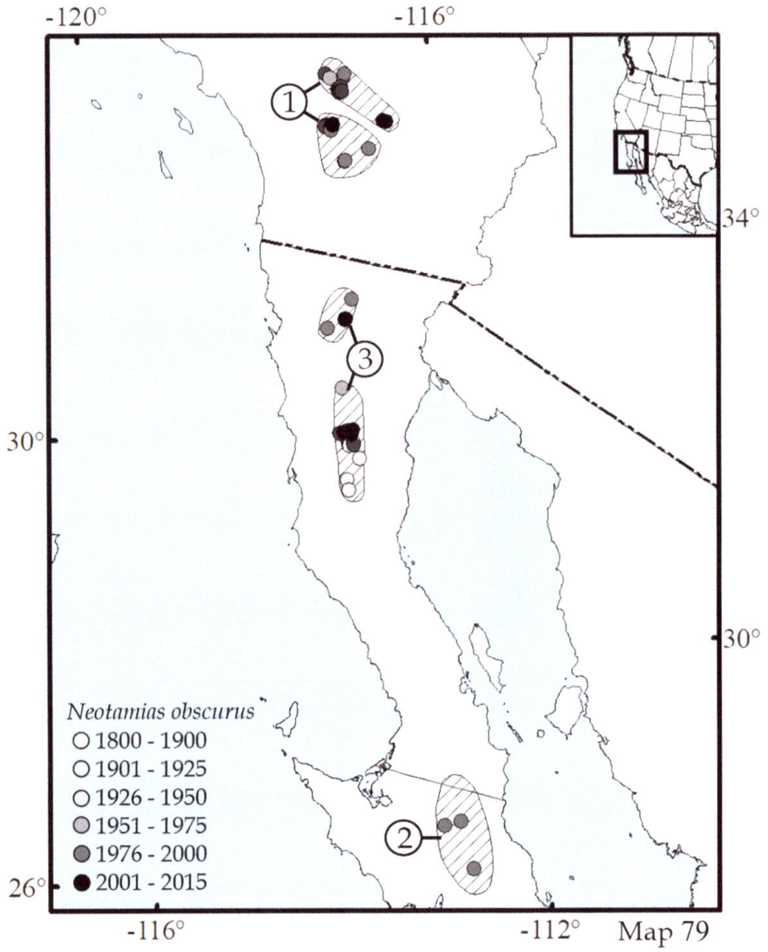


**Map 79** Distribution map of *Neotamias obscurus*

**Additional Literature:** Nelson (1922), Callahan (1976), Blankenship and Bradley (1985), Levenson et al. (1985), Sutton (1992), Best and Granai (1994a), Thorington et al. (2012).

## *Neotamias ochrogenys* (Merriam, 1897)

Yellow-cheeked chipmunk, chichimoco de cachetes amarillos

Monotypic.

**Conservation:** *Neotamias ochrogenys* is listed as Least Concern by the IUCN (Cassola 2016bm).

**Characteristics:** Yellow-cheeked chipmunk is small sized, total length 252.0–277.0 mm and skull length 39.0–41.0 mm. Dorsal pelage with a central dark stripe blackish in many individuals only at the middle and back portions, and not markedly contrasting; in some individuals, very narrow anteriorly or of the same general color as the back; middle stripes similar to the central stripe, but paler brownish and more as a shadow in some individuals; flank stripes mainly absent. Inner paler stripes medium gray or absent, replaced by the general coloration, similar to or slightly paler than the rump; outer stripes similar to the inner paler stripes; fur of underparts whitish. General coloration gray with hairs paler at the tip and external stripes of contrasting shades; ears similar in coloration to the head top; postauricular patches gray, not markedly contrasting; hind feet pale cinnamon; rump similar to the general coloration; facial stripes clear but not markedly contrasting; back of the head medium gray, similar to the nape; tail blackish with hairs paler at the tip. **Baculum L shaped.** Species group *townsendii*.

**Comments:** *Neotamias ochrogenys* was recognized as a distinct species (Sutton and Nadler 1974); it was previously regarded as a subspecies of *N. townsendii* (Merriam 1897). *N. ochrogenys* ranges along the northern coast of California (Map 80) and can be found in sympatry, or nearly so, with *N. sonomae*. *N. ochrogenys* can be differentiated from the other species of the *townsendii* complex (*N. senex*, *N. siskiyou*, and *N. townsendii*) by its larger body size; dark dorsal pelage; tail thinner; and by the structure of its genital bones and its vocalizations. From *N. quadrimaculatus*, by its smaller size; shorter and wider rostrum and the zygomatic width. From *N. sonomae* and *N. merriami*, by its shorter tail; and from *N. siskiyou*, only by the baculum and os clitoridis.

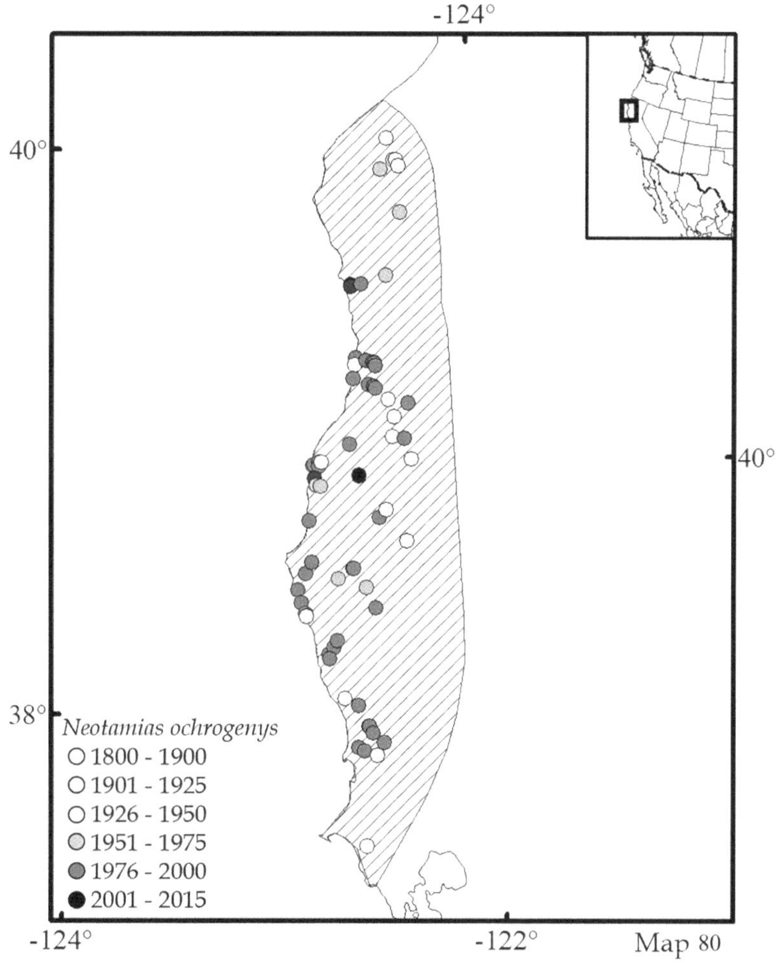


**Map 80** Distribution map of *Neotamias ochrogenys*

**Additional Literature:** Merriam (1897), Howell (1929), Hooper (1944), Adams and Sutton (1968), Sutton (1982), Kain (1985), Gannon et al. (1993).

## ***Neotamias palmeri*** (Merriam, 1897)

Palmer's chipmunk, chichimoco de Spring Mountains

Monotypic.

**Conservation:** *Neotamias palmeri* is listed as under protection by the United States Endangered Species Act (FWS 2022) and as Endangered (B1ab (iii): extent of occurrence less than 5000 km$^2$ in five or less populations and under continuing decline in area, extent and/or quality of habitat) by the IUCN (Lowrey 2016).

**Characteristics:** Palmer's chipmunk is small sized; total length 204.0–233.0 mm and skull length 34.9–36.5 mm. Dorsal pelage with a central dark stripe blackish, narrow and not markedly contrasting, from the nape to the tail base; middle stripes in brownish shades, similar to the flanks, and flank stripes absent. Inner paler stripes medium gray; if absent, area of general coloration, similar or less pale than the rump; outer stripes paler and contrasting with the general coloration; fur of underparts whitish. General coloration gray with a combination of different shades; flanks reddish brown; ears similar in coloration to the head crown; postauricular patches pale gray, not markedly contrasting; hind feet pale reddish gray; rump similar to the general coloration; facial stripes pale, but not markedly contrasting; back of the head slightly darker than the rump coloration; tail blackish with hairs paler at the tip; tail only with blackish hairs distally and cinnamon ventrally. Species group *quadrivittatus*.

**Comments:** *Neotamias palmeri* is endemic to Spring Mountain, Nevada (Map 81), and has allopatric distribution with respect to the other species of *Neotamias*.

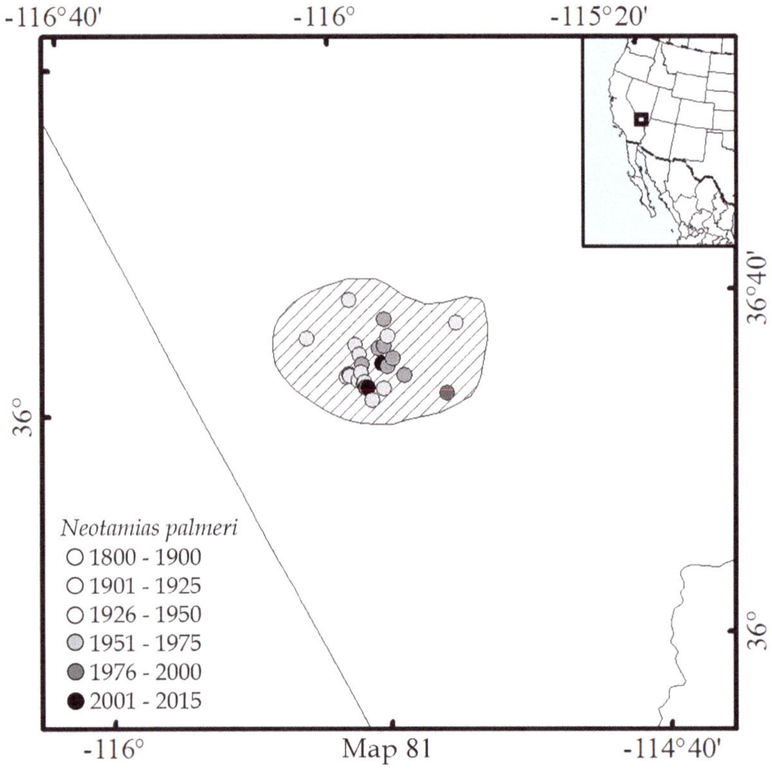


**Map 81** Distribution map of *Neotamias palmeri*

**Additional Literature:** Howell (1929), Hall (1946, 1981), White (1953), Sutton (1992), Best (1993c).

## *Neotamias panamintinus* (Merriam, 1893)

Panamint chipmunk, chichimoco del Valle Panamint

1. *N. p. acrus* (Johnson, 1943). Restricted to Kingston Mountains, southeastern California.
2. *N. p. panamintinus* (Merriam, 1893). Mountain ranges of eastern California and southwestern Nevada.

**Conservation:** *Neotamias panamintinus* is listed as Least Concern by the IUCN (Cassola 2016bn).

**Characteristics:** Panamint chipmunk is small sized; total length 192.0–220.0 mm and skull length 33.0–34.8 mm. Dorsal pelage with a central dark brown stripe, not contrasting and narrow, from the nape to the tail base; middle stripes of similar color to the flanks; flank stripes, if not absent, of similar color to the flanks. Inner paler stripes medium gray, of similar color to the rump, with hairs paler at the tip, not markedly contrasting with the dark stripes; outer stripes paler, with a greater contrast; fur of underparts yellowish white. General coloration grayish in the rump and hips; flanks, shoulders, and sides of the neck reddish brown; ears similar to the head coloration; postauricular patches pale gray contrasting with the reddish brown shoulders and neck flanks; hind feet gray; rump and hips grayish, grizzled in different shades; facial stripes pale, not markedly contrasting; back of the head medium gray with hairs paler at the tip; tail brownish black throughout, fur of underparts cinnamon. Species group *amoenus*.

**Comments:** *Neotamias panamintinus* is restricted to the mountains of eastern California and southwestern Nevada (Map 82). It can be found in sympatry, or nearly so, with *N. merriami*, *N. minimus*, *N. speciosus*, and *N. umbrinus*. *N. panamintinus* can be differentiated from *N. amoenus* by being similar in size but with smaller feet and ears; a paler pelage; paler dorsal stripes and a broader skull. From *N. minimus*, by being more reddish in color; and from *N. speciosus*, by being smaller; ears less pointed; dark facial and dorsal stripes others with few hoary hairs; pale stripes and fur of underparts less ochraceous wash; skull relatively wide, but with a narrow and more pointed rostrum; upper toothrow less divergent anteriorly.

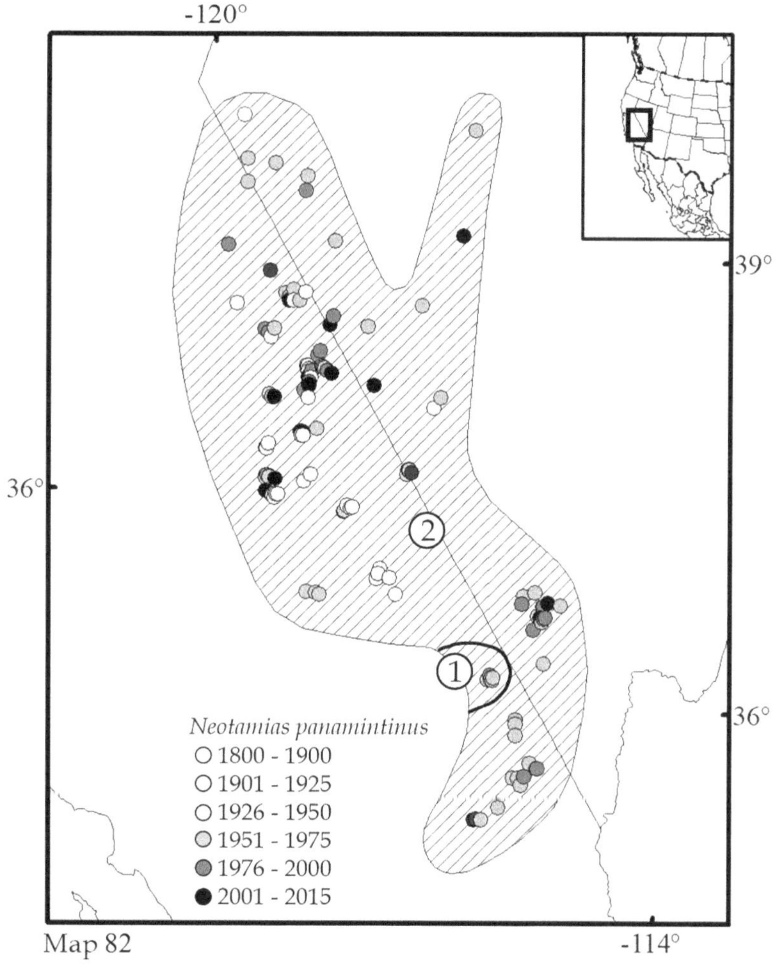


**Map 82** Distribution map of *Neotamias panamintinus*

**Additional Literature:** Howell (1929), Sutton (1992), Best et al. (1994c), Johnson (1943), Callahan (1976), Hirshfeld and Bradley (1977), Levenson (1990), Thorington et al. (2012).

## *Neotamias quadrimaculatus* (Gray, 1867)

Long-eared chipmunk, chichimoco de orejas grandes

Monotypic.

**Conservation:** *Neotamias quadrimaculatus* is listed as Least Concern by the IUCN (Cassola 2016bo).

**Characteristics:** Long-eared chipmunk is small sized; total length 200.0–250.0 mm and skull length 36.3 to 38.5 mm. Dorsal pelage with a central dark stripe blackish from the nape to the tail base, contrasting with the back general coloration; middle stripes similar to the central stripe in coloration and flank stripes absent. Inner paler stripes gray with a mixture of reddish gray, similar to the rump coloration; outer stripes paler; fur of underparts whitish. General coloration grayish on the rump and hips, with hairs paler at the tip; reddish brown in the flanks, shoulders and sides of the neck; back of the ears, outer edge whitish and anterior edge blackish; postauricular patches pale gray, markedly contrasting with the general coloration; hind feet medium gray; rump similar to the general coloration; facial stripes contrasting with the general coloration of the head, darker stripe well marked; back of the head dark grayish with hairs paler at the tip; tail brownish gray and blackish distally, paler at the tip. Species group *townsendii.*

**Comments:** *Neotamias quadrimaculatus* is restricted to Sierra Nevada and western-central Nevada (Map 83). It can be found in sympatry, or nearly so, with *N. alpinus*, *N. amoenus*, *N. merriami*, *N. minimus*, *N. speciosus*, *N. senex*, and *N. townsendii*. *N. quadrimaculatus* can be differentiated from *N. alpinus* by being larger, and from *N. ochrogenys*, *N. senex*, *N. siskiyou*, *N. sonomae*, and *N. townsendii* by its larger size; longer and narrower rostrum, and by its zygomatic width.

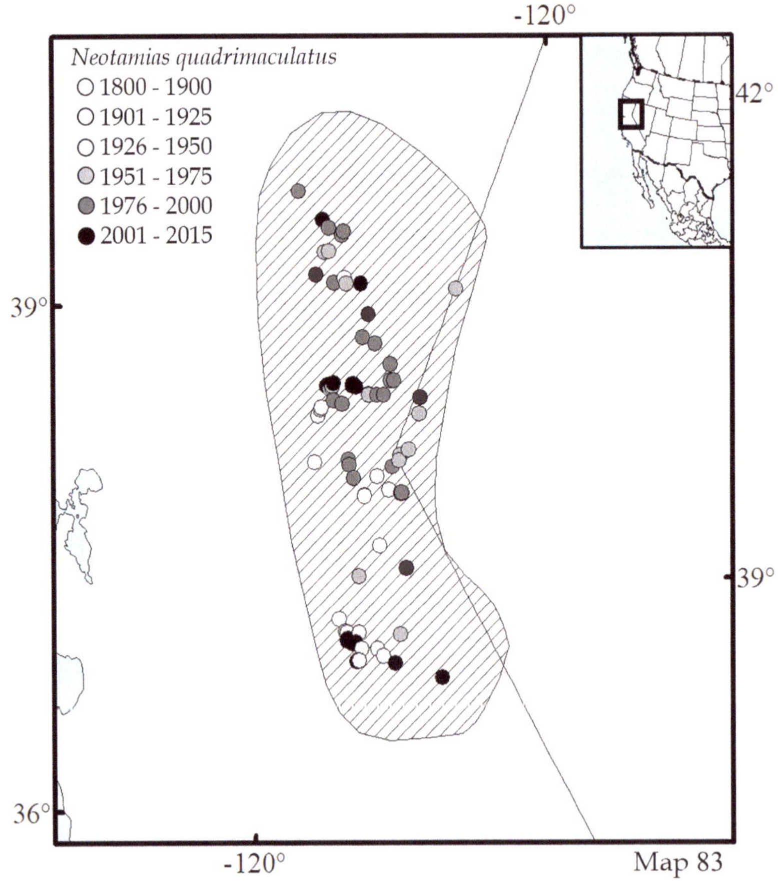


**Map 83** Distribution map of *Neotamias quadrimaculatus*

**Additional Literature:** Grinnell and Storer (1924), Howell (1929), Johnson (1943), White (1953), Burt (1960), Levenson et al. (1985), Jameson and Peeters (1988, 2004), Sutton (1992), Clawson et al. (1994b).

## *Neotamias quadrivittatus* (Say, 1823)

Colorado chipmunk, chichimoco de Colorado

1. *N. q. australis* (Patterson, 1980). Organ Mountains, southern-central New Mexico.
2. *N. q. hopiensis* (Merriam, 1905). Eastern Utah, southern Wyoming, western Colorado, and northeastern Arizona.
3. *N. q. oscuraensis* Sullivan, 1996. Oscura Mountains, central New Mexico.
4. *N. q. quadrivittatus* (Say, 1823). Arizona, Colorado, northern New Mexico, and western Oklahoma.

**Conservation:** *Neotamias quadrivittatus* is listed as Least Concern by the IUCN (Cassola 2016bp).

**Characteristics:** Colorado chipmunk is small sized; total length 197.0–235.0 mm and, skull length 33.5–36.8 mm. Dorsal pelage with a central dark stripe blackish from the nape to the tail base, contrasting with the back general coloration; middle stripes similar to the central stripe in coloration; flank stripes similar to the central stripe in coloration, but thin and short. Inner paler stripes gray with a mixture of pale gray, similar to the neck and rump coloration; outer stripes paler; fur of underparts whitish. General coloration grayish in the shoulders and neck flanks with hair paler at the tip; flanks, hips, and feet ochraceous to cinnamon; back of the ears whitish, outer and anterior parts darker; postauricular patches pale gray, small, similar in color to the outer paler stripe; facial stripes contrasting with the head general coloration; back of the head reddish to cinnamon, sometimes mixed with gray, with hair paler at the tip; tail blackish brown and paler at the tip. Species group *quadrivittatus*.

**Comments:** *Neotamias rufus* was originally described as a subspecies of *N. quadrivittatus*, and ranges in western Utah, northeast Arizona, Colorado, New Mexico, and western Oklahoma (Map 84). It can be found in sympatry, or nearly so, with *N. amoenus*, *N. dorsalis*, *N. minimus*, *N. rufus*, and *N. umbrinus*. *N. quadrivittatus* can be differentiated from *N. amoenus* and *N. minimus* by having the dark outer stripes not well defined and grayish, and the outer lines without a strong contrast. From *N. dorsalis*, by having all the dorsal stripes clearly apparent on the back. From *N. rufus* and *N. umbrinus*, by having the postauricular spot small and grayish, not occupying part of the shoulders, and general coloration in dark gray shades.

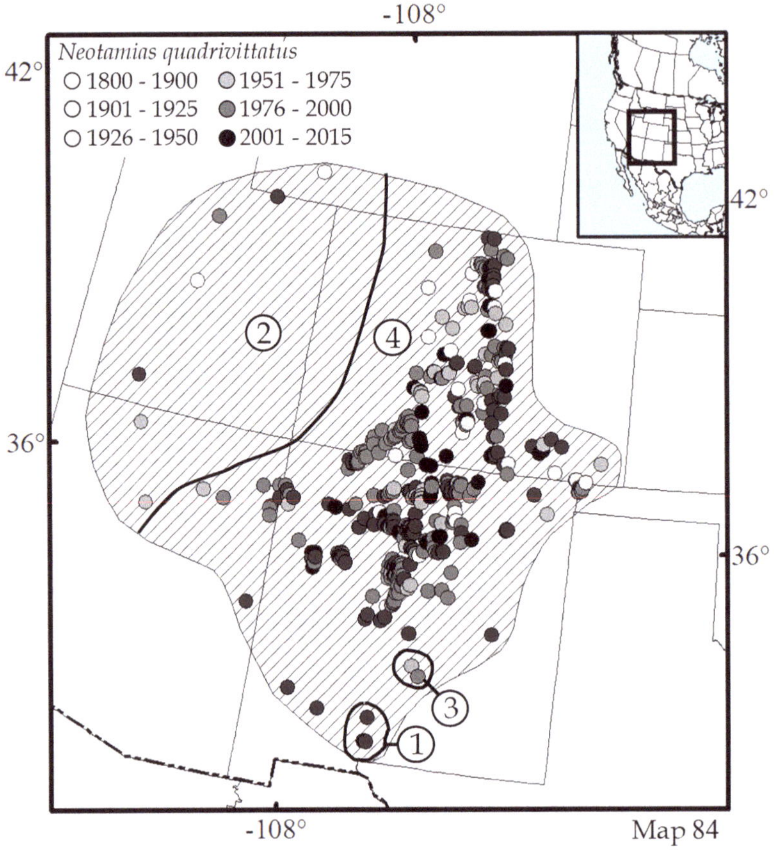


**Map 84** Distribution map of *Neotamias quadrivittatus*

**Additional Literature:** Howell (1929), Hoffmeister and Ellis (1979), Patterson (1984), Levenson et al. (1985), Hoffmeister (1986), Sutton (1992), Best et al. (1994a).

## *Neotamias ruficaudus* (Howell, 1920)

Red-tailed chipmunk, chichimoco de cola roja

1. *N. r. ruficaudus* (Howell, 1920). Montana and central eastern Idaho.
2. *N. r. simulans* (Howell, 1922). Extreme western Montana, southern British Columbia, and Alberta, Idaho, and eastern Washington.

**Conservation:** *Neotamias ruficaudus* is listed as Least Concern by the IUCN (Cassola 2016bq).

**Characteristics:** Red-tailed chipmunk is small sized; total length 223.0–248.0 mm and skull length 34.0–36.2 mm. Dorsal pelage with a central dark stripe blackish from the nape to the rump; middle stripes similar to the central stripe in coloration and flank stripes similar to the central one in coloration. Inner paler stripes dark gray mixed with pale gray, but not contrasting with the dark stripes; outer stripes similar to the inner paler stripe, but paler; fur of underparts whitish. General coloration orangish on the dorsum to dark grayish on the rump and hips, with the hair of a different gray shade at the tip, but not contrasting with the dark or pale stripes; flanks, shoulders, and neck flanks dark reddish brown; back of the ears paler, outer and anterior parts darker; with postauricular patches gray and very small; hind feet dark gray; rump with the general coloration; facial stripes not contrasting with the head general coloration; back of the head dark grayish with hair paler at the tip; tail dark brown grizzle with blackish shades and with hair paler at the tip; fur of underparts reddish with black and pink borders. Species group *quadrivittatus*.

**Comments:** *Neotamias ruficaudus* ranges from southeastern British Columbia and southwestern Alberta to northeastern Washington, northern Idaho, and western Montana (Map 85). It can be found in sympatry, or nearly so, with *N. amoenus* and *N. minimus*. *N. ruficaudus* can be differentiated from *N. amoenus* and *N. minimus* by its larger body and skull and its distinctly reddish tail.

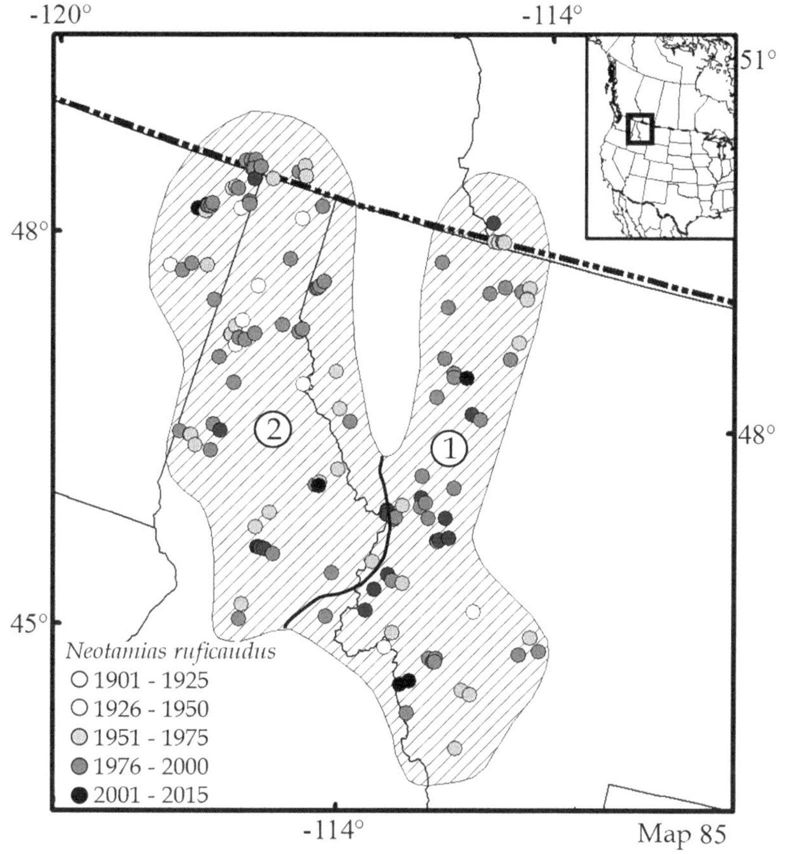


**Map 85** Distribution map of *Neotamias ruficaudus*

**Additional Literature:** Howell (1922, 1929), Ingles (1965), Patterson (1984), Levenson et al. (1985), Patterson and Heany (1987), Sutton (1992), Best (1993e), Thorington et al. (2012).

## *Neotamias rufus* (Hoffmeister and Ellis, 1979)

Hopi chipmunk, chichimoco de los Hopi

Monotypic.

**Conservation:** *Neotamias rufus* is listed as Least Concern by the IUCN (Cassola 2016br).

**Characteristics:** Hopi chipmunk is small sized; total length 198.0–233.0 mm and skull length 33.2–36.7 mm. Dorsal pelage with a central dark stripe brownish from the nape to the tail base and not contrasting with the back general coloration; middle stripes similar to the central stripe, but paler and reddish and flank stripes absent or very small, similar to the flank coloration. Inner paler stripes of similar coloration to the rump and nape, not contrasting with the other stripes and the back general coloration; outer stripes pale gray, close to white in some individuals; fur of underparts whitish. General coloration gray in the rump and hips, with hair of a different shade of gray at the tip, but not contrasting with the dark or pale stripes; flanks, shoulders, and neck reddish brown; ears of similar color to the head top; whitish postauricular patches whitish, of similar color but contrasting with the outer paler stripes; hind feet bright reddish orange; face middle stripes strongly contrasting; back of the head gray with hair paler at the tip; tail with a mix of blackish, reddish brown, and grayish hair, and fur of underparts bright red orange with a black border. Species group *quadrivittatus*.

**Comments:** *Neotamias rufus* was originally described as a subspecies of N. *quadrivittatus. N. rufus* ranges in eastern Utah, western Colorado, and northeastern Arizona (Map 86). It can be found in sympatry, or nearly so, with *N. amoenus*, *N. dorsalis*, *N. minimus*, *N. quadrivittatus*, and *N. umbrinus*. *N. rufus* can be differentiated from *N. amoenus* and *N. minimus* by having the dark outer stripes not well defined and brownish, and the outer lines without a strong contrast. From *N. dorsalis*, by having all the dorsal stripes clearly apparent on the back. From *N. quadrivittatus*, by having the postauricular spot large and brownish, occupying part of the shoulders, and general coloration in brown shades. From *N. umbrinus*, by having the middle dark line in brown shades and not contrasting, and outer paler lines present, but not contrasting with the general coloration.

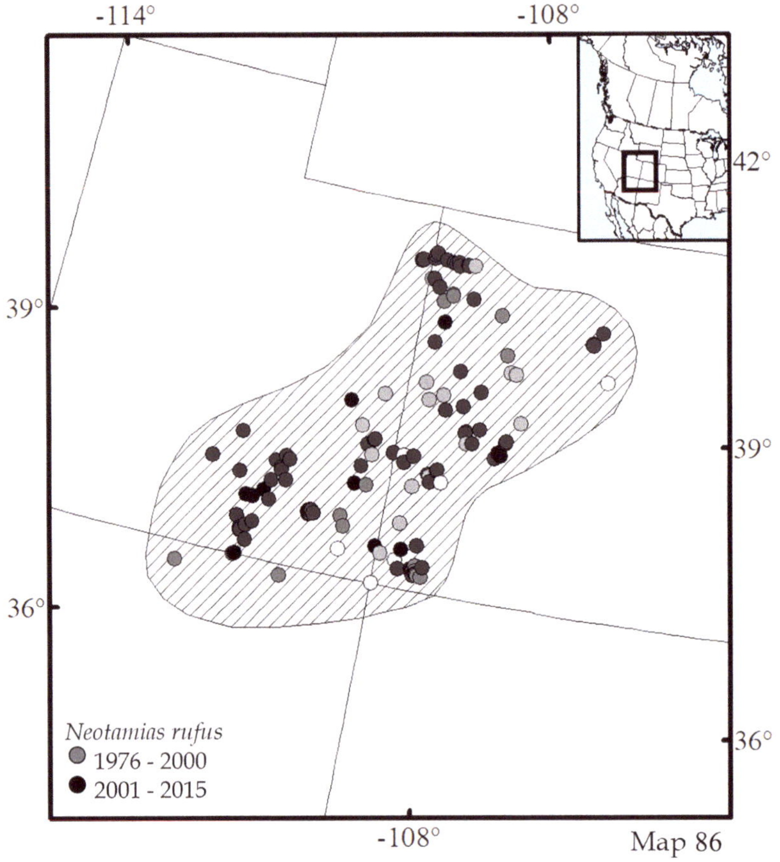


**Map 86** Distribution map of *Neotamias rufus*

**Additional Literature:** Cary (1911), Sutton (1953), Hoffmeister and Ellis (1979), Patterson (1984), Hoffmeister (1986), Sutton (1992), Burt and Best (1994), Thorington et al. (2012).

## ***Neotamias senex*** (Allen, 1890)

Allen's chipmunk, chichimoco de la Sierra Nevada

1. *N. s. pacifica* Sutton and Patterson, 2000. A narrow belt of coastal redwood forests in northwestern California.
2. *N. s. senex* (Allen, 1890). Inland in the Sierra Nevada, Cascade and other ranges in California and Oregon.

**Conservation:** *Neotamias senex* is listed as Least Concern by the IUCN (Cassola 2016bs).

**Characteristics:** Allen's chipmunk is small sized; total length 230.0–260.0 mm and skull length 37.3–40.0 mm. Dorsal pelage with a central dark stripe, black from the nape to the tail base and not contrasting with the back general coloration dark brown; middle and flank stripes similar to the central stripe. Inner paler stripes of similar coloration to those in the rump and nape, which are dark brown; outer stripes paler than the general coloration; fur of underparts reddish brown. General coloration grayish dark brown throughout the body, with only pale reddish areas on the neck flanks; back of the ears paler, outer, and anterior parts darker; postauricular patches whitish, extremely pale; hind feet grayish brown; facial stripes pale, strongly contrasting; head top grayish brown, similar to the general coloration; tail thin grizzled gray with reddish orange shades, with fur of underparts ochraceous, often frosted with buff. Species group *townsendii*.

**Comments:** *Neotamias senex* was recognized as a distinct species (by its ossa genitalia (Sutton and Nadler 1974) and vocalizations (Gannon and Lawlor 1989)), previously regarded as a subspecies of *N. townsendii* (Levenson and Hoffmann 1984; Nadler et al. 1985). *N. senex* ranges from central Oregon south through Sierra Nevada in California (Map 87). It can be found in sympatry, or nearly so, with *N. alpinus*, *N. amoenus*, *N. merriami*, *N. minimus*, *N. quadrimaculatus*, and *N. speciosus*. *N. senex* can be differentiated from *N. alpinus* by being larger. From *N. ochrogenys* by its smaller body size; dorsal pelage paler; tail thicker. From *N. quadrimaculatus*, by its smaller size; zygomatic arches width; and shorter and wider rostrum; and from *N. siskiyou*, only by the baculum and os clitoridis.

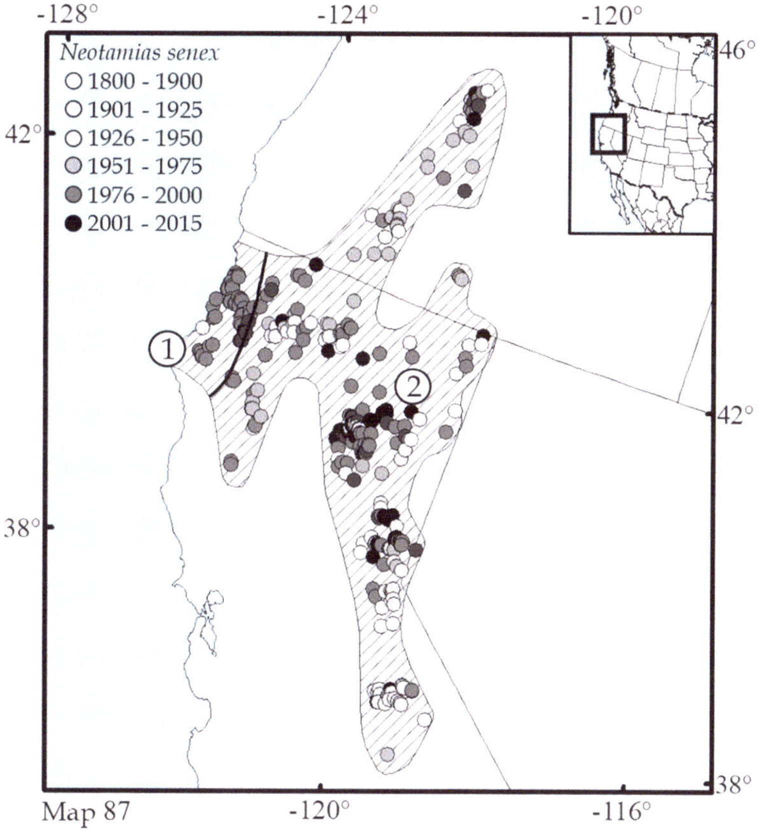


**Map 87** Distribution map of *Neotamias senex*

**Additional Literature:** Howell (1929), Gannon and Lawlor (1989), Sutton (1992), Gannon and Forbes (1995), Thorington et al. (2012).

## *Neotamias siskiyou* (Howell, 1922)

Siskiyou chipmunk, chichimoco de las montañas Siskiyou

1. *N. s. humboldti* Sutton and Patterson, 2000. Inland areas of Oregon and California.
2. *N. s. siskiyou* (Howell, 1922). Coastal areas of Oregon and California.

**Conservation:** *Neotamias siskiyou* is listed as Least Concern by the IUCN (Cassola 2016bt).

**Characteristics:** Siskiyou chipmunk is small sized; total length 250.0–268.0 mm and skull length 39.0 to 41.0 mm. Dorsal pelage with a central dark stripe black from the nape to the tail base, and not contrasting with the back dark brown general coloration; middle stripes similar to the central one, but paler and brownish; flank stripes small and brownish, similar to the flank coloration. Inner paler stripes of similar coloration to that on the flanks and shoulders, which is dark brown; outer stripes paler than the general coloration, **paler on the back**; fur of underparts reddish brown. General coloration grayish dark brown in the rump and hips; shoulders, neck flanks, and flanks dark brownish; outer back of the ears paler and anterior part darker; postauricular patches gray, similar to the outer paler stripe; hind feet dark brown; facial stripes pale, contrasting with the head coloration; back of the head grayish brown, similar to the shoulder coloration; tail with a mixture of blackish and dark brownish hairs, a few hairs paler at the tip. **Tip of the baculum longer than the shaft**. Species group *townsendii*.

**Comments:** *Neotamias siskiyou* was recognized as a distinct species (Sutton and Nadler 1974; Gannon and Lawlor 1989), previously regarded as a subspecies of *N. townsendii* (Howell 1922). *N. siskiyou* ranges in the Siskiyou Mountains, from central Oregon south through northern California coast (Map 88). It can be found in sympatry, or nearly so, with *N. amoenus* and *N. minimus*. *N. siskiyou* can be differentiated from *N. ochrogenys* by its smaller body size; dorsal pelage paler, and tail thicker. From *N. quadrimaculatus* by its smaller size; zygomatic width; shorter and wider rostrum. From *N. ochrogenys* and *N. senex* only by the baculum and os clitoridis.

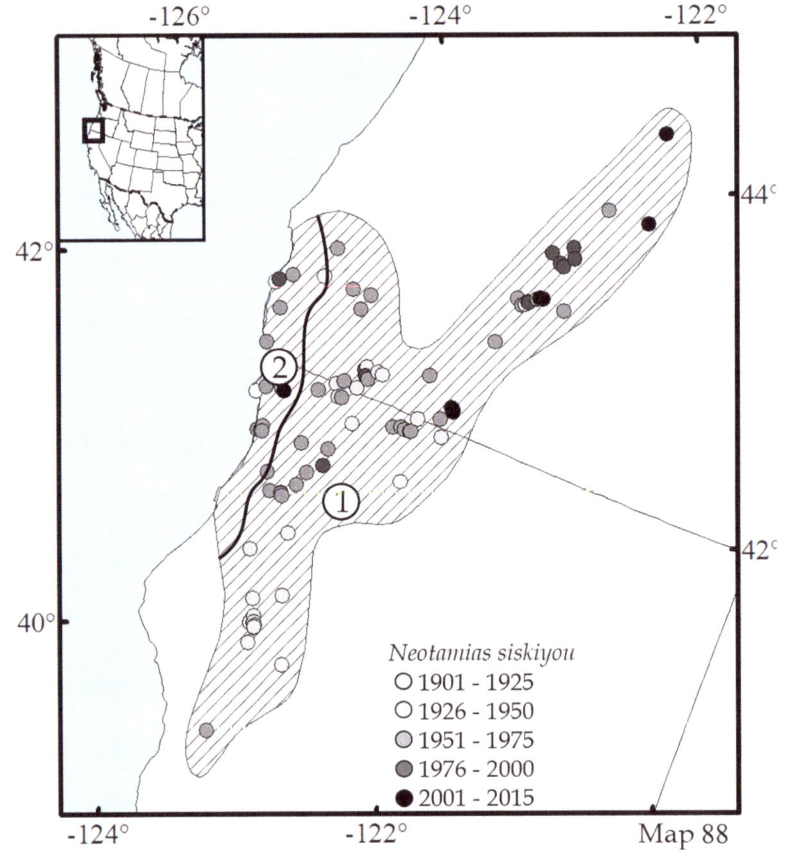


**Map 88** Distribution map of *Neotamias siskiyou*

**Additional Literature:** Howell (1929), Callahan (1980), Levenson et al. (1985), Sutton (1992), Thorington and Hoffmann (2005), Thorington et al. (2012).

## *Neotamias solivagus* (Howell, 1922)

Howell's chipmunks, chichimoco de Coahuila

Monotypic.

**Conservation:** *Neotamias solivagus* (as *Neotamias durangae*) is no listed by the IUCN (Álvarez-Castañeda et al. 2016z) and the Norma Oficial Mexicana (DOF 2019). However, has a smaller distribution area that need to be evaluated.

**Characteristics:** Howell's chipmunk is small sized; total length 221.0–245.0 mm and skull length 35.3–39.5 mm. Dorsal pelage with a central dark stripe blackish from the nape to the tail base, contrasting with the back general coloration; middle stripes brownish to blackish brown and wide; flank stripes very small, paler than the middle stripes. Inner paler stripes forming a narrow stripe dark gray to reddish gray, not contrasting with the central and middle stripes; outer stripes similar to the inner paler ones, but a little paler in some individuals; fur of underparts whitish. General coloration brownish reddish gray; back of the ears whitish, outer, and anterior parts blackish; postauricular patches gray; hind feet pale to medium gray; rump brownish reddish gray; facial stripes contrasting with the head general coloration; tail brownish gray with hair paler at the tip. Species group *quadrivittatus*.

**Comments:** *Neotamias solivagus* is endemic to Coahuila (Map 89), and has allopatric distribution with respect to the other species of *Neotamias*.

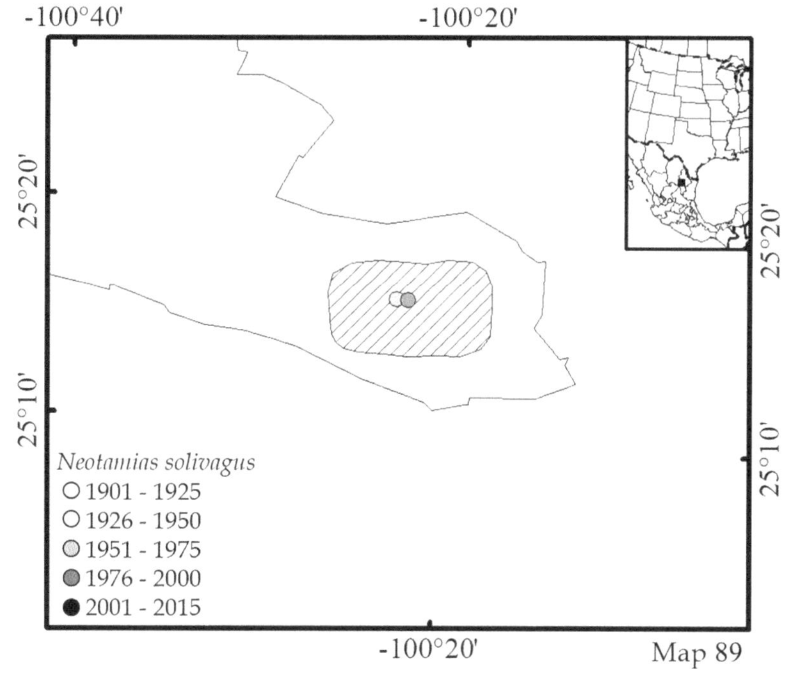


**Map 89** Distribution map of *Neotamias solivagus*

**Additional Literature:** Howell (1922, 1929), Baker (1956), Fleharty (1960), Callahan (1980), Sutton (1992), Best et al. (1993), Thorington et al. (2012).

## ***Neotamias sonomae*** (Grinnell, 1915)

Sonoma chipmunk, chichimoco de Sonoma

1. *N. s. alleni* (Howell, 1922). Restricted to northern San Francisco Bay, central-western California.
2. *N. s. sonomae* (Grinnell, 1915). From northern San Francisco Bay to near the border with Oregon.

**Conservation:** *Neotamias sonomae* is listed as Least Concern by the IUCN (Cassola 2016bu).

**Characteristics:** Sonoma chipmunk is small sized; total length 220.0–270.0 mm and skull length 36.5 to 40.0 mm. Dorsal pelage with a central dark stripe black from the nape to the tail base and not contrasting with the back general coloration, which is dark brown; middle stripes similar to the central stripe; flank stripes only present as a small shadow. Inner paler stripes of similar coloration to that on the rump and shoulders; however, the shoulder area could be more reddish; outer stripes paler than the general coloration, paler on the back; fur of underparts reddish brown. General coloration grayish dark brown in the rump and hips; shoulders, neck flanks, and flanks brownish; back of the ears, outer paler and anterior part darker, but paler than the head top; postauricular patches gray, similar to the outer paler stripe; hind feet dark gray; facial stripes palest, not strongly contrasting with the head coloration; back of the head grayish brown, darker than the shoulder coloration; tail dark brownish dorsally and flanks with blackish hairs, not paler at the tip. **Tip of the baculum less than one-half of the shaft length**. Species group *townsendii*.

**Comments:** *Neotamias sonomae* ranges in northwestern California, from Siskiyou County southward to San Francisco Bay (Map 90). It can be found in sympatry, or nearly so, with *N. amoenus* and *N. ochrogenys*. *N. sonomae* can be differentiated from *N. ochrogenys* and *N. quadrimaculatus* by its longer tail. From *N. quadrimaculatus*, by its smaller size; zygomatic arches width; shorter and wider rostrum; and from *N. siskiyou*, only by the baculum and os clitoridis.

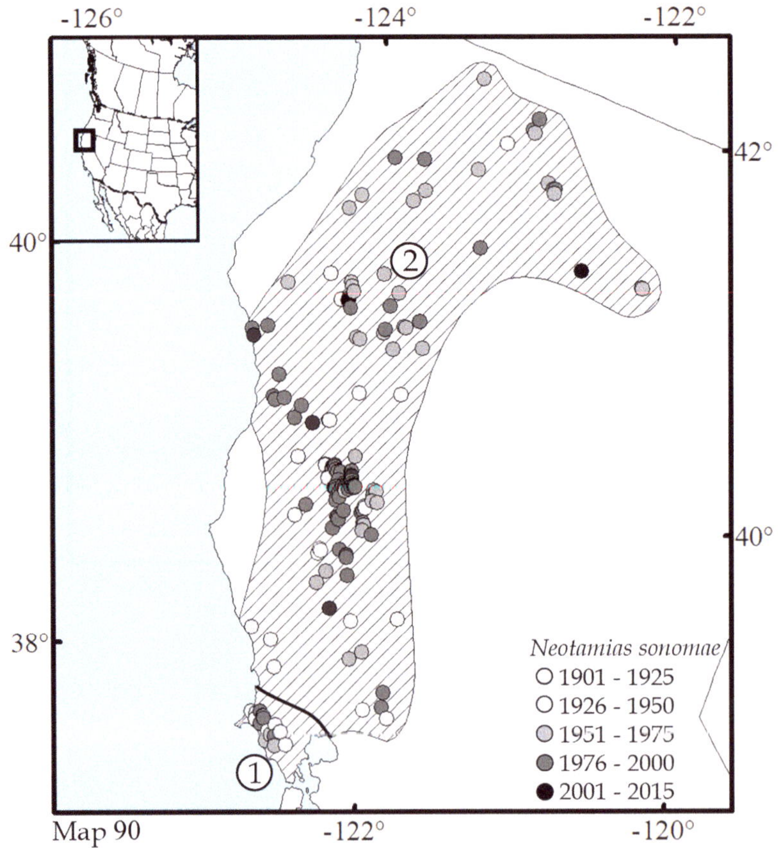


**Map 90** Distribution map of *Neotamias sonomae*

**Additional Literature:** Howell (1922, 1929), Johnson (1943), Patterson (1984), Levenson et al. (1985), Levenson (1990), Sutton (1992), Best (1993b), Thorington et al. (2012).

## ***Neotamias speciosus*** (Merriam, 1890)
Lodgepole chipmunk, chichimoco del pino

1. *N. s. callipeplus* (Merriam, 1893). Restricted to the Mount Pinos, central-western California.
2. *N. s. frater* (Allen, 1890). From Sierra Nevada, Nevada, north through California.
3. *N. s. sequoiensis* (Howell, 1922). From Sierra Nevada, Nevada, south through the Sequoia National Forest, western California.
4. *N. s. speciosus* (Merriam, 1890). From the Sequoia National Forest south through California.

**Conservation:** *Neotamias speciosus* is listed as Least Concern by the IUCN (Cassola 2016bv).

**Characteristics:** Lodgepole chipmunk is small sized; total length 197.0–240.0 mm and skull length 33.5–36.0 mm. Dorsal pelage with a central dark stripe black from the nape to the tail base and contrasting with the back general coloration; middle stripes similar to the central stripe, but brownish, and flank stripes absent. Inner paler stripes of similar coloration to those on the rump and shoulders, with a mixture of paler and gray hairs; outer stripes solid grayish cream, well defined; fur of underparts whitish. General coloration grayish in the rump and hips, with a mixture of different shades of gray; flanks and neck flanks cinnamon to yellowish orange or pale brown; shoulders same as the flanks, but with hairs blackish at the tip; back of the ears lighter, outer and anterior parts darker; postauricular patches pale gray, similar to the outer paler stripes; hind feet gray; facial stripes paler and well developed, contrasting with the dark stripes; back of the head grayish, similar to the nape; tail short with gray hairs dorsally, middle brownish, flanks with blackish hairs, and paler at the tip. Species group *quadrivittatus*.

**Comments:** *Neotamias speciosus* ranges from northern California to the San Bernardino Mountains and extreme western-central Nevada (Map 91). It can be found in sympatry, or nearly so, with *N. alpinus*, *N. amoenus*, *N. dorsalis*, *N. merriami*, *N. minimus*, and *N. panamintinus*. *N. speciosus* can be differentiated from *N. amoenus* by its larger size; longer and narrow ears; sharply contrasting pale and dark stripes; facial stripes less washed with ochraceous shades and dark facial stripe usually darker blackish; skull more massive, with a less pointed rostrum and larger incisive foramina. From *N. panamintinus*, by its larger size; ears more pointed; darker facial and dorsal stripes; pale stripes and fur of the underparts more ochraceous wash. Skull relatively narrow, but with a broader and less pointed rostrum; upper toothrow more divergent anteriorly. From *N. umbrinus*, by its smaller size; longer ears and slightly shorter tail; head crown darker; shoulders brown; with a darker facial stripe, skull shorter and broader, with the incisors shorter and more recurved.

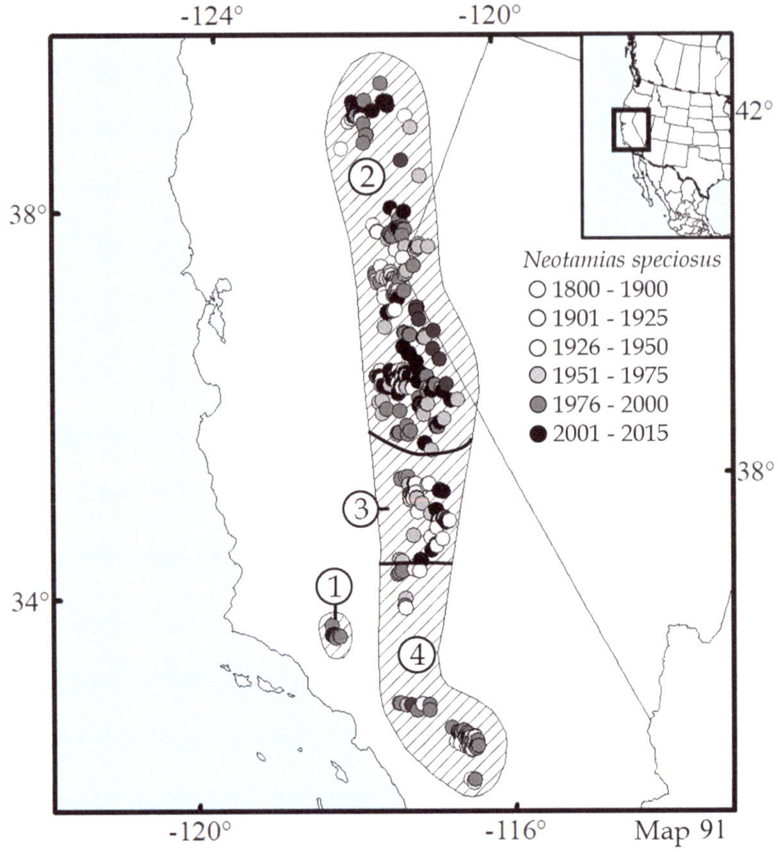


**Map 91** Distribution map of *Neotamias speciosus*

**Additional Literature:** Howell (1929), Johnson (1943), Callahan (1980), Levenson et al. (1985), Sutton (1992), Best et al. (1994b).

## *Neotamias townsendii* (Bachman, 1839)

Townsend's chipmunk, chichimoco del noroeste

1. *N. t. cooperi* (Baird, 1855). From southwest British Columbia south through inland to Oregon.
2. *N. t. townsendii* (Bachman, 1839). From southwest British Columbia south through the coast to Oregon.

**Conservation:** *Neotamias townsendii* is listed as Least Concern by the IUCN (Cassola 2016bw).

**Characteristics:** Townsend's chipmunk is small sized; total length 235.0–265.0 mm and skull length 36.8–40.0 mm. Dorsal pelage and flanks varying from tawny, antique brown, or umber to tawny olive, sayal brown, orange cinnamon, or clay with a central dark stripe black fuscous and wide, from the nape to the rump, not strongly contrasting with the back general coloration; middle and flank stripes similar to the central stripe. Inner paler stripes absent in a wide area where the coloration is the same as in the rump and shoulders, but slightly paler; **outer stripes similar to the inner paler stripes, but narrow and nearly obsolete**; **fur of underparts creamy white to grayish**. General coloration reddish dark brown in the rump, hips, flanks, shoulders, and neck flanks; **outer back part of the ears smoke gray, anterior part fuscous to fuscous black**; post-auricular patches small and gray, not strongly contrasting; hind feet grayish brown; facial stripes brownish gray, not clearly contrasting with the dark stripe; the most contrasting are the upper and lower stripes; back of the head slightly darker than the general body coloration; **tail borders uniformly smoke gray or pale buff, never deep buff**; **upper molars with the posterior or two main cross ridges diverging externally from the anterior protoloph**. Species group *townsendii*.

**Comments:** *Neotamias ochrogenys* was recognized as a distinct species (Sutton and Nadler 1974), previously regarded as a subspecies of *N. townsendii* (Merriam 1897). *N. townsendii* ranges in southern British Columbia and Washington (Map 92). It can be found in sympatry, or nearly so, with *N. amoenus* and *N. quadrimaculatus*. *N. townsendii* can be differentiated from *N. ochrogenys* by its smaller body size; dorsal pelage paler; tail thicker; and from *N. quadrimaculatus*, by its smaller size; zygomatic arches width; shorter and wider rostrum.

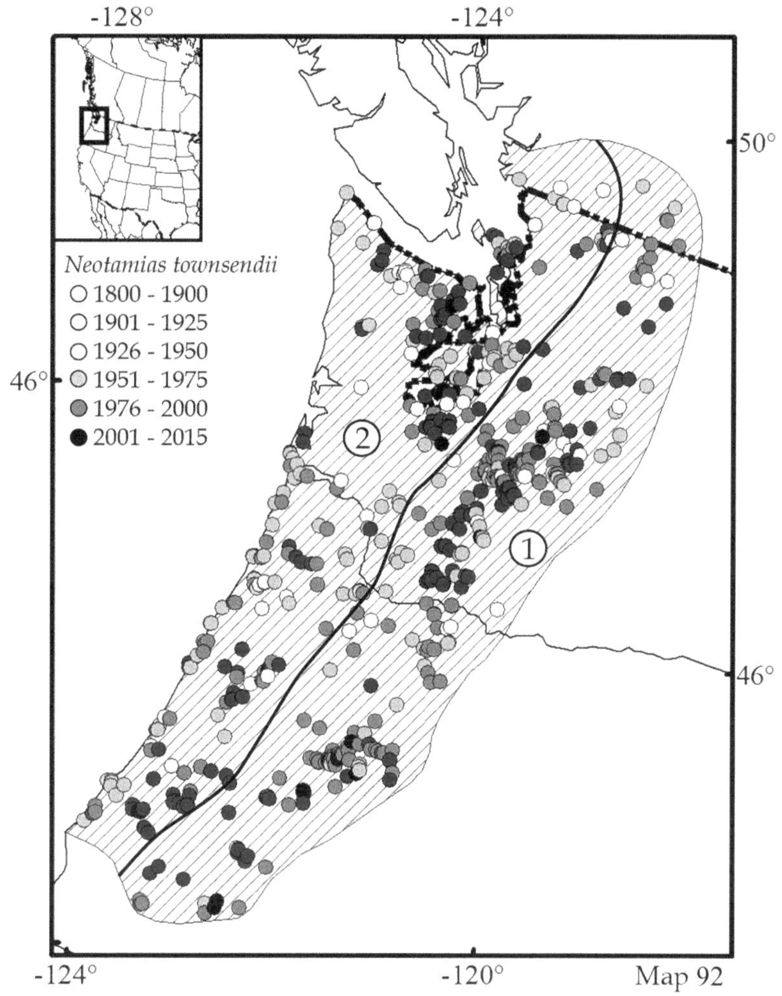


**Map 92** Distribution map of *Neotamias townsendii*

**Additional Literature:** Howell (1929), Maser (1975), Levenson et al. (1985), Sutton (1992, 1993), Thorington and Hoffmann (2005), Thorington et al. (2012).

### *Neotamias umbrinus* (Allen, 1890)

Uinta chipmunk, chichimoco común

1. *N. u. adsitus* (Allen, 1905). Southwestern Utah and northern Arizona.
2. *N. u. fremonti* (White, 1953). Mainly Wyoming.
3. *N. u. inyoensis* (Merriam, 1897). Eastern California, Nevada, and western Utah.
4. *N. u. montanus* (White, 1953). Mainly Colorado.
5. *N. u. nevadensis* (Burt, 1931). Restricted to the Sheep Mountains, southern Nevada.
6. *N. u. sedulus* (White, 1953). Restricted to the Henry Mountains, southeastern Utah.
7. *N. u. umbrinus* (Allen, 1890). Northern Utah, southwestern Wyoming, and southeast Idaho.

**Conservation:** *Neotamias umbrinus* is listed as Least Concern by the IUCN (Cassola 2016bx).

**Characteristics:** Uinta chipmunk is small sized; total length 196.0–243.0 mm and skull length 33.5 to 36.8 mm. Dorsal pelage with a central dark stripe black from the nape to the rump and strongly contrasting with the back general coloration; middle stripes similar to the central stripe, and flank stripes absent. Inner paler stripes of similar coloration to the rump and shoulders, but mixed with pale gray; outer stripes pale gray, strongly contrasting with the dorsal coloration; fur of underparts white to cream. General coloration reddish cinnamon brown in the flanks, shoulders, and neck flanks, grayish on the rump and hips; back of the ears paler, outer and anterior parts darker, similar to the head stripes; postauricular patches paler gray, similar in color to the outer paler stripes; hind feet pale brown; facial stripes paler and well marked, the darker ones not strongly contrasting; back of the head darker than the nape coloration; tail with blackish hairs, orange to reddish distally, paler at the tip. Species group *quadrivittatus*.

**Comments:** *Neotamias umbrinus* ranges from eastern-central California east through southwestern Montana, Wyoming, and northern-central Colorado (Map 93). It can be found in sympatry, or nearly so, with *N. alpinus*, *N. amoenus*, *N. merriami*, *N. minimus*, *N. panamintinus*, *N. quadrivittatus*, and *N. rufus*. *N. umbrinus* can be differentiated from *N. alpinus* by its larger size, and from *N. speciosus*, by being larger; shorter ears; tail slightly longer; back of the head paler; shoulders not brown; facial stripes paler. The skull is longer and narrower, with the incisors longer and less recurved.

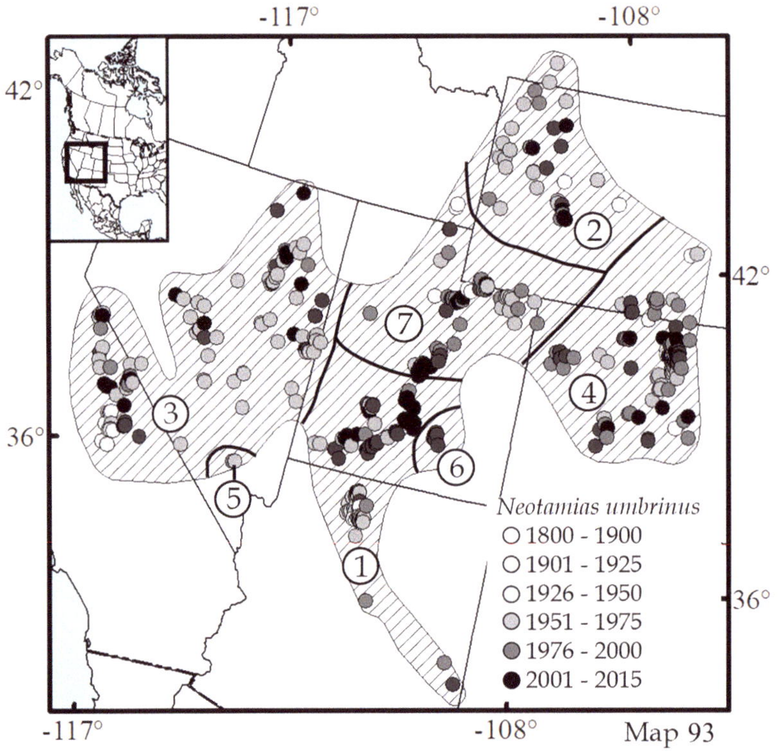


**Map 93** Distribution map of *Neotamias umbrinus*

**Additional Literature:** Howell (1929), Johnson (1943), Sutton (1992), Best et al. (1994a, b, c), Braun et al. (2011a), Thorington and Hoffmann (2005), Thorington et al. (2012), Koprowski et al. (2017).

## *Tamias striatus* (Linnaeus, 1758)

Eastern chipmunk, chichimoco del este

1. *T. s. doorsiensis* Long, 1971. Northeastern Wisconsin.
2. *T. s. fisheri* Howell, 1925. From western New York south through Virginia.
3. *T. s. griseus* Mearns, 1891. Upper Mississippi Valley, from Missouri north through Ontario and Manitoba.
4. *T. s. lysteri* (Richardson, 1829). Southeastern Ontario and Michigan.
5. *T. s. ohioensis* Bole and Moulthrop, 1942. Illinois and Ohio.
6. *T. s. peninsulae* Hooper, 1942. Wisconsin and Michigan.
7. *T. s. pipilans* Lowery, 1943. Louisiana.
8. *T. s. quebecensis* Cameron, 1950. Québec and central Ontario.
9. *T. s. rufescens* Bole and Moulthrop, 1942. Southern Michigan.
10. *T. s. striatus* (Linnaeus, 1758). Southern Ohio to Georgia.
11. *T. s. venustus* Bangs, 1896. Arkansas and Oklahoma.

**Conservation:** *Tamias striatus* is listed as Least Concern by the IUCN (Cassola 2016dt).

**Characteristics:** Eastern chipmunk is small sized; total length 215.0–300.0 mm and skull length 37.4–46.0 mm. Dorsal pelage with the dark central stripe black and narrow, from the nape to the rump, not strongly contrasting with the back general coloration; middle stripes short, only over the flanks, not over the rump or shoulders, similar in color to the central stripe; flank stripes similar to the dark middle ones. **Inner paler stripes absent, wide area of the same coloration as the rump and shoulders;** outer stripes grayish reddish cream, well defined, contrasting with the dorsal coloration; fur of underparts whitish. General coloration reddish, with hairs darker at the tip, giving a grizzled pattern; rump brownish, shoulders grayish; flanks and neck flanks reddish brown; back and outer parts of the ears paler, anterior part darker; hind feet gray; postauricular patches very small, pale gray; facial stripes are the palest; stripes reddish brown not clearly contrasting with the dark stripe, which is of the same color as the head top; back of the head grayish, similar to the nape; tail blackish, with hair paler at the tip; **tail length less than 40% of the head-and-body length;** skull elongated, with a relatively small and weak postorbital process; the zygomatic plate slanting upward at approximately 45° from the occlusal plane; without a temporal foramen in the squamosoparietal suture; lamboidal crest well developed; **upper premolars absent; hypohyal and ceratohyal bones not fused**; **infraorbital foramen perforating the zygomatic plate rather than a canal opening forward on the plate**. Species group *striatus*.

**Comments:** *Tamias striatus* ranges from southern Manitoba east through Nova Scotia, and south through Louisiana to Georgia, but is absent from much of the coastal plain. *T. striatus* was introduced to Newfoundland (Map 94). None of the species of *Neotamias* has sympatric distribution with *T. striatus*. *T. striatus* can be differentiated from the other squirrel species by having nine longitudinal stripes, five darker and four paler, with the paler stripes running between the darker ones. The stripes can vary in the presence, absence, coloration, shade, length, and intensity. From all the species of *Neotamias*, by the absence of the inner paler stripes, with a wide area of the same coloration as in the rump and shoulders.

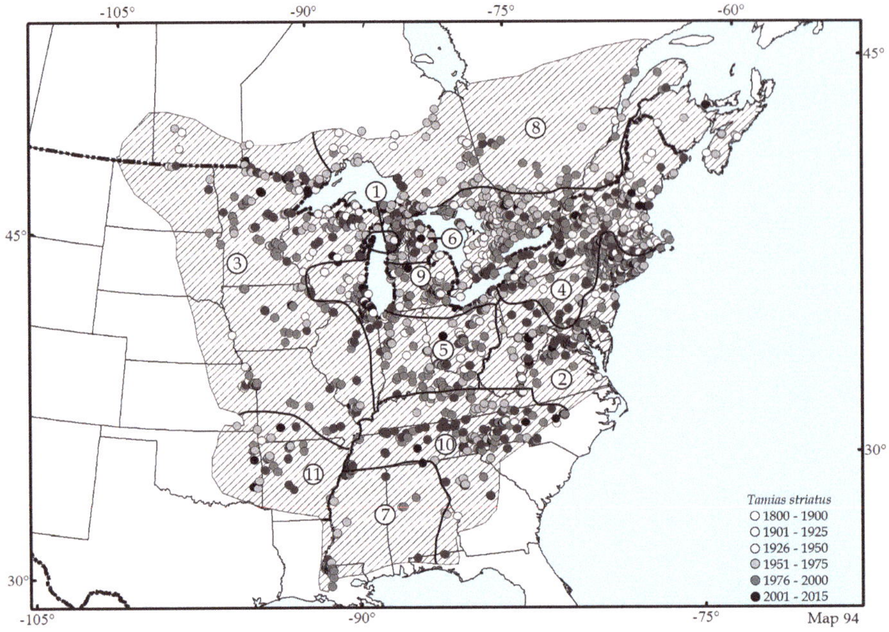


**Map 94** Distribution map of *Tamias striatus*

**Additional Literature:** Howell (1929, 1938), White (1953), Moore (1960), Callahan (1980), Levenson et al. (1985), Snyder (1982), Sutton (1992), Thorington et al. (2012), Koprowski et al. (2017).

## **Suborder Supramyomorpha** D'Elía et al., 2019

This suborder includes anomaluromorphs, castorimorphs, and myomorphs, which are now infraorders. No morphological characteristics are present in all the species of the Supramyomorpha that could be used as diagnostic traits, so the species should be determined based on molecular data. The keys were elaborated based on the review of specimens and the following sources: Lawlor (1979), Hall (1981), Jones and Manning (1992), Álvarez-Castañeda et al. (2017a), and D'Elía et al. (2019).

1. Zygomatic plate oriented more laterally; infraorbital foramen enlarged, narrow, and typically expanded dorsally (Fig. 18); maxillary teeth three or fewer ........ 2

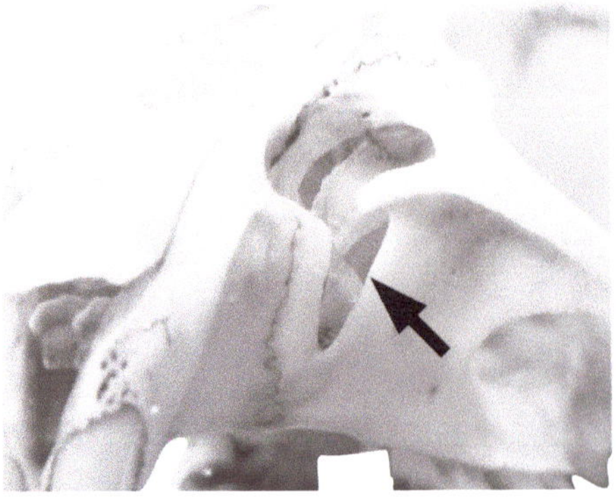

**Fig. 18** Infraorbital foramen enlarged, narrow, and typically expanded dorsally

1a. Zygomatic plate oriented anteriorly; infraorbital foramen small (Fig. 19); maxillary teeth four or five..........................4

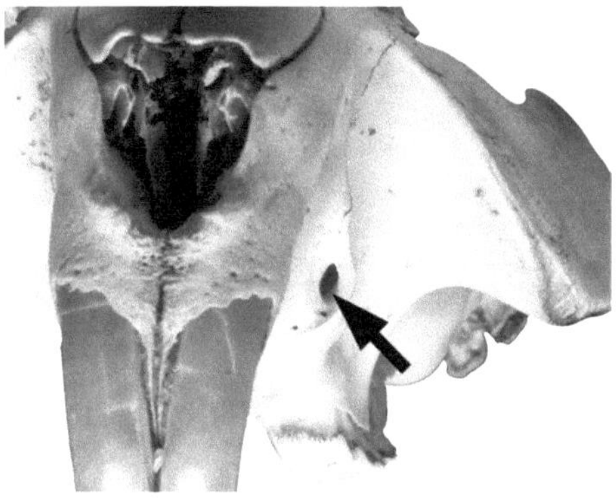

**Fig. 19** Infraorbital foramen small

2. First upper molars with three rows of the cusps arranged longitudinally on the occlusal surface (Fig. 20); three transverse lophs clearly defined.......................................................................................................... family Muridae (p. 429)

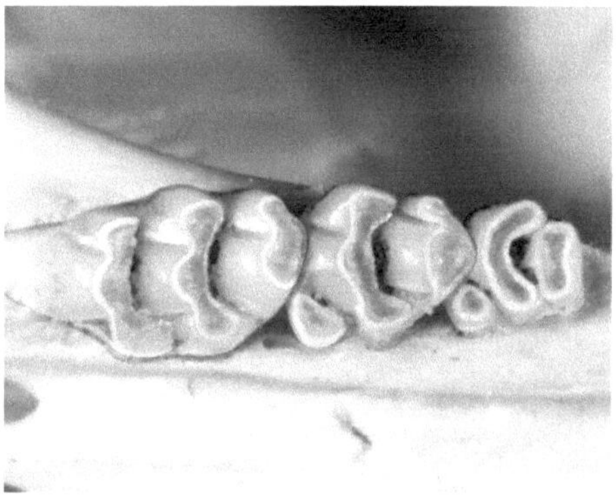

**Fig. 20** First upper molars with three rows of the cusps arranged longitudinally on the occlusal surface

2a. First upper molars with one or two rows of the cusps arranged longitudinally or without cusps, molars flattened on the occlusal surface (Fig. 21); no defined lophs ..........................................................................................................3

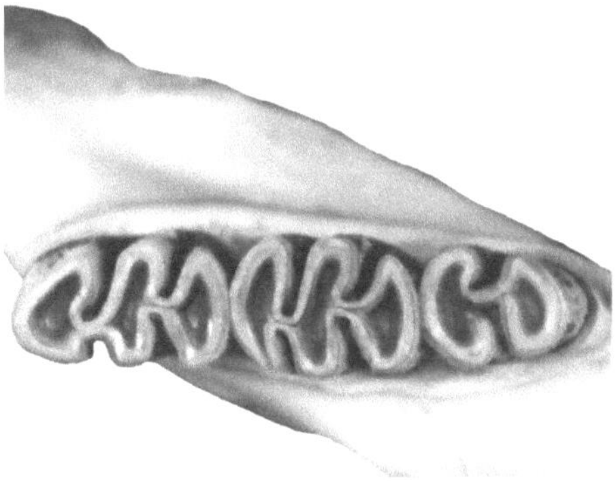

**Fig. 21** First upper molars with one or two rows of the cusps arranged longitudinally or without cusps, molars flattened on the occlusal surface

3. Tail length less than 1.5 times the head-and-body length, either haired or naked and scaly; hind limbs not greatly elongated; infraorbital foramen moderately large.......................................................................... family Cricetidae (p. 139)

3a. Tail length at least 1.5 times the head-and-body length, naked and scaly; hind limbs elongated; infraorbital foramen markedly enlarged ........ family Zapodidae (p. 431)

4. Tail scaly, broad, and flat; hindfoot toes webbed; basioccipital fossa well developed (Fig. 22); skull length greater than 75.0 mm; upper toothrow greater than 18.0 mm (from Alaska east through Labrador Peninsula and south through Sonora, Chihuahua, Coahuila, Nuevo León, and Tamaulipas, except for the Florida Peninsula) ........ family Castoridae; *Castor canadensis* (p. 137)

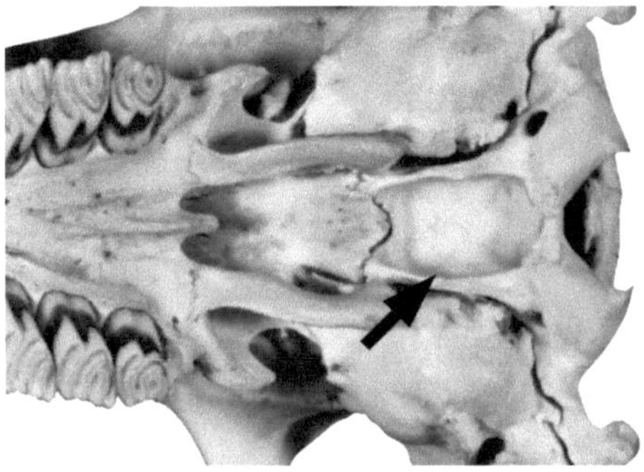

**Fig. 22** Basioccipital fossa well developed

4a. Tail bare or hairy, not scaly, broad, or flat; hindfoot toes not webbed; basioccipital fossa absent (Fig. 23); skull length less than 75.0 mm; upper toothrow less than 18.0 mm ........ 5

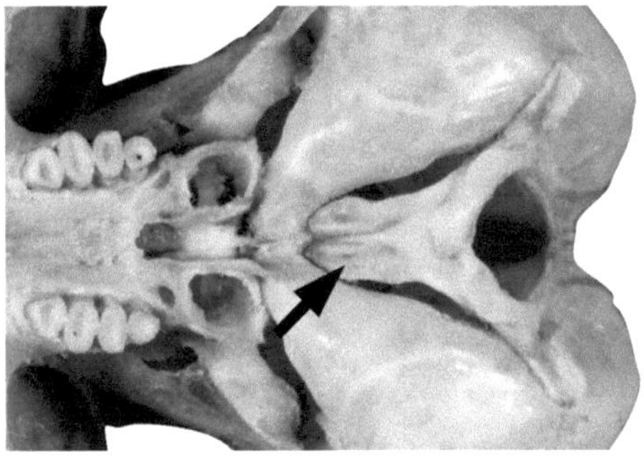

**Fig. 23** Basioccipital fossa absent

5. Tail short and almost naked; forefoot with claws extremely long; rostrum wider than the interorbital constriction; first upper molariforms almost split at the middle (Fig. 24); last upper molars larger than the others ........ family Geomyidae (p. 437)

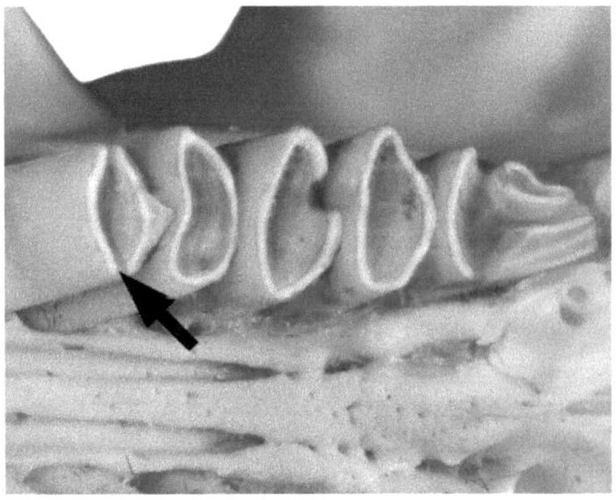

**Fig. 24** First upper molariforms almost split at the middle

5a. Tail long and hairy; forefoot with claws in proportion to its size; rostrum thinner than the interorbital constriction; first upper molariforms not split at the middle (Fig. 25); last upper molars equal or smaller than the others........................................................................................................................................................ family Heteromyidae (p. 492)

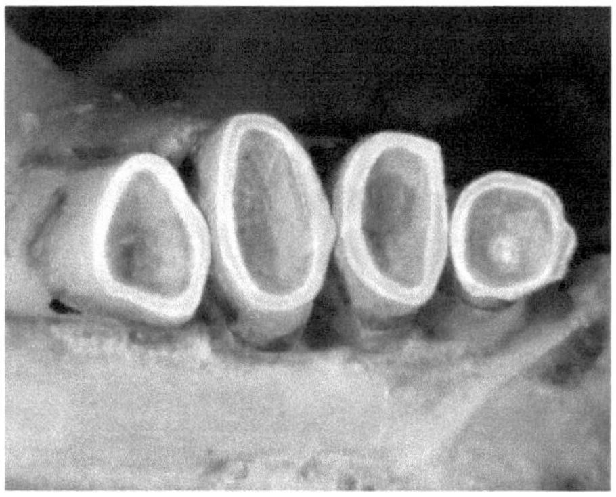

**Fig. 25** First upper molariforms not split at the middle

## **Family Castoridae** Hemprich, 1820

The family includes beavers. It is currently represented by one genus (*Castor*) with two species. They are characterized by t**hick and large** head and body **(95.0–160.0 cm); tail long, wide, flat, mostly naked, and scaly; dorsal pelage brown with grayish hair for protection; hindfoot digits webbed; incisors proportionately large and orange; a distinctive basioccipital depression**. Their natural distribution ranges from all the Mexican states bordering with the United States to central Canada and Alaska. It belongs to the suborder Castorimorpha, superfamily Castoroidea.

## Genus ***Castor*** Linnaeus, 1758

Dorsal pelage varying markedly according to the season and geographic areas, generally rich glossy brown; fur of underparts pale brown to tawny; pelage very dense, underfur fine, overlaid with many coarse guardrails; tail and feet black; body thick and compact, highly modified for aquatic life; limbs short; ears small; hind feet large, pentadactyl, unguiculate, toes webbed; tail broad, flat, nearly hairless and covered with large scales; caudal vertebrae more flattened; rostrum broad and deep; braincase narrow; basioccipital region with a conspicuous pit like a depression; molariforms hypsodont.

### ***Castor canadensis*** Kuhl, 1820

American Beaver, castor

1. *C. c. acadicus* Bailey and Doutt, 1942. Central Québec, Nova Scotia, and New England, from Main to the east.
2. *C. c. baileyi* Nelson, 1927. Southeastern Oregon and northern Nevada.
3. *C. c. belugae* Taylor, 1916. Southern Alaska, central and western Yukon, and extreme northwestern British Columbia.
4. *C. c. caecator* Bangs, 1913. Restricted to Newfoundland.
5. *C. c. canadensis* Kuhl, 1820. From Yukon and western British Columbia east through western Québec and southern Virginia, including Minnesota.
6. *C. c. carolinensis* Rhoads, 1898. From Illinois and Ohio south through Louisiana and east through North Carolina
7. *C. c. concisor* Warren and Hall, 1939. Restricted to central Colorado.

8. *C. c. duchesnei* Durrant and Crane, 1948. Restricted to the Duchesne River, Utah, and Colorado.
9. *C. c. frondator* Mearns, 1897. Central and eastern Arizona, western New Mexico, and northern Sonora.
10. *C. c. idoneus* Jewett and Hall, 1940. Coastal areas of southern Oregon and northern California.
11. *C. c. labradorensis* Bailey and Doutt, 1942. Newfoundland, Labrador Peninsula.
12. *C. c. leucodontus* Gray, 1869. From Southern British Columbia south through southern Oregon and western Idaho.
13. *C. c. mexicanus* Bailey, 1913. From New Mexico, southern Texas, and northern Chihuahua east through northern Tamaulipas.
14. *C. c. michiganensis* Bailey, 1913. Michigan and Wisconsin.
15. *C. c. missouriensis* Bailey, 1919. From Montana to southern Minnesota south through eastern Colorado, Oklahoma, and western Arkansas.
16. *C. c. pallidus* Durrant and Crane, 1948. Restricted to the Raft River, Utah.
17. *C. c. phaeus* Heller, 1909. Restricted to the Chichago Islands and Admiralty Islands, Alaska.
18. *C. c. repentinus* Goldman, 1932. Southeastern Utah, western Arizona, and southeastern California.
19. *C. c. rostralis* Durrant and Crane, 1948. Central and western Colorado.
20. *C. c. sagittatus* Benson, 1933. Southern Yukon and central British Columbia.
21. *C. c. shastensis*, Taylor, 1916. Southwestern Oregon and northern California.
22. *C. c. subauratus* Taylor, 1912. Central California.
23. *C. c. taylori* Davis, 1939. Central and southern Idaho.
24. *C. c. texensis* Bailey, 1905. Southeastern Oklahoma and central Texas.

**Conservation:** *Castor canadensis* is listed as Endangered to Special Protection in the Norma Oficial Mexicana (DOF 2019) and as Least Concern by the IUCN (Cassola 2016g).

**Characteristics:** The same mentioned in the genus.

**Comments:** *Castor canadensis* ranges from Alaska east through the Labrador Peninsula and from Canada south through rivers in northern México, Sonora, Chihuahua, Coahuila, Nuevo León, and Tamaulipas, except for the Florida Peninsula (Map 95). It can be found in sympatry, or nearly so, with different species, and differs from all other species by its larger size; tail paddle-like, flat, nearly hairless, and covered with large scales.

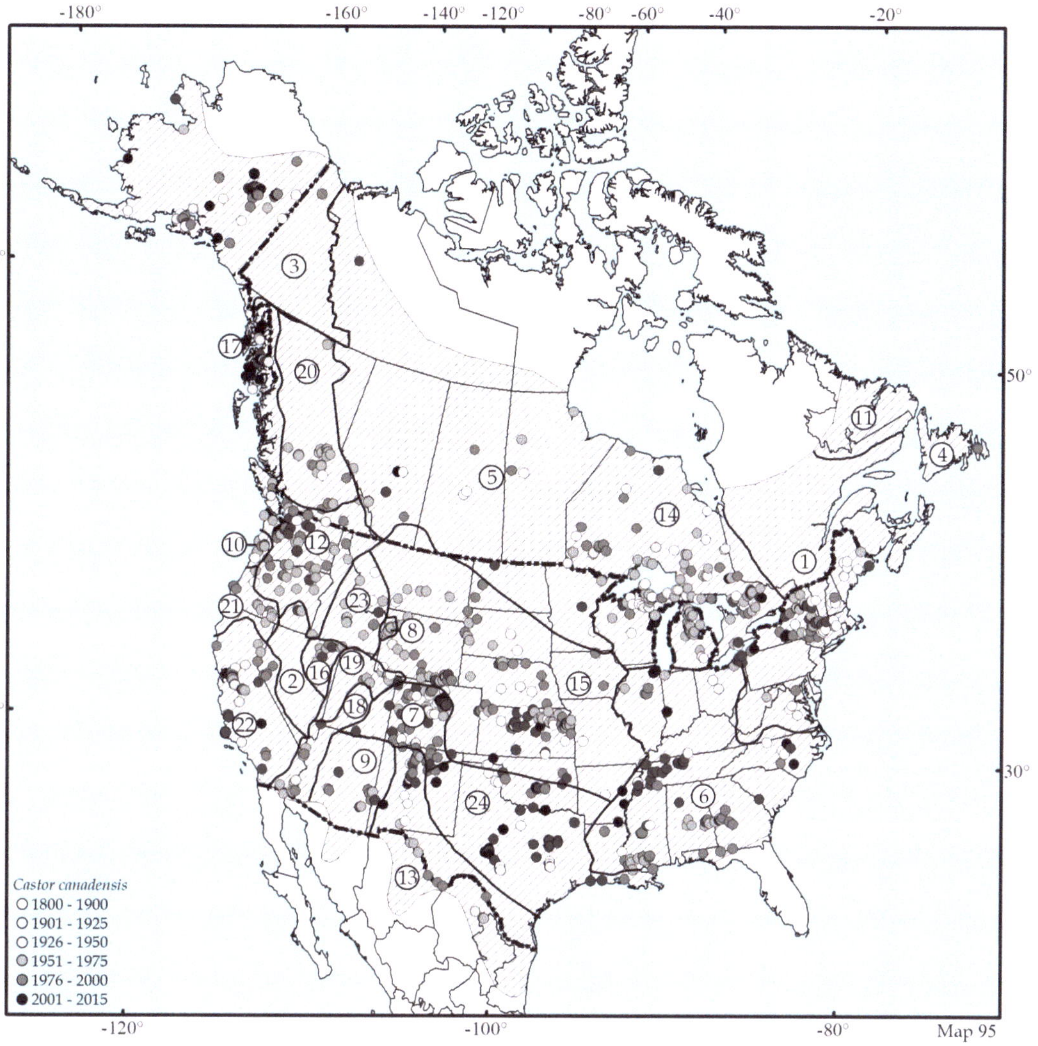


**Map 95** Distribution map of *Castor canadensis*

**Additional Literature:** Taylor (1916), Grinnell et al. (1937a, b), Jenkins and Busher (1979), Hall (1981).

## **Family Cricetidae** Fischer, 1817

The family Cricetidae includes the majority of living rodents in North America. This group is relatively simple, so it was decided not to separate subfamilies in this first version of the keys, seeking to facilitate the identification process. Four subfamilies range in North America: Arvicolinae, Neotominae, Sigmodontinae, and Tylomyinae. The keys were elaborated based on the review of specimens and the following sources: Lawlor (1979), Hall (1981), Jones and Manning (1992), and Álvarez-Castañeda et al. (2017a).

1. Tail length less than 75% of the total length; if it is greater than 75%, total length greater than 300.0 mm; molars with an enamel pattern composed of alternating triangles; zygomatic arches angular; rostrum short .................................................................................................................................................subfamily Arvicolinae (p. 140)

1a. Tail length greater than 75% of the total length, if it is less than 75%, total length less than 150.0 mm; molars without an enamel pattern composed of non-alternating triangles; zygomatic arches smooth; rostrum tapering anteriorly ........2

2. Tail naked; cusps of the upper molars opposite, or nearly so; anterior and lingual cusps of the first upper molars reduced; third upper molars not reduced, if equal to the second upper molars, subsidiary ridges of the upper molars reduced........subfamily Tylomyinae (p. 421)

2a. Tail hairy; cusps of upper molars alternating; anterior and lingual cusps of the first upper molars not reduced; third upper molars reduced........subfamily Neotominae and Sigmodontinae (p. 197)

## **Subfamily Arvicolinae** Gray, 1821

The Arvicolinae includes lemmings and voles, characterized by being small (except *Ondatra*, *Neofiber*, and *Arvicola*), furry, short-tailed, never as long as the head-and-body length; five toes on the each foot, and short-legged; **12 molars show prismatic cusps shaped as alternating triangles and loops, long-crowned and ever-growing (hypsodont) in most genera; postorbital processes absent; outer wall of the infraorbital canal transverse or oblique to the skull long axis and with the front edge emarginate or undercut, not projecting in front of the anterior border of the superior ramus of the maxillary root of the zygomamaxillary root; palatine bones notably thick.** The keys were elaborated based on the review of specimens and the following sources: Miller (1896), Ellerman et al. (1941), Hall and Cockrum (1953), Lawlor (1979), Hall (1981), Verts and Carraway (1984), Jones and Manning (1992), and Álvarez-Castañeda et al. (2017a).

1. Total length greater than 200.0 mm; skull length greater than 40.0 mm; upper incisor width greater than 2.0 mm; zygomatic breadth greater than 25.0 mm........tribe Ondatrini 2

1a. Total length less than 200.0 mm; skull length less than 40.0 mm; upper incisor width less than 2.0 mm; zygomatic breadth less than 25.0 mm........3

2. Tail rounded with thin hair; total length less than 390.0 mm; tail length less than 180.0 mm; hind feet not webbed between the toes; fifth lower molars with shallow reentrant angles; tooth ever-growing (rootless; from Florida north through South Georgia)........*Neofiber alleni* (p. 189)

2a. Tail flattened laterally and with hair; total length greater than 450.0 mm; tail length greater than 200.0 mm; hind feet webbed between the toes; fifth lower molars with deep reentrant angles; molars rooted (from Alaska east through the Labrador Peninsula, Prince Edwards Island, and New Brunswick, and south throughout the United States, except for the Florida Peninsula, in Mexico only in the Colorado River Delta)........*Ondatra zibethicus* (p. 190)

3. Upper incisors grooved; rostrum short, approximately as broad as long or about one-fourth of the skull length........tribe Lemmini *Synaptomys* (p. 185)

3a. Upper incisors smooth; rostrum and longer than broad or more than one-fourth of the skull length........4

4. Upper molariforms non ever-growing (brachyodont) in adults; nasals extending anteriorly beyond the incisors........5

4a. Upper molariforms ever-growing (hypsodont) in adults; nasals not extending anteriorly beyond the incisors........7

5. Lower molars with inner (lingual) reentrant angles, less deep than those on the outer (labial) reentrant angles, not exceeding one-half of the tooth width; cheek-teeth small; upper toothrow somewhat divergent........tribe *Clethrionomyini Clethrionomys* (p. 172)

5a. Lower molars with inner (lingual) reentrant angles much deeper than those on the outer (labial) angles and exceeding one-half of the tooth width; cheek-teeth large; upper toothrow essentially parallel........tribe Phemacomyini 6

6. Tail length less than 50.0 mm........*Phenacomys* (p. 195)

6a. Tail length greater than 50.0 mm........*Arborimus* (p. 191)

7. Squamosal bone with a prominent peg extending anterolaterally into the orbit (Fig. 26); first lower molars with seven closed triangles between the anterior and posterior loops; in winter, dorsal pelage white and fore claws with extra basal growths........tribe Dicrostonychini *Dicrostonyx* (p. 177)

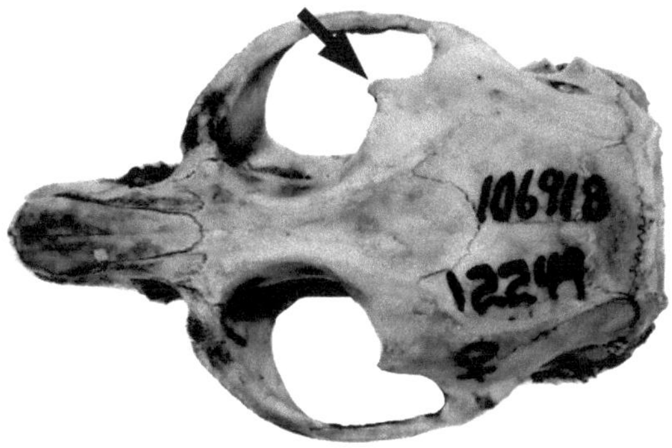

**Fig. 26** Squamosal bone with a prominent peg extending anterolaterally into the orbit

7a. Squamosal bone without a peg extending anterolaterally into the orbit; if present, not prominent (Fig. 27); first lower molars with six or fewer closed triangles between the anterior and posterior loops; in winter, dorsal pelage not white and fore claws without extra basal growths ........................................................................................ tribe Arvicolini 8

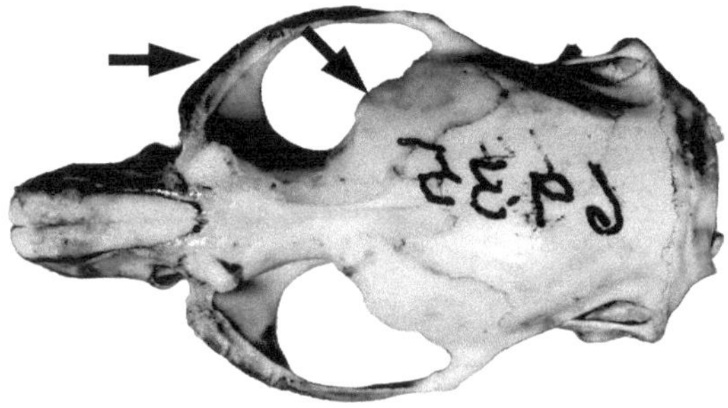

**Fig. 27** Squamosal bone without a peg extending anterolaterally into the orbit; if present, not prominent

8. Skull massive; zygomatic arches broadly divergent, wider anteriorly than posteriorly (Fig. 28); upper toothrow strongly divergent posteriorly; first lower molars with three closed triangles; skull length usually less than 29.0 mm (from Alaska and British Columbia to Hudson Bay, Baffin Island, and Manitoba) .............................tribe Lemmini *Lemmus* (p. 183)

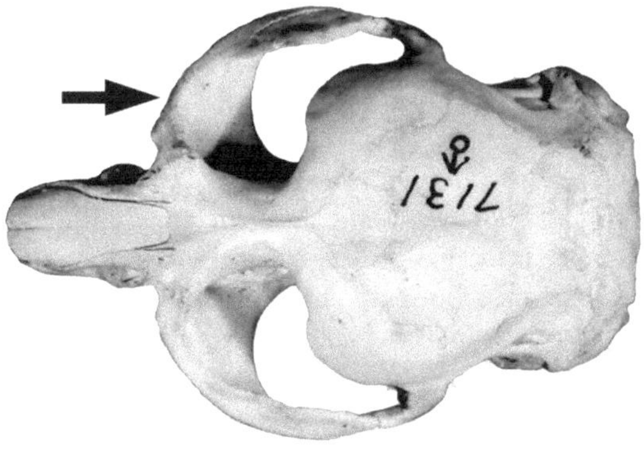

**Fig. 28** Zygomatic arches broadly divergent, wider anteriorly than posteriorly

8a. Skull not massive; zygomatic arches rounded, not wider anteriorly than posteriorly (Fig. 27); upper toothrow only slightly divergent posteriorly; first lower molars with more than three closed triangles; skull length usually greater than 29.0 mm ........................................................................................................................................................9

9. Total length less than 250.0 mm; tail length not exceeding the hindfoot length by 10.0 mm; inner surface of the anterior loop of the first lower molars smooth (from southern Alberta, Saskatchewan, and western North Dakota south through northeastern California, Nevada, Utah, and northern Colorado, with one branch in Washington and Oregon)................ ........................................................................................................................ *Lemmiscus curtatus* (p. 143)

9a. Total length greater than 250.0 mm; tail length usually exceeding the hindfoot length by 10.0 mm; inner surface of the anterior loop of the first lower molars with a distinctly small reentrant fold ........ 10

10. Tail small, almost twice the hindfoot length; first lower molars usually with four closed triangles (from Alaska east through western Mackenzie, and south through northwest British Columbia, including the Aleutian Islands) ........ *Alexandromys oeconomus* (p. 142)

10a. Tail medium sized, more than twice the hindfoot length; first lower molars with five or six closed triangles ........ *Microtus* (p. 144)

## Tribe Arvicolini

### Genus *Alexandromys* Ognev, 1914

The Tundra vole, which ranges in North America, has the **tail short, from 1.8 to 2.4 times the hindfoot length** and bicolored; dorsally darker than the back and whitish ventrally; **first lower molars with four closed triangles and the fifth triangle open and confluent with a short terminal loop.**

#### ***Alexandromys oeconomus*** (Pallas, 1776)

Tundra vole, ratón colicorto nórdico

1. *A. o. amakensis* (Murie, 1930). Restricted to the Amak Island, Alaska.
2. *A. o. elymocetes* (Osgood, 1906). Restricted to the Montague Island, Alaska.
3. *A. o. innuitus* (Merriam, 1900). Restricted to the Saint Lawrence Island, Alaska.
4. *A. o. macfarlani* (Merriam, 1900). Mainly all Alaska, Yukon, and western Mackenzie.
5. *A. o. operarius* (Nelson, 1893). Western and southwestern Alaska.
6. *A. o. popofensis* (Merriam, 1900). Restricted to the Popof Island, Alaska.
7. *A. o. punukensis* (Hall and Gilmore, 1932). Restricted to the Punuk Island, Alaska.
8. *A. o. sitkensis* (Merriam, 1897). Restricted to the Chichagof Island, Alaska.
9. *A. o. unalascensis* (Merriam, 1897). Restricted to the Sanak and Unalaska islands, Alaska.
10. A. *o. yakutatensis* (Merriam, 1900). Restricted to southeastern Alaska, southwestern Yukon, and northwestern British Columbia.

**Conservation:** *Alexandromys oeconomus* (as *Microtus oeconomus*) is listed as Least Concern by the IUCN (Linzey et al. 2016o)

**Characteristics:** Tundra vole is medium sized; total length 160.0–225.0 mm and skull length 22.4–32.5 mm. Dorsal pelage buffy, cinnamon brown to rusty rich; fur of underparts light gray to whitish at the tip with blackish shades at the base, giving a whitish appearance; **tail short, almost two times the hindfoot length, and bicolored, dorsally darker than the back pelage and whitish ventrally**; six plantar tubercles; eight nipples; **first lower molars with four closed triangles and fifth triangle open and confluent with a short terminal loop**; second upper molars with four closed sections; third upper molars with three closed triangles; third lower molars with three transverse loops and no triangles; first lower molars usually with four closed triangles.

**Comments:** *Alexandromys oeconomus* was previously considered within the subgenus *Alexandromys* of *Microtus* (Musser and Carleton 1993, 2005; Conroy and Cook 2000), and was recognized as a distinct genus (Abramson and Lissovsky 2012). *A. oeconomus* is a Holarctic species with a broad range in North America from Alaska east through western Mackenzie and south through southern Northwest Territories and northwestern British Columbia, including the Aleutian Islands (Map 96). It can be found in sympatry with *Microtus abbreivatus*, *M. longicaudus*, *M. pennsylvanicus*, and *M. xanthognathus*. *A. oeconomus* can be differentiated from all the other species of *Microtus* by having the tail shorter, from 1.8 to 2.4 times the hindfoot length, and bicolored, dorsally darker than the back and whitish ventrally; first lower molars with four closed triangles and the fifth triangle open and confluent with a short terminal loop.

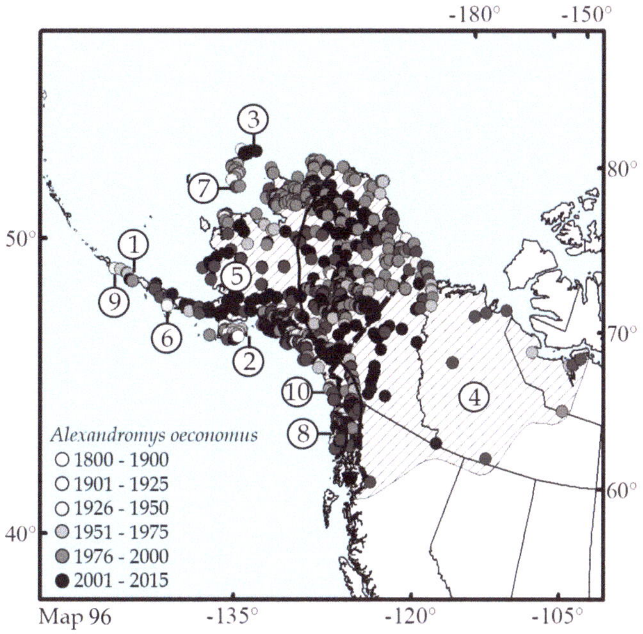


**Map 96** Distribution map of *Alexandromys oeconomus*

**Additional Literature:** Bailey (1900), Modi (1987), Musser and Carleton (2005), Bannikova et al. (2010), Lissovsky and Obolenskaya (2011), Lissovsky et al. (2012).

## Genus *Lemmiscus* Thomas, 1912

Sagebrush vole is medium sized. The genus has only one species in North America. Dorsal pelage pale buffy to ashy gray; fur of underparts silvery white to pale buffy; pelage long, loose, and dense; ears tinged with buff and more than one-half of the hindfoot length. *Lemmiscus* can be differentiated from all the other species of Arvicolinae by having the **third lower molars with four prisms, two terminal transverse loops, and two median triangles**.

### *Lemmiscus curtatus* (Cope, 1868)

Sagebrush vole, chincolo de las artemisas

1. *L. c. curtatus* (Cope, 1868). Southern Nevada and central-eastern California.
2. *L. c. intermedius* (Taylor, 1911). Central-northern Nevada, eastern California, and western Colorado.
3. *L. c. levidensis* Goldman, 1941. Idaho, Wyoming, southern Montana, northern Kansas, and Colorado.
4. *L. c. orbitus* (Dearden and Lee, 1955). Central Colorado.
5. *L. c. pallidus* (Merriam, 1888). Montana, North Dakota, Alberta, and Saskatchewan.
6. *L. c. pauperrimus* (Cooper, 1868). Oregon and Washington.

**Conservation:** *Lemmiscus curtatus* is listed as Least Concern by the IUCN (Cassola 2016al).

**Characteristics:** Sagebrush vole is medium sized; total length 108.0–142.0 mm and skull length 21.5–24.6 mm. Dorsal pelage pale buffy gray to ashy gray; fur of underparts silvery white to pale buffy; pelage long, loose, and dense; ears and nose tinged with buff; ears more than one-half of the hindfoot length; tail indistinctly bicolored and **not exceeding the hindfoot length by 10.0 mm**; six nipples; **third lower molars with four (rather than five) prisms, with two terminal transverse loops and two median triangles**.

**Comments:** *Lemmiscus curtatus* was originally described as *Lagurus curtatus*, but it was considered as a subgenus *Lemmiscus* of *Microtus* (Musser and Carleton 2005); however, it has been as a subgenus of *Lagurus* (Hall 1981; Honacki et al. 1982). *L. curtatus* ranges from southern Alberta, Saskatchewan, and western North Dakota south through northeastern California, Nevada, Utah, and northern Colorado, with one branch in Washington and Oregon (Map 97). All the species of *Lemmiscus* have allopatric distribution.

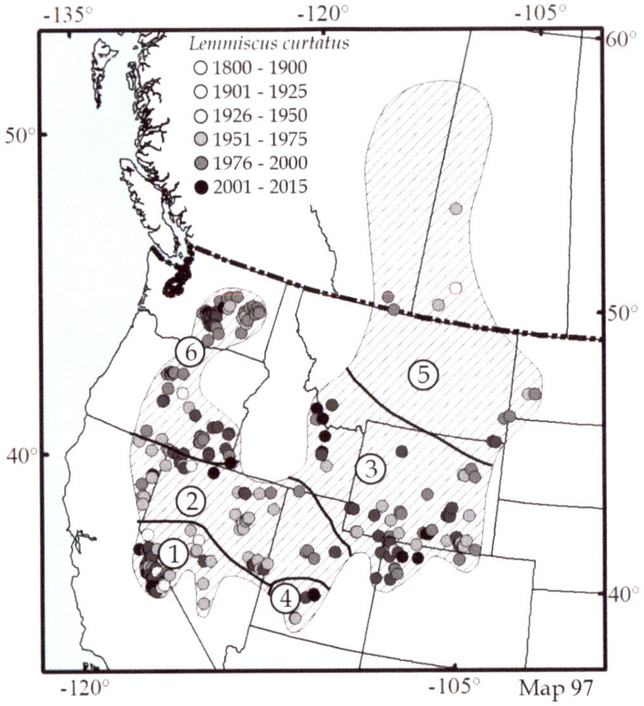


**Map 97** Distribution map of *Lemmiscus curtatus*

**Additional Literature:** Miller (1896), Davis (1939), Carroll and Genoways (1980), Carleton (1980), Modi (1987, 1996), Verts and Carraway (1998).

## Genus *Microtus* Schrank, 1798

Dorsal pelage dark with hair brown at the tip; ventral pelage leaden gray grizzly with red; ears long and mainly naked; feet large; **tail short, almost one-half of the head-and-body length**; five plantar tubercles, one very rudimentary, so they seem to be five; four, six, or eight nipples; **incoming and inner molar angles approximately equal; tooth enamel triangles surrounding the dentine; incisors without a groove**; third upper molars with two closed rounded triangles and one open triangle; second lower molars with one anterior pair of confluent triangles; third lower molars with one posterior loop, **two closed triangles and one anterior loop.** *Microtus* has three subgenera: *Mynomes* (*M. canicaudus*, *M. drummondii*, *M. dukecampbelli*, *M montanus*, *M. ochrogaster*, *M. oregoni*, *M. pennsylvanicus*, and *M. townsendi*), *Pitymys* (*M. guatemalensis*, *M. oaxacensis*, *M. pinetorum*, and *M. quasiater*), and *incertae sedis* (*M. abbreivatus*, *M. californicus*, *M. chrotorrhinus*, *M. longicaudus*, *M. mexicanus*, *M. mogollonensis*, *M. richardsoni*, *M. umbrosus*, and *M. xanthognathus*). The keys were elaborated based on the review of specimens and the following sources: Bailey (1900), Hall (1981), Verts and Carraway (1984), and Álvarez-Castañeda et al. (2017a).

1. Plantar tubercles five; third lower molars with two transverse lobes and one or two closed triangles at the middle.......2

1a. Plantar tubercles five or six; third lower molars with three transverse lobes and without closed triangles at the middle ......................................................................................4

2. Tail length greater than 30% of the total length; third upper molars with two triangles (Oaxaca, México) ........................................................................................................ *Microtus* (*incertae sedis*) *umbrosus* (p. 170)

2a. Tail length less than 30% of the total length; third upper molars with three to five triangles ........................................ 3

3. Lips white; third upper molars with three triangles and third lower molars with two triangles (Chiapas, México, and Guatemala) ........................................................................................ *Microtus* (*Pitymys*) *guatemalensis* (p. 157)

3a. Lips of the same color as the rostrum; third upper molars with five triangles and third lower molars with one triangle (Oaxaca, México) ........................................................................................ *Microtus* (*Pitymys*) *oaxacensis* (p. 158)

4. Third upper molars with two closed triangles ........................................................................................ 5

4a. Third upper molars with three or more closed triangles ........................................................................................ 7

5. Upperparts grayish; tail length greater than 26.0 mm; six nipples (from Alberta, Saskatchewan, and Manitoba south through northern New Mexico, Oklahoma, Arkansas, Tennessee, and West Virginia) ........................................................................................................ *Microtus* (*Mynomes*) *ochrogaster* (p. 152)

5a. Upperparts blackish or brownish; tail length less than 26.0 mm; four nipples ........................................................ 6

6. Hindfoot length generally greater than 15.0 mm; ears generally greater than 12.5 mm; dorsal pelage blackish (from southeastern San Luis Potosí south through central Veracruz, México) ............... *Microtus* (*Pitymys*) *quasiater* (p. 160)

6a. Hindfoot length generally less than 16.0 mm; ears generally less than 12.5 mm; dorsal pelage pale brown (from southern Minnesota, Iowa, Nebraska, Kansas, Oklahoma, and Texas east through the Atlantic Coast, and from central Florida north through Maine) ........................................................................................ *Microtus* (*Pitymys*) *pinetorum* (p. 159)

7. Plantar tubercles five; lateral glands on the flanks or inconspicuous ........................................................................ 8

7a. Plantar tubercles six; lateral glands on the hips (on flanks in *M. xanthognathus*) ........................................................ 9

8. Total length greater than 170.0 mm; tail length greater than 65.0 mm; dorsal pelage grayish to reddish; lateral glands conspicuous on the flanks; third upper molars with three loops of enamel on the lingual side (from British Columbia and Alberta south through Washington and Oregon, south through Montana, Idaho, Wyoming, and Utah) ........................................................................................................ *Microtus* (*incertae sedis*) *richardsoni* (p. 169)

8a. Total length less than 170.0 mm; tail length less than 65 mm; dorsal pelage brown; lateral glands dark or wanting; third upper molars with four loops of enamel on the lingual side (from southwest British Columbia south through northwest California) ........................................................................................ *Microtus* (*Mynomes*) *oregoni* (p. 153)

9. Skull narrow; zygomatic breadth less than 4.5 mm of the braincase breadth; skull with a pronounced median crest; claws enlarged (from northwestern and eastern Alaska south through Yukon, Hall Island and Saint Matthew Island in the Bering Sea, Alaska) ........................................................................ *Microtus* (*incertae sedis*) *abbreviatus* (p. 161)

9a. Skull not narrow; zygomatic breadth greater than 4.5 mm of the braincase breadth; skull without a pronounced median crest; claws not enlarged ........................................................................................................ 10

10. Second upper molars with three closed triangles and a rounded posterior loop ........................................................ 11

10a. Second upper molars with three closed triangles but lacking a posterior loop ........................................................ 13

11. Only occurs in the Field Marsh Island in Waccasassa Bay, Levy County, Florida ........................................................................................................ *Microtus* (*Mynomes*) *dukecampbelli* (p. 149)

11a. Occurs throughout Alaska, most of Canada, northern and eastern United States, to outlier populations into New Mexico and northern Chihuahua ........................................................................................................ 12

12. Occurs from western of the Ontario-Quebec border, Alaska, and British Columbia east through Ontario, and from Canada south through Washington, Idaho, Utah, Kansas, Illinois, Indiana, Ohio, United States, and Chihuahua, México ........................................................................................................ *Microtus* (*Mynomes*) *drummondii* (p. 148)

12a. Occurs from east of the Ontario-Québec border, Ohio, western Kentucky, and North Carolina east through the Atlantic Coast, and from North Carolina north through Newfoundland ................ *Microtus* (*Mynomes*) *pennsylvanicus* (p. 154)

13. Four nipples; zygomatic arches broad; incisive foramina constricted (Fig. 29) ........................................................ 14

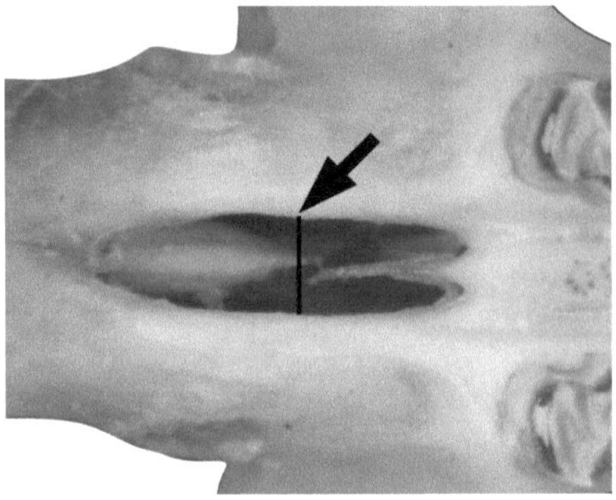

**Fig. 29** Incisive foramina constricted

13a. Eight nipples; zygomatic arches narrow; incisive foramina not constricted (Fig. 30) ........ 15

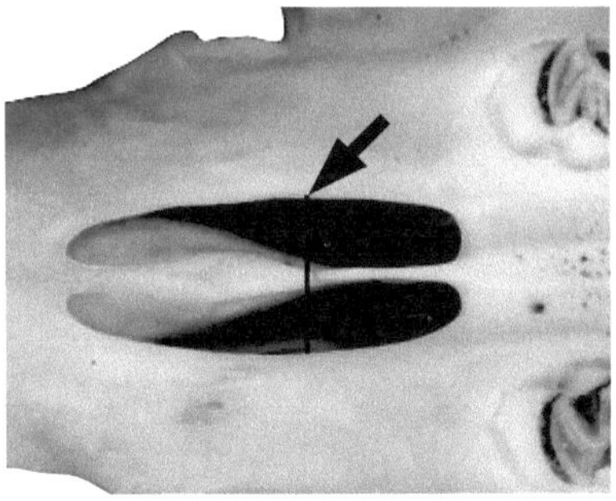

**Fig. 30** Incisive foramina wide

14. Occurs in México (from Chihuahua, Coahuila, and Nuevo León south through Puebla, Veracruz, and Oaxaca by the highlands and Altiplano Central) ........ *Microtus* (*incertae sedis*) *mexicanus* (p. 167)

14a. Occurs in United States (mountainous regions of southern Utah and Colorado, Arizona, New Mexico, and Texas) ........ *Microtus* (*incertae sedis*) *mogollonensis* (p. 168)

15. Nose yellowish; males with a pair of glands on the flanks ........ 16

15a. Nose not yellowish; males with a pair of glands on the hips ........ 17

16. Total length usually greater than 185.0 mm; hindfoot length greater than 23.5 mm; third upper molars with three closed triangles (from central Alaska east through the western coast of Hudson Bay) ........ *Microtus* (*incertae sedis*) *xanthognathus* (p. 171)

16a. Total length usually less than 185.0 mm; hindfoot length less than 23.5 mm; third upper molars with five closed triangles (from the Labrador Peninsula west through western Ontario and northeastern Minnesota and south through the Appalachian Mountains to eastern Tennessee) ........ *Microtus* (*incertae sedis*) *chrotorrhinus* (p. 164)

17. Incisive foramina not constricted posteriorly, wider posteriorly than anteriorly (from southwestern Oregon south through northern Baja California Peninsula) ........ *Microtus* (*incertae sedis*) *californicus* (p. 162)

17a. Incisive foramina constricted posteriorly, wider anteriorly than posteriorly ........ 18

18. Tail length greater than 50% of the head-and-body length (from Alaska and Yukon south through California, Arizona, and New Mexico, and from nearby the Pacific Coast east through Saskatchewan, South Dakota, Wyoming, Colorado, and New Mexico) ........ *Microtus* (*incertae sedis*) *longicaudus* (p. 165)

18a. Tail length less than 50% of the head-and-body length ........ 19

19. Tail unicolor blackish, usually greater than 33% of the head-and-body length; incisive foramina usually greater than 6.0 mm (extreme southwest British Columbia to northwest California, including the Vancouver Island and neighboring islands) ........................................ *Microtus* (*Mynomes*) *townsendii* (p. 156)

19a. Tail usually bicolored, darker dorsally than ventrally, usually less than 33% of the head-and-body length; incisive foramina usually less than 5.0 mm ........................................ 20

20. Tail grayish, usually with a sharp brownish stripe; confined to west of the Cascade Mountain (throughout the Willamette Valley, Oregon, north through Columbia River) ........................................ *Microtus* (*Mynomes*) *canicaudus* (p. 147)

20a. Tail bicolored, but usually not markedly so; east of the Cascade Mountains (from British Columbia south through the Arizona-New Mexico border, and from Washington, Oregon, and California east through Montana, Wyoming, Colorado, and New Mexico) ........................................ *Microtus* (*Mynomes*) *montanus* (p. 150)

## Subgenus *Mynomes* Rafinesque, 1817

Dorsal pelage brown and hoary underneath; chin and feet white; inner toes very short; tail hairy, without scales, depressed or flat; its length one-fifth of the head-and-body length; five or six plantar tubercles; third upper molars with three or more closed triangles; braincase without a pronounced median crest.

### *Microtus canicaudus* Miller, 1897

Gray-tailed vole, ratón colicorto de cola gris

Monotypic.

**Conservation:** *Microtus canicaudus* is listed as Least Concern by the IUCN (Cassola 2016as).

**Characteristics:** Gray-tailed vole is medium sized; total length 140.0–168.0 mm and skull length 24.0–26.2 mm. Dorsal pelage yellowish brown to grayish brown; fur of underparts pale gray at the tip with blackish shades at the base; **tail very short, about 150% of the hindfoot length, and bicolored, darker above;** six plantar tubercles; eight nipples; **palate posterior margin V shaped; incisive foramina not markedly constricted posteriorly;** third upper molars with four loops of enamel on the lingual side.

**Comments:** *Microtus canicaudus* is considered within the subgenus *Mynomes* of *Microtus* (Musser and Carleton 2005), and was previously recognized as a subspecies of *M. montanus* (Hall and Kelson 1951). Electrophoretic and cytological analyses support it at the species level (Modi 1986). This species is endemic throughout the Willamette Valley, Oregon, north through the Columbia River (Map 98). It can be found in sympatry, or nearly so, with *M. californicus*, *M. longicaudus*, *M. oregoni*, *M. richardsoni*, and *M. townsendii*. *M. canicaudus* can be differentiated from all the other species of *Microtus* by having the tail very short, about 150% of the hindfoot length, and bicolored; posterior palate margin V shaped and incisive foramina not markedly constricted posteriorly.

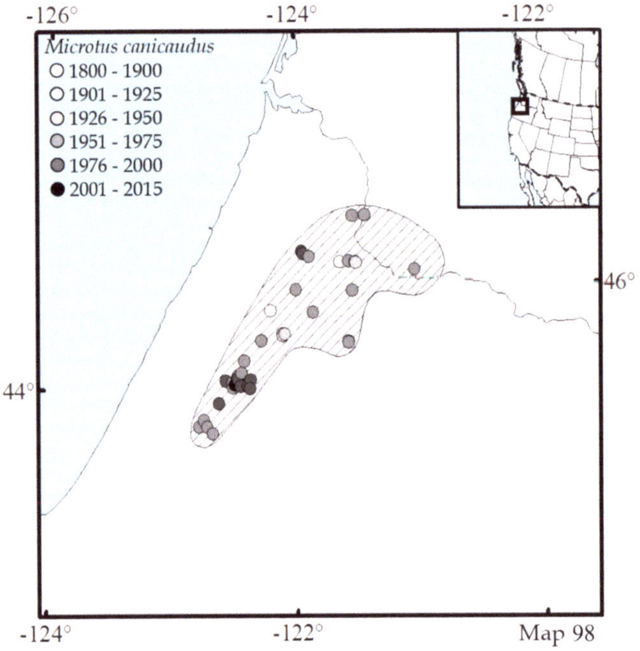


**Map 98** Distribution map of *Microtus canicaudus*

**Additional Literature:** Bailey (1900), Hsu and Johnson (1970), Modi (1987), Verts and Carraway (1987b), Zagorodnyuk (1990).

## ***Microtus drummondii*** (Audubon and Bachman, 1853)

Drummond meadow vole, ratón colicorto del oeste

1. *M. d. admiraltiae* Heller, 1909. Restricted to the Admiralty Island, Alaska.
2. *M. d. alcorni* Baker, 1951. From southwestern Alaska south through British Columbia.
3. *M. d. aphorodemus* Preble, 1902. Restricted to the Hudson Bay western coast.
4. *M. d. aztecus* (Allen, 1893). Restricted to northwestern New Mexico.
5. *M. d. chihuahuensis* Bradley and Cockrum, 1968. Restricted to the Galeana county, Chihuahua.
6. *M. d. drummondii* (Audubon and Bachman, 1853). From the border between Alaska and Yukon east through central Ontario and south through North Dakota and northern Wisconsin.
7. *M. d. finitus* Anderson, 1956. Restricted to southwestern Nebraska and northeastern Colorado.
8. *M. d. funebris* Dale, 1940. Southwestern British Columbia and northern Washington.
9. *M. d. insperatus* (Allen, 1894). From southern Alberta and Saskatchewan southeast through northwestern Wyoming and South Dakota.
10. *M. d. kincaidi* Dalquest, 1941. Restricted to the Moses Lake, California.
11. *M. d. microcephalus* (Rhoads, 1894). Southeastern British Columbia.
12. *M. d. modestus* (Baird, 1858). Southern Colorado and New Mexico.
13. *M. d. pullatus* Anderson, 1956. From Southern Montana and central Idaho south through Utah.
14. *M. d. rubidus* Dale, 1940. Western British Columbia.
15. *M. d. tananaensis* Baker, 1951. Central Alaska.
16. *M. d. uligocola* Anderson, 1956. Restricted to northeastern Colorado.

**Conservation:** *Microtus drummondii chihuahuensis* (as *M. pennsylvanicus chihuahuensis*) is listed as Endangered in the Norma Oficial Mexicana (DOF 2019) and is considered extinct (List et al. 2010). *M. drummondii* (as *M. pennsylvanicus*) is listed as Least Concern by the IUCN (Cassola 2016ay).

**Characteristics:** Drummond meadow vole is medium sized; total length 144.0–198.0 mm and skull length 27.2 ± 0.65 mm. **Dorsal pelage dark blackish brown**, long, and thick; flanks brown to grayish brown; fur of underparts pale gray at the tip with blackish shades at the base; general appearance grayish; tail short, more than twice the hindfoot length, and bicolored; upperparts of limbs dark or grayish brown; six plantar tubercles; eight nipples; **incisive foramen length more than 5.0** mm and constricted posteriorly; first lower molars with five closed triangles; second upper molars with four closed sections; third upper molars with three closed triangles; third lower molars with three transverse loops and no triangles.

**Comments:** *Microtus drummondii* (as *M. pennsylvanicus*) is within the subgenus *Mynomes* of *Microtus* (Musser and Carleton 2005). *M. d. aztecus* was as junior synonym of *M. d. modestus*. *M. drummondii* was recognized as a distinct species (Jackson and Cook 2020); it was previously regarded as a subspecies of *M. pennsylvanicus* (Hollister 1913). *M. drummondii* ranges from Alaska and British Columbia east through Ontario, and from northern Canada south through Washington, Idaho, Utah, Kansas, Illinois, Indiana, Ohio, and Chihuahua, México (Map 99; Jackson and Cook 2020). This species can be found in sympatry, or nearly so, with *M. abbreviatus*, *M. chrotorrhinus*, *M. longicaudus*, *M. mogollonensis*, *M. montanus*, *M. ochrogaster*, *M. oregoni*, *M. pinetorum*, *M. richardsoni*, *M. townsendii*, and *M. xanthognathus*. *M. drummondii* can be differentiated from all the other species of *Microtus* by having the dorsal pelage dark blackish brown; incisive foramen length more than 5.0 mm and constricted posteriorly. From *M. pennsylvanicus*, only using genetic data.

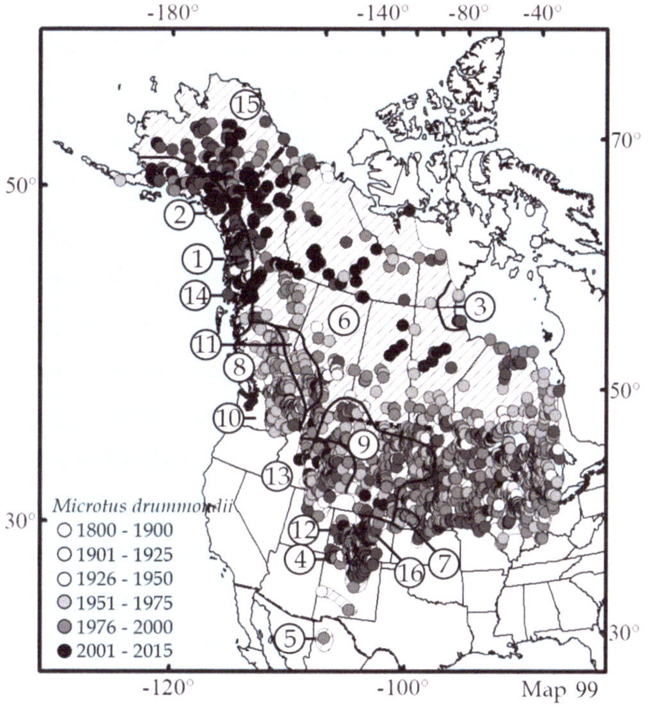


**Map 99** Distribution map of *Microtus drummondii*

**Additional Literature:** Bailey (1900), List et al. (2010).

## *Microtus dukecampbelli* Woods et al., 1982

Florida meadow vole, ratón colicorto de Florida

Monotypic.

**Conservation:** *Microtus dukecampbelli* (as *M. pennsylvanicus dukecampbelli*) is listed as Endangered by the United States Endangered Species Act (FWS 2022), and (as *M. pennsylvanicus*) is listed as Least Concern by the IUCN (Cassola 2016ay).

**Characteristics:** Florida meadow vole is medium sized; total length 178.0–198.0 mm and skull length 30.7–32.0 mm. **Dorsal pelage dark black brown on the back**, long, and thick; **fur of underparts dark gray on the belly**; tail short, more than twice the hindfoot length, and bicolored; six plantar tubercles; eight nipples; incisive foramen length more than 5.5 mm and constricted posteriorly; first lower molars with five closed triangles; second upper molars with four closed sections; third upper molars with three closed triangles; third lower molars with three transverse loops and no triangles.

**Comments:** *Microtus dukecampbelli* was recognized as a distinct species (Jackson and Cook 2020); it was previously regarded as a subspecies of *M. pennsylvanicus* (Woods et al. 1982). This species is only known from the type specimen from Island Field Marsh in Waccasassa Bay, Levy County, Florida (Map 100). It can be differentiated from *M. pennsylvanicus* by being significantly larger in somatic and skull measurements; shorter rostrum, and darker coloration.

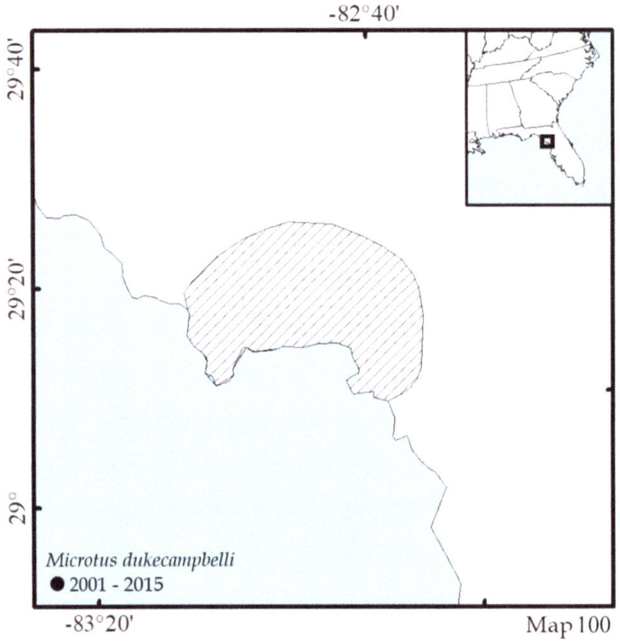


**Map 100** Distribution map of *Microtus dukecampbelli*

**Additional Literature:** Woods (1992).

## *Microtus montanus* (Peale, 1848)

Montane vole, ratón colicorto de las montañas

1. *M. m. amosus* Hall and Hayward, 1941. Central-western Colorado.
2. *M. m. arizonensis* Bailey, 1898. Restricted to Springerville, Apache County, Arizona.
3. *M. m. canescens* Bailey, 1898. Central Washington and British Columbia.
4. *M. m. codiensis* Anderson, 1954. Central-northern Wyoming.
5. *M. m. dutcheri* Bailey, 1898. Restricted to the Whitney Mountains, between Jackass and Monache Meadows.
6. *M. m. fucosus* Hall, 1935. Restricted to the Pahranagat Valley, Nevada.
7. *M. m. fusus* Hall, 1938. Southwestern Kansas and northern New Mexico.
8. *M. m. micropus* Hall, 1935. Central-northern Nevada, southwestern Oregon, southeastern Idaho, and western Colorado.
9. *M. m. montanus* (Peale, 1848). Southern Oregon, western California, and eastern Nevada.
10. *M. m. nanus* (Merriam, 1891). From Oregon east through Wyoming and from Washington south through Colorado and Kansas.
11. *M. m. nevadensis* Bailey, 1898. Restricted to the Ash Meadows, Nevada.
12. *M. m. pratincola* Hall and Kelson, 1951. Western Montana.

13. *M. m. rivularis* Bailey, 1898. Restricted to Saint George, southwestern Colorado.
14. *M. m. undosus* Hall, 1935. Central-western Nevada.
15. *M. m. zygomaticus* Anderson, 1954. Central-northern Kansas.

**Conservation:** *Microtus montanus* is listed as Least Concern by the IUCN (Cassola 2016av).

**Characteristics:** Montane vole is medium sized; total length 140.0–220.0 mm and skull length 25.4–30.2 mm. Dorsal pelage brownish, often with a buffy or grayish wash, **with hair black at the tip**; fur of underparts white to gray, some washed with buffy; tail bicolored and 2.2–2.4 times as long as the hind feet; six plantar tubercles; eight nipples; first lower molars with five closed triangles; third lower molars with three transverse loops and no triangles; **second upper molars with four closed sections;** third upper molars with three closed triangles; **incisive foramina constricted posteriorly**.

**Comments:** *Microtus montanus* is considered within the subgenus *Mynomes* of *Microtus* (Musser and Carleton 2005). *M. canicaudus* was previously recognized as a subspecies of *M. montanus* (Hall and Kelson 1951). Electrophoretic and cytological analyses supported its assignment at the species level (Modi 1986). *M. montanus* ranges from southern-central British Columbia south through the Arizona-New Mexico border, and from Washington, Oregon, and California east through Montana, Wyoming, Colorado, and New Mexico (Map 101). The southernmost population is isolated in the White and Blue mountains in Arizona and New Mexico. *M. montanus* can be found in sympatry, or nearly so, with *M. californicus*, *M. drummondii*, *M. longicaudus*, *M. mogollonensis*, *M. ochrogaster*, and *M. richardsoni*, and can be differentiated from all the other species of *Microtus* by having the tip of the dorsal pelage black; second upper molars with four closed sections, and incisive foramina costricted posteriorly.

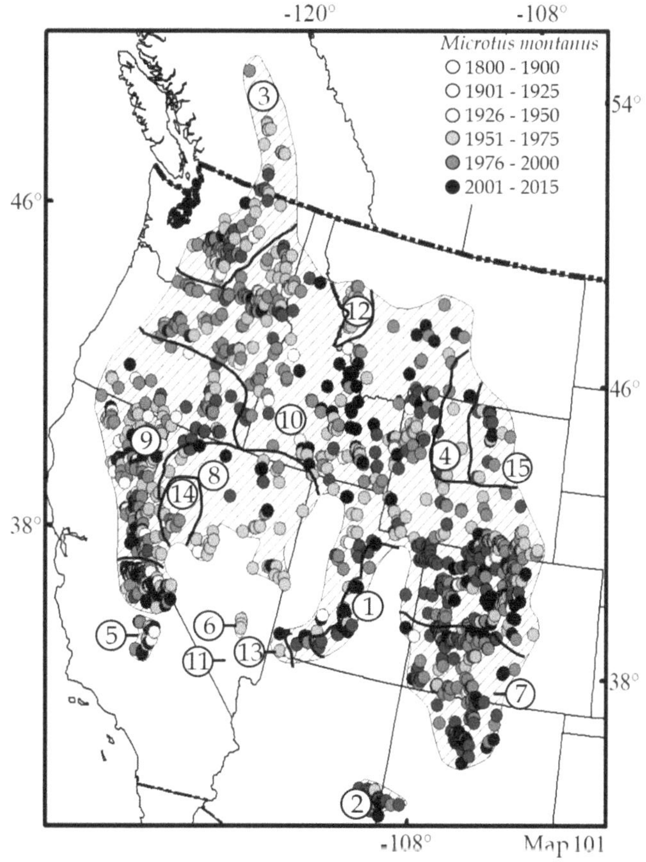


**Map 101** Distribution map of *Microtus montanus*

**Additional Literature:** Bailey (1900), Anderson (1954, 1959), Anderson (1960), Hooper and Hart (1962), Judd et al. (1980), Modi (1987), Moore and Janacek (1990), Zagorodnyuk (1990), Conroy and Cook (2000), Sera and Early (2003).

## ***Microtus ochrogaster*** (Wagner, 1842)

Prairie vole, ratón colicorto de las praderas

1. *M. o. haydenii* (Baird, 1858). From eastern Montana south through northern New Mexico and Oklahoma.
2. *M. o. ludovicianus* Bailey, 1900. Restricted to southeastern Texas and southwestern Louisiana.
3. *M. o. minor* (Merriam, 1888). From eastern Alberta southeast through Iowa and Wisconsin.
4. *M. o. ochrogaster* (Wagner, 1842). From southeastern South Dakota to northwestern Oklahoma east through Michigan to Alabama.
5. *M. o. ohionensis* Bole and Moulthrop, 1942. Restricted to southeastern Pennsylvania, eastern Ohio, western Delaware, and northern West Virginia.
6. *M. o. similis* Severinghaus, 1977. Montana, Wyoming, and western South Dakota.
7. *M. o. taylori* Hibbard and Rinker, 1943. Restricted to southwestern Colorado.

**Conservation:** *Microtus ochrogaster* is listed as Least Concern by the IUCN (Cassola 2016aw). However, *M. o. ludovicianus* is considered extinct.

**Characteristics:** Prairie vole is medium sized; total length 130.0–172.0 mm and skull length 27.6–30.6 mm. Dorsal pelage grayish brown and long; fur of underparts pale gray at the tip with blackish shades at the base; **tail short, about twice the hindfoot length, not bicolored but darker dorsally than ventrally**; ears small; five plantar tubercles; **six nipples**; first lower molars with three closed and two open triangles; second lower molars with an anterior pair of confluent triangles; third lower molars with three transverse loops, the middle loop sometimes constricted or even split into two triangles; third upper molars with two closed triangles; squamosal ridge small; rostrum and nasals short; auditory bullae large and wide.

**Comments:** *Microtus ochrogaster* was within the subgenus *Pedomys* of *Microtus* (Musser and Carleton 1993, 2005; Conroy and Cook 2000). *M. o. ludovicianus* was previously considered a distinct species of *M. ochrogaster* (Lowery 1974) and *M. o. minor* is morphologically different (Musser and Carleton 2005). *M. ochrogaster* ranges throughout the prairie states of the United States and Canada, from eastern-central Alberta, central Saskatchewan, and southern Manitoba south through northern New Mexico, Oklahoma, Arkansas, Tennessee, and West Virginia (Map 102). A disjunctive population (*M. o. ludovicianus*) was found in eastern Texas and western Louisiana, which is apparently extinct. *M. ochrogaster* can be found in sympatry, or nearly so, with *M. drummondii*, *M. longicaudus*, *M. montanus. M. pennsylvanicus*, *M. pinetorum*, and *M. richardsoni*; it can be differentiated from all the other species of *Microtus* by having the tail short, about twice the hindfoot length, not bicolored, but darker dorsally than ventrally.

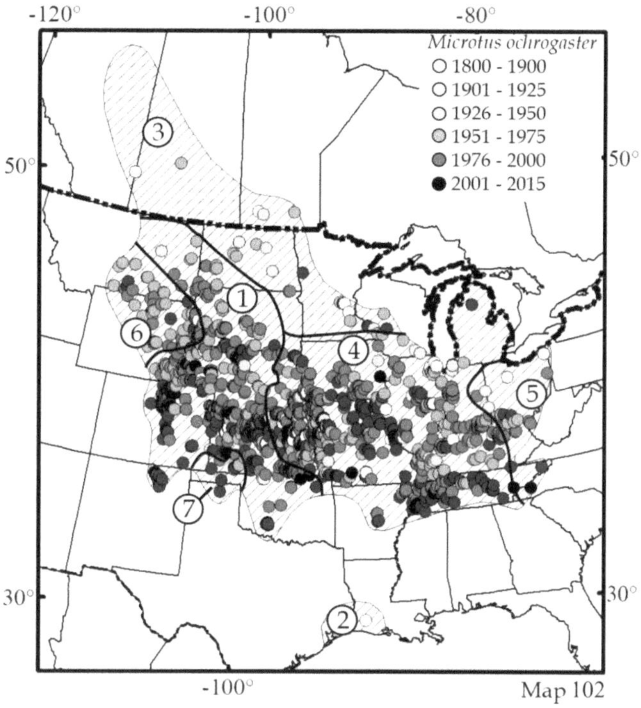


**Map 102** Distribution map of *Microtus ochrogaster*

**Additional Literature:** Bailey (1900), Anderson (1960), Hooper and Hart (1962), Severinghaus (1977), Choate and Williams (1978), Moore and Janacek (1990), Stalling (1990), Zagorodnyuk (1990).

## *Microtus oregoni* (Bachman, 1839)

Creeping vole, ratón colicorto de Oregon

1. *M. o. adocetus* Merriam, 1908. From northern-central California to southcentral Oregon.
2. *M. o. bairdi* Merriam, 1897. From northern-central California to southcentral Oregon.
3. *M. o. oregoni* (Bachman, 1839). From border between British Columbia and Washington southward to Punta Mendocino, California.
4. *M. o. serpens* Merriam, 1897. Restricted to British Columbia.

**Conservation:** *Microtus oregoni* is listed as Least Concern by the IUCN (Cassola 2016ax).

**Characteristics:** Creeping vole is medium sized; total length 129.0–154.0 mm and skull length 20.9–23.4 mm. Dorsal pelage gray, dark brown to black mixed with yellow bisters; fur of underparts pale gray at the tip with blackish shades at the base, general appearance grayish; **tail short, almost twice the hindfoot length and not bicolored, but darker than the back dorsally and paler ventrally;** five plantar tubercles; ears small and blackish; eight nipples; smaller molars, second upper molars with an anterior pair of triangles usually confluent; **third upper molars with three closed triangles**; first lower molars with five closed triangles; third lower molars with three transverse loops.

**Comments:** *Microtus oregoni* is considered within the subgenus *Mynomes* of *Microtus* (Musser and Carleton 2005); it ranges from southwestern British Columbia south through northwestern California (Map 103). *M. oregoni* can be found in sympatry, or nearly so, with *M. californicus*, *M. canicaudus*, *M. drummondii*, *M. longicaudus*, *M. richardsoni*, and *M. townsendii*. *M. oregoni* can be differentiated from all the other species of *Microtus* by having a short tail, almost twice the hindfoot length, and not bicolored, but darker than the back dorsally and lighter ventrally; third upper molars with three closed triangles; first lower molars with five closed triangles and third lower molars with three transverse loops.

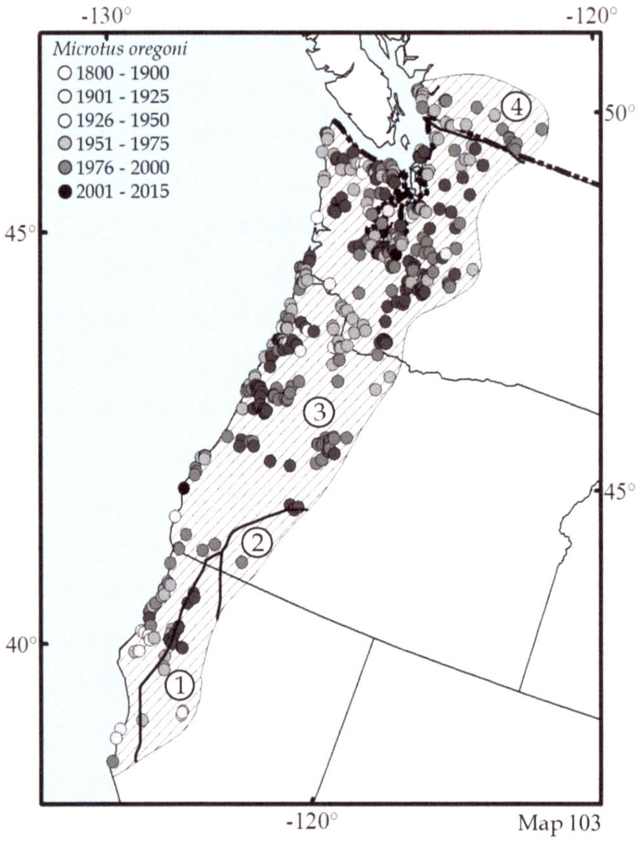


**Map 103** Distribution map of *Microtus oregoni*

**Additional Literature:** Bailey (1900), Anderson (1960), Hooper and Hart (1962), Carraway and Verts (1985), Moore and Janacek (1990), Zagorodnyuk (1990), Modi (1996), Conroy and Cook (2000).

## *Microtus pennsylvanicus* (Ord, 1815)

Meadow vole, ratón colicorto de los pastizales

1. *M. p. acadicus* Bangs, 1897. Restricted to Nova Scotia and the Prince Edward Island.
2. *M. p. breweri* (Baird, 1858). Restricted to the Muskeget Island, Massachusetts.
3. *M. p. copelandi* Youngman, 1967. Restricted to the Grand Manan Island, New Brunswick.
4. *M. p. enixus* Bangs, 1896. Newfoundland and central Québec.
5. *M. p. fontigenus* Bangs, 1896. Western Ontario and southern Québec.
6. *M. p. labradorius* Bailey, 1898. Northern Québec.
7. *M. p. magdalenensis* Youngman, 1967. Restricted to the Grindstone Island, Québec.
8. *M. p. nesophilus* Bailey, 1898. Restricted to the Gull Island, New York.
9. *M. p. nigrans* Rhoads, 1897. All the states around the southern Chesapeake Bay.
10. *M. p. pennsylvanicus* (Ord, 1815). From North Dakota to Northern Kansas east throughout the US Atlantic coast from South Carolina to the north.
11. *M. p. provectus* Bangs, 1908. Restricted to the Brock Island, Rhode Island.
12. *M. p. shattucki* Howe, 1901. Restricted to the Tumble Down Dick and North Haven islands.
13. *M. p. terraenovae* (Bangs, 1894). Restricted to Newfoundland.

**Conservation:** *Microtus pennsylvanicus nesophilus* is considered extinct (Hoffmann and Koeppl 1985; List et al. 2010). *M. p. breweri* is listed as Vulnerable (D2: very small population size with area of occupancy less than 20 km$^2$) by the IUCN (Roach 2020), and (as *M. pennsylvanicus*) is listed as Least Concern by the IUCN (Cassola 2016ay).

**Characteristics:** Meadow vole is medium sized; total length 140.0–195.0 mm and skull length 27.4 ± 0.63 mm. **Dorsal pelage dark blackish brown**, long, and thick; fur of underparts paler gray at the tip with blackish shades at the base; general appearance grayish; flanks brown to grayish brown; **tail short, more than twice the hindfoot length, and bicolored;** upperparts of the limbs dark or grayish brown; six plantar tubercles; eight nipples; incisive foramen length more than 5.0 mm and constricted posteriorly; first lower molars with five closed triangles; second upper molars with four closed sections; third upper molars with three closed triangles; third lower molars with three transverse loops and no triangles.

**Comments:** *Microtus pennsylvanicus* is considered within the subgenus *Mynomes* of *Microtus* (Musser and Carleton 2005). *M. breweri* is an insular species and has been considered the same species as the mainland *M. pennsylvanicus*, electrophoretic differences between the species are considered marginal, likely related to an insular effect (Jones et al. 1986; Modi 1986; Jackson and Cook 2020). However, cranial and dental characters support its assignment as a full species (Baker et al. 2003a; Musser and Carleton 2005). *M. nesophilus* and *M. provectus* were previously regarded as distinct species, but here I am following Musser and Carleton (2005), who have considered them at the subspecies level. *M. pennsylvanicus* ranges from east of the Ontario-Québec border, Ohio, western Kentucky, and North Carolina east through the Atlantic Coast, and from North Carolina north through Newfoundland (Map 104; Jackson and Cook 2020). This species can be found in sympatry, or nearly so, with *M. chrotorrhinus*, *M. ochrogaster*, and *M. pinetorum*. *M. pennsylvanicus* can be differentiated from all the other species of *Microtus* by having the dorsal pelage dark blackish brown; incisive foramen length more than 5.0 mm and constricted posteriorly. From *M. drummondii*, only using genetic data.

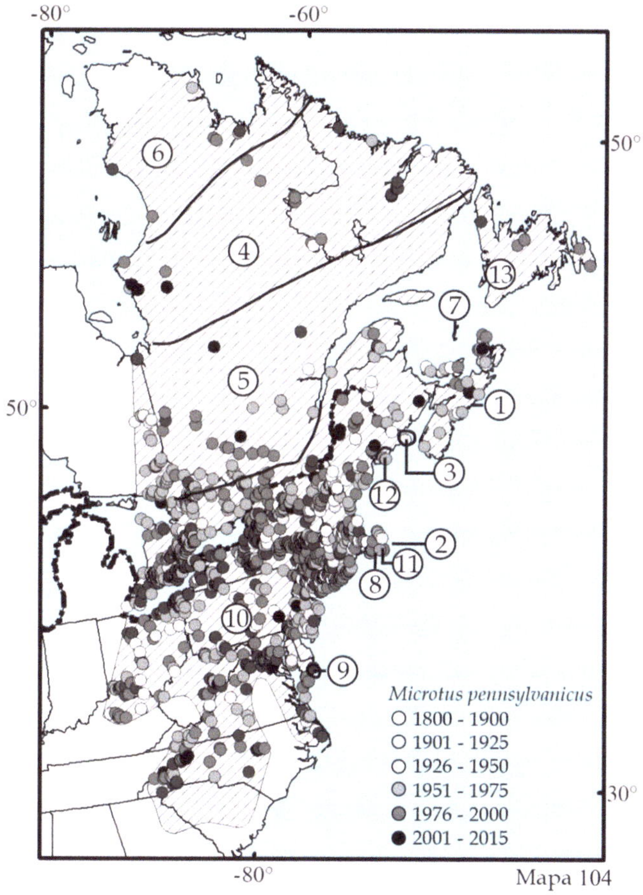


**Map 104** Distribution map of *Microtus pennsylvanicus*

**Additional Literature:** Bailey (1900), Anderson (1960), Tamarin and Kunz (1974), Reich (1981), Hoffmann and Koeppl (1985), Modi (1987), Moore and Janacek (1990), Zagorodnyuk (1990), Jones et al. (1992), Conroy and Cook (2000), Baker et al. (2003a), List et al. (2010).

## ***Microtus townsendii*** (Bachman, 1839)

Townsend's vole, ratón colicorto de la costa noroeste

1. *M. t. cowani* Guiguet, 1955. Restricted to the Triangle Island, British Columbia.
2. *M. t. cummingi* Hall, 1936. Restricted to the Bowen and Texada islands, British Columbia.
3. *M. t. laingi* Anderson and Rand, 1943. Restricted to central-northern Vancouver Island, British Columbia.
4. *M. t. pugeti* Dalquest, 1940. Restricted to the Shaw, San Juan, and Cypress Islands, Washington.
5. *M. t. tetramerus* (Rhoads, 1894). Restricted to the Vancouver, Bunshy, Vargas, Saltspring, and Pender Islands, British Columbia.
6. *M. t. townsendii* (Bachman, 1839). From southern British Columbia south through northwestern California.

**Conservation:** *Microtus townsendii* is listed as Least Concern by the IUCN (Cassola 2016bb).

**Characteristics:** Townsend's vole is medium sized; total length 169.0–235.0 mm and skull length 27.3–29.0 mm. Dorsal pelage dark brownish to blackish brown with a heavy mixture of black-tipped guard hairs; fur of underparts pale gray to grayish brown at the tip, blackish at the base, general appearance grayish; **tail slightly bicolored and large for the genus *Microtus*, average 65.0 mm and less than two and a half times the hindfoot length**; six plantar tubercles; eight nipples; first lower molars with five closed triangles; **second upper molars with four closed sections**; third upper molars with three closed triangles; third lower molars with three transverse loops and no triangles; **incisive foramina long, narrow, and constricted posteriorly**.

**Comments:** *Microtus townsendii* is considered within the subgenus *Mynomes* of *Microtus* (Musser and Carleton 2005). It ranges in coastal and wet areas from British Columbia and Vancouver Island south through Humboldt Bay, California (Map 105). This species can be found in sympatry, or nearly so, with *M. califonicus*, *M. canicaudus***,** *M. drummondii*, *M. longicaudus*, *M. oregoni*, and *M. richardsoni*. *M. townsendii* can be differentiated from all the other species of *Microtus* by having a large bicolored tail with an average of 65.0 mm, less than two-and-a-half times the hindfoot length; second upper molars with four closed sections; incisive foramina long, narrow and constricted posteriorly.

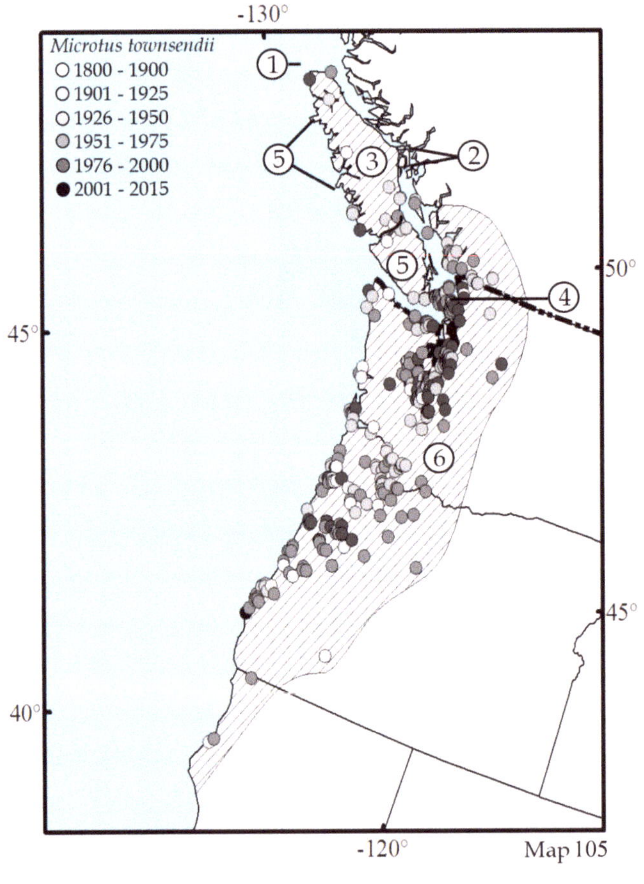


**Map 105** Distribution map of *Microtus townsendii*

**Additional Literature:** Bailey (1900), Anderson (1959), Hooper and Hart (1962), Modi (1987), Cornely and Verts (1988), Zagorodnyuk (1990), Conroy and Cook (2000).

## Subgenus *Pitymys* McMurtrie, 1831

Ears very small; tail short; hair short, dense, and glossy; five plantar tubercles; four nipples, two inguinal pairs; lateral glands on the hips in adult males; skull flat and wide; braincase quadrate; auditory bullae small; molars narrow; third lower molars with two closed triangles; first lower molars with three closed and two open triangles; second lower molars with an anterior pair of confluent triangles; third lower molars with three transverse loops.

### *Microtus guatemalensis* Merriam, 1898

Guatemalan vole, ratón colicorto de Guatemala

Monotypic.

**Conservation:** *Microtus guatemalensis* is listed as Threatened in the Norma Oficial Mexicana (DOF 2019) and as Near Threatened by the IUCN (Matson 2020).

**Characteristics:** Guatemalan vole is medium sized; total length 130.0–155.0 mm and skull length 24.7–28.0 mm. **Dorsal pelage dark brown, some specimens can be near blackish**; fur of underparts leaden gray grizzly with low tawny hairs; pelage long and silky; nose region darker; **tail very short about one and a half times the hindfoot length**; **lips white**; ears large; five plantar tubercles; six nipples; first lower molars with three closed triangles and an interior confluent pair of triangles opening into a terminal loop; **third lower molars with four closed sections including a pair of subequal median triangles; third upper molars with three closed triangles**.

**Comments:** *Microtus guatemalensis* is considered within the subgenus *Pitymys* of *Microtus* (Musser and Carleton 1993, 2005; Conroy and Cook 2000). This species ranges in highland meadows of central Chiapas and central Guatemala (Map 106), and is allopatric with respect to other species of *Microtus. M. guatemalensis* can be differentiated from all the other species of *Microtus* by having the dorsal pelage dark brown, and some specimens can be near blackish; tail very short, about one and a half times the hindfoot length; third lower molars with four closed sections, including a pair of subequal median triangles; third upper molars with three closed triangles.

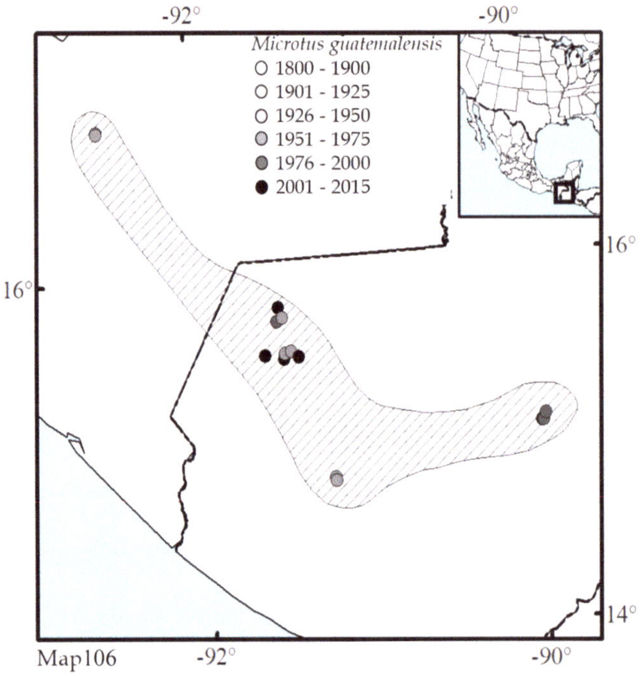


**Map 106** Distribution map of *Microtus guatemalensis*

**Additional Literature:** Bailey (1900), Honacki et al. (1982), Conroy et al. (2001).

## *Microtus oaxacensis* Goodwin, 1966

Tarabundí vole, ratón colicorto de Oaxaca

Monotypic.

**Conservation:** *Microtus oaxacensis* is listed as Threatened in the Norma Oficial Mexicana (DOF 2019) and as Critically Endangered (B1ab (iii); extent of occurrence less than 100 km$^2$ in only one population and under continuing decline in area, extent and quality of habitat) by the IUCN (De Grammont and Cuarón 2018b).

**Characteristics:** Oaxacan vole is medium sized; total length 159.0–163.0 mm and skull length 28.1–30.4 mm. Dorsal pelage blackish brown; fur of underparts leaden gray grizzly, slightly paler gray at the tip; **tail very short,** about less than twice the hindfoot length, and bicolored; pelage long and silky; **foot dorsal region blackish;** five plantar tubercles; six nipples; first lower molars with five closed triangles; second upper molars with four closed sections and third one with three; **third upper molars with five triangles; third lower molars with one medial triangle; skull elongated and more angular than in the other *Microtus* species**.

**Comments:** *Microtus oaxacensis* is considered within the subgenus *Pitymys* of *Microtus* (Musser and Carleton 1993, 2005; Conroy and Cook 2000). This species is restricted to northern-central Oaxaca (Map 107), and is allopatric with respect to other species of *Microtus*. *M. oaxacensis* can be differentiated from all the other species of *Microtus* by having the tail very short, about less than twice the hindfoot length, and bicolored; foot dorsal region blackish; skull elongated and more angular than in the other *Microtus* species; third upper molars with five triangles and third lower molars with one medial triangle.

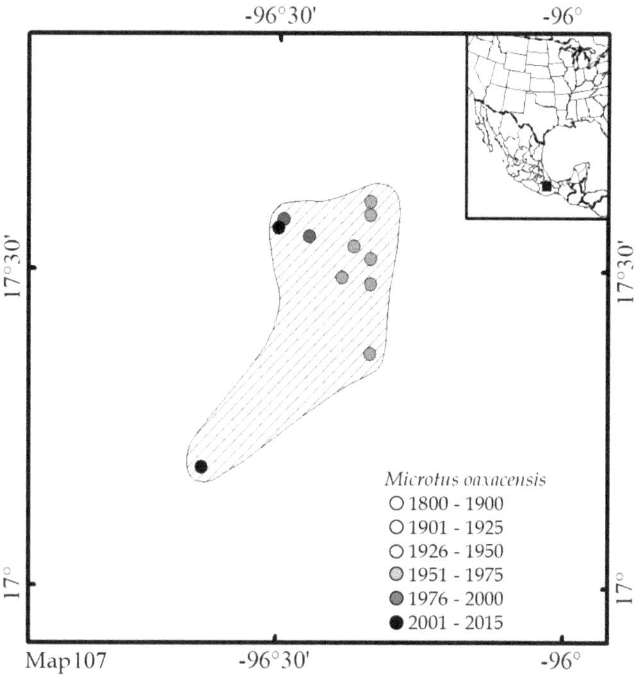


**Map 107** Distribution map of *Microtus oaxacensis*

**Additional Literature:** Goodwin (1966, 1969), Jones and Genoways (1967), Frey and Cervantes (1997b), Conroy et al. (2001).

## *Microtus pinetorum* (Le Conte, 1830)

Woodland vole, ratón colicorto de los bosques

1. *M. p. auricularis* Bailey, 1898. From southern Ohio and Indiana southwest through Mississippi and Central Texas.
2. *M. p. carbonarius* (Handley, 1952). Restricted to western Tennessee and Kentucky, southern Ohio, and southwestern West Virginia.
3. *M. p. nemoralis* Bailey, 1898. From Texas north through Minnesota, west through the Mississippi River, and south through Arkansas.
4. *M. p. parvulus* (Howell, 1916). Restricted to northern Florida.
5. *M. p. pinetorum* (Le Conte, 1830). From southern Virginia southwest through Alabama and Georgia.
6. *M. p. scalopsoides* (Audubon and Bachman, 1841). From Illinois and Virginia north through the Great Lakes and San Lorenzo, and from the Mississippi River east through the Atlantic coast.
7. *M. p. schmidti* (Jackson, 1941). Restricted to central-northern Wisconsin.

**Conservation:** *Microtus pinetorum* is listed as Least Concern by the IUCN (Cassola 2016az).

**Characteristics:** Woodland vole is small sized; total length 102.0–118.0 mm and skull length 25.1–26.6 mm. Dorsal pelage pale reddish brown, sometimes with a cinnamon coloration; fur of underparts pale gray at the tip, with blackish shades at the base, general appearance grayish; **tail shortest of all the species in North America, average 20.0 mm, only slightly larger than the hindfoot length, not bicolored, but dorsally of the same coloration as the back, paler ventrally**; five plantar tubercles; four nipples; first lower molars with three closed and two open triangles; second lower molars with an anterior pair of confluent triangles; third upper molars with two closed triangles; third lower molars with three transverse loops; squamosal ridge small; rostrum and nasals short; auditory bullae small.

**Comments:** *Microtus pinetorum* is considered within the subgenus *Pitymys* of *Microtus* (Musser and Carleton 1993, 2005; Conroy and Cook 2000). It ranges from southern Minnesota, Iowa, Nebraska, Kansas, Oklahoma, and Texas east through the Atlantic Coast, and from central Florida north through Maine (Map 108). This species can be found in sympatry, or nearly so, with *M. chrotorrhinus*, *M. drummondii*, *M. ochrogaster*, and *M. pennsylvanicus*. *M. pinetorum* can be differentiated from all the other species of *Microtus* by having the shortest tail of all species in North America, average 20.0 mm, only slightly larger than the hindfoot length.

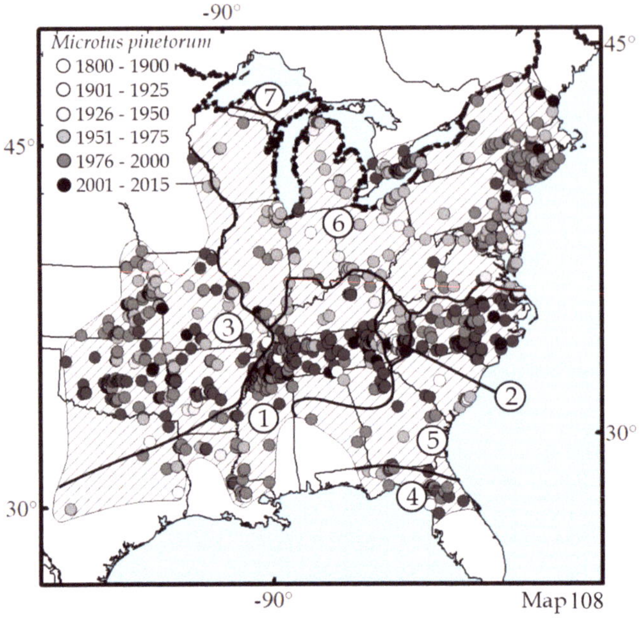


**Map 108** Distribution map of *Microtus pinetorum*

**Additional Literature:** Bailey (1900), Anderson (1960), Hooper and Hart (1962), Goodwin (1966), Smolen (1981), Moore and Janacek (1990), Zagorodnyuk (1990), Modi (1996).

## *Microtus quasiater* (Coues, 1874)

Jalapan pine vole, ratón colicorto de Jalapa

Monotypic.

**Conservation:** *Microtus quasiater* is listed as subject to Special Protection in the Norma Oficial Mexicana (DOF 2019) and as Near Threatened by the IUCN (Álvarez-Castañeda et al. 2019a).

**Characteristics:** Jalapan pine vole is medium sized; total length 114.0–150.0 mm and skull length 23.7–28.0 mm. Dorsal pelage dark brown, sometimes close to blackish brown; fur of underparts small, paler gray at the tip and blackish at the base, general appearance dark grayish; **tail very short, less than one and a half times the hindfoot length, not bicolored but dorsally similar to the back, and paler ventrally;** hair long and silky; ears small; five plantar tubercles; four nipples; first lower molars with three closed and two open triangles; second lower molars with an anterior pair of confluent triangles; third upper molars with two closed triangles; third lower molars with three transverse loops; squamosal ridge small; rostrum and nasals short; auditory bullae small.

**Comments:** *Microtus quasiater* is considered within the subgenus *Pitymys* of *Microtus* (Musser and Carleton 1993, 2005; Conroy and Cook 2000). It ranges on the eastern slopes of Sierra Madre Oriental, from southeastern San Luis Potosí south through central Veracruz, México (Map 109), and is allopatric with respect to other species of *Microtus*. *M. quasiater* can be differentiated from all the other species of *Microtus* by having a tail very small, less than one and a half times the hindfoot length, not bicolored, but dorsal coloration similar to the back and paler ventrally; third upper molars with two closed triangles.

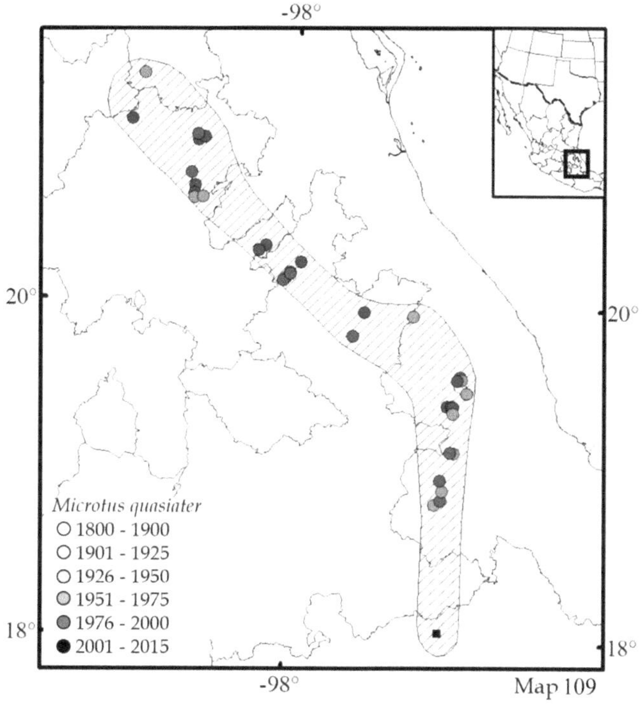


**Map 109** Distribution map of *Microtus quasiater*

**Additional Literature:** Bailey (1900), Hall and Cockrum (1953), Anderson (1960), Moore and Janacek (1990), Cervantes et al. (1994), Conroy et al. (2001), Torres-Flores and López-Wilchis (2018).

## Subgenus ***incertae sedis***

### ***Microtus abbreviatus*** Miller, 1899

Insular vole, ratón colicorto de la Isla San Mateo

1. *M. a. abbreviatus* Miller, 1899. Restricted to the Hall Island, Alaska.
2. *M. a. andersoni* Rand, 1945. Restricted to central-eastern Yukon.
3. *M. a. cantator* Anderson, 1947. Restricted to southeastern Alaska and northwestern British Columbia.
4. *M. a. fisheri* Merriam, 1900. Restricted to the Saint Matthew Island, Alaska.
5. *M. a. miurus* Osgood, 1901. Restricted to southern-central Alaska.
6. *M. a. muriei* Nelson, 1931. Northern Alaska, Yukon, and eastern Mackenzie.
7. *M. a. oreas* Osgood, 1907. Southwestern and central Alaska.

**Conservation:** *Microtus abbreviatus* is listed as Least Concern by the IUCN (Cassola 2016ar).

**Characteristics:** Insular vole is medium sized; total length 171.0–179.0 mm and skull length 25.3–31.7 mm. Dorsal **pelage yellowish brown to dark buff in grizzle**; fur of underparts creamy white to buff at the tip, blackish at the base; **tail very short, about 120% of the hindfoot length, and slightly bicolored, dorsally darker than the back**; six plantar tubercles; eight nipples; first upper molars with five or six closed triangles and with five inner and four outer salient angles; third upper molars with two closed triangles, third triangle open and confluent with the posterior loop.

**Comments:** *Microtus abbreviatus* was considered within the subgenus *Stenocranius* of *Microtus* (Musser and Carleton 1993; Conroy and Cook 2000), but the current treatment is without a subgenus assignation (Musser and Carleton 2005). *M. abbreviatus* is an insular species and was recognized as a distinct species (Jones et al. 1997; Baker et al. 2003a; Musser and Carleton 2005; Weksler et al. 2010); it was previously regarded as a subspecies of *M. miurus* inhabiting mainland Alaska (Conroy and Cook 2000). *M. abbreviatus* ranges from northwestern and eastern Alaska south through Yukon, Hall Island, and Saint Matthew Island, in the Bering Sea, Alaska (Map 110). This species can be found in sympatry, or nearly so, with *M. longicaudus*, *M. pennsylvanicus*, and *M. xanthognathus*. *M. abbreviatus* can be differentiated from all the sympatric species of *Microtus* by having six plantar tubercles; first upper molars with five or six closed triangles and with five inner and four outer salient angles; third upper molars with two closed triangles, with the third triangle open and confluent with the posterior loop.

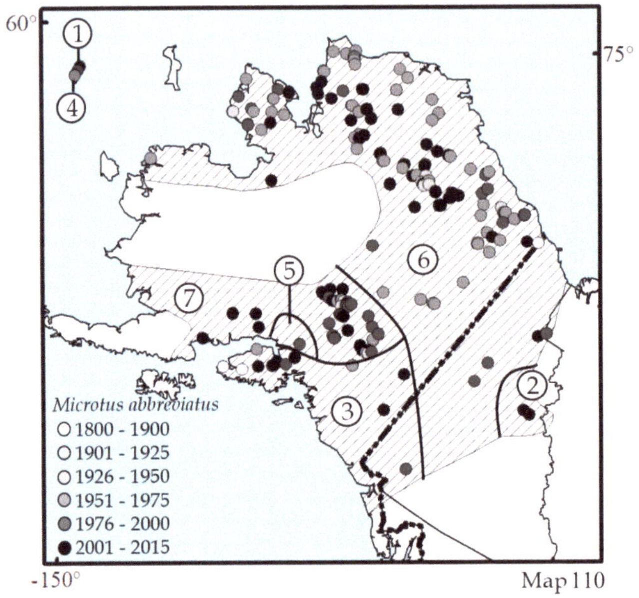


**Map 110** Distribution map of *Microtus abbreviatus*

**Additional Literature:** Bailey (1900), Anderson (1960), Rausch and Rausch (1968), Zagorodnyuk (1990), Musser and Carleton (1993, 2005) Jones et al. (1992), Conroy and Cook (1999a).

## ***Microtus californicus*** (Peale, 1848)

California vole, ratón colicorto de California

1. *M. c. aequivocatus* Osgood, 1928. Restricted to San Telmo Valley, Baja California.
2. *M. c. aestuarinus* Kellogg, 1918. Central Valley, California.
3. *M. c. californicus* (Peale, 1848). Coastal Range from San Luis Obispo to San Francisco Bay, California.
4. *M. c. constrictus* Bailey, 1900. Coastal Range from Manchester to Cape Mendocino, California.
5. *M. c. eximius* Kellogg, 1918. From northern San Francisco Bay through the Klamath Mountain to Southern Oregon.

6. *M. c. grinnelli* Huey, 1931. Restricted to the Sierra de Juárez, Baja California.
7. *M. c. halophilus* von Bloeker, 1937. Restricted to Monterey, California.
8. *M. c. huperuthrus* Elliot, 1903. Restricted to the Sierra San Pedro Mártir, Baja California.
9. *M. c. kernensis* Kellogg, 1918. Tehachapi Mountains and southern of Central Valley, California.
10. *M. c. mariposae* Kellogg, 1918. Eastern slope of the Central Valley, California.
11. *M. c. mohavensis* Kellogg, 1918. Restricted to Victorville, California.
12. *M. c. paludicola* Hatfield, 1935. Restricted to southern San Francisco Bay, California.
13. *M. c. sanctidiegi* Kellogg, 1918. From the México-USA border north through southern Tehachapi Mountains, California.
14. *M. c. sanpabloensis* Thaeler, 1961. Restricted to San Pablo Creek, California.
15. *M. c. scirpensis* Bailey, 1900. Restricted to a spring near Shoshone, California.
16. *M. c. stephensi* von Bloeker, 1932. Restricted to Los Angeles Valley, California.
17. *M. c. vallicola* Bailey, 1898. Restricted to central Sierra Nevada, California.

**Conservation:** *Microtus californicus scirpensis* from California is listed as Endangered by the United States Endangered Species Act (FWS 2022), *M. c. aequivocatus* as Endangered in the Norma Oficial Mexicana (DOF 2019), and (as *M. californicus*) as Least Concern by the IUCN (Álvarez-Castañeda et al. 2016n).

**Characteristics:** California vole is medium sized; total length 149.0–196.0 mm and skull length 26.1–30.7 mm. Dorsal pelage dark blackish brown grizzly with a small area cream, flanks yellowish brown; fur of underparts paler gray at the tip with blackish shades at the base; hair long and wooly; ears hairy and well developed; **feet whitish above; tail bicolored and more than twice** the **hindfoot length**; six plantar tubercles; eight nipples; first lower molars with five closed triangles; second upper molars with four closed sections; third upper molars with three closed triangles; third lower molars with three transverse loops and no triangles; **incisive foramina not constricted posteriorly, wider posteriorly than anteriorly**.

**Comments:** *Microtus californicus* was considered probably within the subgenus *Mynomes* of *Microtus* (Musser and Carleton 1993; Conroy and Cook 2000), but the current treatment is without a subgenus assignation (Musser and Carleton 2005). This species ranges from southwestern Oregon south through the northern Baja California Peninsula, including a large area in California. Two disjunction populations are found in the Mohave Desert and the White Mountain/Panamint ranges in California (Map 111). *M. californicus* can be found in sympatry, or nearly so, with: *M. canicaudus*, *M. longicaudus*, *M. montanus*, *M. oregoni*, *M. richardsoni*, and *M. townsendii.* It can be differentiated from all the other species of *Microtus* by the bicolored tail more than twice the hindfoot length; hindfoot upperparts whitish, and incisive foramina not wider posteriorly than anteriorly.

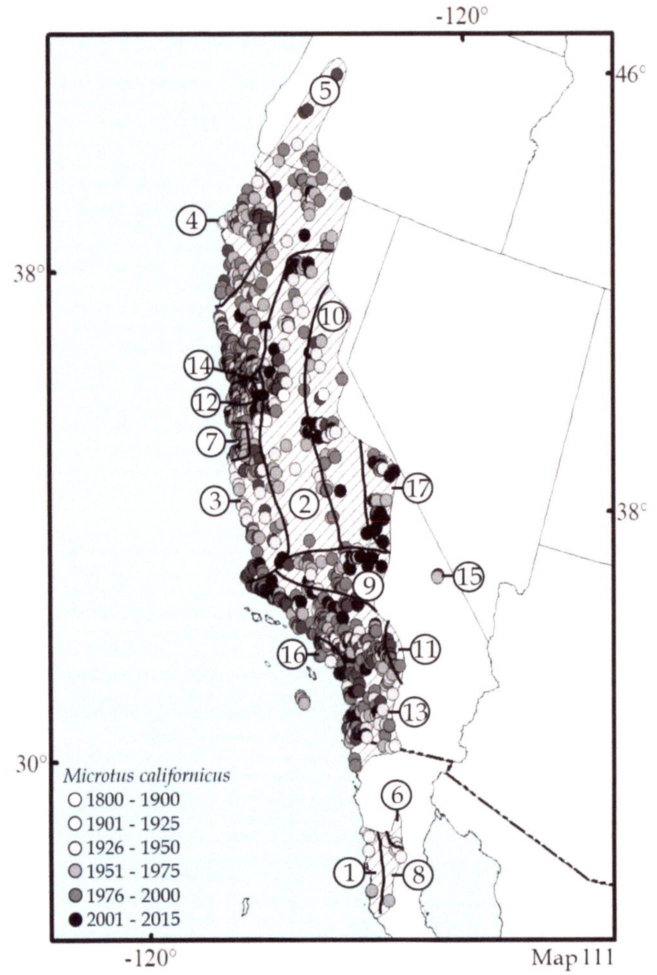


**Map 111** Distribution map of *Microtus californicus*

**Additional Literature:** Bailey (1900), Anderson (1960), Zagorodnyuk (1990), Conroy and Cook (2000), Cudworth and Koprowski (2010).

## *Microtus chrotorrhinus* (Miller, 1894)

Rock vole, ratón colicorto de las rocas

1. *M. c. carolinensis* Komarek, 1932. Smoky Mountains from eastern Tennessee to southern Pennsylvania.
2. *M. c. chrotorrhinus* (Miller, 1894). From Pennsylvania to the Labrador Peninsula.
3. *M. c. ravus* Bangs, 1898. Restricted to the eastern side of the Labrador Peninsula.

**Conservation:** *Microtus chrotorrhinus* is listed as Least Concern by the IUCN (Cassola 2016at).

**Characteristics:** Rock vole is medium sized; total length 140.0–185.0 mm and skull length 25.0–27.7 mm. Dorsal pelage brown with yellowish shades; fur of underparts pale gray at the tip with blackish shades at the base; **tail short but large for the genera, about 225% of the hindfoot length, dark dorsally and paler ventrally but not bicolored**; six plantar tubercles; eight nipples; **incisive foramen length less than 5.0 mm and not markedly tapered posteriorly;** first lower molars with five closed triangles; third lower molars with three transverse loops and no triangles; second upper molars with four closed sections; **third upper molars with five triangles surrounding the dentine**.

**Comments:** *Microtus chrotorrhinus* was considered within the subgenus *Aulacomys* of *Microtus* (Musser and Carleton 1993; Conroy and Cook 2000), but the current treatment is without a subgenus assignation (Musser and Carleton 2005). *M. chrotorrhinus* is known from the Labrador Peninsula west through western Ontario and northeastern Minnesota and south through along the Appalachian Mountains to eastern Tennessee (Map 112). *M. chrotorrhinus* can be found in sympatry, or nearly so, with *M. drummondii*, *M. pennsylvanicus*, and *M. pinetorum*. It can be differentiated from all the other species of *Microtus* by its small tail but large for the genera, about 225% of the hindfoot length, darker in the upperparts and paler ventrally, but not bicolored; incisive foramen length less than 5.0 mm and not markedly tapered posteriorly.

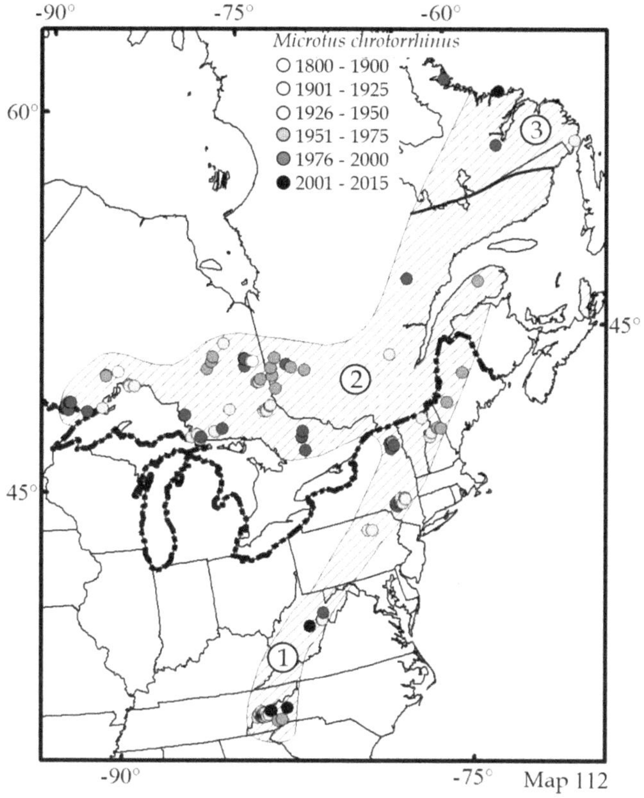


**Map 112** Distribution map of *Microtus chrotorrhinus*

**Additional Literature:** Bailey (1900), Rausch and Rausch (1974), Kirkland and Jannett (1982), Kilpatrick and Crowell (1985), Zagorodnyuk (1990).

## *Microtus longicaudus* (Merriam, 1888)

Long-tailed vole, ratón colicorto de cola larga

1. *M. l. abditus* Howell, 1923. Restricted to coastal areas of Oregon from the central Columbia River.
2. *M. l. alticola* (Merriam, 1890). Northwestern Arizona and eastern Colorado.
3. *M. l. angusticeps* Bailey, 1898. Restricted to coastal areas from Mendocino northward to central Oregon.
4. *M. l. baileyi* Goldman, 1938. Restricted to the Grand Canyon, northern-central Arizona.
5. *M. l. bernardinus* Merriam, 1908. Restricted to the San Bernardino Mountains, California.
6. *M. l. halli* Hayman and Holt, 1941. Northeastern Oregon and southeastern Washington.
7. *M. l. incanus* Lee and Durrant, 1960. Restricted to Mont Henry and Pennell, Utah.
8. *M. l. latus* Hall, 1931. Central Nevada and western Colorado.
9. *M. l. leucophaeus* (Allen, 1894). Restricted to the Graham Mountains, Arizona.
10. *M. l. littoralis* Swarth, 1933. Restricted to islands of northwestern British Columbia.

11. *M. l. longicaudus* (Merriam, 1888). From southern British Columbia south through northern California and east through Montana to New Mexico.
12. *M. l. macrurus* Merriam, 1898. Restricted to the coastal area from the Columbia River to central British Columbia.
13. *M. l. sierrae* Kellogg, 1922. Mountain ranges of western California.
14. *M. l. vellerosus* Allen, 1899. From southern British Columbia and Alberta north through Alaska and Yukon.

**Conservation:** *Microtus longicaudus* is listed as Least Concern by the IUCN (Cassola 2016au).

**Characteristics:** Long-tailed vole is medium sized; total length 162.0–198.0 mm and skull length 26.3–27.7 mm. Dorsal pelage grayish, brownish gray to sepia brown; fur of underparts creamy white to buff at the tip with blackish shades at the base; **tail longest for the genus, about more than three times the hindfoot length**, mono or bicolored; six plantar tubercles; eight nipples; first lower molars with five closed triangles; second upper molars with four closed sections; third upper molars with three closed triangles; third lower molars with three transverse loops and no triangles.

**Comments:** *Microtus longicaudus* was considered within the subgenus *Aulacomys* of *Microtus* (Musser and Carleton 1993; Conroy and Cook 2000), but the current treatment is without a subgenus assignation (Musser and Carleton 2005). *M. coronaries* was previously considered a distinct species relative to *M. longicaudus*; however, it is now recognized as a subspecies of *M. longicaudus* (Jones et al. 1992; Baker et al. 2003a). *M. longicaudus* ranges in western North America, from eastern-central Alaska and Yukon south through southern California, Arizona, and New Mexico, and from the Pacific coast or nearby east through Saskatchewan, South Dakota, Wyoming, Colorado, and New Mexico (Map 113). It can be found in sympatry, or nearly so, with *M. abbreivatus*, *M. californicus*, *M. canicaudus*, *M. drummondii*, *M. mogollonensis*, *M. montanus*, *M. ochrogaster*, *M. oregoni*, *M. richardsoni*, and *M. townsendii*. *M. longicaudus* can be differentiated from all the other species of *Microtus* by having the longest tail for *Microtus*, about more than three times the hindfoot length.

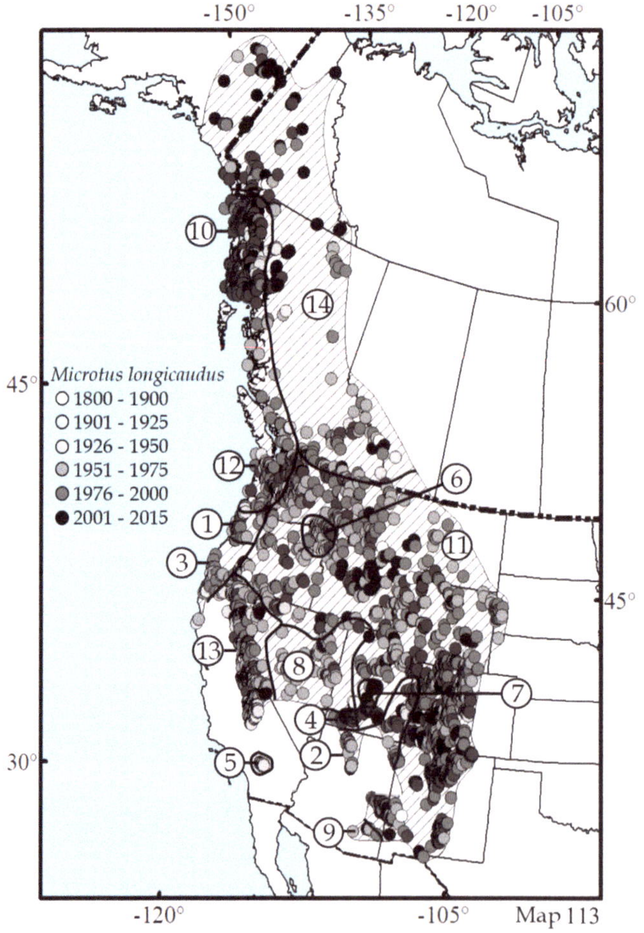


**Map 113** Distribution map of *Microtus longicaudus*

**Additional Literature:** Bailey (1900), Hall (1981), Modi (1987), Smolen and Keller (1987), Moore and Janacek (1990), Zagorodnyuk (1990).

## *Microtus mexicanus* (Saussure, 1861)

Mexican vole, ratón colicorto mexicano

1. *M. m. fulviventer* Merriam, 1898. Restricted to Oaxaca.
2. *M. m. fundatus* Hall, 1948. Restricted to northern Michoacán.
3. *M. m. madrensis* Goldman, 1938. Sierra Madre Occidental from Durango and Chihuahua.
4. *M. m. mexicanus* (Saussure, 1861). Ciudad de México, Estado de México, Hidalgo, Morelos, Puebla, Tlaxcala, and Veracruz.
5. *M. m. neveriae* Hooper, 1955. Restricted to the southeastern Jalisco highlands.
6. *M. m. ocotensis* Álvarez and Hernández-Chávez, 1993. Restricted to Pinal de Amoles, Querétaro.
7. *M. m. phaeus* (Merriam, 1892). From Zacatecas east through San Luis Potosí and south through Jalisco.
8. *M. m. salvus* Hall, 1948. Restricted to the Monte Tancítaro, Michoacán.
9. *M. m. subsimus* Goldman, 1938. From Coahuila, Nuevo León, Tamaulipas, and northeastern San Luis Potosí.

**Conservation:** *Microtus mexicanus* is listed as Least Concern by the IUCN (Álvarez-Castañeda and Reid 2016).

**Characteristics:** Mexican vole is medium sized; total length 128.0–155.0 mm and skull length 23.1–26.9 mm. Dorsal pelage dark to grayish brown, sometimes grizzly with yellow shades, flanks yellowish brown; fur of underparts **yellowish cream with yellowish gray shades at the base**; **tail very short, more than one and a half the hindfoot length, and bicolored**; ears hairy; upperparts of limbs brown or pale gray; six plantar tubercles; **four nipples**; first lower molars with five closed triangles; second upper molars with four closed sections; third upper molars with three closed triangles; third lower molars with three transverse loops and no triangles.

**Comments:** *Microtus mexicanus* has unclear interspecific affinities; it was considered within the subgenus *Microtus* or *Pitymys* (Anderson 1959, 1961; Musser and Carleton 1993; Conroy and Cook 2000; Conroy et al. 2001), but the current treatment is without a subgenus assignation (Musser and Carleton 2005). *M. mogollonensis* was elevated to full species different from *M. mexicanus* (Musser and Carleton 2005). *M. mexicanus* ranges from Chihuahua, Coahuila, and Nuevo León south through Puebla, Veracruz, and Oaxaca by the highlands and the Altiplano Central (Map 114). This species can be found in sympatry, or nearly so, with *M. drummondii*, *M. longicaudus*, and *M. montanus*. *M. oaxacensis*, *M. quasiater*; *M. umbrosus* can be found at short distances in the states of Veracruz or Oaxaca, but it has never been recorded in the same locality. *M. mexicanus* can be differentiated from all the other species of *Microtus* by having the fur of the underparts yellowish cream with yellowish gray shades at the base; tail very short and bicolored, more than one and a half the hindfoot length and with four nipples.

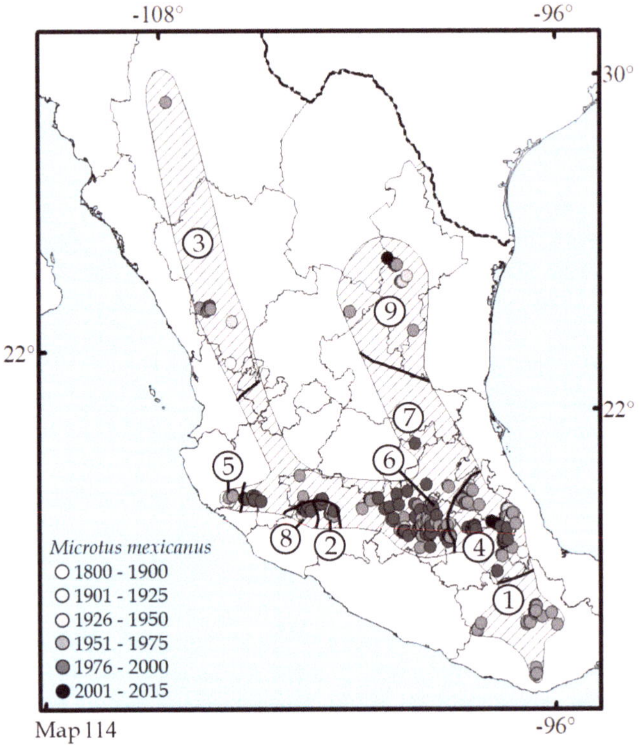


**Map 114** Distribution map of *Microtus mexicanus*

**Additional Literature:** Bailey (1900), Moore and Janacek (1990), Álvarez and Hernández-Chávez (1993), Frey and LaRue (1993), Frey (1999).

## *Microtus mogollonensis* (Mearns, 1890)

Mogollon vole, ratón colicorto de Mogollon

1. *M. m. guadalupensis* Bailey, 1902. Central and southeastern New Mexico and southwestern Texas.
2. *M. m. hualpaiensis* Goldman, 1938. Restricted to the Hualpai Mountains, Arizona.
3. *M. m. mogollonensis* (Mearns, 1890). Western Arizona and eastern New Mexico.
4. *M. m. navaho* Benson, 1934. Restricted to the Navajo Mountains.

**Conservation:** *Microtus mogollonensis* (as *M. mexicanus*) is listed as Least Concern by the IUCN (Álvarez-Castañeda and Reid 2016).

**Characteristics:** Mogollon vole is medium sized; total length 122.0–154.0 mm and skull length 23.6–24.6 mm. Dorsal pelage typically dark brown; ventral pelage buffy to ochraceous gray; six plantar tubercles; **four nipples**; first lower molars with five closed triangles; second upper molars with four closed sections; third upper molars with three closed triangles; third lower molars with three transverse loops and no triangles.

**Comments:** *Microtus mogollonensis* is considered without a subgenus assignation (Musser and Carleton 1993, 2005; Conroy and Cook 2000). However, it is considered a sibling species of *M. mexicanus* with unclear interspecific affinities, considered a member of the subgenus *Microtus* or *Pitymys* (Anderson 1959, 1961; Musser and Carleton 1993; Conroy and Cook 2000; Conroy et al. 2001). Here I followed Musser and Carleton (2005), considering *M. mogollonensis* a valid species different from *M. mexicanus* (Judd et al. 1980; Modi 1987; Frey 1999). *M. mogollonensis* is patchily distributed from the mountains of southern Utah and Colorado south through central Arizona, New Mexico, and a small area in western Texas (Map 115). It can be found in sympatry with, or nearly so, *M. drummondii*, *M. longicaudus*, and *M. montanus*. *M. mogollonensis* can be differentiated from all the other species of *Microtus*, except for *M. mexicanus*, by having the fur of the underparts yellowish cream with yellowish gray shades at the base; tail very short and bicolored, more than one and a half the hindfoot length, and with four nipples.

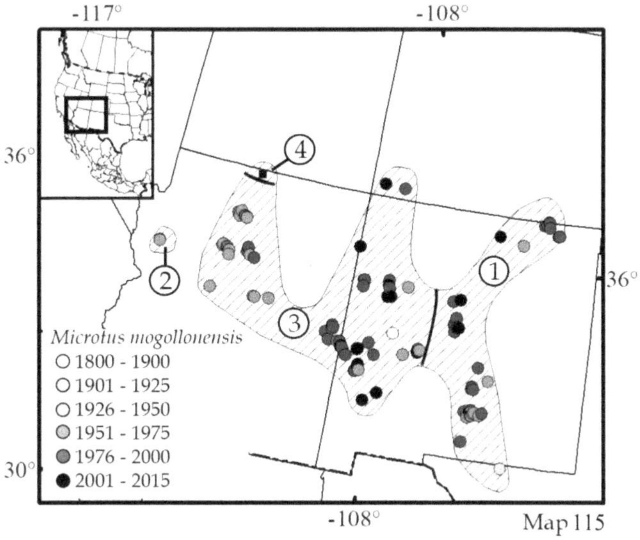


**Map 115** Distribution map of *Microtus mogollonensis*

**Additional Literature:** Bailey (1900), Anderson (1960), Frey and LaRue (1993).

## *Microtus richardsoni* (DeKay, 1842)

North American water vole, ratón colicorto de agua

1. *M. r. arvicoloides* (Rhoads, 1894). From British Columbia south through Oregon.
2. *M. r. macropus* (Merriam, 1891). From the border between British Columbia and Alberta east through Montana and south through Oregon to Wyoming.
3. *M. r. myllodontus* Rasmussen and Chamberlain, 1959. Utah and southeastern Idaho.
4. *M. r. richardsoni* (DeKay, 1842). Eastern British Columbia and western Alberta.

**Conservation:** *Microtus richardsoni* is listed as Least Concern by the IUCN (Cassola 2016ba).

**Characteristics:** North American water vole is medium sized; total length 198.0–274.0 mm and skull length 29.0–34.8 mm. Dorsal pelage grayish sepia, dark sepia to dark reddish brown in a grizzled appearance; fur of underparts pale gray at the tip with blackish shades at the base, general appearance grayish; **tail longer than in all other *Microtus* in North America, average 75.0 mm, more than two and a half times the hindfoot length, and bicolored**; foot large; five plantar tubercles; eight nipples; second upper molars with four closed sections; third upper molars with three closed triangles; third lower molars with three transverse loops; auditory bullae small; first lower molars with five closed triangles.

**Comments:** *Microtus richardsoni* was previously considered within the subgenus *Aulacomys* of *Microtus* (Musser and Carleton 1993, Conroy and Cook 2000), but the current treatment is without a subgenus assignation (Musser and Carleton 2005). This species ranges in two distinct areas. The first is the coastal Mountains from western British Columbia south through the Blue and Cascade mountains of Washington and Oregon. The second is the Rocky Mountains from British Columbia and Alberta south through western Montana, Idaho, Wyoming, and central Utah (Map 116). It can be found in sympatry, or nearly so, with *M. californicus*, *M. canicaudus*, *M. drummondii*, *M. longicaudus*, *M. montanus*, *M. ochrogaster*, *M. oregoni*, *M. townsendii*, and *M. xanthognathus*. *M. richardsoni* can be differentiated from all other species of *Microtus* by having a larger body size; tail on average 75.0 mm, more than two and a half times the hindfoot length.

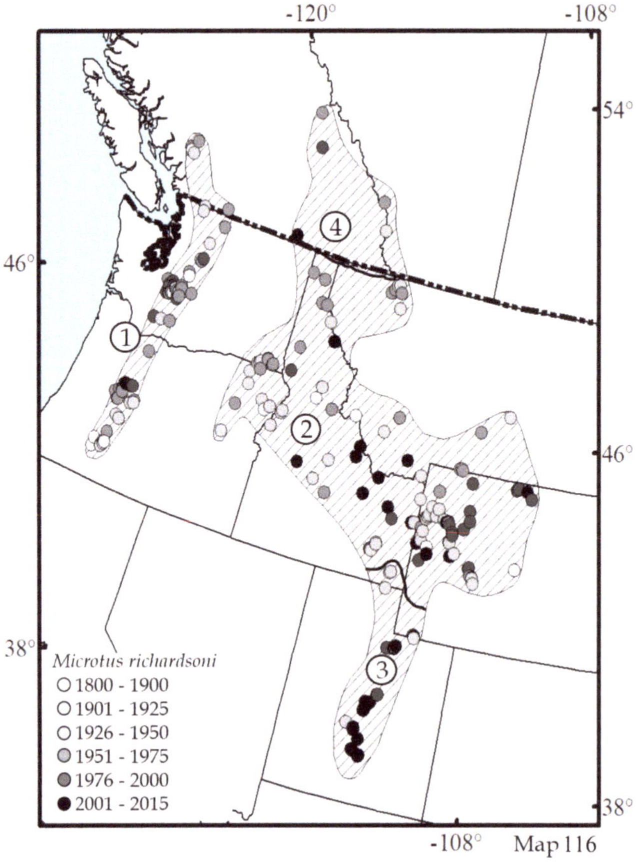


**Map 116** Distribution map of *Microtus richardsoni*

**Additional Literature:** Bailey (1900), Hooper and Hart (1962), Carleton (1980), Hall (1981), Ludwing (1984).

## ***Microtus umbrosus*** Merriam, 1898

Zempoaltepec vole, ratón colicorto del cerro Zempoaltepec

Monotypic.

**Conservation:** *Microtus umbrosus* is listed as subject to Special Protection in the Norma Oficial Mexicana (DOF 2019) and as Endangered (B1ab (i, ii, iii, v) + 2ab (i, ii, iii, v): extent of occurrence less than 5000 km$^2$ in five or less populations and under continuing decline in extent of occurrence, area of occupancy, quality of habitat, and number of mature individuals + area of occupancy less than 500 km$^2$ in five or less populations and under continuing decline in extent of occurrence, area of occupancy, quality of habitat, and number of mature individuals) by the IUCN (de Grammont and Cuarón 2018c).

**Characteristics:** Zempoaltepec vole is medium sized; total length 171.0–187.0 mm and skull length 26.1–28.0 mm. Dorsal pelage dark with brown shades at the tip, general appearance blackish or dark brown; fur of underparts pale gray at the tip and blackish at the base, general appearance dark grayish; **tail long, more than twice the hindfoot length, not bicolored, but dorsally similar to the back and paler ventrally**; ears long and mainly naked; feet large; five plantar tubercles, one very rudimentary, so they seem to be five; four nipples; third upper molars with two closed rounded triangles and one open; second lower molars with an anterior pair of confluent triangles; third lower molars with a posterior loop, two closed triangles and an anterior loop.

**Comments:** *Microtus umbrosus* was considered a member of the subgenus *Microtus* of *Microtus* (Bailey 1900; Anderson 1959). However, genetic analyses suggested that it is related to *M. pinetorum*, a member of the subgenus *Pitymys* (Conroy et al. 2001), but the current treatment is without a subgenus assignation (Musser and Carleton 2005). This species is restricted to the Cerro Zempoaltepec in northern-central Oaxaca (Map 117), and is allopatric with respect to other species of *Microtus*. *M. umbrosus* can be differentiated from all the other species of *Microtus* by having five plantar tubercles; tail length more than 30% of the total length; third lower molars with two closed triangles and third upper molars with two triangles.

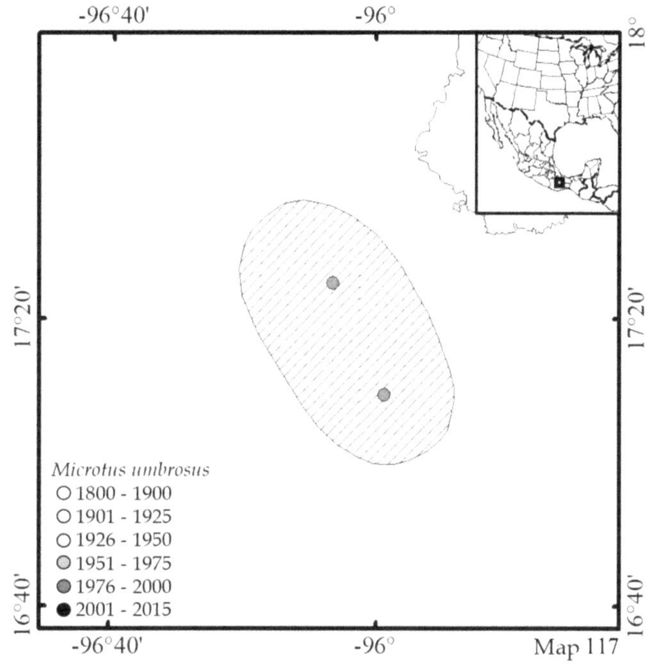


**Map 117** Distribution map of *Microtus umbrosus*

**Additional Literature:** Hall and Cockrum (1953), Musser and Carleton (1993, 2005), Cervantes et al. (1994), Frey and Cervantes (1997a).

## ***Microtus xanthognathus*** (Leach, 1815)
Taiga vole, ratón colicorto de la taiga

Monotypic.

**Conservation:** *Microtus xanthognathus* is listed as Least Concern by the IUCN (Cassola 2016bc).

**Characteristics:** Taiga vole is medium sized; total length 186.0–226.0 mm and skull length 27.7–31.7 mm. Dorsal pelage dark sepia to grayish brown, with black hair on the back; fur of underparts pale gray at the tip with blackish shades at the base and general appearance grayish; **tail long for *Microtus*, almost twice the hindfoot length, and bicolored, black dorsally and dusky gray ventrally; chin with a white spot and cheeks with an ochraceous cinnamon coloration;** six plantar tubercles; eight nipples; first lower molars with five closed triangles; second upper molars with four closed sections; **third upper molars with three closed triangles; third lower molars with three transverse loops and no triangles**; incisive foramina long and narrow.

**Comments:** *Microtus xanthognathus* was considered within the subgenus *Aulacomys* of *Microtus* (Musser and Carleton 1993; Conroy and Cook 2000), but the current treatment is without a subgenus assignation (Musser and Carleton 2005). *M. xanthognathus* ranges from central Alaska east through the Hudson Bay western coast (Map 118). *M. xanthognathus* can be found in sympatry, or nearly so, with *M. abbreivatus*, *M. drummondii*, and *M. richardsoni*. It can be differentiated from all the other species of *Microtus* by having the cheeks with an ochraceous cinnamon coloration; tail almost twice the hindfoot length and bicolored, black dorsally and dusky gray ventrally; third upper molars with three closed triangles, and third lower molars with three transverse loops and no triangles.

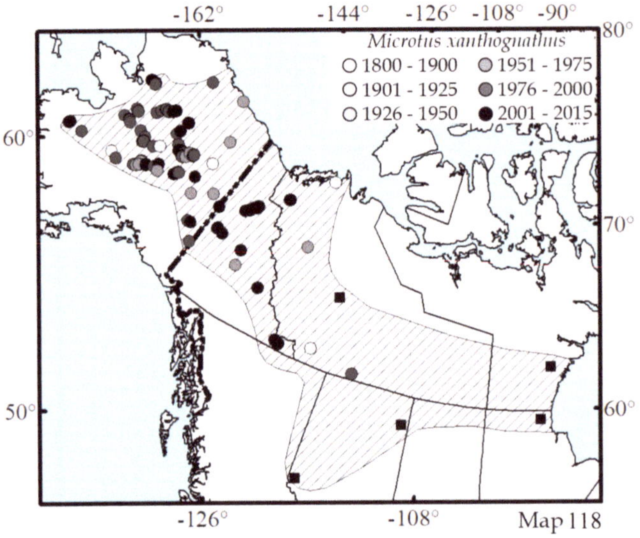


**Map 118** Distribution map of *Microtus xanthognathus*

**Additional Literature:** Bailey (1900), Rausch and Rausch (1974), Youngman (1975), Zagorodnyuk (1990), Conroy and Cook (1999b), Conroy et al. (2001), Musser and Carleton (2005).

## Tribe Clethrionomyini

### Genus *Clethrionomys* Pallas, 1811

Dorsal pelage brown to grayish brown, sometimes with a wide bright reddish to dark rufous stripe in the central back; flanks paler, with blackish hair at the base; fur of underparts whitish with blackish shades at the base; ears hairy and of the same color as the flanks; tail either short or long and sharply or indistinctly bicolored; **molars rooted in adults; most loops of enamel on the upper molars rounded; third upper molars with three loops of enamel on the lingual side; lingual reentrant angles of the lower molars small,** if any is deeper than those on the labial side, **not exceeding one-half of the tooth width;** cheek-teeth small; upper toothrow somewhat divergent. The keys were elaborated based on the review of specimens and the following sources: Bailey (1897), Hall (1981), and Verts and Carraway (1984).

1. Tail short and thick; postpalatal bridge usually incomplete in adults (Fig. 31; from Alaska and northern British Columbia east through the Hudson Bay)........................................................................................ *Clethrionomys rutilus* (p. 176)

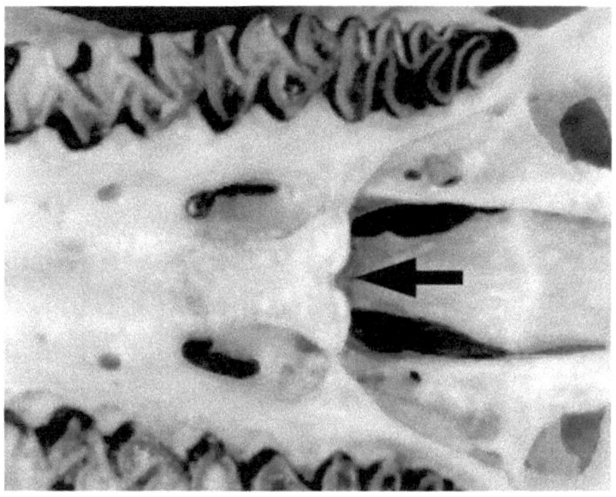

**Fig. 31** Postpalatal bridge usually incomplete in adults

1a. Tail longer and slender; postpalatal bridge always complete in adults (Fig. 32)............................................................2

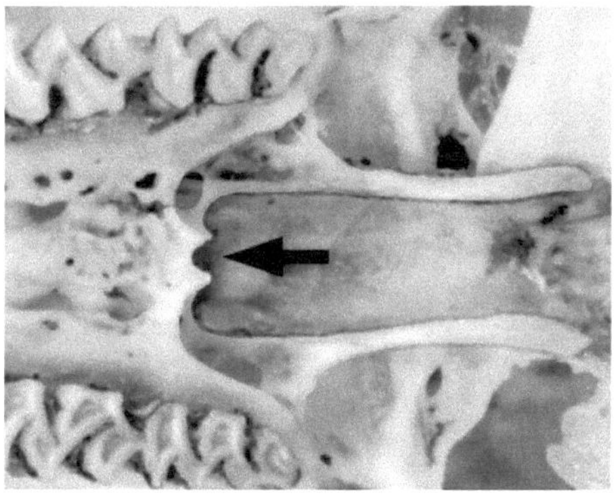

**Fig. 32** Postpalatal bridge always complete in adults

2. Total length usually less than 158.0 mm; tail usually less than 50.0 mm; upperparts distinctly reddish; palate posterior margin without a median spine (from British Columbia and Washington east through the Labrador Peninsula, and south through New Jersey, Georgia, Pennsylvania, Michigan, and south through Arizona and New Mexico) ............................................................................................................................................ *Clethrionomys gapperi* (p. 174)

2a. Total length usually greater than 155.0 mm; tail usually greater than 47.0 mm; upperparts indistinctly reddish; palate posterior margin with a median spine (from the Columbia River, Oregon, south through northwest California) ............................................................................................................................................*Clethrionomys californicus* (p. 173)

## ***Clethrionomys californicus*** (Merriam, 1890)

Western red-backed vole, chincolo de lomo rojo del oeste

1. *C. c. californicus* (Merriam, 1890). From Columbia River south to near the coast through the northern San Francisco Bay.
2. *C. c. mazama* (Merriam, 1897). From the Columbia River south by the Willamette Valley through northern California.
3. *C. c. obscurus* (Merriam, 1897). From the Columbia River south by the Cascade Mountains through northern California.

**Conservation:** *Clethrionomys californicus* (as *Myodes californicus*) is listed as Least Concern by the IUCN (Cassola 2016bd).

**Characteristics:** Western red-backed vole is medium sized; **total length 155.0–160.0 mm and skull length 21.8–23.3 mm**. Dorsal stripe dark brown to dark gray, **central back reddish** but not in the shape of a stripe; **tail long and indistinctly bicolored**; postpalatal bridge always complete in adults; **posterior palate margin with a median spine**.

**Comments:** *Clethrionomys californicus* (Ellerman 1941; Miller and Kellogg 1955; Hall 1981; Kryštufek et al. 2019) was previously assigned to *Myodes californicus* (Kretzoi 1964, 1969; Musser and Carleton 2005). The subspecies *C. californicus occidentalis* and *M. c. caurinus* are now considered subspecies of *C. gapperi* (Johnson and Ostenson 1959; Cowan and Guiguet 1965). *C. californicus* ranges from the Columbia River east through Cascade Range, Oregon, and south through the northern San Francisco Bay, California (Map 119). *C. gapperi* and *C. californicus* are separated by the Columbia River, and *C. californicus* is allopatric with respect to the other species of *Clethrionomys*, from which it can be differentiated by being the largest, with a total length usually greater than 155.0 mm and the tail usually greater than 47.0 mm; back indistinctly reddish; the palate posterior margin has a median spine.

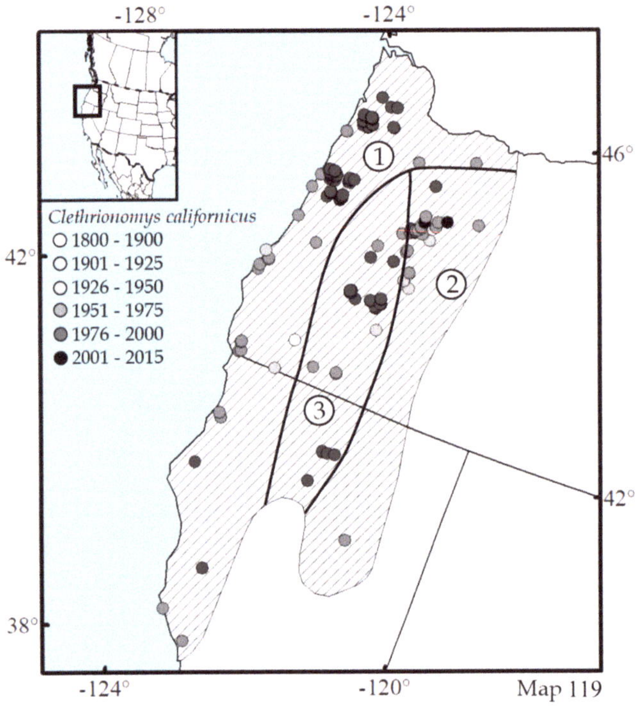


**Map 119** Distribution map of *Clethrionomys californicus*

**Additional Literature:** Alexander and Verts (1992).

## ***Clethrionomys gapperi*** (Vigors, 1830)

Southern red-backed vole, chincolo de lomo rojo del sur

1. *C. g. arizonensis* Cockrum and Fitch, 1952. Restricted to eastern Arizona.
2. *C. g. athabascae* (Preble, 1908). From eastern British Columbia west through western Manitoba and the southern Northwest Territories.
3. *C. g. brevicaudus* (Merriam, 1891). Northeastern Wyoming and central-western South Dakota.
4. *C. g. carolinensis* (Merriam, 1888). Appalachian Mountains, West Virginia.
5. *C. g. cascadensis* Booth, 1945. Cascade Mountain, Washington.
6. *C. g. caurinus* (Bailey, 1898). Western coast of British Columbia.
7. *C. g. galei* (Merriam, 1890). From southern Alberta south through Utah and Colorado.

8. *C. g. gapperi* (Vigors, 1830). From southeastern Manitoba, northeastern Minnesota, and Wisconsin east through the Labrador Peninsula and the northern Saint Lawrence River.
9. *C. g. gaspeanus* Anderson, 1943. Northern New Brunswick, southern bank of the San Lorenzo River.
10. *C. g. gauti* Cockrum and Fitch, 1952. Southern Colorado and northern New Mexico.
11. *C. g. hudsonius* (Anderson, 1940). Western Manitoba, northern Ontario, and western Québec.
12. *C. g. idahoensis* (Merriam, 1891). Southeastern Washington, eastern Oregon, Idaho, western Montana, and Wyoming.
13. *C. g. limitis* (Bailey, 1913). Restricted to western Arizona.
14. *C. g. loringi* (Bailey, 1897). From southern Alberta and Montana east through Manitoba and Minnesota.
15. *C. g. maurus* Kellogg, 1939. Restricted to eastern Kentucky.
16. *C. g. nivarius* (Bailey, 1897). Restricted to the eastern Olympic Peninsula, Washington.
17. *C. g. occidentalis* (Merriam, 1890). Restricted to western Washington.
18. *C. g. ochraceus* (Miller, 1894). From New Hampshire and Vermont northeast through New Brunswick.
19. *C. g. pallescens* Hall and Cockrum, 1940. Restricted to Nova Scotia.
20. *C. g. paludicola* Doutt, 1941. Northeastern Indiana and western Pennsylvania.
21. *C. g. phaeus* (Swarth, 1911). Restricted to Marten Arm, border between British Columbia and Alaska.
22. *C. g. proteus* (Bangs, 1897). Labrador Peninsula.
23. *C. g. rhoadsii* (Stone, 1893). Southern New York and New Jersey.
24. *C. g. rupicola* Poole, 1949. Restricted to Kittatinny Ridge of Berks, Pennsylvania.
25. *C. g. saturatus* (Rhoads, 1894). From northern Washington and Idaho north through British Columbia.
26. *C. g. solus* Hall and Cockrum, 1952. Restricted to the Revillagigedo Island, Alaska.
27. *C. g. stikinensis* Hall and Cockrum, 1952. Restricted to the Stikine River, British Columbia.
28. *C. g. ungava* (Bailey, 1897). Restricted to the northern Labrador Peninsula.
29. *C. g. wrangeli* (Bailey, 1897). Restricted to Sergief Island, Alaska.

**Conservation:** *Clethrionomys gapperi* (as *Myodes californicus*) is listed as Least Concern by the IUCN (Cassola 2016bd).

**Characteristics:** Southern red-backed vole is medium sized; total length 120.0–153.0 mm and skull length 20.9–23.2 mm. Dorsal stripe wide bright chestnut to yellowish brown (occasionally black); **tail long and bicolored**; **postpalatal bridge always complete in adults but without a median spine**.

**Comments:** *Clethrionomys gapperi* (Ellerman 1941; Miller and Kellogg 1955; Hall 1981; Kryštufek et al. 2019) was previously known as *Myodes gapperi* (Kretzoi 1964, 1969; Musser and Carleton 2005). *C. gapperi* includes the subspecies *C. g. occidentalis* and *C. g. caurinus*, previously considered under *C. californicus* (Johnson and Ostenson 1959, Cowan and Guiguet 1965). *C. rutilus* and *C. gapperi* have been considered conspecific (Bee and Hall 1956; Youngman 1975). I follow Carleton (1980) and Musser and Carleton (2005), who considered them different species. *C. gapperi* is widely distributed from British Columbia and Washington east through the Labrador Peninsula and south through New Jersey, the Appalachian Mountains to northern Georgia, Pennsylvania, Michigan, and south through Arizona and New Mexico (Map 120). It is allopatric with respect to the other species of *Clethrionomys*, from which it can be differentiated by having the tail long, bicolored, and dorsally with a wide stripe, bright chestnut to yellowish brown, occasionally black; the posterior palate margin is complete but without a median spine.

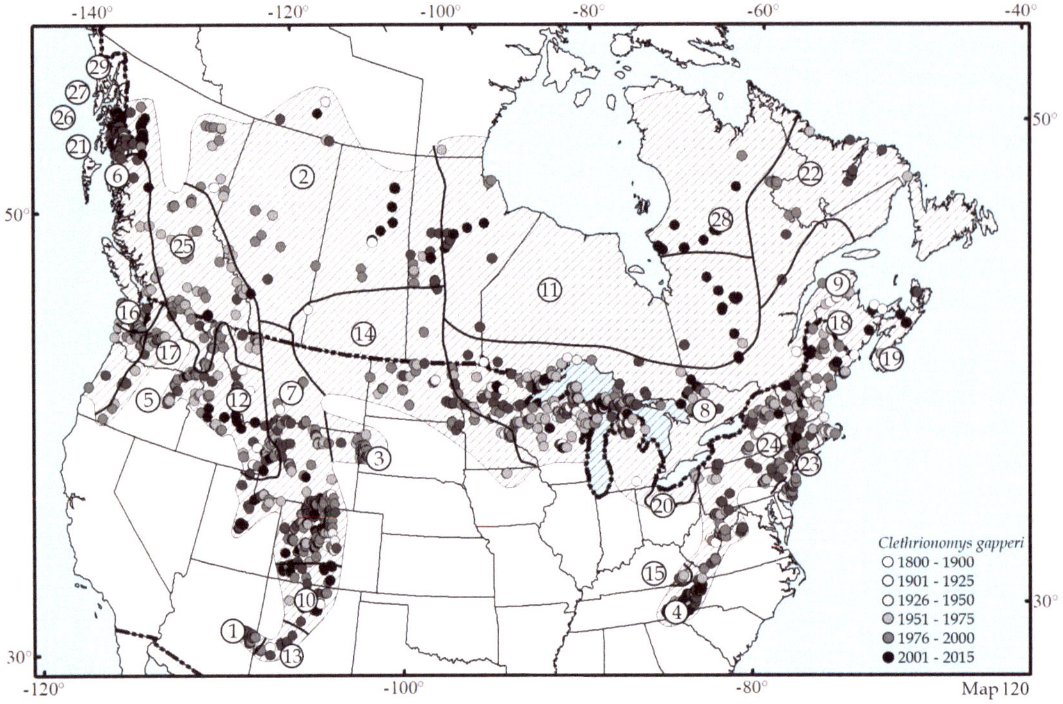


**Map 120** Distribution map of *Clethrionomys gapperi*

**Additional Literature:** Bee and Hall (1956), Youngman (1975), Merritt (1981).

## *Clethrionomys rutilus* (Pallas, 1779)

Northern red-backed vole, chincolo de lomo rojo del norte

1. *C. r. albiventer* Hall and Gilmore, 1932. Restricted to the Saint Lawrence Island, Alaska.
2. *C. r. dawsoni* (Merriam, 1888). From Alaska east through Keewatin, including northern British Columbia and Manitoba.
3. *C. r. glacialis* Orr, 1945. Restricted to the Glacial Bay, Alaska.
4. *C. r. insularis* (Heller, 1910). Restricted to the Hawkins Island, Alaska.
5. *C. r. orca* (Merriam, 1900). Restricted to the southern coast of Alaska.
6. *C. r. platycephalus* Manning, 1957. Restricted to Tuktoyaktuk, Alaska.
7. *C. r. washburni* Hanson, 1952. Eastern Makenzie and western Keewatin.
8. *C. r. watsoni* Orr, 1945. Restricted to the Cape Yakataga, Alaska.

**Conservation:** *Clethrionomys rutilus* (as *Myodes rutilus*) is listed as Least Concern by the IUCN (Linzey et al. 2020).

**Characteristics:** Northern red-backed vole is medium sized; total length 130.0–158.0.0 mm and skull length 21.7–24.5 mm. Dorsal stripe bright reddish to dark rufous; **tail short and sharply bicolored**; **postpalatal bridge usually incomplete in adults**.

**Comments:** *Clethrionomys rutilus* (Miller and Kellogg 1955; Hall 1981; Kryštufek et al. 2019) was previously known as *Myodes rutilus* (Kretzoi 1964, 1969; Musser and Carleton 2005). *C. rutilus* and *C. gapperi* have been considered conspecific (Bee and Hall 1956; Youngman 1975). I follow Carleton (1980) and Musser and Carleton (2005), who considered them different species. *C. rutilus* has a Holarctic distribution, ranging from Alaska east through western coast of the Hudson Bay and south through northern British Columbia (Map 121). It is allopatric with respect to the other species of *Clethrionomys*, from which it can be differentiated in by its short and thick tail; postpalatal bridge usually incomplete in adults.

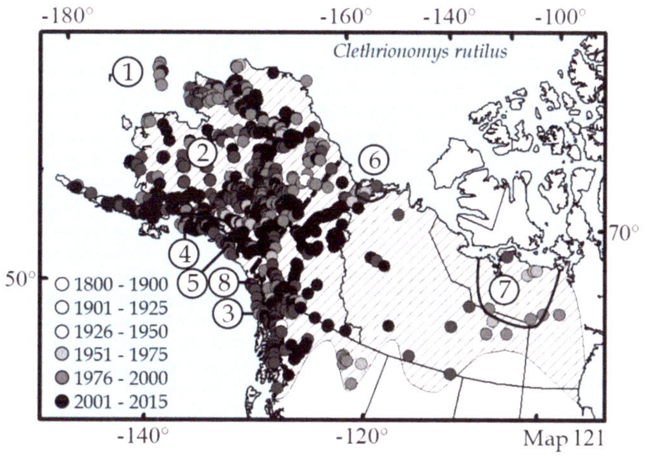


**Map 121** Distribution map of *Clethrionomys rutilus*

**Additional Literature:** Ellerman (1941), Carleton (1981).

## Tribe Dicrostonychini

### Genus *Dicrostonyx* Gloger, 1841

Coloration varying greatly within species; in addition, they have a different summer and winter coloration, so it is not easy to have a diagnostic coloration for each species. In general, all species have different brown shades during summer; **in summer, a dark line can be seen on the back center;** dorsal line clearer in juveniles than in adults; winter coloration white, but with hairs blackish at the base; **tail short; ears reduced; powerful third and fourth feet claws become enlarged in winter, so as to appear double; pointed process projecting forward from the orbit posterior rim.** All the species of *Dicrostonyx* were previously considered a complex regarded as part of *D. torquatus* as a single circumpolar species. The keys were elaborated based on the review of specimens and the following sources: Allen (1919b), Rausch and Rausch (1972), Hall (1981), Krohne (1982), and Jones et al. (1986, 1992).

1. First and second upper molars without an accessory fold at the posterior inner corner; last lower molars usually lacking an accessory antero-inner fold, with an enamel fold (from the eastern coast of the Hudson Bay east through northeastern Labrador, northern Québec, and the Belcher Islands, Canada)........................................*Dicrostonyx hudsonius* (p. 179)

1a. First and second upper molars with a small accessory fold at the posterior inner corner; last lower molars with an accessory anterior inner fold, usually an anterior inner enamel fold........................................2

2. Occurs in Canada........................................3

2a. Occurs in Alaska........................................5

3. Occurs in the Ogilvie Mountains, Yukon Territory, Canada........................................*Dicrostonyx nunatakensis* (p. 180)

3a. Does not occur in the Ogilvie Mountains, Yukon Territory........................................4

4. Nasals slender and pointed posteriorly; interparietal bone broad and square shaped (Canadian Arctic and northern Alaska)........................................*Dicrostonyx groenlandicus* (part, p. 178)

4a. Nasals broader and slightly pointed posteriorly; interparietal bone slender and triangular (restricted to the Hudson Bay western coast, Northwest Territories, Manitoba, and Mackenzie, Canada)........*Dicrostonyx richardsoni* (p. 181)

5. Occurs in the Umnak and Unalaska islands of the Aleutian Archipelago, Alaska, United States........................................ ........................................*Dicrostonyx unalascensis* (p. 182)

5a. Does not occur in the Umnak and Unalaska islands of the Aleutian Archipelago........................................6

6. Dorsal pelage grizzled gray in summer; total length greater than 130.0 mm; tail length greater than 21.0 mm; nasals broader; auditory bullae large (Canadian arctic and northern Alaska) .............*Dicrostonyx groenlandicus* (part, p. 178)

6a. Dorsal pelage reddish-brown in summer; total length less than 135.0 mm; tail length less than 21.0 mm; nasals narrow; auditory bullae small (west of Alaska, the Unimak and Saint Lawrence Islands, Alaska)................................................ ........................................................................................................................... *Dicrostonyx nelsoni* (p. 180)

## ***Dicrostonyx groenlandicus*** (Traill, 1823)

Northern collared lemming, leming de tundra común

1. *D. g. clarus* Handley, 1953. Restricted to the Borden and Melville islands, Franklin.
2. *D. g. groenlandicus* (Traill, 1823). Northeastern Keewatin and western islands, Franklin.
3. *D. g. kilangmiutak* Anderson and Rand, 1945. From Yukon to central Keewatin.
4. *D. g. lentus* Handley, 1953. Restricted to the southern Baffin Island, Franklin.
5. *D. g. rubricatus* (Richardson, 1889). Northern Alaska.

**Conservation:** *Dicrostonyx groenlandicus* is listed as Least Concern by the IUCN (Cassola 2016o).

**Characteristics:** Northern collared lemming is medium sized; total length 129.0–157.0 mm and skull length 23.0–28.0 mm. Dorsal pelage grizzled grayish to grayish brown; fur of underparts pale cinnamon to whitish with a pectoral cinnamon spot in the brown (summer) stage, with no strong contrast between the dorsal and ventral coloration; cinnamon spots on the ears not strongly contrasting with the dorsal coloration; a dark line in the snout center; squamosal bone with a prominent peg extending anterolaterally within the orbit; nasals slender and pointed posteriorly; interparietal bone broad and square shaped.

**Comments:** *Dicrostonyx groenlandicus* was recognized within the *D. torquatus* complex. However, karyotype information elevated *D. groenlandicus* to full species (Rausch and Rausch 1972; Honacki et al. 1982; Krohne 1982; Jones et al. 1986, 1992). *D. kilangmiutak* and *D. rubricatus* (previously known as *Arvicola rubricatus*) were elevated to full species (Baker et al. 2003a), and *D. kilangmiutak* is karyotypically separable from *D. groenlandicus* (Engtrom et al. 1992). However, they are regarded as synonyms of *D. groenlandicus* (Musser and Carleton 2005). *D. g. alascensis* was considered under *D. g. rubricatus*. *D. g. clarus* and *D. g. lentus* are considered synonyms (Borowik and Engtrom 1993). *D. groenlandicus* ranges in the tundra area above the tree line in the Canadian Arctic, northern Alaska, Queen Elizabeth Islands, Southampton, and in the District of Franklin (Map 122), and is allopatric with respect to other species of *Dicrostonyx*.

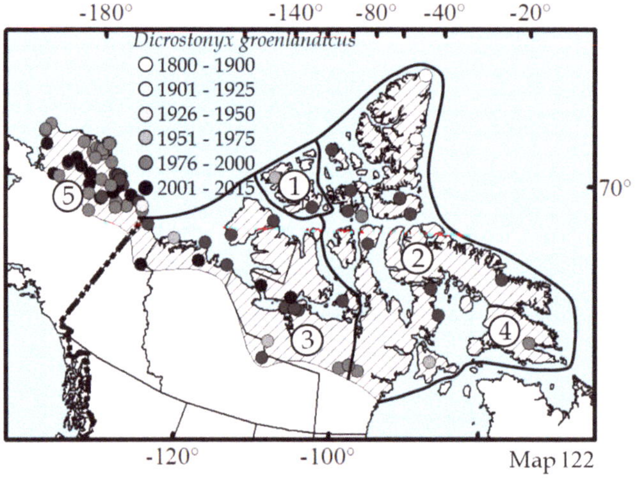


**Map 122** Distribution map of *Dicrostonyx groenlandicus*

**Additional Literature:** Miller (1896), Allen (1919b), Anderson and Rand (1945), Handley (1953), Jarrell and Fredga (1993), Eger (1995), Musser and Carleton (2005).

## ***Dicrostonyx hudsonius*** (Pallas, 1778)

Ungava collared lemming, leming de tundra de Ungava

Monotypic.

**Conservation:** *Dicrostonyx hudsonius* is listed as Least Concern by the IUCN (Cassola 2016p).

**Characteristics:** Ungava collared lemming is medium sized; total length 148.0–160.0 mm and skull length 28.0–32.0 mm. Dorsal pelage buffy gray, blackish dorsal stripe indistinct; fur of underparts pale cinnamon to whitish; strong contrast between the dorsal and ventral coloration, with a cinnamon spot on the ears; a dark line in the snout center; **first and second upper molars without an accessory fold at the posterior inner corner; last lower molars usually lacking an accessory anterior inner fold and an enamel fold**.

**Comments:** *Dicrostonyx hudsonius* was recognized within the *D. torquatus* complex. However, karyotype information considered *D. hudsonius* a full species (Rausch and Rausch 1972; Honacki et al. 1982; Krohne 1982; Jones et al. 1986, 1992). It ranges from the Hudson Bay eastern coast east through northeastern Labrador, northern Québec, and the Belcher Islands, Canada (Map 123). It is allopatric with respect to other species of *Dicrostonyx*, from which it can be differentiated by having the first and second upper molars without an accessory fold at the posterior inner corner; last lower molars usually lacking an accessory anterior inner fold and an enamel fold.

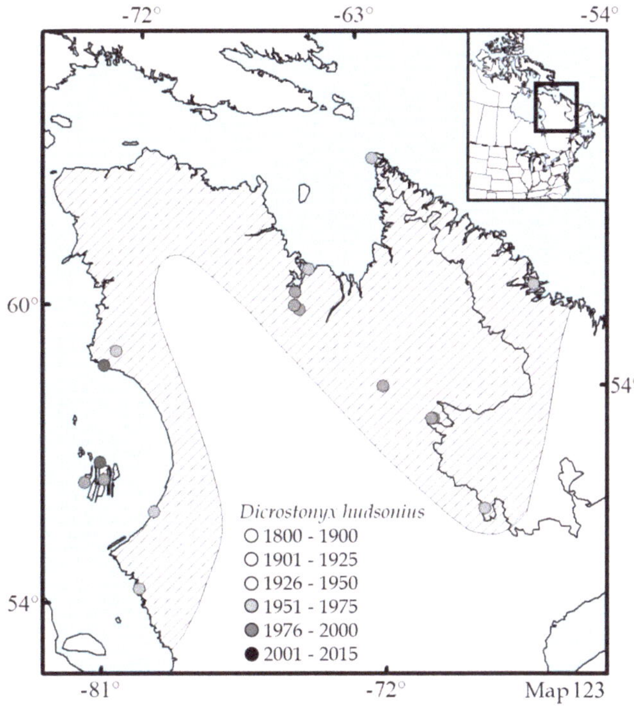


**Map 123** Distribution map of *Dicrostonyx hudsonius*

**Additional Literature:** Miller (1896), Allen (1919b), Anderson and Rand (1945), Borowik and Engtrom (1993), Eger (1995).

## *Dicrostonyx nelsoni* Merriam, 1900

Nelson's collared lemming, leming de tundra de Alaska

Monotypic.

**Conservation:** *Dicrostonyx nelsoni* is listed as Least Concern by the IUCN (Cassola 2016q).

**Characteristics:** Nelson's collared lemming is medium sized; total length 130.0–133.0 mm and skull length 28.2–28.9 mm. Dorsal pelage reddish brown, sometimes with oxide red or intensely cinnamon shades; fur of underparts pale brown to whitish, with a pectoral cinnamon spot poorly differentiated; no contrast between the dorsal and ventral coloration; ear spot brown; dark line in the snout center contrasting with the pale cheek coloration; squamosal bone with a prominent peg extending anterolaterally within the orbit; nasals narrow; supraorbital ridges strong; auditory bullae very small and narrow.

**Comments:** *Dicrostonyx exsul* was elevated to full species (Baker et al. 2003a), but it was considered a junior synonym of *D. nelsoni* (Musser and Carleton 2005). *D. nelsoni* was first recognized within the *D. torquatus* complex and later as a subspecies of *D. groenlandicus*. However, karyotype information recognized *D. nelsoni* as a valid species (Rausch and Rausch 1972; Honacki et al. 1982; Krohne 1982; Jones et al. 1986, 1992). *D. g. peninsulae* was considered a junior synonym of *D. nelsoni*. *D. nelsoni* is only known from western and southwestern costal Alaska, Unimak, and Saint Lawrence Island (Map 124), and is allopatric with respect to other species of *Dicrostonyx*.

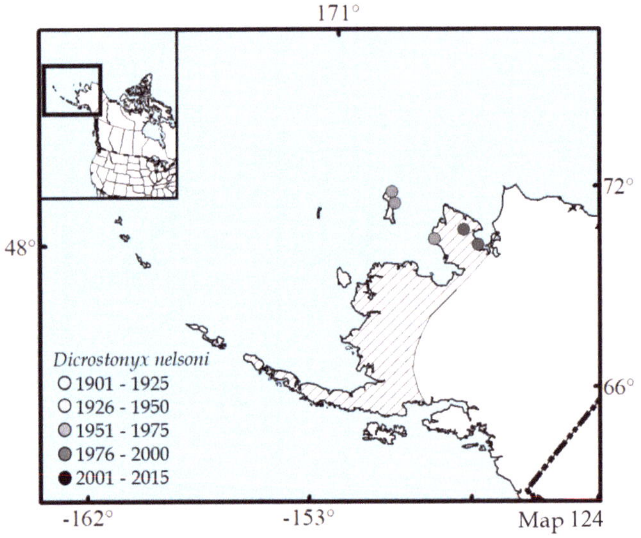


**Map 124** Distribution map of *Dicrostonyx nelsoni*

**Additional Literature:** Youngman (1967), Krohne (1982), Jarrell and Fredga (1993), Eger (1995), Baker et al. (2003a).

## *Dicrostonyx nunatakensis* Youngman, 1967

Ogilvie Mountains collared lemming, leming de tundra del Yukón

Monotypic.

**Conservation:** *Dicrostonyx nunatakensis* is listed as Least Concern by the IUCN (Cassola 2016r).

**Characteristics:** Ogilvie Mountain collared lemming is medium sized; total length 128.0–129.0 mm and skull length 25.3–25.5 mm. Dorsal pelage buffy gray to grayish brown, blackish dorsal stripe present; fur of underparts pale cinnamon to whitish, with a strong contrast between the dorsal and ventral coloration, and with a cinnamon spot on the ears; dark line in the snout center; first and second upper molars with a small accessory fold at the posterior inner corner; last lower molars with an accessory anterior inner fold and usually an anterior inner enamel fold.

**Comments:** *Dicrostonyx nunatakensis* was recognized within the *D. torquatus* complex and later as a subspecies of *D. groenlandicus*. However, karyotype information considered it a valid species (Rausch and Rausch 1972; Honacki et al. 1982; Krohne 1982; Jones et al. 1986, 1992), only known from the Ogilvie Mountains in northern-central Yukon (Map 125). It is allopatric with respect to other species of *Dicrostonyx*. *D. nunatakensis* is morphologically similar to *D. groenlandicus*, so it can only be differentiated by its restricted distribution.

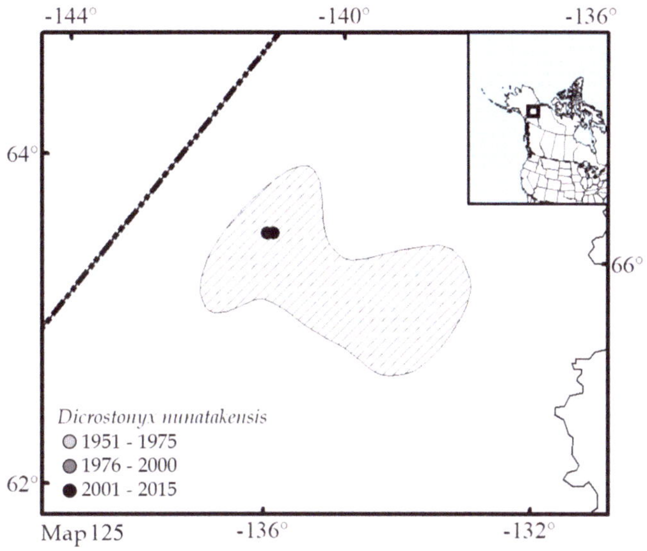


**Map 125** Distribution map of *Dicrostonyx nunatakensis*

**Additional Literature:** Rausch (1977), Honacki et al. (1982), Eger (1995), Jung et al. (2014).

## *Dicrostonyx richardsoni* Merriam, 1900

Richardson's collared lemming, leming de tundra de la Bahía Hudson

Monotypic.

**Conservation:** *Dicrostonyx richardsoni* is listed as Least Concern by the IUCN (Cassola 2016s).

**Characteristics:** Richardson's collared lemming is medium sized; total length 145.0–146.0 mm and skull length 27.5–29.0 mm. Dorsal pelage brown to reddish brown; it is the species with a more reddish coloration; fur of underparts pale brownish to whitish, with no contrast between the dorsal and ventral coloration; with a cinnamon pale spot from the underarms to the pectoral region; ear spot darker brown; a dark line in the snout center; squamosal bone with a prominent peg extending anterolaterally within the orbit; **nasals broad** and slightly pointed posteriorly; **interparietal bone slender and triangular**.

**Comments:** *Dicrostonyx richardsoni* was recognized within the *D. torquatus* complex and later as a subspecies of *D. groenlandicus*. However, karyotype information considered it a valid species (Rausch and Rausch 1972; Krohne 1982; Honacki et al. 1982; Jones et al. 1986, 1992). *D. richardsoni* ranges from the Hudson Bay western coast to the Great Slave Lake, including the Northern Territories, Manitoba, and Mackenzie (Map 126), and is allopatric with respect to other species of *Dicrostonyx*.

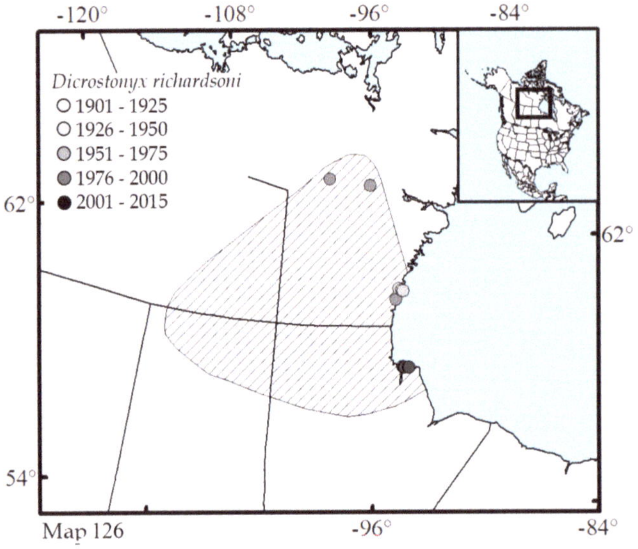


**Map 126** Distribution map of *Dicrostonyx richardsoni*

**Additional Literature:** Allen (1919b), Youngman (1967), Eger (1995).

## *Dicrostonyx unalascensis* Merriam, 1900

Umnak island collared lemming, leming de tundra de la isla Umnak

1. *D. u. stevensoni* Nelson, 1929. Restricted to the Umnak Island, Alaska.
2. *D. u. unalascensis* Merriam, 1900. Restricted to the Unalaska Island, Alaska.

**Conservation:** *Dicrostonyx unalascensis* is listed as Data Deficient by the IUCN (Garibaldi 2019).

**Characteristics:** Umnak Island collared lemming is medium sized; total length 150.0–169.0 mm and skull length 31.2–33.2 mm. Dorsal coloration pale brown to pale grayish brown, appearing opaque; fur of underparts light brown with no contrast between the dorsal and ventral coloration; with no contrasting cinnamon-brown spot on the ears; wide dark line on the snout center; pointed process projecting forward from the orbit posterior rim.

**Comments:** *Dicrostonyx unalascensis* was recognized within the *D. torquatus* complex and later as a subspecies of *D. groenlandicus*. However, karyotypic information considered *D. unalascensis* a valid species (Rausch and Rausch 1972; Krohne 1982; Honacki et al. 1982; Jones et al. 1986, 1992). Only known from the Umnak and Unalaska Islands in the Aleutian Archipelago, Alaska (Map 127), and is allopatric with respect to other species of *Dicrostonyx*.

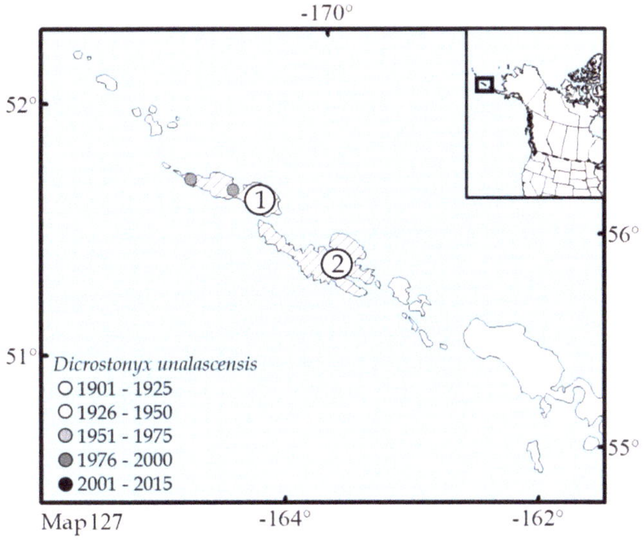


**Map 127** Distribution map of *Dicrostonyx unalascensis*

**Additional Literature:** Jarrell and Fredga (1993), Eger (1995).

# Tribe Lemmini

## Genus *Lemmus* Link, 1795

Dorsal coloration grizzled grayish in summer and grizzled tawny in winter; fur of underparts washed with buff; limbs short; tail not projecting beyond the limbs. *Lemmiscus* can be differentiated from all the other species of Arvicolinae by having the temporal ridges fused into a **sharp median interorbital crest; zygomatic arches broadly divergent, wider anteriorly than posteriorly; toothrows widely divergent posteriorly**. The keys were elaborated based on the review of specimens and the following sources: Davis (1944a), Rausch (1953), Rausch and Rausch (1975), Hall (1981), and Spitsyn et al. (2021).

1. From Yukon west through Alaska, including the Aleutian Islands ........................................*Lemmus nigripes* (p. 183)

1a. From the Yukon–Mackenzie and Alaska-British Columbia borders east through British Columbia and Alberta, Nunavut, and the Baffin Island ........................................ *Lemmus trimucronatus* (p. 184)

### *Lemmus nigripes* (True, 1894)

Western brown lemming, leming pardo del oeste

1. *L. n. alascensis* Merriam, 1900. Restricted to the northern coast of Alaska.
2. *L. n. harroldi* Swarth, 1931. Restricted to the Nunivak Island, Alaska.
3. *L. n. minusculus* Osgood, 1904. Southwestern Alaska.
4. *L. n. nigripes* (True, 1894). Restricted to the Saint George Island, Alaska.
5. *L. n. subarticus* Bee and Hall, 1956. Central-northern Alaska.
6. *L. n. yukonensis* Merriam, 1900. Central Alaska and southwestern Yukon.

**Conservation:** *Lemmus nigripes* (as *L. trimucronatus*) is listed as Least Concern by the IUCN (Cassola 2016am).

**Characteristics:** Western brown lemming is medium sized; total length 120.0–165.0 mm and skull length 28.5–32.0 mm. Dorsal pelage grizzled grayish (summer); fur of underparts washed with buff; limbs short; tail not projecting beyond the limbs; ears markedly reduced; soles almost concealed by hair, with phalanges notably lengthened; temporal ridges fused into a sharp median interorbital crest; **zygomatic arches broadly divergent, wider anteriorly than posteriorly; toothrows widely divergent posteriorly**.

**Comments:** *Lemmus nigripes* was previously considered a subspecies of *L. trimucronatus* (Rausch 1953), and is currently considered a valid species (Spitsyn et al. 2021). *L. trimucronatus* was previously known as *L. sibiricus*; however, Chernyavsky et al. (1993) mentioned they were from different continents, North America and Asia, so they should be considered different species. *L. nigripes* ranges in Asia and northwestern North America from Yukon and west through Alaska, including the Aleutian Islands (Map 128; Fedorov et al. 2003; Abramson and Petrova 2018). *L. nigripes* can be found near the range of *L. trimucronatus*. Both species are cryptic and can only be differentiated by molecular methods and their distribution area.

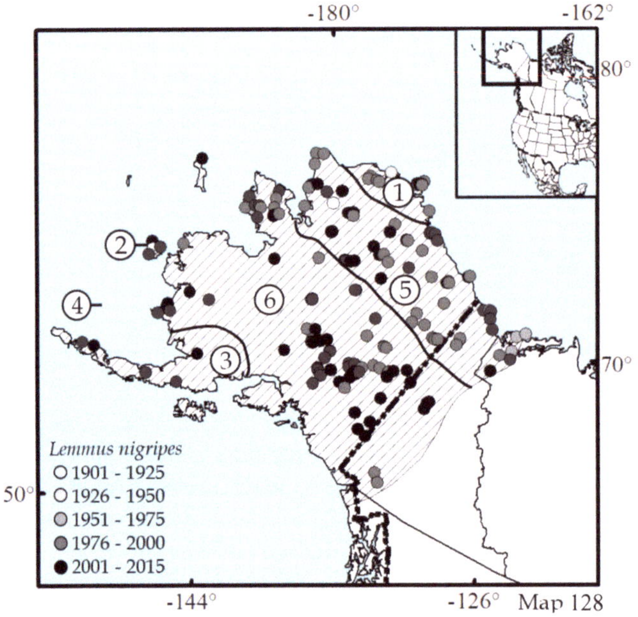


**Map 128** Distribution map of *Lemmus nigripes*

**Additional Literature:** Miller (1896), Davis (1944a), Hall and Cockrum (1953), Rausch (1953), Rausch and Rausch (1975), Batzli (1999), Musser and Carleton (2005).

## *Lemmus trimucronatus* (Richardson, 1825)

Nearctic brown lemming, leming pardo

1. *L. t. helvolus* (Richardson, 1828). Southern Yukon and British Columbia.
2. *L. t. phaiocephalus* Manning and Macpherson, 1958. Restricted to the Bank Island, Franklin.
3. *L. t. trimucronatus* (Richardson, 1825). From Yukon east through the Hudson Bay and the southern Franklin islands.

**Conservation:** *Lemmus trimucronatus* is listed as Least Concern by the IUCN (Cassola 2016am).

**Characteristics:** Brown lemming is medium sized; total length 122.0–160.0 mm and skull length 28.8–31.4 mm. Dorsal pelage grizzled grayish (summer) to grizzled tawny (winter); fur of underparts washed with buff; limbs short; tail not projecting beyond the limbs; ears markedly reduced; soles almost concealed by hair, and phalanges notably lengthened; temporal ridges fused into a sharp median interorbital crest; **zygomatic arches broadly divergent, wider anteriorly than posteriorly; toothrows widely divergent posteriorly**.

**Comments:** *Lemmus trimucronatus* was known as *L. sibiricus*; however, Chernyavsky et al. (1993) mentioned they were from different continents, North America and Asia, so they should be considered different species. *L. trimucronatus* has a broad range in northern North America, from the Yukon–Mackenzie and Alaska-British Columbia borders east through Alberta, Nunavut, and the Baffin Island, including many of the northern islands of northern Canada, in addition to central British Columbia, eastern of the Rocky Mountains in the western side, and western Alberta (Map 129). *L. trimucronatus* can be found near the range of *L. nigripes*. Both species are cryptic and can only be differentiated by molecular methods and their distribution area.

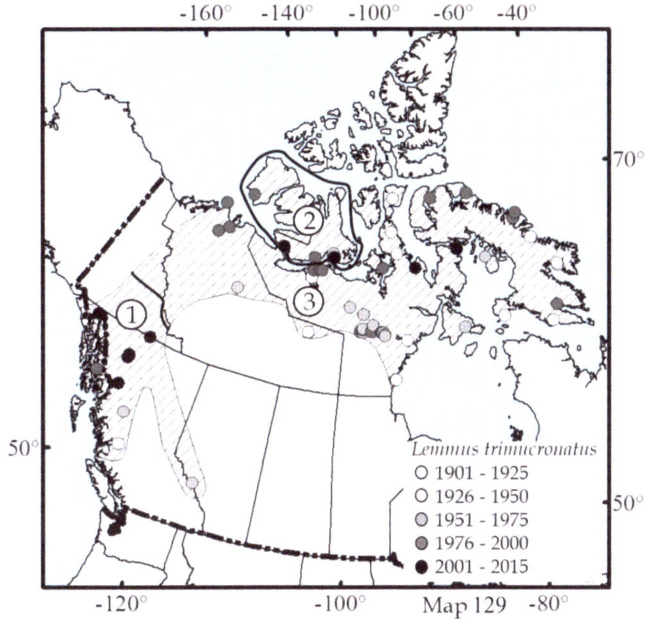


**Map 129** Distribution map of *Lemmus trimucronatus*

**Additional Literature:** Miller (1896), Davis (1944a), Hall and Cockrum (1953), Rausch (1953), Rausch and Rausch (1975), Batzli (1999), Fedorov et al. (2003), Musser and Carleton (2005).

## Género *Synaptomys* Baird, 1857

Dorsal pelage from bright brown to grizzled brown or gray; fur of underparts grizzle with hair pale gray at the tip and plumbeous at the base; nipples six or eight, depending on the species; tail short, slightly longer than the hind feet; six plantar tubercles; **rostrum short, less than 25% of the total skull length; upper incisors longitudinally grooved** (McAllister and Hoffmann 1988). The keys were elaborated based on the review of specimens and the following sources: Howell (1927), Hall (1981), Linzey (1983), and Verts and Carraway (1984).

1. Ear base with a few brighter hairs; eight nipples; lower molars without closed triangles on the labial side; mandibular incisors relatively slender; posterior margin of the palate with a sharply pointed median spine (Fig. 33; from Alaska and Washington east through the Labrador Peninsula and New England, New Brunswick, and south through Montana and Minnesota) ..........subgenus *Mictomys*; *Synaptomys borealis* (p. 186)

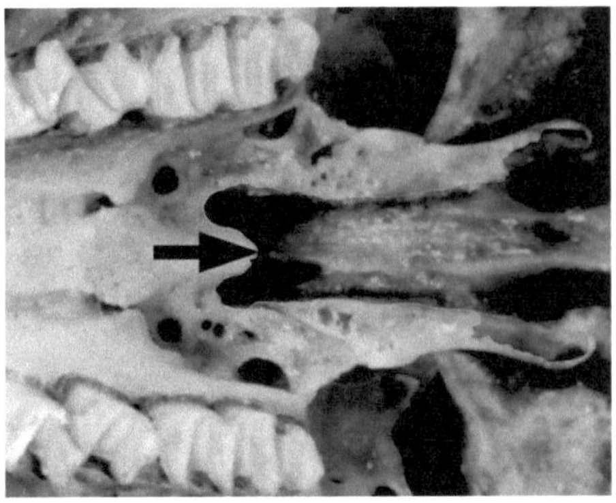

**Fig. 33** Posterior margin of the palate with a sharply pointed median spine

1a. Ear base with brighter hairs; six nipples; lower molars with closed triangles on the labial side; mandibular incisors heavy; posterior margin of the palate with a broad blunt median spine (Fig. 34; from Manitoba, Wisconsin, South Dakota, Nebraska, and Kansas east through the Atlantic coast from southern Québec south through Virginia and North Carolina)subgenus.......................................................................................*Synaptomys*; *Synaptomys cooperi* (p. 188)

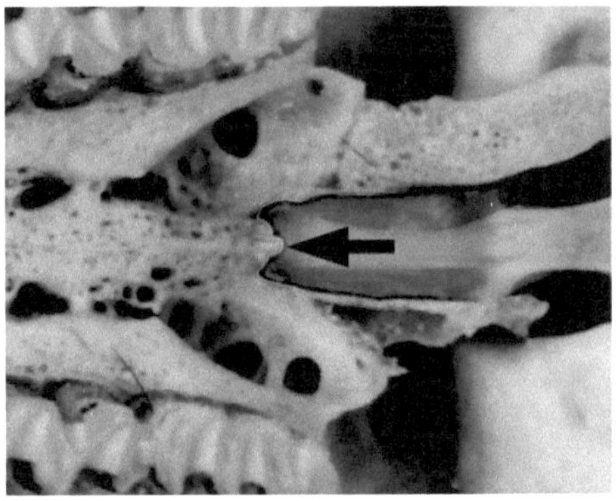

**Fig. 34** Posterior margin of the palate with a broad blunt median spine

## Subgenus *Mictomys* True, 1894

**Eight nipples,** two pairs pectoral and two inguinal; **ear base with few hair appreciably brighter than the rest of the pelage; mandible with molars without close triangles on the outer sides and incisors relatively slender and sharply pointed;** molars almost without outer reentrant angles; **palate with a well-developed posterior spinous process.**

### *Synaptomys borealis* (Richardson, 1828)

Northern bog lemming, leming de pantano del norte

1. *S. b. artemisiae* Anderson, 1932. Restricted to southern British Columbia and northern Washington.
2. *S. b. borealis* (Richardson, 1828). From western Makenzie, eastern British Columbia, and central Alberta east through western Saskatchewan.
3. *S. b. chapmani* Allen, 1903. Southeastern British Columbia, southwestern Alberta, northeastern Washington, and northwestern Montana.
4. *S. b. dalli* Merriam, 1896. From Alaska southeast through southern British Columbia.
5. *S. b. innuitus* (True, 1894). Labrador Peninsula, except the eastern part.
6. *S. b. medioximus* Bangs, 1900. Eastern Labrador Peninsula.

7. *S. b. smithi* Anderson and Rand, 1943. From western Saskatchewan east through eastern Ontario.
8. *S. b. sphagnicola* Preble, 1899. From New Hampshire northeast through New Brunswick.
9. *S. b. truei* Merriam, 1896. Coastal British Columbia and northernmost coast of Washington.

**Conservation:** *Synaptomys borealis* is listed as Least Concern by the IUCN (Cassola 2017d).

**Characteristics:** Okanagan bog lemming is medium sized; total length 118.0–135.0 mm and skull length 23.0–25.8.0 mm. Dorsal pelage grizzled gray to argus brown; fur of underparts light gray at the tip with plumbeous shades at the base; **ear base with a few brighter hairs**; tail bicolored; **nipples eight**; **molariforms without closed triangles on the labial side; lower incisors relatively slender; median spine sharply pointed at the palate posterior margin**.

**Comments:** *Synaptomys borealis* ranges from Alaska and Washington east through the Labrador Peninsula, New England, and New Brunswick, and south through Montana and Minnesota (Map 130), and is allopatric with respect to *S. cooperi*. It can be differentiated from it by its dorsal pelage grizzled gray to argus brown with a few brighter hairs at the ear base; tail bicolored; eight nipples; molariforms lack closed triangles on the labial side; lower incisors relatively slender and with a sharply pointed median spine at the palate posterior margin.

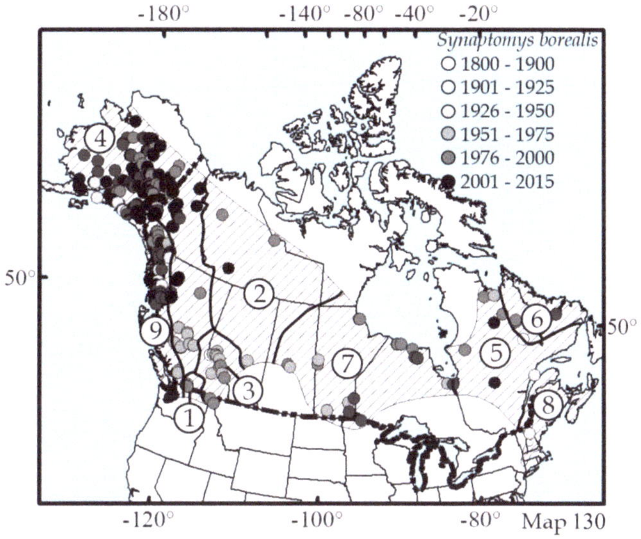


**Map 130** Distribution map of *Synaptomys borealis*

**Additional Literature:** Howell (1927), Hall (1981), Honacki et al. (1982), Conroy and Cook (1999a), Musser and Carleton (2005).

## Subgenus *Synaptomys* Baird, 1857

**Six nipples,** one pair pectoral and two inguinal; hair at the ear base not appreciably brighter than the rest of the pelage; **mandible with molars with close triangles on the outer sides and incisors heavy, outer edges never noticeably worn or prolonged into sharp splinters of the enamel;** molars with outer reentrant angles well developed; **palate with a poorly developed posterior spinous process**.

## *Synaptomys cooperi* Baird, 1857

Southern bog lemming, leming de pantano del sur

1. *S. c. cooperi* Baird, 1858. From Minnesota and Wisconsin northeast through Great Lakes and San Lawrence River.
2. *S. c. gossii* (Coues, 1877). From Nebraska and Kansas west through eastern Illinois.
3. *S. c. helaletes* Merriam, 1896. Northeastern North Carolina and southeastern Virginia.
4. *S. c. kentucki* Barbour, 1956. Northeastern Kentucky.
5. *S. c. paludis* Hibbard and Rinker, 1942. Restricted to the Meade Country State Park, Kansas.
6. *S. c. relictus* Jones, 1958. Restricted to the Rock Creek Fish Hatchery, Nebraska.
7. *S. c. stonei* Rhoads, 1893. From western Kentucky and Tennessee northeast through Massachusetts.

**Conservation:** *Synaptomys cooperi* is listed as Least Concern by the IUCN (Cassola 2016ds).

**Characteristics:** Southern bog lemming is medium sized; total length 118.0–154.0 mm and skull length 23.1–30.6 mm. Dorsal pelage bright chestnut to dark grizzled brown; fur of underparts light gray at the tip with plumbeous shades at the base; in general, there seems to be more contrast between the dorsal and ventral coloration; **ear base with bright hairs**; tail indistinctly bicolored; **six nipples; molariforms with closed triangles on the labial side; lower incisors heavy; median spine at the posterior margin of the palate broadly blunt**.

**Comments:** *Synaptomys cooperi* ranges from Manitoba, Wisconsin, South Dakota, Nebraska, and Kansas east through the Atlantic coast, from southern Québec south through Virginia and North Carolina (Map 131); it is allopatric with respect to *S. borealis*. It can be differentiated from it by its bright chestnut to dark grizzled brown dorsal pelage; ear base with brighter hairs; tail indistinctly bicolored; six nipples; molariforms with triangles on the labial side closed; lower incisors heavy and with a median spine at the posterior palate margin.

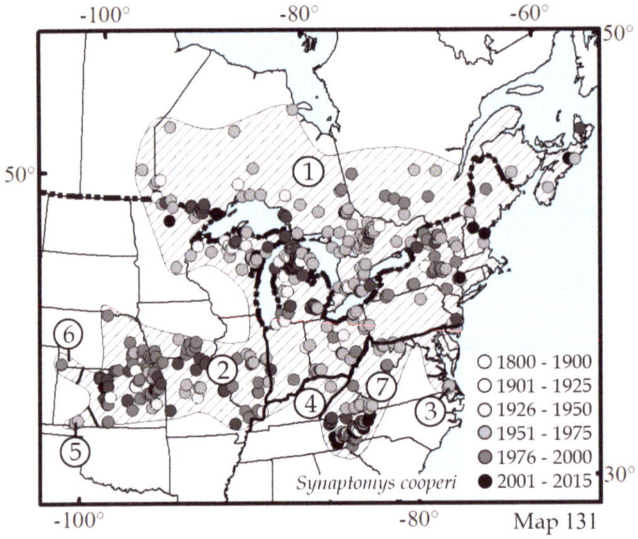


**Map 131** Distribution map of *Synaptomys cooperi*

**Additional Literature:** Howell (1927), Bole and Moulthrop (1942), Wetzel (1955), Connor (1959), Easterla (1968b), Hall (1981), Rose (1981), Linzey (1983), Wilson and Choate (1997), Musser and Carleton (2005), Cook (2017), Rose and Linzey (2021).

# Tribe Ondatrini

## Genus *Neofiber* True, 1884

The Round-tailed muskrats can be differentiated from all other species of the subfamily Arvicolinae by being one of the largest; **total length 300.0–390.0 mm; tail 150.0–180.0 mm and rounded laterally; hind feet not webbed between the toes; fifth lower molars with shallow reentrant angles**.

### *Neofiber alleni* True, 1884

Round-tailed muskrat, rata almizclera de cola redonda

1. *N. a. alleni* True, 1884. Central and eastern Florida.
2. *N. a. apalachicolae* Schwartz, 1953. Northwestern Florida.
3. *N. a. exoristus* Schwartz, 1953. Southeastern Georgia.
4. *N. a. nigrescens* Howell, 1920. Central Florida peninsula.
5. *N. a. struix* Schwartz, 1952. Southern Florida peninsula.

**Conservation:** *Neofiber alleni* is listed as Least Concern by the IUCN (Cassola 2016bg).

**Characteristics:** Round-tailed muskrat is large sized; total length 285.0–381.0 mm and skull length 41.0–51.0 mm. Dorsal pelage dense, brown to blackish brown; **fur of underparts grayish pale buff; tail rounded**, long and sparsely haired; hind feet larger than the forefeet; the postorbital process shell-like and nearly right-angled; molars rootless, all triangles closed; **fifth lower molars with shallow reentrant angles**; third lower molars with one outer fold; third upper molars with two transverse loops and two median triangles.

**Comments:** *Neofiber alleni* ranges in Florida north through South Georgia (Map 132). The only species that can be found in sympatry with the subfamily Arvicolinae is *Microtus pinetorum*. *N. alleni* is the second largest species of the subfamily Arvicolinae, and can be differentiated from *M. pinetorum* by its largest size (total length 300.0–390.0 mm and tail length 150.0–180.0 mm) and the fifth lower molars with shallow reentrant angles.

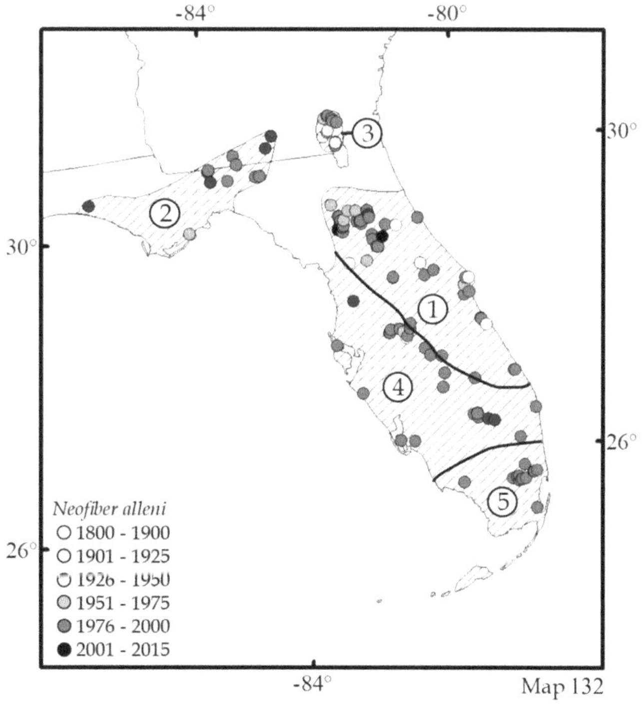


**Map 132** Distribution map of *Neofiber alleni*

**Additional Literature:** Bailey (1900), Howell (1920), Hall and Cockrum (1953), Schwartz (1953), Birkenholz (1972), Musser and Carleton (2005).

## Genus *Ondatra* Link, 1795

These muskrats are large sized, can be differentiated from all other species of the subfamily Arvicolinae by being **the largest, total length 456.0–553.0 mm, skull length 52.2–65.1 mm; tail flattened laterally; hind feet webbed between the toes; fifth lower molars with deep reentrant angles.**

### *Ondatra zibethicus* (Linnaeus, 1766)

Common muskrat, rata almizclera

1. *O. z. albus* (Sabine, 1823). From Saskatchewan to northern Ontario.
2. *O. z. aquilonius* (Bangs, 1899). Northern Labrador Peninsula.
3. *O. z. bernardi* Goldman, 1932. Southern Nevada, western Arizona, eastern California, and northern Baja California.
4. *O. z. cinnamominus* (Hollister, 1910). From southern Alberta and Manitoba southward through New Mexico and Texas.
5. *O. z. goldmani* Huey, 1938. Restricted to southwestern Utah and southeastern Nevada.
6. *O. z. macrodon* (Merriam, 1897). From coastal areas of North Carolina north through southeastern Pennsylvania.
7. *O. z. mergens* (Hollister, 1910). Northern Nevada and northeastern California.
8. *O. z. obscurus* (Bangs, 1894). Restricted to Newfoundland.
9. *O. z. occipitalis* (Elliot, 1903). Coastal areas of southwestern Washington and Oregon.
10. *O. z. osoyoosensis* (Lord, 1863). From southern British Columbia and Alberta south through northern Arizona and New Mexico.
11. *O. z. pallidus* (Mearns, 1890). Arizona and western New Mexico.
12. *O. z. ripensis* (Bailey, 1902). Southern New Mexico, southwestern Texas, and northern Chihuahua.
13. *O. z. rivalicius* (Bangs, 1895). Coastal areas from eastern Texas east through western Florida.
14. *O. z. spatulatus* (Osgood, 1900). From Alaska east through western Saskatchewan.
15. *O. z. zalophus* (Hollister, 1910). Southern Alaska.
16. *O. z. zibethicus* (Linnaeus, 1766). From southeastern Manitoba east through the southern of Labrador Peninsula, and south through Louisiana and Georgia.

**Conservation:** *Ondatra zibethicus* is listed as Least Concern by the IUCN (Cassola 2016ce).

**Characteristics:** Muskrat is large sized; total length 456.0–553.0 mm and skull length 52.2–65.1 mm. Dorsal pelage dark brown; fur of underparts paler; **tail blackish and laterally compressed**, relatively long; top coat bright and hair short and dense; **hind feet webbed**, rigid hair around the digits and longer than in the forelimbs; ears short; molars rooted, all triangles closed; **fifth lower molars with deep reentrant angles**.

**Comments:** *Ondatra zibethicus* ranges from Alaska east through the Labrador Peninsula, Prince Edwards Island, and New Brunswick, and south through all the United States except for the Florida Peninsula; in México, only in the Colorado River delta and northwestern Chihuahua (Map 133). It can be found in sympatry, or nearly so, with mainly all the genera of the subfamily Arvicolinae, from which it can be differentiated by its larger size, greater than 456.0 mm; hind feet webbed, and fifth lower molars with deep reentrant angles.

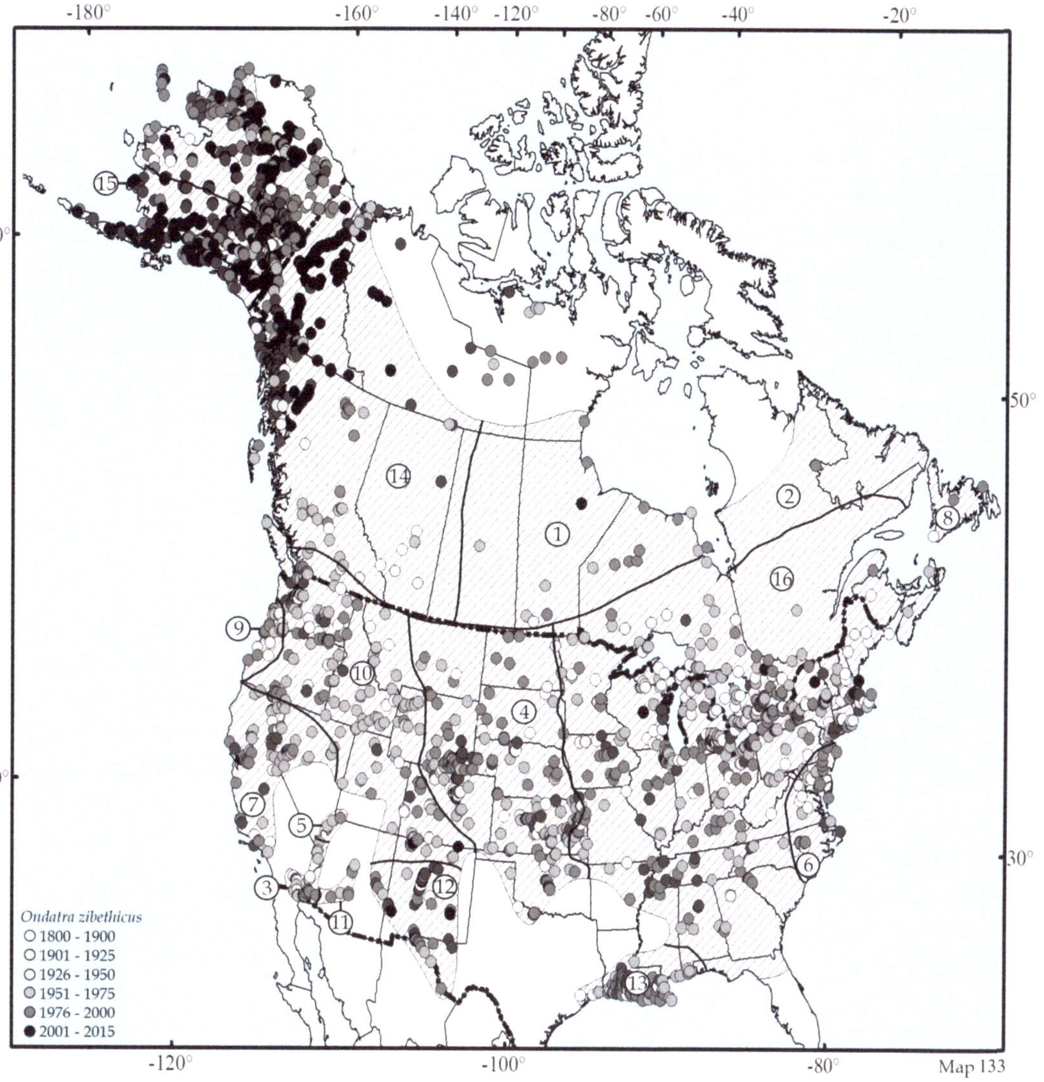


**Map 133** Distribution map of *Ondatra zibethicus*

**Additional Literature:** Hollister (1911), Hall (1955, 1981), Godin (1977), Boyce (1978), Willner et al. (1980), Musser and Carleton (2005).

## Tribe Phemacomyini

### Genus *Arborimus* Taylor, 1915

Coloration ranging between reddish brown with cinnamon shades to dark brown or brownish-gray; tail short, greater than 50.0 mm and from scantily to well haired; ears reduced; **inner (lingual) reentrant angles much deeper than the outer (labial) ones and exceeding one-half of the tooth width in the lower molars; upper molars large and essentially parallel**. *Arborimus* was recognized as a distinct genus (Johnson 1968, 1973; Jones et al. 1997; Musser and Carleton 2005); it was previously regarded as a subspecies of *Phenacomys* (Howell 1926; Hall 1981). The keys were elaborated based on the review of specimens and the following sources: Hall (1981), Verts and Carraway (1984), and Adam and Hayes (1998).

1. Dorsal pelage dark brown; tail slender, distinctly bicolored, and scantly haired; incisive foramina narrow; upper incisors not strongly curved (from the Columbia River south through Humboldt County in California) ........................................................................................................................................ *Arborimus albipes* (p. 192)

1a. Dorsal pelage reddish orange to cinnamon; tail thin, not strongly bicolored, and fairly haired; incisive foramina wide; upper incisors strongly curved ........................................................................................................................................................ 2

2. Zygomatic breadth usually greater than 13.8 mm in males and greater than 14.1 mm in females; total skull length usually greater than 24.8 mm in males and greater than 24.3 mm in females; nasals extending further posteriorly to the maxillary (from Oregon south through California) ..................................................... *Arborimus longicaudus* (p. 193)

2a. Zygomatic breadth usually less than 13.7 mm in males and less than 14.2 mm in females; total skull length usually less than 24.9 mm in males and less than 24.3 mm in females; nasals not extending further posteriorly to the maxillary (from northwest California north through California-Oregon border) ............................................... *Arborimus pomo* (p. 194)

## *Arborimus albipes* (Merriam, 1901)

White-footed vole, chincolo arborícola de patas blancas

Monotypic.

**Conservation:** *Arborimus albipes* is listed as Least Concern by the IUCN (Cassola 2016a).

**Characteristics:** White-footed vole is medium sized; total length 158.0–176.0 mm and skull length 21.2–22.5.0 mm. **Dorsal pelage dark brown or brownish-gray**, sometimes with reddish cinnamon shades on the back; fur of underparts pale gray at the tip with blackish shades at the base; tail scantily haired, sometimes nearly naked at the tip, **distinctly bicolored**, greater than 50.0 mm long and greater than 50% of the head-and-body length; **upper incisors moderately curved**; **incisive foramina narrow**; lower molars with reentrant angles extending much deeper than those on the lingual side; supraorbital ridge poorly defined.

**Comments:** *Arborimus albipes* ranges from the Columbia River, Oregon, south through Humboldt County, California, and also includes the western slope of Cascade Range (Map 134). It is allopatric with respect to other species of *Arborimus*, from which it can be differentiated by its dorsal pelage dark brown; tail slender, distinctly bicolored; upper incisors moderately curved, and incisive foramina narrow.

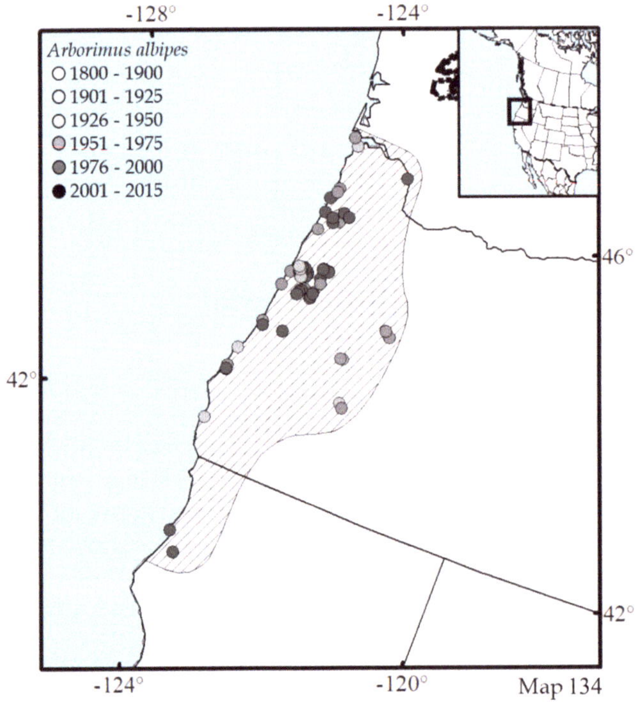


**Map 134** Distribution map of *Arborimus albipes*

**Additional Literature:** Voge and Bern (1949), Johnson and Maser (1982), Carleton and Musser (1984), Verts and Carraway (1995).

## ***Arborimus longicaudus*** (True, 1890)

Red tree vole, chincolo arborícola de las sequoias

1. *A. l. longicaudus* (True, 1890). From central Oregon along the coast south through to San Francisco Bay, California.
2. *A. l. silvicola* Howell, 1921. From the Columbia River along the coast south through central Oregon.

**Conservation:** *Arborimus longicaudus* is listed as Near Threatened by the IUCN (Scheuering 2018).

**Characteristics:** Red tree vole is medium sized; total length 158.0–176.0 mm and skull length 24.8–26.1 mm in males and 24.4–25.4 mm in females. Dorsal pelage reddish, dorsum lacking a median stripe; fur of underparts pale gray with blackish shades at the base; tail thin, well haired, and not distinctly bicolored, greater than 50.0 mm long and greater than 50% of the head-and-body length; six plantar tubercles on the hind feet; upper incisors strongly curved; reentrant angles of the lower molars extending much deeper than those on the lingual side; **supraorbital ridges well developed**.

**Comments:** *Arborimus pomo* is the Californian population previously recognized as *A. longicaudus*. *A. longicaudus* ranges on the western slope of the Cascade Range and from the coastal range of Oregon south through the California-Oregon border (Map 135). It is allopatric with respect to other species of *Arborimus*. *A. longicaudus* can be differentiated from *A. pomo* by its nasals not extending further posteriorly to the maxillary; well-developed supraorbital ridges and greater on average in all measurements.

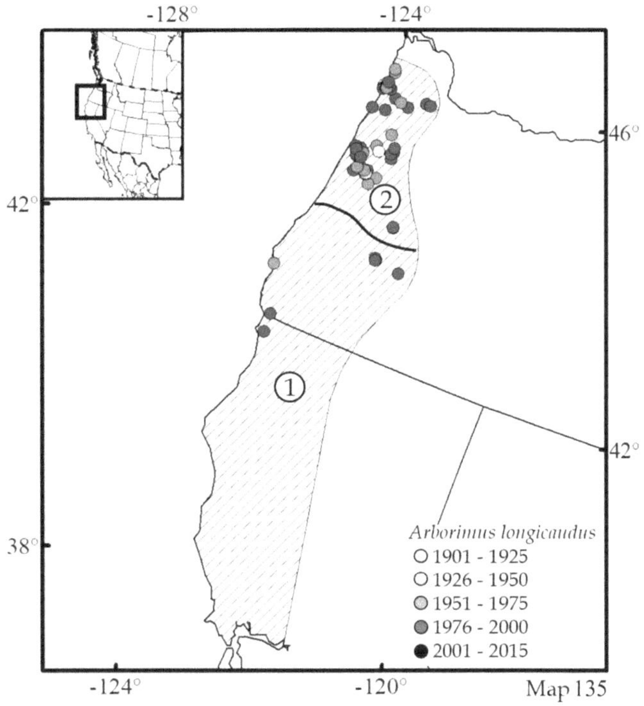


**Map 135** Distribution map of *Arborimus longicaudus*

**Additional Literature:** Voge and Bern (1949), Johnson (1968, 1973), Johnson and Maser (1982), Johnson and George (1991), Hayes (1996).

## *Arborimus pomo* Johnson and George, 1991

Sonoma tree vole, chincolo arborícola de Sonoma

Monotypic.

**Conservation:** *Arborimus pomo* is listed as Near Threatened by the IUCN (Blois and NatureServe 2008).

**Characteristics:** Sonoma tree vole is medium sized; total length 159.0–176.0.0 mm and skull length 24.5–24.9 mm in males and 24.0–24.2 mm in females. Dorsal pelage reddish brown with cinnamon shades; fur of underparts pale gray with blackish shades at the base; tail well haired with dusky-brown shades dorsally, somewhat paler ventrally, greater than 50.0 mm long and greater than 50% of the head-and-body length; six plantar tubercles on the hind feet; upper incisors strongly curved; lower molars with reentrant angles extending much deeper than those on the lingual side; **supraorbital ridges poorly developed**.

**Comments:** *Arborimus pomo* is the Californian population, previously recognized as *A. longicaudus* or *Phenacomys longicaudus*. *A. pomo* was recognized as a distinct species based on chromosome differences, smaller overall size, and certain skull and muscle differences (Johnson and George 1991). This species ranges on coastal areas of northwestern California, from a few miles in northern San Francisco Bay to a few miles north through the south of the California-Oregon border (Map 136), and is allopatric with respect to other species of *Arborimus*. *A. pomo* can be differentiated from *A. longicaudus* by its nasals extending further posteriorly to the maxillary; supraorbital ridges poorly developed and lesser on average in all measurements.

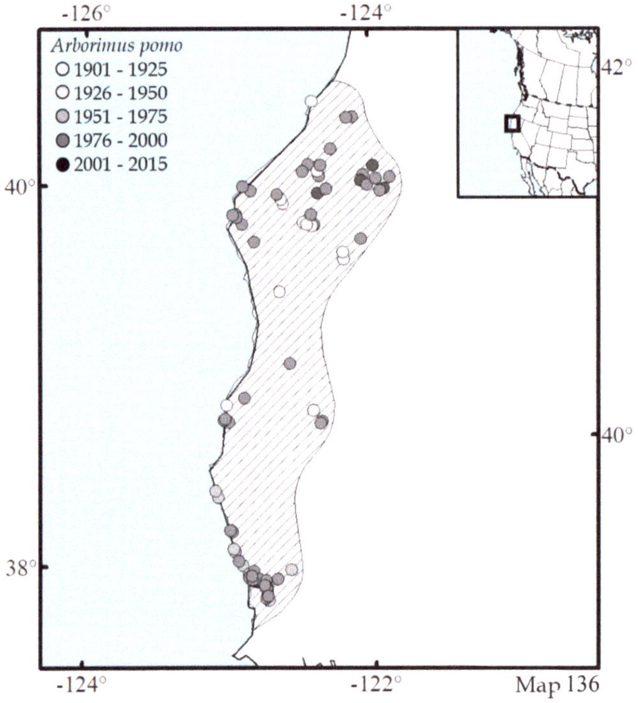


**Map 136** Distribution map of *Arborimus pomo*

**Additional Literature:** Voge and Bern (1949), Jones et al. (1968), Johnson and Maser (1982), Adam and Hayes (1998).

# Género *Phenacomys* Merriam, 1889

Dorsal pelage grizzled dull brown, lacking the yellowish wash; nose and eye rings yellowish to reddish; tail length less than 50.0 mm, generally bicolored, blackish brown dorsally and grayish white ventrally; **five plantar tubercles on the hind feet; inner reentrant angles much deeper than the outer angles and exceeding one-half of the tooth width in the lower molars, upper molars large and essentially parallel**. *Arborimus* was previously considered a subgenus of *Phenacomys* (Howell 1926, Hall 1981), nowadays it is a full genus (Johnson 1973; Musser and Carleton 2005). The keys were elaborated based on the review of specimens and the following sources: Howell (1926), Hall (1981), Verts and Carraway (1984), and McAllister and Hoffmann (1988).

1. Nose and eye rings grayish brown; ear tip lined with pale yellowish hairs; dorsal coloration not yellowish wash; interorbital region flattened in lateral view; median palatal spine not well defined (highly fragmented distribution in British Columbia, Washington, Oregon, Idaho, Montana, Wyoming, California, Nevada, Utah, Colorado, and New Mexico) ........ *Phenacomys intermedius* (p. 195)

1a. Nose and eye rings yellowish to reddish; ear tip tawny or yellowish; dorsal coloration yellowish or tawny wash; interorbital region downward in the lateral view; median palatal spine well defined (from the Yukon Territory east through the Labrador Peninsula and south through near the Canada-United States border) ........ *Phenacomys ungava* (p. 196)

## *Phenacomys intermedius* Merriam, 1889

Western heather vole, campañol del oeste

1. *P. i. celsus* Howell, 1923. Restricted to eastern California and the border with Nevada highlands.
2. *P. i. intermedius* Merriam, 1889. From British Columbia southeast through New Mexico.
3. *P. i. laingi* Anderson, 1942. Restricted to the central coast of British Columbia.
4. *P. i. levis* Howell, 1923. Along the border between British Columbia and Alberta, and northwestern Montana.
5. *P. i. oramontis* Rhoads, 1895. From southwestern British Columbia south through central Oregon.

**Conservation:** *Phenacomys intermedius* is listed as Least Concern by the IUCN (Cassola 2016ct).

**Characteristics:** Western heather vole is medium sized; total length 130.0–156.0 mm and skull length 22.3–25.4 mm. Dorsal pelage dark brown to gray, brown to reddish brown, but lacking the yellowish wash; fur of underparts whitish with blackish shades at the base; nose and eye rings grayish brown; tail thick, generally bicolored, blackish brown dorsally and grayish white ventrally, less than 45.0 mm long and less than 50% of the head-and-body length; **incisors with a groove; interorbital region flattened; median palatal spine not well defined**.

**Comments:** All the subespecies of *Phenacomys ungava* have been considered subspecies of *P. intermedius*. I follow Musser and Carleton (2005), who consider it a different species (Jones et al. 1997; Baker et al. 2003a). *P. intermedius* ranges are highly fragmented in western Canada and the United States. One is located within British Columbia; the second, from south-central British Columbia south through central Oregon; the third, from central British Columbia to Idaho; the fourth, in western Washington; the fifth, in central California; the sixth, in the Nevada–Utah border; the seventh, from Utah and Colorado to New Mexico (Map 137). It is allopatric with respect to other species of *Phenacomys*. It can be differentiated from *Arborimus* by having the smallest tail, less than 50.0 mm. From *P. ungava*, by having the ear tip tawny or yellowish; the interorbital region downward and with a well-developed median spine in the posterior margin of the palate.

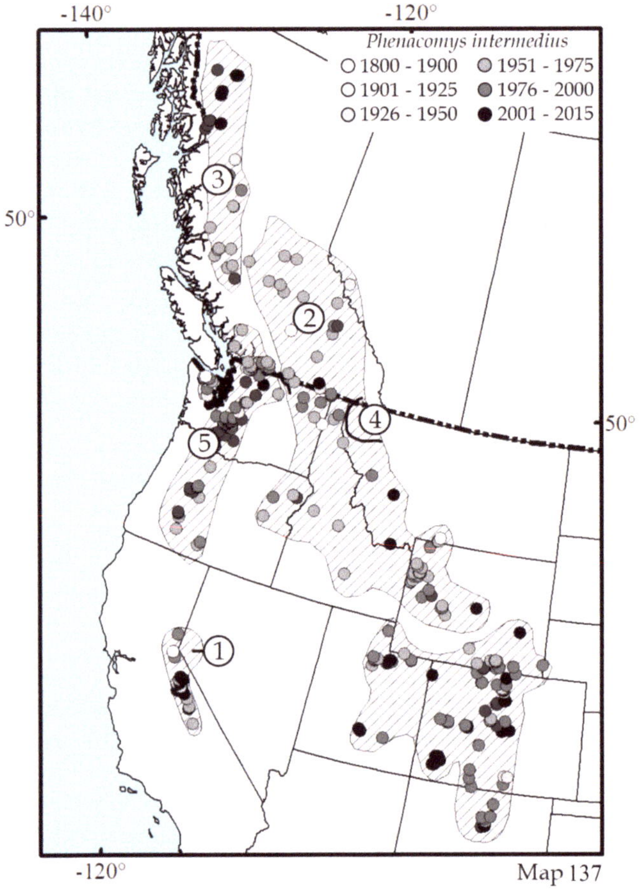


**Map 137** Distribution map of *Phenacomys intermedius*

**Additional Literature:** Anderson (1942, 1947), Voge and Bern (1949), McAllister and Hoffmann (1988).

## *Phenacomys ungava* Merriam, 1889

Eastern heather vole, campañol del este

1. *P. u. crassus* Bangs, 1900. Central-western Labrador Peninsula.
2. *P. u. mackenzii* Preble, 1902. Yukon, Alberta, and Northern Territories
3. *P. u. soperi* Anderson, 1942. Southern Alberta, Saskatchewan, and Manitoba.
4. *P. u. ungava* Merriam, 1889. From Manitoba east through Québec.

**Conservation:** *Phenacomys ungava* is listed as Least Concern by the IUCN (Cassola 2016cu).

**Characteristics:** Eastern heather vole is medium sized, total length 117.0–129.0 mm and skull length 21.3–24.4 mm in males and 24.8–26.1 mm in females. Dorsal pelage grizzled dull brown, lacking a yellowish wash; fur of underparts whitish with blackish shades at the base; nose and eye rings yellowish to reddish; tail generally bicolored, blackish brown dorsally and grayish white ventrally, less than 45.0 mm long and less than 50% of the head-and-body length; **incisors with a groove**; the **interorbital region downward in the lateral view; median palatal spine well defined**.

**Comments:** I follow Musser and Carleton (2005), who consider *Phenacomys ungava* as a distinct species (Jones et al. 1997, Baker et al. 2003a); it was previously regarded as a subspecies of *P. intermedius* (Hall 1981). *P. intermedius celatus* and all other subspecies of *P. ungava* were considered subspecies of *P. intermedius*. *P. ungava* ranges from the Yukon Territory east through the Labrador Peninsula and south through near the Canada–United States border (Map 138), and it is allopatric with respect to other species of *Phenacomys*. It can be differentiated from *Arborimus* by having the smallest tail, less than 50.0 mm. From *P. intermedius*, by having the ear tip with a pale line with yellowish hairs; interorbital region flattened and without a median spine in the posterior margin of the palate.

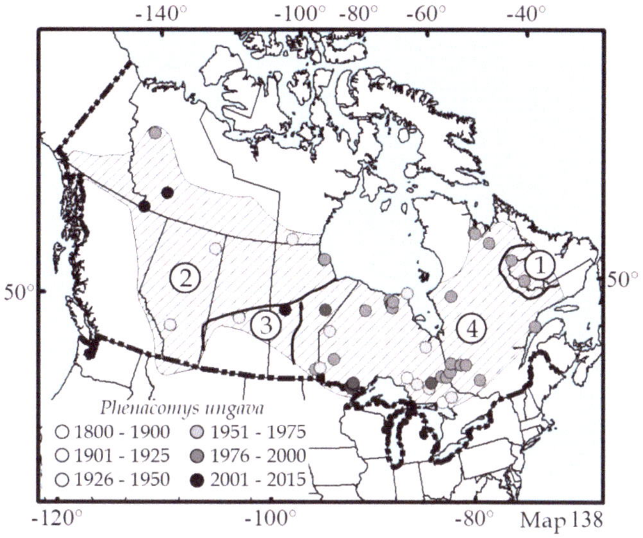


**Map 138** Distribution map of *Phenacomys ungava*

**Additional Literature:** Anderson (1942, 1947), Voge and Bern (1949), Braun et al., (2013).

## **Subfamilies Neotominae** Merriam, 1894 and **Sigmodontinae** Wagner, 1843

1. Occlusal surface of the molars flat-crowned ........ 2

1a. Occlusal surface of the molars with a well-defined cusp pattern ........ 7

2. Occlusal surface of the molars with deep reentrant angles, which may occupy approximately one-half of the tooth width (Fig. 35) ........ tribe Neotomini (p. 208)

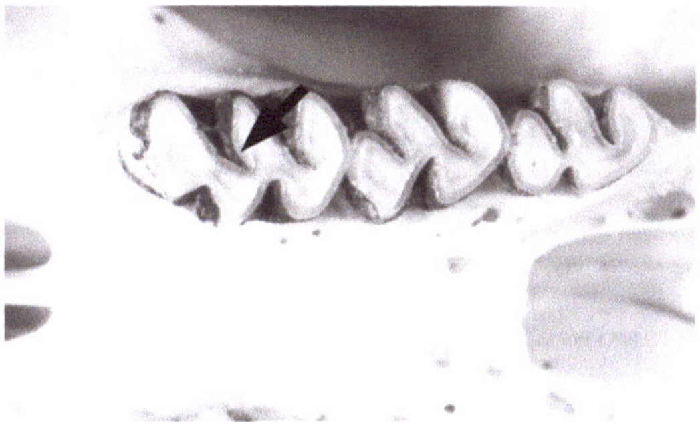

**Fig. 35** The occlusal surface of the molars with deep reentrant angles, which may occupy approximately one-half of the tooth width

2a. Occlusal surface of the molars with reentrant angles but not deep, which may not occupy approximately one-half of the tooth width ........ 6

3. Third lower molars not S shaped ........ 4

3a. Third lower molars S shaped ........ 5

4. Auditory bullae enlarged and inflated, and its main axis approximately parallel to the skull axis (Fig. 36); supraorbital rim with a prominent ridge (endemic to Jalisco and Colima) ........ *Xenomys nelsoni* (p. 385)

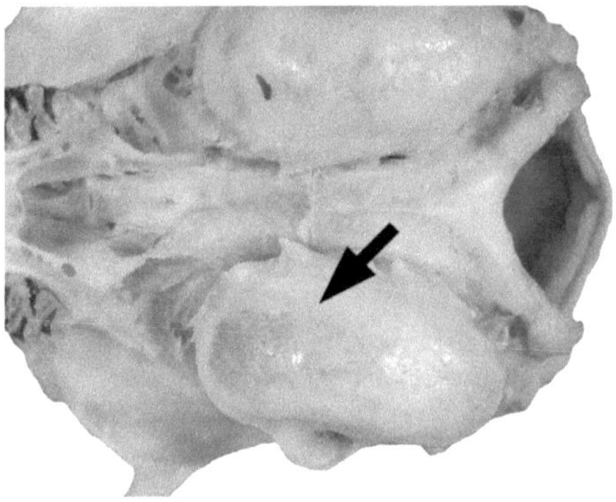

**Fig. 36** Auditory bullae enlarged and inflated, and its main axis approximately parallel to the skull axis

4a. Auditory bullae small, the main axis approximately oblique to the skull axis; no ridges in the supraorbital rim (from the lowlands of Sinaloa south through Oaxaca, including Puebla and Morelos) ........ *Hodomys alleni* (p. 215)

5. Third upper molars with an inner fold and third lower molars 8 shaped (Fig. 37) ........ *Neotoma* (p. 222)

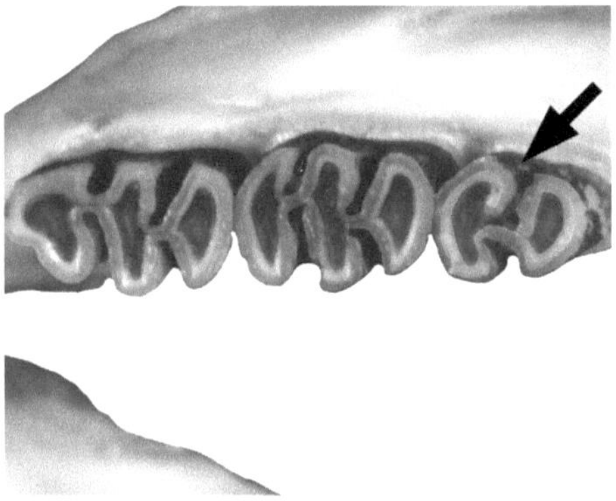

**Fig. 37** The third upper molars with an inner fold and third lower molars are eight shaped

5a. Third upper molars without an inner fold and third lower molars not 8 shaped ........ *Nelsonia* (p. 220)

6. Upper pelage coarsely grizzled; third lower molars S shaped (Fig. 38); supraorbital rim with a prominent ridge extending to the parietal and lambdoidal crest; rostrum short and broad ........ tribe Sigmodontini *Sigmodon* (p. 407)

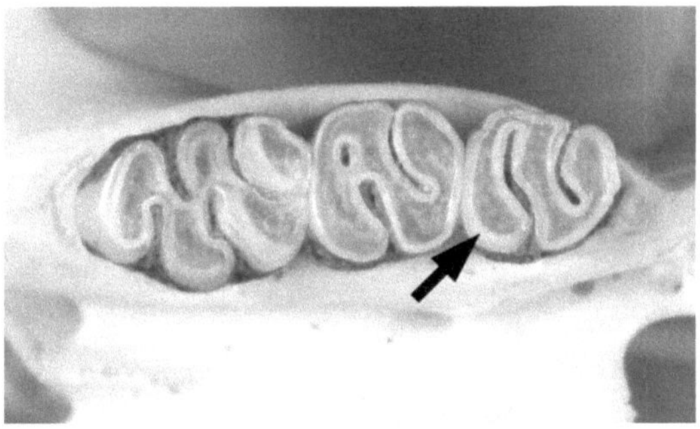

**Fig. 38** The third lower molars are S shaped

6a. Upper pelage not coarsely grizzled; third lower molars not S shaped; supraorbital rim with a ridge but not extending to the parietal and lambdoidal crest; rostrum long and thin (from eastern Michoacán east through central Veracruz) ........ *Neotomodon alstoni* (p. 254)

7. Anterior surface of upper incisors with a groove ........ *Reithrodontomys* (p. 363)

7a. Anterior surface of upper incisors without a groove ........ 8

8. Small sized; head-and-body length less than 144.0 mm; dorsal and lateral coloration dark gray or black; skull length less than 24.0 mm ........ tribe Baiomyini 9

8a. Large sized; head-and-body length greater than 144.0 mm; dorsal and lateral coloration not dark gray; skull length greater than 24.0 mm ........ 10

9. General coloration blackish; posterior extensions of external folds in the upper molars as isolated as deep (Fig. 39; from eastern Oaxaca, central Chiapas south through Guatemala) ........ *Scotinomys teguina* (p. 207)

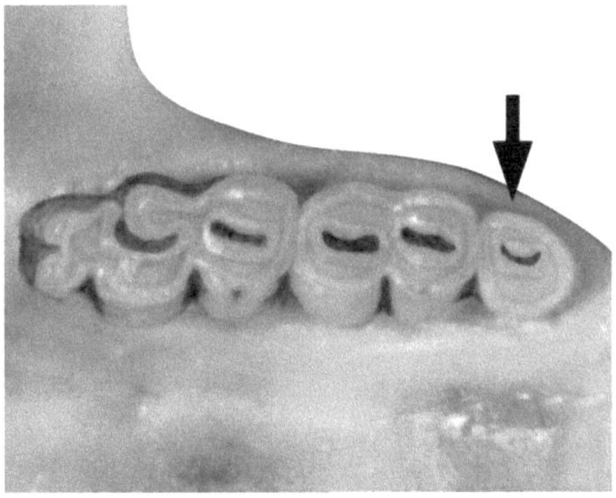

**Fig. 39** Posterior extensions of external folds in the upper molars are as isolated as deep

9a. General coloration brown or dark gray, occasionally blackish; posterior extensions of external folds in the upper molars not as isolated as deep ........ *Baiomys* (p. 203)

10. Tail short, less than the head-and-body length, flat and wide, greater than 4.0 mm in the middle; clearly bicolored; third upper molars reduced (Fig. 40); the coronoid process of mandible large ........ *Onychomys* (p. 256)

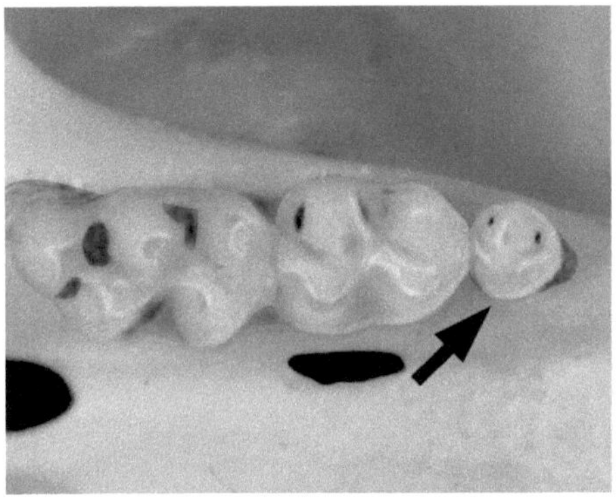

**Fig. 40** Third upper molars reduced

10a. Tail short or long but not flat and wide, less than 4.0 mm wide in the middle; bicolored or unicolor; third upper molars not reduced (Fig. 41); the coronoid process of mandible short ........................ 11

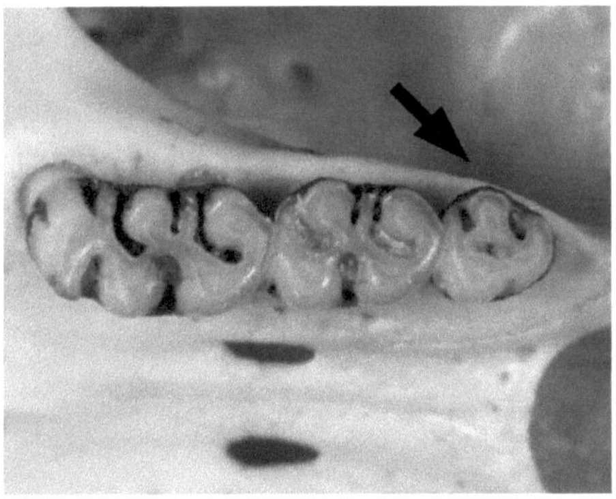

**Fig. 41** Third upper molars not reduced

11. Hind feet wide and relatively long; total length greater than 11% of the total length; palatine with two small foramens on the back (Fig. 42) ........................ 12

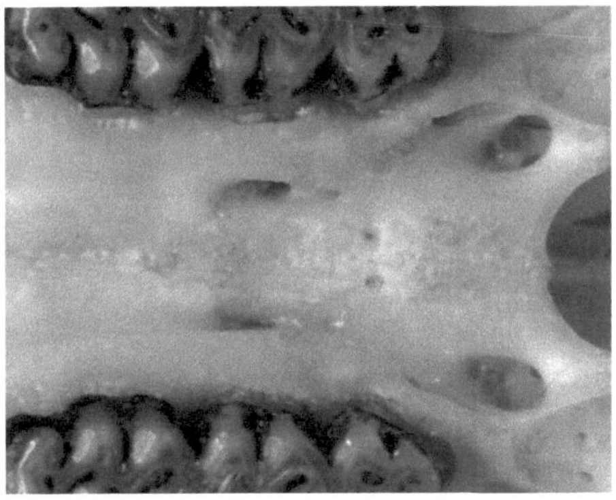

**Fig. 42** Palatine with two small foramens on the back

11a. Hind feet thin and in proportion to the size of specimens; total length less than 11% of the total length; palatine without two small foramens on the back (Fig. 43) ........................ 14

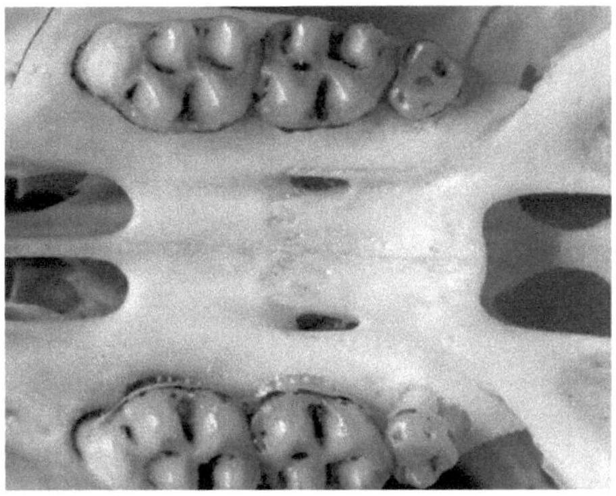

**Fig. 43** Palatine without two small foramens on the back

12. Hind feet less than 25.0 mm; second upper molars with a circular enamel island (Fig. 44); without supraorbital and temporal crests (southern Nuevo León and southern Nayarit south through Central America, including the Yucatán peninsula) ............................................................................................................................................................ *Oligoryzomys fulvescens* (p. 396)

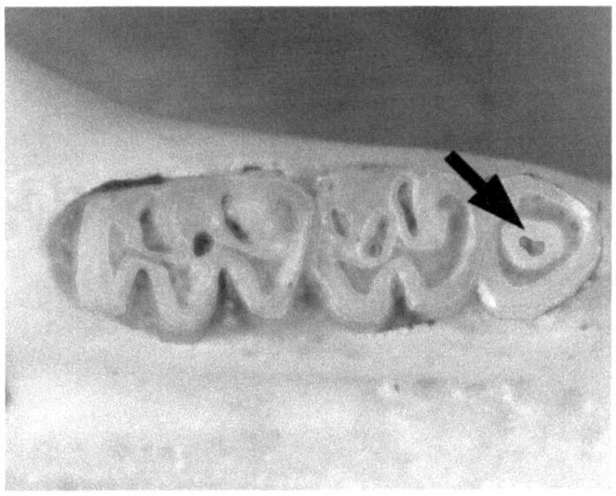

**Fig. 44** Second upper molars with a circular enamel island

12a. Hind feet greater than 25.0 mm; second upper molars with the elliptical enamel island or without it (Fig. 45); supraorbital and temporal crests ........................................................................................................................................ 13

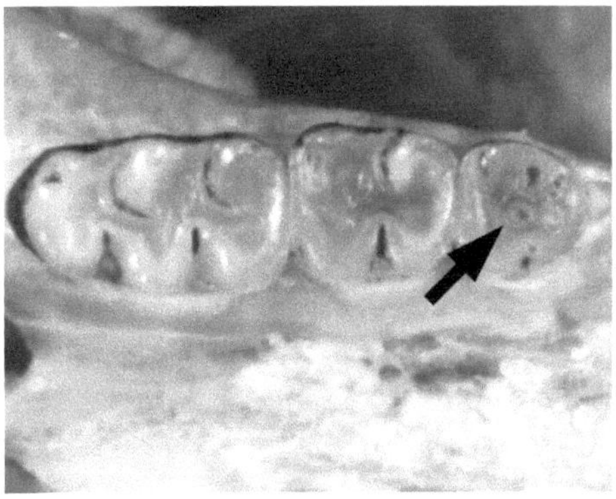

**Fig. 45** Second upper molars with elliptical enamel island or without it

13. Hair of hindfoot digits larger than the respective claw; six mammae (nipples); zygomatic notch shallow (Fig. 46); principal reentrant angles extending farther than halfway across the molar crowns.............................*Handleyomys* (p. 388)

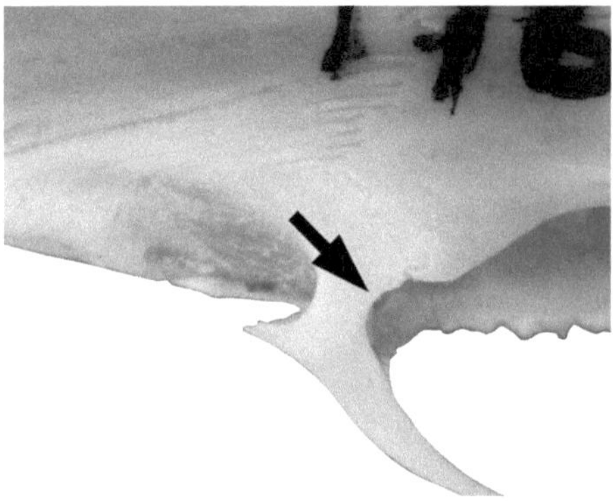

**Fig. 46** Zygomatic notch shallow

13a. Hair of hindfoot digits shorter than the respective claw; eight mammae (nipples); zygomatic notch deep (Fig. 47); principal reentrant angles normally reach less than halfway across the molar crowns ........................ *Oryzomys* (p. 397)

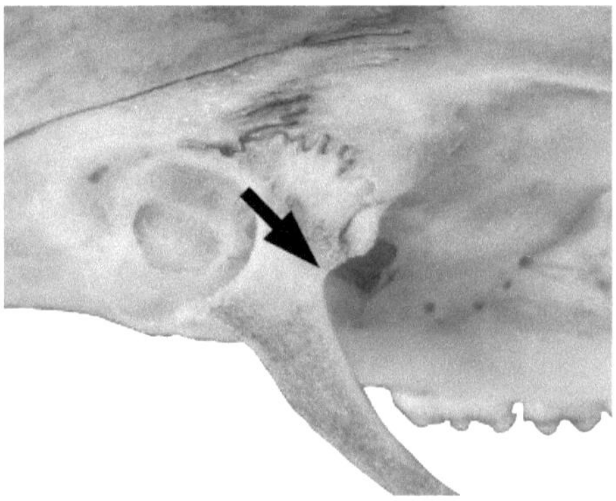

**Fig. 47** Zygomatic notch deep

14. Some digits webbed; braincase markedly flattened ............................................... tribe Ichthyyomyni *Rheomys* (p. 386)

14a. Digits not webbed; braincase not flattened ........................................................................................................ 15

15. Supraorbital rim with a prominent ridge (Fig. 48; from southern Nayarit south through Oaxaca) .................................... ........................................................................................................................ *Osgoodomys banderanus* (p. 262)

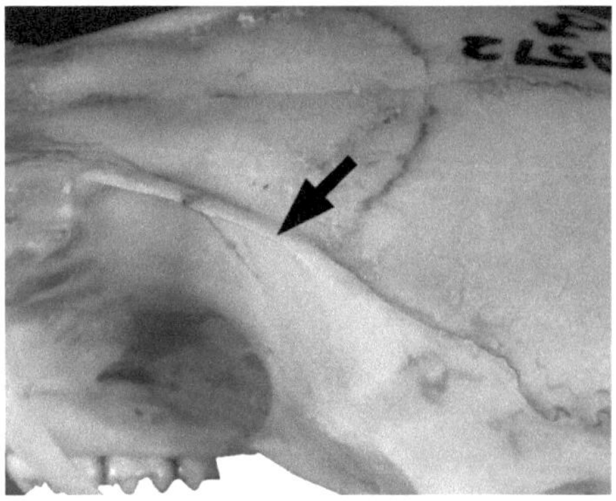

**Fig. 48** Supraorbital rim with a prominent ridge

15a. Supraorbital rim with a small or flat ridge but never prominent ...................................................................... 16

16. Total length greater than 300.0 mm ............................................................................................................... 17

16a. Total length less than 300.0 mm ........ 18

17. First and second lower molars with well-developed enamel islands; first upper molars with an almost circular enamel island ........ *Megadontomys* (p. 216)

17a. First and second lower molars with barely developed enamel islands; first upper molars without a circular enamel island ........ *Peromyscus* (part, p. 263)

(Note: In the genus *Peromyscus*, species are clustered into species groups based on their morphological similarities; they are mentioned to facilitate understanding complex taxonomical genera).

18. Five plantar tubercles on the hind feet; the transverse section of the molars never forming a loop extending to the outer edge (Florida peninsula) ........ *Podomys floridanus* (p. 361)

18a. More than five plantar tubercles on the hind feet; the transverse section of the molars forming a loop extending to the outer edge ........ 19

19. Seven plantar tubercles on the hind feet, one very rudimentary at the base of the fifth digits; hairs clothing the ears of the same color as those of upperparts; almost all stages of wear of first and second lower molars leaving five subtriangular islands of dentine (from southeastern Missouri, eastern Oklahoma, and eastern Texas east through Virginia and central Florida) ........ tribe Ochrotomyini; *Ochrotomys nuttalli* (p. 256)

19a. Six plantar tubercles on the hind feet; ears darker than the upperparts and with few hairs or naked; first and second lower molars without five subtriangular islands of dentine ........ 20

20. Tail with long hair, short, or naked; fifth digits short, smaller than the second and fourth digits; molars relatively small; dorsal profile of the skull concave ........ *Peromyscus* (part, p. 263)

(Note: In the genus *Peromyscus*, species are clustered into species groups based on their morphological similarities; they are mentioned to facilitate understanding complex taxonomical genera.)

20a. Tail without long hair; fifth digits long, similar in size to the second and fourth digits; molars relatively large; dorsal profile of the skull straight ........ *Habromys* (p. 208)

## Tribe Baiomyini

### Genus *Baiomys* True, 1894

**Dorsal pelage blackish sepia to ochraceous buff; fur of underparts slaty gray to white or pale buff;** ears relatively small and slightly rounded; hindfoot soles almost naked and with six tubercles; nasals projecting very slightly ahead of the incisors; coronoid process well developed and strongly curved; anterior palatine foramina long and usually ending posterior to the plane of the front of the first molars; interorbital space relatively wide, usually more than half the width of the widest frontal bones; upper incisors relatively heavy; accessory cusps of the first and second lower molars very small and not obvious in the transverse view until a late stage of wear; inner reentrant angle of the third lower molars relatively small and usually obliterated at an early stage of wear; the **coronoid process of the mandible large, broad, and strongly recurved**. The keys were elaborated based on the review of specimens and the following sources: Packard (1960), Hall (1981), Álvarez-Castañeda et al. (2017a), and Hernández-Canchola and León-Paniagua (2021).

1a. Hindfoot length less than 16.0 mm; skull not slightly convex in lateral view; occipitonasal length less than 19.0 mm; entoglossal process of the basilar rounded or absent (Eje Neovolcánico north through three prongs, the first through the Gulf coast to Texas, the second through the Sierra Madre eastern slope to Arizona and New Mexico, and the third through the Sierra Madre western slope to southern Sonora) ........ *Baiomys taylori* (p. 205)

1. Hindfoot length greater than 16.0 mm; skull slightly convex in lateral view; occipitonasal length greater than 19.0 mm; entoglossal process of the basilar pointed and directed anteriorly ........ 2

2. Shades in buffy brown (from southern Nayarit south through Michoacán) ........ *Baiomys musculus* (p. 205)

2a. Shades in blackish brown (from Guerrero south through Central America) ........ *Baiomys brunneus* (p. 204)

## ***Baiomys brunneus*** (Allen and Chapman, 1897)

Southern pygmy mouse, ratón pigmeo tropical

Monotypic.

**Conservation:** *Baiomys brunneus* (as *B. musculus*) is listed as Least Concern by the IUCN (Reid and Vázquez 2016a).

**Characteristics:** Southern pygmy mouse is very small sized; total length 110.0–123.0 mm and skull length 18.9–20.0 mm. Dorsal pelage dark reddish brown, yellowish ocher to nearly blackish, dorsal hair gray at the base and blackish at the tip; fur of underparts dull cream buff, sometimes grayish white on the throat; flanks usually more or less buffy; feet white or grayish white, tarsal joint slightly dusky; tail much shorter than the head-and-body length and covered with short hairs; palatal fossa anterior margin at the same level or behind the posterior margin of the last upper molars; interorbital breadth 73% of the frontal breadth; skull curved at the frontoparietal suture level in the lateral view, slightly convex.

**Comments:** *Baiomys brunneus* was recognized as a distinct species (Amman and Bradley 2004; Hernández-Canchola and León-Paniagua 2021), previously regarded as a subspecies of *B. musculus* (Russell 1968b) that also included the subspecies *B. m. brunneus*, *B. m. infernatis B. m. nigrescens*, *B. m. pallidus*. *B. brunneus* ranges from Guerrero south through Central America (Map 139). It can be found in sympatry, or nearly so, with *B. musculus* and *B. taylori*. *B. brunneus* can be distinguished from *B. taylori* by having a smaller size in the following measurements: hindfoot length 16.0 mm, occipitonasal length 19.0 mm, zygomatic breadth 10.0 mm, molars less hypsodont, cingular ridges and teeth secondary cusps pronounced. From *B. musculus*, by having the back, face, and ears blackish brown; vibrissae paler; venter darker but paler gray at the tips, becoming darker at the base; forefeet and hind feet sooty-colored to dark brown; tail paler ventrally; nasals toward the midline at the anteriormost point; zygoma less massive; smaller external and skull dimensions, and using genetic data.

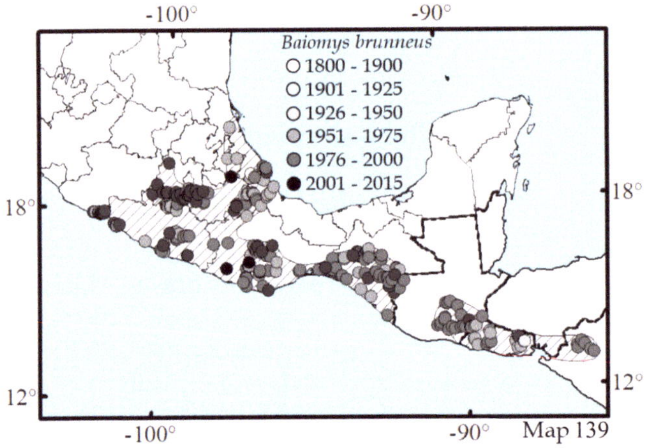


**Map 139** Distribution map of *Baiomys brunneus*

**Additional Literature:** Mearns (1907), Osgood (1909), Miller (1912a), Packard and Montgomery (1978), Yates et al. (1979a), Hall (1981), Stangl et al. (1983), Eshelman and Cameron (1987).

## *Baiomys musculus* (Merriam, 1892)

Jalisco pygmy mouse, ratón pigmeo de Jalisco

Monotypic.

**Conservation:** *Baiomys musculus* is listed as Least Concern by the IUCN (Reid and Vázquez 2016a).

**Characteristics:** Southern pygmy mouse is very small; total length 115.0–135.0 mm and skull length 19.8–20.7 mm. Dorsal pelage dark, olive-brown to buffy brown, hair gray at the base and blackish at the tip; fur of underparts and flanks pale olive-buff to gray, gray at the base and white to buff at the tip; face and head paler than the back with a greater number of buff hairs; throat and chin white through the base; forefeet and hind feet white to gray; tail faintly bicolored, blackish dorsally and whitish ventrally; nasals flared anteriorly; zygoma and zygomatic plate thick.

**Comments:** *Baiomys musculus brunneus*, *B. m. infernatis B. m. nigrescens*, and *B. m. pallidus* were previously considered a subspecies of *B. musculus* (Packard 1960), currently placed within *B. brunneus* (Amman and Bradley 2004; Hernández-Canchola and León-Paniagua 2021). *B. musculus* ranges from southern Nayarit south through Michoacán (Map 140) and can be found in sympatry, or nearly so, with *B. brunneus* and *B. taylori*. *B. musculus* can be distinguished from *B. taylori* by having larger sizes in the following measurement: hindfoot length 16.0 mm, occipitonasal length 19.0 mm, zygomatic breadth 10.0 mm, molars less hypsodont, cingular ridges and teeth secondary cusps pronounced. From *B. brunneus*, by having the back, face, and ears paler, more buff; vibrissae paler; venter paler, whitish to pale olive-buff at the tip; forefeet and hind feet whitish to pale buff; tail paler ventrally; nasals flaring outward; zygoma more massive; larger in external and cranial dimensions, and using genetic data.

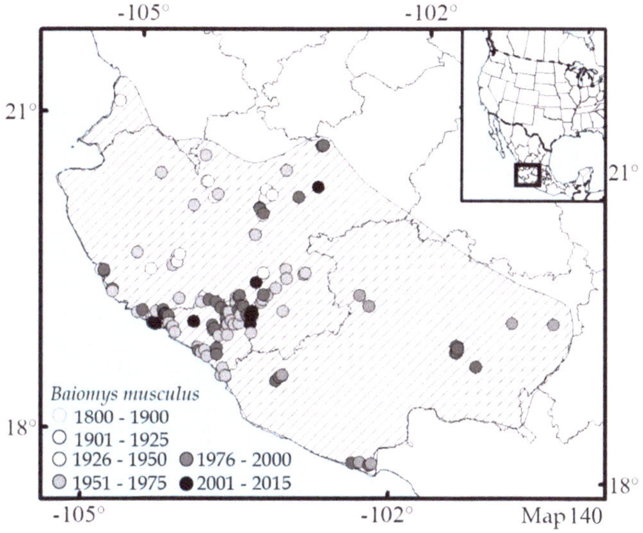


**Map 140** Distribution map of *Baiomys musculus*

**Additional Literature:** Mearns (1907), Osgood (1909), Miller (1912a), Packard and Montgomery (1978), Yates et al. (1979a), Hall (1981), Stangl et al. (1983), Eshelman and Cameron (1987).

## *Baiomys taylori* (Thomas, 1887)

Northern pygmy mouse, ratón pigmeo norteño

1. *B. t. allex* (Osgood, 1904). From southern Nayarit to southern Michoacán.
2. *B. t. analogus* (Osgood, 1909). From eastern Jalisco east through western Veracruz.
3. *B. t. ater* Blossom and Burt, 1942. Southeastern Arizona, southwestern New Mexico, northwestern Sonora, and Chihuahua.
4. *B. t. canutus* Packard, 1960. From southern Sonora south through Nayarit.

5. *B. t. fuliginatus* Packard, 1960. Restricted to southeastern San Luis Potosí.
6. *B. t. paulus* (Allen, 1903). From Chihuahua south through northern Jalisco.
7. *B. t. subater* (Bailey, 1905). Restricted to southeastern Texas.
8. *B. t. taylori* (Thomas, 1887). From western Texas south through northern Veracruz, including northwestern Coahuila.

**Conservation:** *Baiomys taylori* is listed as Least Concern by the IUCN (Timm et al. 2016c).

**Characteristics:** Northern pygmy mouse is the smallest member of the genus; total length 87.0–123.0 mm and skull length 16.8–19.2 mm. Dorsal pelage brown to pale brown with the flanks slightly more reddish; fur of underparts smoke gray washed with cream buff; ears thinly clothed with grayish hairs, of the same coloration as the back; no orbital ring or spot at the base of the whiskers; no lateral line; feet smoke gray; tarsal joints slightly dusky; tail much shorter than the head-and-body length and covered with short hairs, dull dusky dorsally and smoke gray ventrally; nasals short and broad, slightly exceeded by ascending branches of the premaxilla; palatal fossa anterior margin at the same level or behind the second upper molars**; braincase relatively broader; interorbital breadth 88% of the frontal breadth**; **skull in the lateral view not slightly convex; auditory bullae moderate**.

**Comments:** *Baiomys taylori* ranges from Eje Neovolcánico north through three prongs, the first through the Gulf coast to Texas, the second through the Sierra Madre eastern slope to Arizona and New Mexico, and the third through the Sierra Madre western slope to southern Sonora (Map 141). *B. taylori* can be distinguished from *B. musculus* by having less size in the following measurements: hindfoot length 16.0 mm, occipitonasal length 19.0 mm, zygomatic breadth 10.0 mm; in addition, molars more hypsodont, cingular ridges and teeth with secondary cusps reduced or absent, and basihyal entoglossal process reduced or absent.

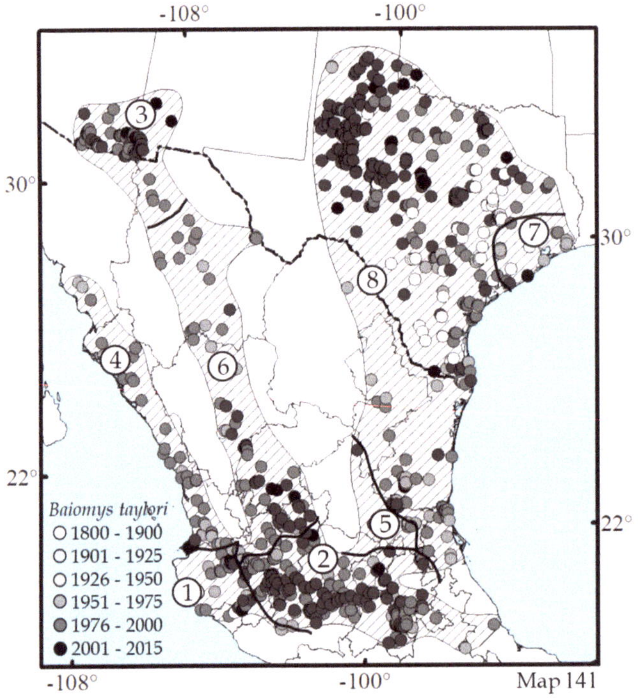


**Map 141** Distribution map of *Baiomys taylori*

**Additional Literature:** Mearns (1907), Osgood (1909), Miller (1912a), Packard (1960), Packard and Montgomery (1978), Yates et al. (1979a), Hall (1981), Stangl et al. (1983), Eshelman and Cameron (1987), Musser and Carleton (2005).

# Genus *Scotinomys* Thomas, 1913

**Small sized and almost unicolored mice, blackish-brown or dark cinnamon**; **six plantar and five palmar tubercles**; tail usually shorter than the head-and-body length and sparsely haired; hind feet narrow; molars hypsodont and narrow (in particular the first upper molars); **molar cusps adapted for piercing and crushing**; glans penis and baculum closely resembling those of specimens of *Baiomys*.

## *Scotinomys teguina* (Alston, 1877)

Short-tailed singing mouse, ratón pigmeo negro

*S. t. teguina* (Alston, 1877). Restricted to Chiapas and Guatemala.

**Conservation:** *Scotinomys teguina* is listed as Special Protection by the Norma Oficial Mexicana (DOF 2019) and as Least Concern by the IUCN (Reid et al. 2016b).

**Characteristics:** Short-tailed singing mouse is small sized; total length 115.0–144.0 mm and skull length 20.0–22.0 mm. **In México, general pelage blackish-brown including feet and tail**; hair moderately long; **eyes small, partially hidden in fur**; ears moderate in size; limbs short; six tubercles on the hind feet and five in the forefeet; six mammae (nipples); tail usually shorter than the head-and-body length and sparsely haired; skull dorsal surface smooth, without prominent ridges; rostrum short, almost of the same breath as the interparietal bone; **optical foramen equal to or larger than the sphenoidal fissure**; **zygomatic notch small or absent, so that the zygomatic plate anterior keel is barely visible, if at all, when the skull is viewed from above**; incisive foramina terminate near the anterior border of the first upper molars; some parts of the posterior extensions of the external folds in the upper molars as isolated as deep.

**Comments:** *Scotinomys teguina* is restricted to the intermediate elevations from eastern Oaxaca and central Chiapas south through Guatemala (Map 142). It can be found in sympatry, or nearly so, with species of *Baiomys*, *Handleyomys*, *Oryzomys*, *Neotoma*, *Tylomys*, *Peromyscus*, and *Reithrodontomys*. *S. teguina* can be distinguished from all other species by having a blackish-brown coloration; eyes small, partially hidden in fur; zygomatic notch small or absent, so that the zygomatic plate anterior keel is barely visible, if at all, when the skull is viewed from above; foramen optical equal to or larger than the sphenoidal fissure.

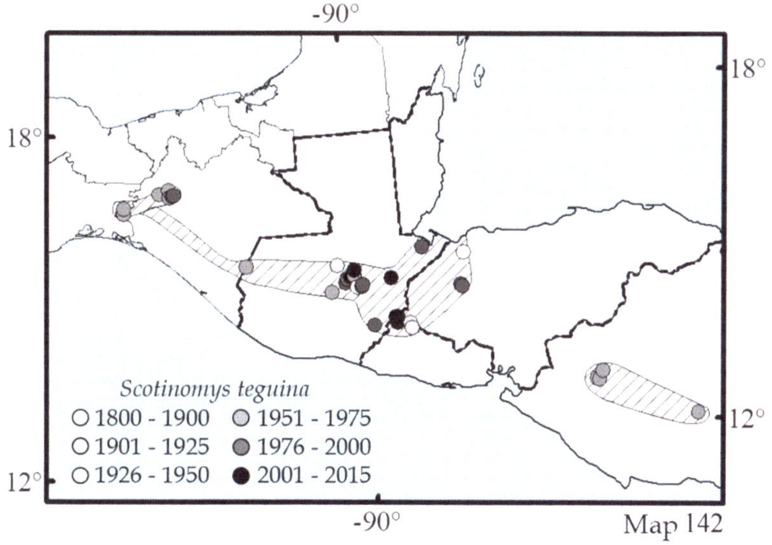


**Map 142** Distribution map of *Scotinomys teguina*

**Additional Literature:** Thomas (1913), Hall and Kelson (1959), Hooper (1972), Hall (1981), Musser and Carleton (2005).

## Tribe Neotomini

### Genus *Habromys* Hooper and Musser, 1964

Dorsal pelage reddish brown to dark brown; fur silky, fine and dense; orbital ring narrow and dusky black; ears wide and long, as much as the hindfoot length; **fifth digits long and similar in size to the second and fourth digits; tail thin and as long as the head-and-body length, covered with hair; molars proportionally larger in relation to the skull**. Previously a subgenus of *Peromyscus* (Hooper and Musser 1964), *Habromys* was elevated to the genus level based on genetic and morphological differences (Carleton 1980). The keys were elaborated based on the review of specimens and the following sources: Osgood (1909), Hall (1981), Carleton (1989), and Álvarez-Castañeda et al. (2017a).

1. Ranges east of the Tehuantepec Isthmus (Chiapas south Central America) ......................... *Habromys lophurus* (p. 212)

1a. Ranges west of the Tehuantepec Isthmus, not in Chiapas .................................................................................... 2

2. Total length less than 170.0 mm; hindfoot length less than 21.0 mm; skull length less than 24.0 mm; upper toothrow length less than 3.6 mm .................................................................................................................................... 3

2a. Total length greater than 170.0 mm; hindfoot length greater than 21.0 mm; skull length greater than 24.0 mm; upper toothrow length greater than 3.6 mm ........................................................................................................................ 4

3. Tail bicolored; rostrum breadth less than 4.0 mm; nasal length greater than 8.3 mm (endemic to southern Estado de México and northern Guerrero) ......................................................................................... *Habromys schmidlyi* (p. 213)

3a. Tail bicolored but no clear boundary between the dorsal and ventral coloration; rostrum breadth greater than 3.9 mm; nasal length less than 8.6 mm (endemic to Estado de México and northeastern Michoacán) .......................................... ........................................................................................................................... *Habromys delicatulus* (p. 209)

4. Total length less than 212.0 mm; skull length less than 26.0 mm; nasal length less than 10.0 mm; toothrow length less than 4.5 mm; zygomatic breadth less than 13.5 mm........................................................................................... 5

4a. Total length greater than 215.0 mm; skull length greater than 28.0 mm; nasal length greater than 10.0 mm; toothrow length greater than 4.5 mm; zygomatic breadth greater than 14.0 mm ............................................................. 6

5. Tail length less than 100.0 mm; skull length less than 25.0 mm; toothrow length less than 3.8 mm; zygomatic breadth less than 12.8 mm; braincase breadth less than 11.8 mm (endemic to Hidalgo and northern Oaxaca)............................ ........................................................................................................................... *Habromys simulatus* (p. 214)

5a. Tail length greater than 100.0 mm; skull length greater than 25.0 mm; toothrow length greater than 3.8 mm; zygomatic breadth greater than 12.8 mm; braincase breadth greater than 11.8 mm (endemic to Sierra de Juárez, northern Oaxaca) ............................................................................................................. *Habromys chinanteco* (p. 208)

6. Tail unicolor; metatarsal regions with a dark spot generally present; skull slightly concave in lateral view (endemic to Cerro Zempoaltépetl, Oaxaca) ..................................................................................... *Habromys lepturus* (p. 211)

6a. Tail usually bicolored; dark spots on the metatarsal region usually absent; skull slightly straight in lateral view (endemic to Sierra de Juárez, Oaxaca)........................................................................................... *Habromys ixtlani* (p. 210)

### *Habromys chinanteco* (Robertson and Musser, 1976)

Chinanteco crested-tailed mouse, ratón arbóreo de Oaxaca

Monotypic.

**Conservation:** *Habromys chinanteco* is listed as **Critically Endangered** (B1ab (iii); extent of occurrence less than 100 km$^2$ in only one population and under continuing decline in area and quality of habitat) by the IUCN (Álvarez-Castañeda et al. 2018b).

**Characteristics:** Chinanteco crested-tailed mouse is medium sized; total length 192.0–212.0 mm and skull length 25.6–26.5 mm. Dorsal pelage grayish brown; lateral line well defined; fur of underparts grayish white; lateral line pale ocher; hair thin and silky; orbital ring narrow and dusky black; hind feet gray brown with whitish digits; tail very thin and as long as the head-and-body length, covered with hair distally, unicolor with a hair tuft at the tip; a spot around the eyes and whiskers dark; skull narrow and nasal bones shorts.

**Comments:** Carleton et al. (2002) considered *Habromys chinanteco* as a possible junior synonym of *H. simulatus*. *H. chinanteco* is only found on Cerro Pelón in the Sierra de Juárez, Oaxaca (Map 143). All species of *Habromys* have allopatric distribution. Species of *Habromys* are very similar to those of *Peromyscus* and can be distinguished by having the fifth digits long, similar in size to the second and fourth digits; molars relatively large, and dorsal skull profile straight.

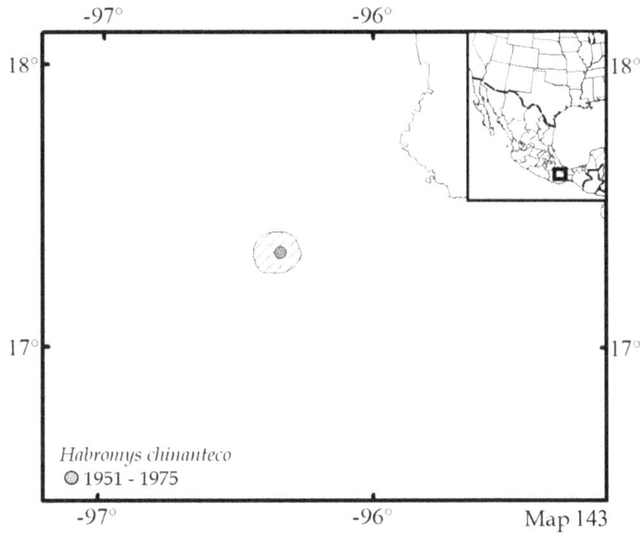


**Map 143** Distribution map of *Habromys chinanteco*

**Additional Literature:** Robertson and Musser (1976), Carleton et al. (2002), Musser and Carleton (2005).

## *Habromys delicatulus* Carleton et al., 2002

Delicate crested-tailed mouse, ratón arbóreo delicado

Monotypic.

**Conservation:** *Habromys delicatulus* is listed as Endangered (B1ab (i, ii, iii, v) + 2ab (i, ii, iii, v): extent of occurrence and quality of habitat, and extent of occurrence less than 5000 km$^2$ severely fragmented and under continuing decline in extent of occurrence, area of occupancy and quality of habitat, and number of subpopulations, and area of occupancy less than 500 km$^2$ and an observed, estimated or projected continuing decline of at least 20% in 5 years, the number of mature individuals less than 250) by the IUCN (Vázquez 2017a).

**Characteristics:** Delicate crested-tailed mouse is small sized, the smallest species of *Habromys*; total length 148.0–163.0 mm and skull length 22.1–23.3 mm. Dorsal pelage brown without dark hair toward the back central part; lateral line yellowish; fur of underparts grayish white; rostrum flanks black-brown to the spot around the eyes; hair silky, fine and dense, lateral line pale ocher; orbital ring narrow and dusky black; ears wide and long, as much as the hindfoot length; hairs on the chin and neck with whitish shades at the base; digits whitish distally; rostrum short; rostrum slender with delicate and fragile zygomatic arches; nasal bones short and narrow; zygomatic plate narrow.

**Comments:** *Habromys delicatulus* is only known from Cañada de La Ermita, Estado de México, and Cerro Garnica, northeastern Michoacán (Map 144). All species of *Habromys* have allopatric distribution. Species of *Habromys* are very similar to those of *Peromyscus* and can be distinguished by having the fifth digits long, similar in size to the second and fourth digits; molars relatively large, and dorsal skull profile straight.

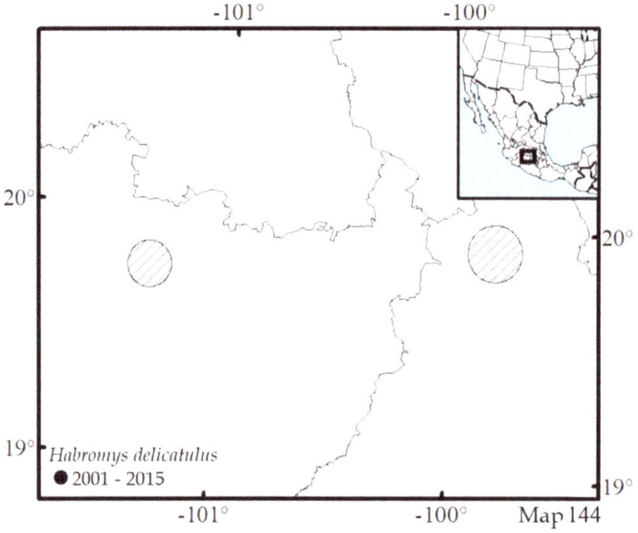


**Map 144** Distribution map of *Habromys delicatulus*

**Additional Literature:** Carleton et al. (2002), León-Paniagua et al. (2007), Musser and Carleton (2005).

## *Habromys ixtlani* (Goodwin, 1964)

Ixtlán crested-tailed mouse, ratón arbóreo de Ixtlán

Monotypic.

**Conservation:** *Habromys ixtlani* is listed as **Critically Endangered** (B1ab (iii); extent of occurrence less than 100 km$^2$ at only one location and under continuing decline in area, extent and quality of habitat) by the IUCN (Álvarez-Castañeda et al. 2018m).

**Characteristics:** Ixtlán crested-tailed mouse is medium sized; total length 210.0–280.0 mm and skull length 29.6–31.8 mm. Dorsal pelage cinnamon brown grizzly with dark shades, mainly on the back; fur of underparts grayish white; flanks and shoulders cinnamon yellowish; ears blackish brown with thin hairs and a whitish edge; hair on the limbs grayish brown with digits whitish; toes and sides of the hind feet whitish, digits top dark; **tail longer than the head-and-body length, evenly bicolored dusky dorsally and paler ventrally;** dark spot around the eyes and whiskers blackish brown; rostrum large; nasal bones long and wide.

**Comments:** *Habromys ixtlani* was recognized as a distinct species (Carleton et al. 2002), previously regarded as a subspecies of *H. lepturus* (Musser 1969). *Habromys ixtlani* is only known from Cerro Machín, in Sierra de Juárez, Oaxaca (Map 145). All species of *Habromys* have allopatric distribution. Species of *Habromys* are very similar to those of *Peromyscus* and can be distinguished from them by having the fifth digits long, similar in size to the second and fourth digits; molars relatively large, and dorsal skull profile straight.

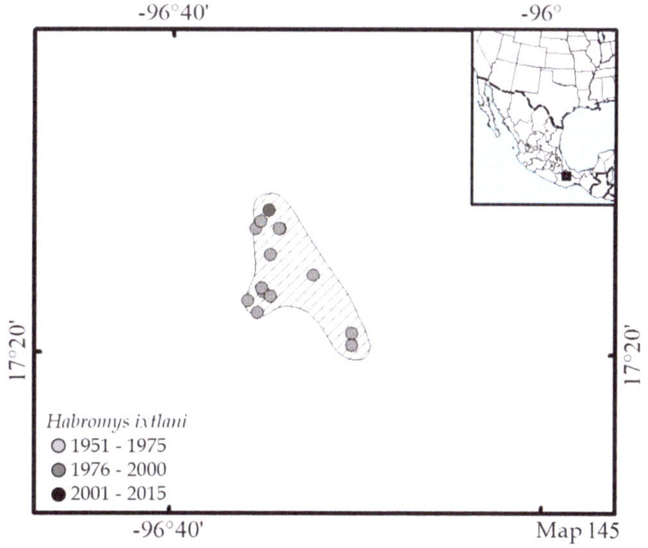


**Map 145** Distribution map of *Habromys ixtlani*

**Additional Literature:** Goodwin (1964), León-Paniagua et al. (2007), Musser and Carleton (2005).

## *Habromys lepturus* (Merriam, 1898)

Zempoaltépetl crested-tailed mouse, ratón arbóreo de Zempoaltépetl

Monotypic.

**Conservation:** *Habromys lepturus* is listed as **Critically Endangered** (B1ab (iii); extent of occurrence less than 100 km$^2$ at only one location and under continuing decline in area and quality of habitat) by the IUCN (Álvarez-Castañeda et al. 2018i).

**Characteristics:** Zempoaltépetl crested-tailed mouse is medium sized; total length 216–262 mm and skull length 28.8–32.6 mm. Dorsal pelage brown palely mixed with cinnamon, with a darker dorsal stripe; fur of underparts white; flanks, shoulders, and head cinnamon to russet, mixed with brownish black; ears thinly clothed with soft brown hairs, scarcely or not at all, edged with paler, with a black hair tuft at the anterior base of the ears; nose side area through the base of the whiskers to and around eye black or brownish black; toes and sides of the hind feet whitish, digits dark at the top; **soles hairy posteriorly; tail about as long as the head-and-body length, rather coarsely haired,** unicolor and slightly penicillate; **with rather long nasal bones, constricted frontal bones; large interparietal bone and relatively large teeth**; rostrum short; nasal long and thin.

**Comments:** *Habromys ixtlani* was recognized as a distinct species (Carleton et al. 2002), previously regarded as a subspecies of *H. lepturus* (Musser 1969). *Habromys lepturus* is only known from Cerro Zempoaltépetl in Oaxaca (Map 146). All species of *Habromys* have allopatric distribution. Species of *Habromys* are very similar to those of *Peromyscus* and can be distinguished by having the fifth digits long, similar in size to the second and fourth digits; molars relatively large, and dorsal skull profile straight.

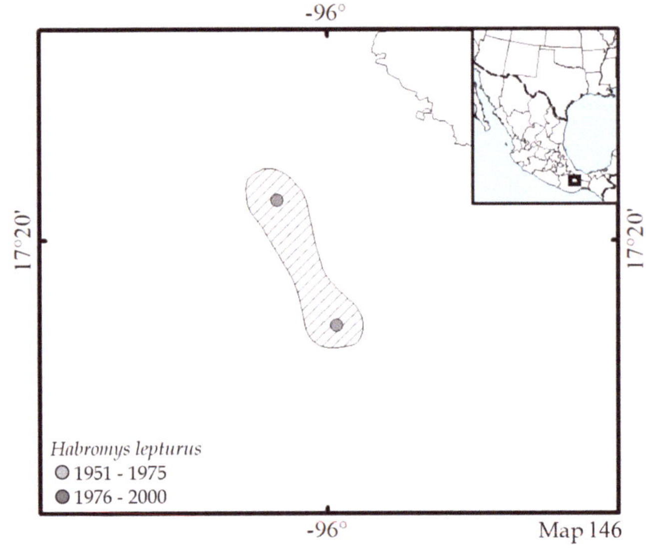


**Map 146** Distribution map of *Habromys lepturus*

**Additional Literature:** Merriam (1898a), Osgood (1909), Hooper (1958), Carleton et al. (2002), León-Paniagua et al. (2007), Musser and Carleton (2005).

## *Habromys lophurus* (Osgood, 1904)

Crested-tailed mouse, ratón arbóreo de cola crestada

Monotypic.

**Conservation:** *Habromys lophurus* is listed as Near Threatened by the IUCN (Reid et al. 2008a).

**Characteristics:** Crested-tailed mouse is medium sized; total length 187.0–230.0 mm and skull length 25.4–28.1 mm. Dorsal pelage from wood brown to beige, with a small dusky area in the middle of the back; fur of underparts whitish, without a pectoral spot; lateral line pale ochraceous buff; orbital ring dusky black, rather narrow but expanded into a distinct spot in front of the eyes; forefeet whitish and hind feet dark brownish to the base of the toes; forearms dusky to the wrists; digits whitish; **tail long, covered with comparatively long soft hairs, terminating in a distinct pencil; pelage soft and "woolly," rather dull and lusterless**; **with a large interparietal bone and short nasals;** rostrum short.

**Comments:** *Habromys lophurus* ranges in the highlands of Chiapas south through Central America (Map 147). All species of *Habromys* have allopatric distribution. Species of *Habromys* are very similar to those of *Peromyscus* and can be distinguished from them by having the fifth digits long, similar in size to the second and fourth digits; molars relatively large, and dorsal skull profile straight.

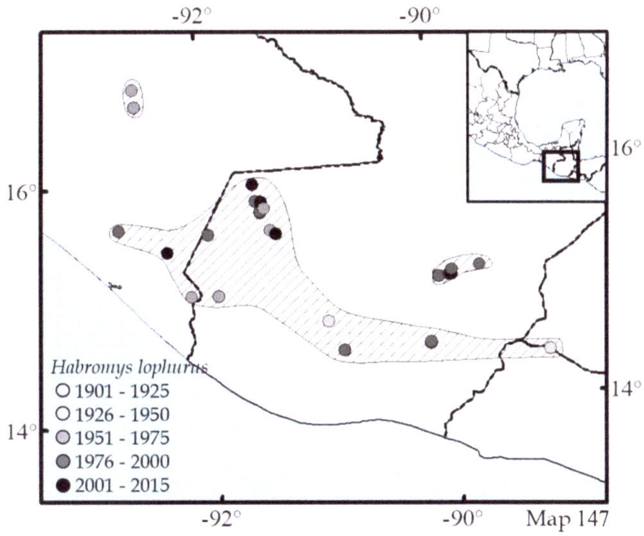


**Map 147** Distribution map of *Habromys lophurus*

**Additional Literature:** Osgood (1904, 1909), Hooper (1958), Robertson and Musser (1976), Carleton et al. (2002), Musser and Carleton (2005).

## *Habromys schmidlyi* Romo-Vázquez et al., 2005

Schmidly crested-tailed mouse, ratón arbóreo de Guerrero

Monotypic.

**Conservation:** *Habromys schmidlyi* is listed as **Critically Endangered** (B1ab (iii); extent of occurrence less than 100 km$^2$ at only one location and under continuing decline in area and quality of habitat) by the IUCN (Álvarez-Castañeda et al. 2018c).

**Characteristics:** Schmidly crested-tailed mouse is small sized; total length 144.0–167.0 mm and skull length 22.8–24.0 mm. Dorsal pelage brown with a darker dorsal stripe; fur of underparts white; lateral line pale ocher; orbital ring narrow and dusky black; digits whitish at the tip; tail with relatively long hair, clearly bicolored (black dorsally and white ventrally) and one hair tuft (about 6.0 mm); rostrum short; nasal bones short and narrow; the zygomatic plate wider than the postpalatal region; mesopterygoidea fossa thin and short.

**Comments:** *Habromys schmidlyi* is only known from Sierra de Taxco on the border of Guerrero and Estado de México (Map 148). All species of *Habromys* have allopatric distribution. Species of *Habromys* are very similar to those of *Peromyscus* and can be distinguished from them by having the fifth digits long, similar in size to the second and fourth digits; molars relatively large, and dorsal skull profile straight.

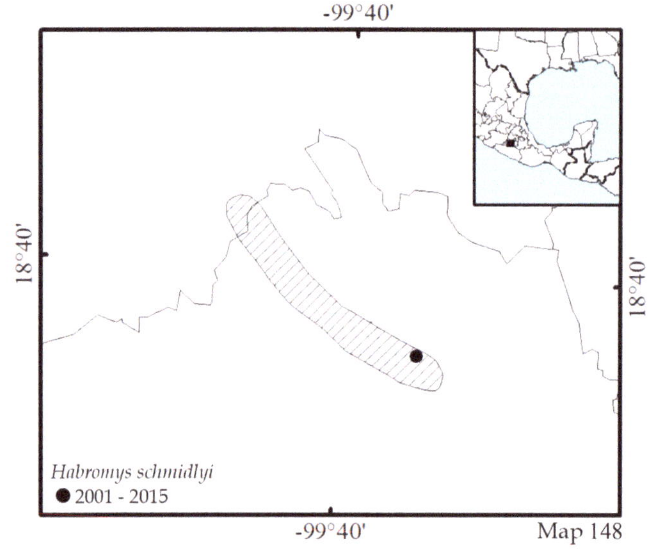


**Map 148** Distribution map of *Habromys schmidlyi*

**Additional Literature:** Romo-Vázquez et al. (2005), León-Paniagua et al. (2007), Musser and Carleton (2005).

## *Habromys simulatus* (Osgood, 1904)

Jico crested-tailed mouse, ratón arbóreo de Xico

Monotypic.

**Conservation:** *Habromys simulatus* is listed as Special Protection by the Norma Oficial Mexicana (DOF 2019) and as Critically Endangered (C2a (i, ii); D: small population size with less than 250 mature individuals under a continuing decline in numbers of mature individuals and population structure with no more than 50 mature individuals or at least 90% in one subpopulation) by the IUCN (Vázquez 2018a).

**Characteristics:** Jico crested-tailed mouse is small sized; total length 178–197 mm and skull length 23.8–25.0 mm. Dorsal pelage brown with a darker dorsal stripe; fur of underparts white, without a pectoral spot; lateral line pale ocher; orbital ring dusky black, rather narrow but expanded into a distinct spot in front of the eyes; forefeet whitish and hind feet dark brownish to the base of the toes; **dark markings of feet and face slightly more intense;** forearms dusky to the wrist; digits whitish; tail covered with hair and crested, bicolored, chiefly brown dorsally and with a narrow white line ventrally, with a hair tuft at the tip; **skull and teeth very small**; braincase inflated and rostrum depressed; **auditory bullae relatively large**; interorbital constriction relatively wide; rostrum short.

**Comments:** Carleton et al. (2002) considered *Habromys chinanteco* as a possible junior synonym of *H. simulatus*. *H. simulatus* is known from two disconnected areas, middle slopes of Sierra Madre Oriental from central Veracruz, Hidalgo, and Puebla (Map 149). All species of *Habromys* have allopatric distribution. Species of *Habromys* are very similar compared to those of *Peromyscus* and can be distinguished from them by having the fifth digits long, similar in size to the second and fourth digits; molars relatively large, and dorsal skull profile straight.

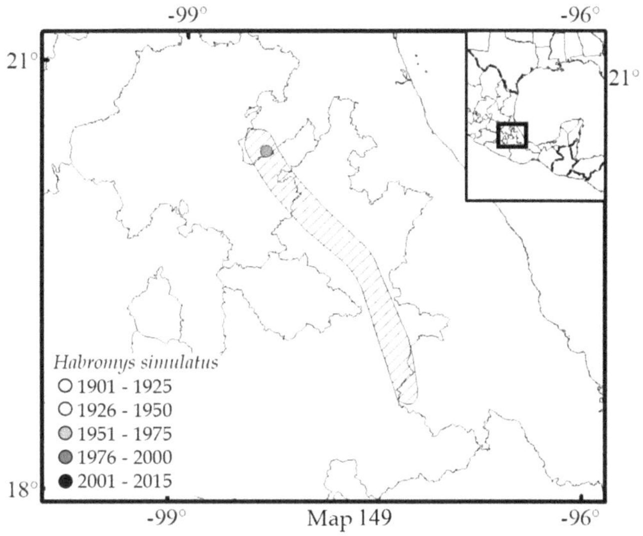


**Map 149** Distribution map of *Habromys simulatus*

**Additional Literature:** Osgood (1904, 1909), Hooper (1958), Musser and Carleton (2005).

## Genus *Hodomys* Merriam, 1894

Dorsal pelage reddish to dark brown; fur of underparts leaden gray with whitish spots sometimes yellowish; few hairs on the tail and shorter than the head-and-body length, without being totally bicolored but darker dorsally; **interorbital region narrow**; no ridges in the supraorbital rim; **incisive foramen extending to the anterior margin of the first maxillary molars**; **auditory bullae smaller in proportion to the skull, and its main axis approximately oblique to the skull axis**; first maxillary molars with two internal and external folds; second upper molars with two external and internal folds; **lower third molars S shaped.**

### *Hodomys alleni* (Merriam, 1892)

Allen's woodrat, rata de campo del occidente

1. *H. a. alleni* (Merriam, 1892). From southern Sinaloa south through Michoacán.
2. *H. a. elatturus* Osgood, 1938. Inland of Michoacán and Guerrero, Estado de México, and Morelos.
3. *H. a. guerrerensis* Goldman, 1938. Restricted to coastal Guerrero.
4. *H. a. vetulus* Merriam, 1894. Southern Puebla, northern Oaxaca, and northeastern Guerrero.

**Conservation:** *Hodomys alleni* is listed as Least Concern by the IUCN (Álvarez-Castañeda et al. 2016o).

**Characteristics:** Allen's woodrat is medium sized; total length 368.0–446.0 mm and skull length 50.2–55.8 mm. The characteristics of the species are the same that those mentioned for the genus.

**Comments:** *Hodomys* has been regarded as a genus (Goldman 1910; Ellerman 1941; Carleton 1980; Edwards and Bradley 2003) and as a subgenus of *Neotoma* (Burt and Barkalow 1942). *H. alleni* ranges from the lowlands of southern Sinaloa south through Oaxaca, including the Río Balsas Basin to Puebla and northern Oaxaca (Map 150). It can be found in sympatry, or nearly so, with *Neotoma mexicana*, *N. phenax*, and *Xenomys nelsoni*. *H. alleni* can be distinguished from all other species by the enamel pattern of the third lower molar chewing-surface not S shaped and incisive foramen extended to the anterior margin of the first maxillary molars.

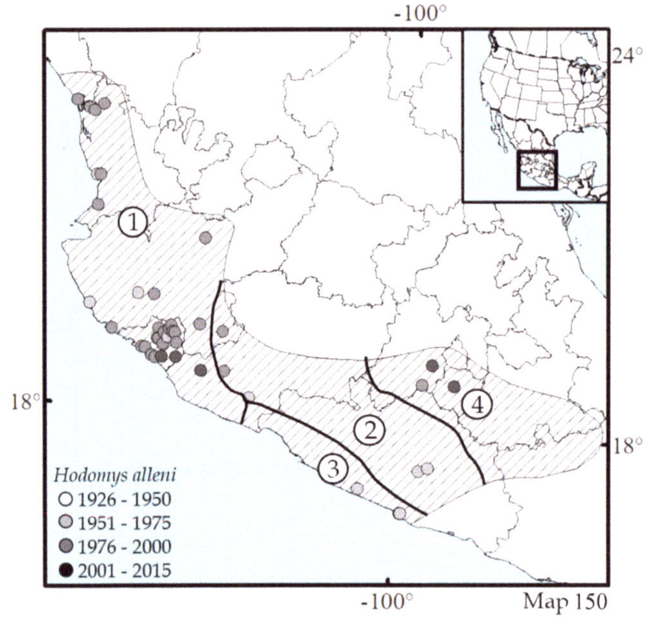


**Map 150** Distribution map of *Hodomys alleni*

**Additional Literature:** Merriam (1894b), Birney and Jones (1972), Genoways and Birney (1974), Musser and Carleton (2005), Bradley et al. (2022b).

## Genus *Megadontomys* Merriam, 1898

Dorsal pelage brown to dark brown; ears and tail long and very scantily haired; hair long, soft, and very dense; skull very elongated, mainly in the rostrum; nasals expanded anteriorly and projecting far beyond the incisors; molars very large and heavy, low, and usually worn flat at an early stage; **first and second upper and lower molars with a supplementary narrow enamel loop on each side, when molars slightly worn, first upper molars with five salient and four reentrant outer angles;** first upper molars with a subcircular enamel island; first and second lower molars prominent and well developed. *Megadontomys* is considered at the genus level (Carleton 1980); it was previously considered a subgenus of *Peromyscus* (Osgood 1909). The keys were elaborated based on the review of specimens and the following sources: Osgood (1909), Hall (1981), Carleton (1989), and Álvarez-Castañeda et al. (2017a).

1. Dorsal surface of the limbs white; ear length greater than 23.0 mm (endemic to Guerrero) ........................................ *Megadontomys thomasi* (p. 219)

1a. Dorsal surface of the limbs dark; ear length less than 23.0 mm ........................................ 2

2. Hair of the flanks ocher brown contrasting with the dark color of the back; upper toothrow length generally greater than 6.5 mm; anterior edge of the zygomatic plate straight or slightly concave (endemic to Hidalgo, Veracruz, and Puebla).... ........................................ *Megadontomys nelsoni* (p. 218)

2a. Hair of the flanks dark and not contrasting with the dark color of the back; upper toothrow length generally equal to or less than 6.5 mm; anterior edge of the zygomatic plate concave (endemic to Sierra de Juárez, Oaxaca)........................ ........................................ *Megadontomys cryophilus* (p. 217)

## *Megadontomys cryophilus* (Musser, 1964)

Oaxacan giant deermouse, ratón gigante de Oaxaca

Monotypic.

**Conservation:** *Megadontomys cryophilus* is listed as Threatened by the Norma Oficial Mexicana (DOF 2019) and as Endangered (B1ab (iii): extent of occurrence less than 5000 km², severely fragmented or known to exist at no more than five locations, and under continuing decline in area, extent, and quality of habitat) by the IUCN (Álvarez-Castañeda 2018a).

**Characteristics:** Oaxacan giant deermouse is similar to members of *Peromyscus* but larger; total length 300.0–331.0 mm and skull length 35.8–37.5 mm. **Dorsal pelage blackish and flanks dark brown**; fur of underparts creamy white but with a greater admixture of slaty undercolor, especially in the neck; dusky areas around the eyes and at base of the vibrissae more extensive; cheeks darker with fewer whitish-tipped hairs; limb dorsal region darker; ears small; rostrum wide; rostrum long and thin; crest in the supraorbital rim well developed, nasals long; zygomatic region widely flaring; auditory bullae small; differentiation of the supraorbital margins of the frontal bones variable; **zygomatic notch deeper; anterior zygomatic plate edge concave**.

**Comments:** *Megadontomys cryophilus* was recognized as a distinct species (Carleton 1989; Vallejo and González-Cózatl 2012; Werbitsky and Kilpatrick 1987); it was previously regarded as a subspecies of *M. thomasi* (Musser 1964). *M. cryophilus* only ranges in highlands of northern Oaxaca from 2400 to 3500 m (Map 151), with allopatric distribution with respect to the other species of *Megadontomys. M. cryophilus* can be distinguished from *M. thomasi* by having the tail slightly longer; ears smaller in all dimensions; upperparts darker with greater black-tipped hairs and with ochraceous hairs with paler ochraceous-buff at the tip along the flanks, with little or no differentiation from the back; zygomatic plate anterior edge convex; braincase slightly less inflated; auditory bullae smaller on average; zygomatic region slightly wider; nasals, lower, and upper molar rows longer.

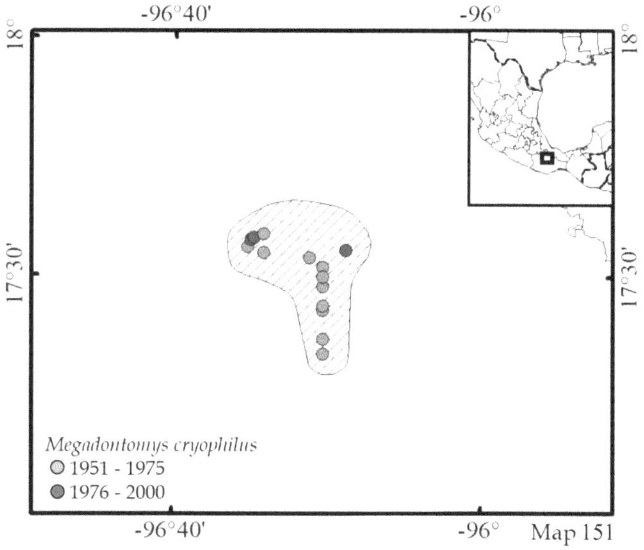


**Map 151** Distribution map of *Megadontomys cryophilus*

**Additional Literature:** Merriam (1898a), Musser and Carleton (2005), Vallejo et al. (2017).

## ***Megadontomys nelsoni*** (Merriam, 1898)

Nelson's giant deermouse, ratón gigante de Veracruz

Monotypic.

**Conservation:** *Megadontomys nelsoni* is listed as Threatened by the Norma Oficial Mexicana (DOF 2019) and as Endangered (B1ab (iii): extent of occurrence less than 5000 km$^2$, severely fragmented or known to exist at no more than five locations, and under continuing decline in area, extent, and quality of habitat) by the IUCN (Álvarez-Castañeda et al. 2018l).

**Characteristics:** Nelson's giant deermouse is similar to members of *Peromyscus* but larger; total length 318.0 mm and skull length 360.5 mm. **Dorsal pelage dark brown contrasting with ocher brown in the flanks**; fur of underparts creamy white; base of the whiskers and around the eyes dusky; ears large; forefoot white; forearms dusky to the wrists; hind feet grayish dusky; toes white, tarsal joint broadly brownish; legs dorsal region whitish; tail unicolor, dusky all around and longer than the head-and-body length; rostrum long and thin; **supraorbital margins elevated**; **anterior zygomatic plate edge straight or slightly concave; without a well developed crest in the supraorbital rim**.

**Comments:** *Megadontomys nelsoni* was recognized as a distinct species (Werbitsky and Kilpatrick 1987; Carleton 1989; Vallejo and González-Cózatl 2012); it was previously regarded as a subspecies of *M. thomasi* (Musser 1964). *M. nelsoni* only ranges in the eastern slopes of Sierra Madre Oriental, from southeastern Puebla, Hidalgo, and central Veracruz (Map 152), with allopatric distribution with respect to the other species of *Megadontomys. M. nelsoni* can be distinguished from *M. thomasi* by having a less massive skull; lacking the supraorbital ridges and a less pronounced posterior reentrant angle in the last lower molars.

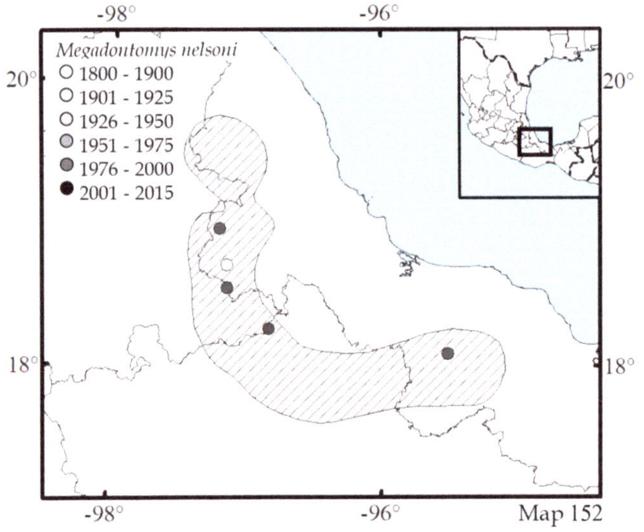


**Map 152** Distribution map of *Megadontomys nelsoni*

**Additional Literature:** Merriam (1898a), Musser and Carleton (2005), Vallejo et al. (2017).

## *Megadontomys thomasi* (Merriam, 1898)

Thomas's giant deermouse, ratón gigante de Guerrero

Monotypic.

**Conservation:** *Megadontomys thomasi* is listed as Special protection by the Norma Oficial Mexicana (DOF 2019) and as Endangered (B1ab (iii): extent of occurrence less than 5000 $km^2$, severely fragmented or known to exist at no more than five locations, and under continuing decline in area, extent, and quality of habitat) by the IUCN (Álvarez-Castañeda and Castro-Arellano 2019a).

**Characteristics:** Thomas's giant deermouse is similar to members of *Peromyscus* but larger; total length 295.0–351.0 mm and skull length 34.7–37.6 mm. **Dorsal pelage reddish brown and dark brown occasionally**; fur of underparts cream-white; blackish spot around the eyes; **limb dorsal region whitish**; **ears large and minutely hairy, appearing almost naked, length greater than 23.0 mm**; **tail longer than the head-and-body length, nearly unicolor and closely covered with short bristly hairs, which do not quite conceal the annulations**; **hindfoot soles naked to the calcaneus**; whiskers large and long, reaching the shoulders; rostrum long and thin; **crest in the supraorbital rim well developed**; interparietal bone very large and broad, subtriangular; incisive foramen very large.

**Comments:** *Megadontomys thomasi* was recognized as a distinct species (Werbitsky and Kilpatrick 1987; Carleton 1989; Vallejo and González-Cózatl 2012); it previously included *M. cryophilus* and *M. nelsoni* as subspecies. *M. thomasi* is endemic to Guerrero (Map 153), with allopatric distribution with respect to the other species of *Megadontomys*. *M. thomasi* can be distinguished from *M. cryophilus* by having the upperparts darker with greater black-tipped hairs and with the hairs along the flanks deep ochraceous-tawny at the tip, with little contrast with the tawny and blackish flanks; anterior edge of the zygomatic plate straight or slightly concave; braincase slightly more inflated, and auditory bullae larger on average; the zygomatic region slightly narrow; nasals and upper molar toothrow shorter. From *M. nelsoni*, by its great massiveness, with prominent supraorbital ridges, and a stronger development of the posterior reentrant angle of the last lower molars.

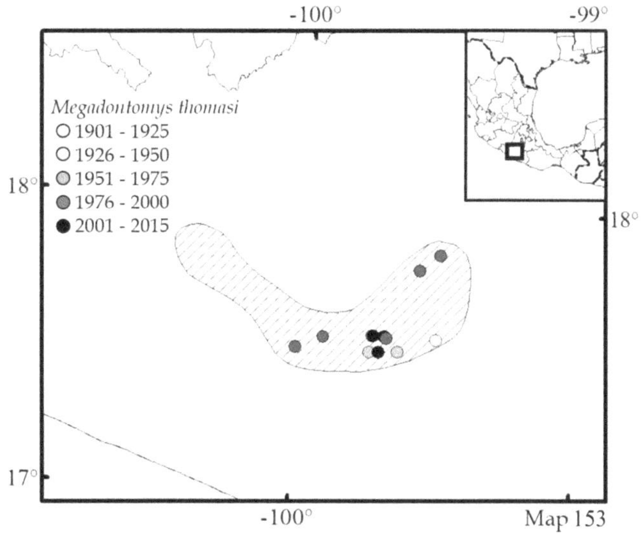


**Map 153** Distribution map of *Megadontomys thomasi*

**Additional Literature:** Merriam (1898a), Musser and Carleton (2005), Vallejo et al. (2017).

# Genus *Nelsonia* Merriam, 1897

Dorsal pelage grayish brown to dark slaty gray, washed with pale shades; flanks tending to be more ochraceous than the back; fur of underparts white, hairs plumbeous basally; tail dark dorsally and paler ventrally; rostrum thin and long; postpalatine foramen approximately at the center between the interpterigoidea foramina and the anterior palatine fossa, separated by a thin septum; interparietal bone well developed; zygomatic plate narrow; palatal bone with no pits; auditory bullae subconical; teeth formed by alternating prisms**; third upper molars with a single external reentrant angle, which nearly separates the tooth into prisms**; **second lower molars with an external and internal reentrant angle; third lower molars with a single internal reentrant angle**. The keys were elaborated based on the review of specimens and the following sources: Hooper (1954), Hall (1981), and Álvarez-Castañeda et al. (2017a).

1. Tail tip white; zygomatic plate narrow and no anterorbital notch (Fig. 49; restricted to the Aguascalientes, Jalisco, Zacatecas, and southern Durango highlands) ................................................................ *Nelsonia neotomodon* (p. 221)

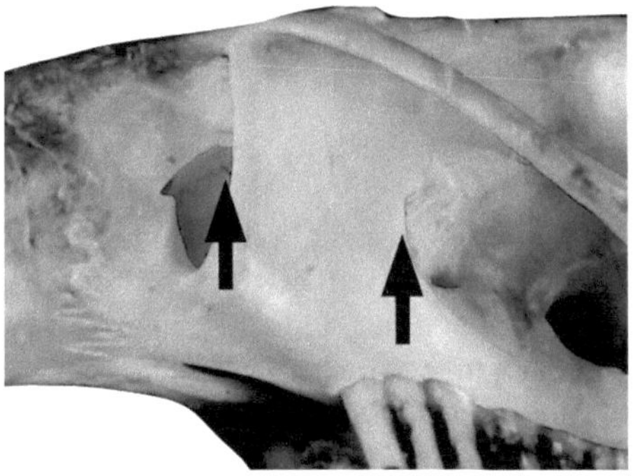

**Fig. 49** Zygomatic plate narrow and no anterorbital notch

1a. Tail tip not white; zygomatic plate wide and anterorbital notch evident (Fig. 50; restricted to the Colima, Jalisco, Michoacán, Estado de México, and Morelos highlands) ...................................................... *Nelsonia goldmani* (p. 220)

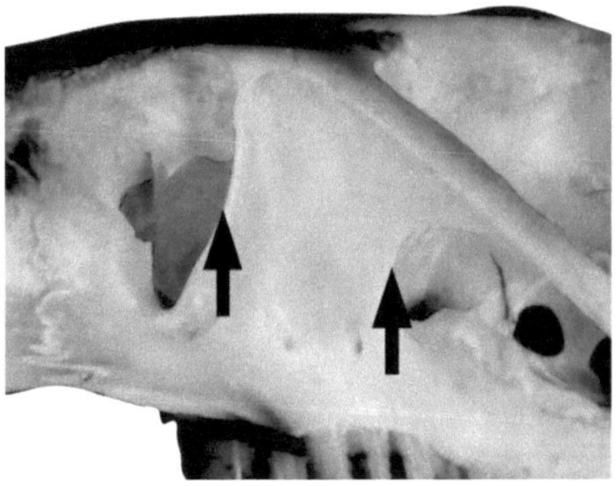

**Fig. 50** Zygomatic plate wide and anterorbital notch evident

## *Nelsonia goldmani* Merriam, 1903

Nelson and Goldman woodrat, rata enana de Michoacán

1. *N. g. cliftoni* Genoways and Jones, 1968. Restricted to Volcán de Colima, Jalisco.
2. *N. g. goldmani* Merriam, 1903. Restricted to Monte Tancítaro and Patanban, Michoacán.

**Conservation:** *Nelsonia goldmani* is listed as Special Protection by the Norma Oficial Mexicana (DOF 2019) and as Endangered (B1ab (iii): extent of occurrence less than 5000 km$^2$, severely fragmented or known to exist at no more than five locations, and under continuing decline in area, extent, and quality of habitat) by the IUCN (Álvarez-Castañeda and Castro-Arellano 2019b).

**Characteristics:** Nelson and Goldman woodrat is medium sized; total length 235.0–255.0 mm and skull length 30.3–33.3 mm. **Dorsal pelage darker with a distinct grayish cast and a fulvous but subdued lateral line; hind feet dusky dorsally; fur of underparts whitish, with leaden gray shades at the base; tail indistinctly bicolored and not white at the tip; zygomatic plate wide with an anterorbital notch.**

**Comments:** *Nelsonia goldmani* was recognized as a distinct species (Engstrom et al. 1992); it was previously regarded as a subspecies of *N. neotomodon* (Hooper 1954). *N. goldmani* ranges on three disjunctive areas in the Eje Neovolcánico, eastern Colima, and southern Jalisco; central Michoacán; central Estado de México and northern Morelos (Map 154), with allopatric distribution with respect to the other species of *Nelsonia*. *N. goldmani* can be distinguished from all the other species of rodents by having the third upper molars with a single external reentrant angle, nearly from the tooth into separated prisms. From *N. neotomodon*, by having relatively broad anterior zygomatic arches and a well developed zygomatic notch.

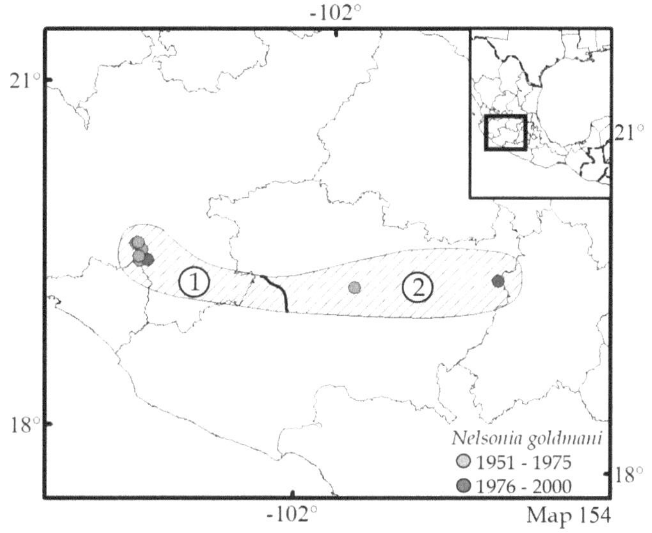


**Map 154** Distribution map of *Nelsonia goldmani*

**Additional Literature:** Merriam (1903), Hooper (1954), Musser and Carleton (2005).

## *Nelsonia neotomodon* Merriam, 1897

Diminutive woodrat, rata enana del oeste

Monotypic.

**Conservation:** *Nelsonia neotomodon* is listed as Special Protection by the Norma Oficial Mexicana (DOF 2019) and as Near Threatened by the IUCN (Álvarez-Castañeda and Castro-Arellano 2008a).

**Characteristics:** Diminutive woodrat is medium sized; total length 233.0–256.0 mm and skull length 31.2–33.0 mm. **Dorsal pelage cinnamon-buff darker dorsal than laterally; fur of underparts whitish, with leaden gray shades at the base; hind feet usually white above; tail distinctly bicolored, usually white at the tip; braincase as the dome; zygomatic plate narrow without an anterorbital notch.**

**Comments:** *Nelsonia neotomodon* is only known from the Sierra Madre Occidental, from southern Durango south through northern Jalisco, Zacatecas, and Aguascalientes. *N. n. goldmani* and *N. n. cliftoni* were previously subspecies of *N. neotomodon* (Hooper 1954), now as *N. goldmani* (Engstrom et al. 1992). *N. neotomodon* is restricted to southern Durango, Zacatecas, Jalisco, and Aguascalientes highlands (Map 155), with allopatric distribution with respect to the other species of *Nelsonia*. *N. neotomodon* can be distinguished from all the other species of rodents by having third upper molars with a single external reentrant angle, nearly from the tooth into separate prisms. From *N. goldmani*, by having relatively narrow anterior zygomatic arches and without a zygomatic notch.

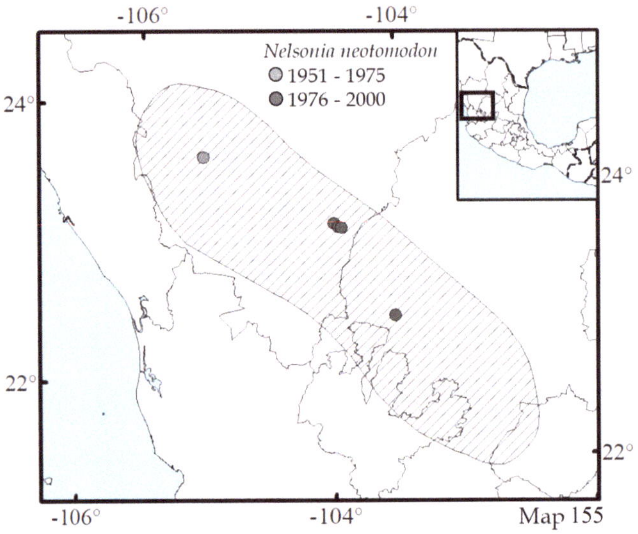


**Map 155** Distribution map of *Nelsonia neotomodon*

**Additional Literature:** Merriam (1903), Musser and Carleton (2005).

## Genus *Neotoma* Say and Ord, 1825

Dorsal pelage pale buff or gray to rich buff or ferruginous, more or less mixed with black; fur of underparts usually white but sometimes pale buff; maxillary toothrow clearly broader anteriorly than posteriorly; first upper molars with the anterointemal reentrant angle varying from deep to shallow or obsolete; **third upper molars with the middle loop undivided by a deepening of the reentrant angles;** interpterygoid fossa varying from wide to narrow; bullae variable in size; frontal bones constricted near or anteriorly to the middle; tail terete, tapering and short-haired; **hind feet naked below along the outer side at least to the tarso-metatarsal joint.** *Neotoma* has three subgenera: *Neotoma*, with four species groups (*floridana* [*N. albigula*, *N. floridana*, *N. goldmani*, *N. leucodon*, *N. magister*, *N. melanura*, *N. nelsoni*, and *N. palatina*]; *lepida* [*N. bryanti*, *N. devia*, *N. fuscipes*, *N. macrotis*, *N. insularis*, *N. lepida*, and *N. stephensi*]; *mexicana* [*N. angustapalata*, *N. ferruginea*, *N. mexicana*, and *N. picta*] and *micropus* [*N. micropus*]); subgenus *Teanopus* (*N. phenax*); and the subgenus *Tenoma* (*N. cinerea*). The keys were elaborated based on the review of specimens and the following sources: Goldman (1932), Hall (1981), Edwards and Bradley (2003), Patton et al. (2007), and Álvarez-Castañeda et al. (2017a).

1. Tail flattened and bushy; soles thickly furred from the heels to the posterior tubercles; first upper molars with a deep anterointernal reentrant angle; third upper molars with an anterior closed triangle and two confluent posterior loops (from the Yukon Territory and Northwest Territories south through California, Arizona, and New Mexico and east through the Dakotas, Nebraska, and Colorado) ........................................ subgenus (*Tenoma*) *Neotoma cinerea* (p. 252)

1a. Tail cylindrical and not bushy; soles not furred from the heels to the posterior tubercle; first upper molars without a deep anterointernal reentrant angle; third upper molars without an anterior closed triangle and two confluent posterior loops ........................................ 2

2. Auditory bullae greatly enlarged (Fig. 51; almost semi-circular in profile but not greatly widened transversely) and situated only slightly obliquely to the long skull axis (endemic to the lowlands of Sonora and Sinaloa) .............................. ........................................................................................................ subgenus (*Teanopus*) *Neotoma phenax* (p. 251)

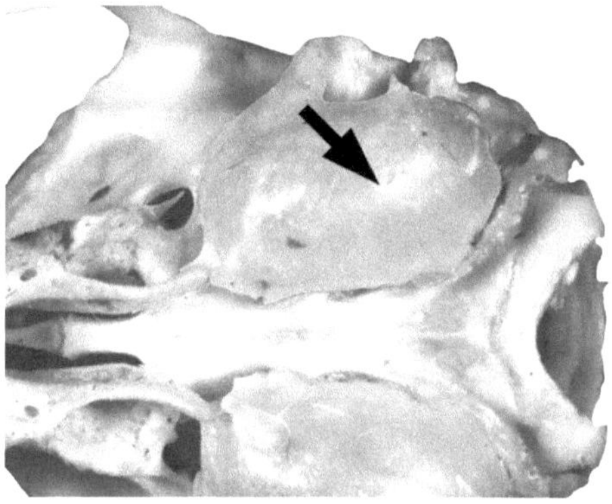

**Fig. 51** Auditory bullae greatly enlarged

2a. Auditory bullae not greatly enlarged and situated obliquely (Fig. 52) ..........................................subgenus (*Neotoma*) 3

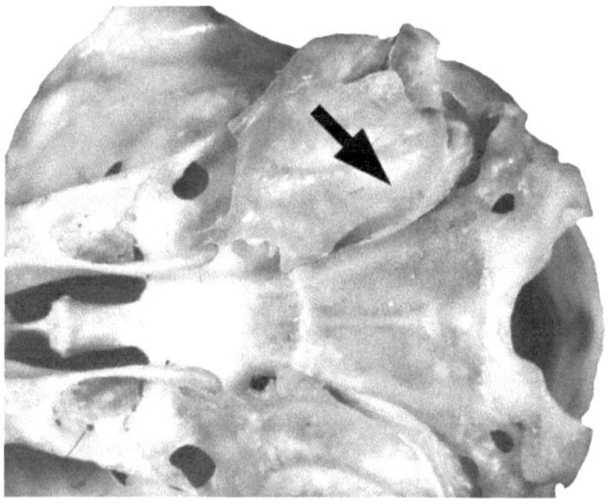

**Fig. 52** Auditory bullae not greatly enlarged and situated obliquely

3. Maxillary toothrow slightly narrower posteriorly than anteriorly; middle lobe of the last upper molars partially or completely divided by an inner reentrant angle ........................................................................................................ 4

3a. Maxillary toothrow much narrower posteriorly than anteriorly; middle lobe of the last upper molars not divided by an inner reentrant angle ........................................................................................................ 5

4. Size and skull large; tail unicolor; interpterygoid fossa narrow; incisive foramina equal to or shorter than the palatal bridge (from the Columbia River, Oregon, south through northern Santa Barbara, California) .................*Neotoma fuscipes* (p. 235)

4a. Size and skull small; tail distinctly bicolored; interpterygoid fossa broad; incisive foramina longer than the palatal bridge (from central and eastern California south through northwestern Baja California) ...................... *Neotoma macrotis* (p. 241)

5. Anterior and internal reentrant angle of the first upper molars extending greater than halfway across the occlusal crown (Fig. 53)........................................................................................................ 6

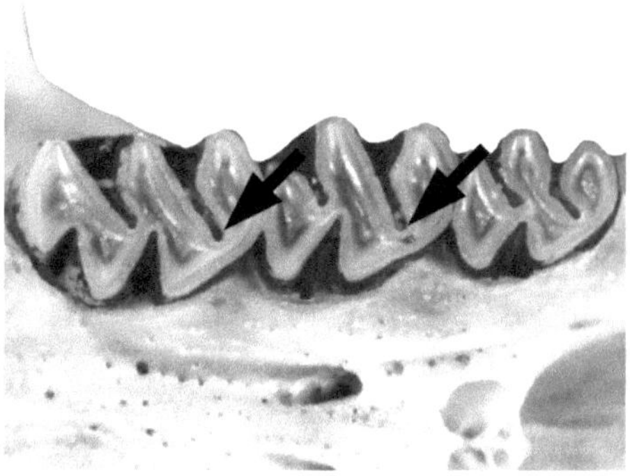

**Fig. 53** Anterior and internal reentrant angle of the first upper molars extending greater than halfway across the occlusal crown

5a. Anterior and internal reentrant angle of the first upper molars extending halfway or less across the occlusal crown (Fig. 54)........10

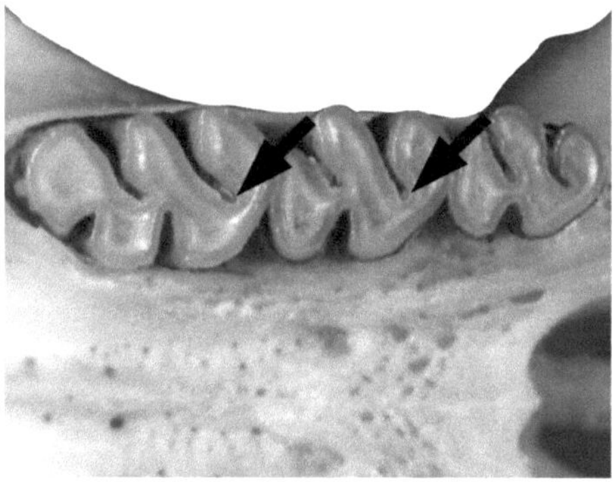

**Fig. 54** Anterior and internal reentrant angle of the first upper molars extending halfway or less across the occlusal crown

6. Tail haired and as long as the head-and-body length; auditory bullae larger in proportion to the skull and inflated ventrally (from eastern Mississippi River, Connecticut, New York, New Jersey, Pennsylvania southwest through Alabama, North Carolina, Maryland, Tennessee, Kentucky, Indiana, and West Virginia)........*Neotoma magister* (p. 243)

6a. Tail with small hair; auditory bullae in proportion to the skull and not ventrally inflated ........7

7. Total length greater than 370.0 mm; sides of the interpterygoid fossa strongly concave near the posterior end of the toothrow (Fig. 55: endemic to southern Tamaulipas and northern San Luis Potosí)..........*Neotoma angustapalata* (p. 229)

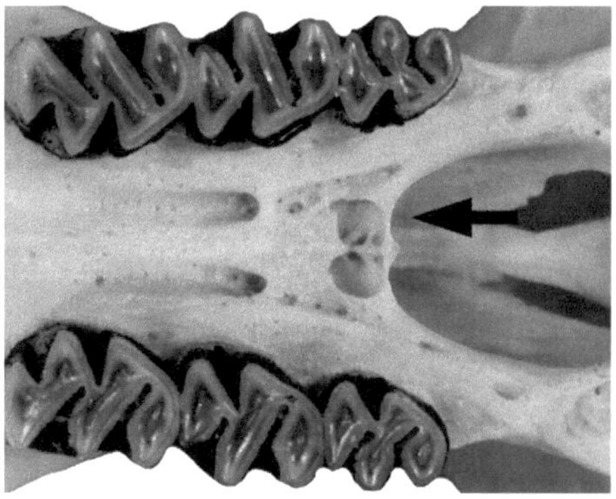

**Fig. 55** Sides of the interpterygoid fossa strongly concave near the posterior end of the toothrow

7a. Total length less than 370.0 mm; sides of the interpterygoid fossa more or less parallel near the posterior end of the toothrow (Fig. 56) ........ 8

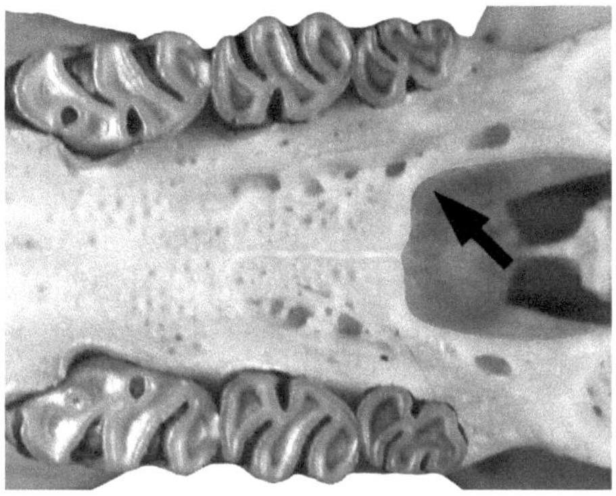

**Fig. 56** Sides of the interpterygoid fossa more or less parallel near the posterior end of the toothrow

8. Ranges Arizona, Utah, and Colorado south through Oaxaca........ *Neotoma mexicana* (p. 245)

8a. Ranges south of the Eje Neovolcánico southward........ 9

9. Endemic to Guerrero........ *Neotoma picta* (p. 249)

9a. Northern Oaxaca and Chiapas (from Oaxaca south through Central America)........ *Neotoma ferruginea* (p. 233)

10. Tail semi-bushy to fully haired ........ 11

10a. Tail not fully haired........ 12

11. Total length less than 310.0 mm; tail semi-bushy and less than 85% of the head-and-body length; auditory bullae larger in proportion to the skull (central Arizona and western New Mexico)........ *Neotoma stephensi* (p. 250)

11a. Total length greater than 310.0 mm; tail fully haired and greater than 85% of the head-and-body length; auditory bullae small in proportion to the skull (from eastern Colorado, Nebraska, and South Dakota east through North Carolina and New York, and south through southern Texas, Gulf coast, and central Florida, including a population in Key Largo)........ *Neotoma floridana* (p. 234)

12. Tail nearly unicolored; ascending branches of the maxillary bone projecting beyond the posterior margin of the nasals (endemic western side of Cofre de Perote and Pico de Orizaba, Veracruz)........ *Neotoma nelsoni* (p. 247)

12a. Tail sharply bicolored; ascending branches of the maxillary bone not projecting beyond the posterior margin of the nasals........ 13

13. Ears small, less than 26.0 mm; sphenopalatine vacuities absent (endemic to Jalisco and western Zacatecas)........ *Neotoma palatina* (p. 248)

13a. Ears large, greater than 26.0 mm; sphenopalatine vacuities present........ 14

14. Palate with a posterior medial spine (Fig. 57); complete posterior half of the septum dividing the anterior palatine foramen (from Southeastern Colorado and southern Kansas south through Coahuila, Nuevo León, Tamaulipas, and San Luis Potosí)........ *Neotoma micropus* (p. 246)

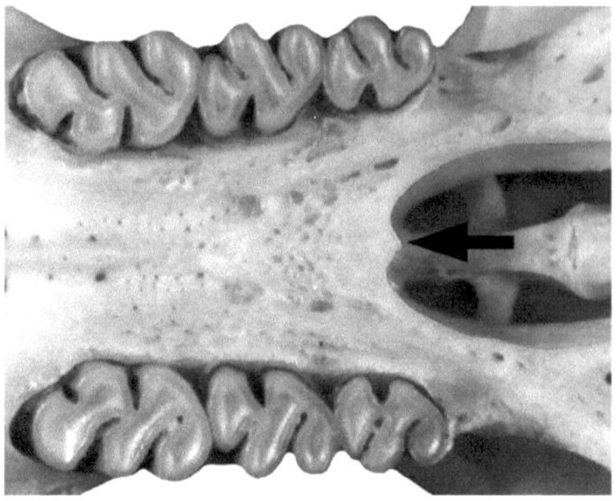

**Fig. 57** Palate with a posterior medial spine

14a. Palate posterior border lacking a medial spine; if spine is present, the posterior half of the septum dividing the anterior palatine foramen is incomplete ........ 15

15a. Hairs on the neck gray at the base, darker than the tips ........ 16

15. Hairs on the neck white at the base; if gray, hind feet greater than 35.0 mm ........ 21

16. Endemic to central area of Altiplanicie Mexicana, from northern Chihuahua and southern Coahuila south through northern San Luis Potosí and Zacatecas ........ *Neotoma goldmani* (p. 236)

16a. Ranges in United States, northwestern Sonora, and the Baja California Peninsula ........ 17

17. Ranges east of the Grande and Conchos rivers, from Colorado to Texas south through Michoacán and Querétaro, Hidalgo, and from Tamaulipas west through Durango ........ *Neotoma leucodon* (part, p. 240)

17a. Ranges west of the Grande and Conchos rivers ........ 18

18. Glans penis with or without a bell shaped extension but not clearly distinguishable at the tip; maxillo-frontal suture joining the lacrimal bone beyond its midpoint (Fig. 58); auditory bullae small relative to the skull size ........ 19

**Fig. 58** Maxillo-frontal suture joining the lacrimal bone beyond its midpoint

18a. Glans penis with a bell shaped extension, clearly distinguishable at the tip; maxillo-frontal suture joins the lacrimal bone behind its midpoint (Fig. 59); auditory bullae inflated relative to the size of the skull ........ 20

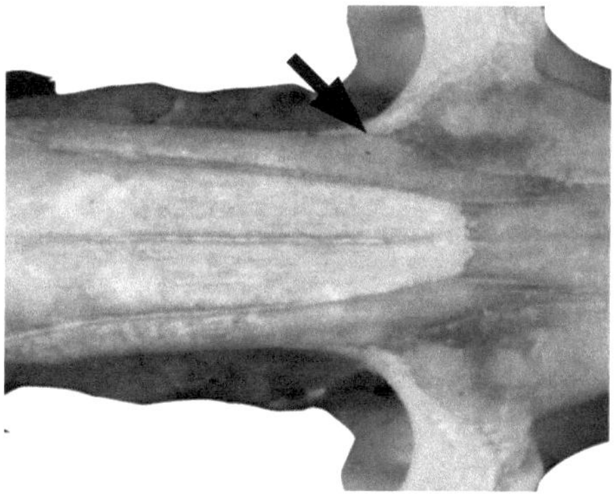

**Fig. 59** Maxillo-frontal suture joining the lacrimal bone behind its midpoint

19. Glans penis thin with elongated baculum and hood, latter with straight, tapered, and a bifurcated fleshy tip; anteroloph of the first upper molars with a deep anteromedial notch (from southern San Francisco Bay south through Baja California Peninsula, the Todos Santos, San Martín, Cedros, Natividad, Margarita, Magdalena, Carmen, Danzante, San José, San Francisco, San Marcos, Coronados, La Partida, and Espíritu Santo Islands. *N. b. anthonyi* and *N. b. bunkeri* considered extinct) ........ *Neotoma bryanti* (p. 230)

19a. Glans penis stout with short baculum and hood, the latter with a straight but a blunt fleshy tip; anteroloph of the first upper molars with a shallow anteromedial notch (endemic to Ángel de la Guardia Island, Baja California) ........ *Neotoma insularis* (p. 237)

20. Frontal bone in contact with the lacrimal bone less than one-half of the maxillary length contact; west of the Colorado River, in eastern California, Nevada, Oregon, Idaho, Utah, and northern Baja California Peninsula ........ *Neotoma lepida* (p. 238)

20a. Frontal in contact with the lacrimal bone much less than the maxilla contact; east of the Colorado River in northern and western Arizona, southeastern California, and northwestern Sonora ........ *Neotoma devia* (p. 232)

21. Ranges east of the Grande and Conchos rivers, from Colorado to Texas south through Michoacán and Querétaro, Hidalgo, and from Tamaulipas west through Durango ........ *Neotoma leucodon* (part, p. 240)

21a. Ranges west of the Grande and Conchos rivers ........ 22

22. Hair on the neck and chest more or less plumbeous basally; tail blackish dorsally, general dorsal coloration darker; rostrum more slender; teeth slightly smaller; auditory bullae smaller (northern Sinaloa, southern Sonora, and a small area in western Chihuahua) ........ *Neotoma melanura* (p. 244)

22a. Hair on the throat and chest pure white to the root; tail grayish brown dorsally, general dorsal coloration pale; rostrum less slender; teeth slightly larger; auditory bullae larger (from Rio Grande and Río Conchos west through southeastern Utah, and from southwestern Colorado west through southeastern California and northeastern Baja California, and south through northern Sinaloa, including Tiburón Island) ........ *Neotoma albigula* (p. 228)

## Subgenus ***Neotoma*** Say and Ord, 1825

Tail cylindrical, tapering, and with short hairs; hind feet naked below along the outer side at least to the tarsometatarsal joint; **first upper molars with the anterointernal angle varying in depth or obsolete; third upper molars with a middle loop either divided or undivided by reentrant angles;** auditory bullae of variable size.

## ***Neotoma albigula*** Hartley, 1894

White-throated woodrat, rata de campo de garganta blanca

1. *N. a. albigula* Hartley, 1894. From western New Mexico, southern Arizona, and southern Nevada south through central Sonora and Chihuahua.
2. *N. a. brevicauda* Durrant, 1934. Restricted to the Grand County, southwestern Utah.
3. *N. a. laplataensis* Miller, 1933. Northeastern Arizona, northwestern New Mexico, southeastern Utah, and southwestern Colorado.
4. *N. a. mearnsi* Goldman, 1915. Restricted to the Pinacate Desert in Arizona and Sonora.
5. *N. a. seri* Townsend, 1912. Restricted to Tiburón Island, Sonora.
6. *N. a. sheldoni* Goldman, 1915. Restricted to northern-central Sonora.
7. *N. a. varia* Burt, 1932. Restricted to Datil (Turner) Island, Sonora.
8. *N. a. venusta* True, 1894. Southeastern California, northeastern Baja California, western Arizona, and Sonora.

**Conservation:** *Neotoma albigula seri* is listed as Threatened and *N. a. varia* as Endangered by the Norma Oficial Mexicana (DOF 2019), and *N. albigula* listed as Least Concern by the IUCN (Lacher and Álvarez-Castañeda 2016).

**Characteristics:** White-necked woodrat is medium sized; total length 317.0–363.0 mm and skull length 41.4–45.6 mm. Dorsal pelage yellowish gray with the flanks ocher; fur of underparts white, with leaden gray shades at the base; **chin, throat, neck, and chest with white hairs through the base**; hair between the hind and forelimbs completely white; feet white; tail shorter than the head-and-body length, with very short hair, **bicolored, grayish brown dorsally and white ventrally**; ears proportionally large; molar toothrow straight; third upper molars with three lobes and two external reentrant angle; anterointernal reentrant angle in the first upper molars shallow; palate concave posteriorly; auditory bullae larger in proportion to the skull, ventrally inflated with an obtuse anterior edge and almost parallel to the longitudinal skull axis. Species group *albigula*.

**Comments:** *Neotoma albigula leucodon* and *N. a. melanura* are now considered full species (Edwards et al. 2001; Bradley and Mauldin 2016). *N. varia* changed from full species (Burt 1932) to *N. albigula* subspecies (Bogan 1997; Álvarez-Castañeda and Ríos 2010). *N. albigula* ranges west of the rivers Grande and Conchos in southeastern Utah and from southwestern Colorado west through southeastern California, western New Mexico, Chihuahua, and northeastern Baja California, and south through central Sonora, including Tiburón Island (Map 156). It can be found in sympatry, or nearly so, with *N. cinerea*, *N. devia*, *N. lepida*, *N. leucodon*, *N. mexicana*, *N. micropus*, *N. phenax*, and *N. stephensi*. *N. albigula* can be distinguished from *N. cinerea* by not having the tail flattened and bushy; hairs of underparts white from the root and first upper molars without a deep anterointernal reentrant angle. From *N. devia* and *N. lepida*, by having a paler coloration; neck and breast hair white at the base; rostrum less slender; nasals narrow and less evenly rounded posteriorly; palate concave posteriorly and auditory bullae proportionally larger. From *N. leucodon*, by having a smaller size; hind feet on average 32.0 mm or less; dorsal coloration paler; rostrum and dentition lighter; supraorbital ridges less developed without projecting frontal shelves. From *N. melanura*, by having a paler coloration; tail grayish brown dorsally; skull more massive; rostrum less slender; teeth and auditory bullae slightly larger. From *N. mexicana*, by having the first upper molars with the anterointernal reentrant angle not deeper, cutting less than halfway across the anterior lobe; skull larger and heavier; rostrum relatively short and more or less slender. From *N. micropus*, by having the palate without a posterior medial spine and posterior septum not complete divided of the palatine foramina. From *N. phenax*, by having the fur of underparts white in particular in the chin, throat, neck, and chest; auditory bullae not distinctively larger in proportion to the skull, without an obtuse anterior edge and not parallel to the skull medial axis. From *N. stephensi*, by having the tail not fully haired; fur of underparts white through the base and supraorbital ridges not continuing forward to the rostrum.

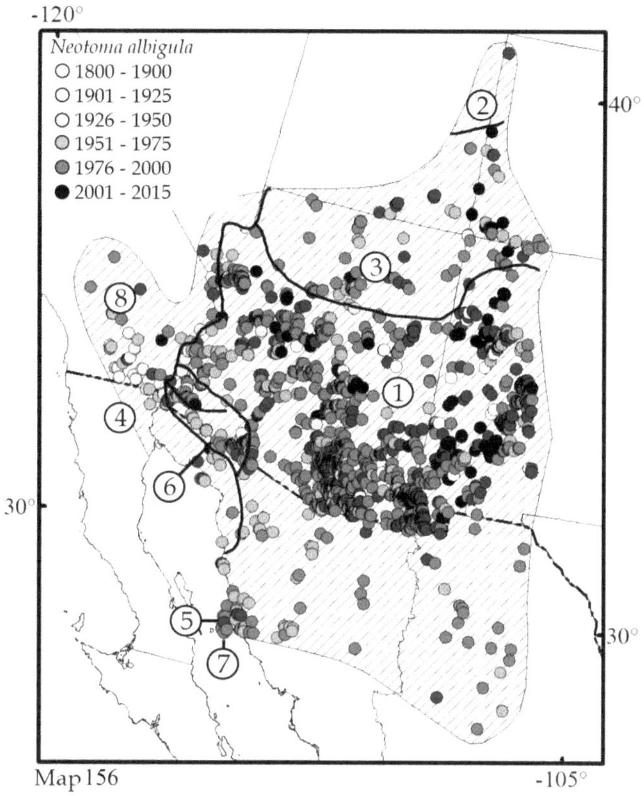


**Map 156** Distribution map of *Neotoma albigula*

**Additional Literature:** Goldman (1910), Hall and Genoways (1970), Macêdo and Mares (1988), Bogan (1997), Álvarez-Castañeda and Cortés-Calva (1999), Álvarez-Castañeda and Ortega-Rubio (2003), Musser and Carleton (2005), Álvarez-Castañeda et al. (2006, 2010b), Bradley and Mauldin (2016).

## *Neotoma angustapalata* Baker, 1951

Tamaulipan woodrat, rata de campo de Tamaulipas

Monotypic.

**Conservation:** *Neotoma angustapalata* is listed as **Near Threatened** by the IUCN (Álvarez-Castañeda and Castro-Arellano 2008b).

**Characteristics:** Tamaulipan woodrat is medium sized; total length 380.0 mm and skull length 33.9 mm. **Dorsal pelage gray or ash gray, head, and cheeks gray; fur of underparts leaden gray grizzly with white** shades and completely white on the throat and groin; tail with sparse hair, blackish dorsally and white ventrally; auditory bullae larger in proportion to the skull; first upper molars with deep anterointernal reentrant angles; **interpterygoidea fossa with concave and wide sides**. Species group *mexicana*.

**Comments:** *Neotoma angustapalata* was considered a possible subspecies of *N. micropus* based on their distribution and morphological similarities (Álvarez 1963; Hooper 1953), but the species status was maintained (Birney 1973; Rogers et al. 2011). *N. angustapalata* ranges are restricted to southwestern Tamaulipas and northeastern San Luis Potosí (Map 157) and has allopatric distribution with respect to other species of *Neotoma*. It can be distinguished from all other species by its deep anterointernal reentrant angle on the first upper molars; not having the maxillovomerine notch in the septum. From *N. leucodon*, by having a larger average skull length (50.22 ± 1.72 *vs* 43.03 ± 2.15), nasal length (24.23 ± 0.75 *vs* 23.14 ± 1.17), and a smaller interorbital constriction (5.92 ± 0.22 *vs* 5.68 ± 0.37; Rogers et al. 2011). From *N. mexicana*, by its larger size, greater than 365.0 mm; fur of underparts darker; tail with scaly appearance and short and sparse hairs; sides of the interpterygoid fossa broader, concave at the posterior end, and a narrower palatine breadth.

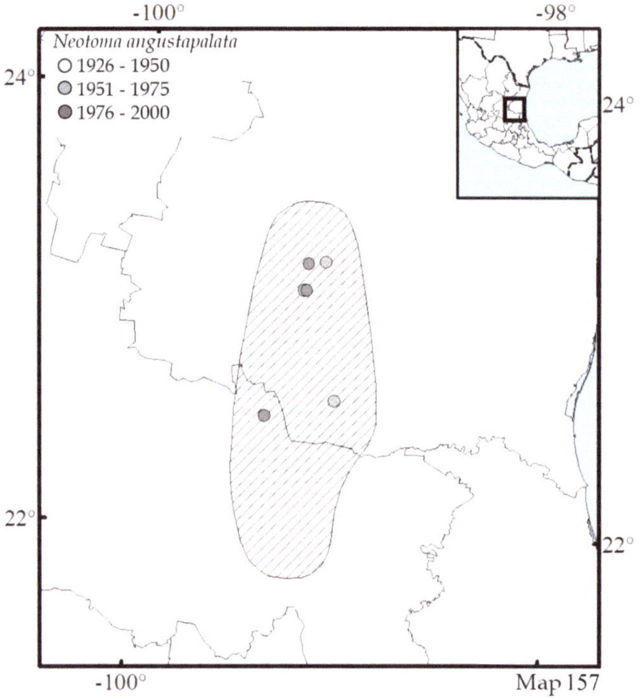


**Map 157** Distribution map of *Neotoma angustapalata*

**Additional Literature:** Baker (1951), Musser and Carleton (2005).

## *Neotoma bryanti* Merriam, 1887

Bryant's woodrat, rata de campo de Baja California

1. *N. b. anthonyi* Allen, 1898. Restricted to Todos Santos Island, Baja California.
2. *N. b. bryanti* Merriam, 1887. From San Felipe and El Rosario south through the Baja California Peninsula, including the Cedros, Magdalena, Margarita, Coronados, Carmen, Danzante, San José, San Francisco, La Partida, and Espíritu Santo Islands.
3. *N. b. intermedia* Rhoads, 1894. From southern San Francisco Bay, California south through El Rosario, Baja California.
4. *N. b. marcosensis* Burt, 1932. Restricted to San Marcos Island, Baja California.
5. *N. b. martinensis* Goldman, 1905. San Martín Island, Baja California.

**Conservation:** *Neotoma bryanti marcosensis* and *N. b. bryanti* are listed as Threatened, *N. b. martinensis* as Endangered, and *N. b. anthonyi* as Extinct by the Norma Oficial Mexicana (DOF 2019) and *Neotoma bryanti* is listed as Least Concern by the IUCN (Álvarez-Castañeda et al. 2017f).

**Characteristics:** Bryant's woodrat is medium sized; total length 284.0–400.0 mm and skull length 36.7–45.2 mm. Dorsal pelage varying depending on the subspecies, from pale brown to dark brown, darker in the back and top of the head; flanks pale yellowish brown, head darker; fur of underparts creamy whitish grizzly with yellowish shades, chest usually crossed by a faint buffy band; throat hair leaden gray through the base, never white; **ears large**; feet white; **tail long, approximately 85% of the head-and-body length; hair bicolored, brownish dorsally and grayish ventrally**; ankle sides usually dusky; feet white; lacrimal contact area with the frontal equal to or greater than the contact area with the maxilla; dentition heavy; **anteroloph of the first upper molars with a deep anteromedial notch; auditory bullae small** in proportion to the skull. Species group *lepida*.

**Comments:** The following subspecies *Neotoma lepida abbreviata* (Espíritu Santo Island), *N. l. latirostra* (Danzante Island), *N. l. marcosensis* (San Marcos Island), *N. l. nudicauda* (Carmen Island), *N. l. perpallida* (San José Island), and *N. l. vicina* (San Francisco Island) are considered *N. bryanti* (Patton et al. 2007). Following Patton et al. (2007) *N. b. bryanti* includes the following taxa: *N. lepida abbreviata*, *N. l. aridicola*, *N. l. arenacea*, *N. bunkeri*, *N. l. felipensis*, *N. l. latirostra*, *N. l. molagrandis*, *N. l. notia*, *N. l. nudicauda*, *N. l. perpallida*, *N. l. pretiosa*, *N. l. ravida*, and *N. l. vicina*. *N. b. intermedia*: *N. l. californica*, *N. l. egressa*, *N. l. gilva*, *N. l. intermedia*, and *N. l. petricola*. *N. bryanti* ranges from the southern San Francisco Bay south through the southern end of the Baja California Peninsula, including the populations of Todos Santos, San Martín, Cedros, Margarita, Magdalena, Carmen, Danzante, San José, San Francisco, Natividad, San Marcos, Coronados, La Partida, and and Espíritu Santo Islands (Map 158). *N. bryanti* can be found in sympatry, or nearly so, with *N. albigula*, *N. fuscipes*, and *N. macrotis*. *N. bryanti* can be distinguished from *N. fuscipes* and *N macrotis* by having the maxillary toothrow much narrower posteriorly than anteriorly and the middle lobe of the last upper molars not divided by an inner reentrant angle. From *N. albigula*, by its darker coloration; neck and chest hair plumbeous at the base; more slenderrostrum more slender; nasals broader and posteriorly more evenly rounded, palate convex posteriorly and auditory bullae smaller. From *N. fuscipes* and *N. macrotis*, by having the tail blackish; hind feet toes dusky; maxillary toothrow much narrower posteriorly than anteriorly and the middle lobe of the last upper molars not divided by an inner reentrant angle. From *N. lepida*, by having the maxillo-frontal suture joining the lacrimal beyond its midpoint and the anteroloph of the first upper molars with a deep anteromedial notch.

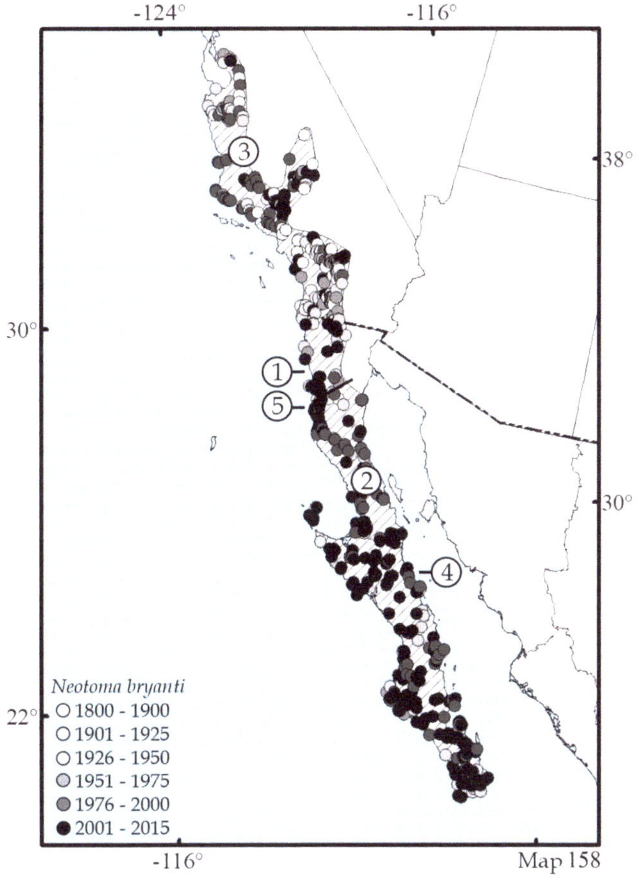


**Map 158** Distribution map of *Neotoma bryanti*

**Additional Literature:** Goldman (1910, 1932), Álvarez Castañeda and Cortés-Calva (1999), Álvarez-Castañeda and Yensen (1999), Cortés-Calva et al. (2001a, b), Verts and Carraway (2002), Álvarez-Castañeda et al. (2008c).

## *Neotoma devia* Goldman, 1927

Arizona woodrat, rata de campo de Arizona

Monotypic.

**Conservation:** *Neotoma devia* is listed as Least Concern by the IUCN (Álvarez-Castañeda et al. 2016c).

**Characteristics:** Arizona woodrat is medium sized; total length 255.0–325.0 mm and skull length 33.4–40.7 mm. Dorsal pelage pale to dark gray. With melanic specimens often found; flanks brown or pale yellowish brown; fur of underparts whitish, with a completely white spot between the hind and fore limbs; ears naked; **tail long, approximately 86% of the head-and-body length,** bicolored; limbs white dorsally; skull angular; contact area of the frontal bone with the lacrimal bone much less than the contact area with the maxilla; **auditory bullae inflated;** sphenoid bone open; **upper molars anteroloph with a shallow to obsolete anteromedial notch**, except in very young individuals. Species group *lepida*.

**Comments:** *Neotoma devia* was recognized as a distinct species (Mascarello 1978; Koop et al. 1985; Patton and Álvarez-Castañeda 2005; Patton et al. 2007); it was previously regarded as a subspecies of *N. lepida* (Goldman 1932), including the subspecies *N. lepida aureotunicata*, *N. l. bensoni*, *N. l. flava*, and *N. l. harteri*. *N. devia* ranges east of the Colorado River in western Arizona, and from southeastern California south through northwestern Sonora (Map 159). It can be found in sympatry, or nearly so, with *N. albigula* and *N. stephensi*. *N. devia* can be distinguished from *N. albigula* by having a darker coloration; hairs on the throat with the root not white; rostrum slender; nasals wider, more evenly rounded posteriorly and auditory bullae smaller. From *N. lepida*, by having a different pattern in the position of the lacrimal bone, in *N. devia* contact area of the frontal bone with the lacrimal bone smaller than the contact area of the maxillary bone with the lacrimal bone. From *N. stephensi*, by having the tail not fully haired and supraorbital ridges that do not continue forward to the rostrum.

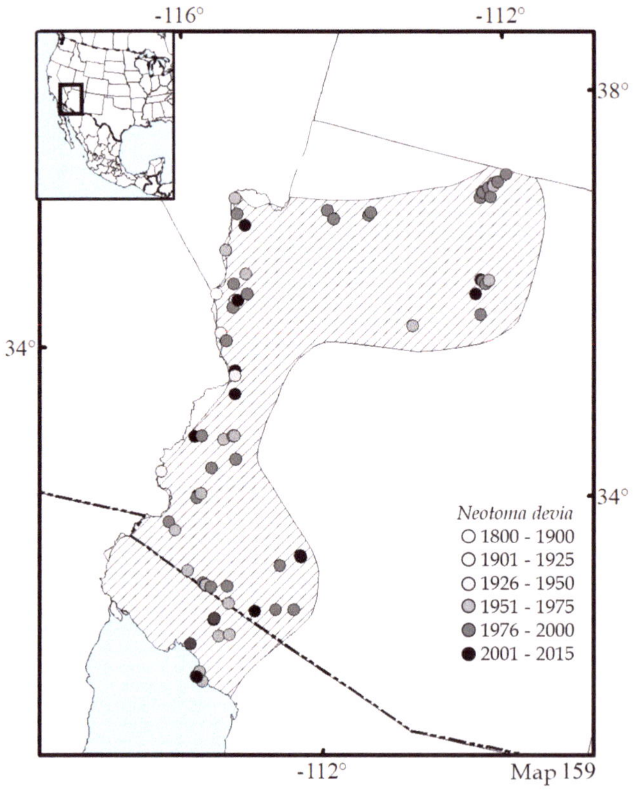


**Map 159** Distribution map of *Neotoma devia*

**Additional Literature:** Goldman (1927), Verts and Carraway (2002), Patton et al. (2007).

## *Neotoma ferruginea* Tomes, 1862

Guatemala woodrat, rata de campo de Guatemala

1. *N. f. chamula* Goldman, 1909. Highlands of central Chiapas.
2. *N. f. isthmica* Goldman, 1904. From the Tehuantepec Isthmus southeast to the lowlands of central and southern Chiapas.
3. *N. f. tropicalis* Goldman, 1904. Western Oaxaca.

**Conservation:** *Neotoma ferruginea* (as *N. mexicana*) is listed as Least Concern by the IUCN (Linzey et al. 2016a).

**Characteristics:** Guatemala woodrat is medium sized; total length 335.0–385.0 mm and skull length 33.0 a 37.5 mm. Dorsal pelage bright rufous to tawny cinnamon ochraceous, back grizzle with black-tipped hairs and darker than the flanks; **fur of underparts white with hair roots white on the neck and chest**; an elongated patch of hair from the chin to the space between the forelegs, pure white from the root to the tip; limbs with the outer surface strongly tinged with dusky, forelegs with the inner surface whitish; hind feet dusky gray; tail indistinctly bicolored, deep dusky dorsally and paler ventrally; skull very large; **nasals emarginated;** frontal bones broad posteriorly; sphenopalatine vacuities small; auditory bullae small. Note: *Neotoma mexicana*, *N. ferruginea*, and *N. picta* can be differentiated mainly by genetic data. Species group *mexicana*.

**Comments:** *Neotoma ferruginea* was reinstated as a distinct species (Edwards and Bradley 2003; Hernández-Canchola et al. 2021a)*;* it was previously regarded as a subspecies of *N. mexicana* (Hall 1955), including the subspecies *N. mexicana chamula*, *N. m. eremita* (part), *N. m. isthmica*, *N. m. picta* (part), and *N. m. tropicalis*. Specimens previously considered as *N. m. eremita* and *N. m. picta* with ranges in Oaxaca should be considered *N. ferruginea. N. ferruginea* ranges in the highlands from Oaxaca and Chiapas south through Central America (Map 160). It can be found in sympatry, or nearly so, with *N. picta*, from which it can be distinguished by its larger size; skull slightly smaller and nasals smaller.

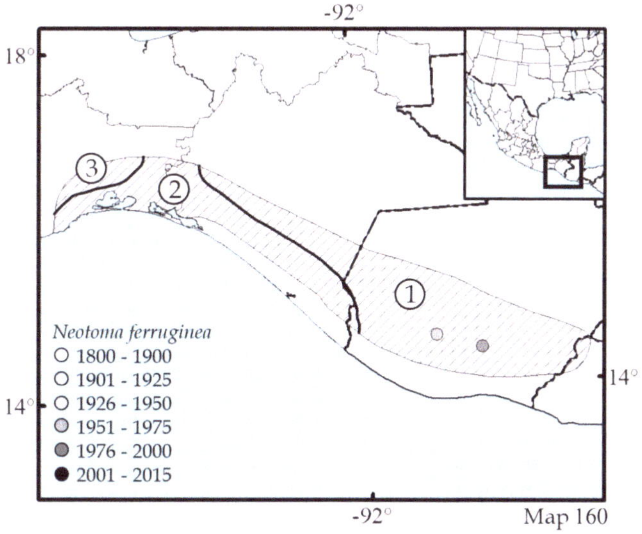


**Map 160** Distribution map of *Neotoma ferruginea*

**Additional Literature:** Ordoñez-Garza et al. (2014).

## ***Neotoma floridana*** (Ord, 1818)

Eastern woodrat, rata de campo del este

1. *N. f. attwateri* Mearns, 1897. Central-eastern Texas.
2. *N. f. baileyi* Merriam, 1894. Northern-central Nebraska and southwestern South Dakota.
3. *N. f. campestris* Allen, 1894. Central-eastern Colorado, southern Nebraska, and central-northwestern Kansas.
4. *N. f. floridana* (Ord, 1818). From central Florida north through Georgia, southern South Carolina, and North Carolina.
5. *N. f. haematoreia* Howell, 1934. Northern Georgia, northwestern South Carolina, and southwestern North Carolina.
6. *N. f. illinoensis* Howell, 1910. From central Louisiana, Arkansas, and Missouri east through southern Illinois, western Kentucky, Tennessee, Alabama, and western Florida.
7. *N. f. osagensis* Blair, 1939. From western Arkansas and Missouri west through central Oklahoma and Kansas.
8. *N. f. rubida* Bangs, 1898. Southwestern Texas, southern Louisiana, southern Mississippi, and southwestern Alabama.
9. *N. f. smalli* Sherman, 1955. Restricted to Key Largo, Florida.

**Conservation:** *Neotoma floridana smalli* is listed as Endangered by the United States Endangered Species Act (FWS 2022) and *Neotoma floridana* is listed as Least Concern by the IUCN (Cassola 2016bz).

**Characteristics:** Eastern woodrat is large in size; total length 311.0–451.0 mm and skull length 39.5–43.0 mm. Dorsal pelage pale cinnamon to buffy gray, back grizzle with blackish shades, pelage coarse; fur of underparts creamy white, with plumbeous shades at the base and axillae creamy buff; head more grayish than the back; posterior area of the nose and flank dark; **ears large;** dusky wash from foot top to the base of the toes**; tail as long as the head-and-body length, color dusky dorsally, slightly paler ventrally, scantily haired;** skull very large, elongated; first upper molars with the anterointernal reentrant angle moderately developed; frontal bones broad, interorbital constriction anterior to the middle part of the frontal bones; palatal bridge shorter than the incisive foramina; sphenopalatine vacuities rather small; auditory bullae small, short, and rounded. Species group *floridana*.

**Comments:** *Neotoma magister* was previously considered a subspecies of *N. floridana* (Schwartz and Odum 1957; Edwards and Bradley 2001). *N. floridana* ranges from eastern Colorado, Nebraska, and South Dakota east through North Carolina and New York, and south through the Gulf coast of Texas and Florida, including a population in Key Largo (Map 161). It can be found in sympatry, or nearly so, with *N. mexicana*, from which it differs by having the first upper molars with the anterointernal reentrant angle not deeper, cutting less than halfway across the anterior lobe; skull larger and heavier; rostrum relatively short and more or less slender. From *N. magister*, by having the interpterygoid fossa not rounded anteriorly and wider; auditory bullae less elongated and not tapered anteriorly; the mastoid process of the squamosal bone broad and first upper molars with the anterointernal reentrant angle shallow.

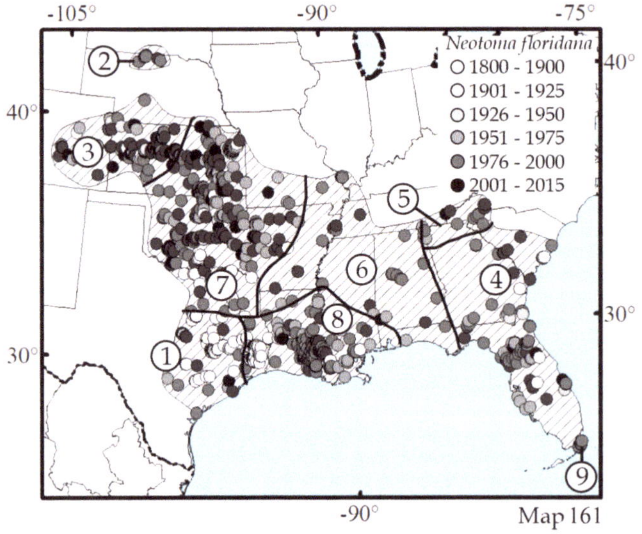


**Map 161** Distribution map of *Neotoma floridana*

**Additional Literature:** Wiley (1980), Musser and Carleton (2005).

## *Neotoma fuscipes* Baird, 1858

Dusky-footed woodrat, rata de campo de patas negras

1. *N. f. annectens* Elliot, 1898. Coast from San Francisco Bay to Monterey Bay, western California.
2. *N. f. fuscipes* Baird, 1858. Klamath Mountains and Sierra Nevada, southern Oregon, and northern California.
3. *N. f. monochroura* Rhoads, 1894. Coast Ranges, Oregon, and northern California.
4. *N. f. perplexa* Hooper, 1938. Coast Range eastern side, central-western California.
5. *N. f. riparia* Hooper, 1938. Diablo Range, central-western California.

**Conservation:** *Neotoma fuscipes riparia* is listed as Endangered by the United States Endangered Species Act (FWS 2022) and *Neotoma fuscipes* is listed as Least Concern by the IUCN (Cassola 2016ca).

**Characteristics:** Dusky-footed woodrat is medium sized; total length 350.0–440.0 mm and skull length 37.5–42.5 mm. Dorsal pelage ochraceous buff, grizzly with darker hair on the back and top of the head; fur of underparts white, with leaden gray shades at the base, neck, chest, and inguinal region white through the base; facial area grayish; ears large and brownish; ankles dusky; forefoot and hindfoot toes white; **hind feet clouded with dusky shades to the toes**; **tail blackish;** skull large; **incisive foramina equal to or shorter than the palatal bridge**; palatal bridge short; auditory bullae large; **maxillary toothrow slightly narrower posteriorly than anteriorly; middle lobe of the last upper molars partially or completely divided by an inner reentrant angle; interpterygoidea fossa narrower**. Species group *fuscipes*.

**Comments:** The subspecies of *Neotoma fuscipes*: *N. f. annectens*, *N. f. bullatior*, *N. f. luciana*, *N. f. macrotis*, *N. f. martirensis*, *N. f. simplex*, and *N. f. streatori* were considered *Neotoma macrotis* (Matocq 2002a, b). *N. fuscipes* ranges from the Columbia River, Oregon south to northern Santa Barbara County, California, including western-central and northern Sierra Nevada (Map 162). It can be found in sympatry, or nearly so, with *N. bryanti* and *N. cinerea*. *N. fuscipes* can be distinguished from *N. bryanti* by having the tail brownish dorsally; hindfoot toes white; maxillary toothrow slightly narrower posteriorly than anteriorly, and middle lobe of the last upper molars partially or completely divided by an inner reentrant angle. From *N. cinerea*, by not having the tail flattened and bushy and first upper molars without a deep anterointernal reentrant angle. From *N. macrotis*, by its larger size and skull; color less gray; tail not bicolored; interpterygoid fossa less broad and incisive foramina equal to or shorter than the palatal bridge.

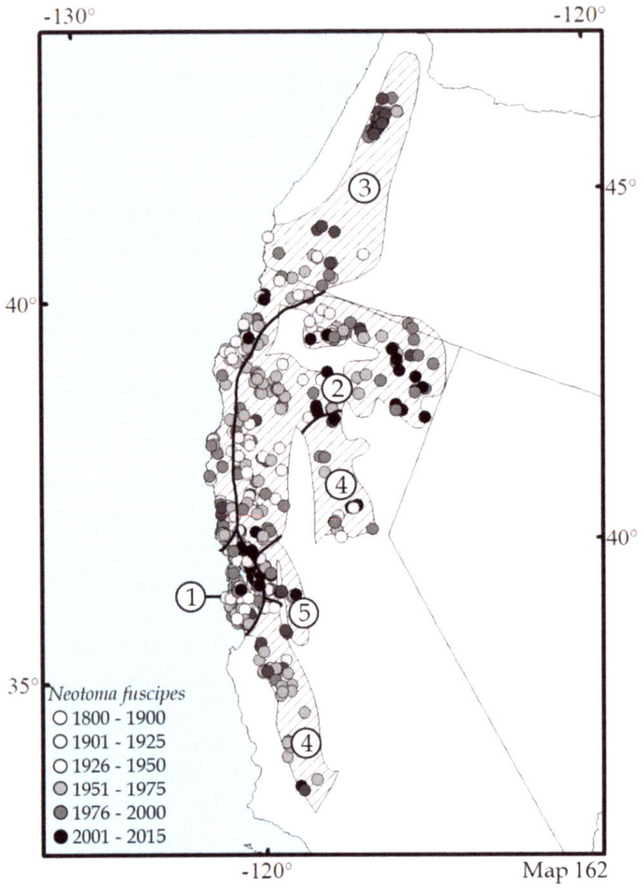


**Map 162** Distribution map of *Neotoma fuscipes*

**Additional Literature:** Goldman (1910), Carraway and Verts (1991a), Musser and Carleton (2005).

## *Neotoma goldmani* Merriam, 1903

Goldman's woodrat, rata de campo del Altiplano

Monotypic.

**Conservation:** *Neotoma goldmani* is listed as Least Concern by the IUCN (de Grammont and Cuarón 2016a).

**Characteristics:** Goldman's woodrat is the smallest in the genus, **total length 265.0–285.0 mm and skull length 36.9–38.4 mm**. Dorsal pelage cream yellow, paler than the head and darker on the back, grizzly with darker hair; fur of underparts white, on the throat plumbeous at the base; limbs white internally; **tail short-haired, blackish dorsally and whitish ventrally**; **premaxilla projections wide posteriorly**; auditory bullae small in proportion to the skull; anterior border of the interparietal bone more convex, posterior angle more developed; molars proportionately large. Species group *lepida*.

**Comments:** *Neotoma goldmani* is known from northern Chihuahua and southern Coahuila south through southern San Luis Potosí and Zacatecas (Map 163). It can be found in sympatry, or nearly so, with *N. leucodon* and *N. mexicana*. *N. goldmani* can be distinguished from all other species of *Neotoma* by its smaller size. From *N. leucodon*, by having the neck hair with the root plumbeous, and skull length less than 39.0 mm. From *N. mexicana*, by having the first upper molars with the antero-internal reentrant angle not deeper, cutting less than halfway across the anterior lobe.

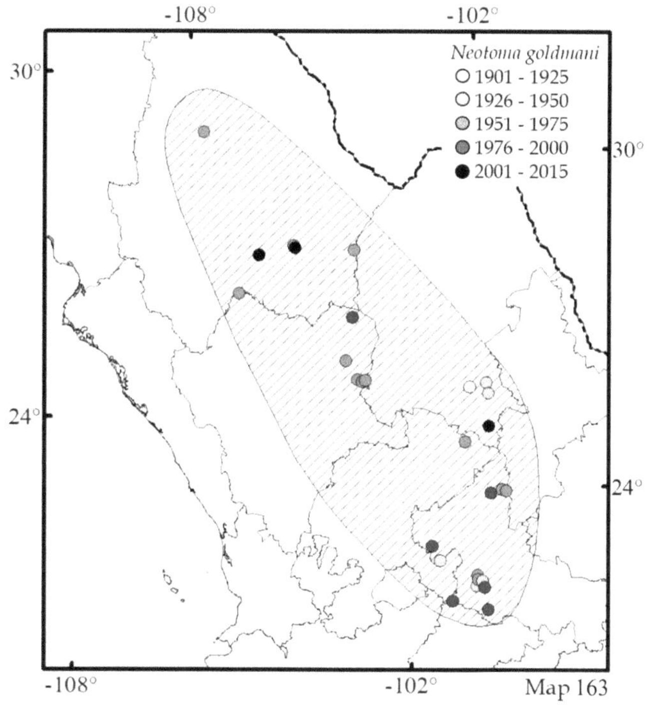


**Map 163** Distribution map of *Neotoma goldmani*

**Additional Literature:** Rainey and Baker (1955), Anderson (1972), Matson and Baker (1986), Hrachovy et al. (1996), Musser and Carleton (2005).

## *Neotoma insularis* Townsend, 1912

Ángel de la Guarda Island woodrat, rata de campo de Isla Ángel de la Guarda

Monotypic.

**Conservation:** *Neotoma insularis* is listed as Endangered by the Norma Oficial Mexicana (DOF 2019) and as Data Deficient by the IUCN (Patton and Álvarez-Castañeda 2017).

**Characteristics:** Ángel de la Guarda Island woodrat is medium sized; total length 287.0–340.0 mm and skull length 37.0–42.3 mm. Dorsal pelage grayish brown to rich creamy buff, darker in the back and top of the head; fur of underparts whitish grizzly with yellowish shades; throat hair leaden gray at the base, never white; spot completely white between the hind feet and forelimbs; ears with slightly pale brown hair; **tail short, approximately 71% of the head-and-body length,** with long hair and bicolored, brownish dorsally and whitish ventrally; limb dorsal region white; **skull short and stocky, with a noticeably short and broad rostrum and squared zygomatic arches**; contact area of the frontal bone with the lacrimal bone much greater than the contact area with the maxillary bone; septum of the incisive foramen with the vomerine portion short and elongated vacuity; **auditory bullae large but narrow**; **anteroloph of the first upper molars with a shallow anteromedial notch**. Species group *lepida*.

**Comments:** *Neotoma insularis* was reinstated as a distinct species by Patton et al. (2007); it was previously regarded as a subspecies of *N. lepida* (Burt 1932; Hall and Kelson 1959; Hall 1981). *N. insularis* is only known from Ángel de la Guarda Island in the Gulf of California, Baja California (Map 164), and cannot be found in sympatry, or nearly so, with any other species of *Neotoma.* It can be distinguished from *N. bryanti* by having approximately less than 75% of the head-and-body length and the maxillo-frontal bone suture joins the lacrimal bone beyond its midpoint.

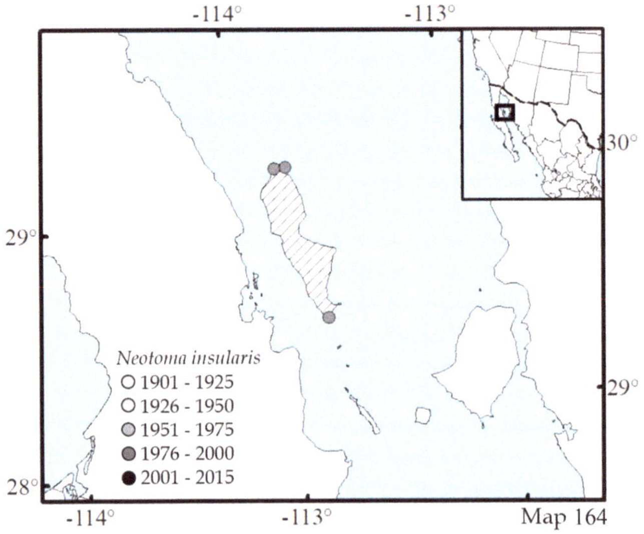


**Map 164** Distribution map of *Neotoma insularis*

**Additional Literature:** Álvarez-Castañeda and Cortés-Calva (1999).

## *Neotoma lepida* Thomas, 1893

Desert woodrat, rata de campo del desierto

1. *N. l. lepida* Thomas, 1893. From southeastern Oregon, eastern and southeastern California to northeastern Baja California east through Nevada, southwestern Idaho, and western Utah.
2. *N. l. marshalli* Goldman, 1939. Restricted to Carrington and Stansbury Island, Great Salt Lake, northwest Utah.
3. *N. l. monstrabilis* Goldman, 1932. Southern Nevada, northern Grand Canyon, Arizona, eastern Utah, and western Colorado.

**Conservation:** *Neotoma lepida* is listed as Least Concern by the IUCN (Lacher et al. 2017)

**Characteristics:** Desert woodrat is medium sized; total length 260.0–326.0 mm and skull length 32.7–40.7 mm. Dorsal pelage yellowish or between buff and cream buff, darker in the back and top of the head; flanks pale or yellowish brown; fur of underparts whitish grizzly with yellowish and plumbeous shades at the base, except on the throat, foreleg inner sides, and small on the pectoral and inguinal regions, white at the base; feet white; **tail short, approximately 80% of the head-and-body length, hair not sharply bicolored;** skull angular, **small**; toothrows and incisors short; contact area of the frontal bone with the lacrimal bone less than one-half of the contact length with the maxillary; incisive foramen extending to the anterior margin of the first maxillary molars; no ridge on the supraorbital edge; **auditory large in proportion to the skull; anteroloph of the first upper molars with a shallow anteromedial notch**. Species group *lepida.*

**Comments:** *Neotoma lepida devia* was elevated as valid species (Patton et al. 2007). The following subspecies, *N. l. abbreviata* (Espíritu Santo Island), *N. l. aridicola*, *N. l. arenacea*, *N. bunkeri* (Cedros Island), *N. l. felipensis*, *N. l. latirostra* (Danzante Island), *N. l. marcosensis* (San Marcos Island), *N. l. molagrandis*, *N. l. notia*, *N. l. nudicauda* (Carmen Island), *N. l. perpallida* (San José Island), and *N. l. vicina* (San Francisco Island), are now allocated to *N. bryanti*; and *N. l. aureotunicata*, *N. l. bensoni*, *N. l. devia*, *N. l. flava*, and *N. l. harteri* are allocated to *N. devia; N. l. sanrafaeli* is considered synonym of *N. lepida monstrabilis*, and *N. l. bella*, *N. l. desertorum*, *N. l. grinnelli*, and *N. l. nevadensis* synonym of *N. l. lepida* (Patton et al. 2007). *N. lepida* ranges in southeastern Oregon, southwestern Idaho, Utah, western Colorado, Arizona northern Grand Canyon, and central California south through the northern Baja California Peninsula (Map 165). It can be found in sympatry, or nearly so, with *N. albigula*, *N. bryanti*, *N. cinerea*, *N. fuscipes*, *N. macrotis*, and *N. mexicana*. *N. lepida* can be distinguished from *N. albigula* by having a darker coloration; hairs on the throat with the root not white; rostrum slender; nasals wider and more evenly rounded posteriorly; bullae smaller. From *N. cinerea*, by lacking a flattened and bushy tail, and first upper molars without a deep anterointernal reentrant angle. From *N. bryanti*, by having the maxillo-frontal bone suture joining the lacrimal bone behind its midpoint; anteroloph of the first upper molars with a shallow anteromedial notch. From *N. devia*, by having a different pattern in the lacrimal bone position, in *N. devia*, the contact area of the frontal bone with the lacrimal bone is less than half the contact area of the maxillary bone with the lacrimal bone. From *N. fuscipes* and *N. macrotis*, by having the maxillary toothrow much narrower posteriorly than anteriorly; middle lobe of the last upper molars not divided by an inner reentrant angle. From *N. insularis*, by having a longer tail, approximately greater than 75% of the head-and-body length, and the maxillo-frontal bone suture joins the lacrimal bone behind its midpoint. From *N. mexicana*, by having the first upper molars with the anterointernal reentrant angle not deeper, cutting less than halfway across the anterior lobe; skull larger and heavier and rostrum relatively short and more or less slender. From *N. stephensi*, by having the tail not fully haired; the supraorbital ridges do not continue forward to the rostrum; skull larger, more angular; zygomatic arches less spreading and squarely shaped; frontal region narrow and less flattened, and braincase less smoothly rounded.

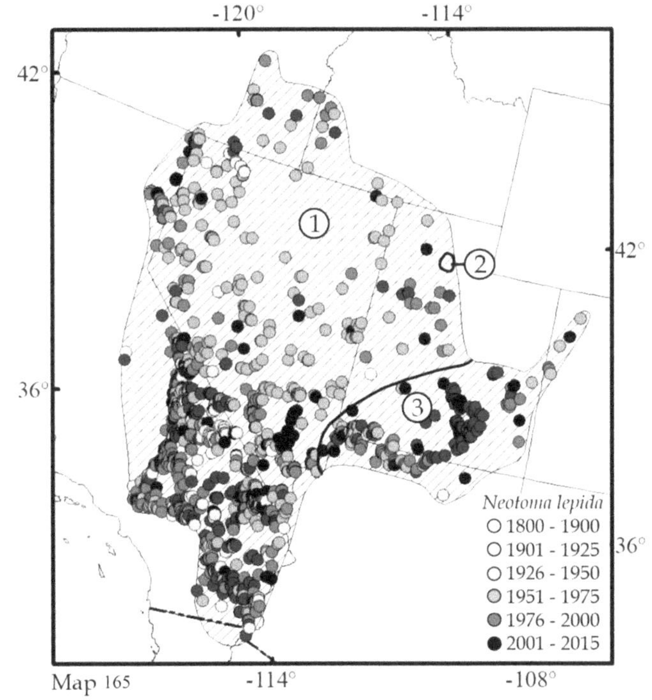


**Map 165** Distribution map of *Neotoma lepida*

**Additional Literature:** Goldman (1910), Verts and Carraway (2002), Musser and Carleton (2005).

## *Neotoma leucodon* Merriam, 1894

White-toothed woodrat, rata de campo de dientes blancos

1. *N. l. durangae* Allen, 1903. Southeastern Chihuahua, Coahuila, and northeastern Durango, Texas, and New Mexico.
2. *N. l. latifrons* Merriam, 1894. Restricted to Queréndaro, Michoacán.
3. *N. l. leucodon* Merriam, 1894. From southeastern Durango, Zacatecas, and San Luis Potosí south through northwestern Michoacán and eastern Hidalgo.
4. *N. l. melas* Dice, 1929. Restricted to Malpais Spring, Lincoln County, central New Mexico.
5. *N. l. robusta* Blair, 1939. Restricted to Jeff Davis County, southwestern Texas.
6. *N. l. subsolana* Álvarez, 1962. Eastern Coahuila, Nuevo León, and southwestern Tamaulipas.
7. *N. l. warreni* Merriam, 1908. From southern Colorado south through Texas, including eastern New Mexico.

**Conservation:** *Neotoma leucodon* is listed as Least Concern by the IUCN (Timm et al. 2016f).

**Characteristics:** White-toothed woodrat is medium sized; total length 333.0–388.0 mm and skull length 38.1–42.5 mm. **Dorsal pelage creamy buff, with blackish hairs grizzle dorsally**, flanks clearest; fur of underparts white, in the pectoral and inguinal regions hair pure white to the roots; hair completely white between the hind limbs and forelimbs; ears proportionally large; tail shorter than the head-and-body length, bicolored, blackish dorsally, and white ventrally; **rostrum and dentition heavy; supraorbital ridges developed, forming slightly projecting frontal shelves**; incisive foramina extended to the anterior margin of the first maxillary molars; auditory bullae larger in proportion to the skull. Species group *albigula*.

**Comments:** *Neotoma leucodon* was a subspecies of *N. albigula* (Goldman 1910) and was elevated to a full species (Edwards et al. 2001). Its ranges from the rivers Grande and Conchos east through, including Colorado, central and eastern New Mexico, and Texas, south through Jalisco, Guanajuato, and Tlaxcala, and from Nuevo León, Tamaulipas, San Luis Potosí and Hidalgo west through central and eastern Durango (Map 166). It can be found in sympatry, or nearly so, with *N. angustapalata*, *N. goldmani*, *N. mexicana*, and *N. micropus*. *N. leucodon* can be distinguished from *N. albigula* by its larger size, hind feet on average greater than 32.0 mm, dorsal coloration darker, rostrum and dentition heavier, and supraorbital ridges more developed, forming slightly frontal shelves. From *N. angustapalata*, by its smaller average size, skull length (50.22 ± 1.72 *vs* 43.03 ± 2.15), nasal length (24.23 ± 0.75 *vs* 23.14 ± 1.17), and least interorbital constriction (5.92 ± 0.22 *vs* 5.68 ± 0.37; Rogers et al. 2011); a shallow anterointernal reentrant angle on the first upper molars. From *N. goldmani*, by its smaller size; neck hair pure white to the roots, and skull length less than 39.0 mm. From *N. mexicana*, by having the first upper molars with the anterointernal reentrant angle not deeper, cutting less than halfway across the anterior lobe; skull larger and heavier and rostrum relatively short and more or less slender. From *N. micropus*, by having the palate without a posterior medial spine and posterior septum which divided the palatine foramina not complete.

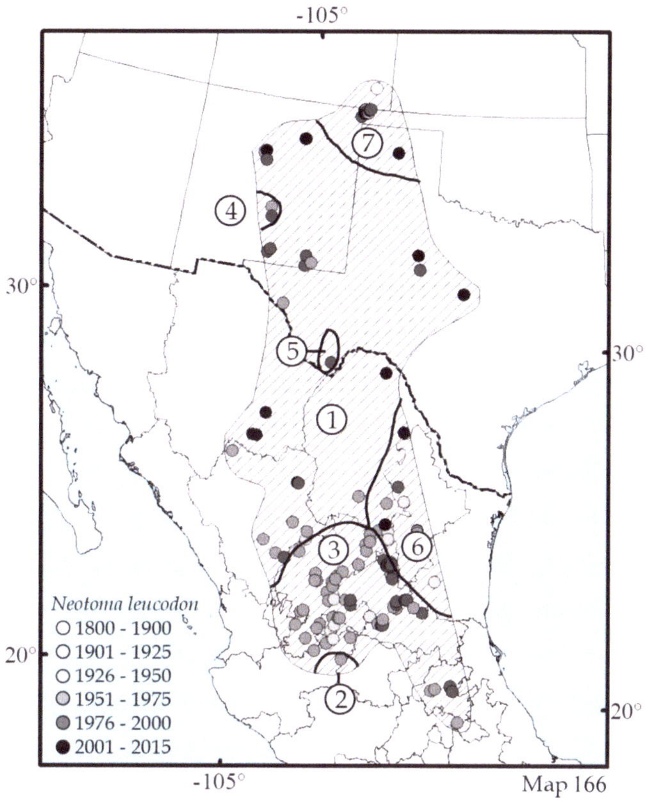


**Map 166** Distribution map of *Neotoma leucodon*

**Additional Literature:** Hall and Genoways (1970), Macêdo and Mares (1988), Rogers et al. (2011).

## *Neotoma macrotis* Thomas, 1893

Big-eared woodrat, rata de campo de orejas grandes

1. *N. m. bullatior* Hooper, 1938. Restricted to the mountains southwestern of Central Valley, California.
2. *N. m. luciana* Hooper, 1938. Coastal mountains, from Monterey to San Luis Obispo area, California.
3. *N. m. macrotis* Thomas, 1893. From San Luis Obispo area, California south through northern Baja California.
4. *N. m. martirensis* Orr, 1934. Northern Baja California, México.
5. *N. m. simplex* True, 1894. South and southeastern mountains of central Valley, California.
6. *N. m. streatori* Merriam, 1894. Sierra Nevada, from the area of Sequoia National Park to Sacramento, California.

**Conservation:** *Neotoma macrotis* is listed as Least Concern by the IUCN (Álvarez-Castañeda and Lacher 2016a).

**Characteristics:** Big-eared woodrat is medium sized; total length 335.0–390.0 mm and skull length 36.1–38.8 mm. Dorsal pelage grayish brown, more or less suffused with buff or ochraceous buff, grizzly with darker hair on the back; fur of under-parts white with leaden gray shades at the base; neck, chest, and inguinal region with white hairs from the root; facial area grayish; ears large; ankles dusky; forefeet white; hind feet clouded with dusky shades to the toes, toes white; **tail distinctly bicolored, brownish black dorsally, whitish ventrally**; skull small; **incisive foramina decidedly longer than the palatal bridge**; palatal bridge short; auditory bullae small, short, and rounded; **maxillary toothrow slightly narrower posteriorly than anteriorly; middle lobe of the last upper molars partially or completely divided by an inner reentrant angle; interpterygoidea fossa broader**. Species group *fuscipes*.

**Comments:** *Neotoma macrotis* was recognized as a distinct species (Matocq 2002a, b); it was previously regarded as a subspecies of *N. macrotis*, including the subspecies (*N. f. annectens*, *N. f. bullatior*, *N. f. luciana*, *N. f. martirensis*, *N. m. simplex*, and *N. f. streatori*). Matocq (2002a, b) does not discuss the subspecific status; for this reason, it is still considered under *N. macrotis*. *N. macrotis* ranges from central and eastern California south through northwestern Baja California (Map 167). It can be found in sympatry, or nearly so, with *N. bryanti* and *N. lepida*. It can be distinguished from them by having the tail brownish dorsally; toes of the hind feet white; maxillary toothrow slightly narrower posteriorly than anteriorly; middle lobe of the last upper molars partially or completely divided by an inner reentrant angle. From *N. fuscipes*, by its smaller size and skull; color grayer; tail distinctly bicolored; interpterygoid fossa much broader and incisive foramina decidedly longer than the palatal bridge.

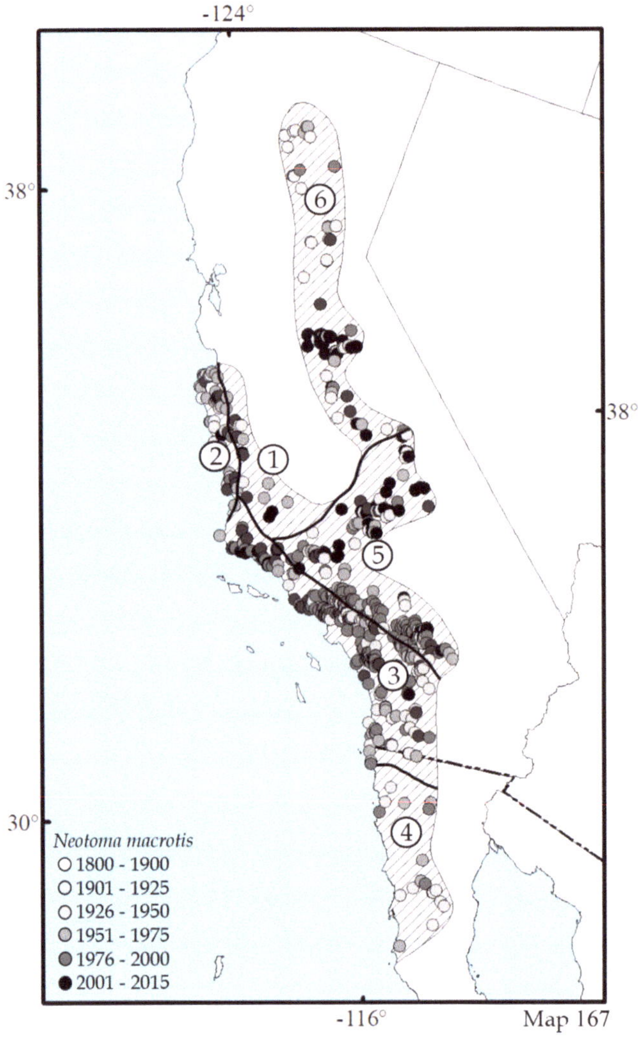


**Map 167** Distribution map of *Neotoma macrotis*

**Additional Literature:** Goldman (1910), Orr (1934), Hall (1981), Carraway and Verts (1991a), Álvarez-Castañeda and Cortés-Calva (1999), Musser and Carleton (2005).

## *Neotoma magister* Goldman, 1904
Appalachian woodrat, rata de campo de los Appalachian

Monotypic.

**Conservation:** *Neotoma magister* is listed as Near Threatened by IUCN (Linzey et al. 2008b).

**Characteristics:** Appalachian woodrat is large sized; **total length 311.0–451.0 mm and skull length 42.7–47.0 mm.** Dorsal pelage from brownish to buffy gray, back grizzle with blackish shades, pelage coarse; fur of underparts white with the hairs white from root, belly sides with hair plumbeous at the base, and axillae creamy buff; head more grayish than the back; nose posterior area and flank dark; **ears large;** foot top dusky wash through the base**; tail fully haired, as long as the head-and-body length**, dark brownish dorsally and slightly paler ventrally; skull very large and elongated; rostrum long; nasals long and narrow, truncate or slightly emarginated posteriorly; no edge on the supraorbital ridge; **auditory bullae larger in proportion to the skull, inflated ventrally with an obtuse anterior edge and almost parallel to the skull medial axis**. Species group *floridana*.

**Comments:** *Neotoma magister* was reinstated as a distinct species by Schwartz and Odum (1957) and Edwards and Bradley (2001); it was previously regarded as a subspecies of *N. floridana* (Burt and Barkalow 1942). *N. magister* has a historical range from western Connecticut, southeastern New York, northern New Jersey, and from central and south Pennsylvania southwest through northeastern Alabama (observed in several cave systems) and northwestern North Carolina, including Maryland, Tennessee, Kentucky, Indiana, and West Virginia (Map 168). *N. magister* is sympatric, or nearly so, with *N. floridana*, from which it differs by its anteriorly rounded, narrower interpterygoid fossa, small sphenopalatine vacuities, more elongated and anteriorly tapered auditory bullae, very broad mastoid process of the squamosal bone, and first upper molars with a deeper anterointernal reentrant angle.

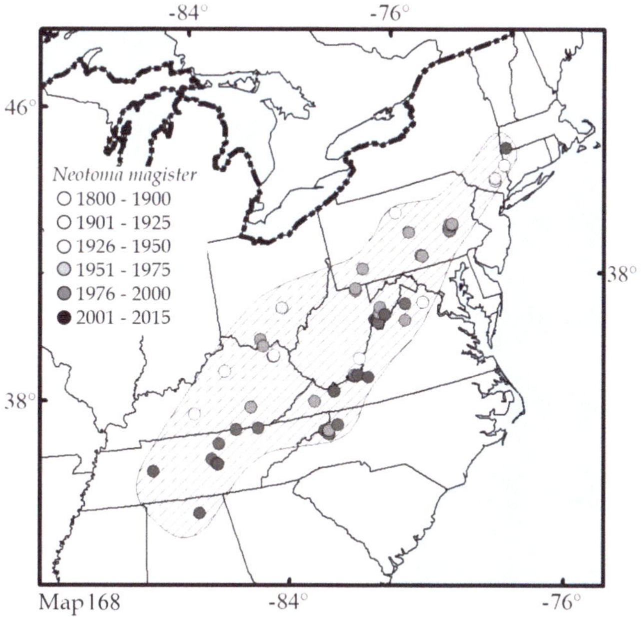


**Map 168** Distribution map of *Neotoma magister*

**Additional Literature:** Goldman (1910), Hall (1981), Musser and Carleton (2005), Castleberry et al. (2006).

## *Neotoma melanura* Merriam, 1894

Dark woodrat, rata de campo negruzca

Monotypic.

**Conservation:** *Neotoma melanura* (as *Neotoma albigula*) is listed as Least Concern by the IUCN (Lacher and Álvarez-Castañeda 2016).

**Characteristics:** Dark woodrat is medium sized; total length 335.0–380.0 mm and skull length 42.0–46.5 mm. Dorsal pelage dark buff and dull buff, **dorsal area darker**; flanks paler; fur of underparts white; central face and outer limb sides pale drab; **throat, neck, and chest with white hair and plumbeous at the base**; feet large and white; tail shorter than the head-and-body length, with hairs very short and bicolored, dark dorsally and whitish ventrally; ears proportionally large; skull long and slender, anterointernal reentrant angle in the first upper molars shallow; palate concave posteriorly; **auditory bullae small in proportion to the skull**, ventrally inflated with an obtuse anterior edge and almost parallel to the skull medial axis. Species group *albigula*.

**Comments:** *Neotoma melanura* was a subspecies of *N. albigula* (Goldman 1905), now elevated to a full species (Bradley and Mauldin 2016). It ranges from northern Sinaloa, southern Sonora, and western Chihuahua (Map 169). It can be found in sympatry, or nearly so, with *N. albigula*, from which it differs by its smaller size, darker coloration, tail black dorsally, skull less massive, slender rostrum, slightly smaller teeth, and relatively smaller auditory bullae.

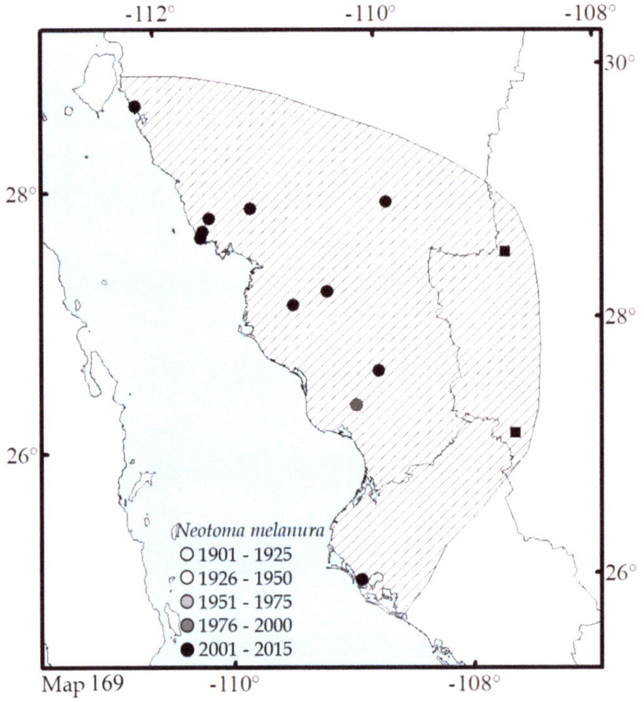


**Map 169** Distribution map of *Neotoma melanura*

**Additional Literature:** Merriam (1894b), Goldman (1910), Macêdo and Mares (1988), Álvarez-Castañeda and Cortés-Calva (1999).

## *Neotoma mexicana* Baird, 1855

Mexican woodrat, rata de campo mexicana

1. *N. m. atrata* Burt, 1939. Restricted to Lincoln County, central-southern New Mexico.
2. *N. m. bullata* Merriam, 1894. Restricted to Santa Catalina Mountains, Pima County, central-southern Arizona.
3. *N. m. distincta* Bangs, 1903. Restricted to Xalapa, central-western Veracruz.
4. *N. m. eremita* Hall, 1955. Northwestern Jalisco and southwestern Nayarit.
5. *N. m. fallax* Merriam, 1894. Central Colorado.
6. *N. m. griseoventer* Dalquest, 1951. Central-eastern San Luis Potosí.
7. *N. m. inopinata* Goldman, 1933. Northeastern Arizona, northwestern New Mexico, southeastern Utah, and southwestern Colorado.
8. *N. m. inornata* Goldman, 1938. Central-western and northwestern Coahuila.
9. *N. m. mexicana* Baird, 1855. From southern Durango and western Zacatecas north through southern Arizona and New Mexico.
10. *N. m. navus* Merriam, 1903. From eastern Durango to western Nuevo León, including northern San Luis Potosí and Zacatecas.
11. *N. m. ochracea* Goldman, 1905. Restricted to Guadalajara, Jalisco.
12. *N. m. pinetorum* Merriam, 1893. Western New Mexico and Central Arizona.
13. *N. m. scopulorum* Finley, 1953. From southern New Mexico to southeastern of Colorado.
14. *N. m. sinaloae* Allen, 1898. From southern Sonora south through northern Nayarit.
15. *N. m. tenuicauda* Merriam, 1892. From southern Nayarit and Zacatecas, and southwestern San Luis Potosí south through central Michoacán.
16. *N. m. torquata* Ward, 1891. From central San Luis Potosí south through eastern Michoacán, Estado de México, Morelos, and Puebla.

**Conservation:** *Neotoma mexicana* is listed as Least Concern by the IUCN (Linzey et al. 2016a).

**Characteristics:** Mexican woodrat is medium sized; total length 326.0–360.0 mm and skull length 42.0–44.2 mm. Dorsal pelage varying widely from pale to dark brown and from reddish yellow to cinnamon, rarely in gray, grizzly with blackish shades that increase toward the back; top of the head paler than the back; flanks yellowish; fur of underparts whitish, with leaden gray shades at the base; **tail moderately long and bicolored**, grayish dorsally and whitish ventrally, skull small and light; rostrum relatively longer and slender; nasals ending posteriorly near the anterior plane of the orbits, only slightly exceeded by the premaxillae; zygomatic arches spreading and squarely shaped, with the sides nearly parallel; **first upper molars with two internal and external notches**; **frontal bones constricted near the middle;** auditory bullae medium, short, and well rounded. Note: *N. mexicana*, *N. ferruginea*, and *N. picta* can be mainly separated by genetic data. Species group *mexicana*.

**Comments:** *Neotoma mexicana ferruginea* (including the subspecies *N. m. chamula*, *N. m. isthmica*, and *N. m. tropicalis*), and *N. m. picta* (with the subspecies *N. m. parvidens*), have been reinstated to the species rank (Edwards and Bradley 2003; Longhofer and Bradley 2006). *N. mexicana* ranges from southeast Utah and central Colorado southwards to the highlands of Oaxaca (Map 170). It can be found in sympatry, or nearly so, with *N. albigula*, *N. cinerea*, *N. floridana*, *N. goldmani*, *N. lepida*, *N. leucodon*, and *N. stephensi*. *N. mexicana* can be distinguished from all the other species of *Neotoma* by having the first upper molars with a deeper anterointernal reentrant angle, cutting more than halfway across the anterior lobe; small and light skull; and relatively long and more slender rostrum. From *N. angustapalata*, by its smaller size; paler fur of underparts; tail not strongly scaly in appearance and with hair; sides of the interpterygoid fossa similar in the anterior and posterior ends; and wider palatine breadth. From *N. cinerea*, by not having the tail flattened and bushy and first upper molars without a deep anterointernal reentrant angle. From *N. stephensi*, by having the tail not fully haired and the supraorbital ridges do not continue forward to the rostrum.

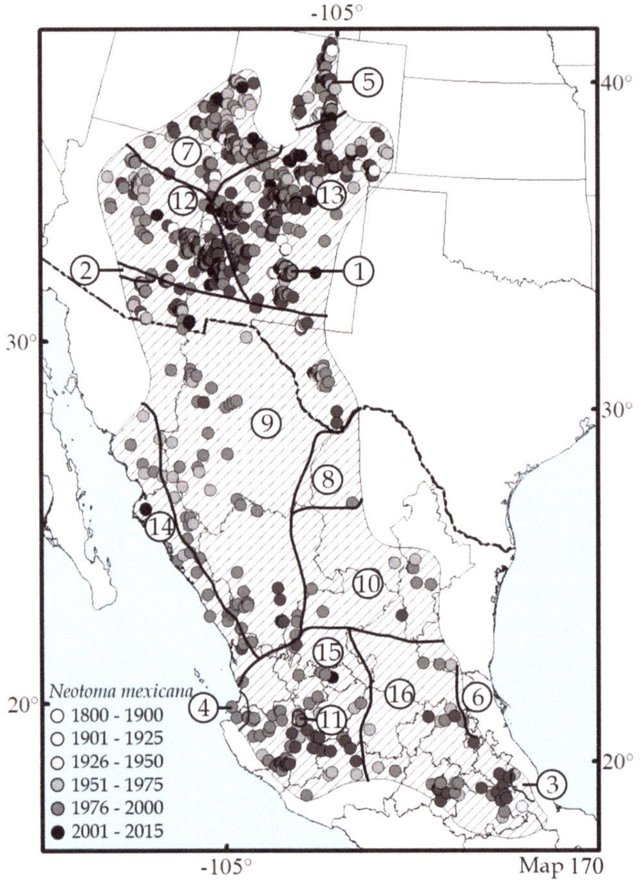


**Map 170** Distribution map of *Neotoma mexicana*

**Additional Literature:** Goldman (1910), Baker (1956), Birney and Jones (1972), Cornely and Baker (1986), Musser and Carleton (2005), Hernández-Canchola et al. (2021a).

## *Neotoma micropus* Baird, 1855

Southern Plain woodrat, rata de campo de las planicies del sur

1. *N. m. canescens* Allen, 1891. From southwestern Kansas, western Oklahoma, and Texas west through New Mexico and northern Chihuahua.
2. *N. m. leucophaea* Goldman, 1933. Restricted to White Sands, southern New Mexico.
3. *N. m. littoralis* Goldman, 1905. Restricted to southeastern Tamaulipas.
4. *N. m. micropus* Baird, 1855. From southwestern Kansas, western Oklahoma, and Texas east through central Kansas, Oklahoma, southern Texas, Coahuila, Nuevo León, and Tamaulipas.
5. *N. m. planiceps* Goldman, 1905. Southeastern San Luis Potosí.

**Conservation:** *Neotoma micropus* is listed as Least Concern by the IUCN (Lacher et al. 2016t).

**Characteristics:** Southern Plain woodrat is medium sized; total length 334.0–411.0 mm and skull length 44.2–52.9 mm. **Dorsal pelage pale gray** to plumbeous gray, rarely gray or brown; fur of underparts white, leaden gray shades at the base, except for the throat, neck, and chest, hair is white through the base; feet white; **tail shorter than the head-and-body length, bicolored, and hair thin,** blackish dorsally and grayish ventrally; **hair short and rather harsh**; **palate with a posterior medial spine; complete posterior half of the septum dividing the anterior palatine foramina;** rostrum heavy; nasals narrower posteriorly; zygomatic arches more widely spreading posteriorly; auditory bullae larger in proportion to the skull, ventrally inflated with an obtuse anterior edge and almost parallel to the skull medial axis.

**Comments:** *Neotoma micropus* ranges from southeastern Colorado and southeastern Kansas south through northern Chihuahua, northeastern Coahuila, Nuevo León, Tamaulipas, and San Luis Potosí (Map 171). It can be found in sympatry, or nearly so, with *N. albigula* and *N. leucodon*. *N. micropus* can be distinguished from *N. albigula* and *N. leucodon* by having the palate with a posterior medial spine and complete posterior half of the septum dividing the anterior palatine foramina. From *N. angustapalata*, by having a shallow anterointernal reentrant angle on the first upper molars and a maxillovomerine notch in the septum.

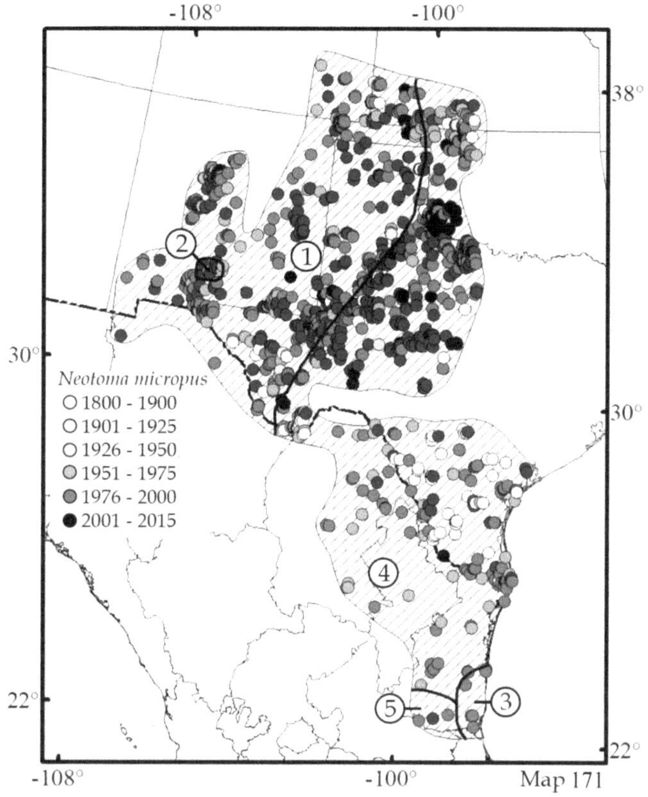


**Map 171** Distribution map of *Neotoma micropus*

**Additional Literature:** Baker (1951), Birney (1973), Braun and Mares (1989), Musser and Carleton (2005).

## *Neotoma nelsoni* Goldman, 1905

Nelson's woodrat, rata de campo de Perote

Monotypic.

**Conservation:** *Neotoma nelsoni* is listed as Critically Endangered (B1ab (iii); extent of occurrence less than 100 km$^2$ at only one location and under continuing decline in area, extent, and quality of habitat) by the IUCN (Álvarez-Castañeda 2018b). The total geographic ranges are estimated to be less than 100 km$^2$ (González-Ruiz et al. 2006).

**Characteristics:** Nelson's woodrat is medium sized; total length 347.0–380.0 mm and skull length 45.0–48.0 mm. Dorsal pelage pale cinnamon grizzly with brown shades, especially in the middle region; fur of underparts white with leaden gray shades at the base, except for the pectoral region where hair is white through the base; top of the head along the back darker brown; cheeks and middle of the face grayish brown; ears large; nose and upper lip blackish; feet white; dorsal foot white; **tail with few hairs and unicolor**, slightly darker than in the dorsum, smoky brown dorsally, slightly paler and grayer ventrally; auditory bullae larger in proportion to the skull, ventrally inflated with an obtuse anterior edge and almost parallel to the skull medial axis; palate with a posterior medial spine; **maxillo-vomerine notch present**; **ascending branches of the maxillary bone projecting beyond the nasal posterior margin**. Species group *albigula*.

**Comments:** *Neotoma nelsoni* is known only from Cofre de Perote, Veracruz (Map 172), and cannot be found in sympatry, or nearly so, with other species of *Neotoma*. It can be distinguished from other species by having the ascending branches of the maxillary bone projecting beyond the nasal posterior margin.

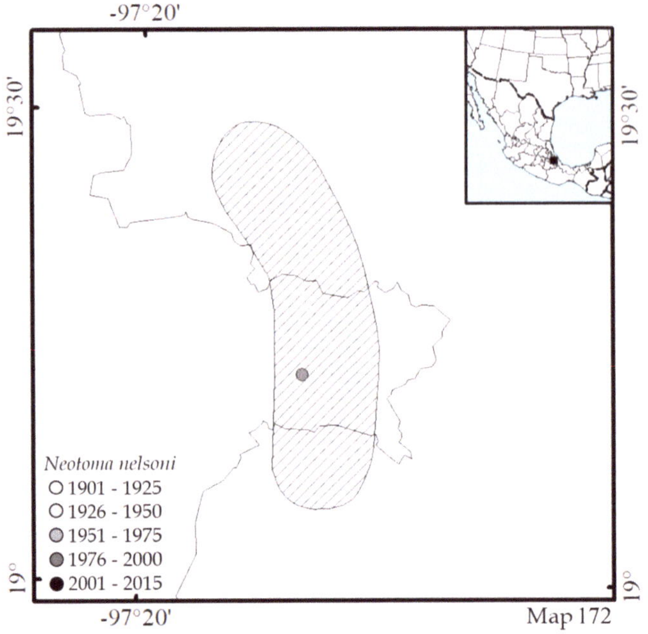


**Map 172** Distribution map of *Neotoma nelsoni*

**Additional Literature:** Goldman (1905, 1910), Musser and Carleton (2005), Álvarez-Castañeda (2018c).

## *Neotoma palatina* Goldman, 1905

Bolaños woodrat, rata de campo de Jalisco

Monotypic.

**Conservation:** *Neotoma palatina* is listed as Vulnerable (B1ab (iii): extent of occurrence less than 20,000 km$^2$, severely fragmented, and under continuing decline in area, extent, and quality of habitat) by the IUCN (Álvarez-Castañeda 2018c).

**Characteristics:** Bolaños woodrat is small sized; total length 326.0–404.0 mm and skull length 36.5–40.5 mm. Dorsal pelage yellowish ocher; fur of underparts white with leaden gray shades at the base, except for the throat, neck, chin, and chest; **ears proportionally small;** cheeks and flanks buffy, becoming much darker on the dorsal region from the abundant admixture of black hairs; muzzle brownish gray; **hair completely white between the hind limbs**; feet white; tail shorter than the body, with very short hair, bicolored, blackish dorsally and soiled whitish ventrally; skull large and robust, **well arched up to the front of the zygomatic arches**; **vomer bone extending backward**; interorbital region narrow; bullae large in proportion to the skull, ventrally inflated with an obtuse anterior edge and almost parallel to the skull medial axis; interpterygoidea narrow and variable in shape; **no sphenopalatine cavities**. Species group *albigula*.

**Comments:** *Neotoma palatina* is only known from eastern-central northern Jalisco and western Zacatecas (Map 173); it is not found in sympatry, or nearly so, with any other species of *Neotoma*. It can be distinguished from other species by having the complete sphenopalatine vacuity closure, with only small irregular vacuities along the palatopterygoid suture, which opens outward into the external pterygoid fossa.

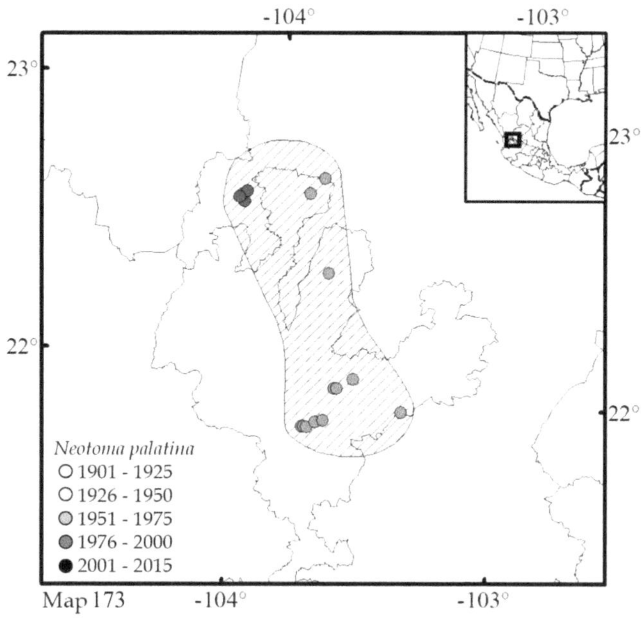


**Map 173** Distribution map of *Neotoma palatina*

**Additional Literature:** Goldman (1905, 1910), Hall and Genoways (1970), Musser and Carleton (2005), Álvarez-Castañeda (2018d).

## *Neotoma picta* Goldman, 1904

Tehuantepec woodrat, rata de campo de Tehuantepec

1. *N. m. parvidens* Goldman, 1904. Coast of southeastern Guerrero and southwestern Oaxaca.
2. *N. p. picta* Goldman, 1904. Guerrero and east Oaxaca.

**Conservation:** *Neotoma picta* (as *N. mexicana*) is listed as Least Concern by the IUCN (Linzey et al. 2016a).

**Characteristics:** Tehuantepec woodrat is medium sized; total length 294.0–395.0 mm and skull length 42.6–48.4 mm. Dorsal pelage rich orange buff to ferruginous, back grizzle with black-tipped hairs; fur of underparts white with white shades at the base on the throat, sometimes in the middle of the breast, and inner sides of the forelegs; head covered with black-tipped hairs; head side brightest; shoulders and flanks moderately darkened; toes white; forefoot yellowish white; hind feet to the toes irregularly clouded with dusky shades; tail indistinctly bicolored, dusky dorsally and paler ventrally; skull very large, elongated; **nasals truncate instead of emarginated posteriorly; frontal bones relatively narrower posteriorly;** frontal bones broad, interorbital bone constriction from the anterior section to the middle; sphenopalatine vacuities small; auditory bullae small, short, and rounded. Note: *Neotoma mexicana*, *N. ferruginea*, and *N. picta* can be separated mainly by genetic data. Species group *mexicana*.

**Comments:** *Neotoma picta* was reinstated as a distinct species by Longhofer and Bradley (2006) and Hernández-Canchola et al. (2021a); it was previously regarded as a subspecies of *N. mexicana* (Hall 1955), including the subspecies *N. m. picta* and *N. m. eremita* with ranges in Guerrero (Map 174). The specimens previously considered *N. m. picta* with ranges in Oaxaca should be considered *N. ferruginea* and those of *N. mexicana eremita* from Guerrero as *N. picta*. It ranges in Sierra Madre of Oaxaca and Guerrero, and can be found in sympatry, or nearly so, with *N. ferruginea*, from which it can be distinguished by its small size; skull slightly larger in size and longer nasals.

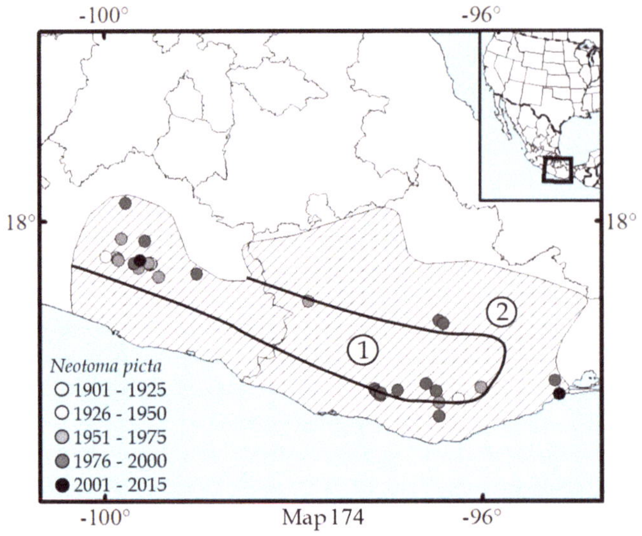


**Map 174** Distribution map of *Neotoma picta*

**Additional Literature:** Goldman (1910), Goodwin (1969).

## *Neotoma stephensi* Goldman, 1904

Stephen's woodrat, rata de campo de Arizona

1. *N. s. relicta* Goldman, 1932. Northeastern Arizona and northwest New Mexico.
2. *N. s. stephensi* Goldman, 1905. Central Arizona and western-central New Mexico.

**Conservation:** *Neotoma stephensi* is listed as Least Concern by the IUCN (Cassola 2016cb).

**Characteristics:** Stephen's woodrat is medium sized; total length 274.0–312.0 mm and skull length 34.0–39.0 mm. Dorsal pelage yellowish to grayish buff, grizzly with darker hair on the back; fur of underparts white to pale yellow with leaden gray shades at the base, except for the throat, chest, and groin where it is white through the base; dusky color extending down from the foot top to one-third of the way below the ankle; **tail semi-bushy pale gray to grayish brown dorsally and paler ventrally**; first upper molars with an anteromedial fold usually absent or shallow; third upper molars with the second loph long and narrow, extending diagonally across the tooth; nasals truncate posteriorly, rarely sharply pointed; auditory bullae larger in proportion to the skull and rounded; **supraorbital ridges remaining lateral as they continue forward to the rostrum**. Species group *lepida*.

**Comments:** *Neotoma stephensi* ranges from central Arizona and western New Mexico (Map 175). It can be found in sympatry, or nearly so, with *N. cinerea*, *N. lepida*, and *N. mexicana*. *N. stephensi* can be distinguished from all other species *Neotoma* by having the supraorbital ridges remaining lateral as they continue forward to the rostrum. Particularly, from *N. albigula*, by having the tail not fully haired; some hairs in the underparts without white shades at the base. From *N. cinerea*, by having the tail semi-bushy, not flattened; first upper molars without a deep anterointernal reentrant angle. From *N. lepida*, by having a smaller skull, less angular; zygomatic arches more spreading and squarely shaped; frontal region broader and more flattened and braincase more smoothly rounded. From *N. mexicana*, by having the first upper molars with the anterointernal reentrant angle not deeper, cutting less than halfway across the anterior lobe, and skull larger and heavier.

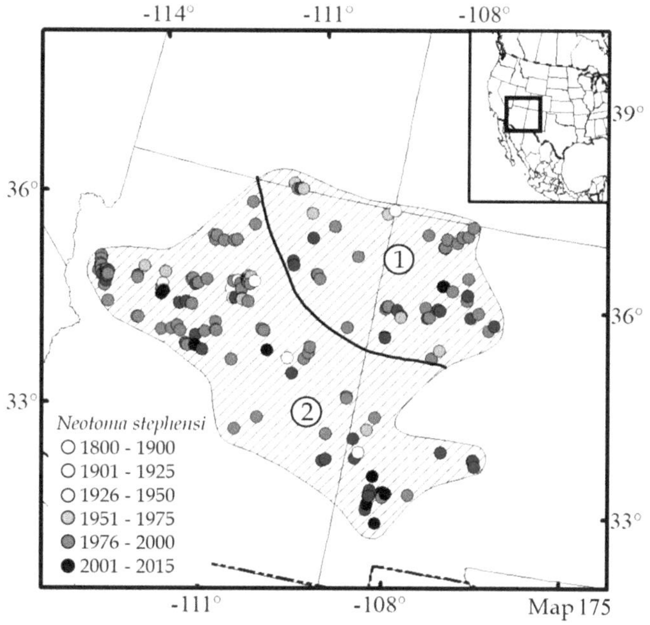


**Map 175** Distribution map of *Neotoma stephensi*

**Additional Literature:** Goldman (1910), Hoffmeister and de la Torre (1960), Hoffmeister (1971), Hall (1981), Jones and Hildreth (1989), Musser and Carleton (2005).

## Subgenus *Teanopus* Merriam, 1903

Tail cylindrical, tapered, and with hair short; hind feet naked below along the outer side at least to the tarsometatarsal joint; third lower molars with a reentrant enamel loop on the inner side passing obliquely forward in front of its mate on the outer side; **auditory bullae larger in proportion to the skull, ventrally inflated with an obtuse anterior edge and almost parallel to the skull medial axis**; sphenoid bone without edge on the supraorbital ridge; lower jaw with a distinct prominence over the roots of the incisors; angle elongated, its lower border strongly inflected and upturned.

### *Neotoma phenax* (Merriam, 1903)

Sonoran woodrat, rata de campo de Sonora

Monotypic.

**Conservation:** *Neotoma phenax* is listed as Special Protection by the Norma Oficial Mexicana (DOF 2019) and as Least Concern by the IUCN (Álvarez-Castañeda 2019c).

**Characteristics:** Sonoran woodrat is large sized; total length 330.0–431.0 mm andskull length 42.7–47.0 mm. Dorsal pelage brownish to buffy gray; fur of underparts grayish-white, chest paler; external limbs dark gray; head more grayish than the back; nose posterior area and flank dark; foot top to the base of the toes dusky wash; tail fully haired, as long as the head-and-body length, dark brownish dorsally, slightly paler ventrally; no ridge on the supraorbital edge; **auditory bullae larger in proportion to the skull, ventrally inflated with an obtuse anterior edge and almost parallel to the skull medial axis**. Species group *phenax*.

**Comments:** *Neotoma phenax* ranges from southern Sonora to northern Sinaloa (Map 176). It can be found in sympatry, or nearly so, with *N. albigula* and *N. mexicana*, from which it can be distinguished by having the fur of underparts grayish-white, particularly, not complete white areas in the chin, throat, neck, and chest; auditory bullae greater, inflated, and larger in proportion to the skull, with an obtuse anterior edge and almost parallel to the skull medial axis.

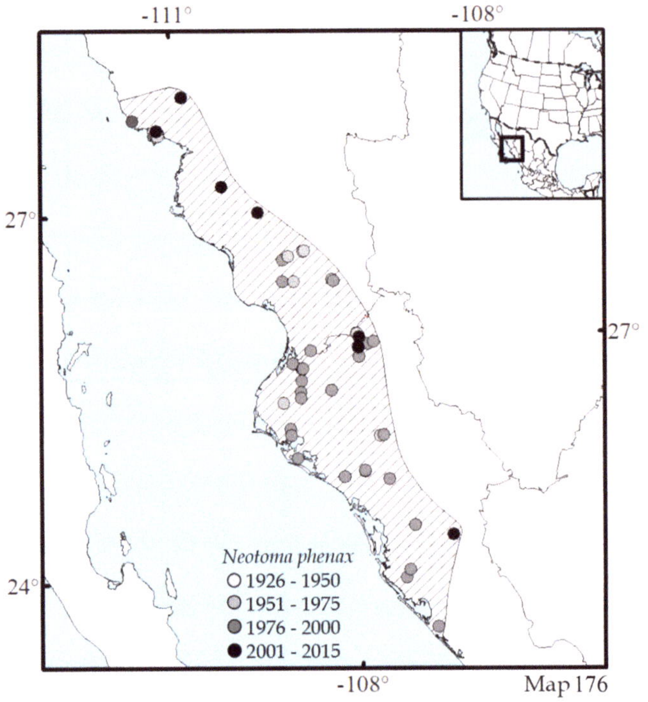


**Map 176** Distribution map of *Neotoma phenax*

**Additional Literature:** Merriam (1903), Burt and Barkalow (1942), Jones and Genoways (1978), Planz et al. (1996), Álvarez-Castañeda and Cortés-Calva (1999), Musser and Carleton (2005).

## Subgenus ***Teonoma*** Gray, 1843

**Tail large and bushy; hindfoot sole normally densely furred from the heels to the posterior tubercles; skull large and angular; temporal ridges prominent, diverging posteriorly to near the anterior border of the interparietal bone, whence they turn abruptly inward and again outward in crossing the interparietal bone to the lambdoid crest**; frontal region narrow, constricted near the middle, somewhat depressed and excavated above along the median line; maxillary arms of the zygomatic broad and heavy; auditory bullae large; **interpterygoid fossa narrow**.

### ***Neotoma cinerea*** Ord, 1815

Bushy-Tailed woodrat, rata de campo de cola peluda

1. *N. c. acraia* (Elliot, 1904). Eastern California, Nevada, and western Utah.
2. *N. c. alticola* Hooper, 1940. Northern Nevada, northeastern California, eastern Oregon, most southern Washington, western and southern Idaho.
3. *N. c. arizonae* Merriam, 1893. Northern Arizona, northwestern New Mexico, eastern Utah, and western Colorado.

4. *N. c. cinerea* (Ord, 1815). Western Colorado, central-western Idaho, Montana, most southwestern Saskatchewan, southern Alberta, and most southeastern British Columbia.
5. *N. c. cinnamomea* Allen, 1895. Southwestern Wyoming and northeastern New Mexico.
6. *N. c. drummondii* (Richardson, 1828). Eastern British Columbia, western Alberta, and Northwest Territories, and southeastern Yukon.
7. *N. c. fusca* True, 1894. Western and northwestern Oregon.
8. *N. c. lucida* Goldman, 1917. Restricted to southern Nevada.
9. *N. c. macrodon* Kelson, 1949. Restricted to central-eastern Utah.
10. *N. c. occidentalis* Baird, 1855. From central Oregon, Washington, northern Idaho, northwestern Montana throughout western British Columbia to southern Yukon, and western Northwest Territories.
11. *N. c. orolestes* Merriam, 1894. Northern-central New Mexico, central Colorado and Wyoming, southern Montana, and southeastern South Dakota.
12. *N. c. pulla* Hooper, 1940. Northern California and southeastern Oregon.
13. *N. c. rupicola* Allen, 1894. Northwestern Colorado, western Nebraska, South Dakota, and North Dakota.

**Conservation:** *Neotoma cinerea* is listed as Least Concern by the IUCN (Cassola 2016by).

**Characteristics:** Bushy-tailed woodrat is **very large sized;** total length 302.0–398.0 mm and skull length 43.0–51.5 mm. Dorsal pelage yellowish gray to ochraceous buff, back much darker by grizzled of dusky hairs; **hair long, thick, and somewhat woolly**; fur of underparts yellowish white but plumbeous basally, chest paler; **ears large; hind feet very large;** fore and hind feet white; **soles thickly furred from the heels to the posterior tubercles; tail moderately long,** almost 80% of the head-and-body length, **bushy, and somewhat distichous,** brownish gray dorsally and white ventrally; skull large, long, and angular; **rostrum elongated**; **incisive foramina long and narrow**; frontal region narrow, depressed, deeply constricted near the middle; **frontal region narrowly constricted and channeled**; temporal ridges prominent; **interorbital region narrow that appears channeled dorsally**; sphenopalatine vacuities absent or present as very narrow slits; auditory bullae larger and elongated; first upper molars with a deep anterointernal reentrant angle; third upper molars with an anterior closed triangle and two confluent posterior loops. Species group *cinerea*.

**Comments:** *Neotoma cinerea* ranges from the Yukon Territory and Northwest Territories south through California, Arizona, and New Mexico and east through the Dakotas, Nebraska, and Colorado (Map 177). It can be found in sympatry, or nearly so, with *N. albigula*, *N. fuscipes*, *N. lepida*, and *N. mexicana*. It can be distinguished from all other species by its tail flattened and bushy; first upper molars with a deep anterointernal reentrant angle; third upper molars with an anterior closed triangle and two confluent posterior loops. From *N. stephensi*, by having the fur of underparts yellowish white, and the supraorbital ridges not continuing forward to the rostrum.

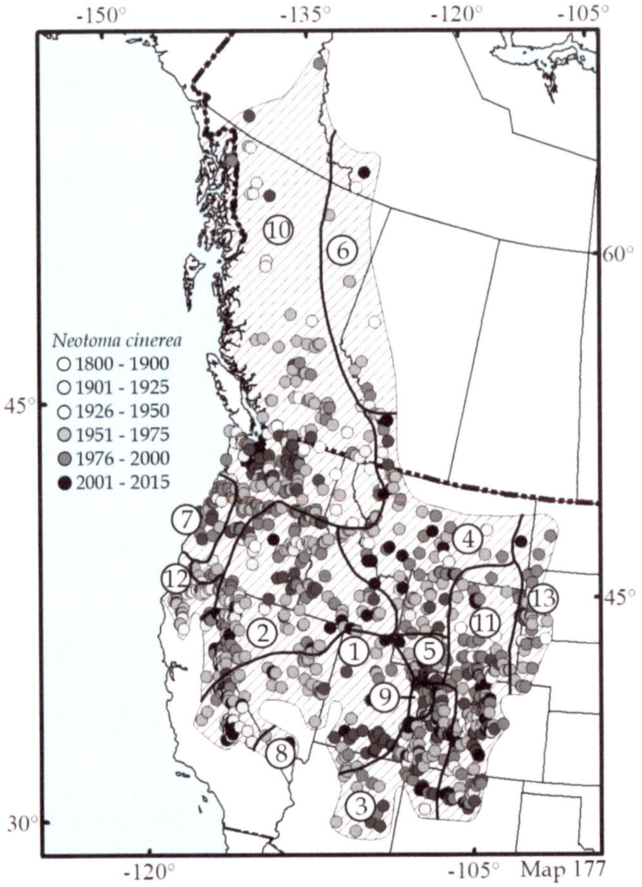


**Map 177** Distribution map of *Neotoma cinerea*

**Additional Literature:** Goldman (1910), Hooper (1940b), Kelson (1952), Hall (1981), Smith (1997), Musser and Carleton (2005).

## Genus *Neotomodon* Merriam, 1898

**Medium sized; ears large and nearly naked; fur soft and dense; with six plantar tubercles and six mammae (nipples)**; incisive foramina long and wide; molars disproportionately large, flat-crowned and heavily enameled; **first and second upper molars similar, with three external loops, two deep external reentrant angles, two internal loops, one shallow internal reentrant angle; third upper molars peglike;** skull broad, with the braincase strongly vaulted and short; zygomata widely spreading with the zygomatic plate extended anteriorly.

### *Neotomodon alstoni* Merriam, 1898

Mexican Volcano mouse, ratón de los Volcanes

1. *N. a. alstoni* Merriam, 1898. From central Michoacán to central Puebla.
2. *N. a. perotensis* Merriam, 1898. Central-eastern Puebla and central-western Veracruz.

**Conservation:** *Neotomodon alstoni* is listed as Least Concern by the IUCN (Álvarez-Castañeda and Castro-Arellano 2016b).

**Characteristics:** Mexican Volcano mouse is medium sized; total length 179.0–227.0 mm and skull length 28.3–31.3 mm. Dorsal pelage soft dense gray to fulvous, occasionally reddish brown; fur of underparts whitish with leaden gray shades to the base and sometimes with a yellowish pectoral spot; ears large, nearly naked; tail relatively short and slightly bicolored; hair dense and soft; six plantar tubercles; six mammae (nipples); braincase round; zygomatic arches short; zygomatic plate extending above; incisive foramen large; supraorbital edge with a prominent ridge; **molar occlusal surface flat-crowned; first and second upper molars similar, each with three external loops and two reentrant angles labially; third lower molars not S shaped.**

**Comments:** *Neotomodon* was reinstated to genus by Carleton (1980, 1989); it was previously regarded as a species of *Peromyscus* (Yates et al. 1979a). *Neotomodon orizabae* was considered a junior synonym of *N. a. perotensis. N. alstoni* is endemic to the Eje Neovolcánico, from eastern Michoacán east through central Veracruz (Map 178). It can be found in sympatry, or nearly so, with different species of *Peromyscus* and *Reithrodontomys. N. alstoni* can be distinguished from the species of *Peromyscus* by its larger size; molar occlusal surface flat-crowned; first and second upper molars similar, each with three external loops and two reentrant angles labially, and third lower molars not S shaped.

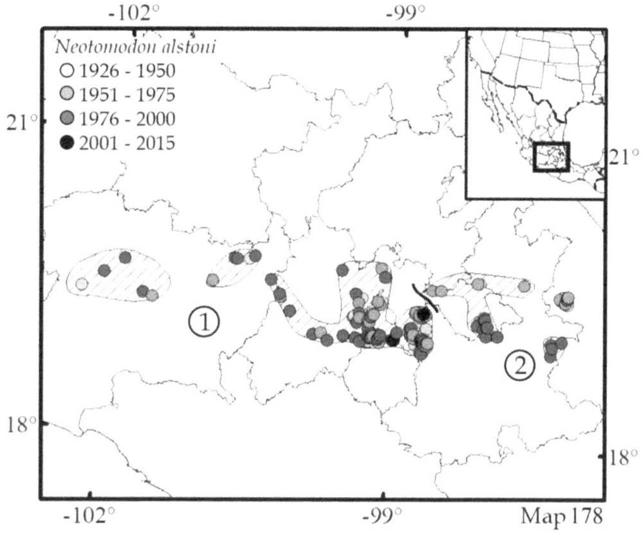


**Map 178** Distribution map of *Neotomodon alstoni*

**Additional Literature:** Merriam (1898a), Williams and Ramírez-Pulido (1984), Stephen et al. (1985), Williams et al. (1985), Musser and Carleton (2005).

## Genus *Ochrotomys* Harlan, 1932

Six plantar tubercles, with a rudimentary one at the base of the fifth digits; six mammae (nipples); molariform teeth relatively wide, with enamel folds much compressed; **tubercles relatively low, tendency for the development of a raised cingulum marked by subsidiary tubercles in inner salient angles on the first and second upper molars**; enamel thicker than in specimens of *Peromyscus*, generally occupying more of the occlusal surfaces of worn teeth; **pattern of the occlusal surfaces in partly worn upper and lower molars much compressed laterally and longitudinally, with five subtriangular islands of dentine in the first and second upper molars; posterior palatine foramen farther back** than in specimens of *Peromyscus*.

## *Ochrotomys nuttalli* (Harlan, 1932)

Golden mouse, ratón dorado

Monotypic.

**Conservation:** *Ochrotomys nuttalli* is listed as Least Concern by the IUCN (Cassola 2016cd).

**Characteristics:** Golden mouse is large sized; total length 170.0–190.0 mm and skull length 17.5–25.5 mm. Dorsal pelage rich tawny ochraceous, nearly clear on the flanks, pale mixed with dusky on the back in fresh pelage; fur of underparts **creamy white suffused with the color of upperparts,** except for the chin and neck**; pelage very soft and thick; hairs clothing the ears of the same color as those of the upperparts;** whiskers mixed brownish and whitish, no dusky spot at the base and no orbital ring; **proximal half of hind feet hairy;** face and head exactly as the sides; six plantar tubercles, with a rudimentary additional tubercle adjacent to the base of the fifth digits; posterior palatine foramina situated farther back and closer to the interpterygoid fossa; **molariforms relatively wide and with enamel folds much compressed, tubercles relatively low; first and second lower molars, in almost all stages of wear, leaving five subtriangular islands of dentine**.

**Comments:** *Ochrotomys* is considered a genus (Hooper 1968); it was previously regarded as a subgenus of *Peromyscus* (Osgood 1909). *O. nuttalli* ranges from southeastern Missouri, eastern Oklahoma, and eastern Texas east through Virginia and central Florida (Map 179). It can be distinguished from *Peromyscus* by having the fur of underparts creamy white suffused with the color of upperparts; pelage very soft and thick; upper molariforms relatively wide, with enamel folds much compressed; tubercles relatively low in the first and second lower molars in almost all stages of wear, leaving five subtriangular islands of dentine.

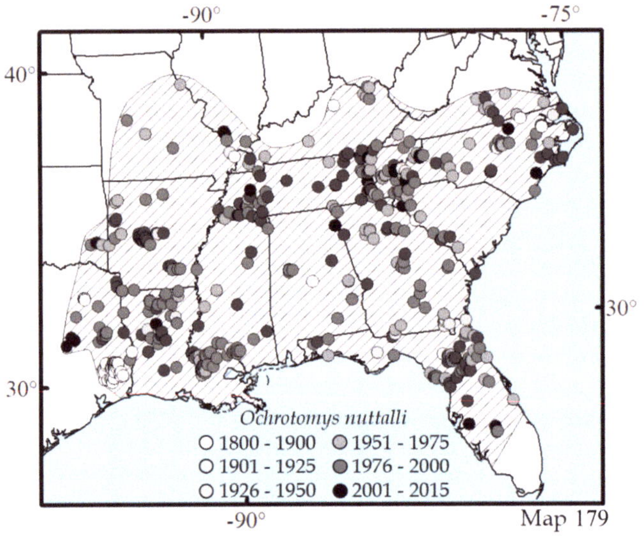


**Map 179** Distribution map of *Ochrotomys nuttalli*

**Additional Literature:** Musser and Carleton (2005).

# Genus *Onychomys* Baird, 1858

**Dorsal pelage brownish or gray to cinnamon or buffy, most intense along the back; underparts white contrasting with the dorsum; tail thick and short, proximal third with the same dorsal coloration, rest and tip whitish and obtuse**; forefeet large with five tubercles; hind feet with four tubercles; **soles with dense hair**; nasals wedge shaped, terminating posteriorly considerable behind the nasal branch end of the premaxillae; **mandible coronoid process well developed;** first and second upper molars large; third molars less than half the size of the second molars; first upper molars with two internal and three external cusps. The keys were elaborated based on the review of specimens and the following sources: Hollister (1914), Hall (1981), and Álvarez-Castañeda et al. (2017a).

1. Tail length less than one-half of the head-and-body length; third lower molars subcircular (Fig. 60); first upper molars less than 50% of the molar series length (from Alberta, Saskatchewan, Manitoba, Washington, and Minnesota south through Sonora, Chihuahua, Coahuila, Nuevo León, and Tamaulipas) .......................... *Onychomys leucogaster* (p. 259)

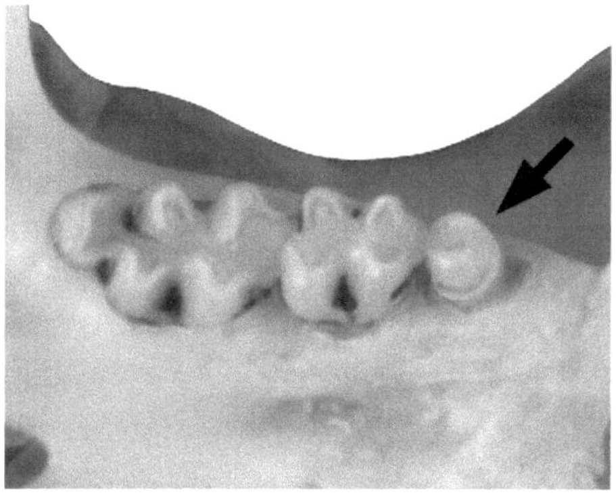

**Fig. 60** Third lower molars subcircular

1a. Tail length greater than one-half of the head-and-body length; third lower molars almost elliptical (Fig. 61); first upper molars greater than 50% of the molar series length .......................................................................................... 2

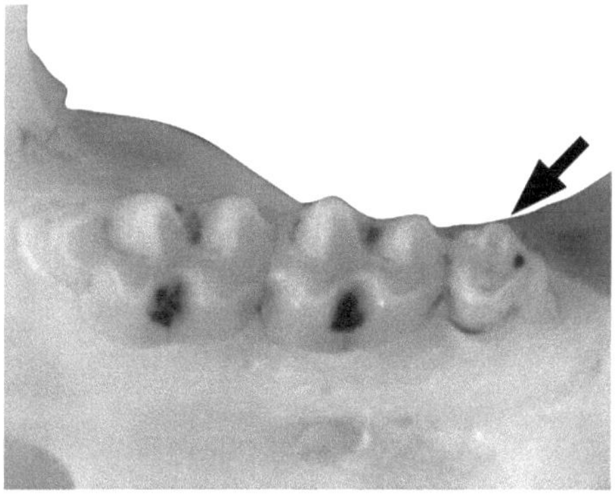

**Fig. 61** Third lower molars almost elliptical

2. Posterior border of the palatine with a "medial projection"; this species is distinguished from *O. arenicola* mainly by the respective karyotypes (from central California, northern Nevada, and southern Arizona south through northern Baja California, Sonora, and central Sinaloa) ................................................................................ *Onychomys torridus* (p. 260)

2a. Posterior border of the palatine without a "medial projection"; this species is distinguished from *O. torridus* mainly by the respective karyotypes (from southern-central New Mexico and west Texas south through Durango, Aguascalientes, and San Luis Potosí) ........................................................................................................ *Onychomys arenicola* (p. 257)

## ***Onychomys arenicola*** Mearns, 1896

Chihuahua grasshopper mouse, ratón chapulinero de Chihuahua

1. *O. a. ater* Anderson, 1972. Restricted to central-western Chihuahua.
2. *O. a. arenicola* Mearns, 1896. From southeast Arizona, central-southern New Mexico, and western Texas south through northern Durango and western Coahuila.
3. *O. a. canus* Merriam, 1904. Southeastern Durango, Zacatecas, western San Luis Potosí, and northeastern Jalisco.
4. *O. a. surrufus* Hollister, 1914. Southeastern Coahuila, Nuevo León, Tamaulipas, and northern and eastern San Luis Potosí.

**Conservation:** *Onychomys arenicola* is listed as Least Concern by the IUCN (Lacher et al. 2016b).

**Characteristics:** Chihuahua grasshopper mouse is small sized; total length 120.0–157.0 mm and skull length 21.5–22.5 mm. Dorsal pelage grab-gray, with a scarce admixture of black-ringed or black-tipped hairs; fur of underparts white as the limb inner part; hind feet with four tubercles on the soles and forefeet with five; feet white; tail short, with two-thirds of the dorsal basal drab, some hairs hoary at the tip, tail white at the tip; ears with a conspicuous white tuff and a dark spot, not black on the anterior ear band; nasals wedge shaped; third upper molars small and almost elliptical; first upper molars more than 50% of the molar series length; **palatine posterior border without a "medial projection"**; phalli longer and rounded at the tip.

**Comments:** *Onychomys arenicola* was recognized as a distinct species (Hinesley 1979; Sullivan et al. 1986); it was previously regarded as a subspecies of *O. torridus* (Mearns 1896; Hollister 1914). The subspecies inhabiting central México were assigned to *O. arenicola* by Riddle (1999) and Lacher et al. (2016b) without a formal review of the specimens. However, Musser and Carleton (2005) did not give a taxonomical assignation to these subspecies. *O. arenicola* ranges from southern-central New Mexico and western Texas south through Durango, Aguascalientes, and San Luis Potosí (Map 180). It can be found in sympatry, or nearly so, with *O. leucogaster*, from which it can be distinguished by its smaller size; paler coloration; longer tail, usually greater than one-half of the head-and-body length; wider interorbital region; teeth lower crowned; first lower molars with the anterior cusp narrow, elongated, and broader transversely; third lower molars markedly reduced, great transverse diameter; crown surface from one-sixth to one-fourth of the second lower molars. From *O. torridus*, by a longer male phallus with a rounded tip. From all other species of rodents, by having the tail short, less than the head-and-body length, flat and wide, more than 4.0 mm wider in the middle, and clearly bicolored.

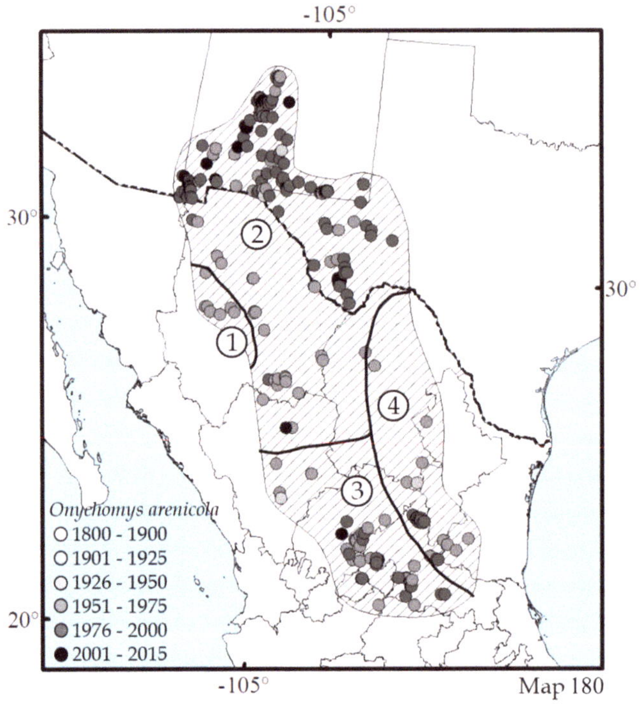


**Map 180** Distribution map of *Onychomys arenicola*

**Additional Literature:** Mearns (1896), Sullivan et al. (1986).

## ***Onychomys leucogaster*** (Wied-Neuwied, 1841)

Northern grasshopper mouse, ratón chapulinero del norte

1. *O. l. albescens* Merriam, 1904. Restricted to central-northern Chihuahua.
2. *O. l. arcticeps* Rhoads, 1898. From central and western South Dakota to Texas east through Wyoming, northern and eastern Colorado, and eastern New Mexico.
3. *O. l. breviauritus* Hollister, 1913. Central and eastern Nebraska, Kansas, and Oklahoma.
4. *O. l. brevicaudus* Merriam, 1891. Western California, Nevada, southern Idaho, and most southwestern Wyoming.
5. *O. l. fuliginosus* Merriam, 1890. Central-northern and northwestern Arizona.
6. *O. l. fuscogriseus* Anthony, 1913. Northeastern California, northeastern Nevada, eastern Oregon, southeastern Washington, and western Idaho.
7. *O. l. leucogaster* (Wied-Neuwied, 1841). From south Manitoba south through southern South Dakota, western Minnesota, and northeastern Iowa.
8. *O. l. longipes* Merriam, 1889. From southern Texas to northern Coahuila, Nuevo León, and Tamaulipas.
9. *O. l. melanophrys* Merriam, 1889. Central Utah and northern Arizona.
10. *O. l. missouriensis* (Audubon and Bachman, 1851). From northeastern Wyoming and northwestern South Dakota north through southern Saskatchewan and Alberta.
11. *O. l. pallescens* Merriam, 1890. Northeastern Arizona, northwestern New Mexico, western Idaho, and central Colorado.
12. *O. l. ruidosae* Stone and Rehn, 1903. Central and southern New Mexico, westernmost Texas, southeastern Arizona, northern Sonora, and Chihuahua.
13. *O. l. utahensis* Goldman, 1939. Western Utah.

**Conservation:** *Onychomys leucogaster* is listed as Least Concern by the IUCN (Timm 2016).

**Characteristics:** Northern grasshopper mouse is small sized; total length 130.0–190.0 mm and skull length 131.0–159.0 mm. Dorsal pelage pale cinnamon or yellowish brown, more intense in the dorsal region; fur of underparts white as in the inner part of the limbs and feet; ears dark gray, with fine silver hairs more abundant toward the edges; hind feet with four tubercles on the soles and forefeet with five; nose wedge shaped; **third upper molars small and almost circular; upper first molars less than 50% of the molar series length**; phalli longer and rounded at the tip.

**Comments:** *Onychomys leucogaster* ranges from Alberta, Saskatchewan, Manitoba, Washington, and Minnesota south through Sonora, Chihuahua, Coahuila, Nuevo León, and Tamaulipas (Map 181). It can be found in sympatry, or nearly so, with *O. arenicola.* It can be distinguished from it by its smaller size; darker coloration; shorter tail; usually less than one-half of the head-and-body length; interorbital region narrow; teeth higher crowned; first anterior cusp of the lower molars more cone shaped, with less indication of an incipient division at the summit into two or three cusplets; third lower molars much larger, subcircular with the transverse and longitudinal diameters subequal, and crown surface from one-half to one-third of the second lower molars. From all the other species of rodents, by having the tail short, less than the head-and-body length, flat and wide, greater than 4.0 mm wider in the middle and clearly bicolored.

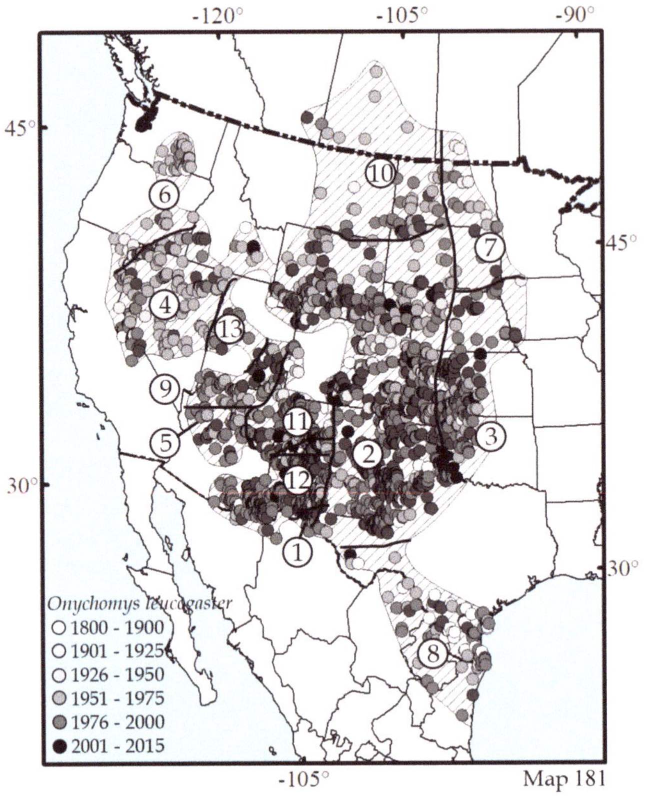


**Map 181** Distribution map of *Onychomys leucogaster*

**Additional Literature:** Hollister (1914), McCarty (1975, 1978), Sullivan et al. (1986), Musser and Carleton (2005).

## *Onychomys torridus* (Coues, 1874)

Southern grasshopper mouse, ratón chapulinero del sur

1. *O. t. clarus* Hollister, 1913. Restricted to Hot Spring Valley, central east California.
2. *O. t. knoxjonesi* Hollander and Willig, 1992. Northern to central Sinaloa.
3. *O. t. longicaudus* Merriam, 1889. Southwestern Utah, northwestern Arizona, Nevada, and central-eastern California.
4. *O. t. macrotis* Elliot, 1903. Restricted to northwestern Baja California.
5. *O. t. pulcher* Elliot, 1904. Southeastern California.
6. *O. t. ramona* Rhoads, 1893. Southwestern California.
7. *O. t. torridus* (Coues, 1874). From western Texas west through southern Arizona and northern Sonora.
8. *O. t. tularensis* Merriam, 1904. Restricted to the southern Central Valley, California
9. *O. t. yakiensis* Merriam, 1904. Southern Sonora and Sinaloa.

**Conservation:** *Onychomys torridus* is listed as Least Concern by the IUCN (Lacher et al. 2016c).

**Characteristics:** Southern grasshopper mouse is small sized; total length 119.0–163.0 mm and skull length 22.8–24.7 mm. Dorsal pelage grayish to pale cinnamon, more intense in the dorsal region; fur of underparts white, including the limb inner part; hind feet with four tubercles on the soles and forefeet with five; nasals wedge shaped; third upper molars small and almost elliptical; first upper molars more than 50% of the molar series length; **palatine posterior border with a "medial projection"; phalli smaller and tip triangular.**

**Comments:** *Onychomys arenicola* was described as a subspecies of *O. torridus* (Mearns 1896) and is considered a distinct species (Hinesley 1979; Sullivan et al. 1986). *O. torridus* ranges from central California, northern Nevada, and southern Arizona south through northern Baja California, Sonora, and central Sinaloa (Map 182), and cannot be found in sympatry, or nearly so, with any other species of *Onychomys*. It can be distinguished from *O. arenicola* by its smaller and triangular shape at the tip. From *O. leucogaster*, by its smaller size; paler coloration; longer tail; usually greater than one-half of the head-and-body length; wider interorbital region; teeth lower crowned; anterior cusp of the first lower molars narrow, elongated and broader transversely; third lower molars much reduced with great transverse diameter and crown surface from one-sixth to one-fourth of the second lower molars. From all other species of rodents, by having the tail short, less than the head-and-body length, flat and wide, greater than 4.0 mm wider in the middle and clearly bicolored.

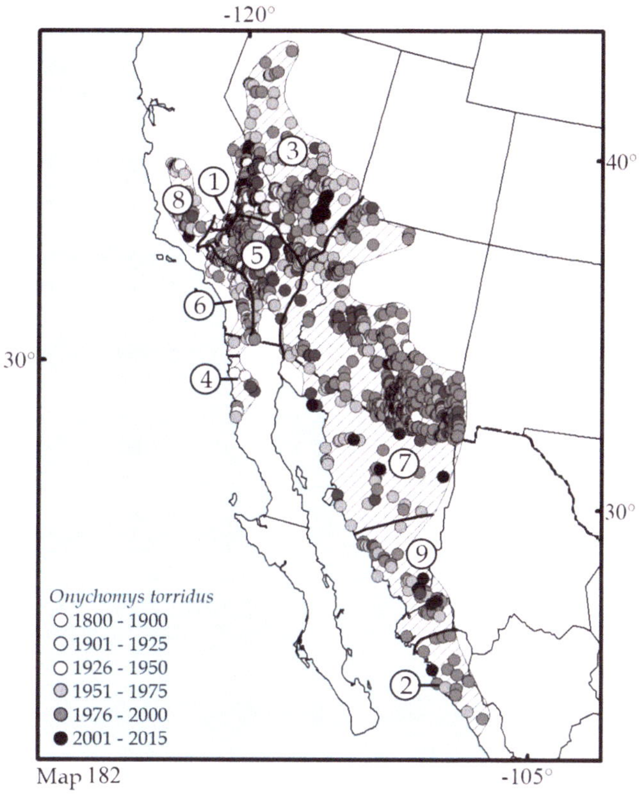


**Map 182** Distribution map of *Onychomys torridus*

**Additional Literature:** McCarty (1975, 1978), Álvarez-Castañeda and Cortés-Calva (1999), Musser and Carleton (2005).

## Genus *Osgoodomys* Hooper and Musser 1964

Tail approximately of the same length as the head-and-body; **pelage soft but rather short;** ears smaller than the hindfoot length; **hindfoot soles naked to the calcaneus**; six plantar tubercles; **skull narrow, supraorbital beads well developed, forming a trenchant shelf above the orbit and bounded on the inner side by a distinct groovelike channel extending from the lacrimal region to or beyond the parietofrontal suture, posterior braincase** so much elongated that more than half of the interparietal bone lies behind the tympanic bullae; lacrimal region swollen; baculum minute, only one-fifth as long as the hind feet; glans penis small, simple, and urethral opening terminal.

## ***Osgoodomys banderanus*** (Allen, 1897)

Michoacán deermouse, ratón de Michoacán

1. *O. b. banderanus* (Allen, 1897). From Nayarit south through Michoacán.
2. *O. b. vicinior* (Osgood, 1904). From eastern Michoacán, Guerrero, and Oaxaca.

**Conservation:** *Osgoodomys banderanus* is listed as Least Concern by the IUCN (Castro-Arellano and Vázquez 2017).

**Characteristics:** Michoacán deermouse is a larger in size; total length 228.0–245.0 mm and skull length 30.6–35.0 mm. Dorsal pelage and flanks ochraceous buff, with a very fine mixture of cinnamon nearly uniformly; fur of underparts creamy with a pectoral patch broad ochraceous buff; lateral line pale yellow; forehead and sides of the head mixed cinnamon and drab gray; orbital ring and spot at the base of the whiskers brown; ears smaller than the hindfoot length; **soles of the hind feet naked to the calcaneus**; six plantar tubercles; tail approximately of the same length as the head-and-body; **pelage soft but rather short**; rostrum long and thin; first and second upper molars with accessory crests between the second and third folds; **supraorbital ridge with a well-developed edge (it can be felt by touching with the finger); a groove on the internal face, running from the lacrimal region to the frontoparietal supraorbital rim**; lacrimal bones well developed; incisive foramen small.

**Comments:** *Osgoodomys* was considered at the genus level by being genetically and morphologically different from *Peromyscus* (Hooper 1968); it was previously regarded as a subgenus of *Peromyscus* (Hooper and Musser 1964). *Osgoodomys banderanus* ranges in the lowlands from southern Nayarit south through Oaxaca and the Río Balsas Basin in Michoacán and Guerrero (Map 183). *O. banderanus* can be found in sympatry, or nearly so, with *Peromyscus aztecus*, *P. hylocetes*, *P. levipes*, *P. labecula*, and *P. perfulvus*. It can be distinguished from all these species by having the supraorbital rim with a prominent, well-developed ridge, forming a trenchant shelf above the orbit and bounded on the inner side by a distinct groove-like channel extending from the lacrimal region to or beyond the parietofrontal suture. From *P. perfulvus*, by having an almost hairless and scaled tail that is not penciled at the tip; feet white; ears blackish; a blackish gray facial mask and rounded supraorbital rim.

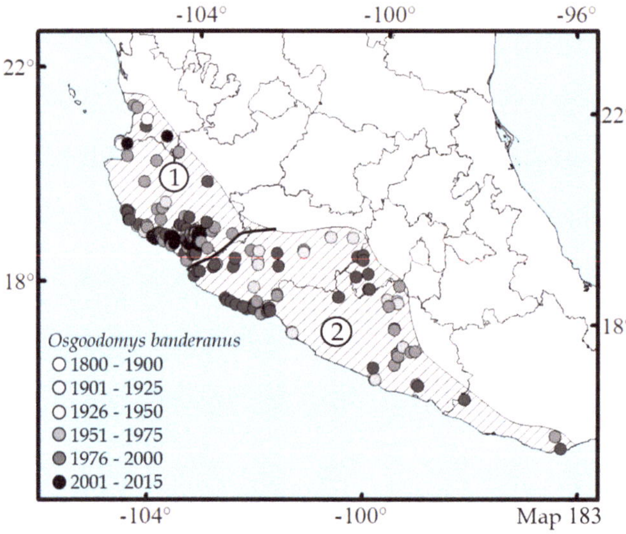


**Map 183** Distribution map of *Osgoodomys banderanus*

**Additional Literature:** Osgood (1909), Carleton (1980), Hall (1981), Musser and Carleton (2005).

# Genus *Peromyscus* Gloger, 1841

Dorsal pelage from rich ochraceous-buffy to blackish brown; fur of underparts usually white but sometimes pale buff or paler than in the dorsum; tail long, at least more than one-third of the total length, often more than one-half; tail with scaly annulations more or less concealed by hair; ears relatively large, membranous, and thinly clothed with hair; soles of the hind feet with six tubercles, hairy proximally or naked medially to the calcaneus; skull with the braincase rather thin-walled, smooth but slightly ridged; interparietal bone well developed; zygomatic bone slender; infraorbital foramen compressed-triangular, bounded on the outside by a broad thin plate; anterior palatine foramen long and separated by a thin bony septum; palate without lateral pits; auditory bullae more or less inflated and obliquely situated, coronoid process short and slightly developed; molar series decreasing in size from the anterior to the posterior part; third upper molars subcircular and usually less than half as large as the second upper molars; upper incisors without grooves. *Peromyscus* has two subgenus and 11 species groups (Table 1). The keys were elaborated based on the review of specimens and the following sources: Osgood (1909), Hall (1981), Carleton (1989), and Álvarez-Castañeda et al. (2017a).

1. Three pairs of mammae (nipples; one pectoral and two inguinal); first and second upper molars without accessory cusps in two principal outer angles (Fig. 62) ........ subgenus *Haplomylomys* 2

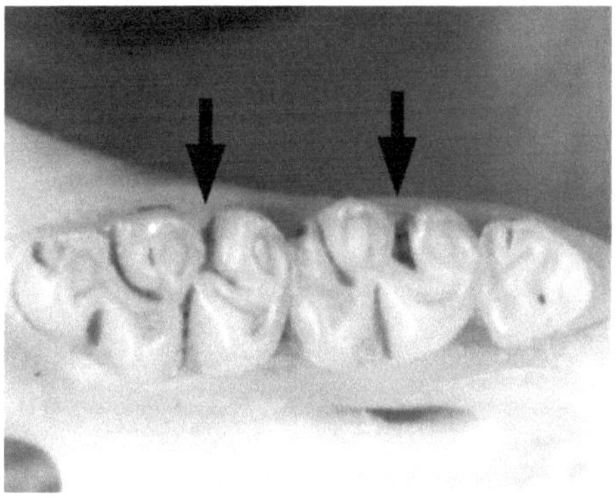

**Fig. 62** First and second upper molars without accessory cusps in two principal outer angles

1a. Two pairs of mammae (nipples; two inguinal); first and second upper molars with accessory cusps in two principal outer angles (Fig. 63) ........ subgenus *Peromyscus* 15

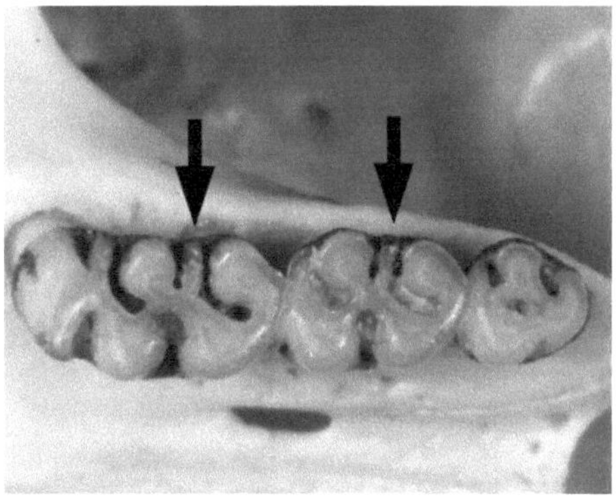

**Fig. 63** First and second upper molars with accessory cusps in two principal outer angles

2. Total length greater than 220.0 mm; tail length usually greater than 120.0 mm; a dark ring around the eyes; skull length usually greater than 28.0 mm (from central California south through the northern Baja California Peninsula) ........ *Peromyscus californicus* (p. 275)

2a. Total length less than 220.0 mm; tail length usually less than 120.0 mm; without a dark ring around the eyes; skull length usually less than 28.0 mm ........ 3

3. Ranges from Coahuila to eastern México ........ 4

3a. Ranges from Coahuila to western México ........ 5

4. Skull length usually greater than 25.9 mm; nasal back V shaped (Fig. 64; endemic to central Coahuila, northernmost Zacatecas, and San Luis Potosí) ........ *Peromyscus hooperi* (p. 284)

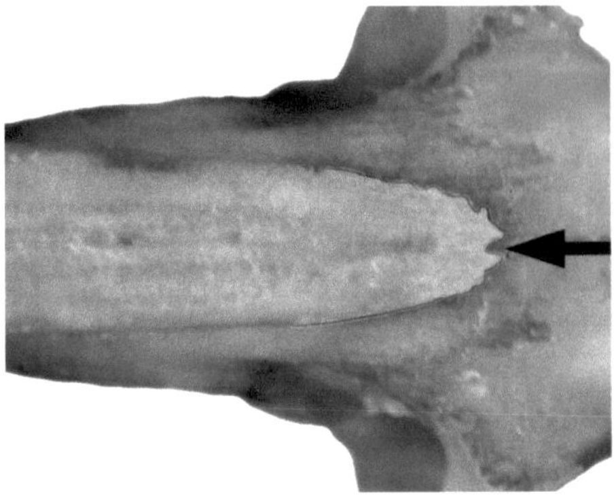

**Fig. 64** Anteroloph of the first upper molars with a deep anteromedial notch

4a. Skull length usually less than 25.7 mm; nasal back not V shaped (Fig. 65; from southwestern Arizona, New Mexico, and Texas south through Sinaloa, Aguascalientes, San Luis Potosí, and Tamaulipas) ........ *Peromyscus eremicus* (part; p. 278)

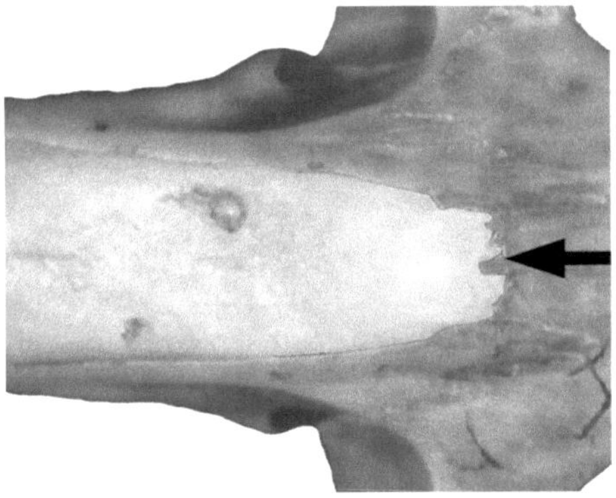

**Fig. 65** Nasal back not V shaped

5. Incisive foramen ending before the anterior margin of the first upper molars (Fig. 66); in lateral view, skull dorsal profile highly arched (endemic to the Ángel de la Guarda, Mejía, Granito and Estanque Islands, Baja California. Extirpated from the three later Islands) ........ *Peromyscus guardia* (p. 283)

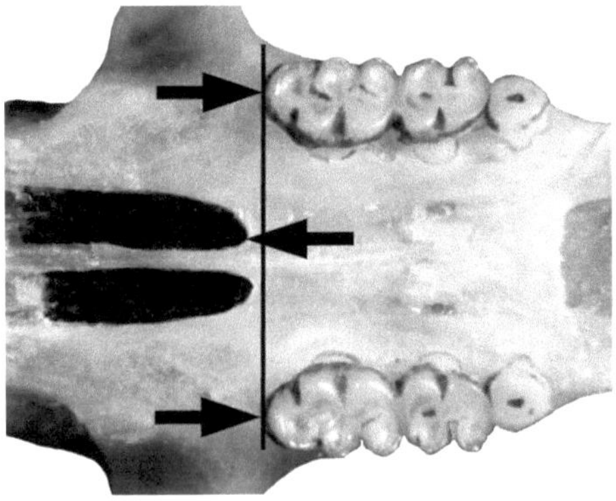

**Fig. 66** Incisive foramen ending before the anterior margin of the first upper molars

5a. Incisive foramen ending at the anterior margin of the first upper molars; in lateral view, skull dorsal profile not highly arched ........ 6

6. Tail length less than the head-and-body length ........ 7

6a. Tail length equal to or greater than the head-and-body length ........ 9

7. Head grayish, clearly contrasting with the dorsal pelage; squamous bone projection wide and tapering toward the nasals; upper toothrow length less than 4.8 mm (endemic to Montserrat Island, Baja California Sur) ........ *Peromyscus caniceps* (part; p. 277)

7a. Head color not contrasting strongly with the rest of the body; squamous bone projection not wide or tapering to the nasals; upper toothrow length greater than 4.0 mm ........ 8

8. Fur of underparts white with cinnamon shades; tail length less than 97.0 mm and equal to or less than the head-and-body length; maxillary extension ending behind the nasals (Fig. 67; endemic to Tortuga Island, Baja California Sur) ........ *Peromyscus dickeyi* (p. 277)

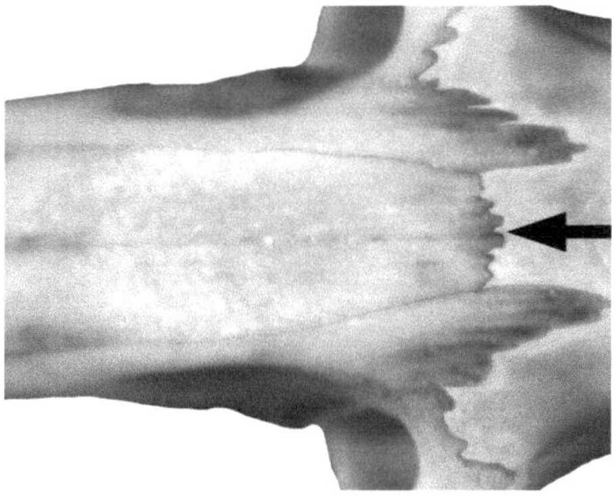

**Fig. 67** Maxillary extension ending behind the nasals

8a. Fur of underparts white, without cinnamon shades; tail length greater than 97.0 mm and greater than the head-and-body length; maxillary extension ending behind at the same level as the nasals (endemic to San Pedro Nolasco Island, Sonora) ........ *Peromyscus pembertoni* (p. 286)

9. Posterior edge of the frontal bone angled (Fig. 68; endemic to San Esteban Island, Baja California) ........ *Peromyscus stephani* (subgenus *Peromyscus* part; p. 354)

**Fig. 68** Posterior edge of the frontal bone angled

9a. Posterior edge of the frontal bone curve (Fig. 69) ........ 10

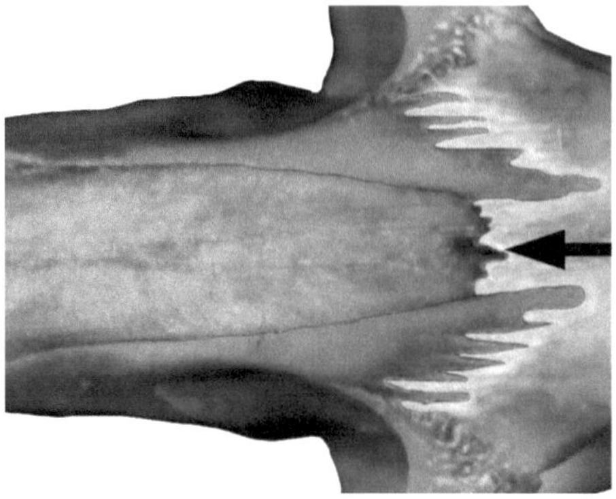

**Fig. 69** Posterior edge of the frontal bone curve

10. Skull flattened; skull length greater than 25.5 mm; upper toothrow length greater than the interorbital width; mastoid width greater than 11.4 mm (from southern Arizona south through Sinaloa) ........ *Peromyscus merriami* (p. 285)

10a. Skull arch shaped; skull length less than 26.2 mm; upper toothrow length less than the interorbital width; mastoid length less than 11.4 mm ........ 11

11. Pelage orange or ocher brown; not ranging in the Baja California Peninsula (from southwestern Arizona, New Mexico, and Texas south through Sinaloa, Aguascalientes, San Luis Potosí, and Tamaulipas) ........ *Peromyscus eremicus* (part; p. 278)

11a. Pelage dark brown; only ranging in the Baja California Peninsula ........ 12

12. Flanks intensely orange; upper toothrow length less than 3.5 mm (from the central Baja California Peninsula southward, including the Carmen and San José Islands, Baja California Sur) ........ *Peromyscus eva* (p. 280)

12a. Flanks brown, occasionally with orange shades but not intense; upper toothrow length greater than 3.5 mm ........ 13

13. Ear length greater than 90% of the hindfoot length; total skull length usually greater than 23.0 mm; an accessory cusp on the second upper molars, in some specimens the accessory cusp can be absent in one tooth (from Oregon, Idaho, and Wyoming south through California, Baja California, Arizona, Sonora, and New Mexico) ........ *Peromyscus crinitus* (part; subgenus *Peromyscus*; p. 301)

13a. Ear length less than 90% of the hindfoot length; total skull length usually less than 23.0 mm; without an accessory cusp on both second upper molars ........ 14

14. Ranges from southwestern California south through La Paz isthmus, Baja California Sur ........ *Peromyscus fraterculus* (p. 281)

14a. Ranges on Cerralvo Island, southeastern Baja California Peninsula ........ *Peromyscus avius* (p. 275)

15. Supraorbital shelf well developed; supraorbital rims with a ridge and straight (in dorsal view, both supraorbital edges are clearly separated and subsequently converge straight to the front of the skull, so that both edges appear wedge shaped) ........ 16

15a. Supraorbital shelf little developed or absent; supraorbital edges without a ridge and almost or completely biconcave (in dorsal view, both are separated on the back and subsequently converge straight to the front of the skull, so that both edges appear hourglass shaped) ........ 39

16. Tail long, usually greater than 110.0 mm and in most specimens greater than 110% of the head-and-body length; tail moderately hairy with long hairs isolated and a terminal tuft of hair about 2.0 mm in length; upperparts generally paler, with orange, ocher, or tawny shades, especially on the flanks; ranging in low and semi-deciduous forest regions but not as described in key 16a ........ 17

16a. Tail short, less than 110.0 mm and less than 110% of the head-and-body length; tail naked o with little hair but shorter, without a terminal tuft of hairs less than 2.0 mm long; upperparts generally blackish, dark gray or dark brown; ranging in humid tropical regions (medium forests and high cloud forests) or oak and conifer forests but not as described in key 16 (except for *P. spicilegus* on the Sierra Madre Occidental west slope and *P. yucatanicus* in the Yucatán Peninsula, found in lowlands) ........ 24

17. Tail length usually greater than 130.0 mm ........ 18

17a. Tail length usually less than 130.0 mm ........ 22

18. Skull length usually less than 27.0 mm; no rims on the supraorbital area (only known from Ciudad Serdán, Puebla) ........ *Peromyscus mekisturus* (p. 329)

18a. Skull length usually greater than 27.0 mm; parietal-temporal ridge well developed on the supraorbital area (Fig. 70) ........ 19

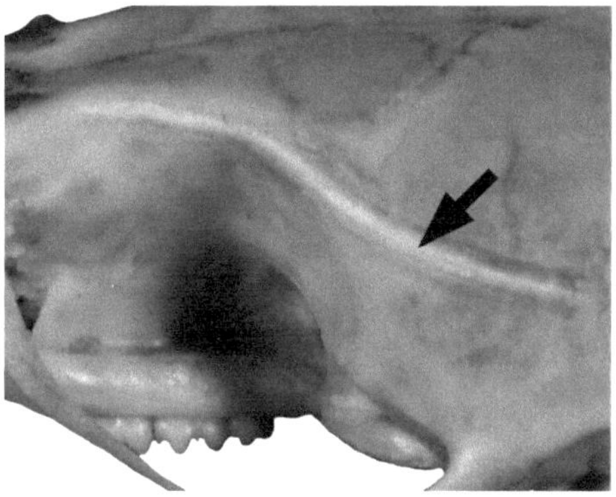

**Fig. 70** Parietal-temporal ridge well developed on the supraorbital area

19. Supraorbital ridges well developed reaching almost the posterior end of the nasals; interorbital constriction narrow; auditory bullae small and not inflated; braincase not domed (from eastern Durango and Nayarit south through northern Michoacán, Guanajuato, and eastern San Luis Potosí) ........ *Peromyscus micropus* (p. 335)

19a. Supraorbital ridges well developed but not reaching the posterior end of the nasals; interorbital constriction not narrow; auditory bullae medium to large and inflated or proportionally inflated; braincase domed or not domed ........ 20

20. Tail relatively short, on average less than 54% of the total length; supraorbital ridges well developed reaching the nasal-premaxillary joint; postorbital constriction narrow; auditory bullae large and inflated; braincase not domed (from Durango, Coahuila, and Nuevo León south through Hidalgo and Estado de México) ........ *Peromyscus zamorae* (p. 359)

20a. Tail relatively large, on average greater than 54% of the total length; supraorbital ridges well developed, not reaching the nasal-premaxillary joint; postorbital constriction wide; auditory bullae medium and proportionally inflated; braincase domed ........ 21

21. Hind feet with the dorsal coloration pale brown, some specimens with a darker brown line over the metatarsals; mastoid area narrow; upper toothrow narrow (Oaxaca, Chiapas, and Guatemala) ........ *Peromyscus leucurus* (p. 323)

21a. Hind feet with the dorsal coloration white or white with brown spots; mastoid area wide; upper toothrow wide (Estado de México, Ciudad de México, Morelos, Guerrero, Puebla, Oaxaca, and southwestern Chiapas) ........ *Peromyscus melanophrys* (p. 331)

22. Metacarpal region whitish; tail dark brown dorsally; skull length usually greater than 28.5 mm (endemic to Santa Catalina Isla, Baja California)........................................................................................... *Peromyscus slevini* (p. 349)

22a. Metacarpal region dark; tail sepia dorsally; skull length usually less than 29.0 mm ....................................................23

23. First and second upper molars usually with ectostylids and mesolophs complete; nasals acute posteriorly, on average with the limits short or the posterior limits of the ascending branches of the premaxillae (coastal lowlands of Jalisco and Colima) ........................................................................................................................ *Peromyscus chrysopus* (p. 297)

23a. First and second upper molars with ectostylids absent and mesolophs absent or short, not reaching the labial border of the tooth; nasals on average exceeding the premaxillae (Río Balsas Basin to Michoacán, Guerrero, and southwestern Estado de México)................................................................................ *Peromyscus perfulvus* (p. 340)

24. Supraorbital rim with a well developed ridge ........................................................................................................25

24a. Supraorbital rim with a poorly developed or absent ridge ..........................................................................................28

25. Tail more or less hairy, hair on the underparts usually white; flanks orange and different from the back; upper toothrow length usually greater than 5.0 mm (endemic to Coalcomán, Michoacán) .............. *Peromyscus winkelmanni* (part; p. 357)

25a. Tail naked, if hair is present ventrally, not white and short; flanks blackish, dark gray or dark brown but never with orange shades; upper toothrow length usually less than 5.0 mm .....................................................................................26

26. Hind feet with the dorsal dark area extending to the base of the toes (endemic to northern Oaxaca)...............................
........................................................................................................................ *Peromyscus melanocarpus* (p. 330)

26a. Hind feet with the dorsal dark area not extending to the base of the toes, at most to the level of the tarsus.................27

27. Pectoral spot ocher; hindfoot length greater than 28.0 mm; upper toothrow length greater than 4.6 mm (central Guerrero and northern-central Oaxaca)................................................................................................ *Peromyscus megalops* (p. 328)

27a. Pectoral spot no ocher; hindfoot length less than 28.0 mm; upper toothrow length less than 4.6 mm (mountains in central Oaxaca) ........................................................................................................ *Peromyscus melanurus* (p. 333)

28. Hindfoot length greater than 30.0 mm; skull length greater than 34.5 mm; upper toothrow length greater than 5.0 mm ......................................................................................................................................................29

28a. Hindfoot length less than 30.0 mm; skull length less than 34.5 mm; upper toothrow length less than 5.0 mm ............30

29. Pelage dark gray; skull length greater than 34.5 mm; rostral length greater than 10.7 mm; upper toothrow length greater than 5.0 mm (mountains of central-northern Chiapas) ................................................. *Peromyscus zarhynchus* (p. 360)

29a. Pelage dark brown; skull length less than 36.0 mm; rostral length less than 11.7 mm; upper toothrow length less than 5.5 mm (mountains of southeastern Chiapas)....................................................*Peromyscus carolpattonae* (part; p. 296)

30. Tail more or less with abundant and short hair, usually bicolored; tail usually white ventrally, contrasting with the dorsal color (endemic to the Yucatán Peninsula) ................................................................... *Peromyscus yucatanicus* (p. 358)

30a. Tail hairless, if hair, very short and isolated, bicolored, or unicolor; ventral tail hair, when present, not contrasting with the back color...........................................................................................................................31

31. Hindfoot length less than 25.0 mm ..........................................................................................................32

31a. Hindfoot length greater than 25.0 mm .......................................................................................................33

32. Ranges in the Eje Neovolcánico and Sierra Madre Occidental ...........................................................................34

32a. Ranges in the Sierra Madre Oriental and Sierra Madre del Sur, Guerrero, Oaxaca, and Chiapas................................35

33. Hindfoot length equal to or greater than 27.0 mm; supraorbital edge weakly crested; rostral length greater than 11.6 mm; upper toothrow length greater than 5.1 mm (endemic to Coalcomán, Michoacán)............................................................................................................................................................*Peromyscus winkelmanni* (part; p. 357)

33a. Hindfoot length less than 27.0 mm; supraorbital edge not crested; rostral length less than 11.7 mm; upper toothrow length less than 5.2 mm (endemic to the Eje Neovolcánico in Colima east through Estado de México) ............................................................................................................................................*Peromyscus hylocetes* (part; p. 313)

34. Upper toothrow length usually greater than 4.5 mm; total length on average 31.5 mm and rostral length on average 10.3 mm (endemic to Eje Neovolcánico Colima east through Estado de México) ............................................................................................................................................*Peromyscus hylocetes* (part; p. 313)

34a. Upper toothrow length usually less than 4.7 mm; total length on average 28.3 mm and rostral length on average 9.4 mm (southeast Sonora and southwest Chihuahua south through northeast Colima and western-central Michoacán).........................................................................................................................................*Peromyscus spicilegus* (p. 352)

35. Dorsal coloration tawny; hind feet with dusky shades; skull short and wide; molars not heavy; auditory bullae large (Hidalgo, Veracruz, and Puebla, and Oaxaca)........................................................................*Peromyscus aztecus* (p. 290)

35a. Dorsal coloration rich tawny; hind feet wholly white just below the tarsal joint to the claws; skull long and rather narrow; molars heavy; auditory bullae small (from the highlands of Chiapas south through Central America)........................................................................................................................................*Peromyscus cordillerae* (p. 300)

36. Hindfoot length greater than 27.0 mm; upper toothrow length greater than 4.6 mm ....................................................37

36a. Hindfoot length less than 27.0 mm; upper toothrow length less than 4.6 mm ..............................................................38

37. Hindfoot length greater than 27.0 mm; dorsal pelage dark gray; tail usually bicolored; upper toothrow length greater than 4.6 mm (restricted to the mountains of southern Chiapas) .........................*Peromyscus carolpattonae* (part; p. 296)

37a. Hindfoot length less than 29.0 mm; dorsal pelage dark brown; tail unicolor and spotted ventrally; upper toothrow length less than 4.9 mm ...............................................................................................................................................83

38. Average tail length 110.0 mm; average skull length 29.4 mm; auditory bullae comparatively large (Fig. 71; from the Pacific coastal plains of Chiapas south through Central America) ....................................*Peromyscus gymnotis* (p. 312)

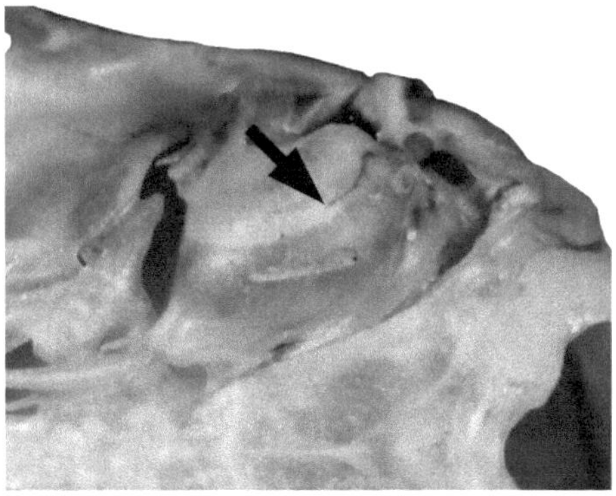

**Fig. 71** Auditory bullae comparatively large

38a. Average tail length 120.0 mm; average skull length 32.0 mm; auditory bullae comparatively small (Fig. 72)..............83

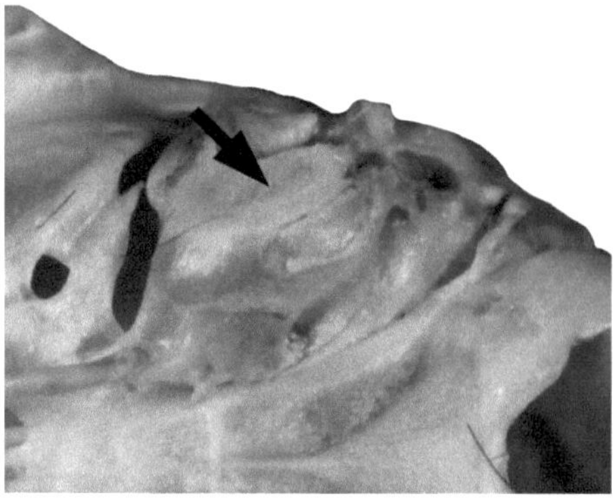

**Fig. 72** Auditory bullae comparatively small

39. Tail length generally less than 80.0 mm and less than 75% of the body-and-head length ............ 40

39a. Tail length generally greater than 80.0 mm and greater than 75% of the body-and-head length ............ 51

40. Ranges restricted to southeastern United States (from Virginia south through northern Florida and west through Louisiana, western Texas, and Arkansas) ............ *Peromyscus gossypinus* (p. 308)

40a. Ranges in different parts of Canada, México, and the United States, except for the southeast ............ 41

41. Tail moderately hairy or hairless, not clearly bicolored; when bicolored, dark dorsally, covering transversely half of the tail (from southern Alberta to Prince Edward Island south through Oaxaca, and Alberta to Arizona and from Chihuahua east through the Atlantic coast, including the northern Yucatán Peninsula and Cozumel Island and excluding the Florida Peninsula) ............ *Peromyscus leucopus* (part; p. 321)

41a. Tail hairy, clearly bicolored, dark dorsally, limited to a narrow stripe covering the vertebral line of the tail ............ 42

42. Dark hair tuft at the base of the ears ............ 43

42a. No dark hair tuft at the base of the ears ............ 44

43. Mastoid oval shaped; shaped mastoid fenestra rounded and visible (from Arizona, Chihuahua, southern Coahuila, and Nuevo León south through Jalisco, Michoacán, Estado de México, Puebla, and central Veracruz) ............ *Peromyscus melanotis* (p. 332)

43a. Mastoid rectangular; shaped mastoid fenestra reduced and elongated (only known from Nahuatzen, Michoacán) ............ *Peromyscus purepechus* (p. 344)

44. Upper toothrow length equal to or greater than 4.0 mm; skull length greater than 27.0 mm (endemic to Santa Cruz Island, Baja California Peninsula) ............ *Peromyscus sejugis* (p. 347)

44a. Upper toothrow length equal to or less than 3.9 mm; skull length usually less than 27.0 mm ............ 45

45. On average, the following measurements less than the figures shown: total length 154.0 mm; hindfoot length 19.0 mm, and tail length 60.0 mm (from northeastern Mississippi to western South Carolina, southward through Alabama, Georgia, and the Florida Peninsula) ............ *Peromyscus polionotus* (p. 341)

45a. On average, the following measurements greater than the figures shown: total length 154.0 mm; hindfoot length 19.0 mm; and tail length 60.0 mm (other areas of the United States, Canada, and México) ............ 46

46. Eastern of the Mississippi River, Minnesota, and Manitoba (from east of the Mississippi River and north through Manitoba east through the Atlantic coast and from the Labrador Peninsula south through Tennessee and North Carolina) ............ *Peromyscus maniculatus* (p. 327)

46a. Western of the Mississippi River, Manitoba, and central Minnesota ............ 47

47. North of 40° ............ 48

47a. South of 40° ............ 49

48. Dorsal tail stripes darker than the rest of the tail but without a strong contrast with the paler sides; skull relatively heavy; skull length on average small (southern Alaska and Yukon south through central Washington and western Oregon) .......... *Peromyscus keeni* (p. 314)

48a. Dorsal tail stripe darker than the rest of the tail, strongly contrasting with the paler sides; skull relatively not heavy; skull length large on average (from west of the Mississippi River northwest through Yukon and south through northern California, northern Sonora, Arizona, New Mexico, Texas, and Arkansas) .......... *Peromyscus sonoriensis* (part; p. 350)

49. Skull length small on average (from the San Francisco Bay area and San Joaquin Valley, extreme western-central region of Nevada south through Baja California Peninsula, including southwestern Arizona and northwestern Sonora) .......... *Peromyscus gambelii* (p. 306)

49a. Skull length large on average (from the northern San Francisco Bay area and San Joaquin Valley and western-central region of Nevada, southwestern Arizona, and northwestern Sonora) .......... 50

50. Tail dorsal stripes darker than the rest of the tail but without a strong contrast with the paler sides (from southern Arizona, New Mexico, southwestern Texas south through Colima, Jalisco, Estado de México, Puebla, Veracruz, and northern Oaxaca, and from eastern-central Chihuahua west through Tamaulipas) .......... *Peromyscus labecula* (p. 317)

50a. Tail dorsal stripes darker than the rest of the tail, strongly contrasting with the paler sides (from west of the Mississippi River northwest through Yukon and south through northern California, northern Sonora, Arizona, New Mexico, Texas, and Arkansas) .......... *Peromyscus sonoriensis* (part; p. 350)

51. Ear length equal to or greater than the hindfoot length; dry ear length greater than 80% of the hindfoot length; auditory bullae large and globe shaped, its length greater than 18% of the skull length .......... 52

51a. Ear length equal to or less than the hindfoot length; dry ear length less than 80% of the hindfoot length; auditory bullae small and not globe shaped, its length less than 18% of the skull length .......... 58

52. Total length greater than 210.0 mm; skull length greater than 29.0 mm; upper toothrow length greater than 4.5 mm .......... 53

52a. Total length less than 210.0 mm; skull length less than 29.0 mm; upper toothrow length less than 4.5 mm .......... 56

53. Hindfoot length less than 24.5 mm; pelage orange yellowish and slightly covered with black; upper toothrow length less than 4.5 mm (from southern Utah and northern-central Colorado southwest through northern Sonora, Chihuahua, and Durango and southeast through Texas, Coahuila, and Nuevo León) .......... *Peromyscus nasutus* (p. 336)

53a. Hindfoot length greater than 24.5 mm; pelage variable but densely covered with black; upper toothrow length greater than 4.5 mm .......... 54

54. Color very dark, chiefly rich blackish brown and black (southern Estado de México, Ciudad de México, Morelos, southeast Puebla, and central Oaxaca) .......... *Peromyscus felipensis* (p. 304)

54a. Color paler chiefly ochraceous buff more or less mixed with dusky shades .......... 55

55. Average size relatively small; pelage dense and glossy; skull with a small braincase (Zacatecas, southeast Coahuila, southwest Durango, and southwestern Nuevo León south through Jalisco, Guanajuato, and San Luis Potosí) .......... *Peromyscus difficilis* (p. 302)

55a. Average size relatively large; pelage loose and dull; skull with a large braincase (Guanajuato, Hidalgo, Michoacán, Estado de México, and Puebla) .......... *Peromyscus amplus* (p. 287)

56. Auditory bullae length equal to or greater than 5.5 mm, greater than 22% of the skull length and generally more than 132% of the upper toothrow length (endemic to Orizaba Basin on Veracruz-Puebla border) .......... *Peromyscus bullatus* (p. 294)

56a. Auditory bullae length less than 5.5 mm, less than 21% of the skull length and generally less than 132% of the upper toothrow length .......... 57

57. North of 32°, except for the Baja California peninsula; skull length on average 27.5 mm (from California, Oregon, Nevada, Utah, Colorado, and Kansas south through Baja California, northern Arizona, southern New Mexico, northern Texas, and southern Baja California Sur) .......... *Peromyscus truei* (p. 356)

57a. South of 32°, except for the Mogollon plateau in New Mexico; skull length on average 28.5 mm (from southwestern New Mexico, Coahuila, and Nuevo León south through central Oaxaca) .......... *Peromyscus gratus* (p. 310)

58. Mesolophid present on the second lower molars (southeastern Kansas, southwestern Missouri, Oklahoma, northwestern Arkansas, and Texas) .......... *Peromyscus attwateri* (p. 289)

58a. Mesolophid absent on the second lower molars .......... 59

59. Total length less than 210.0 mm; tail unicolor, ventral region paler but not contrasting with the dorsum; skull length less than 30.0 mm .......... 60

59a. Total length greater than 210.0 mm; tail clearly bicolored, contrasting coloration between the dorsal and ventral regions; skull length greater than 30.0 mm .......... 81

60. Tail length less than 110.0 mm (Pacific Coastal Plain of Sinaloa and Nayarit) .......... *Peromyscus simulus* (part; p. 348)

60a. Tail length greater than 110.0 mm .......... 61

61. Maxillary length less than 4.0 mm (endemic to Montserrat Island, Baja California Sur) .......... *Peromyscus caniceps* (subgenus *Haplomylomys*, part; p. 277)

61a. Maxillary length greater than 4.0 mm .......... 62

62. Tail length usually less than 96.0 mm; tail with little hair, back hair short and straight; upper toothrow length equal to or less than 4.0 mm (from southern Alberta, eastern Saskatchewan, Montana, Wisconsin, Michigan, Ontario, and New Brunswick southwest through Arizona, Sonora, and Durango, and southeast through Oaxaca, with an isolated population in the northern Yucatán Peninsula and Cozumel Island, east through the Atlantic coast, except for southern Georgia and the Florida Peninsula) .......... *Peromyscus leucopus* (part; p. 321)

62a. Tail length usually greater than 95.0 mm; tail hairy, back hair relatively long and silky; upper toothrow length equal to or greater than 4.0 mm .......... 63

63. Upper toothrow length less than 3.9 mm (Oregon, Idaho, and Wyoming south through California, Baja California, Arizona, Sonora, and New Mexico) .......... *Peromyscus crinitus* (part; p. 301)

63a. Upper toothrow length greater than 4.0 mm .......... 64

64. Tarsus region white .......... 65

64a. Tarsus region as dark as the tibia region .......... 68

65. Hindlimb length equal to or greater than 25.0 mm; total length greater than 210.0 mm; incisive foramen greater than 5.0 mm (restricted to central-western Chihuahua) .......... *Peromyscus polius* (p. 343)

65a. Hindlimb length equal to or less than 24.0 mm; total length less than 210.0 mm; incisive foramen less than 5.0 mm .......... 66

66. Restricted to southern Oklahoma, eastern New Mexico, and Texas .......... *Peromyscus laceianus* (p. 319)

66a. Ranges from Chihuahua, Coahuila, Nuevo León, and Tamaulipas southward .......... 67

67. Dorsal coloration paler and reddish; tarsi paler; total length on average less than 205.0 mm (body length less than 100.0 mm); rostrum, maxillary bone, and interorbital region narrow; auditory bullae small (from Chihuahua and Coahuila south through Nayarit, Jalisco, Guanajuato, Querétaro, and western Hidalgo) .......... *Peromyscus pectoralis* (p. 339)

67a. Dorsal coloration darker and reddish; tarsi darker; total length on average greater than 205.0 mm (body length greater than 100.0 mm); rostrum, maxillary bone, and interorbital region broad; auditory bullae large (from western Tamaulipas south through eastern Hidalgo) .......... *Peromyscus collinus* (p. 298)

68. Island distribution .......... 69

68a. Continental distribution .......... 71

69. Ranges in the Tres Marías Islands (endemic to the Tres Marías Islands, Nayarit) .......... *Peromyscus madrensis* (p. 326)

69a. Ranges in islands of the Sea of Cortez .......... 70

70. Ranges in San Esteban Island (endemic to San Esteban Island, Baja California) .......... *Peromyscus stephani* (part; p. 354)

70a. Ranges in San Pedro Nolasco Island (from western Oregon south through northern Baja California, and from Wyoming, Idaho, Colorado, and western Texas south through Zacatecas and Querétaro, including San Pedro Nolasco Island in the Sea of Cortez) ........ *Peromyscus boylii* (part; p. 292)

71. Ranges in the United States, northwestern or western México........ 72

71a. Ranges in eastern, central, and southern México........ 76

72. Tail not clearly bicolored, often unicolor; upper toothrow length usually less than 4.1 mm (endemic to the lowlands of Sinaloa and Nayarit) ........ *Peromyscus simulus* (part; p. 348)

72a. Tail moderately bicolored; upper toothrow length usually greater than 4.1 mm ........ 73

73. Ranges restricted to the area of Laguna de Chapala in Michoacán and Jalisco........ *Peromyscus sagax* (p. 345)

73a. Ranges from the Sierra Madre Occidental north through southern United States........ 74

74. Ranges restricted to the southern Sierra Madre Occidental in the northwestern Nayarit highlands ........ *Peromyscus carletoni* (p. 295)

74a. Ranges from the central-northern region of Sierra Madre Occidental or Jalisco northward........ 75

75. Back dark; dorsal pelage not contrasting with the flanks; upper toothrow length usually equal to or greater than 4.2 mm; fronto-maxillary suture without teeth (Fig. 73; western slope of Sierra Madre Occidental, Sonora, Durango, and Chihuahua)........ *Peromyscus schmidlyi* (p. 346)

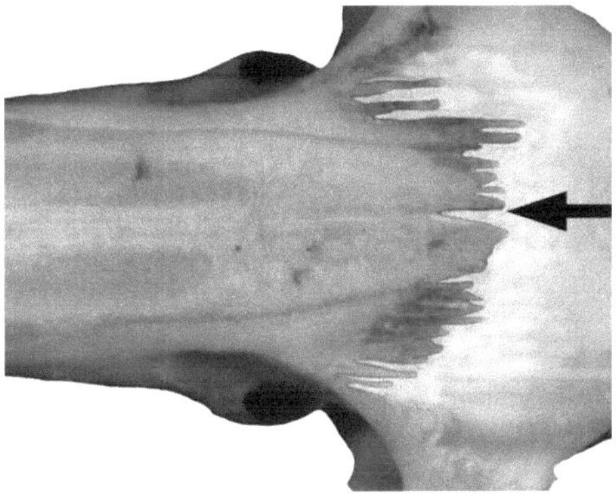

**Fig. 73** Fronto-maxillary suture without teeth

75a. Back pale, usually grizzly gray with black; dorsal pelage strongly contrasting with the flanks; upper toothrow length usually equal to or less than 4.2 mm; fronto-maxillary suture with teeth (Fig. 74; from western Oregon south through northern Baja California, and from Wyoming, Idaho, Colorado, and western Texas south through Zacatecas and Querétaro, including San Pedro Nolasco Island in the Sea of Cortez) ........ *Peromyscus boylii* (part; p. 292)

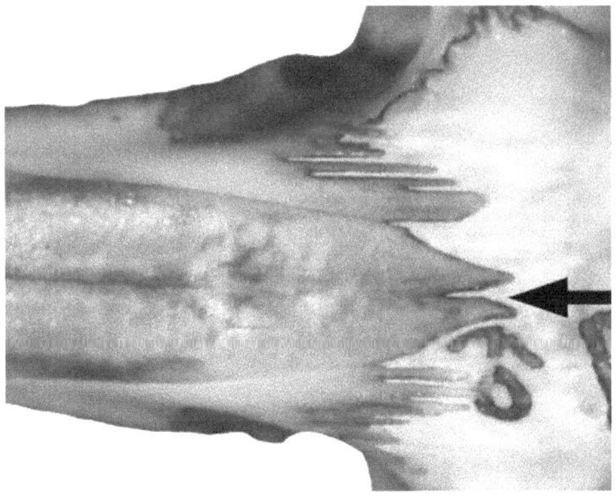

**Fig. 74** Fronto-maxillary suture with teeth

76. Ranges restricted to central-northern Michoacán, southwestern Estado de México, and northern Morelos highlands........ 77

76a. Ranges in eastern, northern, and southern Sierra Madre del Sur........................................................................ 79

77. Ranges restricted to near Lake Zitácuaro, northeastern Michoacán.......................................... *Peromyscus ensinki* (p. 303)

77a. Ranges from highlands of southwestern Jalisco east through Morelos ........................................................... 78

78a. Ranges in highlands of southwestern Jalisco and southeastern Michoacán ................. .*Peromyscus greenbaumi* (p. 311)

78a. Ranges from highlands of central-northern Michoacán east through Morelos................. *Peromyscus kilpatricki* (p. 316)

79. Back pale (usually grizzly gray with black; dorsum strongly contrasting with the flanks); upper toothrow length usually equal to or less than 4.2 mm; fronto-maxillary suture with teeth (Fig. 74; from western Oregon south through northern Baja California, and from Wyoming, Idaho, Colorado, and western Texas south through Zacatecas and Querétaro, including San Pedro Nolasco Island in the Sea of Cortez) ............................................ *Peromyscus boylii* (part; p. 292)

79a. Back dark (dark ochraceous to dark brown; dorsum strongly contrasting with the flanks); upper toothrow length usually equal to or greater than 4.2 mm; fronto-maxillary suture without teeth (Fig. 73)......................................................... 80

80. Dorsal stripe well marked; dorsal coloration usually dark brown or black; ascending branch of the premaxillae not extending beyond the back of the nasals; usually ranges in cloud forest habitats (from southeastern Hidalgo, eastern Puebla, central Veracruz, and central Guerrero south through Central America).................. *Peromyscus beatae* (p. 291)

80a. Dorsal stripe not well defined; dorsal coloration usually ocher; ascending branch of the premaxillae extending beyond the back of the nasals; usually ranges in oak-pine forest habitats (from central Nuevo León and western Tamaulipas south through northeastern Michoacán, Guanajuato, Estado de México, Ciudad de México, Puebla, Tlaxcala, and central Veracruz) ........................................................................................................................ *Peromyscus levipes* (p. 324)

81. Fur of underparts ocher; dorsal coloration dark with ocher; upper toothrow length less than 4.6 mm; nasals not expanded in the anterior part; total length not usually greater than 250.0 mm; skull length less than 32.0 mm (endemic to Tamaulipas and northern San Luis Potosí) ................................................................ *Peromyscus ochraventer* (p. 338)

81a. Fur of underparts pale gray or white; dorsal leaden black; upper toothrow length greater than 4.6 mm; nasals expanded in the anterior part; total length seldom less than 250.0 mm; skull length greater than 32.0 mm .................................... 82

82. Endemic to Hidalgo, Puebla, and Veracruz............................................................................ *Peromyscus furvus* (p. 305)

82a. Endemic to San Luis Potosí and Querétaro ....................................................................... *Peromyscus latirostris* (p. 320)

83. Upperparts and underparts with a paler gray coloration; ring spot around the eyes barely contrasting with the face flanks; small body and skull size; supraorbital bead slightly better developed (Sierra Madre del Sur of Oaxaca on its Pacific slope)..................................................................................................................... *Peromyscus angelensis* (p. 288)

83a. Upperparts and underparts of darker coloration, from reddish to blackish; ring spot around the eyes contrasting with the face flanks; large body and skull size; supraorbital bead slightly developed ................................................................ 84

84. Upperparts cinnamon to darker brown on the back; ring spot around the eyes contrasting with the face flanks, and large on average in body and skull measurements (from Veracruz south through northern Chiapas)........................................ ...................................................................................................................... *Peromyscus mexicanus* (p. 334)

84a. Upperparts dark reddish brown to blackish, darker on the back, with grayish cast; ring spot around the eyes with a greater contrast with the face flanks, and small on average in body and skull measurements (Sierra Madre del Sur of Oaxaca and Veracruz, on its Gulf slope)..................................................................... *Peromyscus totontepecus* (p. 355)

## Subgenus ***Haplomylomys*** Osgood, 1909

Tail always longer than the head-and-body length; **six plantar tubercles; four mammae (nipples); molar accessory tubercles between the outer primary tubercles rudimentary or absent; first and second upper molars usually with three salient and two reentrant outer angles at all stages of wear**; lower molars simple; the coronoid process of the mandible large and elevated; rostral region relatively weak; baculum one-half to two-fifths of the hindfoot length; **glans penis vase shaped;** preputial glands present.

## *Peromyscus avius* Osgood, 1909

Cerralvo island deermouse, ratón de isla Cerralvo

Monotypic.

**Conservation:** *Peromyscus avius* (as *Peromyscus eremicus avius*) is listed as Threatened by the Norma Oficial Mexicana (DOF 2019) and *P. eremicus* is listed as Least Concern by the IUCN (Lacher et al. 2019b). *P. avius* should be listed based on its restricted distribution.

**Characteristics:** Cerralvo island deermouse is medium sized; total length 176.0–208.0 mm andskull length 24.1–25.6 mm. Dorsal pelage ochraceous buff mixed with fine dusky lines, never with orange shades; lateral line, when present, thin and pale yellow; **fur of underparts, cream yellowish, except for the neck and inguinal region**; **ears small,** approximately more than 75% of the hindfoot length; tail longer than the head-and-body length, not bicolored but darker dorsally; interparietal bone rather large; **molars decidedly heavy**. Species group *eremicus*.

**Comments:** *Peromyscus avius* was recognized as a distinct species (Cornejo-Latorre et al. 2017); it was previously regarded as a subspecies of *P. eremicus* (Osgood 1909). *P. avius* is restricted to Cerralvo Island, Baja California Sur (Map 184), with no other species of *Peromycus* found in sympatry. It can be distinguished from other closely related species based on its geographic distribution.

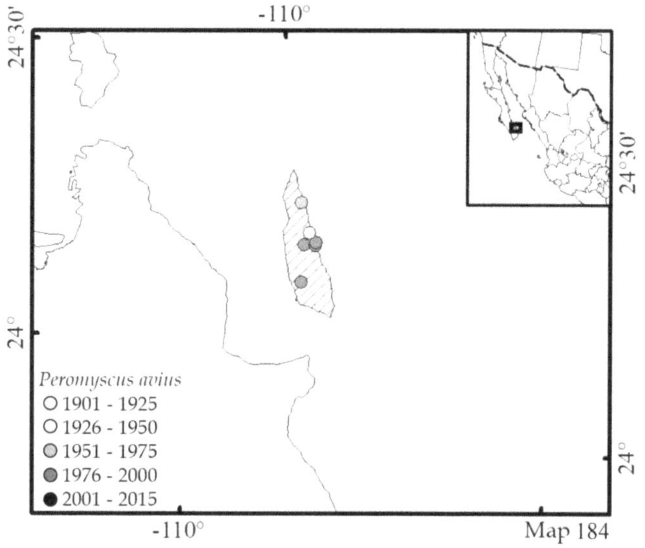


**Map 184** Distribution map of *Peromyscus avius*

**Additional Literature:** Banks (1964), Álvarez-Castañeda and Cortés-Calva (1999).

## *Peromyscus californicus* (Gambel, 1848)

California deermouse, ratón de California

1. *P. c. benitoensis* Grinnell and Orr, 1934. Central-western California.
2. *P. c. californicus* (Gambel, 1848). Coastal areas of western central California.
3. *P. c. insignis* Rhoads, 1895. Southwestern California and northwestern Baja California.
4. *P. c. mariposae* Grinnell and Orr, 1934. Eastern of Central Valley, central-western California.
5. *P. c. parasiticus* (Baird, 1858). Restricted to the southern San Francisco Bay, western-central California.

**Conservation:** *Peromyscus californicus* is listed as Least Concern by the IUCN (Álvarez-Castañeda and Lacher 2016b).

**Characteristics:** California deermouse is large sized; **total length 220.0–285.0** mm and skull length 28.1–32.1 mm. Dorsal pelage grayish brown, with the back with grizzly darker hair and flanks tawny; fur of underparts pale grayish to yellowish brown, with a buffy pectoral spot; flanks ochraceous-tawny; ears very thinly haired, approximately more than 80% of the hindfoot length, with thin hair; **foot soles naked to the end of the calcaneus**; **tail longer than the head-and-body length, unicolor, darker dorsally than ventrally and well haired**; hair long and lax; rostrum thin and long; **second upper molars with accessory cusps in the two principal outer angles**; braincase and auditory bullae inflated; molars robust. Species group *californicus*.

**Comments:** *Peromyscus californicus* ranges from southern San Francisco Bay south through San Quintín valley in northern Baja California, including the western slope of Sierra Nevada (Map 185). It can be found in sympatry, or nearly so, with *P. boylii*, *P. crinitus*, *P. eremicus*, *P. fraterculus*, *P. gambelii*, *P. sonoriensis*, and *P. truei*. *P. californicus* can be distinguished from *P. eremicus* by its larger size, total length greater than 220.0 mm; tail length greater than 117.0 mm, and hindfoot length greater than 25.0 mm; tail well haired and upper parts darker brown. From *P. boylii*, *P. gambelii*, *P. sonoriensis*, and *P. truei*, by having three pairs of mammae (nipples); first and second upper molars without accessory cusps. From *P. gambelii* and *P. sonoriensis*, by having the tail longer than the body-and-head length and with no paravertebral coloration. From *P. crinitus*, by lacking an accessory cusp on the first and second upper molars; the accessory cusp in the first upper molars may be absent in some specimens of *P. crinitus*; tail longer than the head-and-body length, thickly covered with long and soft hairs, terminating in a distinct pencil. From *P. fraterculus*, by its total length greater than 220.0 mm; tail longer than the head-and-body length, commonly indistinctly bicolored and well haired.

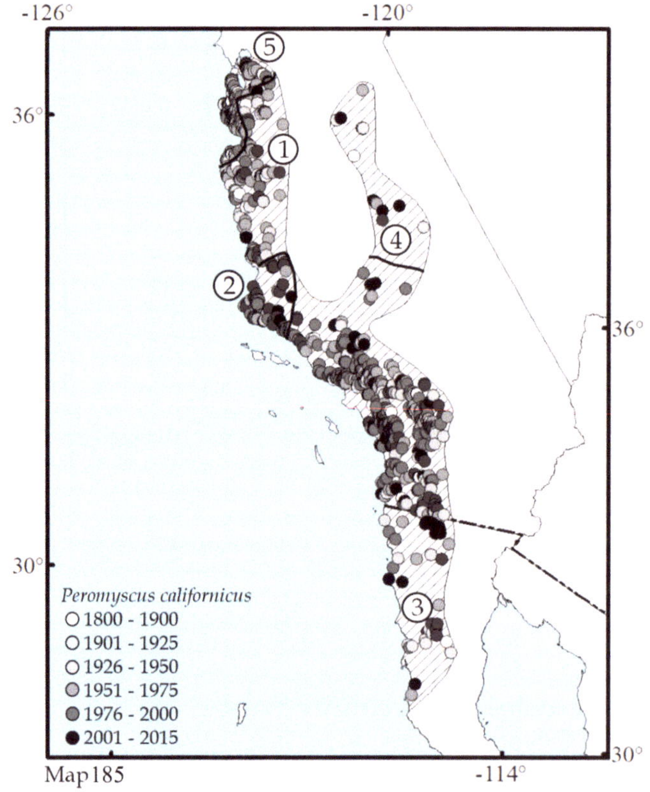


**Map 185** Distribution map of *Peromyscus californicus*

**Additional Literature:** Osgood (1909), Grinnell and Orr (1934), Hooper (1944, 1957), Merritt (1978), Álvarez-Castañeda and Cortés-Calva (1999), Musser and Carleton (2005), Cornejo-Latorre et al. (2017).

## *Peromyscus caniceps* Burt, 1932

Montserrat island deermouse, ratón de Isla Montserrat

Monotypic.

**Conservation:** *Peromyscus caniceps* is listed as Special Protection by the Norma Oficial Mexicana (DOF 2019) and as Critically Endangered (B1ab (v): extent of occurrence less than 100 $km^2$ in only one population and continuing decline in the number of mature individuals) by the IUCN (Álvarez-Castañeda et al. 2018d).

**Characteristics:** Montserrat island deermouse is medium sized; total length 199.0–202.0 mm and skull length 24.5–27.0 mm. Dorsal pelage and flanks yellowish ocher, head gray contrasting with the flanks; lateral line not indistinguishable; fur of underparts whitish grizzly with yellowish shades; ears approximately more than 70% of the hindfoot length; tail shorter than the head-and-body length, unicolor but darker dorsally than ventrally; rostrum thin and long; palatal region greater than the upper toothrow length; auditory bullae small. Species group *eremicus*.

**Comments:** *Peromyscus caniceps* was recognized as a distinct species (Musser and Carleton 2005; Cornejo-Latorre et al. 2017); it was previously regarded as a subspecies of *P. fraterculus* (Hafner et al. 2001). *P. caniceps* is only known from Montserrat Island, Baja California Sur (Map 186), with no other species of *Peromyscus* found in sympatry. It can be distinguished from other closely related species based on its geographic distribution.

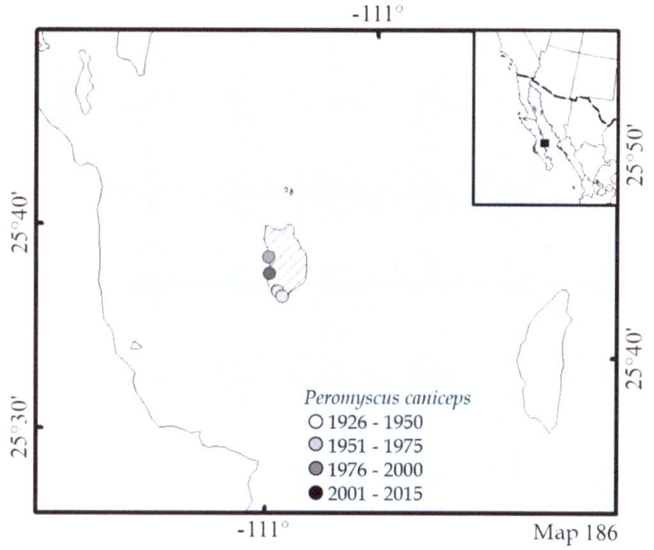


**Map 186** Distribution map of *Peromyscus caniceps*

**Additional Literature:** Burt (1932), Hooper (1968), Avise et al. (1974), Lawlor (1983), Álvarez-Castañeda et al. (1998), Álvarez-Castañeda and Cortés-Calva (1999).

## *Peromyscus dickeyi* Burt, 1932

Tortuga deermouse, ratón de la Isla Tortuga

Monotypic.

**Conservation:** *Peromyscus dickeyi* is listed as Special Protection by the Norma Oficial Mexicana (DOF 2019) and as **Critically Endangered** (B1ac (iv) + 2 ac (iv): extent of occurrence less than 100 $km^2$ in only one population and extreme fluctuations in the number of mature individuals and area of occupancy less than 10 $km^2$, at only a single location and extreme fluctuations in the number of mature individuals) by the IUCN (Álvarez-Castañeda et al. 2008a).

**Characteristics:** Tortuga deermouse is medium sized; total length 186.0–203.0 mm and skull length 25.3–27.6 mm. Dorsal pelage dark grizzly with cinnamon; lateral line thin, pale yellow; fur of underparts whitish grizzly with yellowish shades, some specimens with a pectoral ochraceous spot; lateral line is present along the body; ears approximately more than 75% of the hindfoot length; **tail shorter than the head-and-body length**, not bicolored but darker dorsally; rostrum thin and long; squamous bone projection not wide and not tapered toward the nasals; zygomatic arches nearly parallel-sided; rostrum strong; nasals wide anteriorly; premaxillae projecting well beyond the posterior limits of the nasals. Species group *eremicus*.

**Comments:** *Peromyscus dickeyi* was recognized as a distinct species (Musser and Carleton 2005; Cornejo-Latorre et al. 2017); it was previously regarded as a subspecies of *P. merriami* (Hafner et al. 2001). *P. dickeyi* is only known from Tortuga Island, Baja California Sur (Map 187), with no other species of *Peromyscus* found in sympatry. It can be distinguished from other closely related species based on its geographic distribution.

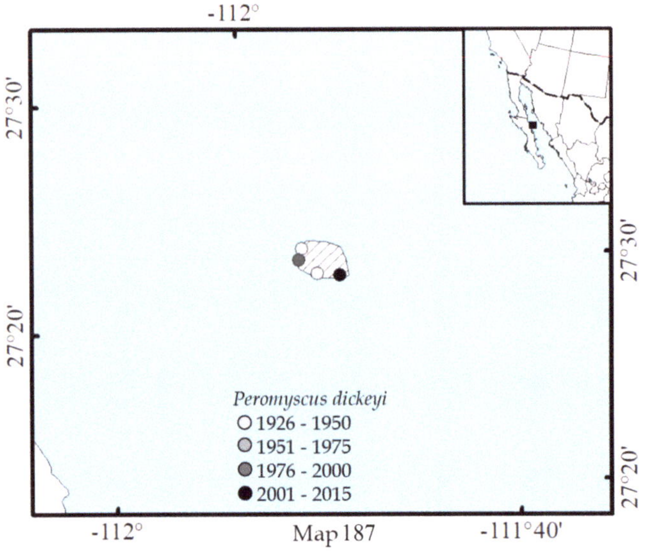


**Map 187** Distribution map of *Peromyscus dickeyi*

**Additional Literature:** Burt (1932), Lawlor (1983), Álvarez-Castañeda and Cortés-Calva (1999), Cortés-Calva and Álvarez-Castañeda (2001).

## *Peromyscus eremicus* (Baird, 1857)

Cactus deermouse, ratón de los cactus

1. *P. e. alcorni* Anderson, 1972. Restricted to central northwestern Chihuahua.
2. *P. e. anthonyi* (Merriam, 1887). Northwestern Chihuahua, southwestern New Mexico, southeastern Arizona, and northeastern Sonora.
3. *P. e. collatus* Burt, 1932. Restricted to Turner Island, western Sonora.
4. *P. e. eremicus* (Baird, 1857). From Arizona and northeastern Sonora east through western Texas and Coahuila.
5. *P. e. interparietalis* Burt, 1932. Restricted to San Lorenzo Island, eastern-central Baja California.
6. *P. e. lorenzi* Banks, 1967. Restricted to Animas Island (San Lorenzo Norte Island), eastern-central Baja California.
7. *P. e. papagensis* Goldman, 1917. Restricted to the Pinacate Desert, northwestern Sonora, and southern Arizona.
8. *P. e. phaeurus* Osgood, 1904. Western Durango, Zacatecas, Aguascalientes, San Luis Potosí, eastern Coahuila, southern Nuevo León, and eastern Tamaulipas.
9. *P. e. pullus* Blossom, 1933. Restricted to Pima County, southern Arizona.
10. *P. e. ryckmani* Banks, 1967. Restricted to Salsipuedes Island, eastern-central Baja California.
11. *P. e. sinaloensis* Anderson, 1972. Southern Sonora, Sinaloa, western Durango, and Nayarit.
12. *P. e. tiburonensis* Mearns, 1897. Restricted to Tiburón Island, western-central Sonora.

**Conservation:** *Peromyscus eremicus collatus*, *P. interparietalis* (including *P. e. lorenzi* and *P. e. ryckmani*), and *P. e tiburonensis* are listed as Threatened by the Norma Oficial Mexicana (DOF 2019) and *P. e. interparietalis*, *P. e. lorenzi*, and *P. e. ryckmani* are listed as Critically Endangered (B1ab (v): extent of occurrence less than 100 km$^2$ in only one population) by the IUCN; *P. eremicus* is listed as Least Concern by the IUCN (Lacher et al. 2019b).

**Characteristics:** Cactus deermouse is medium sized; total length 169.0–218.0 mm and skull length 24.0–26.5 mm. Dorsal pelage gray grizzly with reddish brown shades; head grayish; lateral line thin and pale yellow; fur of underparts whitish grizzly with yellowish shades; ears relatively large and leafy, approximately more than 75% of the hindfoot length, naked or with very thin hair; **hind feet with naked soles**; **tail longer than the head-and-body length and almost naked, finely annulated, and densely covered with short hairs, very slightly or not pencil-like at the tip**; rostrum thin and long; braincase high and slightly concave on the back; nasals relatively wide. Species group *eremicus*.

**Comments:** *Peromyscus eremicus avius* was recognized as a distinct species (Cornejo-Latorre et al. 2017). *P. e. cedrosensis*, *P. e. cinereus*, *P. e. fraterculus*, *P. e. insulicola*, and *P. e. polypolius* were considered subspecies of *P. fraterculus* (Riddle et al. 2000a, b; Cornejo-Latorre et al. 2017). *P. interparietalis* was considered a subspecies of *P. guardia* (Burt 1932; Hafner et al. 2001), previously regarded as a separate species (Banks 1967; Brand and Ryckman 1969); the three subspecies of *P. interparietalis interparietalis*, *P. i. lorenzi*, and *P. i. ryckmani* were recognized as subspecies of *P. eremicus* (Cornejo-Latorre et al. 2017). *P. eremicus* ranges from southwestern Arizona, New Mexico, and Texas south through Sinaloa in the Pacific Coast and in the Altiplano Central to Aguascalientes, San Luis Potosí, and eastern Tamaulipas (Map 188). It can be found in sympatry, or nearly so, with *P. boylii*, *P. californicus*, *P. collinus*, *P. crinitus*, *P. gambelii*, *P. hooperi*, *P. labecula*, *P. laceianus*, *P. melanophrys*, *P. melanotis*, *P. merriami*, *P. pectoralis*, and *P. sonoriensis. P. eremicus* can be distinguished from *P. boylii*, *P. crinitus*, *P. collinus*, *P. gambelii*, *P. laceianus*, *P. melanophrys*, *P. melanotis*, *P. pectoralis*, and *P. sonoriensis* by having three pairs of mammae (nipples); first and second upper molars without accessory cusps. From *P. gambelii*, by having the tail longer than the head-and-body length and without paravertebral coloration. From *P. californicus* by its smaller size, total length less than 220.0 mm; tail length less than 117.0 mm; hindfoot length less than 25.0 mm; tail less well haired and upperparts pale brown. From *P. hooperi*, by its smaller total length; dorsal parts ochraceous or buff; skull and nasal bones relatively shorter; posterior end of the nasal bones truncated and palatine with a short bony protuberance. From *P. merriami*, by its smaller size; less robust skull with a relatively lesser zygomatic breadth; in dorsal view, a shallow zygomatic notch and a smaller infraorbital canal.

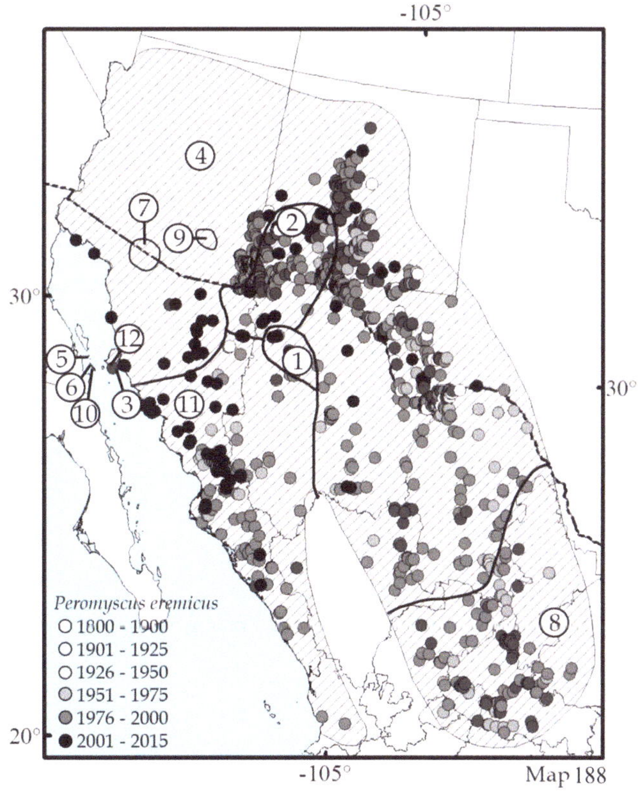


**Map 188** Distribution map of *Peromyscus eremicus*

**Additional Literature:** Burt (1932), Hooper (1957), Commissaris (1960), Hoffmeister and Lee (1963), Lawlor (1971b, 1983), Veal and Caire (1979), Álvarez-Castañeda and Cortés-Calva (1999), Hafner et al. (2001), Musser and Carleton (2005).

## ***Peromyscus eva*** Thomas, 1898

Baja California deermouse, ratón de Baja California

1. *P. e. carmeni* Townsend, 1912. Restricted to Carmen Island, eastern-central Baja California Sur.
2. *P. e. cinereus* Hall, 1931. Restricted to San José Island, central-eastern Baja California Sur.
3. *P. e. eva* Thomas, 1898. Restricted to Baja California Sur.

**Conservation:** *Peromyscus eva carmeni* and *P. e. cinereus* (as *Peromyscus eremicus cinereus*) are listed as Threatened by the Norma Oficial Mexicana (DOF 2019) and *P. eva* is listed as Least Concern by the IUCN (Álvarez-Castañeda and Castro-Arellano 2016a).

**Characteristics:** Baja California deermouse is medium sized; total length 185.0–218.0 mm and skull length 20.9–26.0 mm. **Dorsal pelage with orange shades mainly on the flanks**; flanks orange-yellow; fur of underparts whitish grizzly with yellowish shades; **ears smaller, pale brownish, almost naked, approximately more than 70% of the hindfoot length**; feet white, tarsal joints marked with dusky shades; tail longer than the head-and-body length, dusky dorsally and slightly paler ventrally, often quite uniform blackish all around; rostrum thin and long. Species group *eremicus*.

**Comments:** *Peromyscus eva* was described as a subspecies of *P. eremicus*; based on the morphological differences, it was elevated to a distinct species (Lawlor 1971b). Genetic analyses show that *P. fraterculus cinereus* is a subspecies of *P. eva* (Cornejo-Latorre et al. 2017). *P. eva* ranges from the central Baja California Peninsula southward, including the Carmen and San José Islands, Baja California Sur (Map 189). It can be found in sympatry, or nearly so, with *P. fraterculus*, *P. gambelii*, and *P. truei*. *P. eva* can be distinguished from *P. fraterculus* by having a shorter and slightly harsher pelage, colored with a blend of rufous, buffy, and brown shades; face more grayish; ears slightly smaller ears on average; tail much longer, more than 35 vertebrae; longer rostrum; shallower zygomatic notch; molar toothrow on average greater in length and width. From *P. gambelii* and *P. truei*, by having three pairs of mammae (nipples); first and second upper molars without accessory cusps. From *P. gambelii*, by having the tail longer than the head-and-body length, no paravertebral coloration. *P. eva* can be distinguished from other closely related species based on its geographic distribution.

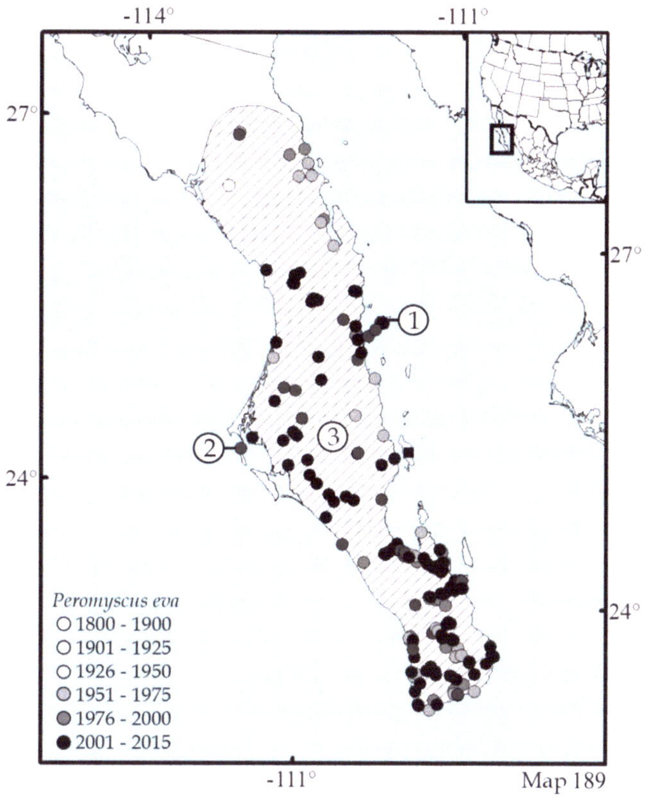


**Map 189** Distribution map of *Peromyscus eva*

**Additional Literature:** Osgood (1909), Lawlor (1983), Álvarez-Castañeda and Cortés-Calva (1999, 2003b), Musser and Carleton (2005).

## *Peromyscus fraterculus* (Miller, 1892)

Northern Baja deermouse, ratón de las Californias

1. *P. f. cedrosensis* Allen, 1898. Restricted to Cedros Island, southwestern Baja California.
2. *P. f. fraterculus* (Miller, 1892). From southwestern California south through northwestern Baja California Sur, including the highlands northern La Paz isthmus.
3. *P. f. insulicola* Osgood, 1909. Restricted to Espiritu Santo Island, southeastern Baja California Sur.
4. *P. f. polypolius* Osgood, 1909. Restricted to Margarita Island, central-western Baja California Sur.
5. *P. f. pseudocrinitus* Burt, 1932. Restricted to Coronados Island, Baja California Sur.

**Conservation:** *Peromyscus fraterculus cedrosensis*, *P. f. insulicola*, *P. f. polypolius*, (all three as *Peromyscus eremicus*) and *P. f. pseudocrinitus* (as *Peromyscus pseudocrinitus*) is listed as Threatened by the Norma Oficial Mexicana (DOF 2019), and *P. fraterculus* is listed as Least Concern by the IUCN (Timm et al. 2016h).

**Characteristics:** Northern Baja deermouse is medium sized; total length 185.0–218.0 mm and skull length 22.3–25.0 mm. Dorsal pelage cinnamon rufous richly sprinkled with black, middle of the dorsum darker, including the flanks, never with orange shades; lateral line, when present, thin and pale yellow; **fur of underparts whitish grizzly with yellowish shades;** head more or less grayish, particularly in the postorbital region; ears approximately more than 75% of the hindfoot length; tail longer than the head-and-body length, not bicolored but darker ventrally; rostrum thin and long. Species group *eremicus*.

**Comments:** *Peromyscus fraterculus* was described as a subspecies of *P. eremicus*; based on genetic differences, it was recognized as a distinct species (Avise et al. 1974; Riddle et al. 2000b). *P. avius* was previously a subspecies but is now considered a distinct species. *P. fraterculus cinereus* is recognized as a subspecies of *P. eva*, and *P. pseudocrinitus*, as a subspecies *P. fraterculus* (Cornejo-Latorre et al. 2017). *P. fraterculus* ranges from southwestern California south through La Paz isthmus in Baja California Sur (Map 190; Álvarez-Castañeda et al. 2010a). It can be found in sympatry, or nearly so, with *P. boylii*, *P. californicus*, *P. crinitus*, *P. eva*, *P. gambelii*, *P. sonoriensis*, and *P. truei*. *P. fraterculus* can be distinguished from *P. boylii*, *P. gambelii*, *P. sonoriensis*, and *P. truei* by its three pairs of mammae (nipples); first and second upper molars without accessory cusps. From *P. gambelii* and *P. sonoriensis*, by having the tail longer than the head-and-body length, without paravertebral coloration. From *P. crinitus*, by lacking an accessory cusp on the first and second upper molars; accessory cusp of the first upper molars may be absent in some specimens of *P. crinitus*; tail longer than the head-and-body length, thickly covered with long, soft hairs, and terminating in a distinct hair tuft. From *P. californicus*, by its smaller size, less than 220.0 mm; tail no indistinctly bicolored and naked. From *P. eremicus*, by its markedly deeper coloration in direct comparison. From *P. eva*, by its larger pelage colored with an admixture of dark brown or black shades; face less grayish; flanks less orange; shorter tail length, 35 vertebrae; toothrow on average shorter and narrower, and zygomatic notch deeper.

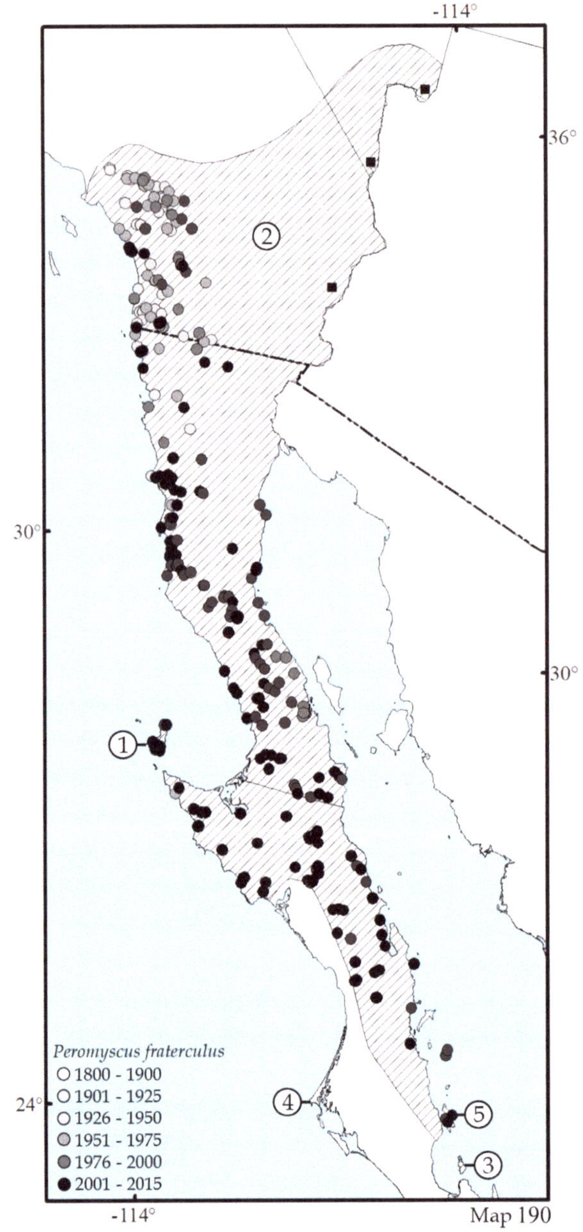


**Map 190** Distribution map of *Peromyscus fraterculus*

**Additional Literature:** Benson (1940), Lawlor (1971b), Veal and Caire (1979), Lawlor (1983), Álvarez-Castañeda (1998), Álvarez-Castañeda and Cortés-Calva (1999, 2003b), Hafner et al. (2001), Álvarez-Castañeda et al. (2008c), Musser and Carleton (2005).

## *Peromyscus guardia* Townsend, 1912

Ángel de la Guarda island deermouse, ratón de la Isla Ángel de la Guarda

1. *P. g. guardia* Townsend, 1912. Restricted to Ángel de la Guarda Island, eastern-central Baja California.
2. *P. g. harbisoni* Banks, 1967. Restricted to Granito Island (Extinct), eastern-central Baja California.
3. *P. g. mejiae* Burt, 1932. Restricted to Mejía Island (Extinct), eastern-central Baja California.

**Conservation:** *Peromyscus guardia* is listed as Endangered by the Norma Oficial Mexicana (DOF 2019) and as Critically Endangered (Possibly Extinct) (B2ab (iv, v): extent of occurrence less than 100 km$^2$ only one population and continuing decline and under continuing decline in the number of subpopulation and number of mature individuals) by the IUCN (Álvarez-Castañeda et al. 2018e).

**Characteristics:** Ángel de la Guarda island deermouse is medium sized; total length 189.0–223.0 mm and skull length 25.5–26.9 mm. Dorsal pelage gray grizzly with reddish brown shades; head grayish; lateral line thin and pale yellow; fur of underparts whitish; ears approximately more than 75% of the hindfoot length, naked or with thin hair; hind feet with six tubercles on the soles; tail longer than the head-and-body length, not bicolored but darker dorsally; rostrum thin and long; nasal posterior edges rounded or pointed; incisive foramina short and not reaching the anterior plane of the first upper molars; braincase high and inflated, slightly concave on the back; rostrum large; auditory bullae large; **nearly complete absence of mesostyles and entostyles on the first upper molars.** Species group *eremicus*.

**Comments:** *Peromyscus interparietalis* was recognized as a subspecies of *P. guardia* (Burt 1932; Hafner et al. 2001); it was previously regarded as a distinct species (Banks 1967; Brand and Ryckman 1969; Cornejo-Latorre et al. 2017). *P. guardia* is restricted to Ángel de la Guarda Archipielago (Ángel de la Guarda, Granito, Mejía, and Estanque Islands), Baja California (Map 191). No other species of *Peromyscus* is found in sympatry. It can be distinguished from *P. fraterculus* by having the skull less arched dorsally, with a longer rostrum and broader interpterygoid fossa; zygomatic arches more compressed anteriorly and auditory bullae larger. *P. guardia* can be distinguished from other closely related species based on its geographic distribution.

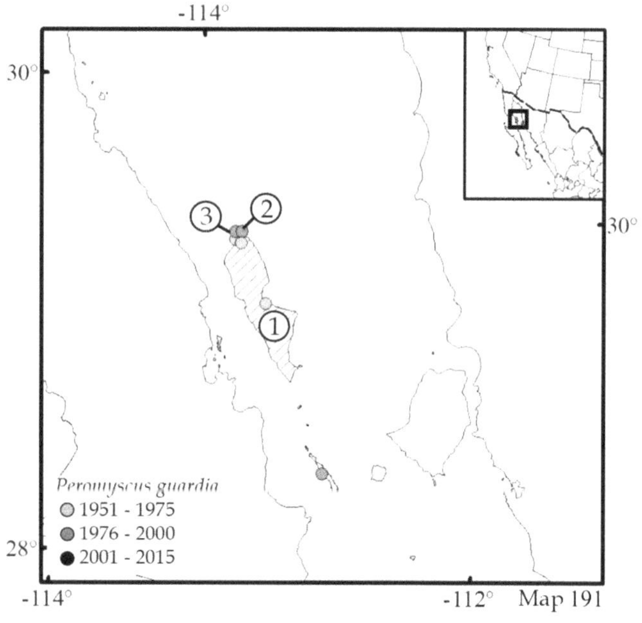


**Map 191** Distribution map of *Peromyscus guardia*

**Additional Literature:** Lawlor (1971a), Lawlor (1983), Álvarez-Castañeda and Cortés-Calva (1999), Rios and Álvarez-Castañeda (2011), Musser and Carleton (2005).

## *Peromyscus hooperi* Lee and Schmidly, 1977

Hooper's deermouse, ratón de Coahuila

Monotypic.

**Conservation:** *Peromyscus hooperi* is listed as Least Concern by the IUCN (Álvarez-Castañeda 2016).

**Characteristics:** Hooper's deermouse is medium sized; total length 172.0–218.0 mm and skull length 25.7–27.5 mm. Dorsal pelage brown grizzly with gray; lateral line slightly marked, yellowish or pale ocher; fur of underparts pale cream; ears long with thin hair; tail longer than the head-and-body length, not bicolored but darker dorsly; rostrum thin and long; **lower frequency of mesolophs and anteroloph absent**. Species group *hooperi*.

**Comments:** *Peromyscus hooperi* ranges from central Coahuila, northernmost Zacatecas, and San Luis Potosí (Map 192). It can be found in sympatry, or nearly so, with *P. difficilis*, *P. eremicus*, *P. labecula*, *P. melanophrys*, and *P. pectoralis*. *P. hooperi* can be distinguished from *P. eremicus* by having a greater total length; pelage of upperparts grayish; skull and nasal bones relatively longer; nasal bone posterior end tapered, V shaped, and palatine spurs without a short bony protuberance. Upperparts of *P. hooperi* are grayish rather than ochraceous or buff as in *P. merriami* and *P. eremicus*. From *P. difficilis*, *P. labecula*, *P. melanophrys*, and *P. pectoralis*, by a lower frequency of mesolophs and anteroloph absent. From *P. labecula*, by having a notably larger tail and skull; less complex occlusal surface of the molars.

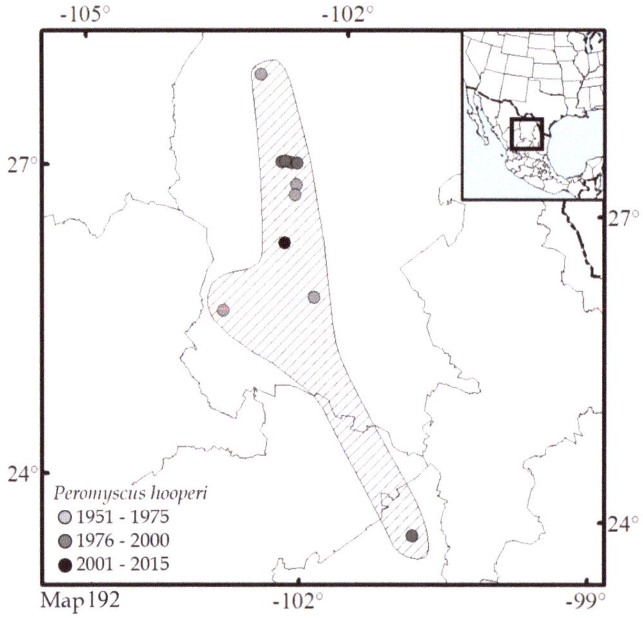


**Map 192** Distribution map of *Peromyscus hooperi*

**Additional Literature:** Lee and Schmidly (1977), Schmidly et al. (1985), Álvarez and Álvarez-Castañeda (1991), Álvarez-Castañeda (2002), Musser and Carleton (2005).

## *Peromyscus merriami* Mearns, 1896

Mesquite deermouse, ratón de los mezquites

1. *P. m. goldmani* Osgood, 1904. Southern Sonora and northern Sinaloa.
2. *P. m. merriami* Mearns, 1896. Southern Arizona and western Sonora.

**Conservation:** *Peromyscus merriami* is listed as Least Concern by the IUCN (Linzey et al. 2016l).

**Characteristics:** Mesquite mouse is medium sized; total length 183.0–223.0 mm and skull length 25.3–27.4 mm. Dorsal pelage grizzly gray with reddish brown shades; head grayish; lateral line pale yellow; fur of underparts whitish grizzly with yellowish shades; ears approximately more than 85% of the hindfoot length, with very thin hair and without the edge whitish; tail cylindrical and longer than the head-and-body length, with short hairs; relatively heavy and rather elongated; rostrum thin and long; **braincase top more flattened than in any other species of *Peromyscus***. Species group *eremicus*.

**Comments:** *Peromyscus dickeyi* was recognized as a distinct species (Musser and Carleton 2005; Cornejo-Latorre et al. 2017); it was previously regarded as a subspecies of *P. merriami* (Hafner et al. 2001). *P. merriami* ranges from southern Arizona south through Sinaloa, including western Chihuahua (Map 193). It can be found in sympatry, or nearly so, with *P. crinitus*, *P. eremicus*, *P. labecula*, *P. leucopus*, and *P. sonoriensis*. *P. merriami* can be distinguished from *P. crinitus* by not having an accessory cusp on the first and second upper molars; the accessory cusp of the first upper molars may be absent in some specimens of *P. crinitus*; tail longer than the head-and-body length, thickly covered with long soft hairs and terminating in a distinct hair tuft. From *P. eremicus*, by its larger size; more robust skull with a relatively greater zygomatic breadth; in dorsal view, a deeper zygomatic notch and a larger infraorbital canal. From *P. labecula*, *P. leucopus*, and *P. sonoriensis*, by having the tail longer than the head-and-body length; no paravertebral coloration; three pairs of mammae (nipples); first and second upper molars without accessory cusps.

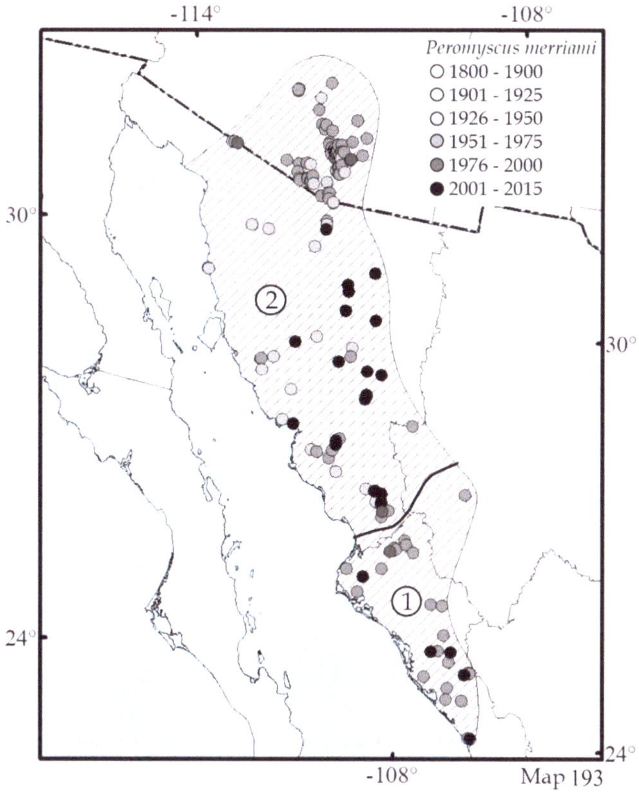


**Map 193** Distribution map of *Peromyscus merriami*

**Additional Literature:** Commissaris (1960), Hoffmeister and Lee (1963), Álvarez-Castañeda and Cortés-Calva (1999).

## ***Peromyscus pembertoni*** Burt, 1932

San Pedro Nolasco deermouse, ratón de la Isla San Pedro Nolasco

Monotypic.

**Conservation:** *Peromyscus pembertoni* is listed as Extinct in the Norma Oficial Mexicana (DOF 2019) and by the IUCN (Álvarez-Castañeda et al. 2017d).

**Characteristics:** San Pedro Nolasco deermouse is medium sized; total length 208.0–213.0 mm and skull length 26.5–28.0 mm. Dorsal pelage pale cinnamon grizzly with dark brown; head paler than the back; lateral line thin and pale yellow; fur of underparts whitish grizzly with yellowish shades; ears approximately more than 75% of the hindfoot length with very thin hair; tail shorter than the head-and-body length, not bicolored and darker dorsally; rostrum thin and long; squamous bone projection not wide, not tapered toward the nasals; maxillary extension ending at the same level as the nasals. Species group *eremicus*.

**Comments:** *Peromyscus pembertoni* was known only from San Pedro Nolasco Island, Sonora (Map 194). *P. pembertoni* can be found in sympatry with *P. boylii*. It can be distinguished from *P. boylii* by having three pairs of mammae (nipples) and first and second upper molars without accessory cusps in the two principal outer angles. From other closely related species, based on its geographic distribution.

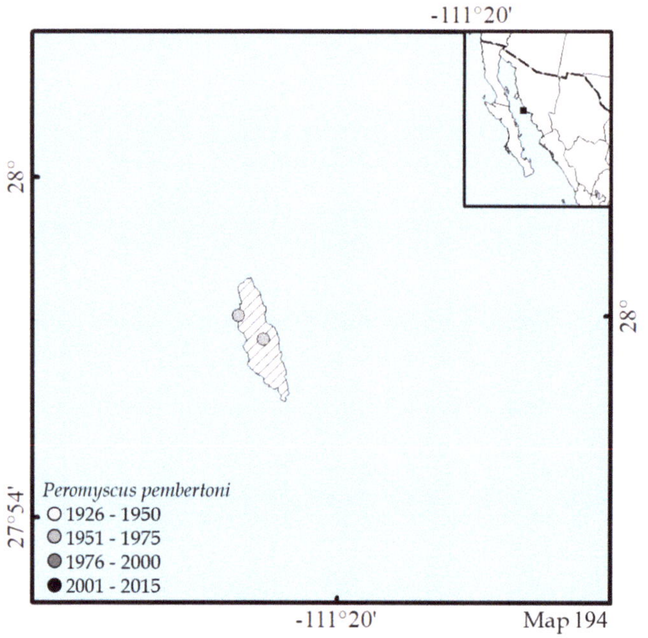


**Map 194** Distribution map of *Peromyscus pembertoni* †

**Additional Literature:** Burt (1932), Lawlor (1983), Álvarez-Castañeda and Cortés-Calva (1999, 2003a), Hafner et al. (2001), Musser and Carleton (2005), Cornejo-Latorre et al. (2017).

## Subgenus ***Peromyscus*** Gloger, 1841

Tail in different sizes in relation to the head-and-body length, **six or four mammae** (nipples; *P. crinitus*, *P.caniceps*, *P. stephani*, and *P. megalops*). **Accessory tubercles present in salient angles of the first and second upper molars** but absent or rudimentary in *P. crinitus*, *P. caniceps*, and *P. stephani*; outer accessory tubercles of the first and second molars only slightly developed. The coronoid process of the mandible usually small and only slightly elevated; rostral region relatively strong; baculum one- to two-thirds as long as the hind feet; **glans penis rod shaped in most species** to vase shaped in *P. hylocetes;* preputial glands rudimentary or absent (Osgood 1909; Linzey and Layne 1969).

### ***Peromyscus amplus*** Osgood, 1904

Highlands rock deermouse, ratón de las rocas de las tierras altas

1. *P. a. amplus* Osgood, 1904. From Michoacán and Guanajuato west through western Veracruz.
2. *P. a. saxicola* Hoffmeister and de la Torre, 1959. Restricted to Querétaro and northern Hidalgo.

**Conservation:** *Peromyscus amplus* (as *Peromyscus difficilis*) is listed as Least Concern by the IUCN (Castro-Arellano and Vázquez 2016c).

**Characteristics:** Highlands rock deermouse is medium sized; total length 235.0–260.0 mm and skull length 28.6–31.8 mm. Dorsal pelage with a clay color produced by the combination of ochraceous buff at the base and a fine peppery mixture of dusky; lateral line rather broad, ochraceous buff; fur of underparts creamy white, with a well-developed ochraceous buff pectoral spot; **ears approximately or equal to the hindfoot length**; tail longer than the head-and-body length, bicolored, dusky brownish dorsally and white ventrally; forefeet and hind feet whitish grizzly, tarsal joint with a small dusky marking; interorbital space narrow; supraorbital border not beaded, never forming any distinct shelf; rostrum and nasals broader and heavier; teeth moderate; **auditory bullae large, well developed**. Species group *truei*.

**Comments:** *Peromyscus amplus* was recognized as a distinct species (Hernández-Canchola et al. 2022); it was previously regarded as a subspecies of *P. difficilis* (Osgood 1909). Its ranges in Guanajuato, Hidalgo, Michoacán, Estado de México, and Puebla (Map 195). It can be found in sympatry, or nearly so, with *P. bullatus*, *P. difficilis*, *P. felipensis*, *P. gratus*, *P. hylocetes*, *P. labecula*, *P. leucopus*, *P. levipes*, *P. mekisturus*, and *P. melanotis*. *P. amplus* can be distinguished from *P. bullatus* by having a larger average size; occipitonasal length greater than 28.6 mm; smaller ears, more or less equal to the hind feet; auditory bullae proportionally less inflated and molars more robust. From *P. difficilis*, by its larger size; dull and reddish coloration; larger braincase, and using genetic data. From *P. felipensis*, by having a smaller skull; coloration much paler, and using genetic data. From *P. gratus*, by having the dorsum often grayish-black; lateral line with ochraceous shades; top of the head often grayish; larger ears, greater than 24.0 mm; a shorter toothrow; auditory bullae proportionally greatly inflated and larger. From *P. labecula*, *P. leucopus*, *P. levipes*, and *P. melanotis*, by its larger size; ears longer than the hind feet; tail longer than the head-and-body length; auditory bullae proportionally greatly inflated. From *P. mekisturus*, by having the tail similar to the head-and-body length or slightly longer.

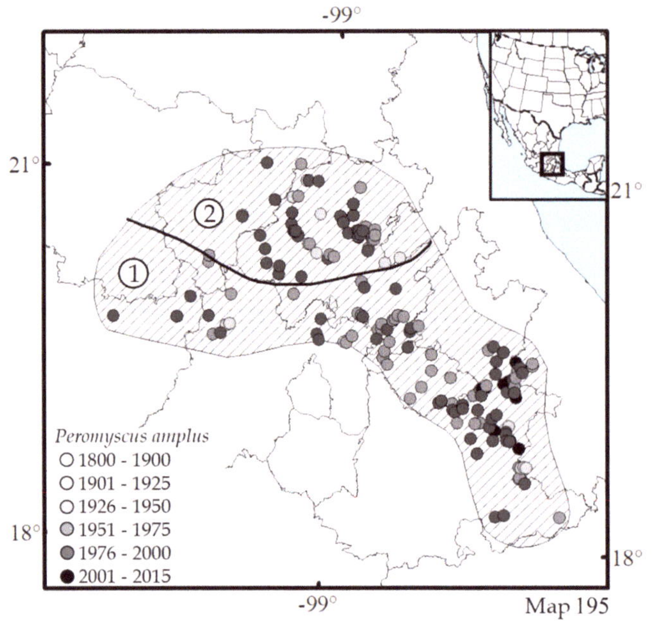


**Map 195** Distribution map of *Peromyscus amplus*

**Additional Literature:** Hooper (1957), Hoffmeister and De la Torre (1961), Zimmerman et al. (1975), Avise et al. (1979), Janecek (1990), DeWalt et al. (1993a), Bradley et al. (1999), Arellano-Meneses et al. (2000), Durish et al. (2004), González-Ruiz and Álvarez-Castañeda (2005), Müdespacher-Ziehl et al. (2005), Musser and Carleton (2005), Fernández et al. (2010).

## ***Peromyscus angelensis*** Osgood, 1904

Angel deermouse, ratón de los ángeles

*P. a. angelensis* Osgood, 1904. Coastal regions of southeast Oaxaca.
*P. a. putlaensis* Goodwin, 1964. Restricted to Putla area, central-western Oaxaca.

**Conservation:** *Peromyscus angelensis* (as *Peromyscus mexicanus*) is listed as Least Concern by the IUCN (Reid and Pino 2016).

**Characteristics:** Angel deermouse is medium sized; total length 218.0–235.0 mm and skull length 28.8–32.4 mm. Dorsal pelage paler gray to darker grayish brown on the back; fur of underparts whitish to cream; flanks with a lateral reddish line; blackish spot on the base of the whiskers; ears dusky brown with the edge faint whitish, approximately 70% of the hindfoot length; forefeet and carpal joints white and proximal one-half of the forearms dusky, overlaid by rufescent shades; hind feet white and tarsal joints dusky brown; tail slightly longer than the head-and-body length, practically naked, with few hairs, and coarsely annulated; rostrum thin and long; supraorbital ridge edge well developed; molariform and auditory bullae proportionally small.

**Comments:** *Peromyscus mexicanus azulensis* was previously considered a subspecies of *P. megalops* (Huckaby 1980). *P. angelensis* was recognized as a distinct species (Pérez-Montes et al. 2022); it was previously regarded as a subspecies of *P. mexicanus* (Huckaby 1980). *P. angelensis* ranges in the Pacific side of the Sierra Madre del Sur in Oaxaca and probably Guerrero (Map 196). It can be found in sympatry, or nearly so, with *P. aztecus*, *P. beatae*, *P. leucopus*, *P. levipes*, *P. megalops*, *P. melanurus*, and *P. tontotepecus*. *P. angelensis* can be distinguished from all other species of *Peromyscus*, except for *P. mexicanus* and *P. tontotepecus*, by having the tail slightly longer than the head-and-body length, practically naked with few hairs, and coarsely annulated; relatively small molars and auditory bullae. From *P. beatae*, and *P. levipes*, by its larger size and tail proportionally longer in relation to the head-and-body. From *P. megalops* and *P. melanurus*, by its smaller size; frontal bone narrow and less distinct; smaller teeth and auditory bullae about the same size. From *P. mexicanus* and *P. totontepecus* by its dorsal pelage paler gray to darker grayish brown and supraorbital ridge edge well developed.

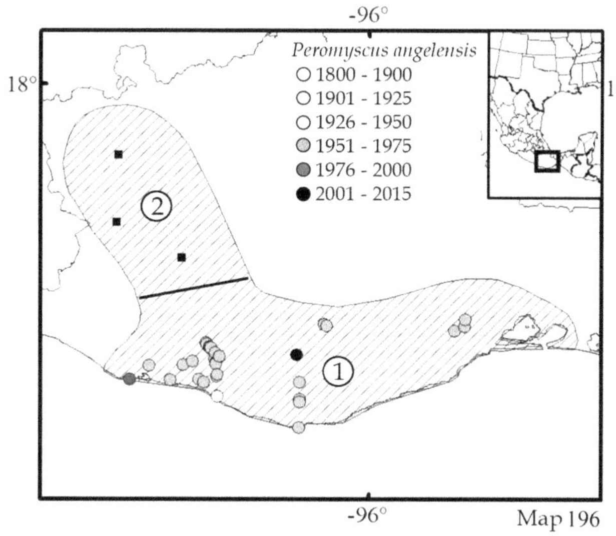


**Map 196** Distribution map of *Peromyscus angelensis*

**Additional Literature:** Osgood (1909), Hooper (1957), Hooper (1968), Musser and Carleton (2005), Trujano-Álvarez and Álvarez-Castañeda (2010).

## *Peromyscus attwateri* Allen, 1895

Texas deermouse, ratón de Texas

Monotypic.

**Conservation:** *Peromyscus attwateri* is listed as Least Concern by the IUCN (Cassola 2016cl).

**Characteristics:** Texas deermouse is medium sized; total length 187.0–218.0 mm and skull length 27.4–30.2 mm. Dorsal pelage darker brown mixed with blackish shades along the medial dorsal area; fur of underparts white with plumbeous shades at the base; flanks cinnamon; ears medium, approximately 85% of the hindfoot length and with very thin hair; tail equal to or longer than the head-and-body length, bicolored, moderately well haired, and generally with a hair tuft at the tip; ankles dark or dusky; braincase elongated and wide but not rounded; zygomatic arches parallel and not converging anteriorly; anterior region of the nasals narrow; infraorbital canal wider; rostrum depressed anteriorly; **first and second upper molars with a mesoloph; first and second lower molars with a long mesolophid**. Species group *truei*.

**Comments:** *Peromyscus attwateri* was reinstated as a distinct species (Schmidly 1973); it was previously regarded as a subspecies of *P. boylii* (Bailey 1906) and within the *truei* species group (Musser and Carleton 2005). *P. attwateri* ranges in southeastern Kansas, southwestern Missouri, Oklahoma, northwestern Arkansas, and Texas (Map 197). It can be found in sympatry, or nearly so, with *P. laceianus*, *P. leucopus*, *P. maniculatus*, *P. pectoralis*, and *P. sonoriensis*. It can be distinguished from all other species of *Peromyscus* by its first and second upper molars with a mesoloph and the lower molars with a longer mesolophid. From *P. leucopus*, *P. maniculatus*, *P. pectoralis*, and *P. sonoriensis*, by having a larger total length and without a dark dorsal narrow stripe covering only the tail vertebral line. From *P. leucopus*, by having larger auditory bullae. From *P. laceianus*, by its larger size, in particular in hindfoot length; darker color with dusky ankles; less sharply bicolored tail; auditory bullae and incisive foramen larger; toothrow larger; posterior region of the nasals rounded; mesopterygoid process larger and more strongly developed.

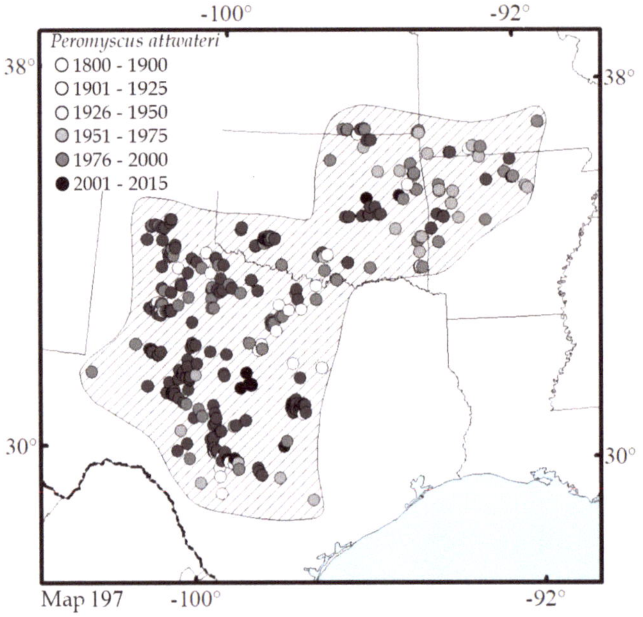


**Map 197** Distribution map of *Peromyscus attwateri*

**Additional Literature:** Osgood (1909), Lee et al. (1972).

## *Peromyscus aztecus* (Saussure, 1860)

Aztec deermouse, ratón azteca

Monotypic.

**Conservation:** *Peromyscus aztecus* is listed as Least Concern by the IUCN (Castro-Arellano and Vázquez 2016b).

**Characteristics:** Aztec deermouse is medium sized; total length 197.0–260.0 mm and skull length 32.7–33.7 mm. Dorsal pelage pale ocher grizzly with pale brown, flanks reddish; fur of underparts light buff; orbital ring blackish; ears dusky, approximately 70% of the hindfoot length; feet white and usually a dusky streak extending from the tarsus to the metatarsus; tail as long as the head-and-body length, bicolored and frequently white at the tip; **supraorbital border angular; angle sharp but not definitely breaded;** interorbital region hourglass-shape; auditive bullae pointed anteriorly, with the braincase anterior half nearly straight. Species group *aztecus*.

**Comments:** *Peromyscus aztecus* was recognized as a distinct species (Álvarez 1961; Carleton 1979, Bradley and Schmidly 1987), with its own species group (Musser and Carleton 2005); it was previously regarded as a subspecies of *P. boylii*. *P. hylocetes* was previously considered a subspecies of *P. aztecus* (Osgood 1909) and is now considered a full species (Sullivan and Kilpatrick 1991; Sullivan et al. 1997). *P. a. cordillerae* (Dickey 1928) and *P. a. hondurensis* (Goodwin 1941) were subspecies of *P. aztecus*, and are now considered subspecies of *P. cordillerae* (Kilpatrick et al. 2021). *P. evides* (Musser 1964) and *P. oaxacensis* (Merriam 1898) were valid species and are now subspecies of *P. aztecus* (Sullivan et al. 1997, 2000; Carleton 1979; Kilpatrick et al. 2021). *P. aztecus* ranges in two disjunctive ranges in Hidalgo, Veracruz, Puebla, and central Oaxaca (Map 198). It can be found in sympatry, or nearly so, with *P. felipensis*, *P. furvus*, *P. gratus*, *P. labecula*, *P. leucopus*, and *P. melanocarpus*. It can be distinguished from *P. felipensis* and *P. gratus* by having the ears smaller than the hind feet; tail not heavily haired and much longer than the head-and-body length; without a pectoral spot; auditory bullae proportionally less inflated and smaller. From *P. furvus*, by its smaller size; tail slightly larger than the head-and-body length; the presence of a lateral line and pectoral spots. From *P. labecula* and *P. leucopus*, by having the tail without a well-defined bicolored pattern and equal to or larger than the head-and-body length. From *P. melanocarpus*, by its smaller size; white carpal and tarsal regions; supraorbital edge present without a well marked and shallow depression, and molars smaller.

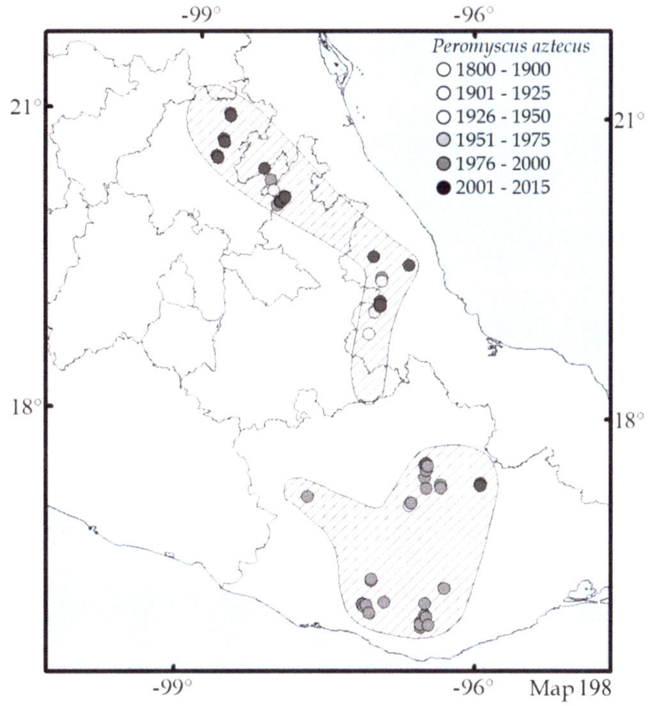


**Map 198** Distribution map of *Peromyscus aztecus*

**Additional Literature:** Hooper (1957), Musser (1964), Vázquez et al. (2001), Musser and Carleton (2005).

## *Peromyscus beatae* Thomas, 1903

Orizaba deermouse, ratón de Orizaba

Monotypic.

**Conservation:** *Peromyscus beatae* is listed as Least Concern by the IUCN (Cassola 2016cm).

**Characteristics:** Orizaba deermouse is medium sized; total length 200.0–234.0 mm and skull length 27.4–29.8 mm. Dorsal pelage brown to dark ocher with a wide darker stripe; occasionally almost black on the back; fur of underparts white to ocher, with a pectoral spot sometimes present; sides with tawny shades; lateral line not sharply marked; orbital ring blackish; ears dusky approximately 85% of the hindfoot length; feet white; tarsal joint sharply marked with dusky shades; tail equal than the head-and-body; skull large for the species group; supraorbital border not angular, almost rounded; auditory bullae large. Species group *boylii*.

**Comments:** *Peromyscus beatae* was reinstated as a species (Bradley and Schmidly 1987; Schmidly et al. 1988); it was previously regarded as a subspecies of *P. boylii levipes* (Álvarez 1961; Hooper 1968) or *P. levipes* (Carleton 1989). *P. beatae* ranges from southeastern Hidalgo, eastern Puebla, central Veracruz, central Guerrero south through Central America (Map 199). *P. beatae* cannot be found in sympatry, or nearly so, with any other species of the *boylii* species group but can be in sympatry, or nearly so, with *P. aztecus*, *P. carolpattonae*, *P. difficilis*, *P. felipensis*, *P. gratus*, *P. labecula*, *P. leucopus*, *P. mekisturus*, *P. megalops*, *P. melanocarpus*, *P. melanophrys*, *P. melanotis*, *P. melanurus*, and *P. mexicanus*. *P. beatae* can be distinguished from *P. carolpattonae*, *P. melanurus*, and *P. mexicanus* by its smaller size; tail slightly larger than the head-and-body; with a tawny lateral line not sharply marked. From *P. mexicanus*, by having the ears with hairs and the tail not coarsely annulated (about 17 annulations/cm). From *P. difficilis*, *P. felipensis*, and *P. gratus*, by having the ears smaller than the hind feet; tail not heavily haired and much longer than the head-and-body length; without a pectoral spot and auditory bullae proportionally greatly inflated. From *P. labecula*, *P. leucopus*, and *P. melanotis*, by having the tail without a well-defined bicolored pattern and equal to or larger than the head-and-body length. From *P. mekisturus*, by having the tail similar than the head-and-body length or slightly longer. From *P. melanophrys*, by its smaller size; without a conspicuous gray patch around the face; tail without annulations and darker at the tip; supraorbital border not sharply angled.

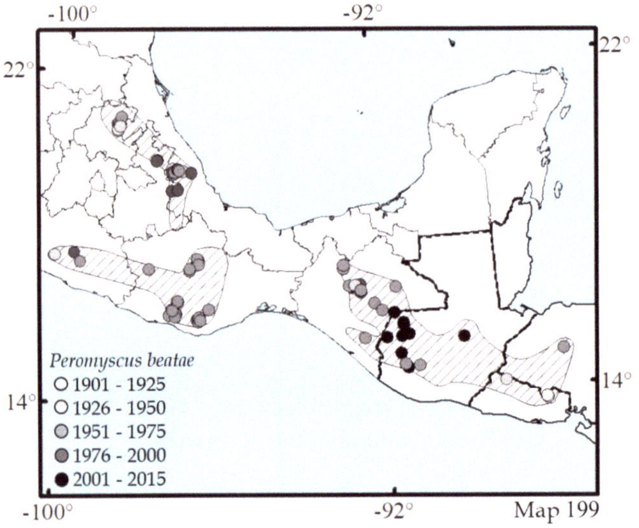


**Map 199** Distribution map of *Peromyscus beatae*

**Additional Literature:** Houseal et al. (1987), Bradley et al. (2000), Musser and Carleton (2005).

## *Peromyscus boylii* (Baird, 1855)

Brush deermouse, ratón de los matorrales

1. *P. b. boylii* (Baird, 1855). Restricted to western Oregon and central-northern California.
2. *P. b. glasselli* Burt, 1932. Restricted to San Pedro Nolasco Island, Sonora.
3. *P. b. rowleyi* (Allen, 1893). From central California and northern Baja California west through southeastern Colorado and western Texas and south through Jalisco, Guanajuato, and Querétaro.
4. *P. b. utahensis* Durrant, 1946. Restricted to Utah and Wyoming.

**Conservation:** *Peromyscus boylii glasselli* is listed as Threatened by the Norma Oficial Mexicana (DOF 2019) and *P. boylii* is listed as Least Concern by the IUCN (Lacher et al. 2016h).

**Characteristics:** Brush deermouse is medium sized; total length 185.0–198.0 mm and skull length 25.8–28.5 mm. Dorsal pelage tawny brown to cinnamon, with the flanks cinnamon orange; fur of underparts whitish with cream, occasionally with an ocher or yellowish pectoral spot; lower face, arms, and lateral line nearly pale ochraceous buff; narrow orbital ring blackish; ears dusky, tuft never containing white hairs but often soft blackish at the base and edged with whitish shades; ears approximately 85% of the hindfoot length with very thin hair; **tail longer than the head-and-body length with hair extending to the vertebral part, with obvious annulations, brown dorsally and whitish ventrally**; **proximal two-fifths of the underside of the hind feet hairy**; limbs white; zygomatic width less anteriorly than posteriory; infraorbital region relatively weak; braincase somewhat-rounded but relatively smaller and less inflated. Species group *boylii*.

**Comments:** The following subspecies of *Peromyscus boylii* were reinstated as distinct species: *P. attwateri* (Schmidly 1973), *P. aztecus* (Álvarez 1961; Carleton 1979; Bradley and Schmidly 1987), *P. beatae* (Schmidly et al. 1988), *P. cordillerae* (Kilpatrick et al. 2021), *P. levipes* (Houseal et al. 1987; Schmidly et al. 1988), *P. madrensis* (Carleton 1977; Carleton et al. 1982), *P. simulus* (Carleton 1977), and *P. spicilegus* (Carleton 1977). The following species were described from populations previously considered *P. boylii: P. carletoni* (Bradley et al. 2014b) was referred to as *P. boylii spicilegus; P. kilpatricki* (Bradley et al. 2016a) to *P. levipes*; and *P. schmidlyi* (Bradley et al. 2004a) to *P. boylii rowleyi*, *P. b. simulus*, and *P. b. spicilegus* (Ordóñez-Garza and Bradley 2011). *P. boylii* ranges from mountainous regions of western Oregon south through northern Baja California, and from Wyoming, Idaho, Colorado, and western Texas south through Zacatecas and Queretaro, including San Pedro Nolasco Island in the Sea of Cortez (Map 200). *P. boylii* cannot be found in sympatry, or nearly so, with any other species of the *boylii* species group, except in northern Querétaro, where it can be found near the range of *P. levipes* and with *P. aztecus*, *P. californicus*, *P. crinitus*, *P. difficilis*, *P. eremicus*, *P. fraterculus*, *P. gambelii*, *P. gratus*, *P. labecula*, *P. laceianus*, *P. leucopus*, *P. levipes*, *P. melanophrys*, *P. melanotis*, *P. nasutus*, *P. polius*, *P. schmidlyi*, *P. sonoriensis*, *P. spicilegus*, and *P. truei*. *P. boylii* can be distinguished from *P. aztecus* and *P. spicilegus* by having a less bright color on the flanks; supraorbital shelf not developed and edges without a ridge and biconcave, both edges appear to have hourglass shape, and auditory bullae less pointed anteriorly and greatly inflated. From *P. californicus*, *P. eremicus*, and *P. fraterculus*, by having two pairs of mammae (nipples); first and second upper molars with accessory cusps. From *P. crinitus*, by its larger size on average, the dorsal coloration paler, and hair without leaden at the base. From *P. difficilis*, *P. gratus*, *P. nasutus*, and *P. truei*, by having the ears smaller than the hind feet; tail less haired and auditory bullae proportionally less inflated. From *P. difficilis*, by its smaller size. From *P. gambelii*, *P. labecula*, *P. leucopus*, *P. melanotis*, and *P. sonoriensis*, by having the tail without a well-defined bicolored pattern and larger than the head-and-body length. From *P. laceianus*, by having the skull slightly larger; rostrum relatively narrow and light; braincase more vaulted; molars slightly larger; ankles dusky; posterior region of the nasals V shaped. From *P. levipes*, by having the dorsal coloration paler tawny brown to cinnamon, usually grizzly gray with black and strongly contrasting with the flanks; upper toothrow length usually equal to or less than 4.2 mm; frontomaxillary suture with teeth. From *P. melanophrys*, by its smaller size not having a conspicuous gray patch around the face; without annulations white at the tail tip and supraorbital border not sharply angled. From *P. polius*, by its smaller size and molars; smaller palatine slits and bullae about the same size but proportionally larger. From *P. schmidlyi*, by being smaller and paler.

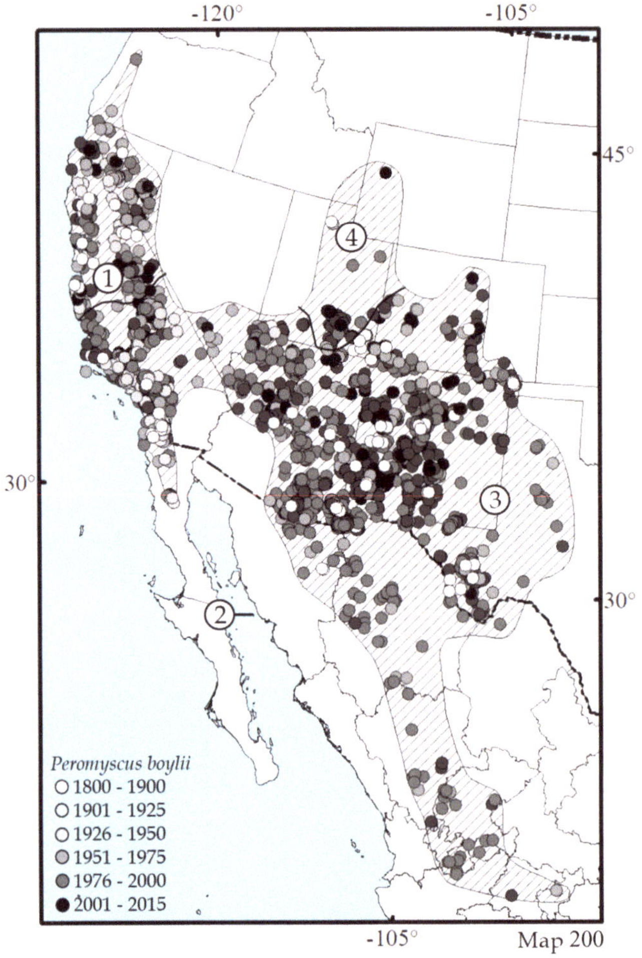


**Map 200** Distribution map of *Peromyscus boylii*

**Additional Literature:** Osgood (1909), Schmidly (1973), Bradley and Schmidly (1987), Álvarez-Castañeda and Cortés-Calva (1999), Bradley et al. (2000), Tiemann-Boege et al. (2000), Musser and Carleton (2005), Kalcounis-Rueppell and Spoon (2009), Bradley et al. (2016a).

## ***Peromyscus bullatus*** Osgood, 1904

Perote deermouse, ratón de Perote

Monotypic.

**Conservation:** *Peromyscus bullatus* is listed as Special Protection by the Norma Oficial Mexicana (DOF 2019) and as **Critically Endangered** (B1ab (iii); extent of occurrence less than 100 $km^2$ at only one location and under continuing decline in area and quality of habitat) by the IUCN (Álvarez-Castañeda 2018d).

**Characteristics:** Perote deermouse is medium sized; total length 178.0–224.0 mm and skull length 27.4–28.6 mm. Dorsal pelage tawny ocher grizzly with dark shades; darker on the back; fur of underparts creamy whitish; **ears approximately more than 115% of the hindfoot length**; **tail shorter than the head-and-body length**; hind and forelimbs whitish; **auditory bullae greatly inflated**. Species group *truei*.

**Comments:** *Peromyscus bullatus* may be a subspecies of *P. truei* (Hooper 1968; Modi and Lee 1984) but is morphologically distinct from *P. difficilis* and *P. gratus* (Carleton 1989; González-Ruiz et al. 2005). *P. bullatus* ranges in the Orizaba Basin, Veracruz-Puebla border (Map 201). It can be found in sympatry, or nearly so, with *P. amplus*, *P. gratus*, *P. labecula*, and *P. melanophrys*. *P. bullatus* can be distinguished from all other species of *Peromyscus* by its larger ears, more than 2.0 mm than the hind feet and auditory bullae proportionally greatly inflated. From *P. amplus*, by its smaller size; occipitonasal length less than 28.6 mm and molars less robust. From *P. gratus*, by having larger ears, greater than 24.0 mm; auditory bullae proportionally more inflated and larger. From *P. labecula*, by having the tail without a well-defined bicolored pattern and equal to or larger than the head-and-body length; ears larger than the hind feet and auditory bullae proportionally greatly inflated. From *P. melanophrys*, by its smaller size and larger ears, more than the hindfoot length.

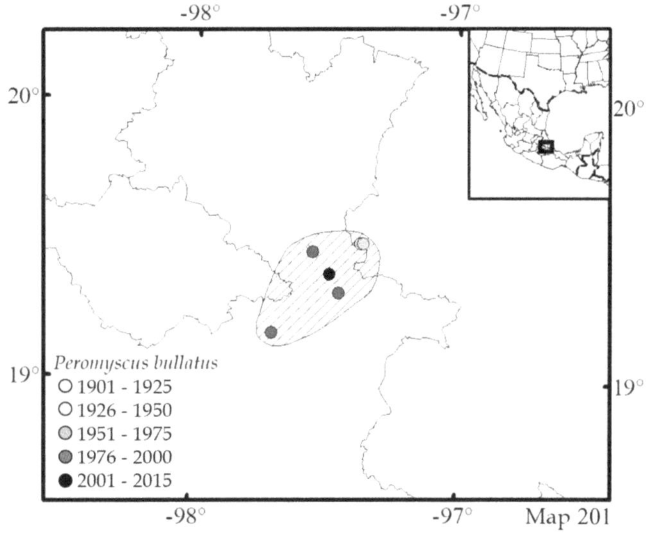


**Map 201** Distribution map of *Peromyscus bullatus*

**Additional Literature:** Osgood (1904, 1909), Hoffmeister (1951), Hooper (1957), González-Ruiz and Álvarez-Castañeda (2005), Musser and Carleton (2005).

## *Peromyscus carletoni* Bradley et al., 2014

Carleton's deermouse, ratón de Nayarit

Monotypic.

**Conservation:** *Peromyscus carletoni* (as *Peromyscus boylii spicilegus*) is listed as Least Concern by the IUCN (Lacher et al. 2016h).

**Characteristics:** Carleton's deermouse is medium sized; total length 171.2–182.0 mm and skull length 24.5–28.7 mm. Dorsal pelage tawny to reddish ocher and hair darker at the tip, concentrating on the back as a line; fur of underparts light buff; orbital ring blackish; lateral line absent; ears dusky, approximately 85% of the hindfoot length; feet white, and usually with a dusky streak extending from the tarsus to the metatarsus; tail bicolored, usually as long as the head-and-body length, and frequently white at the tip; supraorbital border angular; bullae pointing anteriorly, with the braincase anterior half nearly straight. Species group *boylii*.

**Comments:** *Peromyscus carletoni* was recognized as a distinct species (Bradley et al. 2014b); it was previously referred to *P. boylii spicilegus* (Hall 1981). *P. carletoni* is restricted to the highlands of northwestern Nayarit (Map 202) and cannot be found in is sympatry, or nearly so, with any other species of the *boylii* complex but could be found near the range *of P. eremicus*, *P. hylocetes*, *P. labecula*, *P. simulus*, and *P. spicilegus. P. carletoni* can be distinguished from *P. eremicus* by having two pairs of mammas (nipples); first and second upper molars with accessory cusps. From *P. hylocetes* and *P. spicilegus*, by its less bright color on the flanks; supraorbital shelf not developed, edges without a ridge and biconcave; both edges hourglass shaped and auditory bullae less pointed anteriorly and proportionally greatly inflated. From *P. labecula*, by having larger tail in relation to the head-and-body length and without a dark dorsal narrow stripe covering only the vertebral line. From *P. simulus* only with genetic data; however, on average, it has the upper toothrow length greater than 4.1 mm and the tail more bicolor.

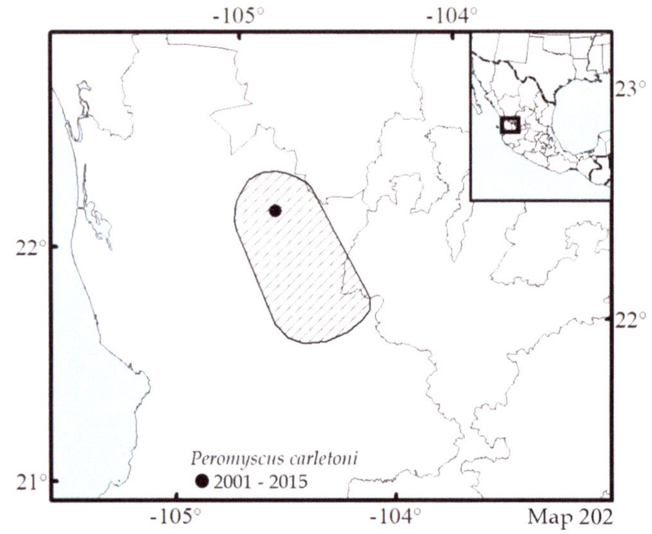


**Map 202** Distribution map of *Peromyscus carletoni*

**Additional Literature:** Osgood (1909).

## *Peromyscus carolpattonae* Álvarez-Castañeda et al., 2019

Carol Patton's deermouse, ratón de El Triunfo

Monotypic.

**Conservation:** *Peromyscus carolpattonae* is not listed in México (DOF 2019) and (as *Peromyscus guatemalensis*) as Least Concern by the IUCN (Vázquez 2016a). The species should be listed because of its restricted distribution.

**Characteristics:** Carol Patton's deermouse is medium sized; total length 257.0–287.0 mm and skull length 33.1–35.8 mm. **Dorsal pelage uniformly grayish brown, basally gray, back central part dark brown or blackish**, brown color paler on the forehead; fur of underparts whitish gray with a pale orange spot on the chest; eye ring black; lateral coat with a strip of orange-brown color that extends to the cheeks; whitish gray outside the fore and hind limbs, darker brown inside the hind limbs, forelegs varying from white to yellowish; toes and fingers yellowish; tail shorter than the head-and-body length, scantily haired and bicolored, brown dark dorsally and yellowish ventrally, tail tip white in some specimens; **hair very long and loose**; **without a definite supraorbital bead and with frontal** bone **constricted**. Species group *mexicanus*.

**Comments:** *Peromyscus carolpattonae* was recognized as a distinct species (Álvarez-Castañeda et al. 2019e), previously referred as *P. guatemalensis* from Chiapas (Huckaby 1980). *P. carolpattonae* is restricted to the southern Chiapas highlands (Map 203) and is sympatric, or nearly so, with *P. beatae*, *P. grandis*, *P. gymnotis*, and *P. mexicanus*. *P. carolpattonae* is very similar to *P. zarhynchus* in its overall smaller external and all craniodental measurements, except for the skull height and basioccipital length, but it is distributed in the southeast (Sierra Madre de Chiapas) and does not co-occur with *P. zarhynchus* (Álvarez-Castañeda et al. 2017a). From *P. beatae*, by its larger size; tail proportionally longer in relation to the head-and-body length and without a lateral line. From *P. grandis*, by having a shorter and more slender skull; with more prominent supraorbital crests; the nasals do not extend beyond the premaxilla, and proportionally larger auditory bullae. From *P. gymnotis*, by its larger external measurements; hindfoot length generally greater than 26.0 mm; upper toothrow length greater than 4.6 mm; supraorbital region developed and smooth, lacking both ledges and beading. From *P. mexicanus*, by lacking well-developed supraorbital crests (Álvarez-Castañeda et al. 2017a).

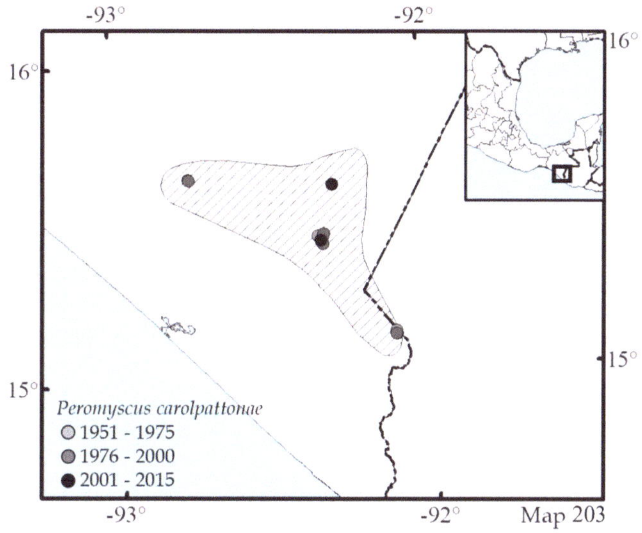


**Map 203** Distribution map of *Peromyscus carolpattonae*

**Additional Literature:** Merriam (1898a), Carleton (1989), Vázquez (2016a, b).

## ***Peromyscus chrysopus*** Hooper, 1955

Coastal tawny deermouse, ratón leonado de la costa

Monotypic.

**Conservation:** *Peromyscus chrysopus* (as *Peromyscus perfulvus*) is listed as Least Concern by the IUCN (Álvarez-Castañeda et al. 2016p).

**Characteristics:** Coastal tawny deermouse is medium sized; total length 288.0–246.0 mm and skull length 27.3–29.0 mm. Dorsal pelage brown, flanks not contrasting with the back, which has long and silky hair; fur of underparts grayish white; ears approximately 80% of the hindfoot length and brown; hind feet white but with a dark coloration on the tarsus, whereas in other species of the group feet are uniformly white; **tail hairy, longer than the head-and-body length, uniformly sepia, slightly paler ventrally than dorsally, and well penciled at the tip;** hind feet white but with coloration on the tarsus; toes buffy; fifth toe relatively long; **first and second upper molars usually with ectostylicls and complete mesolophs; nasals acute posteriorly, their limits short or posterior to the limits of the ascending branches of the premaxilla;** rostrum thin and long; supraorbital area without edges. Species group *melanophrys*.

**Comments:** *Peromyscus chrysopus* was recognized as a distinct species (Castañeda-Rico et al. 2014; López-González et al. 2019); it was previously regarded as a subspecies of *P. perfulvus* (Hooper 1955). It ranges in the coastal lowlands of Jalisco and Colima (Map 204), and can be found in sympatry, or nearly so, with *P. hylocetes*, *P. labecula*, and *P. perfulvus*. *P. chrysopus* can be distinguished from other members of *Peromyscus* by having the tail long, hairy, and uniformly brownish; ears brown instead of blackish; hind feet white but with coloration on the tarsus. From *P. hylocetes*, by its larger size; ears longer than the hind feet; tail longer than the head-and-body length; auditory bullae proportionally greatly inflated. From *P. labecula*, by having the tail without a well-defined dark dorsal narrow stripe covering only the vertebral line and equal to or larger than the head-and-body length. From *P. perfulvus*, by having the first and second upper molars usually with ectostylids and complete mesolophs; nasals acute posteriorly, with their limits short on average, or posterior to the limits of ascending branches of the premaxilla. From *Osgoodomys banderanus*, by having the tail with hair and not scaled or penciled at the tip; the ears are brownish; with a facial mask gray and a sharp supraorbital border.

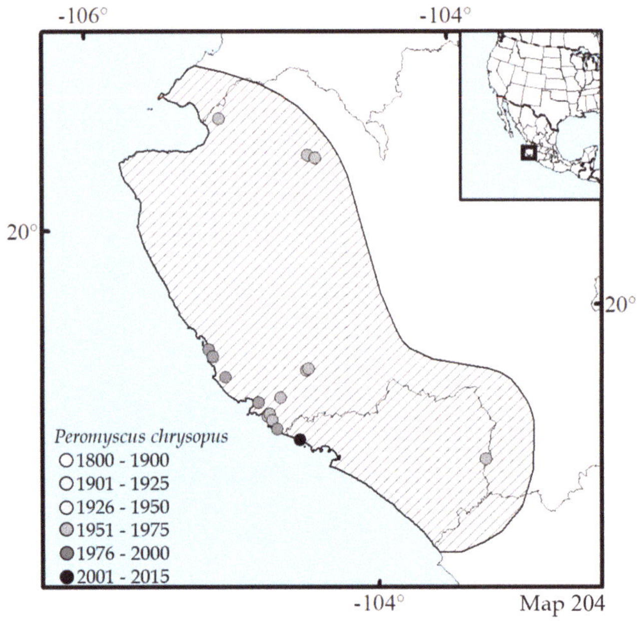


**Map 204** Distribution map of *Peromyscus chrysopus*

**Additional Literature:** Helm et al. (1974), Musser and Carleton (2005), Sánchez-Hernández et al. (2009).

## *Peromyscus collinus* Hooper, 1952

Tamaulipas white-ankled deermouse, ratón de ancas blancas de Tamaulipas

Monotypic.

**Conservation:** *Peromyscus collinus* (as *Peromyscus pectoralis*) is listed as Least Concern by the IUCN (Lacher et al. 2016k).

**Characteristics:** Tamaulipas white ankle deermouse is medium sized; total length 205.0–219.0 mm and skull length 27.0–29.0 mm. Dorsal pelage brown dark with reddish hues; ankles dark; fur of underparts whitish cream; ears approximately 75% of the hindfoot length; tail longer than the head-and-body length, slightly bicolored; **molars clearly smaller**; auditory bullae larger; interparietal bone relatively large. Species group *truei*.

**Comments:** *Peromyscus collinus* was recognized as a distinct species (Hernández-Canchola et al. 2022); it was previously regarded as a subspecies of *P. pectoralis* (Hooper 1952). It ranges from western Tamaulipas south through eastern Hidalgo (Map 205). *P. collinus* and can be found in sympatry, or nearly so, with *P. eremicus*, *P. hooperi*, *P. labecula*, *P. laceianus*, *P. leucopus*, *P. levipes*, *P. mexicanus*, *P. ochraventer*, *P. pectoralis*, and *P. zamorae*. *P. collinus* can be distinguished from *P. hooperi* by its smaller size and the presence of mesolophs and anteroloph. From *P. eremicus*, by having two pairs of mammae (nipples); ear length smaller than the hindfoot length; a lateral line in the flanks; a dark spot around the eyes; a pectoral spot; first and second upper molars with accessory cusps, auditory bullae larger and more orbicular. From *P. labecula* and *P. leucopus*, by having the tail without a well-defined bicolored pattern and equal to or larger than the head-and-body length. From *P. laceianus*, by having darker and reddish upper parts; darker tarsi; larger size with greater total length, tail length, hind foot length, and ear length; braincase narrower and longer; auditory bullae larger. From *P. levipes*, by having a larger tail, ear length equal to or slightly smaller than the hindfoot length, a darker spot around the eyes and a pectoral spot, auditory bullae larger. From *P. mexicanus* and *P. ochraventer*, by having larger ears, equal to or larger than the hindfoot length and auditory bullae of greater size. From *P. pectoralis*, by having the upperparts darker and reddish; darker tarsi; larger external measurements and skull; broader rostrum, maxillary bone, and interorbital region; larger palatine foramen, molar toothrow, and auditory bullae. From *P. zamorae*, by having the ear length equal to or smaller than the hindfoot length; auditory bullae relatively smaller and not globe shaped, its length less than 18% of the skull length.

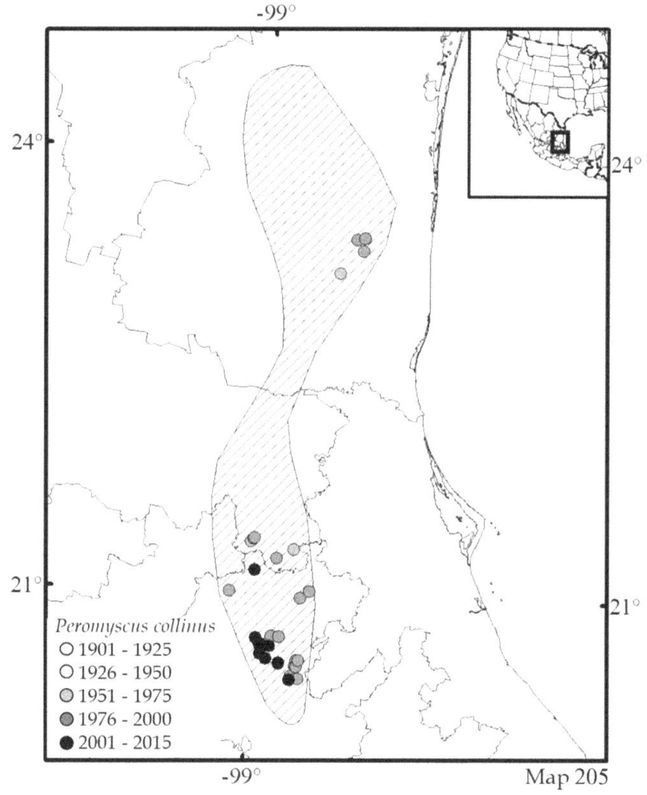


**Map 205** Distribution map of *Peromyscus collinus*

**Additional Literature:** Bailey (1906), Osgood (1909), Schmidly (1972, 1974), Kilpatrick and Zimmerman (1976), Musser and Carleton (2005), Bradley et al. (2015).

## ***Peromyscus cordillerae*** Dickey, 1928

Monte Cacaguatique white-footed Mouse, ratón del Monte Cacaguatique

Monotypic.

**Conservation:** *Peromyscus cordillerae* (as *Peromyscus aztecus*) is listed as Least Concern by the IUCN (Castro-Arellano and Vázquez 2016b).

**Characteristics:** Monte Cacaguatique white-footed mouse is medium sized; total length 200.0–255.0 mm and skull length 30.8–32.3 mm. Dorsal pelage trifle less deep and rich tawny, with a considerable admixture of dusky hairs concentrated chiefly in a broad dorsal band; fur of underparts with a pectoral spot present or absent; orbital ring and spot at the base of the whiskers black and strongly marked; ears proportionally smaller, approximately 60% of the hindfoot length; hind feet wholly white from just below the tarsal joint to the claws; tail as long as the head-and-body length, bicolored but not conspicuously contrasted, and thinly haired; **skull large, long, and rather narrow, with the rostrum and nasals particularly produced;** interorbital region hourglass shaped; auditory bullae small and pointed anteriorly; **molars heavy**. Species group *aztecus*.

**Comments:** *Peromyscus cordillerae* was recognized as a distinct species (Kilpatrick et al. 2021); it was previously regarded as a subspecies of *P. boylii* (Dickey 1928) and later of *P. aztecus* (Carleton 1979). This species has been considered different from *P. aztecus* from central México and was considered *P. oaxacensis* (Hall 1981; Sullivan et al. 1997; Duplechin and Bradley 2014; Bradley et al. 2017). *P. cordillerae* does not include the specimens from Oaxaca (*P. aztecus oaxacensis*) but only those ranging in the highlands of Chiapas, Guatemala, Honduras, and El Salvador (Map 206). It can be found in sympatry, or nearly so, with *P. mexicanus* and *P. zarhynchus*. It can be distinguished from both species by its smaller size. From *P. aztecus*, by having the dorsal coloration rich tawny; hind feet wholly white from just below the tarsal joint to the claws; skull long and rather narrow; molars heavy; auditory bullae small.

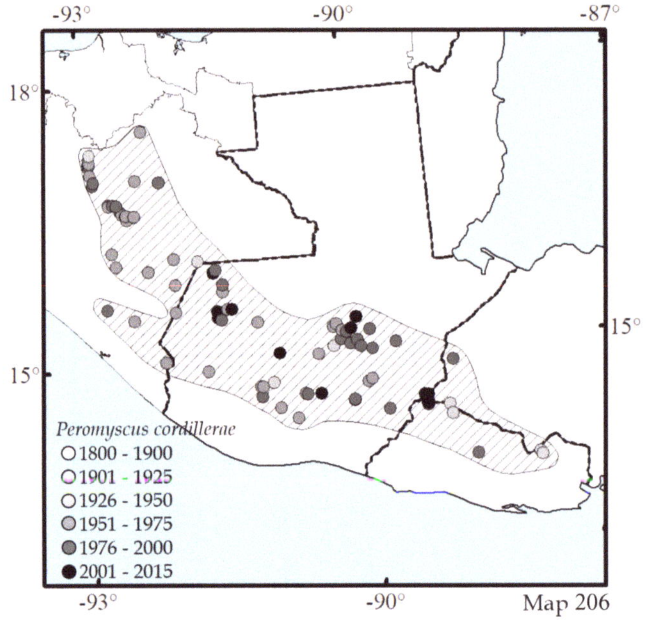


**Map 206** Distribution map of *Peromyscus cordillerae*

**Additional Literature:** Goodwin (1941), Hooper (1957).

## ***Peromyscus crinitus*** (Merriam, 1891)

Canyon deermouse, ratón de las cañadas

1. *P. c. auripectus* (Allen, 1893). Eastern Colorado, northwestern New Mexico, northeastern Arizona, and western Utah.
2. *P. c. crinitus* (Merriam, 1891). Oregon and southwestern Idaho, central and northwestern Nevada, and western California.
3. *P. c. delgadilli* Benson, 1940. Restricted to the Pinacate Desert, northwestern Sonora.
4. *P. c. disparilis* Goldman, 1932. Restricted to northwestern Sonora and southwestern Arizona.
5. *P. c. doutii* Goin, 1944. From southwestern Wyoming south through west Arizona.
6. *P. c. pallidissimus* Huey, 1931. Restricted to San Luis Gonzaga Island, western-central Baja California.
7. *P. c. pergracilis* Goldman, 1939. Eastern Nevada and western Utah.
8. *P. c. stephensi* Mearns, 1897. From southwestern Utah, southern Nevada, and Western California south through western-central Baja California.

**Conservation:** *Peromyscus crinitus pallidissimus* is listed as Threatened by the Norma Oficial Mexicana (DOF 2019) and *P. crinitus* as Least Concern by the IUCN (Lacher and Álvarez-Castañeda 2016).

**Characteristics:** Canyon deermouse is medium sized; total length 161.0–192.0 mm and skull length 23.3–24.5 mm. Dorsal pelage grizzly ocher, brown, and blackish with leaden shades at the base; fur long, loose and silky; fur of underparts paler than in the dorsum and in some cases whitish; **ears large, approximately 90% of the hindfoot length,** with thin hair; **hind feet hairy on the proximal fourth or naked medially to the calcaneus**; **tail longer than the head-and-body length, thickly covered with long and soft hairs, terminating in a distinct pencil; pelage usually long and loose**; rostrum thin and long; coronoid process small, only slightly elevated; anterior breadth of the zygomatic arches less than the braincase breadth; **braincase shallow**; nasal bones not extending beyond the premaxilla; accessory cusps between three principal labial cusps in the first upper molars, rudimentary or lacking in the first lower molars; palate nearly as long as the incisive foramina; **nasals long and slender**. Species group *crinitus*.

**Comments:** *Peromyscus crinitus* was originally considered in the subgenus *Haplomylomys* (Osgood 1909); it is now in the subgenus *Peromyscus* (Hooper and Musser 1964). It ranges from Oregon, Idaho, and Wyoming south through California, Baja California, Arizona, Sonora, and New Mexico (Map 207). It can be found in sympatry, or nearly so, with *P. boylii*, *P. eremicus*, *P. fraterculus*, *P. gambelii*, *P. merriami*, *P. nasutus*, *P. sonoriensis*, and *P. truei*. *P. crinitus* can be distinguished from all the other species by having the hind feet hairy on the proximal fourth or naked medially to the calcaneus, tail longer than the head-and-body length, thickly covered with long and soft hairs, ending in a distinct pencil; shallow braincase and slender nasals. From *P. boylii*, by having on average a smaller size and dorsal pelage more grizzly with leaden shades at the base. From *P. eremicus*, *P. fraterculus*, and *P. merriami*, by having three pairs of mammae; ears more than 90% of the hind-foot length and an accessory cusp or enamel loop on the second upper molars, accessory cusp absent in some specimens. From *P. gambelii* and *P. sonoriensis*, by its larger average size and the tail longer than the head-and-body length, without paravertebral coloration. From *P. nasutus* and *P. truei*, by having the ears smaller than the hind feet, tail not heavily haired, and auditory bullae not proportionally greatly inflated.

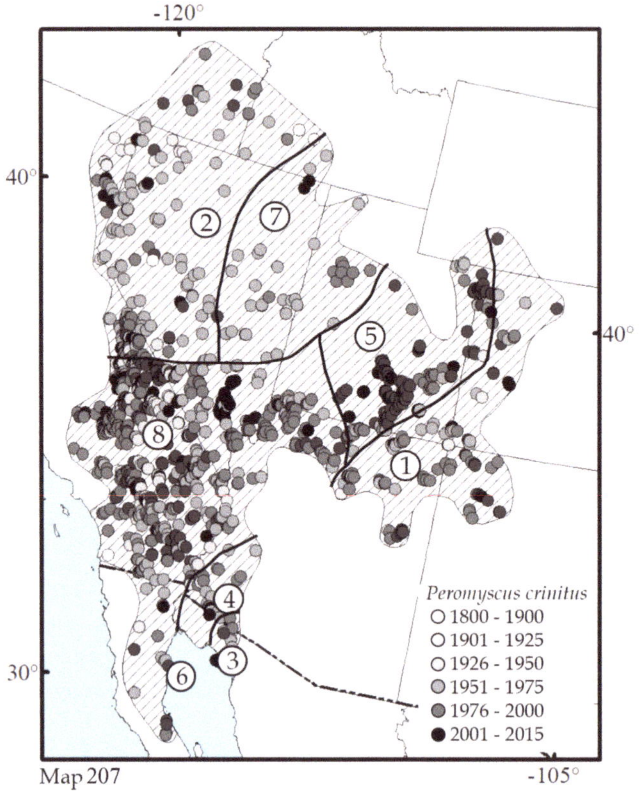


**Map 207** Distribution map of *Peromyscus crinitus*

**Additional Literature:** Davis (1939), Benson (1940), Hall and Hoffmeister (1942), Hall (1946), Hooper (1957), Johnson and Armstrong (1987), Álvarez-Castañeda and Cortés-Calva (1999), Musser and Carleton (2005).

## *Peromyscus difficilis* (Allen, 1891)

Southern rock deermouse, ratón de las rocas del sur

1. *P. d. difficilis* (Allen, 1891). From San Luis Potosí, Querétaro, and Guanajuato, north through southwestern Chihuahua.
2. *P. d. petricola* Hoffmeister and de la Torre, 1959. Southwestern Tamaulipas, northern San Luis Potosí, northwestern Zacatecas, southern Coahuila, and Nuevo León.

**Conservation:** *Peromyscus difficilis* is listed as Least Concern by the (Castro-Arellano and Vázquez 2016c).

**Characteristics:** Southern rock deermouse is medium sized; total length 212.0–255.0 mm and skull length 27.2–30.8 mm. Dorsal pelage ochraceous buff mixed with dusky shades, mainly disposed as fine lines and rather dominating the general effect; flanks of the same coloration as the back; lateral line clear ochraceous buff and usually fairly well defined; **nose, postorbital region, and general facial region distinctly grayish**; a narrow blackish orbital ring; fur of underparts creamy white, usually without a pectoral spot; **ears approximately or equal to the hindfoot length,** thinly haired, bordered with whitish shades, hair tuft at the base about the same color as the surrounding parts; tail longer than the head-and-body length, sharply bicolored, blackish brown dorsally and white ventrally; forefeet and hind feet whitish grizzly, tarsal joint with a small dusky mark; interorbital space narrow; supraorbital border not beaded and seldom markedly sharp-angled, never forming a distinct shelf; nasals quite elongated; teeth moderate; **auditory bullae large and well developed**. Species group *truei*.

**Comments:** *Peromyscus amplus*, *P. felipensis*, and *P. nasutus* were considered subspecies of *P. difficilis* (Osgood 1909; Hoffmeister and De la Torre 1961) which was considered a valid species (Zimmerman et al. 1975; Avise et al. 1979; Hernández-Canchola et al. 2022). Its ranges from Zacatecas, southeast Coahuila, southwest Durango, and southwestern Nuevo León south through Jalisco, Guanajuato, and San Luis Potosí (Map 208). It can be found in sympatry, or nearly so, with *P. beatae*, *P. boylii*, *P. gratus*, *P. labecula*, *P. leucopus*, *P. levipes*, and *P. melanotis*. *P. difficilis* can be distinguished from *P. amplus* by its smaller size; bright and less reddish coloration; smaller braincase; and by using genetic data. From *P. beatae*, *P. boylii*, *P. labecula*, and *P. leucopus*, by its larger size; ears longer than the hind feet; tail longer than the head-and-body length; auditory bullae proportionally greatly inflated. From *P. gratus*, by having the back often grayish-black; lateral line with ochraceous shades; top of the head often grayish; ears larger, greater than 24.0 mm; shorter toothrow; auditory bullae proportionally greatly inflated and larger.

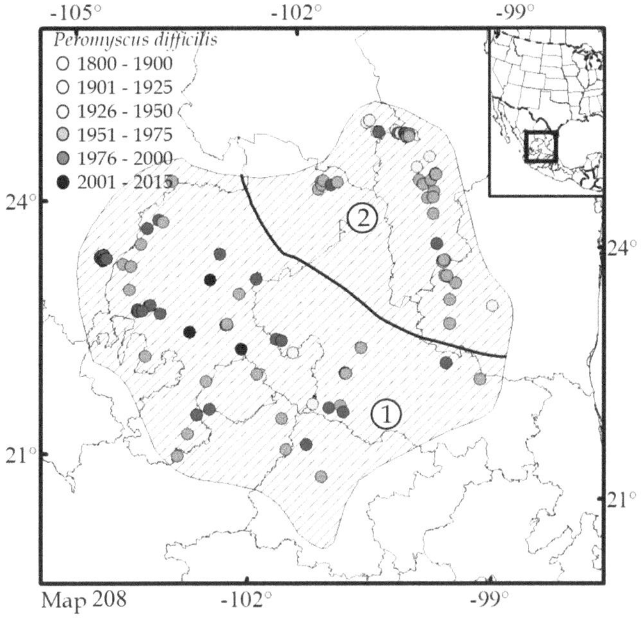


**Map 208** Distribution map of *Peromyscus difficilis*

**Additional Literature:** Hooper (1957), Hoffmeister and De la Torre (1961), Janecek (1990), DeWalt et al. (1993a), Bradley et al. (1999), Arellano-Meneses et al. (2000), González-Ruiz and Álvarez-Castañeda (2005), Musser and Carleton (2005), Fernández et al. (2010).

## ***Peromyscus ensinki*** Bradley et al., 1922

Ensink deermouse, ratón de Zinapécuaro

Monotypic.

**Conservation:** *Peromyscus ensinki* is not listed by the IUCN.

**Characteristics:** Ensink deermouse is medium sized; total length 213.0 mm and skull length 28.2–29.7 mm. Dorsal pelage sepia at the tip and blackish or mouse gray at the base; flanks cinnamon-rufous; fur of underparts dark neutral gray at the base and white at the tip; feet with a dusky grab stripe extending from the ankles to the mid-medapodials; toes white; tail equal than the head-and-body length and slightly bicolored, fuscous dorsally and pale smoke gray ventrally, scantily haired at the base and slightly tufted at the tip; ears approximately 92% of the hindfoot length and deep mouse-gray; vibrissae black; skull elongated, twice longer than wide; rostrum slightly less than one-third of the skull length; nasal bones always larger than the rostral length; molar toothrow length about 15% of the skull length; auditory bullae very long; braincase large and rounded; zygomatic arches nearly parallel. Species group *boylii*.

**Comments:** *Peromyscus ensinki* is only known from 4.8 km S, 3.6 km E Zinapécuaro (the type locality (Map 209)), and cannot be found in sympatry, or nearly so, with any other species of the *boylii* species group but could be found near the range of *P. kilpatricki*, from which it can it be distinguished by its larger size and longer auditory bullae, longer incisive foramen, and wider mesopterygoid fossa, which provide unique morphological features that distinguish it from the other species in the *boylii* group ranging in Michoacán. Only genetic data can be used to separate *P. ensinki* from the other species.

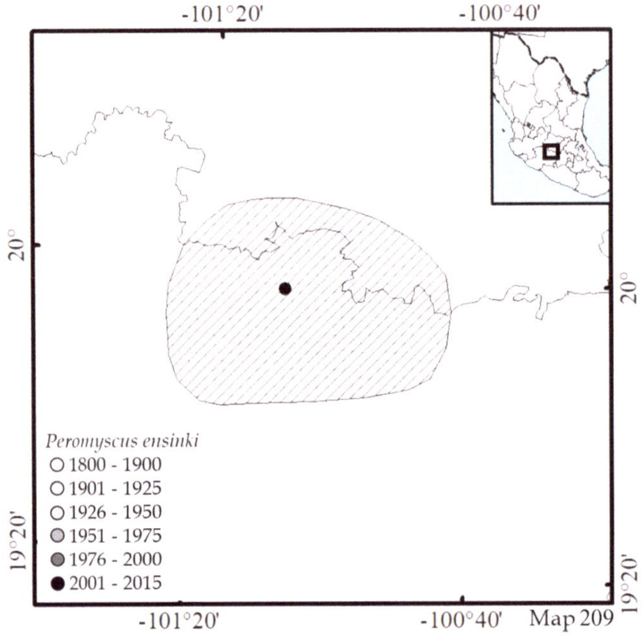


**Map 209** Distribution map of *Peromyscus ensinki*

**Additional Literature:** Bradley et al. (2022a).

## *Peromyscus felipensis* Merriam, 1898

Oaxaca rock deermouse, ratón de las rocas de Oaxaca

Monotypic.

**Conservation:** *Peromyscus felipensis* (as *Peromyscus difficilis*) is listed as Least Concern by the IUCN (Castro-Arellano and Vázquez 2016c).

**Characteristics:** Oaxaca rock deermouse is medium sized; total length 225.0–248.0 mm and skull length 30.4–33.7 mm. Dorsal pelage mixed grayish ochraceous buff and black; general effect on the dorsum nearly black sprinkled with buffy gray; lateral line rather broad, ochraceous buff mixed with dusky shades and seldom very sharply contrasting with the rest of the sides; fur of underparts creamy white with blackish shades at the base, which is never entirely concealed, with a broadly ochraceous buff pectoral spot; **ears approximately or equal to the hindfoot length**; tail longer than the head-and-body length, sharply bicolored, blackish brown dorsally and white ventrally, usually with some mixture of dusky shades on the underside near the base; forefeet and hind feet whitish grizzly, tarsal joint with a small dusky mark; interorbital space narrow; supraorbital border not beaded, never forming a distinct shelf; rostrum and nasals broader and heavier; teeth moderate; **auditory bullae large and well developed but smaller relative to other species of the *truei* group**. Species group *truei*.

**Comments:** *Peromyscus felipensis* was recognized as a distinct species (Hernández-Canchola et al. 2022); it was previously regarded as a subspecies of *Peromyscus difficilis* (Osgood 1909). It ranges from southern Estado de México, Ciudad de México, Morelos, southeast Puebla, and central Oaxaca (Map 210). It can be found in sympatry, or nearly so, with *P. amplus*, *P. aztecus*, *P. beatae*, *P. gratus*, *P. hylocetes*, *P. kilpatricki*, *P. labecula*, *P. leucopus*, and *P. melanophrys*. *P. felipensis* can be distinguished from *P. amplus* by having a larger skull; coloration very dark, rich blackish brown and black; and by using genetic data. From *P. aztecus*, *P. beatae*, *P. hylocetes*, *P. kilpatricki*, *P. labecula*, and *P. leucopus*, its larger size; ears longer than the hind feet; tail longer than the head-and-body length; auditory bullae proportionally greatly inflated. From *P. gratus*, by having the back often grayish-black; lateral line with ochraceous shades; top of the head often grayish; ears larger, greater than 24.0 mm; shorter toothrow; auditory bullae proportionally greatly inflated and larger. From *P. melanophrys*, by having a supraorbital shelf slightly developed or absent and, edges without a ridge and almost or completely biconcave.

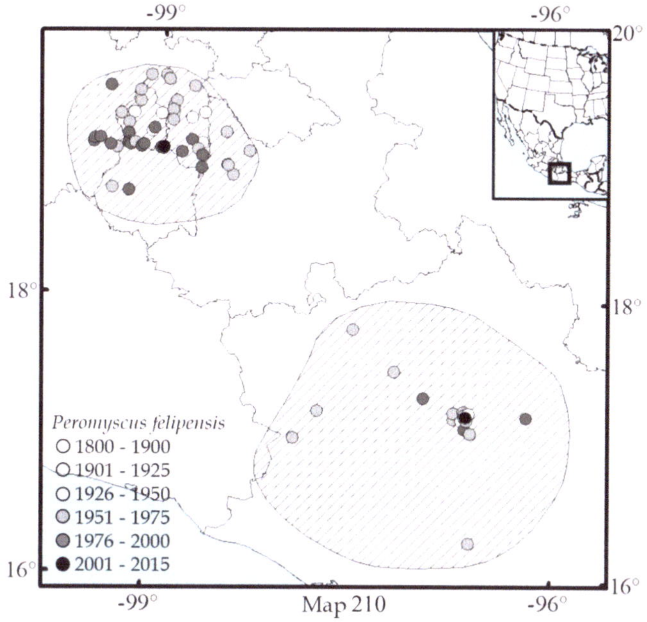


**Map 210** Distribution map of *Peromyscus felipensis*

**Additional Literature:** Hooper (1957), Hoffmeister and De la Torre (1961), Zimmerman et al. (1975), Avise et al. (1979); Janecek (1990), DeWalt et al. (1993a), Bradley et al. (1999), González-Ruiz and Álvarez-Castañeda (2005), Musser and Carleton (2005), Fernández et al. (2010).

## *Peromyscus furvus* Allen and Chapman, 1897

Blackish deermouse, ratón negruzco

Monotypic.

**Conservation:** *Peromyscus furvus* is listed as Data Deficient by the IUCN (Castro-Arellano and Vázquez 2019).

**Characteristics:** Blackish deermouse is large sized; **total length 248.0–282.0 mm and skull length 31.5–36.0 mm**. Dorsal pelage blackish; flanks slightly paler; fur of underparts grayish, darker at the base; eye ring and surrounding area black; flanks snuff brown; feet white; **ears relatively small, nearly naked** and dusky, approximately 75% the hindfoot length; **tail slightly longer than the head-and-body length, very thinly haired, blackish all around or with slight irregular pale markings on the scaly underside**; limbs white, tarsus brown; forefeet and hind feet white; **hindfoot soles naked medially to the calcaneus**; **very broad, inflated anteriorly, and distinctly bell shaped, breadth between the anterior nasals**; palate thick; interorbital region narrow; nasals extending about 2.0 mm beyond the intermaxillar bones; palatine foramina relatively broad. Species group *furvus*.

**Comments:** *Peromyscus latirostris* was recognized as a distinct species (Dalquest 1950, Ávila-Valle et al. 2012). *P. furvus* ranges in Hidalgo, Puebla, and Veracruz (Map 211), and lives in sympatry, or nearly so, with *P. aztecus*, *P. labecula*, *P. latirostris*, *P. leucopus*, *P. levipes*, *P. mexicanus*, and *P. pectoralis*. *P. furvus* can be distinguished from *P. aztecus*, *P. labecula*, *P. leucopus*, *P. levipes*, and *P. pectoralis* by its larger size; tail darker dorsally, naked, with scales, and very broad; breadth across the anterior nasals inflated and distinctly bell shaped. From *P. latirostris*, by its smaller average size and based on genetic differences. From *P. mexicanus*, by having an hourglass-shaped interorbital region; expanded nasals; pectoral mammae (nipples), and tail length slightly longer than the head-and-body length. From *P. ochraventer*, by having larger auditory bullae; a well-developed style in the first molar minor fold; the anterior cingulum of the first upper molars has a smaller lingual cusp and a larger labial cusp, and frequently has a small style anterior to the cleft between the two cusps.

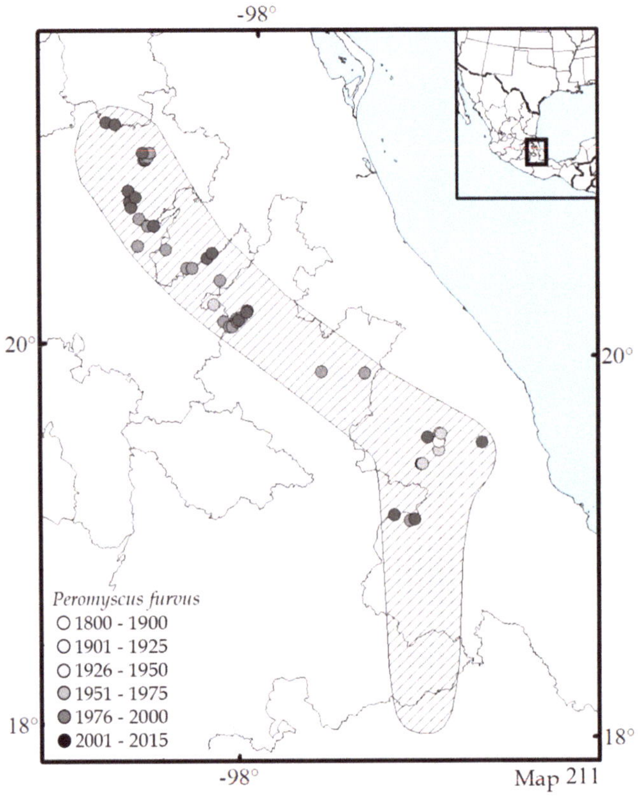


**Map 211** Distribution map of *Peromyscus furvus*

**Additional Literature:** Osgood (1909), Hooper and Musser (1964), Huckaby (1973, 1980), Musser and Carleton (2005), Rogers and Skoy (2011).

## *Peromyscus gambelii* (Baird, 1858)

Pacific common deermouse, ratón común del Pacífico

1. *P. g. anacapae* von Bloeker, 1942. Restricted to Anacapa Island and surrounding islands, California.
2. *P. g. assimilis* Nelson and Goldman, 1931. Restricted to Coronados Island, Baja California.
3. *P. g. catalinae* Elliot, 1903. Restricted to Santa Catalina Island, California.
4. *P. g. cineritius* Allen, 1898. Restricted to San Roque Island, Baja California.
5. *P. g. clementis* Mearns, 1896. Restricted to San Clemente Island, California.

6. *P. g. coolidgei* Thomas, 1898. From central-northern Baja California south through southern Baja California Sur.
7. *P. g. dorsalis* Nelson and Goldman, 1931. Restricted to Natividad Island, Baja California.
8. *P. g. dubius* Allen, 1898. Restricted to Todos Santos Island, Baja California.
9. *P. g. elusus* Nelson and Goldman, 1931. Restricted to the Santa Barbara and Sutil Islands, California.
10. *P. g. exiguus* Allen, 1898. Restricted to San Martín Island, Baja California.
11. *P. g. exterus* Nelson and Goldman, 1931. Restricted to San Nicolas Island, California.
12. *P. g. gambelli* (Baird, 1858). From northern Baja California north through central California.
13. *P. g. geronimensis* Allen, 1898. Restricted to San Gerónimo Island, Baja California.
14. *P. g. hueyi* Nelson and Goldman, 1932. Restricted to Smith Island, Baja California.
15. *P. g. magdalenae* Osgood, 1909. Restricted to Magdalena Island and surrounding mainland areas, Baja California Sur.
16. *P. g. margaritae* Osgood, 1909. Restricted to Margarita Island, Baja California Sur.
17. *P. g. sanctaerosae* von Bloeker, 1940. Restricted to Santa Rosa Island, California.
18. *P. g. santacruzae* Nelson and Goldman, 1931. Restricted to Santa Cruz Island, California.
19. *P. g. streatori* Nelson and Goldman, 1931. Restricted to San Miguel, California.

**Conservation:** *Peromyscus gambelii cineritius* is listed as Endemic (considered extinct by Álvarez-Castañeda and Cortés Calva 1996), *P. g. dorsalis*, *P. g. dubius*, *P. g. exiguus*, *P. g. geronimensis*, *P. g. magdalenae*, and *P. g. margaritae* are listed as Threatened as subspecies of *P. maniculatus* by the Norma Oficial Mexicana (DOF 2019), and *P. maniculatus* is listed as Least Concern by the IUCN (Cassola 2016cq).

**Characteristics:** Pacific common deermouse is medium sized; total length 148.0–195.0 mm and skull length 25.2–28.0 mm. Coloration varying greatly between subspecies, dorsally ochraceous or ochraceous buff mixed with a dusky coloration; fur of underparts creamy white, clearly contrasting with the back; ears dusky, narrowly edged with whitish without a subauricular hair tuft different from the back coloration, approximately 70% of the hind feet; tail shorter than the head-and-body length, bicolored, with a brown to dusky dorsal stripe one-third of the tail diameter; feet whitish; ankles slightly dusky or nearly white; **slightly smaller than in nearby species of the *maniculatus* group**. Species group *maniculatus*.

**Comments:** *Peromyscus gambelii* was reinstated to the species level (Bradley et al. 2019; Greenbaum et al. (2019), including as subspecies all listed above, previously were subspecies of *P. maniculatus* (Osgood 1909) following Bradley et al. (2019) and Greenbaum et al. (2019); populations from central Oregon and eastern-central Washington, considered *P. m. gambelii* (Hall 1981) should be assigned to *P. sonoriensis*. *P. gambelii* ranges from the San Francisco Bay Area and San Joaquin Valley, extreme western-central Nevada south through the Baja California peninsula (Map 212), including southwestern Arizona and northwestern Sonora. It can be found in sympatry, or nearly so, with *P. boylii*, *P. californicus*, *P. crinitus*, *P. eva*, *P. fraterculus*, *P. gambelii*, *P. polius*, *P. sonoriensis*, and *P. truei*. *P. gambelii* can be distinguished from *P. boylii* and *P. polius* by having the tail hairy, shorter than or equal to the head-and-body length, clearly bicolored, with a narrow dark dorsal stripe covering only the vertebral line. From *P. californicus*, *P. eva*, and *P. fraterculus*, by having two pairs of mammae (nipples); first and second upper molars with accessory cusps; tail shorter than the head-and-body length with a paravertebral coloration. From *P. crinitus*, by its smaller average size and the tail shorter than or equal than the head-and-body length and with paravertebral coloration. From *P. melanotis* (extremely similar morphologically), by not having a black preauricular hair tuft; braincase narrower and less rounded; wider interorbital space; prezygomatic notch more prominent; slightly larger auditory bullae, and rostrum shorter and less slender. From *P. sonoriensis* only by using genetic data; however, a slightly larger skull on average. From *P. truei*, by the ears smaller than the hind feet; tail less haired, and auditory bullae proportionally less inflated.

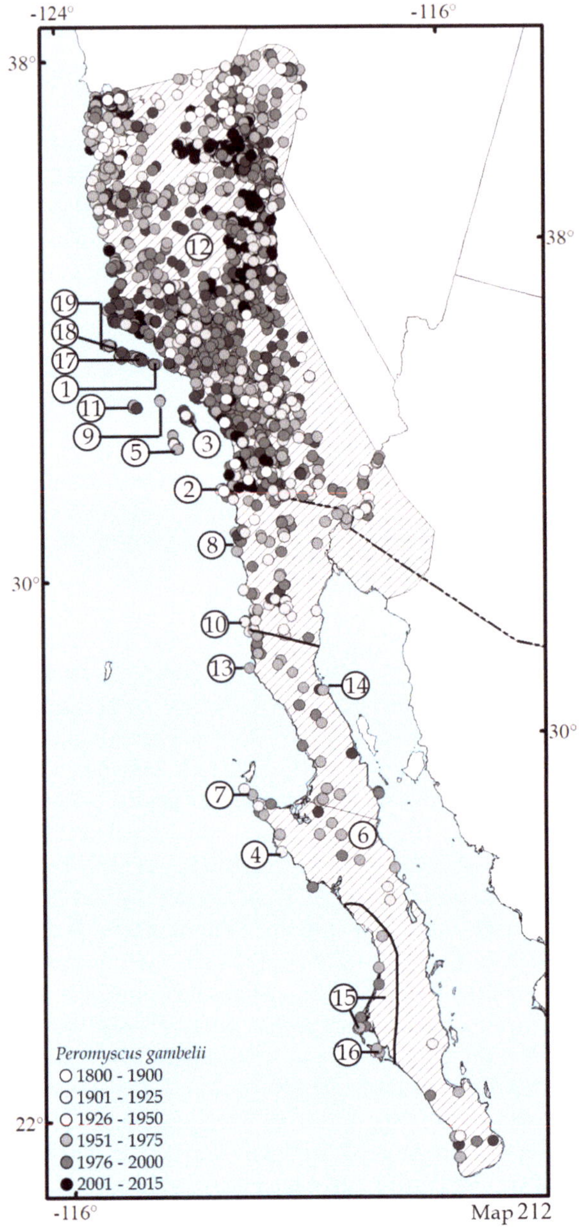


**Map 212** Distribution map of *Peromyscus gambelii*

**Additional Literature:** Lawlor (1983), Álvarez-Castañeda and Cortés-Calva (1999), Álvarez-Castañeda et al. (2008c).

## *Peromyscus gossypinus* (Le Conte, 1853)

Cotton mouse, ratón del algodón

1. *P. g. allapaticola* Schwartz, 1952. Restricted to Key Largo, southern Florida.
2. *P. g. anastasae* Bangs, 1898. Restricted to Cumberland Island in Georgia and Anastasia Island in Florida.
3. *P. g. gossypinus* (Le Conte, 1853). From southeast of Virginia and northeastern Carolina southward, and west through central-eastern Louisiana.
4. *P. g. megacephalus* (Rhoads, 1894). From southeast of Missouri, south Illinois, and northern Tennessee south through western Texas to northwest Georgia.

5. *P. g. palmarius* Bangs, 1896. Restricted from the center to the south of the Florida Peninsula.
6. *P. g. restrictus* Howell, 1939. Known only in the central-western coast of Florida.
7. *P. g. telmaphilus* Schwartz, 1952. Restricted to the southwestern coast of Florida.

**Conservation:** *Peromyscus gossypinus allapaticola* is listed as Endangered by the United States Endangered Species Act (FWS 2022) and *P. gossypinus* is listed as Least Concern by the IUCN (Cassola 2016cn).

**Characteristics:** Cotton mouse is medium sized; total length 175.0–190.0 mm and skull length 27.1–29.9 mm. The species has a wide geographic variation in color. Dorsal pelage from bright rufescent cinnamon to cinnamon rufous, with a dusky mixture in the middle of the back, appearing as a broad stripe from the shoulders to the base of the tail; fur of underparts whitish with a yellowish pectoral spot; orbital ring narrow, slightly widened anteriorly and posteriorly; ears dusky brownish, without all edged whitish; feet white; forearms often dusky or slightly rufescent and dusky; ears approximately 70% of the hindfoot length; **tail shorter than the head-and-body length, not very sharply bicolored, clothed with rather short hairs**; zygomatic bone heavy anteriorly; premaxillae slightly expanded laterally; palatine slits rather broadly open, their outer sides not parallel; interpterygoid fossa broad and square anteriorly. Species group *leucopus*. Note: It is hard to differentiate where it is in sympatry with *P. maniculatus*.

**Comments:** *Peromyscus gossypinus* ranges in the southeastern United States from Virginia south through northern Florida and west through Louisiana, western Texas, and Arkansas (Map 213). It can be found in sympatry, or nearly so, with *P. leucopus*, *P. maniculatus*, *P. polionotus*, and *Podomys floridanus*. *P. gossypinus* can be distinguished from *P. leucopus* by having a darker coloration and teeth clearly larger. From *P. maniculatus*, by having the tail larger to the head-and-body length and without a dark dorsal narrow stripe covering only the vertebral line. From *P. polionotus*, by its larger size; on average, the following measurements are greater than the figures shown: total length 154.0 mm, hindfoot length 19.0 mm, and tail length 60.0 mm. From *P. floridanus*, by having a longer tail, greater than 80% of the head-and-body length; without orange shading on the cheeks, shoulders, and flanks; without hypsodont molars with small accessory tubercles and an interorbital shelf.

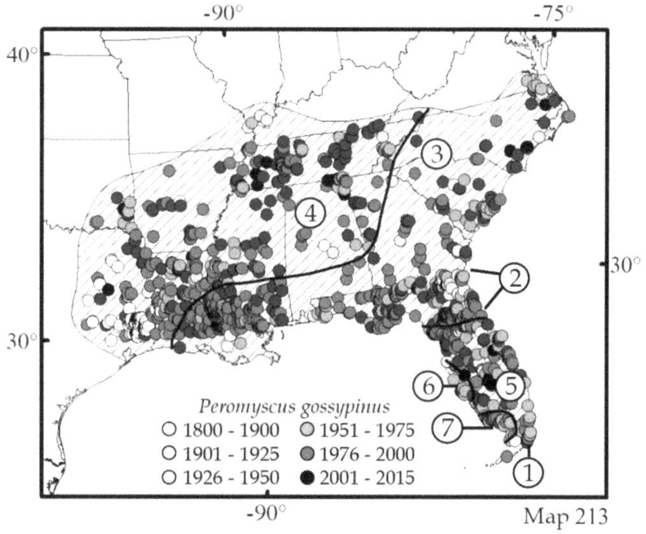


**Map 213** Distribution map of *Peromyscus gossypinus*

**Additional Literature:** Osgood (1909), Wolfe and Linzey (1977), Hall (1981), Musser and Carleton (2005).

## *Peromyscus gratus* Merriam, 1898

Tlalpan deermouse, ratón piñonero mexicano

1. *P. g. erasmus* Finley, 1952. Restricted to near Durango City, Durango.
2. *P. g. gentilis* Osgood, 1904. From southern New Mexico and northwestern Chihuahua south through northern Jalisco and central San Luis Potosí.
3. *P. g. gratus* Merriam, 1898. From northeastern Jalisco to southeastern San Luis Potosí and southeast through Puebla.
4. *P. g. zapotecae* Hooper, 1957. From southeastern Puebla to northern-central Oaxaca.

**Conservation:** *Peromyscus gratus* is listed as Least Concern by the IUCN (Lacher et al. 2016i).

**Characteristics:** Tlalpan deermouse is medium sized; total length 149.0–210.0 mm and skull length 23.9–28.2 mm. Dorsal pelage varying from pale yellowish-brown to brownish black; fur of underparts whitish; hair long and silky; lateral line distinguishable; limbs white; **ears approximately 105% of the hindfoot length;** hind feet white or dusky; six tubercles on the soles**; tail longer than the head-and-body length, covered with short hair**, **bicolored,** with a dorsal stripe slightly darker than the back; **skull shorter and rostrum relatively heavier**; auditory bullae well developed; braincase large rounded; zygomatic arches weak. Species group *truei*. Note: *Peromyscus gratus* and *P. truei* can only be differentiated by their geographic distribution.

**Comments:** *Peromyscus gratus* was reinstated as a distinct species (Modi and Lee 1984), previously considered a subspecies of *P. truei* (Osgood, 1909). *P. sagax* was considered a full species (Bradley et al. 1996), previously a synonym of *P. gratus gratus* (Osgood 1909). *P. gratus* ranges from the Mogollon Plateau in southwestern New Mexico, Coahuila, and Nuevo León south through central Oaxaca (Map 214). It can be found in sympatry, or nearly so, with *P. amplus*, *P. aztecus*, *P. beatae*, *P. boylii*, *P. bullatus*, *P. difficilis*, *P. eremicus*, *P. felipensis*, *P. hylocetes*, *P. kilpatricki*, *P. labecula*, *P. leucopus*, *P. levipes*, *P. megalops*, *P. melanocarpus*, *P. melanophrys*, *P. melanotis*, *P. melanurus*, and *P. truei*. *P. gratus* can be distinguished from *P. amplus*, *P. difficilis*, and *P. felipensis* by having the dorsum often lighter grayish; lateral line without ochraceous shades: top of the head similar to the dorsum; smaller ears, less than 24.0 mm; toothrow proportionally larger; auditory bullae proportionally less inflated and larger. From *P. aztecus*, *P. beatae*, *P. boylii*, *P. hylocetes*, *P. kilpatricki*, *P. labecula*, *P. levipes*, *P. leucopus*, and *P. melanotis*, by having the ears longer than the hind feet; tail longer than the head-and-body length, more heavily haired, and with less pronounced annulations; auditory bullae proportionally greatly inflated. From *P. bullatus*, by having the ear length less than 24.0 mm, equal to the hindfoot length, and auditory bullae proportionally less inflated. From *P. eremicus*, by having two pairs of mammae (nipples); first and second upper molars with accessory cusps. From *P. megalops*, *P. melanocarpus*, and *P. melanurus*, by its smaller size; larger ears; with a dark spot around the eyes and in the pectoral area; auditory bullae proportionally smaller. From *P. melanophrys*, by its smaller size; without a conspicuously gray mask around the face, without annulations, and supraorbital border not sharply angled. From *P. truei*, by having the tail longer than the head-and-body length; ears smaller than the hind feet, and auditory bullae not proportionally greatly inflated (Zimmerman et al. 1975).

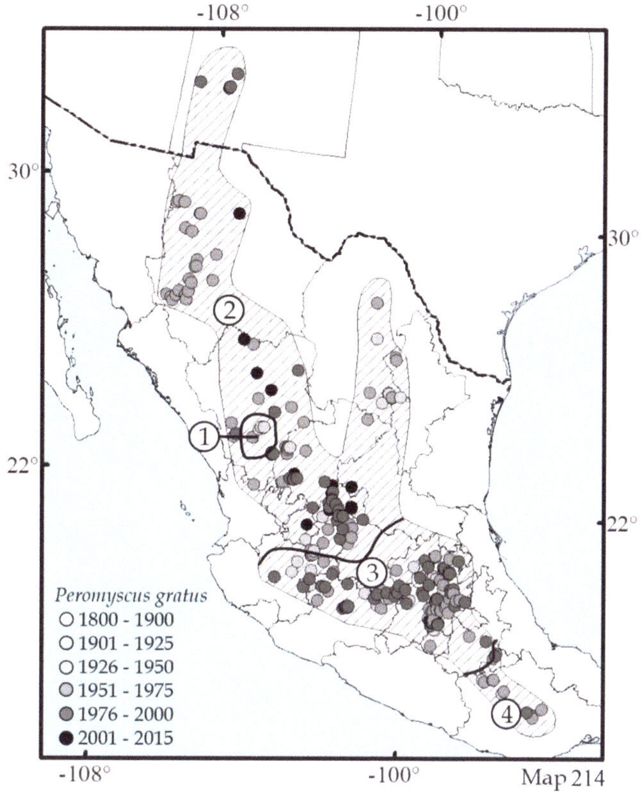


**Map 214** Distribution map of *Peromyscus gratus*

**Additional Literature:** Merriam (1898a), Schmidly (1973), Hoffmeister (1981), Musser and Carleton (2005).

## *Peromyscus greenbaumi* Bradley et al., 2022

Greenbaum deermouse, ratón de Coalcomán

Monotypic.

**Conservation:** *Peromyscus greenbaumi* (as *Peromyscus winkelmanni*) is not listed by the IUCN (Álvarez-Castañeda 2018h).

**Characteristics:** Greenbaum's deermouse is medium sized; mean total length 207.0 mm and skull length 27.7–29.5 mm. Dorsal pelage clove brown and blackish or mouse gray at the base; flanks Dresden brown; fur of underparts dusky neutral gray at the base and white at the tip; feet with a fuscous strip extending from the ankles to the mid-medapodials; toes white; tail as long as the head-and-body length, slightly bicolored, fuscous black dorsally and pale grayish-olive ventrally, scantily haired at the base and slightly tufted at the tip; ears approximately 92% of the hindfoot length and raw umber; and vibrissae black; skull elongated almost twice as long as wide; rostrum on average 41% of the skull length and nasals about 92% of the rostral length; braincase large and rounded; molar toothrow length about 16% of the skull length; interorbital constriction with a smooth outline, not angular; zygomatic arches nearly parallel; auditory bullae large in the *P. boylii* species group. *P. greenbaumi* karyotype $2n = 48$ (Bradley et al. 2022a). Species group *boylii*.

**Comments:** *Peromyscus winkelmanni* was considered present in Sierra de Coalcomán, Michoacán (Álvarez-Castañeda 2005b; Musser and Carleton 2005; Kilpatrick et al. 2021). *P. winkelmanni* is currently restricted to Filo de Caballo in Guerrero (Bradley et al. 2022a). *P. greenbaumi* ranges in Sierra de Coalcomán from southwestern Michoacán norththrough Jalisco along the Colima-Jalisco border between 1600 and 2500 m, in association with pine-oak forest (Map 215). It can be found in sympatry, or nearly so, with *Osgoodomys banderanus*, *P. beatae*, *P. ensinki*, *P. kilpatrick* (Los Reyes, Michoacán), *P. labecula*, *P. levipes*, and *P. megalops*. *P. greenbaumi* can be distinguished from *Osgoodomys banderanus* and *P. megalops* by not having an interorbital bead and ridge. From *P. ensinki* and *P. beatae*, by having its smaller external measurements on average, and skull measurements on average larger in some cranial characters. From *P. labecula*, by its larger size; tail without a well-defined dark dorsal narrow stripe covering only the vertebral line and equal to or larger than the head-and-body length; with a lateral line in the flanks; dark spot around the eyes and supraorbital border sharply angled. From *P. levipes*, by having somatic and skull measurements slightly larger on average.

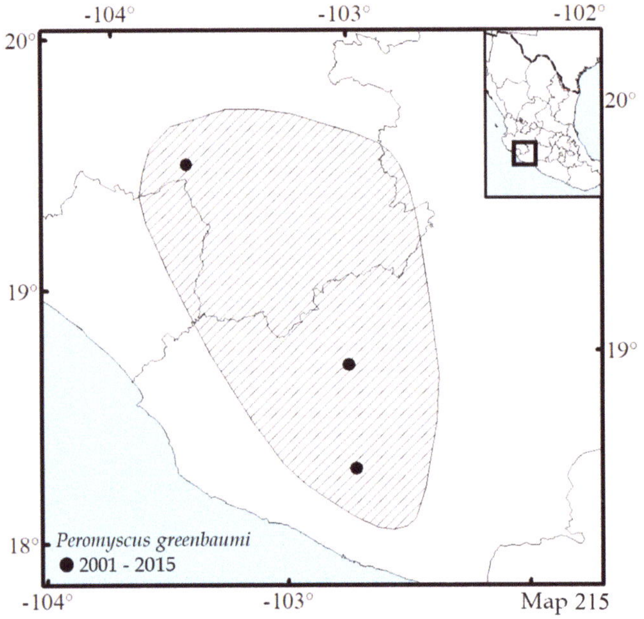


**Map 215** Distribution map of *Peromyscus greenbaumi*

**Additional Literature:** Carleton (1977, 1979), Bradley and Schmidly (1987), Sullivan et al. (1991).

## ***Peromyscus gymnotis*** Thomas, 1894

Naked-eared deermouse, ratón de orejas desnudas

Monotypic.

**Conservation:** *Peromyscus gymnotis* is listed as Least Concern by the IUCN (Vázquez and Reid 2016).

**Characteristics:** Naked-eared deermouse is medium sized; total length 208.0–250.0 mm and skull length 25.9–32.1 mm. Dorsal pelage rich tawny ochraceous to blackish, darker on the back; fur of underparts whitish to cream, often a pectoral region suffused with fulvous shades; **ears moderate size, very scantily haired**, approximately 70% of the hindfoot length; **tail about equal to or slightly shorter than the head-and-body length, scaly annulations slightly finer, nearly unicolor or with slight blotches of yellowish shades on the underparts**; premaxillary bone rather swollen laterally; nasals clearly convex; supraorbital border with a very slight suggestion of a bead. Note: to distinguish *P. gymnotis* from *P. mexicanus* in southern Chiapas, a direct comparison between specimens of the two species is needed. Species group *mexicanus*.

**Comments:** *Peromyscus gymnotis* was recognized as a distinct species (Musser 1971); it was previously regarded as a subspecies of *P. mexicanus* (Osgood 1909). It ranges in the Pacific coastal plain and adjacent foothills from southern Chiapas south through Central America (Map 216). It can be found in sympatry, or nearly so, with *P. carolpattonae*. *P. gymnotis* can be distinguished from *P. carolpattonae* by its smaller average size; hindfoot length generally less than 26.0 mm, upper toothrow length less than 4.6 mm, and with a supraorbital crest well developed.

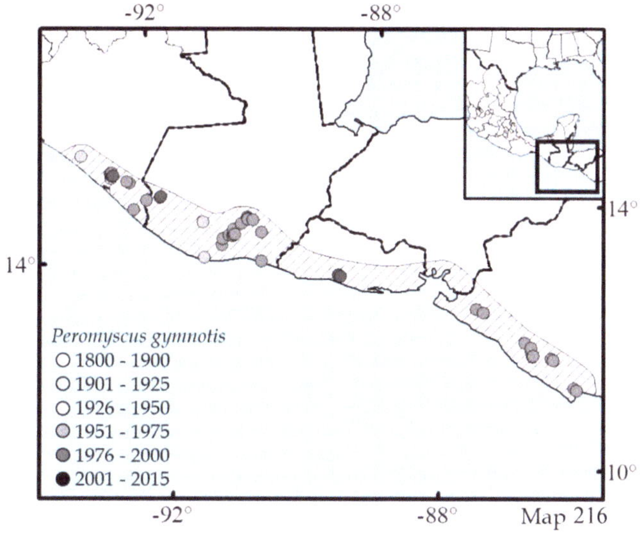


**Map 216** Distribution map of *Peromyscus gymnotis*

**Additional Literature:** Huckaby (1980), Musser and Carleton (2005).

## *Peromyscus hylocetes* Merriam, 1898.

Transvolcanic deermouse, ratón del Eje Neovolcánico

Monotypic.

**Conservation:** *Peromyscus hylocetes* is listed as Least Concern by the IUCN (Vázquez 2016b).

**Characteristics:** Transvolcanic deermouse is medium sized; total length 220.0–238.0 mm and skull length 33.7–34.3 mm. Dorsal pelage pale ocher grizzly, dark mainly on the back central part; flanks tawny; lateral line broad and tawny; fur of underparts light buff; orbital ring blackish; ears dusky; feet white and usually with a dusky streak extending from the tarsus to the metatarsus; **tail usually shorter than the head-and-body length, densely clothed with hair and sharply bicolored**; an angular supraorbital border; braincase anterior half nearly straight; **auditory bullae larger**. Species group *aztecus*. Note: *Peromyscus hylocetes* and *P. spicilegus* in Jalisco need to be directly compared between specimens of both species.

**Comments:** *Peromyscus hylocetes* was reinstated to the species level by Sullivan and Kilpatrick (1991) and Sullivan et al. (1997); it was previously regarded as a subspecies of *P. aztecus* (Osgood, 1909). It ranges along the Eje Neovolcánico, from Colima east through Estado de México (Map 217). It can be found in sympatry, or nearly so, with *P. amplus*, *P. boylii*, *P. carletoni*, *P. gratus*, *P. kilpatricki*, *P. labecula*, *P. leucopus*, *P. levipes*, *P. melanophrys*, *P. melanotis*, and *P. perfulvus*. *P. hylocetes* can be distinguished from *P. amplus* and *P. gratus* by having smaller ears than the hind feet; tail not heavily haired and much longer than the body-and-tail length; without a pectoral spot, and auditory bullae proportionally greatly inflated. From *P. boylii* and *P. levipes*, by having a larger maxillary toothrow, greater than 4.7 mm; brighter color on the sides; an angular supraorbital border; auditory bullae more pointed anteriorly and proportionally less inflated. From *P. carletoni*, by having a bright color on the flanks; supraorbital shelf edges well developed with a ridge and straight, both edges appear wedge shaped, and auditory bullae pointed anteriorly. From *P. labecula* and *P. melanotis*, by having the tail larger than the head-and-body length, without a dark dorsal narrow stripe covering only the vertebral line; an angular supraorbital border and auditory bullae that point anteriorly; braincase anterior half nearly straight. From *P. melanophrys* and *P. perfulvus*, by its smaller size; tail smaller than the head-and-body length and without a dark spot around the eyes.

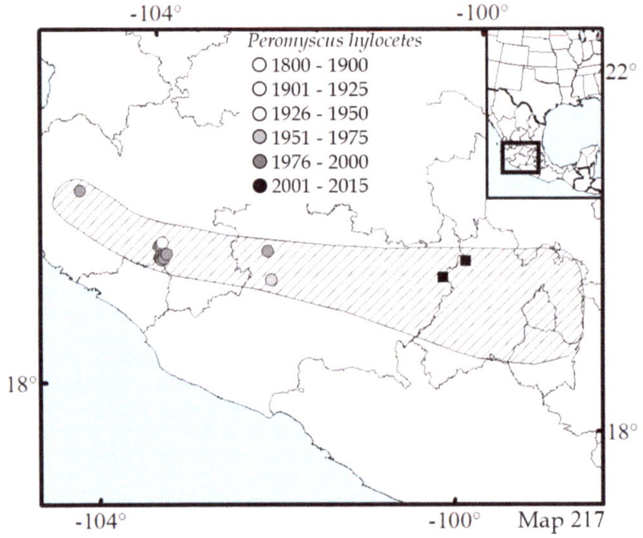


**Map 217** Distribution map of *Peromyscus hylocetes*

**Additional Literature:** Merriam (1898a), Álvarez (1961), Hall (1981), Vázquez et al. (2001), Musser and Carleton (2005).

## *Peromyscus keeni* (Rhoads, 1894)

Northwestern deermouse, ratón del noroeste

1. *P. k. algidus* Osgood, 1909. Restricted to the extreme northwestern British Columbia and southwest Yukon. Note: it can be considered a different species.
2. *P. k. angustus* Hall, 1932. Restricted to the coastal areas of central and southern Vancouver Island and the Bare, Hornby, Newcastle, Parkville, Saltpring, and Sidney Islands.
3. *P. k. balaclavae* McCabe and Cowan, 1945. Restricted to the Balaklava and Hope Islands, British Columbia.
4. *P. k. beresfordi* Guiguet, 1955. Restricted to Beresford Island, British Columbia.
5. *P. k. cancrivorus* McCabe and Cowan, 1945. Restricted to Table Island, British Columbia.
6. *P. k. carli* Guiguet, 1955. Restricted to Cox Island, British Columbia.
7. *P. k. doylei* McCabe and Cowan, 1945. Restricted to Doyle Island, British Columbia.

8. *P. k. georgiensis* Hall, 1938. Restricted to the Bowen, Lasqueti, Sanary, Texada, and Thorman Islands, British Columbia.
9. *P. k. hylaeus* Osgood, 1908. Restricted to the Prince of Wales and Kupreanof Islands, and surrounding mainland, Alaska.
10. *P. k. interdictus* Anderson, 1932. Restricted to the central and northern Vancouver Island, British Columbia.
11. *P. k. isolatus* Cowan, 1935. Restricted to the Pine and Nigei Islands, British Columbia.
12. *P. k. keeni* (Rhoads, 1894). Restricted to the Graham and Moresby Islands, British Columbia.
13. *P. k. macrorhinus* (Rhoads, 1894). Restricted to the coastal area of British Columbia from the central area northward.
14. *P. k. maritimus* McCabe and Cowan, 1945. Known only from Moore Island, British Columbia.
15. *P. k. oceanicus* Cowan, 1935. Restricted to Forrester Island, Alaska.
16. *P. k. oreas* Bangs, 1898. Restricted to central Olimpia peninsula, Washington.
17. *P. k. pluvialis* McCabe and Cowan, 1945. Restricted to Goose Island Group, British Columbia.
18. *P. k. prevostensis* Osgood, 1909. Restricted to the Prevost, Frederick, Hippa, and Marble Islands, British Columbia.
19. *P. k. rubriventer* McCabe and Cowan, 1945. Restricted to the Chatfield, Hecate, Reginald, Ruth, Smythe, and Townsend Islands, British Columbia.
20. *P. k. sartinensis* Guiguet, 1955. Restricted to Sartine Island, British Columbia.
21. *P. k. sitkensis* Merriam, 1897. Restricted to the Baranof, Chichagof, Coronation, Duke, and Warren Islands, Alaska.
22. *P. k. triangularis* Guiguet, 1955. Restricted to Triangle Island, British Columbia.

**Conservation:** *Peromyscus keeni* is listed as Least Concern by the IUCN (Cassola 2016co).

**Characteristics:** Northwestern deermouse is medium sized; **total length 178.0–217.0 mm and skull length 24.2–27.0 mm**. Coloration varying greatly between subspecies, dorsal russet with darker brown; fur of underparts whitish, clearly contrasting with the back; **ears shorter, approximately 65% of the hind feet**; tail on average shorter than the head-and-body length, bicolored with a dark dorsal stripe one-third of the tail diameter; hind feet whitish; **skull heavy in relation to all the *maniculatus*-like**; nasals and rostrum short and thick; posterior nasal endings usually at the same level than the premaxilla. Species group *maniculatus*.

**Comments:** *Peromyscus keeni* was reinstated to the species level (Hogan et al. 1993; Bradley et al. 2019), including the previously subspecies of *P. maniculatus*: *P. m. algidus*, *P. m. angustus*, *P. m. balaclavae*, *P. m. beresfordi*, *P. m. cancrivorus*, *P. m. carli*, *P. m. doylei*, *P. m. georgiensis*, *P. m. hylaeus*, *P. m. interdictus*, *P. m. isolatus*, *P. m. keeni*, *P. m. macrorhinus*, *P. m. maritimus*, *P. m. oreas*, *P. m. oceanicus*, *P. m. pluvialis*, *P. m. prevostensis*, *P. m. rubriventer*, *P. m. sartinensis*, *P. m. sitkensis*, and *P. m. triangularis* (Osgood 1909). The subspecies potentially assigned to *P. keeni* are listed above based on Bradley et al. (2019) and Greenbaum et al. (2019). *P. keeni* ranges from southern Alaska and Yukon south through central Washington and western Oregon, including the islands of British Columbia; it also occurs on most islands in the Queen Charlotte Sound and the Alexander Archipelago (Map 218). *P. keeni* can be found in sympatry, or nearly so, with *P. sonoriensis*, from which it differs by having the heaviest skull; the smallest ears; and the largest tail on average, more or less equal in length as the head-and-body; but mainly by using genetic data.

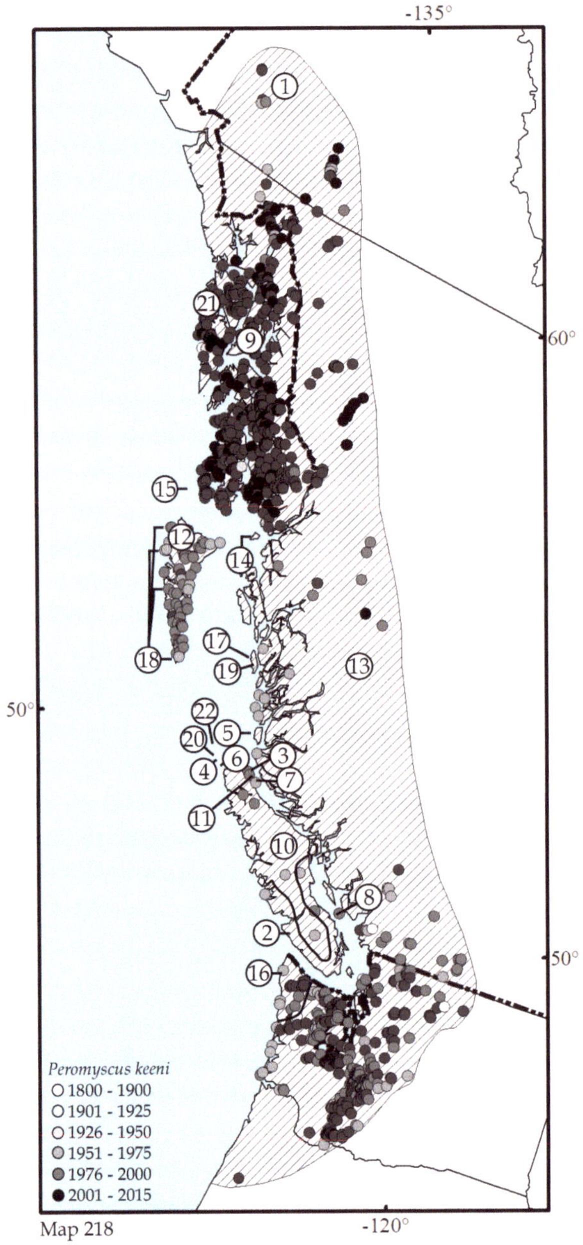


**Map 218** Distribution map of *Peromyscus keeni*

**Additional Literature:** Allard and Greenbaum (1988), Hogan et al. (1993).

## ***Peromyscus kilpatricki*** Bradley et al., 2016

Kilpatrick's deermouse, ratón de Zitácuaro

Monotypic.

**Conservation:** *Peromyscus kilpatricki* (as *Peromyscus levipes*) is listed as Least Concern by the IUCN (Castro-Arellano and Vázquez 2016d).

**Characteristics:** Kilpatrick's deermouse is medium sized; total length 173.0–203.0 mm and skull length 26.0–29.0 mm. Dorsal pelage sepia at the tip and blackish slate at the base, flanks Dresden brown; fur of underparts whitish and blackish slate at the base; ears approximately 85% of the hindfoot length with very thin hair; tail slightly longer than the head-and-body length, occasionally some hair extending to the vertebral part, clove brown dorsally and white ventrally, scantily haired at the base and tufted at the tip; limbs white; feet with a clove brown strip extending slightly past the ankles; toes white; skull elongated, twice as long as wide; braincase slightly rounded; zygomatic arches nearly parallel; auditory bullae medium sized. Species group *boylii*.

**Comments:** *Peromyscus kilpatricki* was historically referred to as *P. levipes*, but karyotypic data and DNA sequence show a different species (Bradley et al. 2016a, b). It ranges in highlands from central Michoacán east through Morelos (Map 219). *P. kilpatricki* cannot be found in sympatry, or nearly so, with any other species of the *boylii* species group but lives in sympatry, or nearly so, with *P. ensinki*, *P. felipensis*, *P. gratus*, *P. hylocetes*, *P. labecula*, *P. levipes*, *P. melanophrys*, *P. melanotis*, *P. leucopus*, and *P. winkelmanni*. *P. kilpatricki* have the coloration and external and cranial measurements similar to *P. levipes*, and can only be differentiated based on DNA analyses. *P. kilpatricki* is on average smaller than *P. beatae*, *P. carletoni*, *P. ensinki*, *P. hylocetes*, *P. levipes*, and *P. winkelmanni* but larger than *P. simulus* and *P. schmidlyi*. *P. kilpatricki* can be distinguished from *P. felipensis* and *P. gratus* by having the ears smaller than the hind feet; tail less haired and auditory bullae proportionally less inflated. From *P. hylocetes* and *P. winkelmanni*, by not having the dark spot around the eyes and a sharply angled supraorbital border. From *P. labecula* and *P. melanotis*, by having the tail without a well-defined narrow stripe covering only the vertebral line and equal to or larger than the head-and-body length. From *P. melanophrys*, by its smaller size; without a conspicuous gray mask around the face and the supraorbital border not sharply angled.

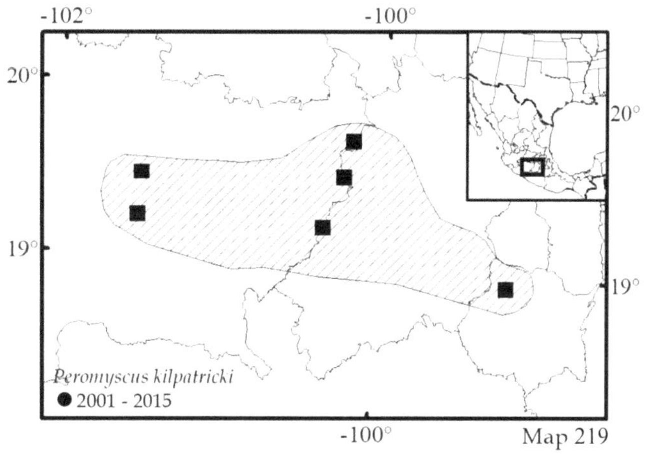


**Map 219** Distribution map of *Peromyscus kilpatricki*

**Additional Literature:** Bradley et al. (2017).

## ***Peromyscus labecula*** Elliot, 1903

Mexican common deermouse, ratón común de México

1. *P. l. blandus* Osgood, 1904. From southern New Mexico, southwestern Texas, eastern-central Chihuahua, and western Tamaulipas south through San Luis Potosí, Guanajuato, and Jalisco.
2. *P. l. fulvus* Osgood, 1904. From central Puebla and Veracruz south through Morelos and central Oaxaca.
3. *P. l. labecula* Elliot, 1903. From southeastern Durango and Zacatecas east through Hidalgo, and south through Estado de México, Michoacán, Colima, Jalisco, and southern Nayarit.

**Conservation:** *Peromyscus labecula* (as *Peromyscus maniculatus*) is listed as Least Concern by the IUCN (Cassola 2016cq).

**Characteristics:** Mexican common deermouse is medium sized; total length 145.0–183.0 mm and skull length 22.5–25.0 mm. Coloration varying greatly between subspecies, dorsal pelage vinaceous buff with dusky shades, occasionally russet, in winter gray and sootier; fur of underparts white; ears dusky, with a white spot at the base, 70% of the hind feet; tail shorter than the head-and-body length, sharply bicolored, with a brown dorsal stripe one-third of the tail diameter dorsally and white ventrally; hind feet whitish; **skull larger and more angular**; braincase arched; frontal bones with the outer edge more curved; interorbital constriction greater; nasals shorter and broad anteriorly; pterygoids short; zygomatic maxillary branch broad and heavy. Species group *maniculatus*.

**Comments:** *Peromyscus labecula* was reinstated to the species level by Bradley et al. (2019) and Greenbaum et al. (2019), including the subspecies *P. l. blandus* and *P. l. fulvus*; it was previously regarded as a subspecies of *P. maniculatus* (Osgood 1909). Specimens from western Chihuahua, *P. sonoriensis rufinus* could be assigned to *P. labecula* (Bradley et al. 2019). *P. labecula* ranges from southern Arizona, New Mexico, and southwestern Texas south through Colima, Jalisco, Estado de México, Puebla, Veracruz, and northern Oaxaca, and from eastern-central Chihuahua west through Tamaulipas (Map 220). *P. labecula* can be found in sympatry, or nearly so, with *P. amplus*, *P. aztecus*, *P. beatae*, *P. boylii*, *P. bullatus*, *P. carletoni*, *P. collinus*, *P. difficilis*, *P. eremicus*, *P. felipensis*, *P. furvus*, *P. gratus*, *P. hooperi*, *P. hylocetes*, *P. kilpatricki*, *P. laceianus*, *P. latirostris*, *P. leucopus*, *P. levipes*, *P. megalops*, *P. mekisturus*, *P. melanocarpus*, *P. melanophrys*, *P. melanotis*, *P. melanurus*, *P. merriami*, *P. mexicanus*, *P. nasutus*, *P. ochraventer*, *P. pectoralis*, *P. perfulvus*, *P. polius*, *P. sagax*, *P. schmidlyi*, *P. simulus*, *P. sonoriensis*, *P. truei*, and *P. winkelmanni*. *P. labecula* can be distinguished from *P. amplus*, *P. bullatus*, *P. collinus*, *P. difficilis*, *P. gratus*, *P. laceianus*, *P. nasutus*, *P. pectoralis*, *P. sagax*, and *P. truei* by having ears smaller than the hind feet; tail less haired and auditory bullae proportionally less inflated. From *P. amplus*, *P. difficilis*, and *P. felipensis*, by its smaller size. From *P. aztecus*, *P. beatae*, *P. boylii*, *P. carletoni*, *P. hylocetes*, *P. kilpatricki*, *P. levipes*, *P. polius*, *P. schmidlyi*, *P. simulus*, and *P. winkelmanni*, by having the tail hairy, shorter than the head-and-body length and clearly bicolored, with a dark dorsum restricted to a narrow stripe covering only the vertebral line without a lateral line in the flanks; without a supraorbital border. From *P. eremicus* and *P. merriami*, by having two pairs of mammae (nipples); first and second upper molars with accessory cusps; tail shorter than the head-and-body length with paravertebral coloration. From *P. furvus*, *P. latirostris*, *P. megalops*, *P. melanocarpus*, *P. melanophrys*, *P. melanurus*, *P. mexicanus*, *P. ochraventer*, and *P. perfulvus*, by its smaller size; smaller tail than the head-and-body length and a proportionally smaller baculum. From *P. melanophrys*, by lacking a conspicuous gray mask around the face; annulations on the tail and the supraorbital border not sharply angled. From *P. hooperi*, by having a notably shorter tail; smaller skull, and more complex occlusal surface of the molars. From *P. leucopus*, it is very hard to differentiate between these two species in some areas; tail with a darker area in the paravertebral area. From *P. melanotis* (extremely similar morphologically), by lacking a black preauricular hair tuft; braincase narrower and less rounded; wider interorbital space; prezygomatic notch more prominent; auditory bullae slightly larger and rostrum shorter and less slender. From *P. sonoriensis*, by its smaller size; shorter tail, less than 75.0 mm; more vinaceous coloration; size and skull larger on average; skull more angular, and mainly using genetic data.

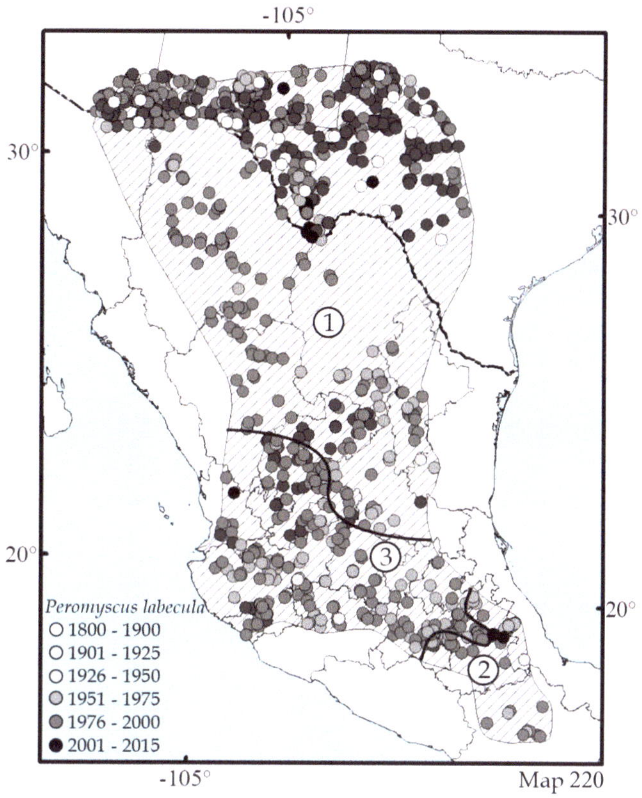


**Map 220** Distribution map of *Peromyscus labecula*

**Additional Literature:** Hooper (1957), Álvarez-Castañeda (1996).

## *Peromyscus laceianus* Bailey, 1906

Northern white-ankled deermouse, ratón de ancas blancas norteño

Monotypic.

**Conservation:** *Peromyscus laceianus* (as *Peromyscus pectoralis*) is listed as Least Concern by the IUCN (Lacher et al. 2016k).

**Characteristics:** Northern white-ankle deermouse is medium sized; total length 175.0–200.0 mm and skull length 27.1–28.2 mm. Dorsal pelage pale ochraceous buff to wood brown, not particularly concentrated on the dorsum; fur of underparts whitish; sides from paler brown to fulvous, not white at the base of the ears; hind feet white, extending to the ankles; tail usually equal to or longer than the head-and-body length, not bicolored pale brown dorsally, white ventrally, and scantily haired; skull large and heavy; auditory bullae medium sized; molars smaller and weaker; braincase relatively elongated; interparietal bone relatively larger. Species group *truei*.

**Comments:** *Peromyscus laceianus* was recognized as a distinct species (Bradley et al. 2015); it was previously regarded as a subspecies of *P. pectoralis* (Bailey 1906). *P. laceianus* is restricted to southcentral Oklahoma, southeastern New Mexico, and Texas (Map 221). It can be found in sympatry, or nearly so, with *P. attwateri*, *P. boylii*, *P. eremicus*, *P. labecula*, *P. leucopus*, *P. melanotis*, *P. nasutus*, *P. pectoralis*, and *P. sonoriensis*. *P. laceianus* can be distinguished from *P. attwateri* by its smaller size, in particular hindfoot length; paler color, with white ankles with dusky markings on the tarsal joints; more sharply bicolored tail; smaller auditory bullae; shorter incisive foramen; posterior margin of nasals truncate; toothrow smaller and shorter; mesopterygoid process poorly developed. From *P. boylii*, by having a slightly smaller skull; rostrum relatively broader and heavier; braincase less vaulted; slightly smaller molars; white ankles; posterior margin of nasals truncate. From *P. eremicus*, by having two pairs of mammae (nipples); ear length smaller than the hindfoot length and having a lateral line in the flank, a dark spot around the eyes; a pectoral spot; first and second upper molars with accessory cusps and auditory bullae larger and more orbicular. From *P. labecula*, *P. leucopus*, *P. melanotis*, and *P. sonoriensis*, by having the tail without a well-defined bicolored pattern and equal to or larger than the head-and-body length. From *P. nasutus*, by having the ear length equal to or smaller than the hindfoot length and auditory bullae smaller and not globe shaped, its length less than 18% of the skull length. From *P. pectoralis*, by a combination of the smallest total length and the greatest skull length, in particular, *P. pectoralis collinus* has a longer skull and a longer total length and from *P. p. pectoralis*, by its smaller skull and total length.

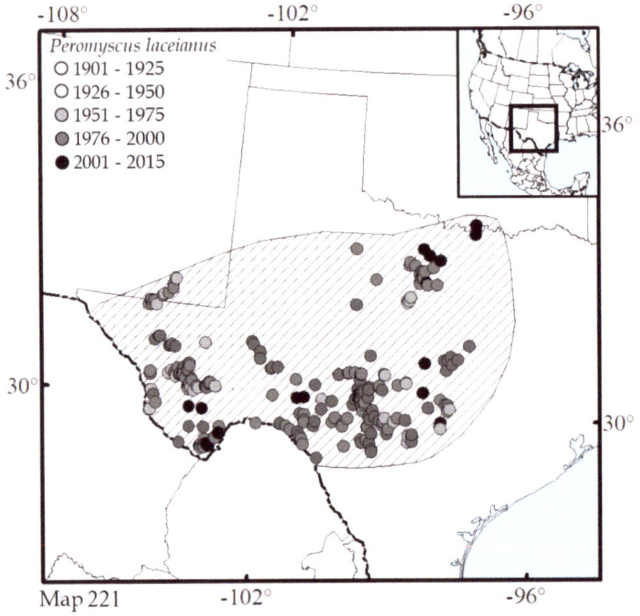


**Map 221** Distribution map of *Peromyscus laceianus*

**Additional Literature:** Osgood (1909), Schmidly (1972), Bradley et al. (2014b).

## *Peromyscus latirostris* Dalquest, 1950

Northern blackish deermouse, ratón negruzco del norte

Monotypic.

**Conservation:** *Peromyscus latirostris* (as *Peromyscus furvus*) is listed as Least Concern by the IUCN (Castro-Arellano and Vázquez 2019).

**Characteristics:** Northern blackish deermouse is medium sized; total length 265.0–278.0 mm and skull length 32.7–37.0 mm. Dorsal pelage dark brown; flanks slightly paler; fur of underparts grayish with darker shades at the base; eye ring and surrounding area black; flanks snuff brown; feet white; ears dusky, approximately 75% of the hindfoot length; **tail naked with scales,** slightly longer than the head-and-body length, **darker dorsally**; limbs white; tarsus brown; forefeet and hind feet white; **skull very broad, inflated anteriorly, and distinctly bell shaped, breadth between the nasal tips**; palate thicker; interorbital region narrow; nasals extending about 2.0 mm beyond the intermaxillar bones; palatine foramina relatively broad. Species group *furvus*.

**Comments:** *Peromyscus latirostris* was recognized as a distinct species (Dalquest 1950, Ávila-Valle et al. 2012); it was previously regarded as a subspecies of *P. furvus*. It ranges along the eastern flanks of Sierra Madre Oriental from southeastern San Luis Potosí to Querétaro (Map 222). It can be found in sympatry, or nearly so, with *P. aztecus*, *P. fulvus*, *P. labecula*, *P. leucopus*, *P. levipes*, *P. mexicanus*, *P. ochraventer*, and *P. pectoralis*. *P. latirostris* can be distinguished from *P. aztecus*, *P. labecula*, *P. leucopus*, *P. levipes*, and *P. pectoralis* by its larger size, tail darker dorsally, naked with scales and very broad, inflated anteriorly, and distinctly bell shaped, breadth the nasals tips. From *P. fulvus*, by its larger average size and based on genetic differences. From *P. mexicanus*, by its hourglass shaped interorbital region; nasals expanded; pectoral mammae (nipples) and tail length slightly longer than the head-and-body length. From *P. ochraventer*, by having larger auditory bullae; a well-developed style in the first molar, with a minor fold anterior to the cingulum of the first upper molars with a smaller lingual cusp, a larger labial cusp and frequently small style anterior to the cleft between the two cusps.

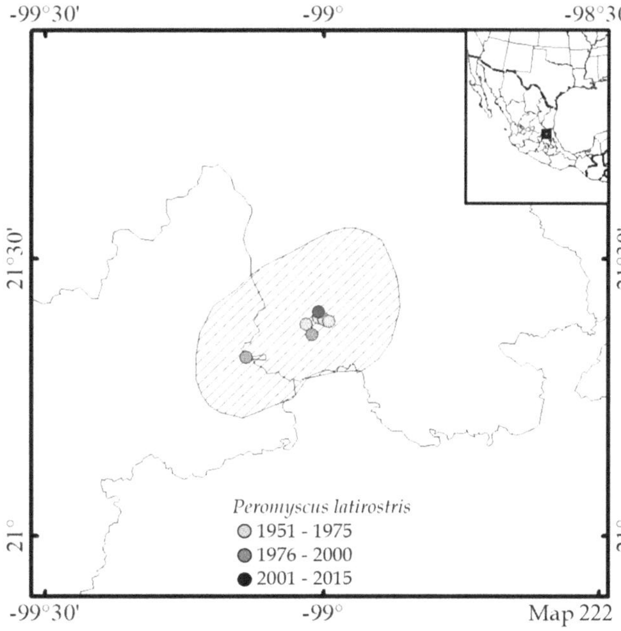


**Map 222** Distribution map of *Peromyscus latirostris*

**Additional Literature:** Hooper and Musser (1964), Huckaby (1973, 1980), Rogers and Skoy (2011).

## ***Peromyscus leucopus*** (Rafinesque, 1818)

White-footed deermouse, ratón de patas blancas

1. *P. l. affinis* (Allen, 1891). From central Oaxaca and southern Veracruz east through western Campeche.
2. *P. l. ammodytes* Bangs, 1905. Restricted to Manomoy Island, Massachusetts.
3. *P. l. aridulus* Osgood, 1909. From southeast Saskatchewan and Alberta south through northwestern Kansas, including eastern Montana and Wyoming and central Dakotas and Nebraska.
4. *P. l. arizonae* (Allen, 1894). Restricted to southeastern Arizona and southwestern New Mexico south through Northern Durango and southwestern Coahuila.

5. *P. l. castaneus* Osgood, 1904. Restricted to the northwestern Yucatán Peninsula.
6. *P. l. caudatus* Smith, 1939. Restricted to southern Nova Scotia and Prince Edward Island.
7. *P. l. cozumelae* Merriam, 1901. Restricted to Cozumel Island, Quintana Roo.
8. *P. l. easti* Paradiso, 1960. Restricted to southeastern Virginia.
9. *P. l. fusus* Bangs, 1905. Restricted to Martha's Vineyard Island and Nantucket Island, Massachusetts.
10. *P. l. incensus* Goldman, 1942. Lowlands of northern and Central Veracruz.
11. *P. l. lachiguiriensis* Goodwin, 1956. Restricted to central-southern Oaxaca.
12. *P. l. leucopus* (Rafinseque, 1818). From Oklahoma and eastern Texas east through Virginia and northern Carolina.
13. *P. l. mesomelas* Osgood, 1904. Highlands of Veracruz and eastern Hidalgo and Puebla.
14. *P. l. novaboracensis* Fisher, 1829. From the eastern Dakotas to northeastern Oklahoma east through Virginia and north through Maine.
15. *P. l. ochraceus* Osgood, 1909. Restricted to central Arizona.
16. *P. l. texanus* (Woodhouse, 1853). From northern-central Texas south through southeastern San Luis Potosí and Tamaulipas.
17. *P. l. tornillo* Mearns, 1896. From southeastern Colorado and southwestern Kansas south through west Texas, New Mexico, and extreme northern Chihuahua.

**Conservation:** *Peromyscus leucopus cozumelae* is listed as Threatened by the Norma Oficial Mexicana (DOF 2019) and *P. leucopus* is listed as Least Concern by the IUCN (Cassola 2016cp).

**Characteristics:** White-footed deermouse is medium sized; total length 130.0–200.0 mm and skull length 24.0–29.5 mm. The species has a strong geographical variation in color. Dorsal pelage pale to dark brown, middle of the back only slightly darker than the remainder of upperparts; fur of underparts whitish with dark at the base; pectoral spot yellowish frequently present; **ears dusky, very narrowly margined with whitish shades, no white spots at the base, medium sized and thinly haired**, approximately 70% of the hindfoot length; without blackish orbital ring; forefeet and hind feet white; upper side of the forearms dusky; ankles brownish; **tail shorter than the head-and-body length,** dusky brownish dorsally and white ventrally; molars and auditory bullae small; zygoma less deeply notched by the infraorbital foramen; palatine slits narrower at either end than the palatine slits, usually ending well beyond the front plane of the first upper molars. Species group *leucopus*. Note: It is hard to differentiate it from *P. maniculatus* where both live in sympatry.

**Comments:** *Peromyscus leucopus* ranges from southern Alberta, eastern Saskatchewan, Montana, Wisconsin, Michigan, Ontario, and New Brunswick southwest through Arizona, Sonora, Coahuila, and Durango, and southeast through Oaxaca, with an isolated population in the northern Yucatán Peninsula, east through the Atlantic coast, from Prince Edward Island, except for southern Georgia and the Florida Peninsula, including the Cozumel Island, Quintana Roo (Map 223). It can be found in sympatry, or nearly so, with *P. amplus*, *P. attwateri*, *P. aztecus*, *P. beatae*, *P. boylii*, *P. collinus*, *P. difficilis*, *P. felipensis*, *P. furvus*, *P. gossypinus*, *P. hylocetes*, *P. labecula*, *P. laceianus*, *P. latirostris*, *P. levipes*, *P. maniculatus*, *P. melanocarpus*, *P. melanophrys*, *P. melanotis*, *P. melanurus*, *P. merriami*, *P. mexicanus*, *P. nasutus*, *P. pectoralis*, *P. polionotus*, *P. polius*, *P. sonoriensis*, *P. truei*, *P. winkelmanni*, and *P. yucatanicus*. *P. leucopus* can be distinguished from *P. amplus*, *P. attwateri*, *P. collinus*, *P. difficilis*, *P. felipensis*, *P. laceianus*, *P. nasutus*, *P. pectoralis*, and *P. truei* by having the ears smaller than the hind feet feet; tail less heavily haired and auditory bullae proportionally less inflated. From *P. aztecus*, *P. beatae*, *P. boylii*, *P. hylocetes*, *P. levipes*, *P. polius*, and *P. winkelmanni*, by having the tail hairy; larger the head-and-body length and not clearly bicolored, with a dark narrow dorsal stripe covering only the vertebral line; without a lateral line and a spot around the eyes. From *P. furvus*, *P. latirostris*, *P. melanocarpus*, and *P. melanurus*, by its smaller size; tail without annulations and the supraorbital border not sharply angled. From *P. gossypinus*, by its paler coloration and clearly smaller teeth. From *P. labecula*, *P. maniculatus*, *P. polionotus*, and *P. sonoriensis*, it is very hard to differentiate it from these species in some places; tail without a darker area in the paravertebral area. From *P. mekisturus*, by having the tail as long as the head-and-body length or slightly longer. From *P. melanotis*, by not having a dark hair tuft at the base of the ears; dorsal pelage similar to the lateral pelage, and nasal length equal to 11.0 mm. From *P. melanophrys*, by its smaller size, without a gray mask around the face and supraorbital border not sharply angled. From *P. merriami*, by having two pairs of mammae (nipples); first and second upper molars with accessory cusps, tail shorter than the body-and-head length. From *P. mexicanus* and *P. yucatanicus*, by its smaller size; tail smaller than the head-and-body length and a proportionally smaller baculum.

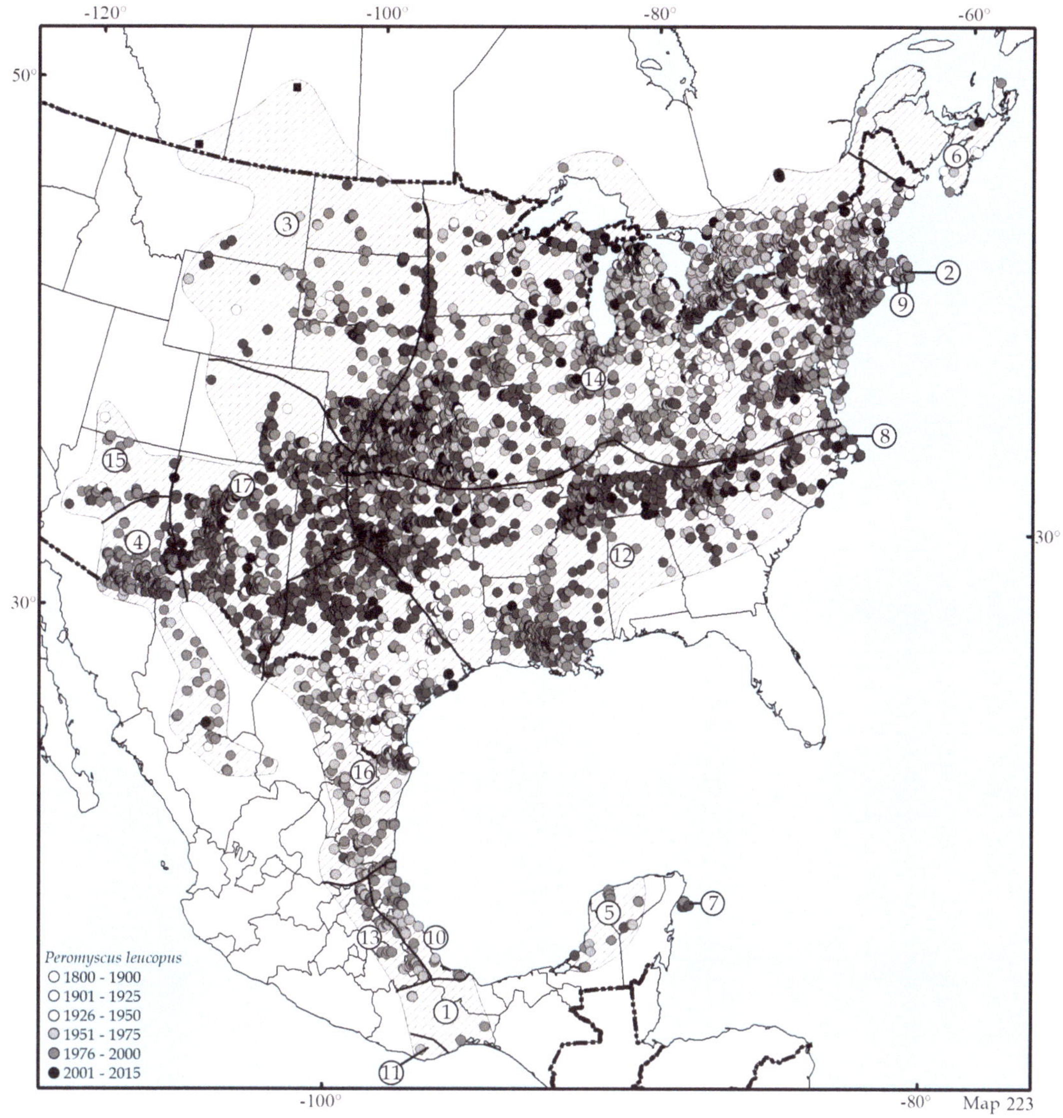


**Map 223** Distribution map of *Peromyscus leucopus*

**Additional Literature:** Merriam (1901a), Osgood (1909), Hooper (1957), Hall (1981), Lackey et al. (1985), Musser and Carleton (2005).

## *Peromyscus leucurus* Thomas, 1894

Tehuantepec deermouse, ratón de Tehuantepec

Monotypic.

**Conservation:** *Peromyscus leucurus* is listed (as *Peromyscus melanophrys*) as Least Concern by the IUCN (Vázquez and Álvarez-Castañeda 2016).

**Characteristics:** Tehuantepec deermouse is large sized; total length 214.0–296.0 mm andskull length 28.7–33.8 mm. Dorsal pelage varying from orange rufous to cinnamon, with a grayish facial mask well marked, fur of underparts grayish-white, dark gray at the base and white at the tip; ankles dark; dorsal hind feet pale to dark brown, with a dorsal dark brown line in some specimens; ears short; tail markedly bicolored and longer than the head-and-body length, dark brown dorsally and grayish ventrally, with a small hair tuft at the posterior end, which can be white; **supraorbital region wide and flat or wide and domed, forming a wide interorbital constriction; auditory bullae medium sized; nasals relatively long; braincase globose; mastoid region narrow;** nasal bones long and tapering or with a U shaped posterior end. Species group *melanophrys*.

**Comments:** *Peromyscus leucurus* was reinstated to the species level by López-González et al. (2019); it was previously regarded as a subspecies of *P. melanophrys* (Osgood 1904). It ranges in Oaxaca, Chiapas, and Guatemala (Map 224), and can be found in sympatry, or nearly so, with *P. difficilis*, *P. gratus*, *P. labecula*, *P. leucopus*, *P. megalops*, *P. melanocarpus*, and *P. melanurus*. *P. leucurus* can be distinguished from all other species, except for *P. melanophrys*, by its black orbital ring and a conspicuous gray mask; tail larger than the head-and-body length, with annulations, covered with short stiff hairs and frequently white at the tip; supraorbital border sharply angled but normally not moulded. From *P. gratus*, *P. labecula*, *P. leucopus*, *P. megalops*, *P. melanocarpus*, and *P. melanurus*, its larger size; tail longer than the head-and-body length; auditory bullae proportionally greatly inflated. From *P. difficilis* and *P. gratus*, by having smaller ears, less than the hindfoot length; auditory bullae not greater in size. From *P. labecula* and *P. leucopus*, by having the tail without a well-defined bicolored pattern and larger than the head-and-body length. From *P. megalops*, *P. melanocarpus*, and *P. melanurus*, by having a darker spot around the eyes and without a pectoral spot. From *P. melanophrys*, by having hind feet with a dorsal pale brown coloration or with a darker brown line over the metatarsals; mastoid area narrow and maxillary toothrow narrow.

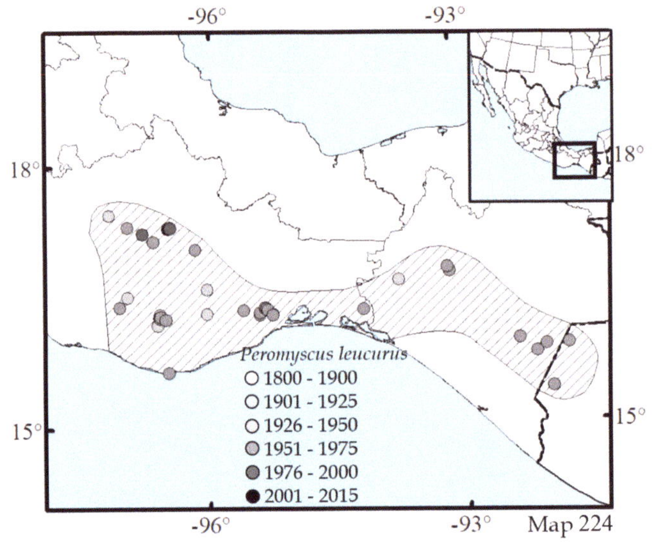


**Map 224** Distribution map of *Peromyscus leucurus*

**Additional Literature:** Baker (1952), Hooper (1955), Carleton (1989), Musser and Carleton (2005).

## *Peromyscus levipes* Merriam, 1898

Nimble-footed deermouse, ratón de patas ágiles

1. *P. l. ambiguus* Álvarez, 1961. Central-southern Nuevo León, Tamaulipas, and northern San Luis Potosí.
2. *P. l. levipes* Merriam, 1898. From southern San Luis Potosí south through northern Puebla and central Morelos.

**Conservation:** *Peromyscus levipes* is listed as Least Concern by the IUCN (Castro-Arellano and Vázquez 2016d).

**Characteristics:** Nimble-footed deermouse is medium sized; total length 184.0–234.0 mm and skull length 22.6–26.2 mm. Dorsal pelage varying by locality but generally ochraceous buff with some dusky shades; fur of underparts white to ocher, with a pectoral spot sometimes present; flanks tawny; lateral line not sharply marked; orbital ring blackish, with a grizzled area between the eyes and the ear base; ears dusky, with a hair tuft soft blackish slate at the anterior base of the ears, edged scarcely with whitish shades; feet white; tarsal joint sharply marked with dusky shades; tail usually shorter than the head-and-body length; **pelage usually longer and softer;** skull relatively large within *Peromyscus*; **second lower molars with the entolophilic present but small**; supraorbital border not angular, almost rounded; auditory bullae larger. Species group *boylii*.

**Comments:** *Peromyscus levipes* was reinstated to the species level by Houseal et al. (1987) and Schmidly et al. (1988); it was previously regarded as a subspecies of *P. boylii* (Osgood 1909). *P. kilpatricki* was historically referred to as *P. levipes*, but karyotype data and the DNA sequence shows that it is a different species (Bradley et al. 2016a, b). It ranges from central Nuevo León and western Tamaulipas, along the Sierra Madre Oriental to northeastern Michoacán, Guanajuato, Estado de México, Ciudad de México, Puebla, Tlaxcala, and central Veracruz (Map 225). *P. levipes* cannot be found in sympatry, or nearly so, with any other species of the *boylii* species group, except for northern Querétaro where it could be found with *P. boylii*; but it could be found in sympatry with *P. amplus*, *P. aztecus*, *P. collinus*, *P. difficilis*, *P. furvus*, *P. gratus*, *P. hylocetes*, *P. kilpatricki*, *P. labecula*, *P. latirostris*, *P. leucopus*, *P. melanocarpus*, *P. melanophrys*, *P. melanotis*, *P. mexicanus*, *P. ochraventer*, *P. pectoralis*, and *P. perfulvus*. *P. levipes* can be differentiated from *P. amplus*, *P. collinus*, *P. difficilis*, *P. gratus*, and *P. pectoralis* by having the ears smaller than the hind feet; tail not heavily haired and much longer than the head-and-body length; without a pectoral spot and auditory bullae proportionally greatly inflated. From *P. aztecus* and *P. hylocetes*, by not having a dark spot around the eyes; auditory bullae relatively larger and without the supraorbital border sharp-angled. From *P. boylii*, by having a dorsal coloration darker ochraceous buff with some dusky shades and without sharp contrasts with the flanks; upper toothrow length usually equal to or greater than 4.2 mm, and fronto-maxillary suture without teeth. From *P. furvus*, *P. latirostris*, *P. ochraventer*, and *P. mexicanus*, by its smaller size; tail slightly larger than the head-and-body length; a tawny lateral line not sharply marked. From *P. mexicanus*, by having the ears with hairs and the tail not coarsely annulated (about 17 annulations/cm). From *P. kilpatricki*, only based on DNA analyses. From *P. labecula*, *P. leucopus*, and *P. melanotis*, by having the tail without a well-defined bicolored pattern and equal to or larger than the head-and-body. From *P. mekisturus*, by not having the tail more than 130% of the head-and-body length. From *P. melanocarpus*, by having white carpal and tarsal regions; the presence of pectoral mammae (nipples); supraorbital edge present without a well marked and shallow depression; smaller size and molars. From *P. melanophrys* and *P. perfulvus*, by its smaller size; without a conspicuous patch of gray around the face; no white annulations at the tail tip and supraorbital border not sharply angled. From *P. ochraventer*, by having the tail more bicolored and rostrum not expanded anteriorly without parallel sides.

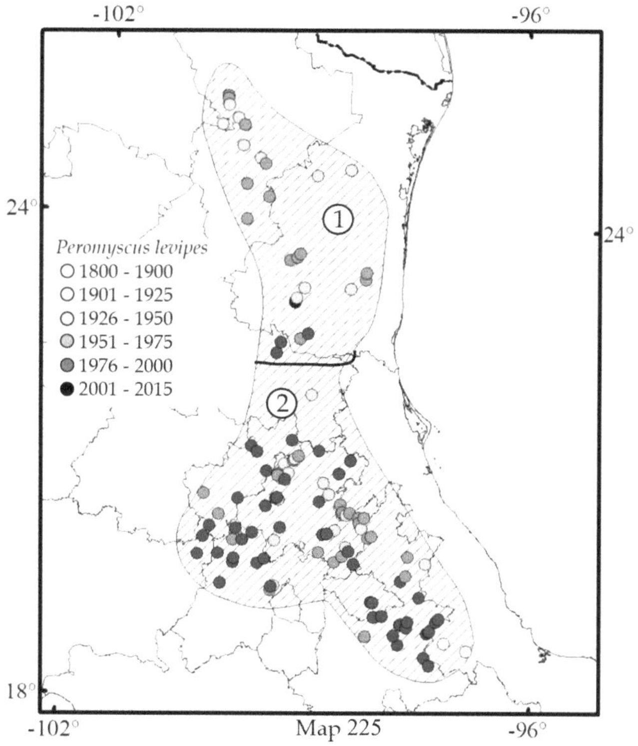


**Map 225** Distribution map of *Peromyscus levipes*

**Additional Literature:** Merriam (1898a), Álvarez-Castañeda (1996), Bradley et al. (2000), Musser and Carleton (2005), Álvarez-Castañeda and González–Ruiz (2009).

## *Peromyscus madrensis* Merriam, 1898

Tres Marías island deermouse, ratón de Isla Tres Marías

Monotypic.

**Conservation:** *Peromyscus madrensis* is listed as Threatened by the Norma Oficial Mexicana (DOF 2019) and as Endangered (B1ab (iv, v) + 2ab (iv, v): extent of occurrence less than 5000 km$^2$ in four populations and under continuing decline in number of subpopulations and number of mature individuals and area of occupancy less than 500 km$^2$ and continuing decline in the number of mature individuals) by the IUCN (Álvarez-Castañeda et al. 2018f).

**Characteristics:** Tres Marías island deermouse is medium sized; total length 210.0–250.0 mm and skull length 28.3–31.5 mm. Dorsal pelage dark yellowish ocher grizzly with dark brown shades, darker on the back; fur of underparts white to yellowish, usually with a pectoral salmon spot; ears approximately 85% of the hindfoot length and of the same color as the body; eyelids darker than the rostrum; tail shorter than the head-and-body length and not bicolored but darker dorsally than ventrally, except for the last third, which is dark all around; hind feet white; upper ankles with a dark spot; skull large, flat, and smoothly rounded; rostrum elongated without supraorbital and superciliary ridges; toothrow and auditory bullae relatively small; **dentition relatively simple with the mesolophid absent and the ectolophid often absent or underdeveloped;** mesoloph typically incomplete. Species group *boylii.*

**Comments:** *Peromyscus madrensis* was reinstated as a distinct species (Carleton 1977; Carleton et al. 1982); it was previously regarded as a subspecies of *P. boylii* (Osgood 1909). *P. madrensis* is restricted to the Islas Marías Archipelago (San Juanito, María Madre, María Magdalena, and María Cleofas), Nayarit (Map 226). It can be distinguished from other closely related species based on its geographic distribution. From all other species of *Peromyscus*, by its large size with a long, scantily haired tail; larger hind feet; relatively small pinnae. From species of the *boylii* species group, the phallus is small and interorbital square. From *P. simulus*, by having the mesoloph more conspicuous. From *P. spicilegus*, the nearest species of the *boylii* group on the mainland is larger with the tail longer; shorter ears; larger braincase and skull clearly broader and flatter.

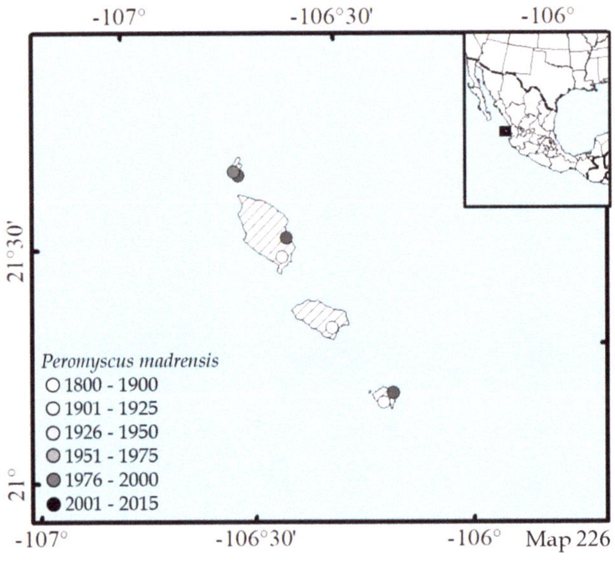


**Map 226** Distribution map of *Peromyscus madrensis*

**Additional Literature:** Merriam (1898a), Hall (1981), Álvarez-Castañeda and Cortés-Calva (1999), Álvarez-Castañeda and Méndez (2005), Musser and Carleton (2005).

## ***Peromyscus maniculatus*** (Wagner, 1845)

Northeastern common deermouse, ratón común del noreste

1. *P. m. abietorum* Bangs, 1896. Restricted to southern Québec, New Brunswick, Nova Scotia, and Maine.
2. *P. m. anticostiensis* Moulthrop, 1937. Restricted to Anticosti Island, Québec.
3. *P. m. argentatus* Copeland and Church, 1906. Restricted to Grand Manan Island, New Brunswick.
4. *P. m. bairdii* (Hoy and Kennicott, 1857). From Manitoba to southeastern Kansas east through eastern Tennessee and Kentucky, west through West Virginia and Pennsylvania, and northwest through Manitoba.
5. *P. m. eremus* Osgood, 1909. Restricted to Grindstone Island, Québec.
6. *P. m. gracilis* (Le Conte, 1855). From southern Ontario and Wisconsin east through southern Québec to Connecticut, and southwestern to northern Georgia.
7. *P. m. maniculatus* (Wagner, 1845). From eastern Manitoba east through eastern Newfoundland.
8. *P. m. nubiterrae* Rhoads, 1896. From western New York south through North Carolina.
9. *P. m. plumbeus* Jackson, 1939. Restricted to eastern Québec.

**Conservation:** *Peromyscus maniculatus* is listed as Least Concern by the IUCN (Cassola 2016cq).

**Characteristics:** Northeastern common deermouse is medium sized; total length 174.0–200.0 mm and skull length 25.0–28.2 mm. Coloration varying greatly between subspecies, dorsum dark brown tinged with fawn, median dorsal region darker; fur of underparts whitish with plumbeous shades at the base clearly contrasting with the back; orbital region and whisker base blackish; ears dusky with pale colors, edges with a preauricular hair tuft with few white hairs, approximately 70% of the hind feet; feet white; **underside of the hind feet hairy, except on the pads and the spaces between them; tail well haired, shorter than the head-and-body length, sharply bicolored with a brownish black dorsal stripe one-third of the tail diameter and distinctly penciled**; hind feet whitish; braincase arched. Species group *maniculatus*.

**Comments:** The following subspecies are considered under *P. gambelii*: *anacapae*, *assimilis*, *catalinae*, *cineritius*, *clementis*, *coolidgei*, *dorsalis*, *dubius*, *elusus*, *exiguous*, *exterus*, *gambelii*, *geronimensis*, *hueyi*, *magdalenae*, *margaritae*, *sanctaerosae*, *santacruzae*, and *streatori*; as subspecies of *P. keeni*: *algidus*, *angustus*, *balaclavae*, *beresfordi*, *cancrivorus*, *carli*, *doylei*, *georgiensis*, *hylaeus*, *interdictus*, *insolatus*, *keeni*, *macrorhinus*, *maritimus*, *oreas*, *oceanicus*, *pluvialis*, *prevostensis*, *rubiventer*, *sartinensis*, *sitkensis*, and *triangularis*; as subspecies of *P. labecula: P. m. blandus*, *P. m. fulvus*, and *P. m. labecula*; and as subspecies of *P. sonoriensis*: *alpinus*, *artemisiase*, *austerus*, *borealis*, *hollisteri*, *inclarus*, *luteus*, *nebrascensis*, *ozarkiarum*, *pallescens*, *rubidus*, *rufinus*, *saturatus*, *saxamans*, and *serratus* (Bradley et al. 2019; Greenbaum et al. 2019). *P. maniculatus* ranges from areas eastern of the Mississippi River north through Manitoba and east through the Atlantic coast from the Labrador Peninsula south through Tennessee and North Carolina (Map 227). Samples from the western side of Great Lakes area and boundaries of provinces of Manitoba and Saskatchewan should be reviewed to propose the boundary between *P. maniculatus* and *P. sonoriensis* in this region. *P. maniculatus* can be found in sympatry, or nearly so, with *P. attwateri*, *P. gossypinus*, *P. leucopus*, and *P. sonoriensis*. *P. maniculatus* can be distinguished from *P. attwateri* and *P. gossypinus* by having a smaller total length, with a dark dorsal narrow stripe covering only the tail paravertebral line. It is very hard to differentiate *P. maniculatus* from *P. leucopus* in some areas; in the former, the tail has a darker area in the paravertebral area. From *P. sonoriensis*, it can be distinguished mainly using genetic data and by having the under side of the hind feet hairy, except on the pads, and with the tail well haired, paravertebral area sharply bicolored and distinctly penciled.

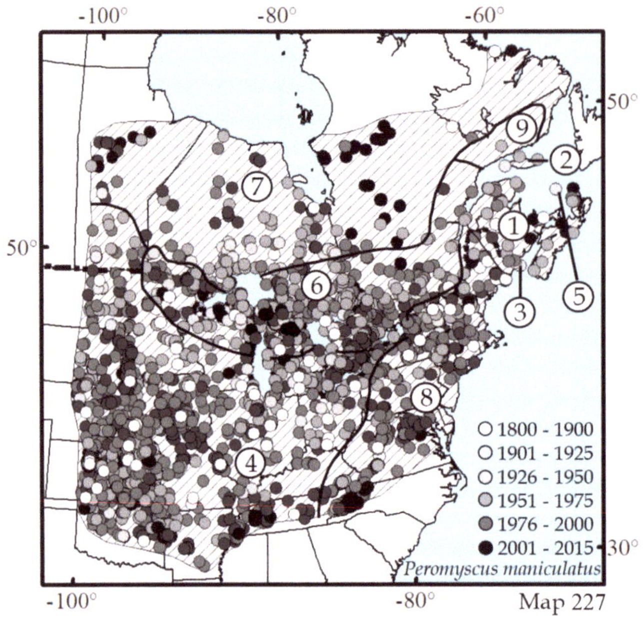


**Map 227** Distribution map of *Peromyscus maniculatus*

**Additional Literature:** Osgood (1909).

## *Peromyscus megalops* Merriam, 1898

Brown deermouse, ratón de rostro ancho

1. *P. m. auritus* Merriam, 1898. Highlands of central-eastern Guerrero and western Oaxaca.
2. *P. m. megalops* Merriam, 1898. Central-southern Oaxaca.

**Conservation:** *Peromyscus megalops* is listed as Least Concern by the IUCN (Castro-Arellano and Vázquez 2016e).

**Characteristics:** Brown deermouse is medium sized; total length 238.0–288.0 mm andskull length 31.2–35.2 mm. Dorsal pelage brown grizzly with blackish brown shades, darker toward the back and paler toward the flanks; fur of underparts creamy to pale yellow; pectoral region and armpits tawny; flanks chiefly rich tawny but with some dusky shades; broadly brown line from the whisker base and around the eyes, and halfway to the base of the ears; forehead brownish; sides of the nose in front of the whiskers grayish cinnamon; nose tip with a tiny whitish spot; ears approximately more than the 50% of the hindfoot length; forefeet and hind feet white; forearms with a narrow dusky line reaching nearly the carpal joint; tarsal joint broadly dusky and sometimes slightly extended on the upper side of the hind feet; **tail longer than the head-and-body length, coarsely haired and irregularly bicolored**; limbs whitish; pelage long and loose; **braincase large and broad with a distinct supraorbital bead**; auditory bullae relatively small; rostrum thin and long; coronoid process small and slightly high; with a groove in the frontal bones. Species group *megalops*.

**Comments:** *Peromyscus melanurus* was recognized as a distinct species (Huckaby 1980); it was previously regarded as a subspecies of *P. megalops* (Osgood 1909). *P. megalops* ranges in the highlands of central Guerrero and northern-central Oaxaca (Map 228). It can be found in sympatry, or nearly so, with *P. beatae*, *P. labecula*, *P. melanocarpus*, *P. melanurus*, *P. melanophrys*, *P. mexicanus*, and *P. winkelmanni*. *P. megalops* can be distinguished from *P. beatae* and *P. winkelmanni* by its larger size; tail proportionally longer than the head-and-body length; large and broad braincase with a distinct supraorbital bead. From *P. beatae*, by lacking a dark spot around the eye. From *P. labecula*, by having the tail without a well-defined dark dorsal narrow stripe covering only the vertebral line and larger than the head-and-body length. From *P. melanocarpus*, by having a dorsal dark area on the hind feet not extending to the base of the toes, at most to the level of the tarsus. From *P. melanurus*, by its larger size; hindfoot length greater than 28.0 mm; pectoral spot ocher; pelage longer; larger skull and less stoutly, and tail nearly unicolor. From *P. melanophrys*, by not having the following characteristics: a dark mask around the eyes; lateral lines, and a distinct supraorbital bead. From *P. mexicanus*, by having a pectoral spot; a distinct supraorbital bead; auditory bullae and molars larger relative to the skull.

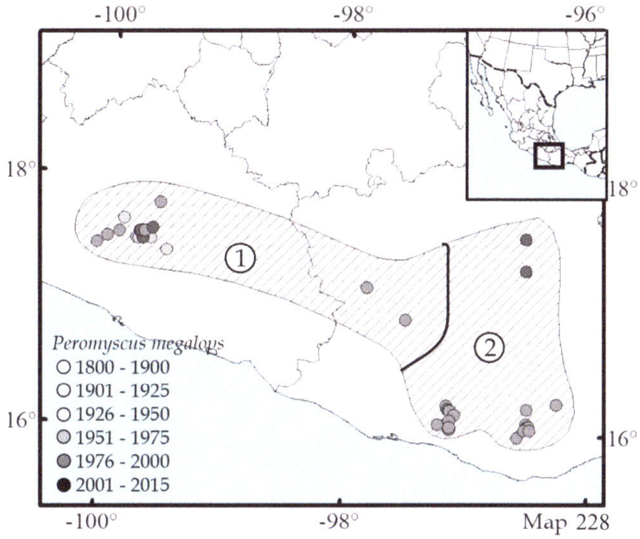


**Map 228** Distribution map of *Peromyscus megalops*

**Additional Literature:** Merriam (1898a), Musser and Carleton (2005).

## *Peromyscus mekisturus* Merriam, 1898

Puebla deermouse, ratón de Puebla

Monotypic.

**Conservation:** *Peromyscus mekisturus* is listed as Threatened in the Norma Oficial Mexicana (DOF 2019) and as Critically Endangered (Possibly Extinct; B1ab (iii); extent of occurrence less than 100 km$^2$ at only one location and under continuing decline in area and quality of habitat) by the IUCN (Álvarez-Castañeda 2018e).

**Characteristics:** Puebla deermouse is medium sized; total length 222.0–249.0 mm and skull length 23.5–25.9 mm. Dorsal pelage grayish-fulvus; fur of underparts buffy-whitish, buffy on the pectoral region and whitish on the chin, lips, and sides of the nose; ears approximately 75% of the hindfoot length; nose gray, with a small whitish fleck at the tip; color ring round eye dusk; **hind feet, except for the toes, chiefly dusky brownish, soles naked, at least medially**; **tail well haired and dusky, about 155% larger than the head-and-body length**; limbs whitish; **skull small; with short nasals and markedly constricted frontal bones;** rostrum short and narrow; zygomatic arches square but narrow, spreading anteriorly, the outer sides strongly convergent anteriorly; interorbital area without a trace of a supraorbital bead; braincase broad and rather flat; interparietal bone narrow; auditory bullae small. Species group *melanophrys*.

**Comments:** Molecular evidence (mtDNA) supported the monophyletic origin of the *melanophrys* species group and questioned that *P. mekisturus* was a valid species (Castañeda-Rico et al. 2014); however, *P. mekisturus* is still considered a full species. *Peromyscus mekisturus* ranges in southeast Puebla, México (Map 229). This species is only known from Ciudad Serdán, Puebla. *P. mekisturus* is sympatric, or nearly so, with *P. amplus*, *P. beatae*, *P. leucopus*, and *P. labecula*. *P. mekisturus* can be distinguished from all other species by having the tail about 155% larger than the head-and-body length, with dusky coloration, and by its restricted distribution, being endemic to the arid land region of Puebla.

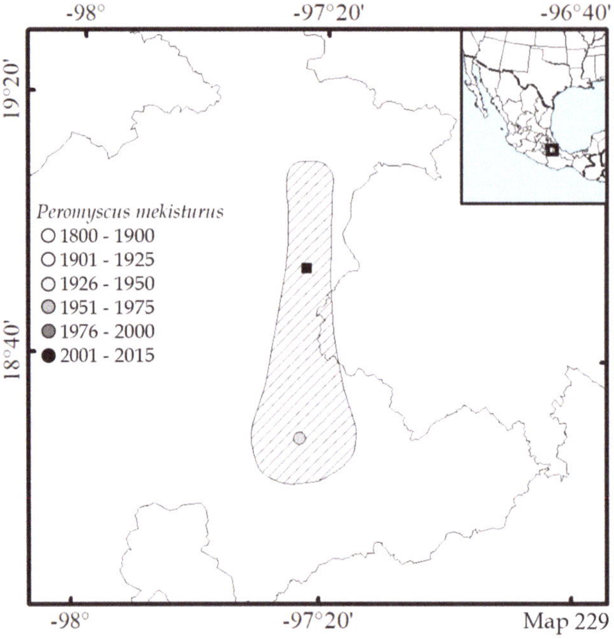


**Map 229** Distribution map of *Peromyscus mekisturus*

**Additional Literature:** Merriam (1898a), Osgood (1909), Musser and Carleton (2005), Álvarez et al. (2021).

## *Peromyscus melanocarpus* Osgood, 1904

Zempoaltepec deermouse, ratón de patas negras

Monotypic.

**Conservation:** *Peromyscus melanocarpus* is listed as Endangered (B1ab (iii): extent of occurrence less than 5000 km$^2$ at five or less locations and under continuing decline in area and quality of habitat) by the IUCN (Álvarez-Castañeda et al. 2018m).

**Characteristics:** Zempoaltepec deermouse is large sized; total length 200.0–263.0 mm and skull length 29.8–34.4 mm. Dorsal pelage dark mummy brown, slightly darker along the midline; **fur of underparts blackish grizzly with white shades**; ears approximately more than half the hindfoot length; **fore and hindfeet dusky brownish to the base of the toes**; tail longer than the head-and-body length, darker dorsally than ventrally, covered with very thin blackish hairs and small ventral spots; **absence of pectoral mammae (nipples);** rostrum thin and long; **supraorbital edge present and well marked, and a shallow depression;** interpterygoid fossa large, extending past to the posterior margin of the third upper molars; auditory bullae relatively small. Species group *megalops*.

**Comments:** *Peromyscus melanocarpus* is restricted to Sierra de Zempoaltepec, Sierra de Juárez, and Sierra Mazateca northern-central Oaxaca (Map 230). It can be found in sympatry, or nearly so, with *P. beatae*, *P. gratus*, *P. labecula*, *P. lepturus*, *P. leucopus*, *P. levipes*, *P. megalops*, *P. melanophrys*, and *P. mexicanus*. *P. melanocarpus* can be distinguished from all other species by having the fore and hindfoot toes with a dusky coloration; absence of pectoral mammae (nipples), supraorbital edge well marked and with a shallow depression. From *P. beatae* and *P. levipes*, by its larger size; without a dark spot around the eyes; tail proportionally longer than the head-and-body; braincase large and broad with a distinct supraorbital bead. From *P. gratus*, by its larger size; without a dark spot around the eyes and a lateral line; ears and auditory bullae larger. From *P. labecula* and *P. leucopus*, by having the tail larger than the head-and-body length and without a well-defined bicolored pattern. From *P. megalops*, by having a dorsal dark area on the hind feet extending to the base of the toes. From *P. melanophrys*, by not having the following characteristics: dark mask around the eyes; lateral line; distinct supraorbital bead. From *P. mexicanus*, by having a pectoral spot; distinct supraorbital bead; larger auditory bullae, and molars in proportion with the skull.

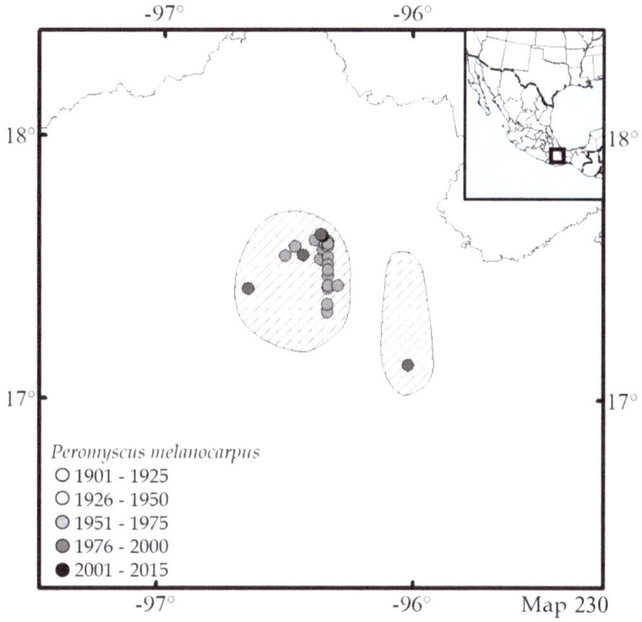


**Map 230** Distribution map of *Peromyscus melanocarpus*

**Additional Literature:** Osgood (1904, 1909), Huckaby (1980), Rickart and Robertson (1985), Musser and Carleton (2005).

## *Peromyscus melanophrys* (Coues, 1874)

Black-eyed deermouse, ratón de máscara

Monotypic.

**Conservation:** *Peromyscus melanophrys* is listed as Least Concern by the IUCN (Vázquez and Álvarez-Castañeda 2016).

**Characteristics:** Black-eyed deermouse is large sized; total length 190.0–297.0 mm and skull length 27.3–32.3 mm. Dorsal pelage varying from orange rufous to cinnamon, with a poorly marked facial mask; fur of underparts grayish, gray at the base and white at the tip in some specimens; feet white to pale brown; ears relatively small; tail markedly bicolored and much longer than the head-and-body length, dark brown dorsally and pale gray ventrally; **supraorbital region wide and flat, subtriangular, with straight, well-developed ridges forming a wide interorbital constriction; braincase globose, with a narrow mastoid region; rostrum and nasals relatively short; auditory bullae medium sized;** nasal bones long with the posterior end blunt and square or U shaped. Species group *melanophrys*.

**Comments:** The subspecies *Peromyscus melanophrys coahuilensis*, *P. m. consobrinus*, *P. m. xenurus*, and *P. m. zamorae* are considered part of the species *P. zamorae*, and *P. m. micropus* of *P. micropus* (López-González et al. 2019). *P. melanophrys* ranges in Estado de México, Ciudad de Mexico, Morelos, Guerrero, Puebla, Oaxaca, and southwestern Chiapas (Map 231) and can be found in sympatry, or nearly so, with *P. beatae*, *P. boylii*, *P. felipensis*, *P. gratus*, *P. labecula*, *P. leucopus*, *P. megalops*, *P. melanocarpus*, and *P. melanurus*. *P. melanophrys* can be distinguished from all other species, except for *P. leucurus* and *P. zamorae*, by its black orbital ring and a conspicuous gray mask; tail larger than the head-and-body length, with annulations covered with short stiff hairs and frequently white at the tip; supraorbital border sharply angled but normally not molded. From *P. beatae*, *P. boylii*, *P. gratus*, *P. labecula*, *P. leucopus*, *P. megalops*, *P. melanocarpus*, and *P. melanurus*, by its larger size; tail longer than the head-and-body length; auditory bullae proportionally greatly inflated. From *P. beatae* and *P. boylii*, by its larger size and the darker spot around the eye. From *P. felipensis* and *P. gratus*, by having smaller ears, less than the hindfoot length; auditory bullae not of larger size. From *P. labecula* and *P. leucopus*, by having the tail larger than the head-and-body length and without a well-defined bicolored pattern. From *P. megalops*, *P. melanocarpus*, and *P. melanurus*, by having a darker spot around the eyes and without a pectoral spot.

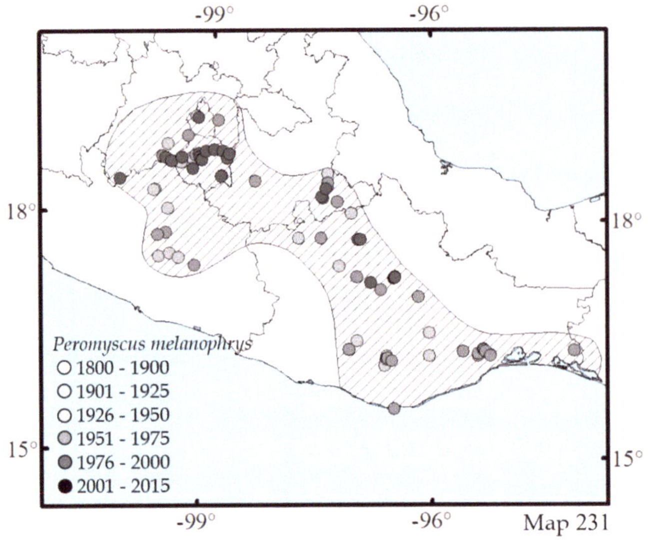


**Map 231** Distribution map of *Peromyscus melanophrys*

**Additional Literature:** Osgood (1909), Hooper (1957), Musser and Carleton (2005), Bradley et al. (2007).

## *Peromyscus melanotis* Allen and Chapman, 1897

Black-eared deermouse, ratón de orejas negras

Monotypic.

**Conservation:** *Peromyscus melanotis* is listed as Least Concern by the IUCN (Álvarez-Castañeda et al. 2016q).

**Characteristics:** Black-eared deermouse is medium sized; total length 140.0–170.0 mm and skull length 25.8–27.5 mm. Dorsal pelage tawny ochraceous to paler yellow brown with the back darker; flank line ocher, very distinct; fur of underparts whitish; **ears dusky brownish to black, with the edges white,** approximately 60% of the hindfoot length; orbital ring very narrow and dusky; whiskers with a small dusky spot at the base; tail well haired, slightly penicillate and shorter than the head-and-body length, bicolored with a sooty brownish dorsal stripe one-third of the tail diameter; limbs whitish; rostrum sharper and nasal bones longer. Species group *maniculatus*.

**Comments:** *Peromyscus melanotis* ranges from Arizona, Chihuahua, southern Coahuila, and Nuevo León south through Jalisco, Michoacán, Estado de México, Puebla, and central Veracruz (Map 232). It can be found in sympatry, or nearly so, with *P. amplus*, *P. beatae*, *P. boylii*, *P. difficilis*, *P. gratus*, *P. hylocetes*, *P. kilpatricki*, *P. labecula*, *P. laceianus*, *P. leucopus*, *P. levipes*, *P. melanophrys*, *P. pectoralis*, *P. perfulvus*, *P. polius*, *P. sagax*, *P. schmidlyi*, and *P. sonoriensis. P. melanotis* can be distinguished from all other species by its small size; short tail and black preauricular hairs at the anterior base of the ears; the preauricular hairs usually form a black hair tuft, but this tuft is very small in many specimens, so it can be missed without careful attention. From *P. amplus*, *P. difficilis*, *P. gratus*, *P. laceianus*, *P. pectoralis*, and *P. sagax*, by having the ears smaller than the hind feet; tail less haired; auditory bullae proportionally less inflated. From *P. beatae*, *P. boylii*, *P. hylocetes*, *P. kilpatricki*, *P. levipes*, *P. polius*, and *P. schmidlyi*, by having the tail hairy, shorter than or equal to the head-and-body length and clearly bicolored, with the dark dorsal narrow stripe covering only the vertebral line; without a lateral line in the flanks and without a supraorbital border. From *P. labecula* and *P. sonoriensis* (extremely similar morphologically), by having a black preauricular hair tuft; braincase broader and more rounded; interorbital space narrow; prezygomatic notch less prominent; slightly smaller auditory bullae; rostrum clearly longer and more slender. From *P. leucopus*, by having a dark hair tuft at the base of the ears; dorsal pelage darker than on the flanks and nasal length equal to 11.0 mm. From *P. melanophrys* and *P. perfulvus*, by its smaller size; without a conspicuous gray mask; supraorbital border not sharply angled.

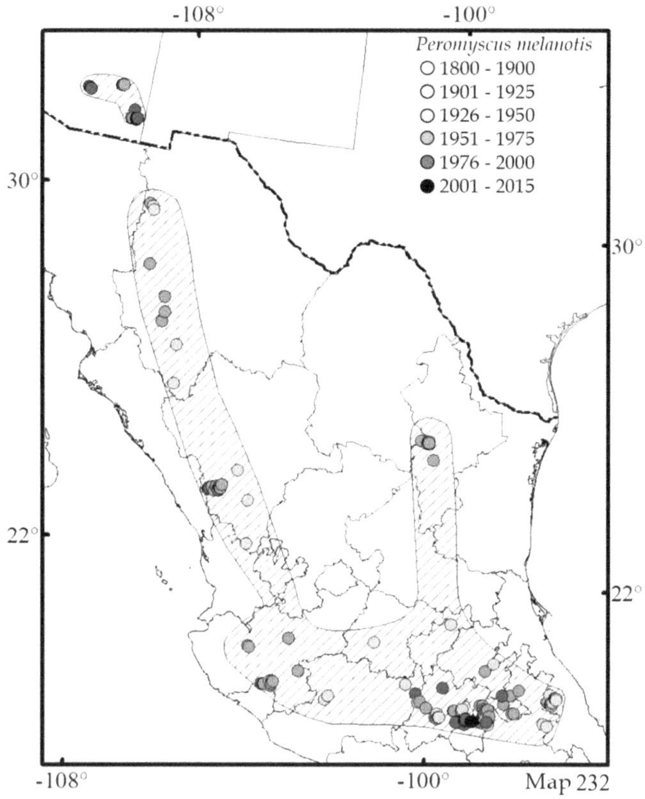


**Map 232** Distribution map of *Peromyscus melanotis*

**Additional Literature:** Osgood (1909), Hooper (1957), Ramírez-Pulido (1969), Avise et al. (1979), Hooper (1947), Álvarez-Castañeda (2005a), Musser and Carleton (2005), Bradley et al. (2019), Greenbaum et al. (2019).

## *Peromyscus melanurus* Osgood, 1909

Black-tailed deermouse, ratón de cola negra

Monotypic.

**Conservation:** *Peromyscus melanurus* is listed as Endangered (B1ab (iii): extent of occurrence less than 5000 km$^2$ at five or less locations and under continuing decline in quality of habitat) by the IUCN (Álvarez-Castañeda et al. 2019b).

**Characteristics:** Black-tailed deermouse is medium sized; total length 238.0–278.0 mm and skull length 29.9–34.3 mm. Dorsal pelage grayish brown to bright tawny ochraceous; fur of underparts creamy to pale yellow, usually without a fulvous pectoral area; flanks chiefly rich tawny but with scarce dusky shades; brown line from the base of the whiskers to around the eyes and halfway to the base of the ears; forehead brownish; sides of the nose in front of the whiskers grayish cinnamon; tip of the nose with a tiny whitish spot; ears approximately more than the 50% of the hindfoot length; forefeet and hind feet white; tarsal joint brown; **tail longer than the head-and-body length, coarsely haired, and unicolor**; limbs whitish; pelage short and loose; **skull with a medium-large and broad braincase, a distinct supraorbital bead, and more stoutly built**; auditory bullae relatively small; rostrum thin and long; coronoid process small and only slightly high; with a groove in the frontal bones. Species group *megalops*.

**Comments:** *Peromyscus melanurus* was recognized as a distinct species (Huckaby (1980); it was previously regarded as a subspecies of *P. megalops* (Osgood 1909). *P. melanurus* ranges in the Pacific slopes of Sierra Madre del Sur of Oaxaca (Map 233). It can be found in sympatry, or nearly so, with *P. beatae*, *P. gratus*, *P. labecula*, *P. leucopus*, *P. levipes*, *P. megalops*, *P. melanophrys*, and *P. mexicanus*. *P. melanurus* can be distinguished, by its larger size; a brown line from the base of the whiskers to around the eyes and halfway to the base of the ears; tail longer than the head-and-body length; coarsely haired; unicolor and with a broad braincase with a distinct supraorbital bead and more stoutly built. From *P. beatae* and *P. levipes*, by its larger size, without a dark spot around the eyes; tail proportionally longer than the head-and-body length; large broad braincase with a distinct supraorbital bead. From *P. gratus*, by its larger size by its larger size; without a dark spot around the eyes and lateral line; ears and smaller auditory bullae. From *P. labecula* and *P. leucopus*, by having the tail without a well-defined bicolored pattern and equal to or larger than the head-and-body length. From *P. megalops*, by its smaller size; hindfoot length less than 28.0 mm; pelage much shorter; smaller skull; tail nearly unicolor. From *P. melanophrys*, by not having the following characteristics: a dark mask around the eyes; lateral lines; and a distinct supraorbital edge. From *P. mexicanus*, by its larger size; frontal bones more distinctly wider; larger teeth and auditory bullae about the same size.

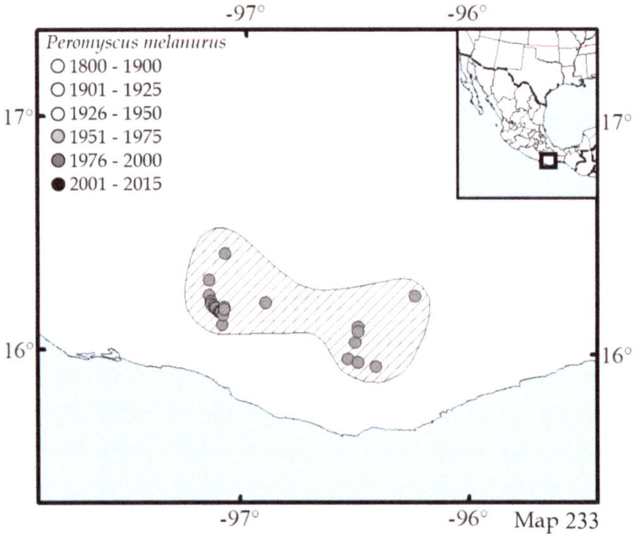


**Map 233** Distribution map of *Peromyscus melanurus*

**Additional Literature:** Musser and Carleton (2005).

## *Peromyscus mexicanus* (Saussure, 1860)

Mexican deermouse, ratón mexicano

1. *P. m. azulensis* Goodwin, 1956. Restricted to Cerro Azul, Tehuantepec Isthmus, Oaxaca.
2. *P. m. mexicanus* (Saussure, 1860). From southeastern San Luis Potosí southward through Veracruz.
3. *P. m. saxatilis* Merriam, 1898. From Chiapas.
4. *P. m. teapensis* Osgood, 1904. From southern Veracruz, Tabasco, and northern Chiapas.

**Conservation:** *Peromyscus mexicanus* is listed as Least Concern by the IUCN (Reid and Pino 2016).

**Characteristics:** Mexican deermouse is medium sized; total length 209.0–239.0 mm and skull length 26.7–32.2 mm. Dorsal pelage cinnamon to darker brown on the back; fur of underparts whitish to cream; pectoral region often yellowish reddish; flanks with a lateral reddish line; blackish spot on the base of the whiskers; ears dusky brown with the edge faint whitish, approximately 70% of the hindfoot length; forefeet and carpal joints white and proximal one-half of the forearms dusky overlaid by rufescent shades; hind feet white and tarsal joints dusky brown; tail slightly longer than the head-and-body length, practically naked with few hairs and coarsely annulated (about 17 annulations/cm); rostrum thin and long; supraorbital ridge edge poorly developed; molariform and auditory bullae proportionally small. Note: to distinguish *P. mexicanus* from *P. gymnotis* in southern Chiapas, a direct comparison between specimens of the two species is needed. Species group *mexicanus*.

**Comments:** *Peromyscus angelensis* and *P. totontepecus* were recognized as a distinct species (Hernández-Canchola et al. 2022; Pérez-Montes et al. 2022); it was previously regarded as a subspecies of *P. mexicanus* (Huckaby 1980). *P. mexicanus* ranges from Veracruz and southern San Luis Potosí south through northern Chiapas (Map 234). It can be found in sympatry, or nearly so, with *P. aztecus*, *P. beatae*, *P. collinus*, *P. furvus*, *P. labecula*, *P. latirostris*, *P. leucopus*, *P. levipes*, *P. ochraventer*, and *P. totontepecus*. *P. mexicanus* can be distinguished from all other species of *Peromyscus*, except for *P. angelensis* and *P. totontepecus*, by having the tail slightly longer than the head-and-body length, practically naked with few hairs and coarsely annulated (about 17 annulations/cm); relatively small molars and auditory bullae. From *P. aztecus*, *P. beatae*, *P. levipes*, *P. labecula*, and *P. leucopus*, by its larger size and tail proportionally longer than the head-and-body length; interorbital areas less hourglass shaped. From *P. collinus*, by having smaller ears; less than the hindfoot length; absence of dark spot around the eyes and pectorals; lateral line and auditory bullae in proportion with the skull. From *P. furvus*, *P. latirostris*, and *P. ochraventer*, by having a beaded or rigged supraorbital border and less developed lophs in the molars. From *P. totontepecus*, by its small size in somatic and skull measurements and general paler coloration.

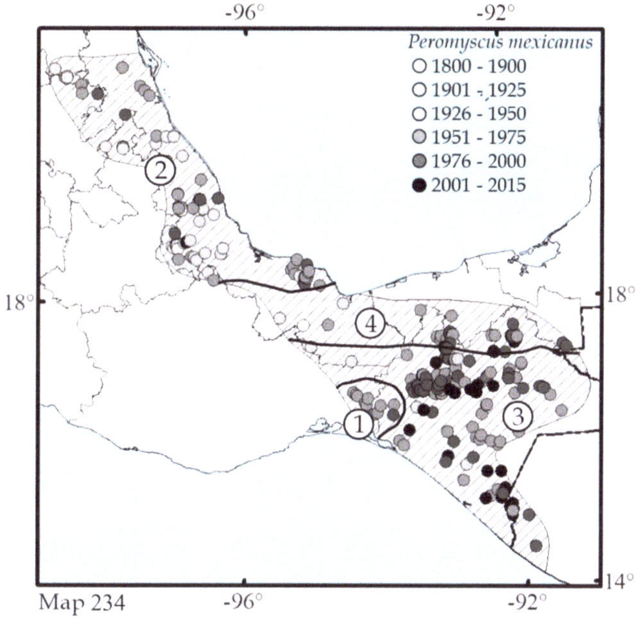


**Map 234** Distribution map of *Peromyscus mexicanus*

**Additional Literature:** Osgood (1909), Hooper (1957), Hooper (1968), Musser and Carleton (2005), Trujano-Álvarez and Álvarez-Castañeda (2010).

## *Peromyscus micropus* Baker, 1952

Small-footed deermouse, ratón de patas pequeñas

Monotypic.

**Conservation:** *Peromyscus micropus* is listed (as subspecies of *Peromyscus melanophrys*) as Least Concern by the IUCN (Vázquez and Álvarez-Castañeda 2016).

**Characteristics:** Small-footed deermouse is medium sized; total length 205.0–280.0 mm, and skull length 27.3–32.1 mm. Dorsal pelage orange rufous to tawny or buff, **with a well marked grayish face mask;** fur of underparts white, sometimes gray at the base; feet white to pale brown, medium sized; ears relatively small; tail markedly bicolored and longer than the head-and-body length, brown dorsally and whitish ventrally; **supraorbital region with straight, well-developed ridges extending from the temporal area to the interorbital constriction and the posterior end of the nasals; frontal region flattened and subtriangular; interorbital constriction and mastoid region narrow; nasals relatively long; rostrum relatively wide; auditory bullae small and not inflated;** nasal bones long and tapering or U shaped at the posterior end. Species group *melanophrys*.

**Comments:** *Peromyscus micropus* was recognized as a distinct species (López-González et al. 2019); it was previously regarded as a subspecies of *P. melanophrys* (Baker, 1952). *P. micropus* ranges from eastern Durango and Nayarit south through northern Michoacán, Guanajuato, and eastern San Luis Potosí (Map 235). It can be found in sympatry, or nearly so, with *P. boylii*, *P. difficilis*, *P. gratus*, *P. labecula*, *P. leucopus*, *P. leucurus*, and *P. melanophrys*. *P. micropus* can be distinguished from all other species, except for *P. melanophrys* and *P. leucurus*, by its black orbital ring and a conspicuous gray mask; tail larger than the head-and-body length, with annulations, covered with short stiff hairs and frequently white at the tip; supraorbital border sharply angled but normally not molded. From *P. boylii*, *P. gratus*, *P. labecula*, and *P. leucopus*, by its larger size; tail longer than the head-and-body length; auditory bullae proportionally greatly inflated. From *P. boylii*, by its larger size and a darker spot around the eyes. From *P. difficilis*, and *P. gratus*, by having smaller ears, less than the hindfoot length; auditory bullae not greater in size. From *P. labecula*, and *P. leucopus*, by having the tail without a well-defined bicolored pattern and larger than the head-and-body length. From *P. melanophrys* and *P. leucurus*, by having the supraorbital ridges well developed, extending from the temporal area to the interorbital constriction and the posterior end of the nasals; interorbital constriction and mastoid region narrow; auditory bullae small and not inflated; braincase not domed.

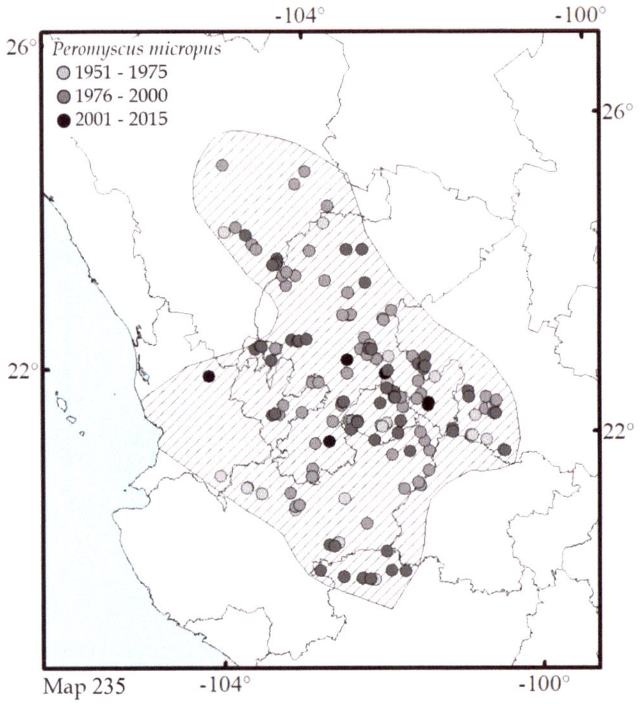


**Map 235** Distribution map of *Peromyscus micropus*

**Additional Literature:** Hooper (1955), Carleton (1989), Musser and Carleton (2005).

## ***Peromyscus nasutus*** (Allen, 1891)

Northern rock deermouse, ratón de las rocas del norte

1. *P. n. griseus* Benson, 1932. Restricted to lava fields in central New Mexico.
2. *P. n. nasutus* (Allen, 1891). From southeast Utah and northern-central Colorado south through Arizona, New Mexico, southern Texas, and central Coahuila.
3. *P. n. penicillatus* Mearns, 1896. Restricted to a strip along the United States–Mexico border from El Paso, Texas to Sierra del Carmen, Coahuila.

**Conservation:** *Peromyscus nasutus* is listed as Least Concern by the IUCN (Lacher et al. 2016j).

**Characteristics:** Northern rock deermouse is medium sized; total length 196.0–216.0 mm and skull length 27.0–28.5 mm. Dorsal pelage from tawny to ocher, more intensely colored on the flanks than on the back; fur of underparts whitish, sometimes with silver hairs; ears equal to or greater than the hindfoot length and naked; tail bicolored, longer than the head-and-body length; **rostrum elongated**; auditory bullae well developed; supraorbital edges biconcave. Species group *truei*.

**Comments:** *Peromyscus nasutus* was reinstated as species (Zimmerman et al. 1975, 1978; Avise et al. 1979); it was previously regarded as a subspecies of *P. difficilis* (Hoffmeister and De la Torre 1961). It ranges from southern Utah and northern-central Colorado southwest through northern Sonora, Chihuahua, and Durango, and southeast through Texas, Coahuila, and Nuevo León (Map 236). It can be found in sympatry, or nearly so, with *P. boylii*, *P. labecula*, *P. laceianus*, *P. leucopus*, *P. sonoriensis*, and *P. truei*. It can be distinguished from *P. boylii*, *P. labecula*, *P. leucopus*, and *P. sonoriensis* by its larger size; ear longer than the hind feet; tail longer than the head-and-body length, auditory bullae proportionally greatly inflated. From *P. laceianus*, by having the ear length equal to or greater than the hindfoot length and larger auditory bullae, globe shaped and its length greater than 18% of the skull length. From *P. truei*, by having the back often grayish-black; lateral line with ochraceous shades; top of the head often grayish; larger ears, greater than 24.0 mm; a shorter toothrow; auditory bullae proportionally greatly inflated and larger.

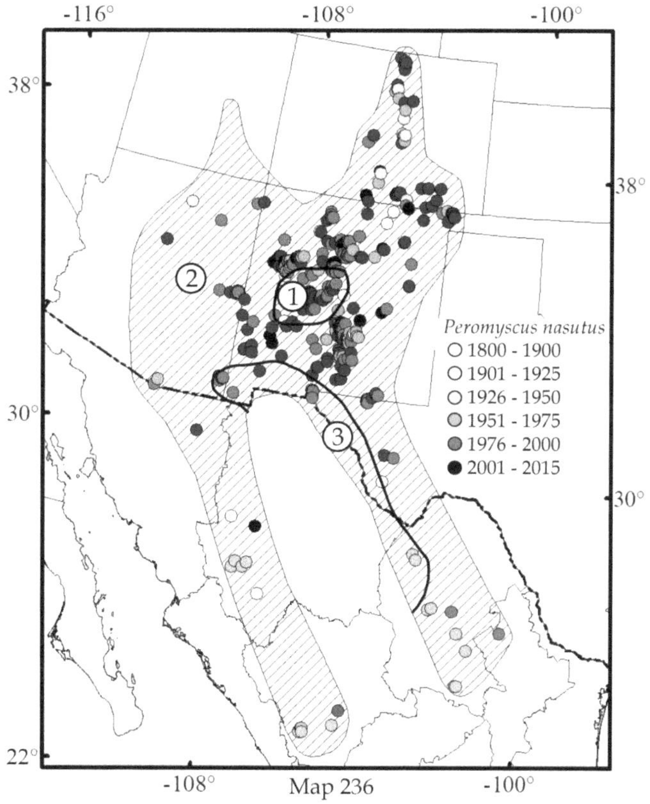


**Map 236** Distribution map of *Peromyscus nasutus*

**Additional Literature:** Osgood (1909), Hooper (1957), Janecek (1990), DeWalt et al. (1993a), Bradley et al. (1999), Musser and Carleton (2005).

## ***Peromyscus ochraventer*** Baker, 1951

Brown belly deermouse, ratón de vientre pardo

Monotypic.

**Conservation:** *Peromyscus ochraventer* is listed as Endangered (B1ab (iii): extent of occurrence less than 5000 $km^2$ at five or less locations and under continuing decline in area and quality of habitat) by the IUCN (Álvarez-Castañeda 2019a).

**Characteristics:** Brown belly deermouse is medium sized; total length 227.0–249.0 mm andskull length 28.7–31.9 mm. Dorsal pelage tawny ocher more intense on the flanks and interspersed with dark shades on the back; **fur of underparts yellowish cinnamon, intense in the pectoral region;** ears approximately 80% of the hindlimb length; cheeks, sides of the neck, shoulders and upper forelegs paler, ochraceous buff to ochraceous orange; tail shorter than the head-and-body length, not clearly bicolored but darker dorsally than ventrally with a flaky appearance; feet white; skull without a beaded or ridged supraorbital border; rostrum expanded anteriorly with almost parallel sides; interorbital area smoothly hourglass shaped; no supraorbital edge; rostrum edges almost parallel. Species group *furvus*.

**Comments:** *Peromyscus ochraventer* is restricted to moist forests of southern Tamaulipas and San Luis Potosí (Map 237). It can be found in sympatry, or nearly so, with *P. collinus*, *P. labecula*, *P. latirostris*, *P. levipes*, and *P. mexicanus*. *P. ochraventer* can be distinguished from all other species by having distinctively brownish underparts; rostrum expanded anteriorly with its sides almost parallel; upper molars with the anterior loph larger. From *P. collinus*, by having less dark upperparts; flanks more ocher; fur of underparts yellowish cinnamon and smaller auditory bullae. From *P. labecula*, by having the tail without a well-defined dark dorsal narrow stripe covering only the vertebral line and equal to or larger than the head-and-body length. From *P. latirostris*, by having a less well-developed style in the minor fold of the first molars. From *P. levipes*, by having the tail less markedly bicolored and the rostrum expanded anteriorly with almost parallel sides. From *P. mexicanus*, by lacking a beaded or rigged supraorbital border and strongly developed lophs that extend uninterruptedly to the respective style.

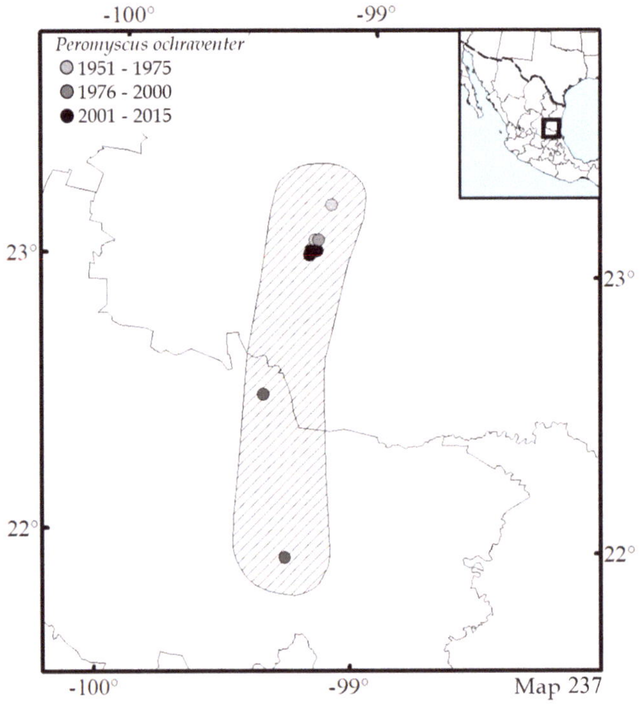


**Map 237** Distribution map of *Peromyscus ochraventer*

**Additional Literature:** Baker (1951), Huckaby (1980), Musser and Carleton (2005), Segura-Trujillo et al. (in press).

## ***Peromyscus pectoralis*** Osgood, 1904

White-ankled deermouse, ratón de ancas blancas

1. *P. p. pectoralis* Osgood, 1904. From southern Chihuahua south through northern Jalisco, central Guanajuato, Querétaro, and Hidalgo.
2. *P. p. zimmermani* Bradley et al., 2015. From northern Coahuila south through northeastern Jalisco and northwestern Guanajuato.

**Conservation:** *Peromyscus pectoralis* is listed as Least Concern by the IUCN (Lacher et al. 2016k).

**Characteristics:** White ankle deermouse is medium sized; total length 185.0–205.0 mm and skull length 25.3–28.0 mm. Dorsal pelage pale ocher to yellowish ocher grizzly with darker shades; fur of underparts whitish cream; some specimens with a yellowish pectoral spot; ears approximately 75% of the hindfoot length; **hindfoot sole somewhat hairy on the proximal third**; soles with six tubercles; tail equal to or longer than the head-and-body length, slightly bicolored; limbs and ankles whitish; braincase less vaulted; **molars clearly smaller**; premaxillae usually ending slightly beyond the even nasals; interparietal bone relatively large. Species group *truei*.

**Comments:** *Peromyscus laceianus* and *P. collinus* were recognized as distinct species (Bradley et al. 2015; Hernández-Canchola et al. 2022); both were previously considered subspecies of *P. pectoralis* (Bailey 1906). The populations of *P. pectoralis laceianus* and *P. p. pectoralis* with ranges in the Altiplanicie Mexicana were described as the new subspecies *P. p. zimmermani* (Bradley et al. 2015). *P. pectoralis* ranges from Río Bravo, northern Chihuahua, and Coahuila south through the Altiplanicie Mexicana and Sierra Madre Oriental to Nayarit, Jalisco, Guanajuato, Querétaro, and western Hidalgo (Map 238). It can be found in sympatry, or nearly so, with *P. amplus*, *P. collinus*, *P. difficilis*, *P. eremicus*, *P. hooperi*, *P. labecula*, *P. latirostris*, *P. leucopus*, *P. levipes*, *P. zamorae*, *P. melanotis*, *P. mexicanus*, and *P. ochraventer*. *P. pectoralis* can be distinguished from *P. amplus* and *P. difficilis* by having the ear length equal to or smaller than the hindfoot length; auditory bullae relatively smaller and not globe shaped, its length less than 18% of the skull length. From *P. collinus*, by having the upperparts lighter and less reddish; tarsi lighter; smaller size; narrow rostrum, maxillary bone, and interorbital region; smaller palatine foramen, molar toothrow, and auditory bullae. From *P. eremicus*, by having two pairs of mammae (nipples); ear length smaller than the hindfoot length; with a lateral line in the flanks; dark spot around the eyes; pectoral spot; first and second upper molars with accessory cusps and auditory bullae larger and more orbicular. From *P. hooperi*, by its smaller size and the presence of mesolophs and anteroloph. From *P. labecula*, *P. leucopus*, and *P. melanotis*, by having the tail without a well-defined bicolored pattern and equal to or larger than the head-and-body length. From *P. levipes*, by having a larger tail, ears equal to or slightly less than the hindfoot length, a darker spot around the eyes and a pectoral spot, auditory bullae larger. From *P. latirostris*, *P. mexicanus*, *P. ochraventer*, and *P. zamorae*, by having larger ears, equal to or larger than the hindfoot length, and auditory bullae greater in size.

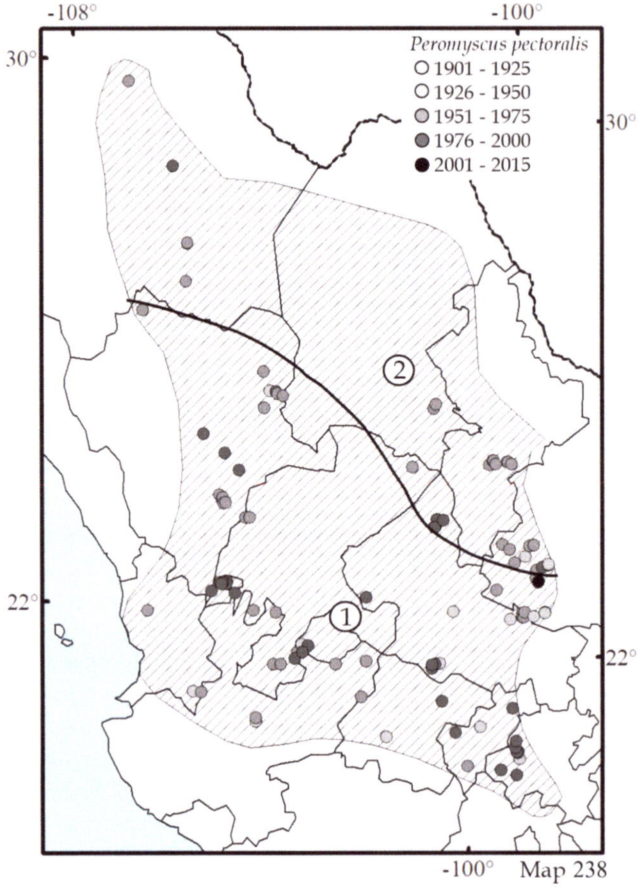


**Map 238** Distribution map of *Peromyscus pectoralis*

**Additional Literature:** Osgood (1909), Schmidly (1972, 1974), Kilpatrick and Zimmerman (1976), Musser and Carleton (2005).

## ***Peromyscus perfulvus*** Osgood, 1945

Tawny deermouse, ratón leonado

Monotypic.

**Conservation:** *Peromyscus perfulvus* is listed as Least Concern by the IUCN (Álvarez-Castañeda et al. 2016p).

**Characteristics:** Tawny deermouse is medium sized; total length 248.0–281.0 mm and skull length 31.9–36.8 mm. Dorsal pelage orange, flanks not contrasting with the back, which have long and silky hair; fur of underparts grayish white; ears approximately 80% of the hindfoot length and brown; hind feet white but with a dark coloration on the tarsi, whereas in other groups of species the feet are uniformly white; **tail longer than the head-and-body length, uniformly sepia, slightly paler ventrally than dorsally, and well penciled at the tip;** hind feet white but with a colored tarsi; toes whitish; fifth toe relatively long; rostrum thin and long; supraorbital edge absent. Species group *melanophrys*.

**Comments:** *Peromyscus chrysopus* was recognized as a distinct species (Castañeda-Rico et al. 2014; López-González et al. 2019); it was previously regarded as a subspecies of *P. perfulvus* (Hooper 1955). *P. perfulvus* ranges in the Río Balsas Basin to Michoacán, Guerrero, and southwestern Estado de México (Map 239). It can be found in sympatry, or nearly so, with *P. chrysopus*, *P. hylocetes*, *P. labecula*, and *P. melanophrys*. *P. perfulvus* can be distinguished from *P. melanophrys* by its tail, longer, hairy, and uniformly brownish; ears brown instead of blackish; hind feet white but with colored tarsi. From *P. hylocetes* and *P. levipes*, by its larger size; ears longer than the hind feet; tail longer than the head-and-body length; auditory bullae proportionally greatly inflated. From *P. labecula*, by having the tail without a well-defined dark dorsal narrow stripe covering only the vertebral line and equal to or larger than the head-and-body length. From *P. chrysopus*, by having the first and second upper molars with ectostylids absent and mesolophs absent or short, not reaching the labial border of teeth; on average, nasals exceeding the premaxilla. From *Osgoodomys banderanus*, by having the tail with hairs, not scaled, and penciled at the tip; ears brownish; with a facial mask gray and a sharp supraorbital border.

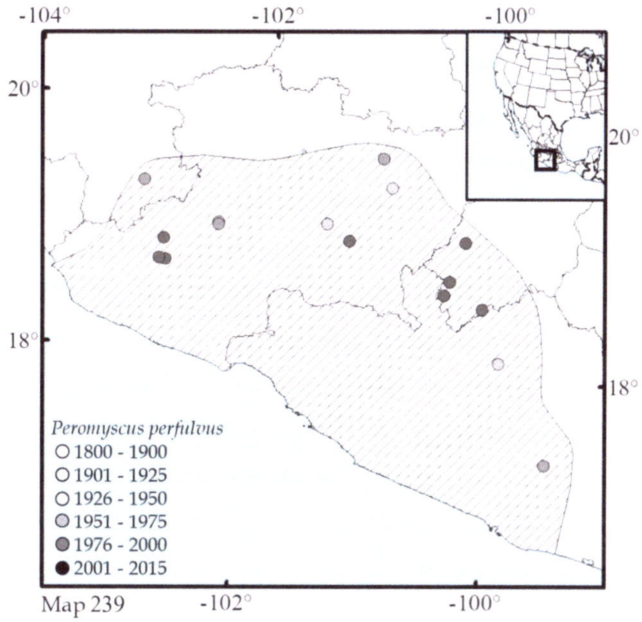


**Map 239** Distribution map of *Peromyscus perfulvus*

**Additional Literature:** Osgood (1945), Álvarez and Hernández-Chávez (1990), Musser and Carleton (2005), Sánchez-Hernández et al. (2009).

## ***Peromyscus polionotus*** (Wagner, 1843)

Beach mouse, ratón del sureste

1. *P. p. albifrons* Osgood, 1909. Restricted to northwestern Florida.
2. *P. p. allophrys* Bowen, 1968. Restricted to the northwestern coast of Florida.
3. *P. p. ammobates* Bowen, 1968. Restricted to the coastal southwest of Alabama.
4. *P. p. colemani* Schwartz, 1954. Eastern Mississippi, Alabama, Georgia, and South and North Carolina.
5. *P. p. decoloratus* Howell, 1939. Restricted to Volucia county, Florida.
6. *P. p. griseobracatus* Bowen, 1968. Restricted to the northwestern coast of Florida.
7. *P. p. leucocephalus* Howell, 1920. Restricted to Santa Rosa Island, Florida.
8. *P. p. lucubrans* Schwartz, 1954. Restricted to central and western South Carolina.

9. *P. p. niveiventris* (Chapman, 1889). Restricted to the central-eastern coast of Florida.
10. *P. p. peninsularis* Howell, 1939. Restricted to the northwestern coast of Florida.
11. *P. p. phasma* Bangs, 1898. Restricted to the type locality, northeastern coast of Florida.
12. *P. p. polionotus* (Wagner, 1843). Restricted to southern Alabama, Georgia, and northwestern Florida.
13. *P. p. rhoadsi* Bangs, 1898. Restricted to central-northern Florida.
14. *P. p. subgriseus* (Chapman, 1893). Restricted to northern-central Florida and southern Georgia.
15. *P. p. sumneri* Bowen, 1968. Restricted to central-northwest Florida.
16. *P. p. trissyllepsis* Bowen, 1968. Restricted to Foster's Island, southern Alabama.

**Conservation:** *Peromyscus polionotus ammobates*, *P. p. phasma*, *P. p. allophrys*, *P. p. trissyllepsis*, and *P. p. peninsularis* are listed as Endangered by the United States Endangered Species Act (FWS 2022) and *Peromyscus polionotus* as Least Concern by the IUCN (Cassola 2016cr).

**Characteristics:** Beach mouse is medium sized; **total length 125.0–137.0 mm** and skull length 22.2–24.1 mm. Dorsal pelage uniformly brownish fawn; **fur of underparts whitish, clearly contrasting with the back and salty gray at the base**; chin and throat with hairs white at the base; sides of the face and orbital region brighter fawn; ears dusky with the edges pale, subauricular hair tuft fawn and whitish, approximately 70% of the hind feet; tail shorter than the head-and-body length, bicolored, brownish black dorsally and white ventrally; hind feet whitish; shorter palatine slits; auditory bullae slightly larger. Species group *maniculatus*.

**Comments:** *Peromyscus polionotus* ranges from northeastern Mississippi to western South Carolina south through Alabama, Georgia, and the Florida Peninsula (Map 240). *P. polionotus* cannot be found in sympatry, or nearly so, or near the range of any other species of the *maniculatus* species group; however, it could be found with *P. gossypinus*, *P. leucopus*, and *Podomys floridanus*. *P. polionotus* can be distinguished from all other species of the *maniculatus* species group by having slightly larger auditory bullae; skull much smaller and total length on average less than 140.0 mm (Osgood 1909; Hall 1981). From *P. gossypinus*, by its smaller size, on average the following measurements less than the following: total length 154.0 mm, hindfoot length 19.0 mm, and tail length 60.0 mm. From *P. leucopus*, it is very hard to differentiate between these species in some areas; tail with a darker paravertebral area. From *Podomys floridanus*, by having a longer tail, greater than 80% of the head-and-body length, and not having orange shading on the cheeks, shoulders, and flanks; without hypsodont molars with small accessory tubercles and without interorbital shelf.

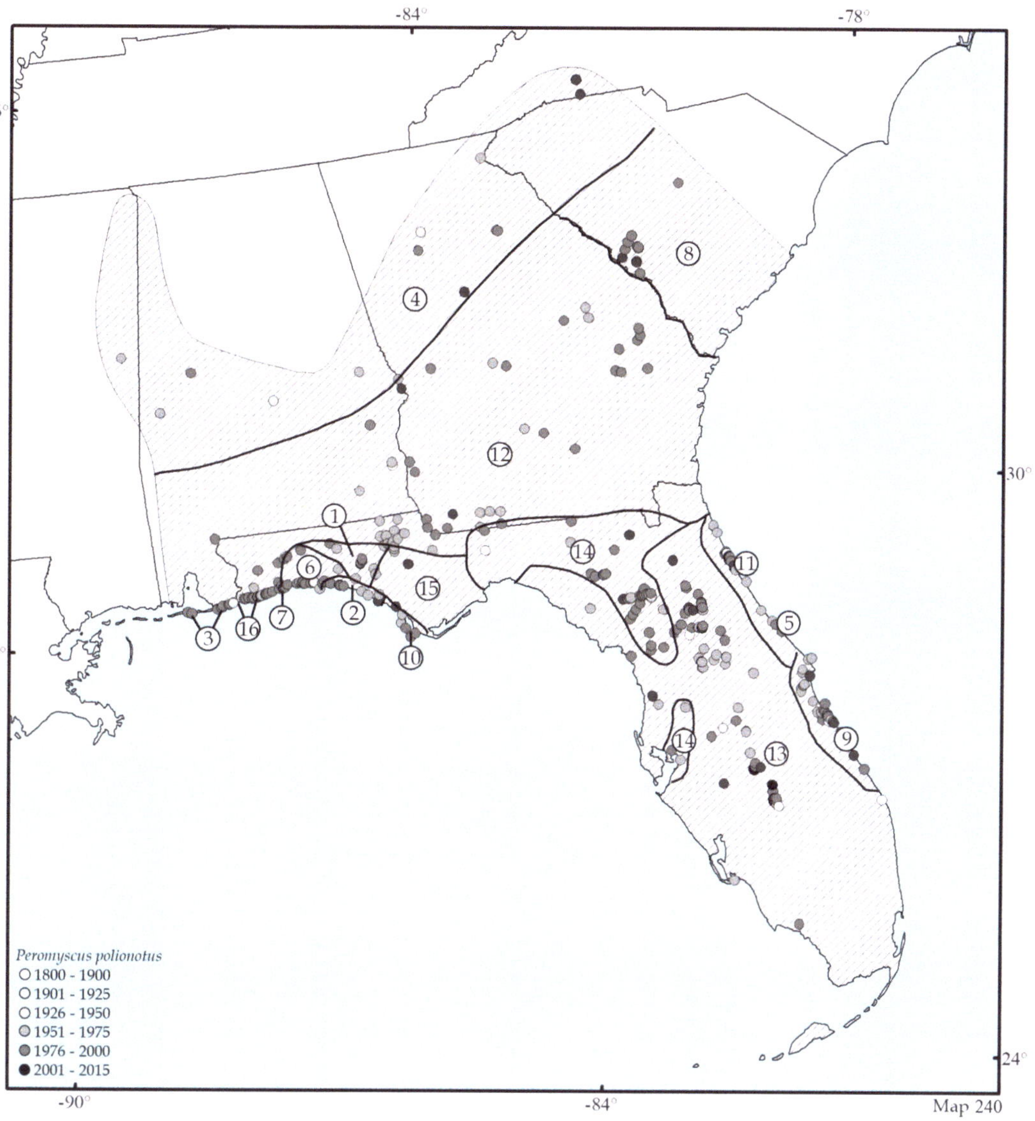


**Map 240** Distribution map of *Peromyscus polionotus*

**Additional Literature:** Musser and Carleton (2005), Bradley et al. (2019), Greenbaum et al. (2019).

## *Peromyscus polius* Osgood, 1904

Chihuahuan deermouse, ratón de Chihuahua

Monotypic.

**Conservation:** *Peromyscus polius* is listed as Near Threatened by the IUCN (Álvarez-Castañeda et al. 2020).

**Characteristics:** Chihuahuan deermouse is medium sized; total length 210.0–234.0 mm and skull length 210.0–234.0 mm. Dorsal pelage grayish broccoli brown; lateral line narrow and with ocher shades; fur of underparts whitish; head slightly more grayish than the body, particularly on the cheeks; orbital ring narrow and dusky; ears approximately 85% of the hind-foot length, grayish dusky, narrow margins with buffy white shades mixed with a grayish and buffy tuft at the base; tail longer than the head-and-body length, bicolored, darker dorsally and whitish ventrally; feet, carpal, tarsal, limbs, wrists, and ankles whitish; skull larger than in *P. b. rowleyi;* palatine slits longer: **molars actually and relatively larger and heavier than those of *P. boylii.*** Species group *boylii.*

**Comments:** *Peromyscus polius* was considered in the *truei* species group (Osgood 1909) and later moved to the *boylii* species group (Hoffmeister 1951). *P. polius* is restricted to western-central Chihuahua (Map 241). It can be found in sympatry, or nearly so, with *P. boylii*, *P. eremicus*, *P. gambelii*, *P. labecula*, *P. leucopus*, and *P. melanotis*. *P. polius* can be distinguished from *P. boylii* by its larger size and molars; palatine foramen longer and auditory bullae about the same size but proportionally smaller. From *P. gambelii*, *P. labecula*, *P. leucopus*, and *P. melanotis*, by having the tail without a well-defined bicolored pattern and equal to or larger than the head-and-body length.

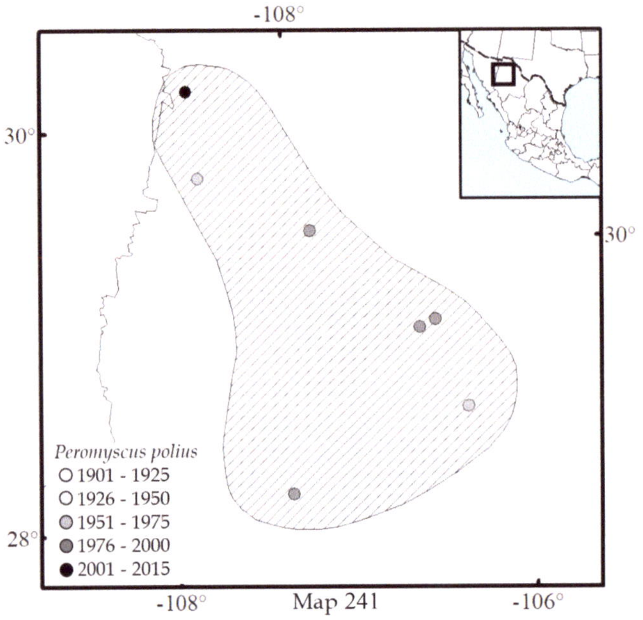


**Map 241** Distribution map of *Peromyscus polius*

**Additional Literature:** Kilpatrick and Zimmerman (1975), Musser and Carleton (2005).

## *Peromyscus purepechus* Léon-Tapia et al., 2020

Purepecha deermouse, ratón purépecha

Monotypic.

**Conservation:** *Peromyscus purepechus* (as *Peromyscus gratus*) is listed as Data Deficient by the IUCN (Lacher et al. 2016i).

**Characteristics:** Purépecha deermouse is medium sized; total length 152.0–194.0 mm and skull length 24.9–28.1 mm. Dorsal pelage Dresden brown with a dark line middorsally; flanks cream-buff; fur of underparts fully white; ears with a cinnamon-brown pelage with black shades; tail well haired, shorter than the head-and-body length, bicolored with a sooty brownish stripe dorsally, one-third of the tail diameter; dresden brown strip extending from the legs to the ankles; toes fully white; mastoid rectangular in shape; mastoid fenestra reduced and elongated. Species group *maniculatus*.

**Comments:** *Peromyscus purepechus* is known only from Nahuatzen, Michoacán (Map 242). It can be found in sympatry, or nearly so, with *P. hylocetes*. *P. purepechus* can be distinguished from *P. hylocetes* by having the tail hairy, shorter than or equal to the head-and-body length and clearly bicolored with the dark dorsal narrow stripe covering only the vertebral line. From *P. labecula* and *P. melanotis*, by having a characteristic combination of ear coloration (black preauricular hairs at the anterior base and ears with cinnamon-brown pelage with black shades); tail length; skull measurements (braincase length: *P. purepechus* = 17.32–19.83 mm vs *P. labecula* = 11.48–18.93 mm, vs *P. melanotis* = 16.22–18.49 mm). Braincase flattened, depth of braincase 8.14–8.8 mm vs 6.97–8.63 mm, and 7.43–8.76 mm; in *P. purepechus*, the mastoid is rectangular and reduced; mastoid fenestra elongated in *P. labecula* vs visibly oval and more rounded in *P. melanotis* and based on genetic data.

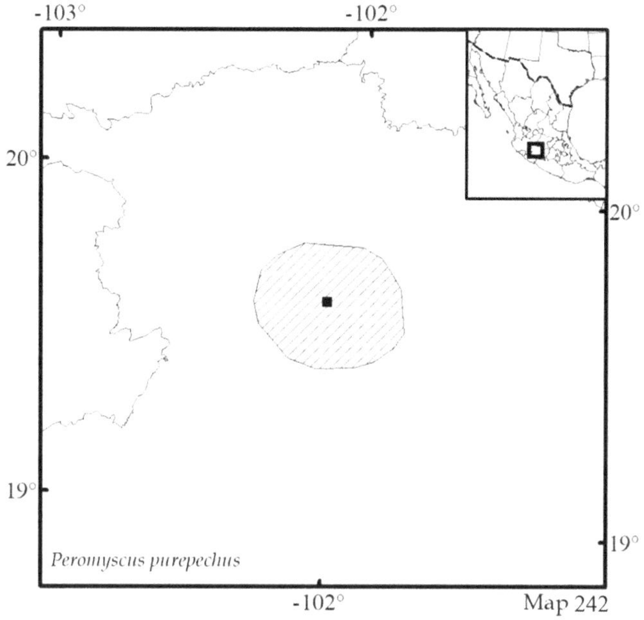


**Map 242** Distribution map of *Peromyscus purepechus*

**Additional Literature:** Léon-Tapia et al. (2020).

## *Peromyscus sagax* Elliot, 1903

Michoacán deermouse, ratón de Sahuayo

Monotypic.

**Conservation:** *Peromyscus sagax* (as *Peromyscus gratus*) is listed as Data Deficient by the IUCN (Lacher et al. 2016i).

**Characteristics:** Michoacán deermouse is medium sized; total length 191.0–220 mm and skull length 24.8–26.9 mm. Dorsal pelage grayish black and buff; fur of underparts whitish with grayish shades; flanks grayish brown and buff with an indistinct lateral line; orbital ring and spot behind the nose black, with a buff spot between the eyes; head sides and shoulders buffy gray; ears large, brown at the base, blackish at the tip, and with narrow white edges, approximately 80% of the hindfoot length; tail equal than the head-and-body length, dusky dorsally and white ventrally; fore and hindfeet whitish; braincase nearly square and broad. Species group *truei*.

**Comments:** *Peromyscus sagax* was reinstated as a distinct species by Bradley et al. (1996); it was previously regarded as a subspecies of *P. truei gratus* (Osgood 1909; Hall 1981). *P. sagax* is known only from a very small area in northwest Michoacán (Map 243). It can be found in sympatry, or nearly so, with *P. labecula* and *P. melanotis*. It can be distinguished from these species by having the tail more heavily haired, with less pronounced annulations and longer than the head-and-body length; auditory bullae proportionally greatly inflated and ears longer than the hind feet.

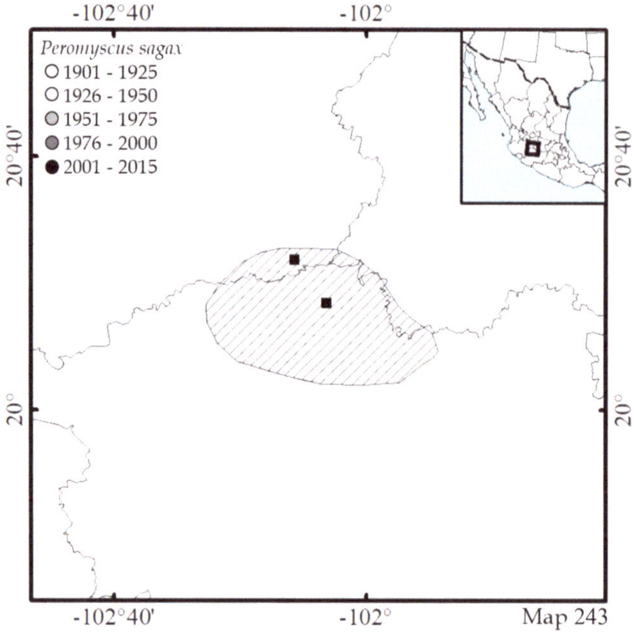


**Map 243** Distribution map of *Peromyscus sagax*

**Additional Literature:** Elliot (1903), Musser and Carleton (2005).

## ***Peromyscus schmidlyi*** Bradley et al., 2004

Schmidly's deermouse, ratón de la Sierra Madre Occidental

Monotypic.

**Conservation:** *Peromyscus schmidlyi* is listed as Least Concern by the IUCN (Álvarez-Castañeda et al. 2017e).

**Characteristics:** Schmidly's deermouse is medium sized; total length 175.0–205.0 mm and skull length 24.2–28.8 mm. Dorsal pelage pale brown with leaden gray shades at the base, flanks reddish cinnamon; fur of underparts whitish with leaden shades at the base; ears approximately 100% of the hindlimb length and dark gray; hind feet mainly whitish; tail longer than the head-and-body length, bicolored, blackish brown dorsally and whitish ventrally, little hair and a tuft at the tip; **surface between the orbital region and the nasals convex in lateral view; postorbital constriction hourglass shaped in dorsal view**. Species group *boylii*.

**Comments:** *Peromyscus schmidlyi* was recognized as a distinct species (Bradley et al. 2004a); it was previously regarded as a subspecies of *P. boylii rowleyi*, *P. b. simulus*, and *P. b. spicilegus* (Ordóñez-Garza and Bradley 2011). *P. schmidlyi* ranges in the highlands from Sonora, Durango, and Sinaloa (Map 244), and cannot be found in sympatry, or nearly so, with any other species of *boylii* complex, but it could be found in simpatry with *P. labecula*, *P. leucopus*, and *P. melanotis*. *P. schmidlyi* can be distinguished from all other species only based on DNA analyses. From *P. b. rowleyi*, by being larger and darker. From *P. spicilegus*, by having the orbital region hourglass shaped and not angular. From *P. labecula* and *P. melanotis*, by being larger in the head-and-body length and without a dark dorsal narrow stripe covering only the vertebral line.

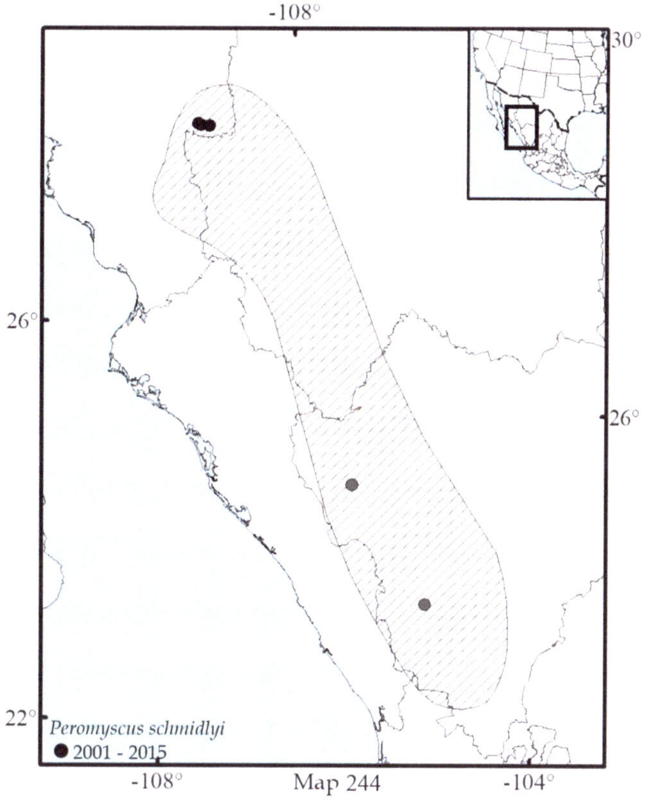


**Map 244** Distribution map of *Peromyscus schmidlyi*

**Additional Literature:** Carleton (1977, 1989), Carleton et al. (1982), Cabrera et al. (2007).

## *Peromyscus sejugis* Burt, 1932

Santa Cruz Island deermouse, ratón de Isla Santa Cruz

Monotypic.

**Conservation:** *Peromyscus sejugis* is listed as Threatened by the Norma Oficial Mexicana (DOF 2019) and as Endangered (B1ab (v) + 2ab (v): extent of occurrence less than 5000 km$^2$ in two populations and under continuing decline in number of mature individuals + area of occupancy less than 500 km$^2$ in five or less populations and under continuing decline in number of mature individuals) by the IUCN (Álvarez-Castañeda 2018f).

**Characteristics:** Santa Cruz Island deermouse is medium sized; total length 160.0–197.0 mm and skull length 26.6–27.4 mm. Dorsal pelage grayish grizzly with hazel shades; fur of underparts whitish; ears approximately 70% of the hindfoot length; tail shorter than the head-and-body length, bicolored with a dark dorsal stripe one-third of the tail diameter; limbs whitish; skull arched toward the sagittal axis; nasals wide; auditory bullae small; palate length greater than the molariform length. Species group *maniculatus*.

**Comments:** *Peromyscus sejugis* closely resembles the *P. maniculatus* species group (Hooper and Musser 1964; Bradley et al. 2019). *P. sejugis* is restricted to the Santa Cruz and San Diego Islands, Baja California Sur (Map 245), no other species of *Peromyscus* is found in sympatry. It can be distinguished from other closely related species based on its geographic distribution.

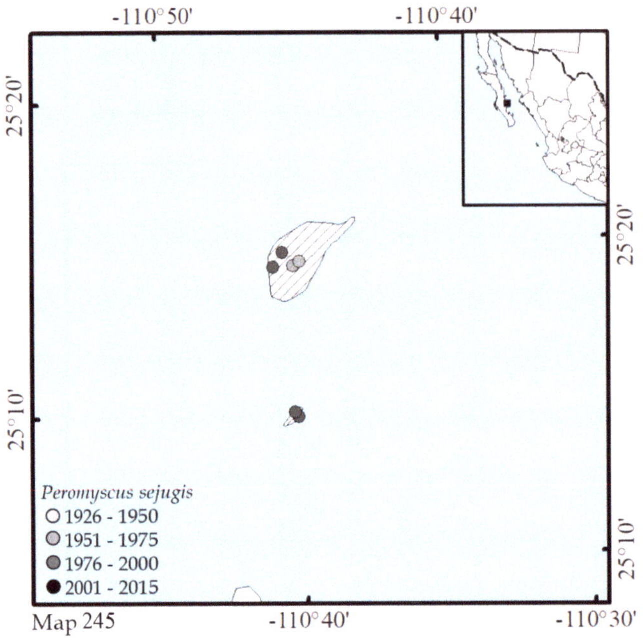


**Map 245** Distribution map of *Peromyscus sejugis*

**Additional Literature:** Burt (1932), Álvarez-Castañeda and Cortés-Calva (1999, 2001), Musser and Carleton (2005).

## *Peromyscus simulus* Osgood, 1904

Nayarit deermouse, ratón de Sinaloa

Monotypic.

**Conservation:** *Peromyscus simulus* is listed as Vulnerable (B1ab (iii, v): Extent of occurrence less than 20,000 km$^2$ at ten or less locations and under continuing decline in area, extent and/or quality of habitat and number of mature individuals) by the IUCN (Álvarez-Castañeda et al. 2018g).

**Characteristics:** Nayarit deermouse is medium sized; total length 191.0–207.0 mm and skull length 26.2–27.4 mm. Dorsal pelage tawny to yellowish ocher and dark at the tip; darker on the back, without a marked lateral line; fur of underparts whitish to whitish cream with grayish shades, pectoral spot frequently present; ears approximately greater than 80% of the hindlimb length; tail smaller than the head-and-body length and **in some specimens blackish all around, not sharply bicolored**; **skull, nasals and rostrum clearly shorter**; parietal narrower and less shelf-like; premaxillae not exceeding the nasals; **zygomatic bone relatively heavy and squared anteriorly**; **molars very small; palate bony and short**. Species group *boylii*.

**Comments:** *Peromyscus simulus* was reinstated to the species level by Carleton (1977); it was previously regarded as a subspecies of *P. boylii* (Osgood 1909; Hall 1981). *P. simulus* is restricted to lowlands of western-central Sinaloa and western Nayarit (Map 246), and cannot be found in sympatry, or nearly so, with any other species of *boylii* complex but could be found in sympatry with *P. carletoni*, *P. labecula*, and *P. leucopus*. *P. simulus* can be distinguished from *P. carletoni* only with genetic data; however, on average it has the upper toothrow length less than 4.1 mm and the tail often unicolor. From *P. labecula* and *P. leucopus*, by having the tail without a well-defined bicolored pattern and equal to or larger than the head-and-body length.

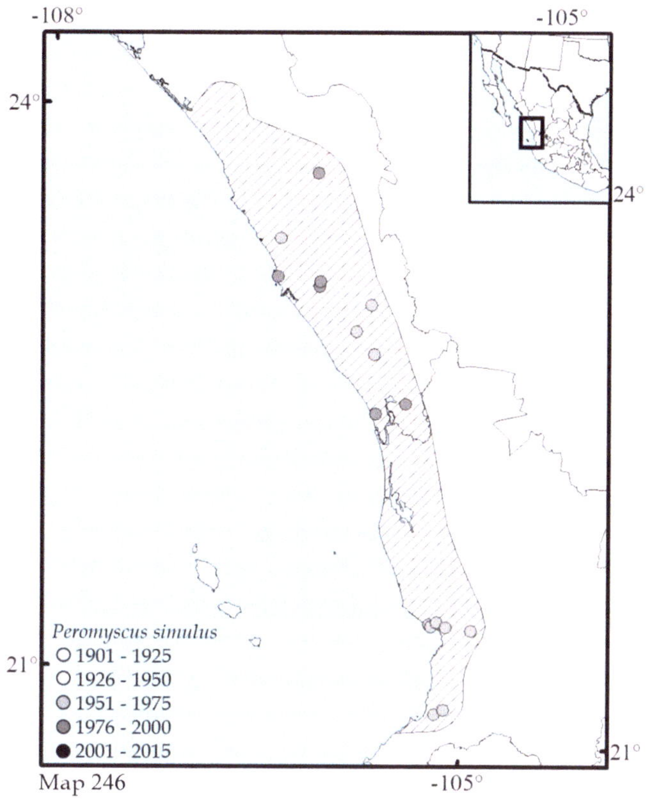


**Map 246** Distribution map of *Peromyscus simulus*

**Additional Literature:** Osgood (1904), Carleton et al. (1982), Roberts et al. (2001), Musser and Carleton (2005).

## *Peromyscus slevini* Mailliard, 1924

Slevin's deermouse, ratón de la Isla Catalina

Monotypic.

**Conservation:** *Peromyscus slevini* is listed as Threatened by the Norma Oficial Mexicana (DOF 2019), and as Critically Endangered (B1ab (v): extent of occurrence less than 100 km$^2$ at only one location and under continuing decline in number of mature individuals) by the IUCN (Álvarez-Castañeda 2019b).

**Characteristics:** Slevin's deermouse is medium sized; total length 208.0–254.0 mm and skull length 29.6–30.7 mm. Dorsal pelage cinnamon reddish grizzly with blackish shades, darker on the back; fur of underparts whitish, grizzly with cinnamon in the pectoral region; ears approximately 75% of the hindfoot length, thin hair; tail longer than the head-and-body length, sepia dorsally and whitish ventrally, with a tuft at the tip; forefoot whitish wrist dark; hind feet whitish but dark from metatarsus; forelegs pale cinnamon; rostrum thin and long; supraorbital edge present but not well marked; interparietal bone rhomboidal; incisive foramen proportionally large and mandible strong. Species group *melanophrys*.

**Comments:** *Peromyscus slevini* was considered one population of *P. fraterculus* (Hafner et al. 2001). However, based on the review of specimens and other analyses, it is still considered a valid species (Musser and Carleton 2005; Álvarez-Castañeda et al. 2010a). *P. slevini* is restricted to Santa Catalina Island, Baja California Sur (Map 247), and no other species of *Peromyscus* could be found in sympatry. However, *P. fraterculus* has been introduced to Santa Catalina Island (Álvarez-Castañeda et al. 2010a). *P. slevini* can be distinguished from *P. fraterculus* by having an accessory cusp or enamel loop on the second upper molars and three pairs of mammae (nipples).

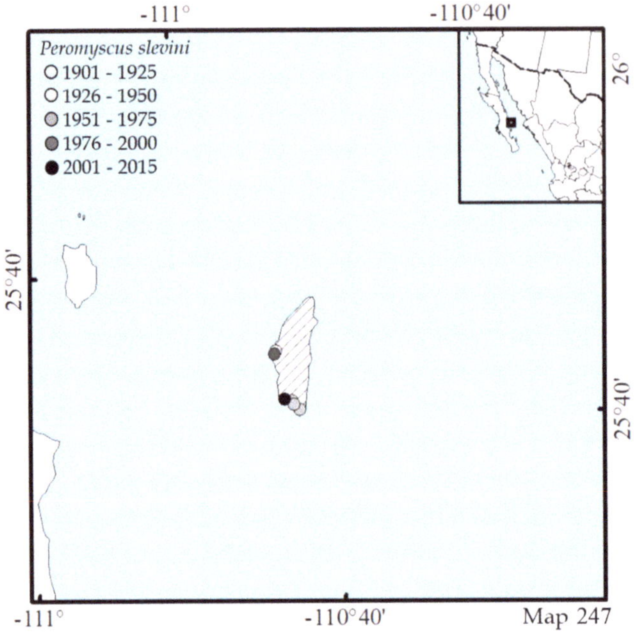


**Map 247** Distribution map of *Peromyscus slevini*

**Additional Literature:** Mailliard (1924), Burt (1932), Hall (1981), Álvarez-Castañeda and Cortés-Calva (1999, 2002a), Álvarez-Castañeda et al. (2006).

## *Peromyscus sonoriensis* (Le Conte, 1853)

North American common deermouse, ratón común de Norteamérica

1. *P. s. alpinus* Cowan, 1937. Restricted to Mountain Revelstoke, southeastern British Columbia.
2. *P. s. artemisiae* (Rhoads, 1894). From southern-central British Columbia southwest through western Wyoming and eastern Idaho.
3. *P. s. austerus* (Baird, 1855). Restricted to a small band in western-central Washington.
4. *P. s. borealis* (Mearns, 1890). From southern Yukon and Mackenzie south through central British Columbia, southern Alberta, and Saskatchewan. Note: this range probably includes a new species.
5. *P. s. hollisteri* Osgood, 1909. Restricted to the Blakeley, Cypress, and San Juan Islands, Washington.
6. *P. s. inclarus* Goldman, 1939. Known only from Fremont Island, Great Salt Lake, Utah.
7. *P. s. luteus* Osgood, 1905. From central South Dakota south through central Texas.
8. *P. s. nebrascensis* (Coues, 1877). From south Alberta and Saskatchewan south through northwestern Texas, including only eastern Colorado and western Dakotas and Nebraska.
9. *P. s. ozarkiarum* Black, 1935. From northeast Oklahoma and southwestern Missouri south through western Arkansas and northern-central Texas.
10. *P. s. pallescens* Allen, 1896. Restricted from north to south central Texas.
11. *P. s. rubidus* Osgood, 1901. From southwestern Washington southward along the Pacific coast to central California.
12. *P. s. rufinus* (Merriam, 1890). From central to central-northern Utah.
13. *P. s. saturatus* Bangs, 1897. Restricted to Saturna Island, British Columbia.
14. *P. s. saxamans* McCabe and Cowan, 1945. Restricted to the Bell, Duncan, Heard, and Hurst Islands, British Columbia.
15. *P. s. serratus* Davis, 1939. Restricted to central Idaho.
16. *P. s. sonoriensis* (Le Conte, 1853). From southeastern Oregon and southern Idaho south through Sonora, including eastern California and western Utah and Arizona.

**Conservation:** *Peromyscus sonoriensis* (as *P. maniculatus*) is listed as Least Concern by the IUCN (Cassola 2016cq).

**Characteristics:** North America common deermouse is medium sized; total length 126.0–176.0 mm and skull length 24.2–27.0 mm. Coloration varying greatly between subspecies, dorsally ochraceous to buff; fur of underparts white; ears dusky, **usually with conspicuous white spots at the anterior base**, rather broadly edged with whitish shades, 70% of the hindfoot length; whisker base with little or no dusky hairs; eyelids sometimes dusky but without an orbital ring; feet and forelegs white, tarsal joints white to buffy, slightly mixed with dusky shades; tail haired, shorter than the head-and-body, more sharply bicolored, with a brown dorsal stripe one-third of the tail diameter above, white ventrally, and more thickly haired; hind feet whitish; braincase arched. Species group *maniculatus*.

**Comments:** *Peromyscus sonoriensis* was reinstated as a distinct species by Bradley et al. (2019) and Greenbaum et al. (2019), including the previous subspecies of *P. maniculatus*: *P. m. alpinus*, *P. m. artemisiae*, *P. m. austerus*, *P. m. borealis*, *P. m. hollisteri*, *P. m. inclarus*, *P. m. luteus*, *P. m. nebrascensis*, *P. m. ozarkiarum*, *P. m. pallescens*, *P. m. rubidus*, *P. m. rufinus*, *P. m. saturatus*, *P. m. saxamans*, *P. m. serratus*, and *P. m. sonoriensis* (Osgood 1909). The populations from central Oregon and eastern-central Washington considered *P. m. gambelii* (Hall 1981) should be assigned to *P. sonoriensis* (Bradley et al. 2019). The subspecies that potentially belong to *P. sonoriensis* are listed above based on Bradley et al. (2019) and Greenbaum et al. (2019). *P. sonoriensis* ranges from west of the Mississippi River northwest through Yukon, including all the south of Manitoba, Saskatchewan, Alberta, and British Columbia south through northern California, northern Sonora, Arizona, New Mexico, Texas, and Arkansas (Map 248). *P. sonoriensis* can be found in sympatry, or nearly so, with *P. attwateri*, *P. boylii*, *P. californicus*, *P. crinitus*, *P. eremicus*, *P. fraterculus*, *P. gambelii*, *P. keeni*, *P. labecula*, *P. laceianus*, *P. leucopus*, *P. maniculatus*, *P. melanotis*, *P. merriami*, *P. nasutus*, and *P. truei*. *P. sonoriensis* can be distinguished from *maniculatus*-like species (*P. gambelii*, *P. labecula*, and *P. maniculatus*) by its larger average size; longer tail, greater than 75.0 mm and with a less vinaceous coloration but mainly using genetic data. From *P. labecula*, by its skull slightly larger. From *P. attwateri*, *P. laceianus*, *P. nasutus*, and *P. truei*, by having the ears smaller than the hind feet; tail less haired and auditory bullae proportionally less inflated. From *P. boylii*, by having the tail hairy, shorter than or equal to the head-and-body length, and clearly bicolored, with the dark dorsal narrow stripe covering only the vertebral line. From *P. californicus*, *P. crinitus*, *P. eremicus*, *P. fraterculus*, and *P. merriami*, by having two pairs of mammae (nipples); first and second upper molars with accessory cusps; tail shorter than the head-and-body length with a paravertebral coloration. From *P. crinitus*, by its smaller average size; tail shorter or similar than the head-and-body length and with a paravertebral coloration. From *P. keeni*, by its lighter skull; larger ears; shorter tail on average, less than the head-and-body length; but mainly using genetic data. From *P. leucopus*, it is very hard to differentiate between these two species in some areas; tail darker in the paravertebral area. From *P. melanotis* (extremely similar morphologically), by lacking a black preauricular hair tuft; braincase narrower and less rounded; wider interorbital space; prezygomatic notch more prominent; slightly larger auditory bullae; rostrum shorter and less slender.

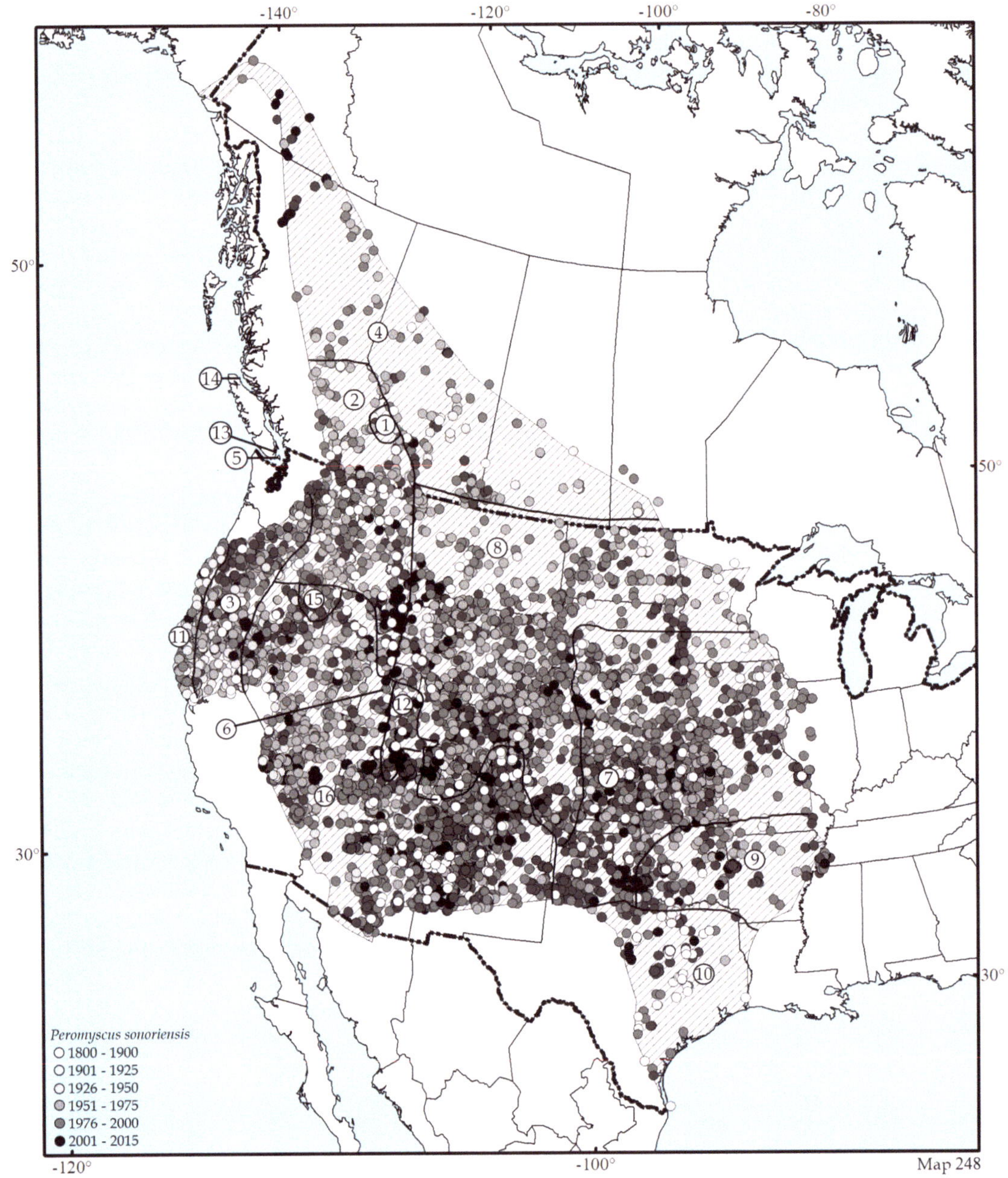


**Map 248** Distribution map of *Peromyscus sonoriensis*

**Additional Literature:** Merriam (1901a), Álvarez-Castañeda and Cortés-Calva (1999), Álvarez-Castañeda (2005a, b).

## ***Peromyscus spicilegus*** Allen, 1897

Gleaning deermouse, ratón de las espigas

Monotypic.

**Conservation:** *Peromyscus spicilegus* (as *Peromyscus boylii spicilegus*) is listed as Least Concern by the IUCN (Lacher et al. 2016h).

**Characteristics:** Gleaning deermouse is medium sized; total length 175.0–232.0 mm andskull length 25.9–30.1 mm. Dorsal pelage tawny to reddish ocher and darker at the tip, concentrating on the back appearing as a line; fur of underparts light buff; orbital ring blackish; lateral line absent; ears dusky, approximately 85% of the hindfoot length; **ears blackish, strongly contrasting with the dorsal rich tawny ochraceous coloration**; feet white and usually with a dusky streak extending from the tarsus to the metatarsus; tail usually as long as the head-and-body length, bicolored and frequently white at the tip; supraorbital border angular; auditory bullae pointed anteriorly, with the anterior half of the braincase nearly straight; **braincase usually more expanded anteriorly, forming an incipient supraorbital shelf**. Species group *aztecus*. Note: *Peromyscus hylocetes* and *P. spicilegus* in Jalisco need to be directly compared between specimens of both species.

**Comments:** *Peromyscus spicilegus* was reinstated to the species level by Carleton (1977); it was previously regarded as a subspecies of *P. boylii spicilegus* (Osgood 1909; Hall 1981) and included in the *P. aztecus* species group (Carleton 1979). It ranges along the western slope of Sierra Madre Occidental, from southeast Sonora and southwestern Chihuahua south through northeastern Colima and western-central Michoacán (Map 249). *P. spicilegus* cannot be found in sympatry, or nearly so, with any other species of the *boylii* complex, but it could be found in sympatry with *P. labecula* and *P. leucopus*. *P. spicilegus* can be distinguished from *P. boylii* and *P. carletoni* by having a bright color on the flanks; supraorbital shelf well developed and edges with a ridge and straight; both edges wedge shaped; auditory bullae pointed anteriorly. From *P. labecula* and *P. leucopus*, by having the tail larger than the head-and-body length; with a lateral line on the flanks, a dark spot around the eyes, and the supraorbital border sharp-angled. From *P. schmidlyi*, by having the orbital region hourglass shaped, angular.

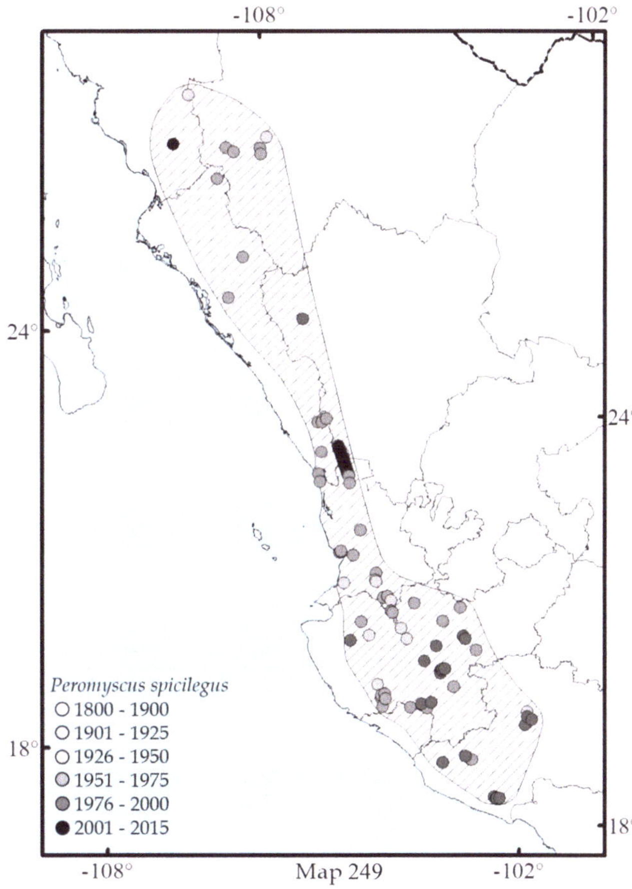


**Map 249** Distribution map of *Peromyscus spicilegus*

**Additional Literature:** Carleton et al. (1982), Roberts et al. (1988), Musser and Carleton (2005), Ordóñez-Garza and Bradley (2011).

## *Peromyscus stephani* Townsend, 1912

San Esteban island deermouse, ratón de Isla San Esteban

Monotypic.

**Conservation:** *Peromyscus stephani* is listed as Threatened by the Norma Oficial Mexicana (DOF 2019) and by the United States Endangered Species Act (FWS 2022), and as **Critically Endangered** (B1ac (iv): extent of occurrence less than 100 km$^2$ at only one location and extreme fluctuations in number of mature individuals) by the IUCN (Álvarez-Castañeda 2018g).

**Characteristics:** San Esteban Island deermouse is medium sized; total length 163.0–210.0 mm and skull length 24.6–27.6 mm. Dorsal pelage grizzly gray with reddish brown shades; head grayish; lateral line pale yellow; fur of underparts whitish; ears approximately more than 85% of the hindfoot length; tail shorter than the head-and-body length and bicolored; fore and hind feet whitish; rostrum thin and long; coronoid process small and slightly high. Species group *boylii*.

**Comments:** *Peromyscus stephani* is restricted to San Esteban Island, Sonora (Map 250). No other species of *Peromyscus* could be found in sympatry with it. *P. stephani* can be distinguished from other closely related species based on its geographic distribution (Hall 1981).

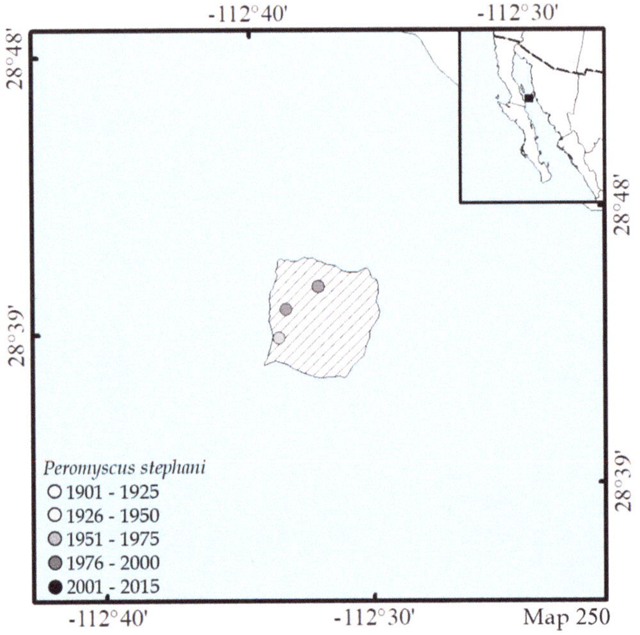


**Map 250** Distribution map of *Peromyscus stephani*

**Additional Literature:** Álvarez-Castañeda and Cortés-Calva (1999).

## *Peromyscus totontepecus* Merriam, 1898

Totontepec deermouse, ratón de Totontepec

Monotipic

**Conservation:** *Peromyscus totontepecus* (as *Peromyscus mexicanus*) is listed as Least Concern by the IUCN (Reid and Pino 2016).

**Characteristics:** Totontepec deermouse is medium sized; total length 240.0–256.0 mm and skull length 30.1–35.2 mm. Dorsal pelage dark reddish brown to blackish, darker on the back, with a grayish cast; fur of underparts whitish to cream; the pectoral region often yellowish reddish; flanks with a lateral reddish line; blackish spot on the base of the whiskers; ears dusky brown with the edge faint whitish, approximately 70% of the hindfoot length; forefeet and carpal joints white and proximal one-half of the forearms dusky overlaid by rufescent shades; hind feet white and tarsal joints dusky brown; tail slightly longer than the head-and-body length, practically naked with few hairs and coarsely annulated; rostrum thin and long; supraorbital ridge with the edge poorly developed; molariform and auditory bullae proportionally small.

**Comments:** *Peromyscus totontepecus* was recognized as a distinct species (Hernández-Canchola et al. 2022; Pérez-Montes et al. 2022); it was previously regarded as a subspecies of *P. mexicanus* (Huckaby 1980). *P. totontepecus* is restricted to the Sierra Madre del Sur of Oaxaca and Veracruz, in the Gulf slope (Map 251). It can be found in sympatry, or nearly so, with *P. angelensis*, *P. aztecus*, *P. beatae*, *P. carolpattonae*, *P. leucopus*, *P. levipes*, and *P. melanocarpus*. *P. totontepecus* can be distinguished from all other species of *Peromyscus*, except for *P. angelensis* and *P. mexicanus*, by having the tail slightly longer than the head-and-body length, practically naked with few hairs and coarsely annulated; relatively small molars and auditory bullae. From *P. angelensis* and *P. mexicanus*, by its large size in somatic and skull measurements and a generally darker coloration. From *P. beatae*, *P. levipes*, and *P. leucopus*, by its larger size and tail proportionally longer than the head-and-body. From *P. melanocarpus*, by having white carpal and tarsal regions; the presence of pectoral mammae (nipples); supraorbital edge present without a well marked or shallow depression; smaller molars.

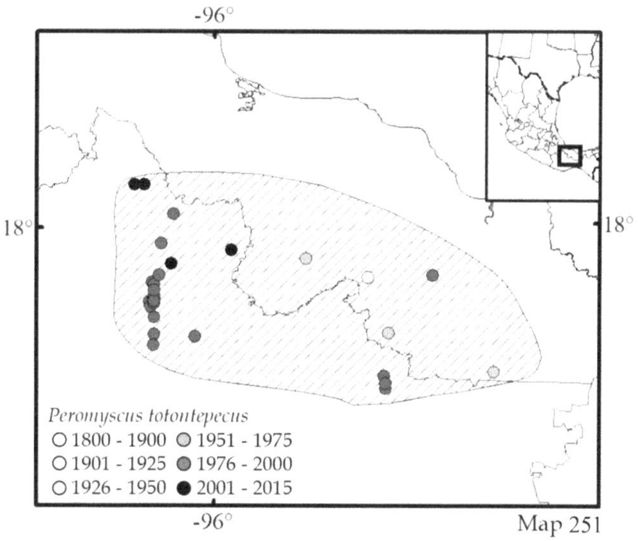


**Map 251** Distribution map of *Peromyscus totontepecus*

**Additional Literature:** Osgood (1909), Hooper (1957), Hooper (1968), Musser and Carleton (2005), Trujano-Álvarez and Álvarez-Castañeda (2010).

## *Peromyscus truei* (Shufeldt, 1885)

Pinyon deermouse, ratón piñonero del norte

1. *P. t. chlorus* Hoffmeister, 1941. Restricted to southern-central California.
2. *P. t. comanche* Blair, 1943. Restricted to central-northern Texas.
3. *P. t. dyselius* Elliot, 1898. Restricted to the southern San Francisco Bay area, California.
4. *P. t. gilberti* (Allen, 1893). From southern-central Oregon south through central California.
5. *P. t. lagunae* Osgood, 1909. Restricted to the southern end of Baja California Sur (no shown in the map).
6. *P. t. martirensis* (Allen, 1893). From southern California to northern Baja California.
7. *P. t. montipinoris* Elliot, 1904. Restricted to southern-central California.
8. *P. t. nevadensis* Hall and Hoffmeister, 1940. From western Nevada to western Utah.
9. *P. t. preblei* Bailey, 1936. Restricted to northern-central Oregon.
10. *P. t. sequoiensis* Hoffmeister, 1941. Coastal area from southwestern Oregon to the San Francisco Bay, California.
11. *P. t. truei* (Shufeldt, 1885). From southeastern Oregon and eastern California east through Colorado, New Mexico, and western Oklahoma.

**Conservation:** *Peromyscus truei* is listed as Least Concern by the IUCN (Cassola 2016es).

**Characteristics:** Pinyon deermouse is medium sized; total length 190.0–200.0 mm and skull length 27.0–30.0 mm. Dorsal pelage pale yellowish-brown to brownish black; fur of underparts whitish; hair long and silky; lateral line distinguishable; limbs white; **ears approximately 100% of the hindfoot length**; **hind feet white or dusky, usually densely haired from the calcaneus to the proximal plantar tubercle**; soles with six tubercles; tail as long as the head-and-body length, bicolored, covered with short hairs, with a dorsal stripe slightly darker than the back; **pelage quite long, loose, and silky**; auditory bullae well developed; braincase large and rounded; zygomatic arches weak. Species group *truei*. Note: *Peromyscus truei* and *P. gratus* can only be differentiated by their geographical distribution.

**Comments:** The subspecies of *Peromyscus truei* from central México were assigned to *P. gratus* by Modi and Lee (1984). It ranges from California, southwestern and central Oregon, Nevada, Utah, Colorado, and Kansas south through Baja California, northern and central Arizona, southern New Mexico, northern Texas (Map 252), and southern Baja California Sur (not shown in the map). It can be found in sympatry, or nearly so, with *P. boylii*, *P. californicus*, *P. crinitus*, *P. eremicus*, *P. eva*, *P. difficilis*, *P. fraterculus*, *P. gambelii*, *P. gratus*, *P. labecula*, *P. leucopus*, *P. maniculatus*, *P. nasutus*, and *P. sonoriensis*. *P. truei* can be distinguished from *P. californicus*, *P. eremicus*, *P. eva*, and *P. fraterculus* by having two pairs of mammae (nipples), and first and second upper molars with accessory cusps. From *P. boylii*, *P. crinitus*, *P. gambelii*, *P. labecula*, *P. leucopus*, *P. maniculatus*, and *P. sonoriensis*, by having the ears longer than the hind feet; tail more heavily haired, with less pronounced annulations and longer than the head-and-body length; auditory bullae proportionally greatly inflated. From *P. californicus*, *P. eremicus*, *P. eva*, and *P. fraterculus*, by having the ear length smaller than the hindfoot length and having a lateral line on the flanks; a dark spot around the eyes; a pectoral spot, auditory bullae larger and more orbicular. From *P. gratus*, by having the tail longer than the body; ear length relatively smaller than the hindfoot length and auditory bullae proportionally less inflated (Zimmerman et al. 1975). From *P. difficilis* and *P. nasutus*, by having the dorsum often grayish-lighter; lateral line without ochraceous shades; top of the head similar to the dorsum; smaller ears, less than 24.0 mm; toothrow proportionally larger; auditory bullae proportionally less inflated and larger.

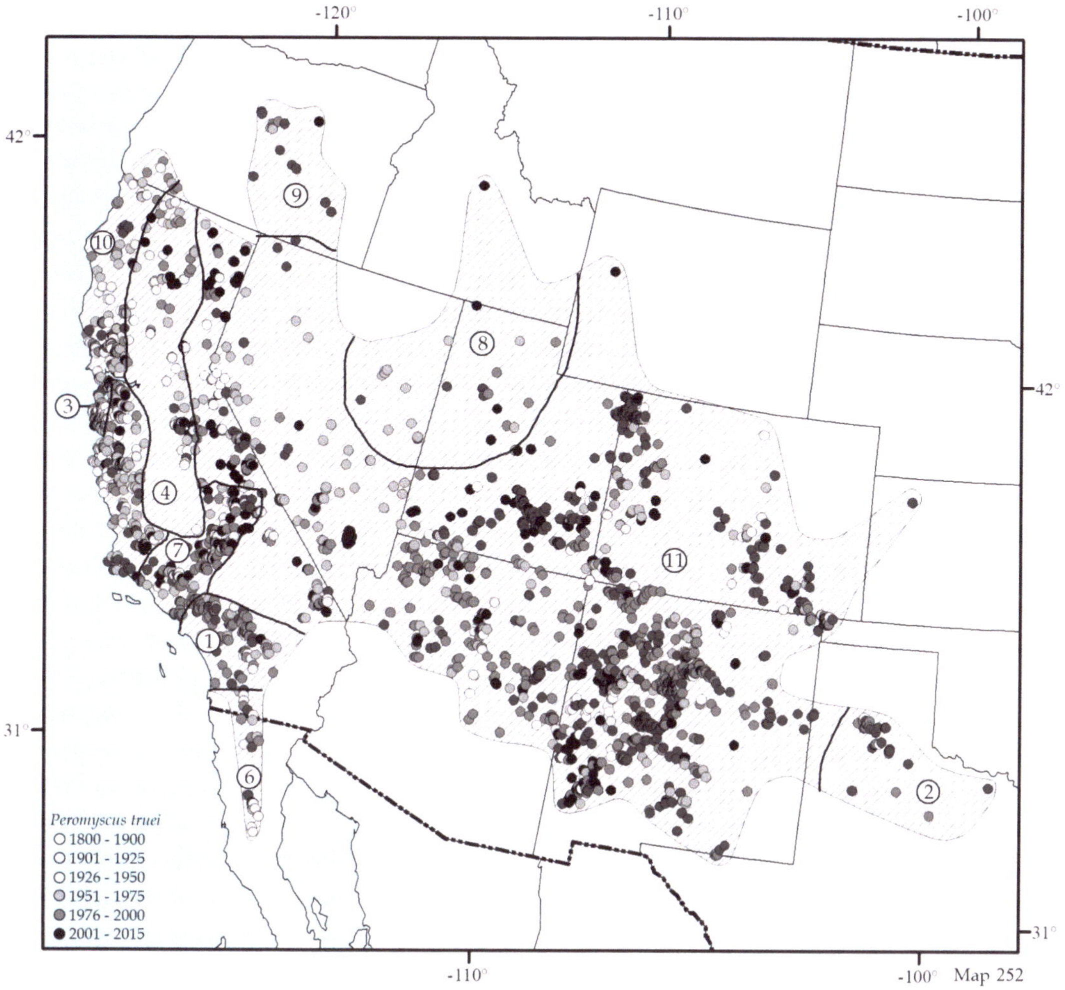


**Map 252** Distribution map of *Peromyscus truei*

**Additional Literature:** Merriam (1898a), Osgood (1909), Hoffmeister (1951, 1981), Schmidly (1973), Álvarez-Castañeda and Cortés-Calva (1999), Musser and Carleton (2005), Kilpatrick et al. (2021).

## ***Peromyscus winkelmanni*** Carleton, 1977

Winkelmann's deermouse, ratón de Guerrero

Monotypic.

**Conservation:** *Peromyscus winkelmanni* is listed as Special Protection by the Norma Oficial Mexicana (DOF 2019) and as Endangered (B1ab (iii): extent of occurrence less than 5000 km$^2$ at one location and under continuing decline in area and quality of habitat) by the IUCN (Álvarez-Castañeda 2018h).

**Characteristics:** Winkelmann's deermouse is medium sized; total length 235.0–265.0 mm and skull length 31.2–31.9 mm. Dorsal pelage tawny grizzly with cheeks blackish and darker; flanks tawny, almost cinnamon; lateral line ocher; fur of underparts paler gray, white at the tip, some specimens have a yellowish pectoral spot; **ears approximately 60% of the hindfoot length and dark**; tail shorter than or equal than the head-and-body length, unicolor, darker ventrally than dorsally; forefeet dark to the metatarsus and then whitish; rostrum thin and long; supraorbital edge present and well marked. Species group *aztecus*.

**Comments:** *Peromyscus winkelmanni* is restricted to Filo de Caballos in Guerrero (Map 253; Smith et al. 1989). Specimens from Sierra de Coalcomán, Michoacán, were considered *P. winkelmanii* and are now regarded as *P. greenbaumi* (Bradley et al. 2022a). *P. winkelmanni* lives in sympatry, or nearly so, with *Osgoodomys banderanus*, *P. kilpatrick*, *P. labecula*, and *P. megalops*. It can be distinguished from *P. hylocetes* and all the other species of *Peromyscus* (except for *O. banderanus* and *P. megalops*) by having an interorbital bead and ridge. From *O. banderanus*, *P. kilpatricki*, and *P. megalops*, by its larger size; it lacks spines on the lappets of glans penis, whereas in many of the other species of *Peromyscus* the spines extend onto the lappets. From *P. labecula*, by its larger size; tail without a well-defined dark dorsal narrow stripe covering only the vertebral line and equal to or larger than the head-and-body length; with a lateral line on the flanks; dark spot around the eyes and supraorbital border sharp-angled.

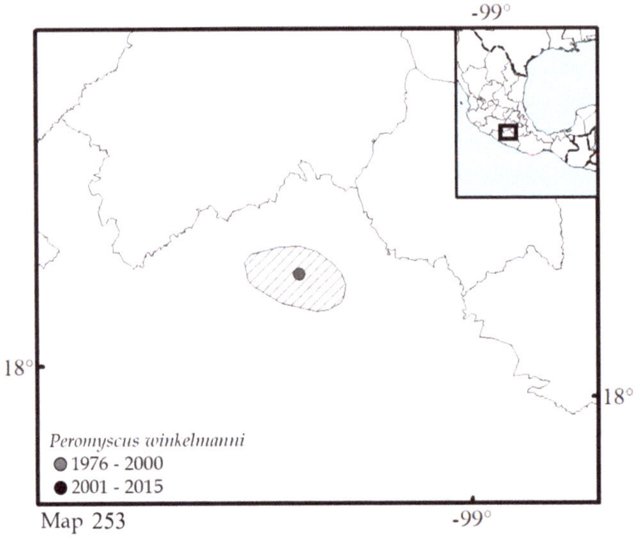


**Map 253** Distribution map of *Peromyscus winkelmanni*

**Additional Literature:** Carleton (1977, 1979), Bradley and Schmidly (1987), Sullivan et al. (1991), Álvarez-Castañeda (2005b), Musser and Carleton (2005), Kilpatrick et al. (2021).

## ***Peromyscus yucatanicus*** Allen and Chapman, 1897

Yucatán deermouse, ratón de Yucatán

1. *P. y. badius* Osgood, 1904. Restricted to the central and southern Yucatán Peninsula, Campeche, Quintana Roo, and northern Guatemala.
2. *P. y. yucatanicus* Allen and Chapman, 1897. Restricted to the northern Yucatán Peninsula, northern Campeche, and Yucatán.

**Conservation:** *Peromyscus yucatanicus* is listed as Least Concern by the IUCN (de Grammont and Cuarón 2016c).

**Characteristics:** Yucatán deermouse is medium sized; total length 181.0–219.0 mm and skull length 25.0–28.7 mm. Dorsal pelage brown grizzly with blackish shades; specimens from the northern Yucatán Peninsula are more yellowish ocher; fur of underparts yellowish white; pure white on the throat and chest; pectoral spot absent; **ears nearly naked**, approximately 75% of the hindfoot length; forefeet white up to the wrists; **hindfoot soles hairy proximally**; **tail thinly haired**, about equal to or slightly shorter than the head-and-body length, dusky dorsally, yellowish ventrally, and frequently blotched with darker pigments; **slight supraorbital bead**, shelf moderately developed; **teeth and auditory bullae small**; nasals not expanded; braincase narrow and elongated. Species group *mexicanus*.

**Comments:** *Peromyscus yucatanicus* ranges in the Yucatán peninsula (Map 254). It can be found in sympatry, or nearly so, with *P. leucopus* and *P. mexicanus*. *P. yucatanicus* can be distinguished from *P. mexicanus* by being smaller, both externally and cranially. From *P. leucopus*, by being larger in most dimensions and with a moderately developed supraorbital ridge.

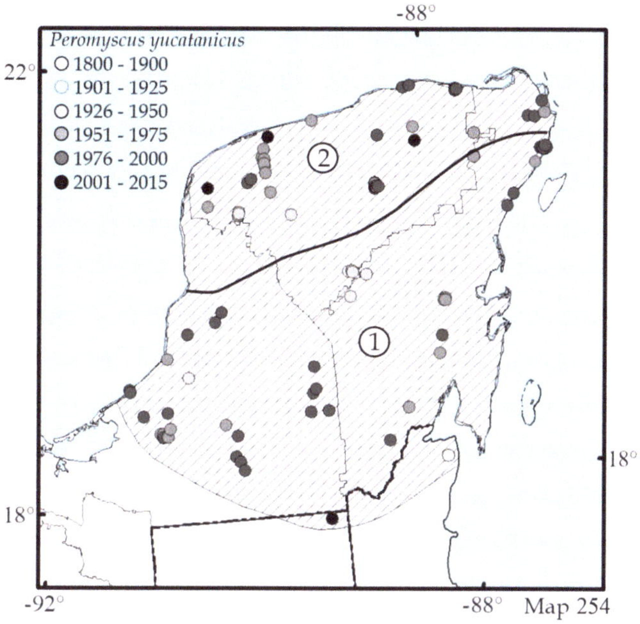


**Map 254** Distribution map of *Peromyscus yucatanicus*

**Additional Literature:** Osgood (1909), Hooper (1957), Lawlor (1965), Huckaby (1980), Young and Jones (1983), Musser and Carleton (2005).

## *Peromyscus zamorae* Osgood, 1904

Zamora deermouse, ratón de Zamora

Monotypic.

**Conservation:** *Peromyscus zamorae* is listed (as subspecies of *Peromyscus melanophrys*) as Least Concern by the IUCN (Vázquez and Álvarez-Castañeda 2016).

**Characteristics:** Zamora deermouse is medium sized; total length 205.0–281.0 mm and skull length 27.7–33.0 mm. Dorsal pelage from olive-tawny with a reddish hue to darker; fur of underparts grayish-white with bases dark gray at the base and white at the tip; ankles dark; feet dorsal coloration white, with the proximal half of foot dusky and the distal half whitish; facial mask around the eyes and over the nose conspicuously gray; ears relatively large; tail markedly bicolored and longer than the head-and-body length, brown dorsally and grayish or whitish ventrally; **supraorbital area with a curved ridge extending from the frontal bones as far as the posterior end of the nasals or to the joint between the nasals and the premaxillary bones; interorbital constriction narrow and curved; the mastoid region wide; auditory bullae large and inflated; rostrum relatively wide and short, with medium-sized nasals**; nasal bones long with the posterior ends tapered or U shaped. Species group *melanophrys*.

**Comments:** *Peromyscus zamorae* was considered at the species level by López-González et al. (2019); it was previously regarded as a subspecies of *P. melanophrys* (Osgood 1904), including the subspecies *P. m. coahuilensis*, *P. m. consobrinus*, and *P. m. xenurus* (López-González et al. 2019). *P. zamorae* ranges from Durango, Coahuila, and Nuevo León south through Hidalgo and Estado de México (Map 255). It can be found in sympatry, or nearly so, with *P. beatae*, *P. boylii*, *P. bullatus*, *P. difficilis*, *P. gratus*, *P. hooperi*, *P. kilpatricki*, *P. labecula*, *P. leucopus P. levipes*, *P. melanotis P. melanophrys*, *P. micropus*, and *P. pectoralis*. *P. zamorae* can be distinguished from all other species, except for *P. melanophrys* and *P. micropus*, by a black orbital ring and a conspicuous gray mask; tail larger than the head-and-body length, with annulations, covered with short stiff hairs and frequently white at the tip; supraorbital border sharply angled but normally not molded. From *P. beatae*, *P. boylii*, *P. bullatus*, *P. gratus*, *P. kilpatricki*, *P. labecula*, *P. leucopus*, *P. levipes*, *P. megalops*, *P. melanotis*, and *P. melanurus*, by its larger size; tail longer than the head-and-body length; auditory bullae proportionally greatly inflated. From *P. beatae*, *P. boylii*, *P. kilpatrick*, and *P. levipes*, by its larger size and a darker spot around the eyes. From *P. bullatus*, *P. difficilis*, *P. gratus*, and *P. pectoralis*, by having smaller ears, less than the hindfoot length; auditory bullae larger. From *P. labecula*, *P. leucopus*, and *P. melanotis*, by having the tail without a well-defined bicolored pattern and larger than the head-and-body length. From *P. melanophrys*, by having a shorter tail, on average less than 54% of the total length; supraorbital ridges reaching the nasal-premaxillary joint; postorbital constriction narrow and curved; auditory bullae large and inflated. From *P. micropus*, by having the supraorbital ridges not reaching the posterior end of the nasals; interorbital constriction narrow and curved; auditory bullae large and inflated.

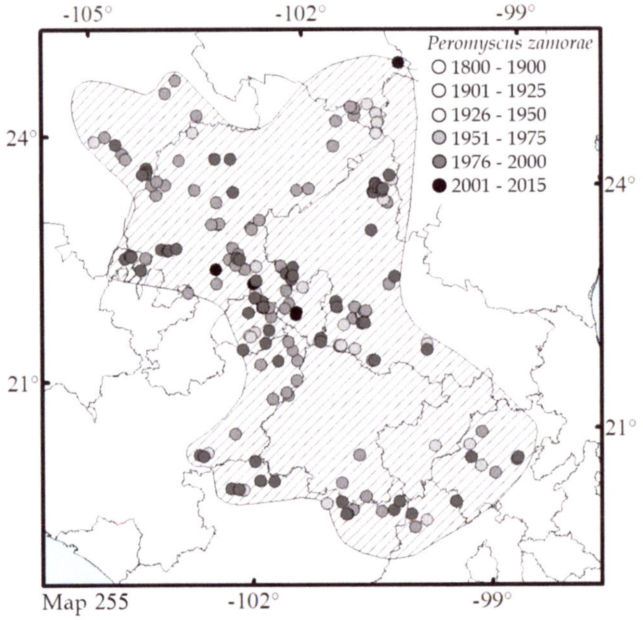


**Map 255** Distribution map of *Peromyscus zamorae*

**Additional Literature:** Baker (1952), Hooper (1955), Carleton (1989), Musser and Carleton (2005).

## *Peromyscus zarhynchus* Merriam, 1898

Chiapan deermouse, ratón de Chiapas

1. *P. z. sancristobalensis* Merriam, 1898. Central and eastern Chiapas.
2. *P. z. zarhynchus* Merriam, 1898. Restricted to the vicinity of Tumbalá, northern Chiapas.

**Conservation:** *Peromyscus zarhynchus* is listed as Special Protection by the Norma Oficial Mexicana (DOF 2019) and as Vulnerable (B1ab (iii): extent of occurrence less than 20,000 km$^2$ at ten or less locations and under continuing decline in area and quality of habitat) by the IUCN (Álvarez-Castañeda et al. 2018j).

**Characteristics:** Chiapan deermouse is the largest member in the genus; **total length 259.0–318.0 mm and skull length 34.5–37.0 mm.** Dorsal pelage dark brown; darker on the back and paler toward the flanks; fur of underparts whitish with grayish shades at the base; pectoral with chestnut color spots, sometimes continuing into the abdomen; lateral line rather broad, clear cinnamon rufous; orbital and anterorbital regions dark blackish brown not very sharply contrasted; ears naked, approximately more than the 60% of the hindfoot length; hind feet long and relatively narrow; feet soiled whitish, tarsal joint broadly brownish and proximal foot slightly brownish; hindfoot soles naked medially to the calcaneus; tail very long, **rather finely scaly, scantily clothed with short hairs**, dusky dorsally and white ventrally; supraorbital edge present and well marked; **rostrum, palatine, and nasals proportionally larger**; shelf of the bony palate rather short; supraorbital border sharp-angled but rarely showing any definite bead; zygomatic bone slightly compressed anteriorly but slightly notched; lower side of the infraorbital plate somewhat produced forward. Species group *mexicanus*.

**Comments:** *Peromyscus zarhynchus* is restricted to the highlands of central and northern Chiapas (Map 256). It can be found in sympatry, or nearly so, with *P. cordillerae*, and *P. mexicanus*. *P. zarhynchus* is very similar to *P. carolpattonae* by its overall larger external and all craniodental measurements, except for the skull height and basioccipital length but does not co-occur (Álvarez-Castañeda et al. 2017a). From *P. cordillerae* and *P. mexicanus*, by its larger size.

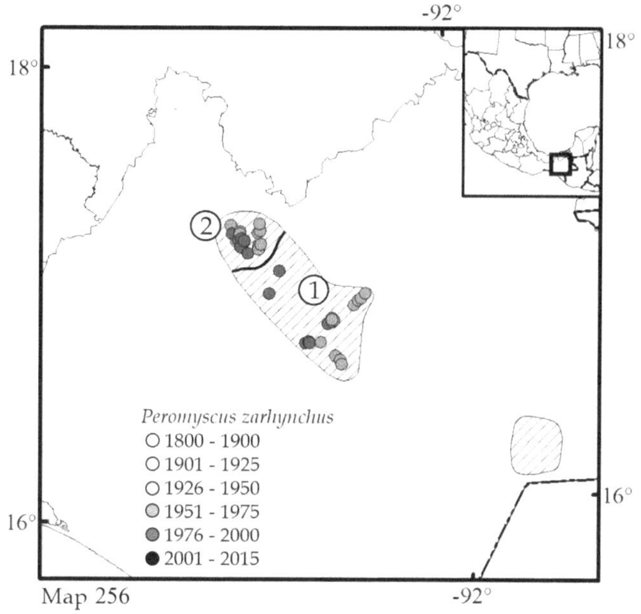


**Map 256** Distribution map of *Peromyscus zarhynchus*

**Additional Literature:** Merriam (1898a), Osgood (1909), Huckaby (1980), Carleton (1989), McClellan and Rogers (1997), Musser and Carleton (2005), Ordoñez-Garza et al. (2010), Lorenzo et al. (2016).

## Genus *Podomys* Chapman, 1889

**Three digital tubercles**; two phalangeal, the latter much reduced in size and subcircular in shape; **six mammae (nipples)**; **accessory tubercles in the salient internal angles of the molars very small**, as seen in the transverse section, never forming a loop extending to the outer tooth edge.

## *Podomys floridanus* Chapman, 1889

Florida deermouse, ratón de Florida

Monotypic.

**Conservation:** *Podomys floridanus* is listed as Near Threatened (B2b (i, ii, iii): extent of occurrence less than 5000 km$^2$, continuing decline the area of extent, continuing decline the area of occurrence, occupancy, and quality of habitat) by the IUCN (Austin and Roach 2019).

**Characteristics:** Florida deermouse is medium sized; total length 178.0–220.0 mm and skull length 28.9–30.6 mm. Dorsal pelage of the head, neck, back, and upper flanks pale ochraceous buff finely mixed with dusky shades, producing a pale grayish-cinnamon effect; lower flanks from the nose to the base of the tail rich ochraceous buff, very palely or not at all mixed with dusky shades; fur of underparts white, although some individuals have a tawny patch on the breast; feet white; ears large and thinly haired, outside of the ears dusky, inside whitish, subauricular hair tuft mixed pale ochraceous buff and dusky; **five plantar tubercles of the hind feet**; tail shorter than the head-and-body length, brown dorsally and white ventrally. **Florida mice has a distinctive skunk-like odor**; **molars slightly more hypsodont; accessory tubercles in salient internal angles of the molars very small, as seen in the transverse section, never forming a loop extending to the outer tooth edge**; supraorbital border rather sharp and shelf-like; posterior end of the nasals slightly exceeding the ascending branches of the premaxillae; palatine slits rather short and expanded.

**Comments:** *Podomys floridanus* was previously classified under the subgenus *Podomys* of *Peromyscus* (Osgood 1909; Hooper 1968); however, genetic and morphometrical analyses suggest that *P. floridanus* should be considered a full species (Carleton 1980, 1989). *P. floridanus* ranges only in the Florida peninsula (Map 257). It can be found in sympatry, or nearly so, with *P. gossypinus* and *P. polionotus*, and differs from these species by having five plantar tubercles on the hind feet; shorter tail, approximately 80% of the head-and-body length; orange shading on the cheeks, shoulders, and flanks; the transverse section of the molars never forming a loop extending to the outer edge; hypsodont molars with small accessory tubercles and an interorbital shelf.

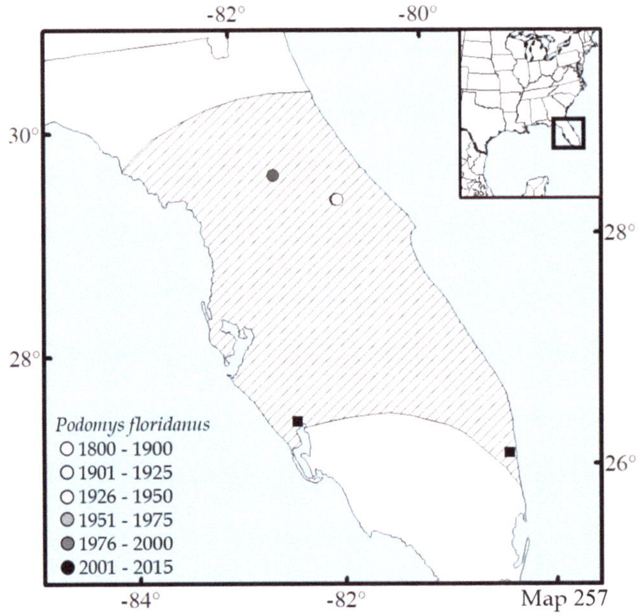


**Map 257** Distribution map of *Podomys floridanus*

**Additional Literature:** Chapman (1894), Layne (1990), Jones and Layne (1993), Musser and Carleton (2005).

# Genus *Reithrodontomys* Giglioli, 1874

Dorsal pelage from rich ochraceous-buffy to dark brown; fur of underparts usually white but sometimes pale buff or paler than in the dorsum; **upper incisors with a deep longitudinal groove near the middle of the tooth;** skull with the braincase smoothly rounded, more or less inflated, without prominent ridges; zygomatic bone slender; outer wall of the interorbital foramen with a broad thin plate; anterior palatine foramina relatively large, forming long narrow slits separated by a thin septum, slightly narrower anteriorly, terminating about at the plane of the anterior border of the toothrow; posterior border of the palate square, often with a slight medial spine, terminating at the plane of the posterior border of the toothrow; pterygoids nearly parallel; auditory bullae more or less inflated, longer than broad and obliquely situated. The descending process of the mandible with a broad flattened plate, strongly deflected inward, the lower portion twisted into a nearly horizontal position and inner margin raised, leaving a distinct depression in the ramus; the coronoid process short; first upper molars with five principal tubercles, one anterior medial one, and two pairs of lateral ones. *Reithrodontomys* has two subgenera, each with two species groups. Subgenus *Aporodon* with *mexicanus* (*R. gracilis* and *R. spectabilis*) and *tenuirostris* (*R. albilabris*, *R. microdon*, *R. tenuirostris*, and *R. wagneri*), and the subgenus *Reithrodontomys* with the species groups: *fulvescens* (*R. fulvescens* and *R. hirsutus*) and *megalotis* (*R. burti*, *R. chrysopsis*, *R. humulis*, *R. megalotis*, *R. montanus*, *R. raviventris*, *R. sumichrasti*, and *R. zacatecae*). Note: The characteristics to distinguish among subgenera correspond to the skull. The key groups are the species of both subgenera. The keys were elaborated based on the review of specimens and the following sources: Howell (1914a), Hooper (1952), Hall (1981), Álvarez-Castañeda et al. (2017a), and Porter et al. (2017).

1. Endemic to Cozumel Island, Quintana Roo; flanks orange more intense; braincase relatively flattened; zygomatic arches broad and strong ........ *Reithrodontomys* (*Aporodon*) *spectabilis* (p. 370)

1a. Ranges in other parts of North America; flanks, if orange, not intense; braincase not flattened; zygomatic arches narrow ........ 2

2. Restricted to salt marshes in vicinity of San Francisco Bay, California ........ *Reithrodontomys* (*Reithrodontomys*) *raviventris* (p. 381)

2a. Ranges in other parts of North America ........ 3

3. Restricted to the Pacific coastal plains of Sonora and Sinaloa; preauricular hair tuft and postauricular area bright buffy; zygomatic notch exceedingly long antero-posteriorly ........ *Reithrodontomys* (*Reithrodontomys*) *burti* (p. 373)

3a. Ranges in other parts of North America ........ 4

4. Restricted to the southeastern United States ........ 5

4a. Ranges in other parts of North America ........ 6

5. Total length greater than 135.0 mm; third upper molars E shaped (Fig. 75) and third lower molars S shaped (from Arizona, Texas, Oklahoma, Kansas, Missouri, Arkansas, Louisiana, and Mississippi south through Central America, except for the Yucatán Peninsula, Tabasco, and the Baja California Peninsula). ........ *Reithrodontomys* (*Reithrodontomys*) *fulvescens* (part; p. 374)

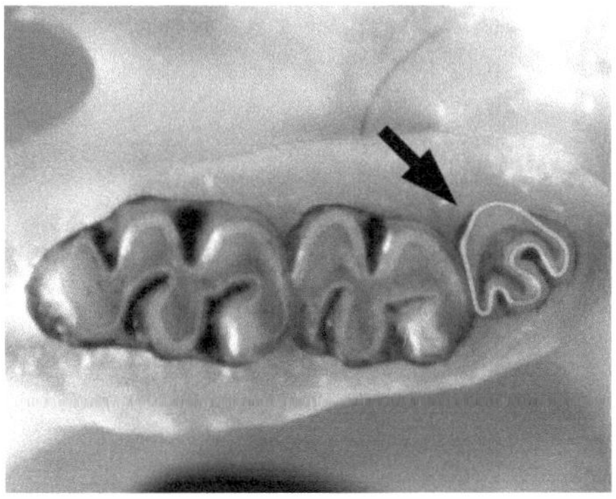

**Fig. 75** Third upper molars E shaped

5a. Total length less than 135.0 mm; third upper and lower molars C shaped (Fig. 76; from Maryland, central Virginia, and southern Ohio south through the center of Florida peninsula and from the Atlantic coast west through west Oklahoma and eastern Texas) ........ *Reithrodontomys* (*Reithrodontomys*) *humulis* (p. 377)

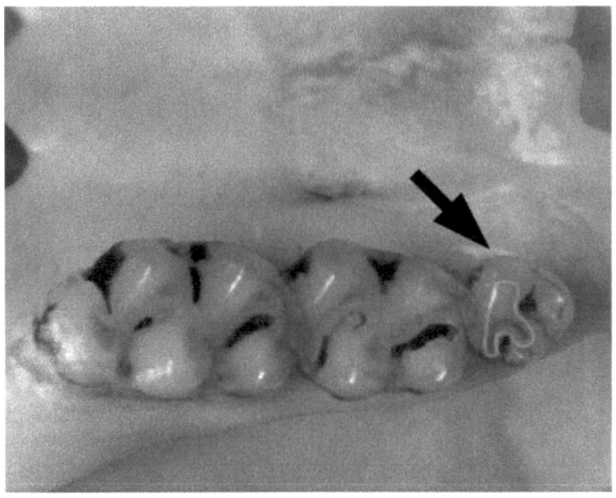

**Fig. 76** Third upper and lower molars C shaped

6. Interorbital constriction near the center of the skull; braincase slightly broader than the anterior zygomatic breadth; third upper molars with the first primary fold at least as long as the second primary fold, each usually extending more than halfway across the crown ........ 7

6a. Interorbital constriction anterior to the center of the skull; braincase notoriously broader than the anterior zygomatic breadth; third upper molars with the first primary fold distinctly shorter than the second primary fold ........ 11

7. Ear length greater than 90% of the hindfoot length; braincase breadth slightly greater than the zygomatic breadth; breadth of the zygomatic plate less than 1.6 mm ........ 8

7a. Ear length less than 90% of the hindfoot length; braincase breadth less than the zygomatic breadth; breadth of the zygomatic plate greater than 1.6 mm ........ 10

8. Dorsal pelage brown or reddish brown with fine grizzly blackish hairs; fur of underparts yellowish (from Chiapas highlands south through Central America) ........ *Reithrodontomys* (*Aporodon*) *microdon* (p. 369)

8a. Dorsal pelage brown or reddish brown with fine grizzly blackish hairs; fur of underparts whitish ........ 9

9. Total length greater than 175.0 mm; auditory bullae larger and oval (Oaxaca highlands) ........ *Reithrodontomys* (*Aporodon*) *albilabris* (p. 365)

9a. Total length less than 175.0 mm; auditory bullae smaller and C-shape rather than spherical (northern Michoacán, Estado de México, Ciudad de México, northern Morelos, and Guerrero) ........ *Reithrodontomys* (*Aporodon*) *wagneri* (p. 372)

10. Total length greater than 204.0 mm; interorbital breadth greater than 4.0 mm; rostral length greater than 9.0 mm (from Chiapas south through Guatemala) ........ *Reithrodontomys* (*Aporodon*) *tenuirostris* (p. 371)

10a. Total length less than 204.0 mm; interorbital breadth less than 4.0 mm; rostral length less than 9.0 mm ........ 12

11. Dorsal coloration of the limbs whitish or ash; skull length less than 22.0 mm, if greater, distance from the anterior foramen to the palatine less than 4.0 mm (from the Yucatán peninsula, Tabasco, and coastal Chiapas south through Central America) ........ *Reithrodontomys* (*Aporodon*) *gracilis* (p. 367)

11a. Dorsal coloration of the limbs ash; skull length greater than 22.0 mm, if smaller, distance from the anterior foramen to the palatine greater than 4.0 mm (from southern Tamaulipas and western-central Michoacán in the highlands south through Central America) ........ *Reithrodontomys* (*Aporodon*) *mexicanus* (p. 368)

12. Third upper molars E shaped and third lower molars S shaped (Fig. 75) ........ 13

12a. Third upper and lower molars C shaped (Fig. 76) ........ 14

13. Tail usually sharply bicolored, paler ventrally than dorsally; hind feet whitish or buffy above; zygomatic breadth less than 11.9 mm; interorbital breadth less than 3.4 mm; supraorbital shelf sometimes slightly elevated (from Arizona, Texas, Oklahoma, Kansas, Missouri, Arkansas, Louisiana, and Mississippi south through Central America, except for the Yucatán Peninsula, Tabasco, and the Baja California Peninsula)........................................................................................................*Reithrodontomys* (*Reithrodontomys*) *fulvescens* (part, p. 374)

13a. Tail unicolor or barely paler ventrally; tarsi and upperparts of the hind feet dusky; zygomatic breadth greater than 11.9 mm; interorbital breadth greater than 3.4 mm; supraorbital shelf trenchant or beaded (Nayarit, Colima, and Jalisco) ........................................................................................... *Reithrodontomys* (*Reithrodontomys*) *hirsutus* (p. 376)

14. Tail length less than 50% of the total length.......................................................................................................15

14a. Tail length greater than 50% of the total length...................................................................................................17

15a. Tail length on average less than the head-and-body length (from southeastern Montana and southwestern North Dakota south through northern Sonora and Chihuahua to northern Durango, and from southeastern Arizona, New Mexico, and Colorado east through the Missouri River)................................*Reithrodontomys* (*Reithrodontomys*) *montanus* (p. 380)

15. Tail length on average greater than the head-and-body length .......................................................................16

16. Tail length about 104% of the head-and-body length; dorsal coloration grayish and fur of underparts white; breadth of the zygomatic plate greater than 1.8 mm (from British Columbia, Saskatchewan, Alberta, and Wisconsin south through Oaxaca, except for southern Baja California peninsula, and from the Pacific coast east through Wisconsin, Indiana, Illinois, Arkansas, Kansas, western Oklahoma, Texas, Tamaulipas, and Veracruz) ........................................................................................... *Reithrodontomys* (*Reithrodontomys*) *megalotis* (part; p. 378)

16a. Tail length about 119% of the head-and-body length; dorsal coloration darker and fur of underparts pinkish yellow; breadth of the zygomatic plate less than 1.8 mm. This species can be distinguished from the others mainly by the karyotypes (from Sierra Madre Occidental in northeastern Chihuahua south through Jalisco and Michoacán)........................ ........................................................................................... *Reithrodontomys* (*Reithrodontomys*) *zacatecae* (p. 383)

17. No buffy hairs on the inner surfaces of the ears; ear length 17.0–19.0 mm; interorbital region strongly constricted, hourglass shaped (from southeastern Jalisco west through western-central Veracruz)............................................................ ...........................................................................................*Reithrodontomys* (*Aporodon*) *chrysopsis* (p. 366)

17a. Some buffy hairs on the inner surfaces of the ears, sometimes only a few and small; ear length less than 18.0 mm; interorbital region broad (from southwest Jalisco and north-central Veracruz south through Central America)...................... ...........................................................................................*Reithrodontomys* (*Reithrodontomys*) *sumichrasti* (p. 382)

## Subgenus *Aporodon* Howell, 1914

**Upper molars with subsidiary enamel loops in the outer primary reentrant angles, in most species, these loops reaching the outer tooth border and appearing as prominent accessory tubercles when viewed in profile**; in other species (*chrysopsis* group) the enamel loops sometimes do not reach the outer tooth border and the accessory tubercles are often absent or markedly reduced.

### *Reithrodontomys albilabris* Merriam, 1901

Oaxaca small-toothed harvest mouse, ratón de dientes pequeños de Oaxaca

Monotypic.

**Conservation:** *Reithrodontomys albilabris* (as *Reithrodontomys microdon*) is listed as Threatened by the Norma Oficial Mexicana (DOF 2019) and as Least Concern by the IUCN (Reid and Vázquez 2016e).

**Characteristics:** Oaxaca small-toothed harvest mouse is small sized; total length 179.0–187.0 mm and skull length 22.0–22.8 mm. **Dorsal pelage brown or reddish brown with fine grizzly blackish hairs**; fur of underparts whitish or yellowish; flanks intensely orange; hair long and thin; **ears orange fuscous**; narrow blackish spot around the eyes; forefoot buffy white with a dusk patch; hind feet very long, dark brown edged with whitish shades; toes white; tail longer than the body, slightly bicolored and with inconspicuous spots; **skull small**; rostrum narrow; palatal foramen ending on the plane of the front molars; interpterygoid fossa very broad; zygomatic arches slender, clearly contracted anteriorly; braincase moderately inflated, tapered and depressed posteriorly; bullae rather large and inflated; third upper and lower molars C shaped. Subgenus *Aporodon*. Species group *tenuirostris*.

**Comments:** *Reithrodontomys albilabris* was recognized as a distinct species (Martínez-Borrego et al. 2022); it was previously regarded as a subspecies of *R. microdon* (Hooper 1950). It ranges in the highlands of northern Oaxaca (Map 258). It can be found in sympatry, or nearly so, with *R. fulvescens*, *R. megalotis*, and *R. sumichrasti*. *R. albilabris* can be distinguished from *R. fulvescens*, *R. megalotis*, *R. mexicanus*, and *R. sumichrasti* by having the third upper molars with the first primary fold distinctly shorter than the second primary fold. From *R. fulvescens*, by not having a pronounced ocher to pale orange lateral liner on the flanks. From *R. mexicanus*, by having larger ears; total length more less equal to the hindfoot length; braincase breadth slightly greater than the zygomatic breadth and the zygomatic plate breadth less than 1.6 mm.

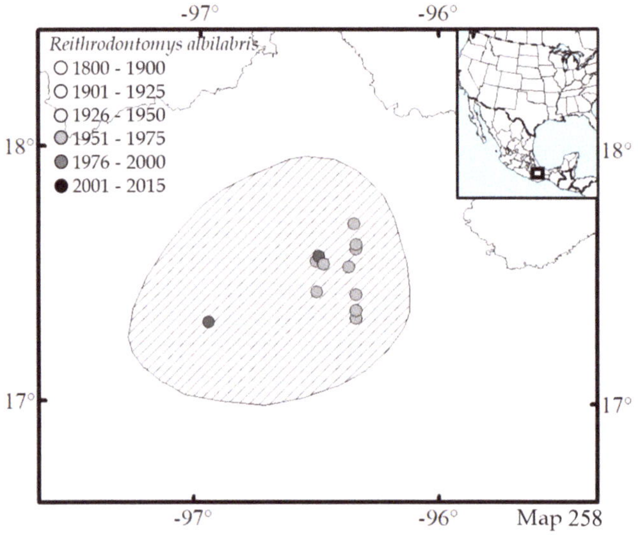


**Map 258** Distribution map of *Reithrodontomys albilabris*

**Additional Literature:** Merriam (1901a), Howell (1914a), Hooper (1952), Spencer and Cameron (1982), Arellano et al. (2003), Musser and Carleton (2005), Bradley (2017).

## ***Reithrodontomys chrysopsis*** Merriam, 1900

Volcano harvest mouse, ratón de las cosechas de los volcanes

1. *R. c. chrysopsis* Merriam, 1900. From southeast Jalisco by Eje Neovolcánico east through Puebla and Tlaxcala.
2. *R. c. perotensis* Merriam, 1901. From Puebla and Tlaxcala by Eje Neovolcánico east through central-western Veracruz.

**Conservation:** *Reithrodontomys chrysopsis* is listed as Least Concern by the IUCN (Álvarez-Castañeda et al. 2016r).

**Characteristics:** Volcano harvest mouse is small sized; total length 170.0–192.0 mm and skull length 23.0–25.1 mm. Dorsal pelage ochraceous buff to orange buff, grizzly with blackish shades, most common on the back; flanks with an indistinct cinnamon medial band; fur of underparts whitish cinnamon; spot around the eyes black; **ears fuscous-black**; **tail long, greater than 90.0 mm and fuscous above, grayish white above;** skull large; auditory bullae very large and moderately inflated; third upper and lower molars C shaped; **braincase subglobular and inflated**; interorbital region constricted; rostrum wider than the interorbital region and long; sphenopalatine bone with large pits; pterygoid fossa relatively narrow. Subgenus *Reithrodontomys*. Species group *megalotis*.

**Comments:** *Reithrodontomys chrysopsis* ranges in the Eje Neovolcánico, from southeastern Jalisco west through western-central Veracruz (Map 259). It can be found in sympatry, or nearly so, with *R. fulvescens*, *R. megalotis*, *R. sumichrasti*, and *R. wagneri*. *R. chrysopsis* can be distinguished from *R. fulvescens*, *R. megalotis*, and *R. sumichrasti* by having the third upper molars with the first primary fold distinctly shorter than the second primary fold. From *R. fulvescens*, by not having a pronounced ocher to pale orange lateral line on the flanks; from *R. megalotis*, by having the braincase breadth greater than 10.7 mm. From *R. wagneri*, by having the second primary fold in the third lower molars faint or absent; third lower molars not resembling the second lower molars in shape; zygomatic plate broader than the mesopterygoid fossa; mesopterygoid fossa not more than 75% as wide as either pterygoid fossa.

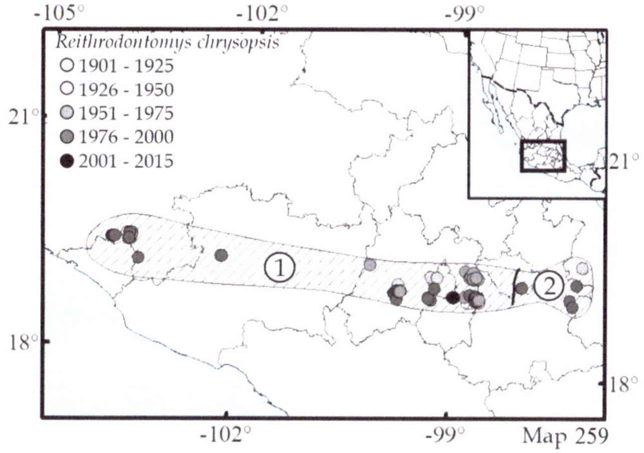


**Map 259** Distribution map of *Reithrodontomys chrysopsis*

**Additional Literature:** Merriam (1900, 1901a), Howell (1914a), Hooper (1952), Spencer and Cameron (1982), Musser and Carleton (2005).

## ***Reithrodontomys gracilis*** Allen and Chapman, 1897

Slender harvest mouse, ratón de las cosechas esbelto

1. *R. g. gracilis* Allen and Chapman, 1897. Yucatán Peninsula, Belize, and northern Guatemala.
2. *R. g. insularis* Jones, 1964. Restricted to Isla del Carmen, Campeche, México.
3. *R. g. pacificus* Goodwin, 1932. Pacific coast, from Chiapas south through Central America.

**Conservation:** *Reithrodontomys gracilis insularis* is listed as Threatened by the Norma Oficial Mexicana (DOF 2019) and as Least Concern by the IUCN (Reid et al. 2016f).

**Characteristics:** Slender harvest mouse is small sized; total length 152.0–192.0 mm and skull length 21.4–22.2 mm. **Dorsal pelage of "peppered" appearance, orange brown**; fur of underparts whitish, sometimes with a slight orange cinnamon shade; flanks of brightest color forming an indistinct lateral line; hair silky and thin; no eye ring is present; ears medium sized, sparsely haired, and dark brown to blackish; ankles fuscous; forefoot buffy white; hind feet grayish white; **tail large,** slender, scaly, and scantily haired, **homogeneous fuscous color**, slightly paler ventrally; skull small; upper incisors strongly recurved; rostrum short and broad; palatal foramen very short; nasals short; **zygomatic arches nearly parallel to the axis of the skull**; **braincase narrow** and moderately flat; bullae rather small; third upper and lower molars C shaped. Subgenus *Aporodon*. Species group *mexicanus*.

**Comments:** *Reithrodontomys gracilis* ranges from the Yucatán peninsula, Tabasco, and coastal Chiapas south through Central America (Map 260). It can be found in sympatry, or nearly so, with *R. fulvescens*, *R. mexicanus*, *R. microdon*, *R. sumichrasti*, and *R. tenuirostris*. *R. gracilis* can be distinguished from *R. fulvescens* and *R. sumichrasti* by having the third upper molars with the first primary fold distinctly shorter than the second primary fold. From *R. mexicanus*, by having the hind feet whitish or dusky above; incisive foramina length less than 4.0 mm; frontal bones broad and flattened in the interorbital area; skull length on average less than 22.0 mm. From *R. microdon*, by having the ears usually more than 3.0 mm smaller than the hind feet; rostrum relatively short and broad; zygomatic arches strong, breadth usually 0.5–1.0 mm broader than the braincase; breadth of the zygomatic plate more than 1.4 mm. From *R. fulvescens*, by not having a pronounced ocher to pale orange lateral line on the flanks. From *R. tenuirostris*, by its smaller size, with total length, interorbital breadth, and rostral length less than 204.0 mm, 4.0 mm, and 9.0 mm, respectively.

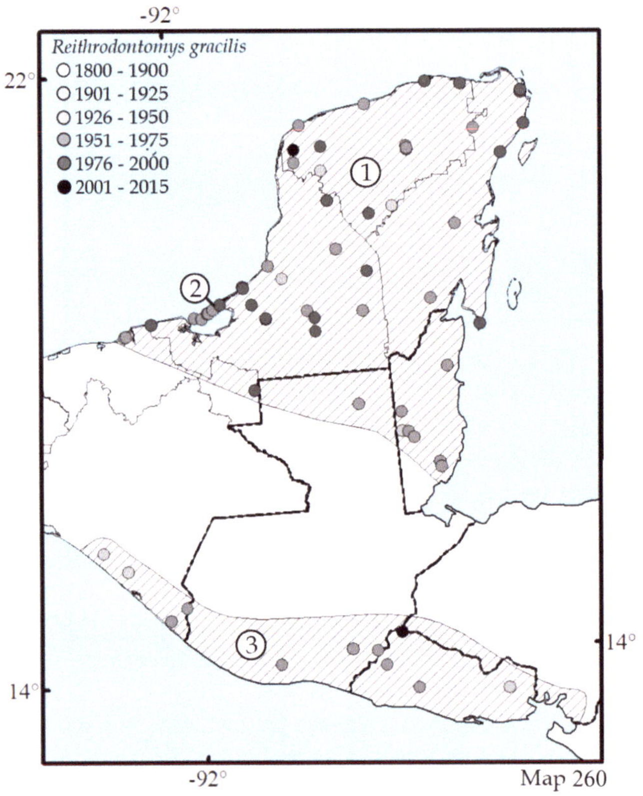


**Map 260** Distribution map of *Reithrodontomys gracilis*

**Additional Literature:** Howell (1914a), Hooper (1952), Spencer and Cameron (1982), Young and Jones (1984), Arellano et al. (2003), Musser and Carleton (2005).

## ***Reithrodontomys mexicanus*** (Saussure, 1860)

Mexican harvest mouse, ratón de las cosechas mexicano

1. *R. m. howelli* Goodwin, 1932. From central Chiapas south through Guatemala.
2. *R. m. mexicanus* (Saussure, 1860). From southern Tamaulipas and eastern San Luis Potosí south through Oaxaca.
3. *R. m. riparius* Hooper, 1955. From southern Michoacán southeast through Oaxaca.
4. *R. m. scansor* Hooper, 1950. Tehuantepec Isthmus and western Chiapas.

**Conservation:** *Reithrodontomys mexicanus* is listed as Least Concern by the IUCN (Delgado et al. 2016a)

**Characteristics:** Mexican harvest mouse is small sized; total length 161.0–203.0 mm and skull length 22.8–23.9 mm. Dorsal pelage tawny to cinnamon-brown, with grizzly blackish shades; fur of underparts whitish or yellowish; **ears large and fuscous or fuscous-black**; indistinct blackish ring around the eyes; fore and hindfeet with upperparts dark brown; **hind-foot digit base dark; tail longer than the head-and-body length, unicolor and dark, tawny dorsally and whitish ventrally; skull short, broad and flat**; rostrum broad; nasals short, ending more or less in line with the ends of the premaxillae; palatal foramen short, ending at or slightly in front of the plane of the first molars; zygomatic arches slender, contracted anteriorly; interpterygoid fossa broad; braincase squarish and depressed posteriorly; auditory bullae small; third upper and lower molars C shaped. Subgenus *Aporodon*. Species group *mexicanus*.

**Comments:** *Reithrodontomys mexicanus* ranges from southern Tamaulipas and western-central Michoacán in the highlands south through Central America (Map 261). It can be found in sympatry, or nearly so, with *R. albilabris*, *R. chrysopsis*, *R. fulvescens*, *R. gracilis*, *R. megalotis*, *R. microdon*, *R. sumichrasti*, *R. tenuirostris*, and *R. wagneri*. *R. mexicanus* can be distinguished from *R. chrysopsis*, *R. fulvescens*, *R. megalotis*, and *R. sumichrasti* by having the third upper molars with the first primary fold distinctly shorter than the second primary fold. From *R. fulvescens*, by not having a pronounced ocher to pale orange lateral line on the flanks. From *R. gracilis*, by having the hind feet dusky above; incisive foramina length greater than 4.0 mm; frontal bones strongly constricted and not markedly flattened; skull length on average greater than 22.0 mm. From *R. albilabris*, *R. microdon*, and *R. wagneri*, by having smaller ears; total length less than 80% of the hindfoot length; skull short, broad, and flat; braincase breadth less than the zygomatic breadth and zygomatic plate breadth greater than 1.6 mm. From *R. tenuirostris*, by its smaller size, with total length, interorbital breadth, and rostral length less than 204.0 mm, 4.0 mm, and 9.0 mm, respectively.

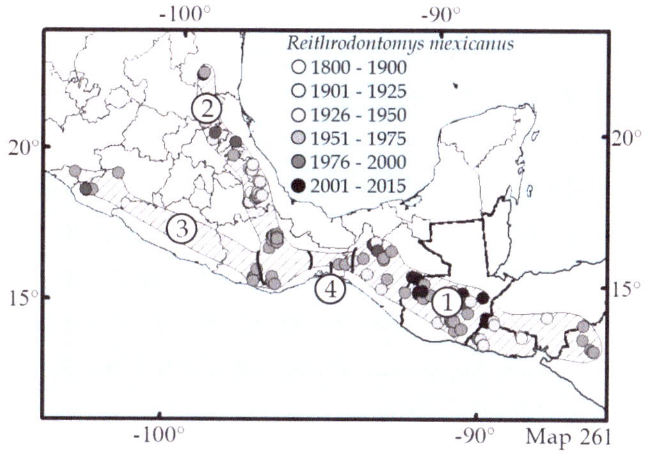


**Map 261** Distribution map of *Reithrodontomys mexicanus*

**Additional Literature:** Howell (1914a), Hooper (1952), Spencer and Cameron (1982), Arellano et al. (2003), Musser and Carleton (2005), Bradley (2017), Martínez-Borrego et al. (2020).

## *Reithrodontomys microdon* Merriam, 1901

Small-toothed harvest mouse, ratón de las cosechas de dientes pequeños

Monotypic.

**Conservation:** *Reithrodontomys microdon* is listed as Threatened by the Norma Oficial Mexicana (DOF 2019) and as Least Concern by the IUCN (Reid and Vázquez 2016e).

**Characteristics:** Small-toothed harvest mouse is small sized; total length 180.0–185.0 mm and skull length 22.4–22.5 mm. **Dorsal pelage brown or reddish brown with fine grizzly blackish hairs**; fur of underparts whitish or yellowish; intense orange lateral line on the flanks; hair long and thin; **ears orange fuscous**; narrow blackish spot around the eyes; forefoot buffy white with a dusky patch; hind feet very long, dark brown edged with whitish shades; toes white; tail longer than the head-and-body length, slightly bicolored, **fuscous dorsally and pale ventrally,** and with inconspicuous spots; **skull small**; rostrum narrow; palatal foramen ending on the plane of the front molars; interpterygoid fossa very broad; zygomatic arches slender, clearly contracted anteriorly; braincase moderately inflated, tapered and depressed posteriorly; bullae rather large and inflated; third upper and lower molars C shaped. Subgenus *Aporodon*. Species group *tenuirostris*.

**Comments:** *Reithrodontomys albilabris* and *R. wagneri* were recognized as a distinct species (Martínez-Borrego et al. 2022); it was previously regarded as a subspecies of *R. microdon* (Merriam 1901a; Hooper 1950). *R. microdon* ranges in central and south Chiapas south through Central America (Map 262). It can be found in sympatry, or nearly so, with *R. fulvescens*, *R. mexicanus*, *R. sumichrasti*, and *R. tenuirostris*. *R. microdon* can be distinguished from *R. fulvescens* and *R. sumichrasti* by having the third upper molars with the first primary fold distinctly shorter than the second primary fold. From *R. fulvescens*, by not having a pronounced ocher to pale orange lateral line on the flanks. From *R. mexicanus*, by having larger ears, total length more less equal to the hindfoot length; braincase breadth slightly greater than the zygomatic breadth, and zygomatic plate breadth less than 1.6 mm.

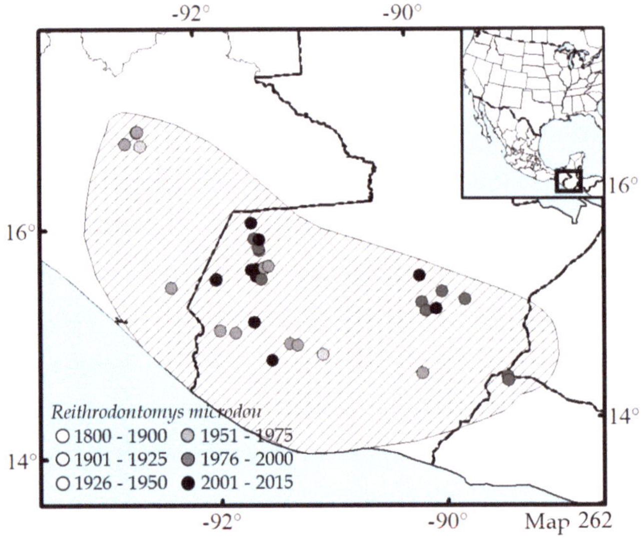


**Map 262** Distribution map of *Reithrodontomys microdon*

**Additional Literature:** Howell (1914a), Hooper (1952), Spencer and Cameron (1982), Arellano et al. (2003), Musser and Carleton (2005), Bradley (2017), Marines-Macías et al. (2021), Martínez-Borrego et al. (2022).

## ***Reithrodontomys spectabilis*** Jones and Lawlor, 1965

Cozumel harvest mouse, ratón de de Isla Cozumel

Monotypic.

**Conservation:** *Reithrodontomys spectabilis* is listed as Threatened by the Norma Oficial Mexicana (DOF 2019) and as Critically Endangered (C2a (i): small population size with less than 250 mature individuals and under continuing decline with less than 50 mature individuals in each subpopulation) by the IUCN (Vázquez et al. 2018).

**Characteristics:** Cozumel harvest mouse is medium sized**; total length 205.0–221.0 mm** and skull length 24.6–26.2 mm. Dorsal pelage brownish ochraceous grizzly with blackish shades; fur of underparts grayish white with the root plumbeous; buff pectoral spot sometimes present; **flanks more intensely orange**; hair silky and thin; ears medium sized and pale brown; tail homogeneous, dark brown dorsally and paler ventrally; **braincase relatively flattened; zygomatic arches broad and strong**; rostrum relatively short and broad; incisive foramina rarely reaching the level of the first upper molars; teeth large; auditory bullae larger but only moderately inflated. Subgenus *Aporodon*. Species group *mexicanus*.

**Comments:** *Reithrodontomys spectabilis* is restricted to the Isla de Cozumel, Quintana Roo (Map 263), and is not in sympatry, or nearly so, with any other species of *Reithrodontomys*. *R. spectabilis* can be distinguished from all other species by having the largest size within *Reithrodontomys;* occlusal surface of the third lower molars similar to the second molars but smaller; the zygomatic plate equal to or slightly wider than the mesopterygoid fossa.

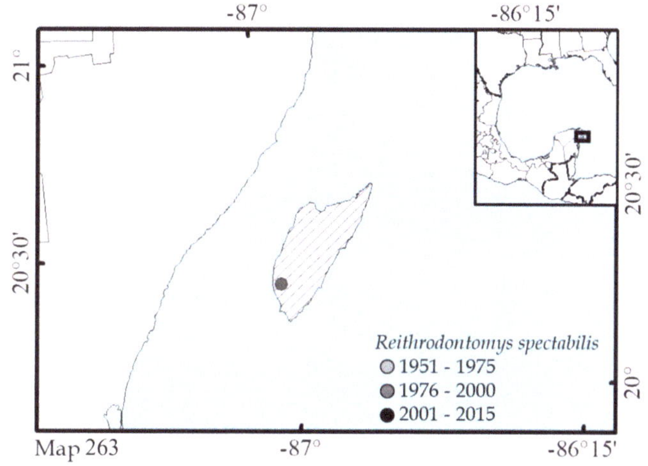


**Map 263** Distribution map of *Reithrodontomys spectabilis*

**Additional Literature:** Jones and Lawlor (1965), Hall (1981), Jones (1982), Spencer and Cameron (1982), Arellano et al. (2003), Musser and Carleton (2005).

## ***Reithrodontomys tenuirostris*** Merriam, 1901

Narrow-nosed harvest mouse, ratón de las cosechas de Guatemala

Monotypic.

**Conservation:** *Reithrodontomys tenuirostris* is listed as Endangered (B1ab (i, iii): extent of occurrence less than 5000 km$^2$, known to exist at no more than five locations, and under continuing decline in area, and quality of habitat) by the IUCN (Vázquez 2019b).

**Characteristics:** Narrow-nosed harvest mouse is small sized; total length 205.0–239.0 mm andskull length 24.5–25.3 mm. Dorsal pelage reddish brown with an orange lateral line; fur of underparts cinnamon yellowish; hair long and slightly shaggy; ears blackish; spot around the eyes narrow and blackish; anterior snout region elongated; **tail long, more than 125% of the head-and-body length**; hind feet long; rostrum long and narrow; nasals tapered to a point posteriorly, ending on a line with the premaxillae; anterior portion of the frontal bones abruptly depressed, forming a shallow sulcus at the posterior end of the nasals; palatal foramen relatively short, not reaching the plane of the first molars; zygomatic arches slender, slightly contracted anteriorly; **braincase broad, markedly inflated, depressed posteriorly**; bullae small and rather flat; third upper and lower molars C shaped. Subgenus *Aporodon*. Species group *tenuirostris*.

**Comments:** *Reithrodontomys tenuirostris* is restricted to the Chiapas and central Guatemala highlands (Map 264). It can be found in sympatry, or nearly so, with *R. fulvescens*, *R. gracilis*, *R. mexicanus*, *R. microdon*, and *R. sumichrasti*. *R. tenuirostris* can be distinguished from *R. fulvescens* and *R. sumichrasti* by having the third upper molars with the first primary fold distinctly shorter than the second primary fold. From *R. fulvescens*, by not having a pronounced ocher to pale orange lateral line on the flanks. From *R. gracilis* and *R. mexicanus*, by its larger size, with total length, interorbital breadth, and rostral length greater than 204.0 mm, 4.0 mm, and 9.0 mm, respectively. From *R. microdon*, by its smaller size and ears blackish and smaller than the hind feet; tail long, more than 125% of the head-and-body length; swollen braincase and narrow rostrum.

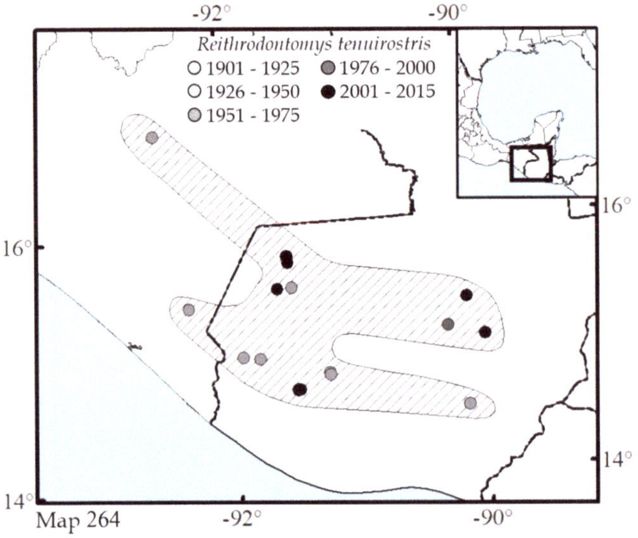


**Map 264** Distribution map of *Reithrodontomys tenuirostris*

**Additional Literature:** Merriam (1901a), Howell (1914a), Hooper (1952), Rogers et al. (1983), Arellano and Rogers (1994), Musser and Carleton (2005), Bradley (2017), Martínez-Borrego et al. (2022).

## ***Reithrodontomys wagneri*** Merriam, 1901

Michoacán small-toothed harvest mouse, ratón de dientes pequeños de Michoacán

1. *R. m. wagneri* Hooper, 1950. Central Michoacán, eastern Estado de México, Ciudad de México, and northern Morelos.
2. *R. m. bakeri* Bradley et al., 2004. Restricted to the montane region of central Guerrero.

**Conservation:** *Reithrodontomys wagneri bakeri* (as *Reithrodontomys bakeri*) is listed as Endangered (B1ab (iii): extent of occurrence less than 5000 km$^2$, known to exist at no more than five locations, and under continuing decline in area and quality of habitat) by the IUCN (Lacher et al. 2018). *Reithrodontomys wagneri* (as *R. microdon*) is listed as Threatened by the Norma Oficial Mexicana (DOF 2019) and as Least Concern by the IUCN (Álvarez-Castañeda et al. 2017a).

**Characteristics:** Michoacán small-toothed harvest mouse is small sized; total length 169.0–173.0 mm and skull length 21.6–22.1 mm. Dorsal pelage ochraceous-tawny with a small proportion of blackish wash dorsally; **fur of underparts whitish or pale pinkish buff, white at the base throughout on the throat**; face side and flanks nearly pure ochraceous-tawny or cinnamon, elsewhere on the body plumbeous-black; hind feet from the ankles to the base of the toes fuscous bordered by whitish; **tail unicolor, fuscous dorsally and pale ventrally, scantily haired at the base and more heavily haired at the tip**; ears fuscous, sparsely sprinkled internally and with brownish or blackish hairs externally; rostrum long and narrow; nasals tapered to a point posteriorly, ending on a line with the premaxillae; anterior portion of the frontal bones abruptly depressed, forming a shallow sulcus at the posterior end of the nasals; palate long (molar row length about 80% of the palate length), **auditory bullae smaller and C shaped rather than spherical;** third upper and lower molars C shaped. Subgenus *Aporodon*. Species group *tenuirostris*.

**Comments:** *Reithrodontomys wagneri* was recognized as a distinct species (Martínez-Borrego et al. 2022); it was previously regarded as a subspecies of a subspecies of *R. microdon* (Hooper 1950). It ranges in northern Michoacán, Estado de México, Ciudad de México, northern Morelos, and Guerrero (Map 265). It can be found in sympatry, or nearly so, with *R. chrysopsis*, *R. fulvescens*, *R. gracilis*, *R. megalotis*, *R. mexicanus*, and *R. sumichrasti*. *R. wagneri* can be distinguished from *R. chrysopsis* by having a second primary fold in the third lower molars well developed; third lower molars resembling but smaller than, the second lower molars; the zygomatic plate small, if any, broader than the mesopterygoid fossa; mesopterygoid fossa approximately as wide as the pterygoids. From *R. fulvescens*, *R. megalotis*, and *R. sumichrasti*, by having the third upper molars with the first primary fold distinctly shorter than the second primary fold. From *R. fulvescens*, by not having a pronounced ocher to pale orange lateral line on the flanks. From *R. mexicanus*, by having larger ears; total length more less equal to the hindfoot length; braincase breadth slightly greater than the zygomatic breadth and zygomatic plate breadth less than 1.6 mm.

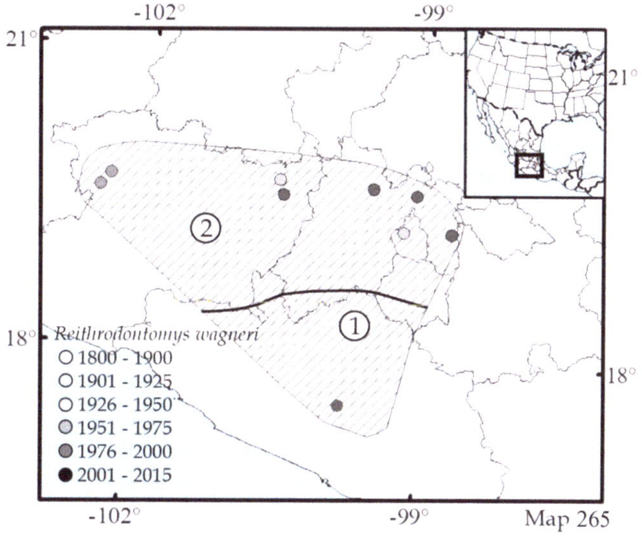


**Map 265** Distribution map of *Reithrodontomys wagneri*

**Additional Literature:** Merriam (1901a), Howell (1914a), Hooper (1952), Spencer and Cameron (1982), Arellano et al. (2003), Musser and Carleton (2005), Bradley et al. (2004b), Bradley (2017).

## Subgenus ***Reithrodontomys*** Giglioli, 1874

**Enamel pattern of the upper molars simple, first and second with two outer reentrant angles each, usually without accessory tubercles.**

### ***Reithrodontomys burti*** Benson, 1939

Sonoran harvest mouse, ratón de las cosechas sonorense

Monotypic.

**Conservation:** *Reithrodontomys burti* is listed as **Data Deficient** by the IUCN (Álvarez-Castañeda et al. 2019e).

**Characteristics:** Sonoran harvest mouse is small sized; total length 116.0–132.0 mm and skull length 19.5–21.3 mm. Dorsal pelage buff to pale cinnamon with blackish shades on the back at the tip; flanks and hip paler; fur of underparts white with leaden gray shades at the base; **cheeks bright buffy**; **preauricular hair tuft and postauricular area bright buffy;** ankles whitish or with a dark line; **pre and post eye areas intensely yellow**; tail faintly bicolored, drab dorsally and whitish ventrally; skull elongated and narrow; **zygomatic notch exceedingly long anteroposteriorly**; third upper and lower molars C shaped; **zygomatic anterior and posterior breadth equal;** interorbital foramen large; nasal proportionally large**.** Subgenus *Reithrodontomys*. Species group *megalotis*.

**Comments:** *Reithrodontomys burti* ranges from western-central Sonora south through western-central Sinaloa (Map 266). *R. burti* is found in sympatry, or nearly so, with *R. fulvescens* and *R. megalotis*, from which it differs by having a preauricular hair tuft and the postauricular area bright buffy; without a pronounced ocher to pale orange lateral line on the flanks and zygomatic notch exceedingly long anteroposteriorly. From all other species of *Reithrodontomys*, by having a preauricular hair tuft and the postauricular area bright buffy; cheeks bright buffy; ankles completely white with a dusky stripe and zygomatic notch exceedingly long anteroposteriorly. In addition, from *R. megalotis*, by having the tail short, less than 95% of the head-and-body length; total length less than 140.0 mm and braincase breadth usually less than 9.8 mm.

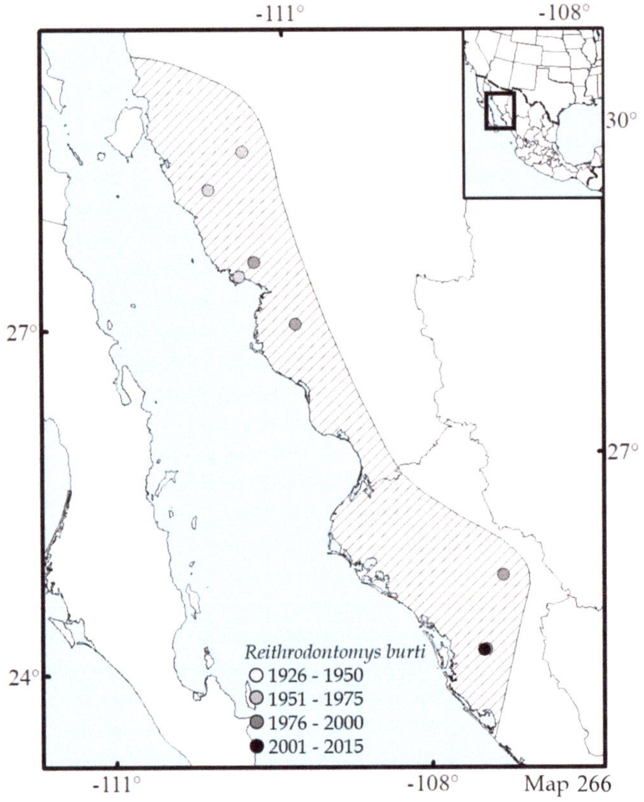


**Map 266** Distribution map of *Reithrodontomys burti*

**Additional Literature:** Benson (1939), Hooper (1952), Spencer and Cameron (1982), Álvarez-Castañeda and Cortés-Calva (1999), Musser and Carleton (2005).

## *Reithrodontomys fulvescens* Allen, 1894

Fulvous harvest mouse, ratón de las cosechas leonado

1. *R. f. amoenus* (Elliot, 1905). Restricted to the Tehuantepec Isthmus, Oaxaca.
2. *R. f. aurantius* Allen, 1895. From southeast Kansas and southwestern Missouri south through Texas to western Mississippi.

3. *R. f. canus* Benson, 1939. From southwestern New Mexico and Texas south through Chihuahua, central Durango, and western Coahuila.
4. *R. f. chiapensis* Howell, 1914. From the Tehuantepec Isthmus to central Chiapas to Honduras.
5. *R. f. difficilis* Merriam, 1901. Central-western Veracruz, eastern Hidalgo, and Puebla.
6. *R. f. fulvescens* Allen, 1894. Southeastern Arizona and northeastern Sonora.
7. *R. f. griseoflavus* Merriam, 1901. From central Durango and Zacatecas south through Jalisco, northwestern Michoacán, and Guanajuato.
8. *R. f. helvolus* Merriam, 1901. From east Guerrero to central and northern Oaxaca.
9. *R. f. infernatis* Hooper, 1950. Western Tlaxcala, southwestern Puebla, and northern Oaxaca.
10. *R. f. intermedius* Allen, 1895. Southeastern Texas, western Coahuila, Nuevo León, and central Tamaulipas.
11. *R. f. laceyi* Allen, 1896. Southwestern Oklahoma and central Texas.
12. *R. f. mustelinus* Howell, 1914. From central and south Michoacán, southern Estado de México southeast through Guerrero and coast of Oaxaca.
13. *R. f. nelsoni* Howell, 1914. Coast of southern Colima and southern Jalisco.
14. *R. f. tenuis* Allen, 1899. From southern Sonora and Chihuahua south through Nayarit and northern Jalisco.
15. *R. f. toltecus* Merriam, 1901. From central Michoacán and southern Guanajuato east through southeastern Hidalgo, Tlaxcala, and southwestern Puebla.
16. *R. f. tropicalis* Davis, 1944. From southern of Tamaulipas and eastern San Luis Potosí south through central-southern Veracruz.

**Conservation:** *Reithrodontomys fulvescens* is listed as Least Concern by the IUCN (Cassola 2016cw).

**Characteristics:** Fulvous harvest mouse is medium sized; total length 145.0–189.0 mm and skull length 20.4–23.6 mm. **Dorsal pelage ochraceous-buff grizzled with black brown; lateral line ocher to pale orange pronounced on the flanks;** fur of underparts whitish or pale yellowish; **ears brown externally and inner surface clothed with ochraceous-tawny shades**; front and sides of the face tinged with grayish shades; feet buffy white; **tail large,** longer than the head-and-body length, bicolored, brown dorsally and grayish white ventrally; **rostrum large**; interpterygoid fossa broad; **incisive foramina slightly longer than the rostrum; braincase elongated**; **third upper molars E shaped and third lower molars S shaped.** Subgenus *Reithrodontomys*. Species group *fulvescens*.

**Comments:** *Reithrodontomys fulvescens* ranges from southern Arizona, Texas, Oklahoma, south of Kansas, southwestern Missouri, Arkansas, Louisiana, and west Mississippi south through Central America (Map 267), excluding the Yucatán Peninsula, Tabasco, and the Baja California Peninsula. *R. fulvescens* is found in sympatry, or nearly so, *R. albilabris*, *R. burti*, *R. chrysopsis*, *R. gracilis*, *R. hirsutus*, *R. humulis*, *R. megalotis*, *R. mexicanus*, *R. microdon*, *R. montanus*, *R. sumichrasti*, *R. tenuirostris*, *R. wagneri*, and *R. zacatecae*. *R. fulvescens* can be distinguished from all other species by having a pronounced ocher to pale orange lateral line on the flanks. From *albilabris*, *R. gracilis*, *R. hirsutus*, *R. mexicanus*, *R. microdon*, and *R. wagneri*, by having the third upper molars with the first primary fold at least as long as the second primary fold, each usually extending more than halfway across the crown. From *R. burti*, *R. chrysopsis*, *R. hirsutus*, *R. humulis*, *R. megalotis*, *R. montanus*, *R. sumichrasti*, *R. tenuirostris*, and *R. zacatecae*, by having a worn occlusal surface of the third lower molars E shaped. From *R. humulis*, by having a reddish coloration, dark mid-dorsal stripe typically present, first primary fold of the third upper molars at least as long as the second primary fold, each usually extending more than halfway across the crown.

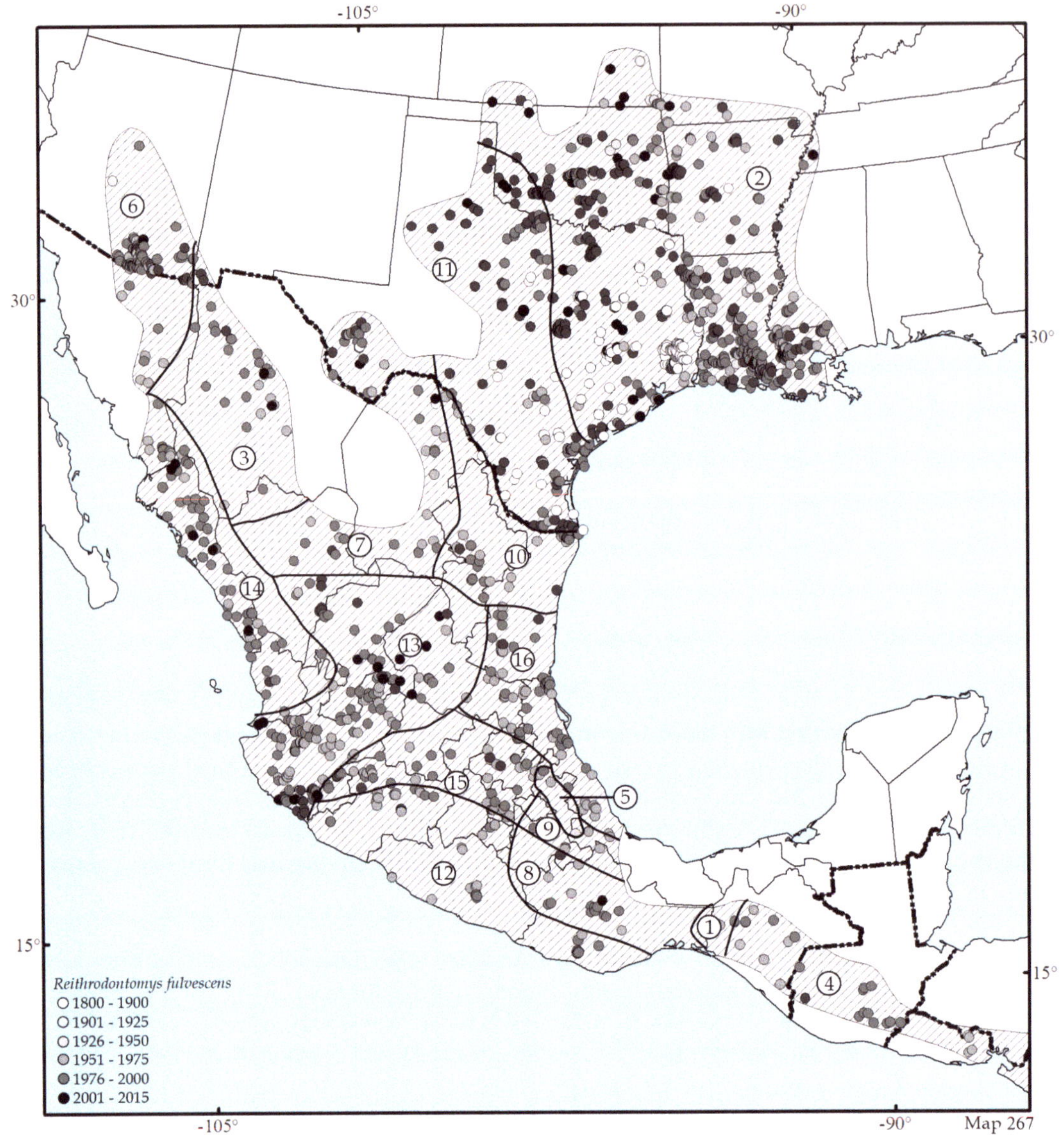


**Map 267** Distribution map of *Reithrodontomys fulvescens*

**Additional Literature:** Merriam (1901a), Howell (1914a), Hooper (1952), Spencer and Cameron (1982), Musser and Carleton (2005), Bradley (2017).

## ***Reithrodontomys hirsutus*** Merriam, 1901

Hairy harvest mouse, ratón de las cosechas leonado

Monotypic.

**Conservation:** *Reithrodontomys hirsutus* is listed as Vulnerable (B1ab (iii): extent of occurrence less than 20,000 km$^2$ at ten or less locations and under continuing decline in area and quality of habitat) by the IUCN (Álvarez-Castañeda et al. 2018h).

**Characteristics:** Hairy harvest mouse is medium sized; **total length 175.0–202.0 mm** and skull length 22.9–24.6 mm. Dorsal ochraceous-buff grizzly with blackish brown shades; fur of underparts grayish white, usually with a distinct tinge of pale buff; flanks brighter than the dorsum; ears brownish; forefoot whitish, washed with pale buff; hind feet grayish white, tinged with dusky shades; ankles fuscous; tail fuscous dorsally and grayish white ventrally; skull large and robust; rostrum and nasals short; ascending branches of the premaxillae extending to the back of the end of the nasals; palatal foramen short and wide; a well-defined supraorbital bead extending back to the parietals; auditory bullae very small; braincase flat and somewhat tapered posteriorly; upper molars with subsidiary enamel loops; third upper and lower molars C shaped. Subgenus *Reithrodontomys*. Species group *fulvescens*.

**Comments:** *Reithrodontomys hirsutus* ranges from southern-central Nayarit south through southwestern Jalisco and northern Colima (Map 268). It can be found in sympatry, or nearly so, with *R. fulvescens*, *R. megalotis*, and *R. sumichrasti*. It differs from these species by having the third upper molars with the first primary fold distinctly shorter than the second primary fold and a worn occlusal surface of the left third lower molars E shaped. From *R. fulvescens*, by not having a pronounced ocher to pale orange lateral line on the flanks.

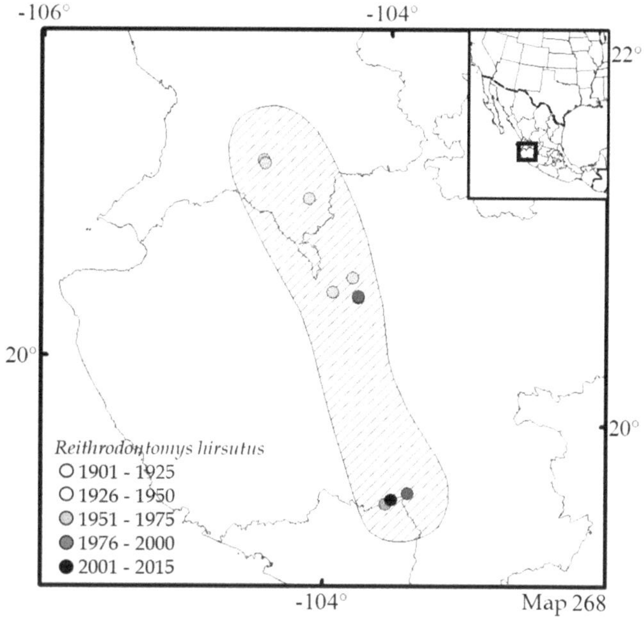


**Map 268** Distribution map of *Reithrodontomys hirsutus*

**Additional Literature:** Merriam (1901a), Howell (1914a), Spencer and Cameron (1982), Musser and Carleton (2005), Bradley (2017).

## ***Reithrodontomys humulis*** (Audubon and Bachman, 1841)

Eastern harvest mouse, ratón de las cosechas del este

1. *R. h. humulis* (Audubon and Bachman, 1841). From eastern Arkansas and Louisiana east through the Atlantic coast, and from Virginia south through south central Florida.
2. *R. h. merriami* Allen, 1895. From western Arkansas and Louisiana west through eastern Oklahoma and Texas.
3. *R. h. virginianus* Howell, 1940. Restricted to southwestern Maryland and eastern Virginia.

**Conservation:** *Reithrodontomys humulis* is listed as Least Concern by the IUCN (Cassola 2016cx).

**Characteristics:** Eastern harvest mouse is small sized**; total length 107.0–128.0 mm and skull length 19.0–21.2 mm.** Dorsal pelage grayer or dark brown to blackish, mixture of blackish brown and cinnamon, usually darkest on the back; distinct medial line black; fur of underparts ashy, usually with a tinge of pale cinnamon; ears fuscous or fuscous-black; feet grayish white; tail bicolored, fuscous dorsally and grayish white ventrally; rostrum short and broad; nasals broad, ending with the premaxillae at the same level; zygomatic arches parallel or slightly expanded posteriorly; palatal foramen broadest in the middle, ending posteriorly about on line with the plane of the first molars; braincase narrow and markedly arched; bullae small and elongated; third upper and lower molars C shaped. Subgenus *Reithrodontomys*. Species group *megalotis*.

**Comments:** *Reithrodontomys humulis* ranges from Maryland, central Virginia, and southern Ohio south through the Florida peninsula, and from the Atlantic coast west through west Oklahoma and eastern Texas (Map 269). It can be found in sympatry, or nearly so, with *R. fulvescens*, *R. megalotis*, and *R. montanus*. *R. humulis* can be distinguished from *R. fulvescens* by having a less reddish coloration, dark mid-dorsal stripe typically absent, first primary fold of the third upper molars distinctly shorter than the second primary fold, extending less than halfway across the crown, a worn occlusal surface of the third lower molars C shaped and without a pronounced ocher to pale orange lateral line on the flanks. From *R. megalotis*, by having a distinct labial ridge, often with cusplets, on the first and second lower molars; major fold and second primary fold in the first and second upper molars tending to coalesce, isolating the anterior from the posterior cusps. From *R. montanus*, by having a smaller average size in the following measurements: total length 114.0 mm; tail length 53.0 mm and hindfoot length 15.0 mm.

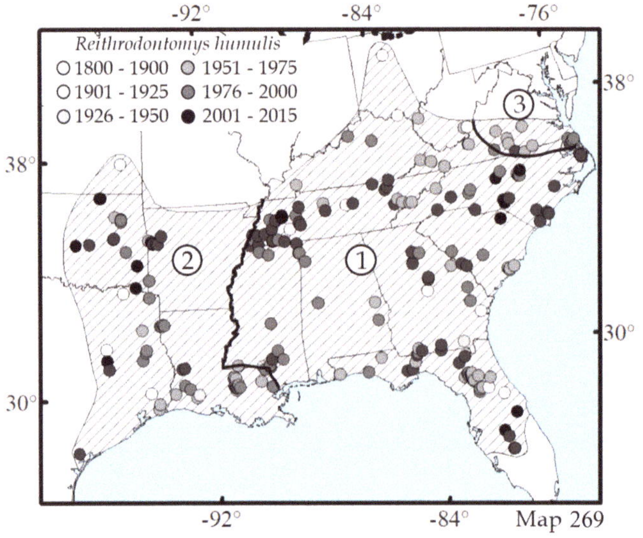


**Map 269** Distribution map of *Reithrodontomys humulis*

**Additional Literature:** Howell (1914a, 1940), Hooper (1952), Spencer and Cameron (1982), Dick (1997), Musser and Carleton (2005).

## ***Reithrodontomys megalotis*** (Baird, 1858)

Western harvest mouse, ratón de las cosechas común

1. *R. m. alticolus* Merriam, 1901. Eastern Guerrero and Oaxaca.
2. *R. m. amoles* Howell, 1914. Restricted to Sierra Gorda, northern Querétaro.
3. *R. m. arizonensis* Allen, 1895. Restricted to the Chiricahua Mountains, southeast of Arizona.
4. *R. m. aztecus* Allen, 1893. From eastern Utah, western Colorado, and southwest of Kansas to northern Arizona, northern New Mexico, and northwestern Texas.
5. *R. m. catalinae* (Elliot, 1904). Restricted to Santa Catalina Island, California.

6. *R. m. distichlis* von Bloeker, 1937. Restricted to the Salinas River, western California.
7. *R. m. dychei* Allen, 1895. From southwest Saskatchewan south through eastern Colorado and east through northeast Arkansas to northeastern Indiana.
8. *R. m. hooperi* Goodwin, 1954. Restricted to central-western Tamaulipas.
9. *R. m. limicola* von Bloeker, 1932. Restricted to Los Angeles Valley, southwestern California.
10. *R. m. longicaudus* (Baird, 1858). From southwestern Oregon southward through western California and northwestern Baja California.
11. *R. m. megalotis* (Baird, 1858). From southern-central British Columbia south through Guanajuato.
12. *R. m. pectoralis* Hanson, 1944. Restricted to Wisconsin.
13. *R. m. peninsulae* (Elliot, 1903). Northern Baja California.
14. *R. m. ravus* Goldman, 1939. Restricted to Stansbury Island, Utah.
15. *R. m. santacruzae* Pearson, 1951. Restricted to Santa Cruz Island, California.
16. *R. m. saturatus* Allen and Chapman, 1897. From Coahuila and Nuevo León south through Puebla and central Veracruz, and west through Michoacán.

**Conservation:** *Reithrodontomys megalotis dycher* is listed as Endangered and *R. m. megalotis* as Special Concern by the Endangered Species Act of Canada, and *R. megalotis* as Least Concern by the IUCN (Cassola 2016cy).

**Characteristics:** Western harvest mouse is small sized; total length 120.0–170.0 mm and skull length 19.6–22.9 mm. Species highly variable in the dorsal pelage, mixed blackish brown and pale ochraceous-buff, darkest in the middle of the back, shading to nearly pure buff on the flanks; **fur of underparts whitish**; **ears usually without dark markings;** ears drab, usually with a hair tuft of ochraceous buff at the base; feet white; tail bicolored, brown dorsally and whitish ventrally**, more than 100% of the total length, and lightly haired**; third upper and lower molars C shaped; braincase broad, flat; palatal foramina long; interpterygoid fossa narrow. Subgenus *Reithrodontomys*. Species group *megalotis*.

**Comments:** *Reithrodontomys zacatecae* was recognized as a distinct species (Hood et al. 1984). *R. megalotis* ranges from British Columbia, Saskatchewan, Alberta, and Wisconsin south through Oaxaca, except for southern Baja California peninsula, and from the Pacific coast east through Wisconsin, Indiana, Illinois, Arkansas, Kansas, western Oklahoma, Texas, Sonora, Chihuahua, Coahuila, Nuevo León, Tamaulipas, and Veracruz (Map 270). It can be found in sympatry, or nearly so, with *R. albilabris*, *R. burti*, *R. chrysopsis*, *R. fulvescens*, *R. hirsutus*, *R. humulis*, *R. mexicanus*, *R. montanus*, *R. raviventris*, *R. sumichrasti*, *R. wagneri*, and *R. zacatecae*. *R. megalotis* can be distinguished from *R. albilabris*, *R. chrysopsis*, *R. hirsutus*, *R. humulis*, *R. mexicanus*, and *R. wagneri* by having the third upper molars with the first primary fold at least as long as the second primary fold, each usually extending more than halfway across the crown. From *R. chrysopsis*, by having the braincase breadth less than 10.7 mm. From *R. burti*, by having a pronounced ocher to pale orange lateral line on the flanks and not having a preauricular hair tuft and a postauricular area bright buffy; the zygomatic notch not exceedingly long anteroposteriorly. From *R. fulvescens*, by having a worn occlusal surface of the third lower molars C shaped, and without a pronounced ocher to pale orange lateral line on the flanks. From *R. humulis*, by having no distinct labial ridge on the first and second lower molars, a major fold and second primary fold in the first and second upper molars meeting but not coalescing, and thus isolating the anterior from the posterior cusps. From *R. montanus*, by having a long tail, more than 90% of the head-and-body length, average of the total length more than 140.0 mm; longer ears; larger braincase, breadth greater than 9.6 mm; zygomatic arches without parallel sides. From *R. raviventris*, by having the dorsal pelage shorter and less dense, varying from pale buff to reddish brown; ears buffy or fuscous and with a broad range, not restricted to salt marshes in vicinity of San Francisco Bay, California. From *R. sumichrasti*, by having the tail shorter than the head-and-body length, usually less than 85.0 mm; breadth of the mesopterygoid fossa less than 1.3 mm; braincase breadth less than 10.7 mm; less than the distance between the posterior palatine foramina. From *R. zacatecae*, by its generally larger size; dorsal fur darker and underparts paler; braincase breadth less than 10.7 mm; mainly by genetic data.

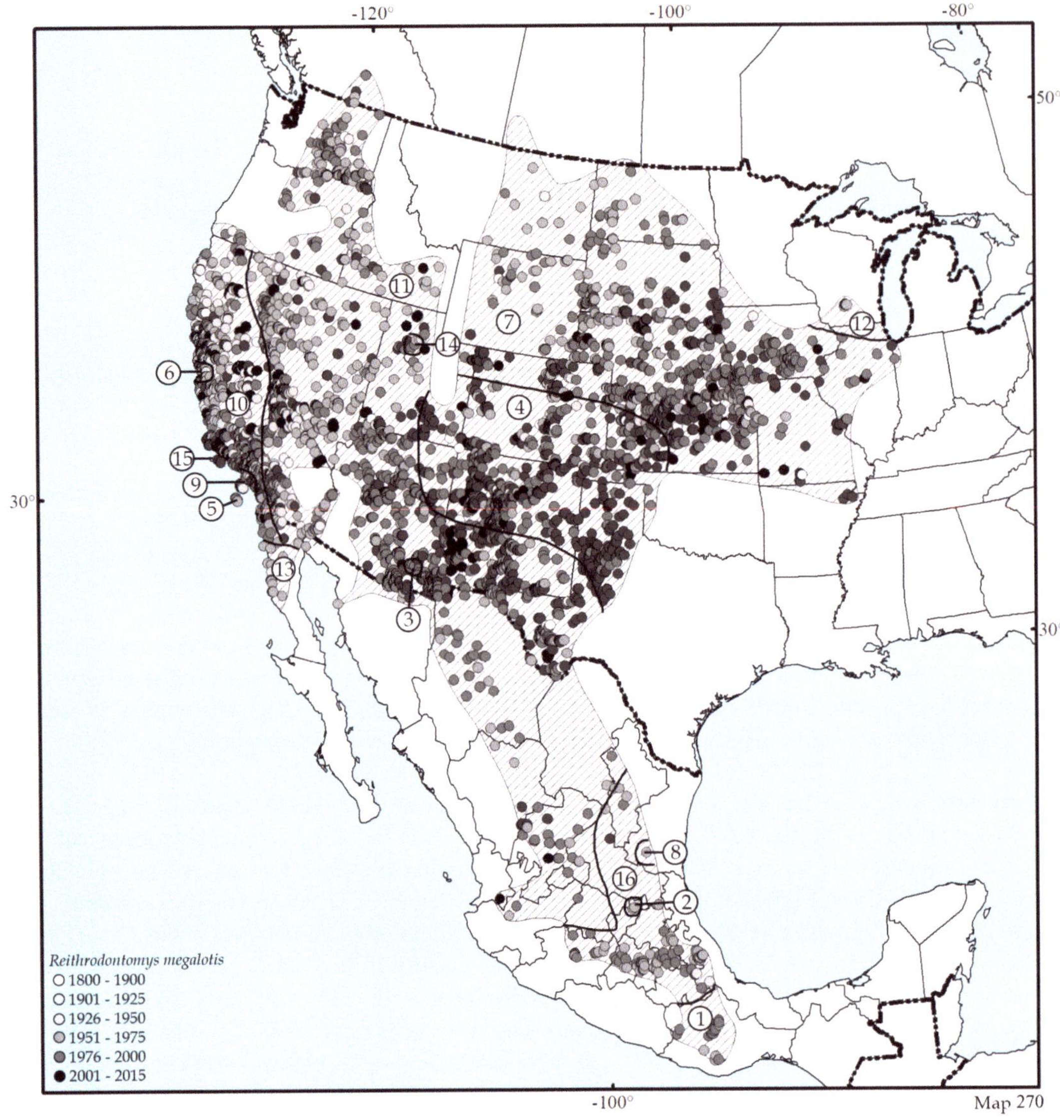


**Map 270** Distribution map of *Reithrodontomys megalotis*

**Additional Literature:** Merriam (1901a), Howell (1914a), Hooper (1952), Webster and Jones (1982b), Spencer and Cameron (1982), Álvarez-Castañeda and Cortés-Calva (1999), Musser and Carleton (2005), Bradley (2017).

## *Reithrodontomys montanus* (Baird, 1855)

Plains harvest mouse, ratón de las cosechas de las planicies

1. *R. m. albescens* Cary, 1903. From southeastern Montana and southwestern North Dakota south through eastern Colorado and western Kansas.
2. *R. m. griseus* Bailey, 1905. From southeastern Nebraska and eastern Kansas south and west through eastern New Mexico, central Texas, and northwest Arkansas.
3. *R. m. montanus* (Baird, 1855). From central-southern Colorado, central New Mexico west through central-southern Arizona and central-northern Sonora, and south through northern Durango.

**Conservation:** *Reithrodontomys montanus* is listed as Least Concern by the IUCN (Lacher et al. 2016l).

**Characteristics:** Plains harvest mouse is small sized; total length 107.0–143.0 mm and skull length 19.1–20.1 mm. Dorsal pelage pale buff to light brown, back dark, contrasting with the flanks and face; no distinct median black line; fur of underparts whitish; **ears small**; feet white; **tail long,** distinctly bicolored, dark brown dorsally and white ventrally; **skull small and narrow**; auditory bullae in proportion to the skull size; third upper and lower molars C shaped. Subgenus *Reithrodontomys*. Species group *megalotis*.

**Comments:** *Reithrodontomys montanus* ranges from southeastern Montana and southwestern North Dakota, south through northern Sonora and Chihuahua to northern Durango, and from southeastern Arizona, New Mexico, and Colorado east through the Missouri River (Map 271). *R. montanus* can be found in sympatry, or nearly so, with *R. fulvescens*, *R. humulis*, and *R. megalotis*. *R. montanus* can be distinguished from *R. fulvescens* by having the worn occlusal surface of the third lower molars C shaped and without a pronounced ocher to pale orange lateral line on the flanks. From *R. humulis*, by having a larger average size in the following measurements: total length 107.0 mm, tail length 47.0 mm, and hindfoot length 14.0 mm. From *R. megalotis*, by having the tail short, less than 95% of the head-and-body length; total length less than 140.0 mm; shorter ears; braincase relatively narrow; breadth usually less than 9.6 mm and zygomatic arches with parallel sides.

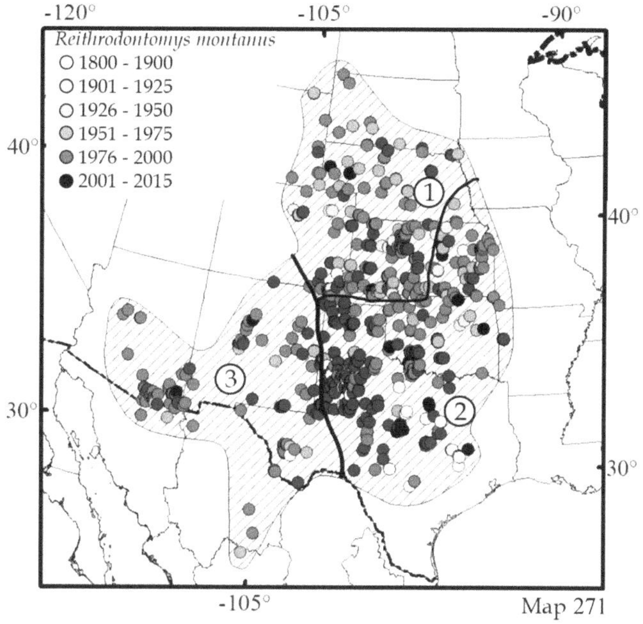


**Map 271** Distribution map of *Reithrodontomys montanus*

**Additional Literature:** Howell (1914a), Hooper (1952), Spencer and Cameron (1982), Wilkins (1986), Musser and Carleton (2005), Bradley (2017).

## ***Reithrodontomys raviventris*** Dixon, 1908

Salt-Marsh harvest mouse, ratón de las cosechas de la Bahía de San Francisco

1. *R. r. halicoetes* Dixon, 1909. Northern San Francisco Bay, California.
2. *R. r. raviventris* Dixon, 1908. Southern San Francisco Bay, California.

**Conservation:** *Reithrodontomys raviventris* is listed as **Endangered** (B1ab (i, ii, iii, iv, v) + 2ab (i, ii, iii, iv, v): extent of occurrence less than 5000 km$^2$ at five or less locations and under continuing decline in extent of occurrence, area of occupancy, and quality of habitat, number of subpopulations, and number of mature individuals and area of occupancy less than 500 km$^2$ at five or less locations and under continuing decline in extent of occurrence, area of occupancy, and quality of habitat, number of subpopulations, and number of mature individuals) by the IUCN (Whitaker and NatureServe 2018).

**Characteristics:** Salt-Marsh harvest mouse is small sized; total length 135.0–162.0 mm and skull length 20.8–21.2 mm. Dorsal pelage mixed black and cinnamon, black predominating on the dorsal area; fur of underparts cinnamon; flanks pale tawny; some individuals with a small white spot on the chin; inner and outer ears black or fuscous, with an ochraceous hair tuft at the anterior base; hind feet from fuscous or brown; toes whitish; front feet sepia, often tinged with buffy white shades; tail slightly paler ventrally; skull longer; rostrum relatively shorter; nasals and palatal foramen shorter; zygomatic arches more widely expanded anteriorly; third upper and lower molars C shaped. Subgenus *Reithrodontomys*. Species group *megalotis*.

**Comments:** *Reithrodontomys raviventris* is restricted to salt marshes in the vicinity of the San Francisco Bay, California (Map 272). *R. raviventris* can be found in sympatry, or nearly so, with *R. megalotis*, from which it differs by having the dorsal pelage long, dense, heavily pigmented, and the ears blackish.

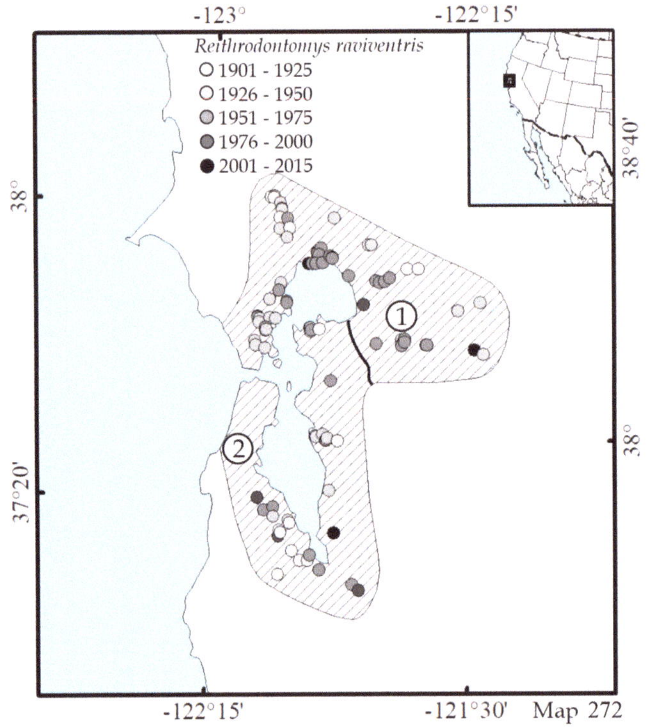


**Map 272** Distribution map of *Reithrodontomys raviventris*

**Additional Literature:** Howell (1914a), Fisler (1965), Shellhammer (1982), Spencer and Cameron (1982), Musser and Carleton (2005), Bradley (2017).

## *Reithrodontomys sumichrasti* (Saussure, 1861)

Sumichrast's harvest mouse, ratón de las cosechas de la montaña

1. *R. s. dorsalis* Merriam, 1901. Highlands of Chiapas and Guatemala.
2. *R. s. luteolus* Howell, 1914. Central Guerrero and Oaxaca.
3. *R. s. nerterus* Merriam, 1901. From south Jalisco and western Michoacán across the Eje Neovolcánico east through western Estado de México.
4. *R. s. sumichrasti* (Saussure, 1861). From Estado de México north through southern San Luis Potosí and south through northern Oaxaca.

**Conservation:** *Reithrodontomys sumichrasti* is listed as Least Concern by the IUCN (Reid et al. 2016c).

**Characteristics:** Sumichrast's harvest mouse is small sized; total length 147.0–206.0 mm and skull length 20.6–24.6 mm. *R. sumichrasti* can be found in two different geographic coloration patterns within its range. In the first pattern, dorsal pelage with a dark coloration, cinnamon with heavy grizzly and blackish dull shades and fur of underparts similar to the dorsal pelage; in the second pattern, a brighter coloration more cinnamon and less grizzly, with blackish dull shades and underparts grayish white. In both color patterns, ears blackish with hair; no dark spot around the eyes; tail of the same size or smaller than the head-and-body length, darker dorsally than ventrally; top of the limbs whitish; auditory bullae small; mesopterygoid fossa broad; paracone of the second upper molars not evenly round posteriorly but distinctly keeled on the posterolabial margin, with the keel projecting to the second primary fold; third upper and lower molars C shaped. Subgenus *Reithrodontomys*. Species group *megalotis*.

**Comments:** *Reithrodontomys sumichrasti* ranges in two separate areas, the first from the highlands of southwestern Jalisco and north-central Veracruz south through Oaxaca; the second from the highlands of central Chiapas south through Nicaragua (Map 273). *R. sumichrasti* can be found in sympatry, or nearly so, with *R. albilabris*, *R. chrysopsis*, *R. fulvescens*, *R. gracilis*, *R. hirsutus*, *R. megalotis*, *R. mexicanus*, *R. microdon*, *R. tenuirostris*, and *R. wagneri*. *R. sumichrasti* can be distinguished from *R. albilabris*, *R. chrysopsis*, *R. gracilis*, *R. mexicanus*, *R. microdon*, and *R. wagneri* by having the third upper molars with the first primary fold at least as long as the second primary fold, each usually extending more than halfway across the crown. From *R. fulvescens* and *R. hirsutus*, by having the worn occlusal surface of the third lower molars C shaped and without a pronounced ocher to pale orange lateral line on the flanks. From *R. megalotis*, by having the tail longer than the head-and-body length, usually greater than 80.0 mm; breadth of the mesopterygoid fossa greater than 1.3 mm; braincase breadth greater than 10.7 mm and greater than the distance between the posterior palatine foramen. From *R. tenuirostris*, by not having the third upper molars with the first primary fold distinctly shorter than the second primary fold.

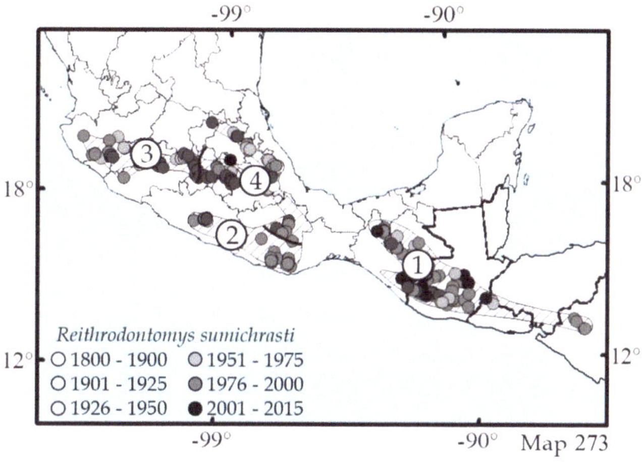


**Map 273** Distribution map of *Reithrodontomys sumichrasti*

**Additional Literature:** Merriam (1901a), Hooper (1952), Hall (1981), Musser and Carleton (2005).

## *Reithrodontomys zacatecae* Merriam, 1901

Zacatecas harvest mouse, ratón de las cosechas de Zacatecas

Monotypic.

**Conservation:** *Reithrodontomys zacatecae* is listed as Least Concern by the IUCN (Álvarez-Castañeda et al. 2016s).

**Characteristics:** Zacatecas harvest mouse is small sized; total length 120.5–130.5 mm and skull length 20.0–20.6 mm. Dorsal pelage ochraceous-buff, heavily mixed with black but not forming a distinct band; fur of underparts cinnamon brown, more intense between the forelegs; flanks and head ochraceous; ears fuscous to dark brown, usually with a blackish patch on the lower inner margin; feet whitish or buffy white; ankles dusky; **tail length approximately 119% of the head-and-body length,** sharply bicolored, dark brown dorsally, grayish white ventrally; skull small; rostrum more slender and tapered at the tip; zygomatic arches slightly narrower anteriorly; nasals tapered to a point posteriorly and ending at the same level than the premaxillae; third upper and lower molars C shaped. Subgenus *Reithrodontomys*. Species group *megalotis*.

**Comments:** *Reithrodontomys zacatecae* was recognized as a distinct species (Hood et al. 1984); it was previously regarded as a subspecies *P. megalotis* (Merriam 1901a). *R. zacatecae* ranges in the Sierra Madre Occidental from southern Chihuahua south through Jalisco and Michoacán (Map 274). *R. zacatecae* can be found in sympatry, or nearly so, with *R. fulvescens* and *P. megalotis*. *R. zacatecae* can be distinguished from *R. fulvescens* by having the worn occlusal surface of the third lower molars C shaped, and without a pronounced ocher to pale orange lateral line on the flanks. From *R. megalotis*, by its smaller size; braincase breadth greater than 10.7 mm; dorsal pelage paler and underparts paler; mainly by genetic data.

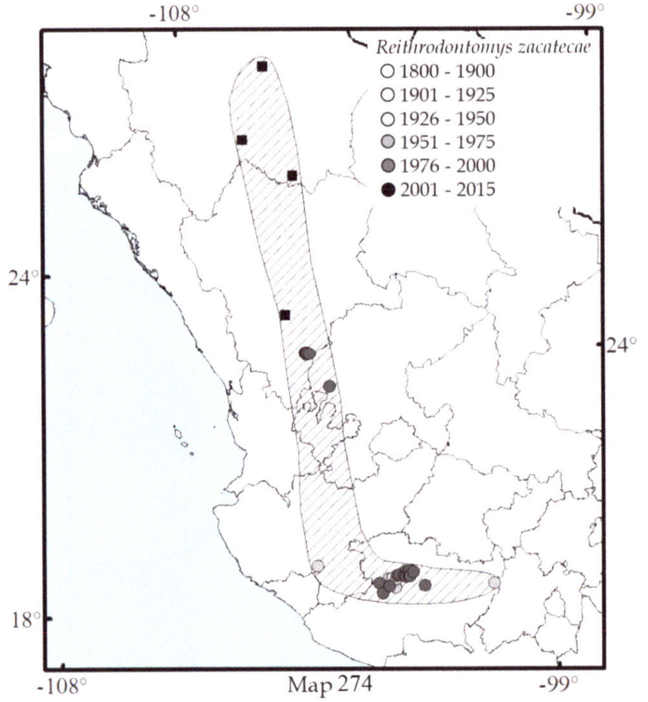


**Map 274** Distribution map of *Reithrodontomys zacatecae*

**Additional Literature:** Howell (1914a), Hooper (1952), Spencer and Cameron (1982), Bell et al. (2001), Musser and Carleton (2005).

## Genus *Xenomys* Merriam, 1892

**Tail cylindrical, tapering, and with short hair; hind feet naked below along the outer side at least to the tarso-metatarsal joint;** enamel pattern of the occlusal surface of the third lower molars S shaped; **lacrimals enlarged, slight postorbital process present; supraorbital shelf markedly elevated and beaded; auditory bullae enlarged and inflated and parallel to the long axis of the skull; paroccipital processes long and stout; interparietal bone enlarged**.

## *Xenomys nelsoni* Merriam, 1892
Magdalena woodrat, rata de Magdalena

Monotypic.

**Conservation:** *Xenomys nelsoni* is listed as Threatened by the Norma Oficial Mexicana (DOF 2019) and as Endangered (A2c: population reduction in the last 10 years of equal to or less than 50% and may not be reversible based on a decline in area of occupancy, extent of occurrence and quality of habitat) by the IUCN (Vázquez 2018c).

**Characteristics:** Magdalena woodrat is medium sized; total length 300.0–333.0 mm and skull length 40.5–44.6 mm. Dorsal pelage cinnamon-brown to pale yellow-brown, hair blackish at the tip; fur of underparts cream white; ears about half as long as the head, nearly naked with inconspicuous hairs; cheeks with a white spot more than halfway to the eyes; whiskers long and blackish, with the distal third grayish; feet white above and darker below; tail shorter than the head-and-body length, conspicuously scaly, monochromatic dorsally dark umber; skull supraorbital rim with a prominent ridge; lacrimal bone enlarged; **auditory bullae enlarged and inflated, the main axis approximately parallel to the main skull axis**; postorbital process present; long and solid paraoccipital process; interparietal bone enlarged, almost as wide as the braincase; lower third molars S shaped; anterior maxillary root of the zygomatic bone strongly notched above.

**Comments:** *Xenomys nelsoni* is patchily distributed in southwestern Jalisco and Colima (Map 275). It can be found in sympatry, or nearly so, with species of *Baiomys*, *Heteromys*, *Hodomys*, *Nyctomys*, *Oryzomys*, *Osgoodomys*, *Reithrodontomys*, and *Sigmodon*. *X. nelsoni* can be distinguished from all other species by its size and by a white spot over each eye and behind each ear.

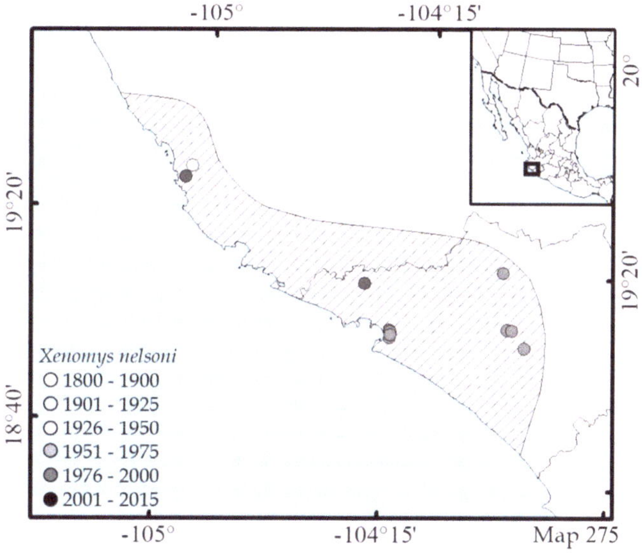


**Map 275** Distribution map of *Xenomys nelsoni*

**Additional Literature:** Merriam (1892b, 1894a), Schaldach (1960), Ceballos (1990), Ceballos and Miranda (2000), Ceballos et al. (2002), Musser and Carleton (2005), González-Ruiz et al. (2006a, b), Fernández (2014), Bradley et al. (2022b).

## Subfamily Sigmodontinae Wagner, 1943

## Tribe Ichthyomyini

### Genus *Rheomys* Thomas, 1906

Dorsal pelage with glossy, grizzled-brownish, and dark brown guard hairs in the fur; fur of underparts distinctly counter shaded, silvery white or pale gray but dark at the base; tail short, equal to or longer than the head-and-body length, dark, unicolor or bicolored depending on the species, dorsal dark brown-black and ventral pale gray-brown or white; **ears small but visible above the unruffled pelage of the head or very small and buried in the head hair**; supraorbital vibrissae absent; rhinarium pigmented; **forefoot with four or fewer separate plantar pads; hind feet proportionately broad with well-developed fringing hairs; braincase flattened; nasal bones long, concealing the incisors and nasal orifices from the dorsal view**; rostrum narrow; nasolacrimal capsules fully exposed in dorsal view; **supraorbital foramina open laterally within the orbital fossa**; **orbicular apophysis of the malleus present**; zygomatic processes of the maxillae and squamosals very delicate; lambdoidal ridges not well developed even in large adults; occipital condyles not produced posteriorly beyond the supraoccipital bone, not visible in the dorsal view; **baculum cartilage tridigitate.** The keys were elaborated based on the review of specimens and the following sources: Hall (1981), and Álvarez-Castañeda et al. (2017a).

1. Total length greater than 265.0 mm; hind feet broad, greater than 36.0 mm; toothrow length greater than 4.6 mm; cutting area of the upper incisors in an inverted V-shape in the frontal view (endemic Oaxaca) .................................................. *Rheomys mexicanus* (p. 386)

1a. Total length less than 265.0 mm; hind feet narrow, less than 36.0 mm; toothrow length less than 4.6 mm; cutting area of the upper incisors not in an inverted V-shape in the frontal view (restricted to Chiapas south through Central America) .................................................. *Rheomys thomasi* (p. 387)

### *Rheomys mexicanus* Goodwin, 1959

Mexican water mouse, ratita nadadora mexicana

Monotypic.

**Conservation:** *Rheomys mexicanus* is listed as Special Protection by the Norma Oficial Mexicana (DOF 2019) and as Endangered (B1ab (iii): extent of occurrence less than 5000 km$^2$ at five or less locations and under continuing decline in area and quality of habitat) by the IUCN (Timm et al. 2018a).

**Characteristics:** Mexican water mouse is medium sized; one of the few species that live in rivers; total length 280.0–302.0 mm and skull length 30.3–32.8 mm. Dorsal pelage blackish, occasionally grizzled, white at the tip; fur of underparts whitish; **tail longer than the head-and-body length, distinctly bicolored, dark brownish or blackish dorsally and pure white ventrally; ears always very small and buried in the head hair; muzzle between the rhinarium and upper lip entirely hairy and undivided by any trace of a naked philtrum; rostrum slender and braincase greatly inflated**; **palate between the molar rows broad**; upper incisors long and slender; molars large (4.8–5.0 mm) and broad (first upper molars, 1.6–1.8 mm) with tall, sharp principal cusps separated by wide reentrant folds; **upper incisors with the cutting area in an inverted V-shape in frontal view**.

**Comments:** *Rheomys mexicanus* is endemic to Oaxaca, México, only known from Guelatao, San José Lachiguiri, Unión Hidalgo, and Totontepec on the Pacific slope of Oaxaca (Map 276; Santos-Moreno et al. 2003). Both species of *Rheomys* are allopatric. *R. mexicanus* can be distinguished from *R. thomasi* by its larger size, greater than 265.0 mm in the head-and-body length, hindfoot length 36.0 mm, and upper toothrow length 4.6 mm; tail longer than the head-and-body length and distinctly bicolored, dark brownish or blackish dorsally and pure white ventrally; ears always very small and buried in the head hair; muzzle between the rhinarium and the upper lip entirely hairy and undivided by any trace of a naked philtrum, slender rostrum and greatly inflated braincase; palate between the molar rows broad.

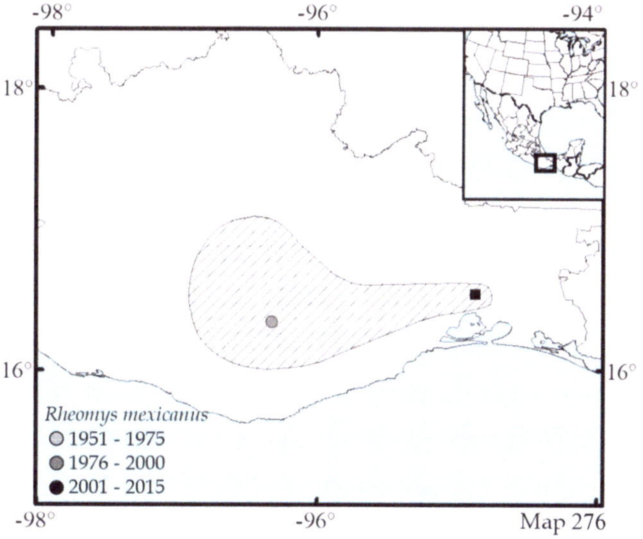


**Map 276** Distribution map of *Rheomys mexicanus*

**Additional Literature:** Goodwin (1959), Voss (1988), Musser and Carleton (2005).

## *Rheomys thomasi* Dickey, 1928

Thomas's water mouse, ratita nadadora de Chiapas

*R. t. chiapensis* Hooper, 1947. Restricted to Chiapas.

**Conservation:** *Rheomys thomasi chiapensis* is listed as Special Protection in the Norma Oficial Mexicana (DOF 2019) and as Near Threatened by the IUCN (Reid et al. 2008c).

**Characteristics:** Thomas's water mouse is medium sized; one of the few species that live in riverbanks, total length 208.0–253.0 mm and skull length 30.0–33.0 mm. Dorsal pelage brown, and black-tipped; fur of underparts grayish white, **tail as long as the head-and-body and not distinctly bicolored, although paler dorsally than ventrally, ears small but always visible above the unruffled head hair, a distinct philtrum usually present**; **rostrum relatively broad and braincase less inflated; palate between the molar rows relatively narrow;** upper incisors relatively short and broad; molars small (4.2–4.5 mm) and narrow (1.4–1.5 mm); lower molar separated by a wide reentrant folds; **upper incisors with the cutting area inverted not V shaped in the frontal view.**

**Comments:** *Rheomys thomasi* ranges from the highlands of Chiapas south through Central America (Map 277). Both species of *Rheomys* are allopatric. *R. thomasi* can be distinguished from *R. mexicanus* by its small size, measuring less than the following: head-and-body length 265.0 mm, hindfoot length 36.0 mm, and upper toothrow 4.6 mm. Tail as long as the head-and-body length and not distinctly bicolored, although paler ventrally than dorsally; ears small but always visible above the unruffled head hair, with a distinct philtrum usually present; relatively broader rostrum and less inflated braincase; palate between the molar rows relatively narrow.

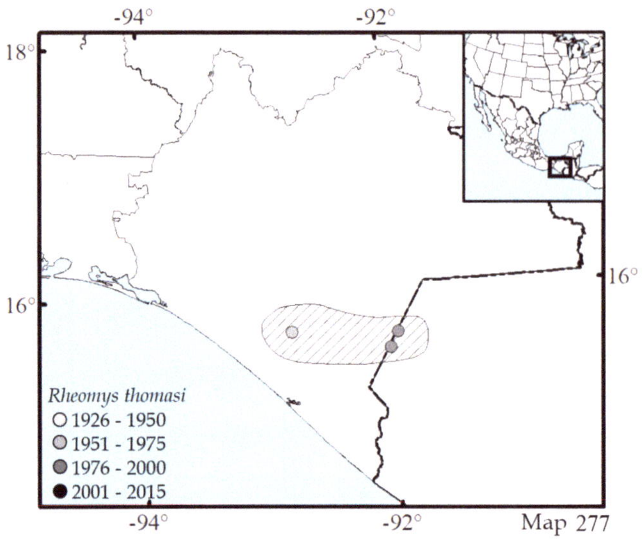


**Map 277** Distribution map of *Rheomys thomasi*

**Additional Literature:** Voss (1988), Musser and Carleton (2005).

## Tribe Oryzomini

### Genus ***Handleyomys*** Voss et al., 2002

Dorsal pelage from rich ochraceous-buffy or ochraceous-tawny to dark-colored shades; pelage short, slightly hispid but not bristly or spiny; ears clothed externally with short fine dusky hairs; first and fifth hindfoot toes notably shorter than the central three; **conspicuous hair tuft of silvery** bristles projecting beyond the tips of the longest claws on the hind feet; eight mammae present; skull narrow with an elongated rostrum; anterior palatine foramina short; supraorbital ridges well developed and interorbital shape cuneate; **posterolateral palatal pits conspicuous and internally perforated**; **shape of the outer margins of both zygomatic arches giving the impression of an elliptical;** auditory bullae small; **principal reentrant angles extending farther across the molar crowns;** satellite rootlets present on the first upper molars. *Handleyomys* has two species groups: *melanotis* (*H. melanotis* and *H. rostratus*) and *chapmani* (*H. alfaroi*, *H. chapmani*, *H. guerrerensis*, *H. rhabdops*, and *H. saturatior*). The keys were elaborated based on the review of specimens and the following sources: Goldman (1918), Hall (1981), and Álvarez-Castañeda et al. (2017a).

1. Ears with inner yellowish or reddish hair ........ 2

1a. Ears with inner black hair ........ 3

2. Hindfoot length greater than 31.0 mm; skull length greater than 30.5 mm; zygomatic breadth greater than 15.0 mm (southern Tamaulipas to Central America, including the Yucatán Peninsula) ........ *Handleyomys rostratus* (p. 394)

2a. Hindfoot length less than 31.0 mm; skull length less than 30.5 mm; zygomatic breadth less than 15.0 mm (from western Jalisco south through Oaxaca) ........ *Handleyomys melanotis* (p. 392)

3. Dorsal hair length greater than 10.0 mm; zygomatic arches wider anteriorly than posteriorly (central Chiapas and Guatemala) ........ *Handleyomys rhabdops* (p. 393)

3a. Dorsal hair length less than 10.0 mm; zygomatic arches wider posteriorly than anteriorly ........ 4

4. Interorbital breadth equal to or greater than 5.0 mm (from southern Tamaulipas and Guerrero south through Central America) ........ *Handleyomys alfaroi* (p. 389)

4a Interorbital breadth less than 5.0 mm ........ 5

5. Dorsal pelage dark and contrasting with the lateral coloration (from northern Chiapas south through Central America) ............................................................ *Handleyomys saturatior* (p. 395)

5a. Dorsal pelage not contrasting with the lateral coloration (from Tamaulipas south through northern Oaxaca) ................ 6

6. Restricted to the Sierra Madre Oriental, from southwestern Tamaulipas south through northwestern Oaxaca ................ ............................................................ *Handleyomys chapmani* (p. 390)

6a. Endemic to Sierra Madre del Sur, Guerrero, and Oaxaca .......................................... *Handleyomys guerrerensis* (p. 391)

## ***Handleyomys alfaroi*** (Allen, 1891)

Alfaro's rice rat, ratón de agua tropical

*H. a. alfaroi* (Allen, 1891). From northeastern Chiapas south through to Central America.

**Conservation:** *Handleyomys alfaroi* is listed as Least Concern by the IUCN (Timm et al. 2016j).

**Characteristics:** Alfaro's rice rat is small sized; total length 205.0–225.0 mm andskull length 24.8–27.6 mm. Dorsal pelage from yellowish dark ocher to tawny or brown, usually grizzly with black hair denser on the back; fur of underparts whitish with yellowish shades, contrasting with the dorsal coloration; tail long, equal to the head-and-body length, with few hairs and paler ventrally; interparietal bone small. Species group *chapmani*.

**Comments:** *Handleyomys* was described as a new genus by Voss et al. (2002); it was previously included within the genus *Oryzomys*. *Handleyomys alfaroi* was recognized within the genus *Handleyomys* by Weksler et al. (2006). *Handleyomys chapmani*, *H. guerrerensis*, *H. rhabdops*, and *H. rostratus* were recognized as full species (Almendra et al. 2014), previously regarded as subspecies of *H. alfaroi*. *Handleyomys alfaroi* ranges in the lowlands from southern Tamaulipas in the Gulf slope and from Guerrero in the Pacific slope south through Central America (Map 278). However, the analyses by Almendra et al. (2018) suggest a possible different species for México, for this reason, *H. alfaroi* will be considered here as ranging only to Central America. *H. alfaroi* can be found in sympatry, or nearly so, with *H. rostratus* and *Oryzomys couesi*; the species of *Oryzomys* can be easily misidentified as *Handleyomys*. *H. alfaroi* can be distinguished from *H. rostratus* by having ears with inner black hair. From the species of *Oryzomys*, by having the hair between the digits larger than the respective claw; six plantar pads; six mammae (nipples); the zygomatic notch shallow; mesopterygoid fossa no fenestrated; and principal reentrant angles extending farther across the molar crowns and auditory bullae large.

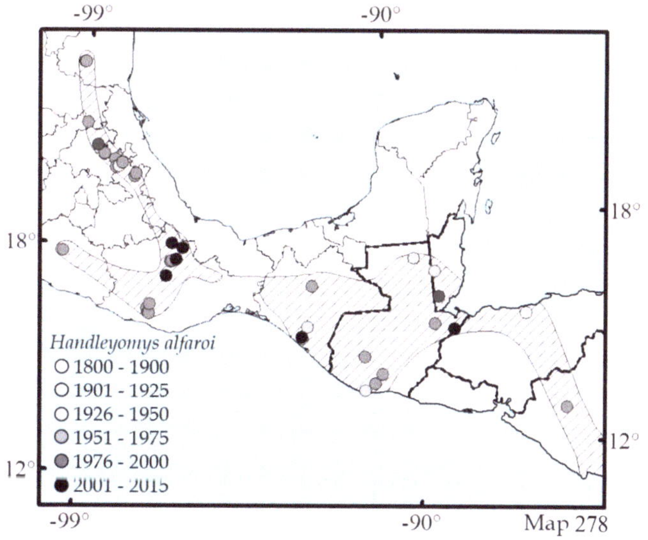


**Map 278** Distribution map of *Handleyomys alfaroi*

**Additional Literature:** Goldman (1918), Hall and Kelson (1959), Haiduk et al. (1979), Hall (1981), Engstrom (1984), Musser and Carleton (2005).

## ***Handleyomys chapmani*** Thomas, 1898

Chapman's rice rat, ratón de agua de la Sierra Madre

1. *H. c. chapmani* Thomas, 1898. From central Veracruz and eastern Puebla south through northern-central Oaxaca.
2. *H. c. dilutior* Merriam, 1901. Central-western Veracruz and eastern Hidalgo.
3. *H. c. huastecae* Dalquest, 1951. Northwestern Veracruz, eastern Hidalgo, and southern Tamaulipas.

**Conservation:** *Handleyomys chapmani caudatus* (as *Oryzomys chapmani caudatus*) is listed as Special Protection by the Norma Oficial Mexicana (DOF 2019) and *H. chapmani* is listed as **Vulnerable** (B2ab (ii, iii): area of occupancy less than 2000 $km^2$ at ten or less locations and under continuing decline in area of occupancy and area, extent and/or quality of habitat) by the IUCN (Vázquez 2018b).

**Characteristics:** Chapman's rice rat is small sized; total length 210.0–222.0 mm and skull length 26.3–27.4 mm. Dorsal pelage grizzled fulvous and black, grizzle black on the top of the head and back central part; flanks in similar shades; fur of underparts whitish with yellowish shades, contrasting with the dorsal coloration; **ears large and blackish**; tail slender, equal to the head-and-body length, blackish dorsally, yellowish ventrally, and terminal all around dark; skull without conspicuous superciliary beads; zygomatic arches rather strongly spreading anteriorly, sides nearly parallel but slightly nearer together anteriorly than posteriorly; rostrum relatively long; nasals blunt posteriorly, ending about on the same plane as the premaxillae; incisive foramina relatively small; braincase dome shaped. Species group *chapmani*.

**Comments:** *Handleyomys* was described as a new genus by Voss et al. (2002); it was previously included within the genus *Oryzomys*. *Handleyomys chapmani* was recognized within the genus *Handleyomys* by Weksler et al. (2006) and Almendra et al. (2014) and as a distinct species (Musser and Carleton 2005); it was previously regarded as a subspecies of *O. alfaroi* (Goldman 1918). *H. chapmani* ranges in the highlands of Sierra Madre Oriental, from southwestern Tamaulipas south through northwestern Oaxaca (Map 279). It can be found in sympatry, or nearly so, with *H. guerrerensis*, *H. melanotis*, *H. rostratus*, *H. saturatior*, *Oryzomys couesi*, and *O. texensis*. The species of *Oryzomys* can be easily misidentified as *Handleyomys*. *H. chapmani* can be distinguished from *H. guerrerensis* only by its distribution and with genetic analysis. From *H. melanotis* and *H. rostratus*, by its smaller size; dark-colored shades with short pelage; conspicuous ears clothed externally and internally with short fine blackish hairs; skull small and delicate in structure and larger auditory bullae. From *H. saturatior*, by having the medial dorsal area not contrasting with the lateral coloration, usually grizzly with few black hairs; fur of underparts in buffy shades. From the species of *Oryzomys*, by having hair between the digits larger than the respective claw; six plantar pads; six mammae (nipples); the zygomatic notch shallow; mesopterygoid fossa no fenestrated; principal reentrant angles extending farther across the molar crowns and auditory bullae large.

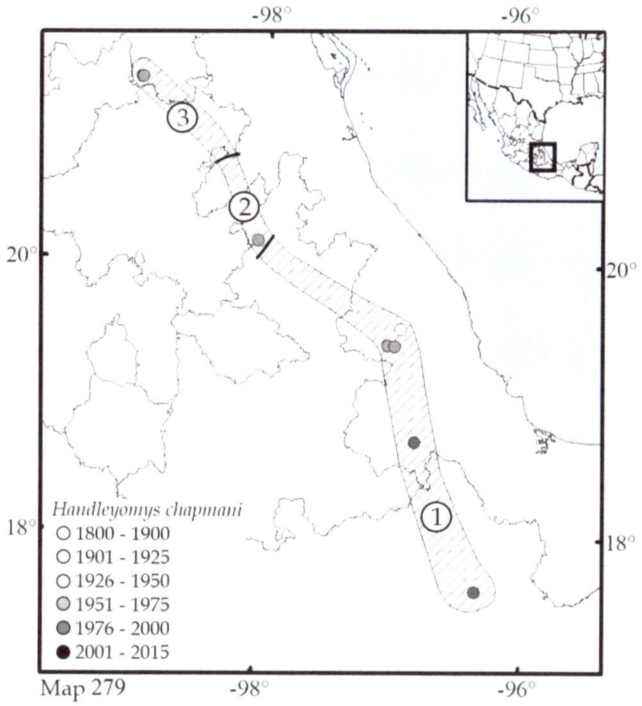


**Map 279** Distribution map of *Handleyomys chapmani*

**Additional Literature:** Merriam (1901b), Goodwin (1969), Haiduk et al. (1979), Musser and Carleton (2005).

## *Handleyomys guerrerensis* Goldman, 1915

Guerrero rice rat, ratón de agua de Guerrero

Monotypic.

**Conservation:** *Handleyomys guerrerensis* (as *Handleyomys alfaroi*) is listed as Least Concern by the IUCN (Timm et al. 2016j).

**Characteristics:** Guerrero rice rat is small sized; total length 210.0 a 227.5 mm and skull length 25.0 a 27.5 mm. Dorsal pelage dark ochraceous-buff to dark ochraceous-tawny, usually grizzly with black hair on the back and top of the head; fur of underparts dull grayish white, plumbeous at the base; outer and inner sides of the ears well clothed with deep, glossy black hairs; cheeks, flanks, and shoulders without conspicuous superciliary beads; zygomatic arches strongly spreading anteriorly, the other sides nearly parallel but slightly nearer together anteriorly than posteriorly; rostrum relatively long; nasals blunt posteriorly, ending about on the same plane as the premaxillae; incisive foramina relatively small, braincase dome shaped. Species group *chapmani*.

**Comments:** *Handleyomys* was described as a new genus by Voss et al. (2002); it was previously included within the genus *Oryzomys*. *Handleyomys guerrerensis* was recognized within the genus *Handleyomys* and as a distinct species (Almendra et al. 2014); it was previously regarded as a subspecies of *O. alfaroi* (Hall and Kelson 1959). *H. guerrerensis* ranges in the Sierra Madre del Sur of Guerrero and southwestern Oaxaca (Map 280) and can be found in sympatry, or nearly so, with *H. chapmani*, *H. melanotis*, *H. rostratus*, *Oryzomys albiventer*, *O. couesi*, and *O. mexicanus*, the species of *Oryzomys* can be easily misidentified as *Handleyomys*. *H. guerrerensis* can be distinguished from *H. chapmani* only by its distribution and with genetic analysis. From *H. melanotis* and *H. rostratus*, by its smaller size; dark-colored forms with short pelage; conspicuous ears clothed externally and internally with short fine blackish hairs; skull small and delicate in structure and larger auditory bullae. From the species of *Oryzomys*, by having hair between the digits larger than the respective claw; six plantar pads; six mammae (nipples); the zygomatic notch shallow; mesopterygoid fossa no fenestrated; principal reentrant angles extending farther across the molar crowns and auditory bullae large.

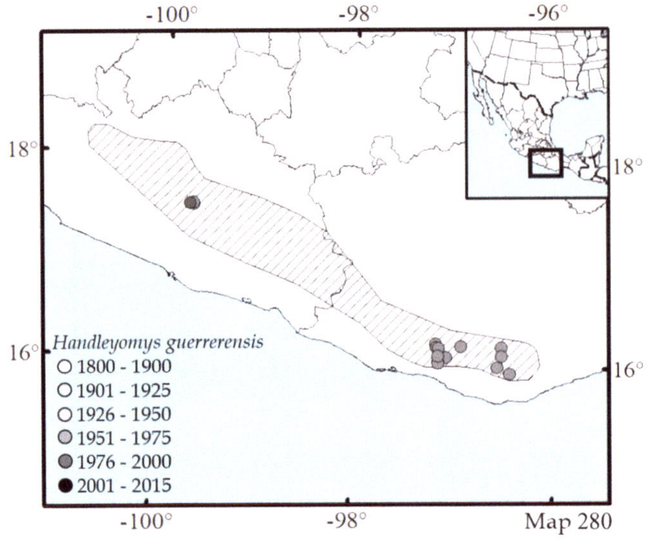


**Map 280** Distribution map of *Handleyomys guerrerensis*

**Additional Literature:** Goldman (1918).

## ***Handleyomys melanotis*** Thomas, 1893

Black-eared rice rat, ratón de agua de orejas negras

1. *H. m. colimensis* Goldman, 1918. From central Jalisco south through Tehuantepec Isthmus, Oaxaca
2. *H. m. melanotis* Thomas, 1893. From southern Sinaloa to central Jalisco.

**Conservation:** *Handleyomys melanotis* is listed as Least Concern by the IUCN (Álvarez-Castañeda et al. 2016u).

**Characteristics:** Black-eared rice rat is medium sized; total length 216.0–277.0 mm andskull length 26.3–28.51 mm. **Dorsal pelage bright intense ochraceous fulvous**, usually grizzly with black shades on the back; fur of underparts whitish yellow with gray shades at the base; **ears large and blackish**; tail long, slightly larger than the head-and-body length, dark brown dorsally and yellowish ventrally; interparietal bone small; incisive foramen small, less than one-half of the palatal length. Species group *melanotis*.

**Comments:** *Handleyomys* was described as a new genus by Voss et al. (2002); it was previously included within the genus *Oryzomys*. *H. melanotis* was recognized within the genus *Handleyomys* by Weksler et al. (2006) and Almendra et al. (2014). *H. rostratus* was previously considered a subspecies of *H. melanotis* (Hooper 1953) but was later elevated to a full species (Engstrom 1984; Musser and Carleton 2005). *H. melanotis* ranges in low to intermediate elevations from western Jalisco south through Oaxaca (Map 281). It can be found in sympatry, or nearly so, with *H. chapmani*, *H. guerrerensis*, *H. saturatior*, *Oryzomys albiventer*, *O. couesi*, *O. mexicanus*, and *O. texensis*; the species of *Oryzomys* can be easily misidentified as *Handleyomys*. *H. melanotis* can be distinguished from *H. chapmani*, *H. guerrerensis*, and *H. saturatior* by having slender individuals; medium sized; rich ochraceous-buffy or ochraceous-tawny coloration; ears large, clothed externally with short, fine dusky hairs and internally rufescent hairs; skull narrow, with an elongated rostrum; short anterior palatine foramina and small auditory bullae. From *H. rostratus*, by having the dorsal pelage bright intense ochraceous fulvous; ears large and blackish; smaller size, with the hind feet and skull length less than 31.0 and 30.5 mm, respectively; the zygomatic breadth greater than 15.0 mm. From the species of *Oryzomys* by having hair between the digits larger than the respective claw; six plantar pads; six mammae (nipples); the zygomatic notch shallow; mesopterygoid fossa no fenestrated; and principal reentrant angles extending farther across the molar crowns and auditory bullae large.

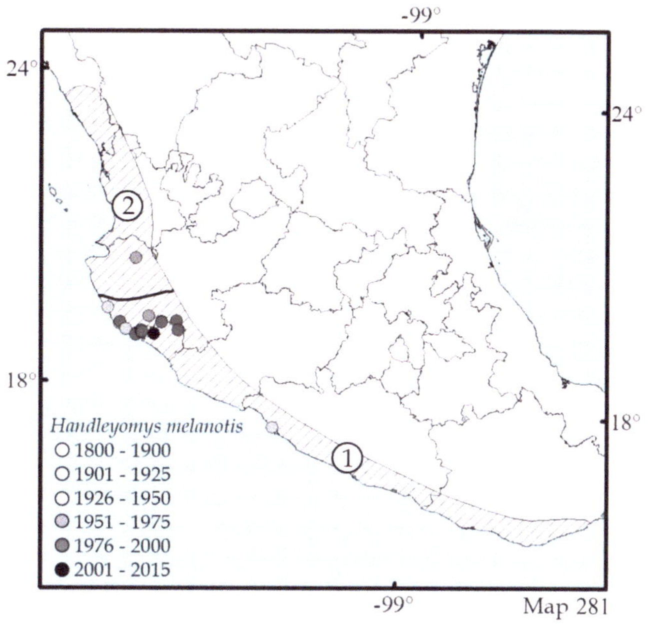


**Map 281** Distribution map of *Handleyomys melanotis*

**Additional Literature:** Merriam (1901b), Goldman (1918), Hooper (1953), Weksler (2003).

## *Handleyomys rhabdops* Merriam, 1901

Highland rice rat, ratón de agua de las tierras altas

Monotypic.

**Conservation:** *Handleyomys rhabdops* is listed as Endangered (B1ab (i, iii): extent of occurrence less than 5000 km$^2$ at five or less locations and under continuing decline in extent of occurrence and area and/or quality of habitat) by the IUCN (Vázquez 2019a).

**Characteristics:** Highland rice rat is medium sized; total length 245.0–270.0 mm and skull length 29.5–32.0 mm. Dorsal pelage very long and woolly, from fulvous to tawny or brown, usually grizzly with black shades on the top of the head and back; fur of underparts soiled whitish; nose and streak from the nose to the eyes blackish; rostrum sides below black streak pale fulvous; **ears thin rather large and blackish**; tail equal to or slightly longer than the head-and-body length, dusky brown dorsally and yellowish ventrally; rostrum long and slender; **incisive foramina short**; **broadly spreading squarish zygomatic arch; interparietal bone small;** zygomatic arches slightly broader anteriorly; braincase large and rounded. Species group *chapmani*.

**Comments:** *Handleyomys* was described as a new genus by Voss et al. (2002); it was previously included within the genus *Oryzomys. Handleyomys rhabdops* was recognized within the genus *Handleyomys* by Weksler et al. (2006) and Almendra et al. (2014) and as a distinct species (Musser and Carleton 2005); it was previously regarded as a subspecies of *O. alfaroi* (Goldman 1918). *H. rhabdops* is restricted to the highlands of central Chiapas and Guatemala (Map 282) and can be found in sympatry, or nearly so, with *H. saturatior* and *Oryzomys couesi;* the species of *Oryzomys* can be easily misidentified as *Handleyomys*. *H. rhabdops* can be distinguished from *H. saturatior* by its larger size, greater than 230.0 mm and zygomatic arches wider posteriorly than anteriorly. From the species of *Oryzomys*, by having hair between the digits larger than the respective claw; six plantar pads; six mammae (nipples); the zygomatic notch shallow; mesopterygoid fossa no fenestrated; principal reentrant angles extending farther across the molar crowns and auditory bullae large.

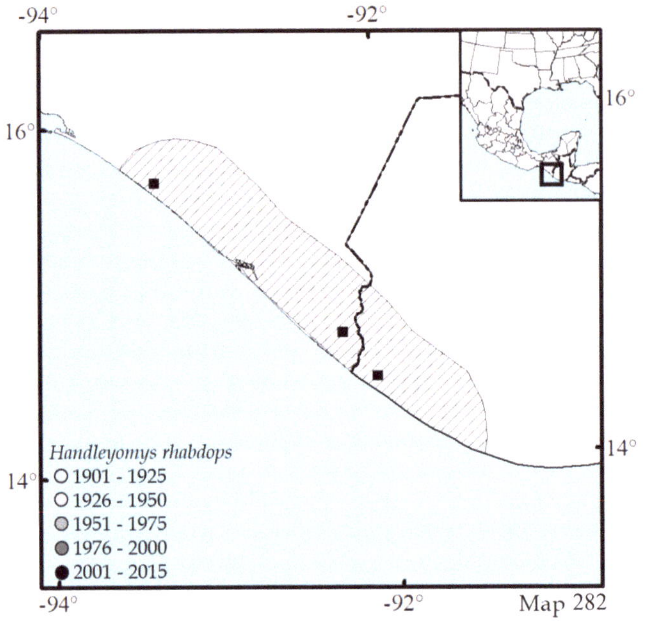


**Map 282** Distribution map of *Handleyomys rhabdops*

**Additional Literature:** Merriam (1901b).

## *Handleyomys rostratus* Merriam, 1901

Long-nosed rice rat, ratón de agua de nariz larga

1. *H. r. carrorum* Lawrence, 1947. Restricted to central south Tamaulipas.
2. *H. r. megadon* Merriam, 1901. From Tabasco to central and southern Yucatán Peninsula.
3. *H. r. rostratus* Merriam, 1901. From southern Tamaulipas south through Tabasco.
4. *H. r. yucatanensis* Merriam, 1901. Restricted to the northern Yucatán Peninsula.

**Conservation:** *Handleyomys rostratus* is listed as Least Concern by the IUCN (Reid and Vázquez 2016b).

**Characteristics:** Long-nosed rice rat is medium sized; total length 218.0–248.0 mm and skull length 30.5–31.5 mm. **Dorsal pelage ochraceous fulvous**, occasionally reddish, usually grizzly with black shades on the back; **fur of underparts buffy white** and gray at the base; **ears large, dark brown**; tail long or slightly larger than the head-and-body length, dark brown dorsally and **irregularly yellowish ventrally; skull large, rather massive, long, and flattened**; **rostrum long and somewhat swollen basally**; nasals large, broad, and flat, ending about the same plane as the premaxillae; incisive foramina medium; superciliary beads moderate; **zygomatic arches narrow but bowed out in the middle**; interparietal bone small. Species group *melanotis*.

**Comments:** *Handleyomys* was described as a new genus by Voss et al. (2002); it was previously included within the genus *Oryzomys*. *Handleyomys rostratus* was recognized within the genus *Handleyomys* by Weksler et al. (2006) and Almendra et al. (2014) and as a distinct species (Engstrom 1984; Musser and Carleton 2005); it was previously regarded as a subspecies of *O. melanotis* (Hooper 1953). *H. rostratus* ranges in the lowlands from southern Tamaulipas to Central America, including the Yucatán Peninsula (Map 283), and can be found in sympatry, or nearly so, with *H. chapmani*, *H. guerrerensis*, *H. melanotis*, *H. saturatior*, and *Oryzomys couesi*; the species of *Oryzomys* can be easily misidentified as *Handleyomys*. *H. rostratus* can be distinguished from *H. chapmani*, *H. guerrerensis*, and *H. saturatior* by being slender; medium sized; rich ochraceous-buffy or ochraceous-tawny; ears large. Clothed externally with short, fine dusky hairs and internally similar in size and rufescent; skull narrow, with an elongated rostrum; short anterior palatine foramen and small auditory bullae. From *H. alfaroi*, by having inner ears with yellowish or reddish hair. From *H. melanotis*, by having the dorsal pelage ochraceous fulvous; ears large and dark brown; larger size with the hind feet and skull length greater than 31.0 mm and 30.5 mm, respectively; the zygomatic breadth less than 15.0 mm. From the species of *Oryzomys*, by having hair between the digits larger than the respective claw; six plantar pads; six mammae (nipples); the zygomatic notch shallow; mesopterygoid fossa not fenestrated; and principal reentrant angles extending farther across the molar crowns and auditory bullae large.

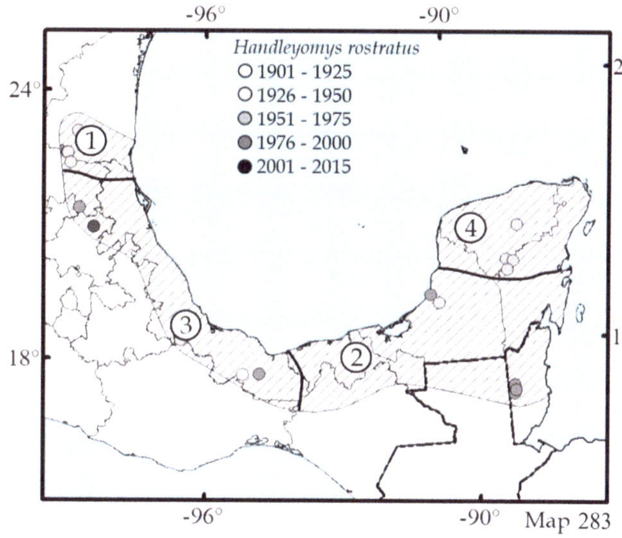


**Map 283** Distribution map of *Handleyomys rostratus*

**Additional Literature:** Merriam (1901b), Goldman (1918), Goodwin (1969).

## *Handleyomys saturatior* Merriam, 1901

Cloud forest rice rat, ratón de agua del bosque de niebla

*H. s. hylocetes* Merriam, 1901. Restricted to the coastal areas of Chiapas.

**Conservation:** *Handleyomys saturatior* is listed as Near Threatened by the IUCN (Reid et al. 2008b).

**Characteristics:** Cloud forest rice rat is small sized; total length 187.0–228.0 mm. Dorsal pelage from yellowish dark to ocher brown, usually grizzly with black hair denser on the back, **slightly "peppered" with fine points of fulvous color**; flanks with paler shades; **fur of underparts soiled buffy, salmon in some specimens, contrasting with the dorsal coloration**; cheeks fulvous; ears and hind feet blackish; tail long, equal to or slightly longer than the head-and-body length, blackish dorsally and pale ventrally; hair on the hind feet digits shorter than the respective claw; skull without conspicuous superciliary beads; interparietal bone small; zygomatic arches with anterior and posterior back wide very similar; braincase dome shaped. Species group *chapmani*.

**Comments:** *Handleyomys* was described as a new genus by Voss et al. (2002); it was previously included within the genus *Oryzomys*. *Handleyomys saturatior* was recognized within the genus *Handleyomys* by Weksler et al. (2006) and Almendra et al. (2014) and as a distinct species (Musser and Carleton 2005), previously as a subspecies of *O. alfaroi* (Goldman 1918). *H. saturatior* is restricted to the cloud forests in the highlands from northern Chiapas south through Central America (Map 284); it can be found in sympatry, or nearly so, with *H. chapmani*, *H. melanotis*, *H. rhabdops*, *H. rostratus*, and *Oryzomys couesi;* the species of *Oryzomys* can be easily misidentified as *Handleyomys*. *H. saturatior* can be distinguished from *H. chapmani* by having the medial dorsal area contrasting with the lateral coloration, usually grizzly with black hair denser on the back, slightly "peppered" with fine points of fulvous shades; fur of underparts soiled buffy, salmon in some specimens. From *H. melanotis* and *H. rostratus*, by its smaller size; dark-colored forms with short pelage; conspicuous ears clothed externally and internally with short fine blackish hairs; skull small and delicate in structure and larger auditory bullae. From *H. rhabdops*, by its smaller size, less than 230.0 mm, and zygomatic arches wider anteriorly than posteriorly. From the species of *Oryzomys*, by having hair between the digits larger than the respective claw; six plantar pads; six mammae (nipples); the zygomatic notch shallow; mesopterygoid fossa no fenestrated; principal reentrant angles extending farther across the molar crowns and auditory bullae large.

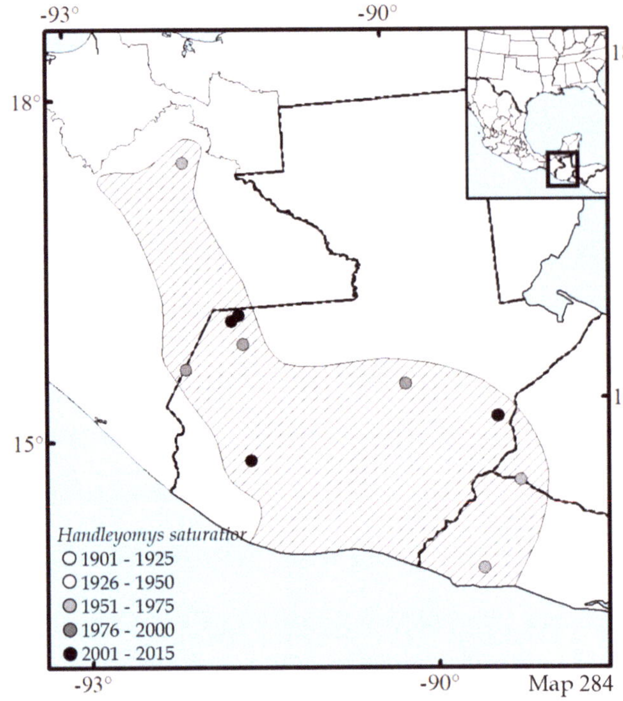


**Map 284** Distribution map of *Handleyomys saturatior*

**Additional Literature:** Merriam (1901b).

## Genus *Oligoryzomys* Bangs, 1900

Very small sized; pelage slightly hispid but not bristly or spiny; hind feet usually less than 25.0 mm; ears larger and coarsely haired; vibrissae short; **four longer hindfoot toes with a conspicuous hair tuft of silvery bristles beyond the tip of the claws**; **tail much longer than the head-and-body**; **eight mammae (nipples) present;** skull broad with a short rostrum; anterior palatine foramina reaching the anterior plane of the first molars; **supraorbital ridges and interorbital shape absent**; **posterolateral palatal pits conspicuous and internally perforated**; **principal reentrant angles usually broader; salient angles formed by worn crowns of tubercles less evenly rounded**; **second upper molars with protocone and hypocone of about equal size**; the inner reentrant angle central in position.

### *Oligoryzomys fulvescens* (Saussure, 1860)

Flavous pigmy rice rat, ratón de agua pigmeo

1. *O. f. engraciae* (Osgood, 1945). Restricted to western Nuevo León and Tamaulipas.
2. *O. f. fulvescens* (Saussure, 1860). From southern Tamaulipas south through Central America, except for the Yucatán Peninsula.
3. *O. f. lenis* (Goldman, 1915). From southern Nayarit to the Tehuantepec Isthmus, Oaxaca.
4. *O. f. mayensis* (Goldman, 1918). Restricted to the Yucatán Peninsula.
5. *O. f. pacificus* (Hooper, 1952). From the southern Tehuantepec Isthmus, Oaxaca, across the coast through Guatemala.

**Conservation:** *Oligoryzomys fulvescens* is listed as Least Concern by the IUCN (Weksler et al. 2016).

**Characteristics:** Flavous pigmy rice rat is small sized; **total length 168.0–198.0 mm and skull length 20.0–22.6 mm**. Dorsal pelage ochraceous-buff to tawny, top of the head, and back moderately with blackish hairs; fur of underparts dull from nearly pure white (lips, throat, and inner sides of hind limbs), generally pale ochraceous-buff on the abdomen and inguinal region; shoulders, cheeks, and flanks more ochraceous but palest; more intensely ochraceous dull yellow on the rump; flanks, back, and head grizzly with dark shades; ears with dark hairs externally and ochraceous-buffy internally; feet white; limbs whitish; hind feet with a hair tuft of silvery hairs projecting beyond the claws of the longest four digits; tail naked, dark brownish dorsally and paler brown or yellowish ventrally, and dusky all around at the tip; **second upper molars with a circular enamel island; without supraorbital and temporal crests**.

**Comments:** *Oligoryzomys fulvescens* ranges in the lowlands from southern Nuevo León and southern Nayarit south through Central America, including the Yucatán peninsula (Map 285). *O. fulvescens* is the only species of *Oligoryzomys* in North America and can be misidentified with *Handleyomys*, *Oryzomys*, and *Reithrodontomys*. *O. fulvescens* can be distinguished from the species of *Handleyomys* and *Oryzomys* by its smaller size; hindfoot length less than 25.0 mm; second upper molars with a circular enamel island and without supraorbital and temporal crests. From *Reithrodontomys*, by lacking a longitudinal groove in the upper incisors, both genera externally similar in shape and size.

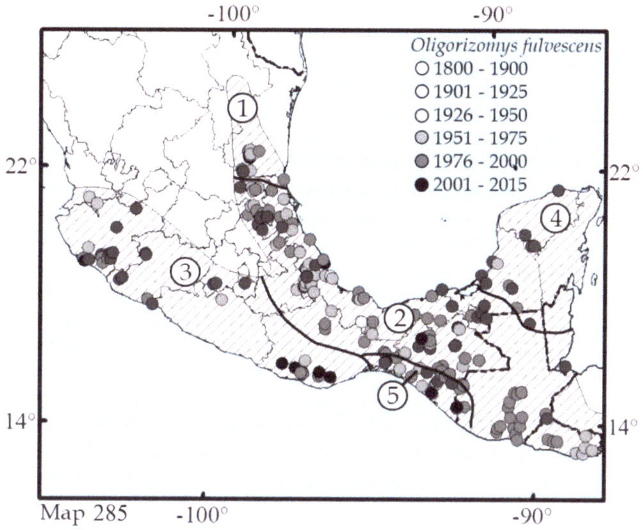


**Map 285** Distribution map of *Oligoryzomys fulvescens*

**Additional Literature:** Merriam (1901b), Goldman (1918), Hall (1981), Dowler and Engstrom (1988), Carleton and Musser (1995), Musser and Carleton (2005).

## Genus *Oryzomys* Baird, 1858

Dorsal pelage brown to dark brown, usually contrasting strongly with paler underparts; pelage slightly hispid but not bristly or spiny; robust form; ears coarsely haired with short vibrissae; first and fifth hindfoot toes notably shorter than the central three toes; **hind feet without a conspicuous hair tuft of silvery bristles beyond the tips of the longest claws**; eight mammae (nipples); skull broad with a short rostrum; **anterior palatine foramina very long**; supraorbital ridges well-developed and interorbital shape cuneate; **posterolateral palatal pits conspicuous and internally perforated**; **auditory bullae large; principal reentrant angles normally reaching less than halfway across the molar crowns**; second upper molars in moderately worn crown crescentic central enamel island extending along the posterior-internal base of the paracone; satellite rootlets present on the first upper molars. In North America, all the species of *Oryzomys* belong to the species group *palustris*. The keys were elaborated based on the review of specimens and the following sources: Goldman (1918), Hall (1981), and Álvarez-Castañeda et al. (2017a).

1. Pelage coloration generally reddish brown, with guard hairs only slightly longer than the body fur; sphenopalatine vacuities absent or weakly expressed as narrow slits near the basisphenoid-presphenoid suture .................................... 2

1a. Pelage coloration predominantly grayish brown, with long glistening guard hairs conspicuously overtopping the body fur; oval sphenopalatine vacuities perforate the walls of the mesopterygoid fossa, exposing much of the presphenoid bone ........................................................................................ 7

2. Dorsal pelage bright fulvous; ears very small and broadly rounded; skull very wide, with a great expansion of the zygomatic arches, especially by an evenly incurved outline of the supraorbital edges (restricted to Valle de México) ........... ........................................................................ *Oryzomys fulgens* (p. 400)

2a. Dorsal pelage, if fulvous, not bright; ears small to large but not broadly rounded; skull narrow or wide, with parallel zygomatic arches ........................................................................................ 3

3. Size relatively small; hindfoot length usually less than 34.0 mm; upper toothrow length usually less than 4.9 mm ........................................................................................ 4

3a. Size relatively large; hindfoot length usually greater than 33.0 mm; upper toothrow length usually greater than 4.7 mm ........................................................................................ 5

4. Coloration of the hind quarters ocher brown; incisive foramina small and broad; zygomatic arches not broadly spreading and squarely shaped (Nuevo León and Tamaulipas south through Central America, including Yucatán Peninsula) ........................................................ *Oryzomys couesi* (p. 399)

4a. Coloration of the hind quarters fulvous brown; incisive foramina long and exceptionally broad; zygomatic arches broadly spreading and squarely shaped (endemic to Los Cabos region, Baja California Sur, extinct) ............................ ........................................................................ *Oryzomys peninsulae* (p. 405)

5. Endemic to Marías Madre Island, Nayarit ........................................................ *Oryzomys nelsoni* (p. 403)

5a. Ranges in mainland México ........................................................................................ 6

6. Fur of underparts white, with the hair root pale gray (central and eastern Jalisco) ............ *Oryzomys albiventer* (p. 398)

6a. Fur of underparts white, with the hair root pale gray, restricted to the Pacific slopes from central Sonora to southeastern Oaxaca ........................................................ *Oryzomys mexicanus* (p. 401)

7. Ranges west of the Mississippi River (west of the Mississippi River in Oklahoma, Arkansas, Texas, Louisiana, and Mississippi) ........................................................ *Oryzomys texensis* (p. 406)

7a. Ranges east of the Mississippi River (eastern of the Mississippi River from Illinois south through Louisiana and east through the Atlantic coast from New Jersey south through the Florida Peninsula) .............. *Oryzomys palustris* (p. 403)

## ***Oryzomys albiventer*** Merriam, 1901

White-bellied rice rat, rata arrocera de vientre blanco

Monotypic.

**Conservation:** *Oryzomys albiventer* (as *O. couesi*) is listed as Least Concern by the IUCN (Linzey et al. 2016n).

**Characteristics:** White-bellied rice rat is medium sized; **total length 276.0–314.0 mm** and skull length 31.5–34.4 mm. Dorsal pelage grizzly ocher with white to tawny shades and grizzly with blackish hair, flanks also grizzly with blackish hair; **fur of underparts white, pale gray at the base;** hindfoot digits with hair shorter than the respective claw; **tail long**, equal to or slightly larger than the head-and-body length, dark brown dorsally and whitish ventrally; **skull large and long**; **nasals large and broad, slightly exceeding the premaxillae**, **supraciliary ridges well developed, nearly straight; molar rows wide and the largest toothrow**; sphenopalatine vacuities small and circular; incisive foramen similar in length to the palate. Species group *palustris*.

**Comments:** *Oryzomys albiventer* was recognized as a distinct species (Carleton and Arroyo-Cabrales 2009); it was previously regarded as a subspecies of *O. couesi* (Goldman 1918). However, Hanson et al. (2010) considered a subspecies of *O. mexicanus* without any analyses or review of specimens: for this reason, *O. albiventer* is still considered a valid species. *O. albiventer* ranges from central and eastern Jalisco at intermediate elevations of 1200–1800 m (Map 286). *O. albiventer* cannot be found in sympatry, or nearly so, with *O. couesi*, *O. mexicanus*, *Handleyomys guerrerensis*, and *H. melanotis*. *Handleyomys* can be easily misidentified as *Oryzomys*. *O. albiventer* can be distinguished from all other species of *Oryzomys* by having the fur of underparts white, pale gray at the base; tail longer; hindfoot length less than 34.0 mm; tail relatively and absolutely long; skull heavily constructed, and molars correspondingly robust. From the species of *Handleyomys*, by having hindfoot digits with hair shorter than the respective claw; five plantar pads; eight mammae (nipples); the zygomatic notch deep; mesopterygoid fossa fenestrated; and principal reentrant angles normally reaching less than halfway across the molar crowns and auditory bullae small.

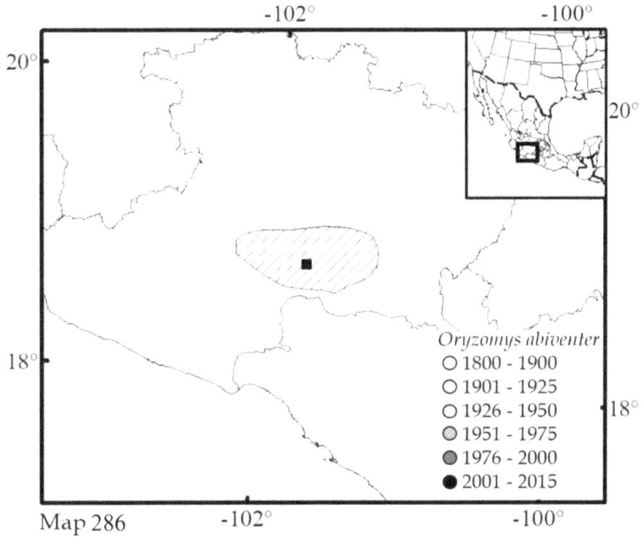


**Map 286** Distribution map of *Oryzomys albiventer*

**Additional Literature:** Merriam (1901b).

## *Oryzomys couesi* (Alston, 1877)

Coues's rice rat, rata arrocera del Golfo

1. *O. c. aquaticus* Allen, 1891. From Rio Grande south through Tamaulipas and eastern Nuevo León.
2. *O. c. aztecus* Merriam, 1901. From central Michoacán and western Hidalgo south through northwestern Oaxaca.
3. *O. c. couesi* (Alston, 1877). From central Veracruz and northern Oaxaca south through Central America, including northern Chiapas and the Yucatán Peninsula.
4. *O. c. cozumelae* Merriam, 1901. Restricted to the Cozumel Island, Quintana Roo.
5. *O. c. crinitus* Merriam, 1901. Valle de México, Estado de México, Ciudad de México, and Hidalgo.
6. *O. c. peragrus* Merriam, 1901. From southern Tamaulipas and San Luis Potosí south through northern Veracruz.

**Conservation:** *Oryzomys couesi cozumelae* is listed as Threatened by the Norma Oficial Mexicana (DOF 2019) and *O. couesi* as Least Concern by the IUCN (Linzey et al. 2016n).

**Characteristics:** Coues's rice rat is medium sized; total length 222.0–294.0 mm and skull length 29.2–32.5 mm. Dorsal pelage ochraceous-buffy to ochraceous-tawny, top of the head and back with darker hair at the tip; fur of underparts light buff to pale ochraceous buff (rarely dull white); **pelage harsh;** cheeks, shoulders, and face sides pale cinnamon-brown; flanks fairly dark, paler toward the back; fore and hindfeet silvery white; ears pale brown; inner surface with larger yellowish hair; hindfoot digits with hair shorter than the respective claw; tail long, equal to or slightly larger than the head-and-body length and mainly naked, brownish dorsally, dull yellowish ventrally and becoming pale brownish toward the tip; **skull light and thin**; sphenopalatine vacuities absent or weakly expressed; braincase narrow; **molar series very long**. Species group *palustris*.

**Comments:** *Oryzomys couesi* was reinstated as a distinct species by Benson and Gehlbach (1979); it was previously regarded as a subspecies of *O. palustris* (Hall 1960). It ranges from Nuevo León and Tamaulipas south through Central America, including Yucatán Peninsula (Map 287; Carleton and Arroyo-Cabrales 2009). However, Hanson et al. (2010) considered that the specimens from central México (subspecies *aztecus*, *crinitus*, and *regillus*, *sensus* Hall 1981) should be considered under *O. mexicanus*, but they do not review specimens from this range; for this reason, the distribution proposed by Carleton and Arroyo-Cabrales (2009) is still considered, based on the review of specimens from the area. *O. couesi* can be found in sympatry, or nearly so, with *O. albiventer*, *O. fulgens*, *O. mexicanus*, *O. texensis*, *Handleyomys alfaroi*, *H. caudatus*, *H. chapmani*, *H. guerrerensis*, *H. melanotis*, *H. rhabdops*, and *H. saturator. Handleyomys* can be easily misidentified as *Oryzomys*. *O. couesi* can be distinguished from *O. albiventer* by not having the fur of underparts white nor pale gray at the base, and tail long. From *O. fulgens*, by having the zygomatic arches approximately in straight lines and the interorbital area narrow; only part of the inner orbit wall forms an even curve; breadth at the posterior end of the olfactory chamber slightly smaller than at the anterior end and frontal premaxillary processes very wide. From *O. mexicanus*, by having a paler coloration; larger molars and toothrow correspondingly larger. However, these characteristics are not obvious; the best way to differentiate it is by genetic analyses. From *O. texensis*, by its coloration generally reddish brown, with the guard hairs only slightly longer than the head-and-body hair; sphenopalatine vacuities absent or weakly expressed as short; narrow slits near the basisphenoid-presphenoid suture. From the species of *Handleyomys*, by having hindfoot digits with hair shorter than the respective claw; five plantar pads; eight mammae (nipples); zygomatic notch deep; mesopterygoid fossa fenestrated; and principal reentrant angles normally reaching less than halfway across the molar crowns and auditory bullae small.

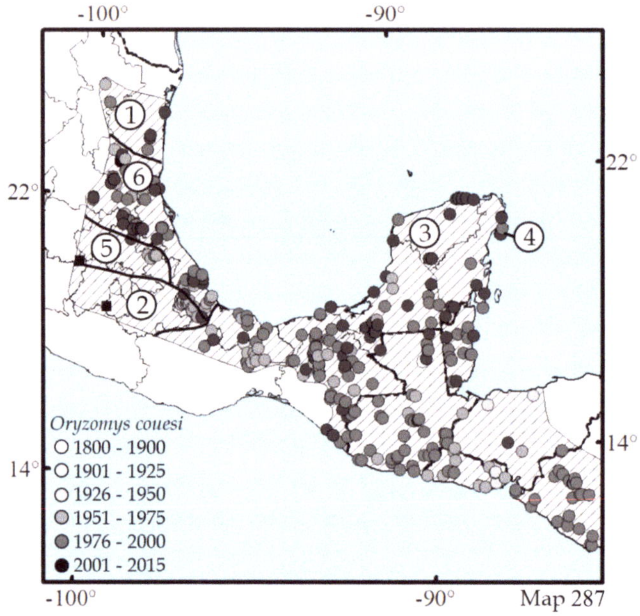


**Map 287** Distribution map of *Oryzomys couesi*

**Additional Literature:** Merriam (1901b), Goldman (1918), Musser and Carleton (2005).

## ***Oryzomys fulgens*** Thomas, 1893

Thomas's rice rat, rata de agua del Valle de México

Monotypic.

**Conservation:** *Oryzomys fulgens* is listed as Threatened by the Norma Oficial Mexicana (DOF 2019). However, all the data consider it extinct. Not listed by the IUCN.

**Characteristics:** Thomas's rice rat is medium sized; total length 311.0 mm and skull length 21.8 mm. **Dorsal pelage bright fulvous, brighter in North America, fur very thick, coarse and woolly**; fur of underparts yellowish, but the lips, chin, throat, and inguinal region whitish; **ears very small, broadly rounded**, thinly haired and of the same color as the head; outer sides of the limbs similar to the back; inner sides whitish; upper surfaces of fore and hind feet thinly clothed with pale silvery-fawn; hindfoot hair digits shorter than the respective claw; tail long, thinly haired, blackish dorsally and yellowish ventrally, darkening towards the tip; **skull very wide; zygomatic arches greatly expanded, especially by evenly incurved outline of the supraorbital edges**; breadth at the posterior end of the olfactory chamber scarcely greater than at the anterior end; nasals broad and flattened; frontal premaxillary processes very narrow and barely attaining the same level as the back of the nasals; sphenopalatine vacuities absent or weakly expressed; anterior palatine foramina large, widely open, their posterior margin just at the level of the front of the first upper molars. Species group *palustris*.

**Comments:** *Oryzomys fulgens* was considered a subspecies of *O. couesi* by Musser and Carleton (2005) without an explanation, when this species has been considered a full species in multiple studies (Goldman 1918; Hall and Kelson 1959; Hall 1981). *O. fulgens* has diagnostic characteristics which distinguish it from the known species (Thomas 1893); for this reason, *O. fulgens* is still considered a valid species until further analyses demonstrating otherwise are available. It was described from "Mexico" (Thomas 1893) and later restricted to the "Valle de México" (Map 288; Goldman 1918). In addition, only the type specimen is known and has an incomplete skull (Goldman 1918). The information of the species is very limited; for this reason, its ranges were considered only in the "Valle de México," the keys were elaborated with the data of the original description (Thomas 1893). *O. fulgens* differs from other species by having a very bright fulvous dorsal pelage, brighter in North America; hair very thick; coarse and woolly, ears very small, broadly rounded; skull very wide; zygomatic arches greatly expanded, especially by the evenly incurved outline of the supraorbital edges.

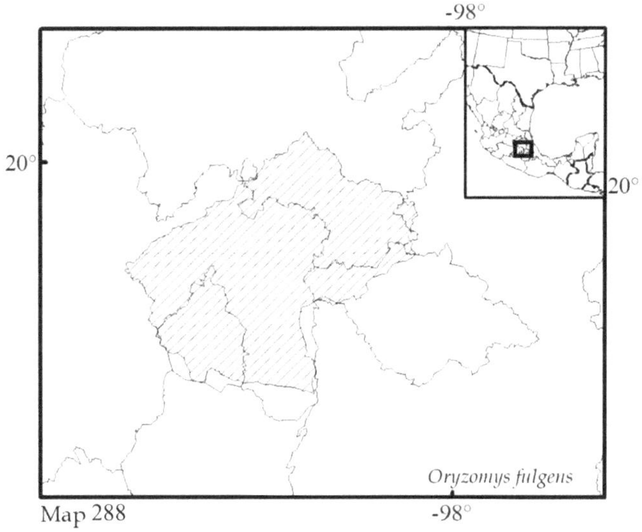


**Map 288** Distribution map of *Oryzomys fulgens (extinct species)*

**Additional Literature:** Merriam (1901b), Hall and Kelson (1959).

## *Oryzomys mexicanus* Allen, 1897

Mexican rice rat, rata arrocera mexicana

1. *O. m. lambi* Burt, 1934. Restricted to the area of Guaymas, Sonora.
2. *O. m. mexicanus* Allen, 1897. Coastal areas from central Sonora to the Tehuantepec Isthmus.
3. *O. m. zygomaticus* Merriam, 1901. Coastal and central Chiapas.

**Conservation:** *Oryzomys mexicanus* (as *O. couesi*) is listed as Least Concern by the IUCN (Linzey et al. 2016n).

**Characteristics:** Mexican rice rat is medium sized; total length 239.0–273.0 mm and skull length 29.8–33.0 mm. **Dorsal pelage ochraceous-buffy; fur of underparts usually nearly pure white**, varying to pale buff, or pale ochraceous-buff; cheeks, shoulders, and lower parts of the flanks warm buff; face, top of the head, and back moderately darkened by grizzle with blackish hairs; ears with outer sides dusky and inner sides with grayish or rusty reddish hairs; feet white; hindfoot digits with hair shorter than the respective claw; tail brownish dorsally, dull yellowish ventrally, and becoming pale brownish toward the tip; sphenopalatine vacuities small and circular; **molars and toothrow smaller**. Species group *palustris*.

**Comments:** *Oryzomys mexicanus* was recognized as a distinct species (Carleton and Arroyo-Cabrales 2009); it was previously regarded as a subspecies of *O. couesi* (Goldman 1918). Hanson et al. (2010) considered *O. albiventer* as a subspecies of *O. mexicanus* without any analysis or review of specimens; for this reason, *O. albiventer* is still considered a full species. *O. mexicanus* ranges in the coastal plain and Pacific slopes from central Sonora to southeastern Oaxaca (Map 289; Carleton and Arroyo-Cabrales 2009), based on a review of specimens from the area. However, Hanson et al. (2010) considered that the specimens from central México (subspecies *aztecus*, *crinitus*, and *regillus*, *sensus* Hall 1981) should be considered under *O. mexicanus* without reviewing specimens from these ranges; for this reason, this work still considers the distribution proposed by Carleton and Arroyo-Cabrales (2009). *O. mexicanus* can be found in sympatry, or nearly so, with *O. albiventer*, *O. couesi*, *Handleyomys guerrerensis*, and *H. melanotis*. *Handleyomys* can be easily misidentified as *Oryzomys*. *O. mexicanus* can be distinguished from *O. albiventer* by not having the fur of underparts white, pale gray at the base; tail long; hindfoot length less than 34.0 mm; skull heavily constructed and molars correspondingly robust. From *O. couesi*, by having a darker coloration; smaller molars and toothrow. However, these characteristics are not obvious, the best way is by genetic analysis. From the species of *Handleyomys*, by having hindfoot digits with hair shorter than the respective claw; five plantar pads; eight mammae (nipples); zygomatic notch deep; mesopterygoid fossa fenestrated; and principal reentrant angles normally reaching less than halfway across the molar crowns and auditory bullae small.

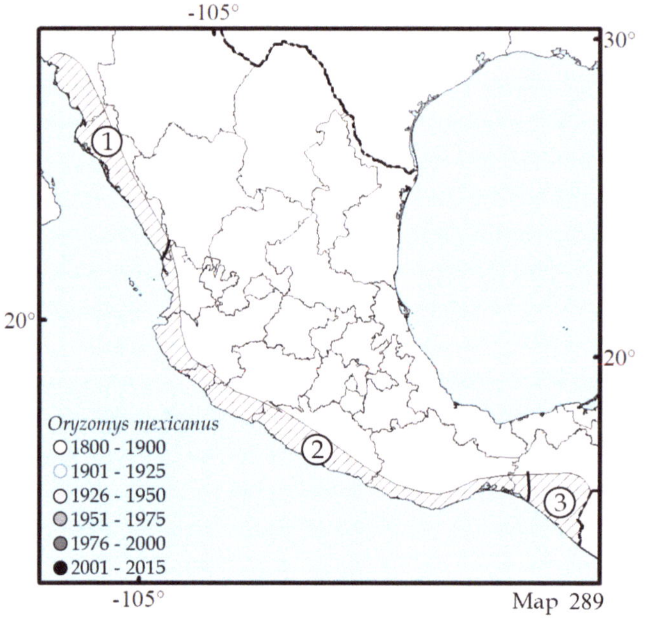


**Map 289** Distribution map of *Oryzomys mexicanus*

**Additional Literature:** Merriam (1901b).

## ***Oryzomys nelsoni*** Merriam, 1898

Nelson's rice rat, rata arrocera de las islas Tres Marías

Monotypic.

**Conservation:** *Oryzomys nelsoni* is listed as extinct by the Norma Oficial Mexicana (DOF 2019) and the IUCN (Timm et al. 2017a).

**Characteristics:** Nelson's rice rat is large sized**; total length 324.0 mm and skull length 34.5–37.8 mm.** Dorsal pelage ochraceous to buff fulvous, more intense in the rump and buff grayish anteriorly, mainly in the face; fur of underparts whitish; fore and hindfeet silvery white; ears pale brown; **hind feet large**; hindfoot digits with hair shorter than the respective claw; **tail proportionally longer** than in near species, equal to or slightly larger than the head-and-body length, brown dorsally and paler ventrally; **skull large and massive; rostrum strongly curved; incisors large and broad**; superciliary beads nearly slight; zygomatic arches rather heavy; sphenopalatine vacuities small and circular; upper incisors large, broad and shortly curved. Species group *palustris*.

**Comments:** *Oryzomys nelsoni* is restricted to María Madre Island, Nayarit (Map 290), and cannot be found in sympatry, or nearly so, with any other species of *Oryzomys*.

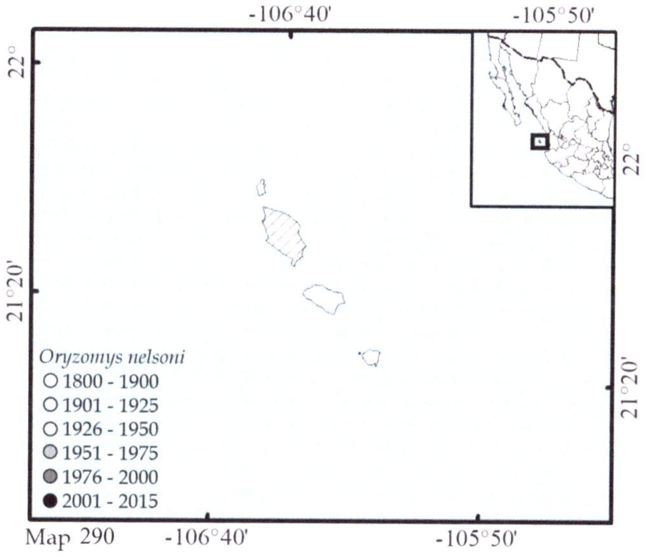


**Map 290** Distribution map of *Oryzomys nelsoni*

**Additional Literature:** Merriam (1901b), Goldman (1918), Álvarez-Castañeda and Méndez (2003), Musser and Carleton (2005), Carleton and Arroyo-Cabrales (2009).

## ***Oryzomys palustris*** Harlan, 1837

Marsh rice rat, rata arrocera de los pantanos

1. *O. p. coloratus* Bangs, 1898. Southern of Florida Peninsula.
2. *O. p. natator* Chapman, 1893. North and center of the Florida Peninsula.
3. *O. p. palustris* (Harlan, 1837). From south Illinois, Kentucky, Tennessee, to Mississippi east through the Atlantic coast from northern Florida to southeastern Pennsylvania and New Jersey.
4. *O. p. planirostris* Hamilton, 1955. Restricted to Pine Island and to a small area on central-western Florida.
5. *O. p. sanibeli* Hamilton, 1955. Restricted to Sanibel Island, Florida.

**Conservation:** *Oryzomys palustris natator* is listed as Endangered by the United States Endangered Species Act (FWS 2022) and as Least Concern by the IUCN (Cassola 2016cf).

**Characteristics:** Marsh rice rat is medium sized; total length 237.0–245.0 mm and skull length 31.0–32.6 mm. Dorsal pelage grayish brown to pale buff darkened; face, top of the head, and back grizzle with blackish hairs; feet whitish; hindfoot digits with hair shorter than the respective claw; ears small; tail short, brownish dorsally, whitish ventrally, and in some specimens dark all around near the tip; rostrum elongated; nasals long and tapering, ending more less as the same plane as the premaxillae; supraciliary bead moderately developed; sphenopalatine foramen oval; foramines perforate mesopterygoid walls; exposing much of the presphenoid bone; zygomatic arches rather narrow; **anterolabial cingulum on the lower third molars absent**. Species group *palustris*.

**Comments:** *Oryzomys palustris* ranges east of the Mississippi River from Illinois south through Louisiana and east through the Atlantic coast from New Jersey south through the Florida Peninsula (Map 291). *O. palustris* cannot be found in sympatry, or nearly so, with any other species of *Oryzomys* or *Handleyomys*, but in the limits of its ranges it can be found in sympatry with *O. texensis*, from which it differs by its darker coloration and with genetic data. From the species of *Handleyomys*, by having hindfoot digits with hair shorter than the respective claw; five plantar pads; eight mammae (nipples); zygomatic notch deep; mesopterygoid fossa fenestrated; and principal reentrant angles normally reaching less than halfway across the molar crowns and auditory bullae small.

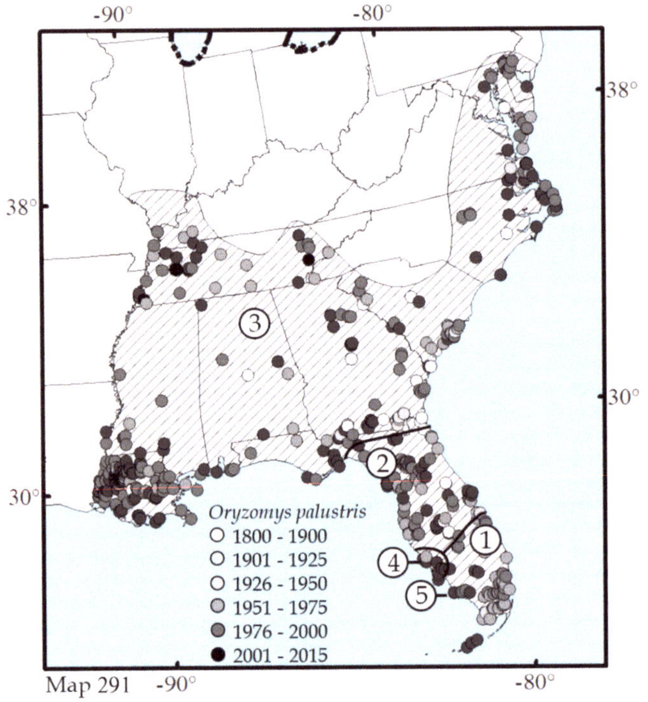


**Map 291** Distribution map of *Oryzomys palustris*

**Additional Literature:** Merriam (1901b), Goldman (1918), Wolfe (1982), Musser and Carleton (2005), Carleton and Arroyo-Cabrales (2009), Hanson et al. (2010), Indorf and Gaines (2013).

## ***Oryzomys peninsulae*†** Thomas, 1897

Lower California rice rat, rata arrocera de Baja California

Monotypic.

**Conservation:** *Oryzomys peninsulae* is listed as extinct by the Norma Oficial Mexicana (DOF 2019) and as Least Concern (as a subspecies of *O. couesi*) by the IUCN (Linzey et al. 2016o). *O. peninsulae* is considered extinct (Álvarez-Castañeda 1994).

**Characteristics:** Lower California rice rat is medium sized; total length 270.0–305.0 mm andskull length 31.5–33.9 mm. Dorsal pelage pale gray to ocher near the tail; fur of underparts whitish, same as the inner part of the limbs; fore and hindfeet silvery white; ears pale brown; **forequarters grayish-brown, hindquarters fulvous brown**; hindfoot digits with hair shorter than the respective claw; tail long, equal to or slightly larger than the head-and-body length, dusky or brown dorsally and solid whitish ventrally; skull broad and massive; premaxillae long reaching far behind the nasals; slight superciliary beads; **upper incisors relatively deep and more curved;** sphenopalatine vacuities small and circular; **incisive foramina long and exceptionally broad; zygomatic arches broadly spreading and squarely shaped; braincase deep**. Species group *palustris*.

**Comments:** *Oryzomys peninsulae* was reinstated as a distinct species by Carleton and Arroyo-Cabrales (2009); it was previously regarded as a subspecies of *O. couesi* (Álvarez-Castañeda and Cortés-Calva 1999) and of *O. palustris* (Hershkovitz 1971). *O. peninsulae* is restricted to the southern Baja California peninsula (Map 292), and cannot be found in sympatry, or nearly so, with any other species of *Oryzomys.*

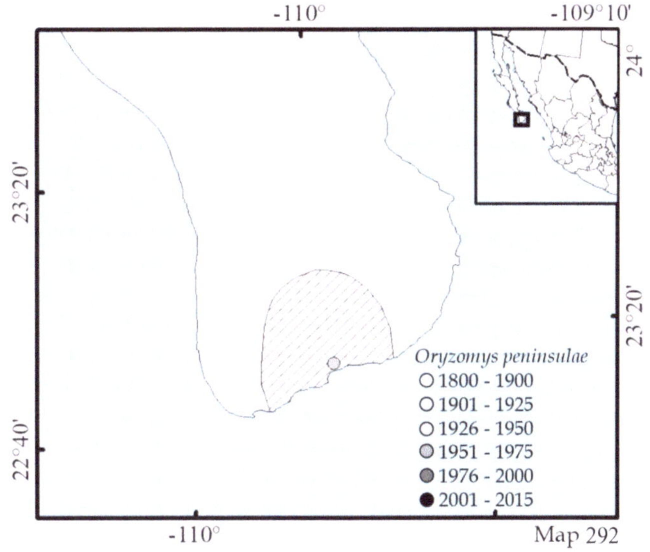


**Map 292** Distribution map of *Oryzomys peninsulae (extinct species)*

**Additional Literature:** Merriam (1901b), Goldman (1918), Álvarez-Castañeda (1994), Carleton and Arroyo-Cabrales (2009).

## ***Oryzomys texensis*** Allen, 1894

Texas rice rat, rata arrocera de Texas

Monotypic.

**Conservation:** *Oryzomys texensis* (as *O. palustris*) is listed as Least Concern by the IUCN (Cassola 2016cf).

**Characteristics:** Texas rice rat is medium sized; total length 226.0–279.0 mm and skull length 30.0–32.1 mm. Dorsal pelage ochraceous-tawny suffusion, darkened on the face, top of the head, and back by overlying blackish hairs; fur of underparts whitish; feet whitish, hindfoot digits with hair shorter than the respective claw; tail long, equal to or slightly larger than the head-and-body length; **skull narrow and less massive**; zygomatic arches less widely spreading; the frontal region narrower; supraorbital borders less projecting; tail brownish dorsally, whitish ventrally, and in some specimens dark all around near the tip; sphenopalatine foramen oval; foramines perforate mesopterygoid walls; exposing much of the presphenoid bone; **anterolabial cingulum on the lower third molars.** Species group *palustris*.

**Comments:** *Oryzomys texensis* was recognized as a distinct species (Hanson et al. 2010); it was previously regarded as a subspecies of *O. palustris* (Goldman 1918; Carleton and Arroyo-Cabrales 2009). *O. texensis* ranges west of the Mississippi River in Oklahoma, Arkansa, Texas, and Louisiana (Map 293). It can be found in sympatry, or nearly so, with *O. couesi*, *O. palustris*, *Handleyomys chapmani*, and *H. melanotis. Handleyomys* species can be easily misidentified as *Oryzomys. O. texensis* can be distinguished from *O. couesi* by its coloration predominantly grayish brown, with long glistening guard hairs conspicuously overtopping the body fur; sphenopalatine vacuities spacious, oval and perforating the mesopterygoid fossa walls, exposing much of the presphenoid bone. From *O. palustris*, by having a paler coloration and with genetic data. From the species of *Handleyomys*, by having hindfoot digits with hair shorter than the respective claw; five plantar pads; eight mammae (nipples); zygomatic notch deep; mesopterygoid fossa fenestrated; and principal reentrant angles normally reaching less than halfway across the molar crowns and auditory bullae small.

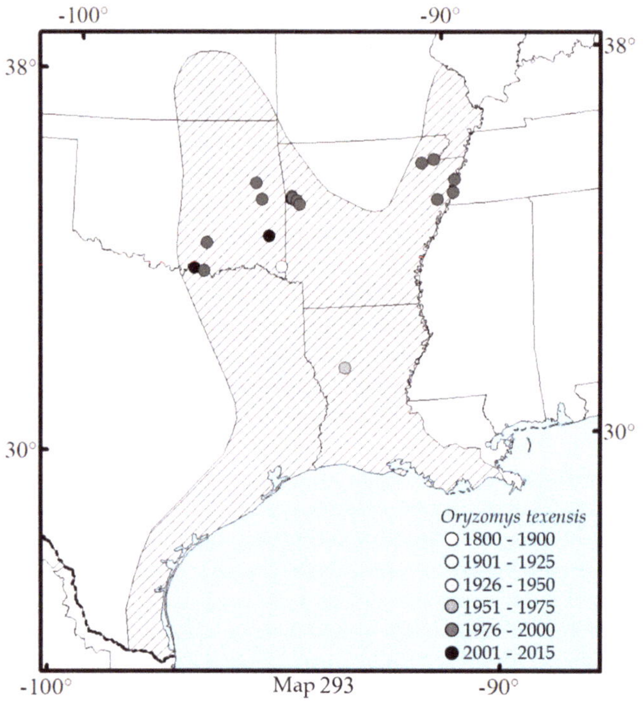


**Map 293** Distribution map of *Oryzomys texensis*

**Additional Literature:** Schmidt and Engstrom (1994).

# Tribe Sigmodontini

## Genus *Sigmodon* Say and Ord, 1825

Dorsal pelage brown to dark brown or salt pepper; fur of underparts white or paler than in the dorsum; **hair short and thick (bushy); ears small; tail shorter than the head-and-body length, showing a series of rings formed by scales with little hair; hindfoot soles blackish;** external hindfoot digits proportionally smaller than the rest; six plantar pads; **ten mammae (nipples);** molariforms flat-crowned with a series of transverse enamel plates; dorsal surface of the rostral premaxillae with a shallow depression, if any; mesopterygoid fossa edges variable across species; supraorbital ridge with the edge well developed and extending to the back of the parietal; interparietal bone wide; **palatal breadth ending at the posterior margin of the last molariforms with a well-defined pit between the middle and last molariforms;** pterygoid fossa deep; zygomatic plate very sharply cut back above, with a process projecting forward on the upper border; bullae moderate; the coronoid process on the mandible well developed. *Sigmodon* has only one subgenus (*Sigmodon*) in North America with two species groups: *fulviventer* (*S. fulviventer*, *S. leucotis*, and *S. ochrognathus*) and *hispidus* (*S. alleni*, *S. arizonae*, *S. hirsutus*, *S. hispidus*, *S. mascotensis*, and *S. toltecus*). The keys were elaborated based on the review of specimens and the following sources: Baker (1969a, b), Hall (1981), and Álvarez-Castañeda et al. (2017a).

1. Tail naked, scaly in appearance, each scale equal to or greater than 0.7 mm in breadth; skull elongated and thin; basioccipital bone large and wide; palatal pits shallow (Fig. 77) .......... 2

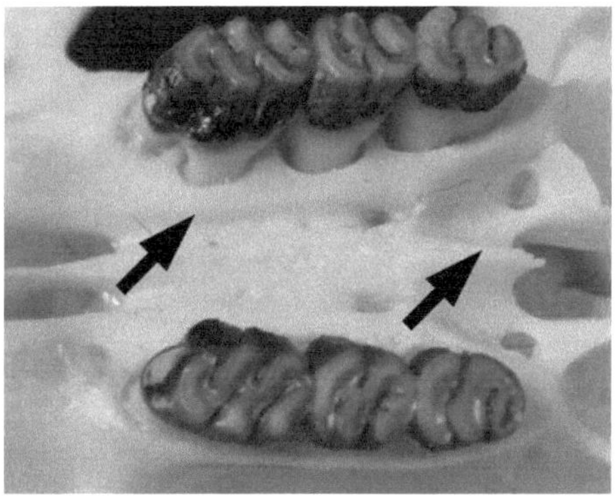

**Fig. 77** Palatal pits shallow

1a. Tail covered with hair, so scales not conspicuous, each scale less than 0.7 mm in breadth; skull short and wide; basioccipital bone large and thin or short and wide; palatal pits deep (Fig. 78) .......... 6

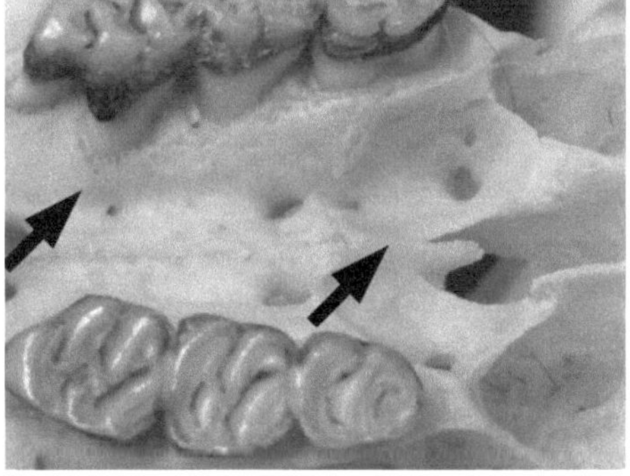

**Fig. 78** Palatal pits deep

2. Hindfoot length on average less than 34.0 mm; minimum distance between the temporal and occipital crest equal to or less than 3.5 mm; lateral nasal margin concave ........ 3

2a. Hindfoot length on average equal to or greater than 34.0 mm; minimum distance between the temporal and occipital crest equal to or greater than 3.5 mm; lateral nasal margin concave or straight ........ 5

3. Dorsal coloration grizzled with hairs on the back paler cream; dorsal color strongly contrasting with the flanks; palatal bone with a depression (north of Río Bravo, from western Arizona and Colorado east through Virginia and the Florida Peninsula and Tamaulipas in the Atlantic southern Sonora to Oaxaca in the Pacific) ........ *Sigmodon hispidus* (p. 415)

3a. Back coloration grizzled with hairs on the back pale orange brownish; dorsal color not strongly contrasting with the flanks; palatal flat; ranges outside of the United States ........ 4

4. Tail length on average less than 120.0 mm; ranges in the Gulf coastal plains of México from Tamaulipas south through the lowlands of Chiapas and northern Guatemala, including the Yucatán Peninsula and Belize) ........ *Sigmodon toltecus* (p. 420)

4a. Tail length greater than 120.0 mm; ranges in the Pacific coastal plains of México, from west of the Tehuantepec Isthmus south through Central America ........ *Sigmodon hirsutus* (p. 414)

5. Palatal ridge present; average total skull length approximately 40.0 mm; oval foramen large (Fig. 79); lateral margin of the nasals concave (from southern Arizona south through Sinaloa and Nayarit, with a disjunctive population along the Colorado River between Nevada, Arizona, and California) ........ *Sigmodon arizonae* (p. 411)

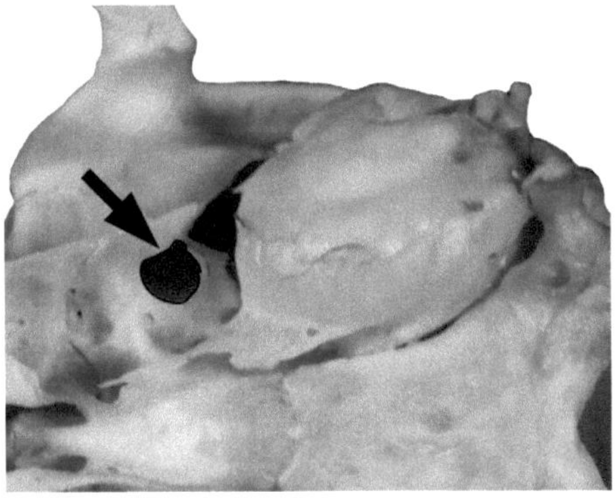

**Fig. 79** Oval foramen large

5a. Palatal ridge absent; average total skull length approximately 36.0 mm; oval foramen small (Fig. 80); lateral margin of the nasals concave or straight (from southern Sinaloa and southwest Zacatecas south through western Chiapas, including Morelos) ........ *Sigmodon mascotensis* (p. 418)

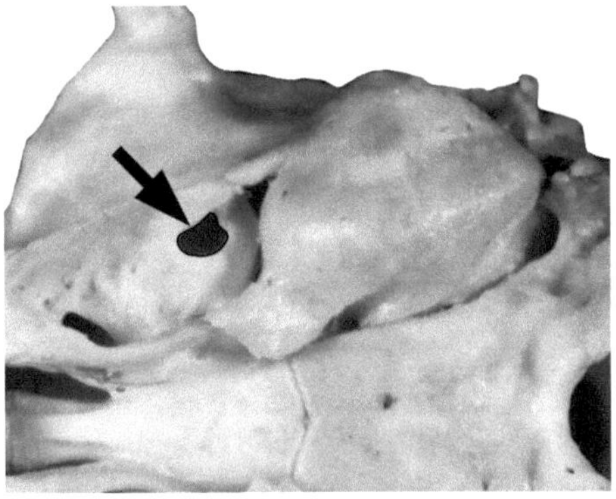

**Fig. 80** Oval foramen small

6. Ear inner coloration whitish, clearly contrasting with the dorsal color; interparietal length less than 2.0 mm; dorsal surface of the premaxillae with a strong depression; edges of the mesopterygoid fossa parallel (Fig. 81; from southwestern Chihuahua, southern Nuevo León and Tamaulipas south through southern Oaxaca) .......*Sigmodon leucotis* (p. 417)

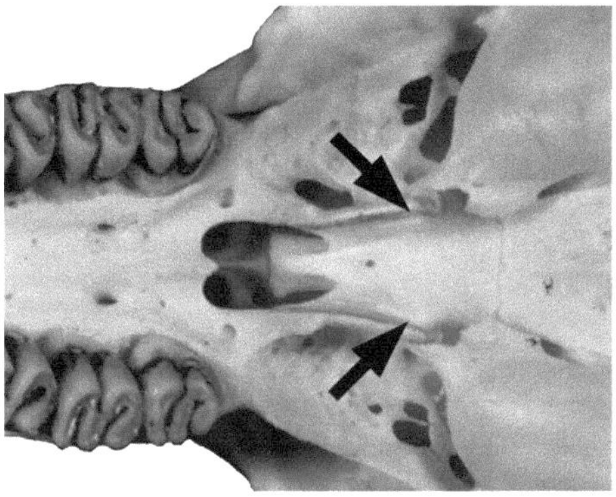

**Fig. 81** Edges of the mesopterygoid fossa parallel

6a. Ear inner coloration not contrasting with the dorsal color; interparietal length greater than 2.0 mm; dorsal surface of the premaxillae with a shallow depression, if any; edges of the mesopterygoid fossa not parallel (Fig. 82).........................7

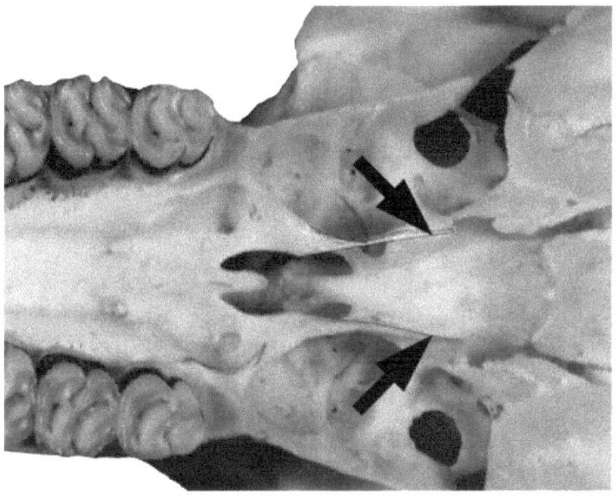

**Fig. 82** Edges of the mesopterygoid fossa not parallel

7. Nose area and eye ring distinctly ochraceous; tail scales less than 0.6 mm; premaxilla-to-condyle length less than 33.2 mm; auditory bullae small and elongated (Fig. 83; from southern Arizona, New Mexico, and Texas south through Durango) ........................................................................................................... *Sigmodon ochrognathus* (p. 419)

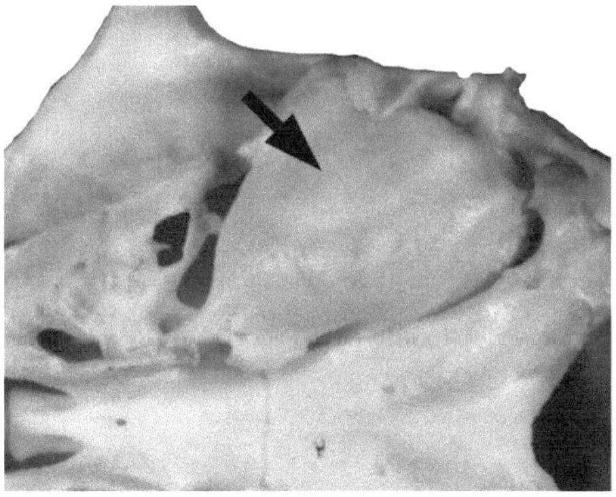

**Fig. 83** Auditory bullae small and elongated

7a. Nose area and eye ring equal to the body color; tail scales greater than 0.6 mm; premaxilla-to condyle-length greater than 33.2 mm; auditory bullae relatively large and wide (Fig. 84) ........ 8

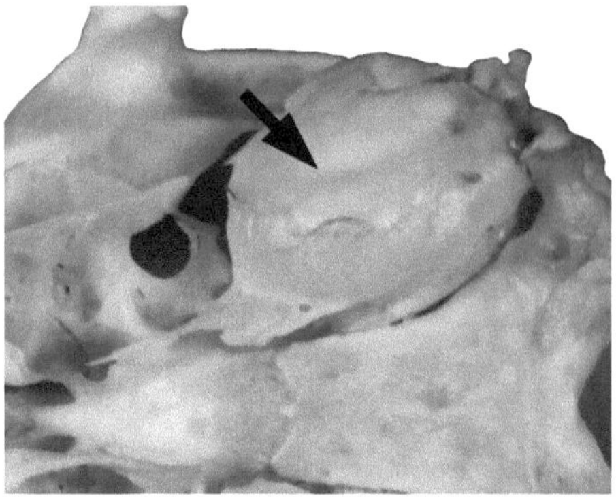

**Fig. 84** Auditory bullae relatively large and wide

8. Dorsal coloration grizzled; fur of underparts yellowish; head-and-body length averaging 179.0 mm; premaxilla-to-condyle length 36.5 mm on average; skull arched, short, and broad; posterior border of the incisive foramen extending back to the anterior border of the first molars (from southeast Arizona, New Mexico, and southwest Texas south through Guanajuato and northwestern Michoacán) ........ *Sigmodon fulviventer* (p. 413)

8a. Dorsal coloration brown; fur of underparts whitish or pale yellow; head-and-body length averaging 168.0 mm; premaxilla-to-condyle length 34.5 mm on average; skull flat, long, and thin; posterior border of the incisive foramen ending before the anterior border of the first molars (from Sinaloa south through Tehuantepec Isthmus) ........ *Sigmodon alleni* (p. 410)

## Subgenus *Sigmodon* Say and Ord, 1825

Cheek teeth tending to become larger antero-posteriorly; supraorbital ridge edge well developed and extending to the parietal back; three pectoral and two inguinal pairs of mammae.

### *Sigmodon alleni* Bailey, 1902

Allen's cotton rat, rata cañera del occidente

Monotypic.

**Conservation:** *Sigmodon alleni* is listed as Vulnerable (A2c + 3c: population reduction in the last 10 years equal to or less than 30% and may not be reversible based on a decline in area of occupancy and quality of habitat, and population reduction of equal to or less than 30% projected to be met within the next 10 years based on a decline in area of occupancy and quality of habitat) by the IUCN (De Grammont and Cuarón 2018d).

**Characteristics:** Allen's cotton rat is medium sized; total length 207.0–228.0 mm and skull length 30.3–31.9 mm. **Dorsal pelage rich brown**; fur of underparts whitish to yellowish; skull short and wide; **incisors strongly recurved**; a crest in the palate; **auditory bullae small in relation to breadth**; palatal pits deep; **basioccipital bone short and narrow;** interparietal length greater than 2.0 mm; mesopterygoid fossa edges not parallel; mandibular angular process round; **mesopterygoid spine small or absent**. Species group *hispidus*.

**Comments:** *Sigmodon planifrons* was recognized as a distinct species (Musser and Carleton (2005), mentioned that the *S. alleni* complex should be reviewed); it was previously regarded as a subspecies of *S. alleni* (Baker 1969a, b; Carleton et al. 1999). Carleton et al. (1999) and León-Paniagua (2017) include *S. planifrons* within *S. alleni*. *S. alleni* ranges from southern Sinaloa south through south Tehuantepec Isthmus (Map 294). It can be found in sympatry, or nearly so, with *S. arizonae*, *S. fulviventer*, *S. hirsutus*, and *S. mascotensis*. *Sigmodon alleni* can be distinguished from all other species of *Sigmodon* by its tail with narrow scales, width equal to or less than 0.65 mm and heavily haired; skull generally short and broad; palatal pits deep; basioccipital one either long and broad or short and broad. From *S. mascotensis*, by having the dorsal pelage uniformly rich brown, occasionally with rufous or cinnamon brown sgades. From *S. fulviventer*, by having the dorsal pelage rich brown; skull with flattened appearance when viewed laterally; bulge of capsular projections for the upper incisors slight, and paraoccipital processes, when viewed from below; incisors slightly hooked rather than straight or curved.

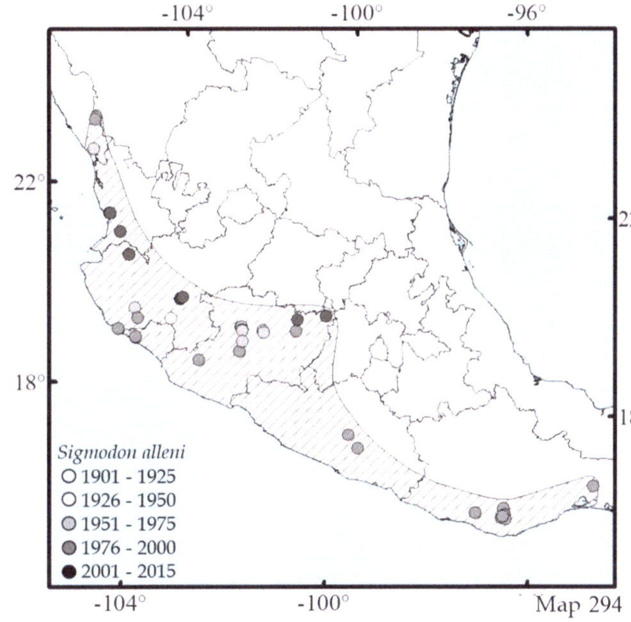


**Map 294** Distribution map of *Sigmodon alleni*

**Additional Literature:** Bailey (1902b), Jimenez (1972), Genoways and Birney (1974), Shump and Baker (1978a), Carroll et al. (2005), Musser and Carleton (2005).

## *Sigmodon arizonae* Mearns, 1890

Arizona cotton rat, rata cañera de Arizona

1. *S. a. arizonae* Mearns, 1890. Central Arizona.
2. *S. a. cienegae* Howell, 1919. From southeastern Arizona south through central Sonora.
3. *S. a. jacksoni* Goldman, 1918. Restricted to Fort Whipple, central Arizona.
4. *S. a. major* Bailey, 1902. From central Sonora south through northern Nayarit, including western Durango.
5. *S. a. plenus* Goldman, 1928. Restricted to the lower Colorado River, southeastern California, western Arizona, and southern Nevada.

**Conservation:** *S. a. arizonae* and S. *a. jacksoni* are believed extinct (Hoffmeister 1986; Arizona Game and Fish Department 2004a, b). *Sigmodon arizonae* is listed as Least Concern by the IUCN (Álvarez-Castañeda et al. 2016ah).

**Characteristics:** Arizona cotton rat is medium sized; total length 247.0–363.0 mm and skull length 27.4–40.4 mm. Dorsal pelage pale buffy gray grizzled with blackish or dark brown shades; fur of underparts pale silver or whitish, blackish at the base; feet with hair grayish to dull brown; tail large, with large "scales" and long hairs, dark dorsally and graded into paler brown or silvery shades ventrally; skull elongated and thin; basioccipital bone large and wide; palatal pits shallow; minimum distance between the temporal and occipital crest averaging greater than or equal to 3.9 mm; nasal lateral margin concave or straight; oval foramen large; **mesopterygoid spine well developed**; **foramen oval and small**. Species group *hispidus*.

**Comments:** *Sigmodon arizonae* was reinstated as a distinct species by Zimmerman (1970); it was previously regarded as a subspecies of *S. hipidus* (Mearns 1890). *S. arizonae* ranges from southern Arizona southward through Sinaloa and Nayarit, with a disjunctive population along the Colorado River between Nevada, Arizona, and California (Map 295). It can be found in sympatry, or nearly so, with *S. alleni*, *S. fulviventer*, *S. hispidus*, and *S. mascotensis*. *S. arizonae* can be distinguished from *S. alleni* and *S. fulviventer* by having the tail with broad scales, with a width equal to or greater than 0.65 mm and sparsely haired; skull generally long and narrow; palatal pits shallow; basioccipital bone long and broad. From *S. hispidus*, by having larger hind feet, greater than 34.0 mm; distance between the temporal and occipital crest on average greater than 3.6 mm; foramen oval diameter greater than three-fourths of the diameter of the third upper molars and crest on the posterior edge of the palate developed; ventral flat surface of the presphenoid wider, with lateral sides not visible. From *S. mascotensis*, by having the dorsal pelage also brown but with darker upperparts and yellowish flanks; foramen oval size averages greater than three-quarters of the diameter of the third upper molars; skull length averages 40.0 mm; an oval-shaped fenestra on the parapterygoid fossa is rare.

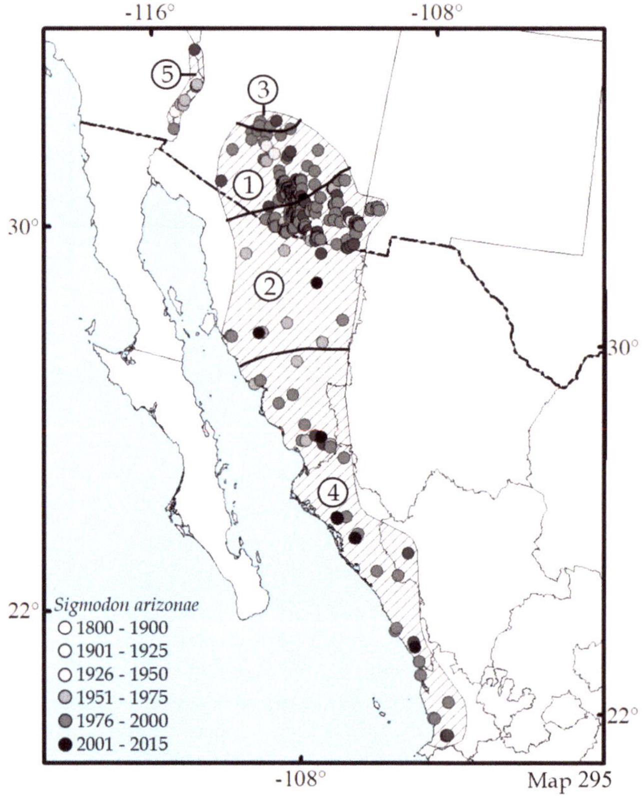


**Map 295** Distribution map of *Sigmodon arizonae*

**Additional Literature:** Bailey (1902b), Zimmerman and Lee (1968), Johnson et al. (1972), Severinghaus and Hoffmeister (1978), Hoffmeister (1986), Álvarez-Castañeda and Cortés-Calva (1999), Arizona Game and Fish Department (2004a, b), Carroll et al. (2005), Musser and Carleton (2005), Gwinn et al. (2011), Martínez-Chapital et al. (2017).

## *Sigmodon fulviventer* Allen, 1889

Tawny-bellied cotton rat, rata cañera de vientre amarillo

1. *S. f. dalquesti* Stangl, 1992. Only known from its type locality in southwestern Texas.
2. *S. f. fulviventer* Allen, 1889. From northern Durango south through northern Jalisco and Guanajuato.
3. *S. f. goldmani* Bailey, 1913. Restricted to Las Palomas area, southwest New Mexico but presumed to be extinct.
4. *S. f. melanotis* Bailey, 1902. Northern Jalisco, Guanajuato, and Michoacán.
5. *S. f. minimus* Mearns, 1894. From central New Mexico and southwestern Arizona south through northern Durango.

**Conservation:** *Sigmodon fulviventer* is listed as Least Concern by the IUCN (Álvarez-Castañeda et al. 2016d).

**Characteristics:** Tawny-bellied cotton rat is medium sized; total length 223.0–270.0 mm and skull length 29.3–34.6 mm. Dorsal pelage grizzled gray, golden gray to yellowish brown; **fur of underparts orange**; flanks slightly paler with whitish hairs; tail brown to blackish brown; skull arched, short, and broad; anterior nasals triangular; palatal pits deep; basioccipital bone large and thin or short and wide; interparietal length greater than 2.0 mm; mesopterygoid fossa with edges not parallel; incisive foramen with the posterior border extending back from the anterior border of the first molars; mandibular angular process rounded. Species group *fulviventer*.

**Comments:** *Sigmodon fulviventer* ranges from southeast Arizona, New Mexico, and southwest Texas south through Guanajuato and northwestern Michoacán (Map 296). It can be found in sympatry, or nearly so, with *S. alleni*, *S. arizonae*, *S. hipidus*, *S. mascotensis*, and *S. ochrognathus*. *S. fulviventer* can be distinguished from *S. arizonae*, *S. hipidus*, and *S. mascotensis* by having the fur of underparts orange; tail with narrow scales with a width equal to or less than 0.65 mm and heavily haired; skull generally short and broad; palatal pits deep; basioccipital bone either long and broad or short and broad. From *S. alleni*, by having the fur of underparts orange; dorsal pelage brown; skull not flattened in appearance when viewed laterally; incisors straight or curved rather than slightly hooked. From *S. ochrognathus*, by having the fur of underparts orange; dorsal pelage with a grizzled coloration; nose area of similar color as the back; without a pronounced medial keel on the basioccipital bone; auditory bullae large and wide; a medial-posterior notch on the interparietal bone and curved paraoccipital processes without distinctive basal notches.

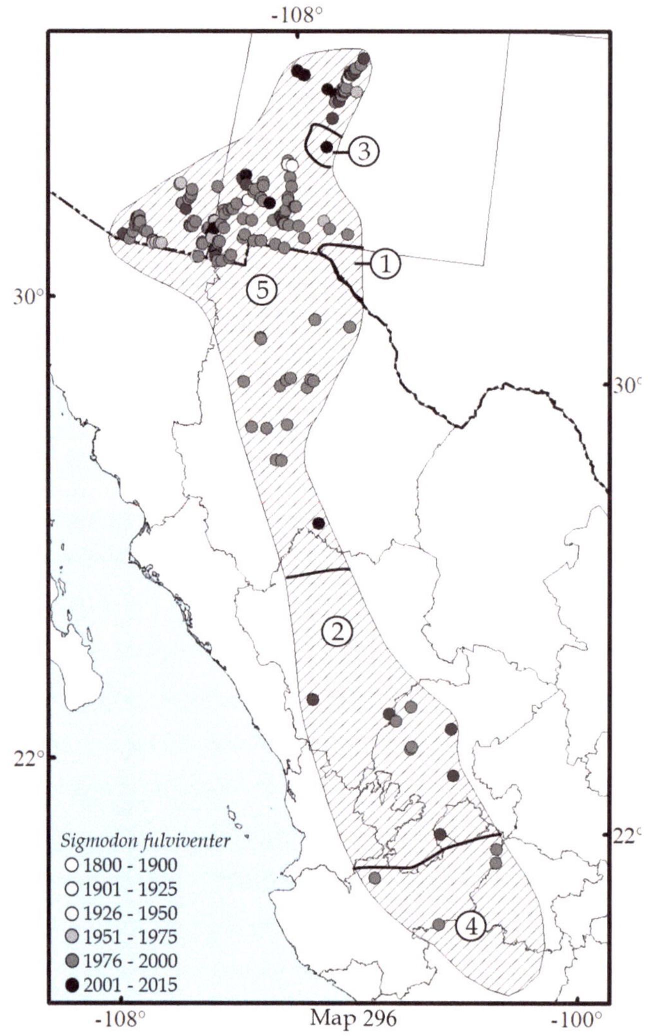


**Map 296** Distribution map of *Sigmodon fulviventer*

**Additional Literature:** Bailey (1902b), Baker (1969a, b), Baker and Shump (1977), Peppers and Bradley (2000), Peppers et al. (2002), Musser and Carleton (2005).

## *Sigmodon hirsutus* Burmeister, 1854

Southern cotton rat, rata cañera del sur

1. *S. h. saturatus* Bailey, 1902 (part). Northern Chiapas, southern Tabasco, northern Guatemala, and southern Belize.
2. *S. h. tonalensis* Bailey, 1902. Restricted to Tehuantepec Isthmus.
3. *S. h. zanjonensis* Goodwin, 1932. Highlands of Chiapas south through Central America.

**Conservation:** *Sigmodon hirsutus* is listed as Least Concern by the IUCN (Delgado et al. 2016b).

**Characteristics:** Southern cotton rat is medium sized; total length 254.0–270.0 mm and skull length 36.3–38.1 mm. Dorsal pelage grizzled blackish to dark brown, with yellowish or grayish shades, flanks slightly paler; fur of underparts light gray to gray, sometimes grizzly with yellowish shades; with a crest in the palate; skull elongated and thin; basioccipital bone large and wide; palatal pits shallow; minimum distance between the temporal and occipital crest averaging less than or equal to 3.2 mm; nasal lateral margin concave. Species group *hispidus*.

**Comments:** *Sigmodon hirsutus* was reinstated as a distinct species by Peppers and Bradley (2000); it was previously regarded as a subspecies of *S. hipidus*. It includes the subspecies *S. h. griseus*, *S. h. hirsutus*, and *S. h. saturatus*; no morphological characteristics can be used to distinguish it from *S. hispidus* or *S. toltecus* (Peppers and Bradley 2000; Carroll et al. 2005). León-Paniagua (2017) still considers it a full species. *S. zanjonensis* was considered a junior species of *S. hirsutus* (Carroll et al. 2005). *S. hirsutus* ranges from the Tehuantepec Isthmus south through Central America (Map 297). It can be found in sympatry, or nearly so, with *S. alleni* and *S. mascotensis*. *S. hirsutus* can be distinguished from *S. alleni* by having the tail with narrow scales equal to or less than 0.65 mm and heavily haired; skull generally short and broad; palatal pits deep; basioccipital bone either long and broad or short and broad. From *S. mascotensis*, by having the distance between the upper and lower ridges from 3.0 to 3.2 mm; hindfoot length 31.0 mm to 32.5 mm; temporal fossa trapezoidal in shape over the posterior half of the parietal and squamosal bones.

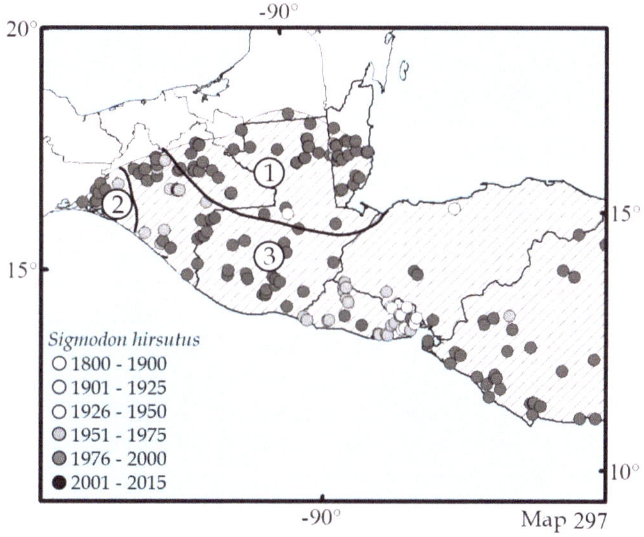


**Map 297** Distribution map of *Sigmodon hirsutus*

**Additional Literature:** Bailey (1902b), Hall and Kelson (1959), Carleton et al. (1999), Peppers et al. (2002), Musser and Carleton (2005), Martínez-Chapital et al. (2017).

## *Sigmodon hispidus* Say and Ord, 1825

Hispid cotton rat, rata cañera espinosa

1. *S. h. alfredi* Goldman and Gardner, 1947. Restricted to southeastern Colorado.
2. *S. h. berlandieri* Baird, 1855. From central Sonora, Coahuila, and Nuevo León south through Estado de México, Tlaxcala, western Puebla, and Oaxaca.
3. *S. h. confinis* Goldman, 1918. Restricted to southwestern Arizona.
4. *S. h. eremicus* Mearns, 1897. Restricted to the Colorado River Delta, southeastern California, southwestern Arizona, northeastern Baja California, and northwestern Sonora (not in the map).
5. *S. h. exsputus* Allen, 1920. Restricted to Big Pine Key, southern Florida.
6. *S. h. floridanus* Howell, 1943. Restricted to the central and southern Florida Peninsula.
7. *S. h. hispidus* Say and Ord, 1825. Missouri, Arkansas, South Carolina, and Florida
8. *S. h. insulicola* Howell, 1943. Restricted to Captiva Island, southwestern Florida.
9. *S. h. komareki* Gardner, 1948. From southern Tennessee and Alabama east through North Carolina, excluding the coastal areas.
10. *S. h. littoralis* Chapman, 1889. From Missouri, Arkansas, and Louisiana east through central Tennessee and western Alabama and along coastal areas to North Carolina.

11. *S. h. spadicipygus* Bangs, 1898. Restricted to the southernmost Florida Peninsula.
12. *S. h. texianus* (Audubon and Bachman, 1853). From southern Nebraska to central Texas, including western Missouri and Arkansas.
13. *S. h. virginianus* Gardner, 1946. Restricted to southern Virginia.

**Conservation:** *Sigmodon hispidus* is listed as Least Concern by the IUCN (Cassola 2016dd).

**Characteristics:** Hispid cotton rat is medium sized; total length 224.0–365.0 mm and skull length 31.0–36.7 mm. Dorsal pelage grizzled blackish to dark brown grizzly with yellowish or grayish shades; flanks slightly paler; fur of underparts whitish, smoky gray or dull buffy brown, sometimes grizzly with yellowish shades; feet dull rusty brown; tail blackish, slightly paler ventrally; **skull generally long and narrow; palatal pits shallow; basioccipital bone long and broad;** mastoid breadth less than 46% of the basal length; crest in the palate; basioccipital bone large and wide; minimum distance between the temporal and occipital crest averaging less than or equal to 3.2 mm; nasal lateral margin concave; **foramen oval in proportion to the skull, third lower molars S shaped.** Species group *hispidus*.

**Comments:** *Sigmodon arizonae* (Zimmerman 1970), *S. hirsutus* (Peppers and Bradley 2000), *S. mascotensis* (Dalby and Lillevick 1969; Zimmerman 1970), and *S. toltecus* (Peppers and Bradley 2000) were previously subspecies of *S. hipidus*. The taxonomic status of *S. h. berlandieri* requires additional analyses for a formal assignment (Carroll et al. 2005). I keep *S. h. berlandieri* as a subspecies until samples from all the ranges of the subspecies are reviewed because two species are observed within its range. No morphological characteristic can be used to distinguish it from *S. hirsutus* or *S. toltecus* (Peppers and Bradley 2000; Carroll et al. 2005). *S. zanjonensis* was reinstated as a distinct species by Carleton et al. (1999); it was previously regarded as a subspecies of *S. hipidus* (Goodwin, 1934). However, genetic analyses of populations of *S. zanjonensis* have reassigned it to a junior species of *S. hirsutus* (Carroll et al. 2005). *S. hispidus* ranges north of Río Bravo, from western Arizona and Colorado east through Virginia, the Florida Peninsula, in the Atlantic south through central Sonora, eastern Chihuahua, Coahuila, Nuevo León, and Tamaulipas to Oaxaca, except central-south Tamaulipas and Veracruz (Map 298). It can be found in sympatry, or nearly so, with *S. arizonae* and *S. ochrognathus*. *S. hispidus* can be distinguished from *S. arizonae* by having smaller hind feet, less than 34.0 mm; distance between the temporal and occipital crest on average less than 3.6 mm; foramen oval diameter less than three-fourths of the diameter of the third upper molars and crest on the posterior of the palate well developed; presphenoid ventral flat surface narrow, with the sides visible. From *S. ochrognathus*, by having the tail with broad scales with a width equal to or greater than 0.65 mm and sparsely haired; skull generally long and narrow; palatal pits shallow; basioccipital bone long and broad.

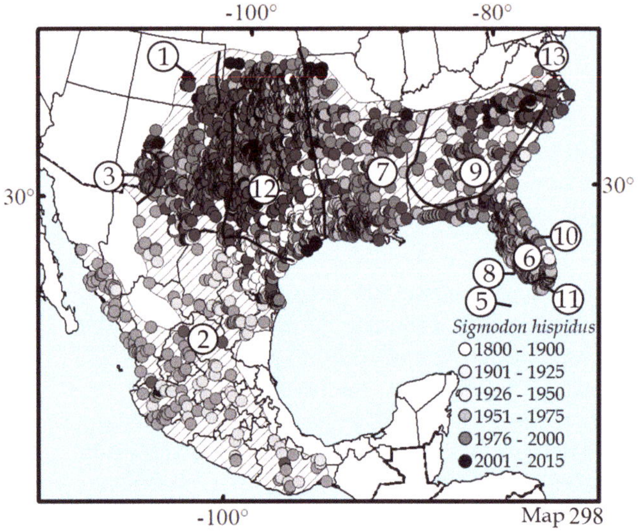


**Map 298** Distribution map of *Sigmodon hispidus*

**Additional Literature:** Bailey (1902b), Hall and Kelson (1959), Easterla (1968c), Anderson (1972), Severinghaus and Hoffmeister (1978), Cameron and Spencer (1981), Hall (1981), Musser and Carleton (2005).

## ***Sigmodon leucotis*** Bailey, 1902

White-eared cotton rat, rata cañera orejas blancas

1. *S. l. alticola* Bailey, 1902. From western Puebla south through northwestern Oaxaca.
2. *S. l. leucotis* Bailey, 1902. From Morelos, Estado de México, and Tlaxcala north through southwestern Chihuahua or southwestern Nuevo León.

**Conservation:** *Sigmodon leucotis* is listed as Least Concern by the IUCN (Álvarez-Castañeda et al. 2016t).

**Characteristics:** White-eared cotton rat is medium sized; total length 230.0–252.0 mm and skull length 31.4–34.1 mm. Dorsal pelage grayish brown; fur of underparts whitish; **ears whitish contrasting with the dorsal coloration**; tail blackish. Brown at the base; skull short and wide; mastoid breadth less than 46% of the basal length; crest in the palate; **prominent depression on the premaxilla**; palatal pits deep; basioccipital bone short or large, wide or narrow; interparietal bone short, less than 2.0 mm in length; mesopterygoid fossa edges parallel; the **mandible angular process hooked**. Species group *fulviventer*.

**Comments:** *Sigmodon leucotis* ranges from southwestern Chihuahua, southern Nuevo León, and Tamaulipas south through southern Oaxaca (Map 299). It can be found in sympatry, or nearly so, with *S. ochrognathus*, from which it differs by having ears with inner whitish color, strongly contrasting with the dorsal coloration; interparietal length less than 2.0 mm; premaxilla dorsal surface with a strong depression; mesopterygoid fossa edges parallel and the mandible angular process hooked.

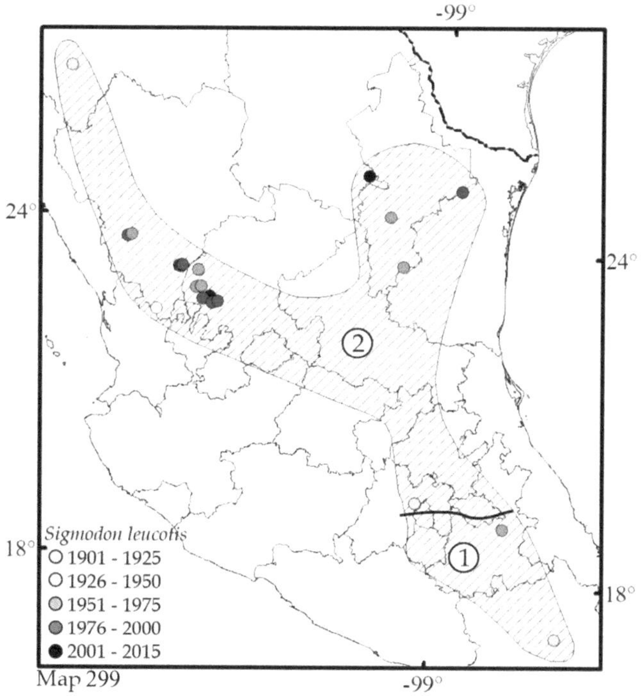


**Map 299** Distribution map of *Sigmodon leucotis*

**Additional Literature:** Bailey (1902b), Hershkovitz (1955), Baker (1969a, b), Zimmerman (1970), Baker and Shump (1977), Shump and Baker (1978b), Musser and Carleton (2005).

## *Sigmodon mascotensis* Allen, 1897

Jalisco cotton rat, rata cañera del oeste

1. *S. m. inexoratus* Elliot, 1903. Restricted to eastern Jalisco and northwestern Michoacán.
2. *S. m. ischyrus* Goodwin 1956. Restricted to Santo Domingo Chontecomatlán, southeastern Oaxaca.
3. *S. m. mascotensis* Allen, 1897. From southern Nayarit south through eastern Chiapas.

**Conservation:** *Sigmodon mascotensis* is listed as Least Concern by the IUCN (Álvarez-Castañeda et al. 2016af).

**Characteristics:** Jalisco cotton rat is medium sized; total length 220.0–314.0 mm and skull length 27.5–37.0 mm. Dorsal pelage grizzled blackish to dark brown grizzly with yellowish or grayish shades; flanks slightly paler; fur of underparts pale to dark gray grizzly with yellowish shades, plumbeous gray at the base; nose area often rich brown; feet dull brownish gray; **hind feet long**; **tail long,** brownish black dorsally and brownish gray ventrally; skull elongated and thin; nasals posteriorly truncate; **nasal lateral margin concave or straight without a crest in the palate**; palatal ridge absent; basioccipital bone large and wide; palatal pits shallow; minimum distance between the temporal and occipital crest averaging greater than or equal to 3.9 mm; oval foramen small; auditory bullae small; interparietal concave-convex; **mesopterygoid spine small**. Species group *hispidus*.

**Comments:** *Sigmodon mascotensis* was reinstated as a distinct species by Dalby and Lillevick (1969) and Zimmerman (1970); it was previously regarded as a subspecies of *S. hipidus* (Bailey 1902b). *S. mascotensis* ranges from the coast of south Sinaloa and southwest Zacatecas south through western Chiapas, including Morelos (Map 300). It can be found in sympatry, or nearly so, with *S. alleni*, *S. arizonae*, *S. fulviventer*, *S. hirsutus*, and *S. toltecus*. *S. mascotensis* can be distinguished from *S. alleni*, *S. fulviventer*, and *S. toltecus* by having the tail with broad scales with a width equal to or greater than 0.65 mm and sparsely haired; skull generally long and narrow; palatal pits shallow; basioccipital bone long and broad. From *S. alleni*, by having the dorsal pelage grayish brown that is uniform throughout. From *S. arizonae*, by having the dorsal pelage grayish brown that is uniform throughout; foramen oval size equal to or less than three-quarters of the diameter of the third upper molars; skull length averages 36.0 mm; oval-shaped fenestra on the parapterygoid fossa is common. From *S. hispidus*, *S. hirsutus*, and *S. toltecus*, by having the distance between the upper and lower ridges greater than 3.9 mm; hindfoot length 34.0–36.0 mm, and temporal fossa bigger and rectangular.

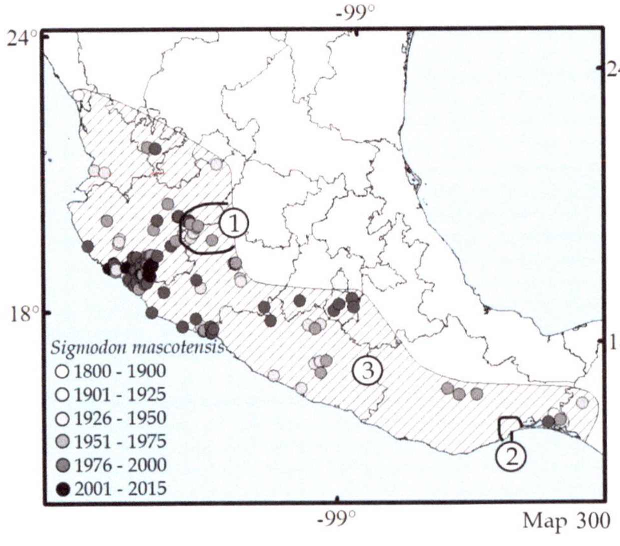


**Map 300** Distribution map of *Sigmodon mascotensis*

**Additional Literature:** Zimmerman and Lee (1968), Severinghaus and Hoffmeister (1978), Elder and Lee (1985), Carleton et al. (1999), Carroll et al. (2005), Musser and Carleton (2005), Martínez-Chapital et al. (2017).

## *Sigmodon ochrognathus* Bailey, 1902

Yellow-nosed cotton rat, rata cañera rostro amarillo

Monotypic.

**Conservation:** *Sigmodon ochrognathus* is listed as Least Concern by the IUCN (Lacher et al. 2016n).

**Characteristics:** Yellow-nosed cotton rat is medium sized; total length 223.0–260.0 mm and skull length 30.2–32.0 mm. **Dorsal pelage with a muddy gray coloration**; fur of underparts whitish; **nose area and eye ring ochraceous**; tail with scales measuring 0.50 mm, contrasting with those of other species measuring 0.75 mm; palatal pits deep; basioccipital bone short or large, wide or narrow; interparietal length greater than 2.0 mm; mesopterygoid fossa edges not parallel; mandible angular process rounded; **pronounced median keel on the basioccipital bone**; **auditory bullae small and elongated; median-posterior notch on the interparietal bone and curved paraoccipital processes with distinctive basal notches**. Species group *hispidus*.

**Comments:** *Sigmodon ochrognathus* ranges from southern Arizona, New Mexico, and Texas south through Durango (Map 301). It can be found in sympatry, or nearly so, with *S. arizonensis*, *S. fulviventer*, *S. hispidus*, and *S. leucotis*. *S. ochrognathus* can be distinguished from *S. arizonensis* and *S. hispidus* by having the tail with narrow scales with a width equal to or less than 0.65 mm and heavily haired; skull generally short and broad; palatal pits deep; basioccipital bone either long and broad or short and broad. From *S. fulviventer*, by having the dorsal pelage with a muddy gray coloration; nose area and eye ring ochraceous; pronounced median keel on the basioccipital; auditory bullae small and elongated; a median-posterior notch on the interparietal; curved paraoccipital processes with distinctive basal notches. From *S. leucotis*, by having the ears in similar color to the top of the head; interparietal length greater than 2.0 mm; premaxillae dorsal surface without a depression; edges of the mesopterygoid fossa not parallel and mandible angular process rounded.

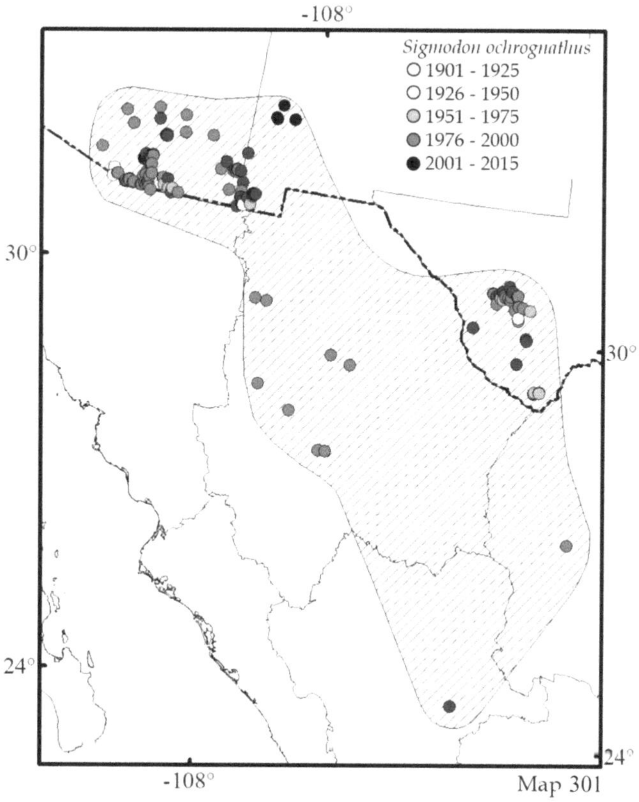


**Map 301** Distribution map of *Sigmodon ochrognathus*

**Additional Literature:** Bailey (1902b), Findley and Jones (1960), Baker (1969a, b), Zimmerman (1970), Baker and Shump (1978), Peppers et al. (2002), Musser and Carleton (2005), Carroll et al. (2005).

## ***Sigmodon toltecus*** (Saussure, 1860)

Toltec cotton rat, rata cañera del Golfo

1. *S. t. microdon* Bailey, 1902. Throughout the Yucatán Peninsula and Belize.
2. *S. t. saturatus* Bailey, 1902 (part). Southeastern Veracruz, Campeche, northern Oaxaca, and Chiapas.
3. *S. t. toltecus* (Saussure, 1860). From southern Tamaulipas to southern-central Veracruz, including western San Luis Potosí, Hidalgo, and Puebla.
4. *S. t. villae* Goodwin, 1958. Restricted to Teopisca, Chiapas.

**Conservation:** *Sigmodon toltecus* is listed as Least Concern by the IUCN (Cassola 2016de).

**Characteristics:** Toltec cotton rat is medium sized; total length 215.0–256.0 mm and skull length 29.3–32.2 mm. Dorsal pelage grizzled in blackish to dark brown grizzly with yellowish or gray shades; flanks slightly paler; fur of underparts pale to dark gray, sometimes grizzly with yellowish shades; skull elongated and thin; basioccipital bone large and wide; palatal pits shallow; minimum distance between temporal and occipital crest averaging less than or equal to 3.2 mm; lateral nasal concave or nearly so. Species group *hispidus*.

**Comments:** *Sigmodon toltecus* was reinstated as a distinct species by Peppers and Bradley (2000); it was previously regarded as a subspecies of *S. hispidus* (Bailey 1902b). It includes the subspecies *S. t. furvus*, *S. t. microdon*, *S. t. saturatus*, *S. t. toltecus*, and *S. t. villae;* no morphological characteristic can be used to distinguish it from *S. hirsutus* or *S. hispidus* (Peppers and Bradley 2000; Carroll et al. 2005). *S. toltecus* ranges in the lowlands from Tamaulipas south through the lowlands of Chiapas and northern Guatemala, including the Yucatán Peninsula and Belize (Map 302). It can be found in sympatry, or nearly so, with *S. alleni*, *S. hispidus*, and *S. mascotensis*. *S. toltecus* can be distinguished from *S. alleni* by having the tail with scales broad with a width equal to or greater than 0.65 mm and sparsely haired, skull generally long and narrow; palatal pits shallow; basioccipital bone long and broad. From *S. mascotensis*, by having the distance between the superior and inferior ridges from 3.0 to 3.2 mm; hindfoot length from 31.0 to 32.5 mm and temporal fossa trapezoidal over the posterior half of the parietal and squamosal bones.

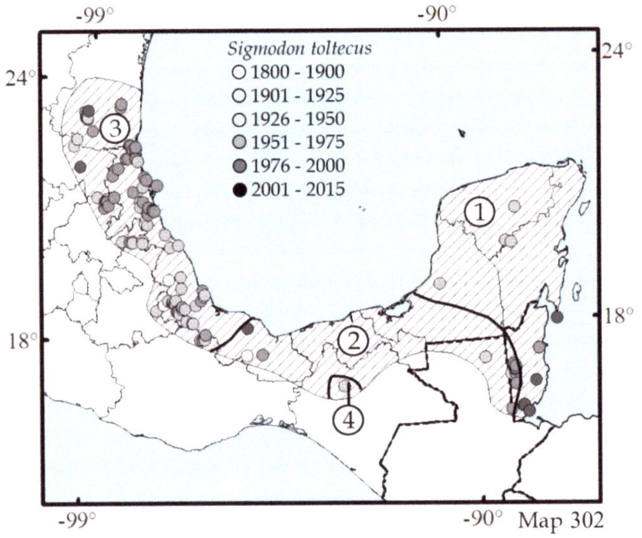


**Map 302** Distribution map of *Sigmodon toltecus*

**Additional Literature:** Hall and Kelson (1959), Hall (1981), Musser and Carleton (2005).

# **Subfamily Tylomyinae** Reig, 1984

1. Tail naked; anterior and lingual cusp of the upper first molars reduced ........ 2

1a. Tail with hair; anterior and lingual cusp of the upper first molars not reduced ........ 3

2. Hindfoot length less than 36.0 mm; pits between the cusps of the upper molars markedly developed; auditory bullae larger; upper toothrow less than 7.8 mm (from northern Chiapas southward, including the Yucatán Peninsula and one record in Guerrero) ........ *Ototylomys* (p. 423)

2a. Hindfoot length greater than 36.0 mm; pits between the cusps of the upper molars not markedly developed; auditory bullae small; upper toothrow greater than 7.8 mm ........ *Tylomys* (p. 426)

3. Auditory bullae not greatly inflated, total length less than 7.0 mm (from southern Jalisco and central Veracruz south through Central America, except for the Yucatán Peninsula) ........ *Nyctomys sumichrasti* (p. 421)

3a. Auditory bullae greatly inflated, total length greater than 7.0 mm (Yucatán Peninsula and northern Belize and Guatemala) ........ *Otonyctomys hatti* (p. 422)

## Genus ***Nyctomys*** Saussure, 1860

Tail long, well haired, and with a terminal hair tuft; **hind feet considerably modified for arboreal life, hallux with a claw**; rostrum short; zygomatic plate narrow and straight anteriorly with a prominent infraorbital foramen; palate broad, ending in front of a line made by the posterior margin of the toothrow, and without lateral pits; **toothrow extremely complex**; braincase and frontal bones broad; supraorbital ridges well developed and extending across the parietals to the occiput; interparietal bone broad and large; **auditory bullae small**.

### ***Nyctomys sumichrasti*** (Saussure, 1860)

Sumichrast's vesper rat, rata del amanecer

1. *N. s. colimensis* Laurie, 1953. Coastal areas of Jalisco and Michoacán.
2. *N. s. pallidulus* Goldman, 1937. Restricted to Oaxaca.
3. *N. s. salvini* (Tomes, 1862). From Chiapas south through Central America.
4. *N. s. sumichrasti* (Saussure, 1860). Restricted to southern Veracruz and northern Oaxaca.

**Conservation:** *Nyctomys sumichrasti* is listed as Least Concern by the IUCN (Samudio et al. 2016).

**Characteristics:** Sumichrast's vesper rat is medium sized; total length 225.0–246.0 mm and skull length 28.5–31.2 mm. **Dorsal pelage tawny to orange**; fur of underparts creamy or white, contrasting with the dorsal coloration; ears small but longer than wide, well haired basally but practically naked otherwise; hind feet short but wide; long digits modified for arboreal life; six plantar tubercles; hallux with a claw; **tail length longer than the head-and-body length, robust and cylindrical with a terminal hair tuft and bone-brown color**; head proportionately large; whiskers long; hair long and silky; supraorbital ridge edge well developed and extending to the occiput; interparietal bone well developed, completely separated from the parietal bone; rostrum short; approximately equal to the braincase breadth; zygomatic plate anterior margin almost perpendicular to the skull axis; failing to reach the anterior plane of the first molars; auditory bullae proportionally to the skull size, less than 6.0 mm.

**Comments:** *Nyctomys sumichrasti* ranges in the lowlands from southern Jalisco and central Veracruz south through South America, except for the Yucatán Peninsula, and with no records in Guerrero (Map 303). It can be found in sympatry, or nearly so, with *Ototylomys* and *Tylomys*. *N. sumichrasti* can be distinguished from *Otonyctomys hatti* by having the auditory bullae proportionally to the skull size, with length less than 7.0 mm. From all the species of *Tylomys*, by having a total length less than 300.0 mm.

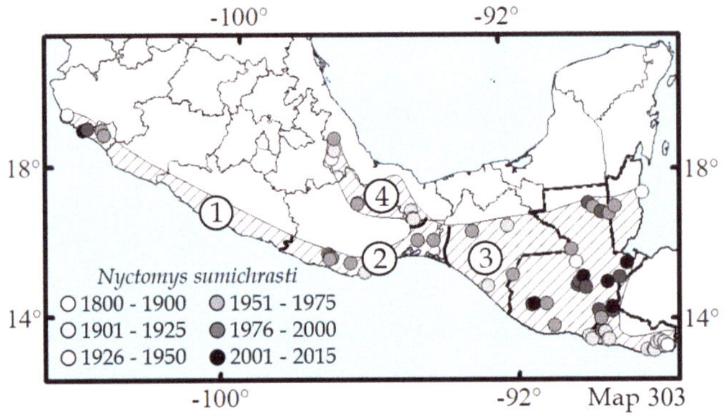


**Map 303** Distribution map of *Nyctomys sumichrasti*

**Additional Literature:** Lee and Elder (1977), Haiduk et al. (1988), Sánchez-Hernández et al. (1999), Hunt et al. (2004), Musser and Carleton (2005).

## Genus *Otonyctomys* Anthony, 1932

Tail long, well haired, and with a terminal hair tuft; hind feet considerably modified for arboreal life **with narrow tarsi; two pairs of inguinal mammae** (nipples); rostrum short; zygomatic plate narrow and **approximately perpendicular to the palatal plane**; palate broad, ending in front of a line made by the posterior margin of the toothrow; without lateral pits; **toothrow relatively small**; braincase and frontal bones broad; supraorbital ridges well developed and extending across the parietals to the occiput; interparietal bone broad and large; **auditory bullae larger**.

### *Otonyctomys hatti* Anthony, 1932

Yucatán vesper rat, rata orejuda trepadora de Yucatán

Monotypic.

**Conservation:** *Otonyctomys hatti* is listed as Threatened by the Norma Oficial Mexicana (DOF 2019) and as Least Concern by the IUCN (Timm et al. 2016l).

**Characteristics:** Yucatán vesper rat is the smallest member within the subfamily Tylominae; total length 196.0–231.0 mm and skull length 24.5–27.0 mm. Dorsal pelage yellowish, tawny or cinnamon grizzly and dark, more intense on the back; flanks with less dark grizzly hair tawny to ochraceous tawny; fur of underparts whitish; hind feet modified for arboreal life, hallux with a claw; **tail dark, covered with hairs of increasing length from base to tip, naked and shiny**; **ears large and naked;** blackish spot at the base of the whiskers and at the anterior margin of the eyes; upper side of the forefeet whitish washed with a warm buff and of the hind feet whitish; metapodials darkened ochraceous-tawny; braincase and frontal bones wide; supraorbital ridge edge well developed and extending to the occiput; interparietal bone well developed separating completely the parietal bone; rostrum short, narrower than the braincase; infraorbital foramen well developed; zygomatic plate anterior margin almost perpendicular to the skull axis; incisive foramen of moderate size, failing to reach the anterior plane of the first molars; **auditory bullae disproportionally large and occupying most of the basicranial region and greatly inflated, maximum length greater than 7.0 mm**.

**Comments:** *Otonyctomys hatti* ranges in the Yucatán Peninsula and a small area of Belize and northeast Guatemala (Map 304). *Otonyctomys* could be found in sympatry, or nearly so, with *Ototylomys*, from which it differs by the disproportionally large auditory bullae, occupying most of the basicranial region and greatly inflated; maximum length greater than 7.0 mm.

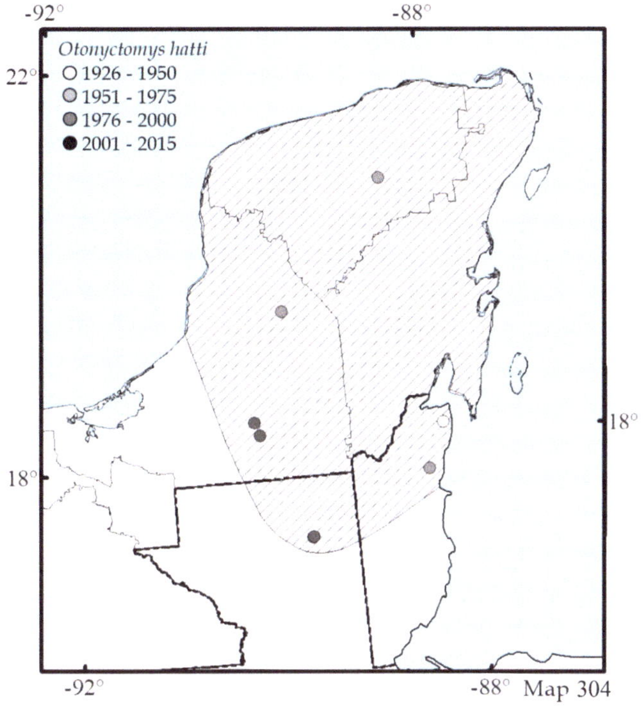


**Map 304** Distribution map of *Otonyctomys hatti*

**Additional Literature:** Anthony (1932), Aranda et al. (1997), Musser and Carleton (2005), MacSwiney et al. (2009a).

## Genus ***Ototylomys*** Merriam, 1901

Dorsal pelage reddish brown, brown to dark brown; flanks paler; fur of underparts white or pale buffy fulvous; tail as long as the head-and-body length, with broad scales, and appearing hairless; ears hairless and leaf-like, thin, and translucent; skull flat; interparietal bone large; superciliary region from a horizontal shelf over the orbits and reaching posterior to the occiput; **auditory bullae large and inflated, as broad anteriorly as posteriorly, without an anterior prolongation, and axes essentially parallel to the main axis of the skull**; anterior maxillary root of the zygomatic arches strongly notched above; **incisive foramen enormous, as broadly open anteriorly as posteriorly**; jaw angle excavated posteriorly, leaving a sharp point projecting backward; the coronoid process reduced, and postcoronoid notch flat and nearly horizontal; **pits between the cusps of the upper molars markedly developed.** The keys were elaborated based on the review of specimens and the following sources: Hall (1981), Álvarez-Castañeda et al. (2017a), and Porter et al. (2017).

1. Hindfoot length greater than 30.0 mm; base of the ears without a white spot of hair; ventral fur dark at the base and tan or reddish-tan at the tip; anterolateral margins of the incisive foramina are tapered medially, beginning near the maxilla-premaxilla suture; edge of the zygomatic plate is concave ventral to zygomatic spine; postglenoid foramina are nearly closed in and reduced to a narrow forward-facing slit (only known Ocozocoautla, Chiapas) .......................................... ........................................................................................................................ *Ototylomys chiapensis* (p. 424)

1a. Hindfoot length generally less than 30.0 mm; base of the ears with a white spot of hair; fur of underparts generally white or cream colored from base to tip; anterolateral margins of the incisive foramina broad and rounded anteriorly; edge of the zygomatic plate straight and vertical; postglenoid foramina distinctly open (from the Yucatán Peninsula and southern Tabasco south through Central America)...................................................................... *Ototylomys phyllotis* (p. 425)

## *Ototylomys chiapensis* Porter et al., 2017

La Pera big-eared climbing rat, rata orejuda trepadora de Chiapas

Monotypic.

**Conservation:** *Ototylomys chiapensis* (as *O. phyllotis*) is listed as Least Concern by the IUCN (Reid and Timm 2016). However, it should be listed as Endangered because of its reduced distribution and the strong deforestation within its ranges.

**Characteristics:** La Pera big-eared climbing rat is large sized; total length 275.0–361.0 mm and skull length 36.4–44.4 mm. Dorsal pelage brown to benzo brown with reddish shades; fur of underparts mottled tan-brown, brown-gray at the base and tan or reddish-tan at the tip, often with white spots (hair all white) on the venter; particularly in the pectoral and inguinal regions; midlateral trends toward a tan coloration on the venter; tail dark brown dorsally and yellowish ventrally; **incisive foramina anterolateral margins tapered medially, beginning near the maxilla-premaxilla suture; zygomatic plate edge concave ventrally to the zygomatic spine; postglenoid foramina nearly closed and reduced to a narrow forward-facing slit**.

**Comments:** *Ototylomys chiapensis* is only known from two localities in northern Ocozocoautla, in the Reserva de la Biosfera Selva El Ocote Zona Sujeta a Conservación Ecológica "La Pera" (Map 305). *O. chiapensis* seem to be allopatric with respect to *O. phyllotis*, from which it differs by having the hindfoot length greater than 30.0 mm; without a white hair spot at base of the ears and with the ventral hairs dark at the base and tan or reddish-tan at the tip; incisive foramina anterolateral margins tapered medially, beginning near the maxilla-premaxilla suture; zygomatic plate edge concave ventral to the zygomatic spine. From all the other species of *Ototylomys*, by having the postglenoid foramina nearly closed and reduced to a narrow forward-facing slit.

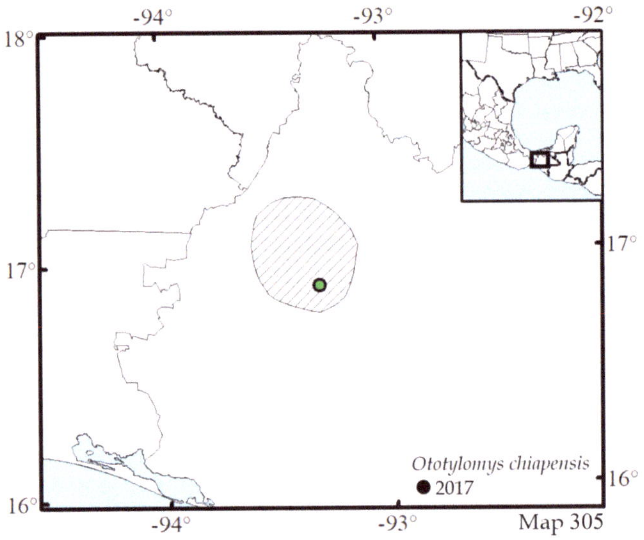


**Map 305** Distribution map of *Ototylomys chiapensis*

**Additional Literature:** Porter et al. (2017).

# ***Ototylomys phyllotis*** Merriam, 1901

Big-eared climbing rat, rata orejuda trepadora de común

1. *O. p. connectens* Sanborn, 1935. Northern Chiapas and western Guatemala.
2. *O. p. phyllotis* Merriam, 1901. Yucatán peninsula including Belize and eastern Guatemala.

**Conservation:** *Ototylomys phyllotis* is listed as Least Concern by the IUCN (Reid and Timm 2016).

**Characteristics:** Big-eared climbing rat is medium sized; total length 236.0–352.0 mm and skull length 32.2–42.3 mm. Dorsal pelage dark grayish brown grizzly with cinnamon to pale brown, sometimes with black-tipped hairs; fur of underparts cream to yellowish white, including the inner side of the fore and hindlegs; ankles and wrists dark; fore and hindfeet whitish; face brownish or grayish dusky above; irregularly yellowish below; **incisive foramina anterolateral margins broad and rounded anteriorly; zygomatic plate edge straight and vertical; postglenoid foramina are distinctly open**.

**Comments:** *Ototylomys phyllotis* ranges from the Yucatán Peninsula and southern Tabasco south through Central America, in addition to the lowlands of eastern Chiapas (Map 306). Hall (1981) recorded *O. p. connectens* from northern Guerrero; however, it is recorded as *Tylomys nudicaudus villai* in the original publication (IB 12221) (Ramírez-Pulido and Sánchez-Hernández 1971). *O. phyllotis* seems to be allopatric with respect to *O. chiapensis*, from which it differs by having the hind-foot length generally less than 30.0 mm; base of the ears with a white spot of hair; fur of underparts generally white or cream from base to tip; incisive foramina anterolateral margins broad and rounded anteriorly; zygomatic plate edge straight and vertical; postglenoid foramina distinctly open.

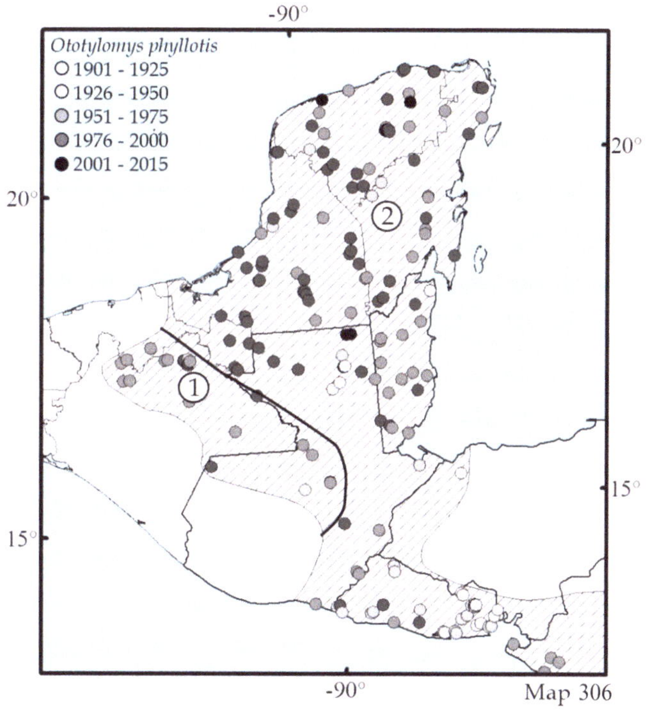


**Map 306** Distribution map of *Ototylomys phyllotis*

**Additional Literature:** Merriam (1901a), Carleton (1980), Lawlor (1969, 1982b), Musser and Carleton (2005), Reid (2009), Porter et al. (2017).

## Genus *Tylomys* Peters, 1866

Dorsal pelage reddish brown to dark brown, paler on the flanks; fur of underparts white or pale buffy fulvous; **inguinal and axillary areas marked with white shades; limbs and feet brown to russet; tail long and naked, proximal two-thirds dark and distal third white; toes white;** braincase elongated and flattened dorsally; upper first molars rectangular; interparietal bone wide; superciliary region with a horizontal shelf over the orbits and reaching posterior to the occiput. The keys were elaborated based on the review of specimens and the following sources: Hall (1981), and Álvarez-Castañeda et al. (2017a).

1. Ears very large, black, and sparsely haired; posterior two-thirds of the tail yellow; auditory bullae larger and broadly rounded anteriorly, without an anterior projection (only know from a region of Tuxtla Gutiérrez, Chiapas)........................................................................................................................................*Tylomys bullaris* (p. 426)

1a. Ears large, not black, and haired; posterior half of the tail yellow or whitish; auditory bullae small, not inflated, and pointed anteriorly, with an anterior projection........................................................................................................2

2. Ears large and nearly naked; tail yellow at the tip; upper toothrow large, heavy, greater than 9.0 mm (Chiapas and Guatemala)........................................................................................................*Tylomys tumbalensis* (p. 428)

2a. Ears small and with few hairs; tail whitish at the tip; upper toothrow small, thin, less than 9.0 mm (from western-central Oaxaca and central Veracruz south through Central America, excluding the Yucatán Peninsula)........................................................................................................................................*Tylomys nudicaudus* (p. 427)

### *Tylomys bullaris* Merriam, 1901

Chiapan climbing rat, rata de árbol de Chiapas

Monotypic.

**Conservation:** *Tylomys bullaris* is listed in México as Threatened in the Norma Oficial Mexicana (DOF 2019) and as Critically Endangered (Possibly Extinct; B1ab (i, ii, iii, iv, v) + 2ab (i, ii, iii, iv, v): extent of occurrence less than 100 km$^2$ at only one location and under continuing decline in extent of occurrence, area of occupancy, quality of habitat, number of subpopulations, and mature individuals + area of occupancy less than 10 km$^2$ at only one location and under continuing decline in extent of occurrence, area of occupancy, quality of habitat, number of subpopulations, and mature individuals) by the IUCN (Álvarez-Castañeda and Castro-Arellano 2019c).

**Characteristics:** Chiapan climbing rat is medium sized; total length 324.0 mm, skull length no measurement of an adult specimen because the only known specimen is a juvenile. Dorsal coloration pale leaden gray (juvenile coloration); upper lip and around the nose whitish; fur of underparts whitish; upper lip and patch on the nose side whitish; **ears very large, black, and sparsely haired; tail long, slender, naked, with the basal third blackish and the distal two-thirds yellow**; toes white; forefoot base brown and hind feet dark brown; interparietal bone broader on the outer side; zygomatic arches spreading more strongly posteriorly and **auditory bullae** larger**, wide and broadly rounded anteriorly, without an anterior prolongation, globose rather than pointed anteriorly**.

**Comments:** *Tylomys bullaris* is only known from the type specimen and type locality in Tuxtla Gutiérrez, Chiapas, México (Map 307). *T. bullaris* seems to be allopatric with respect to *T. nudicaudus* and *T. tumbalensis*, from which it differs by having the ears very large black and sparsely haired; face paler; tail long, slender, naked, and with the basal third blackish and the distal two-thirds yellow; auditory bullae large, inflated, and globose anteriorly, without an anterior projection.

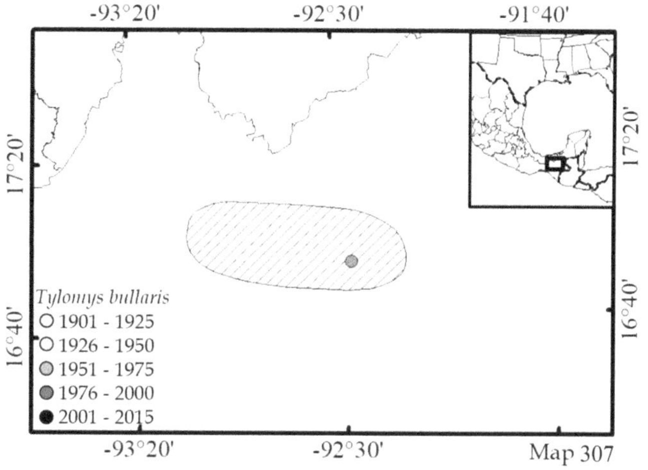


**Map 307** Distribution map of *Tylomys bullaris*

**Additional Literature:** Merriam (1901a), Musser and Carleton (2005), Álvarez-Castañeda and Castro-Arellano (2019a).

## *Tylomys nudicaudus* (Peters, 1866)

Peters's climbing rat, rata de árbol común

*T. n. gymnurus* Villa, 1941. Restricted to Veracruz and northern Oaxaca and northwestern Chiapas.
*T. n. microdon* Goodwin, 1955. Restricted to southern Oaxaca.
*T. n. nudicaudus* (Peters, 1866). Lowlands of central Chiapas south through Central America.
*T. n. villai* Schaldach, 1966. Southwestern Oaxaca and coast of Guerrero.

**Conservation:** *Tylomys nudicaudus* is listed as Least Concern by the IUCN (Vázquez et al. 2017).

**Characteristics:** Peters's climbing rat is medium sized; total length 400.0–500.0 mm and skull length 46.8–49.5 mm. Dorsal pelage reddish brown to pale brown, more intense on the flanks; fur of underparts whitish or pale reddish yellow; groin and armpits stained white; limbs brown; ears large and naked; auditory bullae not markedly inflated.

**Comments:** *Tylomys nudicaudus* ranges from western-central Oaxaca and central Veracruz south through Central America, excluding the Yucatán Peninsula (Map 308). *T. nudicaudus* seems to be allopatric with respect to *T. bullaris* and *T. tumbalensis;* it can be distinguished from *T. bullaris* by having the ears large but not very large; blackish and haired; tail long, slender, naked, with the posterior half yellowish or whitish; auditory bullae not inflated and pointed anteriorly, with an anterior projection. From *T. tumbalensis*, by having a smaller upper toothrow, lighter, less than 9.0 mm; smaller hind feet; ears smaller and haired; tail long, slender, naked, and with the posterior half yellowish or whitish; skull more massive and less slender; rostrum and nasals wide; zygomata bone stronger. Hall (1981) recorded *Ototylomys phyllotis connectens* from northern Guerrero; however, it is recorded as *Tylomys nudicaudus villai* in the original publication (IB 12221) (Ramírez-Pulido and Sánchez-Hernández 1971).

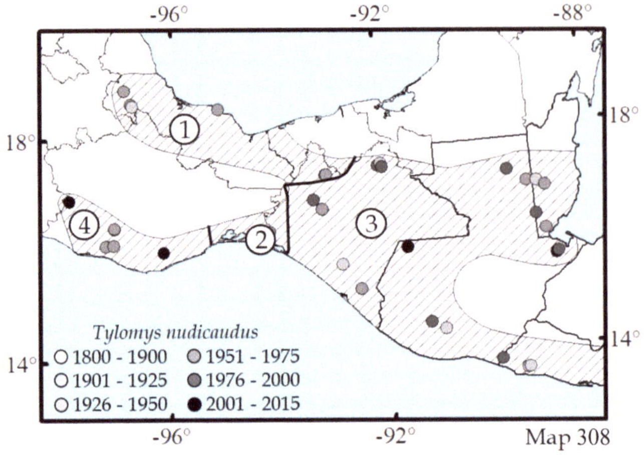


**Map 308** Distribution map of *Tylomys nudicaudus*

**Additional Literature:** Merriam (1901a), Ramírez-Pulido and Sánchez-Hernández (1971), Musser and Carleton (2005), Monroy-Gamboa et al. (Accepted).

## *Tylomys tumbalensis* Merriam, 1901

Tumbalá climbing rat, rata de árbol de Tumbalá

Monotypic.

**Conservation:** *Tylomys tumbalensis* is listed as Special Protection in the Norma Oficial Mexicana (DOF 2019) and as Critically Endangered (Possibly Extinct; B1ab (iii, v): extent of occurrence less than 100 km$^2$ at only one location and under continuing decline in area and quality of habitat and number in mature individuals) by the IUCN (Álvarez-Castañeda and Castro-Arellano 2019d).

**Characteristics:** Tumbalá climbing rat is larger in size; total length 448.0 mm and skull length 49.1 mm. Dorsal pelage dark blackish gray, sides palely washed with brownish shades; underparts, chin, chest and groin hairs white; neck and belly plumbeous, washed with buffy fulvous shades; top of the head blackish through the base of the tail; **tail long, naked and posterior half yellow**; face dark gray, with a dusky ring around the eyes; abdomen leaden gray grizzly with yellowish shades; digits dark brown; **ears large and naked;** skull less massive and more slender; rostrum and nasals more slender; zygomatic arches weaker and less broadly and strongly convex anteriorly; tapering to a point on the sides; auditory bullae not very inflated; **toothrow greater than 9.5 mm**.

**Comments:** *Tylomys tumbalensis* is only known from the type specimen and type locality in Tumbalá, Chiapas, Mexico. However, it is also recorded from El Real and Estación Chajúl of SEDUE, Ocosingo, Chiapas, Chiapas (López-Wilchis and López-Jardines 1998) and La Primavera, Guatemala (Goodwin 1955). Analyses all these data expand the distribution range to the tropical humid forests in central Chiapas and Cuchumantes, Guatemala (Map 309). *T. tumbalensis* is allopatric with respect to *T. bullaris* and *T. nudicaudus.* It differs from *T. bullaris* by having the auditory bullae not inflated and pointed anteriorly, with an anterior projection. From *T. nudicaudus*, by having the upper toothrow larger, heavier, and greater than 9.0 mm; larger hind feet; larger ears, nearly naked; tail long; naked and posterior half yellow. Skull less massive and more slender; rostrum and nasals more slender; zygomatic arches weaker and less broad, with the arms strongly convex anteriorly, tapering to a point on the sides. The type and only specimen is a young-adult.

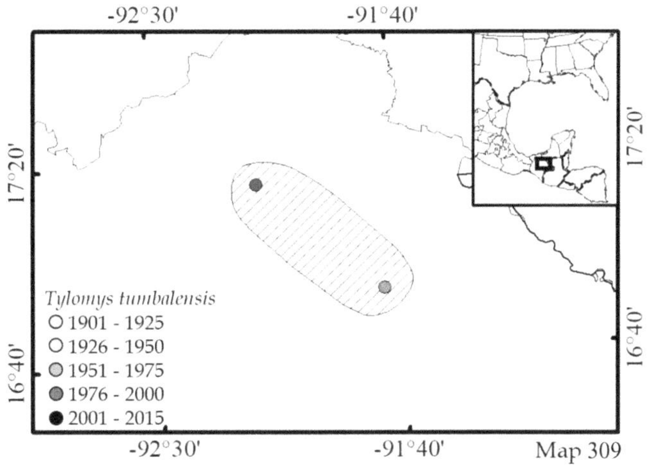


**Map 309** Distribution map of *Tylomys tumbalensis*

**Additional Literature:** Merriam (1901a), Musser and Carleton (2005), Álvarez-Castañeda and Castro-Arellano (2019b).

## **Family Muridae** Illiger, 1811

## (Introduced)

The family Muridae applies to the particular species of Old World rats and mice, some have been introduced into North America. Dorsal pelage uniform grayish-brown to black and underparts paler. Tail long, sparsely haired, and scaly. Molars provided with tubercles (small rounded cusps) arranged in three longitudinal rows; palate extending posteriorly to beyond plane of last molars; palatine slits extending posteriorly to plane of anterior borders of last upper molars; zygomatic bone typically reduced to a splint; infraorbital canal not conspicuously wider in the upper region than in the lower; angular process not markedly distorted outward; cheek-teeth are typically rooted and brachydont. The keys were elaborated based on the review of specimens and the following Hall (1981).

1. Total length less than 250.0 mm; tail length less than 112.0 mm; first upper molars greater than the second and third molars together; occipitonasal length less than 34.0 mm ............................................................ *Mus musculus* (p. 429)

1a. Total length greater than 250.0 mm; tail length greater than 112.0 mm; first upper molars less than the second and third molars together; occipitonasal length greater than 34.0 mm ........................................................................ *Rattus* (p. 430)

## Genus *Mus* Linnaeus, 1758

House mice is medium sized; dorsal pelage gray and light brown to black; fur of the underparts whitish, often with buffy wash; hind foot narrow; fifth digit longer than hallux; **first upper molars greater than the second and third molars together;** supraorbital ridges faint or absent; second upper molar usually with six cusps; third upper molar small; upper incisors compressed and with subapical notch**.**

### ***Mus musculus*** Linnaeus, 1758
House mouse, ratón de casa

**Conservation:** *Mus musculus* is listed as Least Concern and is considered as invasive by the IUCN (Musser et al. 2021).

**Characteristics:** House mice is medium sized; total length 130.0–200.0 mm and skull length 20.1–22.9 mm. Dorsal pelage gray and light brown to black; fur of the underparts whitish, often with buffy wash; tail length similar to the head-and-body length and not sharply bicolored but paler ventrally than dorsally; **ears large rounded**; ears and tail have little hair; snout pointed; hind foot narrow and all the characteristics given for the genus *Mus*.

**Comments:** *Mus musculus* was originally from Asia, Europe, and northern Africa, but through its close association with humans it has been widely introduced across the globe, including many islands, except in the most warm and cold areas (Musser and Carleton, 2005). *M. musculus* can be found in sympatry with many of the rodent species, can be distinguished from all other different form the family Muridae by having the first upper molars with three rows of the cusps arranged longitudinally on the occlusal surface and three transverse lophos clearly defined. From the species of *Rattus* by its smaller size in total length, tail length, and occipitonasal length, with less than 250.0 mm, 112.0 mm, and 34.0 mm, respectively.

**Additional Literature:** Hall (1981), Machólan (1999).

## Genus ***Rattus*** (Fischer, 1803)

Upper parts tawny to black; underparts gray to white; tail sparsely haired and scaled. The keys were elaborated based on the review of specimens and the following (Hall 1981).

1. Tail length equal to or less than that of the head-and-body length; ears small and rounded, length less than the distance between the base of the ears and the posterior margin of the eyes; body hefty; rostrum obtuse; first upper molars without distinct outer notches on the first row of cusps .......... *Rattus norvegicus* (p. 430)

1a. Tail length greater than the head-and-body length; ears large and oval, length greater than the distance between the base of the ears and the posterior margin of the eyes; body elongated; rostrum pointed; first upper molars with distinct outer notches on the first row of cusps .......... *Rattus rattus* (p. 431)

### ***Rattus norvegicus*** (Berkenhout, 1769)
Brown rat, rata café

**Conservation:** *Rattus norvegicus* is listed as Least Concern and is considered as invasive by the IUCN (Ruedas 2013).

**Characteristics:** Brown rat is large sized; total length 316.0–460.0 mm and skull length 41.0–51.5 mm. Dorsal pelage usually reddish or grayish brown to dark gray; underparts are lighter grayish to yellow-white; tail short similar to the head-and-body length and not bicolored, being paler ventrally and darker dorsally; **ears small and rounded**; body hefty; rostrum obtuse; **first upper molars without distinct outer notches on the first row of cusps.**

**Comments:** *Rattus norvegicus* was originally from eastern Asia, but through its close association with humans it has been widely introduced across the globe, including many islands (Musser and Carleton 2005). *R. norvegicus* can be found in sympatry with many of the rodent species, can be distinguished from all other different form the family Muridae by having the first upper molars with three rows of the cusps arranged longitudinally on the occlusal surface and three transverse lophos clearly defined. From *Mus musculus* by its larger size in total length, tail length, and occipitonasal length, with greater than 250.0 mm, 112.0 mm, and 34.0 mm, respectively. From *R. rattus* by its small tail, equal to or less than that of the head-and-body length; ears small and rounded; body hefty and in brown shades; eyes relatively small; rostrum obtuse; skull narrower across braincase; first upper molars without distinct outer notches on the first row of cusps; can weigh twice as much as a *Rattus rattus*.

**Additional Literature:** Hall (1981).

### ***Rattus rattus*** (Linnaeus, 1758)
House rat, rata gris

**Conservation:** *Rattus rattus* is listed as Least Concern and is considered as invasive by the IUCN (Kryštufek et al. 2021).

**Characteristics:** House rat is large sized; total length 325.0–455.0 mm and skull length 39.0–45.4 mm. Dorsal pelage usually black to tawny; underparts are in whitish shades; tail short similar to the head-and-body length and not bicolored, being paler ventrally and darker dorsally; **ears large, elliptical, and naked; body elongated**; rostrum pointed; **first upper molars with distinct outer notches on the first row of cusps.**

**Comments:** *Rattus rattus* was originally from south-eastern Asia, but through its close association with humans it has been widely introduced across the globe, including many islands (Musser and Carleton 2005). *R. rattus* can be found in sympatry with many of the rodent species, can be distinguished from all other different form the family Muridae by having the first upper molars with three rows of the cusps arranged longitudinally on the occlusal surface and three transverse lophos clearly defined. From *Mus musculus* by its larger size in total length, tail length, and occipitonasal length, with greater than 250.0 mm, 112.0 mm, and 34.0 mm, respectively. From *R. norvegicus* by its longer tail, length greater the head-and-body length; ears large and oval; eyes relatively large; body elongated and in grayish shades; rostrum pointed; skull wider across braincase; first upper molars with distinct outer notches on the first row of cusps; can weigh half as a *Rattus norvegicus*.

**Additional Literature:** Hall (1981).

## **Family Zapodidae** Coues, 1875

The family includes the jumping mice. They are characterized by a **hind limbs longer than fore limbs**; **hind feet with five functional digits; tail longer than head-and-body, makes up about 60% of its body length, more or less bicolored and sparsely haired; dorsal pelage with a broad dorsal band yellow-brown, flecked with black**; flanks paler than back; underparts white or suffused with color of flanks; antorbital foramen large and rounded; zygoma depressed; **jugal prolonged and forming suture with lacrimal**; auditory bullae not inflated; fourth upper premolar small or absent; lower premolars absent; **second upper molar with four or more narrow labial reentrant folds**; anterior and posterior cingula of the third upper molar large.

1. Three upper cheek-molars; tail white at the tip ........ *Napaeozapus insignis* (p. 431)

1a. Four upper cheek-molars; tail not white at the tip ........ *Zapus* (p. 433)

## Genus *Napaeozapus* Preble, 1899

**Woodland jumping mouse has a broad brown to black band on the back, much darkened with black shades; flanks buff-yellow to orange with a yellow or red tint, tinged with clay color and scattered with dark guard hairs, with a yellow lateral line;** fur of underparts white; tail sharply bicolored, upperparts brown and fur of underparts paler, **normally white at the tip**; **premolars absent, with a total of three molariforms**; second upper molars about the same size as the first upper molars; with three narrow labial reentrant folds of unequal length on the third upper molars; anterior cingulum of the first upper molars large; incisors orange or yellow, upper incisors grooved; frontal region somewhat swollen; **jugal plates in the zygomatic arches extending dorsally along the maxillary ramus and articulating with the lacrymal bones**; infraorbital foramen large and oval; nasals projecting considerably beyond the incisors. Malaney et al. (2017) reviewed the species of the genus *Napaeozapus*. The revision is not taxonomically oriented and they proposed *N. insignis* and *N. abietorum* as a candidate species to be considered within the genus.

## ***Napaeozapus insignis*** (Miller, 1891)

Woodland jumping mouse, ratón saltarín del bosque

1. *N. i. abietorum* (Preble, 1899). From central western Ontario to central western Québec.
2. *N. i. frutectanus* Jackson, 1919. From southeastern Manitoba, southwestern Ontario, and northern Wisconsin east through northern Michigan.
3. *N. i. insignis* (Miller, 1891). From Southern Ontario, New York, western Pennsylvania, and eastern North Virginia north through eastern Québec and New Brunswick.
4. *N. i. roanensis* (Preble, 1899). From western North Carolina northeast to western Pennsylvania and eastern Ohio.
5. *N. i. saguenayensis* Anderson, 1942. From southcentral Québec west through the southeastern Labrador Peninsula.

**Conservation:** *Napaeozapus insignis* is listed as Least Concern by the IUCN (Cassola 2016bf).

**Characteristics:** Woodland jumping mouse is small sized; total length from 204.0 to 256.0 mm and skull length 19.0–23.3 mm. Pelage tricolored with a distinct broad brown to black band on the back, much darkened with black shades; sides buff-yellow to orange with a yellow or red tint, tinged with clay color and scattered with dark guard hairs; fur of underparts white; **premolars absent, with a total of three molariforms**; second upper molars about the same size as the first upper molars; with three narrow labial reentrant folds of unequal length on the third upper molars; anterior cingulum of the first upper molars large; incisors orange or yellow, upper incisors grooved; frontal region somewhat swollen; **jugal plates in the zygomatic arches extending dorsally along the maxillary ramus and articulating with the lacrimal bones**; infraorbital foramen large and oval; nasals projecting considerably beyond the incisors.

**Comments:** *Napaeozapus insignis* ranges eastern Manitoba east through southern Labrador Peninsula, northern Wisconsin east through Michigan and the Appalachian Mountains highlands from northern Georgia to Southeastern Québec and New Brunswick (Map 310). The most similar species with a sympatric distribution is *Zapus hudsonius*, from which it differs, by having the tail tip white and three molariforms.

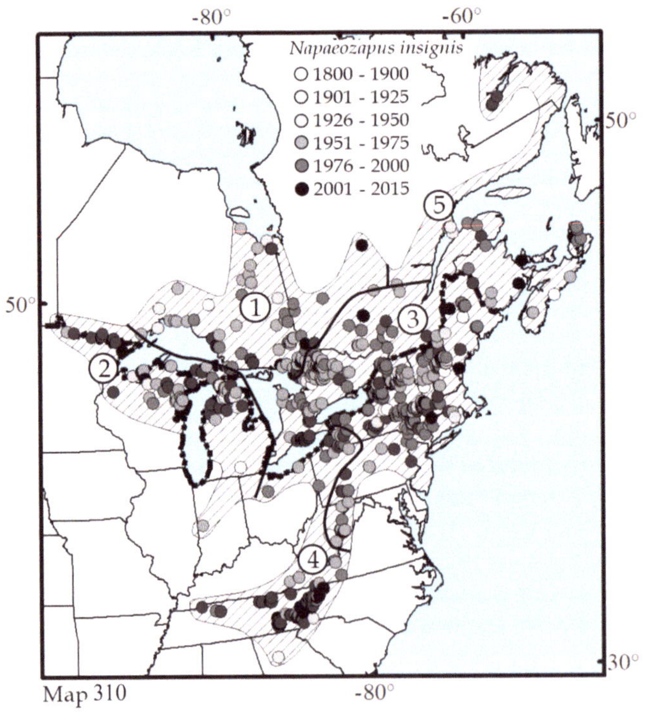


**Map 310** Distribution map of *Napaeozapus insignis*

**Additional Literature:** Whitaker and Wrigley (1972), Meade (1992), Whitaker and Hamilton (1998), Whitaker (1999b), Holden and Musser (2005).

## Genus *Zapus* Coues, 1875

**Pelage tricolored with a distinct broad dorsal band from brown to yellowish brown darkened with brownish black hairs;** sides paler and slightly streaked with brownish-black shades; fur of underparts snow-white or sometimes suffused with yellowish shades but usually separated from the sides by a clear yellowish band; **upper lip with a median groove;** eyes small and midway between the nose and the ears; ears dark with a narrow pale edge and somewhat longer than the surrounding hair; tail long, subcylindrical, greater than 120% of the body length, distinctly bicolored, dark brown dorsally and yellowish white ventrally; hind legs much longer than the forelegs; upper parts of fore and hind feet grayish white; ends of the nasals projecting noticeably beyond the incisors; frontal region moderately swollen; upper incisors grooved; four upper molariforms, the first upper molariforms reduced in size; preorbital foramen large and oval. Malaney et al. (2017) reviewed the genus *Zapus*. The revision is not taxonomically oriented and they proposed candidate species to be considered within the different current species of *Zapus*: within *Z. trinotatus*, they considered two species, the second is *Z. montanus*; within *Z. princeps*, five species, including the new ones: *Z. pacificus Z. okanoganensis*, *Z. oregonus*, and *Z. saltator;* and within *Z. hudsonius*, two species, the new being *Z. luteus*. Because of the absence of taxonomic information and morphological characteristics, these species are not considered in these keys.

1. Flank coloration yellowish orange; pterygoid fossa wide; premolars with a crescentine fold on the occlusal surface; coronoid process of the mandible long and slender (from southwestern British Columbia south through northern San Francisco Bay, California) ........ *Zapus trinotatus* (p. 436)

1a. Flank coloration yellowish but not orange; pterygoid fossa usually narrow; premolars without a crescentine fold on the occlusal surface; coronoid process of the mandible short and broad........ 2

2. Ears with the border slightly white; tail not very distinctly bicolored; incisive foramina greater than 4.6 mm; palatal breath at the third upper molars greater than 4.4 mm (from southern Yukon southwards through central and eastern North Dakota and northwestern South Dakota, and south through eastern-central California and northern-central New Mexico) ........ *Zapus princeps* (p. 434)

2a. Ears without a white border; tail distinctly bicolored; incisive foramina less than 4.6 mm; palatal breath at the third upper molars less than 4.4 mm (from southern Alaska east through the Labrador Peninsula and south through eastern North Carolina, northeastern South Carolina, northern Georgia, Alabama, Mississippi, northeastern Oklahoma, and Colorado) ........ *Zapus hudsonius* (p. 433)

### *Zapus hudsonius* (Zimmermann, 1780)

Meadow jumping mouse, ratón saltarín de la pradera

1. *Z. h. acadicus* (Dawson, 1856). From New York to New Brunswick.
2. *Z. h. alascensis* Merriam, 1897. From southern Alaska to northwestern British Columbia.
3. *Z. h. americanus* (Barton, 1799). From Illinois south through central-eastern Mississippi to eastern North Carolina and east through central New York.
4. *Z. h. campestris* Preble, 1899. Restricted to southern Montana, northern Nebraska, and western North Dakota.
5. *Z. h. canadensis* (Davies, 1798). From central Ontario to central Québec.
6. *Z. h. hudsonius* (Zimmermann, 1780). From central Ontario and northern Wisconsin west through central Alaska.
7. *Z. h. intermedius* Krutzsch, 1954. From southern Manitoba, North Dakota, and western Montana southeast to western Indiana to central Tennessee.
8. *Z. h. ladas* Bangs, 1899. From central Québec east through the Labrador Peninsula.
9. *Z. h. pallidus* Cockrum and Baker, 1950. From southern South Dakota and Nebraska southeast through northwestern Oklahoma and Arkansas.
10. *Z. h. preblei* Krutzsch, 1954. From southwestern Wyoming to northern-central Colorado.
11. *Z. h. tenellus* Merriam, 1897. Restricted to central British Columbia.

**Conservation:** *Zapus hudsonius preblei* is listed as Threatened by the United States Endangered Species Act (FWS 2022). The IUCN listed *Zapus hudsonius* as Least Concern by the IUCN (Cassola 2016eg).

**Characteristics:** Meadow jumping mouse is small sized; **total length 188.0–216.0 mm** and skull length 16.4–19.5 mm. Dorsal pelage ochraceous to dark brown; flanks pale; lateral line ochraceous-buff, indistinct, or absent; fur of underparts white or suffused with ochraceous shades; tail brown to brownish-black dorsally, white to yellowish-white ventrally, and longer than the head-and-body length; **skull narrow in proportion to its length; upper premolars small, with a shallow reentrant fold**; pterygoid fossa narrow; coronoid process of the mandible short and broad; toothrow short.

**Comments:** Previously, *Zapus hudsonius luteus* was treated within *Z. princeps* (Hafner et al. 1981). *Z. hudsonius* ranges from southern Alaska east through the Labrador Peninsula and south through eastern North Carolina, northeastern South Carolina, northern Georgia, Alabama, Mississippi, northeastern Oklahoma, and Colorado (Map 311). It can be found in sympatry, or nearly so, with *Z. trinotatus* in southern British Columbia and with *Z. princeps* in many areas. It differs from *Z. trinotatus* and *Z. princeps* by its smaller size, less than 216.0 mm; narrower skull; smaller premolars; and shorter upper toothrow, on average less than 3.7 mm.

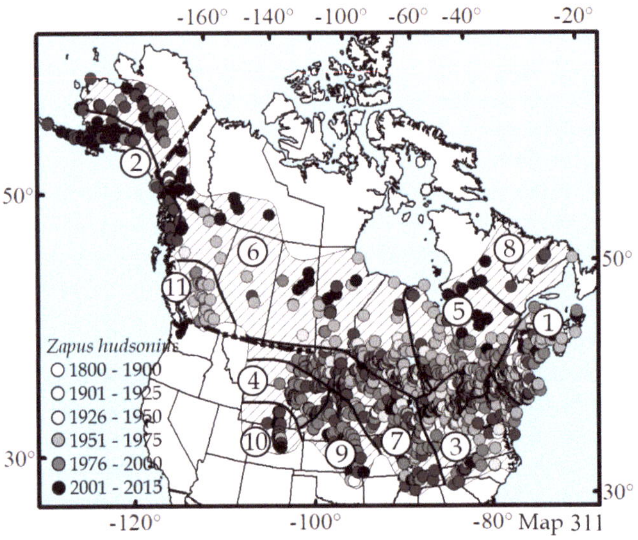


**Map 311** Distribution map of *Zapus hudsonius*

**Additional Literature:** Krutzsch (1954), Whitaker (1972), Hoffmeister (1986), Whitaker and Hamilton (1998), Holden and Musser (2005), Ramey et al. (2005).

## *Zapus princeps* Allen, 1893

Western jumping mouse, ratón saltarín del oeste

1. *Z. p. chrysogenys* Lee and Durrant, 1960. Restricted to Salt Mountains, western Colorado.
2. *Z. p. cinereus* Hall, 1931. Restricted to southern Idaho and central northwestern Utah.
3. *Z. p. curtatus* Hall, 1931. Restricted to northwestern Nevada and a small area in western California.
4. *Z. p. idahoensis* Davis, 1934. From southeastern British Columbia south through central Idaho and from central Wyoming to northern Colorado.
5. *Z. p. kootenayensis* Anderson, 1932. Southern British Columbia and northern Washington.
6. *Z. h. luteus* Miller, 1911. Restricted to central-western Arizona and New Mexico.
7. *Z. p. minor* Preble, 1899. From southern Alberta southeast through central Montana, central North Dakota, southeastern Manitoba, and northwestern South Dakota.
8. *Z. p. oregonus* Preble, 1899. From southern Washington to central Nevada.
9. *Z. p. pacificus* Merriam, 1897. From southern Oregon to central California, including a small area in western Nevada.
10. *Z. p. princeps* Allen, 1893. From central eastern Wyoming to central-northern New Mexico.
11. *Z. p. saltator* Allen, 1899. From southern Yukon to southern British Columbia.
12. *Z. p. utahensis* Hall, 1934. Restricted to central Utah, eastern Idaho, and western Wyoming.

**Conservation:** *Zapus princeps luteus* is listed as Endangered by the United States Endangered Species Act (FWS 2022). *Zapus princeps* is listed as Least Concern by the IUCN (Cassola 2016eh).

**Characteristics:** Western jumping mouse is small sized; total length 216.0–247.0 mm and skull length 19.6–23.0 mm. Dorsal pelage yellowish-gray to salmon-brown and ochraceous; flanks paler than the back; lateral line ochraceous-buff, indistinct, or absent; fur of underparts white, usually suffused with ochraceous shades; tail pale brown to grayish brown dorsally, white to yellowish white ventrally, and longer than the head-and-body length; skull large; nasals projecting beyond the incisors and a large premaxillary plate protruding between the incisors; palate short, with the posterior edge variable in shape; pterygoid fossa moderately narrow; **upper premolars medium sized, with a shallow reentrant fold**; mandible with an angular process more or less in line with the rest of the jaw; coronoid process weak; angular processes extending medially.

**Comments:** *Zapus trinotatus* and *Z. princeps* were formerly considered conspecific, and were regarded as separate species (Baker et al. 2003a; Holden and Musser 2005). *Z. p. luteus* was reconsidered a subspecies of *Z. hudsonius* (Hafner et al. 1981). *Z. princeps* ranges from southern Yukon south through central and eastern North Dakota and northwestern South Dakota, and south through eastern-central California and northern-central New Mexico (Map 312). *Z. princeps* could be found in sympatry with *Z. hudsonius* and *Z. trinotatus*. It differs from *Z. hudsonius* by having larger premolars, a longer rostrum, and a longer baculum. From *Z. trinotatus*, by having smaller premolars; skull narrow and shallow, and a shorter baculum.

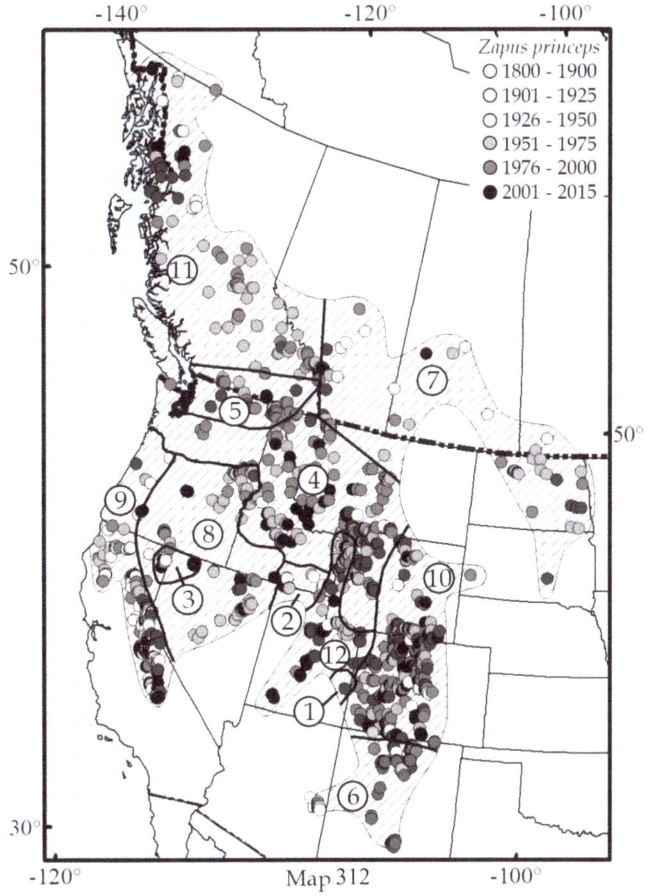


**Map 312** Distribution map of *Zapus princeps*

**Additional Literature:** Krutzsch (1954), Whitaker (1972), Hart et al. (2004).

## *Zapus trinotatus* Rhoads, 1895

Pacific jumping mouse, ratón saltarín del Pacífico

1. *Z. t. eureka* Howell, 1920. Restricted to the central-northern coastal California.
2. *Z. t. montanus* Merriam, 1897. Restricted to central Oregon.
3. *Z. t. orarius* Preble, 1899. Restricted to the northern San Francisco Bay, California.
4. *Z. t. trinotatus* Rhoads, 1895. From southern British Columbia to the northern coast of California.

**Conservation:** *Zapus trinotatus* is listed as Least Concern by the IUCN (Cassola 2016ei).

**Characteristics:** Pacific jumping mouse is small sized; total length from 220.0 to 240.0 mm and skull length 20.8–22.0 mm. Dorsal pelage brighter, of various ochraceous and tawny shades, pattern tricolored, back dark brown to cinnamon-brown, flanks dark orange-brown or ochraceous-buff, occasionally flecked with black; fur of underparts mostly white or diffused with dusky brown shades or of the same color as the flanks; lateral line distinct and bright; belly white, usually suffused with ochraceous shades; ears dark; tail brown dorsally, fur of underparts white to yellowish-white and longer than the head-and-body length; **pterygoid fossa wide,** broad and deep in proportion to its length; **upper premolars large**, with a labial reentrant fold forming a crescentic loop; **coronoid process of the mandible long and slender**.

**Comments:** *Zapus trinotatus* and *Z. princeps* were formerly considered conspecific by some authors but are currently regarded as different species (Baker et al. 2003a; Holden and Musser 2005). *Z. trinotatus* ranges from southwestern British Columbia south through the northern San Francisco Bay, California (Map 313), along the mostly humid west coastal stripe. It can be found in sympatry, or nearly so, with *Z. hudsonius* in southern British Columbia and with *Z. princeps* in many areas. It differs from *Z. princeps* and *Z. hudsonius* by its slightly larger size; markedly distinct separation of colors between the back and underparts; ears usually fringed, of the same color as the back or with pale brown shades; skull broad in relation to its length; pterygoid fossa wide; coronoid process of the mandible long, slender, and divergent from the condyloid process; mandible angle turned in and wide; and premolars with a crescentine fold on the occlusal surface.

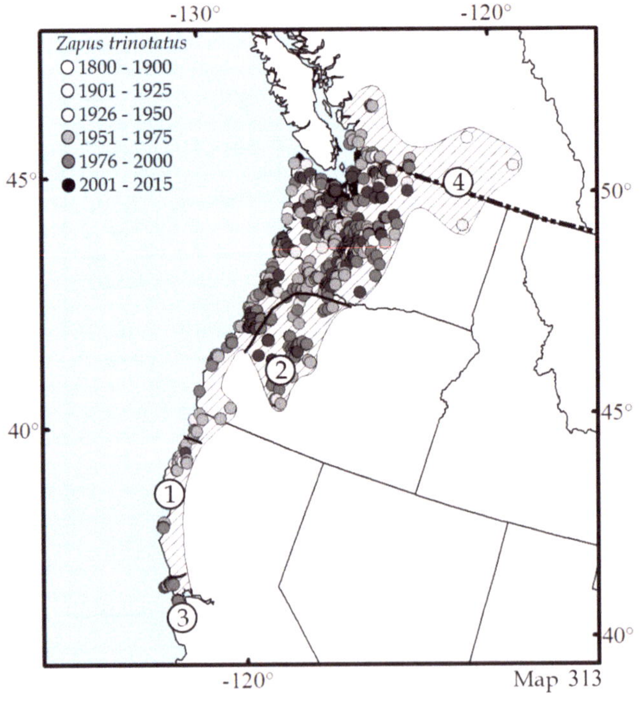


**Map 313** Distribution map of *Zapus trinotatus*

**Additional Literature:** Krutzsch (1954), Whitaker (1972), Gannon (1988, 1999).

# **Family Geomyidae** Bonaparte, 1845

The family Geomyidae includes pocket gophers that are characterized by cylindrical bodies from small to medium sized (150.0–500.0 mm); **ears short; eyes small; forefoot and hindfoot short; tail short; manal claws well developed for digging; cheek pouches external; two holes located in the palate behind the last molar.** Its distribution ranges from Canada south to extreme northwestern Colombia. It belongs to the superfamily Geomyoidea. All living species belong to the subfamily Geomyinae.

NOTE: Geomids have strong sexual dimorphism where males are larger than females. Many specimens in scientific collections tend to be juveniles or young adults. In both cases, these do not exhibit clear identifying features. It is very important to be aware of these two factors in this group because they can be essential for a successful or erroneous identification. For this reason, their distribution areas should also be considered. The keys were elaborated based on the review of specimens and the following sources: Russell (1968a, b), Hall (1981), and Álvarez-Castañeda et al. (2017a).

1. Upper incisor anterior surface smooth (Fig. 85); lower molars bordered by anterior and posterior enamel crests (enamel appears as a stripe of porcelain)............................................................................tribe Thomomyini *Thomomys* (p. 468)

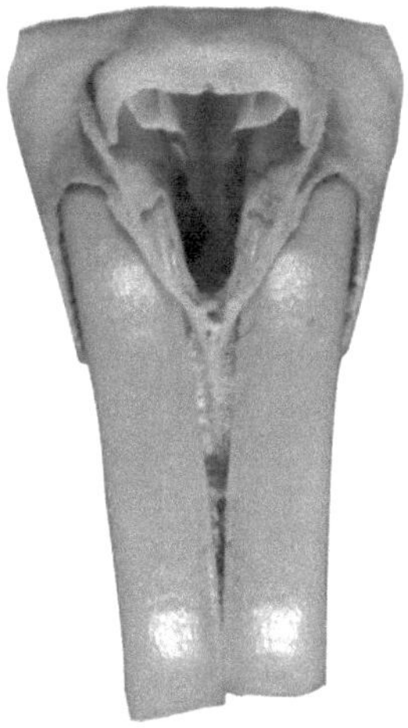

**Fig. 85** Upper incisor anterior surface smooth

1a. Upper incisor anterior surface with one or two grooves (Figs. 86 and 87); lower molars without enamel crests; if present, only posterior ones............................................................................................................................. tribe Geomyini 2

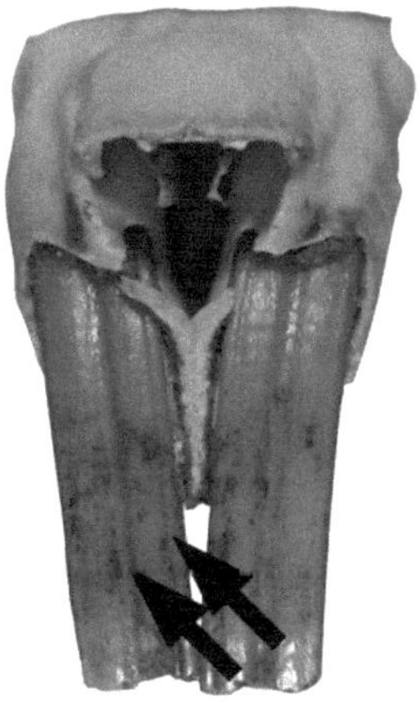

**Fig. 86** Upper incisor anterior surface with two grooves

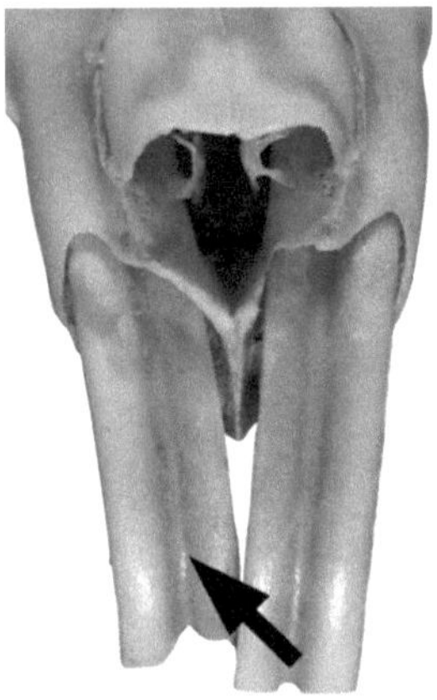

**Fig. 87** Upper incisor anterior surface with one groove

2. Upper incisor anterior surface with two grooves (Fig. 86); occlusal surface of third upper molars longer than wide (Fig. 88).......................................................................................................................3

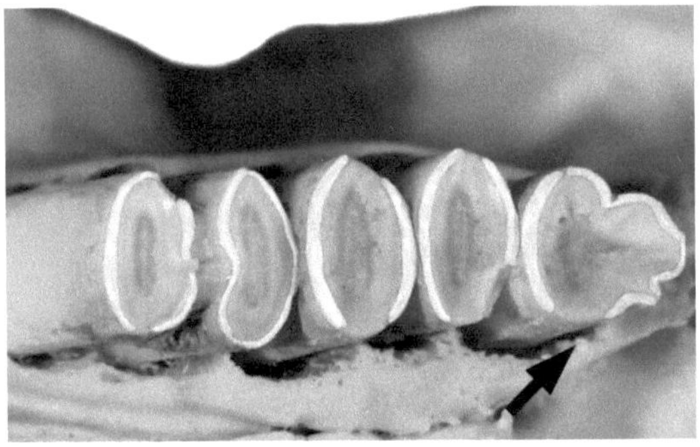

**Fig. 88** Occlusal surface of third upper molars longer than wide

2a. Upper incisor anterior surface with one groove (Fig. 87); occlusal surface of third upper molars not much longer than wide (Fig. 89).......................................................................................................................4

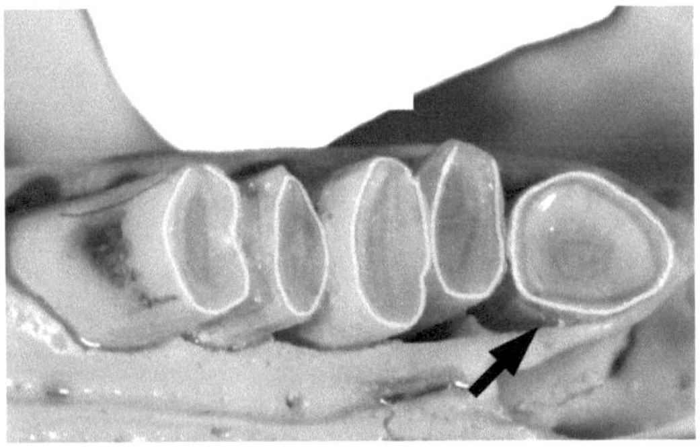

**Fig. 89** Occlusal surface of third upper molars not much longer than wide

3. Upper premolar posterior edge with an enamel plate usually restricted to the posterolingual side; upper third molars clearly consisting of two lobes, longer than wide due to the elongation of the posterior lobe (Fig. 90; endemic to mountains of Michoacán) ........ *Zygogeomys trichopus* (p. 467)

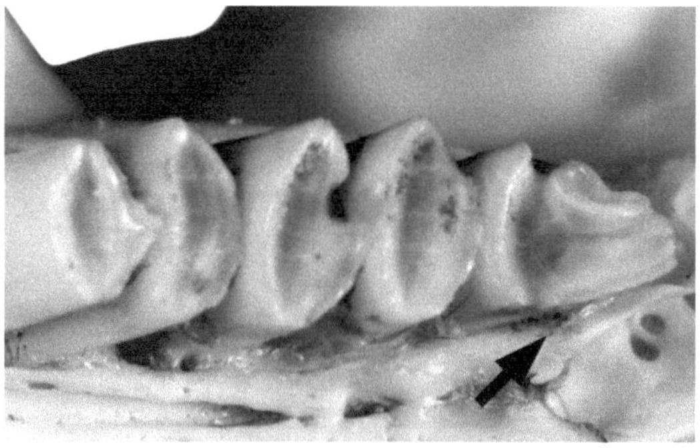

**Fig. 90** Upper third molars clearly consisting of two lobes, longer than wide due to the elongation of the posterior lobe

3a. Upper premolar posterior edge without an enamel plate; upper third molars not clearly consisting of two lobes with a shallow groove in the lingual side; broader than long (Fig. 88) ........ *Geomys* (p. 448)

4. Third upper molars clearly with two lobes (Fig. 89); upper premolars usually with an enamel plate on the back restricted to the lingual region ........ 6

4a. Third upper molars not clearly with two lobes, outer lingual angle not well developed; upper premolars never with an enamel plate in the lingual region ........ 5

5. Sagittal crest absent; zygomatic lateral angles without a plate-shaped expansion (Fig. 91; endemic to Nayarit, Jalisco, and Colima) ........ *Pappogeomys bulleri* (p. 466)

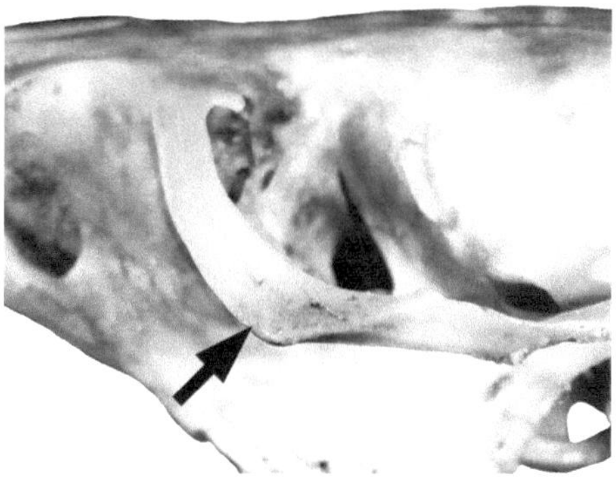

**Fig. 91** Zygomatic lateral angles without a plate-shaped expansion

5a. Sagittal crest present; zygomatic lateral angles with a plate-shaped expansion (Fig. 92; in zygomatic arches parallel to the major skull axis) ........ *Cratogeomys* (p. 441)

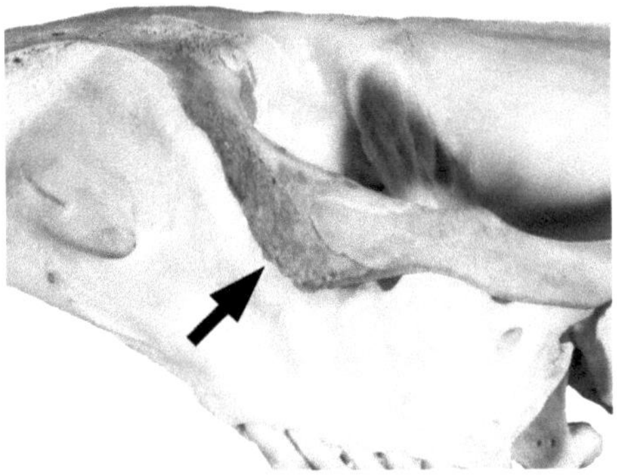

**Fig. 92** Zygomatic lateral angles with a plate-shaped expansion

6. Frontal bone wide and globose; interorbital constriction slightly marked, greater than 11.9 mm (Fig. 93); upper premolars without a posterior enamel plate but sometimes with a small posterolingual plate (Jalisco south through Central America, including southwestern Puebla and central Oaxaca)............................................*Orthogeomys grandis* (p. 464)

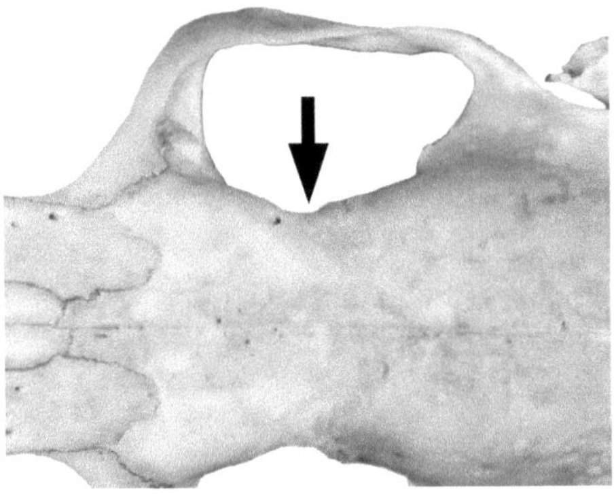

**Fig. 93** Interorbital constriction slightly marked, greater than 11.9 mm

6a. Frontal bone thin and not strongly globose; interorbital constriction well developed, less than 11.9 mm (Fig. 94); upper premolars always with a posterior enamel plate but restricted to the lingual side......................... *Heterogeomys* (p. 462)

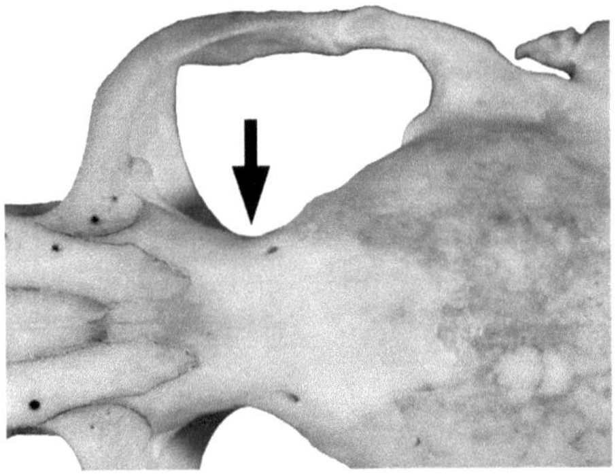

**Fig. 94** Interorbital constriction well developed, less than 11.9 mm

## Tribe Geomyini

### Genus *Cratogeomys* Merriam, 1895

Dorsal pelage in different colors, mainly shades of brown; fur of underparts paler than in the dorsum; tail short and slightly darker dorsally; body compact and cylindrical; hair silky and long, approximately 8.0 mm; skull large and massive; no ridge on squamosal but sagittal crest mainly in adult and old males; interorbital region narrower than rostrum; zygomatic lateral angle with a plate-shaped expansion. The genus *Cratogeomys* can be differentiated from other species of the family Geomyidae by the following characteristics: **rostrum wider than the interorbital region, no zygomatic anterior angle plate-shaped expansion, and third upper molars not clearly showing two lobes**. *Cratogeomys* was previously considered a subgenus of *Pappogeomys* (Russell 1968a; Hall 1981). Honeycutt and Williams (1982), DeWalt et al. (1993b), Demastes et al. (2002) established the current taxonomy. The genus *Cratogeomys* has two species groups: *castanops* (*C. castanops*, *C. fulvescens*, *C. goldmani*, *C. merriami*, and *C. perotensis*) and *fumosus* (*C. fumosus* and *C. planiceps*). *C. gymnurus* is considered a junior synonym of *C. fumosus*. All species of *Cratogeomys* have allopatric distribution. The keys were elaborated based on the review of specimens and the following sources: Russell (1968a, b), Hall (1981), and Álvarez-Castañeda et al. (2017a).

1. Mastoid process generally extends laterally beyond the auditory meatus; breadth across angular jaw processes generally greater than jaw length (including incisors) .......... 2

1a. Mastoid process generally does not extend laterally beyond the auditory meatus; breadth across angular jaw processes generally less than jaw length (including incisors) .......... 3

2. In ventral view, paraoccipital process thick and straight, anterior part of jugal (zygomatic contact) generally wide without a notch (Colima, Jalisco, Michoacán Guanajuato, Querétaro, Estado de México, and Puebla) .......... *Cratogeomys fumosus* (p. 444)

2a. In ventral view, paraoccipital process thin and not completely straight, usually with a notch in the anterior part of jugal (zygomatic contact; Estado de México) .......... *Cratogeomys planiceps* (p. 447)

3. Basioccipital narrow, less than 4.0 mm at its middle; edges of basioccipital straight or sandglass shaped .......... 4

3a. Basioccipital width greater than 4.0 mm at its middle; edges of basioccipital wedge shaped (converging frontward) .......... 5

4. Condylobasal length less than 47.0 mm and palatal length less than 31.5 mm (Durango, Zacatecas, Coahuila, Nuevo León, San Luis Potosí, and Tamaulipas) .......... *Cratogeomys goldmani* (p. 445)

4a. Condylobasal length greater than 47.0 mm and palatal length greater than 31.5 mm (Nazas River and Sierra de Parras northward in Coahuila and Durango north through Colorado, Kansas, Oklahoma, New Mexico, and Texas) .......... *Cratogeomys castanops* (p. 441)

5. Small white patches near the tail base; anterior edge of jugal narrow, less than 2.0 mm (southern Hidalgo, Puebla, and western-central Veracruz) .......... *Cratogeomys perotensis* (p. 447)

5a. Without patches near the tail base; anterior edge of jugal wider and greater than 2.0 mm .......... 6

6. Dorsal pelage pale, beige, or yellowish white; anterior edge of jugal breadth less than 2.5 mm; cranial width less than 26.0 mm (eastern Puebla, western Veracruz, and western Tlaxcala) .......... *Cratogeomys fulvescens* (p. 443)

6a. Dorsal pelage dark gray, dark orange, or chocolate but never beige; anterior edge of jugal breadth greater than 2.5 mm; cranial width greater than 26.0 mm (Estado de México, Ciudad de México, Querétaro, Tlaxcala, Puebla, and Veracruz) .......... *Cratogeomys merriami* (p. 446)

### *Cratogeomys castanops* (Baird, 1852)

Yellow-faced pocket gopher, tuza de cara amarilla

1. *C. c. castanops* (Baird, 1852). From northern Tamaulipas, Coahuila, and Chihuahua north through Colorado, Kansas, Oklahoma, New Mexico, and Texas.
2. *C. c. consitus* Nelson and Goldman, 1934. From northern Tamaulipas, Coahuila, Chihuahua, and Sonora south through the Nazas River and Sierra de Parras, Coahuila, and Durango.

**Conservation:** *Cratogeomys castanops* is listed as Least Concern by the IUCN (Cassola 2016i).

**Characteristics:** Yellow-faced pocket gopher is medium sized; total length 230.0–260.0 mm and skull length 46.0–51.0 mm. Dorsal pelage pale yellowish to dark reddish brown, central part of back grizzly with blackish shades; **yellowish shades generally present**; fur of underparts creamy-whitish to yellowish-ocher; cheeks yellowish; basioccipital width less than 4.0 mm at its middle part and straight or sandglass shaped; zygomatic breadth greater than squamosal breadth; second and third molariforms with no enamel back plate; last upper molariforms semisquare. Species group *castanops*.

**Comments:** The genetic review of the *Cratogeomys castanops* complex (Hafner et al. 2008) considered the subspecies *C. castanops elibatus*, *C. c. goldmani* (part), *C. c. maculatus*, *C. c. peridoneus*, *C. c. planifrons*, *C. c. rubellus*, *C. c. subnubilus*, and *C. c. surculus* (part) as a different species, under the name of *C. goldmani*; the subspecies *C. c. angusticeps*, *C. c. bullatus*, *C. c. convexus*, *C. c. dalquesti*, *C. c. hirtus*, *C. c. parviceps*, *C. c. perplanus*, *C. c. pratensis*, *C. c. simulans*, *C. c. tamaulipensis*, *C. c. torridus*, and *C. c. ustulatus* as junior synonym of *C. castanops castanops;* and the subspecies *C. c. excelsus*, *C. c. goldmani* (part), *C. c. jucundus*, *C. c. perexiguus*, *C. c. sordidulus*, *C. c. subsimus*, and *C. c. surculus* (part) as junior synonyms of *C. castanops consitus*. *C. castanops* ranges from the Nazas River and Sierra de Parras in northern Coahuila and Durango north through Colorado, Kansas, Oklahoma, New Mexico, and Texas (Map 314). *C. castanops* can be distinguished from all other species of *Cratogeomys*, except *C. goldmani*, by its narrow basioccipital bone, less than 4.0 mm at its middle part, and its straight or sandglass-shaped edges and smaller average size. It differs from *C. goldmani* only in the number of chromosomes ($2n = 46$) and its generally larger size, condylobasal length greater than 47.0 mm, and palatal length greater than 31.5 mm.

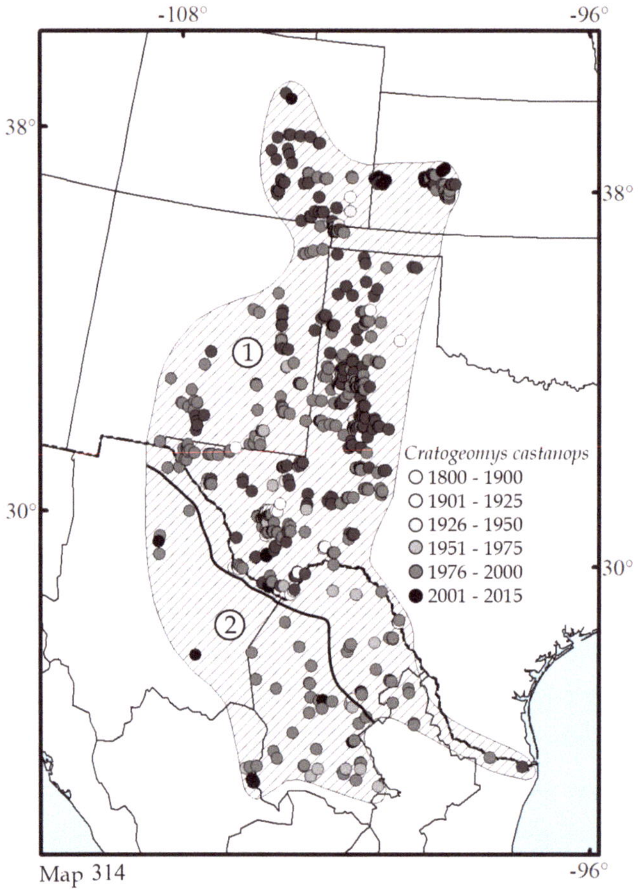


**Map 314** Distribution map of *Cratogeomys castanops*

**Additional Literature:** Russell (1968b), Davidow-Henry et al. (1989), Hafner et al. (2004), Patton (2005b).

## *Cratogeomys fulvescens* Merriam, 1895

Oriental Basin pocket gopher, tuza de la Cuenca Oriental

Monotypic.

**Conservation:** *Cratogeomys fulvescens* is listed as Least Concern by the IUCN (Álvarez-Castañeda et al. 2017b).

**Characteristics:** Oriental Basin pocket gopher is medium sized but is the smallest member of *Cratogeomys*; total length 302.0–327.0 mm and skull length 52.0–58.0 mm. **Dorsal pelage pale, beige, or yellowish white, strong mixture of black-tipped hairs with a salt-pepper appearance**; fur of underparts paler than in the dorsum; basioccipital width greater than 4.0 mm at its middle part, wedge shaped. Species group *castanops*.

**Comments:** *Cratogeomys fulvescens* was recognized as a distinct species (Hafner et al. 2005); previously, it was considered a subspecies of *C. merriami* (Russell 1968b). *C. fulvescens* is distributed in the Oriental Basin, eastern Puebla, western Veracruz, and western Tlaxcala (Hafner et al. 2005; Map 315). It can be distinguished from all other species of *Cratogeomys* by its smaller size; skull width usually less than 26.0 mm; anterior jugal breadth usually from 2.0 to 2.5 mm and dorsal coloration pale, beige, or yellowish white. It differs from *Heterogeomys* species by having the third upper molars with two clear lobes and upper premolars usually with an enamel plate on the back, restricted to the lingual region.

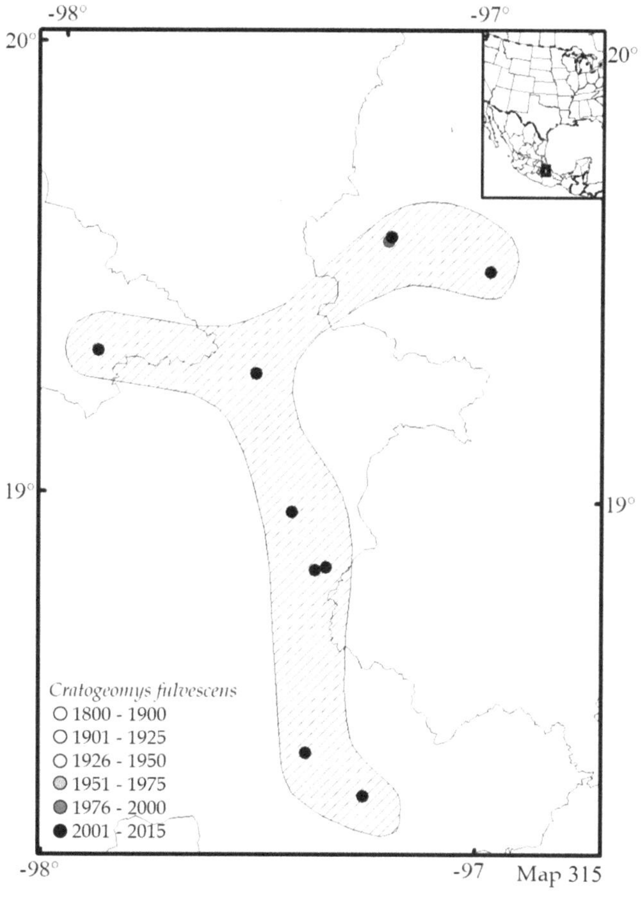


**Map 315** Distribution map of *Cratogeomys fulvescens*

**Additional Literature:** Nelson and Goldman (1934a), Patton (2005b).

## *Cratogeomys fumosus* (Merriam, 1892)

Smoky pocket gopher, tuza del centro de México

1. *C. f. angustirostris* (Merriam, 1903). Restricted to the southwestern central Mexican Plateau.
2. *C. f. fumosus* (Merriam, 1892). Patchily distributed on western Michoacán, eastern slopes of the Sierra Madre del Sur in Jalisco and Colima.
3. *C. f. imparilis* (Goldman, 1939). Patchily distributed in central Michoacán.
4. *C. f. tylorhinus* (Merriam, 1895). Patchily distributed across the southeastern central Mexican Plateau.

**Conservation:** *Cratogeomys fumosus* and *C. neglectus* are listed as Threatened (DOF 2019). Both species were considered to have very small ranges, but the current taxonomy (Hafner et al. 2004) places *neglectus* as a junior synonym of *C. f. fumosus*, thus expanding the range of *C. formosus*. Thus, the Norma Oficial Mexicana conservation status should be reevaluated. The main issue is that both species can be found in sympatry with *Zygogeomys tricopus*, which is under special protection. *C. fumosus* is listed as Least Concern by the IUCN (Álvarez-Castañeda et al. 2016y).

**Characteristics:** Smoky pocket gopher is medium sized; total length 267.0–287.0 mm and skull length 52.0–59.0 mm. Dorsal pelage brown, sometimes pale to dark; fur of underparts paler than in the dorsum; tail short and slightly darker dorsally; **mastoid process extends laterally beyond the auditory meatus**; breadth across angular jaw processes greater than jaw length; paraoccipital process flat and broad when viewed ventrally; anterior jugal part at the suture with zygomatic wide; occipital region of the braincase flattened. Species group *planiceps*.

**Comments:** *Cratogeomys fumosus angustirostris*, *C. f. imparilis*, and *C. f. tylorhinus* were considered subspecies of *C. fumosus*. *C. f. angustirostris* was a junior synonym of *Pappogeomys zinseri* and *P. tylorhinus brevirostris*. *C. f. fumosus* includes *P. gymnurus atratus*, *P. t. gymnurus*, *P. g. inclarus*, *P, z. morulus*, *P. g. tellus*, and *P. t. zodius*; *C. f. tylorhinus* includes *P. neglectus* and *C. t. arvalis* (Hafner et al. 2004). *C. fumosus* is patchily distributed in Colima, Jalisco, Michoacán, Guanajuato, Querétaro, Estado de México, and Puebla (Map 316). *C. fumosus* can be distinguished from *C. merriami* by its mastoid process extending laterally beyond the auditory meatus and breadth across angular jaw processes greater than jaw length, including incisors. From *C. planiceps*, by having the dorsal pelage brown; without a marked black round spot on the ears; paraoccipital process (ventral view) thick, straight, and jugal anterior part wide without a notch. From *Pappogeomys bulleri*, by having a sagittal crest and zygomatic lateral angles with a plate-shaped expansion.

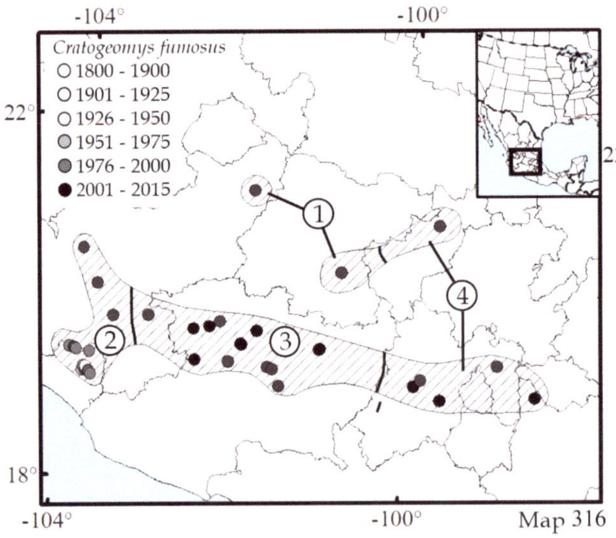


**Map 316** Distribution map of *Cratogeomys fumosus*

**Additional Literature:** Russell (1968b), Cervantes et al. (1993), León et al. (2001), Hafner et al. (2004), Patton (2005b).

## *Cratogeomys goldmani* Merriam, 1895

Goldman's pocket gopher, tuza del Altiplano

1. *C. g. goldmani* Merriam, 1895. From Nazas River and Sierra de Parras, Coahuila, and Durango south through San Luis Potosí and Zacatecas.
2. *C. g. subnubilus* Nelson and Goldman, 1934. From Monterrey, Nuevo León, along western Tamaulipas, and San Luis Potosí.

**Conservation:** *Cratogeomys goldmani* is listed as Least Concern by the IUCN (Álvarez-Castañeda et al. 2016v).

**Characteristics:** Goldman's pocket gopher is medium sized; total length 224.0–270.0 mm and skull length 43.0–51.0 mm. Dorsal pelage pale yellowish to reddish brown, grizzly with blackish shades in the back central part; **fur of underparts whitish, generally ocher**; cheeks more creamy brown; tail short; basioccipital width less than 4.0 mm at its middle part and straight or sandglass shaped; zygomatic breadth greater than squamosal breadth; second and third upper molariforms with no enamel back plate; last upper molariforms semisquare. Species group *castanops*.

**Comments:** The genetic review of the *Cratogeomys castanops* complex (Hafner et al. 2008) considered that the southern subspecies as a different species, under the name of *C. goldmani*. *C. c. goldmani* (part), *C. c. rubellus*, and *C. c. surculus* (part) as junior synonym of *C. g. goldmani*, and *C. castanops elibatus*, *C. c. maculatus*, *C. c. peridoneus*, *C. c. planifrons*, *C. c. subnubilus* as junior synonym of *C. g. subnubilus*. *C. goldmani* ranges from the Nazas River and Sierra de Parras south and from the Sierra Madre Oriental westward, including Durango, Zacatecas, Coahuila, Nuevo León, San Luis Potosí, and Tamaulipas (Map 317). *C. goldmani* can be distinguished from all other species of *Cratogeomys*, except *C. castanops*, by its narrow basioccipital bone, less than 4.0 mm width, and straight or hourglass shaped at its middle part and edges, of smaller average size. From *C. castanops*, by having the number of chromosomes $2n = 42$; smaller size, condylobasal length less than 47.0 mm, and palatal length less than 31.5 mm.

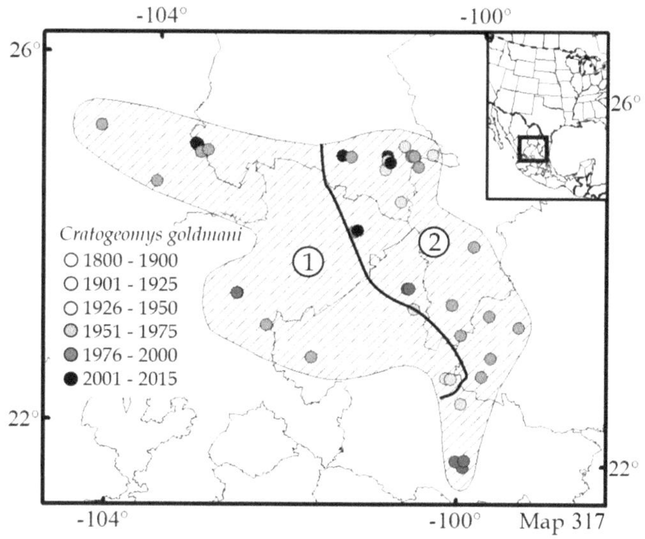


**Map 317** Distribution map of *Cratogeomys goldmani*

**Additional Literature:** Davidow-Henry et al. (1989), Patton (2005b).

## *Cratogeomys merriami* (Thomas, 1893)

Merriam's pocket gopher, tuza del Valle de México

Monotypic.

**Conservation:** *Cratogeomys merriami* is listed as Least Concern by the IUCN (Álvarez-Castañeda et al. 2016ae).

**Characteristics:** Merriam's pocket gopher is medium sized for a pocket gopher but is the largest member of *Cratogeomys*; total length 180.0–253.0 mm in females and 200.0–285.0 mm in males, skull length 58.0 mm. **Dorsal pelage from chocolate brown, chestnut-brown to blackish**; fur of underparts similar but paler yellow to ocher and reddish; melanic specimens have been recorded; basioccipital width greater than 4.0 mm at its middle part and wedge shaped; zygomatic breadth greater than squamosal breadth; jugal anterior edge width greater than 2.5 mm. Species group *castanops*.

**Comments:** Six subspecies of *Cratogeomys merriami* were recognized; *C. m. perotensis* and *C. m. fulvescens* were elevated to full species. *C. m. saccharalis* is not considered valid subspecies and are junior synonyms of *C. merriami* (Hafner et al. 2005). *C. merriami* ranges from Estado de México, Ciudad de México, Querétaro, Tlaxcala, Puebla, and Veracruz (Map 318). However, *C. perotensis* and *C. fulvescens* are geographically very proximal. *C. merriami* can be distinguished from *C. fumosus* and *C. planiceps* by its mastoid process, in general not extending laterally beyond the auditory meatus, or breadth across the angular jaw processes generally less than jaw length, including the incisors. From *C. fulvescens*, by having dorsal pelage in different colors, but never beige, jugal anterior edge width greater than 2.5 mm, and cranial width greater than 26.0 mm.

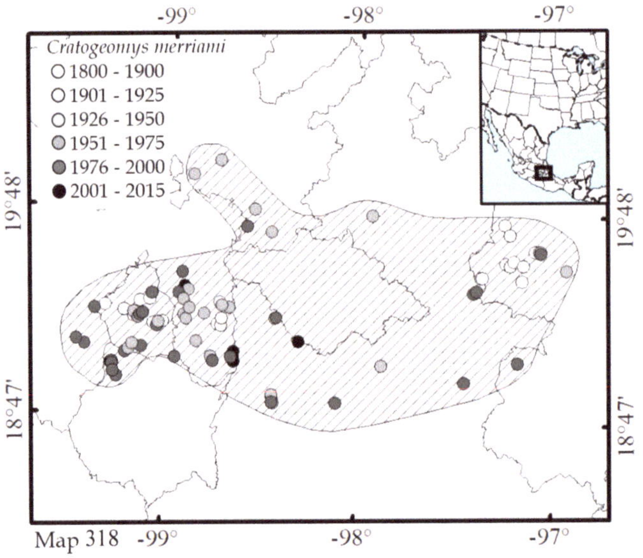


**Map 318** Distribution map of *Cratogeomys merriami*

**Additional Literature:** Patton (2005b).

## *Cratogeomys perotensis* Merriam, 1895

Perote pocket gopher, tuza de Perote

Monotypic.

**Conservation:** *Cratogeomys perotensis* is listed as Least Concern by the IUCN (Álvarez-Castañeda et al. 2017c).

**Characteristics:** Perote pocket gopher is medium sized; total length 310.0 mm and skull length 54.5–58.5 mm. Dorsal pelage brownish, from pale to dark; fur of underparts paler than back. **Individuals commonly possess one or more white patches at the tail base;** basioccipital width greater than 4.0 mm at its middle part and wedge shaped; jugal anterior edge narrow, less than 2.0 mm. Species group *castanops*.

**Comments:** *Cratogeomys perotensis* was recognized as a distinct species (Hafner et al. 2005), previously regarded as a subspecies of *C. merriami* (Russell 1968b). *C. perotensis* ranges from highlands of southern Hidalgo, Puebla, and western-central Veracruz (Map 319). *C. perotensis* can be distinguished from *C. fulvescens* by its small white patches near tail base and jugal anterior edge less than 2.0 mm. From the *Heterogeomys* genus, by having the third upper molars clearly with two lobes and usually the upper premolars with an enamel plate on the back restricted to the lingual region.

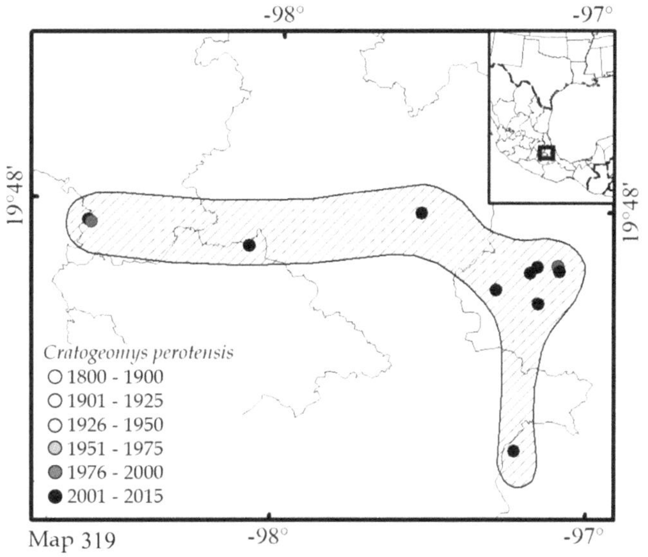


**Map 319** Distribution map of *Cratogeomys perotensis*

**Additional Literature:** DeWalt et al. (1993b).

## *Cratogeomys planiceps* (Merriam, 1895)

Toluca Volcano pocket gopher, tuza del Nevado de Toluca

Monotypic.

**Conservation:** *Cratogeomys planiceps* is listed as Least Concern by the IUCN (Lamoreux 2017).

**Characteristics:** Toluca Volcano pocket gopher is medium sized; total length 372.0 mm and skull length 51.5–66.0 mm. Dorsal pelage chestnut; fur of underparts paler than in the dorsum and leaden gray at the base; pelage similar in legs, rumps, and back; **dorsal limbs whitish; a marked black round spot on the ears**; tail short; **mastoid process extending laterally beyond the auditory meatus**; breadth across the angular jaw processes greater than jaw length; paraoccipital process thin and not straight; jugal anterior part at suture level with zygomatic with a notch. Species group *fumosus*.

**Comments:** *Cratogeomys planiceps* was recognized as a distinct species (Hafner et al. 2004), previously regarded as a subspecies of *C. thylorhinus* (Russell 1968b). *C. planiceps* is restricted to the northern slopes of Volcán de Toluca and Valle de Bravo regions in Estado de México (Hafner et al. 2004; Map 320). It can be distinguished from all other species of *C. merriami* by its mastoid process extending laterally beyond the auditory meatus, and breadth across the angular jaw processes greater than jaw length, including incisors. From *C. fumosus*, by having dorsal pelage chestnut and fur of underparts leaden gray at the base; a marked black round spot on the ears; paraoccipital process (ventral view) thin and not completely straight; usually with a notch in the anterior jugal.

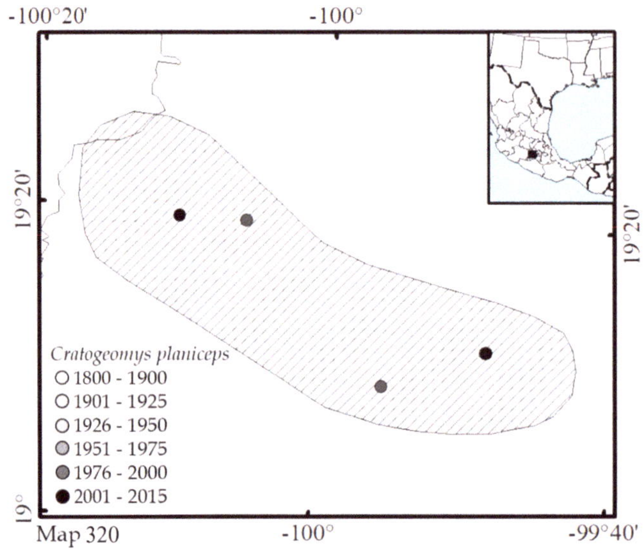


**Map 320** Distribution map of *Cratogeomys planiceps*

**Additional Literature:** DeWalt et al. (1993b), Patton (2005b).

## Genus *Geomys* Rafinesque, 1817

Dorsal pelage in different colors, from pale to dark; fur of underparts paler than in the dorsum; body compact and cylindrical; eyes, ears, and limbs small; claws in limbs well developed; skull large and massive; ridge on the squamosal joining the temporal in adult and old males; interorbital region narrower than rostrum; **middle part of upper incisors anterior surface with two grooves: one large, deep, and medial; second one small and flanked on the inner side**; first upper molars without an enamel plate and larger than the first lower molars; upper molariform not prominently bicolumnar and almost as long as wide; all molariforms elliptical with a small anterior-posterior axis; anterior and posterior margins of molariforms with enamel, other margins with dentin; sagittal crest poorly developed. The genus *Geomys* has four groups of species: *bursarius* (*G. arenarius*, *G. bursarius*, *G. jugossicularis*, *G. knoxjonesi*, *G. lutescens*, and *G. texensis*), *breviceps* (*G. breviceps*), *G. personatus* (*G. attwateri*, *G. personatus*, *G. streckeri*, and *G. tropicalis*), and *pinetis* (*G. pinetis*). No species of *Geomys* are sympatric; however, hybrids can be found in contact zones. The keys were elaborated based on the review of specimens and the following sources: Merriam (1895a), Baker and Williams (1974), Hall (1981), Williams and Cameron (1991), Patton (2005b), and Sudman et al. (2006).

1. Squamosal projection joining jugal not wedge shaped within the zygomatic arches (Fig. 95; restricted to a coastal sand bar of Tamaulipas) ........ 2

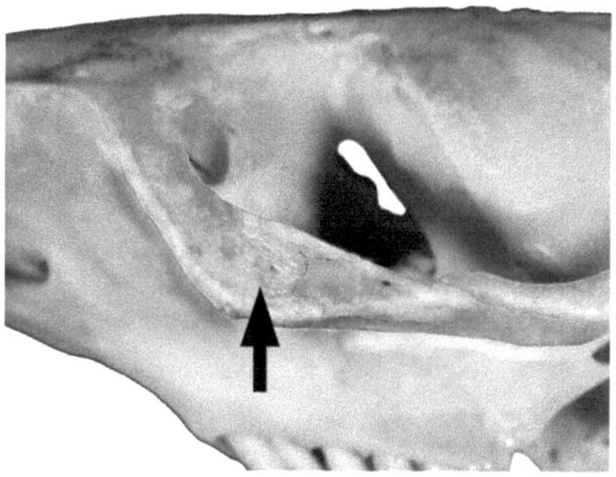

**Fig. 95** Squamosal projection joining jugal not wedge shaped within the zygomatic arches

1a. Squamosal projection ending in a wedge-shape at the middle part of jugal ........ 11

2. In most specimens, anterior and posterior zygomatic breath approximately equal; jugal shorter than or equal to basioccipital, including the exoccipital condyle; posterior end of zygomatic maxillary arm U shaped at the union with jugal ........ 3

2a. In most specimens, anterior zygomatic breath wider than the posterior part; jugal longer than basioccipital, including the exoccipital condyle; posterior end of zygomatic maxillary arm V shaped at its union with jugal (no morphological characteristics can be used to identify the following species but only genetic data) ........ 4

3. Patchy distribution throughout southern Texas and northeastern Tamaulipas, México ........ *Geomys personatus* (p. 457)

3a. Range restricted to the type locality of Carrizo Spring, Texas ........ *Geomys streckeri* (p. 459)

4. Ranges north of 40° latitude and northeast through the Missouri River ........ 5

4a. Ranges south of 40° latitude and west through the Missouri River ........ 6

5. Ranges from eastern Manitoba, North Dakota, and South Dakota east through eastern Illinois and Indiana (eastern North Dakota, South Dakota, Minnesota, Iowa, eastern Missouri, western Wisconsin, central Illinois, and northwestern Indiana) ........ *Geomys bursarius* (p. 453)

5a. Ranges from southern South Dakota south through Oklahoma and Texas (western Wyoming, South Dakota, Nebraska, northeastern Colorado, Kansas, central-western Oklahoma, and northern Texas) ........ *Geomys lutescens* (p. 456)

6. Ranges east of Los Brazos River, Texas and – 99° in Oklahoma ........ 7

6a Ranges west of Los Brazos River, Texas and – 99° in Oklahoma ........ 8

7. Ranges in southern Alabama, Georgia, central and northern Florida ........ *Geomys pinetis* (p. 458)

7a. Ranges in eastern Texas, eastern Oklahoma, central-western Arkansas, and western Louisiana ........ *Geomys breviceps* (p. 452)

8 Ranges in the border area between Colorado, Kansas, New Mexico, and Oklahoma ........ *Geomys jugossicularis* (p. 454)

8a. Ranges in areas other than those mentioned in 8 above ........ 9

9. Southeastern New Mexico and western-central Texas ........ *Geomys knoxjonesi* (p. 455)

9a. Ranges in other areas except southeastern New Mexico and western-central Texas ........ 10

10. Ranges in southeastern Texas, near the coast in basins of rivers Brazos, Colorado, and Guadalupe ........ *Geomys attwateri* (p. 451)

10a. Ranges in central-southern Texas ........ *Geomys texensis* (p. 460)

11. Interparietal bone subquadrate shape; sides of zygomatic arches parallel; edge of premaxilla at the anterior border of palatine foramen wedge shaped; rostrum breadth not much greater than basioccipital length (southern-central New Mexico, southwestern Texas, and northern Chihuahua) ........ *Geomys arenarius* (p. 450)

11a. Interparietal bone triangle shaped; sides of zygomatic arches narrow posteriorly; edge of premaxilla at the anterior palatine foramen subquadrate; rostrum breadth greater than basioccipital length (endemic to southern Tamaulipas, near Tampico) ........ *Geomys tropicalis* (p. 461)

## ***Geomys arenarius*** Merriam, 1895

Desert pocket gopher, tuza del desierto

1. *G. a. arenarius* Merriam, 1895. Southern New Mexico, northern Chihuahua, and southwestern Texas.
2. *G. a. brevirostris* Hall, 1932. Restricted to central-southern New Mexico.

**Conservation:** *Geomys arenarius* is listed as Near Threatened by the IUCN (Lacher et al. 2019a).

**Characteristics:** Desert pocket gopher is small sized; total length 221.0–250.0 mm in females and 244.0–280.0 mm in males, skull length 40.0–43.0 mm. Dorsal pelage sandy brown, yellow-orange, or yellowish; **in general, it shares the ashier coloration of all *Geomys* species**; fur of underparts whitish or yellowish; tail sparsely hairy; **zygomatic arches with a squamous projection ending in a protuberance in jugal middle part;** interparietal bone subquadrate; premaxilla edge at incisor foramen anterior border wedge shaped; mesopterygoid fossa V shaped; interparietal bone subquadrate. Species group *bursarius*.

**Comments:** *Geomys arenarius* was recognized as a distinct species (Jolley et al. 2000), previously regarded as a subspecies of *G. bursarius* (Hafner and Geluso 1983). *G. arenarius* is patchily distributed in southern-central New Mexico, southwestern Texas, and northern Chihuahua. A second isolated population lives in central-eastern New Mexico, in association with deep sandy soils (Schmidly 1977; Davis and Schmidly 1994; Map 321). No other species of *Geomys* are found in sympatry. *G. arenarius* can be distinguished from all other species of groups *bursarius*, *breviceps*, and *pinetis* by having the squamosal branch ends of the zygomatic arches bulge in the middle of the jugal. Chromosomes are $2n = 70$, FN = 88–102.

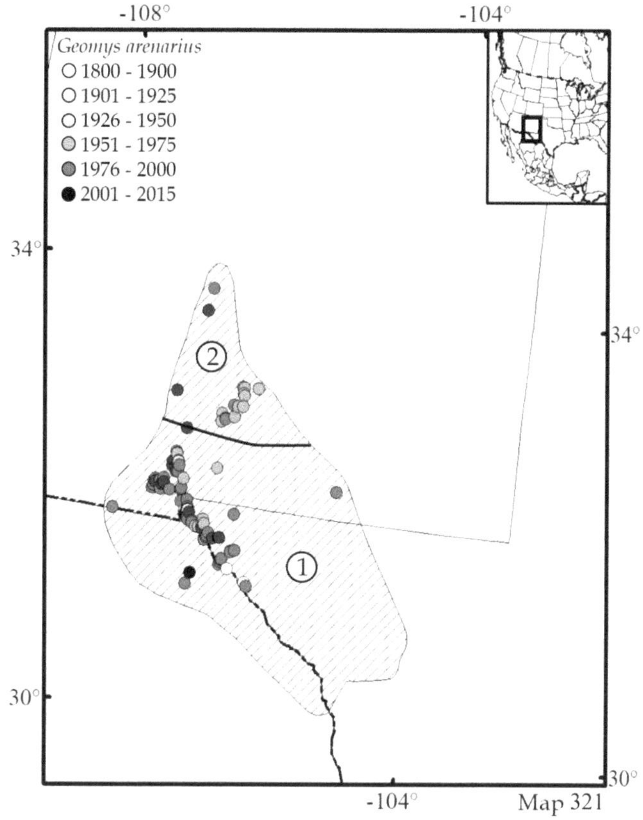


**Map 321** Distribution map of *Geomys arenarius*

**Additional Literature:** Russell (1968a), Stephen and Baker (1974), Williams and Cameron (1991), Patton (1993a, 2005a), Chambers et al. (2009).

## *Geomys attwateri* Merriam, 1895

Attwater's pocket gopher, tuza de Corpus Christi

1. *G. a. ammophilus* Davis, 1940. Restricted to the Brazos River Basin near the coast, central-southeastern Texas.
2. *G. a. attwateri* Merriam, 1895. Restricted to the Colorado and Guadalupe River basins near the coast of southeastern Texas.

**Conservation:** *Geomys attwateri* is listed as Least Concern by the IUCN (Cassola 2016ab).

**Characteristics:** Attwater's pocket gopher is small sized; total length 190.0–235.0 mm and skull length 35.6–41.3 mm. Dorsal pelage pale to dark brown, with back central part generally darker; fur of underparts paler; ears paler than the general coloration and with an inconspicuous darker preauricular spot. Species group *personatus*.

**Comments:** *Geomys attwateri* was recognized as a distinct species (Tucker and Schmidly 1981; Williams and Cameron 1991; Burt and Dowler 1999), previously regarded as a subspecies of *G. bursarius* (Baker and Glas 1951; Patton 1993a). *G. attwateri* is restricted to the basins of rivers Brazos, Colorado, and Guadalupe, in the counties of Aransas, Atascosa, Burleson, Matagorda, Milam, and San Patricio in southeastern Texas, in association with silty clay loam soils with high sand levels (Cameron et al. 1988; Map 322). In the north and east of its ranges, *G. attwateri* can be found in parapatry, or nearly so, with *G. breviceps* (different soil types, high percentage of sandy loam and silt-clay), *G. bursarius* (deep soils of different types), and at the south and west with *G. personatus* (deep sandy soils). *Geomys attwateri* cannot be distinguished from *G. breviceps* using morphological traits but only with cytological and biochemical characteristics (Williams 1999). It can be distinguished from *G. personatus*, by having zygomatic breadth wider anteriorly than posteriorly, jugal longer than basioccipital length, including the exoccipital condyle, and zygomatic maxillary arm posterior end V shaped at the union with the jugal. Chromosomes are $2n = 70$, $FN = 72–74$.

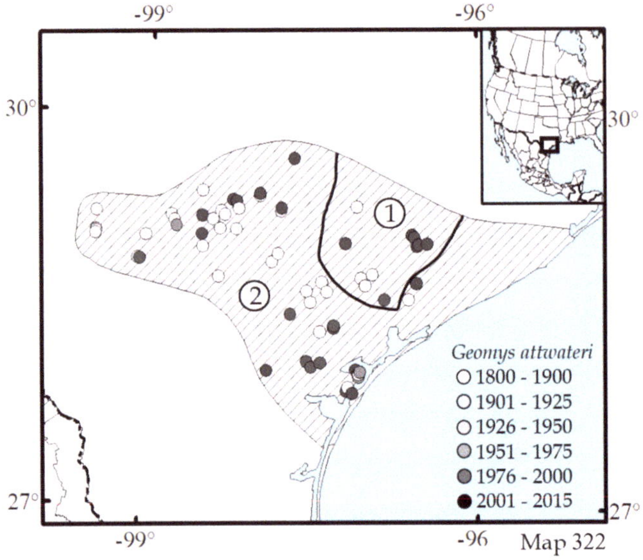


**Map 322** Distribution map of *Geomys attwateri*

**Additional Literature:** Jolley et al. (2000), Sudman et al. (2006), Chambers et al. (2009).

## *Geomys breviceps* Baird, 1855

Baird's pocket gopher, tuza texana del este

1. *G. b. breviceps* Baird, 1855. From southeastern Oklahoma and southwestern Arkansas south through eastern Texas and western Louisiana.
2. *G. b. ozarkensis* Elrod et al., 2000. Northern-central Arkansas.
3. *G. b. sagittalis* Merriam, 1895. Restricted to Galveston Bay, Texas.

**Conservation:** *Geomys breviceps* is listed as Least Concern by the IUCN (Cassola 2016ac).

**Characteristics:** Baird's pocket gopher is small sized; total length 190.0–222.0 mm and skull length 35.6–39.2 mm. Dorsal pelage from pale brown to blackish, highly variable; fur of underparts paler than in the dorsum. Species group *breviceps*.

**Comments:** *Geomys breviceps* was recognized as a distinct species (Baker et al. 2003a; Patton 2005b), previously regarded as a subspecies of *G. bursarius* (Baker and Glas 1951). *G. b. sagittalis* was proposed to be elevated to full species based on its mtDNA differences (Demastes 1994). *G. breviceps* ranges in eastern Texas, eastern Oklahoma, central-western Arkansas, and western Louisiana, associated with different fine sandy loam or clay loam; soil depth greater than 10 cm (Cassola 2016c; Map 323). *G. breviceps* can be found at the south of its ranges in parapatry, or nearly so, with *G. attwateri* (silty clay loam soils with high sand levels) and at the east with *G. bursarius* (soils of different types and depths). It is impossible to distinguish among species morphologically but only using cytological and biochemical characteristics. Karyotypic hybrids between *G. breviceps* and *G. attwateri* with *G. bursarius* have been found. Chromosomes are $2n = 74$, FN = 72.

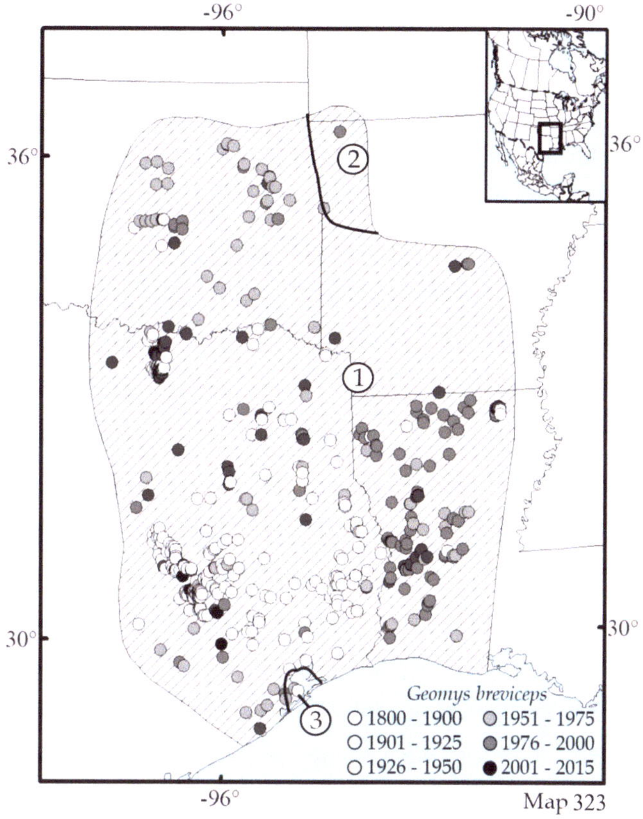


**Map 323** Distribution map of *Geomys breviceps*

**Additional Literature:** Burns et al. (1985), Sulentich et al. (1991), Burt and Dowler (1999), Jolley et al. (2000), Sudman et al. (2006), Chambers et al. (2009).

## *Geomys bursarius* (Shaw, 1800)

Plains pocket gopher, tuza de las Planicies

1. *G. b. bursarius* (Shaw, 1800). From southern Manitoba south through eastern Dakotas, Minnesota, and northwestern Wisconsin.
2. *G. b. illinoensis* Komarek and Spencer, 1931. Central Illinois and central-western Indiana.
3. *G. b. majusculus* Swenk, 1939. From Iowa and northern-central Missouri west through southeastern Nebraska and eastern Kansas.
4. *G. b. missouriensis* McLaughlin, 1958. Eastern Missouri.
5. *G. b. wisconsinensis* Jackson, 1957. Western Wisconsin.

**Conservation:** *Geomys bursarius* is listed as Least Concern by the IUCN (Cassola 2016ad).

**Characteristics:** Plains pocket gopher is small sized; total length 225.0–325.0 mm and skull length 39.0–48.6 mm. Dorsal pelage pale buff brown to different brown to blackish shades; fur of underparts grayish to buff; individuals commonly found with small patches of white hairs; **ears paler at the edges, contrasting with the head color regardless of the general rudimentary coloration**; tail one-fourth of the head-and-body length and sparsely haired; mesopterygoid fossa U shaped. Species group *bursarius*.

**Comments:** The subspecies *Geomys bursarius attwateri*, *G. b. breviceps*, *G. b. jugossicularis*, *G. b. knoxjonesi*, *G. b. lutescens*, and *G. b. texensis* were recognized as a distinct species (Tucker and Schmidly 1981; Heaney and Timm 1983; Baker et al. 1989a; Block and Zimmerman 1991; Sudman et al. 2006). *G. bursarius* is distributed in the Great Plains from Manitoba, eastern North Dakota, South Dakota, Minnesota, Iowa, eastern Missouri, western Wisconsin, central Illinois, and northwestern Indiana, in association with soils of various types and depths (Williams 1999; Map 324). *G. bursarius* can be found in parapatry, or nearly so, at the west with *G. jugossicularis* (sandy loams), *G. lutescens* (deep sandy soils), and *G. knoxjonesis* (deep sandy soils in scrublands), and at the south with *G. breviceps* and *G. texensis*. *Geomys bursarius* cannot be distinguished from *G. attwateri*, *G. breviceps*, *G. jugossicularis*, *G. knoxjonesi*, *G. lutescens*, and *G. texensis* using morphological traits but only with cytological and biochemical characteristics; however, *Geomys bursarius* have ears paler at the edges, contrasting with the head color regardless of the general coloration (Zimmerman 1999). Chromosomes are $2n$ = 70–72, FN = 68–74.

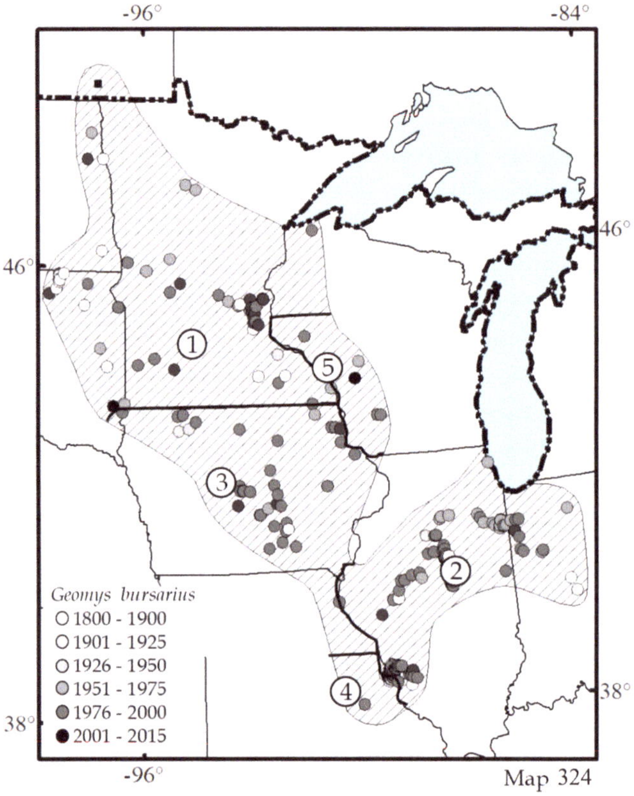


**Map 324** Distribution map of *Geomys bursarius*

**Additional Literature:** Merriam (1895a), Baker and Glas (1951), Russell (1968a), Baker and Genoways (1975), Honeycutt and Schmidly (1979), Sudman et al. (1987), Zimmerman (1999), Chambers et al. (2009), Connior (2011).

## *Geomys jugossicularis* Hooper, 1940

Hall's pocket gopher, tuza de las planicies centrales

1. *G. j. halli* Sudman, Choates and Zimmerman, 1987. From eastern Colorado east through southwestern Kansas, Oklahoma, and northeastern New Mexico.
2. *G. j. jugossicularis* Hooper, 1940. Oklahoma and southwestern Kansas.

**Conservation:** *Geomys jugossicularis* (as *G. bursarius*) is listed as Least Concern by the IUCN (Cassola 2016ad).

**Characteristics:** Hall's pocket gopher is small sized; total length 225.0–325.0 mm and skull length 39.0–48.6 mm. Dorsal pelage yellowish cinnamon; nape and shoulders more ashen-colored; fur of underparts pale buff to whitish at the flanks; ears paler at the outer lower edges; mesopterygoid fossa U shaped. Species group *bursarius*.

**Comments:** *Geomys jugossicularis* was recognized as a distinct species by its high mtDNA divergence (8.1%) from *G. bursarius* (Sudman et al. 2006), previously regarded as a subspecies of *G. bursarius* (Villa and Hall 1949) or *G. lutensis* (Hooper 1940a). *G. jugossicularis* ranges in eastern Colorado and New Mexico, western Kansas, and northwestern Oklahoma, in association with sandy loams (Map 325). It can be found in parapatry, or nearly so, at the east and south of its ranges with *G. bursarius* (different types of soils anddepths) and at the north with *G. lutescens*. The three species can only be distinguished by genetic analyses. Chromosomes are $2n = 70$, FN = 72.

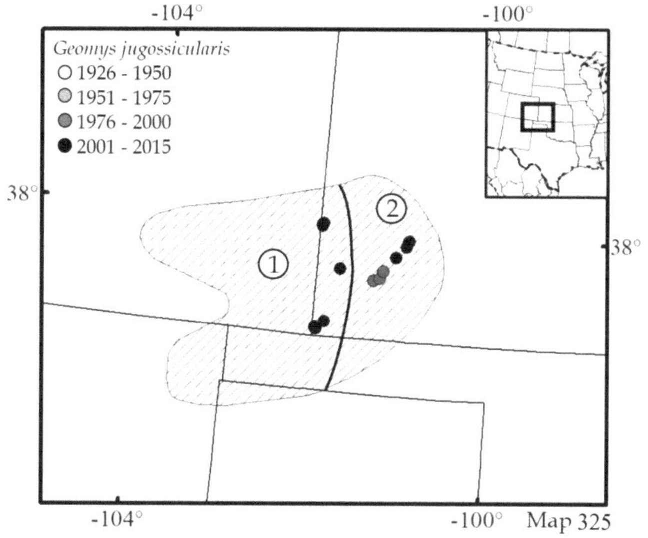


**Map 325** Distribution map of *Geomys jugossicularis*

**Additional Literature:** Chambers et al. (2009).

## *Geomys knoxjonesi* Baker and Genoways, 1975

Jones's pocket gopher, tuza texana del oeste

Monotypic.

**Conservation:** *Geomys knoxjonesi* is listed as Least Concern by the IUCN (Cassola 2016ae).

**Characteristics:** Jones's pocket gopher is small sized; total length 206.0–282.0 mm and skull length 40.9–46.0 mm. Dorsal pelage buff brown to pale yellowish; fur of underparts paler than in the dorsum, some ventral areas covered with almost pure-white hair; lower part of flanks more yellowish; ears paler than the general coloration; feet white; mesopterygoid fossa U shaped. Species group *bursarius*.

**Comments:** *Geomys knoxjonesi* was recognized as a distinct species (Baker et al. 1989b), previously regarded as a subspecies of *G. bursarius* (Baker and Genoways 1975). It ranges in southeastern New Mexico and western-central Texas in association with deep sandy soils in shrublands (Bradley and Baker 1999, Map 326). *G. knoxjonesi* can be found in parapatry, or nearly so, to the north and east with *G. bursarius* (different soil types anddepths) and can only be distinguished using genetic analyses. Chromosomes are $2n = 70$, FN = 68–70. *G. knoxjonesi* is considered of the *bursarius* species group.

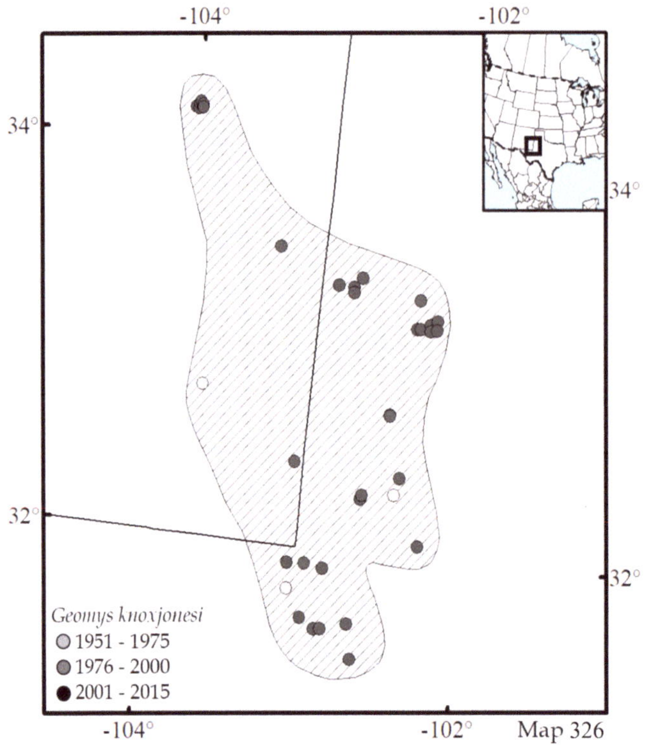


**Map 326** Distribution map of *Geomys knoxjonesi*

**Additional Literature:** Hopton and Cameron (2001), Sudman et al. (2006), Chambers et al. (2009).

## *Geomys lutescens* Merriam, 1890

Sandy hill pocket gopher, tuza arenera

1. *G. l. lutescens* Merriam, 1890. From South Dakota south through central-eastern Colorado and central Kansas.
2. *G. l. industrius* Villa and Hall, 1947. Restricted to southwestern Kansas.
3. *G. l. major* Davis, 1940. From southern Kansas south through eastern New Mexico and northern Texas, including Oklahoma.

**Conservation:** *Geomys lutescens* (as *G. bursarius*) is listed as Least Concern by the IUCN (Cassola 2016ad).

**Characteristics:** Sandy Hill pocket gopher is small sized; total length 225.0–325.0 mm and skull length 39.0–48.6 mm. Dorsal pelage pale brown to brown but generally with buff to pale yellowish shades and back darker than flanks; fur of underparts paler than flanks; back of the head darker; ears paler than the general coloration; tail in some individuals paler than the dorsal pelage; mesopterygoid fossa U shaped. Species group *bursarius*.

**Comments:** *Geomys lutescens* was recognized as a distinct species using morphometric, karyotypic, and electrophoretic methods (Russell 1968a, b; Heaney and Timm 1983; Sudman et al. 1987), previously regarded as a subspecies of *G. bursarius* (Merriam 1890a, b, c). It ranges in western Wyoming, South Dakota, Nebraska, northeastern Colorado, Kansas, central-western Oklahoma, and northern Texas, in association with deep sandy soils (Map 327). *G. lutescens* can be found in parapatry, or nearly so, at the east with *G. bursarius* (different soil types and depths) and at the south with *G. jugossicularis* (sandy loams). The three species can only be distinguished by genetic analyses. Chromosomes are $2n = 72$, FN = 86–98.

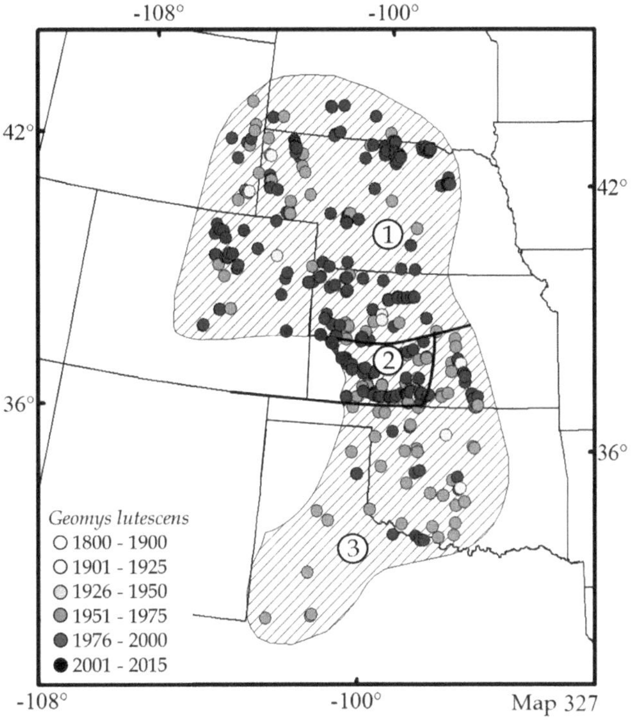


**Map 327** Distribution map of *Geomys lutescens*

**Additional Literature:** Hooper (1940a), Burns et al. (1985), Sudman et al. (2006), Chambers et al. (2009).

## *Geomys personatus* True, 1889

Texas pocket gopher, tuza texana del sur

1. *G. p. davisi* Williams and Genoways, 1981. Restricted to southern Texas.
2. *G. p. fallax* Merriam, 1895. Restricted to southeastern Texas.
3. *G. p. fuscus* Davis, 1940. Restricted to Kinney and Valverde counties, southern Texas.
4. *G. p. maritimus* Davis, 1940. Restricted to Baffin Bay and Flour Bluff, southern Texas.
5. *G. p. megapotamus* Davis, 1940. Southern Texas and northeastern Tamaulipas.
6. *G. p. personatus* True, 1889. Restricted to the Mustang and Padre islands, southern Texas.

**Conservation:** *Geomys personatus* is listed as Threatened in the Norma Oficial Mexicana (DOF 2019) and as Least Concern by the IUCN (Lacher 2016b).

**Characteristics:** Texas pocket gopher is small sized; total length 225.0–305.0 mm in females and 248.0–326.0 mm in males, skull length 44.2–46.8 mm. Dorsal pelage gray-brown sand to pale gray brown; fur of underparts completely whitish cream and gray at the base; back of head with darker hairs; ears proportionally smaller and paler at the edges; squamosal branch of the zygomatic arches not ending in a protuberance in the middle jugal; mesopterygoid fossa U shaped. Species group *personatus*.

**Comments:** *Geomys personatus* has a patchy distribution throughout southern Texas and adjacent northeastern Tamaulipas, occupying deep sandy soils (Lacher 2016b; Map 328). *G. personatus* can be found in parapatry, or nearly so, at the northeast of its range with *G. attwateri* (silty clay loam soils with high sand levels). *G. streckeri* may overlap in range with the subspecies *G. p. fuscus* and *C. p. megapotamus*. *G. personatus* can be distinguished from *G. attwateri* by having the zygomatic breadth wider anteriorly than posteriorly, jugal longer than the basioccipital length, including the exoccipital condyle, and zygomatic maxillary arm at the union with the jugal with posterior end V shaped; from *G. streckeri*, only by genetic analyses. Chromosomes are $2n = 68–70$, FN = 70–76.

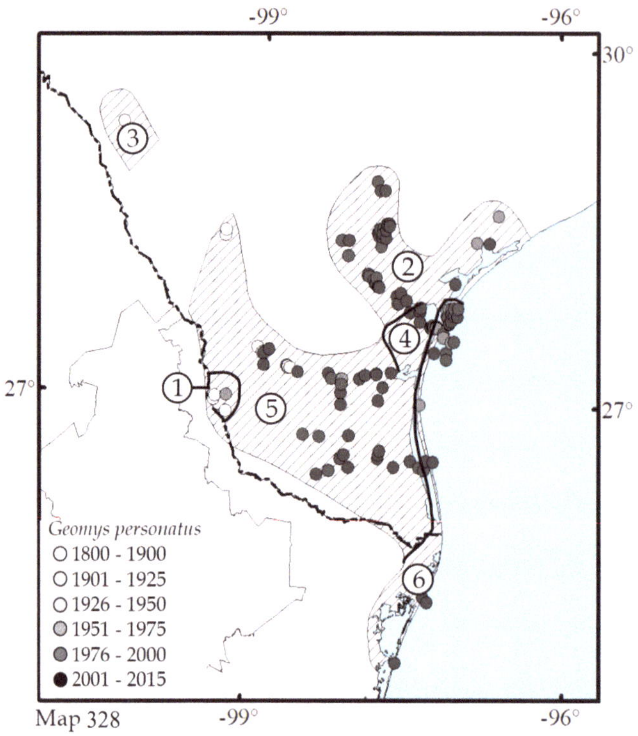


**Map 328** Distribution map of *Geomys personatus*

**Additional Literature:** Davis (1940), Russell (1968a), Williams and Genoways (1981), Stephen (1982), Jolley et al. (2000), Sudman et al. (2006).

## *Geomys pinetis* Rafinesque, 1817

Southeastern pocket gopher, tuza del sureste

1. *G. p. fontanelus* Sherman, 1940. Restricted to eastern Georgia and southwestern South Carolina.
2. *G. p. pinetis* Rafinesque, 1817. Alabama, Florida, Georgia, and South Carolina. From southeastern South Carolina west through south central Alabama and south through central Florida.

**Conservation:** *Geomys pinetis goffi* (Hafner et al. 1998) and *G. p. fontanelus* are listed as Endangered and Possibly Extinct by the Georgia Department of Natural Resources. The IUCN listed *Geomys pinetis* as Least Concern (Cassola 2016af).

**Characteristics:** Southeastern pocket gopher is small sized; total length 215.0–324.0 mm and skull length 38.5–49.7 mm. Dorsal pelage from reddish to grayish brown; fur of underparts paler than in the dorsum; rostrum back part and back of head darker, contrasting with nape and back; rump generally paler; whitish gray coloration from chin to neck. Species group *pinelis*.

**Comments:** *Geomys pinetis* is karyotypically the most divergent taxon within *Geomys* (Qumsiyeh et al. 1988). *G. p. colonus*, *G. p. cumberlandius*, and *G. p. floridanus* were recognized as a distinct species (Hall 1981; Laerm 1981) but not by Patton (1993a, 2005b) and Baker et al. (2003a, b, c). *G. pinetis* ranges in Alabama, Georgia, and Florida, associated with deep sandy soils and open areas in long-leaf pinewood (Cassola 2016af; Map 329). The subspecies *G. p. austrinus*, *G. p. colonus*, *G. p. cumberlandius*, *G. p. floridanus*, *G. p. goffi*, *G. p. mobliensis* are considered junior synonym of *G. p. pinetis* (Williams and Genoway 1980). Sudman et al. (2006) considered that *G. p. mobilensis* may represents a species distinct from *G. pinetis*. No other species of the genus *Geomys* occurs anywhere near its range. *G. pinetis* cannot be distinguished morphologically from any of the other species of *Geomys*. Chromosomes are $2n = 42$, FN = 80.

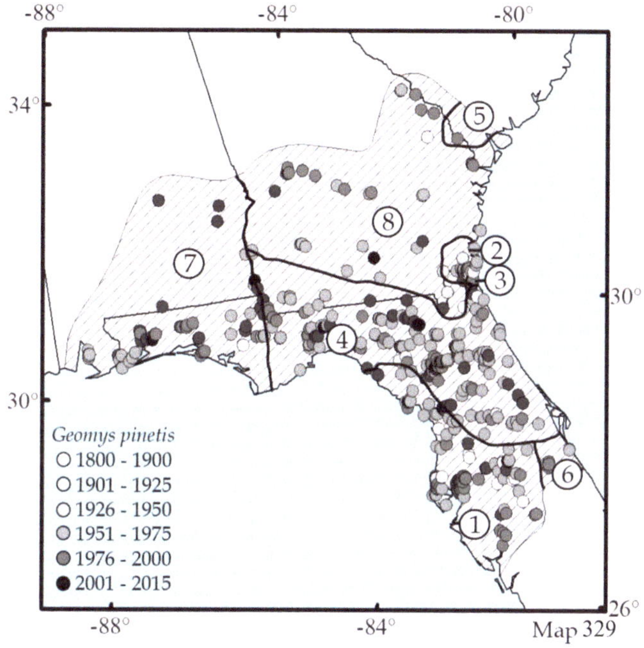


**Map 329** Distribution map of *Geomys pinetis*

**Additional Literature:** Russell (1968a), Pembleton and Williams (1978), Qumsiyeh et al. (1988), Jolley et al. (2000), Sudman et al. (2006), Chambers et al. (2009).

## *Geomys streckeri* Davis, 1943

Strecker's pocket gopher, tuza texana del Carrizo

Monotypic.

**Conservation:** *Geomys streckeri* (as *G. personatus*) is listed as Least Concern by the IUCN (Lacher 2016b), however by its small range need to be evaluated.

**Characteristics:** Strecker's pocket gopher is small sized; total length 220.0–322.0 mm and skull length 44.0–47.0 mm. Dorsal pelage pale buffy without a darker coloration in the back; fur of underparts creamy yellowish, including the flanks; some specimens have the rump paler than the flanks; ears evenly colored. Species group *personatus*.

**Comments:** *Geomys streckeri* was recognized as a distinct species (Chambers et al. 2009), previously regarded as a subspecies of *G. personatus* (Davis 1937). *G. streckeri* has a restricted distribution, known only from its type locality at Carrizo Springs, Texas (Map 330); it is not known to coexist with any other species of *Geomys*. Geographically, the closest species is *G. personatus*; both use the same type of habitat, deep sandy soils in semiarid areas. This species is recognized only using genetic analyses.

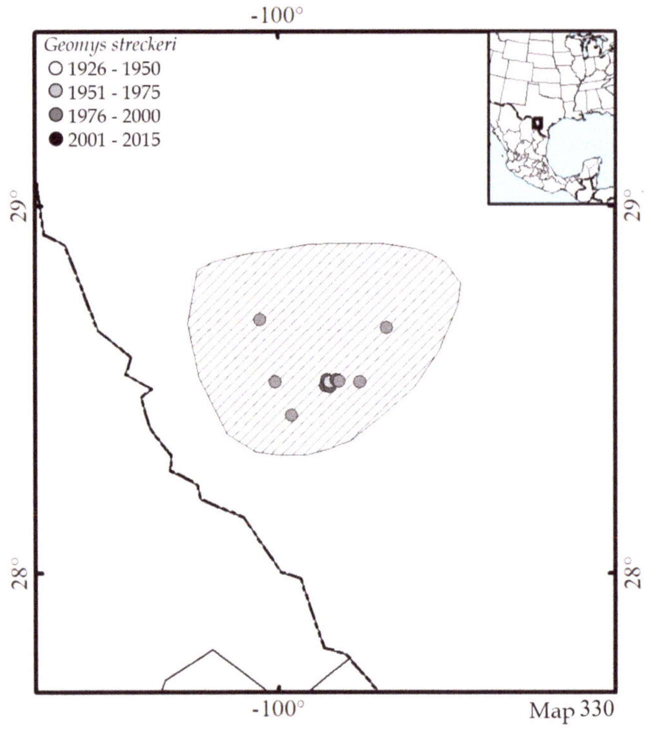


**Map 330** Distribution map of *Geomys streckeri*

**Additional Literature:** Sudman et al. (2006).

## *Geomys texensis* Merriam, 1895

Llano pocket gopher, tuza del centro de Texas

1. *G. t. bakeri* Smolen, Pitts and Bickham, 1993. Restricted to Medina, Uvalde, and Zavala counties, central-south Texas.
2. *G. t. llanensis* Bailey, 1905. Restricted to Gillespie, Kimble, and Zavala counties, central-southern Texas.
3. *G. t. texensis* Merriam, 1895. Restricted to Mason, McCulloch, and San Saba counties, central Texas.

**Conservation:** *Geomys texensis* is listed as Least Concern by the IUCN (Cassola 2016ag).

**Characteristics:** Llano pocket gopher is small sized; total length 180.0–270.0 mm and skull length 35.0–43.2 mm. Dorsal pelage from pale brown to liver brown, including different shades of buffy brown; fur of underparts paler brown and some individuals whitish; feet usually white; temporal region and back of the head darker than the nape; ears paler at the edges with low contrast; mesopterygoid fossa U shaped. Species group *bursarius*.

**Comments:** *Geomys texensis* was recognized as a distinct species (Block and Zimmerman 1991) previously regarded as a subspecies of *G. bursarius* (Wilson and Reeder 1993; Jolley et al. 2000). It ranges only in central-southern Texas, in association with deep soils with brown loamy sand or gravelly sandy loams (Schmidly 2004; Map 331). The only other species with which its ranges overlap at the north is *G. bursarius* (different soil types and depths) and both can only be distinguished using genetic analyses. Chromosomes are $2n = 70$, FN = 68.

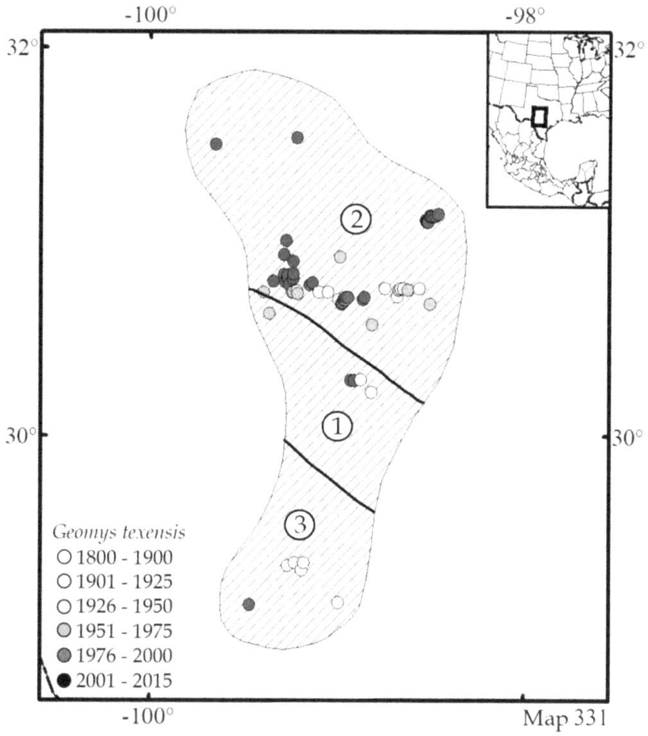


**Map 331** Distribution map of *Geomys texensis*

**Additional Literature:** Cramer and Cameron (2001), Sudman et al. (2006), Chambers et al. (2009).

## *Geomys tropicalis* Goldman, 1915

Tropical pocket gopher, tuza tropical

Monotypic.

**Conservation:** *Geomys tropicalis* is listed as Threatened in the Norma Oficial Mexicana (DOF 2019) and as Endangered (B1ab [ii, iii] + 2ab [ii, iii]: extent of occurrence less than 5000 km$^2$ in five or less populations and under continuing decline in area of occupancy and quality of habitat + area of occupancy less than 500 km$^2$ in five or less populations and continuing decline in area of occupancy and quality of habitat) by the IUCN (Roach 2018a).

**Characteristics:** Tropical pocket gopher is small sized; total length 235.0–250.0 mm in females and 260.0–265.0 mm in males, skull length 41.3–43.1 mm. Dorsal pelage cinnamon to cinnamon buffy without a darker coloration in the central back; fur of underparts paler, in some individuals whitish; ears smaller and evenly colored; zygomatic arches wider at the front than at the back; sagittal crest poorly developed; zygomatic arches with a squamosal branch ending in a protuberance at the middle jugal; mesopterygoid fossa V shaped; interparietal bone triangular. Species group *personatus*.

**Comments:** *Geomys tropicalis* was recognized as a distinct species (Álvarez 1963), previously regarded as a subspecies of *G. personatus* (Goldman 1915). *Geomys tropicalis* is only known from the southeastern Tamaulipas region of coastal México (Baker and Williams 1974; Map 332). *G. tropicalis* can be found in sympatry, or nearly so, at the north of its range with *G. personatus*. However, *G. tropicalis* thrives in coastal deep sandy areas (Roach 2018a). *G. tropicalis* differs by having a squamous wedge-shaped ending projection at the jugal middle part and by its unique karyotype (Davis et al. 1971).

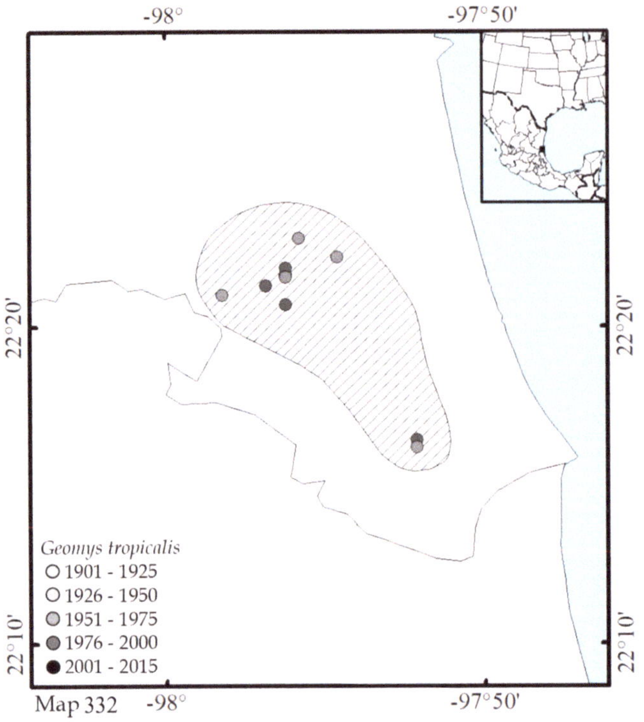


**Map 332** Distribution map of *Geomys tropicalis*

**Additional Literature:** Sudman et al. (2006), Chambers et al. (2009).

## Genus *Heterogeomys* Merriam, 1895

Dorsal pelage from reddish to reddish brown or brown, with hair strong and coarse or soft; fur of underparts paler than in the dorsum; body compact and cylindrical; hair sparse and bristly; skull large and massive; ridge on the squamosal joining the temporal in adult and old males; interorbital region narrower than the rostrum; **middle part of upper incisors anterior surface with one deep groove;** last upper molars semi-lobular, only one labial groove; an enamel plate covers the front and edge of the reentrant angle of first upper and lower molariforms; first lower molariforms with a posterior enamel plate and first upper premolars always with a small plate on the lingual side. The two species of the genus *Heterogeomys* are not sympatric. The keys were elaborated based on the review of specimens and the following sources: Nelson and Goldman (1929), Hall (1981), and Álvarez-Castañeda et al. (2017a).

1. Total length less than 361.0 mm; hindfoot length less than 54.0 mm; hair strong, coarse, short, and sparse (southern Tamaulipas, eastern San Luis Potosí, and Querétaro south through Central America, including the Yucatán Peninsula) .......... *Heterogeomys hispidus* (p. 462)

1a. Total length greater than 361.0 mm; hindfoot length greater than 54.0 mm; hair soft, thick, and woody, almost black (endemic to southeast of Pico de Orizaba, Veracruz) .......... *Heterogeomys lanius* (p. 463)

### *Heterogeomys hispidus* (Le Conte, 1852)

Hispid pocket gopher, tuza gigante tropical

1. *H. h. chiapensis* Nelson and Goldman, 1929. All Chiapas, except northwestern.
2. *H. h. concavus* Nelson and Goldman, 1929. San Luis Potosí, Querétaro, and northwestern Veracruz.
3. *H. h. hispidus* (Le Conte, 1852). Western Veracruz and eastern Puebla.
4. *H. h. isthmicus* Nelson and Goldman, 1929. Restricted to southeastern Veracruz.
5. *H. h. latirostris* Hall and Álvarez, 1961. Restricted to northeastern Veracruz.

6. *H. h. negatus* Goodwin, 1953. Restricted to southern Tamaulipas.
7. *H. h. teapensis* Goldman, 1939. Restricted to southern Tabasco and northwestern Chiapas.
8. *H. h. tehuantepecus* Goldman, 1939. Restricted to northern Oaxaca.
9. *H. h. torridus* Merriam, 1895. Restricted to central Veracruz.
10. *H. h. yucatanensis* Nelson and Goldman, 1929. Yucatán peninsula, Belize, and Guatemala.

**Conservation:** *Heterogeomys hispidus* (as *Orthogeomys hispidus*) is listed as Least Concern by the IUCN (Vázquez et al. 2016c).

**Characteristics:** Hispid pocket gopher is the largest sized; total length 292.0–335.0 mm in females and 309.0–343.0 in males, skull 57.0–59.5 mm in females and 58.5–66.5 mm in males. Dorsal pelage reddish to reddish brown **with hair strong, coarse, short, and extremely sparse**; fur of underparts paler than in the dorsum; tail naked and feet with only sparse hairs; **adult body mass 400–650 g**; third upper molars without two lobes, outer lingual angle not well developed, and upper premolars never with an enamel plate in the lingual region.

**Comments:** *Heterogeomys hispidus* was previously regarded as a species of *Orthogeomys* (Spradling et al. 2016). Its ranges are patchily distributed along the Gulf of México from southern Tamaulipas, eastern San Luis Potosí, and Querétaro south through Central America, including the Yucatán Peninsula, generally at low elevations but can be found at higher elevations in mountains of eastern Querétaro and central Chiapas (Map 333). *H. hispidus* can be distinguished from *H. lanius* mainly by having the dorsal pelage from reddish to reddish-brown with hair strong, coarse, short, and extremely sparse. From species of *Cratogeomys*, by having the third upper molars without two lobes, outer lingual angle not well developed, and upper premolars never with an enamel plate in the lingual region.

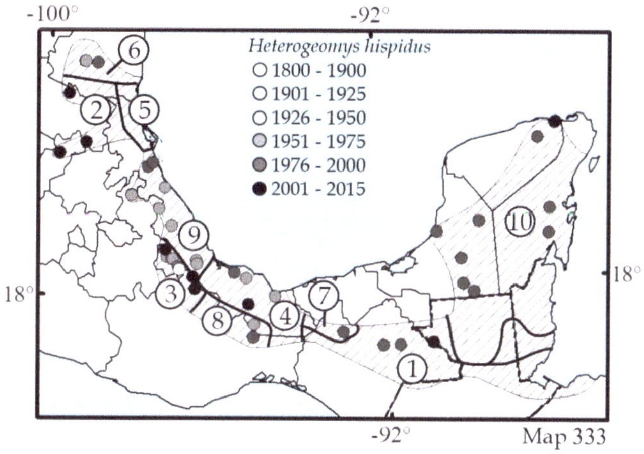


**Map 333** Distribution map of *Heterogeomys hispidus*

**Additional Literature:** Nelson and Goldman (1929), Russell (1968a), Hall (1981), Patton (2005b).

## *Heterogeomys lanius* Elliot, 1905

Big pocket gopher, tuza gigante de Veracruz

Monotypic.

**Conservation:** *Heterogeomys lanius* is listed as Threatened by the Norma Oficial Mexicana (DOF 2019) and as Critically Endangered (Possibly Extinct; B1ab [iii, v]: extent of occurrence less than 100 $km^2$ in only one population and continuing decline in quality of habitat and number of mature individuals) by the IUCN (Vázquez 2017b).

**Characteristics:** Big pocket gopher is large sized; **total length 361.0–460.0 mm and skull length 67.4–75.3 mm**. Dorsal pelage brown with **hair soft, thick, and woody, almost black**; fur of underparts paler than in the dorsum; tail blackish and almost naked; **adult body mass 900 g**.

**Comments:** *Heterogeomys lanius* was previously a species of *Orthogeomys* (Spradling et al. 2016). *H. lanius* is known only from two localities at the southeastern slope of Pico de Orizaba, Veracruz (Map 334). *H. lanius* can be distinguished from *H. hispidus* mainly by its dorsal pelage brown with fur soft, thick, and woody, almost black. From species of *Cratogeomys*, by having the third upper molars without two lobes, the outer lingual angle not well developed, and the upper premolars never with an enamel plate in the lingual region.

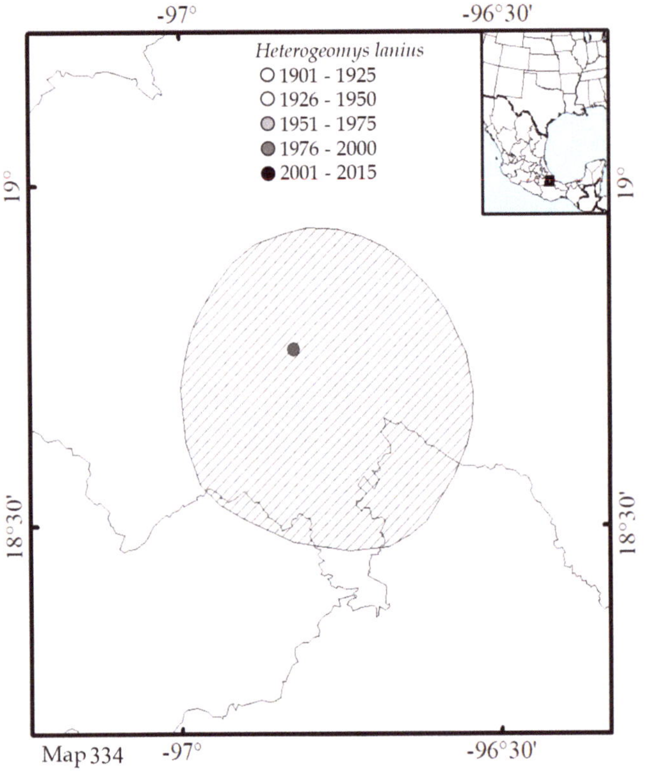


**Map 334** Distribution map of *Heterogeomys lanius*

**Additional Literature:** Nelson and Goldman (1929), Russell (1968a), Hall (1981), Patton (2005b).

## Genus *Orthogeomys* Merriam, 1895

Pelage coarse, in many instances hispid; individuals with so few hairs that they appear naked; **upper incisor with a sulcus slightly medial to the mid-line, touching the middle in some specimens; enamel plate on the posterior wall of upper premolars usually absent, although a small plate restricted to the lingual end of the wall rarely present**; **zygomatic breadth mainly slightly larger than mastoid breadth;** frontal wide and much inflated, no interorbital constriction.

### *Orthogeomys grandis* (Thomas, 1893)

Giant pocket gopher, tuza gigante del Pacífico

1. *O. g. alleni* Nelson and Goldman, 1930. From the southeastern Jalisco coastal plains through central Oaxaca.
2. *O. g. alvarezi* Schaldach, 1966. Restricted to San Gabriel Mixtepec, Oaxaca.
3. *O. g. annexus* Nelson and Goldman, 1933. Restricted to Tuxtla Gutiérrez.

4. *O. g. carbo* Goodwin, 1956. Restricted to the central coast of Oaxaca.
5. *O. g. felipensis* Nelson and Goldman, 1930. Restricted to the Central Valleys of Oaxaca.
6. *O. g. guerrerensis* Nelson and Goldman, 1930. Restricted to the central lowlands of Guerrero.
7. *O. g. huixtlae* Villa R., 1944. Restricted to southeastern Chiapas.
8. *O. g. nelsoni* Merriam, 1895. Restricted to Sierra Norte of Oaxaca.
9. *O. g. scalops* (Thomas, 1894). Restricted to the Isthmus of Tehuantepec, Oaxaca.
10. *O. g. soconuscensis* Villa R., 1949. Restricted to the coast of Chiapas.

**Conservation:** *Orthogeomys cuniculus* is listed as Threatened in the Official Mexican Standard (DOF 2019). The current taxonomical analysis considered *O. cuniculus* as a junior synonym of *Orthogeomys grandis* (Spradling et al. 2016); species with a wide distribution from Jalisco to El Salvador. Given this range, *O. g. cuniculus* should be unlisted from the DOF (2010). The IUCN listed *Orthogeomys grandis* as Least Concern (Vázquez et al. 2016a).

**Characteristics:** Giant pocket gopher **is the largest member of the family**; total length 314.0–390.0 mm in females and 366.0–435.0 mm in males, skull length 63.1–72.0 mm. Dorsal pelage reddish cinnamon to brownish; fur of underparts paler; body compact and cylindrical; hair sparse and bristly; skull large and massive; ridge on the squamosal joining the temporal in adult and old males; rostrum narrower than the interorbital region; **anterior surface of upper incisors with one deep groove**; last upper molars semi-lobular, only one labial groove; an enamel plate covers the front and reentrant angle edge of the first upper and lower molariforms; the first lower molariforms with an enamel plate and the first upper molariforms without a small plate on the lingual side.

**Comments:** *Orthogeomys cuniculus* was considered a junior synonym of *O. grandis* (Spradling et al. 2016). Its ranges are patchily distributed along the Pacific coast of México, from Jalisco south through Central America, including southwestern Puebla and central Oaxaca (Map 335). *O. grandis* can be distinguished from all the other species of the family Geomyidae by its larger size (greater than 350.0 mm), great body mass (greater than 800 g), and the combination of the following characteristics: third upper molars clearly with two lobes; upper premolars without a posterior enamel plate, and upper incisor anterior surface with one groove.

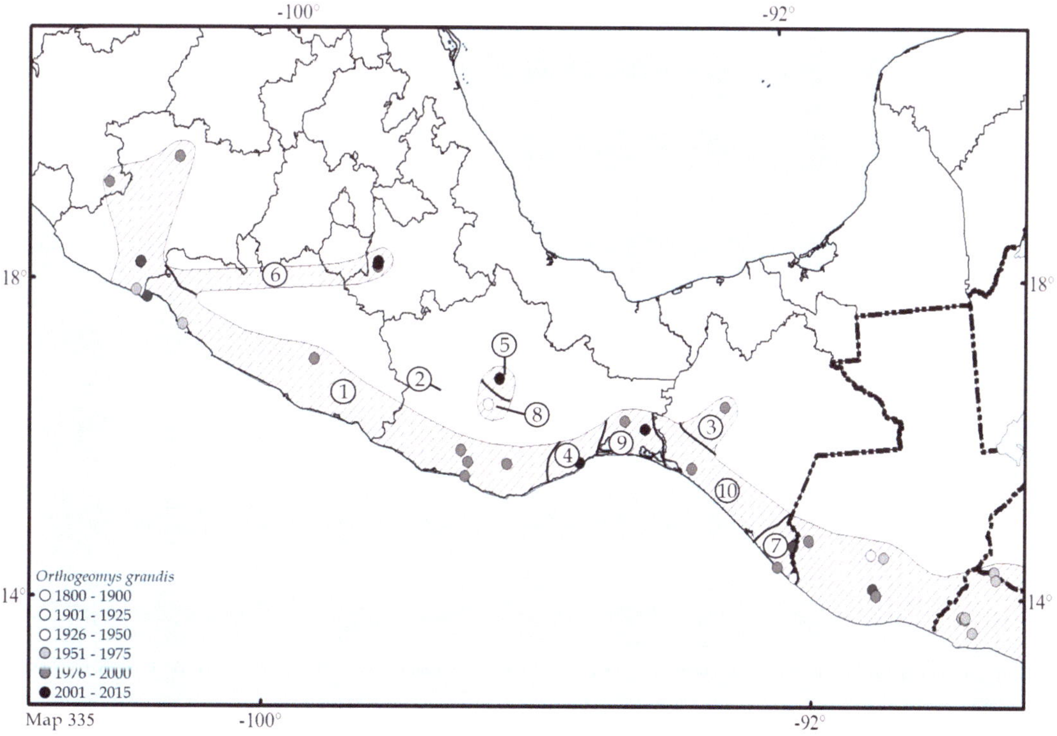


**Map 335** Distribution map of *Orthogeomys grandis*

**Additional Literature:** Nelson and Goldman (1929), Russell (1968a), Patton (2005b).

## Genus *Pappogeomys* Merriam, 1895

**Upper incisors without sulcus;** an **enamel plate on the posterior wall of first upper molars complete; claws on forefeet larger relative to body size.** Pelage long (approx. 10 mm.) and soft, covering all the body.

### *Pappogeomys bulleri* (Thomas, 1892)

Buller's pocket gopher, tuza de Jalisco

1. *P. b. albinasus* Merriam, 1895. Restricted to central Jalisco and southeastern Nayarit.
2. *P. b. alcorni* Russell, 1957. Eastern and central Jalisco.
3. *P. b. bulleri* (Thomas, 1892). Restricted to the Sierra Madre del Sur highlands, Jalisco.
4. *P. b. burti* Goldman, 1939. Restricted to coastal areas and lowlands of Colima and Jalisco.
5. *P. b. nayaritensis* Goldman, 1939. Restricted to lowlands of Nayarit.

**Conservation:** *Pappogeomys bulleri alcorni* is listed as Special Protection in the Norma Oficial Mexicana (DOF 2019). The IUCN lists *P. b. alcorni* as Critically Endangered (A2bc: population reduction in the last 10 years >80% or may not be reversible; data obtained from the index of abundance and decline in area of occupancy; Castro-Arellano and Vázquez 2016a). The current taxonomical analyses confirmed that *P. b. amecensis*, *P. b. flammeus*, *P. b. lagunensis*, and *P. b. lutulentus* need to be considered synonyms under *P. b. alcorni* (Hafner et al. 2009). This taxonomic change expanded the distribution area of *P. b. alcorni* from the type locality in the mountains south of Laguna de Chapala to all eastern and central Jalisco. Based on the above, *P. b. alcorni* should be unlisted from the Norma Oficial Mexicana (DOF 2019) and the IUCN Red List.

**Characteristics:** Buller's pocket gopher is medium sized; total length 200.0–249.0 mm and skull length 39.2–42.9 mm. Dorsal pelage bicolored, pale to dark gray at the base and ocher, tawny, cinnamon at the tip or including blackish shades in melanic specimens; fur of underparts paler than in the dorsum; tail short and slightly darker dorsally; body compact and cylindrical; skull large and massive; **rostrum wider than the interorbital region**; incisors without grooves in the frontal phase; without a zygomatic anterior angle with a plate-shaped expansion.

**Comments:** *Cratogeomys* was considered a subgenus of *Pappogeomys* (Russell 1968a), nowadays a full genus (Honeycutt and Williams 1982; Demastes et al. 2002). The genetic review of the *Pappogeomys bulleri* complex (Hafner et al. 2009) considered that the previous subspecies of *P. bulleri* are junior synonyms of the following subspecies: *P. b. albinasus* includes *P. b. infuscus* and *P. b. nayaritensis* (part); *P. b. bulleri* includes *P. b. amecensis*, *P. b. flammeus*, *P. b. lagunensis*, and *P. b. lutulentus*; and *P. b. burti* includes *P. b. melanurus*. *P. bulleri* is known from Nayarit, Jalisco, and Colima (Map 336), and differs from all the other species of the family Geomyidae by the following characteristics: rostrum wider than the interorbital region; without a zygomatic anterior angle with a plate-shaped expansion, and third upper molars not clearly showing two lobes.

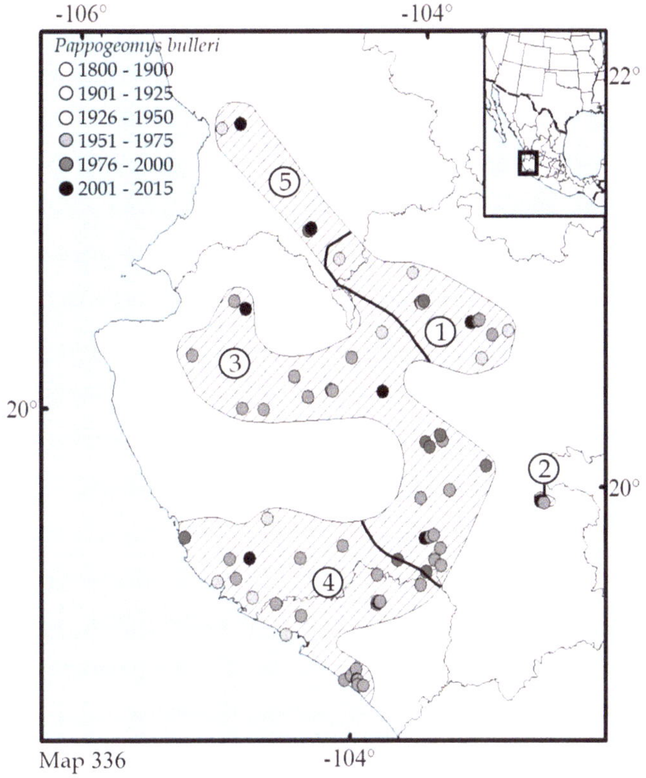


**Map 336** Distribution map of *Pappogeomys bulleri*

**Additional Literature:** Merriam (1895a), Russell (1968b), Hall (1981), DeWalt et al. (1993b), Desmastes et al. (2003), Soler–Frost et al. (2003).

## Genus *Zygogeomys* Merriam, 1895

Body compact and cylindrical; eyes, ears, and limbs small; fore and hindfoot claws well developed; tail short and slightly darker dorsally; **upper incisors bisulcate, a major sulcus on the inner side of the median line and a minor sulcus on the inner convexity**; **third upper molars conspicuously bicolumnar, longer than wide owing to the elongation of the posterior loph**; rostrum narrow relative to its length; maxillary and squamosal roots of the zygomatic arches in contact above the jugal, and anteroextemal angles rounded rather than expanded; zygomata not widely spreading and slender; sagittal crest short but well developed.

### *Zygogeomys trichopus* Merriam, 1895

Michoacán pocket gopher, tuza de Michoacán

1. *Z. t. tarascensis* Goldman, 1938. Only known from Pátzcuaro, Michoacán.
2. *Z. t. trichopus* Merriam, 1895. Only known from Nahuatzen, Cerros Tancítaro, and Patambán, Michoacán.

**Conservation:** *Zygogeomys trichopus* is listed as Endangered by the Official Mexican Standard (DOF 2019) and as Endangered (B1ab [iii, v]: extent of occurrence less than 5000 km$^2$ in five or less populations and continuing decline in quality of habitat and number of mature individuals) by the IUCN (Álvarez-Castañeda et al. 2018k). However, it is considered as pest in the avocado fields, and they are under control.

**Characteristics:** Michoacán pocket gopher is medium sized; total length 292.0–322.0 mm in females and 343.0–346.0 mm in males, skull length 50.0–61.0 mm. **Dorsal pelage grayish-black and silky**; fur of underparts paler than in the dorsum; tail short and slightly darker dorsally; body compact and cylindrical; eyes, ears, and limbs small; fore and hindfoot claws well developed; skull large and massive; rostrum wider than the interorbital region; **upper incisors with two grooves in the middle part of the frontal face**; first upper molars with an enamel plate restricted to the lingual side; **last upper molariforms longer than wider, with two separate prisms on both molar sides**.

**Comments:** *Zygogeomys trichopus* is known from four areas in Michoacán: Nahuatzen, Pátzcuaro, Cerros Tancítaro, and Patambán (Map 337). *Z. trichopus* can be found in sympatry, or nearly so, with *Cratogeomys fumosus*, which is normally more common and abundant. In those areas, the tunnels of *Zygogeomys* are deeper and larger in diameter. *Z. trichopus* can be distinguished from *Cratogeomys* and *Thomomys* by having two grooves in the upper incisor anterior surface; however, the internal grooves are not highly notorious, so their presence is best detected by passing a pencil tip or a thumb nail on the incisor front face.

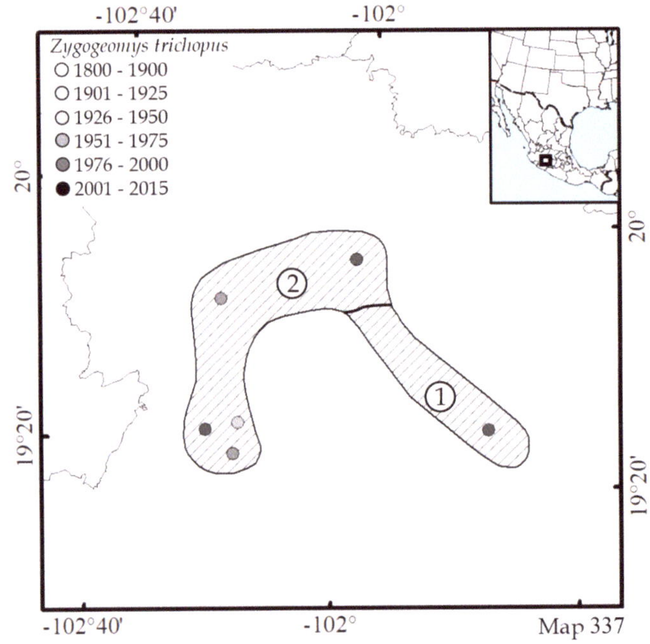


**Map 337** Distribution map of *Zygogeomys trichopus*

**Additional Literature:** Merriam (1895a), Russell (1968a), Hall (1981), Patton (1993a, b, 2005a).

## Tribe Thomomyini

### Genus *Thomomys* Wied-Neuwied, 1839

Dorsal pelage pigmented from almost white to dark color; fur of underparts paler than in the dorsum; tail short and slightly darker dorsally; body compact and cylindrical; eyes, ears, and limbs small; **forefeet small and narrow, with claws well developed**; **six or eight nipples;** skull large and massive; no ridge on the squamosal but a sagittal crest mainly in adult and old males; **anterior surface of the upper incisors smooth, without grooves;** all molariforms monoprismatic and with a tendency to an elliptical shape, with small anterior–posterior axes; **upper incisors slightly procumbent; basitemporal fossa absent.** The genus *Thomomys* included twsubgenera: *Megascapheus* with two species groups *bulbivorus* (*T. bulbivorus*) and *umbrinus* (*T. atrovarius*, *T. bottae*, *T. bulbivorus*, *T. fulvus*, *T. laticeps*, *T. nayarensis*, *T. nigricans*, *T. sheldoni*, *T. townsendii*, and *T. umbrinus*); for the subgenus *Thomomys*, only the species group *talpoides* (*T. clusius*, *T. idahoensis*, *T. mazama*, *T. monticola*, and *T. talpoides*). The keys were elaborated based on the review of specimens and the following sources: Hall (1981), Álvarez-Castañeda (2010), Hafner et al. (2011), Mathis et al. (2013a, b), and Álvarez-Castañeda et al. (2017a).

1. No procumbent upper incisors, root above the fourth upper premolars; sphenoidal fissure absent, in some specimens, a foramen present; angular process not continuous, with a weakly developed flange along the ventral ramus side-............ ..........subgenus *Thomomys* 2

1a Procumbent upper incisors, root between the fourth upper premolars and the first upper molars; sphenoidal fissure present; angular process continuous, with a well-developed flange along the ventral ramus side............ subgenus *Megascapheus* 6

2. Small ears, ear notch height less than 6.9 mm............3

2a. Large ears, ear notch height greater than 6.9 mm............5

3. Dark auricular patch large and extending dorsally; ears without paler edges, similar in color to the auricular patch; posterior braincase portion projecting well beyond the lamboidal crest (southern British Columbia, Alberta, Saskatchewan, and Manitoba south through California, Nevada, Arizona, and New Mexico, and east through North Dakota, South Dakota, Nebraska, and Colorado)............*Thomomys talpoides* (p. 490)

3a. Dark auricular patch very small or absent; ears paler at the edges; posterior braincase portion projecting beyond the lamboidal crest. Restricted to Idaho, Montana, and Wyoming............4

4. Dark auricular patch very small and not extending dorsally; ear edges paler than the auricular patch but similar to the dorsal coloration (eastern and southeastern Idaho, southwestern Montana, western Wyoming, and northern Utah)............ ............*Thomomys idahoensis* (p. 486)

4a. Dark auricular patch absent; ear edges buff-colored and usually paler than the dorsal pelage (restricted to southern-central Wyoming)............*Thomomys clusius* (p. 485)

5. Ear rounded and less than 7.0 mm; subauricular patch dark, five to six times larger than the ears; pale coloration; premaxilar extending 1.5–2.5 mm behind nasals; nasals truncated or round posteriorly (western Washington, western Oregon, and northwestern California)............*Thomomys mazama* (p. 487)

5a. Ears pointed and greater than 7.0 mm; subaricular patch dark, never more than four times larger than ears; dark coloration; premaxilar shorter than nasals or extending at most 0.5 mm behind nasals; nasals V shaped posteriorly (northeastern California and western Nevada)............*Thomomys monticola* (p. 489)

6. Hindfoot length equal to or greater than 34.0 mm............7

6a. Hindfoot length equal to or less than 33.0 mm............8

7. Hindfoot length equal to or greater than 40.0 mm in males and equal to or greater than 38.0 mm in females; dorsal pelage sooty brown; fur of underparts with lead coloration; white throat patch irregular; pterygoids concave on the inner surface and convex on the outer surface; two grooves on the exoccipital (restricted to western Oregon, mainly to Willamette Valley)............*Thomomys bulbivorus* (p. 474)

7a. Hindfoot length less than 40.0 mm in males and less than 38.0 mm in females; dorsal pelage grayish to pale grayish brown; fur of underparts slightly paler; white throat patch absent but present on the chin or feet; pterygoids straight or nearly so; no groove on the exoccipital (northeastern California, northern Nevada, western Oregon, and central Idaho) ............ *Thomomys townsendii* (p. 483)

8. Four pairs of mammae, two pectoral and two inguinal pairs; skull short and wide; nasals slightly tapered and gradually narrowed posteriorly; maxilla-frontal suture concave and lacrimal abuts near the center; zygomatic arches heavy with maxillary expanded to form an external angle near its contact with the jugal............9

8a. Three pairs of mammae, one pectoral and two inguinal pairs; skull not conspicuously short and wide; nasals wedge shaped and tapered posteriorly; maxilla-frontal suture convex and lacrimal abuts in the medial half; zygomatic arches slender and nearly parallel............11

9. Upper incisors usually strongly procumbent (extending beyond the front nasal edge); frontal part of nasals wedge shaped; zygomatic plate thin in side view, almost same height as the jugal (Fig. 96; southern California, and all Baja California Peninsula)............*Thomomys nigricans* (p. 480)

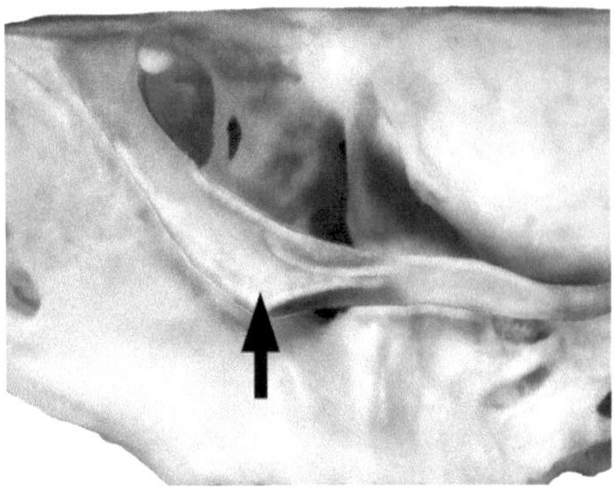

**Fig. 96** Zygomatic plate thin in side view, almost the same height as the jugal

9a. Upper incisors slightly procumbent (usually not extending beyond the front nasal edge); frontal part of nasals slightly rounded; zygomatic plate thick in side view (a widening of the zygomatic bone almost in contact with the jugal), more than twice as high as the jugal (Fig. 97) .......... 10

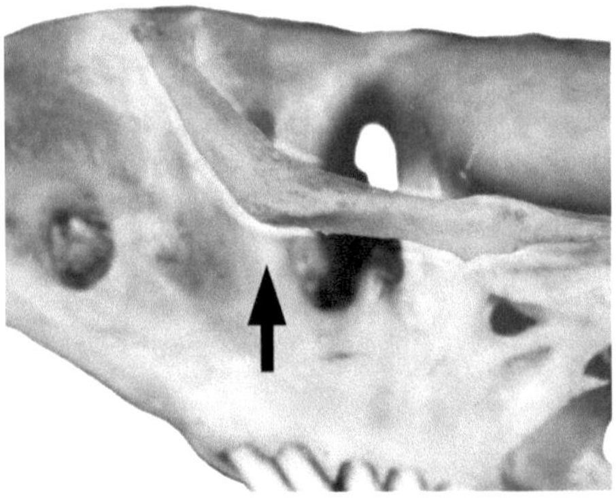

**Fig. 97** Zygomatic plate thick in side view (a widening of the zygomatic bone almost in contact with the jugal), more than twice as high as the jugal

10. Restricted to southern Oregon, California, Nevada, Utah, Colorado, Arizona, New Mexico, southwestern Texas, northwestern Baja California Peninsula, northern-central Sonora, northern Chihuahua, different areas in Coahuila, and northern Zacatecas .......... *Thomomys bottae* (p. 472)

10a. Ranges in southern Oregon and northern California, including the Sacramento Valley and adjacent foothills of the Sierra Nevada, Pacific coast and coastal ranges and the eastern slope of Sierra Nevada .......... *Thomomys laticeps* (p. 478)

11. Restricted to southern Arizona, New Mexico, Texas, Sonora, northwestern Sinaloa, central Chihuahua, and central Coahuila .......... *Thomomys fulvus* (p. 475)

11a. Ranges throughout México .......... 12

12. Hair moderately dense to scarce; dorsal coloration usually of the same shade as flanks and underparts (northeastern Sinaloa south through northwestern Jalisco) .......... *Thomomys atrovarius* (p. 471)

12a. Hair dense; dorsal coloration generally darker than in the flanks, underparts paler (Eje Neovolcánico northward) .......... 13

13. Ranges in the eastern slope of Sierra Madre Occidental, Altiplano Mexicano, and Eje Neovolcánico (southern Arizona south through Veracruz and Puebla) .......... *Thomomys umbrinus* (p. 484)

13a. Ranges in the central region of Sierra Madre Occidental from western Sonora to Nayarit, generally over 2000 m ........................................................................................ 14

14. Back pelage dense; total length generally greater than 180.0 mm; auditory meatus maximum height equal to or greater than 1.5 mm in the dorsal–ventral plane (restricted to western-central Chihuahua southward through western Zacatecas and Jalisco)........................................ *Thomomys sheldoni* (p. 481)

14a. Back pelage not dense; total length generally less than 180.0 mm; auditory meatus maximum height less than 1.5 mm in the dorsal–ventral plane (endemic to Sierra de Nayar, Nayarit)..................................*Thomomys nayarensis* (p. 479)

## Subgenus *Megascapheus* Elliot, 1903

**Rostrum heavy;** procumbent upper incisors, root between fourth upper premolars and first upper molars; based of first lower premolars inclined anteriorly; infraorbital canal openings anterior to the incisive foramina; **sphenoidal fissure open;** anterior enamel plate of the first lower premolars not recurved; the **angular process not broadly continuous with a weakly developed flange along the ventral side of the mandibular ramus; the** anterior enamel plate of the first lower premolars narrow and broadly separated from the lateral enamel plate on the lingual side; chromosome number of living forms from 74 to 82.

### *Thomomys atrovarius* Allen, 1898

Southern pocket gopher, tuza de Sinaloa

1. *T. a. atrovarius* Allen, 1898. From the central Sinaloa coast south through northwestern Jalisco.
2. *T. a. parviceps* Nelson and Goldman, 1934. Restricted to central and northeastern Sinaloa and western Durango.
3. *T. a. simulus* Nelson and Goldman, 1934. Restricted to southeastern Sonora and northeastern Sinaloa.
4. *T. a. sinaloae* Merriam, 1901. Coastal central and northern Sinaloa.

**Conservation:** *Thomomys atrovarius* (as *T. bottae*) is listed as Least Concern by the IUCN (Lacher et al. 2016o).

**Characteristics:** Sinaloa pocket gopher is small sized; total length 210.0–235.0 mm and skull length 36.1–42.0 mm. Dorsal pelage dark-chocolate brown to dull gray-black, some specimens grizzle with hairs paler brown or gray on the flanks; fur of underparts paler than in the dorsum, in grayish shades; tail short and slightly darker dorsally; six nipples and one pair of pectoral mammae; sphenoidal fissure present; suture between the frontal and nasal bones wedge shaped; zygomatic plate thin in side view, almost the jugal height; **upper incisors usually strongly procumbent and extending anteriorly well beyond the nasal ends**. Species group *umbrinus*.

**Comments:** *Thomomys atrovarius* was recognized as a distinct species (Álvarez-Castañeda 2010; Hafner et al. 2011), previously regarded as a subspecies of *T. bottae* (Patton 2005b). *T. umbrinus atrovarius*, *T. u. musculus*, and *T. u. eximius* (part, lowlands) are considered synonyms of *T. atrovarius atrovarius*. *T. umbrinus parviceps* is still considered subspecies as *T. atrovarius parviceps* (Álvarez-Castañeda 2010, Hafner et al. 2011). *T. bottae simulus* and *T. b. sinaloe* have not been assigned to *T. atrovarius*; however, based on their range, I consider them to be populations of *T. atrovarius*. *T. atrovarius* is restricted to the lowlands from northeastern Sinaloa south through northwestern Jalisco, including Nayarit (Map 338). *T. atrovarius* is distinguished from all other *Thomomys* species by its upper incisors usually strongly procumbent (tips of the incisors usually extending anteriorly well beyond the nasal ends) and one pair of pectoral mammae. From *T. umbrinus* and *T. sheldoni*, by having the pelage moderately dense to sparse, sometimes almost naked on the belly; mid-dorsal coloration darker, flanks of the same color as the back or infused with a brownish or grayish wash, and ventral coloration similar to flanks. From *T. nayarensis*, by having the maximum auditory meatus height greater than 1.5 mm in the dorsal–ventral axis.

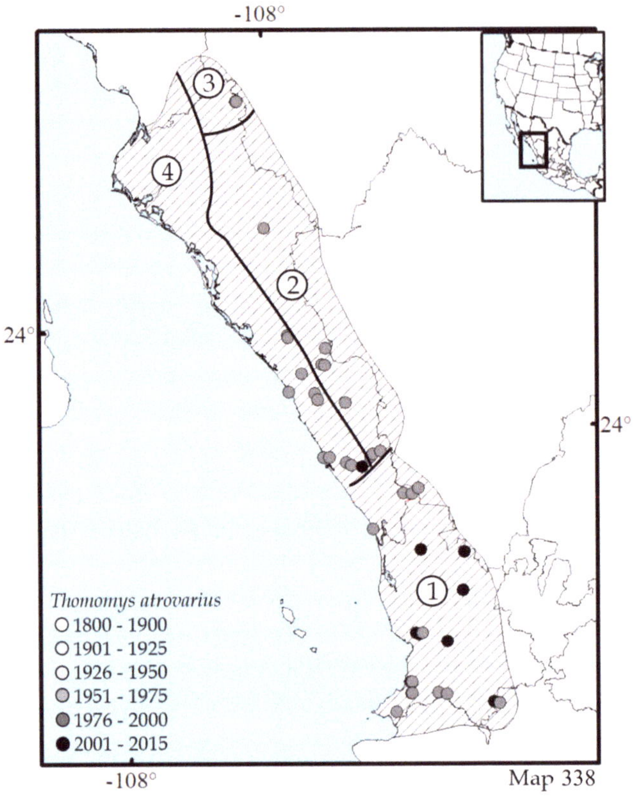


**Map 338** Distribution map of *Thomomys atrovarius*

**Additional Literature:** Nelson and Goldman (1934b), Patton and Smith (1989, 1994), Jones and Baxter (2004).

## *Thomomys bottae* (Eydoux and Gervais, 1836)

Botta's pocket gopher, tuza de abazones del norte

1. *T. b. bottae* (Eydoux and Gervais, 1836). Western coast of California from San Francisco Bay south through Ventura County.
2. *T. b. mewa* Merriam, 1908. Restricted to Maderas County, central California.
3. *T. b. navus* Merriam, 1901. Restricted to Tehama County, northern California.
4. *T. b. pascalis* Merriam, 1901. Restricted to Fresno County, central California.

**Conservation:** *Thomomys bottae* is listed as Least Concern by the IUCN (Lacher et al. 2016o).

**Characteristics:** Botta's pocket gopher is small sized; total length 204.2–233.0 mm and skull length 36.3–40.9 mm. Dorsal pelage almost white or pale yellow to blackish; fur of underparts paler than in the dorsum; tail short and slightly darker dorsally**; eight nipples**; sphenoidal fissure present; suture between the frontal and nasal bones slightly rounded; the zygomatic plate thick in side view and more than twice the jugal height. Species group *umbrinus*.

**Comments:** *Thomomys bottae* was previously considered a junior synonym of *T. umbrinus* (Hall and Kelson 1959; Hall 1981); recognized as a distinct species by Anderson (1966, 1972), Patton and Dingman (1968), Hoffmeister (1969, 1986), Patton (1973), Patton and Smith (1981). The subspecies *Thomomys bottae acrirostratus*, *T. b. agricolaris*, *T. b. awahnee*, *T. b. detumidus*, *T. b. laticeps*, *T. b. leucodon*, *T. b. saxatilis*, and *T. b. silvifugus* were recognized as a distinct species, using molecular data, as *Thomomys laticeps* (Álvarez-Castañeda 2010). *T. b. abbotti*, *T. b. albatus*, *T. b. alticolus*, *T. b. anitae*, *T. b. aphrastus*, *T. b. borjasensis*, *T. b. brazierhowelli*, *T. b. cactophilus*, *T. b. catavinensis*, *T. b. cunicularis*, *T. b. homorus*, *T. b. imitabilis*, *T. b. incomptus*, *T. b. jojobae*, *T. b. juarezensis*, *T. b. litoris*, *T. b. lucidus*, *T. b. magdalenae*, *T. b. martirensis*, *T. b. nigricans*, *T. b. proximarinus*, *T. b. puertae*, *T. b. rhizophagus*, *T. b. ruricola*, *T. b. russeolus*, *T. b. sanctidiegi*, *T. b. siccovallis*, and *T. b. xerophilus* were recognized as populations of *Thomomys nigricans* (Álvarez-Castañeda 2010). *T. b. alpinus*, *T. b. operarius*, *T. b. perpallidus*, *T. b. riparius*, *T. b. absonus*, *T. b. abstrusus*, *T. b. actuosus*, *T. b. albicaudatus*, *T. b. alexandrae*, *T. b. alienus*, *T. b. analogus*, *T. b. angustidens*, *T. b. apache*, *T. b. aridicola*, *T. b. aureiventris*, *T. b. aureus*, *T. b. baileyi*, *T. b. basilicae*, *T. b. birdseyei*, *T. b. bonnevillei*, *T. b. boreorarius*, *T. b. brevidens*, *T. b. camoae*, *T. b. caneloensis*, *T. b. canus*, *T. b. carri*, *T. b. catalinae*, *T. b. cervinus*, *T. b. chrysonotus*, *T. b. cinereus*, *T. b. collinus*, *T. b. collis*, *T. b. comobabiensis*, *T. b. concisor*, *T. b. confinalis*, *T. b. connectens*, *T. b. contractus*, *T. b. convergens*, *T. b. convexus*, *T. b. cultellus*, *T. b. curtatus*, *T. b. depauperatus*, *T. b. depressus*, *T. b. desertorum*, *T. b. desitus*, *T. b. dissimilis*, *T. b. divergens*, *T. b. estanciae*, *T. b. extenuatus*, *T. b. flavidus*, *T. b. fulvus*, *T. b. fumosus*, *T. b. grahamensis*, *T. b. growlerensis*, *T. b. guadalupensis*, *T. b. harquahalae*, *T. b. howelli*, *T. b. hualpaiensis*, *T. b. hueyi*, *T. b. humilis*, *T. b. internatus*, *T. b. lachuguilla*, *T. b. lachuguilla*, *T. b. lachuguilla*, *T. b. lacrymalis*, *T. b. latirostris*, *T. b. latus*, *T. b. lenis*, *T. b. levidensis*, *T. b. limitaris*, *T. b. limpiae*, *T. b. lucrificus*, *T. b. mearnsi*, *T. b. minimus*, *T. b. modicus*, *T. b. morulus*, *T. b. muralis*, *T. b. mutabilis*, *T. b. nanus*, *T. b. nasutus*, *T. b. nesophilus*, *T. b. nicholi*, *T. b. operosus*, *T. b. optabilis*, *T. b. opulentus*, *T. b. osgoodi*, *T. b. paguatae*, *T. b. parvulus*, *T. b. patulus*, *T. b. pectoralis*, *T. b. peramplus*, *T. b. perditus*, *T. b. pervagus*, *T. b. pervarius*, *T. b. phasma*, *T. b. phelleoecus*, *T. b. pinalensis*, *T. b. planirostris*, *T. b. planorum*, *T. b. powelli*, *T. b. proximus*, *T. b. pusillus*, *T. b. retractus*, *T. b. robertbakeri*, *T. b. robustus*, *T. b. rubidus*, *T. b. rufidulus*, *T. b. ruidosae*, *T. b. scotophilus*, *T. b. sevieri*, *T. b. solitarius*, *T. b. spatiosus*, *T. b. stansburyi*, *T. b. sturgisi*, *T. b. suboles*, *T. b. subsimilis*, *T. b. texensis*, *T. b. tivius*, *T. b. toltecus*, *T. b. trumbullensis*, *T. b. tularosae*, *T. b. vanrossemi*, *T. b. vescus*, *T. b. villai*, *T. b. virgineus*, *T. b. wahwahensis*, and *T. b. winthropi*, as populations of *Thomomys fulvus* (Álvarez-Castañeda 2010). Patton and Smith (1990) considered some subspecies as junior synonyms: *T. b. altivallis*, *T. b. angularis*, *T. b. argusensis*, *T. b. diaboli*, *T. b. infrapallidus*, *T. b. lorenzi*, *T. b. neglectus*, *T. b. pallescens*, *T. b. perpes*, *T. b. piutensis*, *T. b. sanctidiegi*, and *T. b. scapterus* as *T. b. bottae*; *T. b. ingens* as *T. b. pascalis; T. b. melanotis*, *T. b. mohavensis*, *T. b. amargosae*, *T. b. oreoecus*, and *T. b. providentialis* as *T. b. perpallidus. T. bottae* ranges from southern Oregon, California, Nevada, Utah, Colorado, Arizona, New Mexico, southwestern Texas, northwestern Baja California, northern-central Sonora, northern Chihuahua, different areas in Coahuila, and northern Zacatecas (Map 339). *T. bottae* can be distinguished from *T. mazama*, *T. monticola*, and *T. talpoides* by not having the sphenoidal fissure. From *T. fulvus*, *T. laticeps*, and *T. nigricans*, only using genetic data. From *T. umbrinus*, by having generally four pairs of mammae (nipples); two pectoral and two inguinal pairs; skull short and wide; nasals slightly tapered and gradually narrowed posteriorly; maxilla-frontal suture concave; zygomatic arches heavy with maxillary expanded to form the external angle near its contact with the jugal.

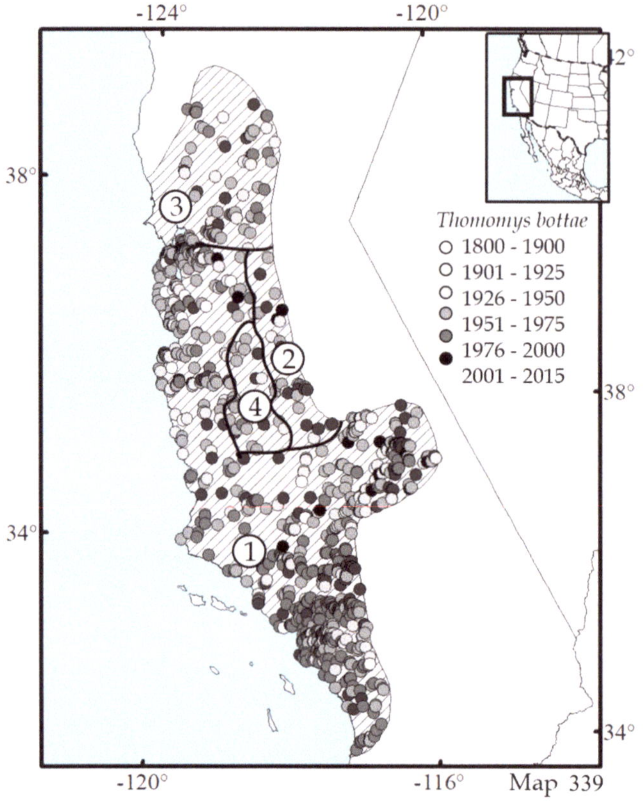


**Map 339** Distribution map of *Thomomys bottae*

**Additional Literature:** Merriam (1895a), Bailey (1915), Russell (1968a), Thaeler (1968a, 1980), Patton (1972), Patton et al. (1984), Patton and Smith (1989, 1991, 1994), Jones and Baxter (2004), Patton (2005b), Hafner et al. (2011).

## *Thomomys bulbivorus* (Richardson, 1829)

Camas pocket gopher, tuza del Valle de Camas

Monotypic.

**Conservation:** *Thomomys bulbivorus* is listed as Least Concern by the IUCN (Cassola 2016dw).

**Characteristics:** Camas pocket gopher is one of the largest members of the genus *Thomomys*; total length 290.0–345.6 mm and skull length 42.8–58.2 mm. Dorsal pelage dark sooty brown; fur of underparts dark lead gray, with an irregular white patch on the throat; ears and nose blackish; eight mammae, two pectoral and two inguinal pairs; skull short and wide; sphenoidal fissure present; **incisors highly procumbent, tip angle distinctly forward**; zygomatic arches usually wider posteriorly; pterygoids convexly inflated and divided by a narrow interpterygoid space. Species group *bulbivorus*.

**Comments:** *Thomomys bulbivorus* has always been a valid species. *T. bulbivorus* is restricted to the Willamette Valley, Oregon (Map 340) and can be distinguished from *T. talpoides* in sympatry, or nearly so, by its larger size, greater than 260.0 mm; sphenoidal fissure present; pterygoids concave on the inner surface and convex on the outer surface; one and two grooves on the exoccipital; incisors highly procumbent and the tip angle distinctly forward.

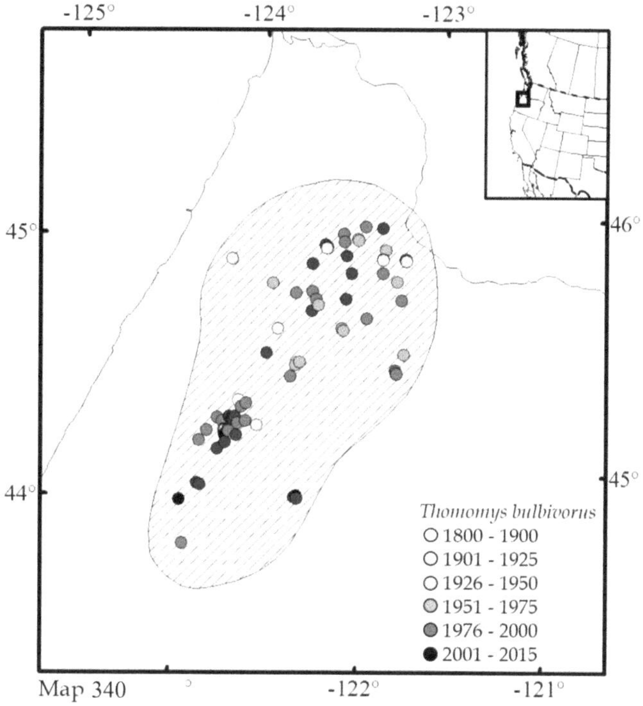


**Map 340** Distribution map of *Thomomys bulbivorus*

**Additional Literature:** Bailey (1915), Russell (1968a), Thaeler (1980), Verts and Carraway (1987a), Patton and Smith (1989, 1994), Carraway and Kennedy (1993), Merriam (1895a), Patton (1999a, 2005b).

## *Thomomys fulvus* (Woodhouse, 1852)

Fulvus pocket gopher, tuza del suroeste

1. *T. f. abstrusus* Hall and Davis, 1935. Restricted to Nye County, southern Nevada.
2. *T. f. actuosus* Kelson, 1951. Restricted to central New Mexico.
3. *T. f. albicaudatus* Hall, 1930. Restricted to Utah County, central Utah.
4. *T. f. alexandrae* Goldman, 1933b. Restricted to Coconino County, northern Arizona.
5. *T. f. alpinus* Merriam, 1897. Restricted to Mount Whitney, Inyo County, western California.
6. *T. f. analogus* Goldman, 1938a. Southeastern Coahuila, southwestern Coahuila, and Nuevo León.
7. *T. f. angustidens* Baker, 1953. Restricted to Sierra del Pino, Coahuila.
8. *T. f. apache* Bailey, 1910. Restricted to Sandoval County, northern New Mexico.
9. *T. f. aureiventris* Hall, 1930. Restricted to Box Elder County, northern Utah.
10. *T. f. aureus* Allen, 1893. Restricted to San Juan County, southwestern Utah.
11. *T. f. basilicae* Benson and Tillotson, 1940. Restricted to central Sonora.
12. *T. f. birdseyei* Goldman, 1937a. Restricted to Washington County, southwestern Utah.
13. *T. f. bonnevillei* Durrant, 1946. Restricted to Juab County, western Utah.
14. *T. f. brevidens* Hall, 1932a. Restricted to Nye County, southern Nevada.
15. *T. f. camoae* Burt, 1937. Restricted to central-southern coast Sonora.
16. *T. f. canus* Bailey, 1910. Restricted to Smoke Creek Desert, Washoe County, northwestern Nevada.
17. *T. f. catalinae* Goldman, 1931. Restricted to Pima County, southern Arizona.
18. *T. f. cervinus* Allen, 1895. Restricted to Maricopa County, southwestern Arizona.
19. *T. f. cinereus* Hall, 1932a. Restricted to Lyon County, Nevada.
20. *T. f. collis* Hooper, 1940. Restricted to Valencia County, western New Mexico.
21. *T. f. concisor* Hall and Davis, 1935. Restricted to Nye County, southern Nevada.
22. *T. f. confinalis* Goldman, 1936. Restricted to Sutton County, southwestern Texas.

23. *T. f. connectens* Hall, 1936. Restricted to Bernalillo County, New Mexico.
24. *T. f. contractus* Durrant, 1946. Restricted to Millard County, western Utah.
25. *T. f. convergens* Nelson and Goldman, 1934. Restricted to the Sonora River delta, Sonora.
26. *T. f. convexus* Durrant, 1939. Restricted to Millard County, western Utah.
27. *T. f. cultellus* Kelson, 1951. Restricted to Mora County, northeastern New Mexico.
28. *T. f. curtatus* Hall, 1932a. Restricted to Nye County, southern Nevada.
29. *T. f. depressus* Hall, 1932a. Restricted to Churchill County, western Nevada.
30. *T. f. desertorum* Merriam, 1901. Restricted to Detrital Valley, Mohave Country, southeastern Arizona.
31. *T. f. dissimilis* Goldman, 1931. Restricted to Garfield County, southern Utah.
32. *T. f. divergens* Nelson and Goldman, 1934. Restricted to Huachinera, eastern Sonora.
33. *T. f. estanciae* Benson and Tillotson, 1939. Restricted to Nácori, eastern Sonora.
34. *T. f. fulvus* (Woodhouse, 1852). From central Arizona to western New Mexico.
35. *T. f. fumosus* Hall, 1932a. Restricted to Nye County, southern Nevada.
36. *T. f. howelli* Goldman, 1936. Restricted to Mesa County, western Colorado.
37. *T. f. humilis* Baker, 1953. Restricted to northern Coahuila.
38. *T. f. internatus* Goldman, 1936. Restricted to Chaffee County, southern Colorado.
39. *T. f. lachuguilla* Bailey, 1902. Southern New Mexico and western Texas.
40. *T. f. lacrymalis* Hall, 1932a. Restricted to Esmeralda County, western Nevada.
41. *T. f. latus* Hall and Davis, 1935. Restricted to White Pine County, eastern Nevada.
42. *T. f. lenis* Goldman, 1942. Restricted to Sevier County, central Utah.
43. *T. f. levidensis* Goldman, 1942. Restricted to Sanpete County, central Utah.
44. *T. f. limpiae* Blair, 1939. Restricted to Jeff Davis County, western Texas.
45. *T. f. lucrificus* Hall and Durham, 1938. Restricted to Churchill County, western Nevada.
46. *T. f. mearnsi* Bailey, 1914. Restricted to the southwest corner of New Mexico.
47. *T. f. minimus* Durrant, 1939. Restricted to Tooele County, northwestern Utah.
48. *T. f. modicus* Goldman, 1931. Restricted to northern Sonora.
49. *T. f. morulus* Hooper, 1940. Restricted to Cibola County, western New Mexico.
50. *T. f. nanus* Hall, 1932a. Restricted to Nye County, southern Nevada.
51. *T. f. nesophilus* Durrant, 1936. Restricted to Antelope Island, Davis County, northern Utah.
52. *T. f. operarius* Merriam, 1897. Restricted to Inyo County, eastern California.
53. *T. f. operosus* Hatfield, 1942. Restricted to Yavapai County, central Arizona.
54. *T. f. optabilis* Goldman, 1936. Restricted to Montrose County, western Colorado.
55. *T. f. opulentus* Goldman, 1935. Restricted to Sierra County, central New Mexico.
56. *T. f. osgoodi* Goldman, 1931. Restricted to Wayne County, central Utah.
57. *T. f. paguatae* Hooper, 1940. Restricted to Valencia County, western central New Mexico.
58. *T. f. peramplus* Goldman, 1931. Restricted to Apache County, northwestern Arizona.
59. *T. f. perditus* Merriam, 1901. From eastern Coahuila and western Nuevo León.
60. *T. f. perpallidus* Merriam, 1886. Restricted to Riverside County, southern California.
61. *T. f. pervagus* Merriam, 1901. Restricted to Rio Arriba and Santa Fe counties, northern New Mexico.
62. *T. f. phelleoecus* Burt, 1933. Restricted to Clark County, southern Nevada.
63. *T. f. pinalensis* Goldman, 1938b. Restricted to Gila County, central Arizona.
64. *T. f. planirostris* Burt, 1931. Restricted to Washington County, southwestern Utah.
65. *T. f. planorum* Hooper, 1940. Restricted to Valencia County, western-central New Mexico.
66. *T. f. powelli* Durrant, 1955. Restricted to Garfield County, southern Utah.
67. *T. f. pusillus* Goldman, 1931. Restricted to Pima County, southern Arizona.
68. *T. f. retractus* Baker, 1953. Restricted to northern Coahuila.
69. *T. f. riparius* Grinnell and Hill, 1936a. Restricted to Riverside County, southern California.
70. *T. f. robertbakeri* Beauchamp-Martin et al., 2019. Restricted to southern-central Texas.
71. *T. f. robustus* Durrant, 1946. Restricted to Tooele County, northwestern Utah.
72. *T. f. sevieri* Durrant, 1946. Restricted to Millard County, western Utah.
73. *T. f. solitarius* Grinnell, 1926. Restricted to Mineral County, western Nevada.
74. *T. f. spatiosus* Goldman, 1938. Restricted to Brewster County, western Texas.
75. *T. f. stansburyi* Durrant, 1946. Restricted to Tooele County, northwestern Utah.
76. *T. f. sturgisi* Goldman, 1938a. From central Coahuila to northwestern Coahuila.
77. *T. f. subsimilis* Goldman, 1933. Restricted to Yuma County, southwestern Arizona.

78. *T. f. texensis* Bailey, 1902. Restricted to southwestern Texas.
79. *T. f. tivius* Durrant, 1937. Restricted to Millard County, western Utah.
80. *T. f. toltecus* Allen, 1893. From southern New Mexico and northern Chihuahua.
81. *T. f. tularosae* Hall, 1932b. Restricted to Otero County, southern New Mexico.
82. *T. f. vanrossemi* Huey, 1934. Restricted to Punta Peñasco, western Sonora.
83. *T. f. vescus* Hall and Davis, 1935. Restricted to Nye County, southern Nevada.
84. *T. f. villai* Baker, 1953. Restricted to northwestern Chihuahua.
85. *T. f. wahwahensis* Durrant, 1937. Restricted to Beaver County, southwestern Utah.
86. *T. f. winthropi* Nelson and Goldman, 1934. Restricted to central Sonora.

**Conservation:** *Thomomys fulvus* (as *T. bottae*) is listed as Least Concern by the IUCN (Lacher et al. 2016o).

**Characteristics:** Fulvus pocket gopher is small sized; total length 199.0–281.0 mm and skull length 36.9–47.2 mm. Dorsal pelage from paler to darker yellow to darker brown; fur of underparts paler than in the dorsum in ocher or yellowish shades; tail short and slightly darker above; six nipples; sphenoidal fissure present; suture between the frontal and nasal bones wedge shaped; the zygomatic plate thin in side view, almost the jugal height. Species group *umbrinus*.

**Comments:** *Thomomys fulvus* was reinstated as a distinct species by Álvarez-Castañeda (2010); previously regarded as a subspecies of *T. umbrinus* (Hall and Kelson 1959; Hall 1981) and *T. bottae* (Patton 1993a, 2005b). The following subspecies: *Thomomys bottae actuosus*, *T. b. alienus*, *T. b. analogus*, *T. b. angustidens*, *T. b. aridicola*, *T. b. baileyi*, *T. b. basilicae*, *T. b. camargensis*, *T. b. camoae*, *T. b. caneloensis*, *T. b. carri*, *T. b. catalinae*, *T. b. cervinus*, *T. b. collinus*, *T. b. collis*, *T. b. comobabiensis*, *T. b. confinalis*, *T. b. convergens*, *T. b. cultellus*, *T. b. desertorum*, *T. b. desitus*, *T. b. divergens*, *T. b. emotus*, *T. b. estanciae*, *T. b. extenuatus*, *T. b. fulvus*, *T. b. grahamensis* *T. b. guadalupensis*, *T. b. hualpaiensis*, *T. b. hueyi*, *T. b. humilis*, *T. b. intermedius*, *T. b. internatus*, *T. b. juntae*, *T. b. lachuguilla*, *T. b. limitaris*, *T. b. limpiae*, *T. b. mearnsi*, *T. b. modicus*, *T. b. morulus*, *T. b. muralis*, *T. b. mutabilis*, *T. b. nelsoni*, *T. b. operosus*, *T. b. opulentus*, *T. b. paguatae*, *T. b. parvulus*, *T. b. patulus*, *T. b. pectoralis*, *T. b. perditus*, *T. b. pervarius*, *T. b. pinalensis*, *T. b. planorum*, *T. b. proximus*, *T. b. pusillus*, *T. b. quercinus*, *T. b. retractus*, *T. b. rubidus*, *T. b. ruidosae*, *T. b. scotophilus*, *T. b. sonoriensis*, *T. b. spatiosus*, *T. b. sturgisi*, *T. b. suboles*, *T. b. texensis*, *T. b. toltecus*, *T. b. tularosae*, *T. b. vanrossemi*, *T. b. villai*, and *T. b. winthropi* are considered subspecies of *T. fulvus* using molecular data (Álvarez-Castañeda 2010). Patton and Smith (1990) considered a many of these subspecies in the "Basin and Range" and "Great Basin" groups. *T. fulvus* was described by Woodhouse (1852; *Geomys fulvus*) as a species different from *Oryctomys* (*Saccophaorus*) *bottae* by Eydoux and Gervais (1836; *Thomomys bottae*) and *Geomys umbrinus* (Richardson 1829; *Thomomys umbrinus*). Patton (1972) showed a change in the karyotype of *T. b. alienus*, which lives within the current ranges of *T. fulvus*, by having a telocentric chromosome "*X*," relative to all other populations examined, currently considered *T. bottae* for having a submetacentric chromosome "X." Hafner et al. (1987), using chromosomal variation, showed that the population of *T. umbrinus* considered by Álvarez-Castañeda (2010) as *T. fulvus* from northern México was different from those in central México in the number of metacentric, submetacentric, subtelocentric, acrocentric, and micro chromosomes, forming different clades; however, Hafner (2017) still considered it as *T. bottae*. Beauchamp-Martin et al. (2019) described a new subspecies under *T. bottae robertbakeri* because they considered all this part of *bottae* as *T. fulvus*, I considered it a subspecies of *T. f. robertbakeri*. Based on morphological data, Hoffmeister (1986) considered some subspecies as junior synonyms: *T. f. chrysonotus*, *T. f. flavidus*, *T. f. harquahalae*, and *T. f. patulus* as *T. f. albatus*; *T. f.* latirostris as *T. f. aureus*; *T. f. hueyi* and *T. f. parvulus*, as *T. f. catalinae*; *T. f. cedrinus*, *T. f. desitus*, *T. f. hualpaiensis*, *T. f. muralis*, and *T. f. suboles*, as *T. f. desertorum*; *T. f. nasutus* and *T. f. mutabilis* as *T. f. fulvus; T. f. alienus*, *T. f. carri. T. f. caneloensis*, *T. f. chicahuae*, *T. f. collinus*, *T. f. extenuatus*, and *T. f. grahamensis*, as *T. f. mearnsi*; *T. f. proximus* as *T. f. modicus*; *T. f. rufidulus* as *T. f. peremplus*; *T. f. absonus*, *T. f. boreorarius*, *T. f. nicholi*, *T. f. trumbullensis*, and *T. f. virgineus* as *T. f. planirostris; T. f. aridicola*, *T. f. comobabiensis*, *T. f. depauperatus*, *T. f. growlerensis*, and *T. f. phasma* as *T. f. pusillus.* Based on morphological data. Beauchamp-Martin et al. (2019) considered: *T. f. lachuguilla* (in part) as *T. f. baileyi*; *T. f. limitaris* and *T. f. pervarius* as *T. f. lachuguilla*; and *T. f. guadalupensis*, *T. f. pectoralis*, *T. f. ruidosae*, and *T. f. scotophilus* as junior synonyms of *T. f. texensis. T. fulvus* ranges in Arizona, New Mexico, Texas, Sonora, northwestern Sinaloa, central Chihuahua, and central Coahuila (revision of the subspecies of Arizona (Hoffmeister 1986), California (Smith and Patton 1990), and the Baja California Peninsula (Trujano-Álvarez and Álvarez-Castaneda 2013) shows that the number of subspecies is overrepresented more in the species of this genus, as is the case of *T. fulvus*. For that reason, I considered that the subspecies taxonomy needs to be reviewed and I do not believe in so great number of subspecies per species, for that reason the subspecies are not mapped; Map 341). *T. fulvus* can be distinguished from *T. mazama*, *T. monticola*, and *T. talpoides* by not having the sphenoidal fissure. From *T. bottae*, *T. nigricans*, and *T. umbrinus*, only using genetic data.

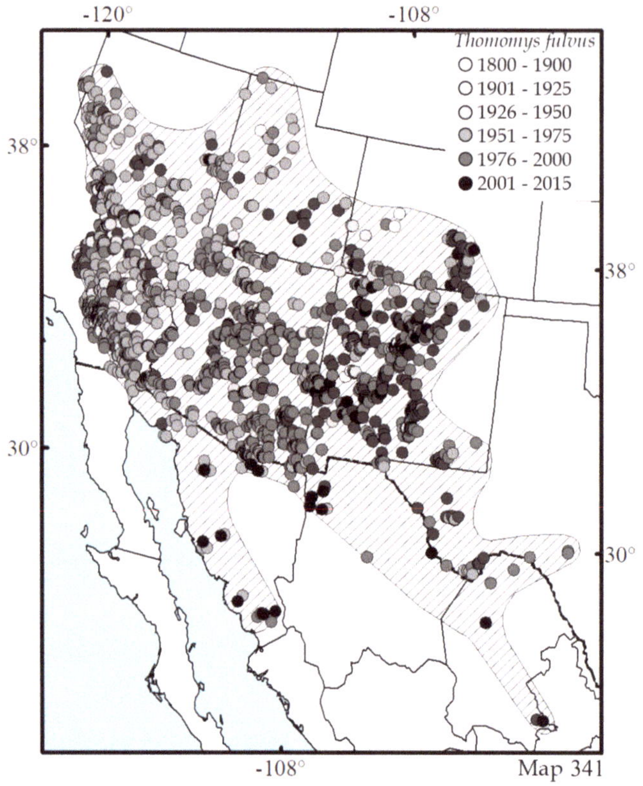


**Map 341** Distribution map of *Thomomys fulvus*

**Additional Literature:** Merriam (1895a), Bailey (1915), Thaeler (1980), Patton and Smith (1989, 1994), Jones and Baxter (2004), Hafner (2017).

## *Thomomys laticeps* Baird, 1855

Northern California pocket gopher, tuza del norte de California

1. *T. l. agricolaris* Grinnell, 1935. Restricted to Yolo County, California.
2. *T. l. awahnee* Merriam, 1908. Southern of Sierra Nevada, California.
3. *T. l. detumidus* Grinnell, 1935. Restricted to Curry County, Oregon.
4. *T. l. laticeps* Baird, 1855. From Humboldt County, California north through southern Oregon.
5. *T. l. leucodon* Merriam, 1897. Highlands around the northern Central Valley, California, and southwestern Oregon.
6. *T. l. saxatilis* Grinnell, 1934. Restricted to Lassen County, California.

**Conservation:** *Thomomys laticeps* (as *T. bottae*) is listed as Least Concern by the IUCN (Lacher et al. 2016o).

**Characteristics:** Northern California pocket gopher is small sized; total length 210.0–227.0 mm and skull length 36.0–40.0 mm. Dorsal pelage almost dark brown to blackish; fur of underparts paler than in the dorsum; tail short and slightly darker above; eight nipples; sphenoidal fissure present; suture between the frontal and nasal bones slightly rounded; zygomatic plate thick in side view, more than twice the jugal height. Species group *umbrinus*.

**Comments:** *Thomomys laticeps* was recognized as a distinct species (Álvarez-Castañeda 2010), previously regarded as a subspecies of *T. bottae* (Patton 2005b). *Thomomys laticeps awahnee*, *T. l. detumidus*, *T. l. laticeps*, and *T. l. leucodon* were considered subspecies of *T. bottae* (Álvarez-Castañeda 2010). Patton and Smith (1990) considered *T. l. minor* and *T. l. silvifugus* as junior synonyms of *T. l. laticeps*, and *T. b. agricolaris* and *T. l. acrirostratus* of *T. b. navus*. *T. laticeps* ranges in Northern California and southern Oregon, including the Sacramento Valley and adjacent foothills of the Sierra Nevada, Pacific coast, and coastal ranges (Map 342). *T. laticeps* can be found in sympatry, or nearly so, with *T. monticola*, *T. mazama*, and *T. talpoides*, from which it can be distinguished by not having the sphenoidal fissure. From *T. talpoides*, by not having the braincase posterior portion projecting well beyond the lamboidal crest; the absence of a narrow flange projecting posteriorly and ventrally at the mandibular angle, and nasals V shaped. From *T. mazama*, by its larger size; a narrow flange projecting posteroventrally from the mandibular angle; a dark subaricular patch one to two times larger than the ears.

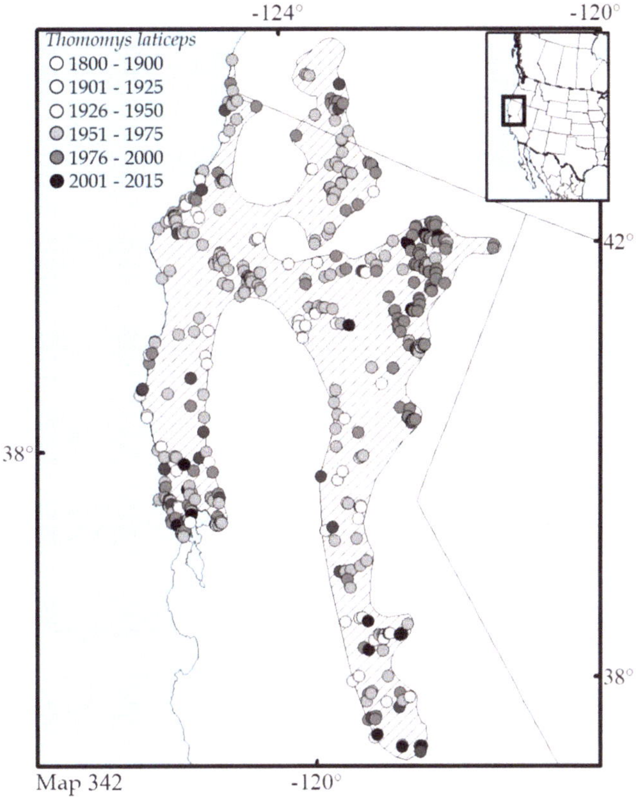


**Map 342** Distribution map of *Thomomys laticeps*

**Additional Literature:** Merriam (1895a), Bailey (1915), Thaeler (1968b, 1980), Patton and Smith (1989, 1994), Jones and Baxter (2004).

## ***Thomomys nayarensis*** Mathis et al., 2013

Nayarit pocket gopher, tuza del Nayar

Monotypic.

**Conservation:** *Thomomys nayarensis* (as *T. bottae*) is listed as Least Concern by the IUCN (Lacher et al. 2016o).

**Characteristics:** Nayarit pocket gopher is small sized; total length 168.0–210.0 mm and skull length 27.0–33.0 mm. Dorsal pelage medium-brown; fur of underparts golden–yellowish brown; side washed with a slightly paler golden shade; tail short and slightly darker dorsally; six nipples; sphenoidal fissure present; suture between the frontal and nasal bones wedge shaped; **zygomatic plate thin in side view, almost the jugal height**. Species group *umbrinus*.

**Comments:** *Thomomys nayarensis* was recognized as a distinct species (Mathis et al. 2013b) from specimens of two localities of Sierra del Nayar, Nayarit (Map 343). *T. nayarensis* can be distinguished from *T. sheldoni* and *T. atrovarius* by having the maximum auditory meatus height less than 1.5 mm in the dorsal–ventral axis and by its generally smaller size. From *T. atrovarius*, by not having sparse pelage; the absence of a darker middorsal line; flanks paler; fur of underparts paler than in the flanks and range in lowlands. From *T. sheldoni*, by having the maximum auditory meatus height shorter and narrower, less than 1.5 mm in the dorsal–ventral axis. From *T. fulvus* and *T. umbrinus*, mainly by the genetic distances; these species are not found in the Sierra Madre Occidental highlands.

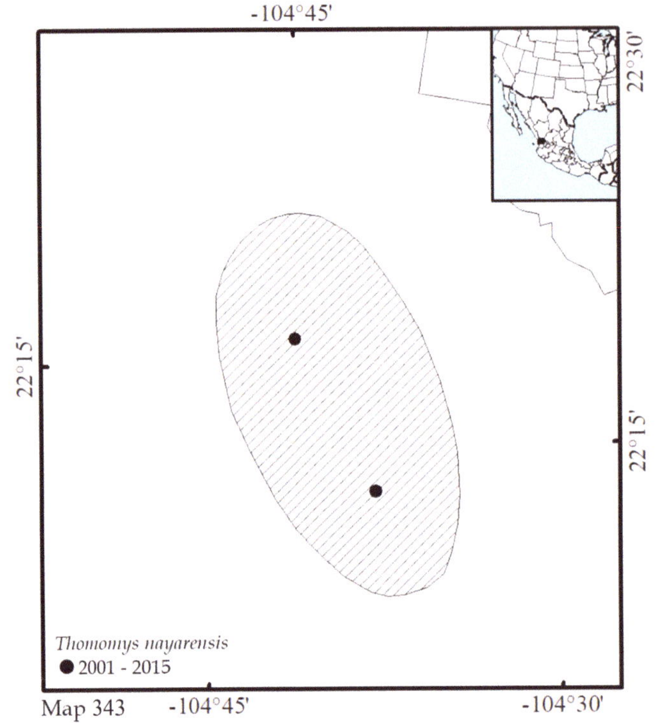


**Map 343** Distribution map of *Thomomys nayarensis*

**Additional Literature:** Álvarez-Castañeda (2010).

## ***Thomomys nigricans*** Rhoads, 1895

California pocket gopher, tuza de Baja California

1. *T. n. anitae* Allen, 1898. From southern Vizcaíno Desert south to the southern tip of Baja California Peninsula.
2. *T. n. martirensis* Allen, 1898. From Sierra Juárez south through the Central Desert, Baja California.
3. *T. n. nigricans* Rhoads, 1895. Restricted to southern California and northwestern Baja California.
4. *T. n. russeolus* Nelson and Goldman, 1909. Restricted to the Vizcaíno Desert, northern Baja California Sur, and southern Baja California.

**Conservation:** *Thomomys nigricans* (as *T. bottae*) is listed as Least Concern by the IUCN (Lacher et al. 2016o).

**Characteristics:** California pocket gopher is small sized; total length 173.0–215.0 mm and skull length 31.0–40.8 mm. Dorsal pelage almost pale yellow to blackish, mainly dark brown; fur of underparts paler than in the dorsum, generally in yellowish shades; tail short and slightly darker dorsally; six nipples and one pair of pectoral mammae; sphenoidal fissure present; suture between the frontal and nasal bones wedge shaped; the zygomatic plate thin in side view, almost the jugal height. Restricted to the Baja California Peninsula and southern California. Species group *umbrinus*.

**Comments:** *Thomomys nigricans* was recognized as a distinct species, as *T. anitae* (Álvarez-Castañeda 2010). Trujano-Álvarez and Álvarez-Castañeda (2013) established the current taxonomy, previously regarded as a subspecies of *T. bottae* (Patton 2005b). *Thomomys bottae anitae*, *T. b. alticolus*, *T. b. imitabilis*, *T. b. incomptus*, *T. b. litoris*, and *T. b. magdalenae* are considered as synonyms of *T. nigricans anitae*. *T. b. abbotti*, *T. b. albatus*, *T. b. aphrastus*, *T. b. brazierhowelli*, *T. b. martirensis*, *T. b. proximarinus*, *T. b. siccovallis*, and *T. b. xerophilus* are considered as synonyms of *T. n. martirensis*. *T. b. cunicularis*, *T. b. jojobae*, *T. b. juarezensis*, *T. b. lucidus*, and *T. b. nigricans*, synonyms of *T. n. nigricans*. *T. b. catavinensis*, *T. b. cactophilus*, *T. b. borjasensis*, *T. b. homorus*, *T. b. rhizophagus*, *T. b. ruricola*, and *T. b. russeolus* are considered synonyms of *T. n. russeolus* (Álvarez-Castañeda 2010; Trujano-Álvarez and Álvarez-Castañeda 2007, 2013). Patton and Smith (1990) considered some subspecies as junior synonyms: *T. b. aderrans*, *T. b. boregoensis*, and *T. b. crassus* as *T. b. albatus*, and *T. b. affinis*, *T. b. cabezonae*, *T. b. jacinteus*, and *T. b. puertae* as *T. n. nigricans*. *T. nigricans* ranges from the San Bernardino Mountains, southern California, south through the entire Baja California peninsula (Map 344), and can be distinguished from *T. bottae* by having one pair of mammae (nipples); upper incisors usually strongly procumbent and nasal frontal part wedge shaped. From *T. fulvus*, it can only be distinguished using genetic analyses. In the contact zone of northeastern Baja California and southeastern California, *T. nigricans* inhabits the highlands and *T. fulvus* the lowlands and Colorado River Valley.

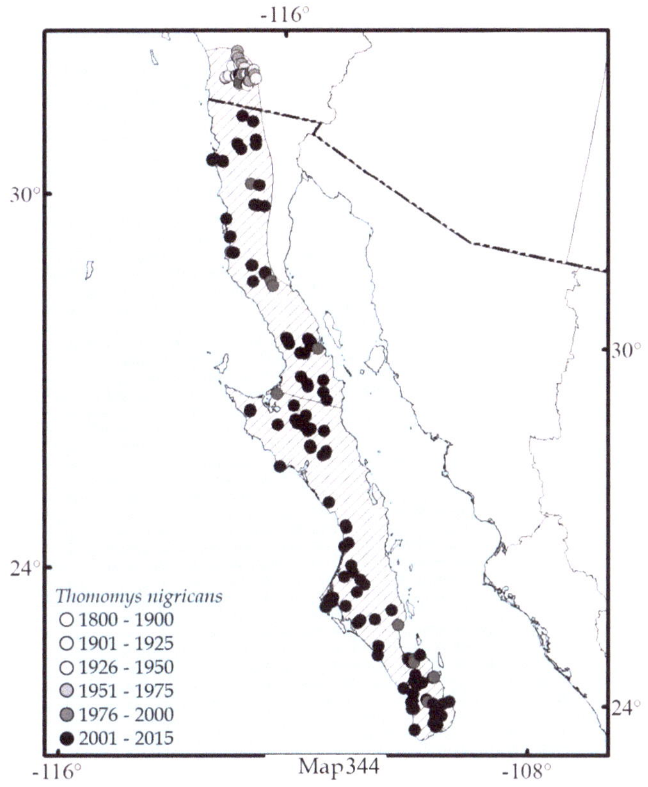


**Map 344** Distribution map of *Thomomys nigricans*

**Additional Literature:** Bailey (1915), Thaeler (1980), Patton and Smith (1989, 1994), Jones and Baxter (2004).

## ***Thomomys sheldoni*** Bailey, 1915

Sheldon's pocket gopher, tuza de la Sierra Madre

1. *T. s. chihuahuae* Nelson and Goldman, 1934. Restricted to the Sierra Madre Occidental highlands, Chihuahua.
2. *T. s. sheldoni* Bailey, 1915. Restricted to the Sierra Madre Occidental highlands, western Durango, northeastern Nayarit, and western Zacatecas.

**Conservation:** *Thomomys sheldoni* (as *T. bottae*) is listed as Least Concern by the IUCN (Lacher et al. 2016o).

**Characteristics:** Sheldoni pocket gopher is small sized; total length 194.0–218.0 mm and skull length 35.0–39.8 mm. Dorsal pelage medium to dark brown, **occasionally with a faint, slightly darker dorsal stripe; fur of underparts golden to yellowish brown with a slightly paler wash of golden brown on the sides; pelage moderately dense**; tail short and slightly darker dorsally; six nipples and one pair of pectoral mammae; sphenoidal fissure present; suture between the frontal and nasal bones wedge shaped; the **zygomatic plate thin in side view, almost jugal height**. Species group *umbrinus*.

**Comments:** *Thomomys sheldoni* was previously considered a different species from *T. umbrinus*, as *T. chihuahuae* (Álvarez-Castañeda 2010). A recent review included the population of *T. u. sheldoni*, which has nomenclatorial priority (Mathis et al. 2013a); for that reason, the species is now known as *T. sheldoni*. *T. u. eximius* (part. highlands) and *T. u. chihuahuae* are considered synonyms of *T. s. chihuahuae*. *T. u. crassidens* and *T. s. sheldoni* are considered synonyms of *T. s. sheldoni*. Hafner et al. (1987) used electromorphic and chromosomal variations to show that the population of *T. umbrinus* from the northern part of Sierra Madre Occidental (Siqueiros and El Vergel) was different from those inhabiting the southern part, mainly based on differences in the number of metacentric, submetacentric, and acrocentric chromosomes. *T. sheldoni* is restricted to the Sierra Madre Occidental highlands from western-central Chihuahua south through western Zacatecas and Jalisco, including western Durango and northeastern Nayarit (Map 345). *T. sheldoni* can be distinguished from *T. atrovarius* by its moderately dense pelage, not sparse; back fur usually darker mid-dorsally, paler flanks and fur of underparts paler than in the flanks, and in association with lowlands. From *T. nayarensis*, by having the maximum auditory meatus height longer and wider, greater than 1.5 mm in the dorsal–ventral axis. From *T. fulvus* and *T. umbrinus* mainly by genetic distances; these species are found in the Altiplano Mexicano but not in the Sierra Madre Occidental highlands.

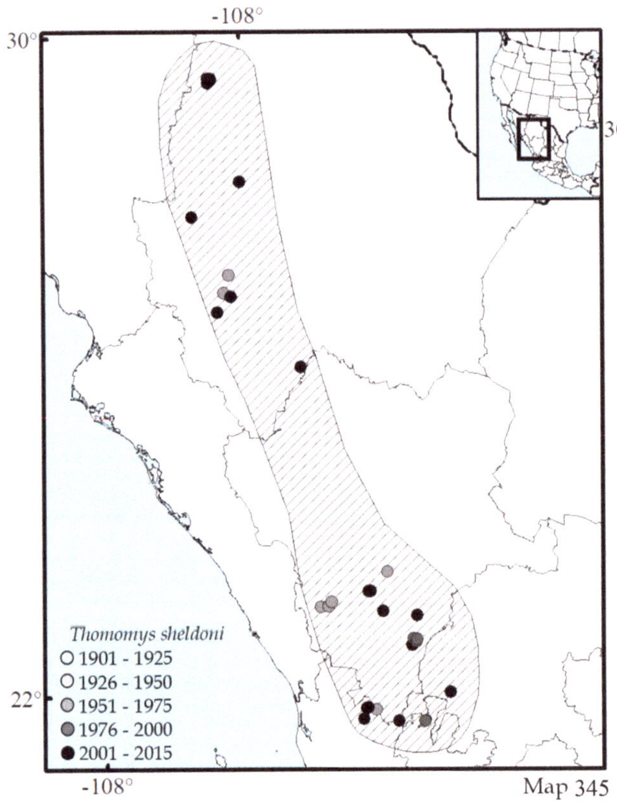


**Map 345** Distribution map of *Thomomys sheldoni*

**Additional Literature:** Bailey (1915), Nelson and Goldman (1934b), Thaeler (1980), Patton and Smith (1989, 1994), Patton (2005b), Mathis et al. (2013a).

## ***Thomomys townsendii*** (Bachman, 1839)

Townsend's pocket gopher, tuza de las montañas del oeste

1. *T. t. nevadensis* Merriam, 1897. Central-northern Nevada, southeastern Oregon, and California.
2. *T. t. townsendii* (Bachman, 1839). Restricted to Snake River, western Idaho, and eastern Oregon.

**Conservation:** *Thomomys townsendii* is listed as Least Concern by the IUCN (Cassola 2016ea).

**Characteristics:** Townsend's pocket gopher is small sized within pocket gophers but is the largest member within *Thomomys*; total length 222.0–289.0 mm and skull length 37.2–46.7 mm. Dorsal pelage grayish to pale grayish brown; fur of underparts slightly paler; rostrum anterior part blackish; postauricular region blackish; chin or feet with a white patch, some specimens with marks above the head; eight mammae, two pectoral and two inguinal pairs; **upper incisors procumbent**; skull relatively larger and flattened dorsally; rostrum and nasals narrow; **first upper molars with a bony protuberance in the orbital region**; sphenoidal fissure present; **pterygoids straight or nearly so; without a groove on the exoccipital**. Species group *umbrinus*.

**Comments:** *Thomomys townsendii* was recognized as a distinct species from *T. bottae* (Thaeler 1968a, b; Patton et al. 1984; Patton and Smith 1989, 1994; Álvarez-Castañeda 2010), with four geographically isolated populations (Álvarez-Castañeda 2010); only two subspecies were recognized: *T. t. nevadensis* (includes *bachmani*, *elkoensis*, and *relictus*) and *T. t. townsendii* (includes *atrogriseus*, *owyhensis*, and *similis*; Smith 1999; Patton 2005b). Molecular analyses (Álvarez-Castañeda 2010) showed that the geographic range is broader than previously considered with four subspecies. *T. bottae brevidens* and *T. b. centralis* are considered synonyms of *T. t. centralis* (Álvarez-Castañeda 2010). *T. b. bachmani*, *T. b. elkoensis*, and *T. b. nevadensis* are considered synonyms of *T. t. nevadensis*. *T. b. townsendii* and *T. b. owyhensis*, of *T. t. townsendii*, and *T. t. similis* was previously a subspecies of *T. bottae*. *T. townsendii* thrives in riverbanks; in Nevada, one population is found adjacent to the Humbolt River and many of its effluents, such as the Reese and Rock Creek rivers, the second in Nevada-Oregon, and in the Quinn River and the Snake River, the first in the Oregon-Idaho area and the second in eastern Idaho. The southern Nevada-Idaho population is associated with the White, Meadow Valley Wash, and Virginia rivers (Map 346). *T. townsendii* differs from *T. bottae* by its relatively larger size, upper incisors usually strongly procumbent, one pair of pectoral nipples; nasal anterior part wedge shaped, and the zygomatic plate thin in side view, almost same height as the jugal, and of relatively larger size.

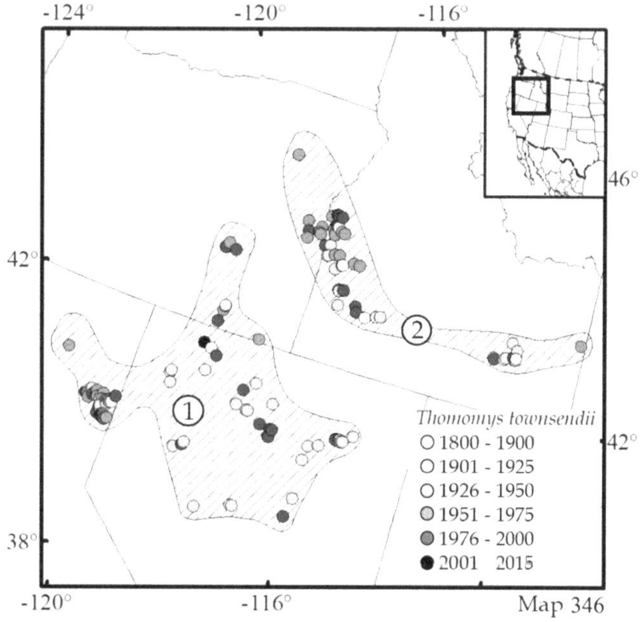


**Map 346** Distribution map of *Thomomys townsendii*

**Additional Literature:** Merriam (1895a), Bailey (1915), Davis (1937), Russell (1968a), Thaeler (1980), Verts and Carraway (2003).

## ***Thomomys umbrinus*** (Richardson, 1829)

Southern pocket gopher, tuza mexicana

1. *T. u. durangi* Nelson and Goldman, 1934. Restricted from southwestern Durango to extreme northwestern Zacatecas.
2. *T. u. goldmani* Merriam, 1901. Central Chihuahua south through central-eastern Durango and southwestern Coahuila.
3. *T. u. intermedius* Mearns, 1897. From southeastern Arizona south through Sonora and northwestern Chihuahua.
4. *T. u. umbrinus* (Richardson, 1829). From eastern-central Zacatecas south through the Eje Neovolcánico in Veracruz.

**Conservation:** *Thomomys umbrinus* is listed as Least Concern by the IUCN (Lacher et al. 2016p).

**Characteristics:** Southern pocket gopher is small sized; total length 174.0–220.0 mm and skull length 27.2–33.5 mm. Dorsal pelage almost white or pale yellow to blackish, **generally with a darkened mid-dorsal stripe**; fur of underparts paler than in the dorsum; tail short and slightly darker dorsally; six nipples; sphenoidal fissure present; suture between the frontal and nasal bones wedge shaped; the **zygomatic plate thin in side view, almost the jugal height**. Species group *umbrinus*.

**Comments:** Hall and Kelson (1959) and Hall (1981) only recognized *Thomomys umbrinus* as a valid species for all the current complex *atrovarius-bottae-fulvus-laticeps-nayarensis-nigricans-sheldoni-umbrinus*, which were previously considered full species (Anderson 1966, 1972; Patton and Dingman 1968; Hoffmeister 1969, 1986; Patton 1973; Patton and Smith 1981; Álvarez-Castañeda 2010). Mathis et al. (2014) reviewed *T. umbrinus* using molecular and morphological data. With the three mtDNA and five nucDNA genes, they found two clades within *T. umbrinus* with a ~15% difference; these clades match those proposed by Hafner et al. (1987) and Álvarez-Castañeda (2010). However, these authors do not recognize the Central Mexican Highlands group as a different species from those of the Eje Neovolcánico. The morphological analyses (Hoffmeister 1986; Hafner et al. 1987) of specimens have reduced the current number of subspecies to four, and consider that all the following subspecies are junior synonyms of *T. u. goldmani*: *T. u. baileyi*, *T. u. camargensis*, *T. u. evexus*, *T. u. juntae*, and *T. u. nelsoni*. *T. u. intermedius*: *T. u. burti*, *T. u. caliginosus*, *T. u. emotus*, *T. u. quercinus*, *T. u. madrensis*, and *T. u. sonoriensis*. *T. u. umbinus*: *T. u. albigularis*, *T. u. arriagensis*, *T. u. atrodorsalis*, *T. u. enixus*, *T. u. martinensis*, *T. u. newmani*, *T. u. orizabae*, *T. u. peregrinus*, *T. u. potosinus*, *T. u. pullus*, *T. u. supernus*, *T. u. tolucae*, *T. u. vulcanius*, and *T. u. zacatecae*. *T. umbrinus* displays a patchy distribution in central México, associated with highlands, from southern Arizona south through Veracruz and Puebla (Map 347). *T. umbrinus* can be distinguished from T. *atrovarius*, *T. bottae*, *T. fulvus*, *T. nayarensis*, and *T. sheldoni* only using genetic data, including both chromosomes and DNA sequences. From *T. bottae*, *T. fulvus*, and *T. nigricans*, it is distinguished by having three pairs of mammae (nipples), one pectoral pair and two inguinal; generally smaller sized with a darkened mid-dorsal stripe; skull not conspicuously short and wide; nasals wedge shaped and narrowing posteriorly; maxilla-frontal suture convex and lacrimal abuts in the medial half, zygomata arch slender and nearly parallel.

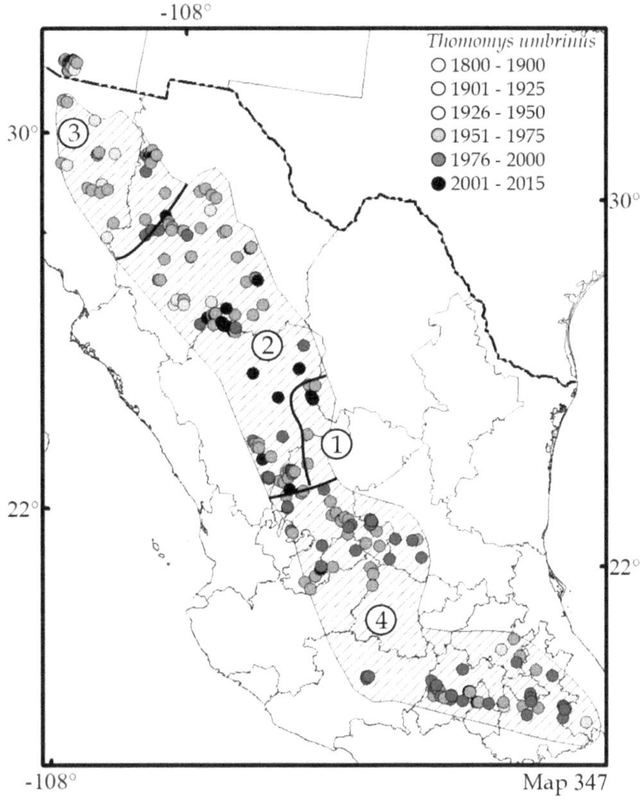


**Map 347** Distribution map of *Thomomys umbrinus*

**Additional Literature:** Merriam (1895a), Bailey (1915), Russell (1968a), Patton (1972), Thaeler (1980), Patton and Smith (1989, 1994), Patton (2005b), Hafner et al. (2007).

## Subgenus *Thomomys* Wied-Neuwied, 1839

**Rostrum slender**; **no p**rocumbent upper incisors, root above the fourth upper premolars; first lower premolar nearly perpendicular to the occlusal surface of toothrow; infraorbital canals opening directly above or slightly posterior to the incisive foramina; **sphenoidal fissure closed** (except for some specimens of *T. clusius*); the anterior enamel plate of the first lower premolars recurved, frequently forming a shallow reentrant angle; the **angular process broadly continuous with a well-developed flange along the ventral side of the mandible ramus; the** anterior enamel plate of the first lower premolars broad and only slightly separated from the posterior enamel plate (rarely continuous) with the lateral enamel plate on the lingual side; diploid numbers from 40 to 60.

### *Thomomys clusius* Coues, 1875

Wyoming pocket gopher, tuza de Wyoming

Monotypic.

**Conservation:** *Thomomys clusius* is listed as Least Concern by the IUCN (Linzey and NatureServe 2017a).

**Characteristics:** Wyoming pocket gopher is small sized; total length 161.0–184.0 mm and skull 28.8–33.0 mm. Dorsal pelage pale yellowish; fur of underparts paler than in the dorsum; **ear fringe with white hairs and without a darker post-auricular patch**; sphenoidal fissure present; rostrum length relatively similar to braincase length; temporal ridges parallel posteriorly; tympanic bullae smaller. Species group *talpoides*.

**Comments:** *Thomomys clusius* was recognized as a distinct species (Thaeler and Hinesley 1979), previously regarded as a subspecies of *T. talpoides* (Bailey 1915). *T. clusius* range is restricted to southern-central Wyoming in Carbon and Sweetwater counties (Map 348). *T. clusius* can be found in sympatry, or nearly so, with *T. talpoides* (Thaeler and Hinesley 1979), and can be distinguished from *T. talpoides* and *T. idahoensis* by its absence of a dark auricular patch; ear edges buff and usually paler than the dorsal pelage (Keinath et al. 2014); fingers with white hairs and approximately 20% smaller in size. From *T. idahoensis*, by its more inflated auditory bullae; smaller baculum, from 10.4 to 13.8 mm vs. 17.8–20.0 mm, and different chromosome number.

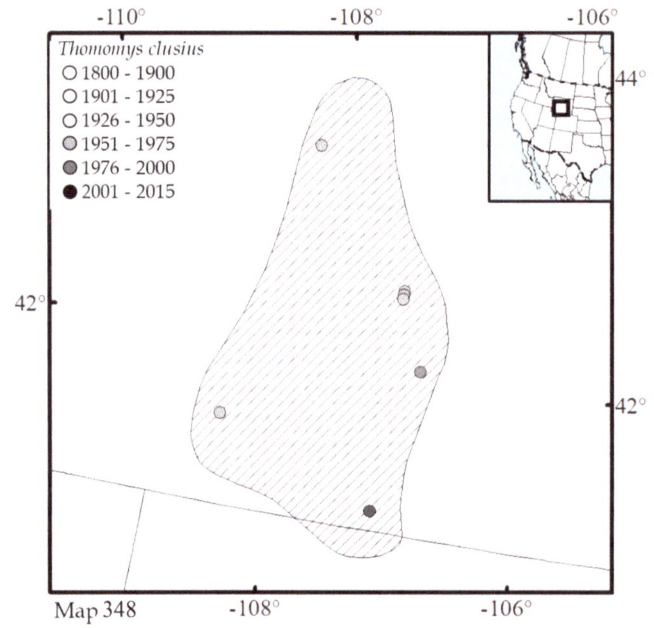


**Map 348** Distribution map of *Thomomys clusius*

**Additional Literature:** Merriam (1895a), Thaeler (1980), Patton and Smith (1989, 1994), Patton (2005b), Cudworth and Grenier (2015).

## *Thomomys idahoensis* Merriam, 1901

Idaho pocket gopher, tuza de Idaho

1. *T. i. confini* Davis, 1937. Restricted to Ravalli County, Montana.
2. *T. i. idahoensis* Merriam, 1901. Southeastern Idaho and southwestern Montana.
3. *T. i. pygmaeus* Merriam, 1901. Southwestern Wyoming and southeastern Idaho.

**Conservation:** *Thomomys idahoensis* is listed as Least Concern by the IUCN (Linzey and NatureServe 2017b).

**Characteristics:** Idaho pocket gopher is small sized; total length 160.0–200.0 mm and skull length 26.7–29.2 mm. Dorsal pelage buff brown to dark brown; hair of underparts paler than in the dorsum; **post-auricular patch absent or very small; ears paler at the edges**; cheeks evenly gray or with darker areas; sphenoidal fissure present; rostrum length relatively similar to braincase length; temporal ridges parallel posteriorly; smaller tympanic bullae. Species group *talpoides*.

**Comments:** *Thomomys idahoensis* was recognized as a distinct species (Thaeler 1972, 1977), previously regarded as a subspecies of *T. talpoides* (Davis 1939; Hall and Kelson 1959; Hall 1981). *Thomomys talpoides confinus* is considered synonym of *T. idahoensis confines*, *T. t. idahoensis* of *T. idahoensis idahoensis*, and *T. t. pygmaeus* of *T. i. pygmaeus* (Thaeler 1972, 1977). *T. idahoensis* has a disjunct range in eastern and southeastern Idaho, southwestern Montana, western Wyoming, and northern Utah (Map 349). Neither of the three known subspecies has a continuous distribution. *T. idahoensis* can be distinguished from *T. talpoides* and *T. clusius* by having the dark auricular patch very small and not extending dorsally; ear edges paler than the auricular patch but similar to the dorsal coloration.

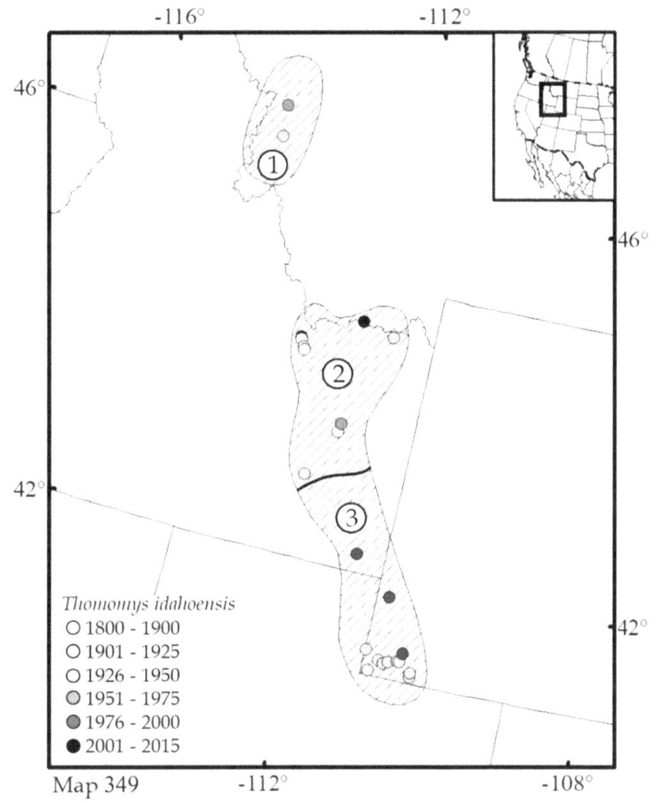


**Map 349** Distribution map of *Thomomys idahoensis*

**Additional Literature:** Bailey (1915), Thaeler (1980), Patton and Smith (1989, 1994), Patton (2005b).

## ***Thomomys mazama*** Merriam, 1897

Western pocket gopher, tuza del oeste

1. *T. m. couchi* Goldman, 1939. Restricted to Mason County, Washington.
2. *T. m. glacialis* Dalquest and Scheffer, 1942. Restricted to Pierce County, Washington.
3. *T. m. helleri* Elliot, 1903. Restricted to Curry County, Oregon.
4. *T. m. hesperus* Merriam, 1901. Coastal area of northwestern Oregon.
5. *T. m. louiei* Gardner, 1950. Restricted to Wahkiakum County, Washington.
6. *T. m. mazama* Merriam, 1897. Central-western Oregon and northern California.
7. *T. m. melanops* Merriam, 1899. Northern Olympia peninsula, Washington.
8. *T. m. nasicus* Merriam, 1897. Restricted to central-western Oregon.
9. *T. m. niger* Merriam, 1901. Restricted to Benton and Lane Counties, Oregon.
10. *T. m. oregonus* Merriam, 1901. Restricted to northwestern Oregon.
11. *T. m. premaxillaris* Grinnell, 1914. Restricted to Tehama County, California.
12. *T. m. pugetensis* Dalquest and Scheffer, 1942. Restricted to Thurston County, Washington.

13. *T. m. tacomensis* Taylor, 1919. Restricted to Tacoma County, Washington.
14. *T. m. tumuli* Dalquest and Scheffer, 1942. Restricted to Thurston County, Washington.
15. *T. m. yelmensis* Merriam, 1899. Restricted to Thurston County, Washington.

**Conservation:** *Thomomys mazama glacialis*, *T. m. pugetensis*, *T. m. tumuli*, and *T. m. yelmensis* are listed as Threatened by the United States Endangered Species Act (FWS 2022) and *T. mazama* as Least Concern by the IUCN (Cassola 2016dx).

**Characteristics:** Western pocket gopher is small sized; total length 194.0–222.0 mm and skull length 30.5–32.3 mm. Dorsal pelage pale yellowish brown to blackish, with reddish shades; fur of underparts paler than in the dorsum, usually with a white spot in the chest; **dark auricular patch five to six times larger than the ears but** small and not extending dorsally; nose and face usually dusky, plumbeous, or blackish; chin, feet, and tail pale gray at the tip, whitish or buff; tail tip naked; slit-like or closed sphenoidal fissure; **wide flange projecting posteroventrally from the mandibular angle;** zygomatic breadth wide compared to skull length; temporal ridges parallel posteriorly; interparietal length shorter than its breadth; interparietal bone longer and narrower with smooth sutures with the frontoparietal bone. Species group *talpoides*.

**Comments:** *Thomomys mazama* was previously considered a subspecies of *T. monticola* (Bailey 1915; Hall and Kelson 1959). *T. mazama* has a patchy range in the highlands of western Washington, western Oregon, and northwestern California (Map 350). *T. mazama* differs from all the other species of *Thomomys* in its smaller size; closed or slit-like sphenoidal fissure; wide flange projecting posteroventrally from the mandibular angle. From *T. laticeps* and *T. bulbivorus*, by having a dark subaricular patch five to six times larger than the ears. From *T. monticola*, by a paler coloration; smaller rounded ears, less than 7.0 mm; three pectoral nipples; zygomatic breadth wider than skull length; temporal ridges parallel posteriorly; interparietal length shorter than its breadth. From *T. talpoides*, by having the interparietal bone longer and narrower, with smooth sutures with the frontoparietal bone.

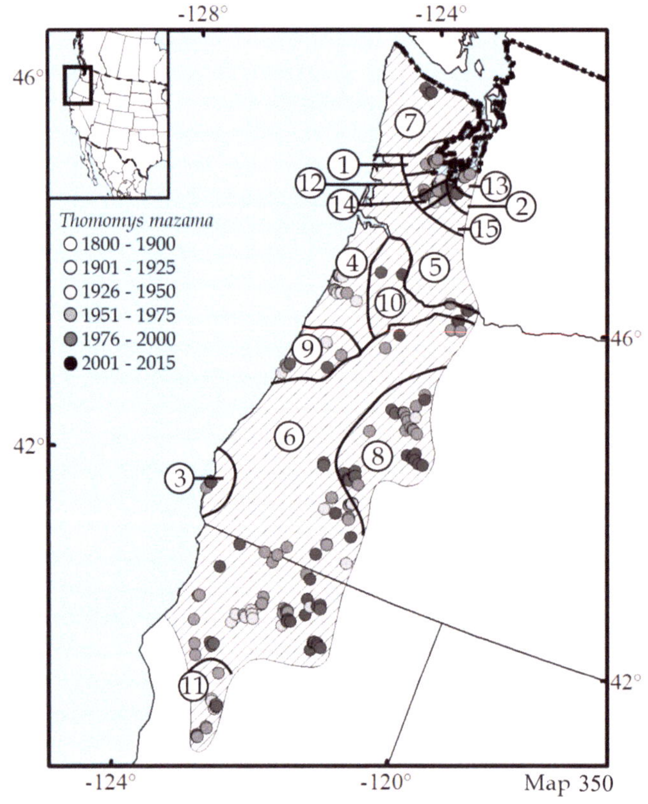


**Map 350** Distribution map of *Thomomys mazama*

**Additional Literature:** Johnson and Benson (1960), Russell (1968a), Thaeler (1980), Patton and Smith (1989, 1994), Verts and Carraway (2003), Patton (2005b), Álvarez-Castañeda (2010).

## ***Thomomys monticola*** Allen, 1893
Mountain pocket gopher, tuza de las montañas

Monotypic.

**Conservation:** *Thomomys monticola* is listed as Least Concern by the IUCN (Cassola 2016dy).

**Characteristics:** Mountain pocket gopher is small sized; total length 190.0–220.0 mm and , skull length 30.2–31.9 mm. Dorsal pelage in different shades of brown; fur of underparts more grayish than the dorsal coloration; **proportionally, the largest ears of all pocket gophers, relatively pointed**; a dark and large auricular patch, approximately three times the ear size; sphenoidal fissure present; rostrum relatively longer than braincase length; temporal ridges diverging posteriorly; interparietal bone longer than wide; tympanic bullae large. Species group *talpoides*.

**Comments:** *Thomomys mazama* was previously considered a subspecies of *T. monticola* (Bailey 1915; Hall and Kelson 1959). *T. monticola* has not received a recent taxonomic review, however studies in process seem to show a complex of many possible species (Patton in literature). *T. monticola* ranges in northern Sierra Nevada, Lassen Peak and Mount Shasta, California, and a small western part of Nevada (Map 351). *T. monticola* can be distinguished from *T. bottae* and *T. laticeps* by an absent sphenoidal fissure. From *T mazama*, by a darker coloration; larger pointed ears greater than 7.0 mm; two pectoral nipples; zygomatic breadth narrower than skull length; temporal ridges diverging posteriorly; interparietal length similar to its breadth. From *T. talpoides*, by having longer and pointed ears; two pairs of pectoral nipples; rostrum relatively longer than braincase length; zygomatic breadth narrower than skull length; temporal ridges diverging posteriorly; interparietal bone longer than wider; tympanic bullae large.

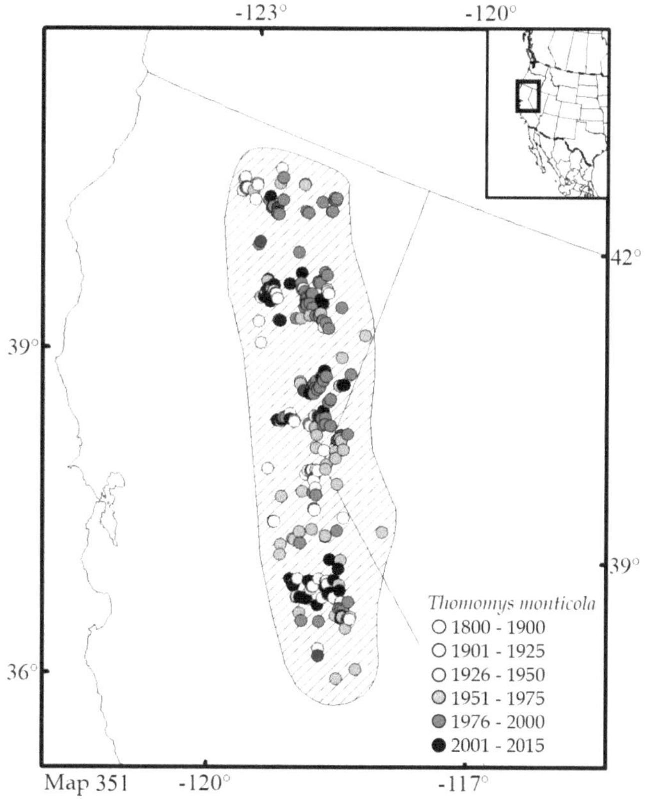


**Map 351** Distribution map of *Thomomys monticola*

**Additional Literature:** Merriam (1895a), Johnson and Benson (1960), Russell (1968a), Bailey (1915), Patton and Smith (1989, 1994), Patton (1999b, 2005b), Thaler (1980), Álvarez-Castañeda (2010).

## *Thomomys talpoides* (Richardson, 1828)

Northern pocket gopher, tuza del norte

1. *T. t. aequalidens* Dalquest, 1942. Restricted to southeastern Oregon.
2. *T. t. agrestis* Merriam, 1908. Restricted to southern-central Colorado.
3. *T. t. andersoni* Goldman, 1939. Restricted to southwestern Alberta.
4. *T. t. attenuatus* Hall and Montague, 1951. Southeastern Wyoming and northeastern Colorado.
5. *T. t. bridgeri* Merriam, 1901. Southeastern Idaho, western and southwestern California.
6. *T. t. bullatus* Bailey, 1914. Southern Saskatchewan, eastern Montana, and northeastern Wyoming.
7. *T. t. caryi* Bailey, 1914. Central-northern Wyoming and southern-central- Montana.
8. *T. t. cheyennensis* Swenk, 1941. Restricted to western Nebraska.
9. *T. t. cognatus* Johnstone, 1955. Restricted to southeastern British Columbia.
10. *T. t. columbianus* Bailey, 1914. Restricted to northern-central Oregon.
11. *T. t. devexus* Hall and Dalquest, 1939. Restricted to central-southeastern Washington.
12. *T. t. douglasii* (Richardson, 1829). Restricted to southwestern Washington.
13. *T. t. durranti* Kelson, 1949. Eastern Idaho and western-central Colorado.
14. *T. t. falcifer* Grinnell, 1926. Restricted to central Nevada.
15. *T. t. fisheri* Merriam, 1901. Western California and eastern Nevada.
16. *T. t. fossor* Allen, 1893. Northeastern Arizona, southwestern Colorado, and northern New Mexico.
17. *T. t. fuscus* Merriam, 1891. Restricted to central Idaho.
18. *T. t. gracilis* Durrant, 1939. Northwestern Idaho, northern, central, and northeastern Nevada.
19. *T. t. immunis* Hall and Dalquest, 1939. Restricted to central-southern Washington.
20. *T. t. incensus* Goldman, 1939. Restricted to southern British Columbia.
21. *T. t. kaibabensis* Goldman, 1938. Restricted to northern-central Arizona.
22. *T. t. kelloggi* Goldman, 1939. Restricted to Park County, Montana.
23. *T. t. levis* Goldman, 1938. Restricted to central Idaho.
24. *T. t. limosus* Merriam, 1901. Restricted to southern Washington.
25. *T. t. loringi* Bailey, 1914. Restricted to central-western Alberta.
26. *T. t. macrotis* Miller, 1930. Restricted to Douglas County, Colorado.
27. *T. t. medius* Goldman, 1939. Restricted to southeastern British Columbia.
28. *T. t. meritus* Hall, 1951. Restricted to central-northern Colorado.
29. *T. t. monoensis* Huey, 1934. Eastern California and western Nevada.
30. *T. t. moorei* Goldman, 1938. Restricted to central Utah.
31. *T. t. nebulosus* Bailey, 1914. Western South Dakota and eastern Wyoming.
32. *T. t. ocius* Merriam, 1901. Southwestern Wyoming, northwestern Colorado, and northeastern Utah.
33. *T. t. oquirrhensis* Durrant, 1939. Restricted to central-northern Utah.
34. *T. t. parowanensis* Goldman, 1938. Restricted to southern Utah.
35. *T. t. pierreicolus* Swenk, 1941. Restricted to western Nebraska.
36. *T. t. pryori* Bailey, 1914. Restricted to southern Montana.
37. *T. t. quadratus* Merriam, 1897. Restricted to southeastern Oregon.
38. *T. t. ravus* Durrant, 1946. Restricted to northeastern Utah.
39. *T. t. relicinus* Goldman, 1939. Restricted to central-southern Idaho.
40. *T. t. retrorsus* Hall, 1951. Restricted to eastern Colorado.
41. *T. t. rostralis* Hall and Montague, 1951. Southern-central Wyoming and central Colorado.
42. *T. t. rufescens* Wied-Neuwied, 1839. Form southwestern Saskatchewan and Manitoba south through South Dakota.
43. *T. t. saturatus* Bailey, 1914. Northern Idaho, northwestern Montana, and southeastern British Columbia.
44. *T. t. segregatus* Johnstone, 1955. Only known from near Wynndel, southeastern British Columbia.
45. *T. t. shawi* Taylor, 1921. Central-southern Washington.
46. *T. t. talpoides* (Richardson, 1828). Western Alberta, central Saskatchewan, central-western Manitoba, and northern-central Montana.
47. *T. t. taylori* Hooper, 1940. Restricted to Valencia County, central New Mexico.

48. *T. t. tenellus* Goldman, 1939. Northwestern Wyoming and southern Montana.
49. *T. t. trivialis* Goldman, 1939. Restricted to central Montana.
50. *T. t. uinta* Merriam, 1901. Restricted to central-northern Utah.
51. *T. t. wallowa* Hall and Orr, 1933. Restricted to northeastern Oregon.
52. *T. t. wasatchensis* Durrant, 1946. Restricted to northern Utah.
53. *T. t. whitmani* Drake and Booth, 1952. Restricted to Walla Walla County, eastern Washington.
54. *T. t. yakimensis* Hall and Dalquest, 1939. Central-southern Washington.

**Conservation:** *Thomomys talpoides* is listed as Least Concern by the IUCN (Cassola 2016dz).

**Characteristics:** Northern pocket gopher is small sized; total length 180.0–253.0 mm and skull length 28.5–36.1 mm. Dorsal pelage rich brown to brownish gray; fur of underparts buff; post-auricular patch grayish black or blackish; ear edges similar in color to the auricular patch and darker than the dorsal coloration; chin often white; tail brown with white shades and almost naked at the tip; sphenoidal fissure closed or slit-like; **braincase posterior portion projecting well beyond the lamboidal crest; the presence of a narrow flange projecting posteriorly and ventrally at the mandibular bone angle**; nasals squarish posteriorly, not V shaped. Species group *talpoides*.

**Comments:** *Thomomys talpoides confinus*, *T. t. idahoensis*, and *T. t. pygmaeus* were recognized as *T. idahoensis* (Thaeler 1972, 1977), and *T. talpoides clusius* as *Thomomys clusius* (Thaeler and Hinesley 1979). Thaeler (1980) suggests that current *T. talpoides* could be divided into 10–12 separate species. *T. talpoides* ranges from southern British Columbia, Alberta, Saskatchewan, and Manitoba south through California, Nevada, Arizona, and New Mexico, and east through North Dakota, South Dakota, Nebraska, and Colorado, and from Washington, Oregon, and California east through Manitoba, eastern North Dakota, and South Dakota, southwestern Nebraska, eastern Colorado, and central New Mexico (the revision of Thaeler (1980) proposing the separation in 10–12 separate species and the reduction of subspecies for other species of the genus (Trujano-Álvarez and Álvarez-Castañeda 2013) shows that the number of subspecies are over represented. I do not believe in so great number of subspecies per species, for that reason the subspecies are nor mapped; Map 352). *T. talpoides* is the most common pocket gopher in the United States (Patton 1999c). *T. talpoides* can be distinguished from all the other species of *Thomomys* by having the braincase posterior portion projecting well beyond the lamboidal crest; a narrow flange projecting posteriorly and ventrally at the mandibular angle; nasals squarish posteriorly, not V shaped. From *T. idahoensis* and *T. clusius*, by having a dark auricular patch larger and extending dorsally, contrasting with the dorsal coloration, ear edges dark, similar to the auricular patch and the dorsal pelage coloration. From *T. monticola*, by having shorter and rounded ears; three pairs of pectoral nipples; rostrum relatively shorter than braincase length; zygomatic breadth wide compared to skull length; temporal ridges nearly parallel; smaller tympanic bullae.

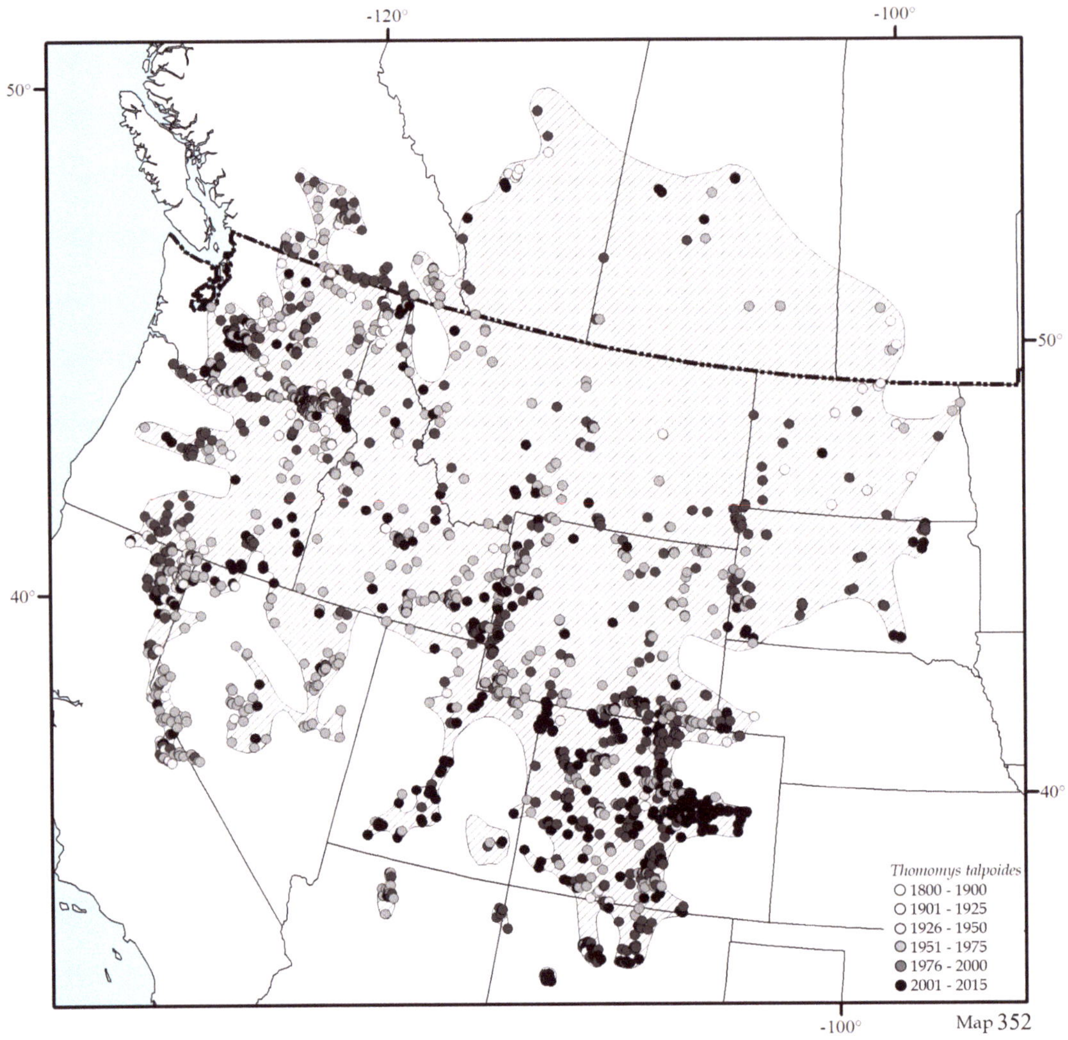


**Map 352** Distribution map of *Thomomys talpoides*

**Additional Literature:** Bailey (1915), Russell (1968a), Thaeler (1980), Patton and Smith (1989, 1994), Verts and Carraway (1999), Patton (2005b), Álvarez-Castañeda (2010).

## **Family Heteromyidae** Gray, 1868

The Family Heteromyidae includes kangaroo rats, kangaroo mice, and both spiny and soft-haired pocket mice characterized by body size between mouse and rat (100.0–500.0 mm in total length); external cheek pouches; forelimbs modified for quadruped or biped saltatorial locomotion; infraorbital canal compressed, completely penetrating against the rostrum, and opening laterally. This family mainly ranges in North America, although the genus *Heteromys* is also distributed in northern South America. It belongs to the superfamily Geomyoidea. All three subfamilies are present in North America: Dipodomyinae, Heteromyinae, and Perognathinae. The keys were elaborated based on the review of specimens and the following sources: Hall (1981) and Álvarez-Castañeda et al. (2017a).

1. Hindlimbs not much longer than forelimbs; auditory bullae not highly developed, without occupying the posterior-dorsal region of the braincase (Fig. 98); lower premolar lobes merging into the labial or lingual side but not in the middle part of the tooth; upper incisors without a longitudinal canal .......................... Subfamily Heteromyinae (p. 523)

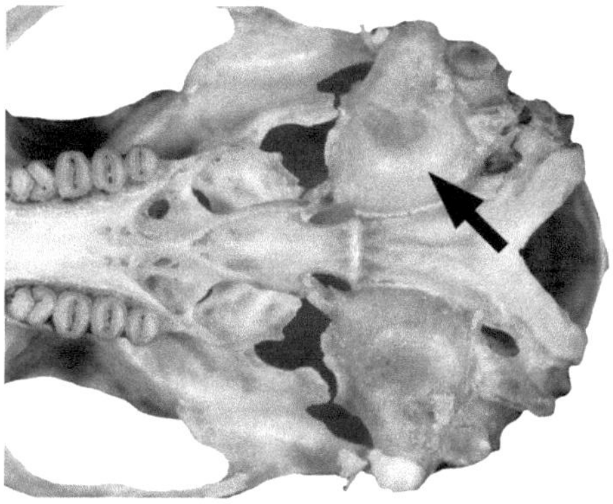

**Fig. 98** Auditory bullae not highly developed, without occupying the posterior-dorsal region of the braincase

1a. Hindlimbs much longer than forelimbs; auditory bullae highly developed, occupying the posterior-dorsal region of the braincase (Figs. 99 and 100); lower premolar lobes merging in the middle part of the tooth; upper incisors with a longitudinal canal ........................................................................ 2

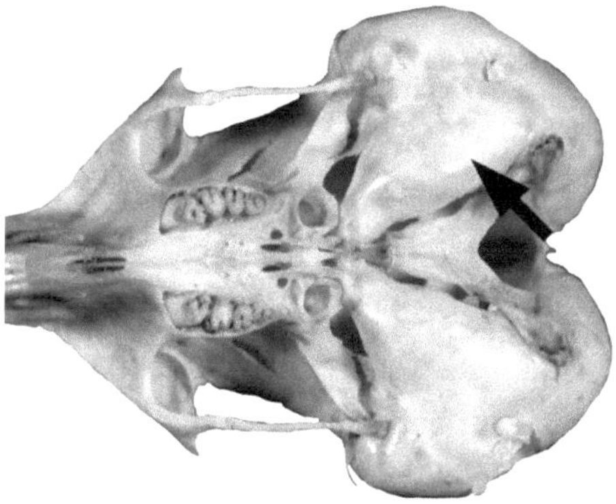

**Fig. 99** Auditory bullae highly inflated; as a result, posterior skull breadth greater than zygomatic breadth

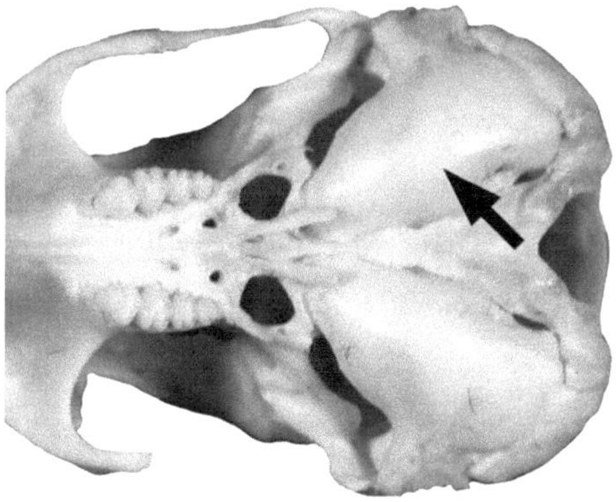

**Fig. 100** Auditory bullae inflated; as a result, posterior skull breadth equal to or slightly greater than zygomatic breadth

2. Forelimbs very short in relation to hindlimbs; tail very long, longer than the head-and-body length and with or without a crest or hair tuft at the tip; auditory bullae highly inflated; as a result, posterior skull breadth greater than zygomatic breadth (Fig. 99); interparietal bone reduced and longer than wide............................Subfamily Dipodomyinae (p. 494)

2a. Forelimbs short relative to hindlimbs; tail generally of equal size as the head-and-body length and with hair as a crest, but not as a hair tuft at the tip; auditory bullae inflated; as a result, posterior skull breadth equal to or slightly greater than zygomatic breadth (Fig. 100); interparietal bone slightly longer than wide.........Subfamily Perognathinae (p. 534)

## **Subfamily Dipodomyinae** Gervais, 1853

The species in this subfamily are characterized by having **forelimbs very small in relation to hindlimbs**; tail very long, longer than the head-and-body length; **auditory bullae highly inflated and, as a result, skull posterior breadth greater than zygomatic breadth**; ventral surface of auditory bullae rarely reaching the level of the grinding surface of molars, never appreciably below that level; interparietal bone reduced and longer than wide; no ethmoid foramen in the frontal bone; center of palate between premolars ridged; pterygoid fossa double; caudal vertebrae with median ventral foramina; calcaneo-navicular or calcaneo-cuneiform articulation present; zygomatic root of maxilla expanded antero-posteriorly; last upper molars small; lophos of upper premolars unite first at or near the tooth center; protoloph usually single-cusped; lophos of upper molars unite progressively from the lingual to the buccal margins; lophos of lower molars unite primitively at the buccal margin, progressively at the tooth center, forming an H-pattern; molariforms brachydont to hypsodont but always rooted; occlusal enamel pattern lost early in life; enamel always complete.

1. 1. Total length greater than 180.0 mm; tail with a greater crest or hair tuft at the tip; auditory bullae length less than the length between the incisive and pterygoid fossa........................*Dipodomys* (p. 494)

2. 1a. Total length less than 180.0 mm; tail without a crest or hair tuft at the tip; auditory bullae length similar to the length between the incisive and pterygoid fossa........................*Microdipodops* (p. 521)

## Genus ***Dipodomys*** Gray, 1841

Dorsal pelage sandy pale to dark brown, flanks similar in color to the dorsum; fur of underparts white; white spots on upper lip, above each eye, and behind each ear; **white line over hindlimbs**; cheeks whitish; eyelids completely blackish; tail length from smaller to larger in relation to the head-and-body length; **white lateral lines extending from the tail base to at least the middle part**; four or five toes on the hindfoot; **hindfoot larger and stronger than the forefoot,** with saltatorial displacement; **head proportionally large relative to body size; auditory bullae highly inflated; as a result, posterior skull breadth greater than zygomatic breadth; interparietal bone reduced, longer than wide.** *Dipodomys* has five species groups: *deserti* (*D. deserti*), *heermanni* (*D. agilis*, *D. californicus*, *D. gravipes*, *D. heermanni*, *D. ingens*, *D. panamintinus*, *D. simulans*, *D. stephensi*, and *D. venustus*), *merriami* (*D. elator*, *D. merriami*, *D. nitratoides*, *D. ornatus*, and *D. phillipsii*), *ordii* (*D. compactus*, *D. microps*, and *D. ordii*), and *spectabilis* (*D. nelsoni*, and *D. spectabilis*). The keys were elaborated based on the review of specimens and the following sources: Setzer (1949), Lidicker (1960a), Hall (1981), and Álvarez-Castañeda et al. (2017a).

1. Hindfood with five toes (the fifth represented only by a nail and located in the mid inner part of the limb)........................2

1a. Hindfood with four toes........................14

2. Head-and-body length (without the tail) greater than 130.0 mm; hindfoot length greater than 49.0 mm; skull breadth across auditory bullae greater than 27.5 mm (San Joaquin Valley, California)........................*Dipodomys ingens* (p. 504)

2a. Head-and-body length (without the tail) less than 130.0 mm; hindfoot length less than 49.0 mm; skull breadth across auditory bullae less than 27.5 mm........................3

3. Total length less than 265.0 mm; hindfoot length less than 40.0 mm........................4

3a. Total length greater than 265.0 mm; hindfoot length greater than 40.0 mm........................5

4. Tail length less than 100% of the head-and-body length and not crested; skull breadth at auditory bullae less than 22.2 mm; interparietal bone rectangular to roundish (restricted to Cameron County, Mustang and Padre islands, Texas, and north Isla Barra Arenosa, Tamaulipas)........................*Dipodomys compactus* (p. 498)

4a. Tail length greater than 100% of the head-and-body length and crested; skull breadth at auditory bullae greater than 22.2 mm; interparietal bone triangular (Alberta, Saskatchewan, southern Washington, Idaho, and North Dakota south through Guanajuato, Querétaro, and Hidalgo, and from Oregon and California east through South Dakota, Nebraska, Kansas, Oklahoma, and Texas)........................*Dipodomys ordii* (p. 511)

5. Ears large, length usually greater than 15.0 mm........................6

5a. Ears small, length usually less than 15.0 mm ........................................................................................................ 7

6. Total length usually less than 315.0 mm, ear length less than 16.0 mm; hindfoot length less than 4.5 mm; nasal length less than 15.0 mm (southwestern California and northwestern Baja California) ............ *Dipodomys agilis* (part; p. 496)

6a. Total length usually greater than 315.0 mm, ear length greater than 16.0 mm; hindfoot length greater than 4.5 mm; nasal length greater than 15.0 mm (along the coastal mountains of western-central California) ........................................................................................................ *Dipodomys venustus* (p. 519)

7. Pelage relatively dark; facial arietiform mark bold black, complete; dorsal dark tail stripe usually at least twice the width of the lateral white stripes ........................................................................................................ 8

7a. Pelage medium-shade or paler; facial arietiform mark weak (dusky patches at whiskers base discontinuous, with a dusky top of the nose); dorsal dark tail stripe less than twice the width of the lateral white stripes ............................ 12

8. Skull "narrow-faced" type; maxillary arches weakly angled and their spread less than 22.0 mm .................................. 9

8a. Skull "broad-faced" type; maxillary arches prominently angled and their spread greater than 22.0 mm ...................... 11

9. Tail thin and ears proportionately smaller; zygomatic arches breadth at the maxilla greater than 55% of skull length (restricted to San Quintín Valley, Baja California, probably extinct) ................................ *Dipodomys gravipes* (p. 502)

9a. Tail wide and ears proportionately large; zygomatic arches breadth at the maxilla less than 55% of skull length........ 10

10. Pelage darker, back dusky cinnamon-buff; smaller auditory bullae, skull breadth across the auditory bullae usually less than 24.5 mm (southwestern California and northwestern Baja California) ................... *Dipodomys agilis* (part; p. 496)

10a. Pelage paler, back near dusky pinkish cinnamon; larger auditory bullae, skull breadth across the auditory bullae usually greater than 24.5 mm (southwestern California south through central Baja California Sur) ........................................................................................................ *Dipodomys simulans* (p. 516)

11. Ears small, length less than 12.8 mm; auditory bullae as viewed from above approaching a globular shape, with all parts of outlines curved (restricted to San Jacinto Valley and adjacent areas, California) .......... *Dipodomys stephensi* (p. 518)

11a. Ears medium sized, length greater than 12.8 mm; auditory bullae, as viewed from above not globular in outline (restricted to central-western California) ........................................................................ *Dipodomys heermanni* (part; p. 503)

12. Skull "narrow-faced" type; maxillary arches weakly angled and their spread less than 22.0 mm (southwestern California and northwestern Baja California) ........................................................................ *Dipodomys agilis* (part; p. 496)

12a. Skull "broad-faced" type; maxillary arches prominently angled and their spread greater than 22.0 mm ...................... 13

13. Small auditory bullae; skull breadth across the auditory bullae less than 24.9 mm; narrowest distance between the auditory bullae greater than 2.0 mm; rostrum wide near its end, usually greater than 4.1 mm (eastern California and western Nevada) ........................................................................................................ *Dipodomys panamintinus* (p. 514)

13a. Larger auditory bullae; skull breadth across the auditory bullae greater than 24.9 mm; narrowest distance between the auditory bullae less than 2.0 mm; rostrum wide near its end, usually less than 4.1 mm (restricted to central-western California) ........................................................................................................ *Dipodomys heermanni* (part; p. 503)

14. Head-and-body length (without the tail) greater than 130.0 mm; hindfoot length greater than 49.0 mm; skull breadth across the auditory bullae greater than 27.5 mm (western Nevada, southeastern California, western Arizona, northeastern Baja California, and northwestern Sonora) ........................................................................ *Dipodomys deserti* (p. 499)

14a. Head-and-body length (without the tail) less than 130.0 mm; hindfoot length less than 49.0 mm; skull breadth across the auditory bullae less than 27.5 mm ........................................................................................................ 15

15. Maxillary arch breadth at the middle greater than 3.9 mm; lower incisors flat on the anterior face, chisel-like (southeastern Oregon, western Idaho, eastern California, Nevada, Utah, western Colorado, and northwestern Arizona) ................ ........................................................................................................ *Dipodomys microps* (p. 507)

15a. Maxillary arch breadth at the middle less 3.9 mm; lower incisors rounded on the anterior face, awl-like .................... 16

16. Restricted to northern-central Texas and southwestern Oklahoma ........................................ *Dipodomys elator* (p. 501)

16a. Ranges in other parts of southern United States and México ........ 17

17. Total length greater than 265.0 mm; hindfoot length usually greater than 40.0 mm; skull length greater than 39.0 mm ........ 18

17a. Total length less than 265.0 mm; hindfoot length usually less than 40.0 mm; skull length less than 39.0 mm ........ 20

18. Total length less than 300.0 mm; skull length less than 40.0 mm; prominent sharp posteroexternal angle of the maxillary arches; first molariform with two lobes in the inner part (southern Oregon and central-northern California) ........ *Dipodomys californicus* (p. 497)

18a. Total length greater than 300.0 mm; skull length greater than 40.0 mm; sharp posteroexternal angle of the maxillary arches; first molariforms with one lobe in the inner part ........ 19

19. Tail tip white, greater than 25% of the tail length; interparietal width greater than 1.5 mm (northeastern Arizona, New Mexico, western Texas, northern Sonora, and Chihuahua, and in Aguascalientes, Jalisco, Zacatecas, and San Luis Potosí) ........ *Dipodomys spectabilis* (p. 517)

19a. Tail tip white, never greater than 25% of the tail length; interparietal width less than 1.5 mm (eastern Chihuahua, northern Coahuila, northwestern Durango, northern Zacatecas and San Luis Potosí, and southern Nuevo León) ........ *Dipodomys nelsoni* (p. 509)

20. Tail dark dorsally, ventral stripes broader than white stripes; nasal and rostrum shorter, nasal length less than 12.4 mm (Central Valley, California) ........ *Dipodomys nitratoides* (p. 510)

20a. Tail white, stripes broader than dark stripes and ventral dark stripes narrower than dorsal stripes; nasal and rostrum longer, nasal length greater than 12.4 mm ........ 21

21. Tail blackish at the tip; black facial lines present to absent; interorbital breadth greater than half of basal length (California, Nevada, Utah, Arizona, New Mexico, and southwestern Texas south through Aguascalientes, Zacatecas, and San Luis Potosí, including the Baja California Peninsula and three islands, Margarita, San José, and Tiburón) ........ *Dipodomys merriami* (p. 505)

21a. Tail white at the tip; black facial lines well marked and earring shaped; interorbital breadth less than half of basal length ........ 22

22. Hindfoot length variable; mastoid breadth: the maxilla breadth ratio generally smaller than 1.08 (Guanajuato, Querétaro, Morelos, Ciudad de México, Hidalgo, Estado de México, Veracruz, Puebla, and northern Oaxaca. Genetic data are needed to differentiate the following species) ........ *Dipodomys phillipsii* (p. 515)

22a. Hindfoot length greater than 37.0 mm; mastoid breadth: the maxilla breadth ratio generally equal to or smaller than 1.08 (Durango, Zacatecas, Guanajuato, Querétaro, and San Luis Potosí) ........ *Dipodomys ornatus* (p. 513)

## ***Dipodomys agilis*** Gambel, 1848

Agile kangaroo rat, rata canguro del sur de California

1. *D. a. agilis* Gambel, 1848. From northwestern Baja California north through Santa Barbara, California.
2. *D. a. perplexus* (Merriam, 1907). Restricted to the San Bernardino Mountain, California.

**Conservation:** *Dipodomys agilis* is listed as Least Concern by the IUCN (Cassola 2016t).

**Characteristics:** Agile kangaroo rat is medium sized; total length 265.0–319.0 mm and skull length 21.3–23.8 mm. Dorsal pelage dark brown; five toes on each hindfoot; pinna medium to large; **dark sole on the hindfoot;** tail well defined dorsally with white ventral stripes and numerous white hairs in the tuft; the maxillary process of zygomatic arches narrow; foramen magnum with an angular dorsal perimeter; suture between each nasal and the premaxilla straight. Species group *heermanni*.

**Comments:** *Dipodomys simulans* was recognized as a distinct species, including *D. antiquaries*, *D. peninsularis*, and *D. paralius* (Best et al. 1986; Williams et al. 1993; Sullivan and Best 1997a), previously regarded as a subspecies of *D. agilis* (Hall 1981). *D. agilis* ranges in southwestern California and northwestern Baja California (Map 353), and can be found in sympatry, or nearly so, with *D. merriami*, *D. simulans*, *D. stephensi*, *D. paramantinus*, and *D. venustus*. *D. agilis* can be distinguished from *D. gravipes* by its smaller size and paler coloration; tail wide at the base, and ears proportionately larger; zygomatic arches breadth at the maxilla level less than 55% of skull length. From *D. heermanni*, by being paler in coloration. From *D. merriami*, by having five toes on each hindfoot. From *D. simulans*, by its darker coloration, back dusky cinnamon-buff; smaller auditory bullae and skull breath across the auditory bullae usually less than 24.5 mm. From *D. panamintinus*, by its much larger ears; darker coloration; tail longer than 155% of the head-and-body length; white tail stripe almost as wide as the dorsal tail stripe at mid-tail. From *D. stephensi*, by its larger ears (mean 17.0 mm); proportionally narrow head, with narrow arietiform spots; tail with many white hairs in the tuft, white dorsal and ventral tail stripes; wide white lateral stripes, auditory bullae not "elongated globose" when viewed from above; foramen magnum with an angular dorsal perimeter; suture between each nasal and premaxilla straight. From *D. venustus*, by its smaller ears and tail; paler coloration and facial markings; rostrum (broad-faced), nasals, and shorter skull, with a narrow zygomatic maxillary arch and lateral side without a pronounced outer angle and lighter incisors.

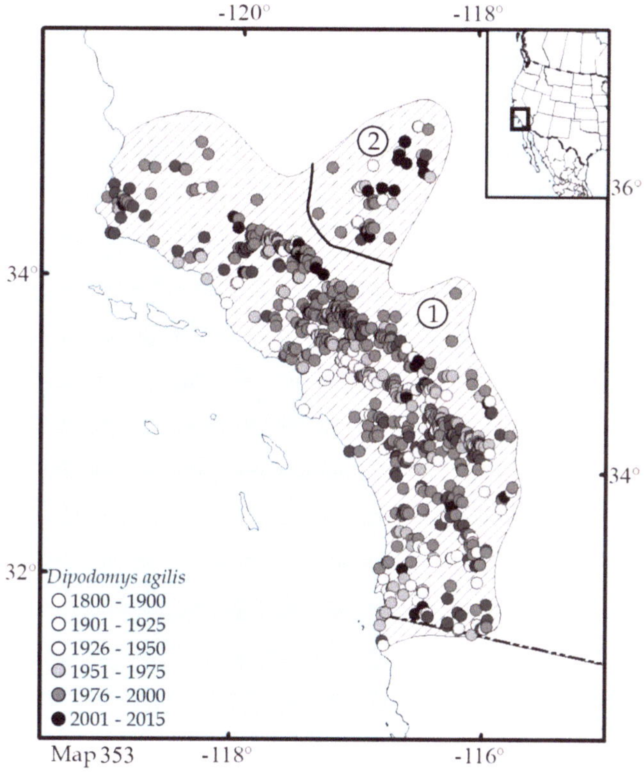


**Map 353** Distribution map of *Dipodomys agilis*

**Additional Literature:** Merriam (1904b), Grinnell (1922), Setzer (1949), Álvarez (1960), Lidicker (1960a, b), Lackey (1967a), Best et al. (1996a), Sullivan and Best (1997b), Patton (2005a).

## ***Dipodomys californicus*** Merriam, 1890

California kangaroo rat, rata canguro de California

1. *D. c. californicus* Merriam, 1890. From central California, from San Francisco Bay north through southern Oregon.
2. *D. c. eximius* Grinnell, 1919. Restricted to Sutter County, California.
3. *D. c. saxatilis* Grinnell and Linsdale, 1929. Restricted to Tehama County, California.

**Conservation:** *Dipodomys californicus* is listed as Least Concern by the IUCN (Cassola 2016u).

**Characteristics:** California kangaroo rat is small sized; total length 260.0–340.0 mm and skull length 37.1–40.2 mm. Dorsal pelage brown and silky; four toes on the hindfoot; **tail with a hair tuft always white at the tip;** four toes (rarely five) on each hindfoot. Species group *heermanni*.

**Comments:** *Dipodomys californicus* was recognized as a distinct species (Fashing 1973; Patton et al. 1976), previously regarded as a subspecies of *D. heermanni* (Grinnell 1922; Hall 1981). *Dipodomys c. eximius* and *D. c. saxatilis* are considered subspecies of *D. californicus* (Williams et al. 1993). *D. californicus* ranges in southern Oregon and central-northern California (Map 354), and can be found in sympatry, or nearly so, with *D. deserti*, *D. heermanni*, and *D. merriami*. *D. californicus* can be distinguished from *D. deserti* by its darker coloration; "broad-faced"; lacks a flatter skull; auditory bullae not more inflated; mastoids do not meet immediately behind the parietals; smaller size, mean total length less than 330.0 mm. From *D. heermanni*, by having four toes on each hindfoot. From *D. merriami*, by its darker coloration; larger size, greater than 260.0 mm, larger hindfoot on average, greater than 40.0 mm, and by the number of toes on each hindfoot.

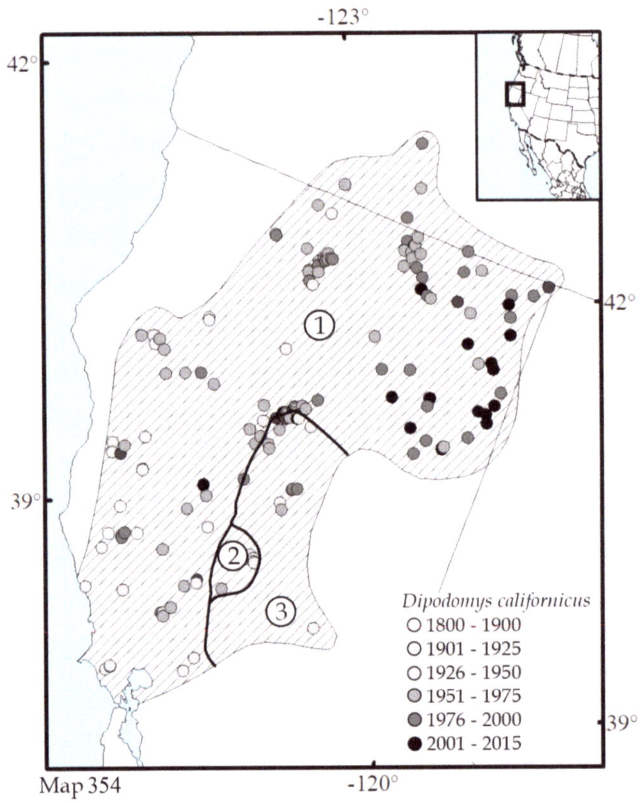


**Map 354** Distribution map of *Dipodomys californicus*

**Additional Literature:** Grinnell and Linsdale (1929), Setzer (1949), Patton et al. (1976), James and James (1984), Kelt (1988a), Patton (2005a).

## *Dipodomys compactus* True, 1889

Gulf Coast kangaroo rat, rata canguro del Golfo

1. *D. c. compactus* True, 1889. Restricted to Padre Island, Texas, and Tamaulipas.
2. *D. c. sennetti* (Allen, 1891). Restricted to Cameron County, southeastern Texas.

**Conservation:** *Dipodomys compactus* is listed as Least Concern by the IUCN (Cassola 2016v).

**Characteristics:** Gulf coast kangaroo rat is medium sized; total length 216.0 a 230.0 mm and skull length 36.3–37.2 mm. Dorsal pelage sandy gray, flanks whitish gray; five toes on each hindfoot; **interparietal bone larger and not pointed posteriorly**. Species group *ordii*.

**Comments:** *Dipodomys compactus* was recognized as a distinct species (Johnson and Sealander 1971; Schmidly and Hendricks 1976; Baumgardner and Schmidly 1981), previously regarded as a subspecies of *D. ordii* (Davis 1942; Johnson and Sealander 1971; Schmidly and Hendricks 1976; Hall 1981). *D. ordii sennetti* is considered a subspecies of *D. compactus* (Williams et al. 1993). *D. compactus* ranges in Cameron County, Barra de Arena, Mustang, and Padre Islands in southern Texas (Map 355). It can be found in sympatry, or nearly so, only with *D. ordii* (Baumgardner and Schmidly 1981), from which it differs by having four toes on the hindfoot and a shorter, non-crested tail; shorter pelage; smaller (shorter and more narrowed) skull, and less inflated auditory bullae.

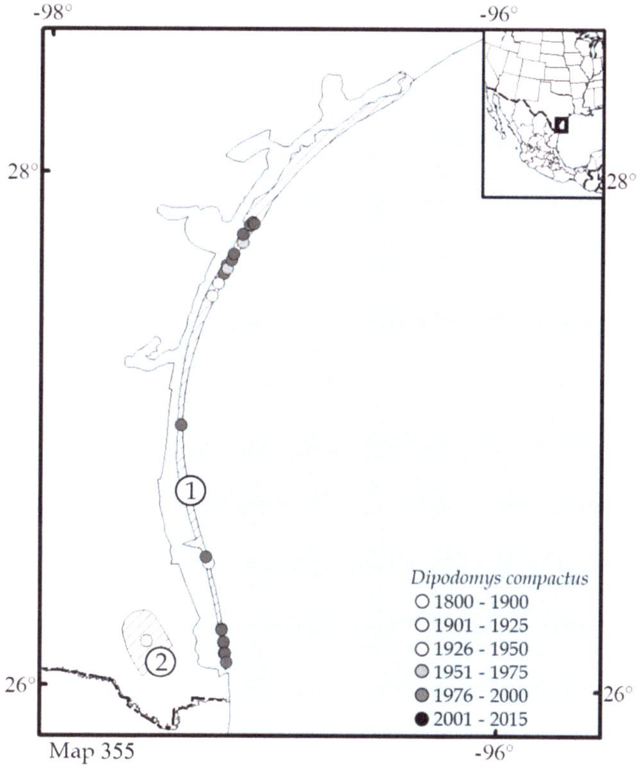


**Map 355** Distribution map of *Dipodomys compactus*

**Additional Literature:** Bailey (1905), Setzer (1949), Baumgardner (1991), Patton (2005a).

## *Dipodomys deserti* Stephens, 1887

Desert kangaroo rats, rata canguro del desierto

1. *D. d. aquilus* Nader, 1965. Restricted to western-central Nevada.
2. *D. d. arizonae* Huey, 1955. Restricted to southern-central Arizona.
3. *D. d. deserti* Stephens, 1887. From central Nevada south through northwestern Sonora and northeastern Baja California.
4. *D. d. sonoriensis* Goldman, 1923. Restricted to western Sonora.

**Conservation:** *Dipodomys deserti* is listed as Least Concern by the IUCN (Álvarez-Castañeda et al. 2016e)

**Characteristics:** Desert kangaroo rat is the largest member of the genus; total length 305.0–378.0 mm and skull length 43.5–48.3 mm. Dorsal pelage brown sand; tail larger than the head-and-body length, without dark lines and with a white hair tuft at the farthest third end, long ridge-like hairs; four toes on each hindfoot; flattest skull within the genus; **greatest inflated auditory bullae within the genus; auditory bullae enlarged; interparietal bone highly reduced; supraoccipital bone visible on the dorsal surface;** mastoids meet immediately behind the parietals, with an inconspicuous spicule between them. Species group *deserti*.

**Comments:** *Dipodomys deserti* is restricted to sandy areas in western Nevada, eastern California, western Arizona, northeastern Baja California, and northwestern Sonora (Map 356). *D. deserti* can be found in sympatry, or nearly so, with *D. californicus*, *D. merriami*, *D. ordii*, *D. panamintinus*, and *D. spectabilis*. *D. deserti* can be distinguished from other species by the flattest skull of the genus and the greatest inflated auditory bullae; mastoids meet immediately behind the parietals, with an inconspicuous spicule between them. From *D. californicus* and *D. ordii*, by its paler coloration; "narrow-faced" and larger size, on average, total length greater than 330.0 mm. From *D. panamintinus* by having the hindfood with five toes. From *D. merriami*, by its paler coloration and larger size. From *D. spectabilis*, by its mastoids meeting immediately behind the parietals, with an inconspicuous spicule (if any) between them.

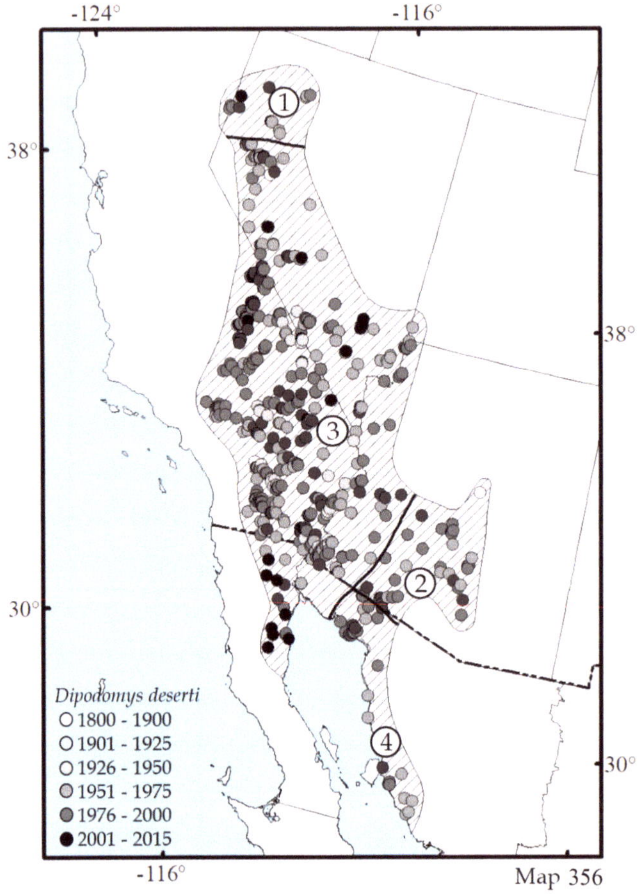


**Map 356** Distribution map of *Dipodomys deserti*

**Additional Literature:** Merriam (1890a), Grinnell (1922), Hall (1946), Setzer (1949), Álvarez (1960), Johnson and Sealander (1971), Brown and Lieberman (1973), Nader (1978), Schnell et al. (1978), Williams et al. (1993), Patton and Álvarez-Castañeda (1999), Patton (2005a).

## *Dipodomys elator* Merriam, 1894

Texas kangaroo rat, rata canguro de Texas

Monotypic.

**Conservation:** *Dipodomys elator* has been considered extirpated from Oklahoma (Caire et al. 1989; Braun et al. 2021), and is listed as Vulnerable (B1ab [i, ii, iii]: Extent of occurrence less than 20,000 km$^2$ with ten or fewer populations and continuing decline in extent of occurrence, area of occupancy, and quality of habitat) by the IUCN (Wahle et al. 2018).

**Characteristics:** Texas kangaroo rat is small sized; total length 260.0–345.0 mm and skull length 36.4–38.8 mm. Dorsal pelage buffy, with blackish shades, flanks grizzled with ocher; tail relatively thick and long, about 160% of the head-and-body length, **with a white hair tuft at the tip**; four toes on each hindfoot; skull broad and rostrum wide, interorbital region narrow; interparietal bone nearly as broad as long; temporal fossa large. Species group *merriami*.

**Comments:** *Dipodomys elator* is restricted to northern-central Texas and southwestern Oklahoma (Map 357), allopatric with respect to all other species of *Dipodomys*. *D. elator* was considered within the *phillipsii* group by Grinnell (1921) and Miller (1924) but not by Davis (1942). *D. elator* differs from *D. phillipsii* and *D. ornatus* by its longer and thicker tail (~162% of the head-and-body length); large body, on average 121.0 mm; wider rostrum; the interorbital region relatively narrower and interparietal region wider.

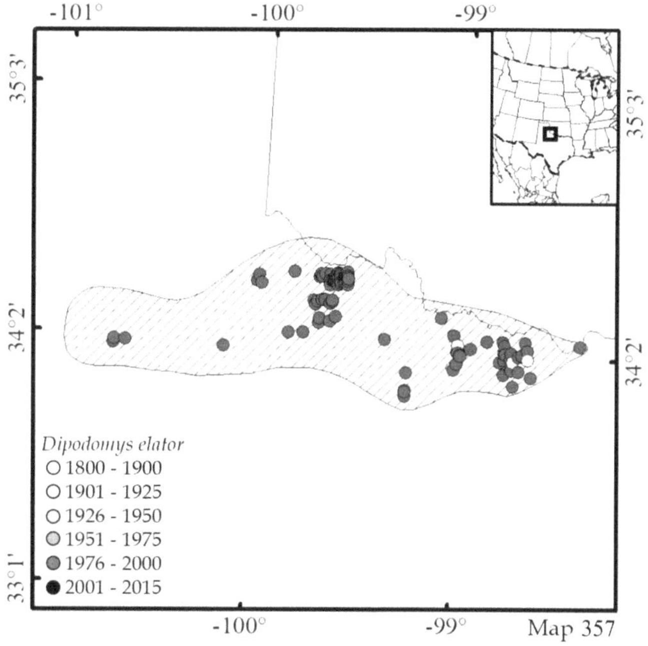


**Map 357** Distribution map of *Dipodomys elator*

**Additional Literature:** Bailey (1905), Davis (1974), Setzer (1949), Jannett (1976), Roberts and Mills (1983), Carter et al. (1985), Best (1987), Hamilton et al. (1987), Williams et al. (1993), Mantooth et al. (2000), Patton (2005a), Braun et al. (2021).

## *Dipodomys gravipes* Huey, 1925

San Quintín kangaroo rat, rata canguro de San Quintín

Monotypic.

**Conservation:** *Dipodomys gravipes* is listed as Extinct by the Norma Oficial Mexicana (DOF 2019) and as Critically Endangered (Possibly Extinct, D: Very small with less than 50 mature individuals) by the IUCN (Álvarez-Castañeda and Lacher 2018).

**Characteristics:** San Quintín kangaroo rat is large sized; total length 286.0–312.0 mm and skull length 22.6–23.1 mm. Dorsal pelage pale brown; tail thin and length greater than the head-and-body length; pinna small; five toes on each hindfoot; **zygomatic arches breadth greater than 55% of skull length**. Species group *heermanni*.

**Comments:** *Dipodomys gravipes* is restricted to San Telmo and San Quintín valleys in northwestern Baja California (Map 358), and can be found in sympatry, or nearly so, with *D. merriami* and *D. simulans*. *D. gravipes* can be distinguished from *D. merriami* by having five toes on each hindfoot and total length greater than 270.0 mm. From *D. agilis* and *D. simulans*, by being larger and heavier; tail thinner, paler, and less sharply bicolored; ears proportionately smaller; hindfoot longer; zygomatic arches breadth at the maxilla level greater than 55% of skull length.

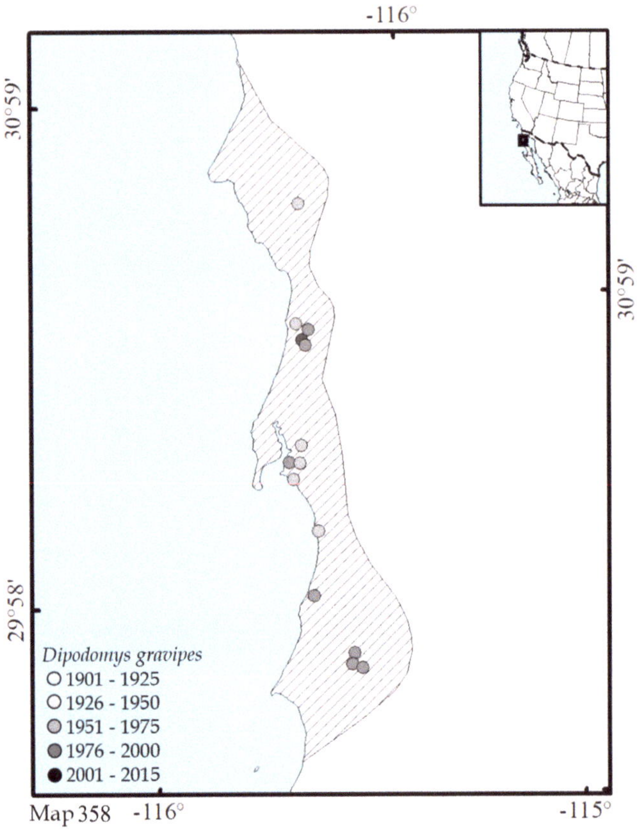


**Map 358** Distribution map of *Dipodomys gravipes*

**Additional Literature:** Huey (1925, 1951), Lackey (1967a), Hall (1981), Best and Schnell (1974), Stock (1974), Best (1978, 1981, 1983a, b), Best and Lackey (1985), Schnell et al. (1978), Williams et al. (1993), Patton and Álvarez-Castañeda (1999), Patton (2005a), Tremor et al. (2019), Cab-Sulub and Álvarez-Castañeda (2020).

# ***Dipodomys heermanni*** LeConte, 1853

Heermann's kangaroo rat, rata canguro del centro de California

Monotypic.

**Conservation:** The subspecies *D. h. morroensis* is listed as Endangered and *D. h. berkeleyensis* as Probably Extinct by the United States Endangered Species Act (FWS 2022). *D. heermanni* is listed as Least Concern by the IUCN (Cassola 2016w).

**Characteristics:** Morro Bay kangaroo rat is medium sized; total length 250.0–313.0 mm and skull length 39.7–40.2 mm. Dorsal pelage strongly tawny-olive throughout, washed with black to ochraceous buff, hair long and silky; tail slightly crested from dusky to blackish; pinna medium sized; five toes on the hindfoot; auditory bullae small to medium; zygomatic arches across maxillary processes prominent. Species group *heermanni*.

**Comments:** Nine subspecies of *Dipodomys heermanni* have been described (*D. h. arenae*, *D. h. berkeleyensis*, *D. h. dixoni*, *D. h. goldmani*, *D. h. heermanni*, *D. h. jolonensis*, *D. h. morroensis*, *D. h. swarthi*, and *D. h. tularensis*), many of them from very few localities and restricted ranges. The genetic and morphometric analyses of specimens from all the recognized subspecies showed very slight genetic and morphometrical variations among populations; under these circumstances, *D. heermanni* should be considered a monotypic species (Benedict et al. 2019). *D. h. californicus*, *D. h. eximius*, and *D. h. saxatilis*, previously subspecies of *D. heermanni* (Hall 1981), are now subspecies of *D. californicus* (Patton et al. 1976). *D. heermanni* ranges from a stripes between Suisun Bay and Lake Tahoe south through a line between Point Conception on the Pacific coast to the foothills of Sierra Nevada and the Tehachapi Mountains (Map 359). *D. heermanni* can be found in sympatry, or nearly so, with *D. agilis*, *D. californicus*, *D. ingens*, *D. nitratoides*, *D. panamintinus*, and *D. venustus*. *D. heermanni* can be distinguished from *D. agilis* by being paler in coloration. From *D. californicus*, *D. ingens*, *D. merriami*, and *D. nitratoides*, by having five toes on each hindfoot. From *D. ingens*, by having the hindfoot length less than 47.0 mm and by being smaller and paler. From *D. panamintinus*, by having the tail longer than 150% of the head-and-body length; tail pale gray or whitish, little hair tuft or no crest at the tip. From *D. venustus*, by its smaller ears and size; nasals and premaxilla narrower; maxillary arch outer angle more developed, and auditory bullae less projected posteriorly.

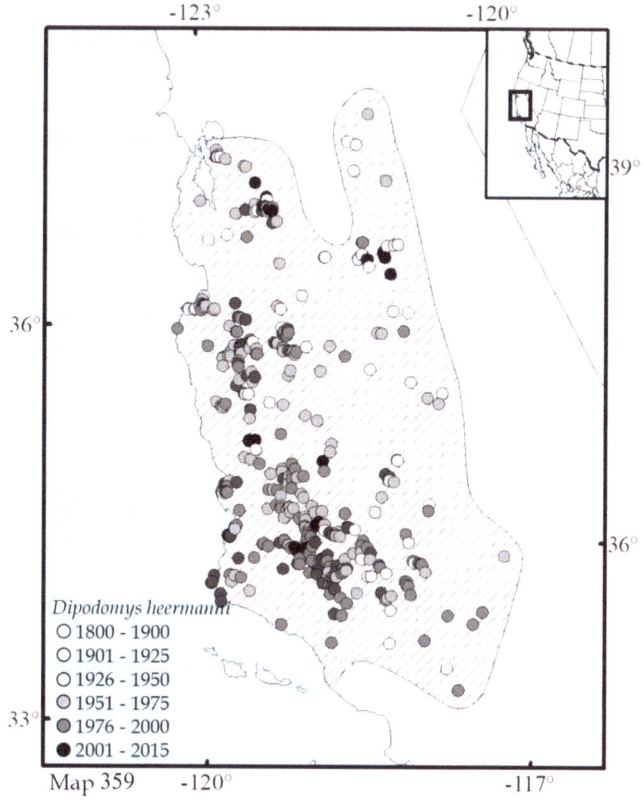


**Map 359** Distribution map of *Dipodomys heermanni*

**Additional Literature:** Grinnell (1922), Burt (1936, 1960), Setzer (1949), Mayer (1952), Lidicker (1960a), Stock (1971, 1974), Homan and Genoways (1978), Kelt (1988b), Williams et al. (1993), Patton (2005a).

## *Dipodomys ingens* (Merriam, 1904)

Giant kangaroo rat, rata canguro gigante

Monotypic.

**Conservation:** *Dipodomys ingens* is listed as Endangered by the United States Endangered Species Act (FWS 2022) and as Endangered (B2ab [i, ii, iii]: area of occupancy less than 500 km$^2$ in five or less populations and continuing decline in extent of occurrence, area of occupancy, and quality of habitat) by the IUCN (Roach 2018c).

**Characteristics:** Giant kangaroo rat is the largest sized; total length 311.0–348.0 mm and skull length 44.2–45.9 mm. Dorsal pelage yellowish brown with scattered blackish and dusky hairs; flanks buff-colored; ears small; tail longer than the head-and-body, thickened, with a dorsal crest of long hairs and a large hair tuft; five toes on each hindfoot; skull large and broad; zygomatic arches broad across the maxillary process. Species group *heermanni*.

**Comments:** *Dipodomys ingens* is restricted to San Joaquin Valley, California (Map 360). It can be distinguished from *D. heermanni* and *D. nitratoides* by being the largest and heaviest kangaroo rat; hindfoot length greater than 47.0 mm; tail length about 128% of the head-and-body length; skull large and broad. From *D. merriami*, *D. nitratoides*, and the other large species like *D. deserti*, *D. nelsoni*, and *D. spectabilis*, by having five toes on each hindfoot.

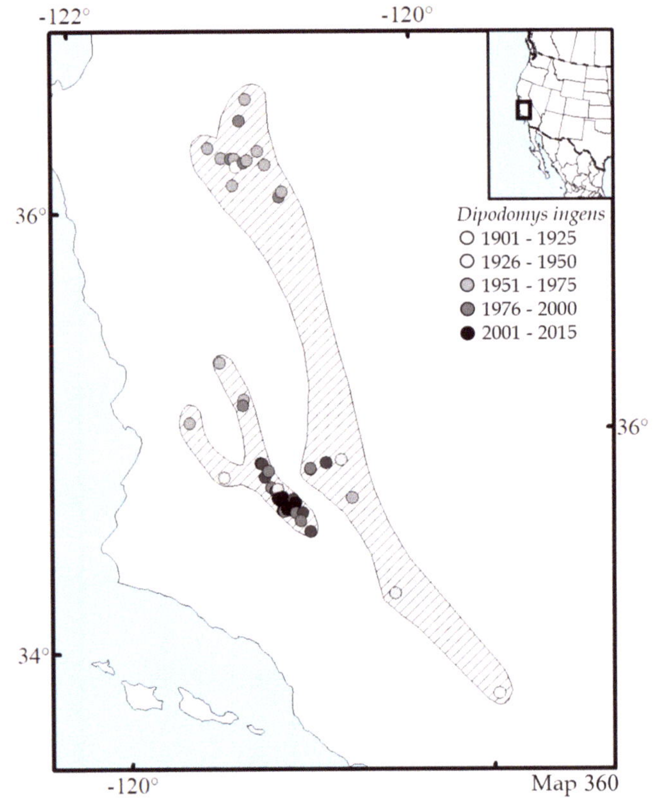


**Map 360** Distribution map of *Dipodomys ingens*

**Additional Literature:** Grinnell (1922, 1932), Shaw (1934), Burt (1936, 1960), Setzer (1949), Lidicker (1960b), Best and Schnell (1974), Stock (1974), Nader (1978), Schnell et al. (1978), Hall (1981), Williams and Kilburn (1991), Williams et al. (1993), Patton (2005a).

## ***Dipodomys merriami*** Mearns, 1890

Merriam's kangaroo rat, rata canguro común de cuatro dedos

1. *D. m. ambiguus* Merriam, 1890. From southern New Mexico and Texas south through Durango, Coahuila, and Nuevo León.
2. *D. m. annulus* Huey, 1951. Restricted to southern Bahía de Los Ángeles, Baja California.
3. *D. m. arenivagus* Elliot, 1904. Restricted to the northeastern coast of Baja California.
4. *D. m. atronasus* Merriam, 1894. Form southern Coahuila and eastern Durango south through San Luis Potosí and Aguascalientes.
5. *D. m. brunensis* Huey, 1951. Restricted to the northeastern coast of Baja California Sur.
6. *D. m. collinus* Lidicker, 1960. Restricted to San Diego County.
7. *D. m. frenatus* Bole, 1936. Southwestern Idaho and northwestern Arizona.
8. *D. m. insularis* Merriam, 1907. Restricted to San José Island, Baja California Sur.
9. *D. m. llanoensis* Huey, 1951. Restricted to the central-western coast of Baja California Sur.
10. *D. m. margaritae* Merriam, 1907. Restricted to Margarita Island, Baja California Sur.
11. *D. m. mayensis* Goldman, 1928. Southern Sonora and northern Sinaloa.
12. *D. m. melanurus* Merriam, 1893. Southern end of Baja California Peninsula.
13. *D. m. merriami* Mearns, 1890. From western Nevada south through Sonora, including eastern California, western and south Arizona.
14. *D. m. mitchelli* Mearns, 1897. Restricted to Tiburón Island, Sonora.
15. *D. m. olivaceus* Swarth, 1929. Southeastern Arizona, southwestern New Mexico, northeastern Sonora, and northwestern Chihuahua.
16. *D. m. parvus* Rhoads, 1894. Restricted to San Bernardino and Riverside Counties, California.
17. *D. m. platycephalus* Merriam, 1907. From the central-western coast of Baja California south through Baja California Sur.
18. *D. m. quintinensis* Huey, 1951. Restricted to San Quintín Valley, Baja California.
19. *D. m. semipallidus* Huey, 1927. Central Baja California.
20. *D. m. trinidadensis* Huey, 1951. Restricted to the Central Desert of Baja California.
21. *D. m. vulcani* Benson, 1934. Restricted to Mohave County, Arizona.

**Conservation:** *Dipodomys merriami parvus* is listed as Endangered by the United States Endangered Species Act (FWS 2022), *D. m. insularis* and *D. m. margaritae* as Endangered by the Norma Oficial Mexicana (DOF 2019). *D. merriami* as Least Concern in the IUCN (Timm et al. 2016e).

**Characteristics:** Merriam's kangaroo rat is small sized; total length 234.0–259.0 mm and skull length 33.9–37.3 mm. Dorsal pelage varies greatly from sandy brown to blackish brown, depending on the substrate color of their habitat; flanks paler, grizzled with yellow and orange; tail length greater than the head-and-body length; four toes on each hindfoot; auditory bullae large; interorbital breadth greater than half of the basal length; wide across the maxillary process of the narrow zygomatic arches. Species group *merriami*.

**Comments:** *Dipodomys insularis* and *D. margaritae* were previously considered valid species (Lidicker 1960a) until the review of *D. merriami* (Álvarez-Castañeda et al. 2009). *D. m. compactus* and *D. m. sennetti* were subspecies of *D. merriami*; now both are considered within *D. compactus* (Johnson and Sealander 1971; Schmidly and Hendricks 1976; Baumgardner and Schmidly 1981). *D. merriami* ranges in southwestern United States and northern México, from California, Nevada, Utah, Arizona, New Mexico, and southwestern Texas south through Aguascalientes, Zacatecas, and San Luis Potosí, including the Baja California Peninsula and three islands, Margarita, San José, and Tiburón (Sonora; Map 361). *D. merriami* can be found in sympatry, or nearly so, with *D. agilis*, *D. deserti*, *D. gravipes*, *D. heermanni*, *D. ingens*, *D. microps*, *D. nelsoni*, *D. ordii*, *D. ornatus*, *D. paramantinus*, *D. phillipsii*, *D. simulans*, *D. spectabilis*, and *D. stephensi*. *D. merriami* can be distinguished from *D. agilis*, *D. compactus*, *D. gravipes*, *D. heermanni*, *D. ingens*, *D. microps*, *D. ordii*, *D. panamintinus*, *D. simulans*, *D. stephensi*, and *D. venustus* by having four toes on each hindfoot. From *D. deserti*, by having a darker coloration; smaller size, total length less than 300.0 mm; without a flattened skull; auditory bullae not greatly inflated, and mastoids that do not meet immediately behind the parietals. From *D. panamintinus*, by having the tail larger than 145% of the head-and-body length; white tail stripe more or less one-half of tail width, and dark ventral stripe extending to the tip. From *D. californicus*, by having a lighter coloration; smaller size, less than 260.0 mm and a small hindfoot, less than 40.0 mm on average. From *D. deserti*, *D. nelsoni*, *D. spectabilis*, and *D. venustus*, by its smaller size. From *D. ornatus* and *D. phillipsii*, by having the tail blackish at the tip and interorbital breadth greater than half of the basal length. From *D. panamintinus*, by having the tail longer than 145% of the head-and-body length, white tail stripe more or less one-half of tail width, and dark ventral stripe extending to the tip. From *D. nitratoides*, *with* which it is sympatric, or nearly so, and morphologically similar, by having paler dorsal and ventral tail stripes, lateral white stripes broader than the darker stripes, and the dark ventral stripe narrower or almost lacking.

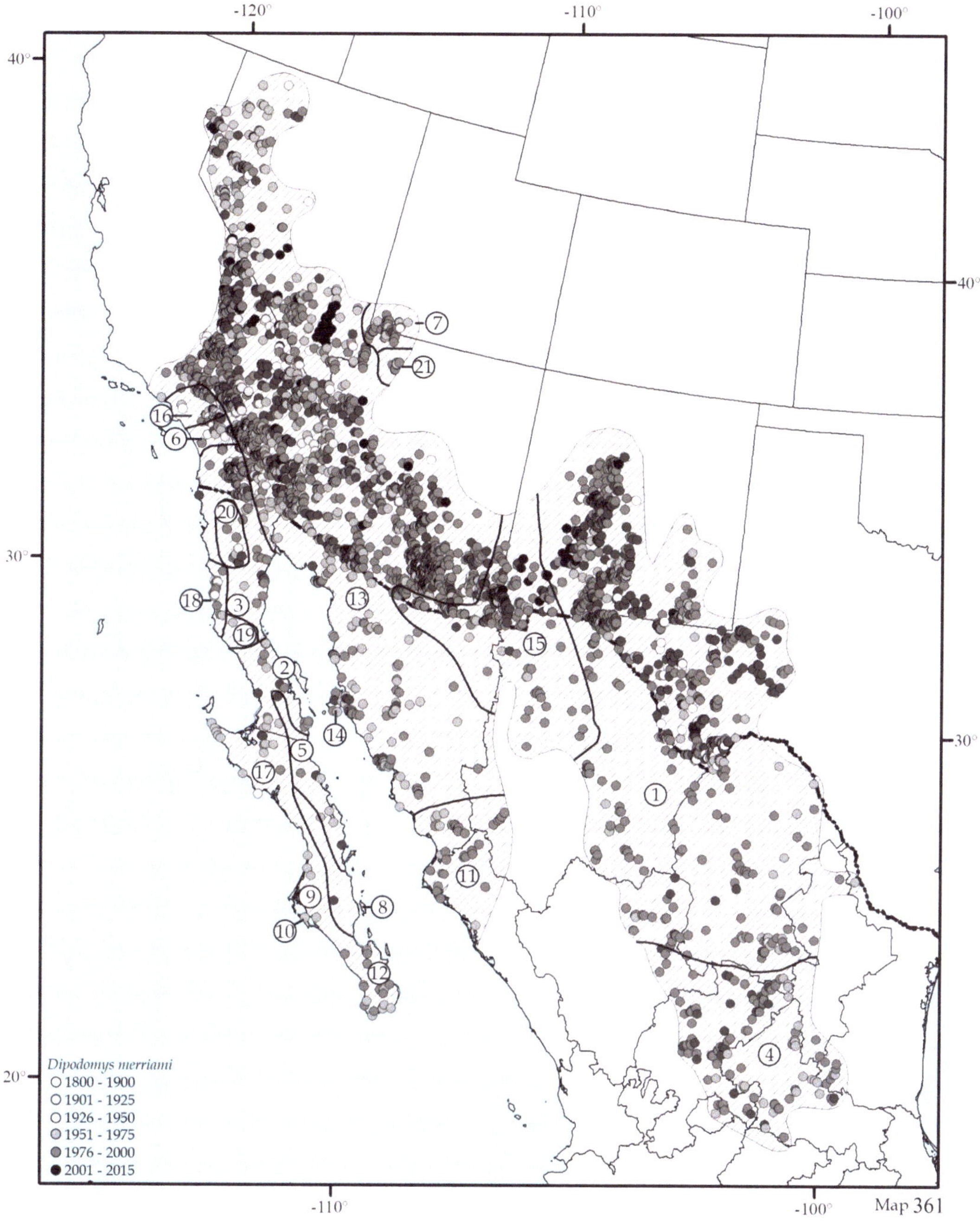


**Map 361** Distribution map of *Dipodomys merriami*

**Additional Literature:** Grinnell (1922), Setzer (1949), Álvarez (1960), Nader (1966), Mazrimas and Hatch (1972), Best and Schnell (1974), Patton et al. (1976), Schnell et al. (1978), Best and Thomas (1991a), Best (1992a), Williams et al. (1993), Patton and Álvarez-Castañeda (1999), Patton (2005a), Álvarez-Castañeda et al. (2008c).

## *Dipodomys microps* (Merriam, 1904)

Chisel-toothed kangaroo rat, rata canguro con incisivos de cincel

1. *D. m. alfredi* Goldman, 1937. Restricted to Gunnison Island, Great Salt Lake, Utah.
2. *D. m. aquilonius* Willett, 1935. Restricted to northeastern Nevada.
3. *D. m. bonnevillei* Goldman, 1937. Restricted to western Utah and western Nevada.
4. *D. m. celsus* Goldman, 1924. Restricted to southeastern Utah and northwestern Arizona.

5. *D. m. centralis* Hall and Dale, 1939. Restricted to central Nevada.
6. *D. m. idahoensis* Hall and Dale, 1939. Restricted to western Idaho.
7. *D. m. leucotis* Goldman, 1931. Restricted to northern Arizona.
8. *D. m. levipes* (Merriam, 1904). Restricted to Inyo County, California.
9. *D. m. microps* (Merriam, 1904). Restricted to the desert area southeastern of Sierra Nevada, California.
10. *D. m. occidentalis* Hall and Dale, 1939. Western and southern Nevada.
11. *D. m. preblei* (Goldman, 1921). Northwestern Nevada and southeastern Oregon.
12. *D. m. russeolus* Goldman, 1939. Restricted to Dolphin Island, Great Salt Lake, Utah.
13. *D. m. subtenuis* Goldman, 1939. Restricted to Carrington Island, Great Salt Lake, Utah.

**Conservation:** *Dipodomys microps* is listed as Least Concern by the IUCN (Cassola 2016x).

**Characteristics:** Chisel-toothed kangaroo rat is small sized; total length 254.0–300.0 mm and skull length 35.2–40.5 mm. Dorsal pelage brown; five toes on each hindfoot; **lower incisors flat on their anterior face, chisel shaped.** Species group *ordii*.

**Comments:** *Dipodomys microps* ranges from southeastern Oregon, western Idaho, eastern California, Nevada, Utah, western Colorado, and northwestern Arizona (Map 362). *D. microps* can be found in sympatry, or nearly so, with *D. ordii*, *D. merriami*, and *D. panamintinus*, and can be distinguished from both (and all other species of kangaroo rats) by its anterior lower incisors flat and chisel-like; skull narrow and constricted, maxillary arches weakly angled; and with dark coloration in the genus. In addition, from *D. ordii*, by having a prominent facemask and dark tail stripes. From *D. merriami* by having five toes on each hindfoot.

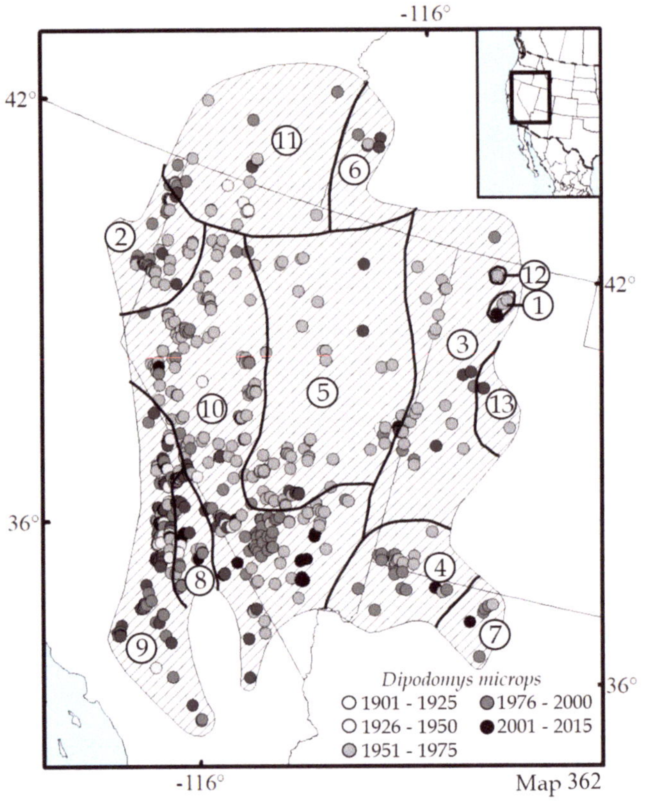


**Map 362** Distribution map of *Dipodomys microps*

**Additional Literature:** Grinnell (1922), Hall and Dale (1939), Setzer (1949), Nader (1966), Hatch et al. (1971), Kenagy (1973), Csuti (1979), Hall (1981), Kenagy and Bartholomew (1985), Hayssen (1991), Williams et al. (1993), Patton (2005a).

## *Dipodomys nelsoni* Merriam, 1907

Nelson's kangaroo rat, rata canguro de cola blanca

Monotypic.

**Conservation:** *Dipodomys nelsoni* is listed as Least Concern by the IUCN (Álvarez-Castañeda et al. 2016w).

**Characteristics:** Nelson's kangaroo rat is large sized; total length 310.0–330.0 mm and skull length 42.0–47.0 mm. Dorsal pelage pale brownish; tail length greater than the head-and-body length; **tail hair tuft at the tip occasionally white, approximately 20.0 mm long and occupying less than 25% of the tail length**; four toes on each hindfoot; **incisors narrow**; **maxillary arch relatively heavy and not flared**; auditory bullae relatively large; supraoccipital and interparietal bones usually small. Species group *spectabilis*.

**Comments:** *Dipodomys nelsoni* was recognized as a distinct species (Anderson 1972; Best 1988a; Matson 1980), previously regarded as a subspecies of *D. spectabilis* (Nader 1966, 1978). *D. nelsoni* ranges from the altiplano Mexicano in eastern Chihuahua, northern Coahuila, northwestern Durango, northern Zacatecas, San Luis Potosí, and southern Nuevo León (Map 363). It can be found in sympatry, or nearly so, with *D. ingens*, *D. ordii*, *D. ornatus*, *D. merriami*, and *D. spectabilis*. *D. nelsoni* can be distinguished from *D. ingens* by having hindfoot length less than 47.0 mm. From *D. ordii*, *D. ornatus*, and *D. merriami*, by its larger size, total length greater than 285.0 mm. Form *D. ordii*, by having five toes in each hindfoot. From *D. spectabilis*, by having less than 25% of the tail end white; slightly smaller size; paler coloration and a narrower maxillary region.

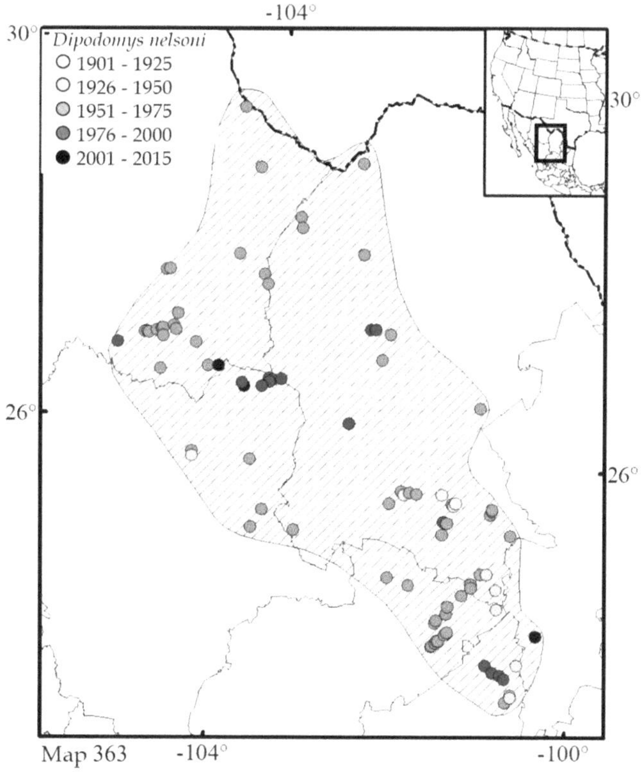


**Map 363** Distribution map of *Dipodomys nelsoni*

**Additional Literature:** Setzer (1949), Baker (1956), Álvarez (1960), Hoffmeister and Nader (1963), Nader (1978), Williams et al. (1993), Patton (2005a).

## *Dipodomys nitratoides* Merriam, 1894

San Joaquín Valley kangaroo rat, rata canguro del Valle de San Joaquín

1. *D. n. brevinasus* Grinnell, 1920. Restricted to the western Central Valley, California.
2. *D. n. exilis* Merriam, 1894. Restricted to the northeastern Central Valley, California.
3. *D. n. nitratoides* Merriam, 1894. Restricted to the eastern Central Valley, California.

**Conservation:** *Dipodomys nitratoides exilis* and *D. n. nitratoides* are listed as Endangered by the United States Endangered Species Act (FWS 2022), and *D. n. exilis* was recorded as extinct by Patton et al. (2019). *D. nitratoides* as **Vulnerable** (B2ab [i, ii, iii, v]: area of occupancy less than 2000 $km^2$ with ten or less populations and continuing decline in extent of occurrence, area of occupancy, quality of habitat, and number of mature individuals) by the IUCN (Roach 2018d).

**Characteristics:** San Joaquin Valley kangaroo rat is small sized; total length 210.0–255.0 mm and skull length 18.0–19.5 mm. Dorsal pelage pale brown to brown, flanks paler grizzled yellowish; tail length greater than the head-and-body length; four toes on each hindfoot; rostrum short with a narrow base and sides nearly parallel; auditory bullae large; interorbital breadth greater than half of the basal length; width across the maxillary process of the zygomatic arches narrow. Species group *merriami*.

**Comments:** *Dipodomys nitratoides* ranges is restricted to the San Joaquin and Cuyama Valleys and the Carrizo Plain in California (Map 364). It can be found in sympatry, or nearly so, with *D. heermanni* and *D. ingens*. *D. nitratoides* can be distinguished from *D. heermanni* and *D. ingens* by having four toes on each hindfoot. From *D. merriami* does not show a sympatric distribution but is morphologically similar and can be distinguished by having dorsal and ventral tail stripes darker and broader than the lateral white stripes.

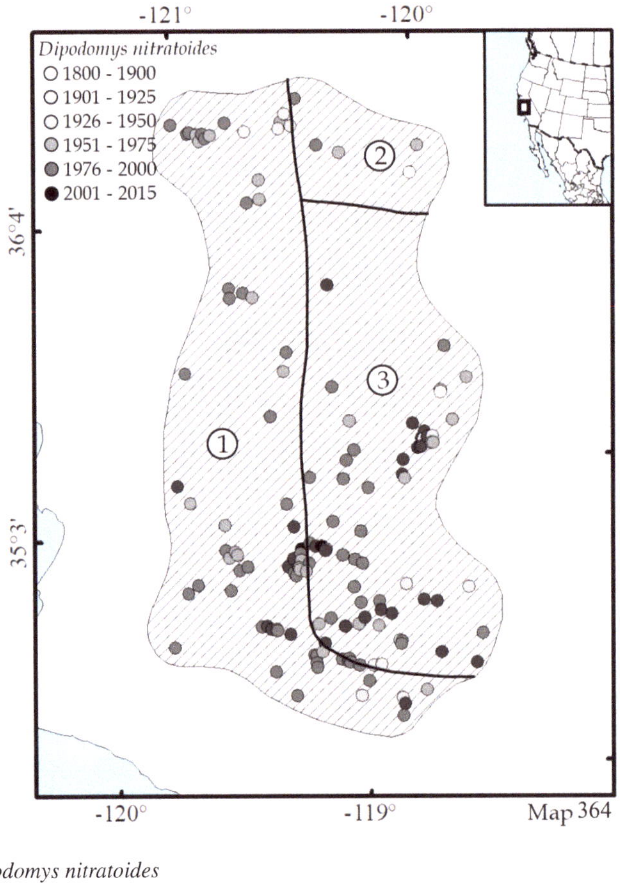


**Map 364** Distribution map of *Dipodomys nitratoides*

**Additional Literature:** Merriam (1894), Grinnell (1922), Hatt (1932), Setzer (1949), Lidicker (1960a), Nader (1966), Johnson and Sealander (1971), Best and Schnell (1974), Hoffmann (1975), Patton et al. (1976), Hall (1981), Best (1991), Williams et al. (1993).

## *Dipodomys ordii* Woodhouse, 1853

Ord's kangaroo rat, rata canguro común de cinco dedos

1. *D. o. attenuatus* Bryant, 1939. Restricted to soutwestern Texas between Chihuahua and Coahuila.
2. *D. o. celeripes* Durrant and Hall, 1939. Northeastern Nevada and northwestern Utah.
3. *D. o. chapmani* Mearns, 1890. Restricted to western Arizona.
4. *D. o. cinderensis* Hardy, 1944. Restricted to southwestern Utah.
5. *D. o. cineraceus* Goldman, 1939. Restricted to northern Utah.
6. *D. o. columbianus* (Merriam, 1894). Southern Washington, Oregon, southern Idaho, northeastern California, and northern Nevada.
7. *D. o. cupidineus* Goldman, 1924. Restricted to northwestern Arizona.
8. *D. o. durranti* Setzer, 1952. Western Coahuila, Nuevo León, and southwestern Tamaulipas.
9. *D. o. evexus* Goldman, 1933. Restricted to central Colorado.
10. *D. o. extractus* Setzer, 1949. Restricted to northern-central Chihuahua.
11. *D. o. fetosus* Durrant and Hall, 1939. Restricted to western Nevada.
12. *D. o. fremonti* Durrant and Setzer, 1945. Restricted to southern Utah.
13. *D. o. idoneus* Setzer, 1949. Southwestern Coahuila and northeastern Durango.
14. *D. o. inaquosus* Hall, 1941. Restricted to central-northern Nevada.
15. *D. o. largus* Hall, 1951. Restricted to Mustang Island, Texas.
16. *D. o. longipes* (Merriam, 1890). Northeastern Arizona, northwestern New Mexico, and southeastern Utah.
17. *D. o. luteolus* (Goldman, 1917). Southeastern Wyoming, South Dakota, northern Nebraska, and northeastern Colorado.
18. *D. o. marshalli* Goldman, 1937. Restricted to Bird Island, Great Salt Lake, Utah.
19. *D. o. medius* Setzer, 1949. Central and southeastern New Mexico.
20. *D. o. monoensis* (Grinnell, 1919). Restricted to western Nevada.
21. *D. o. montanus* Baird, 1855. Central-southern Colorado and central-northern New Mexico.
22. *D. o. nexilis* Goldman, 1933. Central-western Nebraska and central-eastern Utah.
23. *D. o. obscurus* (Allen, 1903). Southern Chihuahua and central Durango.
24. *D. o. oklahomae* Trowbridge and Whitaker, 1940. Restricted to central Oklahoma.
25. *D. o. ordii* Woodhouse, 1853. Southern Arizona and New Mexico, southwestern Texas, northern Sonora, and Chihuahua.
26. *D. o. pallidus* Durrant and Setzer, 1945. Restricted to central-western Utah.
27. *D. o. palmeri* (Allen, 1881). From eastern Durango, Zacatecas, San Luis Potosí south through Tlaxcala.
28. *D. o. panguitchensis* Hardy, 1942. Restricted to central Utah.
29. *D. o. parvabullatus* Hall, 1951. Restricted to Matamoros, Tamaulipas.
30. *D. o. priscus* Hoffmeister, 1942. Southwestern Wyoming and northwestern Colorado.
31. *D. o. pullus* Anderson, 1972. Restricted to central-western Chihuahua.
32. *D. o. richardsoni* (Allen, 1891). Southwestern Colorado, southern Nebraska, Kansas, western Oklahoma, northern Texas, and northeastern New Mexico.
33. *D. o. sanrafaeli* Durrant and Setzer, 1945. Central-eastern Utah and western Colorado.
34. *D. o. terrosus* Hoffmeister, 1942. From northern Wyoming and northwestern South Dakota north through Alberta and Saskatchewan.
35. *D. o. uintensis* Durrant and Setzer, 1945. Restricted to northeastern Utah.
36. *D. o. utahensis* (Merriam, 1904). Restricted to central and northern-central Utah.

**Conservation:** *Dipodomys ordii* is listed as Least Concern by the IUCN (Cassola 2016y).

**Characteristics:** Ord's kangaroo rat is medium sized; total length 208.0–281.0 mm and skull length 35.1–39.3 mm. Dorsal pelage intense orange to dark gray, flanks paler than the dorsum; five toes on each hindfoot; **interparietal bone small and pointed posteriorly**; **lower incisors awl shaped.** Species group *ordii*.

**Comments:** *Dipodomys compactus* was considered a subspecies of *D. ordii* (Johnson and Sealander 1971, Schmidly and Hendricks 1976). *D. ordii* ranges from southern Washington, Idaho, Alberta, Saskatchewan, and North Dakota south through Guanajuato, Querétaro, and Hidalgo, and from Oregon and eastern California east through South Dakota, Nebraska, Kansas, Oklahoma, and Texas (Map 365). It can be found in sympatry, or nearly so, with *D. compactus*, *D. deserti*, *D. microps*, *D. panamintinus*, *D. spectabilis*, *D. merriami*, *D. nelsoni*, and *D. ornatus*. *D. ordii* can be distinguished from *D. compactus* by having a longer and crested tail; larger hair; skull larger and wider; auditory bullae inflated; interparietal bone more acutely pointed posteriorly and triangular. From *D. deserti* by its darker coloration and smaller size, total length less than 330.0 mm. From *D. microps*, by having the incisors not flattened anteriorly and strongly incurved; facemask not prominent and paler tail-stripes and coat. From *D. panamintinus*, by having tail length less than 135% of the head-and-body length; white tail stripe paler, as wide or wider than the ventral dark stripe, never reaching the end of the caudal vertebrae. From *D. spectabilis* and *D. nelsoni*, by its smaller size, total length less than 285.0 mm, and four toes in each hindfoot.

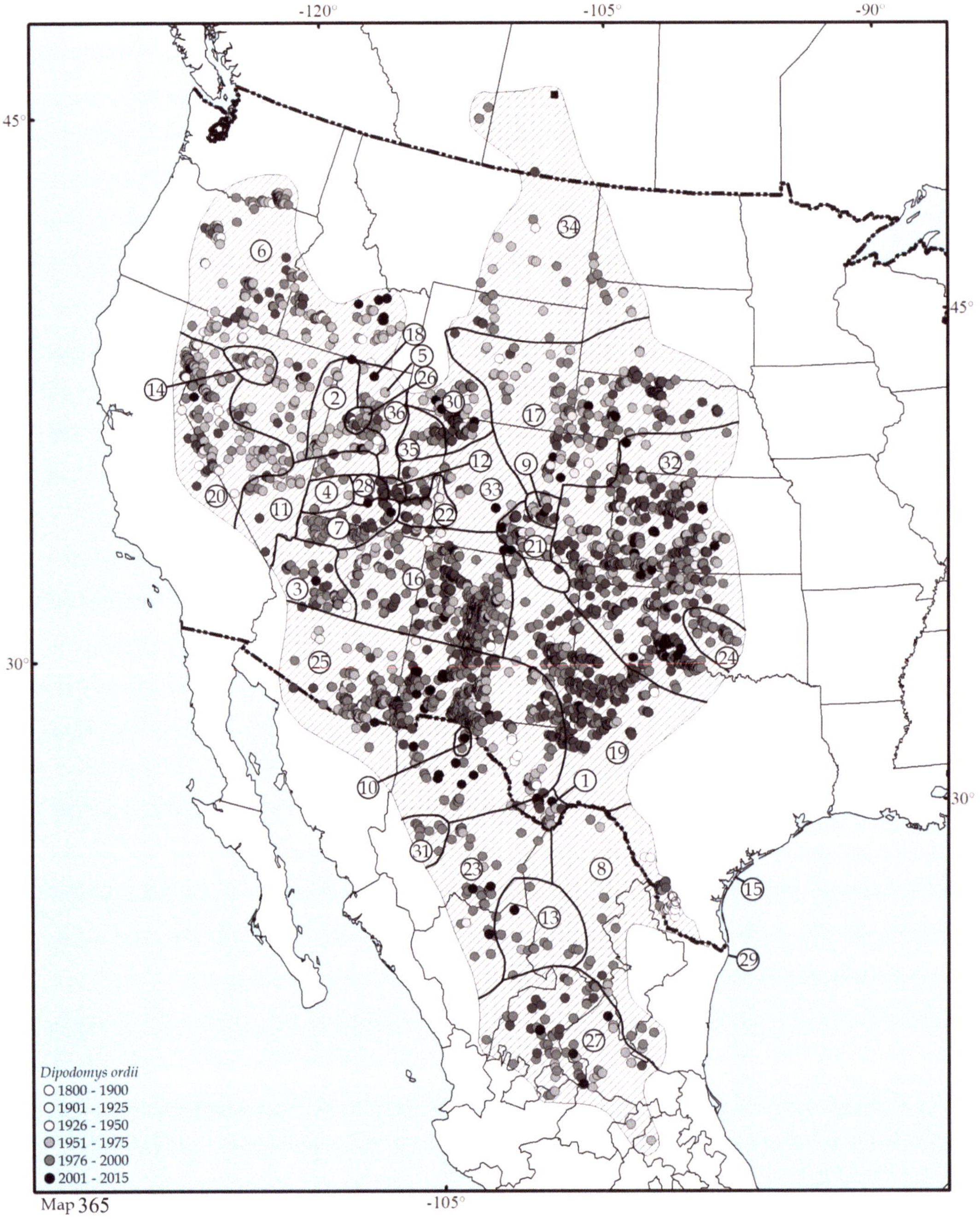


**Map 365** Distribution map of *Dipodomys ordii*

**Additional Literature:** Burt (1936, 1960), Álvarez (1960), Setzer (1949), Jannett (1976), Kennedy et al. (1980), Baumgardner and Schmidly (1981), Garrison and Best (1990), Baumgardner (1991), Williams et al. (1993).

## *Dipodomys ornatus* Merriam, 1894

Plateau kangaroo rat, rata canguro del Altiplano

Monotypic.

**Conservation:** *Dipodomys ornatus* is listed as Least Concern by the IUCN (Hafner 2019).

**Characteristics:** Plateau kangaroo rat is small sized; total length 230.0–290.0 mm and skull length 36.0–40.0 mm. Dorsal pelage pale brown; arietiform spots black and extensive; tail with a white or black hair tuft at the tip; four toes on each hindfoot; auditory bullae relatively small.

**Comments:** *Dipodomys ornatus* was reinstated as a distinct species (Fernández et al. 2012), previously regarded as a subspecies of *D. phillipsii* (Genoways and Jones 1971). It ranges in the Altiplano Mexicano in Durango, Zacatecas, Guanajuato, Querétaro, and San Luis Potosí (Map 366), and is sympatric, or nearly so, with *D. ordii*, *D. merriami*, *D. nelsoni*, and *D. spectabilis*. *D. ornatus* differs from *D. merriami* by having the tail tip white and interorbital breadth less than half of basal length. From *D. nelsoni* and *D. spectabilis*, by its smaller size, total length less than 240.0 mm. From *D. ordii*, by having four toes on each hindfoot. *D. ornatus* can only be differentiated from *D. phillipsii* using genetic data. *D. elator* was considered within the *phillipsii* group, but this was not accepted by Davis (1942); the differences between the two species are that *D. ornatus* (*D. phillipsii*), relative to *D. elator*, has a relatively long and slender tail (~170% of the head-and-body length); smaller body, 105.0 mm on average; rostrum narrow; interorbital region broad, and interparietal region narrow.

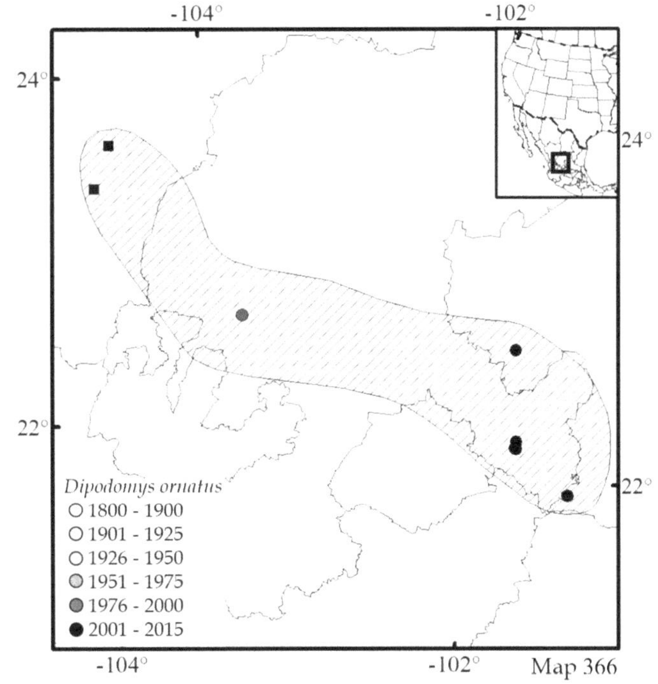


**Map 366** Distribution map of *Dipodomys ornatus*

**Additional Literature:** Setzer (1949), Álvarez (1960), Lidicker (1960b), Best and Schnell (1974), Jones and Genoways (1975), Patton (2005a).

## *Dipodomys panamintinus* (Merriam, 1894)

Panamint kangaroo rat, rata canguro del Valle de Panamint

1. *D. p. argusensis* Huey, 1945. Restricted to Inyo County, California.
2. *D. p. caudatus* Hall, 1946. Restricted to San Bernardino County, California.
3. *D. p. leucogenys* (Grinnell, 1919). Western-central Nevada and eastern-central California.
4. *D. p. mohavensis* (Grinnell, 1918). Central California.
5. *D. p. panamintinus* (Merriam, 1894). Restricted to Inyo County, California.

**Conservation:** *Dipodomys panamintinus* is listed as Least Concern by the IUCN (Cassola 2016z).

**Characteristics:** Panamint kangaroo rat is medium sized; total length 285.0–335.0 mm and skull length 39.0–40.0 mm. Dorsal pelage pale buffy clay-color with pale ochraceous shades; ears small; tail heavily crested and about 58% of the total length, dorsal and ventral stripes pale dusky, lateral stripes white meeting below on distal third; five toes on each hindfoot; **zygomatic arches at maxillary processes very broad,** with prominent rounded angle; auditory bullae small. Species group *heermanni*.

**Comments:** *Dipodomys panamintinus* has discontinuous ranges in eastern California and western Nevada (Map 367). It can be found in sympatry, or nearly so, with *D. agilis*, *D. deserti*, *D. heermanni*, *D. merriami*, *D. microps*, *D. ordii*, *D. nitratoides*, and *D. stephensi*. *D. panamintinus* can be distinguished from *D. agilis*, *D. heermanni*, *D. merriami*, *D. ordii*, and *D. stephensi* by its tail length about 140% of the head-and-body length, white stripe extending to the tip and heavily crested. From *D. agilis*, by having much smaller ears and a paler coloration. From *D. panamintinus*, by having the tail heavily crested and with wide white stripes. From *D. stephensi*, by having hindfoot length usually less than 42.5 mm. From *D. deserti*, *D. merriami* and *D. nitratoides*, by having five toes on each hindfoot. From *D. microps*, by having the incisors curved on their anterior face, not chisel-like. From *D. ordii*, by having the tail length greater than 135% of the head-and-body length.

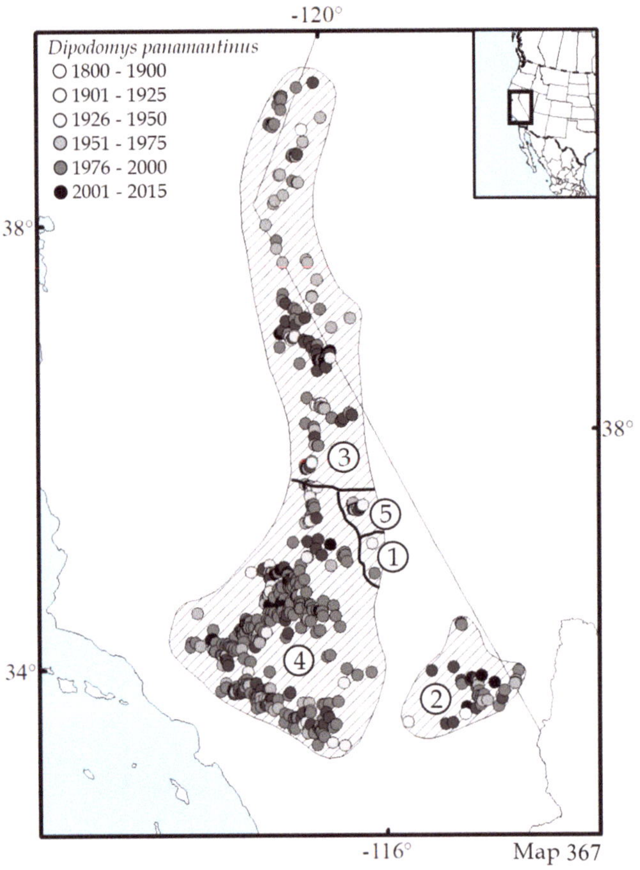


**Map 367** Distribution map of *Dipodomys panamintinus*

**Additional Literature:** Merriam (1894), Grinnell (1922), Hall (1946), Setzer (1949), Beer (1965), Ingles (1965), Nader (1966), Mazrimas and Hatch (1972), Best and Schnell (1974), Patton et al. (1976), Intress and Best (1990), Williams et al. (1993).

## ***Dipodomys phillipsii*** Gray, 1841

Phillips's kangaroo rat, rata canguro del Valle de México

1. *D. p. oaxacae* Hooper, 1947. Southern Puebla and northern Oaxaca.
2. *D. p. perotensis* Merriam, 1894. Puebla and Tlaxcala.
3. *D. p. phillipsii* Gray, 1841. Northern Morelos, Ciudad de México, Estado de México, and Hidalgo

**Conservation:** *Dipodomys phillipsii oaxacae*, *D. p. perotensis*, and *D. p. phillipsii* are listed as Threatened by the Official Mexican Standard (DOF 2019). *D. phillipsii* listed as Least Concern by the IUCN (Álvarez-Castañeda et al. 2016ac).

**Characteristics:** Phillips's kangaroo rat is small sized; total length 263.0–289.0 mm and skull length 39.0–40.0 mm. Dorsal pelage pale brown; four toes on each hindfoot; arietiform spots black and extensive; tail with a white or black hair tuft at the tip; auditory bullae relatively small.

**Comments:** *Dipodomys ornatus* was previously considered a subspecies of *D. phillipsii* (Fernández et al. 2012). *D. phillipsii* ranges in arid lands of central México, Guanajuato, Querétaro, Morelos, Estado de México, Hidalgo, Veracruz, Puebla, and northern Oaxaca (Map 368). *D. phillipsii* differs from the other sympatric species, such as *D. merriami*, by having the tail tip white and interorbital breadth less than half of the basal length. From *D. phillipsii*, genetic data are needed; however, the mastoid breadth: the maxilla breadth ratio is generally smaller than 1.08 mm. From *D. elator*, by having a relatively long and slender tail (~170% of the head-and-body length); small body, mean length 105.0 mm; rostrum narrow; interorbital region broad and interparietal region narrow.

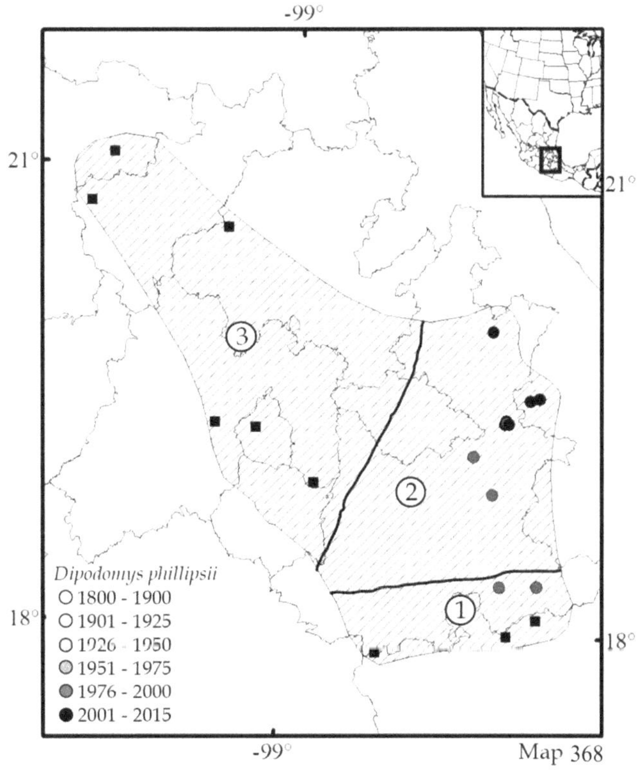


**Map 368** Distribution map of *Dipodomys phillipsii*

**Additional Literature:** Davis (1942), Setzer (1949), Álvarez (1960), Genoways and Jones (1971), Jones and Genoways (1975), Williams et al. (1993), Patton (2005a).

## *Dipodomys simulans* (Merriam, 1904)

Dulzura kangaroo rat, rata canguro de Baja California

1. *D. s. peninsularis* (Merriam, 1907). From Central Desert of Baja California Peninsula south through Baja California Sur.
2. *D. s. simulans* (Merriam, 1904). From southwestern California to Central Desert of Baja California.

**Conservation:** *Dipodomys simulans* is listed as Least Concern by the IUCN (Lacher et al. 2016a).

**Characteristics:** Dulzura kangaroo rat is largesized; total length 270.0–302.0 mm and skull length 38.0–41.5 mm. Dorsal pelage from yellowish to grayish brown, flanks grayish or orange; **tail length thicker than the head-and-body length**, five toes on each hindfoot; **zygomatic arches breadth less than 55% of skull length**. Species group *heermanni*.

**Comments:** *Dipodomys simulans* was recognized as a distinct species, including *D. antiquaries*, *D. peninsularis*, and *D. paralius* (Best et al. 1986; Williams et al. 1993; Sullivan and Best 1997a), previously regarded as a subspecies of *D. agilis* (Hall 1981). *D. simulans* ranges from the most southwestern areas of California south through central Baja California Sur near Magdalena Bay (Map 369). It can be found in sympatry, or nearly so, with *D. agilis*, *D. gravipes*, and *D. merriami*. *D. simulans* can be distinguished from *D. gravipes* by being smaller and paler, having the tail wide, proportionately large ears, and zygomatic arches breadth at the maxilla level less than 55% of skull length. From *D. agilis*, by having a paler dorsal coloration, back coloration near dusky pinkish cinnamon; larger auditory bullae and skull breadth across auditory bullae usually greater than 24.5 mm. From *D. merriami*, by having five toes on each hindfoot.

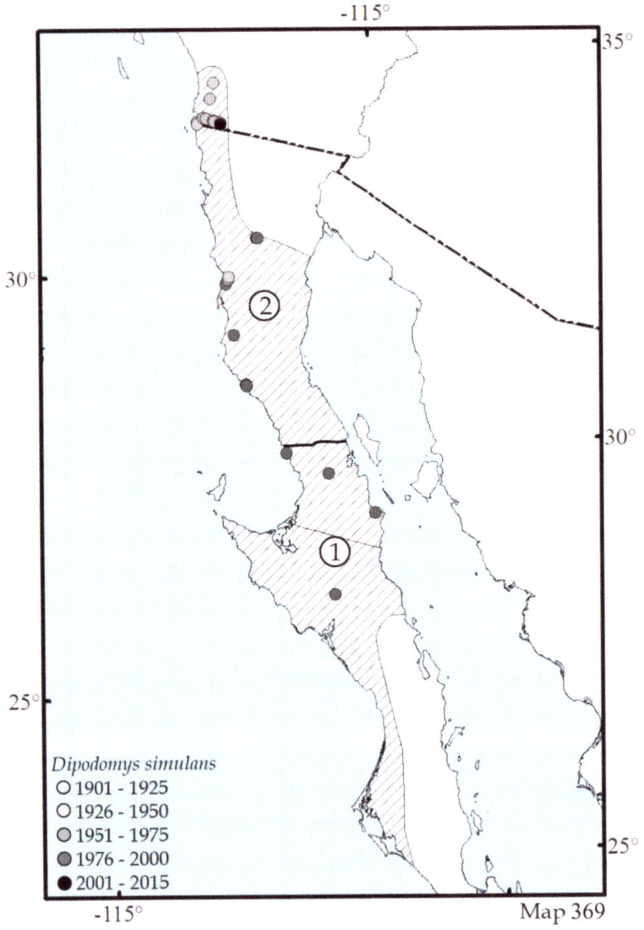


**Map 369** Distribution map of *Dipodomys simulans*

**Additional Literature:** Álvarez (1960), Patton and Álvarez-Castañeda (1999), Patton (2005a), Álvarez-Castañeda et al. (2008c).

## *Dipodomys spectabilis* Merriam, 1890

Banner-tailed kangaroo rat, rata canguro de cola peluda

1. *D. s. baileyi* Goldman, 1923. New Mexico and western Texas.
2. *D. s. clarencei* Goldman, 1933. Northeastern Arizona and northwestern New Mexico.
3. *D. s. cratodon* Merriam, 1907. Zacatecas, San Luis Potosí, and northern Jalisco.
4. *D. s. intermedius* Nader, 1965. Restricted to central Sonora.
5. *D. s. perblandus* Goldman, 1933. Southern Arizona and northern Sonora.
6. *D. s. spectabilis* Merriam, 1890. Southeastern Arizona, southwestern New Mexico, and Chihuahua.
7. *D. s. zygomaticus* Goldman, 1923. Restricted to southern Chihuahua.

**Conservation:** *Dipodomys spectabilis* is listed as Least Concern by the IUCN (Timm et al. 2019).

**Characteristics:** Banner-tailed kangaroo rat is large sized; total length 310.0–349.0 mm and skull length 42.6–48.7 mm. Dorsal pelage pale ochraceous buff, flanks grizzled ocher; tail length greater than the head-and-body length, **tail with a hair tuft always white (40.0 mm) at the tip, occupying greater than 25% of the tail length; tail black throughout before reaching the white tuft;** four toes on each hindfoot; skull large; auditory bullae inflated; interparietal bone variable in size and shape, sometimes fused with the supraoccipital; maxillary arches heavy, with the posterolateral edges slightly slanted or flared out, and anteromedial edge narrow and extending slightly along the premaxilla. Species group *spectabilis*.

**Comments:** *Dipodomys nelsoni* was recognized as a distinct species (Anderson 1972; Matson 1980; Best 1988a), previously regarded as a subspecies of *D. spectabilis* (Nader 1966, 1978). *D. spectabilis* has a disjunct distribution from northeastern Arizona, New Mexico, western Texas, northern Sonora, and Chihuahua, and from Aguascalientes, Jaslisco, Zacatecas, east though San Luis Potosí (Map 370). It can be found in sympatry, or nearly so, with *D. deserti*, *D. ordii*, *D. ornatus*, *D. merriami*, and *D. nelsoni*. *D. spectabilis* can be distinguished from all other *Dipodomys* species by its large and heavy skull; inflated mastoids separated on top by about 3.0 mm, so the interparietal bone has a distinct cuneate shape. From *D. ordii*, *D. ornatus*, and *D. merriami*, by having a larger size, total length greater than 285.0 mm. From *D. nelsoni*, by having a slightly larger size, darker coloration, more than 25% of the tail tip white, and a wide maxillary region. From *D. deserti*, by its mastoids meeting immediately behind the parietals and having an inconspicuous spicule (if any) between them.

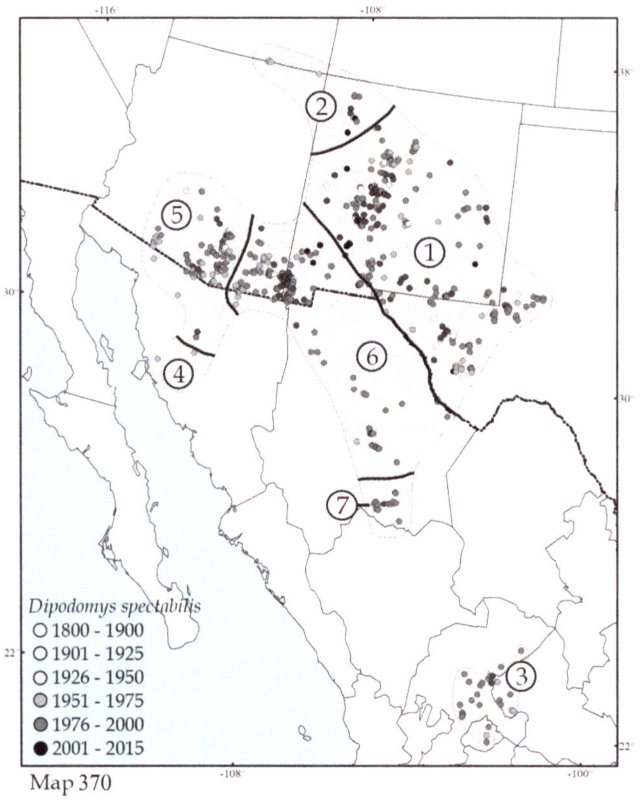


**Map 370** Distribution map of *Dipodomys spectabilis*

**Additional Literature:** Merriam (1890a), Grinnell (1922), Setzer (1949), Dalquest (1953), Baker (1956), Álvarez (1960), Hoffmeister and Nader (1963), Schnell et al. (1978), Williams et al. (1993), Patton (2005a).

## *Dipodomys stephensi* (Merriam, 1907)

Stephens' kangaroo rat, rata canguro de San Jacinto Valley

Monotypic.

**Conservation:** *Dipodomys stephensi* is listed as Endangered by the United States Endangered Species Act (FWS 2022) and as Vulnerable (B1ab [i, ii, iii]: extent of occurrence less than 20,000 km$^2$ with ten or less populations and continuing decline in extent of occurrence, area of occupancy, and quality of habitat) by the IUCN (Roach 2018e).

**Characteristics:** Stephens' kangaroo rat is medium sized; total length 275.0–300.0 mm and skull length 37.8–40.6 mm. Dorsal pelage dark brown; pinna small; tail length 1.45 times the head-and-body length, bicolored, crested, with many hairs in the dorsal and ventral sides, giving the stripes a grizzled appearance, and a few white hairs in the tuft; five toes on each hindfoot; dusky sole on the hindfoot; zygomatic arches at the maxillary processes very broad; auditory bullae small; foramen magnum with a rounded dorsum; suture between each nasal and premaxilla curved. Species group *heermanni*.

**Comments:** *Dipodomys stephensi* is restricted to San Jacinto Valley and adjacent areas in southern California (Map 371), and can be found in sympatry, or nearly so, with *D. agilis* and *D. panamintinus*. *D. stephensi* can be distinguished from *D. agilis* by having shorter ears (average of 15.0 mm), a proportionally wider head, with broader arietiform spots; tail with a few white hairs in the tuft, dorsal and ventral stripes with a grizzled appearance, and lateral white stripes indistinctly framed from the dark stripes; soles on the hindfoot dusky; auditory bullae "elongate globose" when viewed from above; foramen magnum dorsally rounded; suture between each nasal and premaxilla curved. From *D. panamintinus*, by having the hindfoot length usually greater than 42.5 mm; tail length greater than 145% of the head-and-body length; white tail stripe about one-half as wide as the dorsal stripe, and a dark ventral stripe extending to the end of the caudal vertebrae.

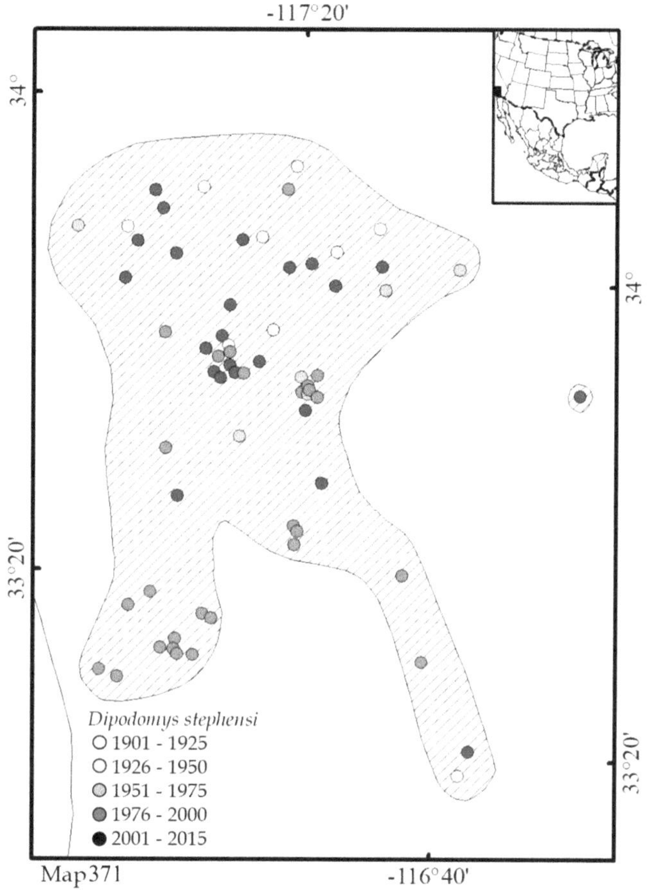


**Map 371** Distribution map of *Dipodomys stephensi*

**Additional Literature:** Grinnell (1922), Burt (1936), Setzer (1949), Lidicker (1960a), Ingles (1965), Lackey (1967a, b), Bleich (1977), Williams et al. (1993), Patton (2005a).

## *Dipodomys venustus* (Merriam, 1904)

Narrow-faced kangaroo rat, rata canguro de rostro angosto

1. *D. v. elephantinus* (Grinnell, 1919). Restricted to San Benito County.
2. *D. v. sanctiluciae* Grinnell, 1919. Western coast of California, from southern Monterey Bay to northern Santa Barbara.
3. *D. v. venustus* (Merriam, 1904). Western coast of California, from Stanford south through Monterey.

**Conservation:** *Dipodomys venustus* is listed as Least Concern by the IUCN (Cassola 2016aa).

**Characteristics:** Narrow-faced kangaroo rat is large sized; total length 293.0–318.0 mm and skull length 39.0–41.6 mm. Dorsal pelage dark brown, with a white spot all around the eye that continues to the ear base; ears large relative to other species in the genus; five toes on each hindfoot; the maxillary process of zygomatic arches narrow. Species group *heermanni*.

**Comments:** *Dipodomys elephantinus* and *D. venustus* may be conspecific with *D. agilis* (Hall 1981; Honacki et al. 1982), but *D. venustus* is a full species by its morphological, genetic, karyotype, and bacular characteristics (Best et al. 1996a); *D. elephantinus* is a subspecies of *D. venustus* (Grinnell 1922). *D. venustus* ranges along the coastal mountains of western-central California, from San Luis Obispo north to the southern San Francisco Bay (Map 372). It can be found in sympatry, or nearly so, with *D. agilis* and *D. heermanni*. *D. venustus* differs from *D. agilis* by having larger ears; a longer tail; darker coloration; bolder facial markings and a proportionally longer rostrum; skull slightly longer; zygomatic maxillary arch broader with a pronounced outer angle; jugal weaker; nasals slightly longer and broader (narrow-faced), and heavier incisors. From D. *heermanni*, by having a darker coloration, larger ears, nasals, and premaxilla broader, maxillary arch outer angle less developed, and auditory bullae projecting more posteriorly.

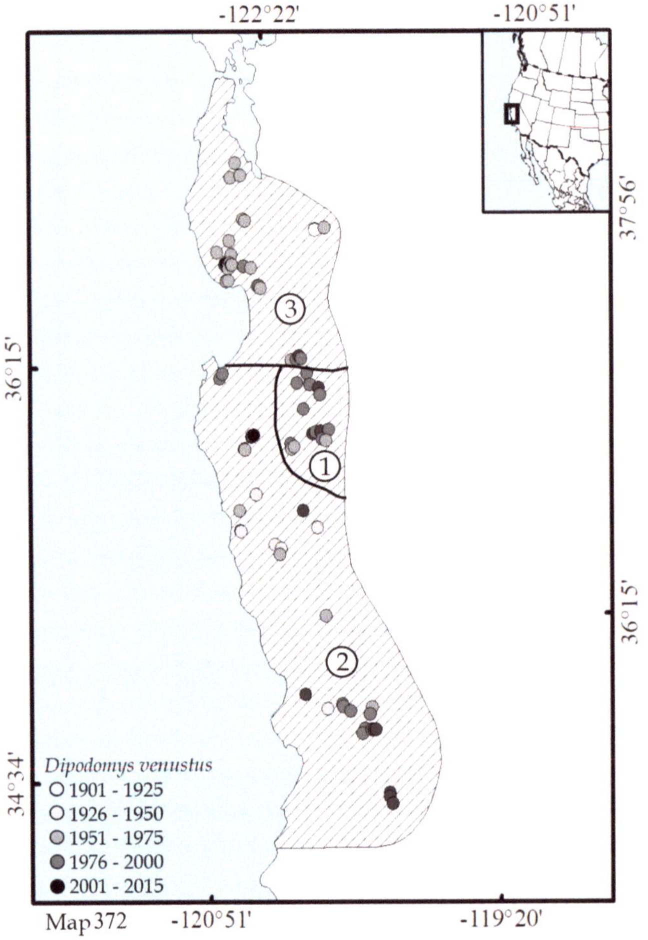


**Map 372** Distribution map of *Dipodomys venustus*

**Additional Literature:** Merriam (1904b), Hatt (1932), Boulware (1943), Best (1986, 1992b), Williams et al. (1993), Best et al. (1996a), Patton (2005a).

## Genus *Microdipodops* Merriam, 1891

Dorsal pelage paler brown, brownish, blackish or grayish; hair of underparts plumbeous or white at the base and white at the tip; tail slightly longer than the head-and-body; **hindfoot soles densely covered with long hair; proportionally, the greatest inflated auditory bullae of any Heteromyidae; ventrally, both bullae contact the glenoid fossa anteriorly, at the basisphenoid symphysis;** molars with an H-pattern; lower premolars with five or six cusps; cervical vertebrae mostly fused. This genus was originally placed in the subfamily Perognathinae (Hall 1941, 1981; Hafner et al. 1979) but now is a member of the Dipodomyinae (Hafner 1982; Hafner and Hafner 1983; Ryan 1989). The keys were elaborated based on the review of specimens and the following sources: Hall (1941, 1981) and Álvarez-Castañeda et al. (2017a).

1. Dorsal pelage dark, brownish, blackish, or grayish; hindfoot length less than 25.0 mm; anterior palatine foramina wide posteriorly, tapering to a sharp point anteriorly (Oregon, California, Nevada, Idaho, and Utah)........................................................ *Microdipodops megacephalus* (p. 521)

1a. Dorsal pelage pale pinkish cinnamon; hindfoot length greater than 25.0 mm; anterior palatine foramina with parallel sides (eastern California, western-central and southern-central Nevada)....................... *Microdipodops pallidus* (p. 522)

### *Microdipodops megacephalus* Merriam, 1891

Dark kangaroo mouse, ratón canguro oscuro

1. *M. m. albiventer* Hall and Durrant, 1937. Restricted to Lincoln County, Nevada.
2. *M. m. ambiguus* Hall, 1941. Restricted to central-northwestern Nevada.
3. *M. m. atrirelictus* Hafner, 1985. Restricted to Riddle, Owyhee County, southwestern Idaho.
4. *M. m. californicus* Merriam, 1901. Restricted to central-western Nevada.
5. *M. m. leucotis* Hall and Durrant, 1941. Restricted to central-western Utah.
6. *M. m. medius* Hall, 1941. Restricted to central Utah.
7. *M. m. megacephalus* Merriam, 1891. Eastern and northeastern Utah.
8. *M. m. nasutus* Hall, 1941. Restricted to Mineral County, Nevada.
9. *M. m. nexus* Hall, 1941. Restricted to northern-central Nevada.
10. *M. m. oregonus* Merriam, 1901. Oregon, northwestern Nevada, and northeastern California.
11. *M. m. paululus* Hall and Durrant, 1941. Restricted to western Utah.
12. *M. m. polionotus* Grinnell, 1914. Restricted to eastern-central California.
13. *M. m. sabulonis* Hall, 1941. Restricted to central-southern Nevada.

**Conservation:** *Microdipodops megacephalus* is listed as Least Concern by the IUCN (Roach 2016).

**Characteristics:** Dark kangaroo mouse is small sized; total length 140.0–177.0 mm and skull length 39.4–40.5 mm. Dorsal pelage brownish, blackish, or grayish, flanks grizzled with ocher; fur of underparts white-tipped and basally plumbeous; **distal half of tail usually darker than the back**; **anterior palatine foramina wide posteriorly and tapering to a sharp point anteriorly**; **premaxillary extending posteriorly nearly as far as the nasals**.

**Comments:** *Microdipodops megacephalus* ranges in the Great Basin of Nevada but extends into small areas in Oregon, California, Idaho, and Utah (Map 373). It can be distinguished from *M. pallidus* by having a darker dorsal coloration that can be brownish, blackish, or grayish; hindfoot less than 25.0 mm; distal half of the tail usually darker than the back; anterior palatine foramina wide posteriorly and tapering to a sharp point anteriorly; premaxilla bone extending posteriorly nearly as far as the nasals. From *Dipodomys*, by its smaller size, less than 180.0 mm; tail without a crest or terminal tuft and absolutely but not proportionally, smaller auditory bullae. From *Chaetodipus* and *Perognathus*, by having forelimbs shorter in relation to the hind legs; interparietal bone slightly longer than wide.

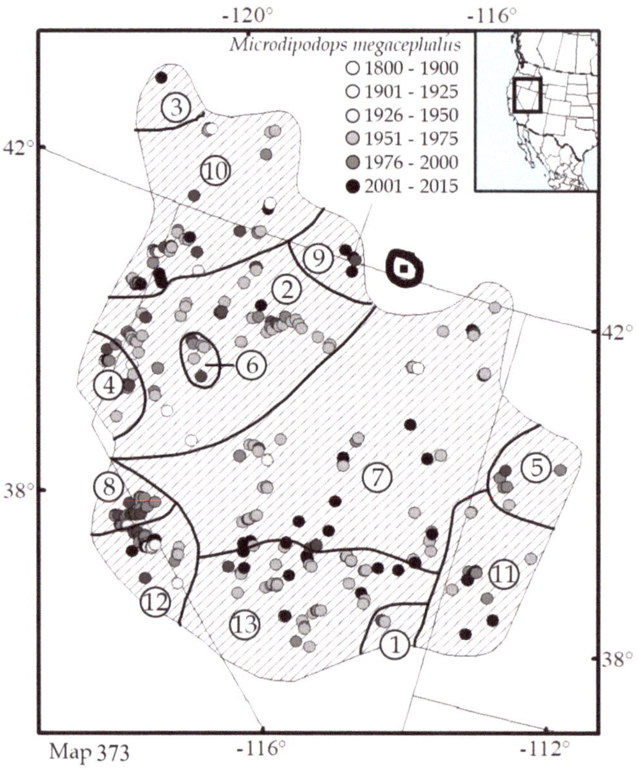


**Map 373** Distribution map of *Microdipodops megacephalus*

**Additional Literature:** Wood (1935), Hall (1946), Hall and Kelson (1959), O'Farrell and Blaustein (1974a), Hafner (1978, 1983), Williams et al. (1993), Patton (2005a).

## *Microdipodops pallidus* Merriam, 1901

Pale kangaroo mouse, ratón canguro pálido

1. *M. p. ammophilus* Hall, 1941. Restricted to central Nevada.
2. *M. p. pallidus* Merriam, 1901. Restricted to western Nevada, one population restricted to Deep Spring Valley, California.
3. *M. p. purus* Hall, 1941. Restricted to central-southern Nevada.
4. *M. p. restrictus* Hafner, 1985. Restricted to Rhodes Salt Marsh, Western Nevada.
5. *M. p. ruficollaris* Hall, 1941. Restricted to central Nevada.

**Conservation:** *Microdipodops pallidus* is listed as Least Concern by the IUCN (NatureServe 2016c).

**Characteristics:** Pale kangaroo mouse is small sized; total length 150.0–173.0 mm and skull length 39.7–40.2 mm. Dorsal pelage pale pinkish cinnamon, flanks grizzled with ocher; fur of underparts white-tipped and basally white; **distal half of the tail usually similar in color to the dorsum and lacking a black tip**; **anterior palatine foramina with parallel sides, of the same front and back width**; **premaxillary bone extending well behind the nasals**.

**Comments:** *M. pallidus* ranges in the Great Basin in western-central and southern-central Nevada, and extends into the northern Mojave Desert basins of eastern California (Map 374). It can be found in sympatry, or nearly so, with *Dipodomys. deserti*, *D. merriami*, *D. microps*, *Chaetodipus formosus*, and *Perognathus longimembris*. *M. pallidus* can be distinguished from *M. megacephalus* by its paler dorsal coloration, pale pinkish cinnamon, and flanks grizzled with ocher; hindfoot greater than 25.0 mm; distal half of the tail usually similar in color to the dorsum and lacking a black tip; anterior palatine foramina with parallel sides, premaxillae extending well behind the nasals. From *Dipodomys*, by its much smaller size, less than 180.0 mm; tail without a crest or hair tuft and smaller auditory bullae. From *Chaetodipus* and *Pergonathus*, by having the front legs shorter than the hindlegs, a much wider head due to greatly inflated bullae; interparietal bone slightly longer than wide.

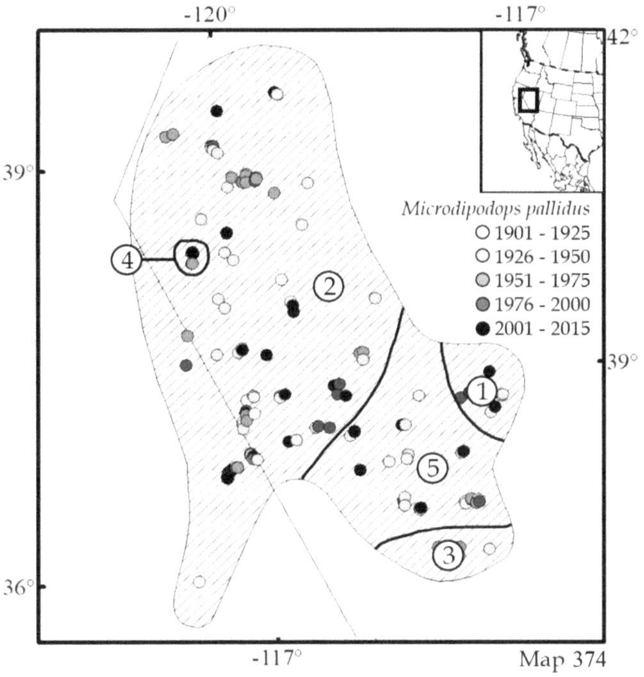


**Map 374** Distribution map of *Microdipodops pallidus*

**Additional Literature:** O'Farrell and Blaustein (1974b), Hafner (1978), Williams et al. (1993), Patton (2005a).

## **Subfamily Heteromyinae** Gray, 1868

The species in this family are characterized by having **hind and forelimbs of similar size; auditory bullae not highly developed without occupying the rear-dorsal region of the skull;** the ventral surface of auditory bullae never reaching the grinding surface of the upper molariforms; ethmoid foramen present in the dorsal part of the orbit; median ventral foramina at the anterior end of the central caudal vertebrae; masseteric crest ending above, not behind, the mental foramen; astragalo-cuboid articulation; lower premolar lobes merging into the labial or lingual side but not in the middle; **upper incisors without a longitudinal canal in frontal view;** protoloph of fourth upper premolars formed by more than one cusp; lophos of upper molars always, and **of lower molars usually, united at the two ends, surrounding the central basin;** molariforms rooted but progressively high-crowned; in later hypsodont kinds, entire crown increasing.

## Genus ***Heteromys*** Desmarest, 1817

Dorsal pelage mouse gray to blackish, sprinkled with slender, ochraceous hairs; a lateral buffy or ochraceous line sometimes present, usually not pronounced; fur of underparts white; limbs of equal size; **pelage hispid, stiff spines mingled with soft slender hairs;** tail usually longer than the head-and-body, not conspicuously crested or penciled at the tip; the auditory region of skull non-inflated and not overlapped by the pterygoids; interpterygoid fossa V shaped, tapering to a rather acute anterior end; molar crowns in early life completely divided by a transverse sulcus into two parallel enamel loops; posterior molars narrower than the premolars. Hafner et al. (2007) proposed *Liomys* as a junior synonym of *Heteromys*; however, Rogers and González (2010) considered *Liomys* as a full genus. The keys were elaborated based on the review of specimens and the following sources: Goldman (1911), Genoways (1973), Hall (1981), and Álvarez-Castañeda et al. (2017a).

1. Tail length greater than the head-and-body length and with little hair; center island in the occlusal surface of molars present in spite of wear; upper premolars with reentrant angles in the anterior lobe; interpterygoid fossa V shaped ..... 2

1a. Tail length shorter than the head-and-body length, with long crest-like hair and a very notorious hair tuft at the tip; center island in the occlusal surface of molars present only in young animals (Fig. 101); upper premolars without reentrant angles in the anterior lobe; interpterygoid fossa U shaped ..... 5

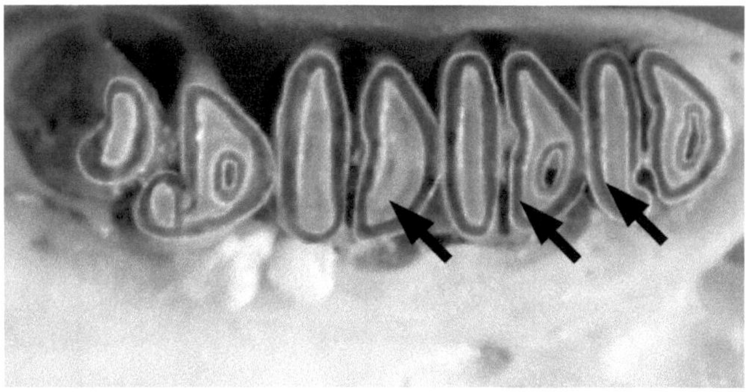

**Fig. 101** Center island in the occlusal surface of molars present only in young animals

2. Hair rough but with soft spines; third upper molars equal to or wider than premolars (endemic to highlands of southern Chiapas) ..... *Heteromys nelsoni* (p. 530)

2a. Hair with abundant spines or bristly as spines; third upper molars narrower than premolars ..... 3

3. Hindfoot soles with hair from the posterior tubercle to the heel (Yucatán Peninsula, northern Guatemala, and Belize) ..... *Heteromys gaumeri* (p. 527)

3a. Hindfoot soles naked ..... 4

4. Total length generally less than 330.0 mm and tail length less than 180.0 mm; lateral line ocher; skull length less than 38.0 mm (southeastern Veracruz, eastern Oaxaca, Chiapas, and from southern Yucatán Peninsula south through Central America) ..... *Heteromys desmarestianus* (p. 526)

4a. Total length generally greater than 330.0 mm and tail length greater than 180.0 mm; no lateral ocher line; skull length greater than 38.0 mm (restricted to Chiapas and Guatemala) ..... *Heteromys goldmani* (p. 528)

5. Hindfoot sole with five tubercles; upper parts of the back grayish brown; lateral line well marked in pink or brown shades; pterygoid process wide (Fig. 102) ..... 6

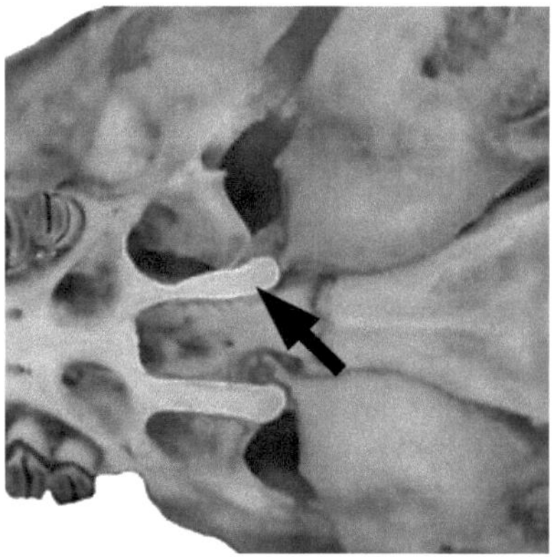

**Fig. 102** Pterygoid process wide

5a. Hindfoot sole with six tubercles; upper parts of the back reddish brown, chocolate brown, or paler; lateral line, if present, in ocher shades; pterygoid process narrow .......................................................... 7

6. Interparietal bone smaller in size, less than 8.6 mm on average, and more triangle shaped (restricted to western Jalisco) .......................................................... *Heteromys bulleri* (p. 525)

6a. Interparietal bone larger in size, greater than 8.6 mm on average, and less triangular and more circular (from Chihuahua and extreme southern Texas southward through Oaxaca, and from Nayarit, Jalisco, Michoacán, and Guerrero east through the Atlantic coast and central Veracruz in the Mexico Gulf coast).......................... *Heteromys irroratus* (p. 529)

7. Lateral line absent; dorsal pelage chocolate brown to paler; interorbital region narrower, less than 23.5% in relation to skull length (from eastern Oaxaca and Chiapas south through Guatemala)........................... *Heteromys salvini* (p. 532)

7a. Lateral line ocher; dorsal pelage reddish brown to paler; interorbital region wide, greater than 23.5% relative to skull length.......................................................... 8

8. Hindfoot length rarely greater than 30.0 mm; skull length less than 32.0 mm (from Sonora and Veracruz south through Guatemala)............................................. *Heteromys pictus* (p. 531)

8a. Hindfoot length rarely less than 30.0 mm; skull length greater than 33.0 mm (endemic to southeastern Jalisco) ........... .......................................................... *Heteromys spectabilis* (p. 533)

## ***Heteromys bulleri*** Thomas, 1893

Manantlán spiny pocket mouse, ratón espinoso de Manantlán

Monotypic.

**Conservation:** *Heteromys bulleri* (as *H. irroratus*) is listed as Least Concern by the IUCN (Castro-Arellano et al. 2016a).

**Characteristics:** Manantlán spiny pocket mouse is medium sized; total length 226.0–255.0 mm and skull length 31.8–33.7 mm. Dorsal pelage grayish brown, sides paler, narrow lateral line brown with contrasting pinky shades; fur of underparts whitish; tail shorter than the head-and-body length; hair silky; **interparietal bone smaller and more triangular or subtriangular;** tip of nasals rounded, interpterygoid fossa U shaped; first lower molariforms with two lophos; metalophos and protolophlophos on the first upper molariforms with three cusps; reentrant angle in the labial side of the first lower molariforms separated by a central valley. Species group *irroratus*.

**Comments:** *Heteromys bulleri* was recognized as a distinct species (Gutiérrez-Costa et al. 2021), previously regarded as a subspecies of *H. irroratus* (Genoways 1973). *H. bulleri* ranges only in highlands of Sierra Madre del Sur in western Jalisco, México, specifically from Sierra Manantlán and Sierra de Juanacatlán (Map 375). It can be found in sympatry, or nearly so, with *H. irroratus*, *H. pictus*, and *H. spectabilis*. *H. bulleri* can be distinguished from *H. pictus* and *H. spectabilis* by having five tubercles on the sole of each hindfoot; back upper parts grayish brown; lateral line well marked in pink or brown shades; pterygoid process wide. From *H. salvini*, by not having a lateral line, and interorbital region narrow. From *H. irroratus*, by its different interparietal shape, which is smaller and more triangular.

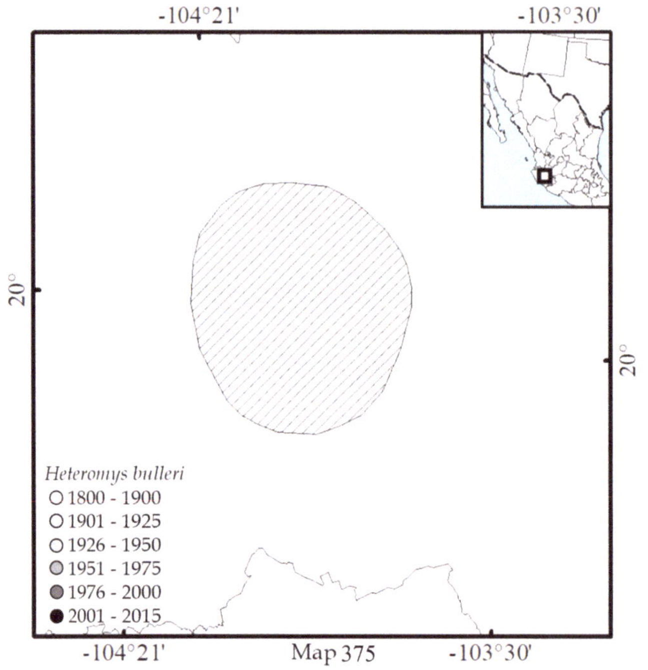


**Map 375** Distribution map of *Heteromys bulleri*

**Additional Literature:** Goldman (1911), Dowler and Genoways (1978), Williams et al. (1993), Patton (2005a), Rogers and Vance (2005).

## *Heteromys desmarestianus* Gray, 1868

Desmarest's spiny pocket mouse, ratón espinoso común

1. *H. d. desmarestianus* Gray, 1868. Restricted to Chiapas.
2. *H. d. griseus* Merriam, 1902. Eastern Oaxaca and western Chiapas.
3. *H. d. temporalis* Goldman, 1911. Restricted to Motzorongo, Veracruz.

**Conservation:** *Heteromys desmarestianus* is listed as Least Concern by the IUCN (Cassola 2016aj).

**Characteristics:** Desmarest's spiny pocket mouse is medium sized; total length 255.0–335.0 mm and skull length 34.2–38.5 mm. **Dorsal pelage grizzled gray to blackish in the middle of the back, interspersed with ocher and increasing sideward,** lateral line pale-ochraceous, though it could be absent or confused with the increased ocher coloration; fur of underparts whitish; tail longer than the head-and-body length, without a conspicuous terminal hair tuft; footpads of the hindfoot naked; hair silky; interpterygoid fossa V shaped; all molariforms of equal shape; three or four lophos in the first lower molariforms. Species group *desmaresianus*.

**Comments:** *Heteromys desmarestianus goldmani* was previously considered a full species (Patton 1993b; Rogers and González 2010). *H. desmarestianus* is within the subgenus *Heteromys* (Goldman 1911; Rogers and González 2010; Hafner et al. 2007). Genetic analyses of all the species of the subgenus *Heteromys* and *Xylomys* found that *H. desmarestianus* is a species complex and proposed the possibility of two additional species for México (Rogers and González 2010). *H. desmarestianus* ranges from southeastern Veracruz, eastern Oaxaca, Chiapas, and the southern Yucatán Peninsula south through Central America (Map 376). It can be found in sympatry, or nearly so, with *H. gaumeri*, *H. irroratus*, *H. nelsoni*, *H. pictus*, and *H. salvini*. *H. desmarestianus* differs from *H. irroratus*, *H. pictus*, and *H. salvini* by having the tail length less than the head-and-body length, with long crest-like hair, and a very notorious hair tuft at the tip; upper premolars without reentrant angles in the anterior lobe; interpterygoid fossa U shaped. From *H. gaumeri*, by having the hindfoot sole naked; lateral line narrow, pale-ochraceous, and tail without a conspicuous terminal hair tuft. From *H. nelsoni*, by having hair with abundant spines and the third upper molars narrower than the premolars. From *H. goldmani*, by its smaller size, total length smaller than 330.0 mm on average; tail less than 180.0 mm and skull less than 38.0 mm; with a lateral ocher line.

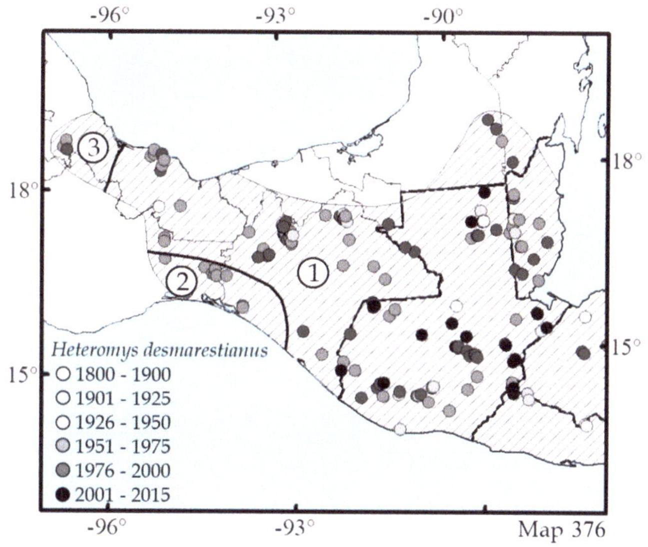


**Map 376** Distribution map of *Heteromys desmarestianus*

**Additional Literature:** Goodwin (1969), Rogers and Schmidly (1982), Schmidt et al. (1989), Rogers (1989, 1990), Williams et al. (1993), Patton (2005a), Hafner et al. (2007).

## *Heteromys gaumeri* Allen and Chapman, 1897

Gaumer's spiny pocket mouse, ratón espinoso de Yucatán

Monotypic.

**Conservation:** *Heteromys gaumeri* is listed as Least Concern by the IUCN (Reid and Vázquez 2016c).

**Characteristics:** Gaumer's spiny pocket mouse is medium sized; total length 295.0–300.0 mm and skull length 33.9–37.5 mm. Dorsal pelage grizzled ocher-yellowish brown to darker; lateral line yellowish ocher; fur of underparts yellowish white; tail longer than the head-and-body length; **footpads of the hindfoot with hair**; hair silky; interpterygoid fossa V shaped; all molariforms of the same shape; three or four lophos in the first lower molariforms. Species group *gaumeri.*

**Comments:** *Heteromys gaumeri* is within the subgenus *Heteromys* (Goldman 1911; Rogers and González 2010). Hafner et al. (2007) do not considered it a subgenus within *Heteromys*. *H. gaumeri* is endemic to the Yucatán Peninsula, northern Guatemala, and Belize (Map 377), and can be found in sympatry, or nearly so, with *H. desmarestianus*. *H. gaumeri* can be distinguished from all other species of *Heteromys* by the presence of hair on the posterior portion of the hindfoot sole. From *H. desmarestianus*, by its smaller size; dorsal pelage grizzle, ocher-yellowish brown and darker; lateral line yellowish, ocher, broad, bright, from the cheeks to the ankles; tail with a conspicuous terminal hair tuft; skull long and angular; supraorbital ridges strongly developed laterally as overhanging shelves; relatively large auditory bullae.

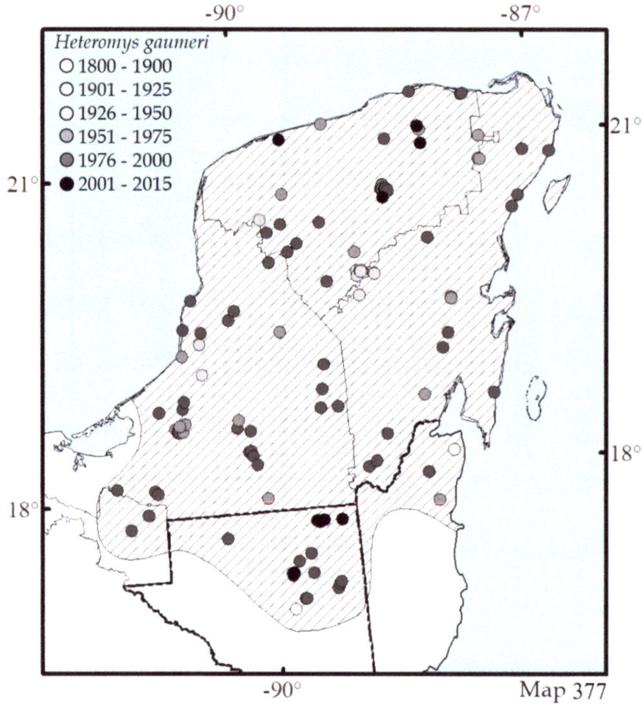


**Map 377** Distribution map of *Heteromys gaumeri*

**Additional Literature:** Genoways (1973), Jones et al. (1974), Engstrom et al. (1987a), Rogers (1989), Schmidt et al. (1989), Williams et al. (1993), Patton (2005a).

## *Heteromys goldmani* Merriam, 1902

Goldman's spiny pocket mouse, ratón espinoso

Monotypic.

**Conservation:** *Heteromys goldmani* is not considered a full species by the IUCN, so it has not been assigned a conservation status.

**Characteristics:** Goldmani's spiny pocket mouse is large sized; total length 300.0–350.0 mm and skull length 34.0–38.1 mm. Dorsal pelage dark; lateral line yellowish ocher; fur of underparts whitish; tail longer than the head-and-body length; footpads of hindfoot naked; hair silky; interpterygoid fossa V shaped; three or four lophos in the first lower molariforms. Species group *desmaresianus*.

**Comments:** *Heteromys goldmani* is within the subgenus *Heteromys* (Goldman 1911; Rogers and González 2010). Hafner et al. (2007) do not consider a subgenus within *Heteromys*. *H. goldmani* was recognized as a distinct species (Rogers and González 2010), previously regarded as a subspecies of *H. desmarestianus* (Patton 1993b; Rogers and González 2010). Hafner et al. (2007) did not mention anything about its taxonomic status. *H. goldmani* has a very restricted distribution on the southern slope of Sierra Madre del Sur in Chiapas and Guatemala (Map 378) and is not found in sympatry, or nearly so, with any other species of *Heteromys*. *H. goldmani* can be distinguished from *H. desmarestianus* by its larger size, total length greater than 330.0 mm on average; tail longer than 180.0 mm, skull greater than 38.0 mm, and without a lateral ocher line.

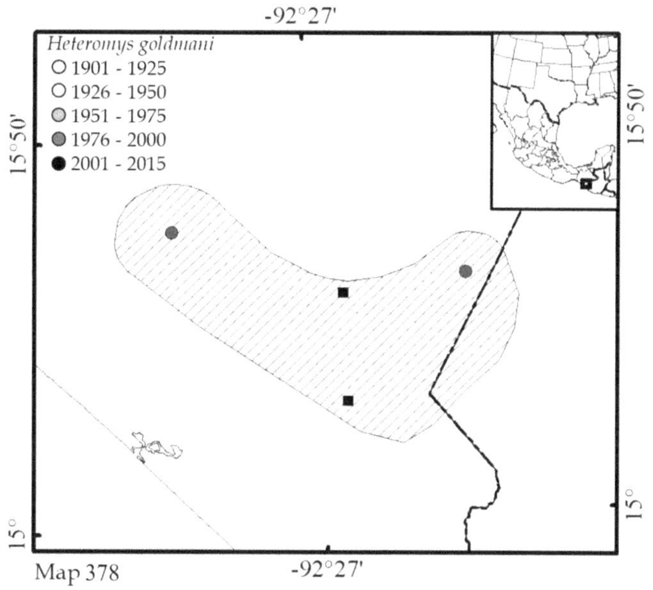


**Map 378** Distribution map of *Heteromys goldmani*

**Additional Literature:** Hall (1981), Rogers and Schmidly (1982), Schmidt et al. (1989), Lorenzo et al. (2019c).

## *Heteromys irroratus* Gray, 1868

Mexican Spiny Mouse, ratón espinoso mexicano

1. *H. i. alleni* Coues, 1881. From southern Chihuahua and southern Nuevo León south through northern Estado de México.
2. *H. i. guerrerensis* (Goldman, 1911). Restricted to central Guerrero.
3. *H. i. irroratus* Gray, 1868. Restricted to Oaxaca.
4. *H. i. jaliscensis* Allen, 1906. Western Jalisco and southwestern Zacatecas.
5. *H. i. texensis* (Merriam, 1902). Southeastern Texas, southward-central Veracruz, including eastern Nuevo León, and eastern San Luis Potosí.
6. *H. i. torridus* (Merriam, 1902). From northern Estado de México south through northern Oaxaca, including eastern Guerrero.

**Conservation:** *Heteromys irroratus* is listed as Least Concern by the IUCN (Castro-Arellano et al. 2016a).

**Characteristics:** Mexican spiny pocket mouse is medium sized; total length 194.0–300.0 mm and skull length 27.3–36.9 mm. **Dorsal pelage grayish brown, sides paler, lateral line narrow, brown with contrasting pinky shades; fur of underparts whitish**; tail shorter than the head-and-body length; hair silky; interpterygoid fossa U shaped; first lower molariforms with two lophos; metalophos and protolophos on the first upper molariforms with three cusps; reentrant angle in the labial side of the first lower molariforms separated by a central valley. Species group *irroratus*.

**Comments:** *Heteromys irroratus* ranges from Chihuahua and extreme southern Texas south through Oaxaca, and from Nayarit, Jalisco, Michoacán, and Guerrero east through the Atlantic coast and central Veracruz in the Mexico Gulf coast (Map 379). *H. irroratus* can be found in sympatry, or nearly so, with *H. bulleri*, *H. pictus*, and *H. spectabilis*. It can be distinguished from *H. bulleri* by its different interparietal shape, which is larger and less triangular. From *H. pictus* and *H. spectabilis*, by having five tubercles on the sole of each hindfoot; back grayish with brown upper parts; lateral line well marked in pink or brown shades; pterygoid process wide. From *H. salvini*, by not having a lateral line and interorbital region narrow. From *H. desmarestianus*, by having the tail length greater than the head-and-body length and with sparse hair; upper premolars with reentrant angles in the anterior lobe and interpterygoid fossa V shaped.

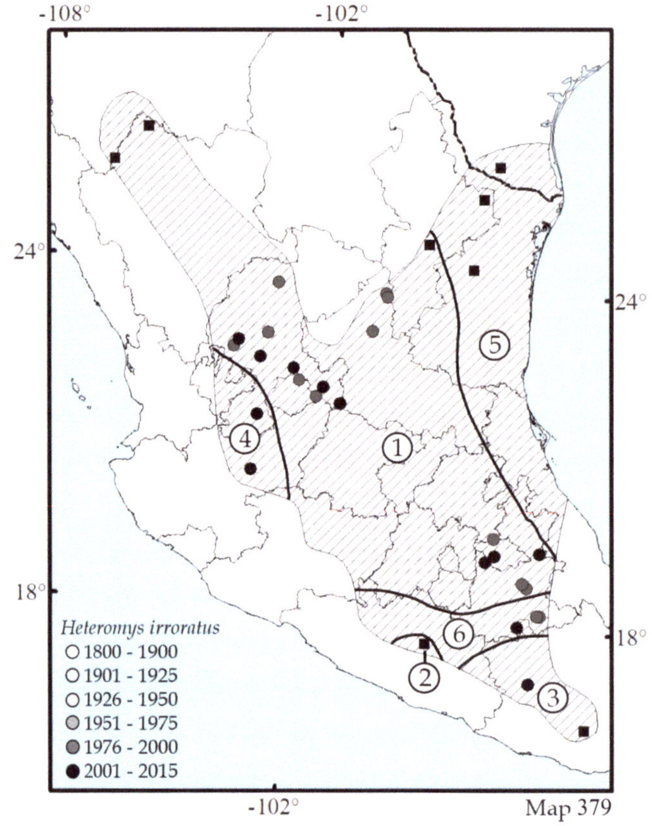


**Map 379** Distribution map of *Heteromys irroratus*

**Additional Literature:** Goldman (1911), Genoways (1973), Dowler and Genoways (1978), Williams et al. (1993), Patton (2005a), Rogers and Vance (2005), Lorenzo et al. (2019c, 2020), Gutiérrez-Costa et al. (2021).

## *Heteromys nelsoni* Merriam, 1902

Nelson's spiny pocket mouse, ratón espinoso de Chiapas

Monotypic.

**Conservation:** *Heteromys nelsoni* is listed as Under Special Protection by the Norma Oficial Mexicana (DOF 2019) and as Endangered (B1ab [i, ii, iii, v]: extent of occurrence less than 5000 km$^2$ in five or less populations and continuing decline in extent of occurrence, area of occupancy, quality of habitat, and number of mature individuals) by the IUCN (Cuarón and Vázquez 2018).

**Characteristics:** Nelson's spiny pocket mouse is large sized; total length 328.0–356.0 mm and skull length 37.7–41.3 mm. Dorsal pelage grizzle grayish to blackish; fur of underparts whitish, contrasting with the dorsal pelage; lateral line absent; tail longer than the head-and-body length; footpads naked; **pelage harsh but spines soft**; interpterygoid fossa V shaped; all molariforms of equal shape; three or four lophos in the first lower molariforms; parietal lateral extension along the lambdoid ridge; **third upper molars equal to or wider than premolars**. Species group *nelsoni*.

**Comments:** *Heteromys nelsoni* is within the subgenus *Xylomys* (Merriam 1902; Rogers and González 2010), but Hafner et al. (2007) considered it a synonym of *Heteromys*. *H. nelsoni* is probably the *Heteromys* species with the smallest range, known only from highlands over 2500 m in southeastern Chiapas and western Guatemala (Map 380). It can be distinguished from all other species of *H. desmarestianus* by its dorsal pelage grizzle from grayish to blackish, lateral line absent, and comparatively soft pelage; more intricate enamel folds in the posterior upper molars; parietal lateral extension greater and lambdoid ridge longer.

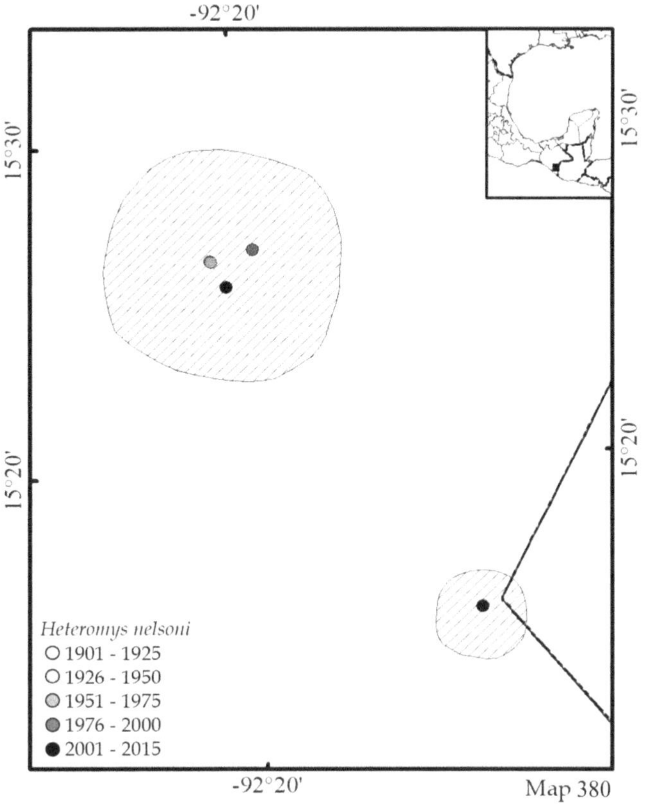


**Map 380** Distribution map of *Heteromys nelson*

**Additional Literature:** Goldman (1911), Hall (1981), Rogers and Schmidly (1982), Rogers and Rogers (1992), Schmidt et al. (1989), Patton (2005a), Rios et al. (2016), Lorenzo et al. (2019c, 2020).

## ***Heteromys pictus*** Thomas, 1893

Western spiny pocket mouse, ratón espinoso pintado

1. *H. p. annectens* Merriam, 1902. From central Guerrero south through central-south Oaxaca.
2. *H. p. hispidus* Allen, 1897. From Sonora south through northern-central Jalisco.
3. *H. p. pictus* Thomas, 1893. Coastal plains from Jalisco and central Veracruz south through Chiapas.
4. *H. p. plantinarensis* (Merriam, 1902). Highlands of western Jalisco, Michoacán, and Guerrero.

**Conservation:** *Heteromys pictus* is listed as Least Concern by the IUCN (Reid and Vázquez 2016d).

**Characteristics:** Painted spiny pocket mouse is small to medium sized; total length 183.0–294.0 mm and skull length 26.0–36.7 mm. **Dorsal pelage speckled with brown, yellow, and ocher; lateral line yellowish ocher and contrasting strongly with the whitish underparts**; hindfoot dorsally whitish; interpterygoid fossa U shaped; first lower molariforms with two lophos; metalophos and protoloph on the first upper molariforms with three cusps; reentrant angle in the labial side of the first lower molariforms separated by a central valley; **skull relatively narrow**. Species group *pictus*.

**Comments:** *Heteromys pictus* ranges in both coasts of México, from Sonora and Veracruz south through Guatemala (Map 381), and can be found in sympatry, or nearly so, with *H. irroratus*, *H. salvini*, and *H. spectabilis. H. pictus* can be distinguished from *H. irroratus* by having six tubercles on the hindfoot soles; upper parts not black grayish brown; a lateral line, if present, in ocher shades; the pterygoid process narrow. From *H. salvini*, by lacking a lateral line, and the interorbital region narrow. From *H. spectabilis*, by having the dorsal pelage speckled with brown, yellow, and ocher, with a lateral line yellowish ocher and contrasting strongly with the whitish underparts; larger size on average; hindfoot length greater than 30.0 mm and skull less than 32.0 mm. From *H. desmarestianus*, by having the tail length greater than the head-and-body length and with sparse hair; upper premolars with reentrant angles in the anterior lobe; interpterygoid fossa V shaped.

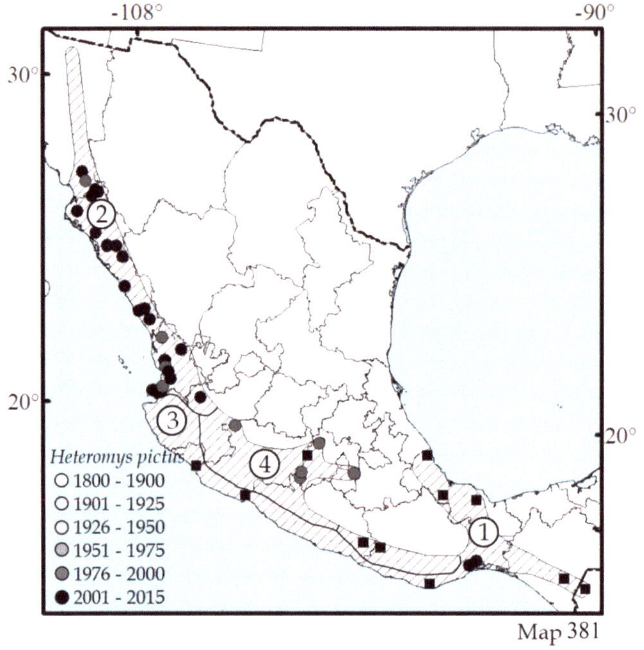


**Map 381** Distribution map of *Heteromys pictus*

**Additional Literature:** Merriam (1902), Goldman (1911), Genoways (1973), Dowler and Genoways (1978), McGhee and Genoways (1978), Hall (1981), Morales and Engstrom (1989), Rogers (1990), Williams et al. (1993), Patton (2005a), Rogers and Vance (2005).

## *Heteromys salvini* Thomas, 1893

Salvin's spiny pocket mouse, ratón espinoso de la costa

*H. s. crispus* (Merriam, 1902). Coastal plains of southeastern Oaxaca and Chiapas.

**Conservation:** *Heteromys salvini* is listed as Least Concern by the IUCN (Vázquez et al. 2016b).

**Characteristics:** Salvin's spiny pocket mouse is the smallest sized of the genus *Heteromys*; total length 185.0–272.0 mm and skull length 28.7–36.4 mm. **Dorsal pelage grayish brown to deep chocolate brown, speckled with dark gray; dominant color interspersed with paler shades of gray; pelage hispid, consisting of stiff spines mingled with slender soft hairs; lateral line absent** and not markedly contrasting with the **whitish underparts**; **tail densely haired and relatively shorter**; dorsal hindfoot whitish; interpterygoid fossa U shaped; metalophos and protoloph on the first upper molariforms with three cusps; reentrant angle in the labial side of the first lower molariforms separated by a central valley. Species group *pictus*.

**Comments:** *Heteromys salvini* is considered different from all the other species of *Liomys* of North America (Genoways 1973) and ranges from eastern Oaxaca and Chiapas south through Guatemala (Map 382). It can be found in sympatry, or nearly so, with *H. desmarestianus* and *H. pictus*. *H. salvini* can be distinguished from *H. pictus* by having a lateral line yellowish ocher, strongly contrasting with the whitish underparts; interorbital region wider. From *H. desmarestianus*, by having the tail length greater than the head-and-body length and with sparse hair; upper premolars with reentrant angles in the anterior lobe and interpterygoid fossa V shaped. From *H. spectabilis*, by its smaller size.

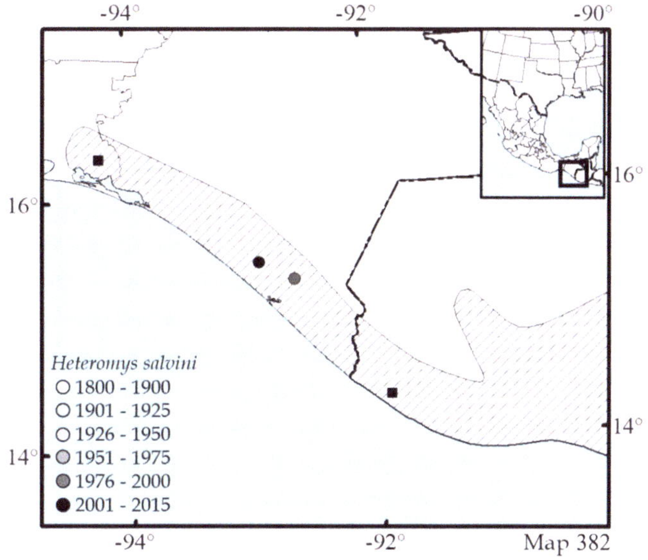


**Map 382** Distribution map of *Heteromys salvini*

**Additional Literature:** Merriam (1902), Goldman (1911), Carter and Genoways (1978), Dowler and Genoways (1978), Williams et al. (1993), Patton (2005a), Rogers and Vance (2005).

## *Heteromys spectabilis* (Genoways, 1971)

Jaliscan spiny pocket mouse, ratón espinoso de Jalisco

Monotypic.

**Conservation:** *Heteromys spectabilis* is listed as Under Special Protection by the Norma Oficial Mexicana (DOF 2019) and as Endangered (B1ab [iii]: extent of occurrence less than 5000 km$^2$ in five or less populations and continuing decline in quality of habitat) by the IUCN (Álvarez-Castañeda and Vázquez 2018).

**Characteristics:** Jaliscan spiny pocket mouse is medium sized; total length 242.0–280.0 mm and skull length 33.0–35.3 mm. Dorsal pelage brown; lateral line yellowish ocher; fur of underparts whitish; tail hairy and shorter than the head-and-body length; interpterygoid fossa U shaped; first lower molariforms with two lophos. Species group *pictus*.

**Comments:** *Heteromys spectabilis* is only known from southeastern Jalisco (Map 383) and can be found in sympatry, or nearly so, with *H. irroratus* and *H. pictus*. *H. spectabilis* can be distinguished from *H. irroratus* by having six tubercles on the soles of each hindfoot; upper parts not black grayish brown and a lateral line, if present, in ocher shades; skull narrower; the pterygoid process narrow and longer. From *H. pictus*, by having the dorsal pelage brown, with a lateral line yellowish ocher, not strongly contrasting with the underparts; smaller size on average; hindfoot length less than 30.0 mm and skull greater than 32.0 mm. From *H. salvini*, by its larger size.

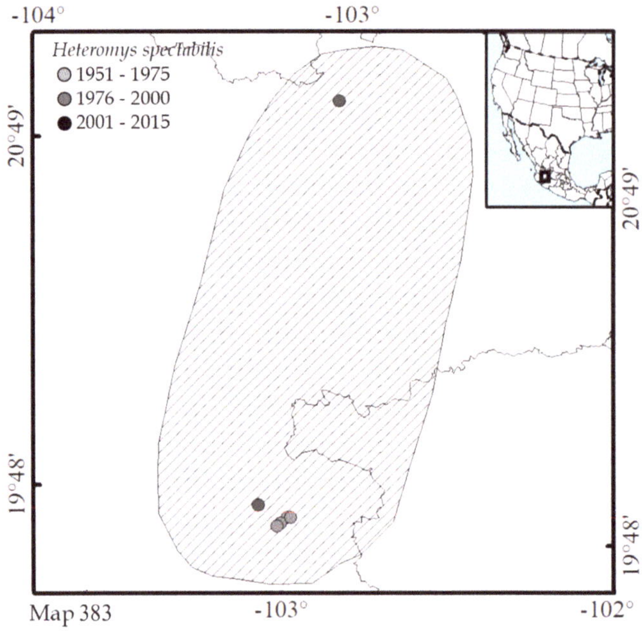


**Map 383** Distribution map of *Heteromys spectabilis*

**Additional Literature:** Goldman (1911), Genoways (1971, 1973), Dowler and Genoways (1978), Williams et al. (1993), Domínguez–Castellanos and Ortega (2003), Patton (2005a), Rogers and Vance (2005).

## Subfamily **Perognathinae** Coues, 1875

The species in this subfamily are characterized by **front legs shorter than hind leg**s but not overly short or small; tail very long, longer than the head-and-body length, usually with a terminal tuft and crested in most *Chaetodipus*; no median ventral foramina in the caudal vertebrae; astragalus articulating with cuboid; auditory bullae inflated and, as a result, posterior skull breadth equal to or slightly greater than zygomatic breadth; the ventral surface of auditory bulla below the grinding surface of the upper molars; interparietal bone variable; palate center between the premolars not ridged; ethmoid foramen in the frontal view; lophos of upper premolars united first at or near the tooth center; protoloph usually single-cusped; lophos of upper molars united progressively at the lingual to buccal margins; lophos of the lower molars united primitively at the buccal margin, progressively at the tooth center, forming an H-pattern; molariforms brachydont to hypsodont but always rooted; enamel pattern lost early in life; enamel always complete; upper incisors smooth or grooved. The keys were elaborated based on the review of specimens and the following sources: Hall (1981) and Álvarez-Castañeda et al. (2017a).

1. Hair soft; mastoid projecting posteriorly beyond the occipital (Fig. 103); interparietal breadth smaller than interorbital breadth, rarely of equal size; skull length less than 25.0 mm ........................................................ *Perognathus* (p. 566)

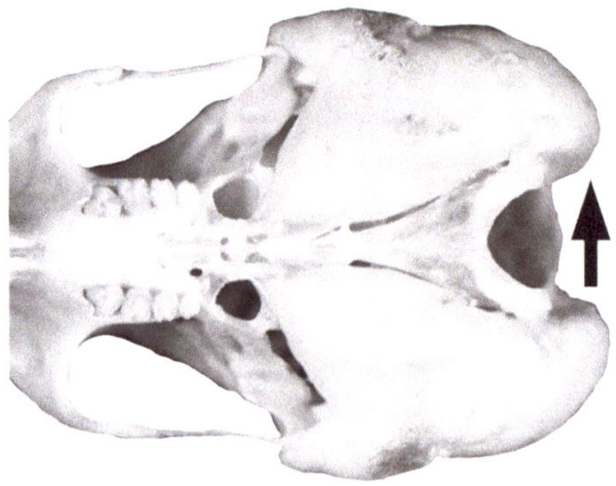

**Fig. 103** Mastoid projecting posteriorly beyond the occipital

1a. Hair coarse, sometimes with stiffened hairs (spines) on the rump; mastoid not projecting posteriorly beyond the occipital (Fig. 104); interparietal breadth equal to or greater than the interorbital breadth; skull length greater than 25.0 mm ........ *Chaetodipus* (p. 535)

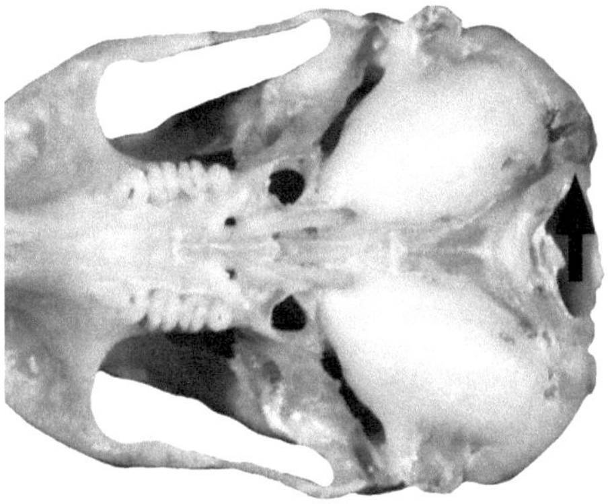

**Fig. 104** Mastoid not projecting posteriorly beyond the occipital

## Genus *Chaetodipus* Merriam, 1889

Dorsal pelage pale gray to dark gray washed with white, ochraceous, yellowish, or tawny shades; fur of underparts white or nearly so; hair generally coarse, some species with spines on the rump and thighs; hindfoot soles naked; mastoids short, not projecting posteriorly beyond the occipital; auditory bullae separated by almost all basisphenoid width; **interparietal wider, breadth equal to or greater than interorbital breadth; supraoccipital with deep lateral indentations**. *Chaetodipus* was previously considered a subgenus of *Perognathus. Chaetodipus* was coined by Merriam's (1889a) and was used (Osgood's 1900) until the revision by Hafner and Hafner (1983) formal splitting of the two subgenus as full genus. The keys were elaborated based on the review of specimens and the following sources: Merriam (1900), Hall (1981), Álvarez-Castañeda et al. (2017a), and Neiswenter et al. (2019).

1. Stiffened hard hairs (spines) lacking ........ 2

1a. Stiffened hard hairs (spines) present on the rump, clearly visible in lateral view ........ 12

2. Tail without a crest of elongated hairs, shorter than the head-and-body length; total length greater than 180.0 mm and the head-and-body length greater than 100.0 mm; skull with a ridge on the supraorbital edge; skull length greater than 28.0 mm (North Dakota and eastern Wyoming south through Jalisco, Guanajuato, Estado de México, Hidalgo and Guanajuato in the western side and through San Luis Potosí and Tamaulipas in the eastern side, and from Colorado, Arizona, and Sonora east through Nebraska, Kansas, Oklahoma, Louisiana, and Tamaulipas) ........ *Chaetodipus hispidus* (p. 553)

2a. Tail with a crest of elongated hairs, longer than the head-and-body length; total length less than 180.0 mm, if greater the head-and-body length less than 100.0 mm; skull without a ridge on the supraorbital edge; skull length less than 28.0 mm ........ 3

3. Posterior margin of auditory bullae projecting slightly beyond the rear plane of the occipital ........ 4

3a. Posterior margin of auditory bullae not projecting beyond the rear plane of the occipital ........ 6

4. Head-and-body length generally less than 90.0 mm; hindfoot length usually less than 26.0 mm; interparietal breadth 5.9 mm or less on average, rarely greater than 6.5 mm (Nevada, Utah, western Arizona, eastern California, and northeastern Baja California) ........ *Chaetodipus formosus* (p. 550)

4a. Head-and-body length generally greater than 90.0 mm; hindfoot length equal to or greater than 26.0 mm; interparietal breadth on average 6.1 mm or greater ........ 5

5. West of lower Colorado River (southeastern California south throughout the Baja California Peninsula, including Montserrat Island) ............ *Chaetodipus rudinoris* (p. 561)

5a. East of lower Colorado River (Arizona to southwestern New Mexico, Sonora to northern Sinaloa) ............ *Chaetodipus baileyi* (p. 542)

6. Ears large, length generally greater than 10.0 mm, rounded at the tip (southern Sonora, southwestern Chihuahua, and Sinaloa) ............ *Chaetodipus artus* (part; p. 541)

6a. Ears small, length generally less than 10.0 mm, pointed ............ 7

7. Interorbital breadth usually less than 5.8 mm (southern Sonora to northern Nayarit) ............ *Chaetodipus pernix* (p. 560)

7a. Interorbital breadth greater than 5.8 mm ............ 8

8. Dorsal pelage dull gray with a faint yellowish line on the head (San Luis Potosí and eastern Zacatecas) ............ *Chaetodipus lineatus* (p. 556)

8a. Dorsal pelage ocher or grayish yellow, sometimes with black hair, without a faint yellowish line on the head ............ 9

9. Tail length considerably greater than the head-and-body length; dorsal pelage reddish brown; tail hair tuft at the tip longer than 16.0 mm; interparietal bone pentagonal ............ 10

9a. Tail length slightly greater than the head-and-body length; dorsal pelage pale yellow; tail hair tuft at the tip less than 16.0 mm; interparietal bone oval ............ 11

10. Nasal-to-occipital length greater than 26.2 mm in males and 25.8 mm in females; nasal length greater than 10.0 mm in males and greater than 9.8 mm in females (southeastern California, extreme southern Nevada and Utah, southern Arizona, southwestern New Mexico, northeastern Baja California, Sonora, western Chihuahua, and northern Sinaloa) ............ *Chaetodipus penicillatus* (p. 558)

10a. Nasal-to-occipital length less than 26.2 mm in males and 25.8 mm in females; nasal length less than 10.0 mm in males and less than 9.8 mm in females (southern New Mexico, and southwestern Texas through Chihuahua, Coahuila, Durango, Zacatecas, Nuevo León, and San Luis Potosí) ............ *Chaetodipus eremicus* (p. 547)

11. Scarce long hairs on the rumps and only backlight visible; mastoid breadth greater than 12.7 mm (endemic to Isla Cerralvo and Los Planes basin, Baja California Sur) ............ *Chaetodipus siccus* (p. 562)

11a. No long hairs as spines on the rumps; mastoid breadth less than 12.7 mm (endemic to the Baja California Peninsula) ............ *Chaetodipus arenarius* (p. 538)

12. Ear length less than 9.0 mm ............ 13

12a. Ear length greater than 9.0 mm ............ 18

13. Distributed in the Baja California Peninsula ............ 14

13a. Not distributed in the Baja California Peninsula ............ 15

14. Lateral line narrow and orange; pelage with marked whitish spines only on the rumps (from southern California south through the central Baja California Peninsula) ............ *Chaetodipus fallax* (p. 548)

14a. Lateral line thin and yellowish or absent; pelage with marked whitish spines on the rump and flanks (from southern Nevada and southeastern California south through the Baja California Peninsula, including Espíritu Santo, San Francisco, San José, Carmen, Coronados, San Marcos, San Lorenzo, Las Ánimas, Ángel de la Guarda, Mejía, Danzante, Magdalena, and Margarita Islands) ............ *Chaetodipus spinatus* (p. 563)

15. Mesopterigoid fossa V shaped; premaxillae extending distally to the same level as the nasals (southern Utah, Arizona, central and southern New Mexico, southwestern Texas, Sonora, northern Chihuahua, and northwestern Coahuila) ............ *Chaetodipus intermedius* (p. 554)

15a. Mesopterigoid fossa U shaped; premaxillae extending distally beyond the posterior border of the nasals ............ 16

16. Relatively paler and smaller; smaller rostrum and larger mastoid; ranges approximately north of 26°; it can be differentiated with genetic analyses only (southeastern New Mexico, western Texas, northern Sonora, and northeastern Chihuahua) .......... *Chaetodipus collis* (p. 545)

16a. Relatively darker and larger; larger rostrum and smaller mastoid; ranges approximately south of 26°, except for a population in southwestern Chihuahua; it can be differentiated with genetic analyses only .......... 17

17. Relatively larger; ranges approximately west of −104°; it can be differentiated with genetic analyses only (southern Chihuahua and northern Durango) .......... *Chaetodipus durangae* (p. 546)

17a. Relatively smaller; ranges approximately east of −104°; it can be differentiated with genetic analyses only (eastern Durango, southern Coahuila, western Nuevo León, Zacatecas, San Luis Potosí, Aguascalientes, and northern Jalisco) .......... *Chaetodipus nelsoni* (p. 557)

18. Distributed in the Baja California Peninsula .......... 19

18a. Not distributed in the Baja California Peninsula .......... 20

19. Few long hairs as spines on the rumps; ear length usually less than 10.0 mm (coastal plains of the southern Baja California Peninsula) .......... *Chaetodipus ammophilus* (p. 537)

19a. Abundant and well-developed spines on the rumps; ear length usually greater than 10.0 mm (from central California south through northern Baja California) .......... *Chaetodipus californicus* (p. 543)

20. Tail with a wide dark dorsal line and few long hairs as a crest; posterior palatal pits large (nearly twice); premaxillae extending beyond the posterior border of the nasals; auditory bullae length averaging less than 6.0 mm (southern Sonora, southwestern Chihuahua, and Sinaloa) .......... *Chaetodipus artus* (part; p. 541)

20a. Tail with a narrow dark dorsal line, with long hair as a crest; posterior palatal pits small; premaxillae extending at the level of the posterior border of the nasals; auditory bullae length greater than 6.5 mm on average (northeastern to southern Sonora and northern Sinaloa) .......... *Chaetodipus goldmani* (p. 551)

## ***Chaetodipus ammophilus*** (Osgood, 1907)

Dalquest's pocket mouse, ratón de bolsas del Cabo

1. *C. a. ammophilus* (Osgood, 1907). Restricted to Margarita Island, Baja California Sur.
2. *C. a. dalquesti* (Roth, 1976). Restricted to the coastal lowlands of the southern part of Baja California Sur.
3. *C. a. sublucidus* (Nelson and Goldman, 1929). Restricted to the coastal plains of La Paz Bay, Baja California Sur.

**Conservation:** *Chaetodipus ammophilus ammophilus* is listed as Threatened and *C. a. dalquesti* as Special Protection by the Norma Oficial Mexicana (DOF 2019). *C. ammophilus* listed as Near Threatened by the IUCN (Roach 2017b).

**Characteristics:** Dalquest's pocket mouse is small sized; total length 148.0–180.0 mm and skull length 24.2–26.0 mm. Dorsal pelage medium gray with little blackish hair interspersed; lateral line usually absent; fur of underparts whitish; tail length greater than the head-and-body length; **coarse hairs on the rumps but not as spines;** skull small and relatively broad; nasals slender; zygomatic arches fragile; braincase vaulted.

**Comments:** *Chaetodipus ammophilus* was described as *C. dalquesti* (Roth 1976). Later, as a junior synonym of *C. arenarius* (Williams et al. 1993), and rested as a valid species (Álvarez-Castañeda and Rios 2011), with *C. arenarius ammophilus* and *C. a. sublucidus* as junior synonyms. *C. ammophilus* has priority over *C. dalquesti*; for this reason, it was established in the current taxonomy (Rios and Álvarez-Castañeda 2013). *C. ammophilus* ranges at the southern end of the Baja California Peninsula, in association with coastal areas from San Evaristo on the Gulf side to Adolfo López Mateos on the Pacific side (Map 384). It can be found in sympatry, or nearly so, with *C. arenarius*, *C. rudinoris*, and *C. spinatus*, and can be distinguished from *C. arenarius* by having very thin stiffened hairs (spines) on the rumps; on average, larger sized; larger ears, greater than 8.0 mm, with a broad black edge. From *C. spinatus*, by not having well-developed thick and stiffened hairs (spines) on the rumps. From *C. rudinoris*, by its smaller size, total length less than 183.0 mm. From *C. siccus*, by having thin spines on the rumps and lacking a brown coloration.

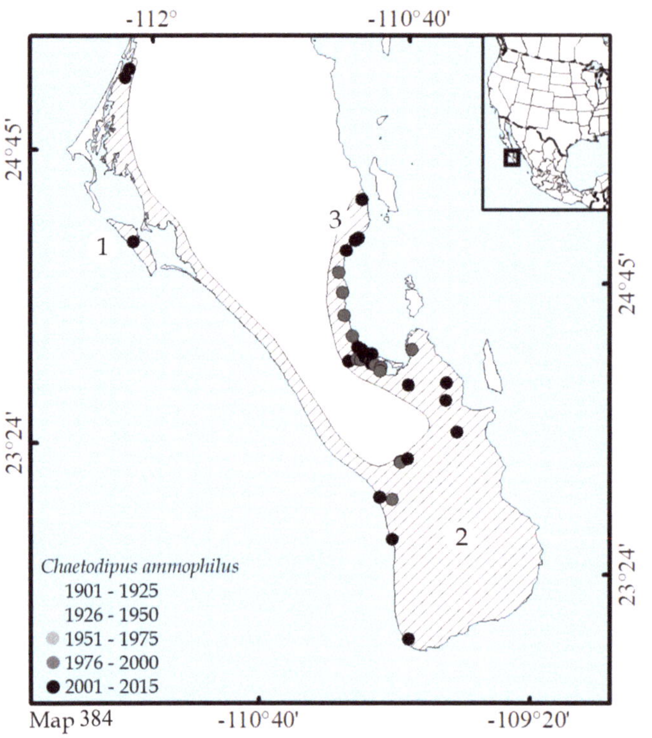


**Map 384** Distribution map of *Chaetodipus ammophilus*

**Additional Literature:** Merriam (1889b), Osgood (1900), Hall (1981), Patton and Álvarez-Castañeda (1999), Riddle et al. (2000b), Patton (2005a), Rios and Álvarez-Castañeda (2010).

## *Chaetodipus arenarius* (Merriam, 1894)

Sand pocket mouse, ratón de bolsas de la arena

1. *C. a. albescens* (Huey, 1926). Restricted to coastal areas of northeastern Baja California.
2. *C. a. albulus* (Nelson and Goldman, 1923). Restricted to Magdalena Island, Baja California Sur.
3. *C. a. ambiguus* (Nelson and Goldman, 1929). From the Central Desert of Baja California south through central Vizcaíno Desert, Baja California Sur.
4. *C. a. arenarius* (Merriam, 1894). Coastal plains from San Ignacio Lagoon and Santa Rosalia southward.
5. *C. a. helleri* (Elliot, 1903). From the Central Desert north through Santo Tomas Bay, Baja California.
6. *C. a. mexicalis* (Huey, 1939). Restricted to the northeastern lowlands of Baja California.
7. *C. a. paralios* (Huey, 1964). Restricted to the lowlands east of Matomí Sierra, from Bahía de Los Ángeles south through El Barril, Baja California.
8. *C. a. ramirezpulidoi* Álvarez-Castañeda and Cortés-Calva, 2004. Restricted to El Mogote Peninsula, La Paz, Baja California Sur.
9. *C. a. sabulosus* (Huey, 1964). Northwestern Baja California Sur, from San Ignacio Lagoon east through Vizcaíno Desert.

**Conservation:** *Chaetodipus arenarius albulus* is listed as Threatened by the Norma Oficial Mexicana (DOF 2019). *C. arenarius* listed as Least Concern by the IUCN (Álvarez-Castañeda and Lacher 2017).

**Characteristics:** Sand pocket mouse is small sized; total length 136.0–182.0 mm, skull length 20.7–23.9 mm. Dorsal pelage yellowish gray, slightly grizzled black; **without spines on the rumps;** lateral line usually absent; fur of underparts whitish; tail darker above and longer than the head-and-body length; skull small and relatively broad; nasals slender; zygomatic arches fragile; braincase vaulted.

**Comments:** *Chaetodipus arenarius siccus* (as understood by Hall 1981; Williams et al. 1993) has received recent revision attention, with Álvarez-Castañeda and Rios (2011) and Rios and Álvarez-Castañeda (2013) elevating both *siccus* and *ammophilus* to species status, the latter including other subspecies historically linked to this species (see those accounts). *C. arenarius* ranges in all sandy areas of the Baja California Peninsula (Map 385) and can be found in sympatry, or nearly so, with *C. ammophilus*, *C. fallax*, *C. penillillatus*, *C. rudinoris*, and *C. spinatus. C. arenarius* can be distinguished from *C. fallax* and *C. spinatus* by lacking stiffened hairs (spines) on the rumps; total length less than 200.0 mm; tail crested and longer than the head-and-body length; without a supraorbital bead; interorbital breadth less than 39% of basilar length. From *C. ammophilus*, by its smaller size on average; ears shorter, less than 8.0 mm, and uniformly pale colored. From *C. fallax*, by *lacking* a well-marked orange lateral line. From *C. formosus*, by *its* smaller size; tail much less crested and upperparts pale gray or brownish; inhabiting sandy areas. From *C. siccus*, by its gray coloration in general and soft pelage. From *C. penicillatus*, by not having a noticeable crest tail and a prominent buffy lateral line, general coloration uniform smoke gray; braincase higher and arched; upper part of foramen magnum constricted by the projecting lateral angles of the margin; the coronoid process more slender and strongly curved backward. From *C. rudinoris*, by its smaller size, total length less than 183.0 mm, and hindfoot less than 26.0 mm. *C. arenarius* is similar to some species of *Perognathus* in pelage texture and total length. Found in sympatry, or nearly so, with *P. longimembris*, from which it can be differentiated by its pelage being less soft fur; mastoid not projecting posteriorly beyond the occipital, and interparietal bone wide, greater than 5.0 mm.

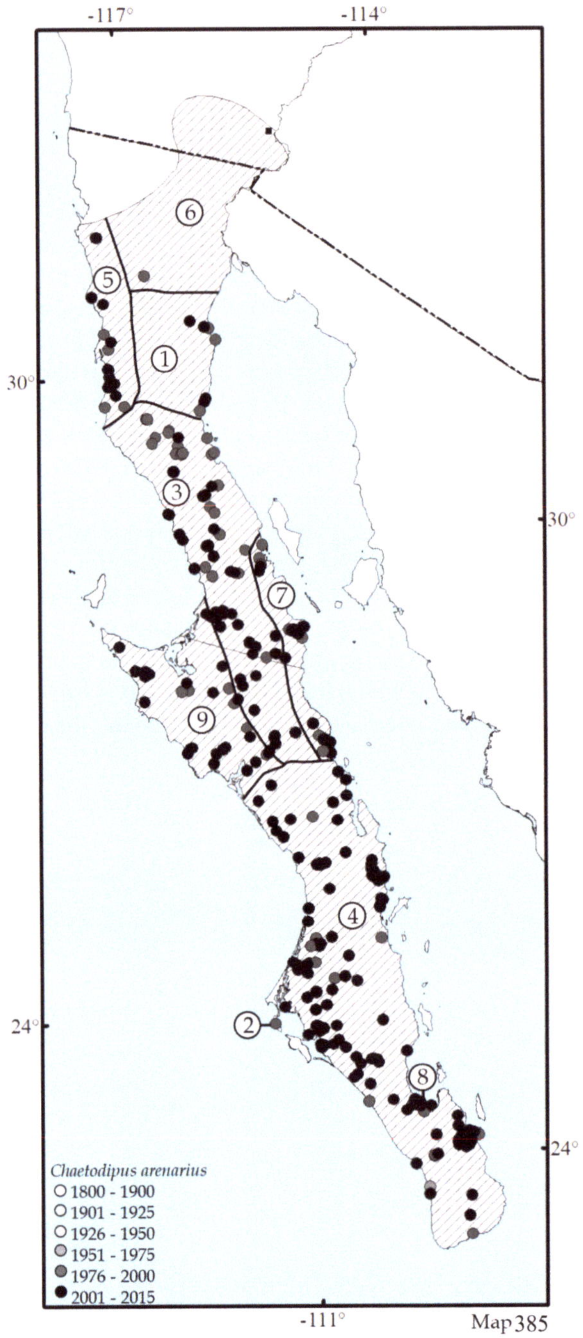


**Map 385** Distribution map of *Chaetodipus arenarius*

**Additional Literature:** Merriam (1889b, 1894), Osgood (1900, 1907), Nelson and Goldman (1923), Huey (1926, 1964), Hatt (1932), Ingles (1965), Roth (1976), Lackey (1991a), Patton and Álvarez-Castañeda (1999), Patton (2005a), Álvarez-Castañeda et al. (2008c).

## *Chaetodipus artus* (Osgood, 1900)

Narrow-skulled pocket mouse, ratón de bolsas de cara delgada

Monotypic.

**Conservation:** *Chaetodipus artus* is listed as Least Concern by the IUCN (Álvarez-Castañeda et al. 2016f).

**Characteristics:** Narrow-skulled pocket mouse is small sized; total length 185.2–193.8 mm and skull length 25.1–27.2 mm. Dorsal pelage brown at the shoulders, darker on the back and blackish on the rumps; lateral line yellow, very distinguishable; fur of underparts whitish; **ears small and rounded at the tip, length generally greater than 10.0 mm**; tail longer than the head-and-body length (approx. 125%) with few hairs as a crest, blackish dorsally and whitish ventrally; hindfoot footpads naked; no stiffened hairs (spines) on the rumps; nasal larger and **posterior margin less than 1.0 mm**; supraoccipital greater than 6.0 mm.

**Comments:** *Chaetodipus artus* was considered conspecific of C. *goldmani* (Hall and Ogilvie 1960), but its specific status became clear when larger samples were studied (Anderson 1964; Patton 1967a, 1969). *C. artus* ranges from Río Mayo and its tributaries in southern Sonora, southwestern Chihuahua, and Sinaloa, in riparian galleries and closed mesic forests (Map 386). This species is sympatric, or nearly so, with *C. baileyi*, *C. goldmani*, and *C. pernix*. It differs from *C. baileyi*, *C. goldmani*, and *C. pernix* by having thick and stiffened hairs (spines) on the rumps (except *C. goldmani*). From C. *pernix*, by its larger size and the interorbital constriction greater than 5.8 mm. From C. *baileyi*, by its smaller size, less than 200.0 mm. From *C. goldmani*, by being slightly smaller on average; tail with less hair and with a broad dorsal stripe; larger posterior palatal pits, nearly twice; supraorbital region broader, greater than 6.0 mm; great premaxillary extension beyond 1.0 mm to the nasal posterior border; mastoid breadth narrower; transverse ridge on the mastoid more strongly marked; auditory bullae smaller, ascending processes of the premaxilla extending posteriorly to the nasals over a distance greater than the lowest breadth of one nasal bone.

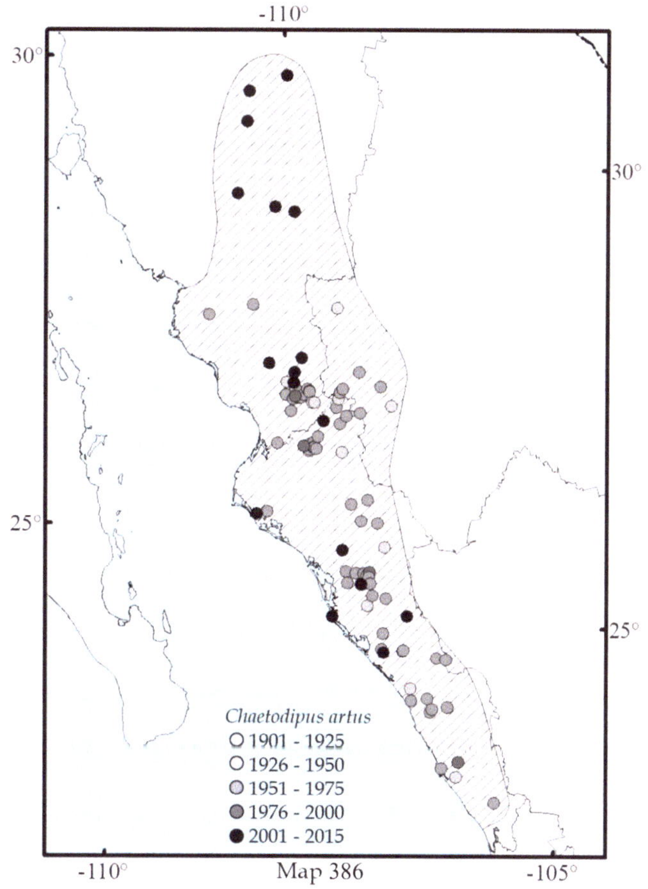


**Map 386** Distribution map of *Chaetodipus artus*

**Additional Literature:** Merriam (1889b), Osgood (1900), Hatt (1932), Burt and Hooper (1941), Anderson (1972), Hall (1981), Best and Lackey (1992a), Best (1993d), Williams et al. (1993), Patton and Álvarez-Castañeda (1999), Patton (2005a).

## ***Chaetodipus baileyi*** (Merriam, 1894)

Bailey's pocket mouse, ratón de bolsas gigantes

1. *C. b. baileyi* (Merriam, 1894). Southern Arizona, southwestern New Mexico in lowlands of Sonora and northwestern Sinaloa.
2. *C. b. domensis* Goldman, 1928. From coastal areas of Sonora north through southwestern Arizona.
3. *C. b. insularis* (Townsend, 1912). Restricted to Isla Tiburón, Sonora.

**Conservation:** *Chaetodipus baileyi insularis* is listed as Endangered by the Norma Oficial Mexicana (DOF 2019). *C. baileyi* is listed as Least Concern by the IUCN (Linzey et al. 2016d).

**Characteristics:** Bailey's pocket mouse is large sized; **total length 196.0–230.0 mm and skull length 29.1–30.6 mm.** Dorsal pelage gray grizzled with yellow; lateral line usually absent; fur of underparts whitish cream; tail bicolored, darker on the back than ventrally, with hair as a crest and a hair tuft at the tip; no stiffened hairs (spines) on the rumps; skull larger and heavily constructed, interparietal bone relatively larger and its width approximately equal to the interorbital breath; auditory bullae barely opposed anteriorly.

**Comments:** *Chaetodipus baileyi extimus*, *C. b. fornicatus*, *C. b. hueyi*, *C. b. mesidios*, and *C. b. rudinoris* were previously considered subspecies of *C. baileyi* (Patton 2005a), the current taxonomy includes these as subspecies of *C. rudinoris* (Riddle et al. 2000a). *C. baileyi* ranges east of the lower Colorado River across southern Arizona to southwestern New Mexico, then through western Sonora to northern Sinaloa (Map 387). It can be found in sympatry, or nearly so, with *C. artus*, *C. goldmani*, *C. hispidus*, *C. intermedius*, *C. pernix*, and *C. penicillatus*. *C. baileyi* can be distinguished from all other species of *Chaetodipus* by its larger size, total length greater than 200.0 mm, and hindfoot greater than 26.0 mm; from *C. artus* and *C. goldmani*, by lacking rump spines. From C. *hispidus*, by having a strongly crested tail with a grayer dorsal coloration, longer than the head-and-body length; without a conspicuous buff-to-ochraceous lateral stripe and larger auditory bullae. From *C. penicillatus*, by the absence of a noticeable crested tail, much larger body and longer hindfoot; interorbital breadth and interparietal breadth nearly equal. From *C. rudinoris*, by its average larger size and a different distribution range.

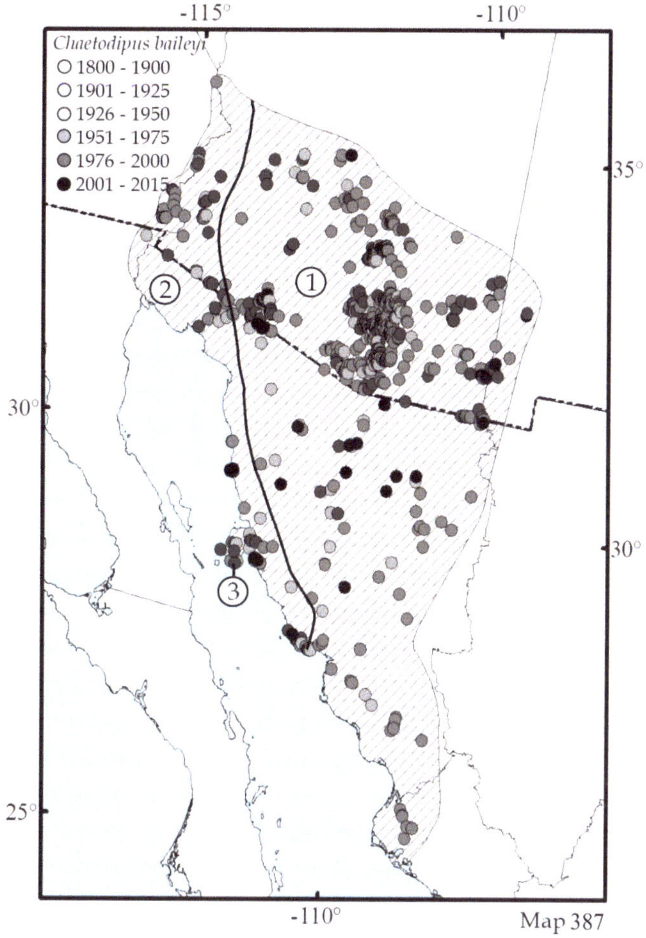


**Map 387** Distribution map of *Chaetodipus baileyi*

**Additional Literature:** Merriam (1889b), Osgood (1900), Burt (1936), Ingles (1947), Patton (1967a, 2005a), Van de Graff (1975), Homan and Genoways (1978), Hoffmeister (1986), Paulson (1988a), Best (1993d), Williams et al. (1993), Patton and Álvarez-Castañeda (1999).

## ***Chaetodipus californicus*** (Merriam, 1889)

California pocket mouse, ratón de bolsas de California

1. *C. c. bensoni* (von Bloeker, 1938). Northwestern part of the Central Valley, California.
2. *C. c. bernardinus* (Benson, 1930). Restricted to the San Bernardino Mountains, California.
3. *C. c. californicus* (Merriam, 1889). Restricted to the central part of San Francisco Bay, California.
4. *C. c. dispa*r (Osgood, 1900). Northeastern part of the Central Valley and from the coastal plains of San Luis Obispo to Santa Monica County, California.
5. *C. c. femoralis* (Allen, 1891). From Santa Monica County, California, to Sierra de Juárez, Baja California.
6. *C. c. marinensis* (von Bloeker, 1938). Coastal plains, from the southern part of San Francisco Bay south through San Luis Obispo, California.
7. *C. c. mesopolius* (Elliot, 1903). Restricted to Sierra San Pedro Mártir, Baja California.
8. *C. c. ochrus* (Osgood, 1904). Southern part of the Central Valley, California.

**Conservation:** *Chaetodipus californicus* is listed as Least Concern by the IUCN (Linzey et al. 2016e).

**Characteristics:** California pocket mouse is medium sized; total length 190.0–235.0 mm and skull length 25.0–28.3 mm. Dorsal pelage sandy brown interspersed with blackish shades; flanks paler; lateral line orange and highly distinguishable; fur of underparts whitish yellow; ears large, approximately one-half of the hindfoot length; tail bicolored with one hair tuft at the tip; **coarse whitish hairs (spines) on the rumps**; braincase arc shaped; mastoid bone reduced; first lower molariforms slightly larger than the last molariforms; **interparietal bone approximately twice as wide as long.**

**Comments:** *Chaetodipus californicus* ranges along the slopes of Sierra Nevada, Coast Ranges, Transverse Range, and Peninsular Ranges, from central California south through northern Baja California (Map 388). *C. californicus* can be found in sympatry, or nearly so, with *C. arenarius*, *C. fallax*, *C. rudinoris*, and *C. spinatus*. It can be distinguished from *C. arenarius* and *C. rudinoris* by having thick and stiffened hairs (spines) on the rump. From *C. fallax*, by having the ears long and with a somewhat squared distal edge; spines not thinner, marked whitish, and without a well marked orange lateral line narrow. From *C. spinatus*, by having spines only on the rump.

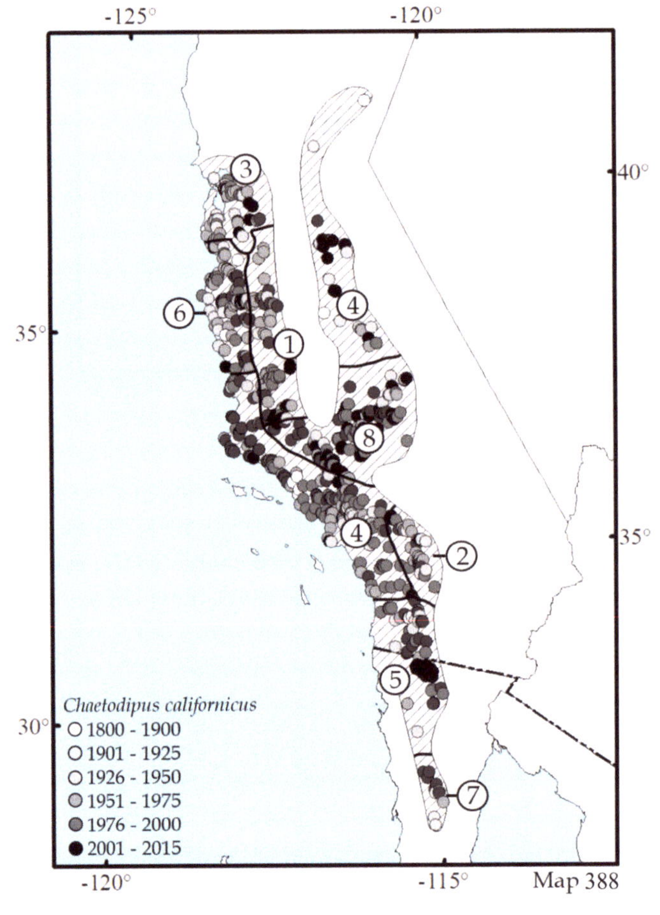


**Map 388** Distribution map of *Chaetodipus californicus*

**Additional Literature:** Merriam (1889b), Osgood (1900), Hall (1981), Best (1993d), Williams et al. (1993), Patton and Álvarez-Castañeda (1999), Patton (2005a).

## *Chaetodipus collis* (Blair, 1938)

Highland coarse-haired pocket mouse, ratón de bolsas de las tierras altas

1. *C. c. collis* (Blair, 1938). Restricted to southwestern Texas and isolated population in southeastern New Mexico.
2. *C. c. mapimiensis* Neiswenter et al. (2019). Restricted to northern Durango, eastern Chihuahua, and western Coahuila.

**Conservation:** *Chaetodipus collis* (as *C. nelsoni*) is listed as Least Concern by the IUCN (Linzey et al. 2016i).

**Characteristics:** Highland coarse-haired pocket mouse is small sized; total length 178.0–186.0 mm and skull length 24.0–27.0 mm. Characteristics very similar to *C. nelsoni* and *C. durangae*. Dorsal pelage orange yellow with sparse grizzled blackish shades; flanks paler; lateral line orange and narrow; fur of underparts whitish; tail longer than the head-and-body length with long hair as a crest and a tuft; **numerous and prominent stiffened hairs (spines) on the rump with distal ends usually dark-colored dorsally**; **white spot at ear base;** hindfoot sole blackish; braincase well arched; interparietal bone straplike; **mesopterigoidea fossa U shaped.**

**Comments:** *Chaetodipus collis* was described by Blair (1938), previously considered a junior synonym of *C. nelsoni canescens* (Borell and Bryant 1942). Neiswenter et al. (2019), using different analyses, considered it a valid species but without morphological differences from *C. durangae* and *C. nelsoni*. However, Baker and Greer (1962) recorded that the specimens from central and southern Durango are darker, have a broad rostrum and large mastoids (currently *C. nelsoni*), and those of northern areas (*C. collis*) are paler, with a smaller rostrum and larger mastoids (Baker 1956). *C. collis* ranges in southeastern New Mexico, western Texas, northern Sonora, and northeastern Chihuahua (Map 389). It can be found in sympatry, or nearly so, with *C. eremicus*, *C. hispidus*, and *C. intermedius*, from which it can be distinguished by its numerous and prominent stiffened hairs (spines) on the rump with distal bristle ends usually darkly dorsally, entire rump spines pale laterally; hindfoot sole blackish. From *C. eremicus*, by its larger size, usually greater than 180.0 mm. From *C. hispidus*, by its smaller size with a relatively long and crested tail. From *C. intermedius*, by its larger size; smaller toothrow length; broad rostrum, and posterior zygomatic narrow.

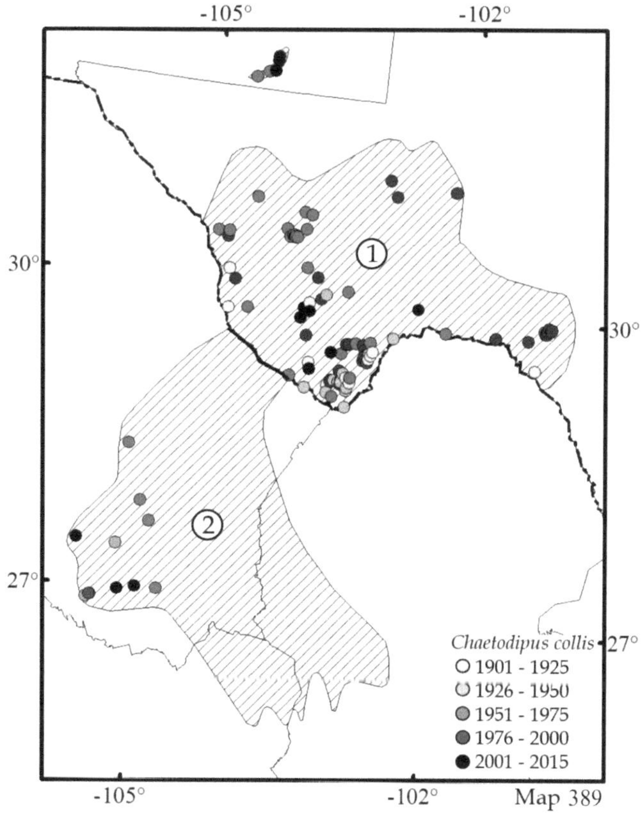


**Map 389** Distribution map of *Chaetodipus collis*

**Additional Literature:** Blair (1938), Anderson (1972), Davis (1974), Wilkins and Schmidly (1979), Findley (1987), Williams et al. (1993), Best (1994b).

## *Chaetodipus durangae* Neiswenter et al., 2019

Durango coarse-haired pocket mouse, ratón de bolsas de Durango

Monotypic.

**Conservation:** *Chaetodipus durangae* (as *C. nelsoni*) is listed as Least Concern by the IUCN (Linzey et al. 2016i).

**Characteristics:** Durango coarse-haired pocket mouse is small sized; total length 203.0–210.0 mm and skull length 24.2–28.5 mm. Characteristics very similar to *C. collis* and *C nelsoni*. Dorsal pelage orange yellow with sparse grizzled blackish shades; flanks paler; lateral line orange and narrow; fur of underparts whitish; tail longer than the head-and-body length with long hair as a crest and a tuft; **numerous and prominent stiffened hairs (spines) on the rump with distal ends usually dark-colored dorsally**; **white spot at ear base;** hindfoot sole blackish; braincase well arched; interparietal bone straplike; **mesopterigoidea fossa U shaped.**

**Comments:** Neiswenter et al. (2019) described *Chaetodipus durangae* as a new species using genetic and morphological analyses; previously, specimens had been assigned to *C. nelsoni nelsoni. C. durangae* ranges are restricted to desert areas of southern Chihuahua and northern Durango (Map 390). It can be found in sympatry, or nearly so, with *C. hispidus*, from which it can be distinguished by its numerous and prominent stiffened hairs (spines) on the rump, with distal ends usually dark dorsally and pale laterally; hindfoot soles blackish.

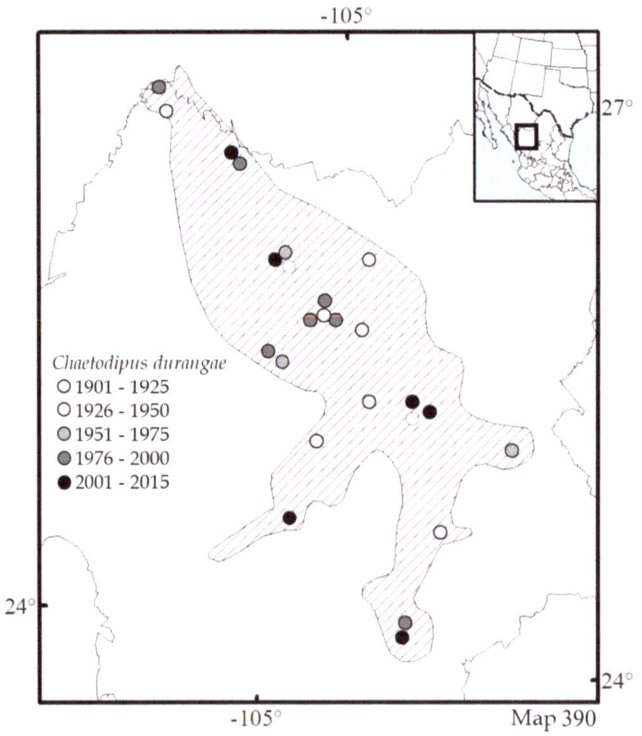


**Map 390** Distribution map of *Chaetodipus durangae*

**Additional Literature:** Anderson (1972), Wilkins and Schmidly (1979), Williams et al. (1993), Best (1994b).

## *Chaetodipus eremicus* (Mearns, 1898)

Chihuahuan pocket mouse, ratón de bolsas de Chihuahua

1. *C. e. atrodorsalis* (Dalquest, 1951). Restricted to eastern San Luis Potosí and southern Nuevo León.
2. *C. e. eremicus* (Mearns, 1898). From southern New Mexico and southwestern Texas, south through San Luis Potosí.

**Conservation:** *Chaetodipus eremicus* is listed as Least Concern by the IUCN (Linzey et al. 2016f).

**Characteristics:** Chihuahuan pocket mouse is small sized; total length 160.0–210.0 mm and skull length 24.2–26.6 mm. Dorsal pelage grayish grizzled brown with dark brown or blackish shades; flanks paler; lateral line orange and narrow, occasionally absent; fur of underparts whitish to yellowish; ears round with a white spot at the base**; tail longer than the head-and-body length with a long, heavily crested hair tuft**; **rump hairs elongated (but with no spines), dark dorsally and pale laterally**; longer hairs silky; mastoid breadth similar to the squamosal breadth; interparietal bone pentagonal, base of the occipital region with rounded corners throughout; auditory bullae widely separated anteriorly.

**Comments:** *Chaetodipus eremicus* was reinstated as a distinct species by Lee et al. (1996), previously regarded as a subspecies of *C. penicillatus* (Osgood 1900; Hoffmeister and Lee 1967; Patton 1993b). *C. eremicus* ranges in the Chihuahuan Desert from southern New Mexico and southwestern Texas south through Chihuahua, Coahuila, Durango, Zacatecas, Nuevo León, and San Luis Potosí (Map 391). It can be found in sympatry, or nearly so, with *C. collis*, *C. intermedius*, *C. lineatus*, and *C. nelsoni*. *C. eremicus* can be distinguished from *C. collis* by the absence of spines and its smaller size, total length usually less than 180.0 mm; hindfoot sole pale pink or white and naked to the heel. From *C. hispidus*, by its crested tail longer than the head-and-body length and without a conspicuous supraorbital bead. From *C. intermedius*, by having elongated hairs (no spines) on the rumps, darker dorsally and paler laterally. From *C. lineatus*, by lacking the dorsal pelage dull gray and without a slightly visible yellowish line on the back but mainly on the head, and flanks more grayish.

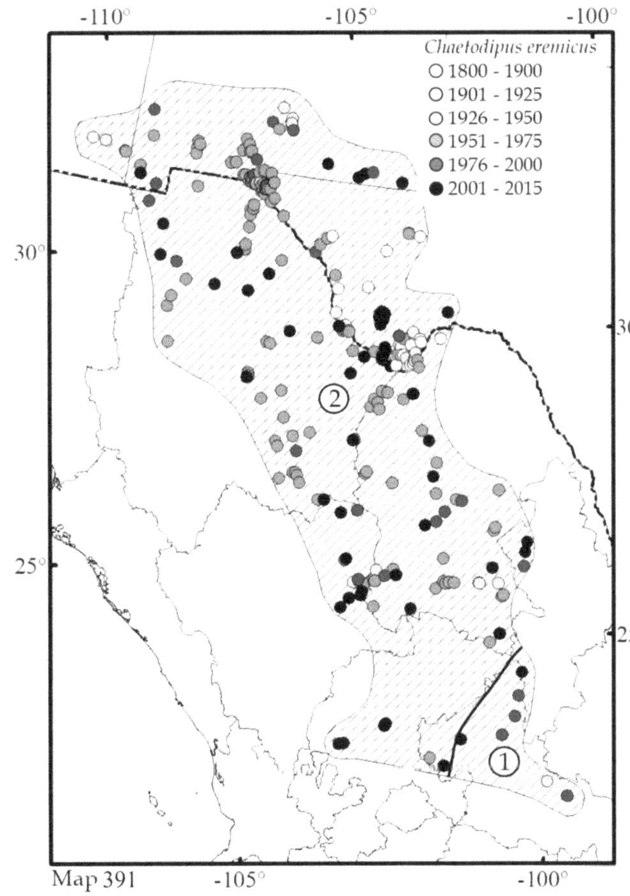


**Map 391** Distribution map of *Chaetodipus eremicus*

**Additional Literature:** Merriam (1889b), Bailey (1931), Dalquest (1951, 1953), Baker (1956), Patton (1969, 2005a), Anderson (1972), Genoways et al. (1977), Wilkins and Schmidly (1979), Best (1993d), Williams et al. (1993), Davis and Schmidly (1994), Manning et al. (1996), Yancey (1997), Mantooth and Best (2005a).

## *Chaetodipus fallax* (Merriam, 1889)

San Diego pocket mouse, ratón de bolsas de San Diego

1. *C. f. anthonyi* (Osgood, 1900). Restricted to Cedros Island, Baja California.
2. *C. f. fallax* (Merriam, 1889). From the southwestern coast of California south through inland central Baja California.
3. *C. f. inopinus* (Nelson and Goldman, 1929). From Bahía del Rosario, Baja California south through Bahía Tortugas, Baja California Sur.
4. *C. f. majusculus* (Huey, 1960). From Santo Tomás south through the coast to Bahía Rosario, Baja California.
5. *C. f. pallidus* (Mearns, 1901). Restricted to southern central California.
6. *C. f. xerotrophicus* (Huey, 1960). Restricted to the islands in southern-central Baja California.

**Conservation:** *Chaetodipus fallax anthonyi* is listed as Threatened by the Norma Oficial Mexicana (DOF 2019) and as Least Concern by the IUCN (Álvarez-Castañeda et al. 2016g).

**Characteristics:** San Diego pocket mouse is small sized; total length 176.0–200.0 mm and skull length 23.9–27.9 mm. Dorsal pelage brown gray grizzle with gray shades; lateral line narrow and orange; fur of underparts whitish or creamy; tail longer than the head-and-body length, with few hairs as a crest; coarse hairs (spines) on the rumps; **this combination of spines present on the rumps and flanks; ears small;** interparietal bone wide with the anterior angle obsolete; braincase arched.

**Comments:** *Chaetodipus anthonyi* from Cedros Island, Baja California, is a subspecies of *C. fallax* (Williams et al. 1993). *C. fallax* ranges from southern California south through the central Baja California Peninsula (Map 392), and can be found in sympatry, or nearly so, with *C. arenarius*, *C. californicus*, *C. rudinoris*, and *C. spinatus*. *C. fallax* can be distinguished from *C. arenarius* and *C. rudinoris* by having thick and very notorious stiffened hairs (spines) on the rumps and a well marked orange narrow lateral line. From *C. californicus* and *C. spinatus*, by having soft hair with marked whitish spines only on the rumps and not on the flanks; well marked orange lateral line, narrow. From *C. californicus* in having the ears short and with rounded distal edge.

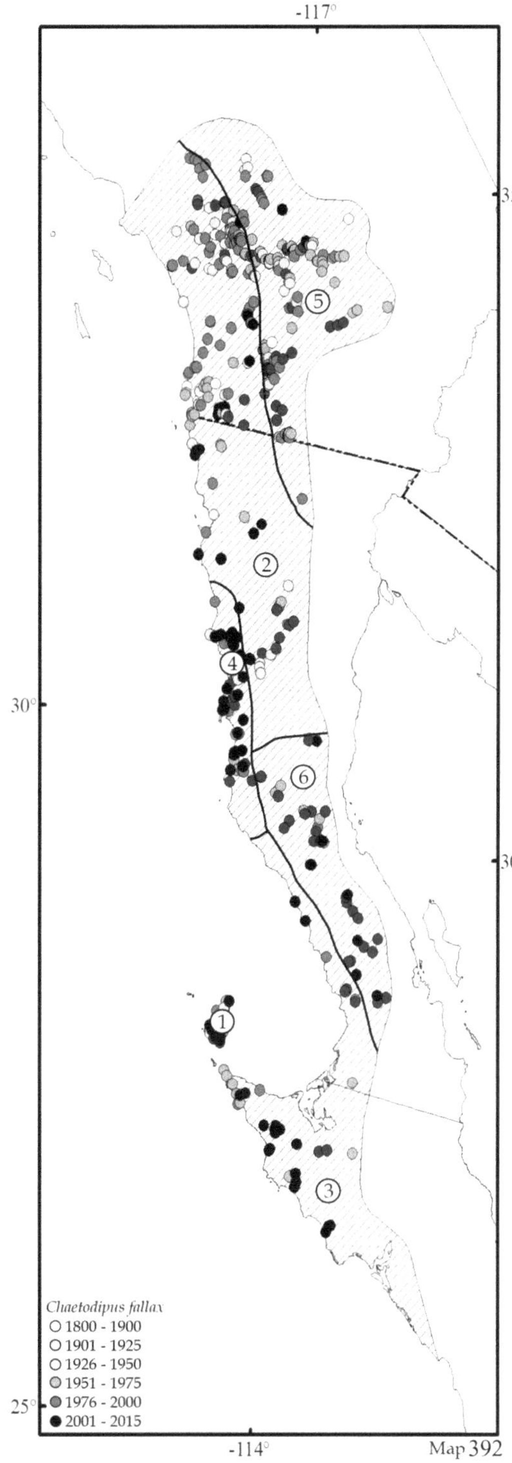


**Map 392** Distribution map of *Chaetodipus fallax*

**Additional Literature:** Merriam (1889b), Osgood (1900), Huey (1960, 1964), Ryan (1968), Caire (1976), Hall (1981), Patton et al. (1981), Best (1993f), Lackey (1996), Patton and Álvarez-Castañeda (1999), Álvarez-Castañeda et al. (2008c), Rios and Álvarez-Castañeda (2010), Camargo et al. (2016).

## *Chaetodipus formosus* (Merriam, 1889)

Long-tailed pocket mouse, ratón de bolsas de cola larga

1. *C. f. cinerascens* (Nelson and Goldman, 1929). Eastern coast of Baja California, from Bahía de Los Ángeles north through San Felipe.
2. *C. f. domisaxensis* (Cockrum, 1956). Central-northern Arizona and central-southern Utah
3. *C. f. formosus* (Merriam, 1889). Southwestern Utah and northwestern Arizona.
4. *C. f. incolatus* (Hall, 1941). Western Utah and eastern Nevada.
5. *C. f. infolatus* (Huey, 1954). From Bahía de Los Ángeles, Baja California south through northeastern Baja California Sur.
6. *C. f. melanocaudus* (Cockrum, 1956). Restricted to northern-central Utah.
7. *C. f. melanurus* (Hall, 1941). Northwestern Nevada and northeastern California.
8. *C. f. mesembrinus* (Elliot, 1904). From southern California south through San Felipe, Baja California.
9. *C. f. mohavensis* (Huey, 1938). Southwestern Utah, northwestern Arizona, southern Nevada, and central-eastern California.

**Conservation:** *Chaetodipus formosus* is listed as Least Concern by the IUCN (Álvarez-Castañeda et al. 2016h).

**Characteristics:** Long-tailed pocket mouse is medium sized; total length 160.0–216.0 mm and skull length 26.6–30.3 mm. Dorsal pelage pale gray, grayish brown to chocolate-dark brown; flanks paler; a narrow orange lateral line in gray specimens; fur of underparts whitish to creamy; thin and silky hair without spines; **large ears, slightly less than the hindfoot**; **tail bicolored, distal half with dense hair and a tuft at the tip, approximately 15.0 mm in length**; no stiffened hairs (spines) on the rumps; interparietal bone markedly wider than long; mastoidal bullae projecting slightly beyond the occipital bone.

**Comments:** *Chaetodipus formosus* was previously placed in the subgenus *Perognathus* (Osgood 1900, Hall 1981) but was subsequently reassigned to *Chaetodipus* (Hafner and Hafner 1983). *C. formosus* ranges from Nevada, Utah, northwestern Arizona, eastern California, and northeastern Baja California (Map 393). It can be found in sympatry, or nearly so, with *C. arenarius*, *C. penicillatus*, and *C. rudinoris*. It can be distinguished from *C. arenarius* by its larger size; tail crested and upperparts darker gray; inhabiting rocky hillsides. From *C. rudinoris*, by having upper parts with yellowish hairs mixed with grayish and weak spines or black hairs on the rump.

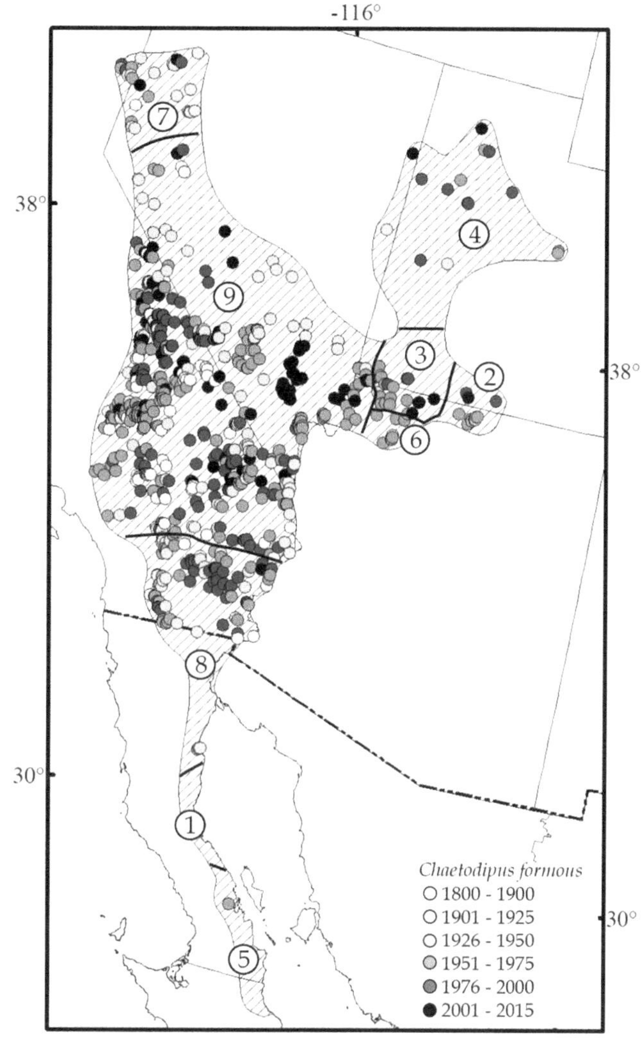


**Map 393** Distribution map of *Chaetodipus formosus*

**Additional Literature:** Merriam (1889b), Osgood (1900), Ingles (1947), Huey (1964), Patton et al. (1981), Best (1993d), Williams et al. (1993), Patton and Álvarez-Castañeda (1999).

## *Chaetodipus goldmani* (Osgood, 1900)

Goldman's pocket mouse, ratón de bolsas Sonorense

Monotypic.

**Conservation:** *Chaetodipus goldmani* is listed as Near Threatened (A2e: population reduction where the causes may not have ceased or may not be understood or may not be reversible, based on the effects of introduced taxa, by the IUCN; Lacher and Álvarez-Castañeda 2019).

**Characteristics:** Goldman's pocket mouse is small sized; total length 139.0–203.0 mm and skull length 23.3–26.8 mm. Dorsal pelage brown at the shoulders, darker on the back, and blackish on the rumps; fur of underparts whitish; lateral line yellow, very distinguishable; **tail longer than the head-and-body length, heavily crested, blackish dorsally**; **antitragal lobe prominent andwide**; **a few stiffened hairs (spines) on the rumps;** skull large and especially robust; nasals larger and posterior margin greater than 1.0 mm; supraoccipital less than 6.0 mm; braincase narrower and higher.

**Comments:** *Chaetodipus goldmani* was considered conspecific *with* C. *artus* (Hall and Ogilvie 1960) but larger samples and multiple types of data unambiguously allowed separating the two as valid species (Anderson 1964; Patton 1967a, 1969). *C. goldmani* ranges from northeastern to southern Sonora and northern Sinaloa, mostly in tropical thorny shrubland vegetation (Map 394). It can be found in sympatry, or nearly so, with *C. artus*, *C. baileyi*, *C. pernix*, and *C. enicillatus. C. goldmani* can be distinguished from *C. artus*, *C. baileyi*, *C. pernix*, and *C. penicillatus* by having thick and stiffened hairs (spines) on the rumps. From *C. artus*, by its larger size on average; tail with more hair and dorsal stripe narrow; supraorbital width narrower, less than 6.0 mm; extension of premaxilla small, less than 1.0 mm to the nasal posterior border; posterior palatal pits smaller; mastoid breadth wide; transverse ridge on the mastoid less strongly marked; larger auditory bullae; ascending processes of the premaxilla extending posteriorly to the nasals, a distance less than the least breadth of one nasal bone. From *C. penicillatus*, by the absence of a noticeable crest tail and a prominently buffy lateral line.

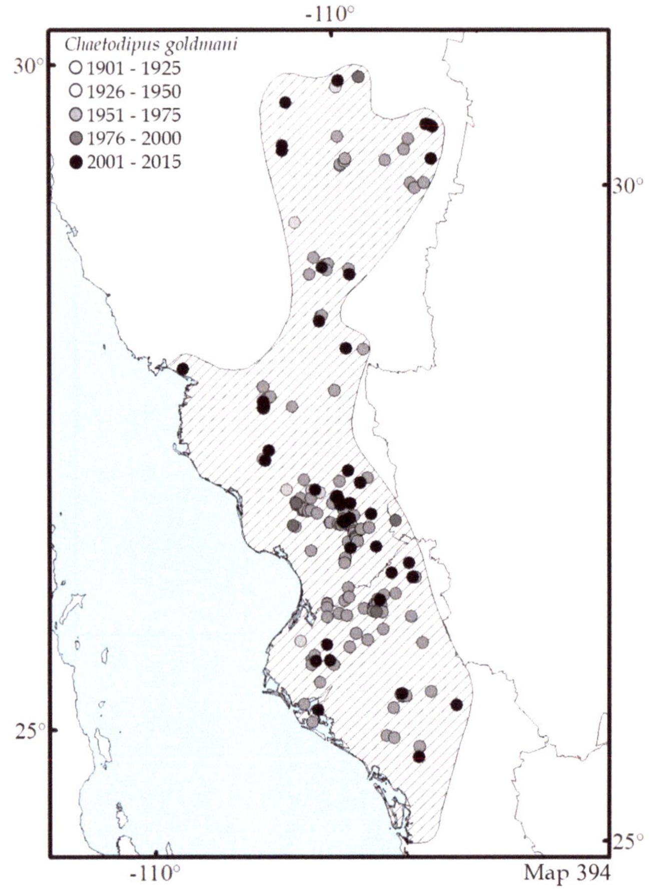


**Map 394** Distribution map of *Chaetodipus goldmani*

**Additional Literature:** Osgood (1900), Hatt (1932), Burt and Hooper (1941), Anderson (1972), Findley (1967), Caire (1978), Straney and Patton (1980), Lackey and Best (1992), Best (1993d), Williams et al. (1993), Patton and Álvarez-Castañeda (1999).

## *Chaetodipus hispidus* (Baird, 1858)

Hispid pocket mouse, ratón de bolsas de los pastizales

1. *C. h. hispidus* (Baird, 1858). From central-western Texas south through Tamaulipas, including eastern Coahuila and eastern Louisiana.
2. *C. h. paradoxus* (Merriam, 1889). From southern North Dakota south through northern Durango, including southeastern Arizona.
3. *C. h. spilotus* (Merriam, 1889). From southeastern Nebraska south through eastern Oklahoma.
4. *C. h. zacatecae* (Osgood, 1900). From southern Coahuila and Durango south through Estado de México and Hidalgo.

**Conservation:** *Chaetodipus hispidus* is listed as Least Concern by the IUCN (Linzey et al. 2016h).

**Characteristics:** Hispid pocket mouse is medium sized, albeit the largest member in the genus; total length 198.0–223.0 mm and skull length 23.4–28.1 mm. Dorsal pelage grizzled brown with orange and blackish shades; flanks paler; lateral line orange and a very distinguishable tail with the back dark and flanks yellowish; **tail length approximately equal to the head-and-body length**; an orange spot around the eyes; no stiffened hairs (spines) on the rumps; skull large; interparietal width greater than the interorbital region; **conspicuous ridge on the supraorbital edge**; auditory bullae less inflated than in other *Chaetodipus* with which it co-occurs; parietal mastoid side short.

**Comments:** Placed in the monotypic subgenus *Burtognathus* by Hoffmeister (1986), *C. hispidus* ranges through the Great Plains of the United States from North Dakota, southeastern Montana, and eastern Wyoming south to Texas, Louisiana, and into northeastern México (Tamaulipas, Nuevo León, and eastern Coahuila); west across Texas, New Mexico, and southeastern Arizona; south along the northern and eastern Sierra Madre Occidental, from northeastern Sonora and Chihuahua south through Jalisco, Estado de México, and Hidalgo (Map 395). *C. hispidus* can be found in sympatry, or nearly so, with *C. baileyi*, *C. collis*, *C. eremicus*, *C. intermedius*, *C. penicillatus*, and *C. nelsoni*. It can be distinguished from all other species of *Chaetodipus* by its non-crested tail equal to or shorter than the head-and-body length; supraorbital bead conspicuous and auditory bullae less inflated than in other *Chaetodipus*. From *C. baileyi*, by its general coloration with buff to ochraceous shades; conspicuous buff-to-ochraceous lateral stripe; auditory bullae smaller. From *C. eremicus*, by its shorter and crested tail, length approximately equal to the head-and-body, and with a conspicuous supraorbital bead. From *C. intermedius*, by its total length usually greater than 180.0 mm and hindfoot length greater than 22.0 mm. From *C. lineatus*, by not having dorsal pelage dull gray and without a slightly visible yellowish line on the back but mainly on the head, sides more grayish. From *C. collis* and *C. nelsoni*, by lacking thick and stiffened hairs (spines) on the rumps. From C. *penicillatus*, by lacking a noticeable crest tail; with a prominent buffy lateral line, and with a conspicuous buff-to-ochraceous lateral stripe. *C. hispidus* is similar in size to *Heteromys irroratus*; it has a coarser pelage, dark gray dorsally, and lacks grooved upper incisors.

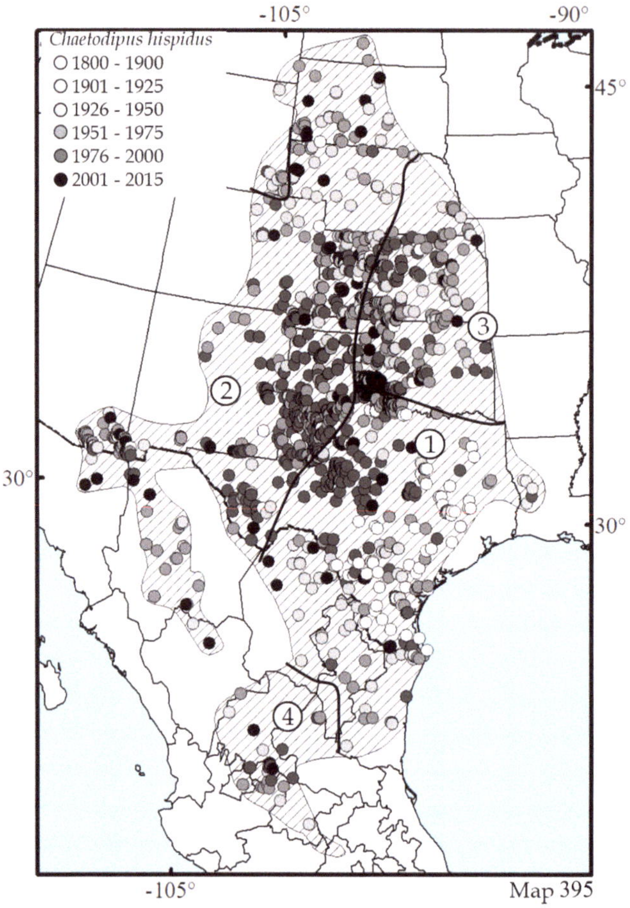


**Map 395** Distribution map of *Chaetodipus hispidus*

**Additional Literature:** Rhoads (1894), Merriam (1889b), Osgood (1900), Elliot (1903), Burt (1936), Glass (1947), Blair (1954), Patton (1967a), Schmidly (1977), Homan and Genoways (1978), Hall (1981), Paulson (1988b), Best (1993d), Williams et al. (1993), Andersen and Light (2012).

## *Chaetodipus intermedius* (Merriam, 1889)

Rock pocket mouse, ratón de bolsas de las rocas

1. *C. i. ater* (Dice, 1929). Restricted to Otero County, New Mexico.
2. *C. i. crinitus* (Benson, 1934). Northern-central western Arizona.
3. *C. i. intermedius* (Merriam, 1889). Central and southern Arizona, central New Mexico, western Texas, northeastern Sonora, central-northern Chihuahua, and western Coahuila.
4. *C. i. lithophilus* (Huey, 1937). From the western coast of Sonora south through Bahía San Jorge.
5. *C. i. minimus* (Burt, 1932). Restricted to Turners Island, Sonora.
6. *C. i. nigrimontis* (Blossom, 1933). Restricted to the Black Mountains, Arizona.
7. *C. i. phasma* (Goldman, 1918). Restricted to northwestern Sonora and southwestern Arizona.
8. *C. i. pinacate* (Blossom, 1933). Restricted to the Pinacate Desert, Sonora, and Arizona.
9. *C. i. rupestris* (Benson, 1932). Restricted to Doña Ana County, New Mexico.
10. *C. i. umbrosus* (Benson, 1934). Restricted to central Arizona.

**Conservation:** *Chaetodipus intermedius minimus* is listed as Threatened by the Norma Oficial Mexicana (DOF 2019). *C. intermedius*, as Least Concern by the IUCN (Linzey et al. 2016h).

**Characteristics:** Rock pocket mouse is small sized; total length 152.0–180.0 mm and skull length 22.7–25.2 mm. Dorsal pelage white to almost black, fur color related to the surrounding soil; paler forms with an orange lateral line very narrow; fur of underparts white to cream, darker forms with a white spot in the chest; ears small, dark at the tip and pale at the base; tail slightly longer than the body, with a dark dorsal part, darker distally, with long hairs dorsally and a hair tuft at the tip (crested); **very small stiffened hairs (spines) on the rumps but larger than other hairs**; braincase well arched; interparietal bone strapped in shape, and touch the auditory bullae; **mesopterigoidea fossa V shaped.**

**Comments:** *Chaetodipus intermedius* ranges from southern Utah, Arizona, central and southern New Mexico, southwestern Texas, Sonora, northern Chihuahua, and northwestern Coahuila (Map 396). It can be found in sympatry, or nearly so, with *C. baileyi*, *C. collis*, *C. eremicus*, *C. hispidus*, and *C. penicillatus*. *C. intermedius* can be distinguished from *C. baileyi*, *C. eremicus*, *C. hispidus*, and *C. penicillatus* by having thick and stiffened hairs (spines) on the rumps. *C. intermedius* has very small stiffened hairs (spines) on the rumps but longer than in other areas. From *C. hispidus*, by having a crested tail longer than the head-and-body length, total length usually less than 180.0 mm and hindfoot length less than 22.0 mm. From *C. collis*, by having less conspicuous rump spines; coarser pelage; smaller size; longer toothrow; narrow rostrum, and posterior zygomatic wide, and from *C. penicillatus*, by its more prominently crested tail; less prominent buffy lateral line; interparietal bone strapped-shape which touches the auditory bullae.

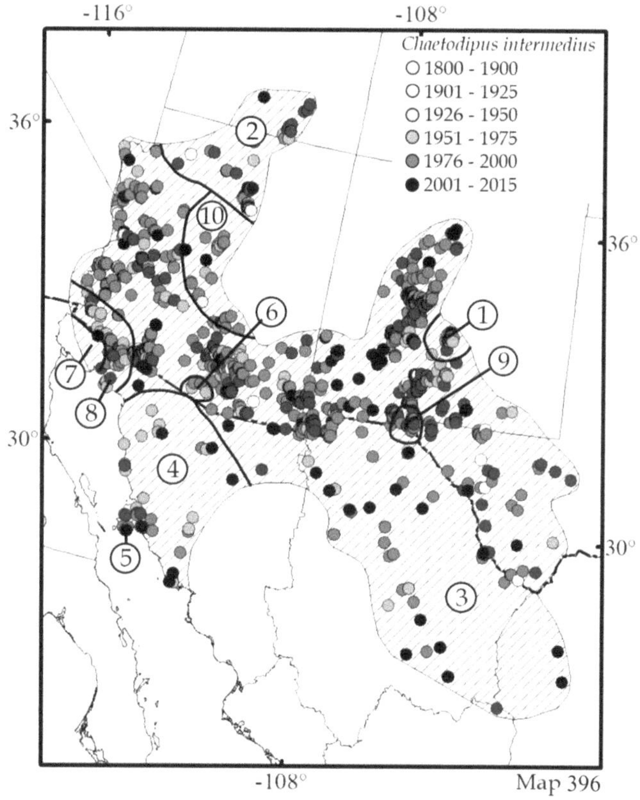


**Map 396** Distribution map of *Chaetodipus intermedius*

**Additional Literature:** Merriam (1889b), Osgood (1900), Hoffmeister (1974), Wilkins and Schmidly (1979), Best (1993d), Williams et al. (1993), Patton and Álvarez-Castañeda (1999).

## *Chaetodipus lineatus* (Dalquest, 1951)

Lined pocket mouse, ratón de bolsas Potosino

Monotypic.

**Conservation:** *Chaetodipus lineatus* is listed as Data Deficient by the IUCN (Vázquez 2017c).

**Characteristics:** Lined pocket mouse is small sized; total length **171.5–176.2 mm and skull length 24.0–25.4 mm. Dorsal pelage dull gray with a slightly visible yellowish line on the back but mainly on the head, flanks more grayish**; lateral line yellowish not very distinguishable; fur of underparts whitish; no stiffened hairs (spines) on the rumps; mastoid breadth similar to squamosal breadth; interparietal bone pentagonal, occipital region base with some rounded corners; auditory bullae widely separated anteriorly**.**

**Comments:** *Chaetodipus lineatus* is endemic to San Luis Potosí and southeastern Zacatecas (Map 397). It can be found in sympatry, or nearly so, with *C. hispidus* and *C. nelson.* It can be distinguished from *C. hispidus* and *C. nelson* by its dorsal pelage dull gray with a slightly visible yellowish line on the back but mainly on the head, and sides more grayish. In addition, from *C. hispidus*, by having a crested tail longer than the head-and-body length, without a conspicuous supraorbital bead. From *C. nelsoni*, by lacking thick and stiffened hairs (spines) on the rumps.

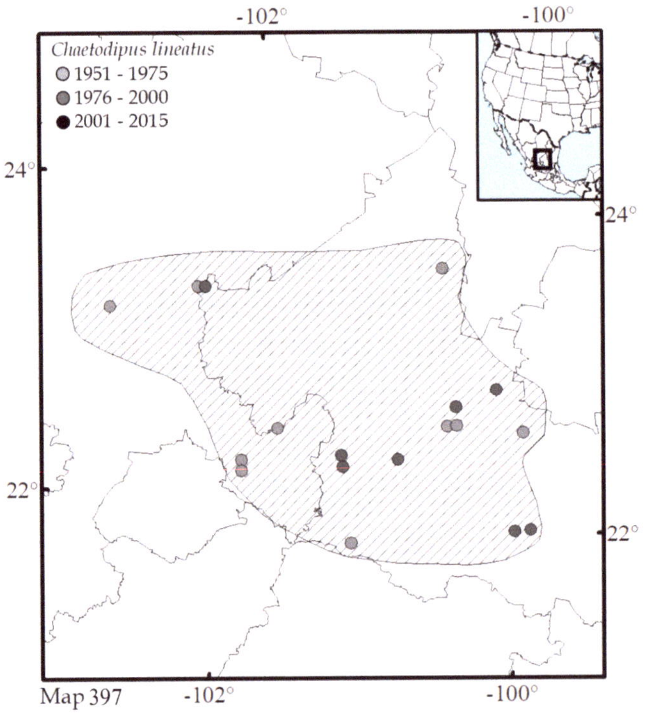


**Map 397** Distribution map of *Chaetodipus lineatus*

**Additional Literature:** Dalquest (1951), Matson and Baker (1986), Best (1993d), Williams et al. (1993).

## *Chaetodipus nelsoni* (Merriam, 1894)

Nelson's pocket mouse, ratón de bolsas del Altiplano

1. *C. n. canescens* (Merriam, 1894). From southern New Mexico and Texas south through southern Sonora, northern Durango, and southern Chihuahua.
2. *C. n. nelsoni* (Merriam, 1894). From northern Durango and southern Chihuahua south through Zacatecas, Jalisco, Michoacán, and San Luis Potosí.

**Conservation:** *Chaetodipus nelsoni* is listed as Least Concern by the IUCN (Linzey et al. 2016i).

**Characteristics:** Nelson's pocket mouse is small sized; total length 181.8–194.0 mm and skull length 24.1–27.8 mm. Characteristics very similar to *C. collis* and *C. durangae*. Dorsal pelage orange yellow with slight grizzled blackish shades; flanks paler; lateral line orange and narrow; fur of underparts whitish; tail longer than the head-and-body length with long hair as a crest and a tuft; **numerous and prominent stiffened hairs (spines) on the rump with distal ends usually dark dorsally**; **white spot at ear base;** hindfoot sole blackish; braincase well arched; interparietal straplike; **mesopterigoidea fossa U shaped.**

**Comments:** Neiswenter et al. (2019) divided *Chaetodipus nelsoni* into three species, describing *C. collis* and *C. durangae*, and limiting *C. nelsoni* to a small area in the southern part of its original range, from eastern Durango, southern Coahuila, western Nuevo León, Zacatecas, San Luis Potosí, Aguascalientes, and northern Jalisco (Map 398). Baker and Greer (1962) noted that specimens from central and southern Durango were darker, with a broader rostrum and larger mastoids (current *C. nelsoni*), than those in the north (*C. collis*), which are paler, with a smaller rostrum and larger mastoids (see also Baker 1956). *C. nelsoni* can be found in sympatry, or nearly so, with *C. eremicus*, *C. hispidus*, and *C. lineatus*, from which it differs by its numerous, prominent rump spines, pale but dark at the tips, and hindfeet sole blackish. From *C. eremicus*, by its larger size, total length usually greater than 180.0 mm. From *C. hispidus*, by its smaller size, with a relatively long crested tail. From *C. lineatus*, by having the dorsal pelage dark-brownish and without a visible yellowish line on the back, present mainly on the head.

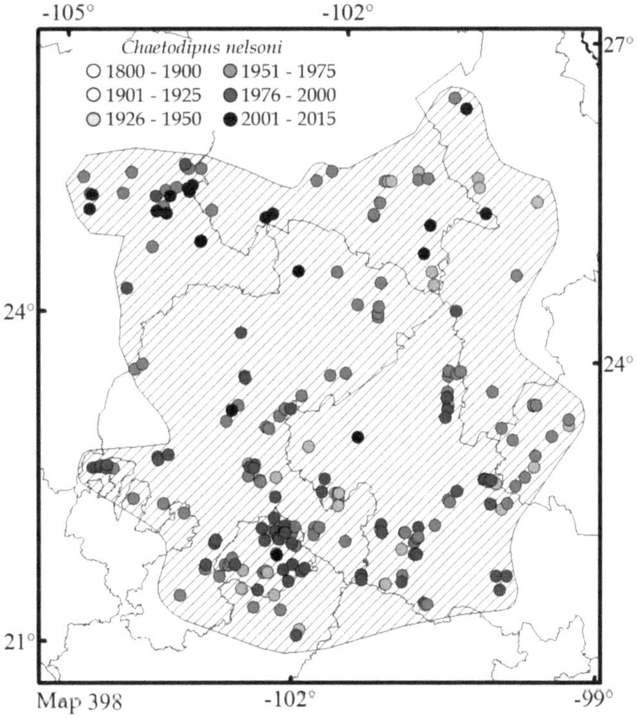


**Map 398** Distribution map of *Chaetodipus nelsoni*

**Additional Literature:** Merriam (1889b), Osgood (1900), Dalquest (1951, 1953), Wilkins and Schmidly (1979), Hall (1981), Williams et al. (1993), Best (1994b), Geluso and Geluso (2015).

## *Chaetodipus penicillatus* (Woodhouse, 1852)

Desert pocket mouse, ratón de bolsas del desierto

1. *C. p. angustirostris* (Osgood, 1900). From northeastern Baja California Sur north through southern California.
2. *C. p. penicillatus* (Woodhouse, 1852). Central and western Arizona.
3. *C. p. pricei* (Allen, 1894). Southern Arizona and Sonora.
4. *C. p. seri* (Nelson, 1912). Restricted to Tiburón Island, Sonora.
5. *C. p. sobrinus* (Goldman, 1939). Southern Nevada and northwestern Arizona.
6. *C. p. stephensi* (Merriam, 1894). Restricted to the Central Valley, California.

**Conservation:** *Chaetodipus penicillatus seri* is listed as Threatened by the Norma Oficial Mexicana (DOF 2019). *C. penicillatus*, as Least Concern by the IUCN (Linzey et al. 2016j).

**Characteristics:** Desert pocket mouse is small sized; total length 162.0–216.0 mm and skull length 24.8–28.2 mm. Dorsal pelage yellowish brown to yellowish gray; flanks paler; lateral line usually barely distinguishable or absent; fur of underparts whitish to creamy; **tail long, with noticeable long black hair as a crest and a large hair tuft at the tip; no stiffened hairs (spines) on the rumps but longer than the rest**; mastoid breadth similar to squamosal breadth; **interparietal bone pentagonal, base of the occipital region with some rounded corners throughout**; auditory bullae very separated anteriorly.

**Comments:** *Chaetodipus penicillatus eremicus* was reinstated as a distinct species by Lee et al. (1996), and *C. p. atrodorsalis* was designated as a subspecies of *C. eremicus*. *C. penicillatus* ranges from eastern California, extreme southern Nevada, and Utah, western and southern Arizona, southwestern New Mexico, northeastern Baja California, Sonora, western Chihuahua, and northern Sinaloa (Map 399). It can be found in sympatry, or nearly so, with *C. arenarius*, C. *baileyi*, *C. goldmani*, *C. hispidus*, *C. intermedius*, *C. pernix*, and *C. rudinoris*. *C. penicillatus* can be distinguished from all other species of *Chaetodipus* by its noticeable crest tail with long hairs and a prominent buffy lateral line. From *C. arenarius*, by having the back coloration yellowish or yellowish-brown; foramen magnum evenly rounded; the coronoid process rising steeply or high above the condyle level. From C. *baileyi* and *C. rudinoris*, by its hindfoot length less than 26.0 mm and interorbital breadth less than the interparietal breadth. From *C. goldmani*, by lacking thick and stiffened hairs (spines) on the rumps. From C. *hispidus*, by lacking a conspicuous buff-to-ochraceous lateral stripe. From *C. intermedius*, by having a noticeable crest tail but smaller; lateral stripe buffy; interparietal bone pentagonal with all angles somewhat rounded and auditory bullae not making contact. From *C. pernix*, by having the mastoid breadth wider than the interorbital breadth; toothrow longer; auditory bullae usually greater than 7.5 mm; angle between the anterior border of the zygomatic and the rostrum less obtuse.

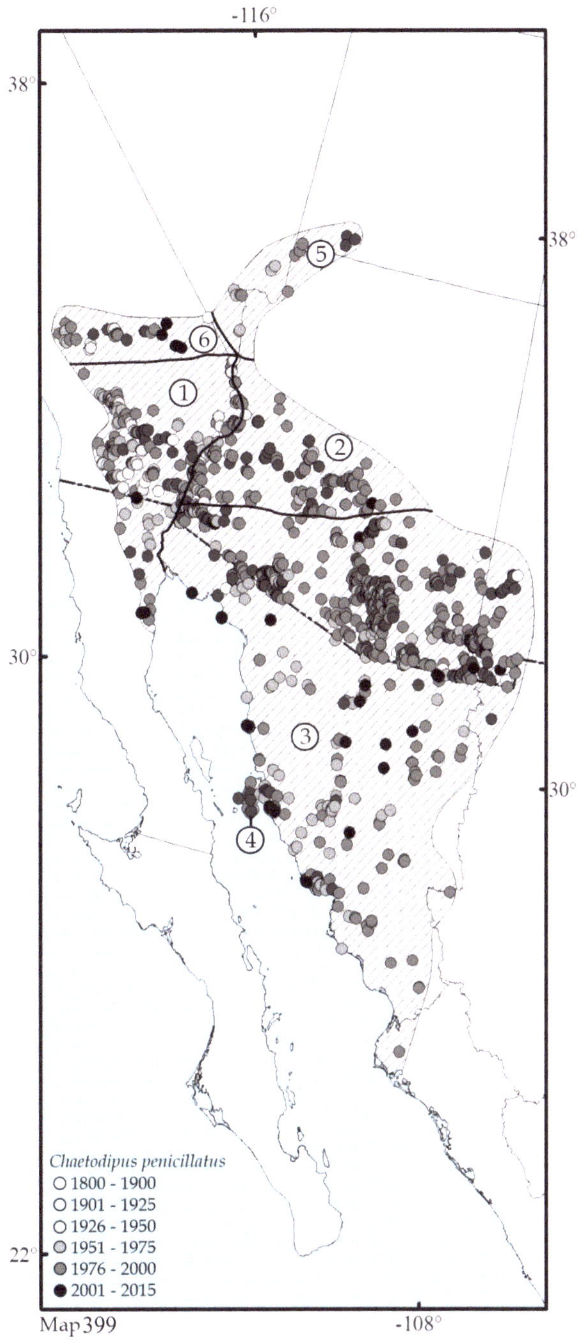


**Map 399** Distribution map of *Chaetodipus penicillatus*

**Additional Literature:** Woodhouse (1852), Merriam (1889b), Osgood (1900), Nelson and Goldman (1929), Hall (1946), Hoffmeister and Lee (1967), Lawlor (1982a), Best (1993d, f), Williams et al. (1993), Patton and Álvarez-Castañeda (1999), Mantooth and Best (2005b).

## ***Chaetodipus pernix*** (Allen, 1898)

Sinaloan pocket mouse, ratón de bolsas de Sinaloa

1. *C. p. pernix* (Allen, 1898). From the Guamuchil area, Sinaloa, south through northern Nayarit.
2. *C. p. rostratus* (Osgood, 1900). From the Guamuchil area, Sinaloa, north through southern Sonora.

**Conservation:** *Chaetodipus pernix* is listed as Least Concern by the IUCN (Álvarez-Castañeda et al. 2016i).

**Characteristics:** Sinaloan pocket mouse is small sized; total length 162.0–175.0 mm and skull length 23.9–24.9 mm. Dorsal pelage brown, profusely lined with blackish hairs and slightly hispid; flanks paler; lateral line yellow, distinguishable; fur of underparts whitish; tail longer than the head-and-body length with few long hairs as a crest; no stiffened hairs (spines) on the rumps; **ears dusky with a minute white spot on the lower margins**; hindfoot soles naked; **nasals relatively broad and flattened**; interparietal bone wider anteriorly than posteriorly; molars small and weak, first lower premolars larger than the last molars.

**Comments:** *Chaetodipus pernix* ranges from the coastal lowlands of southern Sonora south through northern Nayarit (sandy soils; Map 400), and can be found in sympatry, or nearly so, with *C. artus*, *C. baileyi*, *C. goldmani*, and *C. penicillatus*. *C. pernix* can be distinguished from all other species of *Chaetodipus* by its barely developed hairs (spines) on the rumps; tail long, thinly haired, and slightly crested; molars small and weak; lower premolars larger than last molars; nasofrontal suture not emarginated and interparietal bone wide. From *C. artus*, by having smaller ears, pointed, length generally less than 10.0 mm; interorbital constriction less than 5.8 mm. From *C. baileyi* and *C. penicillatus*, by its smaller size, total length less than 200.0 mm; upper parts darker, and nasal bones larger. Specifically from *C. penicillatus*, by lacking a noticeable crest in the tail and a prominent buffy lateral line; mastoid breadth and interorbital breadth narrower; toothrow shorter; mastoid bullae usually less than 7.5 mm; more obtuse angle between the anterior border of the zygomatic and the rostrum.

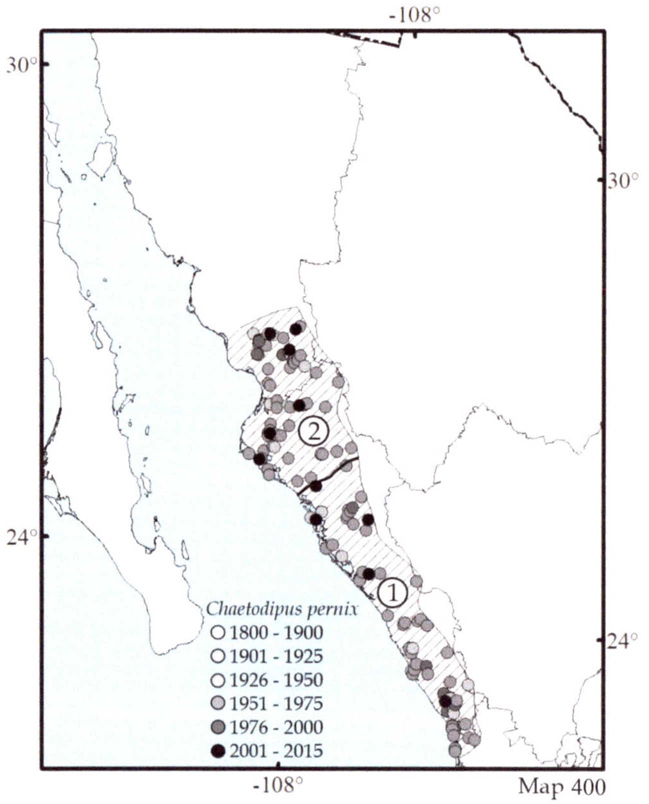


**Map 400** Distribution map of *Chaetodipus pernix*

**Additional Literature:** Allen (1898), Osgood (1900), Hatt (1932), Hoffmeister and Lee (1967), Patton (1967a), Patton et al. (1981), Best (1993d), Best and Lackey (1992b), Williams et al. (1993), Lee et al. (1996), Patton and Álvarez-Castañeda (1999).

## *Chaetodipus rudinoris* (Elliot, 1903)

Baja California pocket mouse, ratón de bolsas de Baja California

1. *C. r. extimus* (Nelson and Goldman, 1930). Central Baja California Sur.
2. *C. r. fornicatus* (Burt, 1932). Restricted to Montserrat Island, Baja California Sur.
3. *C. r. hueyi* (Nelson and Goldman, 1929). Northeastern Baja California and southeastern California.
4. *C. r. mesidios* (Huey, 1964). From the Central Desert of Baja California south through central Baja California Sur.
5. *C. r. rudinoris* (Elliot, 1903). Northern-central Baja California.

**Conservation:** *Chaetodipus rudinoris fornicatus* has been recorded as extirpated from Montserrat Island (Álvarez-Castañeda and Cortés-Calva 2002b; Álvarez-Castañeda and Ortega-Rubio 2003) and listed as *C. baileyi fornicatus* as Endangered by the Norma Oficial Mexicana (DOF 2019) and *C. rudinoris* as Least Concern by the IUCN (Álvarez-Castañeda et al. 2016j).

**Characteristics:** Baja California pocket mouse is large sized; total length 183.0–211.0 mm and skull length 25.3–30.4 mm. Dorsal pelage gray grizzled with yellow shades; lateral line absent; underparts fur whitish; tail with a hair tuft at the tip; no stiffened hairs (spines) on the rumps; skull larger and heavily constructed; interparietal bone relatively large, its breadth approximately equal to interorbital breadth; auditory bullae barely opposed anteriorly.

**Comments:** Riddle et al. (2000a) separated *Chaetodipus rudinoris* from *C. baileyi*. *C. rudinoris* ranges from southeastern California to the southern tip of the Baja California Peninsula, including Montserrat Island, Gulf of California, Baja California Sur (Map 401). It can be found in sympatry, or nearly so, with *C. ammophilus*, *C. arenarius*, *C. californicus*, *C. fallax*, *C. formosus*, *C. penicillatus*, *C. siccus*, and *C. spinatus*. It can be distinguished from *C. californicus*, *C. fallax*, and *C. spinatus* by lacking thick and stiffened hairs (spines) on the rumps. From *C. ammophilus*, *C. arenarius*, *C. fallax*, and *C. siccus*, by its larger size, with total length greater than 183.0 mm and hindfoot length greater than 26.0 mm. From *C. formosus*, by having upperparts without yellowish pelage and lacking weak spines or black hairs on the rump. From *C. penicillatus*, by the absence of a noticeable crest tail and a prominent buffy lateral line.

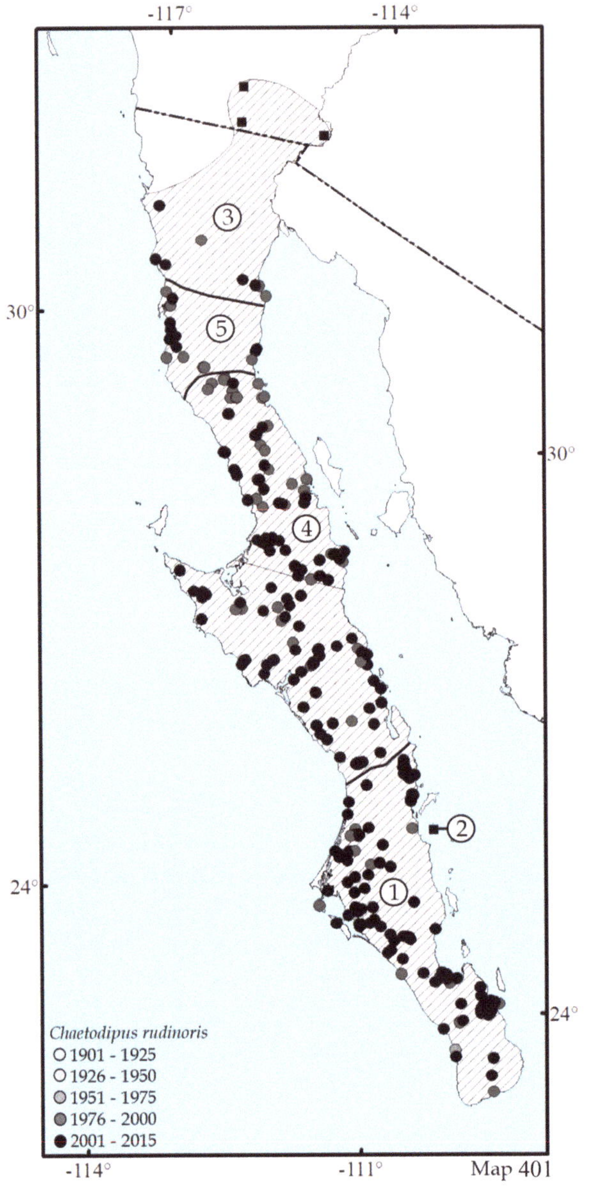


**Map 401** Distribution map of *Chaetodipus rudinoris*

**Additional Literature:** Burt (1936), Ingles (1947), Patton (1967a, 2005a), Homan and Genoways (1978), Paulson (1988a), Álvarez-Castañeda and Ortega-Rubio (2003), Álvarez-Castañeda et al. (2008c).

## *Chaetodipus siccus* (Osgood, 1907)

Cerralvo Island pocket mouse, ratón de bolsas de Isla Cerralvo

(*see addenda, new subspecies *Chaetodipus siccus liae*)

Monotypic.

**Conservation:** *Chaetodipus siccus* (as *C. arenarius siccus*) is listed as Threatened by the Norma Oficial Mexicana (DOF 2019) and as Least Concern by the IUCN (Álvarez-Castañeda and Lacher 2017).

**Characteristics:** Cerralvo Island pocket mouse is small sized; total length 150.0–190.0 mm and skull length 23.0–26.7 mm. Dorsal pelage pale gray, slightly grizzled on the back; lateral line usually absent; fur of underparts whitish; tail longer than the head-and-body length, darker dorsally and paler ventrally; **long scarce hairs on the rumps and only backlight visible;** skull small and relatively broad; nasals slender; zygomatic arches fragile; braincase vaulted.

**Comments:** *Chaetodipus siccus* was recognized as a distinct species (Rios and Álvarez-Castañeda 2011**);** previously, it was considered a subspecies of *C. arenarius* (Williams et al. 1993). *C. siccus* is endemic to Isla Cerralvo and Los Planes Basin, Baja California Sur (Map 402). It can be found in sympatry, or nearly so, with *C. rudinoris* and *C. spinatus. C. siccus* can be distinguished from all other species of *C. spinatus* by lacking well-developed thick and stiffened hairs (spines) on the rumps. From *C. rudinoris*, by its smaller size, with a total length less than 183.0 mm. From *C. ammophilus* and *C. arenarius***,** by having very thin spines on the rumps; general coloration with more brown shades; its size is intermediate between both species.

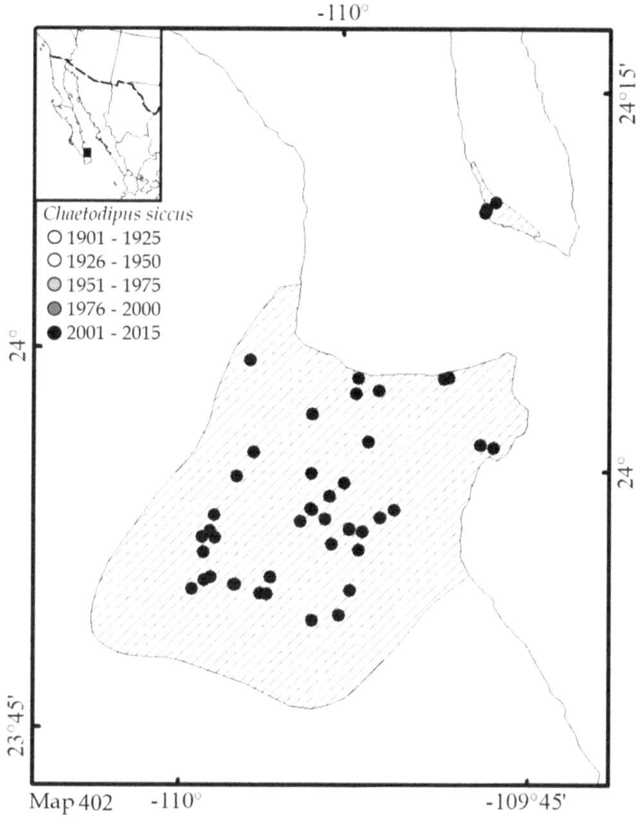


**Map 402** Distribution map of *Chaetodipus siccus*

**Additional Literature:** Merriam (1894), Osgood (1900, 1907), Nelson and Goldman (1929), Hall (1981).

## *Chaetodipus spinatus* (Merriam, 1889)

Spiny pocket mouse, ratón de bolsas espinoso

1. *C. s. broccus* (Huey, 1960). From La Paz Isthmus, Baja California Sur north through the border with Baja California.
2. *C. s. bryanti* (Merriam, 1894). Restricted to San José Island, Baja California Sur.
3. *C. s. evermanni* (Nelson and Goldman, 1929). Restricted to Mejía Island, Baja California.
4. *C. s. guardiae* (Burt, 1932). Restricted to Ángel de la Guarda Island, Baja California.
5. *C. s. lambi* (Benson, 1930). Restricted to Espíritu Santo Island, Baja California Sur.
6. *C. s. latijugularis* (Burt, 1932). Restricted to San Francisco Island, Baja California Sur.
7. *C. s. lorenzi* (Banks, 1967). Restricted to the San Lorenzo and Las Ánimas Islands, Baja California.

8. *C. s. magdalenae* (Osgood, 1907). Restricted to Magdalena Island, Baja California Sur.
9. *C. s. marcosensis* (Burt, 1932). Restricted to San Marcos Island, Baja California Sur.
10. *C. s. margaritae* (Merriam, 1894). Restricted to Margarita Island, Baja California Sur.
11. *C. s. occultus* (Nelson, 1912). Restricted to Carmen Island, Baja California Sur.
12. *C. s. oribates* (Huey, 1960). Central Baja California.
13. *C. s. peninsulae* (Merriam, 1894). From La Paz Isthmus southward in Baja California Sur.
14. *C. s. prietae* (Huey, 1930). From El Rosario, Baja California, south through the border of Baja California.
15. *C. s. pullus* (Burt, 1932). Restricted to Coronados Island, Baja California Sur.
16. *C. s. rufescens* (Huey, 1930). Southern-central Baja California.
17. *C. s. seorsus* (Burt, 1932). Restricted to Danzante Island, Baja California Sur.
18. *C. s. spinatus* (Merriam, 1889). From southern Nevada to San Felipe, Baja California.

**Conservation:** The Norma Oficial Mexicana (DOF 2019) listed many subspecies of *Chaetodipus spinatus* with conservation issues. *C. s. evermanni* as Extinct, *C. s. guardiae*, *C. s. lambi*, *C. s. lorenzi*, *C. s. marcosensis*, *C. s. margaritae*, *C. s. occultus*, *C. s. pullus*, and *C. s. seorsus* as Threatened, and *C. s. bryanti* and *C. s. latijugularis* as Endangered. *C. spinatus* as Least Concern by the IUCN (Linzey et al. 2016k).

**Characteristics:** Spiny pocket mouse is small sized; total length 164.0–225.0 mm and skull length 22.3–25.8 mm. Dorsal pelage grizzle with gray and brown gray shades; lateral line barely distinguishable or absent; fur of underparts whitish to whitish yellow; tail longer than the head-and-body length, with few hairs as a crest; **stiffened hairs (spines) on the rumps, back, and flanks;** skull slender and flattened; interparietal bone broad with anterior angle faintly expressed; mastoids small; supraorbital ridge usually slightly trenchant.

**Comments:** *Chaetodipus spinatus* ranges from southern Nevada and southeastern California south through the Baja California Peninsula, including Espíritu Santo, San Francisco, San José, Carmen, Coronados, San Marcos, San Lorenzo, Las Ánimas, Ángel de la Guarda, Mejía, Danzante, Magdalena, and Margarita Islands (Map 403). It can be found in sympatry, or nearly so, with *C. ammophilus*, *C. arenarius*, *C. fallax*, *C. rudinoris*, and *C. siccus*. It differs from *C. ammophilus*, *C. arenarius*, *C. rudinoris*, and *C. siccus*, by having thick and very notorious stiffened hairs (spines) on the rumps, scattered on the flanks and sides, and often extending to the shoulders. From *C. californicus* and *C. fallax*, by having a pelage with marked whitish spines on the back, flanks, and rumps; lateral line absent or yellowish, very thin.

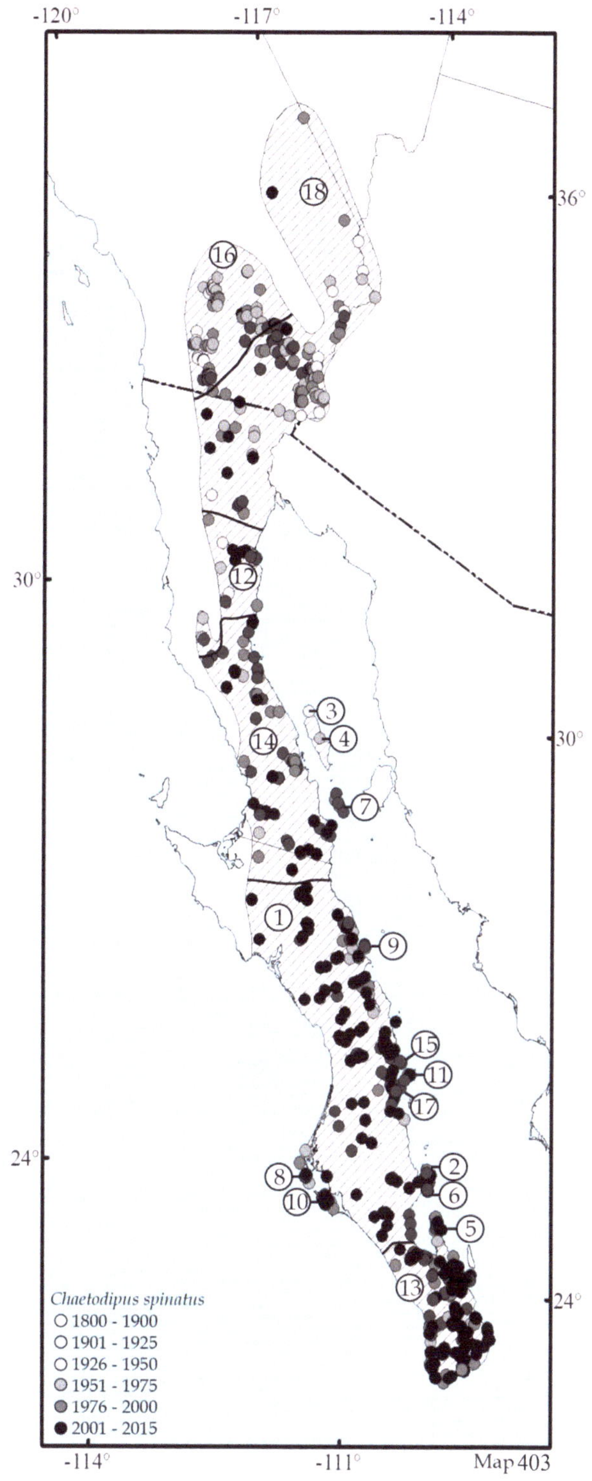


**Map 403** Distribution map of *Chaetodipus spinatus*

**Additional Literature:** Merriam (1889b), Osgood (1900), Hatt (1932), Homan and Genoways (1978), Patton et al. (1981), Lackey (1991b), Best (1993d), Williams et al. (1993), Patton and Álvarez-Castañeda (1999), Álvarez-Castañeda et al. (2008c), Álvarez-Castañeda and Murphy (2014), Álvarez-Castañeda and Nájera-Cortazar (2020).

## Genus *Perognathus* Wied-Neuwied, 1839

Dorsal pelage pale gray to dark gray washed with white, ochraceous, yellowish or tawny shades; fur of underparts white, or nearly so; pale orange spots above the eyes; tail varying in size relative to the head-and-body length but with little fur, bicolored in some species, with a hair tuft at the tip; **hair soft, without spines**; generally with a lateral stripe separating the dorsal and ventral pelage, contrast limited in some species; hindfoot sole moderately hairy; **mastoid bone projecting posteriorly beyond the occipital bone;** ventral auditory bullae meeting, or nearly so, anteriorly; **interparietal narrow (square-shape), breadth equal to or less than the interorbital region breadth; supraoccipital bone without lateral indentations**. *Perognathus* has four species groups: *fasciatus* (*P. fasciatus* and *P. flavescens*), *flavus* (*P. flavus* and *P. nelsoni*), *longimembris* (*P. amplus*, *P*, *inornathus*, and *P. longimembris*), *parvus* (*P. alticola*, *P. mollipilosus* and *P. parvus*). The keys were elaborated based on the review of specimens and the following sources: Merriam (1900), Hall (1981), and Álvarez-Castañeda et al. (2017a).

1. Antitragus lobed; hindfoot length greater than 20.0 mm; tail with a black hair tuft; skull length greater than 24.0 mm ........................................................................................................................................................2

1a. Antitragus not lobed; hindfoot length less than 20.0 mm; tail generally without a hair tuft, small and not black when present; skull length less than 24.0 mm ........................................................................................................4

2. Ears with white or pale yellowish hairs and without a small flap of skin inside; interparietal length greater than 85% of its breadth (only known from the San Bernardino and Tehachapi Mountains in southern California) ............................. ........................................................................................................................ *Perognathus alticola* (p. 567)

2a. Ears with buffy hairs and a small flap of skin inside; interparietal length less than 85% of its breadth...........................3

3. Distributed in British Columbia, Washington, and northeastern Oregon ............................ *Perognathus parvus* (p. 579)

3a. Distributed in central and southern Oregon south across eastern California and east through Idaho west of Columbia and Snake River, southwestern Wyoming, Nevada, Utah, and northern Arizon................. *Perognathus mollipilosus* (p. 578)

4. First lower premolars distinctly larger than the last molars ..........................................................................................5

4a. First lower premolars approximately equal to or less than the last molars ....................................................................6

5. Tail length greater than 52% of total length and with a larger hair tuft; auditory bullae relatively moderate (southeast Oregon and western Utah south through eastern California and from Nevada south along the coast of Sonora, and from southwestern California to the central Baja California Peninsula)............................ *Perognathus longimembris* (p. 575)

5a. Tail length less than 52% of total length and with a small hair tuft; auditory bullae relatively large (restricted to western-central California and the western Mojave Desert) ........................................................ *Perognathus inornatus* (p. 574)

6. Tail tip with or without a small hair tuft; total length greater than 130.0 mm .................................................................7

6a. Tail tip without a hair tuft; total length less than 130.0 mm ..........................................................................................9

7. Pale orange spots above and below the eyes; mastoid breadth greater than 80% of basilar length (from southern North Dakota and eastern Minnesota southwest through northern Arizona, northern Chihuahua, New Mexico, and western Texas)................................................................................................................ *Perognathus flavescens* (part; p. 571)

7a. Without pale orange spots above and below the eyes; mastoid breadth less than 80% of basilar length .........................8

8. Dorsal pelage brown with a small hair tuft at the tip of the tail; interparietal breadth less than rostrum width (disjunct populations in northern Arizona and from western and southern Arizona south to northern Sonora) .............................. ........................................................................................................................ *Perognathus amplus* (p. 568)

8a. Dorsal pelage blackish olive brown, tail without a hair tuft at the tip; interparietal breadth greater than rostrum width (southern Alberta, Saskatchewan, and Manitoba south through northeastern Utah and Colorado) .................................. ...................................................................................................................... *Perognathus fasciatus* (p. 570)

9. Subauricular spots white and postauricular spots absent; a lateral stripe separates the dorsal and ventral coloration; interparietal breadth greater than 4.0 mm; mastoid breadth greater than 80% of basilar length (southern North Dakota and eastern Minnesota southwest through northern Arizona, northern Chihuahua, New Mexico, and western Texas) .................................................................................................................. *Perognathus flavescens* (part; p. 571)

9a. Subauricular spots white to yellowish and postauricular spots large and yellow; without a lateral stripe separating the dorsal and ventral coloration; interparietal breadth less than 4.0 mm; mastoid breadth less than 80% of basilar length....................10

10. Postauricular spots relatively larger; tail length greater than the head-and-body length; interorbital breadth greater than 4.6 mm; rostrum-zygomatic curvature slight (Wyoming and North Dakota southwest through Arizona and New Mexico, and then southeast through Estado de México, Morelos, and Puebla) .................. *Perognathus flavus* (p. 572)

10a. Postauricular spots relatively smaller; tail length less than the head-and-body length; interorbital breadth less than 4.6 mm; rostrum-zygomatic curvature marked (eastern New Mexico, western Oklahoma, and central Texas south through northern Chihuahua, eastern Coahuila, Nuevo León, and Tamaulipas) ............. *Perognathus merriami* (p. 577)

## ***Perognathus alticola*** Rhoads, 1894

White-eared pocket mouse, ratón de abazones de orejas blancas

1. *P. a. alticola* Rhoads, 1894. Only known from the San Bernardino Mountains, San Bernardino County, California.
2. *P. a. inexpectatus* Huey, 1926. Only known from the Tehachapi Mountains, Kern County, California.

**Conservation:** *Perognathus alticola* is listed as Vulnerable (B1ab [i, iii, iv]: Extent of occurrence less than 20,000 $km^2$, severely fragmented and under continued decline in extent of occurrence, quality of habitat, and number of subpopulations) by the IUCN (Naylor and Roach 2017). *P. alticola alticola* in the San Bernardino Mountain is probably extinct.

**Characteristics:** White-eared pocket mouse is large sized; total length 163.0–192.0 mm and skull length 22.1–24.5 mm. Dorsal pelage olivaceous buff to pale wood brown; fur of underparts white; **tail bicolored or tricolored, dorsally of the same color as the back, ventrally white and black at the tip; ear antitragus lobed and covered with white hairs; hindfoot very large, greater than 20.0 mm in length;** skull slightly rounded in dorsal view; auditory bullae well inflated and nearly meeting anteriorly; nasals long; interparietal pentagonal. Species group *parvus*.

**Comments:** *Perognathus alticola* is a distinct species (Best 1994a; Baker et al. 2003a; Patton 2005a, b; Riddle et al. 2014), but some authors have considered it as a subspecies of *P. parvus* (Patton 1993b). Change in spelling from *alticolus* to *alticola* (Erratum 2014). *P. alticola* is only known from two disjunct mountain ranges in southern California, the San Bernardino Mountains, and the Tehachapi Mountains (Map 404). It is sympatric, or nearly so, only with *P. inornatus* and *P. longimembris*. *P. alticola* differs from all the other species of *Perognathus*, except *P. mollipilosus* and *P. parvus*, by having a lobed antitragus on the pinna and covered with white hairs; hindfoot length greater than 20.0 mm. From *P. mollipilosus*, by having white hairs on the ear. From *P. inornatus*, by *its* larger size; total length greater than 21.0 mm on average; tail with a greater hair tuft at the tip; relatively greater mastoid breadth and narrower interparietal bone.

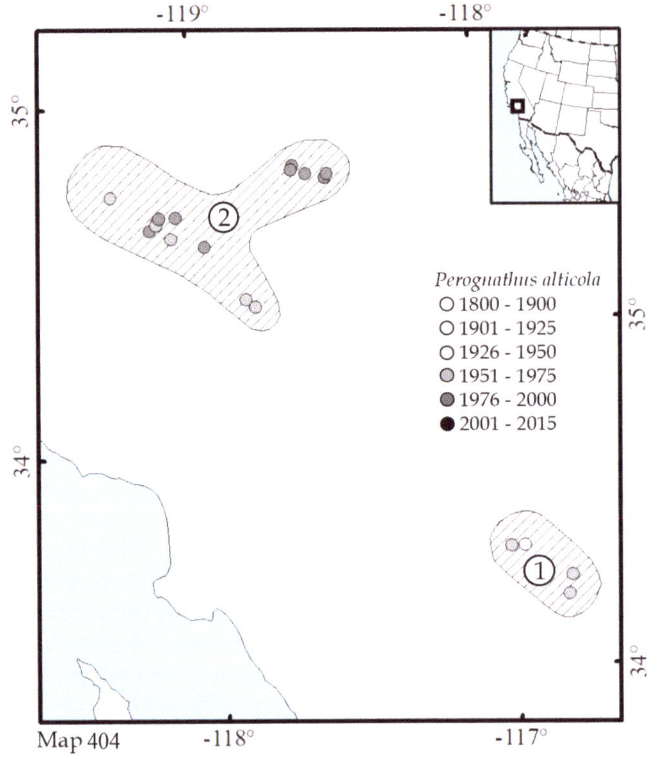


**Map 404** Distribution map of *Perognathus alticola*

**Additional Literature:** Williams et al. (1993), Best (1993c).

## ***Perognathus amplus*** Osgood, 1900

Arizona pocket mouse, ratón de abazones de Arizona

1. *P. a. amplus* Osgood, 1900. Restricted to Yavapai County, Arizona.
2. *P. a. cineris* Benson, 1933. Restricted to Coconino County, Arizona.
3. *P. a. pergracilis* Goldman, 1932. Restricted to Mojave County, Arizona.
4. *P. a. taylori* Goldman, 1932. Restricted to southern-central Arizona.

**Conservation:** *Perognathus amplus amplus* is listed as Special Protection by the Official Mexican Standard (DOF 2019) but is not listed as of conservation concern by the United States Endangered Species Act (FWS 2022) and is considered as Least Concern by the IUCN (Álvarez-Castañeda et al. 2016k).

**Characteristics:** Arizona pocket mouse is medium sized; total length 127.0–179.0 mm and skull length 20.3–25.8 mm. Dorsal pelage yellowish to pale ocher, grizzled blackish, varying according to the population; lateral stripe yellow, well defined; pale orange spots above and below the eyes; large orange spot at the ear back; fur of underparts whitish; **tail greater than the head-and-body length, with a hair tuft at the tip (4.0 mm)**; nasals and rostrum slender; interparietal bone reduced; mastoid bone highly developed; first lower molars equal to or smaller than the last molars; zygomatic arches narrower anteriorly; auditory bullae proportionally smaller than in other species of *Perognathus*. Species group *longimembris*.

**Comments:** Seven subspecies were recognized in *Perognathus amplus* (*P. a. ammodytes*, *P. a. amplus*, *P. a. cineris*, *P. a. jacksoni*, *P. a. pergracilis*, *P. a. rotundus*, and *P. a. taylori*; Hall and Kelson 1959; Hall 1981); limited morphological analyses reduced that number to four (Hoffmeister 1986; Patton 2005a). Genetic analyses showed only three major groups: *P. a. amplus* (including *P. a. jacksoni* and *P. a. pergracilis*), *P. a. cineris* (including *P. a. ammodytes*), and *P. a. taylori* (including *P. a. rotundus*; McKnight 1995, 2005). *P. amplus* occurs south and east of the Colorado River in Arizona, with disjunct populations (subspecies *cineris*) north of the Mogollon Rim in the western Colorado Plateau and the remainder of the range south of the Mogollon Rim across western and southern Arizona south to northern Sonora (Map 405). It is sympatric, or nearly so, with *P. flavescens*, *P. flavus*, and *P. longimembris*. *P. amplus* is stiffened distinct from *P. longimembris*. From *P. flavus*, it differs by having the postauricular patch paler; tail with a hair tuft at the tip and larger than the head-and-body length; hindfoot length greater than 18.5 mm on average. From *P. longimembris*, by its smaller body size and hindfoot size; tail smaller and without a hair tuft at the tip; shorter nasal bones and narrow interorbital width; first lower premolars distinctly larger than the last molars.

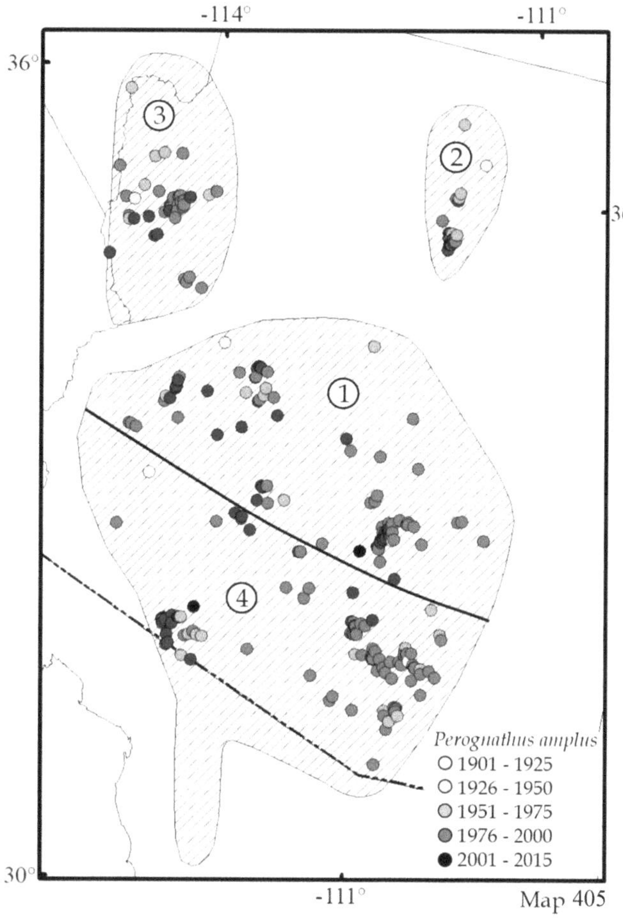


**Map 405** Distribution map of *Perognathus amplus*

**Additional Literature:** Best (1993f), Williams et al. (1993), McKnight (1995, 2005), Patton and Álvarez-Castañeda (1999).

## ***Perognathus fasciatus*** Wied-Neuwied, 1839

Olive-backed pocket mouse, ratón de abazones de coloración verdosa

1. *P. f. callistus* Osgood, 1900. Southwestern Wyoming, northwestern Colorado, and northeastern Utah.
2. *P. f. fasciatus* Wied-Neuwied, 1839. From southern Saskatchewan and Manitoba south through northern Nebraska.
3. *P. f. infraluteus* Thomas, 1893. Restricted to central Colorado.
4. *P. f. litus* Cary, 1911. Restricted to central Wyoming.
5. *P. f. olivaceogriseus* Swenk, 1940. From southeastern Alberta south through Wyoming and Nebraska.

**Conservation:** *Perognathus fasciatus* is listed as Least Concern by the IUCN (Cassola 2016ch).

**Characteristics:** Olive-backed pocket mouse is small sized; total length 130.0–143.5 mm and skull length 21.8–23.6 mm. **Dorsal pelage from grayish to buffy-olivaceous grizzled with black hairs**; buffy lateral line and sides bright; fur of underparts white to buffy; postauricular spots buffy; skull small with a vaulted braincase; **interparietal bones generally pentagonal and wider than the rostrum**; auditory bullae barely in contact anteriorly; first lower molars equal to or smaller than the last molars. Species group *fasciatus*.

**Comments:** *Perognathus fasciatus* ranges from the central Great Plains from southern Alberta, Saskatchewan, and Manitoba, south through northeastern Utah and Colorado (Map 406). It is sympatric, or nearly so, with *P. flavescens* and *P. flavus*, which are very similar and hard to differentiate; however, *P. fasciatus* has the dorsal coloration black olive brown and the lower premolars equal to or slightly smaller than the last molars. From *P. flavus*, it differs by having the postauricular patch paler; tail longer than the head-and-body length, with a hair tuft at the tip; auditory bullae less inflated, and a wider interbullae region.

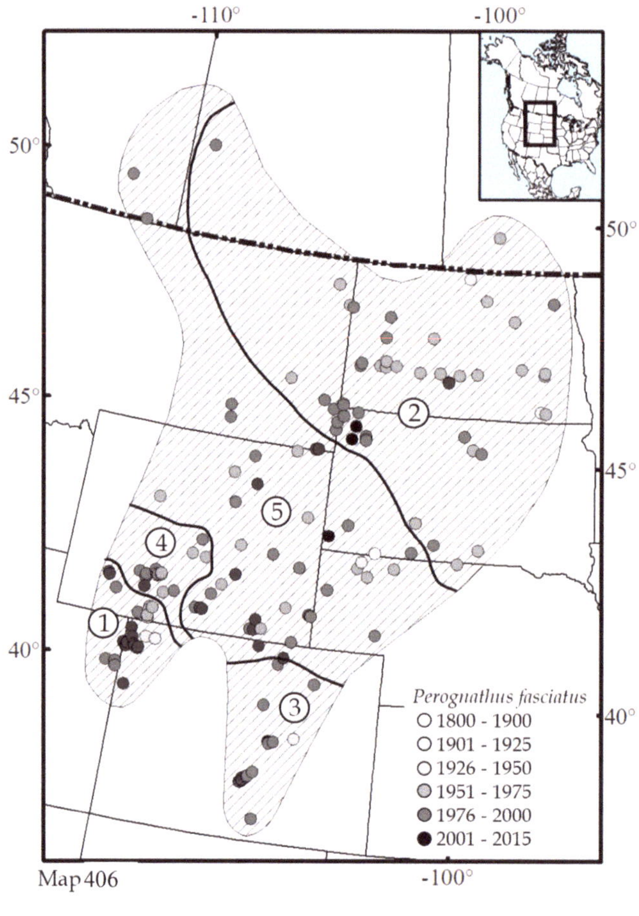


**Map 406** Distribution map of *Perognathus fasciatus*

**Additional Literature:** Merriam (1889b), Blair (1940), Genoways and Jones (1972), Williams and Genoways (1979), Hall (1981), Manning and Jones (1988a), Williams et al. (1993), Neiswenter and Riddle (2011).

## ***Perognathus flavescens*** Merriam, 1889

Plains pocket mouse, ratón de abazones de las planicies

1. *P. f. apache* Merriam, 1889. Northeastern Arizona, western and southern New Mexico, and northern Chihuahua.
2. *P. f. caryi* Goldman, 1918. Eastern Utah and western Colorado.
3. *P. f. cleomophila* Goldman, 1918. Restricted to central Arizona.
4. *P. f. cockrumi* Hall, 1954. Central Kansas and central Oklahoma.
5. *P. f. copei* Rhoads, 1894. Eastern New Mexico and northern Texas.
6. *P. f. flavescens* Merriam, 1889. Southern North Dakota, central and southwestern Wyoming, western Kansas and Nebraska, and eastern Colorado.
7. *P. f. gypsi* Dice, 1929. Restricted to central New Mexico.
8. *P. f. melanotis* Osgood, 1900. Restricted to Casas Grandes, Chihuahua.
9. *P. f. perniger* Osgood, 1904. From North Dakota and Wisconsin south through southeastern Nebraska.
10. *P. f. relictus* Goldman, 1938. Central New Mexico and southern Colorado.

**Conservation:** *Perognathus flavescens* is listed as Least Concern by the IUCN (Lacher et al. 2016e).

**Characteristics:** Plains pocket mouse is small sized; total length 113.0–140.0 mm and skull length 22.0–24.0 mm. Dorsal pelage varying widely according to the subspecies, from orange brown to grayish brown, grizzled with blackish shades; orange lateral stripe well defined; **pale orange spots above and below the eyes, and at the ear back**; fur of underparts whitish; tail greater than the head-and-body length, bicolored, dark dorsally and pale ventrally; skull short and broad, with long nasals; first lower molars equal to or smaller than the last molars; **mastoid breadth greater than 80% of the basilar length**; auditory bullae in contact anteriorly in some populations. Species group *fasciatus*.

**Comments:** *Perognathus apache* was previously considered a full species (Merriam 1889a; Osgood 1900; Hoffmeister 1986); it is currently accepted as the subspecies *P. flavescens apache* (Williams 1978b; Patton 2005a). *P. flavescens* ranges from southern North Dakota and eastern Minnesota southwest through northern Arizona, northern Chihuahua, New Mexico, and western Texas (Map 407). It is sympatric, or nearly so, with *P. fasciatus*, *P. flavus*, and *P. merriami*. *P. flavescens* differs from other species of *Perognathus* by having relatively smaller ears and postauricular patches less conspicuous than in other species; tail short, less than the head-and-body length, without a hair tuft; auditory bullae closely abutting anteriorly and medium sized relative to other species; interparietal bone relatively broad and strap-like. *P. fasciatus* is very similar and hard to differentiate; however, *P. flavescens* has a buffy dorsal coloration, frequently stronger, overlaid with black to olive gray hair, and the lower premolars are distinctly smaller than the last molars. *P. flavescens* differs from *P. flavus* generally by having a paler dorsal coloration without black-tipped guard hairs; postauricular patch small and not stronger, contrasting with the dorsal coloration; on average, 18% of the total length and with a longer tail; smaller auditory bullae (bulla length 37% of skull length, on average); wider interparietal bone, greater than 4.0 mm on average. From *P. merriami*, by its larger size and longer tail; wider interparietal and larger in all cranial dimensions, on average.

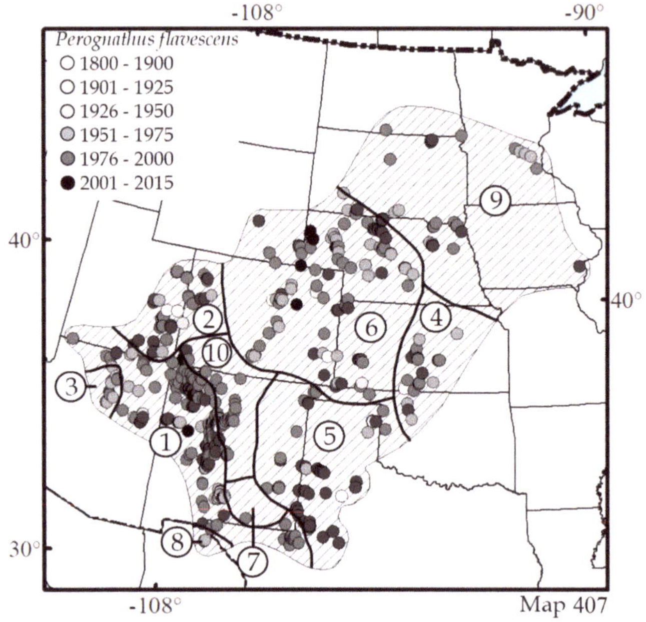


**Map 407** Distribution map of *Perognathus flavescens*

**Additional Literature:** Bailey (1929), Benson (1933), Reed and Choate (1986), Jones et al. (1991), Williams et al. (1993), Monk and Jones (1996).

## *Perognathus flavus* Baird, 1855

Silky pocket mouse, ratón de abazones de pelo sedoso

1. *P. f. bimaculatus* Merriam, 1889. Restricted to western Coconino and northern Yavapai counties, Arizona.
2. *P. f. bunkeri* Cockrum, 1951. Restricted to eastern Colorado, western Kansas, and Oklahoma.
3. *P. f. flavus* Baird, 1855. From southeastern Arizona, New Mexico, and northwestern Texas south through northwestern Durango.
4. *P. f. fuliginosus* Merriam, 1890. Restricted to central Coconino Country, Arizona.
5. *P. f. fuscus* Anderson, 1972. Restricted to the Cuauhtémoc area, Chihuahua.
6. *P. f. goodpasteri* Hoffmeister, 1956. Restricted to the southern Navajo and Apache counties, Arizona.
7. *P. f. hopiensis* Goldman, 1932. Northeastern Arizona, southeastern Utah, southwestern Colorado, and northwestern New Mexico.
8. *P. f. medius* Baker, 1954. From Durango and southern Coahuila south through Guanajuato.
9. *P. f. mexicanus* Merriam, 1894. From Querétaro south through Morelos and Puebla.
10. *P. f. pallescens* Baker, 1954. Western Coahuila, southeastern Chihuahua, and eastern Durango.
11. *P. f. parviceps* Baker, 1954. Restricted to eastern Jalisco.
12. *P. f. piperi* Goldman, 1917. Restricted to eastern Montana and eastern South Dakota.
13. *P. f. sanluisi* Hill, 1952. Restricted to southern-central Colorado.
14. *P. f. sonoriensis* Nelson and Goldman, 1934. Restricted to the coast of Sonora.

**Conservation:** *Perognathus flavus* is listed as Least Concern by the IUCN (Lacher et al. 2016f).

**Characteristics:** Silky pocket mouse is very small sized; total length 100.0–122.0 mm and skull length 19.7–24.1 mm. Dorsal pelage finely lined with black on ochraceous buff; lateral stripe yellowish, barely evident or absent; large pale orange spots above and below the eyes and **behind the ears**; fur of underparts whitish; tail length greater than the head-and-body length, dusky or buffy dorsally and whitish ventrally, **naked and without a hair tuft at the tip**; interparietal bone similar in length and width; auditory bullae not in contact anteriorly; first lower molars equal to or smaller than the last molars. Species group *flavus*.

**Comments:** *Perognathus flavus gilvus* and *P. f. merriami* have been elevated to a separate species, *P. merriami* (Anderson 1972; Lee and Engstrom 1991). Wilson (1973) considered them to be conspecific. *P. flavus* ranges from the central and southern portion of the Great Plains and intermountain areas of Wyoming and North Dakota southwest through Arizona and New Mexico, and then southeast through Estado de México and Puebla (Map 408). A disjunct population occurs along the central coast of Sonora. *P. flavus* is sympatric, or nearly so, with *P. amplus*, *P. flavescens*, and *P. merriami*. *P. flavus* differs from *P. amplus* and *P. fasciatus* by having conspicuous postauricular patches that contrast with the dorsal coloration; tail without a hair tuft and shorter than the head-and-body length. In addition, from *P. amplus*, by its hindfoot length less than 18.5 mm on average. From *P. fasciatus*, by having the auditory bulla more inflated and with the interbullar region narrower. From *P. flavescens*, by having a darker dorsal coloration with multiple black-tipped guard hairs; postauricular patch larger and contrasting with the dorsal coloration; 18% smaller in total length, on average, with a shorter tail; larger auditory bulla (bulla length 40% of occipitonasal length, on average); narrow interparietal bone, less than 3.6 mm width on average. From *P. longimembris*, by its smaller average total and hindfoot size, tail shorter and non-penciled, nasal bones shorter, and a narrower interorbital region. *P. flavus* is hard to distinguish from *P. merriami* in the sympatric areas of southwestern New Mexico and western Texas. Comparing the two species, *P. flavus* is slightly smaller, mainly in all external and skull measurements, with a longer and coarser pelage and a darker, more-contrasting mid-dorsal pinkish coloration; larger postauricular spots and ears; larger and more inflated auditory bullae; narrower interorbital and interparietal width, and shorter nasals.

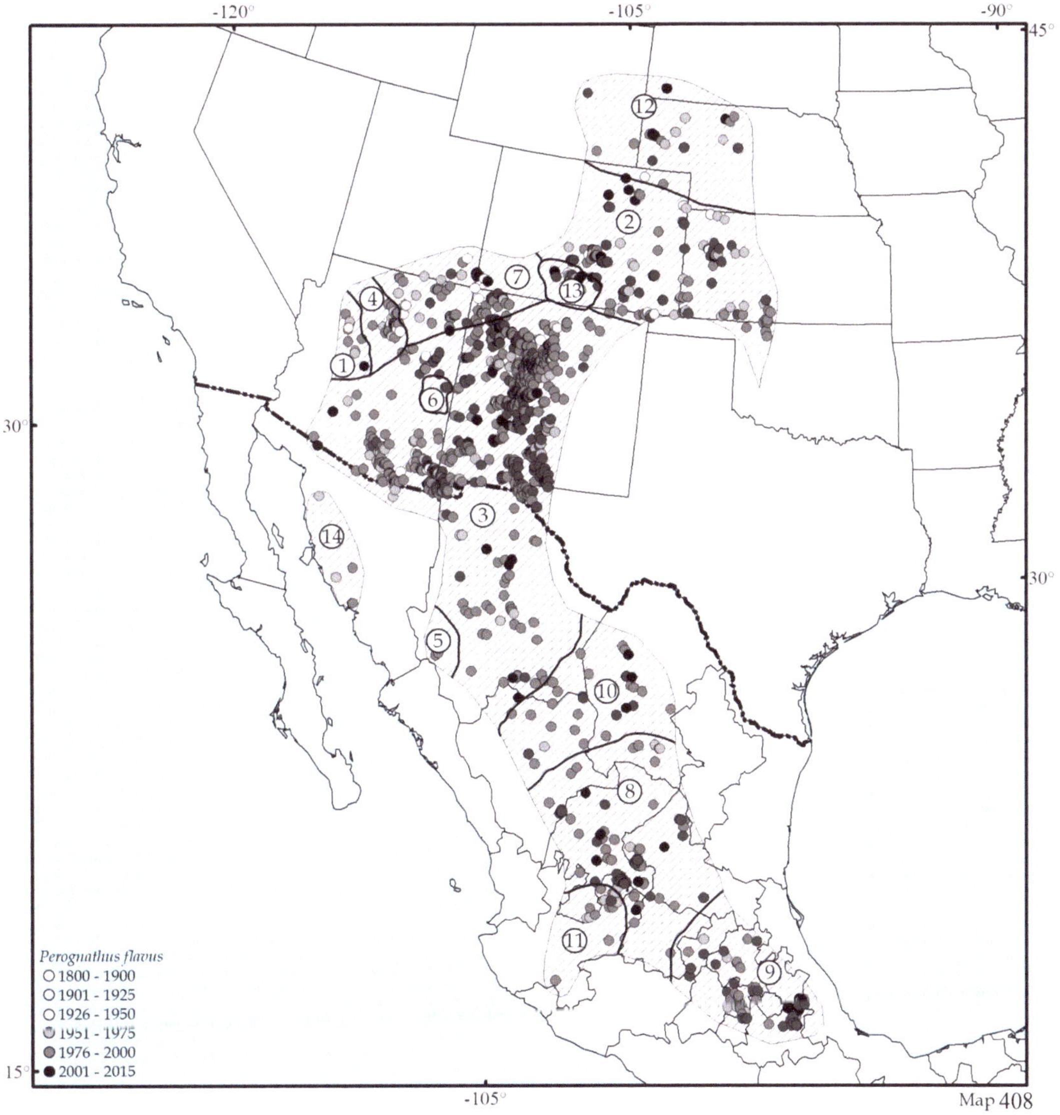


**Map 408** Distribution map of *Perognathus flavus*

**Additional Literature:** Osgood (1900), Bailey (1905), Baker (1954), Baker and Greer (1962), Webster (1968), Williams (1978b), Hoffmeister (1986), Best (1993f), Williams et al. (1993), Best and Skupski (1994a, b), Patton and Álvarez-Castañeda (1999), Neiswenter and Riddle (2010).

## *Perognathus inornatus* Merriam, 1889

San Joaquín pocket mouse, ratón de abazones de San Joaquín

1. *P. i. inornatus* Merriam, 1889. Restricted to the Central Valley, California.
2. *P. i. neglectus* Taylor, 1912. Restricted to Kern County, California.
3. *P. i. sillimani* von Bloeker, 1937. Restricted to the coastal areas of central California.

**Conservation:** *Perognathus inornatus* is listed as Least Concern by the IUCN (Cassola 2016ci).

**Characteristics:** San Joaquin pocket mouse is medium sized; total length 117.0–171.0 mm and skull length 20.1–25.2 mm. Dorsal pelage ochraceous buff to pinkish, overlaid with grizzle black hairs; lateral line moderately well marked; fur of underparts white; tail bicolored; skull short and broad, with short nasals; interparietal bone small and square; first lower molars equal to or smaller than the last molars; auditory bullae opposed anteriorly. Species group *longimembris*.

**Comments:** *Perognathus inornatus* include a species complex (Williams et al. 1993). *P. inornatus* has ranges restricted to western-central California in the Sacramento, San Joaquin, Salinas, and Cuyama valleys and the Carrizo Plain, and extends eastward into the Mojave Desert (Map 409). *P. inornatus* is sympatric, or nearly so, with *P. alticola*, *P. longimembris*, and *P. mollipilosus but* not with *P. amplus*. *P. inornatus* is hard to differentiate from *P. amplus* and *P. longimembris*. From *P. longimembris*, by its larger size and relatively shorter tail with a smaller hair tuft at the tip; interorbital region relatively narrower but within the ranges of *P. longimembris*. From *P. alticola* and *P. mollipilosus*, by not having a lobed antitragus on the pinna; its smaller mean size, less than 210.0 mm; tail with a smaller hair tuft at the tip; relatively greater mastoid bullae and mastoid breadth, and narrower interparietal bone.

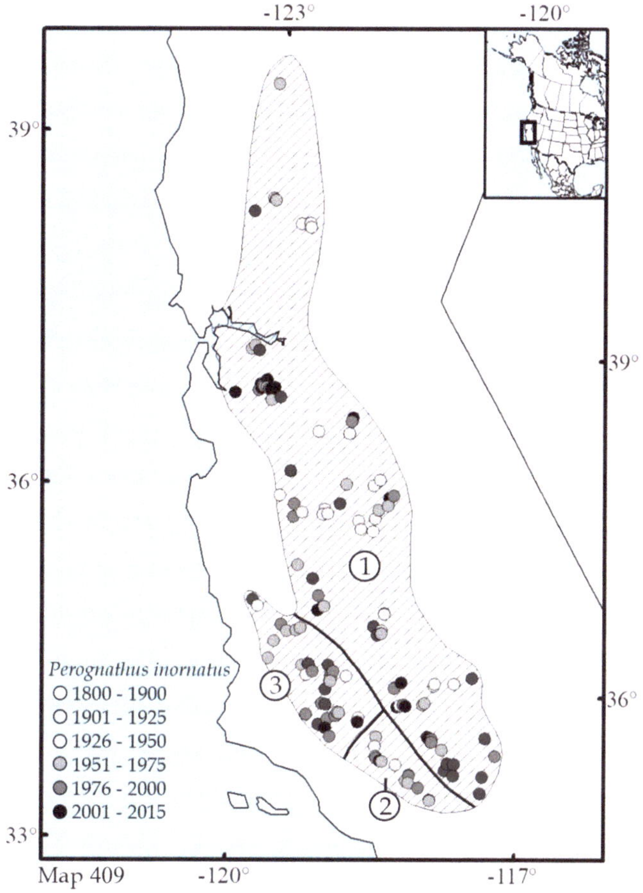


**Map 409** Distribution map of *Perognathus inornatus*

**Additional Literature:** Merriam (1889b), Osgood (1918), Hall (1981), Best (1993c, f), McKnight (2005).

## ***Perognathus longimembris*** (Coues, 1875)

Little pocket mouse, ratón de abazones pequeño

1. *P. l. aestivus* Huey, 1928. Restricted to the San Rafael Valley south to Valle de la Trinidad, Baja California.
2. *P. l. arcus* Benson, 1935. Restricted to southern-central Utah.
3. *P. l. arenicola* Stephens, 1900. Eastern California and north-eastern Baja California.
4. *P. l. arizonensis* Goldman, 1931. Restricted to northwestern Arizona and southern-central Utah.
5. *P. l. bangsi* Mearns, 1898. From central San Bernardino south through eastern San Diego County, California.
6. *P. l. bombycinus* Osgood, 1907. Southwestern Arizona, southeastern California, northwestern Sonora, and northeastern Baja California.
7. *P. l. brevinasus* Osgood, 1900. From Los Angeles, southwestern San Bernardino, and south through northern-central Baja California.
8. *P. l. cantwelli* von Bloeker, 1932. From coastal Los Angeles County south to the border of México including northwestern Baja California.
9. *P. l. gulosus* Hall, 1941. Restricted to western Utah and western Nevada.
10. *P. l. kinoensis* Huey, 1935. Restricted to western Sonora.
11. *P. l. longimembris* (Coues, 1875). Central valley of California and southern part of the Sierra Nevada.
12. *P. l. nevadensis* Merriam, 1894. Northern Nevada, southeastern Oregon, and northeastern California.
13. *P. l. panamintinus* Merriam, 1894. Restricted to central and western Nevada and eastern California.
14. *P. l. pimensis* Huey, 1937. Restricted to southern-central Arizona.
15. *P. l. psammophilus* von Bloeker, 1937. Coastal range of western California.
16. *P. l. salinensis* Bole, 1937. Restricted to the Saline Valley, Inyo County, California.

17. *P. l. tularensis* Richardson, 1937. Restricted to Tulare County, California
18. *P. l. venustus* Huey, 1930. Restricted to the San Agustín area, Baja California.
19. *P. l. virginis* Huey, 1939. Southeastern Nevada, northwestern Arizona, and southwestern Utah.

**Conservation:** *Perognathus longimembris pacificus* is listed as Endangered by the United States Endangered Species Act (FWS 2022). *P. longimembris* as Least Concern by the IUCN (Cassola 2016cj).

**Characteristics:** Little pocket mouse is small sized; total length 110.0–151.0 mm and skull length 18.6–23.1 mm. Dorsal pelage pale to dark grayish brown; lateral stripe orange, either distinguishable or not; **whitish spots behind the ears**; fur of underparts whitish; tail longer than the head-and-body length, bicolored **with a long hair tuft at the tip (3.5 mm)**; **first lower molariforms distinctly larger than the last molars**. Species group *longimembris*.

**Comments:** *Perognathus longimembris internationalis* was designated as a synonym of *P. l. brevinasus* and *P. l. pacificus* as *P. l. cantwelli* (Patton and Fisher 2023). *Perognathus longimembris* ranges from southeastern Oregon to western Utah, then south through Nevada, Arizona, Sonora, and northern Baja California Sur (Map 410). It is sympatric, or nearly so, with *P. alticola*, *P. inornatus*, and *P. amplus*. *P. longimembris* is very hard to differentiate from *P. amplus* and *P. inornatus*. It differs from *P. flavus* by its larger average total and hindfoot size; tail longer and with a hair tuft at the tip; larger nasal and wider interorbital bones. From *P. inornatus*, by its smaller size and relatively larger tail with a greater hair tuft at the tip; interorbital region relatively wider. *Chaetodipus arenarius* is similar to *Perognathus* in pelage type and size and can be differentiated from it by its soft pelage; mastoid projecting posteriorly beyond the occipital; interparietal bone narrow, less than 5.0 mm.

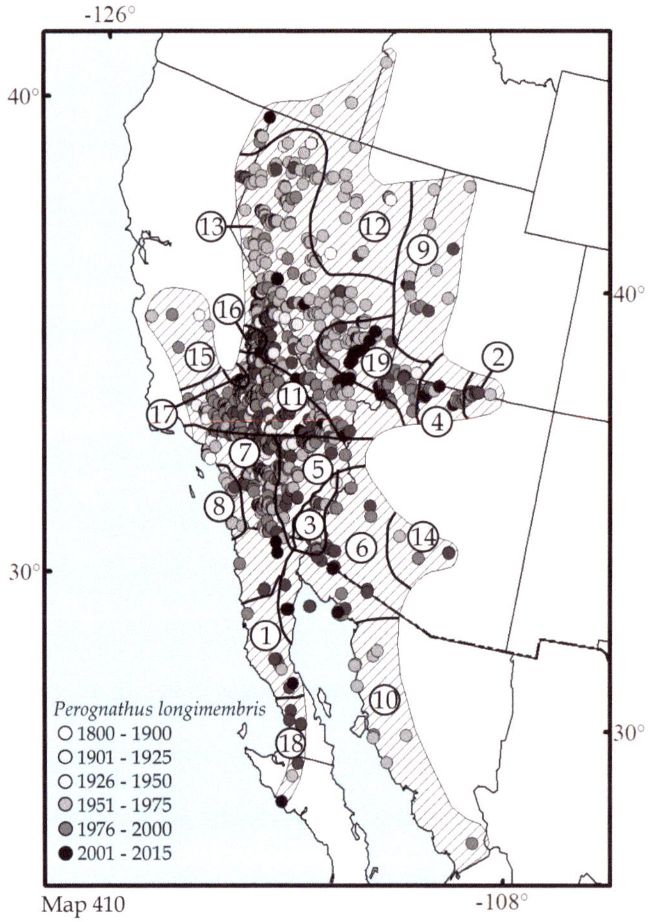


**Map 410** Distribution map of *Perognathus longimembris*

**Additional Literature:** Merriam (1889b), Osgood (1918), Ingles (1965), Williams et al. (1993), Patton and Álvarez-Castañeda (1999), McKnight (2005), Álvarez-Castañeda et al. (2008c), Laabs et al. (2022).

## *Perognathus merriami* Allen, 1892

Merriam's pocket mouse, ratón de abazones del este

1. *P. m. gilvus* Osgood, 1900. Eastern New Mexico, western Texas, northern Chihuahua, and northwestern Coahuila.
2. *P. m. merriami* Allen, 1892. From central Texas south through Tamaulipas, including eastern Coahuila.

**Conservation:** *Perognathus merriami* is listed as Least Concern by the IUCN (Lacher et al. 2016g).

**Characteristics:** Merriam's pocket mouse is very small sized; total length 105.0–120.0 mm and skull length 19.1–21.7 mm. **Dorsal pelage grizzled in yellowish to yellow orange,** with a slight blackish tinge produced by black tips on on the longest hairs; lateral stripe yellowish, indistinct or absent; pale yellow spots above and below the eyes and **small behind the ears**; postauricular spot small and buffy, subauricular spot white; fur of underparts whitish; tail length less than the head-and-body length, **with few hairs and without a hair tuft at the tip**; first lower molars equal to or smaller than the last molars. Species group *flavus*.

**Comments:** Wilson (1973) considered *Perognathus merriami* to be conspecific with *P. flavus* (see that account). *P. merriami* ranges from eastern New Mexico, western Oklahoma, and central Texas south through northern Chihuahua, eastern Coahuila, Nuevo León, and Tamaulipas (Map 411); it is sympatric, or nearly so, with *P. flavus* and *P. flavescens*. *P. merriami* is difficult to distinguish from *P. flavus* in the sympatric areas of southwestern New Mexico and western Texas. Compared with *P. flavus*, *P. merriami* is slightly larger in nearly all external and skull measurements; it has a shorter, coarser, and paler pelage with a less-contrasting mid-dorsal yellowish or yellowish-orange coloration; postauricular spots and ears smaller; auditory bullae smaller and less inflated; greater interorbital and interparietal widths and nasal length. From *P. flavescens*, it differs by its smaller size and shorter tail; nearly parallel zygomatic arches; interparietal bone narrower, and smaller average skull measurements.

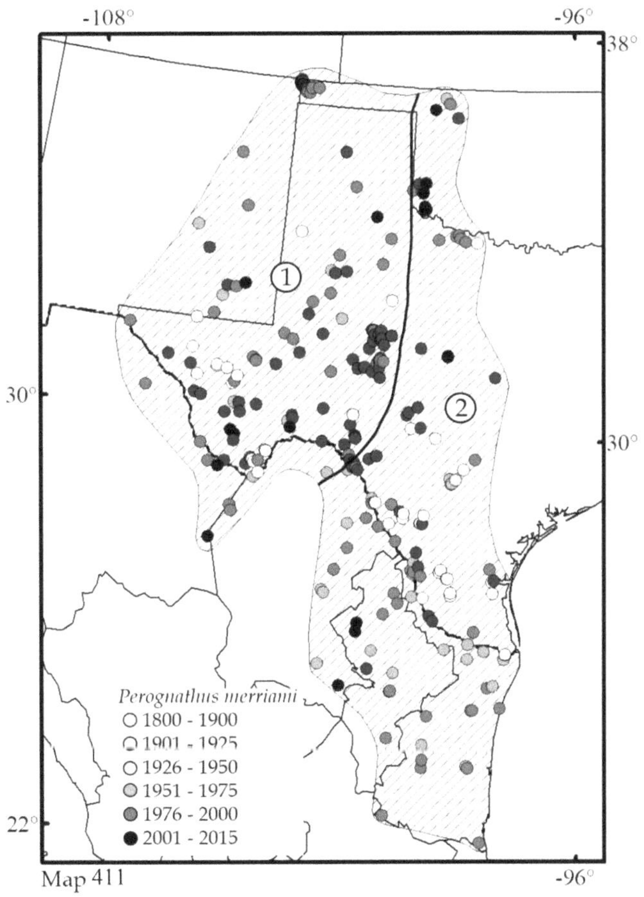


**Map 411** Distribution map of *Perognathus merriami*

**Additional Literature:** Osgood (1900), Bailey (1905), Davis and Robertson (1944), Al-Uthman (1952), Tamsitt (1954), Álvarez (1963), Davis (1974), Dalquest and Horner (1984), Williams et al. (1993), Best and Skupski (1994b).

## *Perognathus mollipilosus* Coues, 1875

Great Basin pocket mouse, ratón de abazones del Great Basin

1. *P. m. bullatus* Durrant and Lee, 1956. Restricted to central Utah.
2. *P. m. clarus* Goldman, 1917. Eastern Idaho, northeastern Utah, and southwestern Wyoming.
3. *P. m. idahoensis* Goldman, 1922. Restricted to central-southern Idaho.
4. *P. m. mollipilosus* Coues, 1875. Central-southern Oregon and eastern California.
5. *P. m. olivaceus* Merriam, 1889. From eastern California to western Utah and southern Idaho
6. *P. m. trumbullensis* Benson, 1937. Southern Utah and northern Arizona.
7. *P. m. xanthonotus* Grinnell, 1912. Restricted to canyons on the southeastern Sierra Nevada in Kern County, California.

**Conservation:** *Perognathus mollipilosus* (as *P. parvus*) is listed as Least Concern by the IUCN (Cassola 2016ck).

**Characteristics:** Great Basin pocket mouse is large sized (within *Perognathus*); total length 148.0–199.5 mm and skull length 22.0–25.0 mm. Dorsal pelage pinkish buff to ochraceous buff, slightly grizzle with blackish shades; hair of underparts white to buffy; lateral line olivaceous; tail longer, bicolored, dark dorsally and with a hair tuft at the tip; **ear antitragus lobed and not clothed with white hairs; hindfoot very long, greater than 20.0 mm**; skull slightly rounded in dorsal view; auditory bullae well inflated and close to meeting anteriorly; nasals long; interparietal pentagonal. Species group *parvus*.

**Comments:** Genetic analyses by Riddle et al. (2014) split the traditional *Perognathus parvus* (Hall 1981) into two separate species, restricting *P. parvus* from central and western Oregon to eastern Idaho, with the remaining range of traditional *P. parvus* occupied by *P. mollipilosus*. These authors, however, did not evaluate its subspecies status. The current study, therefore, retains all the subspecies mapped within the range of *P. mollipilosus* (Map 412). *P. xanthonotus*, originally described as a separate species, is listed as a subspecies herein (following Williams et al. 1993; Jones et al. 1997; Patton 2005). *P. alticola* has been regarded as a subspecies of *P. mollipilosus* (Patton 1993b, sensu *P. parvus*). The majority of the literature about *P. mollipilosus* refers to it as *P. parvus*. *P. mollipilosus* ranges from central and southern Oregon south to Idaho west of the Columbia and Snake River, southwestern Wyoming, eastern California, Nevada, Utah, and northern Arizona (Map 412). It is sympatric, or nearly so, with *P. flavus*, *P. flavescens*, *P. formosus*, and *P. longimembris*. *P. mollipilosus* can be found in sympatry with *P. parvus* only in a small area, differing from it by genetic data. *P. mollipilosus* differs from all the other species of *Perognathus*, except *P. alticola* and *P. parvus*, by having a lobed antitragus on the pinna; hindfoot length greater than 20.0 mm; tail distinctly bicolored, neither crested nor conspicuously tufted, and darkdorsally. It lacks the white hairs on the ears characteristic of *P. alticola*. From *P. inornatus*, by the combination of lobed versus non-lobed antitragus and larger average total size; total length greater than 210.0 mm on average; tail with a greater hair tuft at the tip and relatively greater mastoid breadth. *P. mollipilosus* differs from *P. alticola* by lacking white on the ear, is smaller in overall size, tail with a shorter/longer tuft on the tip, mastoid breadth greater or lesser, and interparietal wider or narrower.

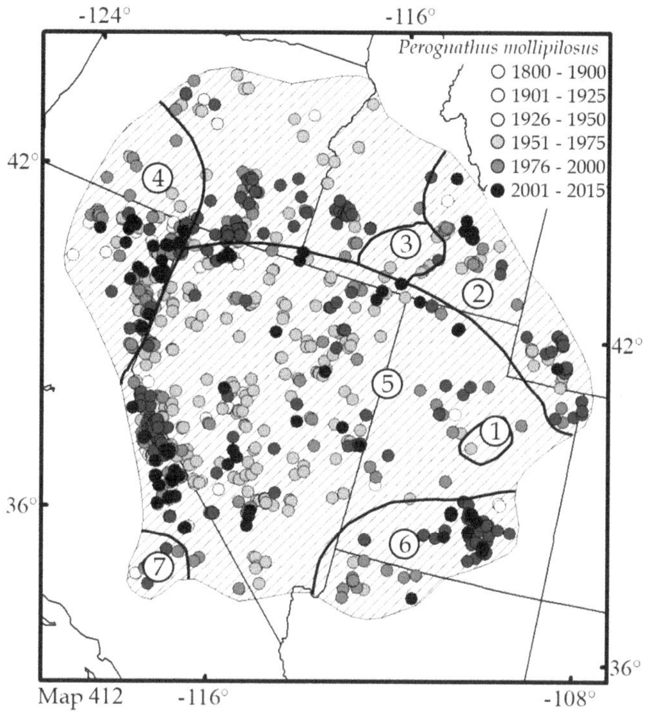


**Map 412** Distribution map of *Perognathus mollipilosus*

**Additional Literature:** Osgood (1900), Hall (1946), Durant (1952), Cowan and Guiguet (1956), Iverson (1967), Blair et al. (1968), Hall (1981), McLaughlin (1984), Hoffmeister (1986), Verts and Kirkland (1988), Best (1993c).

## *Perognathus parvus* (Peale, 1848)

Columbia Plateau pocket mouse, ratón de abazones de Columbia

1. *P. p. columbianus* Merriam, 1894. Restricted to central and southeast Washington.
2. *P. p. laingi* Anderson, 1932. Restricted to Southern British Columbia.
3. *P. p. lordi* (Gray, 1868). Restricted to northern and eastern Washington.
4. *P. p. parvus* (Peale, 1848, part). Restricted to northernmost Oregon.
5. *P. p. yakimensis* Broadbrooks, 1954. Restricted to central-southern Washington.

**Conservation:** *Perognathus parvus* (including *P. mollipilosus*) is listed as Least Concern by the IUCN (Cassola 2016ck).

**Characteristics:** Columbia Plateau pocket mouse is large sized (within *Perognathus*); total length 148.0–199.5 mm and skull length 22.0–25.0 mm. Dorsal pelage pinkish buff to ochraceous buff, slightly grizzled with blackish shades; fur of underparts white to buffy; lateral line olivaceous; tail longer, bicolored, dark dorsally and with a hair tuft at the tip; **ear antitragus lobed and not clothed with white hairs; hindfoot very large, greater than 20.0 mm;** skull slightly rounded in dorsal view; auditory bullae well inflated and close to meeting anteriorly; nasals long; interparietal pentagonal. Species group *parvus*.

**Comments:** Riddle et al. (2014), based on molecular data, restricted *P. parvus* to southern British Columbia, Washington, and northeastern Oregon (Map 413). The remaining part of the range of *P. parvus* (sensu lato) is now allocated to *P. mollipilosus* (see above). As Riddle et al. (2014) do not deal with the nomenclature of the subspecies, herein I kept all the previous subspecies within the recognized range of *P. parvus* (sensu stricto). *P. parvus* is sympatric, or nearly so, with *P. mollipilosus* only in a small area, differing from it only by genetic data. *P. parvus* differs from all the other species of *Perognathus*, except *P. alticola* and *P. mollipilosus*, by having a lobed antitragus on the pinna.

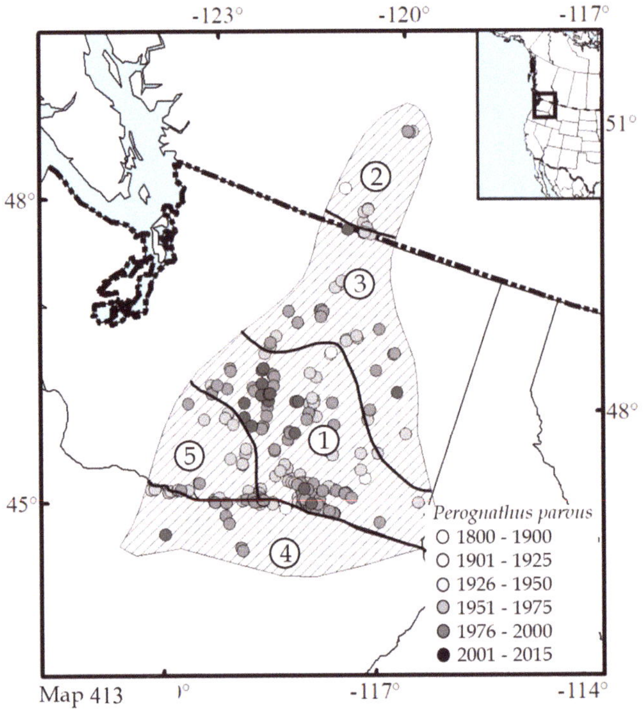


**Map 413** Distribution map of *Perognathus parvus*

**Additional Literature:** Osgood (1900), Hall (1946), Durant (1952), Cowan and Guiguet (1956), Iverson (1967), Blair et al. (1968), Hall (1981), McLaughlin (1984), Hoffmeister (1986), Verts and Kirkland (1988), Williams et al. (1993), Patton (2005a).

## References


Abramov, A. V. 1999. A taxonomic review of the genus *Mustela* (Mammalia, Carnivora). Zoosystematica Rossica 8:357–364.

Abramov, A.V. 2000a. The taxonomic status of the Japanese weasel, *Mustela itatsi* (Carnivora, Mustelidae). Zoologicheskii Zhurnal 79:80–88.

Abramov, A. 2000b. A taxonomic review of the genus *Mustela* (Mammalia, Carnivora). Zoosystematica Rossica 8:357–364.

Abramov, A.V. 2016. *Gulo gulo*. The IUCN Red List of Threatened Species 2016:e.T9561A45198537.

Abramson, N. I., and A. A. Lissovsky. 2012. Subfamily Arvicolinae. Pp. 127–141, *in* The mammals of Russia: A taxonomic and geographic reference (Pavlinov, I. Y., and A. A. Lissovsky, eds.). Moscow: KMK Scientific Press.

Abramson, N. I., and T. V. Petrova. 2018. Genetic analysis of type material of the Amur lemming resolves nomenclature issues and creates challenges for the taxonomy of true lemmings (*Lemmus*, Rodentia: Cricetidae) in the eastern Palearctic. Zoological Journal of the Linnean Society 182:465–477.

Abreu, E., S. Pavan, T. Nunes, T. Mirian, D. Wilson, A. Percequillo, and J. Maldonado. 2020a. Museomics of tree squirrels: a dense taxon sampling of mitogenomes reveals hidden diversity, phenotypic convergence, and the need of a taxonomic overhaul. BMC Evolutionary Biology 20:77.

Abreu-Jr E. F., S. E. Pavan, M. T. N. Tsuchiya, D. E. Wilson, A. R. Percequillo, and J. E. Maldonado. 2020b. Spatiotemporal diversification of Tree Squirrels: is the South American invasion and speciation really that recent and fast? Frontiers in Ecology and Evolution 8.

Abril, V. V., E. A. G. Carnelossi, S. González, and J. M. B Duarte. 2010. Elucidating the evolution of the red brocket deer *Mazama americana* complex (Artiodactyla, Cervidae). Cytogenetic and Genome Research 128:177–187.

Adams, D. R., and D. A. Sutton. 1968. A description of the baculum and os clitoris of *Eutamias townsendii ochrogenys*. Journal of Mammalogy 49:764–768.

Adams, J. K. 1989. *Pteronotus davyi*. Mammalian Species 346:1–5.

Adam, M. D., and J. P. Hayes. 1998. *Arborimus pomo*. Mammalian Species 593:1–5.

Adam, P. J. 2004. *Monachus tropicalis*. Mammalian Species 747:1–9.

Agnarsson, I., M. Kuntner, and L. J. May-Collado. 2010. Dogs, cats, and kin: A molecular species-level phylogeny of Carnivora. Molecular Phylogenetics and Evolution 54:726–745.

Aguirre, L., H. Mantilla, B. Miller, and L. Dávalos. 2015. *Chiroderma salvini*. The IUCN Red List of Threatened Species 2015:e.T4666A22037356.

Akins, J. B., M. L. Kennedy, G. D. Schnell, C. Sánchez-Hernández, M. de la L. Romero-Almaraz, M. C. Wooten, and T. L. Best. 2007. Flight speeds of three species of Neotropical bats: *Glossophaga soricina*, *Natalus stramineus*, and *Carollia subrufa*. Acta Chiropterologica 9:477–482.

Alexander, L. F. 1996. A morphometric analysis of geographic variation within *Sorex monticolus* (Insectivora: Soricidae). Miscellaneous Publications of the University of Kansas Natural History Museum 88:1–54.

Alexander, L. F., and B. J. Verts. 1992. *Clethrionomys californicus*. Mammalian Species 406:1–6.
Allard, M. W., and I. F. Greenbaum. 1988. Morphological variation and taxonomy of chromosomally differentiated *Peromyscus* from the Pacific Northwest. Canadian Journal of Zoology 66:2734–2739.
Allen, G. M. 1916. Bats of the genus *Corynorhinus*. Bulletin of the Museum of Comparative Zoology 60:331–356.
Allen, G. M. 1919a. Notes on the synonymy and nomenclature of the smaller spotted cats of tropical America. Bulletin of the American Museum of Natural History 41:341–419.
Allen, G. M. 1919b. The American collared lemmings. Bulletin Museum Comparative Zoology 62:509–543.
Allen, G. M. 1930. The walrus in New England. Journal of Mammalogy 11:139–145.
Allen, G. M. 1942. Extintct and vanishing mammals of the western hemisphere. American Committee for International Wildlife Protection, Special Publication 11:1–61.
Allen, J. A. 1890. A review of some of the North American ground squirrels of the genus *Tamias*. Bulletin of the American Museum of Natural History 3:45–116.
Allen, J. A. 1893. On a collection of mammals from San Pedro Martir region of Lower California. Bulletin American Museum Natural History 5:181–202.
Allen, J. A. 1897. Description of a new vespertilionine bat from Yucatan. Bulletin American Museum of Natural History 9:231–232.
Allen, J. A. 1898. Descriptions of new mammals from western Mexico and Lower California. Bulletin of the American Museum of Natural History 10:143–158.
Allen, J. A. 1900. List of bats collected by Mr. H. H. Smith in the Santa Marta region of Colombia, with descriptions of new species. Bulletin of the American Museum of Natural History 13:87–94.
Allen, J. A. 1901. A preliminary study of the North American opossum of the Genus *Didelphis*. Bulletin of the American Museum of Natural History 14:149–188.
Allen, J. A. 1903. List of mammals collected by Mr. J. H. Batty in New Mexico and Durango, with descriptions of new species and subspecies. Bulletin of the American Museum of Natural History 19:587–612.
Allen, J. A. 1904a. New forms of the mountain goat (*Oreamnos*). Bulletin of the American Museum of Natural History 20:19–21.
Allen, J. A. 1904b. Mammals from the District of Santa Marta, Colombia, collected by Mr. Herbert H. Smith, with field notes by Mr. Smith. Bulletin of the American Museum of Natural History 20:407–468.
Allen, J. A. 1904c. The *Tamandua* anteaters. Bulletin of the American Museum of Natural History 20:385–398.
Allen, J. A. 1904d. New bats from tropical America, with notes on the species of *Otopterus*. Bulletin of the American Museum of Natural History 20:227–237.
Allen, J. A. 1906. Mammals from the States of Sinaloa and Jalisco, Mexico, collected by J. H. Batty during 1904 and 1905. Bulletin of the American Museum of Natural History 22:191–262.
Allen, J. A. 1913. Ontogenetic and other variations in muskoxen, with a systematic review of the muskox group, recent and extinct. Memoryes of the American Museum of Natural History, new series 1:101–226.
Allen, J. A. 1915a. Review of the south american Sciuridae. Bulletin of the American Museum of Natural History 34:147–309.
Allen, J. A. 1915b. Notes on American deer of the genus *Mazama*. Bulletin of the American Museum of Natural History 34:521–553.
Allen, J. A. 1919. Notes on the synonymy and nomenclature of the smaller spotted cats of tropical America. Bulletin of the American Museum of Natural History 1:341–419.
Allen, J. A. 1939. Bats. Harvard University Press. Cambridge, U.S.A.
Allen, P. J. 2004. *Monachus tropicalis*. Mammalian Species 747:1–9.
Al-Uthman, H. S. 1952. Geographical variations in the Merriam pocket mouse (*Perognathus merriami*) in three biotic provinces in Texas. M. A. thesis. The University of Texas. Austin, U.S.A.
Almazán-Catalán, J. A., A. Taboada-Salgado, C. Sánchez-Hernández, M. de la L. Romero-Almaraz, Q. Jiménez-Salmerón, and E. Guerrero-Ibarra. 2009. Registros de murciélagos para el estado de Guerrero, México. Acta Zoológica Mexicana (n. s.) 25:177–185.
Almendra, A. L., D. S. Rogers, and F. X. González-Cózatl. 2014. Molecular phylogenetics of the *Handleyomys chapmani* complex in Mesoamerica. Journal of Mammalogy 95:26–40.
Almendra, A. L., F. X. González-Cózatl, M. D. Engstrom, and D. S. Rogers. 2018. Evolutionary relationships and climatic niche evolution in the genus *Handleyomys* (Sigmodontinae: Oryzomyini). Molecular Phylogenetics and Evolution 128:12–25.
Alonso-Mejía, A., and R. A. Medellín. 1991. *Micronycteris megalotis*. Mammalian Species 376:1–6.
Alonso-Mejía, A., and R. A. Medellín. 1992. *Marmosa mexicana*. Mammalian Species 421:1–4.
Álvarez, J., M. R. Willig, J. K. Jones, Jr. and W. D. Webster. 1991. *Glossophaga soricina*. Mammalian Species 379:1–7.
Álvarez, T. 1960. Sinopsis de las especies mexicanas del género *Dipodomys*. Revista de la Sociedad Mexicana de Historia Natural 21:391–424.
Álvarez, T. 1961. Taxonomic status of some mice of the *Peromyscus boylii* group in eastern Mexico, with description of a new subspecies. University of Kansas Publications of the Museum of Natural History 14:111–120.
Álvarez, T. 1963. The recent mammals of Tamaulipas, Mexico. University of Kansas Publications, Museum of Natural History 14:363–473.
Álvarez, T. 1968. Notas sobre una coleccion de mamíferos de la región costera del Río Balsas entre Michoacán y Guerrero. Revista de la Sociedad Mexicana de Historia Natural 29:21–35.
Álvarez, T., and S. T. Álvarez-Castañeda. 1990. Cuatro nuevos registros de murciélagos (Chiroptera) del estado de Chiapas. Anales de la Escuela Nacional de Ciencias Biológicas 55:157–161.
Álvarez, T., and S. T. Álvarez-Castañeda. 1991. Análisis de la fauna de roedores del área de El Cedral, San Luis Potosí México. Anales del Instituto de Biología, Universidad Nacional Autónoma de México, Serie Zoología 62:169–180.
Álvarez, T., and S. T. Álvarez-Castañeda 1996. Aspectos biológicos y ecológicos de los murciélagos de Ixtapan del Oro, México. Pp. 169–182, *in* Contribution in Mammalogy: a memorial volume Honoring Dr. J. Knox Jones, Jr. (Genoways, H. H., and R. Baker, eds.). Special Publications, Museum Texas Tech University. Lubbock, U.S.A.
Álvarez, T., and C. E. Aviña. 1964. Nuevos registros en México de la familia Molossidae. Revista de la Sociedad Mexicana de Historia Natural 25:243–254.
Álvarez, T., and L. González-Quintero. 1970. Análisis polínico del contenido gástrico de murciélagos Glossophaginae de México. Anales de la Escuela Nacional de Ciencias Biológicas 18:137–165.

Álvarez, T., and N. González-Ruiz. 2000. Variación geográfica de *Saccopteryx bilineata* (Chiroptera: Emballonuridae) en México, con descripción de una nueva subespecie. Anales de la Escuela Nacional de Ciencias Biológicas 46:305–316.

Álvarez, T., and J. J. Hernández-Chávez. 1990. Cuatro nuevos registros del ratón de campo *Peromyscus* (Rodentia: Muridae) en el Estado de México, México. Anales de la Escuela Nacional de Ciencias Biológicas 33:163–173.

Álvarez, T., and J. J. Hernández-Chávez. 1993. Taxonomía del metorito *Microtus mexicanus* en el centro de México con la descripción de una nueva subespecie. Pp. 137–156, *in* Avances en el estudio de los mamíferos de México (Medellin, R. A., and G. Ceballos, eds.). Publicaciones Especiales. Asociación Mexicana de Mastozoología, Distrito Federal, México.

Álvarez, T., and A. Ocaña. 1999. Sinopsis de restos arqueozoológicos de vertebrados terrestres. Basada en informes del Laboratorio de Paleozoología del INAH. Instituto de Antropología e Historia, Colección Científica 386:1–108.

Álvarez, T., and O. Polaco. 1984. Estudio de los mamíferos capturados en La Michilía, sureste de Durango, México. Anales de la Escuela Nacional de Ciencias Biológicas 28:99–148.

Álvarez, T., and J. Ramírez-Pulido. 1968. Descripción de una nueva subspecie de *Spermophilus adocetus* (Rodentia, Sciuridae) de Michoacán, México y estado taxónomico de S. *a. arceliae* (Villa R., 1942). Revista de la Sociedad Mexicana de Historia Natural 29:181–189.

Álvarez, T., and J. Ramírez-Pulido. 1972. Notas acerca de murciélagos mexicanos. Anales de la Escuela Nacional de Ciencias Biológicas 19:167–178.

Álvarez, T., and N. Sánchez-Casas. 1997a. Notas sobre la alimentación de *Musonycteris* y *Choeroniscus* (Mammalia: Phyllostomidae) en México. Revista Mexicana de Mastozoología 2:113–115.

Álvarez, T., and N. Sánchez-Casas. 1997b. Contribución al conocimiento de los mamíferos, excepto Chiroptera y Rodentia, de Michoacán, México. Anales de la Escuela Nacional de Ciencas Biológicas 42:47–74.

Álvarez, T., S. T. Álvarez-Castañeda, and J. C. López-Vidal. 1994. Claves para los murciélagos de México. Publicación Especial, Centro de Investigaciones Biológicas de Baja California Sur y Escuela Nacional de Ciencias Biológicas, Instituto Politécnico Nacional. La Paz, México.

Álvarez, T., J. Arroyo-Cabrales, and M. González-Escamilla. 1987. Mamíferos (excepto Chiroptera) de la costa de Michoacán. Anales de la Escuela Nacional de Ciencias Biologicas 31:13–62.

Álvarez-Castañeda, S. T. 1994. Current status of the rice rat, *Oryzomys couesi peninsularis*. The Southwestern Naturalist 39:99–100.

Álvarez-Castañeda, S. T. 1996. Los mamíferos de Morelos. Centro de Investigaciones Biológicas del Noroeste, S. C. La Paz, México.

Álvarez-Castañeda, S. T. 1998. *Peromyscus pseudocrinitus*. Mammalian Species 601:1–3.

Álvarez-Castañeda, S. T. 2001. *Peromyscus sejugis*. Mammalian Species 658:1–3.

Álvarez-Castañeda, S. T. 2002. *Peromyscus hooperi*. Mammalian Species 709:1–3.

Álvarez-Castañeda, S. T. 2005a. *Peromyscus melanotis*. Mammalian Species 764:1–4.

Álvarez-Castañeda, S. T. 2005b. *Peromyscus winkelmanni*. Mammalian Species 765:1–3.

Álvarez-Castañeda, S. T. 2007. Systematics of the antelope ground squirrel (*Ammospermophilus*) from islands adjacent to the Baja California Peninsula. Journal of Mammalogy 88:1160–1169.

Álvarez-Castañeda, S. T. 2010. Phylogenetic structure of the *Thomomys bottae-umbrinus* complex in North America. Molecular Phylogenetics and Evolution 54:671–679.

Álvarez-Castañeda, S. T. 2016. *Peromyscus hooperi*. The IUCN Red List of Threatened Species 2016:e.T16667A22361329.

Álvarez-Castañeda, S. T. 2018a. *Megadontomys cryophilus*. The IUCN Red List of Threatened Species 2018a:e.T12940A22353517.

Álvarez-Castañeda, S. T. 2018b. *Neotoma nelsoni*. The IUCN Red List of Threatened Species 2018:e.T14592A22372306.

Álvarez-Castañeda, S. T. 2018c. *Neotoma palatina*. The IUCN Red List of Threatened Species 2018:e.T14593A22370476.

Álvarez-Castañeda, S. T. 2018d. *Peromyscus bullatus*. The IUCN Red List of Threatened Species 2018:e.T16653A22361454.

Álvarez-Castañeda, S. T. 2018e. *Peromyscus mekisturus*. The IUCN Red List of Threatened Species 2018:e.T16675A22362990.

Álvarez-Castañeda, S. T. 2018f. *Peromyscus sejugis*. The IUCN Red List of Threatened Species 2018:e.T16688A22364100.

Álvarez-Castañeda, S. T. 2018g. *Peromyscus stephani*. The IUCN Red List of Threatened Species 2018:e.T16692A22362173.

Álvarez-Castañeda, S. T. 2018h. *Peromyscus winkelmanni*. The IUCN Red List of Threatened Species 2018:e.T16695A22362540.

Álvarez-Castañeda, S. T. 2019a. *Peromyscus ochraventer*. The IUCN Red List of Threatened Species 2019:e.T16683A22363628.

Álvarez-Castañeda, S. T. 2019b. *Peromyscus slevini*. The IUCN Red List of Threatened Species 2019:e.T16690A143643850.

Álvarez-Castañeda, S. T. 2019c. *Neotoma phenax*. The IUCN Red List of Threatened Species 2019:e.T14594A22370414.

Álvarez-Castañeda, S. T., and T. Álvarez. 1991. Los murciélagos de Chiapas. Instituto Politécnico Nacional, Escuela Nacional de Ciencias Biológicas, Distrito Federal, México.

Álvarez-Castañeda, S. T., and M. Bogan. 1997. *Myotis milleri*. Mammalian Species 561:1–3.

Álvarez-Castañeda, S. T., and M. Bogan. 1998. *Myotis peninsularis*. Mammalian Species 573:1–2.

Álvarez-Castañeda, S. T., and P. Cortés-Calva. 2003. *Peromyscus eva*. Mammalian species 738:1–3.

Álvarez-Castañeda, S. T., and I. Castro-Arellano. 2008a. *Nelsonia neotomodon*. The IUCN Red List of Threatened Species 2008:e.T14487A4438546.

Álvarez-Castañeda, S. T., and I. Castro-Arellano. 2008b. *Neotoma angustapalata*. The IUCN Red List of Threatened Species 2008:e.T14583A4446470.

Álvarez-Castañeda, S. T., and E. Rios. 2010. A phylogenetic analysis of *Neotoma varia* (Rodentia: Cricetidae), a rediscovered, endemic, and threatened rodent from Datil Island, Sonora, Mexico. Zootaxa 2647:51–60.

Álvarez-Castañeda, S. T., and I. Castro-Arellano. 2016a. *Peromyscus eva*. The IUCN Red List of Threatened Species 2016:e.T16660A22360342.

Álvarez-Castañeda, S. T., and I. Castro-Arellano. 2016b. *Neotomodon alstoni*. The IUCN Red List of Threatened Species 2016:e.T14600A22345400.

Álvarez-Castañeda, S. T., and I. Castro-Arellano. 2019a. *Megadontomys thomasi*. The IUCN Red List of Threatened Species 2019:e.T12942A22353461.

Álvarez-Castañeda, S. T., and I. Castro-Arellano. 2019b. *Nelsonia goldmani*. The IUCN Red List of Threatened Species 2019:e.T14486A22338391.

Álvarez-Castañeda, S. T., and I. Castro-Arellano. 2019c. *Tylomys bullaris*. The IUCN Red List of Threatened Species 2019:e.T22570A22340265.

Álvarez-Castañeda, S. T., and I. Castro-Arellano. 2019d. *Tylomys tumbalensis*. The IUCN Red List of Threatened Species 2019:e.T22575A22340781.

Álvarez-Castañeda, S. T., y P. Cortés Calva. 1996. Anthropogenic extinction of the endemic deer mouse *Peromyscus maniculatus cineritius* on San Roque Island, Baja California Sur, México. Southwestern Naturalist 41:459–461.

Álvarez-Castañeda, S. T., and P. Cortés-Calva. 1999. Familia Muridae. Pp. 445–566, *in* Mamíferos del noroeste de México (Álvarez-Castañeda, S. T., and J. L. Patton, eds.). Centro de Investigaciones Biológicas del Noroeste, La Paz, México.

Álvarez-Castañeda, S. T., and P. Cortés-Calva. 2001. *Peromyscus sejugis*. Mammalian Species 658:1–2.
Álvarez-Castañeda, S. T., and P. Cortés-Calva. 2002a. *Peromyscus slevini*. Mammalian Species 705:1–2.
Álvarez-Castañeda, S. T., and P. Cortés-Calva. 2002b. Extirpation of bailey's pocket mouse, *Chaetodipus baileyi fornicatus* (Heteromyidae: Mammalia), from isla Montserrat, Baja California Sur, México. Western American Naturalist 62:496–497.
Álvarez-Castañeda, S. T., and P. Cortés-Calva. 2003a. *Peromyscus pembertoni*. Mammalian Species 734:1–2.
Álvarez-Castañeda, S. T., and P. Cortés-Calva. 2003b. *Peromyscus eva*. Mammalian Species 738:1–3.
Álvarez-Castañeda, S. T., and P. Cortés-Calva. 2011. Genetic evaluation of the Baja California rock squirrel *Otospermophilus atricapillus* (Rodentia: Sciuridae). Zootaxa 3138:35–51.
Álvarez-Castañeda, S. T., and N. González-Ruiz. 2009. *Peromyscus levipes*. Mammalian Species 824:1–6.
Álvarez-Castañeda, S. T., and N. González-Ruiz. 2008. Análisis preliminar de las relaciones filogenéticas entre los grupos de especies del género *Peromyscus*. Pp. 5–26, *in* Avances en el Estudio de los Mamíferos de México (Lorenzo, C., E. Espinoza, and J. Ortega, eds.). Publicaciones Especiales, Vol. II, Asociación Mexicana de Mastozoología, A. C., Distrito Federal, México.
Álvarez-Castañeda, S. T., and T. Lacher. 2016a. *Neotoma macrotis*. The IUCN Red List of Threatened Species 2016:e.T14597A22370553.
Álvarez-Castañeda, S. T., and T. Lacher. 2016b. *Peromyscus californicus*. The IUCN Red List of Threatened Species 2016:e.T16654A22361553.
Álvarez-Castañeda, S. T., and T. Lacher. 2017. *Chaetodipus arenarius*. The IUCN Red List of Threatened Species 2017:e.T92459293A123795361.
Álvarez-Castañeda, S. T. and T. Lacher. 2018. *Dipodomys gravipes*. The IUCN Red List of Threatened Species 2018:e.T6676A22227742.
Álvarez-Castañeda, S. T. and W. López-Forment. 1995. Datos sobre los mamíferos del área aledaña a Palpan, Morelos, México. Anales del Instituto de Biología, Universidad Nacional Autónoma de México, Serie Zoología 66:123–133.
Álvarez-Castañeda, S. T., and C. Lorenzo. 2016. Genetic evidence supports *Sylvilagus mansuetus* (Lagomorpha: Leporidae) as a subspecies of *S. bachmani*. Zootaxa. 4196:289–295.
Álvarez-Castañeda, S. T., y C. Lorenzo. 2017. Phylogeography and phylogeny of *Lepus californicus* (Lagomorpha: Leporidae) from Baja California Peninsula and adjacent islands. The Biological Journal of the Linnean Society 121:15–27.
Álvarez-Castañeda, S. T., and L. Méndez. 2003. *Oryzomys nelsoni*. Mammalian Species 735:1–2.
Álvarez-Castañeda, S. T., and L. Méndez. 2005. *Peromyscus madrensis*. Mammalian Species 774:1–3.
Álvarez-Castañeda, S. T., and R. W. Murphy. 2014. The endemic insular and peninsular species *Chaetodipus spinatus* (Mammalia, Heteromyidae) breaks patterns for Baja California. PlosOne 9:e116146.
Álvarez-Castañeda, S. T., and L. A. Nájera-Cortazar. 2020. Does islands populations change in size and form in relation to mainland counterparts? Journal of Mammalogy 101:373–385.
Álvarez-Castañeda, S. T., and A. Ortega-Rubio. 2003. Current status of rodents on islands in the Gulf of California. Biological Conservation 109:157–163.
Álvarez-Castañeda, S. T., and J. L. Patton. 1999. Mamíferos del Noroeste Mexicano I. Centro de Investigaciones Biológicas del Noroeste. La Paz, México 1:1–583.
Álvarez-Castañeda, S. T., and J. L. Patton. 2000. Mamíferos del Noroeste Mexicano II. Centro de Investigaciones Biológicas del Noroeste. La Paz, México 2:584–873.
Álvarez-Castañeda, S. T., and F. Reid. 2016. *Microtus mexicanus*. The IUCN Red List of Threatened Species 2016:e.T13443A115113184.
Álvarez-Castañeda, S. T., and E. Rios. 2010. A phylogenetic analysis of *Neotoma varia* (Rodentia: Cricetidae), a rediscovered, endemic, and threatened rodent from Datil Island, Sonora, Mexico. Zootaxa 2647:51–60.
Álvarez-Castañeda, S. T., and E. Rios. 2011. Revision of *Chaetodipus arenarius* (Rodentia: Heteromyidae). The Zoological Journal of the Linnean Society 160:213–228.
Álvarez-Castañeda, S. T., and E. Vázquez. 2018. *Heteromys spectabilis*. The IUCN Red List of Threatened Species 2018:e.T12077A22225458.
Álvarez-Castañeda, S. T., and E. Yensen. 1999. *Neotoma bryanti*. Mammalian Species 619:1–3.
Álvarez-Castañeda, S. T., G. Arnaud, and E. Yensen. 1996. *Spermophilus atricapillus*. Mammalian Species 521:1–3.
Álvarez-Castañeda, S. T., P. Cortés-Calva, and C. Gómez-Machorro. 1998. *Peromyscus caniceps*. Mammalian Species 602:1–3.
Álvarez-Castañeda, S. T., P. Cortés-Calva, L. Méndez, and A. Ortega-Rubio. 2006. Sea of Cortes island development call for mitigation. BioSciences 56:825–829.
Álvarez-Castañeda, S. T., I. Castro-Arellano, T. Lacher, and E. Vázquez. 2008a. *Peromyscus dickeyi*. The IUCN Red List of Threatened Species 2008:e.T16657A6238543.
Álvarez-Castañeda, S. T., I. Castro-Arellano, T. Lacher, E. Vázquez, and J. Arroyo-Cabrales. 2008b. *Peromyscus polius*. The IUCN Red List of Threatened Species 2008:e.T16686A6289329.
Álvarez-Castañeda, S. T., E. Rios, P. Cortés-Calva, N. González-Ruiz, and C. G. Suárez-Gracida. 2008c. Los mamíferos de las Reservas de El Valle de los Cirios y El Vizcaíno. Centro de Investigaciones Biológicas del Noroeste, S. C - Comisión Nacional para el Conocimiento y Uso de la Biodiversidad. Distrito Federal, México.
Álvarez-Castañeda, S. T., W. Z. Lidicker, and E. Rios. 2009. Revision of the *Dipodomys merriami* complex in the Baja California Peninsula, México. Journal of Mammalogy 90:992–1008.
Álvarez-Castañeda, S. T., G. Arnaud, P. Cortés-Calva, and L. Méndez. 2010a. Invasive migration of a mainland rodent to Santa Catalina Island and its effect on the endemic species *Peromyscus slevini*. Biological Invasions 12:437–439.
Álvarez-Castañeda, S. T., A. Gutiérrez, and M. De la Paz. 2010b. Rediscovery of the *Neotoma* population on Datil [Turner] Island, Sonora, México. Western North American Naturalist 70:437–440.
Álvarez-Castañeda, S. T., T. Álvarez, and N. González-Ruiz. 2015. Guía para la identificación de los mamíferos de México en campo y laboratorio/ Keys for identifying Mexican mammals in the field and in the laboratory. Centro de Investigaciones Biológicas del Noroeste, S. C. y Asociacion Mexicana de Mastozoología, A. C. Guadalajara, México.
Álvarez-Castañeda, S. T., I. Castro-Arellano, and T. Lacher. 2016a. *Neotamias merriami*. The IUCN Red List of Threatened Species 2016:e. T21358A22269203.
Álvarez-Castañeda, S. T., I. Castro-Arellano, and T. Lacher. 2016b. *Neotamias obscurus*. The IUCN Red List of Threatened Species 2016:e. T21359A22268421.
Álvarez-Castañeda, S. T., I. Castro-Arellano, and T. Lacher. 2016c. *Neotoma devia*. The IUCN Red List of Threatened Species 2016:e. T14586A22371569.

Álvarez-Castañeda, S. T., I. Castro-Arellano, and T. Lacher. 2016d. *Sigmodon fulviventer*. The IUCN Red List of Threatened Species 2016:e. T20212A22355460.

Álvarez-Castañeda, S. T., I. Castro-Arellano, and T. Lacher. 2016e. *Dipodomys deserti*. The IUCN Red List of Threatened Species 2016:e. T6686A22228301.

Álvarez-Castañeda, S. T., I. Castro-Arellano, and T. Lacher. 2016f. *Chaetodipus artus*. The IUCN Red List of Threatened Species 2016:e. T4327A115067951.

Álvarez-Castañeda, S. T., I. Castro-Arellano, and T. Lacher. 2016g. *Chaetodipus fallax*. The IUCN Red List of Threatened Species 2016:e. T4330A22226385.

Álvarez-Castañeda, S. T., I. Castro-Arellano, and T. Lacher. 2016h. *Chaetodipus formosus*. The IUCN Red List of Threatened Species 2016:e. T4331A22226290.

Álvarez-Castañeda, S. T., I. Castro-Arellano, and T. Lacher. 2016i. *Chaetodipus pernix*. The IUCN Red List of Threatened Species 2016:e. T4337A22225579.

Álvarez-Castañeda, S. T., I. Castro-Arellano, and T. Lacher. 2016j. *Chaetodipus rudinoris*. The IUCN Red List of Threatened Species 2016:e. T136837A22225520.

Álvarez-Castañeda, S. T., I. Castro-Arellano, and T. Lacher. 2016k. *Perognathus amplus*. The IUCN Red List of Threatened Species 2016:e. T16633A22224910.

Álvarez-Castañeda, S. T., I. Castro-Arellano, T. Lacher, and E. Vázquez. 2016l. *Sciurus oculatus*. The IUCN Red List of Threatened Species 2016:e.T20017A22246721.

Álvarez-Castañeda, S. T., I. Castro-Arellano, T. Lacher, and E. Vázquez. 2016m. *Notocitellus annulatus*. The IUCN Red List of Threatened Species 2016:e.T20479A22265951.

Álvarez-Castañeda, S. T., I. Castro-Arellano, T. Lacher, and E. Vázquez. 2016n. *Microtus californicus*. The IUCN Red List of Threatened Species 2016:e.T13427A22349460.

Álvarez-Castañeda, S. T., I. Castro-Arellano, T. Lacher, and E. Vázquez. 2016o. *Hodomys alleni*. The IUCN Red List of Threatened Species 2016:e.T10211A22379623.

Álvarez-Castañeda, S. T., I. Castro-Arellano, T. Lacher, and E. Vázquez. 2016p. *Peromyscus perfulvus*. The IUCN Red List of Threatened Species 2016:e.T16685A22364160.

Álvarez-Castañeda, S. T., I. Castro-Arellano, T. Lacher, and E. Vázquez. 2016q. *Peromyscus melanotis*. The IUCN Red List of Threatened Species 2016:e.T16678A115136970.

Álvarez-Castañeda, S. T., I. Castro-Arellano, T. Lacher, and E. Vázquez. 2016r. *Reithrodontomys chrysopsis*. The IUCN Red List of Threatened Species 2016:e.T19404A22385521.

Álvarez-Castañeda, S. T., I. Castro-Arellano, T. Lacher, and E. Vázquez. 2016s. *Reithrodontomys zacatecae*. The IUCN Red List of Threatened Species 2016:e.T19419A115151808.

Álvarez-Castañeda, S. T., I. Castro-Arellano, T. Lacher, and E. Vázquez. 2016t. *Sigmodon leucotis*. The IUCN Red List of Threatened Species 2016:e.T20215A22355853.

Álvarez-Castañeda, S. T., I. Castro-Arellano, T. Lacher, and E. Vázquez. 2016u. *Handleyomys melanotis*. The IUCN Red List of Threatened Species 2016:e.T15606A22328750.

Álvarez-Castañeda, S. T., I. Castro-Arellano, T. Lacher, and E. Vázquez. 2016v. *Cratogeomys goldmani*. The IUCN Red List of Threatened Species 2016:e.T136647A22216690.

Álvarez-Castañeda, S. T., I. Castro-Arellano, T. Lacher, and E. Vázquez. 2016w *Dipodomys nelsoni*. The IUCN Red List of Threatened Species 2016:e.T6690A22228791.

Álvarez-Castañeda, S. T., A. D. Cuarón, and P. C. de Grammont. 2016x. *Cryptotis peregrina*. The IUCN Red List of Threatened Species 2016:e. T136550A22284600.

Álvarez-Castañeda, S. T., T. Lacher, and E. Vázquez. 2016y. *Cratogeomys fumosus*. The IUCN Red List of Threatened Species 2016:e. T16026A22216828.

Álvarez-Castañeda, S. T., T. Lacher, and E. Vázquez. 2016z. *Neotamias durangae*. The IUCN Red List of Threatened Species 2016:e. T21357A22268753.

Álvarez-Castañeda, S. T., T. Lacher, and E. Vázquez. 2016aa. *Xerospermophilus perotensis*. The IUCN Red List of Threatened Species 2016:e. T20489A22264586.

Álvarez-Castañeda, S. T., T. Lacher, and E. Vázquez. 2016ab. *Neotamias bulleri*. The IUCN Red List of Threatened Species 2016:e. T21356A103309807.

Álvarez-Castañeda, S. T., T. Lacher., and E. Vázquez. 2016ac. *Dipodomys phillipsii*. The IUCN Red List of Threatened Species 2016:e. T92463503A22228991.

Álvarez-Castañeda, S. T., T. Lacher, and E. Vázquez. 2016ad. *Callospermophilus madrensis*. The IUCN Red List of Threatened Species 2016:e. T20485A22263330.

Álvarez-Castañeda, S.T., T. Lacher, and E. Vázquez. 2016ae. *Cratogeomys merriami*. *The IUCN Red List of Threatened Species* 2016:e. T16028A22217011.

Álvarez-Castañeda, S. T., I. Castro-Arellano, T. Lacher, E. Vázquez, and J. Arroyo-Cabrales. 2016af. *Sigmodon mascotensis*. The IUCN Red List of Threatened Species 2016:e.T20216A115157887.

Álvarez-Castañeda, S. T., J. Matson, N. Woodman, P. C. de Grammont, and G. A. Hammerson. 2016ag. *Sorex ornatus*. The IUCN Red List of Threatened Species 2016:e.T41408A115184758.

Álvarez-Castañeda, S. T., I. Castro-Arellano, and T. Lacher. 2016ah. *Sigmodon arizonae*. The IUCN Red List of Threatened Species 2016:e. T20211A22355540.

Álvarez-Castañeda, S. T., T. Álvarez, and N. González-Ruiz. 2017a. Keys for identifying Mexican Mammals. The Johns Hopkins University Press. Baltimore, U.S.A.

Álvarez-Castañeda, S. T., T. Lacher, and E. Vázquez. 2018a. *Cynomys mexicanus*. The IUCN Red List of Threatened Species 2018:e. T6089A22260873.

Álvarez-Castañeda, S. T., T. Lacher, and E. Vázquez. 2018b. *Habromys chinanteco*. The IUCN Red List of Threatened Species 2018:e. T9608A22376453.

Álvarez-Castañeda, S. T., T. Lacher, and E. Vázquez. 2018c. *Habromys schmidlyi*. The IUCN Red List of Threatened Species 2018:e. T136616A22376358.

Álvarez-Castañeda, S. T., T. Lacher, and E. Vázquez. 2018d. *Peromyscus caniceps*. The IUCN Red List of Threatened Species 2018:e. T16655A22361697.

Álvarez-Castañeda, S. T., T. Lacher, and E. Vázquez. 2018e. *Peromyscus guardia*. The IUCN Red List of Threatened Species 2018:e. T16664A22359973.

Álvarez-Castañeda, S. T., T. Lacher, and E. Vázquez. 2018f. *Peromyscus madrensis*. The IUCN Red List of Threatened Species 2018 e.T16671A22361002.

Álvarez-Castañeda, S. T., T. Lacher, and E. Vázquez. 2018g. *Peromyscus simulus*. The IUCN Red List of Threatened Species 2018:e. T16689A22362390.

Álvarez-Castañeda, S. T., T. Lacher, and E. Vázquez. 2018h. *Reithrodontomys hirsutus*. The IUCN Red List of Threatened Species 2018:e. T19409A22386368.

Álvarez-Castañeda, S. T., T. Lacher, I. Castro-Arellano, and E. Vázquez. 2018i. *Habromys lepturus*. The IUCN Red List of Threatened Species 2018:e.T9609A22376889.

Álvarez-Castañeda, S. T., T. Lacher, E. Vázquez, and J. Arroyo-Cabrales. 2018j. *Peromyscus zarhynchus*. The IUCN Red List of Threatened Species 2018:e.T16697A22362848.

Álvarez-Castañeda, S. T., T. Lacher, E. Vázquez, and J. Arroyo-Cabrales. 2018k. *Zygogeomys trichopus*. The IUCN Red List of Threatened Species 2018:e.T23323A22216603.

Álvarez-Castañeda, S. T., T. Lacher, E. Vázquez, J. Arroyo-Cabrales, and P. C. de Grammont. 2018l. *Megadontomys nelsoni*. The IUCN Red List of Threatened Species 2018:e.T12941A22353381.

Álvarez-Castañeda, S. T., T. Lacher, E. Vázquez, J. Arroyo-Cabrales, and P. C. de Grammont. 2018m. *Peromyscus melanocarpus*. The IUCN Red List of Threatened Species 2018:e.T16676A22363072.

Álvarez-Castañeda, S. T., E. Vázquez, I. Castro-Arellano, and T. Lacher. 2018n. *Habromys ixtlani*. The IUCN Red List of Threatened Species 2018:e.T136582A22376638.

Álvarez-Castañeda, S. T., A. D. Cuarón, and P. C. Grammont. 2018o. *Cryptotis phillipsii*. The IUCN Red List of Threatened Species 2018:e. T136639A22283893.

Álvarez-Castañeda, S. T., T. Lacher, and E. Vázquez. 2019a. *Microtus quasiater*. The IUCN Red List of Threatened Species 2019:e. T13453A22348115.

Álvarez-Castañeda, S. T., T. Lacher, and E. Vázquez. 2019b. *Peromyscus melanurus*. The IUCN Red List of Threatened Species 2019:e. T16679A22363320.

Álvarez-Castañeda, S. T., T. Lacher, and E. Vázquez. 2019c. *Peromyscus pseudocrinitus*. The IUCN Red List of Threatened Species 2019:e. T16687A143642485.

Álvarez-Castañeda, S. T., T. Lacher, E. Vázquez, and J. Arroyo-Cabrales. 2019d. *Peromyscus sagax*. The IUCN Red List of Threatened Species 2019:e.T136710A22362792.

Álvarez-Castañeda, S. T., I. Castro-Arellano, T. Lacher, E. Vázquez, and J. Arroyo-Cabrales. 2019e. *Reithrodontomys burti*. The IUCN Red List of Threatened Species 2019:e.T19403A22385577.

Álvarez-Castañeda, S. T., C. Lorenzo, C. A. Segura-Trujillo, and S. G. Pérez-Consuegra. 2019f. Two new species of *Peromyscus* from Chiapas, Mexico, and Guatemala. Pp. 543–558, *in* From field to laboratory: a memorial volume in honor of Robert J. Baker (Bradley, R. D., H. H. Genoways, D. J. Schmidly, and L. C. Bradley, eds.). Special Publications, Museum of Texas Tech University. Lubbock, U.S.A.

Álvarez-Castañeda, S. T., T. Lacher, E. Vázquez, and J. Arroyo-Cabrales. 2020. *Peromyscus polius*. The IUCN Red List of Threatened Species 2020:e.T16686A22364219.

Álvarez-Castañeda, S. T., C. A. Segura-Trujillo, and N. González-Ruiz. 2021. *Peromyscus mekisturus*. Mammalian Species, in press.

American Society of Mammalogy. 2022. ASM's Mammal Diversity Database. https://www.mammaldiversity.org/ Accessed on 27 December 2022.

Amman, B. R., and R. D. Bradley. 2004. Molecular evolution in *Baiomys* (Rodentia: Sigmodontinae): evidence for a genetic subdivision in *B. musculus*. Journal of Mammalogy 85:162–166.

Ammerman, L. K., D. N. Lee, and R. S. Pfau. 2016. Patterns of Genetic Divergence among *Myotis californicus*, *M. ciliolabrum*, and *M. leibii* Based on Amplified Fragment Length Polymorphism. Acta Chiropterologica 18:337–347.

Amori, G., F. Chiozza, C. Rondinini, and L. Luiselli. 2011. Worldwide conservation hotspots for Soricomorpha focusing on endemic island taxa: an analysis at two taxonomic levels. Endangered Species Research 15:143–149.

Amstrup, S. C. 2003. Polar bear, *Ursus maritimus*. Pp. 587–610, *in* Wild Mammals of North America: Biology, Management, and Conservation (Feldhamer, G. A., B. C. Thomson, and J. A. Chapman, eds). John Hopkins University Press. Baltimore, USA.

Andersen, J. M., Y. F. Wiersma, and G. Stenson. 2009. Movement patterns of hooded seals (*Cystophora cristata*) in the Northwest Atlantic Ocean during the post-mout and pre-breeding seasons. Journal of Northwest Atlantic Fisheries Science 42:1–11.

Andersen, K. 1906a. On the bats of the genera *Micronycteris* and *Glyphonycteris*. Annals and magazine of Natural History 18:50–65.

Andersen, K. 1906b. Brief diagnoses of a new genus and ten new forms of stenodermatous bats. Annals and magazine of Natural History 18:419–423.

Andersen, K. 1908. A monograph of the chiropteran genera *Uroderma, Enchisthenes,* and *Artibeus*. Proceedings of the Zoological Society of London 204–319.

Anderson, A. E., and O. C. Wallmo. 1984. *Odocoileus hemionus*. Mammalian Species 219:1–9.

Anderson, E. 1977. Pleistocene Mustelidae (Mammalia, Carnivora) from Fairbanks, Alaska. Bulletin of the Museum of Comparative Zoology 148:1–21.

Anderson, R. M. 1934. *Sorex palustris brooksi*, a new water shrew from Vancouver Island. The Canadian Field Naturalist 48:134.

Anderson, R. M. 1947. Catalogue of Canadian Recent mammals. National Museum of Canada Bulletin, Biological Series 102:1–238.

Anderson, R. M. 1942. Canadian voles of the genus *Phenacomys* with description of two new Canadian subspecies. Canadian Field-Naturalist 56:56–60.

Anderson, R. M., and A. L. Rand. 1945. The Varying Lemming (Genus *Dicrostonyx*) in Canada. Journal of Mammalogy 26:301–306.

Andersen, J. J., and J. E. Light. 2012. Phylogeography and subspecies revision of the hispid pocket mouse, *Chaetodipus hispidus* (Rodentia: Heteromyidae). Journal of Mammalogy 93:1195–1215.

Anderson, S. 1954. Subspeciation in the meadow mouse, *Microtus montanus*, in Wyoming and Colorado. University of Kansas Publications, Museum of Natural History 7:489–506.

Anderson, S. 1959. Distribution, variation, and relationships of the montane vole, *Microtus montanus*. University of Kansas Publications, Museum of Natural History 9:415–511.

Anderson, S. 1960. The baculum of Microtine rodents. University of Kansas Publications, Museum of Natural History 12:181–216.

Anderson, S. 1961. Mammals of Mesa Verde National Park, Colorado. University of Kansas Publications, Museum of Natural History 14:29–67.

Anderson, S. 1962. Tree squirrels (*Sciurus colliaei* group) from western Mexico. American Museum Novitates 2093:1–13.

Anderson, S. 1964. The systematic status of *Perognathus artus* and *Perognathus goldmani* (Rodentia). American Museum Novitates 2184:1–27.

Anderson, S. 1966. Taxonomy of gophers, especially *Thomomys* in Chihuahua, Mexico. Systematic Zoology 15:189–198.

Anderson, S. 1969. *Macrotus waterhousii*. Mammalian Species 1:1–4.

Anderson, S. 1972. Mammals of Chihuahua: taxonomy and distribution. Bulletin of the American Museum of Natural History 148:149–410.

Anderson, S. and A. S. Gaunt. 1962. A classification of the white-sided jackrabbits of Mexico. American Museum Novitates 2088:1–16.

Anderson, S., and C. E. Nelson. 1965. A systematic revision of *Macrotus* (Chiroptera). American Museum Novitates 2212:1–39.

Angerbjörn, A., and M. Tannerfeldt. 2014. *Vulpes lagopus*. The IUCN Red List of Threatened Species 2014:e.T899A57549321.

Anthony, H. E. 1923. Mammals from Mexico and South America. American Museum Novitates 54:1–10.

Anthony, H. E. 1932. A new genus of rodents from Yucatan. American Museum Novitates 586:1–3.

Appollonio, M. 1999. Dama dama. *in* The Atlas of European Mammals (Mitchell-Jones, A. J., G. Amori, W. Bogdanowicz, B. Kryštufek, P. J. H. Reijnders, F. Spitzenberger, M. Stübbe, J. B. M. Thissen, V. Vohralík, and J. Zima, eds). Academic Press, London, UK.

Aranda, M., J. E. Escobedo, and C. Pozo. 1997. Registros recientes de *Otonyctomys hatti* (Rodentia: Muridae) en Quintana Roo, México. Acta Zoológica Mexicana 72:63–65.

Arbogast, B. S. 1999. Mitochondrial DNA phylogeography of the New World flying squirrels (*Glaucomys*): implications for Pleistocene biogeography. Journal of Mammalogy 80:142–155.

Arbogast, B. S., R. A. Browne, and P. D. Weigl. 2001. Evolutionary genetics and Pleistocene biogeography of North American Tree squirrels (*Tamasciurus*). Journal of Mammalogy 82:302–319.

Arbogast, B. S., K. I. Schumacher, N. J. Kerhoulas, A. L. Bidlack, J. A. Cook, and G. J. Kenagy. 2017. Genetic data reveal a cryptic species of New World flying squirrel: *Glaucomys oregonensis*. Journal of Mammalogy 98:1027–1041.

Arcangeli, J., J. E. Light, and F. A. Cervantes 2018. Molecular and morphological evidence of the diversification in the gray mouse opossum, *Tlacuatzin canescens* (Didelphimorphia), with description of a new species. Journal of Mammalogy 99:138–158.

Arellano, E., and D. S. Rogers. 1994. *Reithrodontomys tenuirostris*, Mammalian Species 477:1–3.

Arellano-Meneses, A. G., L. A. Hernandez-Carbajal, I. E. Lira Galera, G. Ruiz-Guzman, and C. Mudespacher-Ziehl. 2000. Karyotypical studies on *Peromyscus difficilis amplus* (Rodentia: Muridae). Cytologia 65:25–28.

Arellano, E., D. S. Rogers, and F. A. Cervantes. 2003. Genic differentiation and phylogenetic relationships among tropical harvest mice (*Reithrodontomys*: Subgenus Aporodon). Journal of Mammalogy 84:129–143.

Arita, H. T. 1999. Southern long-nosed bat/*Leptonycteris curasoae*. Pp. 76–78, *in* The Smithsonian book of North American mammals (Wilson, D. E., and S. Ruff, eds.). Smithsonian Institution Press, Washington, U.S.A.

Arita, H. T., and S. R. Humphrey. 1988. Revision taxonómica de los murciélagos magueyeros del género *Leptonycteris* (Chiroptera: Phyllostomidae). Acta Zoológica Mexicana 29:1–60.

Arita, H. T., J. G. Robinson, and K. H. Redford. 1990. Rarity in neotropical forest mammals and its ecological correlates. Conservation Biology 4:181.

Armitage, K. B., J. F. Downhower, and G. E. Svendsen. 1976. Seasonal changes in weights of marmots. The American Midland Naturalist 96:36–51.

Armitage, K. B., D. W. Johns, and D. C. Andersen. 1979. Cannibalism among yellow-bellied marmots. Journal of Mammalogy 60:205–206.

Arizona Game and Fish Department. 2004a. *Sigmodon arizonae arizonae*. Arizona Game and Fish Department Heritage Management Data System. http://www.azgfd.com/w_c/edits/documents/Sigmaraz.D.pdf, accessed 15 December 2008.

Arizona Game and Fish Department. 2004b. *Sigmodon arizonae jacksoni*. Arizona Game and Fish Department Heritage Management Data System. http://www.azgfd.com/w_c/edits/documents/Sigmarja.D.pdf, accessed 15 December 2008.

Armstrong, D. M. 1971. Notes on variation in *Spermophilus tridecemlineatus* (Rodentia, Sciuridae) in Colorado and adjacent states, and description of a new subspecies. Journal of Mammalogy 52:528–536.

Armstrong, D. M. 1972. Distribution of mammals in Colorado. Museum of Natural History University of Kansas Monographs 3:1–415.

Armstrong, D. M., and J. K. Jones, Jr. 1971a. Mammals from the Mexican state of Sinaloa. I. Marsupialia, Insectivora, Edentata, Lagomorpha. Journal of Mammalogy 52:747–757.

Armstrong, D. M., and J. K. Jones, Jr. 1971b. *Sorex merriami*. Mammalian Species 2:1–2.

Armstrong, D. M., and J. K. Jones, Jr. 1972a. *Megasorex gigas*. Mammalian Species 16:1–2.

Armstrong, D. M., and J. K. Jones, Jr. 1972b. *Notiosorex crawfordi*. Mammalian Species 17:1–5.

Armstrong, D. M., J. P. Fitzgerald, and C. A. Meaney. 2011. Mammals of Colorado. University Press of Colorado. Colorado, U.S.A.

Arroyo-Cabrales, J. 2016. *Rhogeessa genowaysi*. The IUCN Red List of Threatened Species 2016:e.T19680A21989676.

Arroyo-Cabrales, J., and T. Álvarez. 1990. Restos óseos de murciélagos procedentes de las excavaciones en las grutas de Loltún. Colección científica: Serie prehistoria del Instituto Nacional de Antropología e Historia 194:1–103.

Arroyo-Cabrales, J., and S. T. Álvarez-Castañeda. 2015a. *Artibeus hirsutus*. The IUCN Red List of Threatened Species 2015:e.T2131A21996678.

Arroyo-Cabrales, J., and S. T. Álvarez-Castañeda. 2015b. *Nyctinomops femorosaccus*. The IUCN Red List of Threatened Species 2015:e.T14994A22010542.

Arroyo-Cabrales, J., and S. T. Álvarez-Castañeda. 2017a. *Corynorhinus rafinesquii*. The IUCN Red List of Threatened Species 2017:e.T17600A21976905.

Arroyo-Cabrales, J., and S. T. Álvarez-Castañeda. 2017b. *Corynorhinus townsendii*. The IUCN Red List of Threatened Species 2017:e. T17598A21976681.

Arroyo-Cabrales, J., and S. T. Álvarez-Castañeda. 2017c. *Euderma maculatum*. The IUCN Red List of Threatened Species 2017:e.T8166A22028573.

Arroyo-Cabrales, J., and S. T. Álvarez-Castañeda. 2017d. *Idionycteris phyllotis*. The IUCN Red List of Threatened Species 2017:e. T10790A21990019.

Arroyo-Cabrales, J., and S. T. Álvarez-Castañeda. 2017e. *Lasiurus xanthinus*. The IUCN Red List of Threatened Species 2017:e.T41532A22004260.

Arroyo-Cabrales, J., and S. T. Álvarez-Castañeda. 2017f. *Myotis austroriparius*. The IUCN Red List of Threatened Species 2017:e. T14147A22059907.

Arroyo-Cabrales, J., and S. T. Álvarez-Castañeda. 2017g. *Myotis ciliolabrum*. The IUCN Red List of Threatened Species 2017:e.T14153A22058110.

Arroyo-Cabrales, J., and S. T. Álvarez-Castañeda. 2017h. *Myotis evotis*. The IUCN Red List of Threatened Species 2017:e.T14157A22059133.

Arroyo-Cabrales, J., and S. T. Álvarez-Castañeda. 2017i. *Myotis keenii*. The IUCN Red List of Threatened Species 2017:e.T14171A22055579.

Arroyo-Cabrales, J., and P. C. de Grammont. 2017a. *Antrozous pallidus*. The IUCN Red List of Threatened Species 2017:e.T1790A22129152.

Arroyo-Cabrales, J., and P. C. de Grammont. 2017b. *Myotis thysanodes*. The IUCN Red List of Threatened Species 2017:e.T14206A22063246.

Arroyo-Cabrales, J., and J. K. Jones. 1988a. *Balantiopteryx plicata*. Mammalian Species 301:1–4.

Arroyo-Cabrales, J., and J. K. Jones. 1988b. *Balantiopteryx io* and *Balantiopteryx infusca*. Mammalian Species 313:1–3.

Arroyo-Cabrales, J., and S. Ospina-Garcés. 2015. *Musonycteris harrisoni*. The IUCN Red List of Threatened Species 2015:e.T14003A22099002.

Arroyo-Cabrales, J., and S. Ospina-Garcés. 2016a. *Myotis findleyi*. The IUCN Red List of Threatened Species 2016:e.T14159A22058800.

Arroyo-Cabrales, J., and S. Ospina-Garcés. 2016b. *Myotis planiceps*. The IUCN Red List of Threatened *Species* 2016:e.T14191A22066742.

Arroyo-Cabrales, J., and S. Ospina-Garcés. 2016c. *Myotis sodalis*. The IUCN Red List of Threatened Species 2016:e.T14136A22053184.

Arroyo-Cabrales, J., and S. Ospina-Garcés. 2016d. *Myotis peninsularis*. The IUCN Red List of Threatened Species 2016:e.T14189A22066405.

Arroyo-Cabrales, J., and S. Ospina-Garcés. 2016e. *Myotis vivesi*. The IUCN Red List of Threatened Species 2016:e.T14209A22069146.

Arroyo-Cabrales, J., and S. Ospina-Garcés. 2016f. *Rhogeessa mira*. The IUCN Red List of Threatened Species 2016:e.T19683A22007311.

Arroyo-Cabrales, J., and R. D. Owen. 1996. Intraspecific variation and phenetic affinities of *Dermanura hartii*, with reapplication of the specific name *Enchisthenes hartii*. Pp. 67–81, *in* Contributions in mammalogy: a memorial volume honoring Dr. J. Knox Jones, Jr. (Genoways, H. H., and R. J. Baker, eds.). The Museum, Texas Tech University, Lubbock, U.S.A.

Arroyo-Cabrales, J., and R. D. Owen. 1997. *Enchisthenes hartii*. Mammalian Species 546:1–4.

Arroyo-Cabrales, J., and S. Perez. 2017. *Myotis californicus*. The IUCN Red List of Threatened Species 2017:e.T14150A22061366.

Arroyo-Cabrales, J., and O. J. Polaco. 1997. *Rhogeessa mira*. Mammalian Species 550:1–2.

Arroyo-Cabrales, J., and F. Reid. 2016. *Platyrrhinus helleri*. The IUCN Red List of Threatened Species 2016:e.T88159886A88159952.

Arroyo-Cabrales, J., R. R. Hollander, and J. K. Jones, Jr. 1987. *Choeronycteris mexicana*. Mammalian Species 291:1–5.

Arroyo-Cabrales, J., R. A. van den Bussche, K. Haiduk-Sigler, R. K. Chesser, and R. J. Baker. 1997. Genic variation in island populations of *Natalus stramineus* (Chiroptera: Natalidae). Occasional Papers, Museum of Texas Tech University 1771: 1–9.

Arroyo-Cabrales, J., E. K. V. Kalko, R. K. Laval, J. E. Maldonado, R. A. Medellín, O. J. Polaco, and B. Rodríguez-Herrera. 2005. Rediscovery of the Mexican flat–headed bat *Myotis planiceps* (Vespertilionidae). Acta Chiropterologica 7:309–314.

Arroyo-Cabrales, J., O. J. Polaco, D. E. Wilson, and A. L. Gardner. 2008. Nuevos registros de murciélagos para el estado de Nayarit, México. Revista Mexicana de Mastozoología 12:141–162.

Arroyo-Cabrales, J., Miller, B., F. Reid, A. D. Cuarón, and P. C. de Grammont. 2015a. *Mimon cozumelae*. The IUCN Red List of Threatened Species 2015:e.T136561A21991024.

Arroyo-Cabrales, J., S. T. Álvarez-Castañeda, A. D. Cuarón, and P. C. de Grammont. 2015b. *Glossophaga morenoi*. The IUCN Red List of Threatened Species 2015:e.T9276A22108155.

Arroyo-Cabrales, J., B. Miller, F. Reid, A. D. Cuarón, and P. C. de Grammont. 2015c. *Centronycteris centralis*. The IUCN Red List of Threatened Species 2015:e.T136350A22023809.

Arroyo-Cabrales, J., Miller, B., F. Reid, A. D. Cuarón, and P. C. de Grammont. 2016. *Lasiurus borealis*. The IUCN Red List of Threatened Species 2016:e.T11347A22121017.

Arroyo-Cabrales, J., Miller, B., F. Reid, A. D. Cuarón, and P. C. de Grammont. 2017. *Myotis auriculus*. The IUCN Red List of Threatened Species 2017:e.T14145A22060698.

Ashley, M., J. Norman, and L. Stross. 1996. Phylogenetic analysis of the Perissodactyla family Tapiridae using mitochondrial cytochrome c oxidase (COII) sequences. Journal of Mammalian Evolution 3:315–326.

Audet, D., M. D. Engstrom, and M. B. Fenton. 1993. Morphology, Karyology, and echolocation calls of *Rhogeessa* (Chiroptera: Vespertilionidae) from the Yucatán Peninsula. Journal of Mammalogy 74:498–502.

Audet, A. M., C. B. Robbins, and S. Larivière. 2002. *Alopex lagopus*, Mammalian Species 713:1–10.

Aune, K., D. Jørgensen, and C. Gates. 2017. *Bison bison*. The IUCN Red List of Threatened Species 2017:e.T2815A123789863.

Aurioles-Gamboa, D., and F. J. Camacho-Ríos. 2007. Diet and feeding overlap of two otariids, *Zalophus californianus* and *Arctocephalus townsendi*: Implications to survive environmental uncertainty. Aquatic Mammals 33:315–326.

Aurioles-Gamboa, D., F. Elorriaga-Verplancken, and C. J. Hernández-Camacho. 2010. Guadalupe fur seal population status on the San Benito Islands, Mexico. Marine Mammal Science 26:402–408.

Austin, J., and N. Roach. 2019. *Podomys floridanus*. The IUCN Red List of Threatened Species 2019:e.T17830A22339074.

Astúa, D., D. Lew, L. P. Costa, and R. Pérez-Hernandez. 2021. *Didelphis marsupialis* (amended version of 2016 assessment). The IUCN Red List of Threatened Species 2021:e.T40501A197310576.

Aurioles-Gamboa, D. 2015. *Arctocephalus townsendi*. The IUCN Red List of Threatened Species 2015:e.T2061A45224420.

Aurioles-Gamboa, D., and J. Hernández-Camacho. 2015. *Zalophus californianus*. The IUCN Red List of Threatened Species 2015:e. T41666A45230310.

Ávila-Flores, R., J. J. Flores-Martínez, and J. Ortega. 2002. *Nyctinomops laticaudatus*. Mammalian Species 697:1–6.

Ávila-Valle, Z. A., A. Castro-Campillo, L. León-Paniagua, I. H. Salgado-Ugalde, A. G. Navarro-Sigüenza, B. E. Hernández-Baños, and J. Ramírez-Pulido. 2012. Geographic variation and molecular evidence of the blackish deer mouse complex (*Peromyscus furvus*, Rodentia: Muridae). Mammalian Biology 77:166–177.

Avise J. C., M. H. Smith, R. K. Selander, T. E. Lawlor, and P. R. Ramsey. 1974. Biochemical polymorphism and systematics in the genus *Peromyscus*. V. Insular and mainland species of the subgenus *Haplomylomys*. Systematic Zoology 23:226–238.

Avise, C., J. Smith, and R. K. Selander. 1979. Biochemical polymorphism and systematics in the genus *Peromyscus* VII. Geographic differentiation in members of the *truei* and *maniculatus* species group. Journal of Mammalogy 60:177–192.

Bailey, B. 1929. Mammals of Sherburne County, Minnesota. Journal of Mammalogy 10:153–164.

Bailey, V. 1897. Revision of the American voles of the genus *Evotomys*. Proceedings of the Biological Society of Washington 11:113–138.

Bailey, V. 1900. Revision of American voles of the genus *Microtus*. North American Fauna 17:1–88.

Bailey, V. 1902a. Seven new mammals from western Texas. Proceedings of the Biological Society of Washington 15:117–120.

Bailey, V. 1902b. Synopsis of the North American species of *Sigmodon*. Proceedings of the Biological Society of Washington 15:101–116.

Bailey, V. 1905. Biological survey of Texas. North American Fauna, 25:1–222.

Bailey, V. 1915. Revision of the Pocket Gophers of the Genus *Thomomys*. North American Fauna 39:1–136.

Bailey, V. 1931. Mammals of New Mexico. North American Fauna 53:1–412.

Bailey, V. 1936. The mammals and life zones of Oregon. North American Fauna 55:1–416.

Baird, A. B., D. M. Hills, J. C. Patton, and J. W. Bickham. 2008. Evolutionary history of the genus *Rhogeessa* (Chiroptera: Vespertilionidae) as revaluated by mitochondrial DNA sequences. Journal of Mammalogy 89:744–754.

Baird, A. B., D. M. Hillis, J. C. Patton, and J. W. Bickham. 2009. Speciation by monobrachial centric fusions: A test of the model using nuclear DNA sequences from the bat genus *Rhogeessa*. Molecular Phylogenetics and Evolution 50:256–267.

Baird, A. B. 2010. Genetic identification of cryptic species: Anexample in *Rhogeessa*. Pp. 22–23, *in* Molecular approaches in natural resource conservation and management (DeWoody, J. A., J. W. Bickham, C. H. Michler, K. M. Nichols, O. E. Rhodes, Jr., and K. E. Woeste, eds.). Cambridge University Press. Cambridge, U.S.A.

Baird, A. B., M. R. Marchán-Rivadeneira, S. G. Pérez, and R. J. Baker. 2012. Morphological analysis and description of two new species of *Rhogeessa* (Chiroptera: Vespertilionidae) from the neotropics. Occasional Papers of the Museum, Texas Tech University 307:1–25.

Baird, A. B., J. K. Braun, M. A. Mares, J. C. Morales, J. C. Patton, C. Q. Tran, and J. W. Bickham. 2015. Molecular systematic revision of tree bats (Lasiurini): doubling the native mammals of the Hawaiian Islands. Journal of Mammalogy 96:1255–1274.

Baird, A. B., J. Braun, M. Engstrom, B. Lim, M. Mares, J. Patton, and J. Bickham. 2021. On the utility of taxonomy to reflect biodiversity: the example of *Lasiurini* (Chiroptera: Vespertilionidae). Therya 12:283.

Baker, R. H. 1951. Two new moles (genus *Scalopus*) from Mexico and Texas. University of Kansas Publications, Museum of Natural History 5:17–24.

Baker, R. H. 1952. Geographic range of *Peromyscus melanophrys*, with description of new subspecies. University of Kansas Publications, Museum of Natural History 5:251–258.

Baker, R. H. 1954. The silky pocket mouse (*Perognathus flavus*) of Mexico. University of Kansas Publications, Museum of Natural History 7:339–347.

Baker, R. H. 1955. A new species of bat (genus *Myotis*) from Coahuila, Mexico. Proceedings of the Biological Society of Washington 68:165–166.

Baker, R. H. 1956. Mammals of Coahuila, México. University of Kansas Publications, Museum of Natural History 9:125–335.

Baker, R. H. 1969a. Cotton rat of the *Sigmodon fulviventer* group. Pp. 177–232, *in* Contribution in Mammalogy (Jones Jr., J. K., ed.). Miscellaneous Publication, Museum of Natural History, University of Kansas 51:1–428.

Baker, R. H. 1969b. Cotton rats of the *Sigmodon fulviventer* group. University of Kansas, Museum of Natural History, Miscellaneous Publications 51: 177–232.

Baker, R.H. 1983. Michigan mammals. Michigan State University Press. Lasing, U.S.A.

Baker, R. H. 1984a. Origin, classification, and distribution of the white-tailed deer. Pp. 1–18, *in* White-tailed deer: ecology and management (Halls, L. K. ed.), Stackpole Books. Harrisburg, U.S.A.

Baker, R. H., and B. P. Glas. 1951. The taxonomic status of the Pocket Gopher *Geomys bursarius* and *Geomys breviceps*. Proceedings of the Biological Society of Washington 64:55–58.

Baker, R. H., and J. K. Greer. 1962. Mammals of the Mexican state of Durango. Publications of the Museum, Michigan State University, Biological Series 2:25–154.

Baker, R. H., and K. A. Shump. 1977. *Sigmodon fulviventer*. Mammalian Species 94:1–4.

Baker, R. H., and K. A. Shump. 1978. *Sigmodon ochrognathus*. Mammalian Species 97:1–2.

Baker, R. H., R. G. Webb, and P. Dalby. 1967. Notes on reptiles and mammals from southern Zacatecas. The American Midland Naturalist 77:223–226.

Baker, R. H., and H. J. Stains. 1955. A new long-eared *Myotis* (*Myotis evotis*) from northeastern Mexico. University of Kansas Publications, Museum of Natural History 9:81–84.

Baker, R. J. 1979. Karyology. Pp. 107–155, *in* Biology of bats of the New World family Phyllostomatidae Part III (Baker, R. J., J. K. Jones, Jr., and D. C. Carter, eds.). Special Publications of the Museum, Texas Tech University 16:1–441.

Baker, R. J. 1984b. A sympatric cryptic species of mammal: a new species of *Rhogeessa* (Chiroptera: Vespertilionidae). Systematic Zoology 33:178–183.

Baker, R. J., and C. L. Clark. 1987. *Uroderma bilobatum*. Mammalian Species 279:1–4.

Baker, R. J., and H. H. Genoways. 1975. A new subspecies of *Geomys bursarius* (Mammalia: Geomyidae) from Texas and New Mexico. Occasional Papers, The Museum Texas Tech University 29:1–18.

Baker, R. J., and H. H. Genoways. 1978. Zoogeography of Antillean bats. Pp. 53–97, *in* Zoogeography in the Caribbean (Gill, F. B., ed.). Special Publication, Academy of Natural Sciences of Philadelphia 13:1–128.

Baker, R. J., and T. C. Hsu. 1970. Chromosomes of the desert shrew, *Notiosorex crawfordi* (Coues). The Southwestern Naturalist 14:448–449.

Baker, R. J., and J. K. Jones, Jr. 1972. *Tadarida aurispinosa* in Sonora, Mexico. The Southwestern Naturalist 17:308–309.

Baker, R. J., and V. R. McDaniel. 1972. A new subspecies of *Uroderma bilobatum* (Chiroptera, Phyllostomatidae) from Middle America. Occasional Papers, the Museum Texas Tech University 7:1–4.

Baker, R. J., and J. L. Patton. 1967. Karyotypes and karyotypic variation of North American vespertilionid bats. Journal of Mammalogy 41:270–286.

Baker, R. J., and C. Ward. 1967. Distribution of bats in southeastern Arkansas. Journal of Mammalogy 48:130–132.

Baker, R. J., and S. J. Williams. 1974. *Geomys tropicalis*. Mammalian Species 35:1–4.

Baker, R. J., H. H. Genoways, S. J. Williams, and J. W. Warner. 1973. Cytotypes and morphometrics of two phyllostomatid bats, *Micronycteris hirsuta* and *Vampyressa pusilla*. Occasional Papers of the Museum, Texas Tech University 17: 1–10.

Baker, R. J., J. W. Bickham, and M. J. Arnold. 1985. Chromosomal evolution in *Rhogeessa* (Chiroptera: Vespertilionidae): possible speciation by centric fusions. Evolution 39:233–243.

Baker, R. J., S. K. Davis, R. D. Bradley, M. J. Hamilton, and R. A. Van Den Busseh. 1989a. Ribosomal-DNA, mitochondrial-DNA, chromosomal, and allozymic studies on a contact zone in the pocket gopher, *Geomys*. Evolution 43:63–75.

Baker, R. J., C. S. Hood, and R. L. Honeycutt. 1989b. Phylogenetic relationships and classification of the higher categories of the New World bat family Phyllostomidae. Systematic Zoology 38:228–238.

Baker, R. J., C. A. Porter, J. C. Patton, and R. A. Van Den Bussche. 2000. Systematics of bats of the family Phyllostomidae based on RAG2 DNA sequences. Occasional Papers, The Museum of Texas Tech University 202:1–16.

Baker, R. J., S. Solari, and F. G. Hoffmann. 2002. A new Central American species from the *Carollia brevicauda* complex. Occasional Papers of the Museum, Texas Tech University 217:1–12.

Baker, R. J., L. C. Bradley, R. D. Bradley, J. W. Dragoo, M. D. Engstrom, R. S. Hoffman, C. A. Jones, F. Reid, D. W. Rice, and C. Jones. 2003a. Revised checklist of North American mammals north of Mexico, 2003. Occasional Papers, Museum of Texas Tech University 229:1–23.

Baker, R. J., M. B. O'Neill, and L. R. McAliley. 2003b. A new species of desert shrew, *Notiosorex*, based on nuclear and mitochondrial sequence data. Occasional Papers of the Museum, Texas Tech University 222:1–12.

Baker, R. J., S. R. Hoofer, C. A. Porter, and R. A. Van Den Bussche. 2003c. Diversification among the New World leaf–nosed bats: an evolutionary hypothesis and classification inferred from digenomic congruence of DNA sequence. Occasional Papers, The Museum of Texas Tech University 230:1–32.

Baker, R. J., M. M. Mcdonough, V. J. Swier, P. A. Larsen, J. P. Carrera, and L. K. Ammerman. 2009. New species of bonneted bat, genus *Eumops* (Chiroptera: Molossidae) from lowlands of western Ecuador and Peru. Acta Chiropterologica 11:1–13.

Banbury, J. L., and G. S. Spicer. 2007. Molecular systematics of chipmunks (*Neotamias*) inferred by mitochondrial control region sequences. Journal of Mammalian Evolution 14:149–162.

Bangs, O. 1896. A review of the squirrels of eastern North America. Proccedings of the Biological Society of Washington 10:145–167.

Banks, R. C. 1964. The mammals of Cerralvo Island, Baja California. Transactions of the San Diego Society of Natural History 13:397–404.

Banfield, A. W. F. 1961. Notes on mammals of the Kluane Game Santuary. Yukon Territory. Natural Museum Canada Bulletin 172:128–135.

Banfield, A. W. F. 1974. The mammals of Canada. University of Toronto Press. Toronto, Canada.

Banfield, A. W. F. 1987. The mammals of Canada. University of Toronto Press. Ontario, Canada.

Banfield, A. W. F. 1962a. A revision of the reindeer and Caribou, genus *Rangifer*. Bulletin Natural Museum Canada 177, Biological series 66:1–137.

Banfield, A. W. F. 1962b. Notes on the mammals of Pelee Island, Ontario. Bulletin Natural Museum Canada 183:118–122.

Bannikova, A. A., V. S. Lebedev, A. A. Lissovsky, V. Matrosova, N. I. Abramson, E. V. Obolenskaya, and A. S. Tesakov. 2010. Molecular phylogeny and evolution of the Asian lineage of vole genus *Microtus* (Rodentia: Arvicolinae) inferred from mitochondrial cytochrome b sequence. Biological Journal of the Linnean Society 99:595–613.

Barbee, R. W., and C. M. Fugler. 1977. Varation in the three species of the chiropteran genus *Carollia* of north-western Amazonia. Elisha Mitchell Science Society Journal 93:101.

Barbour, R. W., and W. H. Davis. 1969. Bats of America. University Press of Kentucky. Lexington, U.S.A.

Barbour, R. W., and W. H. Davis. 1970. The status of *Myotis occultus*. Journal of Mammalogy 51:150–151.

Barbour, R. W., and W. H. Davis. 1974. Mammals of Kentucky. University Press of Kentucky. Lexington, U.S.A.

Barkalow, F. S., Jr., and M. Shorten. 1973. The world of the gray squirrel. J. B. Lippincott Company. Philadelphia, U.S.A.

Barnes, L. G., D. P. Domning, and C. E. Ray. 1985. Status of studies on fossil marine mammals. Marine Mammal Science 1:15–53.

Barquez, R., and M. Diaz. 2015. *Eumops perotis*. The IUCN Red List of Threatened Species 2015:e.T8247A97207171.

Barquez, R., and M. Diaz. 2016a. *Lasiurus ega*. The IUCN Red List of Threatened Species 2016:e.T11350A22119259.

Barquez, R., and M. Diaz. 2016b. *Myotis keaysi*. The IUCN Red List of Threatened Species 2016:e.T14170A22056048.

Barquez, R., B. Rodriguez, B. Miller, and M. Diaz. 2015a. *Eumops auripendulus*. The IUCN Red List of Threatened Species 2015:e.T8241A97206888.

Barquez, R., S. Perez, B. Miller, and M. Diaz. 2015b. *Artibeus lituratus*. The IUCN Red List of Threatened Species 2015:e.T2136A21995720.

Barquez, R., S. Perez, B. Miller, and M. Diaz. 2015c. *Carollia perspicillata*. The IUCN Red List of Threatened Species 2015:e.T3905A22133716.

Barquez, R., S. Perez, B. Miller, and M. Diaz. 2015d. *Chrotopterus auritus*. The IUCN Red List of Threatened Species 2015:e.T4811A22042605.

Barquez, R., S. Perez, B. Miller, and M. Diaz. 2015e. *Desmodus rotundus*. The IUCN Red List of Threatened Species 2015:e.T6510A21979045.

Barquez, R., S. Perez, B. Miller, and M. Diaz. 2015f. *Diaemus youngi*. The IUCN Red List of Threatened Species 2015:e.T6520A21982777.

Barquez, R., S. Perez, B. Miller, and M. Diaz. 2015g. *Glossophaga soricina*. The IUCN Red List of Threatened Species 2015:e.T9277A22107768.

Barquez, R., S. Perez, B. Miller, and M. Diaz. 2015h. *Noctilio albiventris*. The IUCN Red List of Threatened Species 2015:e.T14829A22019978.

Barquez, R., S. Perez, B. Miller, and M. Diaz. 2015i. *Noctilio leporinus*. The IUCN Red List of Threatened Species 2015:e.T14830A22019554.

Barquez, R., S. Perez, B. Miller, and M. Diaz. 2015j. *Phyllostomus discolor*. The IUCN Red List of Threatened Species 2015:e.T17216A22136476.

Barquez, R., B. Rodriguez, B. Miller, and M. Diaz. 2015k. *Molossus molossus*. The IUCN Red List of Threatened Species 2015:e.T13648A22106602.

Barquez, R., B. Rodriguez, B. Miller, and M. Diaz. 2015l. *Molossus rufus*. The IUCN Red List of Threatened Species 2015:e.T13644A22107969.

Barquez, R., B. Rodriguez, B. Miller, and M. Diaz. 2015m. *Nyctinomops laticaudatus*. The IUCN Red List of Threatened Species 2015:e.T14995A22011208.

Barquez, R., M. Diaz, and J. Arroyo-Cabrales. 2015n. *Nyctinomops macrotis* (errata version published in 2016). The IUCN Red List of Threatened Species 2015:e.T14996A97207443.

Barquez, R., M. Diaz, E. Gonzalez, A. Rodriguez, S. Incháustegui, and J. Arroyo-Cabrales. 2015o. *Tadarida brasiliensis*. The IUCN Red List of Threatened Species 2015:e.T21314A22121621.

Barquez, R., S. Perez, B. Miller, and M. Diaz. 2016a. *Eptesicus brasiliensis*. The IUCN Red List of Threatened Species 2016:e.T7916A22114459.

Barquez, R., S. Perez, B. Miller, and M. Diaz. 2016b. *Eptesicus furinalis*. The IUCN Red List of Threatened Species 2016:e.T7927A22118013.

Barquez, R., M. Diaz, R. Samudio Jr, and J. Arroyo-Cabrales. 2016c. *Myotis albescens*. The IUCN Red List of Threatened Species 2016:e.T14140A22049892.

Barquez, R. M., N. P. Giannini, and M. A. Mares. 1993. Guide to the bats of Argentina. Oklahoma Museum of Natural History, University of Oklahoma, Norman, U.S.A.

Barquez, R., B. Lim, B. Rodriguez, B. Miller, and M. Diaz. 2015. *Peropteryx macrotis*. The IUCN Red List of Threatened Species 2015:e.T16709A22101100.

Bateman, G. C., and T. A. Vaughan. 1974. Night activities of mormoopid bats. Journal of Mammalogy 55:45–65.

Barthelmes, E. L. 2016. Family Eretizontidae (New World Porcupines). Pp. 372–397, in Handbook of the mammals of the world 6. Lagomorpha and Rodentia I (Wilson, D. E., T. E. Lacher, Jr., and R. A. Mittermeier, eds.). Lynx editions. Barcelona, Spain.

Bartholomew, G. A., and R. A. Boolootlan. 1960. Numbers and population structure of pinnipeds on the California Chanel Islands. Journal of Mammalogy 41:366–375.

Barragán, F., C. Lorenzo, A. Morón, M. A. Briones-Salas, and S. López. 2010. Bat and rodent diversity in a fragmented landscape on the Isthmus of Tehuantepec, Oaxaca, México. Tropical Conservation Science 3:1–16.

Barry, R. 2018. Sylvilagus obscurus. Pp. 149–152, in Pikas, Rabbits and Hares of the World (Smith, A., C. H. Johnston, P. C. Alves, and K. Hackländer, eds.). The Johns Hopkins University Press/IUCN/SSC. Baltimore, U.S.A.

Barry, R., and H. C. Lanier. 2019. *Sylvilagus obscurus*. The IUCN Red List of Threatened Species 2019:e.T41301A45192437.

Bartels, M. A., and P. D. Thompson. 1993. *Spermophilus lateralis*. Mammalian Species 440:1–8.

Bartig, J. L., T. L. Best, and S. L. Burt. 1993. *Tamias bulleri*. Mammalian Species 438:1–4.

Batzli, G. O. 1999. Brown lemming/*Lemmus sibiricus*. Pp. 653–654, *in* The Smithsonian Book of North American Mammals (Wilson, D. E., and S. Ruff, eds.). Smithsonian Institution Press. Washington, U.S.A.

Batzli, G. O., and S. T. Sobaski. 1980. Distribution, abundance, and foraging patterns of ground squirrels near Atkasook, Alaska. Arctic and Alpine Research 12:501–510.

Baumgardner, G. D., K. T. Wilkins, and D. J. Schmidly. 1977. Noteworthy additions to the bat fauna of the Mexican states of Tamaulipas (San Carlos Mountains) and Queretaro. Mammalia 41:237–238.

Baumgardner, G. D. 1991. *Dipodomys compactus*. Mammalian Species 369:1–4.

Baumgardner, G. D., and D. J. Schmidly. 1981. Systematics of the southern races of two species of kangaroo rats (*Dipodomys compactus* and *D. ordii*). Occasional Papers, The Museum, Texas Tech University 73:1–27.

Baumgartner, L. L. 1943. Pelage studies of fox squirrels (*Sciurus niger rufiventer*). The American Midland Naturalist 29:588–590.

Beauchamp-Martin, S. L., F. B. Stangl, Jr., D. J. Schmidly, R. D. Stevens, and R. D. Bradley. 2019. Systematic review of Botta's pocket gopher (*Thomomys bottae*) from Texas and southeastern New Mexico, with description of a new taxon. Pp. 515–542, *in* From field to laboratory: a memorial volume in honor of Robert J. Baker (Bradley, R. D., H. H. Genoways, D. J. Schmidly, and L. C. Bradley, eds.). Special Publications, Museum of Texas Tech University. Lubbock, U.S.A.

Bee, J. W., and E. R. Hall. 1956. Mammals of northern Alaska on the Arctic Slope. Miscelaneous Publications Museum Natural History, University of Kansas 8:1–309.

Beer, J. R. 1965. The interparietal in kangaroo rats. The Southwestern Naturalist 10:145–150.

Beever, E. A. 2018. *Lepus othus*. Pp. 204–205, *in* Pikas, Rabbits and Hares of the World (Smith, A., C. H. Johnston, P. C. Alves, and K. Hackländer, eds.). The Johns Hopkins University Press/IUCN/SSC. Baltimore, U.S.A.

Beever, E. A., and J. French. 2018. *Sylvilagus nuttallii*. Pp. 147–149, *in* Pikas, Rabbits and Hares of the World (Smith, A., C. H. Johnston, P. C. Alves, and K. Hackländer, eds.). The Johns Hopkins University Press/IUCN/SSC. Baltimore, U.S.A.

Beever, E. A., D. E. Brown, and C. Lorenzo. 2018a. *Lepus californicus*. Pp. 170–173, *in* Pikas, Rabbits and Hares of the World (Smith, A., C. H. Johnston, P. C. Alves, and K. Hackländer, eds.). The Johns Hopkins University Press/IUCN/SSC. Baltimore, U.S.A.

Beever, E. A., D. E. Brown, and J. Berger. 2018b. *Lepus townsendii*. Pp. 218–220, *in* Pikas, Rabbits and Hares of the World (Smith, A., C. H. Johnston, P. C. Alves, and K. Hackländer, eds.). The Johns Hopkins University Press/IUCN/SSC. Baltimore, U.S.A.

Bekoff, M. 1977. *Canis latrans*. Mammalian Species 79:1–9.

Belant, J., D. Biggins, D. Garelle, R. G. Griebel, and J. P. Hughes. 2015. *Mustela nigripes*. The IUCN Red List of Threatened Species 2015:e.T14020A45200314.

Belcher, R. L., and T. E. Lee, Jr. 2002. *Arctocephalus townsendi*. Mammalian Species 700:1–5.

Belk, M. C., and H. D. Smith. 1991. *Ammospermophilus leucurus*. Mammalian Species 368:1–8.

Bell, D. M., M. J. Hamilton, C. W. Edwards, L. E. Wiggins, R. Muñiz Martínez, R. E. Strauss, R. D. Bradley, and R. J. Baker. 2001. Patterns of karyotypic megaevolution in *Reithrodontomys*: evidence from a cytochrome-*b* phylogenetic hypothesis. Journal of Mammalogy 82:81–91.

Bello-Gutiérrez, J., R. Reyna-Hurtado, and W. Jorge. 2010. Central American red brocket deer *Mazama temama* (Kerr 1792). Pp. 166– 171, *in* Neotropical Cervidology: Biology and Medicine of Latin American Deer (Duarte, J. M. B., and S. González, eds.). Funep/IUCN. Jaboticabal, Brazil and Gland, Switzerland.

Bello, J., R. Reyna, and J. Schipper. 2016. *Mazama temama*. The IUCN Red List of Threatened Species 2016:e.T136290A22164644.

Benedict, F. A. 1957. Hair structure as a generic character in bats. University of California Publications in Zoology, 59:285–547.

Benedict, R. A., J. D. Druecker, and H. H. Genoways. 1999. New records and habitat information for *Sorex merriami* in Nebraska. Great Basin Naturalist 59:285–287.

Benedict, R. A., H. H. Genoways, and J. R. Choate. 2006. Taxonomy of the short-tailed shrews (genus *Blarina*) in Florida. Occasional Papers, Museum of Texas Tech University 251:1–19.

Benedict, B. D., A. A. Castellanos, and J. E. Light. 2019. Phylogeographic assessment of the Heermann's kangaroo rat (*Dipodomys heermanni*). Journal of Mammalogy 100:72–91.

Beneski, Jr., J. T., and D. W. Stinson. 1987. *Sorex palustris*. Mammalian Species 296:1–6.

Benson, D. L., and F. R. Gehlbach. 1979. Ecological and taxonomical notes on the rice rat (*Oryzomys couesi*) in Texas. Journal of Mammalogy 60:225–228.

Benson, S. B. 1933. Concealing coloration among some desert rodents of the southwestern United States. University of California Publications in Zoology 40:1–7.

Benson, B. B. 1939. Descriptions and records of harvest mice (Genus *Reithrodontomys*) from Mexico. Proceedings of the Biological Society of Washington 52:147–150.
Benson, S. B. 1940. New subspecies of the canyon mouse (*Peromyscus crinitus*) from Sonora, Mexico. Proceedings Biological Society of Washington 53:1–4.
Benson, S. B. 1947. Description of a mastiff bat (genus *Eumops*) from Sonora, Mexico. Proceedings of the Biological Society of Washington 60:133–134.
Benson, S. B., and R. M. Bond. 1939. Notes on *Sorex merriami* Dobson. Journal of Mammalogy 20:348–351.
Bertram, G. C. L., and C. K. R. Bertram. 1964. Manatees in the Guianas. Zoologica 49:115–120.
Berta, A., and M. Churchill. 2011. Pinniped taxonomy: review of currently recognized species and subspecies, and evidence used for their description. Mammal Review 42:207–234.
Best, T. L. 1978. Variation in kangaroo rats (genus *Dipodomys*) of the *heermanni* group in Baja California, Mexico. Journal of Mammalogy 59:160–175.
Best, T. L. 1981. Bacular variation in kangaroo rats (genus *Dipodomys*) of the *heermanni* group in Baja California, Mexico. The Southwestern Naturalist 25:529–534.
Best, T. L. 1982. Relationships between ecogeographic and morphologic variation of the agile kangaroo rat (*Dipodomys agilis*) in Baja California, Mexico. Bulletin of the Southern California Academy of Sciences 80:60–69.
Best, T. L. 1983a. Morphologic variation in the San Quintin kangaroo rat (*Dipodomys gravipes* Huey 1925). The American Midland Naturalist 109:409–413.
Best, T. L. 1983b. Intraspecific variation in the agile kangaroo rat *(Dipodomys agilis)*. Journal of Mammalogy 64:426–436.
Best, T. L. 1986. *Dipodomys elephantinus*. Mammalian Species 255:1–4.
Best, T. L. 1987. Sexual dimorphism and morphometric variation in the Texas kangaroo rat (*Dipodomys elator* Merriam 1894). The Southwestern Naturalist 32:53–59.
Best, T. L. 1988a. *Dipodomys nelsoni*. Mammalian Species 326:1–4.
Best, T. L. 1988b. *Dipodomys spectabilis*. Mammalian Species 311:1–10.
Best, T. L. 1991. *Dipodomys nitratoides*. Mammalian Species 381:1–7.
Best, T. L. 1992a. *Dipodomys margaritae*. Mammalian Species 400:1–3.
Best, T. L. 1992b. *Dipodomys venustus*. Mammalian Species 403:1–4.
Best, T. L. 1993a. *Tamias palmeri*. Mammalian Species 443:1–6.
Best, T. L. 1993b. *Tamias sonomae*. Mammalian Species 444:1–5.
Best, T. L. 1993c. *Perognathus inornatus*. Mammalian Species 450:1–5.
Best, T. L. 1993d. *Chaetodipus lineatus*. Mammalian Species 451:1–3.
Best, T. L. 1993e. *Tamias ruficaudus*. Mammalian Species 452:1–7.
Best, T. L. 1993f. Patterns of morphologic and morphometric variation in heteromyid rodents. Pp. 197–235, *in* Biology of the Heteromyidae (Genoways, H. H., and J. H. Brown, eds.). Special Publication, The American Society of Mammalogists 10:1–719.
Best, T. L. 1994a. *Perognathus alticolus*. Mammalian Species 463:1–4.
Best, T. L. 1994b. *Chaetodipus nelsoni*. Mammalian Species 484:1–6.
Best, T. L. 1995a. *Sciurus variegatoides*. Mammalian Species 500:1–6.
Best, T. L. 1995b. *Spermophilus mohavensis*. Mammalian Species 509:1–7.
Best, T. L. 1995c. *Sciurus oculatus*. Mammalian Species 498:1–3.
Best, T. L. 1995d. *Sciurus nayaritensis*. Mammalian Species 492:1–5.
Best, T. L. 1995e. *Sciurus deppei*. Mammalian Species 505:1–5.
Best, T. L. 1995f. *Sciurus colliaei*. Mammalian Species 497:1–4.
Best, T. L. 1995g. *Sciurus alleni*. Mammalian Species 501:1–4.
Best, T. L. 1995h. *Spermophilus annulatus*. Mammalian Species 508:1–4.
Best, T. L. 1995i. *Spermophilus adocetus*. Mammalian Species 504, 20:1–4.
Best, T. L. 1996. *Lepus californicus*. Mammalian Species 530:1–10.
Best, T. L., and G. Ceballos. 1995. *Spermophilus perotensis*. Mammalian Species 507:1–3.
Best, T. L., and N. J. Granai. 1994a. *Tamias obscurus*. Mammalian Species 472:1–6.
Best, T. L., and N. J. Granai. 1994b. *Tamias merriami*. Mammalian Species 476:1–9.
Best, T. L., and T. H. Henry. 1993. *Lepus alleni*. Mammalian Species 424:1–8.
Best, T. L., and T. H. Henry. 1994a. *Lepus arcticus*. Mammalian Species 457:1–9.
Best, T. L., and T. H. Henry. 1994b. *Lepus othus*. Mammalian Species 458:1–5.
Best, T. L., and J. A. Lackey. 1985. *Dipodomys gravipes*. Mammalian Species 236:1–4.
Best, T. L., and J. A. Lackey. 1992a. *Chaetodipus artus*. Mammalian Species 418:1–3.
Best, T. L., and J. A. Lackey. 1992b. *Chaetodipus pernix*. Mammalian Species 420:1–3.
Best, T. L., and J. B. Jennings. 1997. *Myotis leibii*. Mammalian Species 547:1–6.
Best, T. L., and S. Riedel. 1995. *Sciurus arizonensis*. Mammalian Species 496:1–5.
Best, T. L., and G. D. Schnell. 1974. Bacular variation kangaroo rats (genus *Dipodomys*). The American Midland Naturalist 91:257–270.
Best, T. L., and M. P. Skupski. 1994a. *Perognathus flavus*. Mammalian Species 471:1–10.
Best, T. L., and M. P. Skupski. 1994b. *Perognathus merriami*. Mammalian Species 473:1–7.
Best, T. L. and H. H. Thomas. 1991a. *Dipodomys insularis*. Mammalian Species 374:1–3.
Best, T. L., and H. H. Thomas. 1991b. *Spermophilus madrensis*. Mammalian Species 378:1–2.
Best, T. L., R. M. Sullivan, J. A. Cook, and T. Yates. 1986. Chromosomal, genetic and morphologic variation in the agile kangaroo rat, *Dipodomys agilis* (Rodentia: Heteromyidae). Systematic Zoology 35:311–324.
Best, T. L., C. L. Lewis, K. Caesar, and A. S. Titus. 1990a. *Ammospermophilus interpres*. Mammalian Species 365:1–6.
Best, T. L., A. S. Titus, K. Caesar, and C. L. Lewis. 1990b. *Ammospermophilus harrisii*. Mammalian Species 366:1–7.

Best, T. L., A. S. Titus, C. L. Lewis, and K. Caesar. 1990c. *Ammospermophilus nelsoni*. Mammalian Species 367:1–7.
Best, T. L., J. L. Bartig, and S. L. Burt. 1992. *Tamias canipes*. Mammalian Species 411:1–5.
Best, T. L., S. L. Burt, and J. L. Bartig. 1993. *Tamias durangae*. Mammalian Species 437:1–4.
Best, T. L., S. L. Burt, and J. L. Bartig. 1994a. *Tamias quadrivittatus*. Mammalian Species 466:1–7.
Best, T. L., R. G. Clawson, and J. A. Clawson. 1994b. *Tamias speciosus*. Mammalian Species 478:1–9.
Best, T. L., R. G. Clawson, and J. A. Clawson. 1994c. *Tamias panamintinus*. Mammalian Species 468:1–7.
Best, T. L., R. K. Chesser, D. A. McCullough, and G. D. Baumgardner. 1996a. Genic and Morphometric Variation in Kangaroo Rats, Genus *Dipodomys*, from Coastal California. Journal of Mammalogy 77:785–800.
Best, T. L., W. M. Kiser, and P. W. Freeman. 1996b. *Eumops perotis*. Mammalian Species 534:1–8.
Best, T. L., W. M. Kiser, and J. C. Rainey. 1997. *Eumops glaucinus*. Mammalian Species 551:1–6.
Best, T. L., J. L. Hunt, L. A. McWilliams, and K. G. Smith. 2001. *Eumops hansae*. Mammalian Species 687:1–3.
Best, T. L., J. L. Hunt, L. A. Williams, and K. G. Smith. 2002. *Eumops auripendulus*. Mammalian Species 708:1–5.
Bhatnagar, K. P. 1978. Breech presentation in the hairy-legged vampire, *Diphylla ecaudata*. Journal of Mammalogy 59:864–866.
Bininda-Emonds, O. R. P., J. L. Gittleman, and A. Purvis. 1999. Building large trees by combining phylogenetic information: A complete phylogeny of the extant Carnivora (Mammalia). Biological Reviews 74:143–175.
Birkenholz, D. E. 1972. *Neofiber alleni*. Mammalian Species 15:1–4.
Birney, E. C., and H. H. Genoways. 1973. Chromosomes of *Spermophilus adocetus* (Mammalia; Sciuridae), with comments on the subgenetic affinities of the species. Experimentia 29:228–229.
Birney, E. C., and J. K. Jones, Jr. 1972. Woodrats (genus *Neotoma*) of Sinaloa, Mexico. Transactions of the Kansas, Academy of Science 74:197–211.
Birney, E. C. 1973. Systematics of three species of woodrats (genus *Neotoma*) in central North America. Miscellaneous Publications of the Museum of Natural History, University of Kansas 58:1–173.
Birney, E. C., J. B. Bowles., R. M. Timm, and S. L. Williams. 1974. Mammalian distributional records in Yucatán and Quintana Roo, with comments on reproduction, structure, and status of peninsular populations. Bell Museum Natural History, University of Minnesota, Occasional Papers 13:1–25.
Bisbal, E. F. J. 1991. Distribución y taxonomía del venado Matacán (*Mazama* sp) en Venezuela. Acta Biológica Venezuelica 13:89–104.
Blair, W. F. 1938. Two new Pocket-mice from Western Texas. Occasional Papers of the Museum of Zoology, University of Michigan 381:1–3.
Blair, W. F. 1940. Two cases of abnormal coloration in mammals. Journal of Mammalogy 21:461–462.
Blair, W. F. 1954. Mammals of the mesquite plains biotic district in Texas and Oklahoma, and speciation in the central grasslands. Texas Journal of Science 6:235–264.
Blair, W. F., A. P. Blair, P. Brodkorb, F. R. Cagle, and G. A. Moore. 1968. Vertebrates of the United States. Second ed. McGraw-Hill Publishing Co., New York.
Blair, F. W., A. P. Blair, P. Brodkorb, F. R. Cagle, and G. A. Moore. 1958. Vertebrates of the United States. McGraw-Hill. New York, U.S.A.
Blanco, J. C. 1998. Mamíferos de España. Planeta. Madrid, España.
Blankenship, D. J., and G. L. Bradley. 1985. Electrophoretic comparison of two southern California chipmunks (*Tamias obscurus* and *Tamias merriami*). Bulletin of the Southern California Academy of Sciences 84:48–50.
Bleich, V. C. 1977. *Dipodomys stephensi*. Mammalian Species 73:1–3.
Block, S. B., and E. G. Zimmerman. 1991. Allozymic variation and systematics of plains pocket gophers (*Geomys*) of south-central Texas. The Southwestern Naturalist 36:29–36.
Blois, J. and NatureServe (Clausen, M. K., G. Hammerson, and S. Cannings). 2008. *Arborimus pomo*. The IUCN Red List of Threatened Species 2008:e.T2018A9174950.
Blood, B. R., and M. K. Clark. 1998. *Myotis vivesi*. Mammalian Species 588:1–5.
Bobrinskii, N. A., B. A. Kuznetsov, and A. P. Kuzyakin. 1944. Opredelitl' mlekopitayushchikh SSSR [Guide to the mammals of the U.S.S.R.]. Sovietskaya Nauka, Moscow, Russia.
Bock, A. 2020. *Lepus europaeus* (Lagomorpha: Leporidae), Mammalian Species 52:125–142.
Bogan, M. A. 1974. Identification of *Myotis californicus* and *M. leibii* in southwestern North America. Proceedings of the Biological Society of Washington 87:49–56.
Bogan, M. A. 1978. A new species of *Myotis* from the Islas Tres Marias, Nayarit, Mexico, with comments on variation in *Myotis nigricans*. Journal of Mammalogy 59:519–530.
Bogan, M. A. 1997. On the status of *Neotoma varia* from Isla Datil, Sonora. Pp. 81–88, *in* Life among the muses: papers in honor of James S. Findley (Yates, T. L., W. L. Gannon, and D. Wilson, eds.). Special Publication, the Museum of Southwestern Biology 3:1–290.
Bogan, M. A. 1999. Family Vespertilionidae. Pp. 139– 181, *in* Mamíferos del noroeste de México (Álvarez-Castañeda, S. T., and J. L. Patton, eds.). Centro de Investigaciones Biológicas del Noroeste, S. C., La Paz, México.
Boitani, L., M. Phillips, and Y. Jhala. 2018. *Canis lupus* (errata version published in 2020). The IUCN Red List of Threatened Species 2018:e. T3746A163508960.
Bogdanowicz, W., S. Kasper, and R. D. Owen. 1998. Phylogeny of Plecotine bats: reevaluation of morphological and chromosomal data. Journal of Mammalogy 79:78–90.
Bole, B. P., Jr., and P. N. Moulthrop. 1942. The Ohio Recent mammal collection in the Cleveland Museum of Natural History. Science Publications, Cleveland Museum Natural History 5:83–181.
Bonvicino, C. R., F. C. Almeida, and R. Cerqueira. 2000. The karyotype of *Sphiggurus villosus* (Rodentia: Erethizontidae). Studies on Neotropical Fauna and Environment 35:81–83.
Bonvicino, C. R., V. Penna-Firme, and E. Braggio. 2002. Molecular and Karyologic Evidence of the Taxonomic Status of *Coendou* and *Sphiggurus* (Rodentia: Hystricognathi). Journal of Mammalogy 83:1071–1076.
Bonner, W. N. 1981. Grey seal *Halichoerus grypus* Fabricius. 1791. Pp. 111–144, *in* Handbook of marine mammals (Ridgway, S. H., and R. Harrison, eds). Academic Press. London, United Kingdom.
Borell, A. E., and M. D. Bryant. 1942. Mammals of the Big Bend area of Texas. University of Californian Publications in Zoology 48:1–61.
Borowik, O. A., and M. D. Engstrom. 1993. Chromosomal evolution and biogeography of collared lemmings (*Dicrostonyx*) in the eastern and High Arctic of Canada. Canadian Journal of Zoology 71:1481–1493.

Boskovic, R., K. M. Kovacs, M. O. Hammill, and B. N. White. 1996. Geographic distribution of mitochondrial DNA haplotypes in grey seals (*Halichoerus grypus*). Canadian Journal of Zoology 74:1787–1796.

Boulware, J. T. 1943. Two new subspecies of kangaroo rats (genus *Dipodomys*) from southern California. University of California Publications in Zoology 46:391–396.

Boveng, P. 2016. *Phoca largha*. The IUCN Red List of Threatened Species 2016:e.T17023A45229806.

Bowen, D. 2016. *Halichoerus grypus*. The IUCN Red List of Threatened Species 2016:e.T9660A45226042.

Bowles, J. B. 1975. Distribution and biogeography of mammals of lowa. Special Publication Museum of Texas Tech University 9:1–184.

Bowyer, R. T., and D. M. Leslie. 1992. *Ovis dalli*. Mammalian Species 393:1–7.

Boyce, M. S. 1978. Climatic variability and body size variation in the muskrats (*Ondatra zibethicus*) of North America. Oecologia 36:1–19.

Boyce, M. S. 1980. First record of the fringe-tailed bat, *Myotis thysanodes*, from southeastern Wyoming. The Southwestern Naturalist 25:114–115.

Boyeskorov, G. 1999. New data on moose (*Alces*, Artiodactyla) Systematic. Säugetierkundliche Mitteilungen 44:3–13.

Bradbury, J. W., and S. L. Vehrencamp. 1976. Social organization and foraging in emballonurid bats: I. Field studies. Behavioral Ecology and Sociobiology 1:337–381.

Bradley, R. D. 2017. Genus *Reithrodontomys*. Pp. 367–383, *in* Handbook of the mammals of the world: rodents II (Wilson, D. E., T. E. Lacher, and R. A. Mittermeier, eds.). Lynx Edicions, Barcelona, Spain.

Bradley, R. D., and R. J. Baker. 1999. *Geomys knoxjonesi*. Pp. 486–487, *in* The Smithsonian Book of North American Mammals (Wilson, D. E., and S. Ruff, eds.). Smithsonian Institution Press. Washington, U.S.A.

Bradley, R. D., and M. R. Mauldin. 2016. Molecular data indicate a cryptic species in *Neotoma albigula* (Cricetidae: Neotominae) from northwestern Mexico. Journal of Mammalogy 97:187–199.

Bradley, R. D., and D. J. Schmidly. 1987. The glans penes and bacula in Latin American taxa of the *Peromyscus boylii* group. Journal of Mammalogy 68:595–616.

Bradley, R. D., D. J. Schmidly, and R. D. Owen. 1989. Variation in the glans penes and bacula among latin american populations of the *Peromyscus boylii* species complex. Journal of Mammalogy 70:712–725.

Bradley, R. D., D. J. Schmidly, and C. W. Kilpatrick. 1996. The relationships of *Peromyscus sagax* to the *P. boylii* and *P. truei* species groups in Mexico based on morphometric, karyotipic, and allozymic data. Pp. 95–106, *in* Contributions in Mammalogy: a memorial volume honoring Dr. J. K. Jones, Jr. (Genoways, H. H., and R. J. Baker, eds.). Museum Texas Tech University. Lubbock, U.S.A.

Bradley, R. D., D. J. Schmidley, and C. Jones. 1999. The northern rock mouse, *Peromyscus nasutus* (Mammalia: Rodentia), from Davis Mountain Texas. Occasional papers, Museum of Texas Tech University 190:1–3.

Bradley, R. D., I. Tiemann-Boege, C. W. Kilpatrick, and D. J. Schmidly. 2000. Taxonomic status of *Peromyscus boylii sacarensis*: interferences from DNA sequences of the mitochondrial cytochrome–*b* gene. Journal of Mammalogy 81:875–884.

Bradley, R. D., D. S. Carroll, M. L. Haynie, R. Muñiz-Martínez, M. J. Hamilton, and W. L. Kilpatrick. 2004a. A new species of *Peromyscus* from western Mexico. Journal of Mammalogy 85:1184–1193.

Bradley, R. D., F. Mendez-Harclerode, M. J. Hamilton, and G. Ceballos. 2004b. A new species of *Reithrodontomys* from Guerrero, Mexico. Occasional Papers of the Museum of Texas Tech University 231:1–12.

Bradley, R. D., N. D. Durish, D. S. Rogers, J. R. Miller, M. D. Engstrom, and C. W. Kilpatrick. 2007. Toward a molecular phylogeny for *Peromyscus*: evidence from mitochondrial cytochrome–*b* sequences. Journal of Mammalogy 88:1146–1159.

Bradley, R. D., L. K. Ammerman, R. J. Baker, L. C. Bradley, J. A. Cook, R. C. Dowler, C. Jones, D. J. Schmidly, F. B. Stangl, Jr., R. A. Van Den Bussche, and B. Würsig. 2014a. Revised checklist of North American mammals north of Mexico, 2014. Museum of Texas Tech University 327:1–27.

Bradley, R. D., N. Ordóñez-Garza, C. G. Sotero-Caio, H. M. Huynh, C. W. Kilpatrick, L. I. Iñiguez-Dávalos, and D. J. Schmidly. 2014b. Morphometric, karyotypic, and molecular evidence for a new species of *Peromyscus* (Cricetidae: Neotominae) from Nayarit, Mexico. Journal of Mammalogy 95:176–186.

Bradley, R. D., D. J. Schmidly, B. R. Amman, R. N. Platt II, K. M. Neumann, H. M. Huynh, R. Muñiz-Martínez, C. López-González, and N. Ordóñez-Garza. 2015. Molecular and morphometric data reveal multiple species in *Peromyscus pectoralis*. Journal of Mammalogy 96:446–459.

Bradley, R. D., N. Ordoñez-Garza, G. Ceballos, D. S. Rogers, and D. J. Schmidly. 2016a. A new species in the *Peromyscus boylii* species group (Cricetidae: Neotominae) from Michoacán, México. Journal of Mammalogy 97:154–165.

Bradley, R. D., M. Nuñez-Tabares, T. J. Soniat, S. Kerr, R. W. Raymond, and N. Ordóñez-Garza. 2016b. Molecular systematics and phylogeography of *Peromyscus nudipes* (Cricetidae: Neotominae). Pp. 201–214, *in* Contributions in Natural History: a memorial volume in honor of Clyde Jones (Manning, R. M., J. R. Goetze, and F. D. Yancey, II, eds.). Special Publications, Museum of Texas Tech University 65:1–273.

Bradley, R. D., N. Ordóñez-Garza, and L. C. Bradley. 2017. Oaxacan deermouse *Peromyscus oaxacensis*. Pp. 392, *in* Handbook of the Mammals of the World, Volume 7, Rodents II (Wilson, D. E., T. E. Lacher, Jr., and R. A. Mittermeier, eds.). Lynx Edicions. Barcelona, Spain.

Bradley, R. D., J. Q. Francis, R. N. Platt II , T. J. Soniat, D. Alvarez, and L. L. Lindsey. 2019. Mitochondrial DNA Sequence Data Indicate Evidence for Multiple Species within *Peromyscus maniculatus*. Special Publications of the Museum of Texas Tech University 70:1–59.

Bradley, R. D., N. Ordóñez-Garza, C. W. Thompson, E. A. Wright, G. Ceballos, C. W. Kilpatrick, and D. J. Schmidly. 2022a. Two new species of *Peromyscus* (Cricetidae: Neotominae) from the Transverse Volcanic Belt of Mexico. Journal of Mammalogy 103:255–274.

Bradley, R. D., C. W. Edwards, L. L. Lindsey, J. R. Bateman, M. N. B. Cajimat, M. L. Milazzo, C. F. Fulhorst, M. D. Matocq, and M. R. Mauldin. 2022b. Reevaluation of the phylogenetic relationships among Neotomini rodents (*Hodomys*, *Neotoma*, and *Xenomys*) and comments on the woodrat classification. Journal of Mammalogy 103:1221–1236.

Bradley, W. G., and L. Baker. 1965. Skull measurements of desert bighorn sheep from the Desert Game Range. Transactions of the Desert Bighorn Council 9:70–74.

Bradley, W. G., and L. R. Baker. 1967. Range of variation in Nelson bighorn sheep from the Desert Game Range and its taxonomic significance. Transactions of the Desert Bighorn Council 11:114–140.

Brand, L. R., and R. E. Ryckman. 1969. Biosystematics of *Peromyscus eremicus*, *P. guardia*, and *P. interparietalis*. Journal of Mammalogy 50:501–513.

Brandborg, S. M. 1955. Life history and management of the mountain goat in Idaho. Wildlife Bulletin of the Idaho Department Fish and Game 2:1–142.

Brandon-Jones, D. 2006. Apparent confirmation that *Alouatta villosa* (Gray, 1845) is a senior synonym of *A. pigra* Lawrence, 1933 as the species-group name for the black howler monkey of Belize, Guatemala, and Mexico. Primate Conservation 21:41–43.
Braun, J. K., and M. A. Mares. 1989. *Neotoma micropus*. Mammalian Species 330:1–9.
Braun, J. K., S. B. Gonzalez-Perez, G. M. Street, J. M. Mook, and N. J. Czaplewski. 1988. *Phenacomys ungava* (Rodentia: Cricetidae). Mammalian Species 45:18–29.
Braun, J. K., Q. D. Layman, and M. A. Mares. 2009. *Myotis albescens*. Mammalian Species 846:1–9.
Braun, J. K., A. A. Johnson, and M. A. Mares. 2011a. *Tamias umbrinus* (Rodentia: Sciuridae). Mammalian Species 43:216–227.
Braun, J. K., T. S. Eaton, and M. A. Mares. 2011b. *Marmota caligata* (Rodentia: Sciuridae). Mammalian Species 43:155–171.
Braun, J. K., S. B. Gonzalez-Perez, G. M. Street, J. M. Mook, and N. J. Czaplewski. 2013. *Phenacomys ungava* (Rodentia: Cricetidae). Mammalian Species 45:18–29.
Braun, J. K., B. Yang, S. B. González-Pérez, and M. A. Mares. 2015. *Myotis yumanensis* (Chiroptera: Vespertilionidae). Mammalian Species 47:1–14.
Braun, J. K., S. B. C. Coyner, and M. A. Mares. 2021. Modern extirpation of the Texas kangaroo rat, *Dipodomys elator*, in Oklahoma: changing land use and climate over a century of time as the road to eventual extinction. Therya 12:177–186.
Braun, J. K., S. B. Gonzalez-Perez, G. M. Street, J. M. Mook, and N. J. Czaplewski. 2013. *Phenacomys ungava* (Rodentia: Cricetidae). Mammalian Species 899:18–29.
Briggs, K. T., and V. G. Morejohn. 1976. Dentition, cranial morphology and evolution in elephant seals. Mammalia 40:199–222.
Briones-Salas, M., and V. Sánchez-Cordero. 2004. Mamíferos. Pp. 423–447, *in* Biodiversidad de Oaxaca (García-Mendoza, A. J., M. J. Ordoñez, and M. Briones-Salas eds.). Instituto de Biología, UNAM, Fondo Oaxaqueño para la Conservación de la Naturaleza, World Wildlife Fund. Distrito Federal, México.
Brito, D., D. Astúa, D. Lew, and N. de la Sancha. 2021. *Metachirus nudicaudatus* (amended version of 2015 assessment). The IUCN Red List of Threatened Species 2021:e.T40509A197311536.
Broadbooks, H. E. 1965. Ecology and distribution of the pikas of Washington and Alaska. The American Midland Naturalist 73:299–335.
Brook, S. M., J. Pluháček, R. Lorenzini, S. Lovari, M. Masseti, O. Pereladova, and S. Mattioli. 2018. *Cervus canadensis*. The IUCN Red List of Threatened Species 2018:e.T55997823A142396828.
Brown, D. E. 1984. Arizona's tree squirrels. Arizona Game and Fish Department. Phoenix, U.S.A.
Brown, D. E. 1989. The ocelot. Pp. 421–433, *in* Audubon wildlife report (Chandler, W. J., ed.). Harcourt Brace Jovanovich. San Diego, U.S.A.
Brown, D. E., and A. T. Smith. 2019. *Lepus callotis*. The IUCN Red List of Threatened Species 2019:e.T11792A45177499.
Brown, D. E., R. D. Babb, C. Lorenzo, and M. M. Altemus. 2014. Ecology of the Antelope Jackrabbit (*Lepus alleni*). The Southwestern Naturalist 59:577–589.
Brown, D. E., G. Beatty, J. E. Brown, and A. T. Smith. 2018a. History, status, and population trends of cottontail rabbits and jackrabbits in the western United States. Western Wildlife 5:16–42.
Brown, D. E., C. Lorenzo, and M. Altemus. 2018b. *Lepus alleni*. Pp. 159–163, *in* Pikas, Rabbits and Hares of the World (Smith, A., C. H. Johnston, P. C. Alves, and K. Hackländer, eds.). The Johns Hopkins University Press/IUCN/SSC. Baltimore, U.S.A.
Brown, D. E., C. Lorenzo, and M. B. Traphagen. 2018c. *Lepus callotis*. Pp. 173–176, *in* Pikas, Rabbits and Hares of the World (Smith, A., C. H. Johnston, P. C. Alves, and K. Hackländer, eds.). The Johns Hopkins University Press/IUCN/SSC. Baltimore, U.S.A.
Brown, D. E., C. Lorenzo, and S. T. Álvarez-Castañeda. 2019. *Lepus californicus*. The IUCN Red List of Threatened Species 2019:e.T41276A45186309.
Brown, D. E., M. B. Traphagen, C. Lorenzo, and M. Gomez-Sapiens. 2018d. Distribution, status and conservation needs of the white-sided jackrabbit, *Lepus callotis* (Lagomorpha). Revista Mexicana de Biodiversidad 89:313–324.
Brown, J. H., and G. A. Lieberman. 1973. Resource utilization and coexistence of seed-eating desert rodents in sand dune habitats. Ecology 54:788–797.
Brown, R. H., H. H. Genoways, and J. K. Jones, Jr. 1971. Bacula of some Neotropical bats. Mammalia 35:456–464.
Brown, R. J., and R. L. Rudd. 1981. Chromosomal comparisons within the *Sorex ornatus-S. vagrans* complex. Wassman Journal Biology 39:30–35.
Brunet, A. K., R. M. Zink, K. M. Kramer, R. C. Blackwell-Rago, S. L. Farrell, T. V. Line, and E. C. Birney. 2002. Evidence of introgression between masked shrews (*Sorex cinereus*), and prairie shrews (*Sorex haydeni*), in Minnesota. The American Midland Naturalist 147:116–122.
Brunner, S. 2004. Fur seals and sea lions (Otariidae): identification of species and taxonomic review. Systematics and Biodiversity 1:339–439.
Bryant, W. E. 1945. Phylogeny of Neartic Sciuridae. The American Midland Naturalist 33:257–390.
Bucher, J. E., and S. R. Hoffmann.1980. *Caluromys derbianus*. Mammalian Species 140:1–4.
Buck, C. L., and B. M. Barnes. 1999. Temperatures of hibernacula and changes in body composition of Arctic ground squirrels over winter. Journal of Mammalogy 80:1264–1276.
Buechner, H. K. 1960. The bighorn sheep in the United States, its past, present and future. Wildlife Monographs 4:1–174.
Burn, D. M., and A. M. Doroff. 2005. Decline in sea otter (*Enhydra lutris*) populations along the Alaska Peninsula, 1986–2001. Fishery Bulletin 103:270–279.
Burns, J. C., J. R. Choate, and E. G. Zimmerman. 1985. Systematic relationships of Pocket Gophers (genus *Geomys*) on the central Great Plains. Journal of Mammalogy 66:102–118.
Burns, J. J. 1981a. Bearded seal *Erignathus barbatus* Erxleben, 1777. Pp. 145–170, *in* Handbook of marine mammals (Ridgway, S. H., and R. Harrison, eds.). Academic Press. New York, U.S.A.
Burns, J. J. 1981b. Ribbon seal *Phoca fasciata* Zimmerman, 1783. Pp. 89–109, *in* Handbook of marine mammals (Ridgway, S. H., and R. Harrison, eds). Academic Press. New York, U.S.A.
Burns, J. J., and F. H. Fay. 1972. Comparative morphology of the skull of the ribbon seal, *Histriophoca fasciata*, with remarks on systematics of Phocidae. Journal of Zoology 161:363–394.
Burt, M. S., and R. C. Dowler. 1999. Biochemical systematics of *Geomys breviceps* and two chromosomal races of *Geomys attwateri* in eastern Texas. Journal of Mammalogy 80:799–809.
Burt, S. L., and T. L. Best. 1994. *Tamias rufus*. Mammalian Species 460:1–6.

Burt, W. H. 1932. Descriptions of heretofore unknown mammals from islands in the Gulf of California, Mexico. Transactions of the San Diego Society of Natural History 7:161–182.

Burt, W. H. 1936. A study of the baculum in the genera *Perognathus* and *Dipodomys*. Journal of Mammalogy 17:145–156.

Burt, W. H. 1960. Bacula of North American mammals. Miscellaneous Publications of thc Museum of Zoology, University of Michigan 113:1–76.

Burt, W. H., and F. S. Barkalow. 1942. A comparative study of the bacula of wood rats (subfamily Neotominae). Journal of Mammalogy 23:287–297.

Burt, W. H., and R. P. Grossenheider. 1964. A field guide to the mammals. The Riverside Press, Cambridge, U.S.A.

Burt, W. H., and E. T. Hooper. 1941. Notes on mammals from Sonora and Chihuahua, Mexico. Occasional Papers of the Museum of Zoology, University of Michigan 430:1–7.

Burt, W. H., and R. A. Stirton. 1961. The mammals of El Salvador. Miscelaneous Publication of Museum Zoology, University Michigan 117:1–69.

Burton, K. L., and M. D. Engstrom. 2001. Species diversity of bats (Mammalia: Chiroptera) in Iwokrama Forest, Guyana, and the Guianan subregion: implications for conservation. Biodiversity and Conservation 10:613–657.

Buskirk, S. W., L. F. Ruggiero, K. B. Aubry, K. B. Pearson, J. R. Squires, K. S. Mckelvey, G. M. Koehler, and C. J. Krebs. 2000. Comparative ecology of *Lynx* in North America. Pp. 397–418, *in* Ecology and Conservation of *Lynx* in the United States (Ruggiero, L. F., K. B. Aubry, S. W. Buskirk, G. M. Koehler, C. J. Krebs, K. S. McKelvey, and J. R. Squires, eds). University Press of Colorado. Boulder, U.S.A.

Byrd, G. V., and N. Norvell. 1993. Status of the Pribilof shrew based on summer distribution and habitat use. Northwestern Naturalist 74:49–54.

Cab-Sulub, L., and S. T. Álvarez-Castañeda. 2020. Analysis of the remaining habitat of an endemic species rediscovered. Mammalian Biology 100:307–314.

Cabrera, A. 1919. Genera *Mammalium*. Monotremata. Marsupialia. Museo Nacional de Ciencias Naturales. Madrid, Spain.

Cabrera, H., S. T. Álvarez-Castañeda, N. González-Ruíz, and J. P. Gallo-Reynoso. 2007. Distribution and natural history of Schmidly's Deermouse (*Peromyscus schmidlyi*). The Southwestern Naturalist 52:620–623.

Cáceres, M. C., and R. M. R. Barclay. 2000. *Myotis septentrionalis*. Mammalian Species 634:1–4.

Cáceres, N. C., and A. P. Carmignotto. 2006. *Caluromys lanatus*, Mammalian Species 803:1–6.

Caire, W. 1976. Phenetic relationships of pocket mice in the subgenus *Chaetodipus* (Rodentia: Heteromyidae). Journal of Mammalogy 57:375–378.

Caire, W. 1978. The distribution and zoogeography of the mammals of Sonora, Mexico. Ph.D. dissertation. The University of New Mexico. Albuquerque, U.S.A.

Caire, W., R. K. Laval, M. L. Laval, and R. Clawson. 1979. Notes on the ecology of *Myotis keenii* (Chiroptera, Vespertilionidae) in eastern Missouri. The American Midland Naturalist 102:404–407.

Caire, W., J. D. Taylor, B. P. Glass, and M. A. Mares. 1989. Mammals of Oklahoma. University of Oklahoma Press. Norman, U.S.A.

Caire, W., J. E. Vaughan, and V. E. Diersing. 1978. First record of *Sorex arizonae* (Insectivora: Soricidae) from Mexico. The Southwestern Naturalist 23:532–533.

Calahorra-Oliart, A., S. M. Ospina-Garcés, and L. León-Paniagua. 2021. Cryptic species in *Glossophaga soricina* (Chiroptera: Phyllostomidae): do morphological data support molecular evidence? Journal of Mammalogy 102:54–68.

Callahan, J. R. 1976. Systematics and biogeography of the *Eutamias obscurus* complex (Rodentia: Sciuridae). Ph.D. dissertation. The University of Arizona. Tucson, U.S.A.

Callahan, J. R. 1977. Diagnosis of *Eutamias obscurus* (Rodentia: Sciuridae). Journal of Mammalogy 58:188–201.

Callahan, J. R. 1980. Taxonomic status of *Eutamias bulleri*. The Southwestern Naturalist 25:1–8.

Callahan, J. R., and D. R. Davis. 1976. Desert chipmunks. The Southwestern Naturalist 21:127–130.

Caldwell, D. K. 1961. The harbor seal in South Carolina. Journal of Mammalogy 42:425.

Caldwell, D. K., and M. C. Caldwell. 1969. The harbor seal, *Phoca vitulina concolor*, in Florida. Journal of Mammalogy 50:379–380.

Camargo, I., and S. T. Álvarez-Castañeda. 2018. *Sorex ornatus juncensis* (Mammalia: Soricomorpha) rediscovering after 90 years without records, and taxonomical comments. Mammalia 83:193–197.

Camargo, I., and S. T. Álvarez-Castañeda. 2020. A new species and three subspecies of the desert shrew (*Notiosorex*) from the Baja California peninsula and California. Journal of Mammalogy 101:872–886.

Camargo, I., E. Rios, and S. T. Álvarez-Castañeda. 2016. Geographical extension of *Chaetodipus fallax* (Rodentia: Heteromyidae) in the Baja California peninsula. Mammalia 81:315–318.

Camargo, I, S. T. Álvarez-Castañeda, P. D. Polly, J. D. Stuhler, and J. E. Maldonado. 2021. Molecular phylogenetic and taxonomic status of the large-eared desert shrew *Notiosorex evotis* (Eulipotyphla: Soricidae), with the designation of a neotype. Journal of Mammalogy 103:1422–1330.

Cameron, G. N., and S. R. Spencer. 1981. *Sigmodon hispidus* Mammalian Species 158:1–9.

Cameron, G. N., S. R. Spencer, B. D. Eshelman, L. R. Williams, and M. J. Gregory. 1988. Activity and burrow structure of Attwater's pocket gopher (*Geomys attwateri*). Journal of Mammalogy 69:667–677.

Cano-Sánchez, E., F. Rodriguez-Gomez, L. A. Ruedas, K. Oyama, L. Leon-Paniagua, A. Mastretta-Yanes, and A. Velazquez. 2022. Using Ultraconserved Elements to Unravel Lagomorph Phylogenetic Relationships. Journal of Mammalian Evolution 29:395–411.

Carleton, M. D. 1977. Interrelationships of populations of the *Peromyscus boylii* species group (Rodentia, Muridae) in western Mexico. Occasional Papers, Museum of Zoology, University of Michigan 675:1–47.

Carleton, M. D. 1979. Taxonomic status and relationships of *Peromyscus boylii* from El Salvador. Journal of Mammalogy 60:280–296.

Carleton, M. D. 1980. Phylogenetic relationships in neotomine-peromyscine-rodents (Muroidea) and a reappraisal of the dichotomy within New World Cricetinae. Miscellaneous Publications, Museum of Zoology, University of Michigan 157:93–108.

Carleton, M. D. 1981. A survey of gross stomach morphology in Microtinae (Rodentia: Muroidea): Zeitschrift ftir Saugetierkunde 46:93–108.

Carleton, M. D. 1989. Systematics and evolution in avance study *Peromyscus*. Pp. 7–141, *in* Advances in the study of *Peromyscus* (Kirkland Jr., G. L., and J. L. Lane, eds.). Texas Tech University Press. Texas, U.S.A.

Carleton, M. D., and J. Arroyo-Cabrales. 2009. Review of the *Oryzomys couesi* complex (Rodentia: Cricetidae: Sigmodontinae) in western Mexico. Pp. 93–127, *in* Systematic Mammalogy: contributions in honor of Guy G. Musser (Voss. R. S., and M. D. Carleton, eds.). Bulletin of the American Museum of Natural History 331:1–450.

Carleton, M. D., and G. G. Musser. 1984. Muroid rodents. Pp. 289–379, *in* Orders and families of recent mammals of the world (Anderson, S., and J. K. Jones, Jr., eds.) John Wiley and Sons. New York, U.S.A.

Carleton, M. D., and G. G. Musser. 1995. Systematic studies of oryzomyine rodents (Muridae: Sigmodontinae): definition and distribution of *Oligoryzomys vegetus* (Bangs, 1902). Proceedings of the Biological Society of Washington 108:338–369.
Carleton, M. D., and S. Olson. 1999. Amerigo Vespucci and the rat of Fernando de Noronha: A new genus and species of Rodentia (Muridae: Sigmodontinae) from a volcanic island off Brazil's continental shelf. American Museum Novitates 3256:59.
Carleton, M. D., D. E. Wilson, A. L. Gardner, and M. A. Bogan. 1982. Distribution and systematics of *Peromyscus* (Mammalia: Rodentia) of Nayarit, Mexico. Smithsonian Contributions in Zoology 352:1–46.
Carleton, M. D., R. D. Fisher, and A. L. Gardner. 1999. Identification and distribution of cotton rats, genus *Sigmodon* (Muridae: Sigmodontinae) of Nayarit, Mexico. Proceedings of the Biological Society of Washington 112:813–856.
Carleton, M. D., O. Sánchez, and G. Urbano-Vidales. 2002. A new species of *Habromys* (Muroidea: Neotominae) from Mexico, with generic review of species definitions and remarks on diversity patterns among Mesoamerica small mammals restricted to humid montane forest. Proceedings of the Biological Society of Washington 115:488–553.
Carraway, L. N. 1985. *Sorex pacificus*. Mammalian Species 231:1–5.
Carraway, L. N. 1987. Analysis of characters for distinguishing *Sorex trowbridgii* from sympatric *S. vagrans*. The Murrelet 68:29–30.
Carraway, L. N. 1988. Records of reproduction in *Sorex pacificus*. The Southwestern Naturalist 33:479–480.
Carraway, L. N. 1990. A morphologic and morphometric analysis of the "*Sorex vagrans* species complex" in the Pacific coast region. Special Publications of the Texas Tech University Museum 32:1–76.
Carraway, L. N. 2007. Shrews (Eulypotyphla: Soricidae) of Mexico. Monographs of the Western North American Naturalist 3:1–91.
Carraway, L. N., and P. K. Kennedy. 1993. Genetic variation in *Thomomys bulbivorus*, an endemic to the Willamette Valley, Oregon. Journal of Mammalogy 74:952–962.
Carraway, L. N., and B. J. Verts. 1985. *Microtus oregoni*. Mammalian Species 233:1–6.
Carraway, L. N., and B. J. Verts. 1991a. *Neotoma fuscipes*. Mammalian Species 386:1–10.
Carraway, L. N., and B. J. Verts. 1991b. *Neurotrichus gibbsii*. Mammalian Species 387:1–7.
Carraway, L. N., and B. J. Verts. 1993. *Aplodontia rufa*. Mammalian Species 431:1–10.
Carraway, L. N., and B. J. Verts. 1994. *Sciurus griseus*. Mammalian Species 474:1–7.
Carraway, L. N., and B. J. Verts. 1996. *Aplodontia rufa*. Mammalian Species 431:1–10.
Carraway, L. N., and R. M. Timm. 2000. Revision of the extant taxa of the genus *Notiosorex* (Mammalia: Insectivora: Soricidae). Proceeding Biological Society of Washington 113:302–318.
Carraway, L. N., L. F. Alexander, and B. J. Verts. 1993. *Scapanus townsendii*. Mammalian Species 434:1–7.
Carroll, L. E., and H. H. Genoways. 1980. *Lagurus curtatus*. Mammalian Species 124:1–6.
Carroll, D. S., L. L. Peppers, and R. D. Bradley. 2005. Molecular systematics and phylogeography of the *Sigmodon hispidus* species group. Pp. 87–100, *in* Contribuciones mastozoológicas en homenaje a Bernardo Villa (Sánchez-Cordero, V., and R. A. Medellín, eds.). Instituto de Biología e Instituto de Ecología, Universidad Nacional Autónoma de México y Comisión Nacional para el Conocimiento y Uso de la Biodiversidad. Distrito Federal, México.
Carter, D. C., and W. B. Davis. 1961. *Tadarida aurispinosa* (Peale) (Chiroptera: Molossidae) in North America. Proceedings of the Biological Society of Washington 74:161–165.
Carter, D. C., and P. G. Dolan. 1978. Catalogue of type specimens of Neotropical bats in selected European museums. Special Publications, The Museum, Texas Tech University 15:1–136.
Carter, C. H., and H. H. Genoways. 1978. *Liomys salvini*. Mammalian Species 84:1–5.
Carter, D. C., and J. K. Jones, Jr. 1978. Bats from the Mexican state of Hidalgo. Occasional Papers, The Museum, Texas Tech University 54:1–12.
Carter, D. C., W. D. Webster, J. K. Jones, Jr., C. Jones, and R. D. Suttkus. 1985. *Dipodomys elator*. Mammalian Species 232:1–3.
Cary, M. 1911. A biological survey of Colorado. North American Fauna 33:1–256.
Caso, A., T. de Oliveira, and S. V. Carvajal. 2015. *Herpailurus yagouaroundi*. The IUCN Red List of Threatened Species 2015:e.T9948A50653167.
Cassola, F. 2016a. *Arborimus albipes*. The IUCN Red List of Threatened Species 2016:e.T2017A22389204.
Cassola, F. 2016b. *Blarina brevicauda*. The IUCN Red List of Threatened Species 2016:e.T41451A115187102.
Cassola, F. 2016c. *Blarina carolinensis*. The IUCN Red List of Threatened Species 2016:e.T41452A115187223.
Cassola, F. 2016d. *Blarina hylophaga*. The IUCN Red List of Threatened Species 2016:e.T41453A115187348.
Cassola, F. 2016e. *Callospermophilus lateralis*. The IUCN Red List of Threatened Species 2016:e.T42468A22265474.
Cassola, F. 2016f. *Callospermophilus saturatus*. The IUCN Red List of Threatened Species 2016:e.T42562A22262657.
Cassola, F. 2016g. *Castor canadensis*. The IUCN Red List of Threatened Species 2016:e.T4003A22187946.
Cassola, F. 2016h. *Condylura cristata*. The IUCN Red List of Threatened Species 2016:e.T41458A115187740.
Cassola, F. 2016i. *Cratogeomys castanops*. The IUCN Red List of Threatened Species 2016:e.T16025A115131062.
Cassola, F. 2016j. *Cryptotis goldmani*. The IUCN Red List of Threatened Species 2016:e.T41371A22285527.
Cassola, F. 2016k. *Cryptotis mexicana*. The IUCN Red List of Threatened Species 2016:e.T41374A22286065.
Cassola, F. 2016l. *Cynomys gunnisoni*. The IUCN Red List of Threatened Species 2016:e.T42453A115189620.
Cassola, F. 2016m. *Cynomys leucurus*. The IUCN Red List of Threatened Species 2016:e.T42454A22261371.
Cassola, F. 2016n. *Cynomys ludovicianus*. The IUCN Red List of Threatened Species 2016:e.T6091A115080297.
Cassola, F. 2016o. *Dicrostonyx groenlandicus*. The IUCN Red List of Threatened Species 2016:e.T42618A115195764.
Cassola, F. 2016p. *Dicrostonyx hudsonius*. The IUCN Red List of Threatened Species 2016:e.T42619A115195917.
Cassola, F. 2016q. *Dicrostonyx nelsoni*. The IUCN Red List of Threatened Species 2016:e.T42620A22331765.
Cassola, F. 2016r. *Dicrostonyx nunatakensis*. The IUCN Red List of Threatened Species 2016:e.T6567A22332006.
Cassola, F. 2016s. *Dicrostonyx richardsoni*. The IUCN Red List of Threatened Species 2016:e.T42622A115196066.
Cassola, F. 2016t. *Dipodomys agilis*. The IUCN Red List of Threatened Species 2016:e.T6684A22228553.
Cassola, F. 2016u. *Dipodomys californicus*. The IUCN Red List of Threatened Species 2016:e.T42599A115193510.
Cassola, F. 2016v. *Dipodomys compactus*. The IUCN Red List of Threatened Species 2016:e.T6685A115083122.
Cassola, F. 2016w. *Dipodomys heermanni*. The IUCN Red List of Threatened Species 2016:e.T42600A22227931.
Cassola, F. 2016x. *Dipodomys microps*. The IUCN Red List of Threatened Species 2016:e.T42603A22227645.

Cassola, F. 2016y. *Dipodomys ordii*. The IUCN Red List of Threatened Species 2016:e.T6691A115083268.
Cassola, F. 2016z. *Dipodomys panamintinus*. The IUCN Red List of Threatened Species 2016:e.T42604A115193637.
Cassola, F. 2016aa. *Dipodomys venustus*. The IUCN Red List of Threatened Species 2016:e.T42605A22227166.
Cassola, F. 2016ab. *Geomys attwateri*. The IUCN Red List of Threatened Species 2016:e.T136380A22217970.
Cassola, F. 2016ac. *Geomys breviceps*. The IUCN Red List of Threatened Species 2016:e.T136840A115212715.
Cassola, F. 2016ad. *Geomys bursarius*. The IUCN Red List of Threatened Species 2016:e.T42588A115192675.
Cassola, F. 2016ae. *Geomys knoxjonesi*. The IUCN Red List of Threatened Species 2016:e.T136258A22218149.
Cassola, F. 2016af. *Geomys pinetis*. The IUCN Red List of Threatened Species 2016:e.T42589A115192878.
Cassola, F. 2016ag. *Geomys texensis*. The IUCN Red List of Threatened Species 2016:e.T9062A22217724.
Cassola, F. 2016ah. *Glaucomys sabrinus*. The IUCN Red List of Threatened Species 2016:e.T39553A22256914.
Cassola, F. 2016ai. *Glaucomys volans*. The IUCN Red List of Threatened Species 2016:e.T9240A115091392.
Cassola, F. 2016aj. *Heteromys desmarestianus*. The IUCN Red List of Threatened Species 2016:e.T47804700A115400334.
Cassola, F. 2016ak. *Ictidomys tridecemlineatus*. The IUCN Red List of Threatened Species 2016:e.T42564A22263122.
Cassola, F. 2016al. *Lemmiscus curtatus*. The IUCN Red List of Threatened Species 2016:e.T42624A115196202.
Cassola, F. 2016am. *Lemmus trimucronatus*. The IUCN Red List of Threatened Species 2016:e.T136712A115211700.
Cassola, F. 2016an. *Marmota broweri*. The IUCN Red List of Threatened Species 2016:e.T42455A22258026.
Cassola, F. 2016ao. *Marmota flaviventris*. The IUCN Red List of Threatened Species 2016:e.T42457A115189809.
Cassola, F. 2016ap. *Marmota monax*. The IUCN Red List of Threatened Species 2016:e.T42458A115189992.
Cassola, F. 2016aq. *Marmota olympus*. The IUCN Red List of Threatened Species 2016:e.T42459A22257452.
Cassola, F. 2016ar. *Microtus abbreviatus*. The IUCN Red List of Threatened Species 2016:e.T13425A22350031.
Cassola, F. 2016as. *Microtus canicaudus*. The IUCN Red List of Threatened Species 2016:e.T42625A22348218.
Cassola, F. 2016at. *Microtus chrotorrhinus*. The IUCN Red List of Threatened Species 2016:e.T42626A115196387.
Cassola, F. 2016au. *Microtus longicaudus*. The IUCN Red List of Threatened Species 2016:e.T42627A115196586.
Cassola, F. 2016av. *Microtus montanus*. The IUCN Red List of Threatened Species 2016:e.T42630A22346732.
Cassola, F. 2016aw. *Microtus ochrogaster*. The IUCN Red List of Threatened Species 2016:e.T42631A115196932.
Cassola, F. 2016ax. *Microtus oregoni*. The IUCN Red List of Threatened Species 2016:e.T42632A115197199.
Cassola, F. 2016ay. *Microtus pennsylvanicus*. The IUCN Red List of Threatened Species 2016:e.T13452A115114123.
Cassola, F. 2016az. *Microtus pinetorum*. The IUCN Red List of Threatened Species 2016:e.T42633A115197344.
Cassola, F. 2016ba. *Microtus richardsoni*. The IUCN Red List of Threatened Species 2016:e.T42634A115197660.
Cassola, F. 2016bb. *Microtus townsendii*. The IUCN Red List of Threatened Species 2016:e.T13487A115114983.
Cassola, F. 2016bc. *Microtus xanthognathus*. The IUCN Red List of Threatened Species 2016:e.T42628A22348489.
Cassola, F. 2016bd. *Myodes californicus*. The IUCN Red List of Threatened Species 2016:e.T42616A115195295.
Cassola, F. 2016be. *Myodes gapperi*. The IUCN Red List of Threatened Species 2016:e.T42617A115195411.
Cassola, F. 2016bf. *Napaeozapus insignis*. The IUCN Red List of Threatened Species 2016:e.T42612A115194392.
Cassola, F. 2016bg. *Neofiber alleni*. The IUCN Red List of Threatened Species 2016:e.T14520A22356567.
Cassola, F. 2016bh. *Neotamias alpinus*. The IUCN Red List of Threatened Species 2016:e.T42568A22266784.
Cassola, F. 2016bi. *Neotamias amoenus*. The IUCN Red List of Threatened Species 2016:e.T42569A115190467.
Cassola, F. 2016bj. *Neotamias canipes*. The IUCN Red List of Threatened Species 2016:e.T21364A22266973.
Cassola, F. 2016bk. *Neotamias cinereicollis*. The IUCN Red List of Threatened Species 2016:e.T42570A22267056.
Cassola, F. 2016bl. *Neotamias minimus*. The IUCN Red List of Threatened Species 2016:e.T42572A115190804.
Cassola, F. 2016bm. *Neotamias ochrogenys*. The IUCN Red List of Threatened Species 2016:e.T42573A22267475.
Cassola, F. 2016bn. *Neotamias panamintinus*. The IUCN Red List of Threatened Species 2016:e.T42574A22267550.
Cassola, F. 2016bo. *Neotamias quadrimaculatus*. The IUCN Red List of Threatened Species 2016:e.T42575A22267619.
Cassola, F. 2016bp. *Neotamias quadrivittatus*. The IUCN Red List of Threatened Species 2016:e.T42576A115191051.
Cassola, F. 2016bq. *Neotamias ruficaudus*. The IUCN Red List of Threatened Species 2016:e.T42577A22268024.
Cassola, F. 2016br. *Neotamias rufus*. The IUCN Red List of Threatened Species 2016:e.T42578A115191185.
Cassola, F. 2016bs. *Neotamias senex*. The IUCN Red List of Threatened Species 2016:e.T42579A22268269.
Cassola, F. 2016bt. *Neotamias siskiyou*. The IUCN Red List of Threatened Species 2016:e.T42580A22268201.
Cassola, F. 2016bu. *Neotamias sonomae*. The IUCN Red List of Threatened Species 2016:e.T42581A115191316.
Cassola, F. 2016bv. *Neotamias speciosus*. The IUCN Red List of Threatened Species 2016:e.T42582A115191427.
Cassola, F. 2016bw. *Neotamias townsendii*. The IUCN Red List of Threatened Species 2016:e.T42584A115191888.
Cassola, F. 2016bx. *Neotamias umbrinus*. The IUCN Red List of Threatened Species 2016:e.T42585A115192014.
Cassola, F. 2016by. *Neotoma cinerea*. The IUCN Red List of Threatened Species 2016:e.T42673A115200351.
Cassola, F. 2016bz. *Neotoma floridana*. The IUCN Red List of Threatened Species 2016:e.T42650A115199202.
Cassola, F. 2016ca. *Neotoma fuscipes*. The IUCN Red List of Threatened Species 2016:e.T14587A22371665.
Cassola, F. 2016cb. *Neotoma stephensi*. The IUCN Red List of Threatened Species 2016:e.T42651A115199398.
Cassola, F. 2016cc. *Neurotrichus gibbsii*. The IUCN Red List of Threatened Species 2016:e.T41468A115188045.
Cassola, F. 2016cd. *Ochrotomys nuttalli*. The IUCN Red List of Threatened Species 2016:e.T42674A115200634.
Cassola, F. 2016ce. *Ondatra zibethicus*. The IUCN Red List of Threatened Species 2016:e.T15324A22344525.
Cassola, F. 2016cf. *Oryzomys palustris*. The IUCN Red List of Threatened Species 2016:e.T42675A115200837.
Cassola, F. 2016cg. *Parascalops breweri*. The IUCN Red List of Threatened Species 2016:e.T41469A115188181.
Cassola, F. 2016ch. *Perognathus fasciatus*. The IUCN Red List of Threatened Species 2016:e.T42608A115194044.
Cassola, F. 2016ci. *Perognathus inornatus*. The IUCN Red List of Threatened Species 2016:e.T42609A22224670.
Cassola, F. 2016cj. *Perognathus longimembris*. The IUCN Red List of Threatened Species 2016:e.T16636A115135230.
Cassola, F. 2016ck. *Perognathus parvus*. The IUCN Red List of Threatened Species 2016:e.T42610A115194207.
Cassola, F. 2016cl. *Peromyscus attwateri*. The IUCN Red List of Threatened Species 2016:e.T42652A115199528.

Cassola, F. 2016cm. *Peromyscus beatae*. The IUCN Red List of Threatened Species 2016:e.T136323A22364310.
Cassola, F. 2016cn. *Peromyscus gossypinus*. The IUCN Red List of Threatened Species 2016:e.T42653A115199668.
Cassola, F. 2016co. *Peromyscus keeni*. The IUCN Red List of Threatened Species 2016:e.T135164A115204632.
Cassola, F. 2016cp. *Peromyscus leucopus*. The IUCN Red List of Threatened Species 2016:e.T16669A115136270.
Cassola, F. 2016cq. *Peromyscus maniculatus*. The IUCN Red List of Threatened Species 2016:e.T16672A22360898.
Cassola, F. 2016cr. *Peromyscus polionotus*. The IUCN Red List of Threatened Species 2016:e.T42654A115199876.
Cassola, F. 2016cs. *Peromyscus truei*. The IUCN Red List of Threatened Species 2016:e.T16694A115137578.
Cassola, F. 2016ct. *Phenacomys intermedius*. The IUCN Red List of Threatened Species 2016:e.T42636A115197827.
Cassola, F. 2016cu. *Phenacomys ungava*. The IUCN Red List of Threatened Species 2016:e.T42637A115198018.
Cassola, F. 2016cv. *Poliocitellus franklinii*. The IUCN Red List of Threatened Species 2016:e.T41787A22265037.
Cassola, F. 2016cw. *Reithrodontomys fulvescens*. The IUCN Red List of Threatened Species 2016:e.T19407A115150749.
Cassola, F. 2016cx. *Reithrodontomys humulis*. The IUCN Red List of Threatened Species 2016:e.T42678A115201061.
Cassola, F. 2016cy. *Reithrodontomys megalotis*. The IUCN Red List of Threatened Species 2016:e.T19410A115151098.
Cassola, F. 2016cz. *Scapanus orarius*. The IUCN Red List of Threatened Species 2016:e.T41474A115188698.
Cassola, F. 2016da. *Scapanus townsendii*. The IUCN Red List of Threatened Species 2016:e.T41475A22322352.
Cassola, F. 2016db. *Sciurus carolinensis*. The IUCN Red List of Threatened Species 2016:e.T42462A22245728.
Cassola, F. 2016dc. *Sciurus nayaritensis*. The IUCN Red List of Threatened Species 2016:e.T20015A115155124.
Cassola, F. 2016dd. *Sigmodon hispidus*. The IUCN Red List of Threatened Species 2016:e.T20213A115157685.
Cassola, F. 2016de. *Sigmodon toltecus*. The IUCN Red List of Threatened Species 2016:e.T136559A115209897.
Cassola, F. 2016df. *Sorex arcticus*. The IUCN Red List of Threatened Species 2016:e.T41385A115182930.
Cassola, F. 2016dg. *Sorex bendirii*. The IUCN Red List of Threatened Species 2016:e.T41389A115183051.
Cassola, F. 2016dh. *Sorex cinereus*. The IUCN Red List of Threatened Species 2016:e.T41392A115183208.
Cassola, F. 2016di. *Sorex fumeus*. The IUCN Red List of Threatened Species 2016:e.T41396A22312838.
Cassola, F. 2016dj. *Sorex haydeni*. The IUCN Red List of Threatened Species 2016:e.T41399A115183736.
Cassola, F. 2016dk *Sorex hoyi*. The IUCN Red List of Threatened Species 2016:e.T41400A115183871.
Cassola, F. 2016dl. *Sorex longirostris*. The IUCN Red List of Threatened Species 2016:e.T41401A115184004.
Cassola, F. 2016dm. *Sorex merriami*. The IUCN Red List of Threatened Species 2016:e.T41403A115184201.
Cassola, F. 2016dn. *Sorex nanus*. The IUCN Red List of Threatened Species 2016:e.T41406A115184594.
Cassola, F. 2016do. *Sorex palustris*. The IUCN Red List of Threatened Species 2016:e.T41410A115184897.
Cassola, F. 2016dp. *Sorex tenellus*. The IUCN Red List of Threatened Species 2016:e.T41419A22318690.
Cassola, F. 2016dq. *Sorex trowbridgii*. The IUCN Red List of Threatened Species 2016:e.T41421A115185589.
Cassola, F. 2016dr. *Sorex ugyunak*. The IUCN Red List of Threatened Species 2016:e.T41423A115185865.
Cassola, F. 2016ds. *Synaptomys cooperi*. The IUCN Red List of Threatened Species 2016:e.T42639A115198182.
Cassola, F. 2016dt. *Tamias striatus*. The IUCN Red List of Threatened Species 2016:e.T42583A115191543.
Cassola, F. 2016du. *Tamiasciurus douglasii*. The IUCN Red List of Threatened Species 2016:e.T42586A115192165.
Cassola, F. 2016dv. *Tamiasciurus hudsonicus*. The IUCN Red List of Threatened Species 2016:e.T42587A115192299.
Cassola, F. 2016dw. *Thomomys bulbivorus*. The IUCN Red List of Threatened Species 2016:e.T42594A22216513.
Cassola, F. 2016dx. *Thomomys mazama*. The IUCN Red List of Threatened Species 2016:e.T21810A115163696.
Cassola, F. 2016dy. *Thomomys monticola*. The IUCN Red List of Threatened Species 2016:e.T42596A22216069.
Cassola, F. 2016dz. *Thomomys talpoides*. The IUCN Red List of Threatened Species 2016:e.T42597A115193142.
Cassola, F. 2016ea. *Thomomys townsendii*. The IUCN Red List of Threatened Species 2016:e.T42598A115193369.
Cassola, F. 2016eb. *Urocitellus armatus*. The IUCN Red List of Threatened Species 2016:e.T42463A22264746.
Cassola, F. 2016ec. *Urocitellus beldingi*. The IUCN Red List of Threatened Species 2016:e.T42464A22264836.
Cassola, F. 2016ed. *Urocitellus columbianus*. The IUCN Red List of Threatened Species 2016:e.T42466A22265632.
Cassola, F. 2016ee. *Urocitellus parryii*. The IUCN Red List of Threatened Species 2016:e.T20488A22262403.
Cassola, F. 2016ef. *Urocitellus richardsonii*. The IUCN Red List of Threatened Species 2016:e.T42561A22262546.
Cassola, F. 2016eg. *Zapus hudsonius*. The IUCN Red List of Threatened Species 2016:e.T42613A115194664.
Cassola, F. 2016eh. *Zapus princeps*. The IUCN Red List of Threatened Species 2016:e.T42614A115195084.
Cassola, F. 2016ei. *Zapus trinotatus*. The IUCN Red List of Threatened Species 2016:e.T23192A115167834.
Cassola, F. 2017a. *Sciurus aberti*. The IUCN Red List of Threatened Species 2017:e.T42461A22245623.
Cassola, F. 2017b. *Sorex rohweri*. The IUCN Red List of Threatened Species 2017:e.T136282A22317740.
Cassola, F. 2017c. *Sorex sonomae*. The IUCN Red List of Threatened Species 2017:e.T41418A22318770.
Cassola, F. 2017d. *Synaptomys borealis*. The IUCN Red List of Threatened Species 2017:e.T42638A22377185.
Cassola, F. 2018. *Marmota caligata*. The IUCN Red List of Threatened Species 2018:e.T42456A122560084.
Cassola, F. 2019. *Sorex preblei*. The IUCN Red List of Threatened Species 2019:e.T41413A117935777.
Castañeda-Rico, S., L. León-Paniagua, E. Vázquez-Domínguez, and A. G. Navarro-Sigüenza. 2014. Evolutionary diversification and speciation in rodents of the Mexican lowlands: the *Peromyscus melanophrys* species group. Molecular Phylogenetics and Evolution 70:454–463.
Castleberry, S. B., M. T. Mengak, and W. M. Ford. 2006. *Neotoma magister*. Mammalian Species 789:1–5.
Castro-Arellano, I., and E. Vázquez. 2016a. *Pappogeomys bulleri*. The IUCN Red List of Threatened Species 2016:e.T92474664A22217245.
Castro-Arellano, I., and E. Vázquez. 2016b. *Peromyscus aztecus*. The IUCN Red List of Threatened Species 2016:e.T16651A22361906.
Castro-Arellano, I., and E. Vázquez. 2016c. *Peromyscus difficilis*. The IUCN Red List of Threatened Species 2016:e.T16658A115135855.
Castro-Arellano, I., and E. Vázquez. 2016d. *Peromyscus levipes*. The IUCN Red List of Threatened Species 2016:e.T16670A115136642.
Castro-Arellano, I., and E. Vázquez. 2016e. *Peromyscus megalops*. The IUCN Red List of Threatened Species 2016:e.T16674A115136745.
Castro-Arellano, I., and E. Vázquez. 2017. *Osgoodomys banderanus*. The IUCN Red List of Threatened Species 2017:e.T15629A22359053.
Castro-Arellano, I., and E. Vázquez. 2019. *Peromyscus furvus*. The IUCN Red List of Threatened Species 2019:e.T16661A22359896.
Castro-Arellano, I., H. Zarza, and R. A. Medellín. 2000. *Philander opossum*. Mammalian Species 638:1–8.

Castro-Arellano, I., R. Timm, and S. T. Álvarez-Castañeda. 2016a. *Heteromys irroratus*. The IUCN Red List of Threatened Species 2016:e. T12074A22225187.

Castro-Arellano, I., R. Timm, J. Matson, and N. Woodman. 2016b. *Notiosorex evotis*. The IUCN Red List of Threatened Species 2016:e. T136273A115205277.

Ceballos, G. 1990. Comparative natural history of small mammals from tropical forest in western Mexico. Journal of Mammalogy 71:263–266.

Ceballos, G., and R. A. Medellín. 1988. *Diclidurus albus*. Mammalian Species 316:1–4.

Ceballos, G., and A. Miranda. 2000. A field guide to the mammals from the Jalisco coast. Fundación Ecológica de Cuixmala-Universidad Nacional Autónoma de México. Distrito Federal, México.

Ceballos-G, G., and D. E. Wilson. 1985. *Cynomys mexicanus*. Mammalian Species 248:1–3.

Ceballos, G., H. Zarza, and M. A. Steele. 2002. *Xenomys nelsoni*. Mammalian Species 704:1–3.

Cerqueira, R. 1985. The distribution of *Didelphis* in the South America (Polyprotodontia, Didelphidae). Journal of Biogeography 12:135–145.

Cervantes, F. A. 1993. *Lepus flavigularis*. Mammalian Species 423:1–3.

Cervantes F. A., and L. Guevara. 2010. Rediscovery of the critically endangered Nelson's small-eared shrew (*Cryptotis nelsoni*), endemic to Volcán San Martín, Eastern México. Mammalian Biology 75:451–454.

Cervantes, F. A., and C. Lorenzo. 1997. *Sylvilagus insonus*. Mammalian Species 568:1–4.

Cervantes, F. A., C. Lorenzo, and R. S. Hoffmann. 1990. *Romerolagus diazi*. Mammalian Species 360:1–7.

Cervantes, F. A., V. J. Sosa, J. Martínez, R. M. González, and R. C. Dowler. 1993. *Pappogeomys tylorhinus*. Mammalian Species 433:1–4.

Cervantes, F. A., C. Lorenzo, J. Vargas, and T. Holmes. 1992. *Sylvilagus cunicularius*. Mammalian Species 412:1–4.

Cervantes, F. A., J. Martínez, and R. M. González. 1994. Karyotype of the Mexican tropical voles *Microtus quasiater* and *M. umbrousus* (Arvicolinae: Muridae). Acta Theriologica 39:373–377.

Cervantes, F. A., S. T. Álvarez-Castañeda, B. VillaRamírez, C. Lorenzo, A. Rojas, A. L. Colmenares, and J. Vargas. 1996. Natural history of the insular black Jackrabbit (*Lepus insularis*) from Espiritu Santo Island, Mexico. The Southwestern Naturalist 41:186–189.

Cervantes, F. A., J. P. Ramírez-Silva, A. Marín, and G. L. Portales. 1999a. Allozyme variation of cottontail rabbits (*Sylvilagus*) from Mexico. Zeitschrift für Saügetierkunde 64:356–362.

Cervantes, F. A., C. Lorenzo, and O. G. Ward. 1999b. Chromosomal relationships among spiny pocket mice, *Liomys* (Heteromydae), from Mexico. Journal of Mammalogy 80:823–832.

Cervantes, F. A., C. Lorenzo, V. Farías, and J. Vargas. 2008. *Lepus flavigularis*. The IUCN Red List of Threatened Species 2008:e.T11790A3306162.

Chambers, R. R., P. D. Sudman, and R. D. Bradley. 2009. A phylogenetic assessment of Pocket Gophers (*Geomys*): evidence from nuclear and mitochondrial genes. Journal of Mammalogy 90:537–547.

Chapman, F. M. 1894. Remarks on certain land mammals from Florida, with a list of the species known to occur in the state. Bulletin of the American Museum of Natural History 6:333–346.

Chapman, J. A. 1974. *Sylvilagus bachmani*. Mammalian Species 34:1–4.

Chapman, J. A. 1975a. *Sylvilagus transitionalis*. Mammalian Species 55:1–4.

Chapman, J. A. 1975b. *Sylvilagus nuttallii*. Mammalian Species 56:1–3.

Chapman, J. A. and G. Ceballos. 1990. Chapter 5: The Cottontails. Pp. 95–110, *in* Rabbits, hares and pikas: status survey and conservation action plan (Chapman, J. A., and J. C. Flux, eds). IUCN. Gland, Switzerland.

Chapman, J. A., and G. A. Feldhamer. 1981. *Sylvilagus aquaticus*. Mammalian Species 151:1–4.

Chapman, J. A., and G. Willner. 1978. *Sylvilagus audubonii*. Mammalian Species 106:1–4.

Chapman, J. A., and G. Willner. 1981. *Sylvilagus palustris*. Mammalian Species 153:1–3.

Chapman, J. A., J. G. Hockman, and M. M. Ojeda. 1980. *Sylvilagus floridanus*. Mammalian Species 136:1–8.

Chapman J. A., K. L. Cramer, N. J. Dippenaar, and T. J. Robinson. 1992. Systematics and biogeography of the New England cottontail, *Sylvilagus transitionalis* (Bangs, 1895), with the description of a new species from the Appalachian Mountains. Proceedings of the Biological Society of Washington 105:841–866.

Chapskii, K. K. 1975. Obosnavanie dvukh novykh podvidov nastoyashchikh tyulenei semeistva Phocidae. Trudy Zoologischeskogo Instituta Akademii Nauk SSSR 53:282–333.

Chesser, R. K. 1983. Cranial variation among population of the black tailed prairie dog in New Mexico. Occasional Papers, the Museum Texas Tech University 84:1–13.

Chernyavsky, F. B., N. I. Abramson, A. A. Tsvetkova, E. M. Anbinder, and L. P. Kurysheva. 1993. On systematics and zoogeography of true lemming of the genus *Lemmus* (Rodentia, Cricetidae) of Beringia. Zoologicheskii Zhurnal 72:111–122.

Christman, G. N. 1971. The mountain bison. American Western 8:44–47.

Choate, J. R. 1969. Taxonomic status of the shrew, *Notiosores* (*Xenosorex*) *phillipsii* Schaldach, 1966 (Mammalia: Insectivora). Proccedings of the Biological Society of Washington 82:469–476.

Choate, J. R. 1970. Systematics and zoogeography of Middle Americas shrews of the genus *Cryptotis*. University of Kansas Publications, Museum of Natural History 19:195–317.

Choate, J. R. 1973. *Cryptotis mexicana*. Mammalian Species 28:1–3.

Choate, J. R., and E. D. Fleharty. 1974. *Cryptotis goodwini*. Mammalian Species 44:1–3.

Choate, J. R., and S. L. Williams. 1978. Biogeographic interpretation of variation within and among populations of the prairie vole, *Microtus ochrogaster*. Occasional Papers, The Museum, Texas Tech University 49:1–25.

Choromanski-Norris J. F., K. Fritzell, and A. B. Sargeant. 1986. Seasonal activity cycle and weight changes of the Franklin's ground squirrel. American Midland Naturalist 116:101–107.

Choromanski-Norris, J., E. K. Fritzell, and A. B. Sargeant. 1989. Movements and habitat use of Franklin's ground squirrels in duck-nesting habitat. Journal of Wildlife Management 53:324–331.

Clare, E. L., A. M. Adams, A. Z. Maya-Simões, J. L. Eger, P. D. N. Hebert, and M. B. Fenton. 2013. Diversification and reproductive isolation: cryptic species in the only New World high-duty cycle bat, *Pteronotus parnellii*. BMC Ecology and Evolution 13:26. https://doi.org/10.1186/1471-2148-13-26.

Clark, J. L. 1964. The great arc of wild sheep. University of Oklahoma Press. Norman, U.S.A.

Clark, T. W., E. Anderson, C. Douglas, and M. Strickland. 1987. *Martes americana*. Mammalian Species 289:1–8.

Clark, T. W., R. S. Hoffmann, and C. F. Nadler. 1971. *Cynomys leucurus*. Mammalian Species 7:1–4.

Clawson, R. G., J. A. Clawson, and T. L. Best. 1994a. *Tamias alpinus*. Mammalian Species 461:1–6.

Clawson, R. G., J. A. Clawson, and T. L. Best. 1994b. *Tamias quadrimaculatus*. Mammalian Species 469:1–6.

Cloutier, D., and D. W. Thomas. 1992. *Carollia perspicillata*. Mammalian Species 417:1–9.

Clutton-Brock, J., G. B. Corbert, and M. Hills. 1976. A review of the family Canidae, with a classification by numerical methods. Bulletin of the British Museum (Natural History). Zoology 29:117–199.

Coates-Estrada, R., and A. Estrada. 1986. Manual de identificación de campo de los mamíferos de la Estación de la Biología "Los Tuxtlas." Instituto de Biología, Universidad Nacional Autónoma de México, México. Distrito Federal, México.

Cockrum, E. L. 1961. The recent mammals of Arizona: their taxonomy and distribution. University of Arizona Press. Tucson, U.S.A.

Cockrum, E. L. and A. L. Gardner. 1960. Underwood's mastiff bat in Arizona. Journal of Mammalogy 41:510–511.

Cockrum, E. L., and G. Van R. Bradshaw. 1963. Notes on mammals from Sonora, Mexico. American Museum Novitates 2138:1–9.

Corbet, G. B. 1978. The mammals of the Palaearctic region: A taxonomic review. British Museum (Natural History). London, U.K.

Cole, F. R., and D. E. Wilson. 2006. *Leptonycteris yerbabuenae*. Mammalian Species 797:1–7.

Cole, F. R., and D. E. Wilson. 2009. *Urocitellus canus* (Rodentia: Sciuridae). Mammalian Species 834:1–8.

Cole, F. R., and D. E. Wilson. 2010. *Microtus miurus* (Rodentia: Cricetidae). Mammalian Species 42:75–89.

Colella, J. P., E. J. Johnson, and J. A. Cook. 2018. Reconciling molecules and morphology in North American *Martes,* Journal of Mammalogy 99:1323–1335.

Colella, J. P., R. E. Wilson, S. L. Talbot, and J. A. Cook. 2019. Implications of introgression for wildlife translocations: the case of North American martens. Conservation Genetics, 20:153–166.

Colella, J. P., L. M. Frederick, S. L. Talbot, and J. A. Cook. 2021. Extrinsically reinforced hybrid speciation within Holarctic ermine (*Mustela* spp.) produces an insular endemic. Diversity and Distributions 27:747–762.

Collins, A. C. 2008. The taxonomic status of spider monkeys in the twenty-first century. Pp. 50–78, *in* Spider Monkeys: behavior, ecology and evolution of the genus *Ateles* (Campbell, C. J. ed.). Cambridge, United Kingdom.

Collins, P. W. 1982. Origin and differentiation of the island fox: a study of evolution in insular populations. M. A. thesis, The University of California. Santa Barbara, U.S.A.

Collins, P. W. 1993. Taxonomic and biogeographic relationships of the island fox (*Urocyon littoralis*) and gray fox (*Urocyon cinereoargenteus*) from western North America. Pp. 351–390, *in* Proceedings of the third Channel Islands symposium: recent advances in California Islands research (Hochberg, F. G., ed.). Santa Barbara Museum of Natural History, Santa Barbara, U.S.A.

Collins, A., and J. Dubach. 2000. Phylogenetic relationships among spider monkeys (*Ateles*) haplotypes based on mitochondrial DNA variation. International Journal of Primatology 21:381–420.

Coltman, D. W., G. Stenson, M. O. Hammill, T. Haug, S. Davis, and T. L. Fulton. 2007. Panmictic population structure in the hooded seal (*Cystophora cristata*). Molecular Ecology 16:1639–1648.

Commissaris, L. R. 1960. Morphological and ecological differentiation of *Peromyscus merriami* from southern Arizona. Journal of Mammalogy 41:305–310.

Committee on Taxonomy. 2014. List of marine mammal species and subspecies. Available at: www.marinemammalscience.org, accessed November 2019.

Connior, M. B. 2011. *Geomys bursarius* (Rodentia: Geomyidae). Mammalian Species 43:104–117.

Connor, P. F. 1959. The bog lemming *Synaptomys cooperi* in southern New Jersey. Publication Museum Michigan State University, Biological Series 1:161–248.

Conroy, C. J., and J. A. Cook. 1999a. MtDNA evidence for repeated pulse of speciation within arvicoline and murid rodents. Journal Mammalian Evolution 6:221–245.

Conroy, C. J., and J. A. Cook. 1999b. *Microtus xanthognathus*. Mammalian Species 627:1–5.

Conroy, C. J., and J. A. Cook. 2000. Molecular systematics of a Holartic rodents (*Microtus*: Muridae). Journal of Mammalogy 81:344–359.

Conroy, C. J., Y. Hortelano, F. A. Cervantes, and J. A. Cook. 2001. The phylogenetic position of southern relictual species of *Microtus* (Muridae: Rodentia) in North America. Mammalian Biology 66:332–344.

Conway, M. C., and C. G. Schmitt. 1978. Record of the Arizona shrew *(Sorex arizonae)* from New Mexico. Journal of Mammalogy 59:631.

Cook, J. A. 2017. Southern bog lemming *Synaptomys cooperi*. Pp. 291–292, *in* Handbook of the mammals of the world, vol. 7 (Wilson, D. E., T. E. Lacher, Jr., and R. A. Mittermeier, eds.). Lynx Edicions, Barcelona, Spain.

Cooke, B. D., J. F. C. Flux, and N. Bonino. 2018. Introduced lagomorphs. Pp. 13–17, *in* Lagomorphs: Pikas, Rabbits, and Hares of the World (Smith, A. T., C. H. Johnston, P. Alves, and K. Hackländer, K., eds). Johns Hopkins University Press, Baltimore, U.S.A.

Coonan, T., K. Ralls., B. Hudgens., B. Cypher, and C. Boser. 2013. *Urocyon littoralis*. The IUCN Red List of Threatened Species 2013:e. T22781A13985603.

Corbet, G. B., and J. E. Hill. 1980. A world list of mammalian species. British Museum (Natural History). London, United Kingdom.

Cornejo-Latorre, C., P. Cortés-Calva, and S. T. Álvarez-Castañeda. 2021. *Peromyscus fraterculus* (Rodentia: Cricetidae). Mammalian Species 53(1008):112–124.

Cornejo-Latorre, C., P. Cortés-Calva, and S. T. Álvarez-Castañeda. 2017. The evolutionary history of the subgenus *Haplomylomys* (Cricetidae: *Peromyscus*). Journal of Mammalogy 98:1627–1640.

Cornely, J. E., and B. J. Verts. 1988. *Microtus townsendii*. Mammalian Species 325:1–9.

Cornely, J. E., and R. J. Baker. 1986. *Neotoma mexicana*. Mammalian Species 262:1–7.

Cornely, J. E., L. N. Carraway, and B. J. Verts. 1992. *Sorex preblei*. Mammalian Species 416:1–3.

Cortés-Calva, P., and S. T. Álvarez-Castañeda. 2001. *Peromyscus dickeyi*. Mammalian Species 659:1–2.

Cortés-Calva, P., S. T. Álvarez-Castañeda, and E. Yensen. 2001a. *Neotoma anthonyi*. Mammalian Species 663:1–3.

Cortés-Calva, P., E. Yensen, and S. T. Álvarez-Castañeda. 2001b. *Neotoma martinensis*. Mammalian Species 657:1–3.

Cortés-Calva, P., J. P. Gallo-Reynoso, J. Delgadillo-Rodríguez, C. Lorenzo, and S. T. Álvarez-Castañeda. 2013. The effect of feral dogs and other alien species on native mammals of Isla de Cedros, Mexico. Natural Areas Journal 33:466–473.

Cortés-Ortiz, L., E. Bermingham, C. Rico, E. Rodríguez-Luna, I. Sampaio, and M. Ruiz-García. 2003. Molecular systematics and biogeography of the Neotropical monkey genus, *Alouatta*. Molecular Phylogenetics and Evolution 26:64–81.

Cortés-Ortiz, L, T. F. Duda, D. Canales-Espinosa, F. García-Orduña, E. Rodríguez-Luna, and E. Bermingham. 2007. Hybridization in large-bodied New World primates. Genetics 176:2421–2425.

Cortés-Ortiz, L., L. Aguiar, I. Agostini, M. A. Kelaita, F. E. Silva, and J. C. Bicca-Marques. 2015. Hybridization in howler monkeys: Current understanding and future directions. Pp. 107–131, *in* Howler Monkeys: adaptive radiation, systematics, and morphology (Kowalewski, M. M., P. A. Garber, L. Cortés-Ortiz, B. Urbani, and D. Youlatos, eds.). Springer. New York, USA.

Cortes-Ortíz, L., D. Canales-Espinosa, F. M. Cornejo, D. Guzman-Caro, A. Link, P. Moscoso, P. Méndez-Carvajal, E. Palacios, V. Rodríguez, M. Rosales-Meda, D. Solano, K. Williams-Guillén, and S. de la Torre. 2020a. *Alouatta palliata*. The IUCN Red List of Threatened Species 2020:e.T39960A17925090.

Cortes-Ortíz, L., D. Canales-Espinosa, L. K. Marsh, R. A. Mittermeier, P. Méndez-Carvajal, M. Rosales-Meda, D. Solano, and K. Williams-Guillén. 2020b. *Ateles geoffroyi*. The IUCN Red List of Threatened Species 2020:e.T2279A17929000.

Cortes-Ortíz, L., M. Rosales-Meda, L. K. Marsh, and R. A. Mittermeier. 2020c. *Alouatta pigra*. The IUCN Red List of Threatened Species 2020:e.T914A17926000.

Costa, L. P., and J. L. Patton. 2006. Diversidade e limites geográficos e sistemáticos de marsupiais brasileiros. Pp. 321–341, *in* Os marsupiais do Brasil (Cáceres, N. C., and E. L. A. Monteiro-Filho, eds.). Campo Grande, Editora UFMS.

Cowan, I. M. 1936. Distribution and variation in deer (Genus *Odocoileus*) of the Pacific coastal region of North America. California Fish and Game 22:155–246.

Cowan, I. M. 1940. Distribution and variation in the native sheep of North America. American Midland Naturalist 24:505–580.

Cowan, I. M. 1956. What and where are the mule and black-tailed deer? Pp. 334–359, *in* The deer of North America (Taylor, W. P. ed.). The Stackpole Co., Harrisburg and The Wildlife Management Institution, Washington, U.S.A.

Cowan, I. M., and C. J. Guiguet. 1956. The mammals of British Columbia. British Columbia Provincial Museum of Natural History and Anthropology Handbook, Victoria, British Columbia 11:1–413.

Cowan, I. M., and C. J. Guiguet. 1965. The mammals of British Columbia. Third Edition. British Columbia Province Museum, Handbook 11:1–414.

Cowan, I. M., and W. McCrory. 1970. Variation in the mountain goat, *Oreamnos americanus* (Blainville). Journal of Mammalogy 51:60–73.

Craig, B. A., and J. E. Reynolds III. 2004. Determination of manatee population trends along the Atlantic coast of Florida using a Bayesian approach with temperature-adjusted aerial survey data. Marine Mammal Science 20:386–400.

Cramer, M. J., and G. N. Cameron. 2001. *Geomys texensis*. Mammalian Species 679:1–3.

Cramer, M. J., M. R. Willig, and C. Jones. 2001. *Trachops cirrhosus*. Mammalian Species 656:1–6.

Crase, F. T. 1973. New size records for the western gray squirrel. The Murrelet 54:20–21.

Cronin, M. A. 1992. Intraspecifi c mitochondrial DNA variation in North American cervids. Journal of Mammalogy 73:70–82.

Cronin, M. A., J. L. Bodkin, B. E. Ballachey, J. A. Estes, and J. C. Patton. 1996. Mitochondrial-DNA variation among subspecies and populations of Sea otters (*Enhydra lutris*). Journal of Mammology 77:546–557.

Csuti, B. A. 1979. Patterns of adaptation and variation in the Great Basin Kangaroo rat (*Dipodomys microps*). University of California Publications in Zoology 111:1–69.

Cuarón, A. D., and E. Vázquez. 2018. *Heteromys nelsoni*. The IUCN Red List of Threatened Species 2018:e.T10009A22223445.

Cuarón, A. D., and P. C. de Grammont. 2017. *Cryptotis goodwini*. The IUCN Red List of Threatened Species 2017:e.T48269679A123794650.

Cuarón, A. D., and P. C. de Grammont. 2018a. *Cryptotis magna*. The IUCN Red List of Threatened Species 2018:e.T5766A22285160.

Cuarón, A. D., and P. C. de Grammont. 2018b. *Sorex sclateri*. The IUCN Red List of Threatened Species 2018:e.T20394A22316927.

Cuarón, A. D., I. J. March, and P. M. Rockstroh. 1989. A second armadillo (*Cabassous centralis*) for the faunas of Guatemala and Mexico. Journal of Mammalogy 70:870–871.

Cuarón, A. D., P. C. de Grammont, and K. McFadden. 2016a. *Procyon pygmaeus*. The IUCN Red List of Threatened Species 2016:e.T18267A45201913.

Cuarón, A. D., P. C. de Grammont, N. Woodman, and J. Matson. 2016b. *Cryptotis mayensis*. The IUCN Red List of Threatened Species 2016:e.T136488A22284758.

Cuarón, A. D., F. Reid, J. F. González-Maya, and K. Helgen. 2016c. *Galictis vittata*. The IUCN Red List of Threatened Species 2016:e.T41640A45211961.

Cuarón, A. D., F. Reid, K. Helgen, and J. F. González-Maya. 2016d. *Eira barbara*. The IUCN Red List of Threatened Species 2016:e.T41644A45212151.

Cuarón, A. D., J. F. González-Maya, K. Helgen, F. Reid, J. Schipper, and J. W. Dragoo. 2016e. *Mephitis macroura*. The IUCN Red List of Threatened Species 2016:e.T41634A45211135.

Cuarón, A. D., K. Helgen, and F. Reid. 2016f. *Conepatus semistriatus*. The IUCN Red List of Threatened Species 2016:e.T41633A45210987.

Cuarón, A. D., K. Helgen, and F. Reid. 2016g. *Spilogale gracilis*. The IUCN Red List of Threatened Species 2016:e.T136797A45221721.

Cuarón, A. D., K. Helgen, F. Reid, J. Pino, and J. F. González-Maya. 2016h. *Nasua narica*. The IUCN Red List of Threatened Species 2016:e.T41683A45216060.

Cuarón, A. D., P. C. de Grammont, N. Woodman, and J. Matson. 2017. *Sorex saussurei*. The IUCN Red List of Threatened Species 2017:e.T41416A22317311.

Cuarón, A. D., P. C. de Grammont, and J. Matson. 2018. *Sorex stizodon*. The IUCN Red List of Threatened Species 2018:e.T20395A22316681.

Cudworth, N. L., and J. Koprowski. 2010. *Microtus californicus* (Rodentia: Cridetidae). Mammalian Species 42:230–243.

Cudworth, N. L., and M. B. Grenier. 2015. *Thomomys clusius* (Rodentia: Geomyidae). Mammalian Species 47:57–62.

Culver, M., W. E. Johnson, J. Pecon-Slattery, and S. J. O'Brien. 2000. Genomic ancestry of the American puma (*Puma concolor*). Journal of Heredity 91:186–197.

Currier, M. J. P. 1983. *Felis concolor*. Mammalian Species 200:1–7.

Cypher, B., and R. List. 2014. *Vulpes macrotis*. The IUCN Red List of Threatened Species 2014:e.T41587A62259374.

Czaplewski, N. J. 1983. *Idionycteris phyllotis*. Mammalian Species 208:1–4.

Czech, H. A., A. A. Bohlman, and W. B. Sutton. 2017. New Records for *Sorex hoyi* (American pygmy shrew) in Alabama. Southeastern Naturalist 16:464–472.

Czernay, S. 1987. Spiesshirsche und Pudus. A. Ziemsen Verlag. Wittenberg Lutherstadt, Germany.

Da Cunha-Tavares, V., and C. A. Mancina. 2008. *Phyllops falcatus* (Chiroptera: Phyllostomidae). Mammalian Species 811: 1–7.

Dalby, P. L., and H. A. Lillevick. 1969. Taxonomic analysis of electrophoretic blood serum patterns in the cotton rat, *Sigmodon*. Publications of the Museum Michigan State University, Biological Series 4:65–104.

Dalquest, W. W. 1947. Notes on the natural history of the bat, *Myotis yumanensis*, in California, with a description of a new race. The American Midland Naturalist 38:224–247.

Dalquest, W. W. 1948. Mammals of Washington. University of Kansas Museum of Natural History Publications 2:1–444.

Dalquest, W. W. 1950. Record of mammals from San Luis Potosi. Occasional Papers of the Museum of Zoology, Louisiana State University 23:1–15.

Dalquest, W. W. 1951. Six new mammals from the state of San Luis Potosi, Mexico. Journal of the Washington Academy of Sciences 41:361–364.

Dalquest, W. W. 1953. Mammals of the Mexican state of San Luis Potosi. Louisiana State University Press. Baton Rouge, U.S.A.

Dalquest, W. W., and N. V. Horner. 1984. Mammals of north central Texas. Midwestern State University Press, Wichita Falls, U.S.A.

Dalquest, W. W., and D. R. Orcutt. 1942. The biology of the least shrew-mole, *Neurotrichus gibbsii minor*. The American Midland Naturalist 27:387–401.

Dalquest, W. W., and J. H. Roberts. 1951. Behavior of young grisons in captivity. The American Midland Naturalist 46:359–366.

Dalquest, W. W., H. J. Werner, and J. H. Roberts. 1952. The facial glands of a fruit eating bat, *Artibeus jamaicensis* Leach. Journal of Mammalogy 33:102–103.

Dávalos, L. M. 2006. The geography of diversification in the mormoopids (Chiroptera: Mormoopidae). Biological Journal of the Linnean Society 88:101–118.

Davalos, L., J. Molinari, H. Mantilla-Meluk, C. Medina, J. Pineda, and B. Rodriguez. 2016. *Pteronotus personatus*. The IUCN Red List of Threatened Species 2016:e.T18709A115145223.

Davalos, L., J. Molinari, B. Miller, and B. Rodriguez. 2018. *Peropteryx kappleri*. The IUCN Red List of Threatened Species 2018:e. T16707A22100544.

Davalos, L., J. Molinari, H. Mantilla-Meluk, C. Medina, J. Pineda, and B. Rodriguez. 2019. *Mormoops megalophylla*. The IUCN Red List of Threatened Species 2019:e.T13878A22086060.

Davidow-Henry, B. R., J. K. Jones, Jr. and R. R. Hollander. 1989. *Cratogeomys castanops*. Mammalian Species 338:1–6.

Davis, B. L., S. L. Williams, and G. Lopez. 1971. Chromosomal studies of Geomys. Journal of Mammalogy 52:617–620.

Davis, C. S., I. Stirling, C. Strobeck, and D. W. Coltman. 2008. Population structure of ice-breeding seals. Molecular Biology 17:3078–3094.

Davis, W. B. 1937. Variations in Townsend pocket gophers. Journal of Mammalogy 18:145–158.

Davis, W. B. 1939. The recent mammals of Idaho. Caxton Printers. Caldwell, U.S.A.

Davis, W. B. 1940. Distribution and variation of pocket gophers (genus *Geomys*) in the southwestern United States. Bulletin of the Texas Agricultural Experiment Station 590:1–38.

Davis, W. B. 1942. The systematic status of four kangaroo rats. Journal of Mammalogy 23:328–333.

Davis, W. B. 1944a. Geographic variation in brown lemmings (Genus *Lemmus*). The Murrelet 25:19–25.

Davis, W. B. 1944b. Notes on Mexican mammals. Journal of Mammalogy 25:370–403.

Davis, W. B. 1959. Taxonomy of the eastern pipistrel. Journal of Mammalogy 40:521–531.

Davis, W. B. 1965. Review of the *Eptesicus brasiliensis* complex in Middle America with the description of a new subspecies from Costa Rica. Journal of Mammalogy 46:229–240.

Davis, W. B. 1966a. Review of South American bats of the genus *Eptesicus*. The Southwestern Naturalist 11:245–274.

Davis, W. B. 1966b. The mammals of Texas. Bulletin Texas Parks andWildlife Departament 41:1–267.

Davis, W. B. 1968. Revision of the genus *Uroderma*. Journal of Mammalogy 49:676–698.

Davis, W. B. 1969. A review of the small fruit bats (genus *Artibeus*) of Middle America. Part I. The Southwestern Naturalist 14:15–29.

Davis, W. B. 1970. A review of the small fruit bats (genus *Artibeus*) of Middle America. Part II. The Southwestern Naturalist 14:389–402.

Davis, W. B. 1974. The mammals of Texas. Bulletin Texas Parks and Wildlife Deptament 41:1–294.

Davis, W. B. 1980. New *Sturnira* (Chiroptera: Phyllostomidae) from Central and South America, with key to currently recognized species. Occasional Papers, Museum of Texas Tech University 70:1–5.

Davis, W. B. 1984. Review of the large fruit-eating bats of the "*Artibeus lituratus*" complex (Chiroptera: Phyllostomidae) in Middle America. Occasional Papers, Museum of Texas Tech University 93:1–16.

Davis, W. B., and D. C. Carter. 1962. Notes on Central American bats with description of a new subspecies of *Mormoops*. The Southwestern Naturalist 7:64–74.

Davis, W. B., and D. C. Carter. 1978. A review of the round-eared bats of the *Tonatia silvicola* complex, with description of three new taxa. Occasional Papers, Museum of Texas Tech University 53:1–13.

Davis, W. B., and A. L. Gardner. 2008. Genus *Eptesicus* Rafinesque, 1820. Pp. 440–450, *in* Mammals of South America. Vol. 1: Marsupials, xenarthrans, shrews, and bats (Gardner, A. L., ed.). University of Chicago Press. Chicago, U.S.A.

Davis, W. H., and C. L. Rippy. 1968. Distribution of *Myotis lucifugus* and *Myotis austroriparius* in the southeastern United States. Journal of Mammalogy 49: 113–117.

Davis, W. B., and L Robertson, Jr. 1944. The mammals of Culberson County, Texas. Journal of Mammalogy 25:254–273.

Davis, W. B., and D. J. Schmidly. 1994. The mammals of Texas. Texas Parks and Wildlife Press. Austin, U.S.A.

Davis, W. B., D. C. Carter, and R. H. Pine. 1964. Noteworthy records of Mexican and Central American bats. Journal of Mammalogy 45:375–387.

Dawson, N. G., J. P. Colella, M. P. Small, K. D. Stone, S. L. Talbot, and J. A. Cook. 2017. Historical biogeography sets the foundation for contemporary conservation of martens (genus *Martes*) in northwestern North America, Journal of Mammalogy 98:715–730.

de Grammont, P. C., and A. Cuarón. 2016a. *Neotoma goldmani*. The IUCN Red List of Threatened Species 2016:e.T14588A115122856.

de Grammont, P. C., and A. Cuarón. 2016b. *Notocitellus adocetus*. The IUCN Red List of Threatened Species 2016:e.T20477A22265744.

de Grammont, P. C., and A. Cuarón. 2016c. *Peromyscus yucatanicus*. The IUCN Red List of Threatened Species 2016:e.T16696A22362477.

de Grammont, P. C., and A. Cuarón. 2016d. *Sciurus alleni*. The IUCN Red List of Threatened Species 2016:e.T20004A22248517.

de Grammont, P. C., and A. D. Cuarón. 2018a. *Tamiasciurus mearnsi*. The IUCN Red List of Threatened Species 2018:e.T21378A22250725.

de Grammont, P. C., and A. D. Cuarón. 2018b. *Microtus oaxacensis*. The IUCN Red List of Threatened Species 2018:e.T13449A22346873.

de Grammont, P. C., and A. D. Cuarón. 2018c. *Microtus umbrosus*. The IUCN Red List of Threatened Species 2018:e.T42635A22346455.

de Grammont, P. C., and A. D. Cuarón. 2018d. *Sigmodon alleni*. The IUCN Red List of Threatened Species 2018:e.T115591937A22355693.

de Grammont, P. C., and J. Matson. 2018. *Sorex macrodon*. The IUCN Red List of Threatened Species 2018:e.T20392A22314529.

de Grammont, P. C., A. Cuarón, and E. Vázquez. 2016. *Sciurus colliaei*. The IUCN Red List of Threatened Species 2016:e.T20007A22248115.

de la Sancha, N., R. Pérez-Hernandez, L. P. Costa, D. Brito, and N. Cáceres. 2016. *Philander opossum*. The IUCN Red List of Threatened Species 2016:e.T40516A22176779.

de la Torre, L. 1956. The dental formula of the bats of the genus *Diaemus*. Proceedings of the Biological Socociety of Washington 69:191–192.

de la Torre, J. A., and R. A. Medellín. 2010. *Pteronotus personatus*. Mammalian Species 42:244–250.

de Oliveira, T. G. 1998a. *Herpailurus yagouaroundi*. Mammalian Species 578:1–6.

de Oliveira, T. G. 1998b. *Leopardus wiedii*. Mammalian Species 579:1–6.

de Oliveira, T., A. Paviolo, J. Schipper, R. Bianchi, E. Payan, and S. V. Carvajal. 2015. *Leopardus wiedii*. The IUCN Red List of Threatened Species 2015:e.T11511A50654216.

de Villa-Meza, A., R. Ávila-Flores, A. D. Cuarón, and D. Valenzuela-Galván. 2011. *Procyon pygmaeus* (Carnivora: Procyonidae). Mammalian Species 43:87–93.

Decher, J., and J. R. Choate. 1995. *Myotis grisescens*. Mammalian Species 510:1–7.

Decker, D. M., and W. C. Wozencraft. 1991. Phylogenetic analysis of recent procyonid genera. Journal of Mammalogy 72:42–55.

D'Elía, G., P. Fabre, and E. P. Lessa. 2019. Rodent systematics in an age of discovery: recent advances and prospects. Journal of Mammalogy 100:852–871.

Delgado, C., D. Tirira, M. Gómez-Laverde, J. Matson, and R. Samudio. 2016a. *Reithrodontomys mexicanus*. The IUCN Red List of Threatened Species 2016:e.T19411A115151358.

Delgado, C., M. Aguilera, R. Timm, and R. Samudio. 2016b. *Sigmodon hirsutus*. The IUCN Red List of Threatened Species 2016:e.T136426A115207583.

Delibes-Mateos, M., R. Villafuerte, B. D. Cooke, and P. C. Alves. 2018. *Oryctolagus cuniculus* (Linnaeus, 1758). Pp. 99–104, *in* Lagomorphs: Pikas, Rabbits, and Hares of the World (Smith, A. T., C. H. Johnston, P. Alves, and K. Hackländer, K., eds). Johns Hopkins University Press, Baltimore, U.S.A.

DeMaster, D. P., and I. Stirling. 1981. *Ursus maritimus*. Mammalian Species 145:1–7.

Demastes, J. W. 1994. Systematics and zoogeography of the mer rouge pocket gopher (*Geomys breviceps breviceps*) based on cytochrome-b sequences. The Southwestern Naturalist 39:276–280.

Demastes, J. W., T. A. Spradling, M. S. Hafner, D. J. Hafner, and D. L. Reed. 2002. Systematics and phylogeography in pocket gophers in the genera *Cratogeomys* and *Pappogeomys*. Molecular Phylogenetics and Evolution 22:144–154.

Demboski, J. R., and J. A. Cook. 2001. Phylogeography of the dusky shrew, *Sorex monticolus* (Insectivora, Soricidae): insight into deep and shallow history in northwestern North America. Molecular Ecology 10:1227–1240.

Demboski, J. R., and J. A. Cook. 2003. Phylogenetic diversification within the *Sorex cinereus* group (Soricidae). Journal of Mammalogy 84:144–158.

Desmastes, J. W., A. L. Butt, M. S. Hafner, and J. E. Light. 2003. Systematics of a rare species of pocket gopher *Pappogeomys alcorni*. Journal of Mammalogy 84:753–761.

Deutsch, C. J., J. P. Reid, R. K. Bonde, D. E. Easton, H. I. Kochman, and T. J. O'Shea. 2003. Seasonal movements, migratory behavior, and site fidelity of West Indian manatees along the Atlantic coast of the United States. Wildlife Monographs 51:1–77.

Deutsch, C. J., C. Self-Sullivan, and A. Mignucci-Giannoni. 2008. *Trichechus manatus*. The IUCN Red List of Threatened Species 2008:e.T22103A9356917.

DeWalt, T. S., E. G. Zimmerman, and J. V. Planz. 1993a. Mitochondrial-DNA phylogeny of species of the *boylii* and *truei* group of the genus *Peromyscus*. Journal of Mammalogy 74:352–362.

DeWalt, T. S., P. D. Sudman, M. S. Hafner, and S. K. Davis. 1993b. Phylogenetic relationships of pocket gophers (*Cratogeomys* and *Pappogeomys*) based on mitochondrial DNA cytochrome b sequences. Molecular Phylogenetics and Evolution 2:193–204.

Dice, L. R. 1937. Mammals of the San Carlos Mountains and vicinity. University of Michigan Studies, Scientific Series 12:245–268.

Dick, T. S. 1997. *Reithrodontomys humulis*. Mammalian Species 565:1–6.

Dickerson, B. R., R. R. Ream, S. N. Vignieri, and P. Bentzen. 2010. Population structure as revealed by mtDNA and microsatellites in northern fur seals, *Callorhinus ursinus*, throughout their range. PLoS ONE:e10671.

Dickey, D. R. 1928. Five new mammals of the genus *Peromyscus* from El Salvador. Proceedings of the Biological Society of Washington 41:1–6.

Diersing, V. E. 1980. Systematics and Evolution of the Pygmy Shrews (Subgenus *Microsorex*) of North America. Journal of Mammalogy 61:76–101.

Diersing, V. E. 1981. Systematic status of *Sylvilagus brasiliensis* and *S. insonus* from North America. Journal of Mammalogy 62:539–556.

Diersing, V. E. 2019. Taxonomic revision of the long-tailed shrew, *Sorex dispar* Batchelder, 1911, from the Appalachian Region of North America, with the description of a new subspecies. Journal of Mammalogy 100:1837–1846.

Diersing, V. E., and D. F. Hoffmeister. 1977. Revision of the shrews *Sorex merriami* and a description of a new species of the subgenus *Sorex*. Journal of Mammalogy 58:321–333.

Diersing, V. E., and D. F. Hoffmeister. 1981. Distribution and systematics of the Masked Shrew (*Sorex cinereus*) in Illinois. Natural History Miscellanea 213:1–11.

Dobson, G. E. 1878. Catalogue of the Chiroptera of the British Museum. British Museum. London, England.

Dobson, G. E. 1885. Notes on species of Chiroptera in the collection of the Genoa Civic Museum, with descriptions of new species. Annali del Museo Civico di Storia Naturale di Genova 22:16–19.

DOF (Diario Oficial de la Federación). 2010. NOM-059-SEMARNAT-2010, Protección ambiental-Especies nativas de México de flora y fauna silvestres-Categorías de riesgo y especificaciones para la inclusión, exclusión o cambio-Lista de especies en riesgo. Diario Oficial de la Federación, 30 de diciembre de 2010. Ciudad de México. México.

DOF (Diario Oficial de la Federación). 2019. Modificación del anexo normativo III. lista de especies en riesgo de la Norma Oficial Mexicana NOM-059-SEMARNAT-2010, Protección ambiental-Especies nativas de México de flora y fauna silvestres-Categorías de riesgo y especificaciones para la inclusión, exclusión o cambio-Lista de especies en riesgo. Diario Oficial de la Federación, 14 de noviembre de 2019. Ciudad de México, México.

Dokuchaev, N. E.1994. Siberian shrew Sorex minutissimus found in Alaska. Zoologichesky Zhurnal 73:254–256.

Dokuchaev, N. E. 1997. A new species of shrew (Soricidae, Insectivora) from Alaska. Journal of Mammalogy 78: 811–17.

Dolan, J. M., Jr. 1963. Beitrag zur systematischen Gliederung des Tribus Rupicaprini Simpson, 1945. Zeitschrift für zoologische Systematik und Evolutionsforschung 1:311–407.
Dolan, P. G. 1989. Systematics of Middle American mastiff bats of the genus *Molossus*. Special Publications of the Museum, Texas Tech University 29:1–71.
Dolan, P. G., and D. C. Carter. 1977. *Glaucomys volans*. Mammalian Species 78:1–6.
Dolan, P. G., and D. C. Carter. 1979. Distributional notes and records for Middle American Chiroptera. Journal of Mammalogy 60:644–649.
Dolan, P. G., and R. L. Honeycutt. 1978. Systematic and evolutionary implications of genic variation in the mastiff bat, *Eumops* (Chiroptera: Molossidae). Bat Research News 19:72.
Domínguez-Castellanos, Y., and J. Ortega. 2003. *Liomys spectabilis*. Mammalian Species 718:1–3.
Domning, D. 2016. *Hydrodamalis gigas*. The IUCN Red List of Threatened Species 2016:e.T10303A43792683.
Domning, D. P., and H. Furusawa. 1995. Summary of taxa and distribution of Sirenia in the North Pacific Ocean. *Island Arc* 3: 506–512.
Domning, D. P., J. Thomason, and D. G. Corbett. 2007. Steller's sea cow in the Aleutian Islands. Marine Mammal Science 23:976–983.
Donkin, R. A. 1985. The peccary-with observations on the introduction of pigs to the New World. Transactions of the American Philosophical Society 75:1–152.
Doroff, A., and A. Burdin. 2015. *Enhydra lutris*. The IUCN Red List of Threatened Species 2015:e.T7750A21939518.
Dowler, R. C., and M. D. Engstrom. 1988. Distributional records of mammals from the southwestern Yucatan peninsula of Mexico. Annals Carnegie Museum 57:159–166.
Dowler, R. C., and H. H. Genoways. 1978. *Liomys irroratus*. Mammalian Species 82:1–6.
Dragoo, J. W. 2009. Family Mephitidae (skunks). Pp. 532–563, *in* Handbook of the mammals of the world, vol. 1 Carnivores (Wilson, D. E., and R. A. Mittermeier, eds.). Lynx Edicions, CI and IUCN. Barcelona, Spain.
Dragoo J. W., and R. L. Honeycutt 1997. Systematics of Mustelid-like Carnivores. Journal of Mammalogy 78:426–443.
Dragoo, J. W., and S. R. Sheffield. 2009. *Conepatus leuconotus* (Carnivora: Mephitidae). Mammalian Species 827:1–8.
Dragoo, J. W., J. R. Choate, T. L. Yates, and T. P. O'Farrell. 1990. Evolutionary and taxonomic relationships among North American arid-land foxes. Journal of Mammalogy 71:318–332.
Dragoo, J. W., R. L. Honeycutt, and D. J. Schmidly. 2003. Taxonomic status of the white-backed hog-nosed skunks, genus *Conepatus* (Carnivora: Mephitidae). Journal of Mammalogy 84:159–176.
Drew, R. E., J. G. Hallett, K. B. Aubry, K. W. Cullings, S. M. Koepf, and W. J. Zielinski. 2003. Conservation genetics of the fisher (*Martes pennanti*) based on mitochondrial DNA sequencing. Molecular Ecology 12:51–62.
Driscoll, C. A., M. Menotti-Raymond, A. L. Roca, K. Hupe, W. E. Johnson, E. Geffen, E. H. Harley, M. Delibes, D. Pontier, A. C. Kitchener, N. Yamaguchi, S. J. O'Brien, and D. W. Macdonald. 2007. The Near Eastern origin of cat domestication. Science 317:519–523.
Duckworth, J. W., N. S. Kumar, M. Anwarul Islam, H. Sagar Baral, and R. Timmins. 2015. Axis axis. The IUCN Red List of Threatened Species 2015:e.T41783A22158006.
Durish, N. D., K. E. Halcomb, C. W. Kilpatrick, and R. D. Bradley. 2004. Molecular systematics of the Peromyscus truei species group. Journal of Mammalalogy 85:1160–1169.
Durant, S. D. 1952. Mammals of Utah. University of Kansas Publications, Museum of Natural History 6:1–549.
Duplechin, R. M., and R. D. Bradley. 2014. *Peromyscus oaxacensis* Merrian, 1898 Oaxacan deermouse. Pp. 383–384, *in* Mammals of México (Ceballos, G., ed.). John Hopkins University Press. Baltimore, U.S.A.
Easterla, D. A. 1965. The spotted bat in Utah. Journal of Mammalogy 46:665–668.
Easterla, D. A. 1968a. First records of the pocketed free-tailed bat for Texas. Journal of Mammalogy 49:515–516.
Easterla, D. A. 1968b. Range extension of the southern bog lemming in Arkansas. The Southwestern Naturalist 13:364.
Easterla, D. A. 1968c. Hispid cotton rat north of the Missouri River. The Southwestern Naturalist 13:364–365.
Easterla, D. A. 1968d. First records of *Blarina brevicauda* minima in Missouri and Arkansas. The Southwestern Naturalist 13:448–449.
Easterla, D. A. 1970. First records of the spotted bat in Texas and notes on its natural history. The American Midland Naturalist 83:306–308.
Easterla, D. A. 1971. Notes on young and adults of the spotted bat, *Euderma maculatum.* Journal of Mammalogy 52:475–476.
Edelman, A. J. 2003. *Marmota olympus*. Mammalian Species 736:1–5.
Edelman, A. J. 2019. *Sylvilagus obscurus* (Lagomorpha: Leporidae). Mammalian Species 51:128–135.
Edwards, R. L. 1963. Observations on the small mammals of the southeastern shore of Hudson Bay. Canadian Field Naturalist 77:1–12.
Edwards, C. W., and R. D. Bradley. 2001. Molecular phylogenetics of the *Neotoma floridana* species group. Journal of Mammalogy 82:791–798.
Edwards, C. W., and R. D. Bradley. 2002. Molecular Systematics and Historical Phylobiogeography of the *Neotoma mexicana* Species Group. Journal of Mammalogy 83:20–30.
Edwards, C. W., and R. D. Bradley. 2003. Molecular systematics of the genus *Neotoma*. Molecular Phylogenetics and Evolution 25:489–500.
Edwards, C. W., C. F. Fulhorst, and R. D. Bradley. 2001. Molecular phylogenetics of the *Neotoma albigula* species group: further evidence of a paraphyletic assemblage. Journal of Mammalogy 82:267–279.
Eger, J. L. 1974. A new subspecies of the bat *Eumops auripendulus* (Chiroptera: Molossidae), from Argentina and Eastern Brazil. Life Sciences Occasional Papers, Royal Ontario Museum 25:1–8.
Eger, J. L. 1977. Systematic of the genus *Eumops* (Chiroptera: Molossidae). Life Sciences Contributions, Royal Ontario Museum 110:1–69.
Eger, J. L. 1990. Patterns of geographic variation in the skull of Neartic ermine (*Mustela erminea*). Canadian Journal of Zoology 68:1241–1249.
Eger, J. L. 1995. Morphometric variation in the Nearctic collared lemming (*Dicrostonyx*). Journal of Zoology of London 235:143–161.
Eger, J. L. 2008. Family Molossidae P. Gervais, 1856. Pp. 399–439, *in* Mammals of South America. Vol. 1: Marsupials, xenarthrans, shrews, and bats (Gardner, A. L., ed.). University of Chicago Press. Chicago, U.S.A.
Egoscue, H. J. 1979. *Vulpes velox*. Mammalian Species 122:1–5.
Eisenberg, J. F. 1989. Mammals of the Neotropics: The northern Neotropics, Panama, Columbia, Venezuela, Guyana, Suriname, French Guiana. University of Chicago. Chicago, U. S. A.
Eisenberg, J. F., and K. H. Redford. 1999. Mammals of the Neotropics, 3. The central Neotropics. University of Chicago Press. Chicago, U.S.A.
Eizirik, E., T. Haag., A. S. Santos, F. M. Salzano, L. Silveira, F. C. C. Azevedo, and M. M. Furtado. 2008. Jaguar conservation genetics. Cat News Special Issue 4:31–34.

Elder, F. F. B., and M. R. Lee. 1985. The chromosomes of *Sigmodon ochrognathus* and *S. fulviventer* suggest a realignment of *Sigmodon* species groups. Journal of Mammalogy 66:511–518.

Elizalde-Arellano, C., J. C. López-Vidal, E. Quhart, J. I. Campos-Rodríguez, and R. Hernández-Arciga. 2010. Nuevos registros y extensiones de distribución de mamíferos para Guanajuato, México. Acta Zoológica Mexicana (n. s.) 26:73–98.

Elizalde-Arellano, C., J. C. López-Vidal, E. Uría-Galicia, H. Montellano-Rosales, J. Arroyo-Cabrales, and R. A. Medellín. 2008. Citologia vaginal y ciclo estral de *Diphylla ecaudata*. Pp. 253–268, *in* Avances en el estudio de los mamíferos de México (Lorenzo, C., E. Espinoza, and J. Ortega, eds.). Volumen II. Asociación Mexicana de Mastozoología, A. C. y el Colegio de la Frontera Sur. San Crístóbal de las Casas, México.

Elizalde-Arellano, C., J. C. López-Vidal, J. Arroyo-Cabrales, R. A. Medellín, and J. W. Laundré. 2007. Food sharing behavior in the hairy–legged vampire bat *Diphylla ecaudata*. Acta Chiropterologica 9:314–319.

Ellerman, J. R. 1940. The families and genera of living rodents. Volume I. Rodents other than Muridae. British Museum (Natural History), London, United Kingdom.

Ellerman, J. R. 1941. The families and genera of living rodents. British Museum (Natural History), London 2:1–690.

Ellerman, J. R., and T. C. S. Morrison-Scott. 1951. Checklist of Palaearctic and indian mammals 1758 to 1946. British Museum (Natural History), London, United Kingdom.

Ellerman, J. R., R. W. Hayman, and G. W. C. Holt. 1941. The families and genera of living rodents, with a list of names from (1758–1936). Brtish Museum, Natural history London 2:1–690.

Elliot, D. G. 1901. A synopsis of the mammals of North America and the adjacent seas. Field Columbian Museum Publications, Zoology Series 2:1–522.

Elliot, D. G. 1903. A list of a collection of Mexican mammals with descriptions of some apparently new forms. Field Columbian Museum Publications, Zoology Series 71:141–522.

Elliot, D. G. 1905. A checklist of mammals of the North American continent, the West Indes and the neighboring seas. Field Columbian Museum Publications Zoological series 6:1–192.

Elliott, C. L., and J. T. Flinders. 1991. *Spermophilus columbianus*. Mammalian Species 372:1–9.

Emmons, L. H. 1997. Mammals of the Rio Urucuti Basin, south central Chuquisaca, Bolivia. Pp. 30–33, *in* A rapid assessment of the humid forest of south central Chuquisaca, Bolivia (Schulenberg, T. S., and K. Awbrey, eds.). Conservation International, RAP Working Papers 8, Washington, U.S.A.

Emmons, L. H. 2005. A Revision of the Genera of Arboreal Echimyidae (Rodentia: echimyidae, Echimyinae), with descriptions of two new genera. Pp. 247–310, *in* Mammalian Diversification: From Chromosomes to Phylogeography (A Celebration of the Career of James L. Patton) (Lacey, E. A., and P. Myers). University of California Publications in Zoology.

Emmons, L. 2016a. *Cuniculus paca*. The IUCN Red List of Threatened Species 2016:e.T699A22197347.

Emmons, L. 2016b. *Dasyprocta punctata*. The IUCN Red List of Threatened Species 2016:e.T89497686A78319610.

Emmons, L. 2016c. *Erethizon dorsatum*. The IUCN Red List of Threatened Species 2016:e.T8004A22213161.

Emmons, L. H., and F. Feer. 1990. Neotropical rainforest mammals. A field guide. University of Chicago Press. Chicago, U.S.A.

Emmons, L. H., and F. Feer. 1997. Neotropical rainforest mammals, a field guide. 2nd ed. University of Chicago Press.

Engstrom, M. D. 1984. Chromosomal, genic, and morphological variation in the *Oryzomys melanotis* species group. PhD. Dissertation, Texas A & M University, College Station, U.S.A.

Engstrom, M. D., and D. E. Wilson. 1981. Systematics of *Antrozous dubiaquercus* (Chiroptera: Vespertilionidae), with comments on the status of *Bauerus* Van Gelder. Annals Carnegie Museum 50:371–383.

Engstrom, M. D., H. H. Genoways, and P. K. Tucker. 1987a. Morphological variation, karyology, and systematic relationships of *Heteromys gaumeri* (Rodentia: Heteromyidae). Pp. 289–303, *in* Studies in Neotropical mammalogy: essays in honor of Philip Hershkovitz (Patterson, B. D., and R. M. Timm, eds.). Fieldiana: Zoological, new series 39:1–506.

Engstrom, M. D., T. E. Lee, and D. E. Wilson. 1987b. *Bauerus dubiaquercus*. Mammalian Species 282:1–3.

Engstrom, M. D., O. Sánchez-Herrera, and G. Urbano-Vidales. 1992. Distribution, geographic variation, and systematic relationships within *Nelsonia* (Rodentia: Sigmodontinae). Proceedings of the Biological Society of Washington 105:867–881.

Erratum. 2014. Erratum and Nomenclature Committee Opinion on use of *Perognathus alticola* versus *P. alticolus*. Journal of Mammalogy 95:910.

Ernest, K. A., and M. A. Mares. 1987. *Spermophilus tereticaudus*. Mammalian Species 274:1–9.

Escobedo-Morales, L. A., L. León-Paniagua, J. Arroyo-Cabrales, and F. Greenaway. 2006. Distributional records for mammals from Chiapas, México. The Southwestern Naturalist 51:269–272.

Escobedo-Morales, L. A., S. Mandujano, L. E. Eguiarte, M. A. Rodrguez-Rodrguez, and J. E. Maldonado. 2016. First phylogenetic analysis of mesoamerican brocket deer *Mazama pandora* and *Mazama temama* (Cetartiodactyla: Cervidae) based on mitochondrial sequences: implications on neotropical deer evolution. Mammalian Biology: Zeitschrift für Säugetierkunde 81:303–313.

Escobedo-Morales, L. A., L. León-Paniagua, E. Martínez-Meyer, and S. Mandujano. 2023. Reevaluation of the status of the Central American brocket deer *Mazama temama* (Artiodactyla: Cervidae) subspecies based on morphological and environmental evidence. Journal of Mammalogy 104:333–346.

Eshelman, B. D., and G. N. Cameron. 1987. *Baiomys taylori*. Mammalian Species 285:1–7.

Eshelman, B. D., and C. S. Sonnemann. 2000. *Spermophilus armatus*. Mammalian Species 637:1–6.

Espinoza, J., C. Lorenzo, and E. Rios. 2011. Variación morfológica y morfométrica de *Heteromys desmarestianus* en Chiapas, México. Therya 2:139–154.

Estes, J. A. 1980. *Enhydra lutris*. Mammalian Species 133:1–8.

Esteva, M., A. F. Cervantes, S. V. Bryant, and J. A. Cook. 2010. Molecular phylogeny of long-tailed shrews (genus *Sorex*) from Mexico and Guatemala. Zootaxa 2615:47–65.

Fabre, P.-H., J. L. Patton, and Y. L. R. Leite. 2016. Family Echimyidae (hutias, South American spiny-rats and coypu). Pp. 552–641, *in* Handbook of the Mammals of the World. Vol 6. Lagomorphs and Rodents I (Wilson, D. E., T. E. Lacher, Jr, and R. A. Mittermeier, eds.). Lynx Edicions. Barcelona, Spain.

Faries, K. M., T. V. Kristensen, J. Beringer, J. D. Clark, D. White, Jr, and L. S. Eggert. 2013. Origins and genetic structure of black bears in the Interior Highlands of North America. Journal of Mammalogy 94:369–377.

Fashing, N. J. 1973. Implications of karyotypic variation in the kangaroo rat, *Dipodomys heermanni*. Journal of Mammalogy 54:1018–1020.

Faulhaber, C. A., and A. T. Smith. 2008. *Sylvilagus palustris*. The IUCN Red List of Threatened Species 2008:e.T41303A10435830.

Fauteaux, D., G. Lupien, F. Fabianek, J. Gagnon, M. Séguy, and L. Imbeau. 2014. An illustrated key to the mandibles of small mammals of eastern Canada. Canadian Field Naturalist 128:25–37.

Fay, F. H. 1981. Walrus *Odobenus rosmarus* (Linneaus, 1758). Pp. 1–23, *in* Handbook of Marine Mammals. The walrus, sea lions, fur seals, and sea otter (Ridgway, S. H., and R. Harrison, eds.). Academic Press. San Diego, U.S.A.

Fay, F. H. 1982. Ecology and biology of the Pacific walrus, *Odobenus rosmarus divergens* Illiger. North American Fauna 74:1–279.

Fay, F. H. 1985. *Odobenus rosmarus*. Mammalian Species 238:1–7.

Feijó, A., B. Patterson, and P. Cordeiro-Estrela. 2018. Taxonomic revision of the long-nosed armadillos, Genus *Dasypus* Linnaeus, 1758 (Mammalia, Cingulata). PLoS ONE 13:e0195084.

Feijó, A., J. F. Vilela, J. Cheng, M. A. A. Schetino, R. T. F. Coimbra, C. R. Bonvicino, F. R. Santos, B. D. Patterson, and P. Cordeiro-Estrela. 2019. Phylogeny and molecular species delimitation of long-nosed armadillos (*Dasypus*: Cingulata) supports morphology-based taxonomy. Zoological Journal of the Linnean Society 186:813–825.

Fellers, G. M., W. Z. Lidicker Jr., A. Linzey, and NatureServe. 2016. *Aplodontia rufa*. The IUCN Red List of Threatened Species 2016:e. T1869A115057269.

Feldhamer, G. A. 1993. Habitat partitioning, body size, and timing of parturition in pygmy shrews and associated soricids. Journal of Mammalogy 74:403–411.

Felten, V. H. 1956. Fledermäuse (Mammalia, Chiroptera) aus El Salvador. Senckenbergiana Biologica 37:179–212.

Fenton, M. B., and R. M. R. Barclay. 1980. *Myotis lucifugus*. Mammalian Species 142:1–8.

Fernández, J. A. 2012. Phylogenetics and biogeography of the microendemic rodent *Xerospermophilus perotensis* (Perote ground squirrel) in the Oriental Basin of Mexico. Journal of Mammalogy 93:1431–1439.

Fernández, J. A. 2014. Mitochondrial phylogenetics of a rare Mexican endemic Nelson's woodrat, *Neotoma nelsoni* (Rodentia: Cricetidae), with comments on its biogeographic history. The Southwestern Naturalist 59:81–90.

Fernández, J. A., F. García-Campusano, and M. S. Hafner. 2010. *Peromyscus difficilis*. Mammalian Species 867:220–229.

Fernández, J. A., F. A. Cervantes, and M. S. Hafner. 2012. Molecular systematics and biogeography of the Mexican endemic kangaroo rat, *Dipodomys phillipsii* (Rodentia: Heteromyidae). Journal of Mammalogy 93:560–571.

Ferrell, C. S., and D. E. Wilson. 1991. *Platyrrhinus helleri*. Mammalian Species 373:1–5.

Fertl, D., A. J. Schiro, G. T. Regan, C. A. Beck, N. Adimey, L. Price-May, A. Amos, G. A. J. Worthy, and R. Crossland. 2005. Manatee occurrence in the northern Gulf of Mexico, west of Florida. Gulf and Caribbean Research 17:69–94.

Festa-Bianchet, M. 2020a. *Ovis canadensis*. The IUCN Red List of Threatened Species 2020:e.T15735A22146699.

Festa-Bianchet, M. 2020b. *Ovis dalli*. The IUCN Red List of Threatened Species 2020:e.T39250A22149895.

Festa-Bianchet, M. 2022. *Oreamnos americanus*. The IUCN Red List of Threatened Species 2022:e.T42680A211860282. Accessed on 11 August 2022.

Findley, J. S. 1955a. Taxonomy and distribution of some American shrews. University of Kansas Publications, Museum of Natural History 7:613–618.

Findley, J. S. 1955b. Speciation of the wandering shrew. University of Kansas Publications, Museum of Natural History 9:1–68.

Findley, J. S. 1967. A black population of the Goldman pocket mouse. The Southwestern Naturalist 12:191–192.

Findley, J. S. 1972. Phenetic relationships among bats of the genus *Myotis*. Systematic Zoology 21:31–52.

Findley, J. S. 1987. The natural history of New Mexican mammals. University of New Mexico Press, Albuquerque, U.S.A.

Findley, J. S., and C. Jones. 1960. Geographic variation in the yellow-nosed cotton rat. Journal of Mammalogy 41:462–469.

Findley, J. S., and C. Jones. 1967. Taxonomic relationships of bats of the species *Myotis fortidens*, *M. lucifugus*, and *M. occultus*. Journal of Mammalogy 48:429–444.

Findley, J. S., A. H. Harris, D. E. Wilson, and C. Jones. 1975. Mammals of New Mexico. University of New Mexico Press. Albuquerque, U.S.A.

Fisler, G. F. 1965. Adaptations and speciation in harvest mice of the marshes of San Francisco Bay. University of California Publications in Zoology 77:1–108.

Fish, F. E., B. R. Blood, and B. D. Clark. 1991. Hydrodynamics of the feet of fish-catching bats: influence of the wáter surface on drag and morphological design. Journal of Experimental Zoology 258:164–173.

Fitch, J. H., and K. A. Shump. 1979. *Myotis keenii*. Mammalian Species 121:1–3.

Fitch, J. H., K. A. Shump, A. U. Shump. 1981. *Myotis velifer*. Mammalian Species 149:1–5.

Fleharty, E. D. 1960. The status of the gray-necked chipmunk in New Mexico. Journal of Mammalogy 41:235–242.

Fleming, T. H., E. T. Hooper, and D. E. Wilson. 1972. Three Central American bat communities: structure, reproductive cycles, and movement patterns. Ecology 53:655–670.

Flyger, V., and J. E. Gates. 1982. Fox and gray squirrels. Pp. 209–229, *in* Wild mammals of North America (Chapman, J. A., and G. A. Feldhamer, eds.). Johns Hopkins University Press. Baltimore, U.S.A.

Flynn, L. J., L. L. Jacobs, Y. Kimura, and E. H. Lindsay. 2019. Rodent Suborders. Fossil Imprint 75:292–298.

Folkow, L. P., E. S. Nordoy, and A. S. Blix. 2010. Remarkable development of diving performance and migrations of hooded seals (*Cystophora cristata*) during their first year of life. Polar Biology 33:433–441.

Fonseca, R. M., S. R. Hoofer, C. A. Porter, C. A. Cline, D. A. Parish, F. G. Hoffmann, and R. J. Baker 2007. Morphological and molecular variation within Little big-eared bats of the genus M*icronycteris* (Phyllostomidae: Micronycterinae) from San Lorenzo, Ecuador. Pp. 721–746, *in* The quintessential naturalist: Honoring the life and legacy of Oliver P. Pearson (Kelt, D. A., E. P. Lessa, J. Salazar-Bravo, and J. L. Patton, eds.). University of California Publications in Zoology 134:1–981.

Forsten, A., and P. M. Youngman. 1982. *Hydrodamalis gigas*. Mammalian Species 165:1–3.

Ford, L. S., and R. S. Hoffmann. 1988. *Potos flavus*. Mammalian Species 321:1–9.

Francis, C. M., and N. B. Simmons. 2022. On behalf of the G. B. T. W. G. of the IUCN SSC Bat Specialist Group. On the taxonomy of *Myotis lucifugus*.

Franzmann, A. W. 1978. Moose. Pp. 67–82, *in* Big game of North America: ecology and management (Schmidt, J. L. and D. L. Gilbert, eds.). Stackpole Books. Harrisburg, U.S.A.

Franzmann, A. W. 1981. *Alce alce*. Mammalian Species 154:1–7.
Franzmann, A. W., *et al.* 1978. Alaskan moose measurements and weights and measurement-weight relationships. Canadian Journal of Zoology 56:298–306.
Frase, B. A., and R. S. Hoffmann. 1980. *Marmota flaviventris*. Mammalian Species 135:1–8.
Freeman, P. W. 1981. A multivariate study of the family Molossidae (Mammalia: Chiroptera): morphology, ecology, evolution. Fieldiana Zoology (New Series) 7:1–173.
French, T. W., and G. L. Kirkland Jr., 1983. Taxonomy of the Gaspe shrew, *Sorex gaspensis*, and the rock shrew, *S. dispar*. Canadian Field-Naturalist 97:75–78.
Frey, J. K. 1999. Mogollon vole/*Microtus mogollonensis*. Pp. 634–635, *in* The Smithsonian book of North American Mammals (Wilson, D. E., and S. Ruff, eds.). Smithsonian Institution Press. Washington, U.S.A.
Frey, J. K., and F. A. Cervantes. 1997a. *Microtus umbrosus*. Mammalian Species 555:1–3.
Frey, J. K., and F. A. Cervantes. 1997b. *Microtus oaxacensis*. Mammalian Species 556:1–3.
Frey, J. K., and C. T. LaRue. 1993. Note on the distribution of the Mogollon vole (*Microtus mogollonensis*) in New Mexico and Arizona. The Southwestern Naturalist 38:176–178.
Frey, J. K., and D. W. Moore. 1990. Status of Hayden's shrew (*Sorex haydeni*) in Kansas. The Southwestern Naturalist 35:84–86.
Frick, W. F., J. P. Hayes, and P. A. Heady, III. 2008. Island biogeography of bats in Baja California, Mexico: Patterns of bat species richness in a near-shore archipelago. Journal of Biogeography 35:353–364.
Frick, W. F., J. P. Hayes, P. A. Heady, III, and J. P. Haynes. 2009. Facultative nectar-feeding behavior in a gleaning insectivorous bat (*Antrozous pallidus*). Journal of Mammalogy 90:1157–1164.
Fritzell, E. K., and K. J. Haroldson. 1982. *Urocyon cinereoargenteus*. Mammalian Species 189:1–8.
Fujita, M. S., and T. H. Kunz. 1984. *Pipistrellus subflavus*. Mammalian Species 228:1–6.
Fumagalli, L., P. Taberlet, D. T. Stewart, L. Gielly, J. Hausser, and P. Vogel. 1999. Molecular phylogeny and evolution of *Sorex* shrews (Soricidae: Insectivora) inferred from mitocondrial DNA sequence data. Molecular Phylogenetics and Evolution 11:222–235.
FWS. 2022. U.S. Fish and Wildlife Service. https://www.fws.gov/. Accessed on 10 September 2022.
Galindo-Leal, C., and G. Zuleta. 1997. The distribution, habitat, and conservation status of the Pacific water shrew, *Sorex bendirii*, in British Columbia. Canadian-Field Naturalist 111:422–428.
Gallina, S., and H. Lopez Arevalo. 2016. *Odocoileus virginianus*. The IUCN Red List of Threatened Species 2016:e.T42394A22162580.
Gallo-Reynoso, J. P. 1994. Factors affecting the population status of the Guadalupe fur seal, *Arctocephalus townsendi* (Merriam, 1897), at Isla Guadalupe, Baja California, México. PhD thesis. University of California, Santa Cruz, U.S.A.
Gallo-Reynoso, J. P., and A. L. Figueroa-Carranza. 1996. Size and weight of Guadalupe fur seals. Marine Mammal Science 12:318–321.
Gallo-Reynoso, J. P., A. L. Figueroa-Carranza, I. D. Barba-Acuña, D. Borjes-Flores, and I. J. Pérez-Cossío. 2020. Steller Sea Lions (*Eumetopias jubatus*) Along the Western Coast of Mexico. Aquatic Mammals 46:411–416.
Gannon, W. L. 1988. *Zapus trinotatus*. Mammalian Species 315:1–5.
Gannon, W. L. 1999. *Zapus trinotatus*. Pp. 669–670, *in* The Smithsonian book of North American mammals (Wilson, D. E., and S. Ruff, eds.). Smithsonian Institution Press. Washington, U.S.A.
Gannon, W. L., and R. B. Forbes. 1995. *Tamias senex*. Mammalian Species 502:1–6.
Gannon, W. L., and T. E. Lawlor. 1989. Variation of the chip vocalization of three species of Townsend chipmunks (genus *Eutamias*). Journal of Mammalogy 70:740–753.
Gannon, M. R., M. R. Willig, and J. K. Jones, Jr. 1989. *Sturnira lilium*. Mammalian Species 333:1–5.
Gannon, W. L., R. B. Forbes, and D. E. Kain. 1993. *Tamias ochrogenys*. Mammalian Species 445:1–4.
Gentry, A., J. Clutton-Brock, and C. P. Groves. 2004. The naming of wild animal species and their domestic derivatives. Journal of Archaeological Science 31:645–651.
García, M., C. Jordan, G. O'Farril, C. Poot, N. Meyer, N. Estrada, R. Leonardo, E. Naranjo, Á. Simons, A. Herrera, C. Urgilés, C. Schank, L. Boshoff, and M. Ruiz-Galeano. 2016. *Tapirus bairdii*. The IUCN Red List of Threatened Species 2016:e.T21471A45173340.
García-Estrada, C., A. Damon, C. Sánchez-Hernández, L. Soto-Pinto, and G. Ibarra-Núñez. 2006. Bat diversity in montane rainforest and shaded coffee under different management regimes in southeastern Chiapas, Mexico. Biological Conservation 132:351–361.
García-Perea, R. 1992. New data on the systematics of *Lynxes*. Cat News 16:15–16.
García-Rodríguez, R. 1998. Phylogeography of the West Indian manatee (*Trichechus manatus*): how many populations and how many taxa? Molecular Ecology 7:1137–1149.
Gardner, A. L. 1963. Nota acerca de la distribución de los murciélagos en México. Revista de la Sociedad Mexicana de Historia Natural 24:41–44.
Gardner, A. L. 1965. New bat records from the Mexican state of Durango. Proceedings of the Western Foundation Vertebrate Zoology 1:101–106.
Gardner, A. L. 1973. The systematic of the genus *Didelphis* (Marsupialia: Didelphidae) in North and Middle America. Special Publications Museum Texas Tech University 4:1–81.
Gardner, A. L. 1977. Feeding habits. Pp. 293–350, *in* Biology of bats of the New World family Phyllostomatidae, Part 2 (Baker, R. J., J. K. Jones, Jr., and D. C. Carter, eds.). Special Publications Museum, Texas Tech University 13:1–364.
Gardner, A. L. 1982. Virginia oposum. Pp. 3–36, *in* Wild mammals of North America (Chapman, J. A., and G. A. Feldhamer, eds.). Johns Hopkins Press. Baltimore, U.S.A.
Gardner, A. L. 1986. The taxonomic status of *Glossophaga morenoi* Martinez y Villa, 1938 (Mammalia: Chiroptera: Phyllostomidae). Proceedings of the Biological Society of Washington 99:489–492.
Gardner, A. L. 1993. Order Didelphimorphia. Pp. 15–23, *in* Mammal species of the world: a taxonomic and geographic reference. (Wilson, D. E., and D. M. Reeder, eds.). Second edition. Smithsonian Institution Press. Washington, U.S.A.
Gardner, A. L. 2005a. Order Didelphiomorpha. Pp. 3–18, *in* Mammal species of the world: a taxonomic and geographic reference (Wilson, D. E., and D. M. Reeder, eds.). Third edition. Johns Hopkins University Press. Baltimore, U.S.A.
Gardner, A. L. 2005b. Order Cingulata. Pp. 94–99, *in* Mammal species of the world: a taxonomic and geographic reference (Wilson, D. E., and D. M. Reeder, eds.). Third edition. Johns Hopkins University Press. Baltimore, U.S.A.
Gardner, A. L. 2005c. Order Pilosa. Pp. 100–103, *in* Mammal species of the world: a taxonomic and geographic reference (Wilson, D. E., and D. M. Reeder, eds.). Third edition. Johns Hopkins University Press. Baltimore, U.S.A.

Gardner, A. L. 2007a. Suborder Vermilingua Illiger, 1811. Pp. 168–177, *in* Mammals of South America. Vol. 1. Marsupials, xenarthrans, shrews, and bats (Gardner, A. L. ed.). University of Chicago Press. Chicago, U.S.A.

Gardner, A. L. 2007b. Tribe Sturnirini [Miller, 1907]. Pp. 363–376, *in* Mammals of South America. Vol. 1. Marsupials, xenarthrans, shrews, and bats (Gardner, A. L., ed.). University of Chicago Press. Chicago, U.S.A.

Gardner, A. L. 2008. Mammals of South America. Vol. 1: marsupials, xenarthrans, shrews, and bats. University of Chicago Press. Chicago, U.S.A.

Gardner, A. L., and D. C. Carter. 1972. A review of the Peruvian species of *Vampyrops* (Chiroptera: Phyllostomatidae). Journal of Mammalogy 53:72–82.

Gardner, A. L., and V. L. Naples. 2007. Genus *Tamandua* Gray, 1825. Pp. 173–177, *in* Mammals of South America. Vol. 1. Marsupials, xenarthrans, shrews, and bats (Gardner, A. L. ed.). University of Chicago Press, Chicago, U.S.A.

Gardner, A. L., and J. L. Patton. 1972. New species of *Philander* (Marsupialia: Didelphidae) and *Mimon* (Chiroptera: Phyllostomidae) from Peru. Occasional Papers of the Museum of Zoology, Louisiana State University 43:1–12.

Gardner, A. L., and J. Ramírez-Pulido. 2020. Type localities of Mexican land mammals, with comments on taxonomy and nomenclature. Special Publications of the Museum of Texas Tech University 73: 1–134.

Gardner, A. L., R. K. LaVal, and D. E. Wilson. 1970. The distributional status of some Costa Rican bats. Journal of Mammalogy 51:712–729.

Garibaldi, A. 2019. *Dicrostonyx unalascensis*. The IUCN Red List of Threatened Species 2019:e.T39974A22331447.

Garrison, T. E. and T. L. Best. 1990. *Dipodomys ordii*. Mammalian Species 353:1–10.

Garshelis, D. L., B. K. Scheick, D. L. Doan-Crider, J. J. Beecham, and M. E. Obbard. 2016. *Ursus americanus*. The IUCN Red List of Threatened Species 2016:e.T41687A114251609.

Gatica-Colmina, A., B. Navarrete-Laborde, A. Ortiz-González, and O. C. Rosas-Rosas. 2014. Nuevo registro de distribución del puerco espín del norte *Erethizon dorsatum* en Chihuahua, México. Acta Zoológica Mexicana 30:399–402.

Geist, V. 1971. Mountain sheep: a study in behavior and evolution. University Chicago Press. Chicago, U.S.A.

Geist, V. 1998. Deer of the world: their evolution, behavior, and ecology. Stackpole Books. Mechanicsburg, U.S.A.

Geffen, E., A. Mercure, D. J. Girman, D. W. Macdonald, and R. K. Wayne. 1992. Phylogenetic relationships of the foxlike canids: mitochondrial DNA restriction fragment, site and cytochrome b sequence analyses. Journal of Zoology (London) 228:27–39.

Gelatt, T., and K. Sweeney. 2016. *Eumetopias jubatus*. The IUCN Red List of Threatened Species 2016:e.T8239A45225749.

Gelatt, T., R. Ream, and D. Johnson. 2015. *Callorhinus ursinus*. The IUCN Red List of Threatened Species 2015:e.T3590A45224953.

Geluso, K. 2009. Distributional Records for Seven Species of Mammals in Southern New Mexico. Occasional Papers Museum of Texas Tech University Number 287:1–7.

Geluso, K. N., and K. Geluso. 2015. Distribution and natural history of Nelson's pocket mouse (*Chaetodipus nelsoni*) in the Guadalupe Mountains in southwestern New Mexico. Occasional Papers Museum of Texas Tech University 332:1–20.

Genoways, H. H. 1971. A new species of spiny pocket mouse (genus *Liomys*) from Ialisco, Mexico. Occasional Papers of the Museum of Natural History, University of Kansas 5:1–7.

Genoways, H. H. 1973. Systematic and evolutionary relationships of spiny pocket mice, genus *Liomys*. Special Publications Museum Texas Tech University 5:1–368.

Genoways, H. H., and R. J. Baker. 1996. A new species of the genus *Rhogeessa,* with comments on geographic distribution and speciation in the genus. Pp. 83–87, *in* Contributions in mammalogy: a memorial volume honoring Dr. J. Knox Jones, Jr. (Genoways, H. H., and R. J. Baker, eds.). Museum of Texas Tech University, Lubbock, U.S.A.

Genoways, H. H., and E. C. Birney. 1974. *Neotoma alleni*. Mammalian Species 41:1–4.

Genoways, H. H., and J. H. Brown. 1993. Biology of the Heteromyidae. American Society of Mammalogists, Special Publication 10:1–719.

Genoways, H. H., and J. R. Choate. 1972. A multivariate analysis of systematic relationships among populations of the short-tailed shrews (genus *Blarina*) in the southeast. Systematic Zoology 21:106–116.

Genoways, H. H., and J. R. Choate. 1998. Natural history of the southern short-tailed shrew, *Blarina carolinensis*. Occasional Papers of the Museum of Southwestern Biology 8:1–43.

Genoways, H. H., and J. K. Jones, Jr. 1967. Notes on distribution and variation in the Mexican big-eared bat, *Plecotus phyllotis*. The Southwestern Naturalist 12:477–480.

Genoways, H. H., and J. K. Jones, Jr. 1969. Taxonomic status of certain long-eared bats (genus *Myotis*) from the southwestern United States and Mexico. The Southwestern Naturalist 14:1–13.

Genoways, H. H., and J. K. Jones, Jr. 1971. Systematics of southern banner-tailed kangaroo rats of the *Dipodomys phillipsii* group. Journal of Mammalogy 52:265–287.

Genoways, H. H., and J. K. Jones, Jr. 1972. Mammals from southwestern North Dakota. Occasional Papers Museum, Texas Tech University 6:1–36.

Genoways, H. H., and J. K. Jones, Jr. 1975. Annotated checklist of mammals of the Yucatan Peninsula, Mexico. IV. Carnivora, Sirenia, Perissodactyla, Artiodactyla. Occasional Papers The Museum, Texas Tech University 26:1–22.

Genoways, H. H., and S. L. Williams. 1979. Records of bats (Mammalia, Chiroptera) from Suriname. Annals of the Carnegie Museum 48:323–335.

Genoways, H. H., R. J. Baker, and J. E. Corney. 1977. Mammals of the Guadalupe Mountains National Park, Texas. Pp. 271–332, *in* Biological investigations of the Guadalupe Mountains National Park, Texas (Genoways, H. H., and R. J. Baker, eds.). National Park Service, Proceedings and Transactions Series Number 4:1–442.

Gentry, A. 2006. Mammal Species of the World. A taxonomic and geographic reference. 2005. D. E. Wilson & D.M. Reeder (Eds.) Ed. 3, 2 vols. Baltimore, Johns Hopkins University Press: A nomenclatural review. Bulletin of Zoological Nomenclature 63:215–219.

George, S. B. 1986. Evolution and historical biogeography of sorycine shrews. Systematic Zoology 35:153–162.

George, S. B. 1988. Systematics, historical biogeography, and evolution of the genus *Sorex*. Journal of Mammalogy 69:443–461.

George, S. B. 1989. *Sorex trowbridgiis*. Mammalian Species 337:1–5.

George, S. B., and J. D. Smith. 1991. Inter- and intraspecific variation among coastal and island populations of *Sorex monticolus* and *Sorex vagrans* in the Pacific Northwest. Pp. 75–91, *in* The biology of the Soricidae (Findley, J. S., and T. L. Jones, eds.). The Museum of Southwestern Biology, University of New Mexico, Albuquerque, U.S.A.

George, S. B., R. Choate, and H. H. Genoways. 1981. Distribution and taxonomic status of *Blarina hylophaga* Elliot (Insectivora: Soricidae). Annals of the Carnegie Museum 50:493–513.

George, S. B., J. R. Choate, and H. H. Genoways. 1986. *Blarina bravicauda*. Mammalian Species 261:1–9.

Giannatos, G., J. Herrero, and S. Lovari. 2007. *Capra hircus (Europe assessment)*. The IUCN Red List of Threatened Species 2007:e. T136383A4283792.

Gier, H. T. 1975. Ecology and social behavior of the coyote. Pp. 247–262, *in* The wild canids (Fox, M. W., ed.), Van Nostrand Reinhold. New York, U.S.A.

Gilbert, D. A., N. Lehman, S. J. O'Brien, and R. K. Wayne. 1990. Genetic fingerprinting reflects population differentiation in the California Channel Island fox. Nature 344:764–767.

Gill, A. E. and E. Yensen. 1992. Biochemical differentiation in the Idaho ground squirrel, *Spermophilus brunneus* (Rodentia: Sciuridae). Great Basin Naturalist 52:155–159.

Gillihan, S. W., and K. R. Foresman. 2004. *Sorex vagrans*. Mammalian Species 744:1–5.

Girón, L. E., J. G. Owen, and M. E. Rodríguez. 2010. Van Gelder's Bat (*Bauerus dubiaquercus*) from El Salvador, Central America. The Southwestern Naturalist 55:585–587.

Glass, B. P. 1947. Geographic variation in *Perognathus hispidus*. Journal of Mammalogy 28:174–179.

Glass, B. P., and R. J. Baker. 1968. The status of the name *Myotis subulatus* Say. Proceedings of the Biological Society of Washington 81:257–260.

Glass, B. P., and C. M. Ward. 1959. Bats of the genus *Myotis* from Oklahoma. Journal of Mammalogy 40:194–201.

Global Invasive Species Database. 2022a. Species profile: *Capra hircus*. Downloaded from http://www.iucngisd.org/gisd/speciesname/Capra+hircus on 20–12–2022.

Global Invasive Species Database. 2022b. Species profile: *Axis axis*. Downloaded from http://www.iucngisd.org/gisd/speciesname/Axis+axis on 20–12–2022.

Global Invasive Species Database. 2022c. Species profile: *Dama dama*. Downloaded from http://www.iucngisd.org/gisd/speciesname/Dama+dama on 20–12–2022.

Global Invasive Species Database. 2022d. Species profile: *Sus scrofa*. Downloaded from http://www.iucngisd.org/gisd/species.php?sc=73 on 20–12–2022.

Global Invasive Species Database. 2022e. Species profile: *Canis lupus*. Downloaded from http://www.iucngisd.org/gisd/speciesname/Canis+lupus on 20–12–2022.

Godin, A. J. 1977. Wild mammals of New England. Johns Hopkins University Press, Baltimore, U.S.A.

Goldman, E. A. 1905. Twelve new wood rats of the genus *Neotoma*. Proceedings of the Biological Survey of Washington 18:27–34.

Goldman, E. A. 1910. Revision of the wood rat of the genus *Neotoma*. North American Fauna 31:1–124.

Goldman, E. A. 1911. Revision of the spiny pocket mice (genera *Heteromys* and *Liomys*) North American Fauna 34:1–70.

Goldman, E. A. 1913. Descriptions of new mammals from Panama and Mexico. Smithsonian Miscellaneous Collection 60:1-20.

Goldman, E. A. 1915. Five new mammals from Mexico and Arizona. Proceedings of the Biological Society of Washington 28:133–137.

Goldman, E. A. 1918. The rice rats of North America (genus *Oryzomys*). North American Fauna 43:1–100.

Goldman, E. A. 1920. Mammals of Panama. Smithonian Miscellaneous Collections 69:1–306.

Goldman, E. A. 1927. A new woodrat from Arizona. Proceedings of the Biological Society of Washington 40:205–206.

Goldman, E. A. 1932. Revision of the wood rats of *Neotoma lepida* group. Journal of Mammalogy 13:59–67.

Goldman, E. A. 1937. The wolves of North America. Journal of Mammalogy 18:37–45.

Goldman, E. A. 1943. The races of the ocelot and margay in Middle America. Journal of Mammalogy 24:372–385.

Goldman, E. A. 1944. The wolves of North America. Pp. 387–636, *in* Classification of wolves. The American Wildlife Institution. Washington, U.S.A.

Goldman, E. A. 1945. A new pronghorn antelope from Sonora. Proceedings of the Biological Society of Washington 58:3–4.

Goldman, E. A. 1950. Raccoons of North and Middle America. North American Fauna 60:1–153.

Goldman, E. A. 1951. Biological investigations in Mexico. Smithsonian Miscellaneous Collection 115:1–476.

Goldman, E. A., and R. T. Moore. 1945. The biotic provinces of Mexico. Journal of Mammalogy 26:347–360.

Golley, F. B. 1962. Mammals of Georgia: a study of their distribution and functional role in the ecosystem. University Georgia Press. Athens, U.S.A.

Gompper, M. E. 1995. *Nasua narica*. Mammalian Species 487:1–10.

Gompper, M, and D. Jachowski. 2016. *Spilogale putorius*. The IUCN Red List of Threatened Species 2016:e.T41636A45211474.

Gongora, J., R. Reyna-Hurtado, H. Beck, A. Taber, M. Altrichter, and A. Keuroghlian. 2011. *Pecari tajacu*. The IUCN Red List of Threatened Species 2011:e.T41777A10562361.

Gonzalez, E., R. Barquez, and B. Miller. 2016a. *Lasiurus blossevillii*. The IUCN Red List of Threatened Species 2016:e.T88151055A22120040.

Gonzalez, E., R. Barquez, and J. Arroyo-Cabrales. 2016b. *Lasiurus cinereus*. The IUCN Red List of Threatened Species 2016:e.T11345A22120305.

González, C. A. L., D. E. Brown, and J. P. Gallo-Reynoso. 2003. The ocelot *Leopardus pardalis* in north-western Mexico: ecology, distribution and conservation status. Oryx 37:358–364.

González, S., J. E. Maldonado, J. Ortega, A. C. Talarico, L. Bidegaray-Batista, J. E. Garcia, and J. M. B. Duarte. 2009. Identification of the endangered small red brocket deer (*Mazama bororo*) using noninvasive genetic techniques (Mammalia, Cervidae) Molecular Ecology Resources. 2009:754–758.

González-Cózatt, F. X., R. M. Vallejo, and F. A. Cervantes. 2007. Avances en el estudio de la sistemática de lagomorfos utilizando marcadores moleculares: filogenia del género *Sylvilagus* basada en secuencias del gen 16s. Pp. 31–46, *in* Tópicos de sistemática, biogeografía, ecología y conservación de mamíferos (Sánchez-Rojas, G., and A. Rojas-Martínez, eds.). Universidad Autónoma del Estado de Hidalgo. Pachuca, México.

González-Ruiz, N., and S. T. Álvarez-Castañeda. 2005. *Peromyscus bullatus*. Mammalian Species 770:1–3.

González-Ruíz, N., S. T. Álvarez-Castañeda, and T. Álvarez. 2004. Distribution, taxonomy, and conservation status of the perote mouse *Peromyscus bullatus* (Rodentia: Muridae) in México. Biodiversity and Conservation 14:3423–3436.

González-Ruiz, N., J. Ramírez-Pulido, and H. H. Genoways. 2006a. Geographic distribution, taxonomy, and conservation status of Nelson's woodrat (*Neotoma nelsoni*) in Mexico. The Southwestern Naturalist 51: 112–126.

González-Ruiz, N., S. T. Álvarez-Castañeda, and J. Ramírez-Pulido. 2006b. *in litt. Neotoma nelsoni* [with keys of Genus *Neotoma*] (Rodentia: Cricetidae). Mammalian Species.

González-Ruiz, N., J. Ramírez-Pulido, and J. Arroyo-Cabrales. 2011. A new species of mastiff bat (Molossidae: *Molossus*) from Mexico. Mammalian Biology 76:461–469.

Goodwin, G. G. 1932. New records and sorne observations on Connecticut mammals. Journal of Mammalogy 13:36–40.

Goodwin, G. G. 1934. Mammals collected by A. W. Anthony in Guatemala, 1924–1928. Bulletin of the American Museum of Natural History 68:1–60.

Goodwin, G. G. 1941. A new *Peromyscus* from western Honduras. American Museum Novitates 1121:1.

Goodwin, G. G. 1942a. A summary of recognizable species of *Tonatia*, with descriptions of two new species. Journal of Mammalogy 23:204–209.

Goodwin, G. G. 1942b. Mammals of Honduras. Bulletin of the American Museum of Natural History 79:107–195.

Goodwin, G. G. 1946. Mammals of Costa Rica. Bulletin of the American Museum of Natural History 87:271–474.

Goodwin, G. G. 1955. New tree-climbing rats from Mexico and Colombia. American Museum Novitates, American Museum of Natural History 1738:1–5.

Goodwin, G. G. 1958. Bats of the genus *Rhogeessa*. American Museum Novitates 1923:1–17.

Goodwin, G. G. 1959. Descriptions of some new mammals. American Museum Novitates 1967:1–8.

Goodwin, G. G. 1964. A new species and a new subspecies of *Peromyscus* from Oaxaca, Mexico. American Museum Novitates 2183:1–8.

Goodwin, G. G. 1966. A new species of vole (Genus *Microtus*) from Oaxaca, Mexico. American Museum Novitates 2243:1–4.

Goodwin, G. G. 1969. Mammals from the state of Oaxaca, Mexico, in the American Museum of Natural History. Bulletin of the American Museum of Natural History 141:1–270.

Goodwin, G. G., and A. M. Greenhall. 1961. A review of the bats of Trinidad and Tobago: descriptions, rabies infections, and ecology. Bulletin of the American Museum of Natural History 122:187–302.

Gordon, K. 1943. The natural history and behavior of the western chipmunk and the mantled ground squirrel. Oregon State College Studies in Zoology 5:1–104.

Gosselin, J., and F. Boily. 1994. Unusual southern occurrence of a juvenile bearded seal, *Erignathus barbatus*, in the St. Lawrence estuary, Canada. Marine Mammal Science 10:480–483.

Graham, R. W., and M. W. Graham. 1994. Late Quaternary distribution of *Martes* in America. Pp. 26–58, *in* Martens, sables, and fishers: biology and conservation (Buskirk, S. W., A. S. Harestad, M. G. Raphael, and R. A. Powell, eds.). Cornell University Press. Ithaca, New York.

Graham, R. 2001. "Comment on "Skeleton of extinct North American sea mink (*Mustela macrodon*)" by Mead et al.". Quaternary Research 56:419–421.

Green, J. S., and J. T. Flinders. 1980. *Brachylagus idahoensis*. Mammalian Species 125:1–4.

Green, M. M. 1943. The three pelages of the smoky shrew. Canadian Field Naturalist 57:96.

Greenbaum, I. F., R. L. Honeycutt, and S. E. Chirhart. 2019. Taxonomy and phylogenetics of the *Peromyscus maniculatus* species group. Pp. 559–576, *in* From field to laboratory: A memorial volume in honor of Robert J. Baker (Bradley, R. D., H. H. Genoways, D. J. Schmidly, and L. C. Bradley, eds.). Special Publications, Museum of Texas Tech University 71:1–911.

Greenhall, A. M., and W. A. Schutt. 1996. *Diaemus youngi*. Mammalian Species 533:1–7.

Greenhall, A. M., G. Joermann, and U. Schmidt. 1983. *Desmodus rotundus*. Mammalian Species 202:1–6.

Greenhall, A. M., U. Schmidt, and G. Joermann. 1984. *Diphylla ecaudata*. Mammalian Species 227:1–3.

Gregorin, R., G. L. Capusso, and V. R. Furtado. 2008. Geographic distribution and morphological variation in *Mimon bennettii* (Chiroptera, Phyllostomidae). Iheringia, Série Zoologia 98:404–411.

Grinnell, J. 1913. A distributional list of the mammals of California. Proceeding of the California Academy of Sciences 3:265–390.

Grinnell, H. W. 1918. A synopsis of the bats of California. University of California Publications in Zoology 17:223–404.

Grinnell, J. 1921, Revised List of the Species in the Genus *Dipodomys*. Journal of Mammalogy 2:94–97.

Grinnell, J. 1922. A geographical study of the kangaroo rats of California. University California Publications in Zoology 24:1–124.

Grinnell, J. 1932. Habitat relations of the giant kangaroo rat. Journal of Mammalogy 13:305–320.

Grinnell, J. 1933. Mammals of California. Proceedings of the California Academy of Sciences 3:71–234.

Grinnell, J., and J. Dixon. 1918. Natural History of the ground squirrels of California. Monthly Bulletin of the State Commission of Horticulture 7:597–708.

Grinnell, J., and J. M. Linsdale. 1929. A new kangaroo rat from the upper Sacramento Valley, California. University of California Publications in Zoology 30:453–459.

Grinnell, J., and R. T. Orr. 1934. Systematic review of the *californicus* group of the rodent genus *Peromyscus*. Journal of Mammalogy 15:210–220.

Grinnell, J., and T. I. Storer. 1924. Animal life in the Yosemite: an account of the mammals, birds, reptiles, and amphibians in a cross-section of the Sierra Nevada. University of California Press. Berkeley, U.S.A.

Grinnell, J., J. Dixon, and J. M. Linsdale. 1937a. Fur-bearing mammals of California. University California Press, Berkeley 2:377–777.

Grinnell, J., J. B. Dixon, and J. M. Linsdale. 1937b. Fur-bearing mammals of California. University of California Press, Berkeley 1:1–1375.

Groves, C. P. 1981. Systematic relationships in the Bovini (Artiodactyla, Bovidae). Zeitschrift für Zoologisches Systematik und Evolutionsforschung 19:264–278.

Groves, C. P. 1986. The taxonomy, distribution and adaptations of recent equids. Pp. 11–65, *in* Equids in the Ancient World (Meadow, R. H., and H. P. Uepermann, eds.). Dr Ludwig Reichert Verlag. Wiesbaden, Germany.

Groves, C. P. 2005. Order Primates. Pp. 111–184, *in* Mammal species of the world. A taxonomic and geographic reference (Wilson, D. E., and D. A. M. Reeder, eds.). Third edition. The Johns Hopkins University Press. Baltimore, U.S.A.

Groves, C. 2006. The genus *Cervus* in eastern Eurasia. European Journal of Wildlife Research 52:14–22.

Groves, C., and P. Grubb. 2011. Ungulate taxonomy. The Johns Hopkins University Press. Baltimore, U.S.A.

Grubb, P. 1993. Order Artiodactyla. Pp. 377–414, *in* Mammal species of the world. A taxonomic and geographic reference (Wilson, D. E., and D. A. M. Reeder, eds.). Third edition. The Johns Hopkins University Press. Baltimore, U.S.A.

Grubb, P. 2005. Order Artiodactyla. Pp. 637–722, *in* Mammal species of the world. A taxonomic and geographic reference (Wilson, D. E., and D. A. M. Reeder, eds.). Third edition. The Johns Hopkins University Press. Baltimore, U.S.A.

Grzimek, B. 1975. Grzimek's Animal Life Encyclopaedia, Mammals, I-IV. Van Nostrand Reinhold. New York, U.S.A.

Guerrero, J. A., E. de Luna, and C. Sánchez-Hernández. 2003. Morphometrics in the quantification of character state identity for the assessment of primary homology: an analysis of character variation of the genus *Artibeus* (Chiroptera: Phyllostomidae). Biological Journal of the Linnean Society 80:45–55.

Guerrero, J. A., E. de Luna, and D. González. 2004. Taxonomic status of *Artibeus jamaicensis triomylus* inferred from molecular and morphometric data. Journal of Mammalogy 85:866–874.

Guerrero, J. A., J. Ortega, D. González, and J. E. Maldonado. 2008. Molecular phylogenetics and taxonomy of the fruit-eating bats of the genus *Artibeus* (Chiroptera: Phyllostomidae). Pp. 125–146, *in* Avances en el estudio de los mamíferos de México, Volumen II (Lorenzo, C., E. Espinoza, and J. Ortega eds.). Asociacin Mexicana de Mastozoologia, A. C. y el Colgio de la Forntera Sur. San Cristóbal de las Casas, México.

Guevara, L. 2017. They can dig it: semifossorial habits of the Mexican small-eared shrew (Mammalia: *Cryptotis mexicanus*). Revista Mexicana de Biodiversidad 88:1003–1005.

Guevara, L. 2023. A new species of small-eared shrew (Soricidae, Cryptotis) from El Triunfo Biosphere Reserve, Chiapas, Mexico. Journal of Mammalogy 104:546–561.

Guevara, L., and F. Cervantes. 2010. Rediscovery of the critically endangered Nelson's small-eared shrew (*Cryptotis nelsoni*), endemic to Volcán San Martín, Eastern México. Mammalian Biology 75:451–454.

Guevara, L., and F. Cervantes. 2013. Molecular systematics of small-eared shrews (Soricomorpha, Mammalia) within *Cryptotis mexicanus* species group from Mesoamérica. Acta Theriologica 59:233–242.

Guevara, L., and V. Sánchez-Cordero. 2018. New records of a critically endangered shrew from Mexican cloud forests (*Soricidae*, *Cryptotis nelsoni*) and prospects for future field research. Biodiversity Data Journal 6:e26667.

Guevara, L., V. Sánchez-Cordero, L. León-Paniagua, and N. Woodman. 2014a. A new species of small-eared shrew (Mammalia, Eulipotyphla, *Cryptotis*) from the Lacandona rain forest, Mexico. Journal of Mammalogy 95:739–753.

Guevara, L., C. Lorenzo, S. Ortega-García, and V. Sánchez-Cordero. 2014b. Noteworthy records of an endemic shrew from Mexico (Mammalia, Soricomorpha, *Crytotis griseoventris*), with comments on taxonomy. Mammalia 78:405–408.

Guevara, L., F. Cervantes, and V. Sánchez-Cordero. 2015. Riqueza, distribución y conservación de los topos y las musarañas (Mammalia, Eulipotyphla) de México. Therya 6:43–68.

Guggisberg, C. A. W. 1975. Wild cats of the world. Taplinger Publishing Company. New York, U.S.A.

Guilday, J. E. 1968. Grizzly bears from eastern North America. The American Midland Naturalist 79:247–250.

Gunn, A. 2016. *Rangifer tarandus*. The IUCN Red List of Threatened Species 2016:e.T29742A22167140.

Gunn, A., and M. Forchhammer. 2022. *Ovibos moschatus*. The IUCN Red List of Threatened Species 2022:e.T29684A22149286.

Gureev, A. A. 1979. Fauna SSSR. Mlekopitajuschie [Mammalia], Tom. 4, Vyp. 2. Nasekomojadnye (Mammalia, Insectivora). Nauka, Russia.

Gurnell, J. C. 1987. The natural history of squirrels. Facts on File. New York, U.S.A.

Gustavsson, I., and C. O. Sundt. 1968. Karyotypes in five species of deer (*Alces alces* L., *Capreolus capreolus* L., *Cervus elaphus* L., *Cervus nippon nippon* Temm. and *Dama dama* L.). Hereditas 60:233–247.

Gutiérrez, E. E., S. A. Jansa, and R. S. Voss. 2010. Molecular systematics of mouse opossums (Didelphidae: *Marmosa*): assessing species limits using mitochondrial DNA sequences, with comments on phylogenetic relationships and biogeography. American Museum Novitates 3692:1–22.

Gutiérrez, E. E., K. M. Helgen, M. M. McDonough, F. Bauer, M. T. Hawkins, L. A. Escobedo-Morales, B. D. Patterson, and J. E. Maldonado. 2017. A gene-tree test of the traditional taxonomy of American deer: the importance of voucher specimens, geographic data, and dense sampling. ZooKeys. 2017:87.

Gutiérrez-Costa, M. A., F. X. González-Cózatl, M. M. Ramírez-Martínez, L. I. Iñiguez-Dávalos, and D. S. Rogers. 2021. Molecular data suggest that *Heteromys irroratus bulleri* should be recognized as a species-level taxon. Therya 12:139–148.

Gwinn, R. N., G. H. Palmer, and J. L. Koprowski. 2011. *Sigmodon arizonae*. Mammalian Species 43:149–154.

Hacklander, K., and S. Schai-Braun. 2018. *Lepus europaeus* Pallas, 1778 European Hare. Pp. 187–190, *in* Lagomorphs: Pikas, Rabbits, and Hares of the World (Smith, A. T., C. H. Johnston, P. C. Alves, and K. Hackländer, eds). Johns Hopkins University Press. Baltimore, USA.

Hacklander, K., and S. Schai-Braun. 2019. *Lepus europaeus*. The IUCN Red List of Threatened Species 2019:e.T41280A45187424.

Hafner, D. J. 1984. Evolutionary relationships of the Nearctic Sciuridae. Pp. 3–23, *in* The biology of ground-dwelling squirrels: annual cycles, behavioral ecology, and sociality (Murie, J. O., and G. R. Michener, eds.). University of Nebraska Press. Lincoln, U.S.A.

Hafner, D. J. 1992. Speciation and persistence of a contact zone in Mojave Desert ground squirrels, subgenus *Xerospermophilus*. Journal of Mammalogy 73:770–778.

Hafner, D. 2019. *Dipodomys ornatus*. The IUCN Red List of Threatened Species 2019:e.T92464382A92464420.

Hafner, D. J., and K. N. Geluso. 1983. Systematic relationships and historical zoogeography of the desert pocket gopher, *Geomys arenarius*. Journal of Mammalogy 64:405–413.

Hafner, D. J., and A. T. Smith. 2010. Revision of the subspecies of the American pika, *Ochotona princep*s (Lagomorpha: Ochotonidae). Journal of Mammalogy 91:401–417.

Hafner, D. J., and T. L. Yates. 1983. Systematic status of the Mohave ground squirrel, *Spermophilus mohavensis* (subgenus *Xerospermophilus*). Journal of Mammalogy 64:397–404.

Hafner, D. J., J. C. Hafner, and M. S. Hafner. 1979. Systematic status of Kangaroo Mice, genus *Microdipodops*: morphometric, chromosomal, and protein analyses. Journal of Mammalogy 60:1–10.

Hafner, D. J., K. E. Petersen, and T. L. Yates. 1981. Evolutionary relationships of jumping mice (genus *Zapus*) of the southwestern United States. Journal of Mammalogy 62:501–512.

Hafner, D. J., B. R. Riddle, and S. T. Álvarez-Castañeda. 2001. Evolutionary relationships of white-footed mice (*Peromyscus*) on islands in the Sea of Cortez, México. Journal of Mammalogy 82:775–790.

Hafner, D. J., M. S. Hafner, G. L. Hasty, T. A. Spradling, and J. W. Demastes. 2008. Evolutionary relationships of pocket gophers (*Cratogeomys castanops* species group) of the Mexican Altiplano. Journal of Mammalogy 89:190–208.

Hafner, J. C. 1978. Evolutionary relationships of kangaroo mices, genus *Microdipodops*. Journal of Mammalogy 59:354–366.

Hafner, J. C., and M. S. Hafner. 1983. Evolutionary relationships of Heteromyid rodents. Great Basin Naturalist, Memories 7:3–29.

Hafner, J. C., J. E. Light, D. J. Hafner, M. S. Hafner, E. Reddington, D. S. Rogers, and B. R. Riddle. 2007. Basal clades and molecular systematic of heteromyid rodents. Journal of Mammalogy 88:1129–1145.

Hafner, M. S. 1982. A biochemical investigation of geomyoid systematics (Mammalia: Rodentia). Zeitschrift für Zoologische Systematik und Evolutionsforschung 20:118–130.

Hafner, M. 2017. Familia Geomyidae Pp. 234-269, *in* Handbook of the Mammals of the World – Volume 6, Lagomorphs and Rodents I (Wilson, D. E., T. E. Lacher, Jr, and R. A. Mittermeier). Lynx Editions, Barcelona, Spain.

Hafner, M. S., J. C. Hafner, J. L. Patton, and M. F. Smith. 1987. Macrogeographic patterns of genetic differentiation in the Pocket Gopher *Thomomys umbrinus*. Systematic Zoology 36:18–34.

Hafner, M. S., T. A. Spradling, J. E. Light, D. J. Hafner, and J. R. Demboski. 2004. Systematic revision of pocket gophers of the *Cratogeomys gymnurus* species group. Journal of Mammalogy 85:1170–1183.

Hafner, M. S., J. E. Light, D. J. Hafner, S. V. Brant, T. A. Spradling, and J. W. Demastes. 2005. Cryptic species in the Mexican pocket gopher *Cratogeomys merriami*. Journal of Mammalogy 86:1095–1108.

Hafner, M. S., D. J. Hafner, J. W. Demastes, G. L. Hasty, J. E. Light, and T. A. Spradling. 2009. Evolutionary relationships of pocket gophers of the genus *Pappogeomys* (Rodentia: Geomyidae). Journal of Mammalogy 90:47–56.

Hafner, M. S., A. R. Gates, V. Mathis, J. W. Demastes, and D. J. Hafner. 2011. Redescription of the pocket gopher *Thomomys atrovarius* from the Pacific coast of mainland Mexico. Journal of Mammalogy 92:1367–1382.

Hahn, W. L. 1905. A new bat from Mexico. Proceedings of the Biological Society of Washington 18:247–248.

Haiduk, M. W., C. Sánchez-Hernández, and R. J. Baker. 1988. Phylogenetic relationships of *Nyctomys* and *Xenomys* to other cricetine genera based on data from G-banded chromosomes. The Southwestern Naturalist 33:397–403.

Haiduk, M. W., J. W. Bickham, and D. J. Schmidly. 1979. Karyotypes of six species of *Oryzomys* from Mexico and Central America. Journal of Mammalogy 60:610–615.

Hall, A. 2002. Gray seal *Halichoerus grypus*. Pp. 522–524, *in* Encyclopedia of Marine Mammals (Perrin, W. F., B. Wursig, and J. G. M. Thewissen, eds.). Academic Press, San Diego, U.S.A.

Hall, E. R. 1941. Revision of the rodent genus *Microdipodops*. Field Mususem of Natural History, Zoological Series 27:233–277.

Hall, E. R. 1946. Mammals of Nevada. University of California Press. Berkeley, U.S.A.

Hall, E. R. 1951a. American weasley. University of Kansas Publications, Museum of Natural History 4:1–466.

Hall, E. R. 1951b. Asynopsis of the North American Lagomorpha. University of Kansas Publications, Museum of Natural History 5:119–202.

Hall, E. R. 1955. Handbook of the mammals of Kansas. University of Kansas Museum of Natural History, Miscellaneous Publications 7:1–303.

Hall, E. R. 1960. *Oryzomys couesi* only subspecifically differences from the marsh rice rat *Oryzomys palustris*. The Southwestern Naturalist 5:171–173.

Hall, E. R. 1981. The mammals of North America. Second edition. John Wiley and Sons. New York, U.S.A.

Hall, E. R. 1984. Geographic variation among brown and grizzly bears (*Ursus arctos*) in North America. Special Publications, Museum of Natural History, University of Kansas 13:1–16.

Hall, E. R., and E. L. Cockrum. 1953. A synopsis of the North American microtine rodents. University of Kansas Publications, Museum of Natural History 5:373–498.

Hall, E. R., and F. H. Dale. 1939. Geographic races of the kangaroo rat, *Dipodomys microps*. Occasional Papers of the Museum of Zoology, Louisiana State University 4:47–62.

Hall, E. R., and W. W. Dalquest 1950. Synopsis of the American Bat of the genus *Pipistrellus*. University of Kansas Publications, Museum of Natural History 1:591–602.

Hall, E. R., and D. F. Hoffmeister. 1942. Geographic variation in the canyon mouse, *Peromyscus maniculatus*. Journal of Mammalogy 23:51–65.

Hall, E. R., and H. H. Genoways. 1970. Taxonomy of the *Neotoma albigula*-group of woodrats in central Mexico. Journal of Mammalogy 51:504–516.

Hall, E. R., and R. M. Gilmore. 1932. New mammals from St. Lawrence Island, Bering Sea, Alaska. University of California Publications in Zoology 38:391–404.

Hall, E. R. and R. M. Gilmore. 1934. *Marmota caligata broweri*, a new marmot from northern Alaska. Canadian Field-Naturalist 48:57–59.

Hall, E. R., and K. R. Kelson. 1951. A new subspecies of *Microtus montanus* from Montana and comments on *Microtus canicaudus* Miller. University of Kansas Publications, Museum of Natural History 5:73–79.

Hall, E. R., and K. R. Kelson. 1959. The mammals of North America. The Ronald Press Co. New York, U.S.A.

Hall, E. R., and M. B. Ogilvie. 1960. Conspecificity of two pocket mice, *Perognathus goldmani* and *P. artus*. University of Kansas Publications, Museum of Natural History 9:513–518.

Hallett, J. G. 1978. *Parascalops breweri*. Mammalian Species 98:1–4.

Hallett, J. G. 1999. Hairy-tailed mole *Parascalops breweri*. Pp. 62–63, *in* The Smithsonian book of North American Mammals (Wilson, D. E., and S. Ruff, eds.). Smithsonian Institution Press. Washington, U.S.A.

Halls, L. K. 1978. White-tailed deer. Pp. 43–65, *in* Big game of North America: ecology and management (Schmidt, J. L., and D. L. Gilbert, eds.). Stackpole Books. Harrisburg, U.S.A.

Halls, L. K. 1984. White-tailed deer: ecology and management. Stackpole Books. Harrisburg, U.S.A.

Hamilton, M. J., R. K. Chesser, and T. L. Best. 1987. Genetic Variation in the Texas Kangaroo Rat, *Dipodomys elator* Merriam. Journal of Mammalogy 68:775–781.

Hamilton, W. J. 1940. The biology of the smoky shrew (*Sorex fumeus fumeus* Miller). Zoologica 25:473–492.

Hammerson, G. A. 2016. *Sorex pacificus*. The IUCN Red List of Threatened Species 2016:e.T41409A22318109.

Hammerson, G. A., J. Matson, F. Reid, and N. Woodman. 2019. *Sorex neomexicanus*. The IUCN Red List of Threatened Species 2019:e. T136608A22319242.

Hanák, V., and V. Mazák. 1991. Enciclopedia de los Animales, Mamíferos de todo el Mundo. Susaeta. Madrid, España.

Handley, C. O. 1953. Three new lemmings (*Dicrostonyx*) from Arctic America. Journal of Washington Academic Science 43:197–200.

Handley, C. O., Jr. 1955. A new species of free-tailed bat (genus *Eumops*) from Brazil. Proceedings of the Biological Society of Washington 68:177–178.

Handley, C. O., Jr. 1959. A revision of American bats of the genera *Euderma* and *Plecotus*. Proceedings of the United States National Museum 110:95–246.
Handley, C. O., Jr. 1960. Description of new bats from Panama. Proceedings of the United States National Museum 3442:459–479.
Handley, C. O., Jr. 1966. Description of new bats (*Chiroderma* and *Artibeus*) from Mexico. Anales del Instituto de Biología, Universidad Nacional Autónoma de México 36:297–301.
Handley, C. O., Jr. 1980. Mammals of the Smithsonian Venezuelan Project. Brigham Young University, Science Bulletin, Biological Series 20:1–91.
Handley, C. O., Jr. 1991. The identity of *Phyllostoma planirostre* Spix, 1823 (Chiroptera: Stenodermatinae). Pp. 12–17, *in* Contributions to mammalogy in honor of Karl F. Koopman (Griffiths, T. A., and D. Klingener, eds.). Bulletin of the American Museum of Natural History 206:1–432.
Hansen, C. G. 1980. Physical characteristics. Pp. 52–63, *in* The desert bighorn: its life history, ecology, and management (Monson, G., and L. Sumner, eds.). University of Arizona Press. Tucson, U.S.A.
Hansen, R. M. 1954. Molt patterns in ground squirrels. Proceedings of the Utah Academy of Sciences, Arts and Letters 31:57–60.
Hanson, J. D., J. L. Indorf, V. J. Swier, and R. D. Bradley. 2010. Molecular divergence within the *Oryzomys palustris* complex: evidence for multiple species. Journal of Mammalogy 91:336–347.
Hardy, R. 1950. A new tree squirrel from central Utah Proceedings of the Biological Society of Washington 63:13–14.
Harington, C. R. 1966. A polar bear's life. Report on Polar Bears. Lectures presented at the Eighth Annual Meeting of the Washington Area Associates. Pp. 3–7. Arctic Institute of North America Research Paper 34. Calgary, Canada.
Harper, F. 1961. Land and fresh-water mammals of the Ungava Peninsula. University of Kansas Museum Natural History, Misclaneous Publications 27:1–178.
Harrington, C. R. 1966. Extralimital occurrences of walruses in the Canadian Artic. Journal of Mammalogy 47:506–513.
Harris, A. H. 1974. *Myotis yumanensis* in interior southwestern North America, with comments on *Myotis lucifugus*. Journal of Mammalogy 55:589– 607.
Harris, C. J. 1968. Otters; a study of the recent Lutrinae. Weinfield and Nicolson. London, United Kingdom.
Harris, D. E., B. Lelli, and G. Jakush. 2002. Harp seal records from the southern Gulf of Maine: 1997–2001. Northeastern Naturalist 9:331–340.
Harris, W. P., Jr. 1937. Revision of *Sciurus variegatoides,* a species of Central American squirrel. Miscellaneous Publications of the Museum of Zoology, University of Michigan 38:1–39.
Harrison, D. L. 1975. *Macrophyllum macrophyllum*. Mammalian Species 62:1–3.
Harrison, R. G., S. M. Bogdanowicz, R. S. Hoffmann, E. Yensen, and P. W. Sherman. 2003. Phylogeny and evolutionary history of the ground squirrels (Rodentia: Marmotinae). Journal of Mammalian Evolution 10:249–276.
Hart, E. B. 1992. *Tamias dorsalis*. Mammalian Species 399:1–6.
Hart, E. B., M. C. Belk, E. Jordan, and M. W. Gonzalez. 2004. *Zapus princeps*. Mammalian Species 749:1–7.
Hartman, G. D., and T. L. Yates. 1985. *Scapanus orarius*. Mammalian Species 253:1–5.
Hartman, G. D. 1999. Star-nosed mole *Condylura cristata*. Pp. 65–67, *in* The Smithsonian book of North American Mammals (Wilson, D. E., and S. Ruff, eds.). Smithsonian Institution Press. Washington, U.S.A.
Hassanin, A., and E. J. P. Douzery. 2003. Molecular and morphological phylogenies of Ruminantia and the alternative position of the Moschidae. Systematic Biology 52:206–228.
Hassanin, A., F. Delsuc, A. Ropiquet, C. Hammer, B. Jansen Van Vuuren, C. Matthee, M. Ruiz-Garcia, F. Catzeflis, V. Areskoug, T. T. Nguyen, A. Couloux. 2012. Pattern and timing of diversification of Cetartiodactyla (Mammalia, Laurasiatheria), as revealed by a comprehensive analysis of mitochondrial genomes. Comptes Rendus Biologies 335:32–50.
Hatch, F. T., E. J. Ridley, and J. A. Mazrimas. 1971. Some *Dipodomys* species: ecologic and taxonomic features, estrous cycle, and breeding attempts. U.S. Atomic Energy Commission, Topical Report, Lawrence Livermore Laboratory UCRL-51140:1–25.
Hartman, D. S. 1971. Behavior and ecology of the Florida manatee, *Trichechus manatus latirostris* (Harlan), at Crystal River, Citrus County. Ph.D. dissertation. Cornell University, Cornell, U.S.A.
Hatt, R. T. 1932. The vertebral columns of ricochetal rodents. Bulletin of the American Museum of Natural History 63:599–738.
Hawes, M. L. 1975. Ecological adaptations in two species of shrews. Ph. D. dissertation. University of British Columbia. Vancouver, Canada.
Hayes, J. P. 1996. *Arborimus longicaudus*. Mammalian Species, 532:1–5.
Hayssen, V. 1991. *Dipodomys microps*. Mammalian Species 389:1–9.
Hayssen, V., F. Miranda, and B. Pasch. 2011. *Cyclopes didactylus* (Pilosa: Cyclopedidae). Mammalian Species 44(895):51–58.
Hayssen, V., J. Ortega, A. Morales-Leyva, and N. Martínez-Méndez. 2013. *Cabassous centralis* (Cingulata: Dasypodidae). Mammalian Species 45(898):12–17.
Heaney, L. R. 1983. *Sciurus granatensis* (ardilla roja, ardilla chisa, red-tailed squirrel). Pp. 489–490, *in* Costa Rican natural history (Janzen, D. H. ed.). The University of Chicago Press, Chicago, U.S.A.
Heaney, L. R., and R. M. Timm. 1983. Relationships of pocket gophers of the genus *Geomys* from the central and northern Great Plains. University of Kansas Publications, Museum of Natural History, Miscellaneous Publications 74:59.
Heckeberg, N. S. 2020. The systematics of the Cervidae: a total evidence approach. PeerJ, 8, e8114.
Heckeberg, N. S., D. Erpenbeck, G. Wörheide, and G. E. Rössner. 2016. Systematic relationships of five newly sequenced cervid species. PeerJ. 4:e2307.
Heffelfinger, J. R. 2000. Status of the name *Odocoileus hemionus crooki* (Mammalia: Cervidae). Proceedings of the Biological Society of Washington 113:319–333.
Helgen, K. 2016. *Conepatus leuconotus*. The IUCN Red List of Threatened Species 2016:e.T41632A45210809.
Helgen, K., and F. Reid. 2016a. *Martes americana*. The IUCN Red List of Threatened Species 2016:e.T41648A45212861.
Helgen, K., and F. Reid. 2016b. *Mephitis mephitis*. The IUCN Red List of Threatened Species 2016:e.T41635A45211301.
Helgen, K., and F. Reid. 2016c. *Mustela frenata*. The IUCN Red List of Threatened Species 2016:e.T41654A45213820.
Helgen, K., and F. Reid. 2016d. *Taxidea taxus*. The IUCN Red List of Threatened Species 2016:e.T41663A45215410.
Helgen, K., and F. Reid. 2018. *Martes pennanti*. The IUCN Red List of Threatened Species 2018:e.T41651A125236220.
Helgen, K. M., and D. E. Wilson. 2003. Taxonomic status and conservation relevance of the raccoons (*Procyon* spp.) of the West Indies. Journal of Zoology, London 259:69–76.

Helgen, K. M., and D. E. Wilson. 2005. A systematic and zoogeographic overview of the raccoons of Mexico and Central America. Pp. 221–236, *in* Contribuciones mastozoológicas en homenaje a Bernardo Villa (Sánchez-Cordero, V., and R. A. Medellín, eds.). Instituto de Biología e Instituto de Ecología, Universidad Nacional Autónoma de México y Comisión Nacional para el Conocimiento y Uso de la Biodiversidad. Distrito Federal, México.

Helgen, K., F. Reid, and R. Timm. 2016a. *Spilogale angustifrons*. The IUCN Red List of Threatened Species 2016:e.T136636A45221538.

Helgen, K., R. Kays, and J. Schipper. 2016b. *Potos flavus*. The IUCN Red List of Threatened Species 2016:e.T41679A45215631.

Helgen, K., A. D. Cuarón., J. Schipper, and J. F. González-Maya. 2016c. *Spilogale pygmaea*. The IUCN Red List of Threatened Species 2016:e. T41637A45211592.

Helgen, K. M., F. R. Cole, L. E. Helgen, and D. E. Wilson. 2009. Generic revision in the Holartic ground squirrel genus *Spermophilus*, Journal of Mammalogy 90:270–305.

Helm, J. D., III, C. Sánchez-Hernández, and R. H. Baker. 1974. Observaciones sobre los ratones de las marismas, *Peromyscus perfulvus* Osgood (Rodentia Cricetidae). Anales del Instituto de Biología, Universidad Nacional Autónoma de México, Serie Zoología 45:141–146.

Hemmer, H. 1978. The evolutionary systematics of living Felidae: Present status and current problems. Carnivore 1:71–79.

Hennings, D., and R. S. Hoffmann. 1977. A review of the taxonomy of the *Sorex vagrans* species complex from western North America. Occasional papers of the Museum of Natural History, the University of Kansas 68:1–35.

Hensley, A. P., and K. T. Wilkins. 1988. *Leptonycteris nivalis*. Mammalian Species 307:1–4.

Henttonen, H., B. Sheftel, M. Stubbe, R. Samiya, J. Ariunbold, V. Buuveibaatar, S. Dorjderem, Ts. Monkhzul, M. Otgonbaatar, and M. Tsogbadrakh. 2016. *Sorex minutissimus*. The IUCN Red List of Threatened Species 2016:e.T29666A115171049.

Heptner, V. G., K. K. Chapskii., V. A. Arsen'ev, and V. E. Sokolov. 1996. Mammals of the Soviet Union. Smithsonian Institution Libraries and National Science Foundation.

Herd, R. M. 1983. *Pteronotus parnellii*. Mammalian Species 209:1–5.

Herd, R. M., and M. B. Fenton. 1983. An electrophoretic, morphological, and ecological investigation of a putative hybrid zone between *Myotis lucifugus* and *Myotis yumanensis* (Chiroptera: Vespertilionidae). Canadian Journal of Zoology 61:2029–2050.

Hermanson, J. W. and T. J. O'Shea. 1983. *Antrozous pallidus*. Mammalian Species 213:1–8.

Hernández-Camacho, J. 1977. Notas para una monografia de *Potos flavus* (Mammalia: Carnivora) en Colombia. Caldasia 11:147–181.

Hernández-Camacho, J., and G. A. Cadena. 1978. Notas para la revisión del género *Lonchorhina* (Chiroptera, Phyllostomidae). Caldasia 12:199–251.

Hernández-Canchola, G., and L. León-Paniagua. 2020. *Sturnira parvidens* (Chiroptera: Phyllostomidae). Mammalian Species 52:57–70.

Hernández-Canchola, G., and L. León-Paniagua. 2021. About the specific status of *Baiomys musculus* and *B. brunneus*. Therya 12:291–301.

Hernández-Canchola, G., L. León-Paniagua, and J. A. Esselstyn. 2021a. Mitochondrial DNA indicates paraphyletic relationships of disjunct populations in the *Neotoma mexicana* species group. Therya 12:411–421.

Hernández-Canchola, G., J. Ortega, and L. León-Paniagua. 2021b. *Sturnira hondurensis* (Chiroptera: Phyllostomidae). Mammalian Species 53:23–34.

Hernández-Canchola, G., L. León-Paniagua, and J. A. Esselstyn. 2022. Mitochondrial DNA and other lines of evidence clarify species diversity in the *Peromyscus truei* species group (Cricetidae: Neotominae). Mammalia.

Hernández-Meza, B., Y. Domínguez-Castellanos, and J. Ortega. 2005. *Myotis keaysi*. Mammalian Species 785:1–3.

Hernández-Mijangos, L. A., R. Gálvez-Mejía, M. Díaz-Negrete, and C. M. Cruz-Durante. 2008. Nuevas localidades en la distribución de murciélagos filostóminos (Chiroptera: Phyllostomidae) en Chiapas, México. Revista Mexicana de Mastozoología 12:163–169.

Herrera, N. D., K. C. Bell, C. M. Callahan, E. Nordquist, B. A. J. Saver, J. Sullivan, J. R. Demobski, and J. M. Good. 2022. Genomic resolution of cryptic species diversity in chipmunks. Evolution 76:2004–2019.

Herron, M. D., T. A. Castoe, and C. L. Parkinson. 2004. Sciurid phylogeny and the paraphyly of Holarctic ground squirrels (*Spermophilus*). Molecular Phylogenetics and Evolution 31:1015–1030.

Hershkovitz, P. 1950. Mammals of northern Colombia. Preliminary report no 6: Rabbits (Leporidae), with notes on the classification and distribution of the South American forms. Proceedinsg of the United States National Museum 100:327–375.

Hershkovitz, P. 1951. Mammals from British Honduras, Mexico, Jamaica and Haiti. Fieldiana: Zoology 31:547–569.

Hershkovitz, P. 1954. Mammals of northern Colombia, preliminary report No. 7. Tapir (genus *Tapirus*) with a systematic review of the American species. Proceedings of the United States National Museum 103:165–496.

Hershkovitz, P. 1955. South American marsh rats genus *Holochilus* with a summary of Sigmodont rodents. Fieldiana: Zoology, Chicago Natural History Museum 37:639–673.

Hershkovitz, P. 1971. A new rice rat of the *Oryzomys palustris* group (Cricetinae, Muridae) from northwestern Colombia, with remarks on distribution. Journal of Mammalogy 52:700–709.

Hershkovitz, P. 1981. *Philander* and four-eyed opossums once again. Proceedings of the Biological Society of Washington 93:943–946.

Hershkovitz, P. 1997. Composition of the family Didelphidae Gray, 1821 (Didelphoidea: Marsupialia), with a review of the morphology and behavior of the included four-eyed pouched opossums of the genus *Philander* Tiedemann, 1808. Fieldiana, Zoology, New Series 86:1–103.

Hibbard, C. W. 1950. Mammals of the Rexroad formation from Fox Canyon, Meade County, Kansas. Contribution Museum of Paleontolgy, University of Michigan 8:113–192.

Hicks, S. A., and S. M. Carr. 1997. Are there two species of pine marten in North America? Genetic and evolutionary relationship within *Martes*. Pp. 15–28, *in* Martes: taxonomy, ecology, techniques, and management (Proulx, G., H. N. Bryant, and P. M. Woodard, eds.). Proceedings of the Second International Martes Symposium. Provincial Museum of Alberta. Edmonton, Canada.

Hidalgo-Mihart, M. G., L. Cantú-Salazar., A. González-Romero, and C. A. López-Gonzalez. 2004. Historical and present distribution of coyote (*Canis latrans*) in Mexico and Central America. Journal of Biogeography 31:2025–2038.

Higdon, J. W., O. R. P. Bininda-Emonds, R. M. D. Beck, and S. H. Ferguson. 2007. Phylogeny and divergence of the pinnipeds (Carnivora: Mammalia) assessed using a multigene dataset. BMC Evolutionary Biology 7:216.

Hildebrand, M. 1952. The integument in Canidae. Journal of Mammalogy 33:419–428.

Hildebrand, M. 1954. Comparative morphology of the body skeleton in recent Canidae. University California Publications in Zoology 52:399–470.

Hillman, C. N., and T. W. Clark. 1980. *Mustela nigripes*. Mammalian Species 126:1–3.

Hilton, C. D., and T. L. Best. 1993. *Tamias cinereicollis*. Mammalian Species 436:1–5.

Hinesley, L. L. 1979. Systematics and distribution of two chromosome forms in the Southern Grasshopper Mouse, Genus *onychomys*. Journal of Mammalogy 60:117–128.

Hirshfeld, J. R., and W. G. Bradley.1977. Growth and development of two species of chipmunks: e*utamias panamintinus* and *E. palmeri*. Journal of Mammalogy 58:44–52.

Hoffmann, W. M. 1975. Geographic variation and taxonomy of *Dipodomys nitratoides* from the California San Joaquin Valley. M. A. thesis, California State University, Fresno.

Hoffmann, F. G., and R. J. Baker 2001. Systematics of bats of the genus *Glossophaga* (Chiroptera: Phyllostomidae) and phylogeography in *G. soricina* based on the cytochrome-*b* gene. Journal of Mammalogy 82:1092–1101.

Hoffmann, F. G., and R. J. Baker 2003. Comparative phylogeography of short-tailed bats (*Carollia*: Phyllostomidae). Molecular Ecology 12:3403–3414.

Hoffmann, F. G., S. R. Hoofer, and R. J. Baker 2008. Molecular dating of the diversification of Phyllostominae bats based on nuclear and mitochondrial DNA sequences. Molecular Phylogenetics and Evolution 49:653–658.

Hoffmann, F. G., J. G. Owen, and R. J. Baker. 2003. mtDNA perspective of chromosomal diversification and hybridization in Peters' tent-making bat (*Uroderma bilobatum*: Phyllostomidae). Molecular Ecology 12:2981–2993.

Hoffmann, M., and C. Sillero-Zubiri. 2021. *Vulpes vulpes*. The IUCN Red List of Threatened Species 2021:e.T23062A193903628.

Hoffmann, R. S., and R. D. Fisher. 1978. Additional Distributional Records of Preble's shrew (*Sorex preblei*). Journal of Mammalogy 59:883–884.

Hoffmann, R. S., and J. W. Koeppl. 1985. Zoogeography. Pp. 85–115, *in* Biology of New World Microtus (Tamarin, R. H. ed.). American Society of Mammalogists. Shippensburg, Pennsylvania.

Hoffmann, R. S., and D. L. Pattie. 1968. A guide to Montana mammals: identification, habitat, distribution and abundance. University of Montana. Missoula, U.S.A.

Hoffmann, R. S., and R. S. Peterson. 1967. Systematics and zoogeography of *Sorex* in the Bering Strait area. Systematic Zoology 16:126–136.

Hoffmann, R. S., and A. T. Smith. 2005. Order Lagomorpha. Pp. 185–211, *in* Mammal Species of the World (Wilson, D. E., and D. M. Reeder, eds.). John Hopkins University Press. Baltimore, U.S.A.

Hoffmann, R. S., and R. J. Thorington. 2005. Family Sciuridae. Pp. 754–818, *in* Mammal Species of the World (Wilson, D. E., and D. M. Reeder, eds.). John Hopkins University Press. Baltimore, U.S.A.

Hoffmann, R. S., J. W. Koeppel, and C. F. Nadler. 1979. The relationships of the Amphiberingian marmots (Mammalia: Sciuridae). Occasional Papers of the Museum of Natural History, University of Kansas 83:1–56.

Hoffmann, R. S., P. L. Wright, and F. E. Newby. 1969. The distribution of some mammals in Montana. I. Mammals other than bats. Journal of Mammalogy 50:579–604.

Hoffmann, R. S., C. G. Anderson, R. W. Thorington, Jr., and L. R. Heaney. 1993. Family Sciuridae. Pp. 419–465, *in* Mammal Species of the World, a Taxonomic and Geographic Reference (Wilson, D., and D. M. Reeder, eds.), Smithsonian Institution Press. Washington, U.S.A.

Hoffmeister, D. F. 1951. A taxonomic and evolutionary study of the pinon mouse, *Peromyscus truei.* Illinois Biological Monographs 21:1–104.

Hoffmeister, D. F. 1956. Mammals of the Graham (Pinaleno) Mountains, Arizona. The American Midland Naturalist 55:257–288.

Hoffmeister, D. F. 1969. The species problem in the *Thomomys bottae-Thomomys umbrinus* complex of gophers in Arizona. Miscellaneous Publications, Museum of Natural History, University of Kansas 51:75–91.

Hoffmeister, D. F. 1970. The seasonal distribution of bats in Arizona: a case for improving mammalian range maps. The Southwestern Naturalist 15:11–22.

Hoffmeister, D. F. 1971. Mammals of Grand Canyon. University of Illinois Press. Urbana, U.S.A.

Hoffmeister, D. F. 1974. The taxonomic status of *Perognathus penicillatus minimus* Burt. The Southwestern Naturalist 19:213–214.

Hoffmeister, D. F. 1977. Noteworthy range extensions of mammals in northem Mexico and Arizona. The Southwestern Naturalist 22:150–151.

Hoffmeister, D. F. 1981. *Peromyscus truei*. Mammalian Species 161:1–5.

Hoffmeister, D. F. 1986. Mammals of Arizona. The University of Arizona Press and the Arizona Game and Fish Department. Tucson, U.S.A.

Hoffmeister, D. E., and V. E. Diersing. 1978. Review of the tassel-eared squirrels of the subgenus *Otosciurus*. Journal of Mammalogy 59:402–413.

Hoffmeister, D. F., and L. S. Ellis. 1979. Geographic variation in *Eutamias quadrivitattus* with comments on the taxonomy of other Arizonan chipmunks. The Southwestern Naturalist 24:655–666.

Hoffmeister, R. G., and D. F. Hoffmeister. 1991. The hyoid in North American squirrels, Sciuridae, with remarks on associated musculature. Anales del Instituto de Biolofía, Universidad Nacional Autónoma de México, Serie Zoología 62:219–234.

Hoffmeister, D. F., and P. H. Krutzsch. 1955. A new subspecies of *Myotis evotis* (H. Allen) from southeastern Arizona and Mexico. Chicago Academy of Sciences, Natural History Miscellaneous 151:1–4.

Hoffmeister, D. F., and M. R. Lee. 1963. Taxonomic review of cottontails, *Sylvilagus floridanus* and *Sylvilagus nuttallii*, in Arizona. The American Midland Naturalist 70:138–148.

Hoffmeister, D. F., and M. R. Lee. 1967. Revision of the pocket mice *Perognathus penicillatus*. Journal of Mammalogy 48:361–380.

Hoffmeister, D. F., and I. A. Nader. 1963. Distributional notes on Arizona mammals. Transactions of Illinois State Academy of Science 56:9293.

Hoffmeister, D. F., and L. de la Torre. 1960. A revision of the Wood Rat *Neotoma stephensi*. Journal of Mammalogy 41:476–491.

Hoffmeister, D. F. and L. de la Torre. 1961. Geographic Variation in the Mouse *Peromyscus difficilis*. Journal of Mammalogy 42:1–13.

Hogan, K. M., M. C. Hedin, H. S. Koh, S. K. Davis, and I. F. Greenbaum. 1993. Systematics and taxonomic implications of karyotypic, electrophoretic, and mitochondrial-DNA variation in *Peromyscus* from the Pacific Northwest. Journal of Mammalogy 74:819–831.

Hoisington-Lopez, J. L., L. P. Waits, and J. Sullivan. 2012. Species limits and integrated taxonomy of the Idaho ground squirrel (*Urocitellus brunneus*): genetic and ecological differentiation. Journal of Mammalogy 93:589–604.

Holden, M. E., and G. G. Musser. 2005. Family Dipodidae. Pp. 871–893, *in* Mammal Species of the World (Wilson, D. E., and D. M. Reeder, eds.). The Johns Hopkins University Press. Baltimore, U.S.A.

Hollister, N. 1911. A systematic synopsis of the muskrats. North American Fauna 32:1–47.

Hollister, N. 1913. Three new subspecies of Grasshopper mice. Proceedings of the Biological Society of Washington 26:215–216.

Hollister, N. 1914. A systematic account of the grasshopper mice. Proceedings of the United States Natural History Museum 47:427–488.

Hollister, N. 1915. The genera and subgenera of raccoons and their allies. Proceedings United States National Museum 49:143–150.

Hollister, N. 1916. A systematic account of the prairie dogs. North American Fauna 40:1–37.

Holloway, G. I., and R. M. R. Barclay. 2001. *Myotis ciliolabrum*. Mammalian Species 670:1–5.

Homan, J. A., and H. H. Genoways. 1978. An analysis of hair structure and its phylogenetic implications among heteromyid rodents. Journal of Mammalogy 59:740–760.

Honacki, J. H., K. E. Kinman, and J. W. Koeppl (eds.). 1982. Mammal species of the world: a taxonomic and geographic reference. Allen Press, Inc. and The Association of Systematics Collections. Lawrence, U.S.A.

Honeycutt, R. L., and D. J. Schmidly. 1979. Chromosomal and morphological variation in the plains pocket gopher, *Geomys bursarius* in Texas and adjacent states. Occasional Papers, The Museum, Texas Tech University 58:1–54.

Honeycutt, R. L., and S. L. Williams, 1982. Genic differentiation in pocket gophers of the genus *Pappogeomys*, with comments on intergeneric relationships in the subfamily Geominae. Journal of Mammalogy 63:208–217.

Hood, C. S., and A. L. Gardner. 2008. Family Emballonuridae Gervais, 1856. Pp. 188–207, *in* Mammals of South America. Vol. 1. Marsupials, xenarthrans, shrews, and bats (Gardner, A. L., ed.). University of Chicago Press. Chicago, U.S.A.

Hood, C. S., and J. K. Jones, Jr. 1984. *Noctilio leporinus*. Mammalian Species 216:1–7.

Hood, C. S., and J. Pitocchelli. 1983. *Noctilio albiventris*. Mammalian Species 197:1–5.

Hood, C. S., L. W. Robbins, R. J. Baker, and H. S. Shellhammer. 1984. Chromosomal studies and evolutionary relationships of an endangered species, *Reithrodontomys raviventris*. Journal of Mammalogy 65:655–667.

Hoofer, S. R., and R. A. Van den Bussche. 2001. Phylogenetic relationships of plecotine bast and allies based on mitochondrial ribosomal sequences. Journal of Mammalogy 82:131–137.

Hoofer, S. R., and R. A. Van den Bussche. 2003. Molecular phylogenetics of the chiropteran family Vespertilionidae. Acta Chiropterologica 5:1–63.

Hoofer, S. R., R. A. Van Den Bussche, and I. Horáček. 2006. Generic status of the American pipistrelles (Vespertilionidae) with description of a new genus. Journal of Mammalogy 87:981–992.

Hoofer, S. R., S. A. Reeder, E. W. Hansen, and R. A. Van Den Bussche. 2003. Molecular phylogenetics and taxonomic review of noctilionoid and vespertilionoid bats (Chiroptera: Yangochiroptera). Journal of Mammalogy 84, 809–821.

Hoogland, J. L. 1995. The black-tailed prairie dog: social life of a burrowing mammal. The University of Chicago Press. Chicago, U.S.A.

Hoogland, J. L. 1996. *Cynomys ludovicianus*. Mammalian Species 535:1–10.

Hooper, E. T. 1940a. A new race of pocket gopher of the species *Geomys lutescens* from Colorado. Occasional Papers of the Museum of Zoology, University of Michigan 420:1–3.

Hooper, E. T. 1940b. Geographical variation in bushy-tailed woodrats. University of California Publications in Zoology 42:407–424.

Hooper, E. T. 1944. San Francisco Bay as a factor influencing speciation in rodents. Miscellaneous Publications, Museum of Zoology, University of Michigan 59:1–89.

Hooper, E. T. 1947. Notes on Mexican mammals. Journal of Mammalogy 28:40–57.

Hooper E. T. 1950. A new subspecies of harvest mouse (*Reithrodontomys*) from Chiapas, Mexico. Journal of the Washington Academy of Sciences 40:418–419.

Hooper, E. T. 1952. A systematic review of the harvest mice (genus *Reithrodontomys*) of Latin America. Miscellaneous Publications of the Museum of Zoology, University of Michigan 77:1–255.

Hooper, E. T. 1953. Notes on mammals of Tamaulipas, Mexico. Occasional Papers of the Museum of Zoology, University of Michigan 544:1–12.

Hooper, E. T. 1954. A synopsis of the cricetine rodents genus *Nelsonia*. Occasional Papers Museum Zoology, University of Michigan 550:1–12.

Hooper, E.T. 1955. Notes on mammals of western Mexico. Occasional Papers of the Museum of Zoology University of Michigan 565:1–26.

Hooper, E. T. 1957. Dental patterns in mice of the genus *Peromyscus*. Miscellaneous Publications, Museum of Zoology, University of Michigan 99:1–59.

Hooper, E. T. 1958. The male phallus in mice of the genus *Peromyscus*. Miscellaneous Publications, Museum of Zoology, University of Michigan 105:1–24.

Hooper, E. T. 1968. Classification. Pp. 27–74, *in* Biology of *Peromysus* (Rodentia) (King, J. A. ed.). Special Publications American Society of Mammalogists 2:1–593.

Hooper, E. T. 1972. A synopsis of the rodent genus *Scotinomys*. Occasional Papers of the Museum of Zoology, University of Michigan 665:1–32.

Hooper, E. T., and B. S. Hart. 1962. A synopsis of recent North American microtine rodents. Miscellaneous Publications of the Museum of Zoology, University of Michigan 120:1–68.

Hooper, E. T., and G. G. Musser. 1964. Notes on classification of the rodents of the genus *Peromyscus*. Occasional Papers of the Museum of Zoology, University of Michigan 635:1–13.

Hope, A. G., E. Waltari, N. E. Dokuchaev, S. Abramov, T. Dupal, A. Tsvetkova, H. Henttonen, S. O. MacDonald, and J. A. Cook. 2010. High-latitude diversification within Eurasian least shrews and Alaska tiny shrews (Soricidae). Journal of Mammalogy 91:1041–1057.

Hope, A. G., K. A. Speer, J. R. Demboski, S. L. Talbot, and J. A. Cook. 2012. A Climate for speciation: rapid spatial diversification within the *Sorex cinereus* complex of shrews. Molecular Phylogentic and Evolution 64:671–684.

Hope, A. G., N. Panter, J. A. Cook, S. L. Talbot, and D. W. Nagorsen. 2014. Multilocus phylogeography and systematic revision of North American water shrews (genus: *Sorex*). Journal of Mammalogy 95: 722–738.

Hope, A. G., J. L. Malaney, K. C. Bell, F. Salazar-Miralles, A. S. Chavez, B. R. Barber, and J. A. Cook. 2016. Revision of widespread red squirrels (genus: *Tamiasciurus*) highlights the complexity of speciation within North American forests. Molecular Phylogenetics and Evolution 100:170–182.

Hope, A. G., R. B. Stephens, S. D. Mueller, V. V. Tkach, and J. R. Demboski. 2020. Speciation of North American pygmy shrews (Eulipotyphla: Soricidae) supports spatial but not temporal congruence of diversification among boreal species. Biological Journal of the Linnean Society 129:41–60.

Hopton, M. E., and G. N. Cameron. 2001. *Geomys knoxjonesi*. Mammalian Species 672:1–3.

Houseal, T. W., I. F. Greenbaum, D. J. Schmidly, S. A. Smith, and K. M. Davis. 1987. Karyotypic variation in *Peromyscus boylii* from Mexico. Journal of Mammalogy 68:281–296.

Howard, W. E. 1949. A means to distinguish skulls of coyotes and domestic dogs. Journal of Mammalogy 30:169–171.

Howell, A. B. 1926. Vole of the genus *Phenacomys*. North American Fauna 48:1–66.

Howell, A. B. 1927. Revision of the American lemming mice (Genus *Synaptomys*). North American Fauna 50:1–38.

Howell, A. H. 1906. Revision of the skunks of the genus *Spilogale*. North American Fauna 26:1–55.
Howell, A. H. 1914a. Revision of the harvest mice (genus *Reithrodontomys*). North American Fauna 36:1–97.
Howell, A. H. 1914b. Ten new marmots from North America. Proceedings of the Biological Society of Washington, 27:1318.
Howell, A. H. 1915a. Descriptions of a new genus and seven new races of flying squirrels. Procceding of the Biological Society Washington 28:109–114.
Howell, A. H. 1915b. Revision of the American marmots. United States Department of Agriculture Bureau of Biological Survey, North American Fauna 37:1–80.
Howell, A. H. 1918. Revision of the flying squirrels. North American Fauna 44:1–64.
Howell, A. H. 1920. Description of a new race of the Florida water-rat (*Neofiber alleni*). Journal of Mammalogy 1:79–80.
Howell, A. H. 1922. Diagnoses of seven new chipmunks of the genus *Eutamias*, with a list of the American species. Journal of Mammalogy 3:178–185.
Howell, A. H. 1924. Revision of the American pikas. North American Fauna 47:1–57.
Howell, A. H. 1925. Preliminary descriptions of five new chipmunks from North America. Journal of Mammalogy 6:51–54.
Howell, A. H. 1929. Revision of the American chipmunks (genera *Tamias* and *Eutamias*). North American Fauna 52:1–64.
Howell, A. H. 1938. Revision of the North American ground squirrels, with a classification of the North American Sciuridae. North American Fauna 56:1–256.
Howell, A. H. 1940. A new race of the harvest mouse (*Reithrodontomys*) from Virginia. Journal of Mammalogy 21:346.
Hrachovy, S. K., R. D. Bradley, and C. Jones. 1996. *Neotoma goldmani*. Mammalian Species 545:1–3.
Hsu, T. C., and M. L. Johnson. 1970. Cytological distinction between *Microtus montanus* and *Microtus canicaudus*. Journal of Mammalogy 51:824–826.
Huchon, D., F. M. Catzeflis, and E. J. P. Douzery. 2000. Variance of molecular datings, evolution of rodents and the phylogenetic affinities between Ctenodactylidae and Hystricognathi. Proceedings of the Royal Society B: Biological Sciences, 267:393–402.
Huckaby, D. G. 1973. Biosystematics of the *Peromyscus mexicanus* group (Rodentia). Ph.D. dissertation, University of Michigan, Ann Arbor.
Huckaby, D. G. 1980. Species limits in the *Peromyscus mexicanus* group (Mammalia: Rodentia: Muroidea). Contributions in Science, Natural History Museum of Los Angeles County 326:1–24.
Hückstädt, L. 2015. *Mirounga angustirostris*. The IUCN Red List of Threatened Species 2015:e.T13581A45227116.
Huey, L. M. 1925. Three new kangaroo rats of the genus *Dipodomys* from Lower California. Proccedings of the Biological Society of Washington 38:83–84.
Huey, L. M. 1951. The kangaroo rats (*Dipodomys*) of Baja California, Mexico. Transactions of the San Diego Society of Natural History 11:205–256.
Huey, L. M. 1926. The description of a new subspecies of *Perognathus* from Lower California with a short discussion of the taxonomic position of other peninsular members of this genus. Proceedings of the Biological Society of Washington 39:67–70.
Huey, L. M. 1960. Comments on the pocket mouse, *Perognathus fallax*, with descriptions of two new races from Baja California, Mexico. Transactions of the San Diego Society of Natural History 12:413–420.
Huey, L. M. 1964. The mammals of Baja California, Mexico. Transactions of the San Diego Society of Natural History 13:85–168.
Humphrey, S. R. 1974. Zoogeography of the nine-banded armadillo (*Dasypus novemcinctus*) in the United States. Bio-Science 24:457–462.
Humphrey, S. R., and T. H. Kunz. 1976. Ecology of a Pleistocene relict, the western big-eared bat *(Plecotus townsendii),* in the southern Great Plains. Journal of Mammalogy 57:470–494.
Hundertmark, K. 2016. *Alces alces*. The IUCN Red List of Threatened Species 2016:e.T56003281A22157381.
Hundertmark, K. J., G. F. Shields, I. G. Udina, R. T. Bowyer, A. A. Danilkin, and C. C. Schwartz. 2002. Mitochondrial phylogeography of moose (*Alces alces*): late Pleistocene divergence and population expansion. Molecular Phylogenetics and Evolution 22:375–387.
Hundertmark, K. J., R. T. Bowyer, G. F. Shields, and C. C. Schwartz. 2003. Mitochondrial phylogeography of moose (*Alces alces*) in North America. Journal of Mammalogy 84:718–728.
Hunt, J. L., L. A. McWilliams, T. L. Best, and K. G. Smith. 2003. *Eumops bonariensis*. Mammalian Species 733:1–5.
Hunt, J. L., J. E. Morris, and T. L. Best. 2004. *Nyctomys sumichrasti*. Mammalian Species 754:1–6.
Hunter, L. 2015. Wild Cats of the World. Bloomsbury Publishing. New York, U.S.A.
Hurtado, N., and V. Pacheco. 2014. Phylogenetic analysis of the genus *Mimon* Gray, 1847 (Mammalia, Chiroptera, Phyllostomidae) with description of a new genus. Therya 5:751–791.
Hurtado, N., and G. D'Elía. 2018. Taxonomy of the genus *Gardnerycteris* (Chiroptera: Phyllostomidae). Acta Chiropterologica 20:99–115.
Husar, S. L. 1978. *Trichechus manatus*. Mammalian Species 93:1–5.
Husson, A. M. 1962. The bats of Suriname. Zoologische Verhandelingen, Rijksmuseum van Natuurlijke Historie te Leiden 58:1–282.
Husson, A. M. 1978. The mammals of Suriname. Zoologische Monographieën van het Rijksmuseum van Natuurlijke Historie 2:1–569.
Hutterer, R. 2005. Order Soricomorpha. Pp. 220–311, *in* Mammal species of the world. A taxonomic and geographic reference (Wilson, D. E., and D. A. M. Reeder, eds.). Third edition. The Johns Hopkins University Press. Baltimore, U.S.A.
Hwang, Y. T., and S. Larivière. 2001. *Mephitis macroura*. Mammalian Species 686:1–3.
Indorf, J. L., and M. S. Gaines. 2013. Genetic divergence of insular marsh rice rats in subtropical Florida. Journal of Mammalogy 94:897–910.
Ingles, L. G. 1947. Mammals of California and its coastal waters. Stanford Univ. Press, Stanford, California.
Ingles, L. G. 1965. Mammals of the Pacific states. Stanford University Press. Stanford, U.S.A.
Intress, C., and T. L. Best. 1990. *Dipodomys panamintinus*. Mammalian Species 354:1–7.
Itoo, T., H. Katoh, K. Wada, K. Shimazaki, and K. Arai. 1977. Report of an ecological survey of the Steller sea lion in the coastal waters of Hokkaido, northern Japan, 1975–1976 winter. Geiken Tsushin 305:1–18.
IUCN SSC Antelope Specialist Group. 2016. *Antilocapra americana*. The IUCN Red List of Threatened Species 2016:e.T1677A115056938.
Iudica, C. A. 2000. Systematic revision of the Neotropical fruit bats of the genus *Sturnira*: a molecular and morphological approach. Ph.D. dissertation. University of Florida. Gainsville, U.S.A.
Ivanitskaya, E. Y., A. I. Kozlovskii, N. V. Orlov, Y. M. Kovalskaya, and M. I. Baskevich. 1986. New data on karyotypes of common shrews (*Sorex*, Soricidae, Insectivora) in fauna of the USSR. Zoologicheskii Zhurnal 65:1228–1236.

Iverson, S. L. 1967. Adaptations to arid environments in *Perognathus parvus* (Peale). Unpubl. Ph.D. dissert., Univ. British Columbia, Vancouver.
Jackson, D. J., and J. A. Cook. 2020. A precarious future for distinctive peripheral populations of meadow voles (*Microtus pennsylvanicus*). Journal of Mammalogy 101:36–51.
Jackson, H. H. T. 1914. New moles of the genus *Scalopus*. Proceedings of the Biological Society of Washington 27:19–21.
Jackson, H. H. T. 1915. A review of the American moles. North American Fauna 38:1–100.
Jackson, H. H. T. 1925. The *Sorex arcticus* and *Sorex arcticus cinereus* of Kerr. Journal of Mammalogy 6:55–56.
Jackson, H. H. T. 1928. A taxonomical revision of the American long-tailed shrew (genera *Sorex* and *Microsorex*). North American Fauna 51:1–238.
Jackson, H. H. T. 1933. Five new shrew of the genus *Cryptotis* from Mexico and Guatemala. Proceedings of the Biological Society of Washington 46:79–82.
Jackson, H. H. T. 1951. Classification of the races of coyote. Pp. 227–341, *in* The clever coyote (Young, S. P., and H. H. T. Jackson, eds.). North American Wildlife Management Institute. Washington, U.S.A.
Jackson, H. H. T. 1961. Mammals of Wisconsin. The University of Wisconsin Press. Madison, U.S.A.
James, A. H., and D. K. James. 1984. *Dipodomys californicus* in Sierra Valley, Plumas County, California. California Fish and Game 70:58–64.
Jameson, E. W., and H. Peeters. 1988. California mammals. University of California Press. Berkeley, U.S.A.
Jameson, E. W., and H. Peeters. 2004. Mammals of California. University of California Press. Berkeley, U.S.A.
Janecek, L. L. 1990. Genetic variation in the *Peromyscus truei* group (Rodentia: Cricetidae). Journal of Mammalogy 71:301–308.
Jannett, F. J., Jr. 1976. Bacula of *Dipodomys ordii compactus* and *Dipodomys elator*. Journal of Mammalogy 57:382–387.
Jannett, F.J., Jr., and R. J. Oehlenschlager. 1994. Range extension and first Minnesota records of the Smokey Shrew *Sorex fumeus*. The American Midland Naturalist 131:364–365.
Jarrell, G. H., and S. O. MacDonald. 1989. Checklist to the mammals of Alaska. University of Alaska Museum. Fairbanks, U.S.A.
Jarrell, G. H., and K. Fredga. 1993. How many kinds of Lemmings? A taxonomic overview. Pp. 46–57, *in* The biology of Lemming (Stenseth, N. C., and R. A. Ims, eds.). Linnean Society Symposium series 15. Academic Press. London, United Kingdom.
Jenkins, S. H., and P. E. Busher. 1979. *Castor canadensis*. Mammalian Species 120:1–8.
Jenkins, S. H., and B. D. Eshelman. 1984. *Spermophilus beldingi*. Mammalian Species 221:1–8.
Jennings, J. B., T. L. Best, S. E. Burnett, and J. C. Rainey. 2002. *Molossus sinaloae*. Mammalian Species 691:1–5.
Jiménez, J. J. 1972. Comparative post-natal growth in five species of the *Sigmodon* II. Cranial character relationships. Revista de Biología Tropical 20:5–27.
Jiménez-Guzmán, A. 1968. Nuevos registros de murciélagos para Nuevo León, México. Anales de Instuto de Biología, Universidad Nacional Autónoma de México, Serie Zoología 39:133–144.
Jiménez-Guzmán, A., S. Contreras-Arquieta, and M. A. Zuñiga-Ramos. 1994. Historia de la Mastofauna de Nuevo Leon, Mexico y su bibliografia. Publicaciones Biológicas, Facultad de Ciencias Biológicas, Universidad Autónoma de Nuevo León 2:1–39.
Johnson, D. H. 1943. Systematic review of the chipmunks (genus *Eutamias*) of California. University of California Publications in Zoology 48:63–147.
Johnson, D. H. 1951. The water shrews of the Labrador Peninsula. Proceedings of the Biological Society of Washington 64:109–116.
Johnson, D. W., and D. M. Armstrong. 1987. *Peromyscus crinitus*. Mammalian Species 287:1–8.
Johnson, M. L. 1968. Application of blood protein electrophoretic studies to problems in mammalian taxonomy. Systematic Zoology 17:23–30.
Johnson, M. L. 1973. Characters of the heather vole, *Phenacomys*, and the red tree vole, *Arborimus*. Journal of Mammalogy 54:239–244.
Johnson, M. L., and S. B. Benson. 1960. Relationships of the pocket gophers of the *Thomomys mazama-talpoides* complex in the Pacific Nortthwest. The Murrelet 41:17–22.
Johnson, M. L., and C. W. Clanton. 1954. Natural history of *Sorex merriami* in Washington State. The Murrelet 35:1–4.
Johnson, M. L., and S. B. George. 1991. Species limits within the *Arborimus longicaudus* species-complex (Mammalia: Rodentia) with a description of a new species of California. Contributions in Science, Natural History Museum of Los Angeles County 429:1–16.
Johnson, M. L., and C. Maser. 1982. Generic relationships of *Phenacomys albipes*. Northwest Science 56:17–19.
Johnson, M. L., and B. T. Ostenson. 1959. Comments on the nomenclature of some mammals of the Pacific Northwest. Journal of Mammalogy 40:574–576.
Johnson, M. L., and T. L. Yates. 1980. A new Townsend's mole (*Scapanus townsendii*) from the State of Washington. Occasional Papers, The Museum, Texas Tech University 63:1–6.
Johnson, W. E., and R. K. Sealander. 1971. Protein variation and systematics in kangaroo rats (genus *Dipodomys*). Systematic Zoology 20:377–405.
Johnson, W. E., R. K. Sealander, M. H. Smith, and Y. J. Kim. 1972. Biochemical genetics of sibling species of the cotton rat (*Sigmodon*). Studies in Genetics VII. University of Texas Publications 7213:297– 305.
Johnson, W. E., E. Eizirik, J. Pecon-Slattery, W. J. Murphy, A. Antunes, E. Teeling, and S. J. O'Brien. 2006. The late Miocene radiation of modern Felidae: a genetic assessment. Science 311:73–77.
Jolley, T. W., R. L. Honeycutt, and R. D. Bradley. 2000. Phylogenetic relationships of pocket gophers (genus *Geomys*) based on the mitochondrial 12s rRNA gene. Journal of Mammalogy 81:1025–1034.
Jones, C. 1977. *Plecotus rafinesquii*. Mammalian Species 69:1–4.
Jones, C., and C. N. Baxter. 2004. *Thomomys bottae*. Mammalian Species 7421–14.
Jones, C., and N. J. Hildreth. 1989. *Neotoma stephensi*. Mammalian Species 328:1–3.
Jones, C., and J. N. Layne. 1993. *Podomys floridanus*. Mammalian Species 427:1–5.
Jones, C., and R. W. Manning. 1989. *Myotis austroriparius*. Mammalian Species 332:1–3.
Jones, C., and R. D. Suttkus. 1971. Wing loading in *Plecotus rafinesquii*. Journal of Mammalogy 52:458–460.
Jones, C., R. S. Hoffman, D. W. Rice, M. D. Engstrom, D. Bradley, D. J. Schmidly, C. Jones, and R. J. Baker. 1997. Revised checklist of North American mammals north of Mexico, 1997. Occasional Papers, The Museum, Texas Tech University 173:1–20.
Jones, J. K., Jr. 1964. Bats from western and southern Mexico. Transactions of Kansas Academy of Science 67:509–516.
Jones, J. K., Jr. 1966. Bats from Guatemala. University ok Kansas Publication, Museum of Natural History 16:439–472.
Jones, J. K., Jr. 1977. *Rhogeessa gracilis*. Mammalian Species 76:1–2.

Jones, J. K., Jr. 1978. A new bat of the genus *Artibeus* from the Lesser Antillean Island of St. Vincent. Occasional Papers, The Museum, Texas Tech University 51:1–6.

Jones, J. K., Jr. 1982. *Reithrodontomys spectabilis*. Mammalian Species 193:1.

Jones, J. K., Jr., and T. Álvarez. 1962. Taxonomic status of the free-tailed bat, *Tadarida yucatanica* Miller. University of Kansas Publications, Museum of Natural History 9:125–133.

Jones, J. K., Jr., and T. Álvarez. 1964. Additional records of mammals from the Mexican state of San Luis Potosí. Journal of Mammalogy 45:302–303.

Jones, J. K., Jr., and J. Arroyo-Cabrales. 1990. *Nyctinomops aurispinosus*. Mammalian Species 350:1–3.

Jones, J. K., Jr., and E. C. Birney. 1988. Handbook of mammals of the north-central states. The University of Minnesota Press. Minneapolis, U.S.A.

Jones, J. K., Jr., and D. C. Carter. 1976. Annotated checklist with keys to subfamilies and genera. Pp. 7–38, *in* Biology of bats of the New World family Phyllostomatidae, Part I (Baker, R. J., J. K. Jones, Jr., and D. C. Carter, eds.). Special Publications, The Museum, Texas Tech University 10:1–218.

Jones, J. K., Jr., and J. R. Choate. 1978. Distribution of two species of long-eared bats of the genus *Myotis* on the northern Great Plains. Prairie Naturalist 10:49–52.

Jones, J. K., Jr., and H. H. Genoways. 1967. Notes on the Oaxacan vole, *Microtus oaxacensis* Goodwin, 1966. Journal of Mammalogy 48:320–321.

Jones, J. K., and H. H. Genoways. 1975. *Dipodomys phillipsii*. Mammalian Species 51:1–3.

Jones, J. K., and H. H. Genoways. 1978. *Neotoma phenax*. Mammalian Species 108:1–3.

Jones, J. K., Jr., and J. A. Homan. 1974. *Hylonycteris underwoodi*. Mammalian Species 32:1–2.

Jones, J. K., Jr., and C. S. Hood. 1993. Synopsis of South American bats of the family Emballonuridae. Occasional Papers, The Museum, Texas Tech University, 155:1–32.

Jones, J. K., Jr., and T. E. Lawlor. 1965. Mammals from Isla Cozumel, Mexico, with description of a new species of harvest mouse. University of Kansas, Publications Museum of Natural History 16:409–419.

Jones, J. K., Jr., and R. R. Johnson. 1967. Sirenians. Pp. 366–373, *in* Recent mammals of the world-a synopsis of families (Anderson, S., and J. K. Jones eds.), Ronald Press Co. New York, U.S.A.

Jones, J. K., Jr., and R. W. Manning. 1992. Illustrated key to skulls of genera of North American land mammals. Texas Tech University Press. Lubbock, U.S.A.

Jones, J. K., Jr., T. Álvarez, and M. R. Lee. 1962. Noteworthy mammals of Sinaloa, Mexico. University of Kansas Publications, Museum of Natural History 14:145–159.

Jones, J. K., Jr., J. D. Smith, and T. Alvarez. 1965. Notes on bats from the cape región of Baja California. Transactions of the San Diego Society of Natural History 14:53–56.

Jones, J. K., Jr., H. H. Genoways, and L. C. Watkins. 1970. Bats of the genus *Myotis* from western Mexico, with a key to the species. Transactions of the Kansas Academy of Sciences 73:409–418.

Jones, J. K., Jr., J. D. Smith, and R. W. Turner. 1971. Noteworthy records of bats from Nicaragua, with a checklist of the chiropteran fauna of the country. Occasional Papers Museum Natural History, University of Kansas 2:1–35.

Jones, J. K., Jr., J. R. Choate, and A. Cadena. 1972. Mammals from the Mexican state of Sinaloa. 2. Chiroptera. Occasional Papers Museum of Natural History, University of Kansas 6: 1–29.

Jones, J. K., Jr., J. D. Smith, and H. H. Genoways. 1973. Annotated checklist of mammals of the Yucatan Peninsula, Mexico: I. Chiroptera. Occasional Papers, The Museum, Texas Tech University 13:1–31.

Jones, J. K., Jr., H. H. Genoways, and T. E. Lawlor. 1974. Annotated checklist of mammals of the Yucatan Peninsula, Mexico II. Rodentia. Occasional Papers, The Museum, Texas Tech University 22:1–24.

Jones, J. K., Jr., D. C. Carter, H. H. Genoways, R. S. Hoffmann, and D. W. Rice. 1982. Revised checklist of North American mammals north of Mexico. Occasional Papers, The Museum, Texas Tech University 80:1–21.

Jones, J. K., Jr., D. C. Carter, and W. D. Webster. 1983a. Records of mammals from Hidalgo, Mexico. The Southwestern Naturalist 28:378–380.

Jones, J. K., Jr., D. M. Armstrong, R. S. Hoffman, and C. Jones. 1983b. Mammals of the northern Great Plains. University of Nebraska Press. Lincoln, U.S.A.

Jones, J. K., Jr., D. C. Carter, H. H. Genoways, R. S. Hoffmann, D. W. Rice, and C. Jones. 1986. Revised checklist of North American mammals north of Mexico. 1986. Occasional Papers, The Museum, Texas Tech University 107:1–22.

Jones, J. K., Jr., J. Arroyo-Cabrales, and R. D. Owen. 1988. Revised checklist of bats (Chiroptera) of Mexico and Central America. Occasional Papers, The Museum, Texas Tech University 120:1–34.

Jones, J. K., Jr., R. W. Manning, and J. R. Geotze. 1991. Noteworthy records of seven species of small mammals from westcentral Texas. Occasional Papers, The Museum, Texas Tech University 143:1–4.

Jones, J. K., Jr., R. S. Hoffmann, D. W. Rice, C. Jones, R. J. Baker, and M. D. Engstrom. 1992. Revised checklist of North American mammals north of Mexico, 1991. Occasional Papers, The Museum, Texas Tech University 146:1–23.

Juárez-López, R., M. Pérez-López, Y. Bravata-de la Cruz, A. J. de la Cruz, F. M. Contreras-Moreno, D. Thornton, and M. G. Hidalgo-Mihart. 2017. Range extension of the northern Naked-Tailed Armadillo (*Cabassous centralis*) in southern Mexico. Western North American Naturalist 77:398–403.

Judd, S. R., S. P. Cross, and S. Pathak. 1980. Non-Robertsonian chromosomal variation in *Microtus montanus*. Journal of Mammalogy 61:109–113.

Jung, T. S., B. G. Slough, D. W. Nagorsen, and P. M. Kukka. 2014. New records of the Ogilvie Mountains Collared Lemming (*Dicrostonyx nunatakensis*) in central Yukon. The Canadian Field-Naturalist 128:265–268.

Junge, J. A., and R. S. Hoffman. 1981. An annotated key to the long-tailed shrews (genus *Sorex*) of the United States and Canada, with notes on the Middle American *Sorex*. Occasional Papers of the Museum of Natural History, the University of Kansas 94:1–48.

Junge, J. A., R. S. Hoffmann, and R. W. DeBry. 1983. Relationships within the Holarctic *Sorex arcticus-Sorex tundrensis* species complex. Acta Theriologica 28:339–350.

Kain, D. E. 1985. The systematic status of *Eutamias ochrogenys* and *Eutamias senex* (Rodentia: Sciuridae). M. A. thesis, Humboldt State University. Arcata, U.S.A.

Kalcounis-Rueppell, M. C., and T. R. Spoon. 2009. *Peromyscus boylii* (Rodentia: Cricetidae). Mammalian Species 838:1–14.

Kays, R. 2018. *Canis latrans*. The IUCN Red List of Threatened Species 2018:e.T3745A163508579.
Keinath, D. A., H. R. Griscom, and M. D. Andersen. 2014. Habitat and distribution of the Wyoming pocket gopher (*Thomomys clusius*). Journal of Mammalogy 95:803–813.
Keith, J. O. 1965. The Abert squirrel and its dependence on Ponderosa pine. Ecology 46:150–163.
Kellnhauser, R. T. 1983. The acceptance of *Lontra* Gray for the New World river otters. Canadian Journal of Zoology 61:278–279.
Kellogg, R. 1956. What and where are the whitetails? Pp. 31–55, *in* The deer of North America (Taylor, W. P. ed.), The Stackpole Co. Harrisburg, U.S.A.
Kelly, P. A. 2018. *Sylvilagus bachmani*. Pp. 122–125, *in* Pikas, Rabbits and Hares of the World (Smith, A., C. H. Johnston, P. C. Alves, and K. Hackländer, eds.). The Johns Hopkins University Press/IUCN/SSC. Baltimore, U.S.A.
Kelly, M., D. Morin, and C. A. Lopez-Gonzalez. 2016. *Lynx rufus*. The IUCN Red List of Threatened Species 2016:e.T12521A50655874.
Kelly, P. A., C. Lorenzo, and S. T. Álvarez-Castañeda. 2019. *Sylvilagus bachmani*. The IUCN Red List of Threatened Species 2019:e. T41302A45192710.
Kelson, K. R. 1952. Comments on the taxonomy and geographic distribution of some North American woodrats (genus *Neotoma*). University of Kansas Publications, Museum of Natural History 5:233–242.
Kelt, D. A. 1988a. *Dipodomys californicus*. Mammalian Species 324:1–4.
Kelt, D. A. 1988b. *Dipodomys heermanni*. Mammalian Species 323:1–7.
Kenagy, G. J. 1973. Daily and seasonal patterns of activity and energetics in a heteromyid rodent community. Ecology 54:1201–1219.
Kenagy, G. J., and G. A. Bartholomew. 1985. Seasonal reproductive patterns in five coexisting California desert rodent species. Ecological Monographs 55:371–397.
Kennedy, M. L., M. L. Beck, and T. L. Best. 1980. Intraspecific Morphologic Variation in Ord's Kangaroo Rat, *Dipodomys ordii*, from Oklahoma. Journal of Mammalogy 61:311–319.
Kennedy, M. L., T. L. Best, and M. J. Harvey. 1984. Bats of Colima, Mexico. Mammalia 48:397–408.
Kennedy, M. L., P. K. Kennedy, M. A. Bogan, and J. L. Waits. 2002. Taxonomic assessment of the black bear (*Ursus americanus*) in the eastern United States. The Southwestern Naturalist 47:335–347.
Kenneth, T. W. 1987. *Lasiurus seminolus*. Mammalian Species 280:1–5.
Kenyon, K. W. 1960. A ringed seal from the Pribiof Islands, Alaska. Journal of Mammalogy 41:520–521.
Kenyon, K. W. 1977. Caribbean monk seal extinct. Journal of Mammalogy 58:97–98.
Keuling, O., and K. Leus. 2019. *Sus scrofa*. The IUCN Red List of Threatened Species 2019:e.T41775A44141833.
Keuroghlian, A., A. Desbiez, R. Reyna-Hurtado, M. Altrichter, H. Beck, A. Taber, and J. M. V. Fragoso. 2013. *Tayassu pecari*. The IUCN Red List of Threatened Species 2013:e.T41778A44051115.
Kitchner, A. C., *et al.* 2017. A revised taxonomy of the Felidae. The final report of the Cat Classification Task Force of the IUCN/SSS Cat Specialist Group. Cat News Special Issue 11. Stämpfli Publikationen AG, Bern, Switzerland.
Kilpatrick, C. W., and K. L. Crowell. 1985. Genic variation of the rock mole, *Microtus chrotorrhinus*. Journal of Mammalogy 66:94–101.
King, C. M. 1983a. *Mustela erminea*. Mammalian Species 195:1–8.
King, C. M. 1989. The natural history of weasels and stoats. Christopher Helm Publishers. London, United Kingdom.
King, J. E. 1954. The otariid seals of the Pacific coast of America. Bulletin of the British Museum (Natural History) Zoology 2:311–337.
King, J. E. 1956. The monk seals genus *Monachus*. Bulletin of the British Museum (Natural History), Zoology Series 3:203–256.
King, J. E. 1966. Relationships of the hooded and elephant seals (genera *Cystophora* and *Mirounga*). Journal of Zoology, London 148:385–398.
King, J. E. 1983b. Seals of the world. Second edition. British Museum (Natural History). London, United Kingdom.
King, S. R. B., L. Boyd, W. Zimmermann, and B. E. Kendall. 2015. *Equus ferus*. The IUCN Red List of Threatened Species 2015:e. T41763A97204950.
Kinlaw, A. 1995. *Spilogale putorius*. Mammalian Species 511:1–7.
Kilpatrick, C. W., and E. G. Zimmerman. 1976. Biochemical variation and systematics of *Peromyscus pectoralis*. Journal of Mammalogy 57:506–522.
Kilpatrick, C. W., N. Pradhan, and R. W. Norris. 2021. A re-examination of the molecular systematics and phylogeography of taxa of the *Peromyscus aztecus* species group, with comments on the distribution of *P. winkelmanni*. Therya 12:331–346.
Kirkland, G. L. 1977. A Re-examination of the subspecific status of the Maryland Shrew, *Sorex cinereus fontinalis* Hollister. Proceedings of the Pennsylvania Academy of Science 51:43–46.
Kirkland Jr., G. L., 1981. *Sorex dispar* and *Sorex gaspensis*. Mammlian Species 155:1–4.
Kirkland Jr., G. L., and F. J. J. Jannett. 1982. *Microtus chrotorrhinus*. Mammalian Species 180:1–5.
Kirkland Jr., G. L., and H. M. van Duesen. 1979. The shrews of the *Sorex dispar* group: *Sorex dispar* Batchelder and *Sorex gaspensis* Anthony and Goodwin. American Museum Novitates 2675:1–21.
Kirkland, G. L., and J. M. Levengood. 1987. First record of the Maryland Shrew (*Sorex fontinalis*) from West Virginia. Proceedings of the Pennsylvania Academy of Science 61:35–37.
Kirkpatrick, R. D., A. M. Cartwright, J. C. Brier, and E. J. Spicka. 1975. Additional mammal records from Belize. Mammalia 39:330–331.
Kiser, W. M. 1995. *Eumops underwoodi*. Mammalian Species 516:1–4.
Kivett, V. K., J. O. Murie, and A. L. Steiner. 1976. A comparative study of scent-gland location and related behavior in some northwestern Nearctic ground squirrel species (Sciuridae): an evolutionary approach. Canadian Journal of Zoology 54:1294–1306.
Koepfli, K. P., and R. K. Wayne. 1998. Phylogenetic relationships of otters (Carnivora: Mustelidae) based on mitochondrial cytochrome b sequences. Journal of Zoology, London 246:401–416.
Koepfli, K. P., K. A. Deer, G. J. Slater, C. Begg, K. Begg, L. Grassman, M. Lucherini, G. Veron, and R. K. Wayne. 2008. Multigene phylogeny of the Mustelidae: Resolving relationships, tempo and biogeographic history of a mammalian adaptive radiation. BMC Biology 6:10.
Koop, B. F., R. J. Baker, and J. T. Mascarello. 1985. Cladistical analysis of chromosomal evolution withing the genus *Neotoma*. Occasional Papers, The Museum, Texas Tech University 96:1–9.
Koopman, K. F. 1971. The systematic and historical status of the Florida *Eumops* (Chiroptera, Molossidae). American Museum Novitates 2478: 1–6.

Koopman, K. F. 1978. Zoogeography of Peruvian bats with special emphasis on the role of the Andes. American Museum Novitates 2651:1–33.
Koopman, K. F. 1982. Biogeography of the bats of South America. Pp. 273–300, *in* Mammalian biology of South America (Mares, M. A., and H. H. Genoways, eds.). Special Publication, Pymatuning Laboratory of Ecology, University of Pittsburgh. Pittsburgh, U.S.A.
Koopman, K. F. 1984. Bats. Pp, 145–186, *in* Orders and families of Recent mammals of the world (Anderson, S., and J. K. Jones, Jr., eds.). John Wiley and Sons. New York, U.S.A.
Koopman, K. F. 1993. Order Chiroptera. Pp. 137–242, *in* Mammal species of the world (Wilson, D. E., and D. M. Reeder, eds.). Second ed. Smithsonian Institution Press. Washington, U.S.A.
Koopman, K. F. 1994. Chiroptera: systematics. Handbook of zoology: a natural history of the phyla of the animal kingdom. VIII. Mammalia. Walter de Gruyter. New York, U. S. A.
Koprowski, J. L. 1994a. *Sciurus carolinensis*. Mammalian Species 480:1–9.
Koprowski, J. L. 1994b. *Sciurus niger*. Mammalian Species 479:1–9.
Koprowski, J. L. 2017. *Ammospermophilus nelsoni*. The IUCN Red List of Threatened Species 2017:e.T1149A22251492.
Koprowski, J., Roth, L., N. Woodman, J. Matson, L. Emmons, and F. Reid. 2016. *Sciurus deppei*. The IUCN Red List of Threatened Species 2016:e.T20001A115154440.
Koprowski, J., Roth, L., F. Reid, N. Woodman, R. Timm, and L. Emmons. 2017. *Sciurus aureogaster*. The IUCN Red List of Threatened Species 2017:e.T20006A22248035.
Koprowski, J. L., E. A. Goldsteins, K. R. Bennett, and C. Pereira Mendes. 2016a. Family Sciuridae (tree, flying and ground squirrels, chipmunks, marmots and prairie dogs). Pp. 648–836, *in* Handbook of the Mammals of the world 6. Lagomorpha and Rodents. (Wilson, D. E., T. E. Lacher, Jr., and R. A. Mittermeier, eds.). Lynx Edicions, IUCN. Barcelona, Spain.
Koprowski, J. L., N. Ramos, B. S. Pasch, and C. A. Zugmeyer. 2006. Observations on the ecology of the endemic Mearns's squirrel (*Tamiasciurus mearnsi*). Southwestern Naturalist 51:426–430.
Koprowski, J. L., M. A. Steele, and N. Ramos-Lara. 2016b. *Tamiasciurus mearnsi* (Rodentia: Sciuridae). Mammalian Species 48:66–72.
Kovacs, K. M. 2009. Bearded seal *Erignathus barbatus*. Pp. 97–101, *in* Encyclopedia of Marine Mammals (Perrin, W. F., B. Wursig, and J. G. M. Thewissen, eds.). Academic Press. San Diego, U.S.A.
Kovacs, K. M. 2015. *Pagophilus groenlandicus*. The IUCN Red List of Threatened Species 2015:e.T41671A45231087.
Kovacs, K. M. 2016a. *Erignathus barbatus*. The IUCN Red List of Threatened Species 2016:e.T8010A45225428.
Kovacs, K. M. 2016b. *Cystophora cristata*. The IUCN Red List of Threatened Species 2016:e.T6204A45225150.
Kovacs, K. M., and D. M. Lavigne. 1986. *Crystophora cristata*. Mammalian Species 258:1–9.
Kovacs, K. M., A. Aguilar, D. Aurioles, V. Burkanov, C. Campagna, N. J. Gales, T. Gelatt, S. D. Goldsworthy, S. J. Goodman, G. J. G. Hofmeyr, T. Härkönen, L. Lowry, L. Lydersen, J. Schipper, T. Sipilä, C. Southwell, D. Thompson, and F. Trillmich. 2012. Global threats to pinnipeds. Marine Mammal Science 28:414–436.
Krebs, C. J., and D. L. Murray. 2018. *Lepus americanus*. Pp. 163–165, *in* Pikas, Rabbits and Hares of the World (Smith, A., C. H. Johnston, P. C. Alves, and K. Hackländer, eds.). The Johns Hopkins University Press/IUCN/SSC. Baltimore, U.S.A.
Kretzoi, M. 1964. Über einige homonyme und synonyme Säugetiernamen. Vertebrata Hungarica 6:131–138.
Kretzoi, M. 1969. Skizze einer Arvicoliden-Phylogenie. Vertebrata Hungarica 12:111–121.
Krohne, D. T. 1982. The Karyotype of *Dicrostonyx hudsonius*. Journal of Mammalogy 63:174–176.
Krumbiegel, I. 1980. Die unterartliche Trennung des Bisons, *Bison bison* (Linné, 1788), und seine Rückzüchtung. Säugetierkundliche Mitteilungen 28:148–160.
Krutzsch, P. H. 1954. North American jumping Mice (Genus *Zapus*). University of Kansas Publications, Museum of Natural History 7:349–472.
Kryštufek, B., and V. Vohralik. 2013. Taxonomic revision of the Palaearctic rodents (Rodentia). Part 2. Sciuridae: *Urocitellus, Marmota* and *Sciurotamias*. Lynx 44:27–138.
Kryštufek, B., A. S. Tesakov, V. S. Lebedev, A. A. Bannikova, N. I. Abramson, and G. Shenbrot. 2019. Back to the future: the proper name for red-backed voles is *Clethrionomys tilesius* and not *Myodes pallas* Mammalia 84:1–4.
Kryštufek, B., L. Palomo, R. Hutterer, G. Mitsainas, and N. Yigit. 2021. *Rattus rattus* (amended version of 2016 assessment). The IUCN Red List of Threatened Species 2021: e.T19360A192565917.
Kumirai, A., and J. K. Jones, Jr. 1990. *Nyctinomops femorosaccus*. Mammalian Species 349:1–5.
Kunz, T. H. 1982. *Lasionycteris noctivagans*. Mammalian Species 172:1–5.
Kunz, T. H., and R. A. Martin. 1982. *Plecotus townsendii*. Mammalian Species 175:1–6.
Kurta, A. 1995. Mammals of the Great Lakes region. Revised edition. University of Michigan Press. Ann Arbor, U.S.A.
Kurta, A., and R. H. Baker. 1990. *Eptesicus fuscus*. Mammalian Species 356:1–10.
Kurta, A., and G. C. Lehr. 1995. *Lasiurus ega*. Mammalian Species 515:1–7.
Kurten, B., and E. Anderson. 1980. Pleistocene mammals of North America. Columbia University Press. New York, U.S.A.
Kuwayama, R., and T. Ozawa. 2000. Phylogenetic relationships among European red deer, wapiti, and sika deer inferred from mitochondrial DNA sequences. Molecular Phylogenetics and Evolution 15:115–123.
Kwiecinski, G. G. 1998. *Marmota monax*. Mammalian Species 591:1–8.
Kwiecinski, G. G. 2006. *Phyllostomus discolor*. Mammalian Species 801:1–11.
Kyle, C. J., C. Strobeck, R. D. Weir, H. Davis, and N. J. Newhouse. 2004. Genetic structure of sensitive and endangered northwestern badger populations (*Taxidea taxus taxus* and *T. t. jeffersonii*). Journal of Mammalogy 85:633–639.
Laabs, D. M., M. L. Allaback, and D. R. Mitchell. 2022. Distribution, morphology, and karyotype of San Joaquin pocket mice from the western Mojave Desert. California Fish and Wildlife Journal 108:e10.
Lacher, T. E., Jr. 2016a. Family Caviidae (Cavies, Capibaras, and Maras). Pp. 406–439, *in* Handbook of the mammals of the world 6. Lagomorpha and Rodentia I (Wilson, D. E., T. E. Lacher, Jr., and R. A. Mittermeier, eds.). Lynx Edicions. Barcelona, Spain.
Lacher, T. 2016b. *Geomys personatus*. The IUCN Red List of Threatened Species 2016:e.T9055A22218317.
Lacher, T., and S. T. Álvarez-Castañeda. 2016. *Neotoma albigula*. The IUCN Red List of Threatened Species 2016:e.T14582A115122696.
Lacher, T., and S. T. Álvarez-Castañeda. 2019. *Chaetodipus goldmani*. The IUCN Red List of Threatened Species 2019:e.T4332A22226182.
Lacher, T., R. Timm, and S. T. Álvarez-Castañeda. 2016a. *Dipodomys simulans*. The IUCN Red List of Threatened Species 2016:e.T136630A115210884.

Lacher, T., R. Timm, and S. T. Álvarez-Castañeda. 2016b. *Onychomys arenicola*. The IUCN Red List of Threatened Species 2016:e. T15337A115127153.
Lacher, T., R. Timm, and S. T. Álvarez-Castañeda. 2016c. *Onychomys torridus*. The IUCN Red List of Threatened Species 2016:e. T15339A115127523.
Lacher, T., R. Timm, and S. T. Álvarez-Castañeda. 2016d. *Otospermophilus variegatus*. The IUCN Red List of Threatened Species 2016:e. T20495A22263993.
Lacher, T., R. Timm, and S. T. Álvarez-Castañeda. 2016e. *Perognathus flavescens*. The IUCN Red List of Threatened Species 2016:e. T16634A115134864.
Lacher, T., R. Timm, and S. T. Álvarez-Castañeda. 2016f. *Perognathus flavus*. The IUCN Red List of Threatened Species 2016:e. T16635A115135061.
Lacher, T., R. Timm, and S. T. Álvarez-Castañeda. 2016g. *Perognathus merriami*. The IUCN Red List of Threatened Species 2016:e. T16637A115135383.
Lacher, T., R. Timm, and S. T. Álvarez-Castañeda. 2016h. *Peromyscus boylii*. The IUCN Red List of Threatened Species 2016:e. T16652A115135521.
Lacher, T., R. Timm, and S. T. Álvarez-Castañeda. 2016i. *Peromyscus gratus*. The IUCN Red List of Threatened Species 2016:e. T16663A115136134.
Lacher, T., R. Timm, and S. T. Álvarez-Castañeda. 2016j. *Peromyscus nasutus*. The IUCN Red List of Threatened Species 2016:e. T16682A115137189.
Lacher, T., R. Timm, and S. T. Álvarez-Castañeda. 2016k. *Peromyscus pectoralis*. The IUCN Red List of Threatened Species 2016:e. T16684A115137332.
Lacher, T., R. Timm, and S. T. Álvarez-Castañeda. 2016l. *Reithrodontomys montanus*. The IUCN Red List of Threatened Species 2016:e. T19413A115151509.
Lacher, T., R. Timm, and S. T. Álvarez-Castañeda. 2016m. *Sciurus griseus*. The IUCN Red List of Threatened Species 2016:e.T20011A115154716.
Lacher, T., R. Timm, and S. T. Álvarez-Castañeda. 2016n. *Sigmodon ochrognathus*. The IUCN Red List of Threatened Species 2016:e. T20217A115157990.
Lacher, T., R. Timm, and S. T. Álvarez-Castañeda. 2016o. *Thomomys bottae*. The IUCN Red List of Threatened Species 2016:e. T21799A115163311.
Lacher, T., S. T. Álvarez-Castañeda, and R. Timm. 2016p. *Thomomys umbrinus*. The IUCN Red List of Threatened Species 2016:e. T21800A115163507.
Lacher, T., R. Timm, and S. T. Álvarez-Castañeda. 2016q. *Xerospermophilus spilosoma*. The IUCN Red List of Threatened Species 2016:e. T42563A22262899.
Lacher, T., R. Timm, and S. T. Álvarez-Castañeda. 2016r. *Xerospermophilus tereticaudus*. The IUCN Red List of Threatened Species 2016:e. T20493A22264318.
Lacher, T., S. T. Álvarez-Castañeda, and R. Timm. 2016s. *Neotamias dorsalis*. The IUCN Red List of Threatened Species 2016:e. T42571A115190634.
Lacher, T., S. T. Álvarez-Castañeda, and R. Timm. 2016t. *Neotoma micropus*. The IUCN Red List of Threatened Species 2016:e.T14591A115123286.
Lacher, T., R. Timm, and S. T. Álvarez-Castañeda. 2017. *Neotoma lepida*. The IUCN Red List of Threatened Species 2017:e. T116988741A123797359.
Lacher, T., D. Hafner, and R. Timm. 2019a. *Geomys arenarius*. The IUCN Red List of Threatened Species 2019:e.T9054A22218401.
Lacher, T., R. Timm, and S. T. Álvarez-Castañeda. 2019b. *Peromyscus eremicus*. The IUCN Red List of Threatened Species 2019:e. T16659A143641683.
Lackey, J. A. 1967a. Biosystematics of *heermanni* group kangaroo rats in southern California. Transactions of the San Diego Society of Natural History 14:313–344.
Lackey, J. A. 1967b. Growth and development of *Dipodomys stephensi*. Journal of Mammalogy 48:624–632.
Lackey, J. A. 1991a. *Chaetodipus arenarius*. Mammalian Species 384:1–4.
Lackey, J. A. 1991b. *Chaetodipus spinatus*. Mammalian Species 385:1–4.
Lackey, J. A. 1996. *Chaetodipus fallax*. Mammalian Species 517:1–6.
Lackey, J. A., and T. L. Best. 1992. *Chaetodipus goldmani*. Mammalian Species 419:1–5.
Lackey, J. A., D. G. Huckaby, and B. G. Ormiston. 1985. *Peromyscus leucopus*. Mammalian Species 247:1–10.
Lacoste, K. N., and G. B. Stenson. 2000. Winter distribution of harp seals (*Phoca groenlandica*) off eastern Newfoundland and southern Labrador. Polar Biolgy 23:805–811.
Lamoreux, J. 2017. *Cratogeomys planiceps*. The IUCN Red List of Threatened Species 2017:e.T136249A22216740.
Lanier, H., and D. Hik. 2016. *Ochotona collaris*. The IUCN Red List of Threatened Species 2016:e.T41257A45182533.
Lanier, H. C. 2018. *Ochotona collaris*. Pp. 36–39, *in* Pikas, Rabbits and Hares of the World (Smith, A., C. H. Johnston, P. C. Alves, and K. Hackländer, eds.). The Johns Hopkins University Press/IUCN/SSC. Baltimore, U.S.A.
Lanier, H. C., and C. Nielsen. 2019. *Sylvilagus aquaticus*. The IUCN Red List of Threatened Species 2019:e.T41296A45190578.
Laerm, J. 1981. Systematic status of the Cumberland Island pocket gopher, *Geomys cumberlandius*. Brimleyana 6:141–161.
Larivière, S. 1999a. *Mustela vison*. Mammalian Species 608:1–9.
Larivière, S. 1999b. *Lontra longicaudis*. Mammalian Species 609:1–5.
Larivière, S. 2001. *Ursus americanus*. Mammalian Species 647:1–11.
Larivière, S. and M. Pasitschniak-Arts. 1996. *Vulpes vulpes*. Mammalian Species 537:1–11.
Larivière, S., and L. R. Walton. 1997. *Lynx rufus*. Mammalian Species 563:1–8.
Larivière, S., and L. R. Walton. 1998. *Lontra canadensis*. Mammalian Species 587:1–8.
Larrison, E. J., and D. R. Johnson. 1981. Mammals of Idaho. University Press of Idaho, Moscow, U.S.A.
Larry, G. M. 1978. *Chironectes minimus*. Mammalian Species 109:1–6.

Larsen, P. A., S. R. Hoofer, M. C. Bozeman, S. C. Pedersen, H. H. Genoways, C. J. Phillips, D. E. Pumo, and R. J. Baker. 2007. Phylogenetics and phylogeography of the *Artibeus jamaicensis* complex based on cytochrome-*b* DNA sequences. Journal of Mammalogy 88:712–727.

Larsen, P. A., M. R. Marchán-Rivadeneira, and R. J. Baker. 2010. Taxonomic status of Andersen's fruit-eating bat (*Artibeus jamaicensis aequatorialis*) and revised classification of *Artibeus* (Chiroptera: Phyllostomidae). Zootaxa 2648:45–60.

Lassieur, S., D. E. Wilson. 1989. *Lonchorhina aurita*. Mammalian Species 347:1–4.

Latch, E. K., J. R. Heffelfinger, J. A. Fike, and O. E. Rhodes. 2009. Species-wide phylogeography of North American mule deer (*Odocoileus hemionus*): cryptic glacial refugia and postglacial recolonization. Molecular Ecology 18:1730–1745.

Laundré, J. W. 2018. *Sylvilagus audubonii*. Pp. 120–122, *in* Pikas, Rabbits and Hares of the World (Smith, A., C. H. Johnston, P. C. Alves, and K. Hackländer, eds.). The Johns Hopkins University Press/IUCN/SSC. Baltimore, U.S.A.

Laundré, J., and T. W. Clark. 2003. Managing puma hunting in the western United States: Through a metapopulation approach. Animal Conservation 6:159–170.

LaVal, R. K. 1967. Records of bats from the southeastern United States. Journal of Mammalogy 48:645–648.

LaVal, R. K. 1970. Infraspecific relationships of bats of the species *Myotis austroriparius*. Journal of Mammalogy 51:542–552.

LaVal, R. K. 1973a. A revision of the Neotropical bats of the genus *Myotis*. Natural History Museum, Los Angeles County Science Bulletin 15:1–54.

LaVal, R. K. 1973b. Systematic of the genus *Rhogeessa* (Chiroptera: Vespertilionidae). Occasional Papers fo the Museum of Natural History, University of Kansas 19:1–47.

LaVal, R. K., and M. L. LaVal. 1980. Ecological studies and management of Missouri bats, with emphasis on cavedwelling species. Terrestrial series, Missouri Deptament of Conservation, Jefferson City 8:1–53.

Lavariega, M. C., N. Martín-Regalado, y R. M. Gómez-Ugalde. 2012. Mamíferos del centro-occidente de Oaxaca, México. Therya 3:349–370.

Lavoie, M., A. Renard, and S. Larivière. 2019. *Lynx canadensis* (Carnivora: Felidae). Mammalian Species 51:136–154.

Lawlor, T. E. 1982a. The evolution of body size in mammals: evidence for insular populations in Mexico. American Naturalist 119:54–72.

Lawlor, T. E. 1982b. *Ototylomys phyllotis*. Mammalian Species 181:1–3.

Lawlor, T. E. 1965. The Yucatan deer mouse, *Peromyscus yucatanicus*. Univerity Kansas Publication, Museum Natural History 16:421–438.

Lawlor, T. E. 1969. A systematic study of the rodent genus *Ototylomys*. Journal of Mammalogy 50:28–42.

Lawlor, T. E. 1971a. Evolution of *Peromyscus* on Northern Island in the Gulf of California, Mexico. Transactions of the San Diego Society of Natural History 16:91–124.

Lawlor, T. E. 1971b. Distribution and relationships of six species of *Peromyscus* in Baja California and Sonora, Mexico. Occasional Papers of the Museum of Zoology, University of Michigan 661:1–22.

Lawlor, T. E. 1979. Handbook to the orders and families of living mammals. Mad River Press. Eureka, U.S.A.

Lawlor, T. E. 1983. The mammals. Pp. 265–284, *in* Island biogeography in the Sea of Cortez (Case, T. J. and M. L. Cody, eds.). University of California Press. Berkeley, U.S.A.

Lawrence, B., and W. H. Bossert. 1967. Multiple character analysis of *Canis lupus, latrans, andfamiliaris,* with a discussion of the relationships of *Canis niger.* American Zoology 7:223–232.

Layne, J. N. 1970. Climbing behavior of *Peromyscus floridanus* and *Peromyscus gossypinus*. Journal of Mammalogy 51:580–591.

Layne, J. N. 1990. The Florida mouse. Pp. 1–21, *in* Burrow associates of the gopher tortoise (Dodd, Jr., C. K., R. E. Ashton, Jr., R. Franz, and E. Wester, eds.). Eighth Annual Meeting, Gopher Tortoise Council, Florida Museum of Natural History, University of Florida. Gainesville, U.S.A.

Lazcano-Barrero, M. A., and J. M. Packard. 1989. The occurrence of manatees (*Trichechus manatus*) in Tamaulipas, Mexico. Marine Mammal Science 5:202–205.

Lazell, J. D., Jr. 1981. Field and taxonomic studies of tropical American raccoons. National Geographic Society Research Reports 13:381–385.

Le Boeuf, B. J., K. W. Kenyon, and B. Villa-Ramírez. 1986. The Caribbean monk seal is extinct. Marine Mammal Science 2:70–72.

Lee, D. S., J. D. Funderburg, Jr., and M. K. Clark. 1982. A distributional survey of North Carolina mammals. Occasional Papers, North Carolina Biological Survey 1982 10:1–70.

Lee, H. S., and D. F. Hoffmeister. 2003. *Sorex arizonae*. Mammalian Species 732:1–3.

Lee, M. R., and F. B. Elder. 1977. Karyotypes of eight especies of Mexican rodents (Muridae). Journal of Mammalogy 58:479–487.

Lee, M. R., and D. F. Hoffmeister. 1963. Status of certain fox squirrels in Mexico and Arizona. Proceedings of the Biological Society of Washington 76:181–189.

Lee, M. R., and D. J. Schmidly. 1977. A new species of *Peromyscus* (Rodentia: Muridae) from Coahuila, Mexico. Journal of Mammalogy 58:263–268.

Lee, M. R., D. J. Schmidly, and C. H. Huheey. 1972. Chromosomal variation in certain populations of *Peromyscus boylii* and its systematic implications. Journal of Mammalogy 53:697–707.

Lee, T. E., Jr., and M. D. Engstrom. 1991. Genetic variation in the silky pocket mouse (*Perognathus flavus*) in Texas and New Mexico. Journal of Mammalogy 72:273–285.

Lee, T. E., Jr., B. R. Riddle and P. L. Lee. 1996. Speciation in the desert pocket mouse (*Chaetodipus penicillatus* Woodhouse). Journal of Mammalogy 77:58–68.

Lent, P. C. 1988. *Ovibos moschatus*. Mammalian Species 302:1–9.

León, L., T. V. Monterrubio, and M. S. Hafner. 2001. *Cratogeomys neglectus*. Mammalian Species 685:1–4.

León-Paniagua, L., A. G. Navarro-Sigüenza, B. E. Hernández-Baños, and J. C. Morales. 2007. Diversification of the arboreal mice of the genus *Habromys* (Rodentia: Cricetidae: Neotominae) in the Mesoamerican highlands. Molecular Phylogenetics and Evolution 42:653–664.

León-Paniagua, L. 2017. Family Cricetidae. Pp. 321–334, *in* Handbook of mammals of the world, volume 7, Rodent II (Wilson D. E., T. E. Lacher, Jr., and R. A. Mittermeier, eds.). Lyx Edicions. Barcelona, Spain.

Léon-Tapia, M. A., J. A. Fernández, Y. Rico, F. A. Cervantes, and A. Espinosa de los Monteros. 2020. A new mouse of the *Peromyscus maniculatus* species complex (Cricetidae) from the highlands of central Mexico. Journal of Mammalogy 101:1117–1132.

Leopold, A. S. 1959. Wildlife of Mexico: the game birds and mammals. University of California Press, Berkeley, U.S.A.

Lessa, E., and J. L. Patton, 1989. Structural constraints, recurrent shapes, and allometry in pocket gophers (genus *Thomomys*). Biological Journal of the Linnean Society 36:349–363.

Levenson, H. R. 1990. Sexual size dimorphism in chipmunks. Journal of Mammalogy 71:161–170.

Levenson, H. and R. S. Hoffmann. 1984. Systematic relationships among taxa in the Townsend chipmunk group. Southwestern Naturalist 29:157–168.

Levenson, H. R., R. S. Hoffman, C. F. Nadler, L. Deutsch, and S. D. Freeman. 1985. Systematics of the holartic chipmunks (*Tamias*). Journal of Mammalogy 66:219–242.

Levin, E. Y., and V. Flyger. 1971. Uroporphyrinogen III cosynthetase activity in the fox squirrel (*Sciurus niger*). Science 174:59–60.

Lewis, T. H. 1983. The anatomy and histology of the rudimentary eye of *Neurotrichus*. Northwest Science 57:8–15.

Lewis, S. E., and D. E. Wilson. 1987. *Vampyressa pusilla*. Mammalian Species 292:1–5.

Lewis-Oritt, N., C. A. Porter, and R. J. Baker. 2001. Molecular systematics of the family Mormoopidae (Chiroptera) based on cytochrome *b* and recombination activating gene 2 sequences. Molecular Phylogenetics and Evolution 20:426–436.

Li, X., A. C. Tzika, Y. Liu, K. V. Doninck, Q. Zhu, and M. C. Milinkovitch. 2010. Preliminary genetic status of the spotted seal *Phoca largha* in Liaodong Bay (China) based on microsatellite and mitochondrial DNA analyses. Trends in Evolutionary Biology 2:33–38.

Lidicker, W. Z., Jr. 1960a. An analysis of intraspecific variation in the kangaroo rat *Dipodomys merriami*. University California Publication in Zoology 67:125–217.

Lidicker, W. Z., Jr. 1960b. The baculum of *Dipodomys ornatus* and its implications for superspecific groupings of kangaroo rats. Journal of Mammalogy 41:495–499.

Lidicker, W. Z., Jr. 1960c. A new subspecies of the cliff cipmunk from Central Chihuahua. Proceedings of the Biological Society of Washington 73:267–274.

Lim, B. K. 1987. *Lepus townsendii*. Mammalian Species 288:1–6.

Lim, B. K., W. A. Pedro, and F. C. Passos. 2003. Differentiation and species status of the Neotropical yellow-eared bats *Vampyressa pusilla* and *V. thyone* (Phyllostomidae) with a molecular phylogeny and review of the genus. Acta Chiropterologica 5:15–29.

Lim, B. K., M. D. Engstrom, N. B. Simmons, and J. M. Dunlop. 2004. Phylogenetics and biogeography of least sac-winged bats *(Balantiopteryx)* based on morphological and molecular data. Mammalian Biology 69:225– 237.

Lim, B. 2015. *Balantiopteryx io*. The IUCN Red List of Threatened Species 2015:e.T2532A22030080.

Lim, B., and B. Miller. 2016. *Rhynchonycteris naso*. The IUCN Red List of Threatened Species 2016:e.T19714A22010818.

Lim, B. K., B. Miller, F. Reid, J. Arroyo-Cabrales, A. D. Cuarón, and P. C. de Grammont. 2016a. *Balantiopteryx plicata*. The IUCN Red List of Threatened Species 2016:e.T2533A22029659.

Lim, B. K., B. Miller, F. Reid, J. Arroyo-Cabrales, A. D. Cuarón, and P. C. de Grammont. 2016b. *Diclidurus albus*. The IUCN Red List of Threatened Species 2016:e.T6561A21986615.

Linares, O. J. 1986. Murciélagos de Venezuela. Cuadernos Lagoven. Caracas, Venezuela.

Lindqvist, C., L. Bachmann, L. W. Andersen, E. W. Born, U. Arason, K. M. Kovacs, C. Lydesren, A. V. Abramov, and Ø. Wiig. 2009. The Laptev Sea walrus *Odobenus rosmarus laptevi*: an enigma revisted. Zoologica Scripta 38:113–127.

Lindsay, S. L. 1981. Taxonomic and biogeographic relationships of Baja California chickarees (*Tamiasciurus*). Journal of Mammalogy 62:673–682.

Lindsay, S. L. 1982. Systematic relationships of parapatric tree squirrel species (*Tamiasciurus*) in the Pacific Northwest. Canadian Journal of Zoology 60:2149–2156.

Ling, J. K., and M. M. Bryden. 1992. *Mirounga leonina*. Mammalian Species 391:1–8.

Linzey, A. V. 1983. *Synaptomys cooperi*. Mammalian Species 210:1–5.

Linzey, A. V., and J. L. Layne. 1969. Comparative morphology of the male reproductive tract in the rodent genus *Peromyscus* (Muridae). American Museum Novitates 2355:1–47.

Linzey, A. V., and NatureServe. 2017a. *Thomomys clusius*. The IUCN Red List of Threatened Species 2017:e.T42595A22216284.

Linzey, A. V., and NatureServe. 2017b. *Thomomys idahoensis*. The IUCN Red List of Threatened Species 2017:e.T21809A22215570.

Linzey, A.V., R. Timm, S. T. Álvarez-Castañeda, I. Castro-Arellano, and T. Lacher. 2008a. *Sciurus arizonensis*. The IUCN Red List of Threatened Species 2008:e.T20005A9132802.

Linzey, A. V. and NatureServe (Hammerson, G., J. C. Whittaker, and Norris, S. J.) 2008b. *Neotoma magister*. The IUCN Red List of Threatened Species 2008:e.T14581A4446084.

Linzey, A. V., J. Matson, and S. Pérez. 2016a. *Neotoma mexicana*. The IUCN Red List of Threatened Species 2016:e.T14590A115123126.

Linzey, A. V., R. Timm, S. T. Álvarez-Castañeda, I. Castro-Arellano, and T. Lacher. 2016b. *Ammospermophilus leucurus*. The IUCN Red List of Threatened Species 2016:e.T42452A115189458.

Linzey, A. V., R. Timm, S. T. Álvarez-Castañeda, I. Castro-Arellano, and T. Lacher. 2016c. *Ictidomys mexicanus*. The IUCN Red List of Threatened Species 2016:e.T20487A22262744.

Linzey, A. V., R. Timm, S. T. Álvarez-Castañeda, I. Castro-Arellano, and T. Lacher. 2016d. *Chaetodipus baileyi*. The IUCN Red List of Threatened Species 2016:e.T4328A115068087.

Linzey, A. V., R. Timm, S. T. Álvarez-Castañeda, I. Castro-Arellano, and T. Lacher. 2016e. *Chaetodipus californicus*. The IUCN Red List of Threatened Species 2016:e.T4329A115068220.

Linzey, A. V., R. Timm, S. T. Álvarez-Castañeda, I. Castro-Arellano, and T. Lacher. 2016f. *Chaetodipus eremicus*. The IUCN Red List of Threatened Species 2016:e.T136606A115210279.

Linzey, A. V., R. Timm, S. T. Álvarez-Castañeda, I. Castro-Arellano, and T. Lacher. 2016g. *Chaetodipus hispidus*. The IUCN Red List of Threatened Species 2016:e.T4333A115068352.

Linzey, A. V., R. Timm, S. T. Álvarez-Castañeda, I. Castro-Arellano, and T. Lacher. 2016h. *Chaetodipus intermedius*. The IUCN Red List of Threatened Species 2016:e.T4334A115068570.

Linzey, A. V., R. Timm, S. T. Álvarez-Castañeda, I. Castro-Arellano, and T. Lacher. 2016i. *Chaetodipus nelsoni*. The IUCN Red List of Threatened Species 2016:e.T4335A115068711.

Linzey, A. V., R. Timm, S. T. Álvarez-Castañeda, I. Castro-Arellano, and T. Lacher. 2016j. *Chaetodipus penicillatus*. The IUCN Red List of Threatened Species 2016:e.T4336A115068852.

Linzey, A.V., R. Timm, S. T. Álvarez-Castañeda, I. Castro-Arellano, and T. Lacher. 2016k. *Chaetodipus spinatus*. The IUCN Red List of Threatened Species 2016:e.T4338A115068992.

Linzey, A. V., R. Timm, S. T. Álvarez-Castañeda, I. Castro-Arellano, and T. Lacher. 2016l. *Peromyscus merriami*. The IUCN Red List of Threatened Species 2016:e.T16680A115137090.

Linzey, A. V., R. Timm, L. Emmons, and F. Reid. 2016m. *Sciurus niger*. The IUCN Red List of Threatened Species 2016:e.T20016A115155257.

Linzey, A. V., R. Timm, N. Woodman, J. Matson, and R. Samudio. 2016n. *Oryzomys couesi*. The IUCN Red List of Threatened Species 2016:e.T15592A115128044.

Linzey, A. V., S. Shar, D. Lkhagvasuren, R. Juškaitis, B. Sheftel, H. Meinig, G. Amori, and H. Henttonen. 2016o. *Microtus oeconomus*. The IUCN Red List of Threatened Species 2016:e.T13451A115113894.

Linzey, A. V., R. Timm, S. T. Álvarez-Castañeda, I. Castro-Arellano, and T. Lacher. 2019. *Sciurus arizonensis*. The IUCN Red List of Threatened Species 2019:e.T20005A22247935.

Linzey, A. V., H. Henttonen, B. Sheftel, and N. Batsaikhan. 2020. *Myodes rutilus*. The IUCN Red List of Threatened Species 2020:e.T4975A164372228.

Lira-Torres, I., and M. Briones-Salas. 2011. Impacto de la ganadería extensiva y caceria de susbsistencia sobre abundancia relativa de mamíferos en la Selva Zoque, Oaxaca, México. Therya 2:217–244.

Lira-Torres, I., and M. Briones-Salas. 2012. Abundancia relativa y patrones de actividad de los mamíferos de los Chimalapas, Oaxaca, México. Acta Zoológica Mexicana (n. s.) 28:566–585.

Lissovsky, A. A., and E. V. Obolenskaya. 2011. The structure of craniometrical diversity of grey voles *Microtus* subgenus *Alexandromys*. Proceedings of the Zoological Institute RAS 315:461–477.

Lissovsky, A. A., T. V. Petrova, S. P. Yatsentyuk, F. N. Golenishchev, N. I. Putincev, I. V. Kartavtseva, I. N. Sheremetyeva, and N. I. Abramson. 2012. Multilocus phylogeny and taxonomy of East Asian voles *Alexandromys* (Rodentia, Arvicolinae). Zoologica Scripta 47:9–20.

List, R., O. R. W. Pergams, J. Pachecho, J. Cruzado, and G. Ceballos. 2010. Genetic divergence of *Microtus pennsylvanicus chihuahuensis* and conservation implications of marginal population extinctions. Journal of Mammalogy 91:1093–1101.

Litvaitis, J., and H. C. Lanier. 2019. *Sylvilagus transitionalis*. The IUCN Red List of Threatened Species 2019:e.T21212A45181534.

Litvaitis, M. K., W. Lee, J. A. Litvaitis, and T. D. Kocher. 1997. Variation in the mitochondrial DNA of the *Sylvilagus* complex occupying the northeastern United States. Canadian Journal of Zoology 75:595–605.

Livaitis, J. A., M. Litvaitis, A. Kovach, and H. Kilpatrick. 2018. *Sylvilagus transitionalis*. Pp. 156–157, *in* Pikas, Rabbits and Hares of the World (Smith, A., C. H. Johnston, P. C. Alves, and K. Hackländer, eds.). The Johns Hopkins University Press/IUCN/SSC. Baltimore, U.S.A.

Long, C. A. 1965. The mammals of Wyoming. University of Kansas Publications, Museum of Natural History 14:493–758.

Long, C. A. 1972. Taxonomic revision of the North American Badger, *Taxidea taxus*. Journal of Mammalogy 53:725–759.

Long, C. A. 1973. *Taxidea taxus*. Mammalian Species 26:1–4.

Long, C. A. 1974. *Microsorex hoyi* and *Microsorex thompsoni*. Mammalian Species 33:1–4.

Long, C. A. 1999a. Pygmy shrew, *Sorex hoyi*. Pp. 25–26, *in* The Smithsonian Book of North American Mammals (Wilson, D. E., and S. Ruff, eds.). Smithsonian Institution Press. Washington, U.S.A.

Long, C. A. 1999b. American badger, *Taxidea taxus*. Pp. 177–179, *in* The Smithsonian book of North American mammals (Wilson, D. E., and S. Ruff, eds.). Washington, U.S.A.

Long, C. A., and R. S. Hoffmann.1992. *Sorex preblei* from the Black Canyon, first record for Colorado. The Southwestern Naturalist 37:318–319.

Long, C. A., and C. F. Long. 1965. Dental abnormalities in North American badgers, genus *Taxidea*. Transactions of the Kansas Academy of Sciences 68:145–155.

Longhofer, L. K., and R. D. Bradley. 2006. Molecular systematics of the genus *Neotoma* based on DNA sequences from intron 2 of the Alcohol dehydrogenase gene. Journal of Mammalogy 87:961–970.

López-González, C. 1998. Systematics and biogeography of the bats of Paraguay. Ph.D. dissertation, Texas Tech University, Lubbock, U.S.A.

López-González, C. 2003. Murciélagos (Chiroptera) del Estado de Durango, México: Composición, distribución y estado de conservación. Vertebrata Mexicana 13:15–23.

López-González, C., and D. F. García-Mendoza. 2006. Murciélagos de la Sierra Tarahumara, Chihuahua, México. Acta Zoológica Mexicana (n. s.) 22:109–135.

López-González, C., and S. J. Presley. 2001. Taxonomic status of *Molossus bondae* J. A. Allen, 1904 (Chiroptera: Molossidae), with description of a new subspecies. Journal of Mammalogy 82:760–774.

López-González, C., D. F. García-Mendoza, J. C. López-Vidal, and C. Elizalde-Arellano. 2019. Multiple lines of evidence reveal a composite of species in the plateau mouse, *Peromyscus melanophrys* (Rodentia, Cricetidae). Journal of Mammalogy 100:1583–1598.

López-González, C., S. J. Presley, R. D. Owen, and M. R. Will. 2001. Taxonomic status of *Myotis* (Chiroptera: Vespertilionidae) in Paraguay. Journal of Mammalogy 82:138–160.

López-Wilchis, R., and J. López-Jardines. 1998. Los mamíferos de México depositados en colecciones de Estados Unidos y Canadá. Universidad Autónoma Metropolitana, Distrito Federal, México.

López-Wilchis, R., L. M. Guevara-Chumacero, N. P. Ángeles, J. Juste, C. Ibáñez, and I. D. L. A. Barriga-Sosa. 2012. Taxonomic status assessment of the Mexican populations of funnel-eared bats, genus *Natalus* (Chiroptera: Natalidae). Acta Chiropterologica 14:305–316.

López-Wilchis, R., J. W. Torres-Flores, J. Arroyo-Cabrales. 2020. *Natalus mexicanus* (Chiroptera: Natalidae). Mammalian Species 52:27–39.

Lorenzini, R., and L. Garofalo. 2015. Insights into the evolutionary history of *Cervus* (Cervidae, tribe Cervini) based on Bayesian analysis of mitochondrial marker sequences, with first indications for a new species. Journal of Zoological Systematics and Evolutionary Research 53:340–349.

Lorenzo, C., and S. T. Álvarez-Castañeda. 2011. *Sylvilagus mansuetus*. The IUCN Red List of Threatened Species 2011:e.T21210A9257211.

Lorenzo, C., and D. E. Brown. 2019. *Lepus alleni*. The IUCN Red List of Threatened Species 2019:e.T41272A45185265.

Lorenzo, C., and F. A. Cervantes. 1995. The G-banded karyotype of the Tapeti rabbit (*Sylvilagus brasiliensis*) from Chiapas, Mexico. Revista Sociedad Mexicana Historia Natural 46:1–5.

Lorenzo, C., and C. H. Johnston. 2019. *Lepus insularis*. The IUCN Red List of Threatened Species 2019:e.T11794A45177986.

Lorenzo, C., and H. C. Lanier. 2019a. *Sylvilagus cunicularius*. The IUCN Red List of Threatened Species 2019:e.T21211A45181292.

Lorenzo, C., and H. C. Lanier. 2019b. *Sylvilagus graysoni*. The IUCN Red List of Threatened Species 2019:e.T21206A45180643.

Lorenzo, C., P. Cortés-Calva, G. Ruiz-Campos, and S. T. Álvarez-Castañeda. 2013. Current distributional status of two subspecies of *Sylvilagus bachmani* in the Baja California Peninsula, Mexico. Western North American Naturalist 73:219–223.

Lorenzo, C., F. A. Cervantes, and M. A. Aguilar. 1993. The karyotypes of some Mexican cottontail rabbits of the genus *Sylvilagus*. Pp. 129–136, *in* Avances en el estudio de los mamíferos de México (Medellín, R. A., and G. Ceballos, eds.). Publicación especial. Asociación Mexicna de Mastozoología, A. C. Distrito Federal, México.

Lorenzo, C., T. Rioja, A. Carrillo, and F. A. Cervantes. 2008. Population fluctuations of *Lepues flavigularis* (Lagomorpha: Leporidae) at Tehuantepec Isthmus, Oaxaca, Mexico. Acta Zoológica Mexicana (n. s.) 24:207–220.

Lorenzo, C., A. Carrillo-Reyes, M. Gómez-Sánchez, A. Vélazquez, and E. Espinoza. 2011. Diet of the endangered Tehuantepec jackrabbit, *Lepus flavigularis*. Therya 2:67–76.

Lorenzo, C., D. E. Brown, S. Amirsultan, and M. García. 2014. Evolutionary history of the antelope jackrabbit, *Lepus alleni*. Journal of the Arizona-Nevada Academy of Sciences 45:70–75.

Lorenzo, C., T. Rioja-Paradela, and A. Carrillo-Reyes. 2015. State and knowledge and conservation of endangered and critical endangered lagomorphs worldwide. Therya 6:11–30.

Lorenzo, C., S. T. Álvarez-Castañeda, S. G. Pérez-Consuegra, and J. L. Patton. 2016. Revision of the Chiapan deer mouse, *Peromyscus zarhynchus*, with the description of a new species. Journal of Mammalogy 97:910–918.

Lorenzo, C., T. Rioja-Paradela, A. Carrillo-Reyes, M. de la Paz-Cuevas, and S. T. Álvarez-Castañeda. 2018a. *Sylvilagus mansuetus*. Pp. 145–147, *in* Pikas, Rabbits and Hares of the World (Smith, A., C. H. Johnston, P. C. Alves, and K. Hackländer, eds.). The Johns Hopkins University Press/IUCN/SSC. Baltimore, U.S.A.

Lorenzo, C., T. Rioja-Paradela, A. Carrillo-Reyes, M. de la Paz-Cuevas, F. A. Cervantes, and S. T. Álvarez-Castañeda. 2018b. *Lepus insularis*. Pp. 198–200, *in* Pikas, Rabbits and Hares of the World (Smith, A., C. H. Johnston, P. C. Alves, and K. Hackländer, eds.). The Johns Hopkins University Press/IUCN/SSC. Baltimore, U.S.A.

Lorenzo, C., J. Vázquez, L. Rodríguez-Martínez, A. Bautista, A. García-Méndez, and F. A. Cervantes. 2018c. *Sylvilagus cunicularius*. Pp. 131–135, *in* Pikas, Rabbits and Hares of the World (Smith, A., C. H. Johnston, P. C. Alves, and K. Hackländer, eds.). The Johns Hopkins University Press/IUCN/SSC. Baltimore, U.S.A.

Lorenzo, C., J. P. Ramírez-Silva, F. A. Cervantes, and R. Ferrera-Muro. 2018d. *Sylvilagus graysoni*. Pp. 142–144, *in* Pikas, Rabbits and Hares of the World (Smith, A., C. H. Johnston, P. C. Alves, and K. Hackländer, eds.). The Johns Hopkins University Press/IUCN/SSC. Baltimore, U.S.A.

Lorenzo, C., F. A. Cervantes, J. Vargas, and R. Ferrera-Muro. 2018e. *Sylvilagus insonus*. Pp. 144–145, *in* Pikas, Rabbits and Hares of the World (Smith, A., C. H. Johnston, P. C. Alves, and K. Hackländer, eds.). The Johns Hopkins University Press/IUCN/SSC. Baltimore, U.S.A.

Lorenzo, C., T. Rioja-Paradela, A. Carrillo-Reyes, E. C. Sáintiz-López, and J. Bolaños. 2018f. *Lepus flavigularis*. Pp. 191–193, *in* Pikas, Rabbits and Hares of the World (Smith, A., C. H. Johnston, P. C. Alves, and K. Hackländer, eds.). The Johns Hopkins University Press/IUCN/SSC. Baltimore, U.S.A.

Lorenzo, C., D. E. Brown, and H. C. Lanier. 2019a. *Sylvilagus insonus*. The IUCN Red List of Threatened Species 2019:e.T21207A45180771.

Lorenzo, C., J. Bolaños-Citalán, D. Navarrete-Gutiérrez, J. A. Pérez-López, and L. Guevara. 2019b. In search of shrews of Chiapas: analysis of their distribution and conservation. Therya 10:121–129.

Lorenzo, C., T. Rioja-Paradela, A. Carrillo-Reyes, E. Sántiz-López, and J. Bolaños-Citalán. 2019c. Projected impact of global warming on the distribution of two pocket mouse species with implications on the conservation of *Heteromys nelsoni* (Rodentia: Heteromyidae). Revista de Biologia Tropical 67:1210–1219.

Lorenzo, C., J. E. Bolaños-Citalán, and O. G. Retana-Guiascón. 2020. Rediscovery of *Heteromys nelsoni* in its type locality after over a century. Mammalian 84:6–9.

Lotze, J., and S. Anderson 1979. *Procyon lotor*. Mammalian Species 119:1–8.

Loureiro, L. O., M. D. Engstrom, and B. K. Lim. 2020. Single nucleotide polymorphisms (SNPs) provide unprecedented resolution of species boundaries, phylogenetic relationships, and genetic diversity in the mastiff bats (*Molossus*). Molecular Phylogenetics and Evolution 143:106690.

Loughlin, T. R., M. A. Perez, and R. L. Merrick. 1987. *Eumetopias jubatus*. Mammalian Species 283:1–7.

Loughlin, T. R., G. A. Antonelis, J. D. Baker, A. E. York, C. W. Fowler, R. L. DeLong, and H. W. Braham. 1994. Status of the northern fur seal population in the United States during 1992. Pp. 9–28, *in* Fur Seal Investigations 1992 (Sinclair, E. H., ed.). United States Department of Commerce, NOAA Technical Memorandum NMFS-AFSC 45:9–28.

Loughry, J., C. McDonough, and A. M. Abba. 2014. *Dasypus novemcinctus*. The IUCN Red List of Threatened Species 2014:e.T6290A47440785.

Lowery, G. H., Jr. 1974. The mammals of Louisiana and its adjacent waters. Louisiana State University Press, Baton Rouge, U.S.A.

Lowrey, C. 2016. *Neotamias palmeri*. The IUCN Red List of Threatened Species 2016:e.T21355A22267875.

Lowry, L. 2016a. *Histriophoca fasciata*. The IUCN Red List of Threatened Species 2016:e.T41670A45230946.

Lowry, L. 2016b. *Odobenus rosmarus*. The IUCN Red List of Threatened Species 2016:e.T15106A45228501.

Lowry, L. 2016c. *Phoca vitulina*. The IUCN Red List of Threatened Species 2016:e.T17013A45229114.

Lowry, L. 2016d. *Pusa hispida*. The IUCN Red List of Threatened Species 2016:e.T41672A45231341.

Lluch, B. C. 1965. Further notes on the biology of the manatee. Anales del Instituto Nacional de Investigaciones en Biologia Pesquera, México 1:405–419.

Luckett, W. P., and J.-L. Hartenberger. 1985 Evolutionary relationships among rodents: a multidisciplinary analysis. Plenum Press. New York, U.S.A.

Ludwing, D. R. 1984. *Microtus richardsoni*. Mammalian Species 223:1–6.

Lyon, M. W., Jr. 1903. Classification of the hares and their allies. Smithsonian Miscellaneous Collections 45:321–463.

Lyon, M. W., Jr. 1936. Mammals of Indiana. American Midland Naturalist 17:1–384.

Macdonald, D. W., N. Yamaguchi, A. C. Kitchener, M. Daniels, K. Kilshaw, and D. Driscoll. 2010. The Scottish wildcat: On the way to cryptic extinction through hybridisation: past history, present problem, and future conservation. Pp. 471–491, *in* Biology and Conservation of Wild Felids (Macdonald D. W., and A. J. Loveridge, eds). Oxford University Press. Oxford, UK.

MacDonald, S. O., and J. A. Cook. 2009. Recent mammals of Alaska. University of Alaska Press. Fairbanks, U.S.A.

MacDonald, S. O., and C. Jones. 1987. *Ochotona collaris*. Mammalian Species 281:1–4.

MacClintock, D. 1970. Squirrels of North America. Van Nostrand Reinhold Company, New York, U.S.A.

Macêdo, R. H., and M. A. Mares. 1988. *Neotoma albigula*. Mammalian Species 310:1–7.

Macholán, M. 1999. *Mus musculus*. Academic Press, London, UK.
MacSwiney, G. M. C., B. Bolívar-Cimé, F. M. Clarke, and P. A. Racey. 2009a. Insectivorous bat activity at cenotes in the Yucatan Peninsula, Mexico. Acta Chiropterologica 11:139–147.
MacSwiney, G. M. C., S. Hernández-Betancourt, and R. Avila-Flores. 2009b. *Otonyctomys hatti.* Mammalian Species 825:1–5.
Mailliard, J. 1924. A new deer mouse (*Peromyscus slevini*) from the Gulf of California. Proceedings of the California Academy of Sciences 12:1219–1222.
Malaney, J. L., J. R. Demboski, and J. A. Cook. 2017. Integrative species delimitation of the widespread North American jumping mice (Zapodinae). Molecular Phylogenetics and Evolution 114:137–152.
Maldonado, J. E., C. Vilá, and R. K. Wayne. 2001. Tripartite genetic subdivisions in the Ornate Shrew (*Sorex ornatus*). Molecular Ecology 10:127–147.
Maldonado, J. E., F. Hertel, and C. Vilá. 2004. Discordant patterns of morphological variation in genetically divergent populations of Ornate Shrews (*Sorex ornatus*). Journal of Mammalogy 85:886–896.
Maldonado, J. E., S. Young, L. H. Simons, S. Stone, L. D. Parker, and J. Ortega. 2015. Conservation genetics and phylogeny of the Arizona shrew in the "Sky Islands" of the Southwestern United States. Therya 6:401–420.
Manning, R. W. 1993. Systematics and evolutionary relationships of the long-eared myotis, *Myotis evotis* (Chiroptera: Vespertilionidae). Special Publications, The Museum, Texas Tech University 37:1–58.
Manning, R. W., and J. K. Jones, Jr. 1988a. *Perognathus fasciatus*. Mammalian Species 303:1–4.
Manning, R. W., and J. K. Jones, Jr. 1988b. A new subspecies of fringed *Myotis*, *Myotis thysanodes*, from the northwestern coast of the United States. Occasional Papers of the Museum, Texas Tech University 123:1–6.
Manning, R. W., and J. K. Jones, Jr. 1989. *Myotis evotis*. Mammalian Species 329:1–5.
Manning, R. W., F. D. Yancey II, and C. Jones. 1996. Nongeographic variation and natural history of two sympatric species of pocket mice, *Chaetodipus nelsoni* and *Chaetodipus eremicus,* from Brewster County, Texas. Pp. 191–195, *in* Contributions in mammalogy: a memorial volume honoring Dr. J. Knox Jones, Jr. (Genoways, H. H., and R. J. Baker, eds.). Museum of Texas Tech University. Lubbock, U.S.A.
Manning, R. W., M. R. Heaney, M. Sagot, and R. J. Baker. 2014. Noteworthy record of Crawford's desert shrew (*Notiosorex crawfordi*) from southern Nevada. The Southwestern Naturalist 59:145–147.
Manning, T. H. 1971. Geographical variation in the polar bear *Ursus maritimus* Phipps. Canadian Wildlife Service. Ottawa, Canada.
Mantilla-Meluk, H. 2014. Defining Species and Species Boundaries in Uroderma (Chiroptera: Phyllostomidae) with a Description of a New Species. Occasional Papers, The Museum, Texas Tech University 325:1–25.
Mantilla-Meluk, H., and R. J. Baker. 2010. New species of *Anoura* (Chiroptera: Phyllostomidae) from Colombia, with systematic remarks and notes on the distribution of the *A. geoffroyi* complex. Occasional Papers, The Museum, Texas Tech University 292:1–24.
Mantilla-Meluk, H., and J. Muñoz-Garay. 2014. Biogeography and taxonomic status of *Myotis keaysi pilosatibialis* LaVal 1973 (Chiroptera: Vespertilionidae). Zootaxa 3793:60–70.
Mantooth, S. J., and T. L. Best. 2005a. *Chaetodipus eremicus*. Mammalian Species 768:1–3.
Mantooth, S. J., and T. L. Best. 2005b. *Chaetodipus penicillatus*. Mammalian Species 767:1–7.
Mantooth, S. J., C. Jones, and R. D. Bradley. 2000. Molecular systematics of *Dipodomys elator* (Rodentia: Heteromyidae) and its phylogeographic implications. Journal of Mammalogy 81:885–894.
Mantooth, S. J., D. J. Hafner, R. W. Bryson, and B. R. Riddle. 2013. Phylogeographic diversification of antelope squirrels (*Ammospermophilus*) across North American deserts. Biological Journal of the Linnean Society 109:949–967.
Manville, R. H., and P. G. Favour, Jr. 1960. Southern distribution of the Atlantic walrus. Journal of Mammalogy 41:499–503.
Manville, R. H., and S. P. Young. 1965. Distribution of Alaskan mammals. United States Bureau of Sport Fisheries and Wildlife, Circular 211:1–45.
Mares, M. A., R. A. Ojeda, and R. M. Barquez. 1989. Guide to the mammals of Salta Province, Argentina. University of Oklahoma Press. Norman, U.S.A.
Marchán-Rivadeneira, M. R., P. A. Larsen, C. J. Phillips, R. E. Strauss, and R. J. Baker. 2012. On the association between environmental gradients and skull size variation in the great fruit-eating bat, *Artibeus lituratus* (Chiroptera: Phyllostomidae). Biological Journal of the Linnean Society 105:623–634.
Marcot, J. D. 2007. Molecular phylogeny of terrestrial artiodactyls. conflicts and resolution. Pp. 4–18, *in* The Evolution of Artiodactyls (Prothero, D., and S. E. Foss, eds). Johns Hopkins University Press. Baltimore, U.S.A.
Marcy, A. E., S. Fendorf, J. L. Patton, and E. A. Hadly. 2013. Morphological adaptations for digging and climate-impacted soil properties define Pocket Gopher (*Thomomys* spp.) distributions. PLoS ONE 8(5):e64935.
Marshall, L. G. 1978. *Chironectes minimus*. Mammalian Species 109:1–6.
Marines-Macías, T., P. Colunga-Salas, and L. León-Paniagua. 2021. *Reithrodontomys microdon* (Rodentia: Cricetidae). Mammalian Species 54(1020):1–8.
Martin, G. M. 2016. *Marmosa mexicana*. The IUCN Red List of Threatened Species 2016:e.T40504A22173751.
Martin, G. M. 2017. *Tlacuatzin canescens*. The IUCN Red List of Threatened Species 2017:e.T12813A22177663.
Martin, C. O., and D. J. Schmidly. 1982. Taxonomic review of the pallid bat, *Antrozous pallidus* (Le Conte). Special Publications, The Museum, Texas Tech University 18:1–48.
Martínez, L., and B. Villa. 1938. Contribuciones al conocimiento de los murciélagos de México I. Anales del Instituto de Biología, Universidad Nacional Autónoma de México 9:339–360.
Martínez-Borrego, D., E. Arellano, F. X. González-Cózatl, and D. S. Rogers. 2020. *Reithrodontomys mexicanus* (Rodentia: Cricetidae). Mammalian Species 52:114–124.
Martínez-Borrego, D., E. Arellano, F. X. González-Cózatl, I. Castro-Arellano, L. León-Paniagua, and D. S. Rogers. 2022. Molecular systematics of the *Reithrodontomys tenuirostris* group (Rodentia: Cricetidae) highlighting the *Reithrodontomys microdon* species complex. Journal of Mammalogy 103:29–44.
Martínez-Chapital, S. T., G. D. Schnell, C. Sánchez-Hernández, and M. L. Romero-Almaraz. 2017. *Sigmodon mascotensis* (Rodentia: Cricetidae). Mammalian Species 49:109–118.
Maser, C. 1975. Characters useful in identifying certain western Oregon mammals. Northwest Science 49:158–159.

Mascarello, J. T. 1978. Chromosomal, biochemical, mensural, penile, and cranial variation in desert woodrats (*Neotoma lepida*). Journal of Mammalogy 59:477–495.

Masseti, M., and D. Mertzanidou. 2008. *Dama dama*. The IUCN Red List of Threatened Species 2008:e.T42188A10656554.

Mathis, V. L., M. S. Hafner, D. J. Hafner, and J. W. Demastes. 2013a. Resurrection and redescription of the pocket gopher *Thomomys sheldoni* from the Sierra Madre Occidental of Mexico. Journal of Mammalogy 94:544–560.

Mathis, V. L., M. S. Hafner, D. J. Hafner, and J. W. Demastes. 2013b. *Thomomys nayarensis*, a new species of pocket gopher from the Sierra del Nayar, Nayarit, Mexico. Journal of Mammalogy 94:983–994.

Mathis, V. L., M. S. Hafner, and D. J. Hafner. 2014. Evolution and phylogeography of the *Thomomys umbrinus* species complex (Rodentia: Geomyidae). Journal of Mammalogy 95:754–771.

Matocq, M. D. 2002a. Morphological and molecular analysis of a contact zone in the *Neotoma fuscipes* species complex. Journal of Mammalogy 83:866–883.

Matocq, M. D. 2002b. Phylogeographical structure and regional history of the dusky-footed woodrat, *Neotoma fuscipes*. Molecular Ecology 11: 229–242.

Matson, J. O. 1975. *Myotis planiceps*. Mammalian Species 60:1–2.

Matson, J. O. 1980. The status of banner-tailed kangaroo rats, genus *Dipodomys,* from central Mexico. Journal of Mammalogy 61:563–566.

Matson, J. 2018. *Sorex pribilofensis*. The IUCN Red List of Threatened Species 2018:e.T20391A22314622.

Matson, J. 2020. *Microtus guatemalensis*. The IUCN Red List of Threatened Species 2020:e.T13432A22350117.

Matson, J. O., and R. H. Baker. 1986. Mammals of Zacatecas. Special Publications, The Museum, Texas Tech University 24:1–88.

Matson, J., and P. C. de Grammont. 2018. *Sorex milleri*. The IUCN Red List of Threatened Species 2018:e.T20397A22316066.

Matson, J. O., and N. Ordóñez-Garza. 2017. The taxonomic status of Long-tailed shrews (Mammalia: genus *Sorex*) from Nuclear Central America. Zootaxa 4236:461–483.

Matson, J. O., and D. R. Patten. 1975. Notes on some bats from the state of Zacatecas, Mexico. Contributions in Science, Los Angeles County Museum of Natural History 263:1–12.

Matson, J., N. Woodman, and F. Reid. 2016a. *Sorex vagrans*. The IUCN Red List of Threatened Species 2016:e.T41425A115186125.

Matson, J., N. Woodman, I. Castro-Arellano, and P. C. de Grammont. 2016b. *Sorex montícola*. The IUCN Red List of Threatened Species 2016:e. T41405A194054902.

Matson, J., N. Woodman, I. Castro-Arellano, and P. C. de Grammont. 2016c. *Scalopus aquaticus*. The IUCN Red List of Threatened Species 2016:e.T41471A115188304.

Matson, J., N. Woodman, I. Castro-Arellano, and P. C. de Grammont. 2016d. *Scapanus latimanus*. The IUCN Red List of Threatened Species 2016:e.T41473A115188559.

Matson, J., P. C. de Grammont, and I. Castro-Arellano. 2017a. *Sorex emarginatus*. The IUCN Red List of Threatened Species 2017:e. T41395A22312750.

Matson, J., N. Woodman, I. Castro-Arellano, and P. C. de Grammont. 2017b. *Sorex mediopua*. The IUCN Red List of Threatened Species 2017:e. T136656A22318332.

Matson, J., N. Woodman, I. Castro-Arellano, and P. C. de Grammont. 2017c. *Sorex ventralis. The IUCN Red List of Threatened Species* 2017:e. T41426A22315839.

Matson, J., N. Woodman, I. Castro-Arellano, and P. C. de Grammont. 2017d. *Sorex veraecrucis*. The IUCN Red List of Threatened Species 2017:e. T136811A22319172.

Matson, J., N. Woodman, I. Castro-Arellano, and P. C. de Grammont. 2017e. *Sorex veraepacis*. The IUCN Red List of Threatened Species 2017:e.T41427A22315421.

Matson, J., A. D. Cuarón, and P. C. de Grammont. 2018. *Cryptotis nelsoni*. The IUCN Red List of Threatened Species 2018:e.T136389A22284939.

Matson, J., N. Woodman, I. Castro-Arellano, and P. C. de Grammont. 2019. *Sorex ixtlanensis*. The IUCN Red List of Threatened Species 2019:e. T136339A22319335.

Mattson, D. J. 1998. Diet and morphology of extant and recently extinct northern bears. Ursus 10:479–96.

Mattson, D. J., and T. Merrill. 2002. Extirpations of grizzly bears in the contiguous United States, 1850–2000. Conservation Biology 16:1123–1136.

Matyushkin, E. N. 1979. Rysi Golarktiki. Sbornik Trudov Zoologicheskovo Muzeya MGU 18:76–161.

Mayer, W. V. 1952. The hair of California mammals with keys to the dorsal guard hairs of California mammals. The American Midland Naturalist 48:480–512.

Mayer, J. J., and R. M. Wetzel. 1987. *Tayassu pecari*. Mammalian Species 293:1–7.

Mayer, J. J., and I. L. Brisbin, Jr. 2009. Wild Pigs: Biology, Damage, Control Techniques and Management Archived 24 October 2014 at the Wayback Machine, Savannah River National Laboratory Aiken, South Carolina, SRNL-RP-2009-00869.

Mazrimas, J. A., and F. T. Hatch. 1972. A possible relationship between satellite DNA and the evolution of kangaroo rat species (genus *Dipodomys*). Nature New Biology 240:102–105.

McAllister, J. A., and R. S. Hoffmann. 1988. *Phenacomys intermedius*. Mammalian Species 305:1–8.

McBee, K., and R. J. Baker. 1982. *Dasypus novemcinctus*. Mammalian Species 162:1–9.

Mccabe, R. E., and R. T. Mccabe. 1984. Of slings and arrows: an historical retrospection. Pp. 19–72, *in* White-tailed deer: ecology and management (L. K. Halls, ed.). Stackpole Books. Harrisburg, U.S.A.

McCarty, R. 1975. *Onychomys torridus*. Mammalian Species 59:1–5.

McCarty, R. 1978. *Onychomys leucogaster*. Mammalian Species 87:1–6.

McCay, T. S. 2001. *Blarina carolinensis*. Mammalian Species 673:1–7.

McCleery, R., C. Faulhaber, and A. Sovie. 2018. *Sylvilagus palustris*. Pp. 152–154, *in* Pikas, Rabbits and Hares of the World (Smith, A., C. H. Johnston, P. C. Alves, and K. Hackländer, eds.). The Johns Hopkins University Press/IUCN/SSC. Baltimore, U.S.A.

McClellan, D. A., and D. S. Rogers. 1997. *Peromyscus zarhynchus*. Mammalian Species 562:1–3.

McDaniel, V. R., M. J. Harvey, R. Tumlison, and K. N. Paige. 1982. Status of the small-footed bat, *Myotis leibii leibii,* in the southern Ozarks. Proceedings of the Arkansas Academy of Sciences 36:92–94.

McDonald, J. N. 1981. North American bison: their classification and evolution. University of California Press. Berkeley, U.S.A.

McDonald, R. A., A. V. Abramov, M. Stubbe, J. Herrero, T. Maran, A. Tikhonov, P. Cavallini, A. Kranz, G. Giannatos, B. Krytufek, and F. Reid. 2019. *Mustela nivalis*. The IUCN Red List of Threatened Species 2019:e.T70207409A147993366.

McDonough, M. M., L. K. Ammerman, R. M. Timm, H. H. Genoways, P. A. Larsen, and R. J. Baker. 2008. Speciation within bonneted bats (Genus *Eumops*): the complexity of morphological, mitochondrial, and nuclear data sets in systematics. Journal of Mammalogy 89:1306–1315.

McFadden, K. W., and S. Meiri. 2013. Dwarfism in insular carnivores: a case study of the Pygmy Raccoon. Journal of Zoology 289:213–221.

McGowan, C., W. S. Davidson, and L. A. Howes. 1999. Genetic analysis of an endangered Pine Marten (*Martes americana*) population from Newfoundland using randomly amplified polymorphic DNA markers. Canadian Journal of Zoology 77:661–666.

McGhee, M. E., and H. H. Genoways. 1978. *Liomys pictus*. Mammalian Species 83:1–5.

McGrew, J. C. 1979. *Vulpes macrotis*. Mammalian Species 123:1–6.

Mcgrath, G. 1987. Relationships of nearctic tree squirrels of the genus Sciurus. Ph.D. dissert., University of Kansas, Lawrence.

McKenna, M., and S. K. Bell. 1997. Classification of mammals above species level. Columbia University Press. New York, U.S.A.

McKnight, M. L. 1995. Mitochondrial DNA phylogeography of *Perognathus amplus* and *Perognathus longimembris* (Rodentia: Heteromyidae): a possible mammalian ring species. Evolution 49:816–826.

McKnight, M. L. 2005. Phylogeny of the *Perognathus longimembris* species group based on mitochondrial cytochrome–*b*: how many species? Journal of Mammalogy 86:826–832.

McLaughlin, C. A. 1984. Protrogomorph, Sciuromorph, Castorimorph, Myomorph (Geomyoid, Anomaluroid, Pedetoid, and Ctenodactyloid) rodents. Pp. 267–288, *in* Orders and families of recent mammals of the World (Anderson, S., and J. K. Jones, Jr., eds.). John Wiley & Sons. New York, U.S.A.

McLean, B. S. 2018. *Urocitellus parryii* (Rodentia: Sciuridae). Mammalian Species 50:84–99.

McLean, B. S., D. J. Jackson, and J. A. Cook. 2016. Rapid divergence and gene flow at high latitudes shape the history of Holarctic ground squirrels (*Urocitellus*). Molecular Phylogenetics and Evolution 102:174–188.

McLellan, B. N., M. F. Proctor, D. Huber, and S. Michel. 2017. *Ursus arctos*. The IUCN Red List of Threatened Species 2017:e.T41688A121229971.

McLellan, L. J. 1984. A morphometric analysis of *Carollia* (Chiroptera, Phyllostomidae). American Museum Novitates 2791:1–35.

McLeod, B. A., T. R. Frasier, and Z. Lucas. 2014. Assessment of the extirpated Maritimes Walrus using morphological and ancient DNA analyses. PLoS ONE 9:e.99569.

McManus, J. J. 1942. *Didelphis virginiana*. Mammalian Species 40:1–6.

McTaggart Cowan, I. 1938. Geographic distribution of color phases of the red fox and black bear in the Pacific Northwest. Journal of Mammalogy 19:202–206.

Mead, J. I., A. E. Spiess, and K. D. Sobolik. 2000. Skeleton of extinct North American Sea Mink (*Mustela macrodon*). Quaternary Research 53:247–262.

Meade, L. 1992. New distributional records for selected species Kentucky mammals. Transactions of the Kentucky Academy of Science 53:127–132.

Meagher, M. 1986. *Bison bison*. Mammalian Species 266:1–8.

Mearns, E. A. 1890. Description of supposed new species and subspecies of mammals from Arizona. Bulletin of the American Museum of Natural History 2:277–307.

Mearns, E. A. 1896. Preliminary diagnoses of new mammals from the mexican border of the United States. Proceedings of the United States Natural History Museum 18:443–447.

Mearns, E. A. 1907. Mammals of the Mexican boundary of the United States. Pt. I, families Didelphidae to Muridae. Bulletin United States National Museum 56:1–530.

Mech, L. D. 1974. *Canis lupus*. Mammalian Species 37:1–6.

Medellín, R. A. 1988. Prey of *Chrotopterus auritus*, with notes on feeding behavior. Journal of Mammalogy 69:841–844.

Medellín, R. A. 1989. *Chrotopterus auritus*. Mammalian Species 343:1–5.

Medellín, R. 2016a. *Leptonycteris nivalis*. The IUCN Red List of Threatened Species 2016:e.T11697A22126172.

Medellín, R. 2016b. *Leptonycteris yerbabuenae*. The IUCN Red List of Threatened Species 2016:e.T136659A21988965.

Medellín, R. A., and H. T. Arita. 1989. *Tonatia evotis* and *Tonatia silvicola*. Mammalian Species 334:1–5.

Medellín, R. A., D. E. Wilson, and L. Navarro. 1985. *Micronycteris brachyotis*. Mammalian Species 251:1–4.

Medellín, R. A., G. Ceballos, and H. Zarza. 1989. *Spilogale pygmaea*. Mammalian Species 600:1–3.

Medellín, R. A., G. Cancino, M. A. Clemente, and R. O. Guerrero. 1992. Noteworthy records of three mammals from Mexico. The Southwestern Naturalist 37:427–429.

Medellín, R. A., H. T. Arita, and O. Sánchez. 1997. Identificacion de los murcielagos de México, clave de campo. Asociación Mexicana de Mastozoología, A. C., México, Publicaciones Especiales 2:1–83.

Medellín, R. A., A. L. Gardner, and J. M. Aranda. 1998. The taxonomic status of the Yucatán brown brocket, *Mazama pandora* (Mammalia: Cervidae). Proceedings of the Biological Society of Washington 111:1–14.

Medellín, R. A., M. Equihua, and M. Amín. 2000. Bat diversity and abundance as indicators of disturbance in Neotropical rainforests. Conservation Biology 14:1666–1675.

Medellín, R. A., H. T. Arita, and O. Sánchez. 2008. Identificación de los murciélagos de México, clave de campo, segunda edición. Instituto de Ecología, Universidad Nacional Autónoma de México – Comisión para el Conocimiento y Uso de la Biodiversidad. Distrito Federal, México.

Meiri, M., P. Kosintsev, K. Conroy, S. Meiri, I. Barnes, and A. Lister. 2018. Subspecies dynamics in space and time: a study of the red deer complex using ancient and modern DNA and morphology. Journal of Biogeography 45:367–380.

Meisfjord, J., and R. C. Sundt. 1996. Genetic variation between populations of the harp seal, *Phoca groenlandica*. Journal of Marine Science 53:89–95.

Mejenes-López, S. de M. A., M. Hernández-Bautista, J. Barragán-Torres, and J. Pacheco-Rodríguez. 2010. Los mamíferos en el estado de Hidalgo, México. Therya 1:161–188.

Mengak, M. T. 1987. Abundance and distribution of shrews in western South Carolina. Brimleyana 13:63–66.

Melo-Ferreira, J., F. A. Seixas, E. Cheng, L. S. Mills, and P. C. Alves. 2014. The hidden history of the snowshoe hare, *Lepus americanus*: extensive mitochondrial DNA introgression inferred from multilocus genetic variation. Molecular Ecology 23: 4617–4630.

Menu, H. 1984. Révision du statut de *Pipistrellus subflavus* (F. Cuvier, 1832). Proposition d'un taxon generique noveau: *Perimyotis nov. gen.* Mammalia 48:409–416.

Mercure, A., K. Ralls, K. P. Koepfli, and R. K. Wayne. 1993. Genetic subdivisions among small canids: mitochondrial DNA, differentiation of swift, kit, and arctic foxes. Evolution 47:1313–1328.
Merriam, C. H. 1884. Mammals of the Adirondack region, northeastern New York. Vol. I and II. Transactions Linnaean Society, New York. New York, U.S.A.
Merriam, C. H. 1889a. Preliminary revision of North American pocket mice (genera *Perognathus* et *Cricetodipus* auct.) with description of new species and subspecies and a key to the known forms. North American Fauna 1:1–29.
Merriam, C. H. 1889b. Revision of the North American Pocket Mice. North American Fauna 1:1–29.
Merriam, C. H. 1889c. Description of a new spermophile from southern California. North American Fauna 2:15–16.
Merriam, C. H. 1890a. Descriptions of three new kangaroo rats, with remarks on the identity of *Dipodomys ordii* of Woodhouse. North American Fauna 4:41–49.
Merriam, C. H. 1890b. Description of a new prairie dog from Wyoming. North American Fauna 4:33–35.
Merriam, C. H. 1890c. Description of twenty-six new species of North American mammals. North American Fauna 4:1–55.
Merriam, C. H. 1891. Results of a biological reconnaissance of south-central Idaho. North American Fauna 5:1–416.
Merriam, C. H. 1892a. Description of a new prairie dog (*Cynomys mexicanus*) from Mexico. Proceedings of the Biological Society of Washington 7:157–158.
Merriam, C. H. 1892b. Description of a new genus and species of murine rodent *(Xenomys nelsoni)* from the state of Colima, western Mexico. Proceedings of the Biological Society of Washington 7:159–163.
Merriam, C. H. 1893. Descriptions of eight new ground squirrels of the genera *Spermophilus* and *Tamias* from California, Texas, and Mexico. Proceedings of the Biological Society of Washington 8:129–138.
Merriam, C. H. 1894a. A new subfamily of murine rodents -the Neotominae- with description of a new genus and species and a synopsis of the known forms. Proceedings of the Academy Natural Sciences of Philadelphia 14:225–252.
Merriam, C. H. 1894b. Abstract of a study of the American wood rats, with descriptions of fourteen new species and subspecies of the genus *Neotoma*. Proceedings of the Biological Survey of Washington 9:117–128.
Merriam, C. H. 1894c. Preliminary descriptions of eleven new kangaroo rats of the genera *Dipodomys* and *Perodipus*. Proceedings of the Biological Society of Washington 9:109–116.
Merriam, C. H. 1895a. Monographyc revision of the pocket gophers Family Geomyidae (exclusive of the species *Thomomys*). North American Fauna 8:1–258.
Merriam, C. H. 1895b. Revision of the shrews of the American genera *Blarina* and *Notiosorex*. North American Fauna 10:1–34.
Merriam, C. H. 1895c. Synopsis of the American shrews of the genus *Sorex*. North American Fauna 10:57–124.
Merriam, C. H. 1897. Notes on the chipmunks of the genus *Eutamias* occurring west of the east base of the Cascade-Sierra system, with descriptions of new forms. Proceedings of the Biological Society of Washington 11:189–212.
Merriam, C. H. 1898a. Description of twenty new species and new subgenus of *Peromyscus* from Mexico and Guatemala. Proceedings of the Biological Society of Washington 12:115–125.
Merriam, C. H. 1898b. Mammals of Tres Marias Islands off western Mexico. Proceedings of the Biological Society of Washington 12:13–19.
Merriam, C. H. 1900. Descriptions of twenty-six new mammals from Alaska and British North America. Proceedings of the Washington Academy of Sciences 2:13–30.
Merriam, C. H. 1901a. Seven new mammals from Mexico, including a new genus of rodents. Proceedings of the Washington Academy of Sciences 3:559–563.
Merriam, C. H. 1901b. Synopsis of the rice rats (genus *Oryzomys*) of the United States and Mexico. Proceedings of the Washington Academy of Sciences 3:273–295.
Merriam, C. H. 1901c. Two new bighorns and a new antelope from Mexico and the United States. Proceedings of the Biological Society of Washington 14:31–32.
Merriam, C. H. 1901d. A new brocket from Yucatan. Proceedings of the Biological Society of Washington 14:105–106.
Merriam, C. H. 1901e. Six new mammals from Cozumel Island, Yucatan. Proceedings of the Biological Society of Washington 14:99–104.
Merriam, C. H. 1902. Twenty new pocket mice *(Heteromys* and *Liomys)* from Mexico. Proceedings of the Biological Society of Washington 15:41–50.
Merriam, C. H. 1903. Four new mammals, including a new genus (*Teanopus*), from Mexico. Proceedings of the Biological Society of Washington 16:79–82.
Merriam, C. H. 1904a. Two new squirrels of the *aberti* group. Proceedings of the Biological Society of Washington 17:129–130.
Merriam, C. H. 1904b. New and little known kangaroo rats of the genus *Perodipus*. Proceedings of the Biological Society of Washington, 17:139–145.
Merriam, C. H. 1918. Review of the grizzly and big brown bears of North America (genus *Ursus*). North American Fauna 41:1136.
Merritt, J. F. 1978. *Peromyscus californicus*. Mammalian Species 85:1–6.
Merritt, J. F. 1981. *Clethrionomys gapperi*. Mammalian Species 146:1–9.
Michener, G. R., and J. W. Koeppl. 1985. *Spermophilus richardsonii*. Mammalian Species 243:1–8.
Mies, R., A. Kurta, and D. G. King. 1996. *Eptesicus furinalis*. Mammalian Species 526:1–7.
Miller, B., and B. Rodriguez. 2016a. *Myotis elegans*. The IUCN Red List of Threatened Species 2016:e.T14156A115121563.
Miller, B., and B. Rodriguez. 2016b. *Lasiurus intermedius*. The IUCN Red List of Threatened Species 2016:e.T11352A115101697.
Miller, B., F. Reid, J. Arroyo-Cabrales, A. D. Cuarón, and P. C. de Grammont. 2015a. *Carollia sowelli*. The IUCN Red List of Threatened Species 2015:e.T136268A22003903.
Miller, B., F. Reid, J. Arroyo-Cabrales, A. D. Cuarón, and P. C. de Grammont. 2015b. *Carollia subrufa*. The IUCN Red List of Threatened Species 2015:e.T3906A22133926.
Miller, B., F. Reid, J. Arroyo-Cabrales, A. D. Cuarón, and P. C. de Grammont. 2015c. *Trachops cirrhosus*. The IUCN Red List of Threatened Species 2015:e.T22029A22042903.
Miller, B., F. Reid, J. Arroyo-Cabrales, A. D. Cuarón, and P. C. de Grammont. 2015d. *Dermanura phaeotis*. The IUCN Red List of Threatened Species 2015:e.T83683287A21997769.

Miller, B., F. Reid, J. Arroyo-Cabrales, A. D. Cuarón, and P. C. de Grammont. 2016a. *Eumops underwoodi*. The IUCN Red List of Threatened Species 2016:e.T8248A22025754.

Miller, B., F. Reid, J. Arroyo-Cabrales, A. D. Cuarón, and P. C. de Grammont. 2016b. *Hylonycteris underwoodi*. The IUCN Red List of Threatened Species 2016:e.T10598A22036808.

Miller, B., F. Reid, J. Arroyo-Cabrales, A. D. Cuarón, and P. C. de Grammont. 2016c. *Glossophaga commissarisi*. The IUCN Red List of Threatened Species 2016:e.T9273A22108801.

Miller, B., F. Reid, J. Arroyo-Cabrales, A. D. Cuarón, and P. C. de Grammont. 2016d. *Glossophaga leachii*. The IUCN Red List of Threatened Species 2016:e.T9274A128959800.

Miller, B., F. Reid, J. Arroyo-Cabrales, A. D. Cuarón, and P. C. de Grammont. 2016e. *Artibeus jamaicensis*. The IUCN Red List of Threatened Species 2016:e.T88109731A21995883.

Miller, B., F. Reid, J. Arroyo-Cabrales, A. D. Cuarón, and P. C. de Grammont. 2016f. *Centurio senex*. The IUCN Red List of Threatened Species 2016:e.T4133A22009493.

Miller, B., F. Reid, J. Arroyo-Cabrales, A. D. Cuarón, and P. C. de Grammont. 2016g. *Rhogeessa tumida*. The IUCN Red List of Threatened Species 2016:e.T19685A22006890.

Miller, B., F. Reid, J. Arroyo-Cabrales, A. D. Cuarón, and P. C. de Grammont. 2016h. *Eptesicus fuscus*. The IUCN Red List of Threatened Species 2016:e.T7928A22118197.

Miller, B., F. Reid, J. Arroyo-Cabrales, A. D. Cuarón, and P. C. de Grammont. 2016i. *Molossus sinaloae*. The IUCN Red List of Threatened Species 2016:e.T13650A22106433.

Miller, F. L., R. H. Russell, and A. Gunn. 1977. Distributions, movements, and numbers of Peary caribou and muskoxen on western Queen Elizabeth Islands, Northwest Territories, 1972–74. Canadian Wildlife Service Report Series 40:1–55.

Miller, G. S., Jr. 1896. The genera and subgenera of voles and lemmings. North American Fauna 12:1–85.

Miller, G. S., Jr. 1897. Revision of the North American bats of the family Vespertilionidae. North American Fauna 13:1–135.

Miller, G. S., Jr. 1900. Note on *Micronycteris brachyotis* (Dobson) and *M. microtis* Miller. Proceedings of the Biological Society of Washington 13:154–155.

Miller G. S., Jr. 1902. Twenty new American bats. Proceedings of the Academy Natural Scieces Philadelphia 54:389–412.

Miller, G. S., Jr. 1907. The families and genera of bats. Bulletin of the United States National Museum 57:1–282.

Miller, G. S., Jr. 1912a. List of North American land mammals in the United States National Museum, 1911. Bulletin of the United States Natural Museum 79:1–455.

Miller, G. S. 1912b. Catalogue of the mammals of Western Europe (Europe exclusive of Russia) in the collection of the British Museum. William Clowes and Sons. London, United Kingdom.

Miller, G. S., Jr. 1914. The generic name of the collared peccaries. Proceedings of the Biological Society of Washington 27:215.

Miller, G. S., and R. Kellogg. 1955. List of North American Recent Mammals. Bulletin of the United States Natural Museum 205:1–954.

Miller, J. S., Jr. 1913. Revision of the bats of the genus *Glossophaga*. Proceeding of the United States National Museum 46:413–429.

Miller, J. S., Jr., and G. M. Allen. 1928. The American bats of the genus *Myotis* and *Pizonyx*. Bulletin of the United States Natural Museum 144:1–209.

Millions, D. G., and B. J. Swanson. 2007. Impact of natural and artificial barriers to dispersal on the population structure of bobcats. Journal of Wildlife Management 71:96–102.

Mills, L., and A. T. Smith. 2019. *Lepus americanus*. The IUCN Red List of Threatened Species 2019:e.T41273A45185466.

Milner, J., C. Jones, and J. K. Jones, Jr. 1990. *Nyctinomops macrotis*. Mammalian Species 351:1–4.

Miranda, F., D. A. Meritt, D. G. Tirira, and M. Arteaga. 2014. *Cyclopes didactylus*. The IUCN Red List of Threatened Species 2014:e. T6019A47440020.

Miranda, F. R., D. M. Casali, F. A. Perini, F. A. Machado, and F. R. Santos. 2018. Taxonomic review of the genus *Cyclopes* Gray, 1821 (Xenarthra: Pilosa), with the revalidation and description of new species. Zoological Journal of the Linnean Society 183:687–721.

Modi, W. S. 1986. Karyotypic differentiation among two sibling species pairs of New World microtine rodents. Journal of Mammalogy 67:159–165.

Modi, W. S. 1987. Phylogenetic analyses of chromosomal banding patterns among the Nearctic Arvicolidae (Mammalia: Rodentia). Systematic Zoology 36:109–136.

Modi, W. S. 1996. Phylogenetic history of LINE-1 among arvicolid rodents. Molecular Biology and Evolution 13:633–641.

Modi, W. S., and M. R. Lee. 1984. Systematic implications of chromosomal banding analyses of populations of *Peromyscus truei* (Rodentia: Muridae). Proceedings of the Biological Society of Washington 97:716–723.

Moehlman, P. D., F. Kebede, and H. Yohannes. 2015. *Equus africanus*. The IUCN Red List of Threatened Species 2015:e.T7949A45170994.

Moehrenschlager, A., and M. Sovada. 2016. *Vulpes velox*. The IUCN Red List of Threatened Species 2016:e.T23059A57629306.

Montiel-Parra, G., A. L. Carlos-Delgado, R. Paredes-León, and T. M. Pérez. 2019. Epizoic arthropods of the Mexican Shrew, *Sorex oreopolus* (Mammalia: Soricidae). Therya 10:33–37.

Monk, R. R., and J. K. Jones, Jr. 1996. *Perognathus flavescens*. Mammalian Species 525:1–4.

Monroy-Gamboa, A. G. 2021. The ghost mammals from Mexico and their implications. Therya 12:477–486.

Monroy-Gamboa, A. G., S. T. Álvarez-Castañeda, and A. L. Trujano-Álvarez. Accepted. *Tylomys nudicaudus*. Mammalian Species.

Monson, G., and L. Sumner. 1980. The desert bighorn: its life history, ecology, and management. University of Arizona Press. Tucson, U. S. A.

Moore, C. M., and P. W. Collins. 1995. *Urocyon littoralis*. Mammalian Species 489:1–7.

Moore, C. M., and L. L. Janacek. 1990. Genic relationships among North American *Microtus* (Mammalia: Rodentia). Annals of Carnegie Museum 59:249–259.

Moore, J. C. 1953. Distribution of marine mammals to Florida waters. The American Midland Naturalist 49:117–158.

Moore, J. C. 1960. The relationships of the gray squirrel; *Sciurus carolinensis*, to its nearest relatives. Proceedings of the Annual Conference of the Southeastern Association of Game and Fish Commissioners 13:356–363.

Moore, S. E., and E. I. Barrowclough. 1984. Incidental sighting of a ribbon seal (*Phoca fasciata*) in the western Beaufort Sea. Arctic 37:290.

Morales, J. C., and M. D. Engstrom. 1989. Morphological variation in the painted spiny pocket mouse, *Liomys pictus* (Family Heteromyidae), from Colima and southern Jalisco, México. Royal Ontario Museum, Life Sciences, Occasional Papers 38:1–16.

Morales-Malacara, J. B., and R. López-Wilchis. 1990. Epizoic fauna of *Plecotus mexicanus* (Chiroptera: Vespertilionidae) in Tlaxcala, Mexico. Journal of Medical Entomology 27:440–445.
Moreno-Valdez, A., R. L. Honeycutt, and W. E. Grant. 2004. Colony dynamics of *Leptonycteris nivalis* (Mexican long-nosed bat) related to flowering *Agave* in Northern Mexico. Journal of Mammalogy 85:453–459.
Morales-Vela, B., and L. D. Olivera-Gómez. 1997. Distribución del manati (*Trichechus manatus*) en la costa norte y centro-norte del estado de Quintana Roo, México. Anales del Instituto de Biología, Serie Zoology 68:153–164.
Morales-Vela, B., J. A. Padilla-Saldivar, and A. A. Mignucci-Giannoni. 2003. Status of the manatee (*Trichechus manatus*) along the northern and western coasts of the Yucatán Peninsula, México. Caribbean Journal of Science 39:42–49.
Mowat, G., D. C. Heard, and C. J. Schwarz. 2013. Predicting grizzly bear density in western North America. PLoS ONE 8:e82757.
Müdespacher-Ziehl, C., R. Espiritu-Mora, M. Martínez-Coronel, and S. Gaona. 2005. Chromosomal studies of two populations of *Peromyscus difficilis felipensis* (Rodentia: Muridae). Cytologia 70:243–248.
Murie, J. O. 1973. Population characteristics and phenology of a Franklin ground squirrel (*Spermophilus franklinii*) colony in Alberta, Canada. The American Midland Naturalist 90:334–340.
Murray J. L., and G. L. Gardner. 1997. *Leopardus pardalis*. Mammalian Species 548:1–10.
Musser, G. G. 1964. Notes on geographic dlstributlon, habitat, and taxonomy of some mexican mammals. Occasional Papers of the Museum of Zoology, University of Michigan 636:1–22.
Musser, G. G. 1968. A systematic study of the Mexican and Guatemalan gray squirrel, *Sciurus aureogaster* F. Cuvier (Rodentia: Sciuridae). Miscellaneous Publications of the Museum of Zoology, University of Michigan 137:1–112.
Musser, G. G. 1969. Notes on *Peromyscus* (Muridae) of Mexico and Central America. American Museum Novitates 2357:1–23.
Musser, G. G. 1970. Identity of the type-specimens of *Sciurus aureogaster* F. Cuvier and *Sciurus nigrescens* Bennett (Mammalia, Sciuridae). American Museum Novitates 2438:1–19.
Musser, G. G. 1971. Peromyscus allophylus Osgood: a synonym of *Peromyscus gymnotis* Thomas (Rodentia, Muridae). American Museum Novitates 2453:1–10.
Musser, G. G., and M. D. Carleton. 1993. Family Muridae. Pp. 501–755, *in* Mammal Species of the World, (Wilson, D. E., and D.M. Reeder, eds.). Second Edition. Smithsonian Institution Press, Washington and London.
Musser, G. G., and M. D. Carleton. 2005. Superfamily Muroidea. Pp. 894–1534, *in* Mammal species of the world. A taxonomic and geographic reference (Wilson, D. E., and D. A. M. Reeder, eds.), Third edition. The Johns Hopkins University Press. Baltimore, U.S.A.
Musser, G., R. Hutterer, B. Kryštufek, N. Yigit, and G. Mitsainas. 2021. *Mus musculus* (amended version of 2016 assessment). The IUCN Red List of Threatened Species 2021: e.T13972A197519724.
Mycroft, E. E., A. B. A. Shafer, and D. T. Stewart. 2011. Cytochrome-b sequences variation in Water Shrews (*Sorex palustris*) from eastern and western North America. Northeastern Naturalist 18:497–508.
Nader, I. A. 1966. Roots of teeth as a generic character in the kangaroo rats, *Dipodomys*. Bulletin of the Biological Research Centre, Baghdad 2:62–69.
Nader, I. A. 1978. Kangaroo rats: intraspecific variation in *Dipodomys spectabilis* Merriam and *Dipodomys deserti* Stephens. Illinois Biological Monographs 49:1–116.
Nadler, C. F. 1966. Chromosomes and systematics of American ground squirrels of the subgenus *Spermophilus*. Journal of Mammalogy 47:579–596.
Nadler, C. F. 1968. The chromosomes of *Spermophilus townsendi* (Rodentia: Sciuridae) and report of a new subspecies. Cytogenetics 7: 144–157.
Nadler, C. F., and D. A. Sutton. 1967. Chromosomes of some squirrels (Mammalia-Sciuridae) from the genera *Sciurus* and *Glaucomys*. Experientia 23:249–251.
Nadler, C. F., R. S. Hoffmann, and K. R. Greer. 1971. Chromosomal divergence during evolution of ground squirrel populations (Rodentia: *Spermophilus*). Systematic Zoology 20:298–305.
Nadler, C. F., R. S. Hoffmann, and A. Woolf. 1973. G-Band patterns as chromosomal markers, and the interpretation of chromosomal evolution in wild sheep *(Ovis)*. Experientia 29:117–119.
Nadler, C. F., R. S. Hoffmann, N. N. Vorontsov, J. W. Koeppl, L. Deutsch, and R. I. Sukernik. 1982. Evolution in ground squirrels. II. Biochemical comparisons in Holarctic populations of *Spermophilus*. Zeitschrift für Säugetierkunde 47:198–214.
Nadler, C. F., E. I. Lyapunova, R. S. Hoffmann, N. N. Vorontsov, L. L. Shaitorova, and Y. M. Borisov. 1984. Chromosomal evolution in Holarctic ground squirrels. II. Giemsa band homologies of chromosomes, and the tempo of evolution. Zeitschrift für Saügetierkunde 49:78–90.
Nagorsen, D. W. 1987. *Marmota vancouverensis*. Mammalian Species 270:1–5.
Nagorsen, D. W. 1996. Opossums, shrews and moles of British Columbia. UBC Press, Vancouver, British Columbia, Canada.
Nagorsen, D. W. 2002. An identification manual to the small mammals of British Columbia. Ministry of Sustainable Resource Management, Ministry of Water, Land and Air Protection, and Royal British Columbia Museum. Victoria, Canada.
Nagorsen, D. W., N. Panter, and A. G. Hope. 2017. Are the western water shrew (*Sorex navigator*) and American water shrew (*Sorex palustris*) morphologically distinct? Canadian Journal of Zoology 95:727–736.
Nájera-Cortazar, L. A., S. T. Álvarez-Castañeda, and E. de Luna. 2015. Phylogeny and geometric morphometrics analyses of *Myotis peninsularis*. Acta Chiropterologica 17:37–47.
Naranjo, E. J., and E. Espinoza-Medinilla. 2001. Los mamíferos de la Reserva Ecológica Huitepec, Chiapas, México. Revista Mexicana de Mastozoología 5:58–67.
Nash, D. J., and R. N. Seaman. 1977. *Sciurus aberti*. Mammalian Species 80:1–5.
Nason, E. S. 1948. Morphology of hair of eastern North American bats. American Midland Naturalist 39:345–361.
NatureServe. 2016a. *Urocitellus townsendii*. The IUCN Red List of Threatened Species 2016:e.T20476A112212554.
NatureServe. 2016b. *Urocitellus washingtoni*. The IUCN Red List of Threatened Species 2016:e.T20475A112211642.
NatureServe (Morefield, J.). 2016c. *Microdipodops pallidus*. The IUCN Red List of Threatened Species 2016:e.T42607A115193914.
NatureServe. 2017. *Sorex lyelli*. The IUCN Red List of Threatened Species 2017:e.T41402A22313470.
Navarrete, D., and J. Ortega. 2011. *Tamandua mexicana*. Mammalian Species 43:56–63.
Navarro L. D., and D. E. Wilson. 1982. *Vampyrum Spectrum*. Mammalian Species 184:1–4.

Navarro L. D., T. Jiménez, and J. Juárez. 1990. Los mamíferos de Quintana Roo. Pp. 371–450, *in* Diversidad biológica en la Reserva de la Biósfera de Sian Ka'an Quintana Roo, México (Navarro D., and J. G. Robinson, eds.). Centro de Investigaciones de Quintana Roo and Program of Studies in Tropical Conservation, University of Florida. Quintana Roo, México.

Naylor, L., and N. Roach. 2016. *Sorex jacksoni*. The IUCN Red List of Threatened Species 2016:e.T20390A22314744.

Naylor, L., and N. Roach. 2017. *Perognathus alticola*. The IUCN Red List of Threatened Species 2017: e.T16631A22224213.

McNab, B. K. 1969. The economics of temperature regulation in Neotropical bats. Compo Biochemetri Physiology 31:227–268.

Neiswenter, S. A., and B. R. Riddle. 2010. Diversification of the *Perognathus flavus* species group in emerging arid grasslands of western North America, Journal of Mammalogy 91:348–362.

Neiswenter, S. A., and B. R. Riddle. 2011. Landscape and climatic effects on the evolutionary diversification of the *Perognathus fasciatus* species group, Journal of Mammalogy 92:982–993.

Neiswenter, S. A., D. J. Hafner, J. E. Light, G. D. Cepeda, K. C. Kinzer, L. F. Alexander, and B. R. Riddle. 2019. Phylogeography and taxonomic revision of Nelson's pocket mouse (*Chaetodipus nelsoni*). Journal of Mammalogy 100:1847–1864.

Nelson, E. W. 1898. Descriptions of new squirrels from Mexico and Central America. Proceedings of the Biological Sciety of Washington 12:145–156.

Nelson, E. W. 1899. Mammals of the Tres Marias Islands. North American Fauna 14:15–20.

Nelson, E. W. 1909. The rabbits of North America. North American Fauna 29:1–278.

Nelson, E. W. 1912. A new subspecies of pronghorn antelope from Lower California. Proceedings of the Biological Society of Washington 25:107–108.

Nelson, E. W. 1922. Lower California and its natural resources. National Academy of Sciences, First Memories 16:1–194.

Nelson, E. W., and E. A. Goldman. 1909. Eleven new mammals from Lower California. Proceedings of the Biological Society of Washington 22:23 –28.

Nelson, E. W., and E. A. Goldman. 1923. Six new pocket mice from Lower California and notes on the status of several described species. Proceedings of the Biological Society of Washington 42:103–112.

Nelson, E. W., and E. A. Goldman. 1929. Four new pocket gophers of the genus *Heterogeomys* from Mexico. Proceedings of the Biological Society of Washington 42:147–152.

Nelson, E. W., and E. A. Goldman. 1931. Three new Raccoons from Mexico and Salvador. Proceedings of the Biological Society of Washington 44:17–22.

Nelson, E. W., and E. A. Goldman. 1932. Two new cacomistles from Mexico, with remarks on the genus *Jentinkia*. Journal of the Washington Academy of Sciences 22:484–488.

Nelson, E. W., and E. A. Goldman. 1934a. Revision of the pocket gophers of the genus *Cratogeomys*. Proceedings of the Biological Society of Washington 47:135–154.

Nelson, E. W., and E. A. Goldman. 1934b. Pocket gophers of the genus *Thomomys* of the Mexican mainland and bordering territory. Journal of Mammalogy 15:105–124.

Nelson, T. W., and K. Shump, Jr. 1978. Cranial variation and size allometry in *Agouti paca* from Ecuador. Journal of Mammalogy 59:387–394.

Nielsen C. K., and L. K. Berkman 2018a. *Sylvilagus aquaticus*. Pp. 117–200, *in* Pikas, Rabbits and Hares of the World (Smith, A., C. H. Johnston, P. C. Alves, and K. Hackländer, eds.). The Johns Hopkins University Press/IUCN/SSC. Baltimore, U.S.A.

Nielsen C. K., and L. K. Berkman. 2018b. *Sylvilagus floridanus*. Pp. 137–140, *in* Pikas, Rabbits and Hares of the World (Smith, A., C. H. Johnston, P. C. Alves, and K. Hackländer, eds.). The Johns Hopkins University Press/IUCN/SSC. Baltimore, U.S.A.

Nielsen, C., and H. C. Lanier. 2019. *Sylvilagus floridanus*. The IUCN Red List of Threatened Species 2019:e.T41299A45191626.

Nielsen, C., D. Thompson, M. Kelly, and C. A. Lopez-Gonzalez. 2015. *Puma concolor*. The IUCN Red List of Threatened Species 2015:e. T18868A97216466.

Nowak, R. M. 1991. Walker's mammals of the world. Fifth edition. The Johns Hopkins University Press. Baltimore, U.S.A.

Nowak, R. M. 1999. Walker's mammals of the world. Sixth edition. The Johns Hopkins University Press. Baltimore, U.S.A.

Nowak, R. M. 2005. Walker's Carnivores of the World. Johns Hopkins University Press, Baltimore, U.S.A. and London, UK.

Nowak, R. M., and J. L. Paradiso. 1983. Walker's mammals of the world. Fourth Edition. Johns Hopkins University Press, Baltimore, U.S.A.

Oaks, E. C., P. J. Young, G. L. Kirkland, and D. F. Schmidt. 1987. *Spermophilus variegatus*. Mammalian Species 272:1–8.

O'Corry-Crowe, G. M., and R. L. Westlake. 1997. Molecular investigations of spotted seals (*Phoca largha*) and harbor seals (*P. vitulina*), and their relationship in areas of sympatry. Pp. 291–304, *in* Molecular genetics of marine mammals (Dizon, A. E., S. J. Chivers, and W. F. Perrin, eds). The Society of Marine Mammalogy, Lawrence, U.S.A.

O'Corry-Crowe, G., and C. Bonin. 2009. The biogeography and population structure of spotted seals (*Phoca largha*) as revealed by mitochondrial and microsatellite DNA. Draft Report to the National Marine Mammal Laboratory. Harbor Branch Oceanographic Institute, Florida Atlantic University, Fort Pierce, U.S.A.

O'Farrell, M. J., and A. R. Blaustein. 1974a. *Microdipodops megacephalus*. Mammalian Species 46:1–3.

O'Farrell, M. J., and A. R. Blaustein. 1974b. *Microdipodops pallidus*. Mammalian Species 47:1–2.

O'Farrell, M. J., and E. H. Studier. 1980. *Myotis thysanodes*. Mammalian Species 137:1–5.

O'Gara, B. W. 1969. Horn casting by female pronghorns. Journal of Mammalogy 50:373–375.

O'Gara, B. W. 1978. *Antilocapra americana*. Mammalian Species 90:1–7.

O'Gara, P. W. 1990. The pronghorn (*Antilocapra americana*). Pp. 231– 264, *in* Horns, pronghorns, and antlers (Bubenik, G. A., and A. B. Bubenik, eds.). Springer-Verlag. New York, U.S.A.

Ohdachi, S. D., M. Hasegawa, M. A. Iwasa, P. Vogel, T. Oshida, L.-K. Lin, and H. Abe. 2006. Molecular phylogenetics of soricid shrews (Mammalia) based on mitochondrial cytochrome–*b* gene sequences: with special reference to the Soricinae. Journal of Zoology 270:177–191.

Ojasti, J., and O. J. Linares. 1971. Adiciones a la fauna de murciélagos de Venezuela con notas sobre las especies del género *Diclidurus*. Acta Biológica Venezolana 7:421–441.

Ojeda, R., C. Bidau, and L. Emmons. 2016. *Myocastor coypus*. The IUCN Red List of Threatened Species 2016:e.T14085A121734257.

Okhotina, M. V. 1983. A taxonomic revision of *Sorex arcticus* Kerr, 1972. (Soricidae, Insectivora). Zoologeskii Zhurnal 62:409–417.

Olguín-Monroy, H. C., L. León-Paniagua, U. M. Samper-Palacios, and V. Sánchez-Cordero. 2008. Mastofauna de la region de los Chimalapas, Oaxaca, México. Pp. 165–216, *in* Avances en el estudio de los mamíferos de México, Volumen II. (Lorenzo, C., E. Espinoza, and J. Ortega, eds.). Asociación Mexicana de Mastozoología, A. C. Distrito Federal, México.

Oliveira T. G. 1998a. *Herpailurus yagouaroundi*. Mammalian Species 578:1–6.

Oliveira T. G. 1998b. *Leopardus wiedii*. Mammalian Species 579:1–6.

Oliver, W. L. R. 1976. The management of yapoks (*Chironectes minimus*) at Jersey Zoo, with observations on their behavior. Annual Reptort Jersey Wildlife Preservation Trust 13:32–36.

Ordóñez-Garza, N., and R. D. Bradley. 2011. *Peromyscus schmidlyi* (Rodentia: Cricetidae). Mammalian Species 43:31–36.

Ordoñez-Garza, N., J. O. Matson, R. E. Strauss, R. D. Bradley, and J. Salazar-Bravo. 2010. Patterns of phenotypic and genetic variation in three species of endemic Mesoamerican *Peromyscus* (Rodentia: Cricetidae). Journal of Mammalogy 91:848–859.

Ordoñez-Garza, N., C. W. Thompson, M. K. Unkefer, C. W. Edwards, J. G. Owen, and R. D. Bradley. 2014. Systematics of the *Neotoma mexicana* species group (Mammalia: Rodentia: Cricetidae) in Mesoamerica: new molecular evidence on the status and relationships of *N. ferruginea* Tomes, 1862. Proceedings of the Biological Society of Washington 127:518–532.

Orr, R. T. 1934. Two new woodrat from Lower California, Mexico. Proceedings Biological Society Washington 47:109–112.

Orr, R. T. 1940. The rabbits of California. Occasional Papers California Academy of Sciences 19:1–227.

Ortega, J., and H. T. Arita. 1997. *Mimon bennettii*. Mammalian Species 549:1–4.

Ortega, J., and I. Alarcón-D. 2008. *Anoura geoffroyi*. Mammalian Species 818:1–7.

Ortega, J., and I. Castro-Arellano. 2001. *Artibeus jamaicensis*. Mammalian Species 662:1–9.

Ortega, J., M. Tschapka, T. P. González-Terrazas, G. Suzán, and R. A. Medellín. 2009. Phylogeography of *Musonycteris harrisoni* along the pacific coast of Mexico. Acta Chiropterologica 11:259–270.

Ortega Reyes, J., D. G. Tirira, M. Arteaga, and F. Miranda. 2014. *Tamandua mexicana*. The IUCN Red List of Threatened Species 2014:e. T21349A47442649.

Ortega, J., B. Vite-de León, A. Tinajero-Espitia, and J. A. Romero-Meza. 2008. *Carollia subrufa*. Mammalian Species 823:1–4.

Ortiz-Martínez T., V. Rico-Gray, and E. Martínez-Meyer. 2008. Predicted and verified distributions of *Ateles geoffroyi* and *Alouatta palliata* in Oaxaca, Mexico. Primates 49:186–19.

Osgood, W. H. 1900. Revision of the pocket mice of the genus *Perognathus*. North American Fauna 18:1–73.

Osgood, W. H. 1904. Thirty new mice of the genus *Peromyscus* from Mexico and Guatemala. Proceedings of the Biological Society of Washington 17:55–77.

Osgood, W. H. 1907. Four new pocket mice. Proceedings of the Biological Society of Washington 20:19–22.

Osgood, W. H. 1909. Revision of the mice of the genus *Peromyscus*. North American Fauna 28:33–252.

Osgood, W. H. 1913. Two new mouse opossums from Yucatan. Proceedings of the Biological Society of Washington 26:175–176.

Osgood, W. H. 1918. The status of *Perognathus longimembris* (Coues). Proceedings of the Biological Society of Washington, 31:95–96.

Ostroff, A. C., and E. J. Finck. 2003. *Spermophilus franklinii*. Mammalian Species 724:1–5.

Owen J. G., and R. S. Hoffmann. 1983. *Sorex ornatus*. Mammalian Species 212:1–5.

Owen, J. G., D. J. Schmidly, and W. B. Davis. 1984. A morphometric analysis of three species of *Carollia* (*Chiroptera*, Glossophaginae) from Middle America. Mammalia 48:85–93.

Owen, J. G., R. J. Baker, and S. L. Williams. 1996. Karyotypic variation in spotted skunks (Carnivora: Mustelidae: *Spilogale*) from Texas, Mexico and El Salvador. Texas Journal of Science 48:119–122.

Packard, R. L. 1960. Speciation and evolution of the pygmy mice, genus *Baiomys*. University of Kansas Publications, Museum of Natural History 9:579–670.

Packard, R. L., and F. W. Judd. 1968. Comments on some mammals from western Texas. Journal of Mammalogy 49:535–538.

Packard, R. L., and J. B. Montgomery, Jr. 1978. *Baiomys musculus*. Mammalian Species 102:1–3.

Palmer, F. G. 1937. Geographic variation in the mole *Scapanus latimanus*. Journal of Mammalogy 18:280–314.

Paradiso, J. L. 1967. A review of the wrinkle-faced bats (*Centurio senex* Gray) with a description of a new subspecies. Mammalia 31:595–604.

Paradiso, J. L., and R. M. Nowak. 1972. *Canis rufus*. Mammalian Species 1972:1–4.

Pasitschniak-Arts, M. 1993. *Ursus arctos*. Mammalian Species 439:1–10.

Pasitschniak-Arts, M., and S. Larivière. 1995. *Gulo gulo*. Mammalian Species 499:1–10.

Patterson, B. D. 1980a. Montane mammalian biogeography in New Mexico. The Southwestern Naturalist 25:33–40.

Patterson, B. D. 1980b. A new subspecies of *Eutamias quadrivittatus* (Rodentia: Sciuridae) from the Organ Mountains, New Mexico. Journal of Mammalogy 61:455–464.

Patterson, B. D. 1984. Geographic variation and taxonomy of Colorado and hopi chipmunks (genus *Eutamias*). Journal of Mammalogy 65:442–456.

Patterson, B. D., and L. R. Heany. 1987. Preliminary analysis of geographic variation in red-tailed chipmunks (*Eutamias ruficaudus*). Journal of Mammalogy 68:782–791.

Patterson, B. D., and Norris, R. W. 2016. Towards a uniform nomenclature for ground squirrels: the status of the Holarctic chipmunks. Mammalia 80:241–251.

Patton, J. L. 1967a. Chromosome studies of certain pocket mice, genus *Perognathus* (Rodentia: Heteromyidae). Journal of Mammalogy 48:27–37.

Patton, J. L. 1967b. Chromosomes and evolutionary trends in the pocket mouse subgenus *Perognathus* (Rodentia: Heteromyidae). The Southwestern Naturalist 12:429–438.

Patton, J. L. 1969. Chromosome evolution in the pocket mouse, *Perognathus goldmani* Osgood. Evolution 23:645–662.

Patton, J. L. 1972. Patterns of geographic variation in karyotype in the Pocket Gopher, *Thomomys bottae* (Eydoux and Gervais). Evolution 26:574–586.

Patton, J. L. 1973. An analysis of natural hybridization between the pocket gophers *Thomomys bottae* and *Thomomys umbrinus*, in Arizona. Journal of Mammalogy 54:561–584.

Patton, J. L. 1993a. Family Geomyidae. Pp. 469–476, *in* Mammal species of the world, a taxonomic and geographic reference (Wilson, D. E., and D. M. Reeder, eds.). Second edition. Smithsonian Institution Press, Washington, U.S.A.

Patton, J. L. 1993b. Family Heteromidae. Pp. 477–486, *in* Mammal species of the world, a taxonomic and geographic reference (Wilson, D. E., and D. M. Reeder, eds.). Second edition. Smithsonian Institution Press. Washington, U.S.A.

Patton, J. L. 1999a. *Thomomys bulbivorus*. Pp. 468–469, *in* The Smithsonian Book of North American Mammals (Wilson, D. E., and S. Ruff, eds.). Smithsonian Institution Press. Washington, U.S.A.

Patton, J. L. 1999b. *Thomomys monticola*. Pp. 473–474, *in* The Smithsonian Book of North American Mammals (Wilson, D. E., and S. Ruff, eds.). Smithsonian Institution Press. Washington, U.S.A.

Patton, J. L. 1999c. *Thomomys talpoides*. Pp. 474–477, *in* The Smithsonian Book of North American Mammals (Wilson, D. E., and S. Ruff, eds.). Smithsonian Institution Press. Washington, U.S.A.

Patton, J. L. 2005a. Family Heteromyidae. Pp. 844–858, *in* Mammal species of the world. A taxonomic and geographic reference (Wilson, D. E., and D. A. Reeder, eds.). Third edition. The Johns Hopkins University Press. Baltimore, U.S.A.

Patton, J. L. 2005b. Family Geomyidae. Pp. 859–870, *in* Mammal species of the world. A taxonomic and geographic reference (Wilson, D. E., and D. A. Reeder, eds.). Third edition. The Johns Hopkins University Press. Baltimore, U.S.A.

Patton, J. L., and S. T. Álvarez-Castañeda. 1999. Family Heteromyidae. Pp. 351–443, *in* Mamíferos del Noroeste Mexicano (Álvarez-Castañeda, S. T., and J. L. Patton, eds.). Centro de Investigaciones Biológicas del Noroeste, S. C. La Paz, México.

Patton, J. L., and S. T. Álvarez-Castañeda. 2005. Phylogeography on the desert woodrat, *Neotoma lepida*, with comments on systematics and biogeographic history. Pp. 375–388, *in* Contribuciones mastozoologicas en honor de Bernanrdo Villa (Sánchez-Cordero, V., and R. Medellín, eds.). Instituto de Biología, Universidad Nacional Autónoma de México. Distrito Federal, México.

Patton, J. L., and S. T. Álvarez-Castañeda. 2017. *Neotoma insularis*. The IUCN Red List of Threatened Species 2017:e.T116989038A119112253.

Patton, J. L., and M. N. Da Silva. 1997. Definition of species of pouched four-eyed opossum (Didelphidae, *Philander*). Journal of Mammalogy 78:90–102.

Patton, J. L., and R. E. Dingman. 1968. Chromosome studies of pocket gophers, genus *Thomomys*. I. The specific status of *Thomomys umbrinus* (Richardson) in Arizona. Journal of Mammalogy 49:1–13.

Patton, J. L., and L. H. Emmons. 2015. Family Dasyproctidae Bonaparte, 1838. Pp. 733–762, *in* Mammals of South America Volume 2: Rodents (Patton, J. L., U. F. G. Pardiñas, and G. D'Elía, eds). University of Chicago Press. Chicago, U.S.A.

Patton, J. L., and R. N. Fisher. 2023. Taxonomic reassessment of the Little pocket mouse, *Perognathus longimembris* (Rodentia, Heteromyidae) of southern California and northern Baja California. Therya 14:131–160.

Patton, J. L., and A. L. Gardner. 1971. Parallel evolution of multiple sex-chromosome systems in the phyllostomatid bats, *Carollia* and *Choeroniscus*. Experientia 27:105–106.

Patton, J. L., and M. F. Smith. 1981. Molecular evolution in *Thomomys*: Phyletic systematics, paraphyly, and rates of evolution. Journal of Mammalogy 62:493–500.

Patton, J. L., and M. F. Smith. 1989. Genetic structure and the genetic and morphologic divergence among pocket gopher species (genus *Thomomys*). Pp. 284–304, *in* Speciation and its consequences (Otte, D., and J. A. Endler, eds.). Sinauer Associates Incorporated. Sunderland, U.S.A.

Patton, J. L., and M. F. Smith. 1990. The evolutionary dynamics of the Pocket Gopher *Thomomys bottae*, with emphasis on California populations. University of California Publications, zoology 123.

Patton, J. L., and M. F. Smith. 1991. Molecular evolution in *Thomomys*: phyletic systematics, paraphyly, and rates of evolution. Journal of Mammalogy 62:493–500.

Patton, J. L., and M. F. Smith. 1994. Paraphyly, polyphyly, and the nature of species boundaries in pocket gophers (Genus *Thomomys*). Systematic Biology 43:11–26.

Patton, J. L., H. MacArthur, and S. Y. Yang. 1976. Systematic relationships of the four-toed populations of *Dipodomys heermanni*. Journal of Mammalogy 57:159–163.

Patton, J. L., S. W. Sherwood, and S. Y. Yang. 1981. Biochemical systematics of chaetodipine pocket mice, genus *Perognathus*. Journal of Mammalogy 62:477–492.

Patton, J. L., M. F. Smith, R. D. Price, and R. A. Hellenthal. 1984. Genetics of hybridization between the pocket gophers *Thomomys bottae* and *Thomomys townsendii* in northeastern California. The Great Basin Naturalist 44:431–440.

Patton, J. L., D. G. Huckaby, and S. T. Álvarez-Castañeda. 2007. The evolutionary history and a systematic revision of woodrats of the *Neotoma lepida* Group. University of California Publications in Zoology 135:1–411.

Patton, J. L., D. F. Williams, P. A. Kelly, B. L. Cypher, and S. E. Phillips. 2019. Geographic variation and evolutionary history of *Dipodomys nitratoides* (Rodentia: Heteromyidae), a species in severe decline. Journal of Mammalogy 100:1546–1563.

Paulson, D. D. 1988a. *Chaetodipus baileyi*. Mammalian Species 297:1–5.

Paulson, D. D. 1988b. *Chaetodipus hispidus*. Mammalian Species 320:1–4.

Pavan, A. C., and G. Marroig. 2016. Integrating multiple evidences in taxonomy: species diversity and phylogeny of mustached bats (Mormoopidae: *Pteronotus*). Molecular Phylogenetics and Evolution 103:184–198.

Pavan, A. C., and Y. G. Marroig. 2017. Timing and patterns of diversification in the Neotropical bat genus *Pteronotus*. Mormoopidae. Molecular Phylogenetics and Evolution 108:61–69.

Pavan, A. C., and V. da C. Tavares. 2020. *Pteronotus gymnonotus* (Chiroptera: Mormoopidae). Mammalian Species 52:40–48.

Paviolo, A., P. Crawshaw, A. Caso, T. de Oliveira, C. A. Lopez-Gonzalez, M. Kelly, C. De Angelo, and E. Payan. 2015. *Leopardus pardalis*. The IUCN Red List of Threatened Species 2015:e.T11509A97212355.

Pavlinov, I. Y., and A. A. Lissovsky. 2012. The mammals of Russia: A Taxonomic and Geographic Reference. KMK Scientific Press Ltd. Moscow, Russia.

Pech-Canché, J. M., C. MacSwiney, and E. Estrella 2010. Importancia de los detectores ultrasónicos para mejorar los inventarios de murciélagos Neotropicales. Therya 1:227–234.

Pelton, M. R. 2003. Black bear (*Ursus americanus*). Pp. 547–555, *in* Wild mammals of North America: biology, management, and conservation (Feldhamer, G. A., B. C Thompson, and J. A. Chapman, eds). Johns Hopkins University Press. Baltimore, U.S.A.

Pembleton, E. F., and S. L. Williams. 1978. *Geomys pinetis*. Mammalian Species 86:1–3.

Peppers, L., and R. D. Bradley. 2000. Cryptic species in *Sigmodon hispidus*: evidence from DNA sequence. Journal of Mammalogy 81:332–343.

Peppers, L. L., D. S. Carroll, and R. D. Bradley. 2002. Molecular systematics of the genus *Sigmodon* (Rodentia: Muridae): evidence from the mitochondrial cytochrome *b* gene. Journal of Mammalogy 83:396–407.

Pérez, E. M. 1992. *Agouti paca*. Mammalian Species 404:1–7.
Perez, S., P. C. de Grammont, and A. D. Cuarón. 2017. *Myotis fortidens*. The IUCN Red List of Threatened Species 2017:e.T14161A22056846.
Pérez, S. G., M. R. Jolón, J. E. Mérida, and A. J. Andino-Madrid. 2019. First record of the shrew *Cryptotis lacandonensis* (Eulipotyphla: Soricidae) for Guatemala. Therya 10:187–193.
Pérez-Hernandez, R., D. Brito, T. Tarifa, N. Cáceres, D. Lew, and S. Solari. 2016a. *Chironectes minimus*. The IUCN Red List of Threatened Species 2016:e.T4671A22173467.
Pérez-Hernandez, R., D. Lew, and S. Solari. 2016b. *Didelphis virginiana*. The IUCN Red List of Threatened Species 2016:e.T40502A22176259.
Pérez-Irineo, G., and A. Santos-Moreno. 2012. Diversidad de mamíferos terrestres de talla grande y media de la selva caducifolia del noreste de Oaxaca, México. Revista Mexicana de Biodiversidad 83:164–169.
Pérez-Montes, L. E., S. T. Álvarez-Castañeda, and C. Lorenzo. 2023. Current status of the *Peromyscus mexicanus* complex in Oaxaca, México. Therya 14:85–98.
Perini, F. A., C. A. M. Russo, and C. G. Schrago. 2010. The evolution of South American endemic canids: a history of rapid diversification and morphological parallelism. Journal of Evolutionary Biology 23:311–322.
Perry, E. A., G. B. Stenson, S. E. Bartlett, W. S. Davidson, and S. M. Carr. 2000. DNA sequence analysis identifies genetically distinguishable populations of harp seals (*Phoca groenlandicus*) in the northwest and northeast Atlantic. Marine Biology 137:53–58.
Peters, S. L., B. K. Lim, and M. D. Engstrom. 2002. Systematics of dog-faced bats (*Cynomops*) based on molecular and morphometric data. Journal of Mammalogy 83:1097–1110.
Peterson, A. T., L. Canseco-Márquez, J. L. Contreras-Jiménez, G. Escalona-Segura, O. Flores-Villela, J. García-López, B. Hernández-Baños, C. A. Jiménez-Ruíz, L. León-Paniagua, S. Mendoza-Amaro, A. G. Navarro-Sigüenza, V. Sánchez-Cordero, and D. E. Willard. 2004. A preliminary biological survey of Cerro Piedra Larga, Oaxaca, Mexico: Birds, mammals, reptiles, amphibians, and plants. Anales del Instituto de Biología, Universidad Nacional Autónoma de México, Serie Zoología 75:439–466.
Peterson, R. L. 1966. The mammals of eastern Canada. Oxford University Press. Toronto, Canada.
Peterson, R. S., and A. Symansky. 1963. First record of the Gaspé shrew from New Bnmswick. Journal of Mammalogy 44:278–279.
Peterson, R. S., C. L. Hubbs, R. L. Gentry, and R. L. Delong. 1968. The Guadalupe fur seal: habitat, behavior, population size, and field identification. Journal of Mammalogy 49:665–675.
Pfeiffer, C. J., and G. H. Gass. 1963. Note on the longevity and habits of captive *Cryptotis parva*. Journal of Mammalogy 44:427–428.
Phillips, C. J. 1971. The dentition of glossophagine bats: development, morphological characteristics, variation, pathology, and evolution. Miscellaneous Publications Museum of Natural History, University of Kansas 54:1–138.
Phillips, C. J., and J. K. Jones, Jr. 1971. A new species of the long-nosed bat, *Hylonycteris underwoodi*, from Mexico. Journal of Mammalogy 52:77–80.
Phillips, C. J., and J. K. Jones, Jr. 1969. Dental abnormalities in North American bats. 1. Emballonuridae, Noctilionidae, and Chilonycteridae. Transactions of the Kansas Academy of Sciences 71:509–520.
Phillips, C. J., G. W. Grimes, and G. L. Forman. 1977. Oral biology. Pp. 121–246, *in* Biology of bats of the New World family Phyllostomatidae, Part II (Baker, R. J., J. K. Jones, Jr., and D. C. Carter, eds.), Special Publications Museum, Texas Tech University 3:1–364.
Phillips, M. 2018. *Canis rufus* (errata version published in 2020). The IUCN Red List of Threatened Species 2018:e.T3747A163509841.
Phuong, M. A., M. C. W. Lim, D. R. Wait, K. C. Rowe, and C. Moritz. 2014. Delimiting species in the genus *Otospermophilus* (Rodentia: Sciuridae), using genetics, ecology, and morphology. Biological Journal of the Linnean Society 113:1136–1151.
Piaggio, A. J., and G. S. Spicer. 2001. Molecular phylogeny of the chipmunks inferred from mitochondrial cytochrome *b* and cytochrome oxidase II gene sequences. Molecular Phylogenetics and Evolution 20:335–350.
Piaggio, A. J., and S. L. Perkins. 2005. Molecular phylogeny of North American long-eared bats (Vespertilionidae: *Corynorhinus*); inter-and intra-specific relationships inferred from mitochondrial and nuclear DNA sequences. Molecular Phylogenetics and Evolution 37:762–775.
Piaggio, A. J., E. W. Valdez, M. A. Bogan, and G. S. Spicer. 2002. Systematics of *Myotis occultus* (Chiroptera: Vespertilionidae) inferred from sequences of two mitochondrial genes. Journal of Mammalogy 83:386–395.
Pine, R. H. 1972. The bats of the genus *Carollia*. Technical Monography, Texas A&M University, Texas Agriculture Experimental Station 8:1–125.
Pine, R. H. 1973. Anatomical and nomenclatural notes on opossums. Proceedings of the Biological Society of Washington 86:391–402.
Pine, R. H., D. C. Carter, and R. K. Laval. 1971. Status of *Bauerus* Van Gelder and its relationships to other nyctophiline bats. Journal of Mammalogy 52:663–669.
Pineda, J., and B. Rodriguez. 2015. *Eumops hansae*. The IUCN Red List of Threatened Species 2015:e.T8245A22026314.
Pino, J., R. Samudio Jr, J. F. González-Maya, and J. Schipper. 2020. *Bassariscus sumichrasti*. The IUCN Red List of Threatened Species 2020:e. T2613A166521324.
Pinto Sandoval, E. D., L. D. Rola, J. A. Morales-Donoso, S. Gallina, R. Reyna-Hurtado, and J. M. Barbanti Duarte. 2022. Integrative analysis of *Mazama temama* (Artiodactyla: Cervidae) and designation of a neotype for the species. Journal of Mammalogy 103:447–458.
Pitra, C., J. Fickel, E. Meijaard, and C. P. Groves. 2004. Evolution and phylogeny of Old World deer. Molecular Phylogenetics and Evolution 33:880–895.
Pizzimenti, J. J. 1975. Evolution of the prairie dog genus *Cynomys*. Occasional Papers Museum Natural History, University of Kansas 39:1–73.
Pizzimenti, J. J. 1976a. Genetic divergence and morphological convergence in the prairie dogs, *Cynomys gunnisoni* and *Cynomys leucurus*. I. Morphological and ecological analyses. Evolution 30:345–366.
Pizzimenti, J. J. 1976b. Genetic divergence and morphological convergence in the prairie dogs, *Cynomys gunnisoni* and *Cynomys leucurus*. II. Genetic analyses. Evolution 30:367–379.
Pizzimenti, J. J., and G. D. Collier. 1975. *Cynomys parvidens*. Mammalian Species 52:1–3.
Pizzimenti, J. J., and R. S. Hoffmann. 1973. *Cynomys gunnisoni*. Mammalian Species 25:1–4.
Pizzimenti, J. J., and C. F. Nadler. 1972. Chromosomes and serum proteins of the Utah prairie dog, *Cynomys parvidens* (Sciuridae). The Southwestern Naturalist 17:279–286.
Planz, J. V., E. G. Zimmerman, T. A. Spradling, and D. R. Akins. 1996. Molecular phylogeny of the *Neotoma floridana* species group. Journal of Mammalogy 77:519–535.
Plumpton, D., and J. K. Jones, Jr. 1992. *Rhynchonycteris naso*. Mammalian Species 413:1–5.

Pocock, R. I. 1920. On the external and cranial characters of the European badger (*Meles*) and of the American badger (*Taxidea*). Proceedings of the Zoological Society of London 1920:423–436.
Pocock, R. I. 1921. The external characters and classification of the Procyonidae. Proceedings of the Zoological Society of London 1921:389–422.
Pocock, R. I. 1928. The structure of the auditory bulla in the Procyonidae and the Ursidae with a note on the bulla of *Hyaena*. Proceedings of the Zoological Society of London 1928:963–974.
Pocock, R. I. 1941. The races of the ocelot and the margay. Field Museum of Natural History Zoological Series 27:319–369.
Poglayen-Neuwall, I. 1989. Procyonids. Pp. 450–468, *in* Grzimek's Encyclopedia of Mammals (Parker, S., ed.). McGraw-Hill. New York, U.S.A.
Poglayen-Neuwall, I., and D. E. Toweill. 1988. *Bassariscus astutus*. Mammalian Species 327:1–8.
Polaco, O. J., and R. Muñiz-Martínez. 1987. Los murciélagos de la costa de Michoacán, México. Anales de la Escuela Nacional de Ciencias Biológicas 31:63–89.
Polziehn, R. O., and C. A. Strobeck. 2002. Phylogenetic comparison of red deer and wapiti using mitochondrial DNA. Molecular Phylogenetics and Evolution 22:342–356.
Poole, K. G. 2003. A review of the Canada lynx, *Lynx canadensis*, in Canada. Canadian Field Naturalist 117:360–376.
Ports, M. A., and S. B. George. 1990. *Sorex preblei* in the northern Great Basin. Great Basin Naturalist 50:93–95.
Porter, C. A., N. E. Beasley, N. Ordóñez-Garza, L. L. Lindsey, D. S. Rogers, N. Lewis-Rogers, J. W. Sites, and R. D. Bradley. 2017. A new species of big-eared climbing rat, genus *Ototylomys* (Cricetidae: Tylomyinae), from Chiapas, Mexico. Journal of Mammalogy 98:1310–1329.
Porter, C. A., S. R. Hoofer, C. A. Cline, F. G. Hoffmann, and R. J. Baker. 2007. Molecular phylogenetics of the phyllostomid bat genus *Micronycteris* with descriptions of two new subgenera. Journal of Mammalogy 88:1205–1215.
Powell, R. A. 1981. *Martes pennanti*. Mammalian Species 156:1–6.
Powell, R. A. 1993. The Fisher: Life History, Ecology, and Behavior. University of Minnesota Press. Minneapolis, U.S.A.
Powell, J. A., and G. B. Rathbun. 1984. Distribution, and abundance of manatees along the northern coast of the Gulf of Mexico. Northeast Gulf Science 7:1–97.
Presley, S. J. 2000. *Eira barbara*. Mammalian Species 636:1–6.
Prestrud, P. 1991. Adaptations by the arctic fox (*Alopex lagopus*) to the polar winter. Arctic 44:132–138.
Puckett, E. E., P. D. Etter, E. A. Johnson, and L. S. Eggert. 2015. Phylogeographic analyses of American black bears (*Ursus americanus*) suggest four glacial refugia and complex patterns of postglacial admixture. Molecular Biology and Evolution 32:2338–2350.
Quakenbush, L., J. Citta, and J. Crawford. 2009. Biology of the spotted seal (*Phoca largha*) in Alaska from 1962 to 2008. Report to National Marine Fisheries Service. Fairbanks, Alaska.
Quigley, H., R. Foster, L. Petracca, E. Payan, R. Salom, and B. Harmsen. 2017. *Panthera onca*. The IUCN Red List of Threatened Species 2017:e. T15953A123791436.
Qumsiyeh, M. B., C. Sanchez-Hernandez, S. K. Da-Vis, J. C. Patton, and R. J. Baker. 1988. Chromosomal evolution in *Geomys* as revealed by G- and C-band analysis. The Southwestern Naturalist 33:1–13.
Rachlow, J., Becker, P. A., and L. Shipley. 2016. *Brachylagus idahoensis*. The IUCN Red List of Threatened Species 2016:e.T2963A45176206.
Rachlow, J. L., P. A. Becker, and L. A. Shipley. 2018. *Brachylagus idahoensis*. Pp. 87–90, *in* Pikas, rabbits and hares of the world (Smith, A., C. H. Johnston, P. C. Alves, and K. Hackländer, eds.). The Johns Hopkins University Press/IUCN/SSC. Baltimore, U.S.A.
Rafinesque, C. S. 1817. Descriptions of seven new genera of North American quadrupeds. The American Monthly Magazine and Critical Review 2:44–46.
Rainey, D. G., and R. H. Baker. 1955. The pigmy woodrat, *Neotoma goldmani*, its distribution and systematic position. University of Kansas Publications, Museum of Natural History 7:619–624.
Ralls, K., J. Ballou, and R. L. Brownell, Jr. 1983. Genetic diversity in California sea otters: theoretical considerations and management implications. Biological Conservation 25:209–232.
Raman, T. R. S. 2013. The Chital (*Axis axis* Erxleben). University Press.
Ramey, C. A., and D. J. Nash. 1976. Geographic variation in Abert's squirrel (*Sciurus aberti*). The Southwestern Naturalist 21:135–139.
Ramey II, R. R., H.-P. Liu, C. W. Epps, L. M. Carpenter, and J. D. Wehausen. 2005. Genetic relatedness of the Preble's meadow jumping mouse (*Zapus hudsonius preblei*) to nearby subspecies of *Z. hudsonius* as inferred from variation in cranial morphology, mitochondrial DNA and microsatellite DNA: implications for taxonomy and conservation. Animal Conservation 8:329–346.
Ramírez-Chaves, H. E., M. Alarcón Cifuentes, E. A. Noguera-Urbano, W. A. Pérez., M. M. Torres-Martínez, P. A. Ossa-López, F. A. Rivera-Páez, and D. M. Morales-Martínez. 2023. Systematics, morphometrics, and distribution of *Eptesicus fuscus miradorensis*, with notes on baculum morphology and natural history. Therya 14:299–311.
Ramírez-Pulido, J. 1969. Contribución al conocimiento de los mamíferos del Parque Nacional "Lagunas de Zempoala", Morelos, México. Anales del Instituto de Biología, Universidad Nacional Autónoma de México, Serie Zoología 40:253–290.
Ramírez-Pulido, J., and C. Sánchez-Hernández. 1971. *Tylomys nudicaudus* from the Mexican States of Puebla and Guerrero. Journal of Mammalogy 52:481.
Ramírez-Pulido, J., and C. Sánchez-Hernández. 1972. Regurgitaciones de lechuzas, procedentes del cañon del Zopilote, Guerrero, México. Revista de la Sociedad Mexicana de Historia Natural 33:107–112.
Ramírez-Pulido, J., R. López Wilchis, C. Müdespacher and I. E. Lira. 1983. Lista y bibliografia reciente de los mamíferos de México. Universidad Autónoma Metropolitana, Iztapalapa. Distrito Federal, México.
Ramírez-Pulido, J., M. C. Brition, A. Perdomo, and A. Castro. 1986. Guia de los mamiferos de México. Universidad Autonoma Metropolitana. Distrito Federal, Méxco.
Ramírez-Pulido, J., N. González-Ruiz, A. L. Gardner, and J. Arroyo-Cabrales. 2014. List of recent land mammals of Mexico, 2014. Museum of Texas Tech University Special Publications 63:1–69.
Randi, E., N. Mucci, M. Pierpaoli, and E. Douzery. 1998. New phylogenetic perspectives on the Cervidae (Artiodactyla) are provided by the mitochondrial cytochrome b gene. Proceedings of the Royal Society of London. Series B: Biological Sciences 265:793–801.
Rausch, R. L. 1953. On the status of some Arctic mammals. Arctic 6:91–148.
Rausch, R. E. 1977. [On the zoogeography of some Beringian mammals.] In Advances in modern theriology: 162–177. Academy of Sciences USSR, Moscow.

Rausch, R. L., and V. R. Rausch. 1968. On the biology and systematic position of *Microtus abbreviatus* Miller, a vole endemic to the St. Matthew Islands, Bering Sea. Zeitschrift für Säugetierkunde 33:65–99.
Rausch, R. L., and V. R. Rausch. 1971. The somatic chromosomes of some North American marmots (Sciuridae), with remarks on the relationships of *Marmota broweri* Hall and Gilmore. Mammalia 35:85–101.
Rausch, R. L., and V. R. Rausch. 1972. Observations on chromosomes of *Dicrostonyx torquatus stevensoni* Nelson and chromosomal diversity in varying lemmings. Zeitschrift für Säugetierkunde 33:65–69.
Rausch, V. R., and R. L. Rausch. 1974. The chromosome complement of the yellow-cheeked vole *Microtus xanthognathus* (Leach). Canadian Journal Genetic and Cytology 16:267–272.
Rausch, R. L., and V. R. Rausch. 1975. Relationships of the red-backed vole, *Clethrionomys rutilus* (Pallas) in North America: Karyotypes of the subspecies *dawsoni* and *albiventer*. Systematic Biology 24:163–170.
Rausch, R. L., and V. R. Rausch. 1997. Evidence for specific independence of the shrew (Mammalia: Soricidae) of St. Paul Island (Pribilov Islands, Bering Sea). Zeitschrift für Säugetierkunde 62:193–202.
Rausch, R. L., J. E. Feagin, and V. R. Rausch. 2007. *Sorex rohweri* sp. nov. (Mammalia, Soricidae) from northwestern North America. Mammalian Biology 72:93–105.
Reddell, J. R. 1968. The hairy-legged vampire, *Diphylla ecaudata,* in Texas. Journal of Mammalogy 49:769.
Redford, K. H., and J. F. Eisenberg. 1992. Mammals of the Neotropics. The southern cone: Chile, Argentina, Uruguay, Paraguay. The University of Chicago Press, Chicago, U.S.A.
Redondo, R. A. F., L. P. S. Brina, R. F. Silva, A. D. Ditchfield, and F. R. Santos. 2008. Molecular systematics of the genus *Artibeus* (Chiroptera: Phyllostomidae). Molecular Phylogenetics and Evolution 49:44–58.
Reducker, D. W., T. L. Yates, and I. F. Greenbaum. 1983. Evolutionary affinities among southwestern long-eared Myotis (Chiroptera: Vespertilionidae). Journal of Mammalogy 64:666–677.
Reed, C. A. 1951. Locomotion and appendicular anatomy in three soricoid insectivores. The American Midland Naturalist 45:513–671.
Reed, E. K. 1955. Bison beyond the Pecos. Texas Journal of Sciences 7:130–135.
Reed, K. M., and J. R. Choate. 1986. Natural history of the plains pocket mouse in agriculturally disturbed sandsage prairie. Prairie Naturalist 18:79–90.
Reich, L. M. 1981. *Microtus pennsylvanicus*. Mammalian Species 159:1–8.
Reid, F. A. 1997. A field guide to the mammals of Central America and southeast Mexico. Oxford University Press. New York, U.S.A.
Reid, F. A. 2006. A field guide to mammals of North America north of Mexico. Peterson Field Guides. Houghton Mifflin Company. New York, U.S.A.
Reid, F. A. 2009. A field guide to the mammals of Central America and Southeast Mexico. Oxford University Press. New York, U.S.A.
Reid, F. 2016a. *Sciurus variegatoides*. The IUCN Red List of Threatened Species 2016:e.T20024A22246448.
Reid, F. 2016b. *Sorex maritimensis*. The IUCN Red List of Threatened Species 2016:e.T136779A22312357.
Reid, F., and J. Pino. 2016. *Peromyscus mexicanus*. The IUCN Red List of Threatened Species 2016:e.T16681A22363818.
Reid, F., and R. Timm. 2016. *Ototylomys phyllotis*. The IUCN Red List of Threatened Species 2016:e.T15666A115129272.
Reid, F., and E. Vázquez. 2016a. *Baiomys musculus*. The IUCN Red List of Threatened Species 2016:e.T2465A115062118.
Reid, F., and E. Vázquez. 2016b. *Handleyomys rostratus*. The IUCN Red List of Threatened Species 2016:e.T15612A115128466.
Reid, F., and E. Vázquez. 2016c. *Heteromys gaumeri*. The IUCN Red List of Threatened Species 2016:e.T10007A22223646.
Reid, F., and E. Vázquez. 2016d. *Heteromys pictus*. The IUCN Red List of Threatened Species 2016:e.T12075A22225256.
Reid, F., and E. Vázquez. 2016e. *Reithrodontomys microdon*. The IUCN Red List of Threatened Species 2016:e.T19412A22386904.
Reid, F., E. Vázquez, and L. Emmons. 2008a. *Habromys lophurus*. The IUCN Red List of Threatened Species2008:e.T9610A13004630.
Reid, F., E. Vázquez, L. Emmons, and T. McCarthy. 2008b. *Handleyomys saturatior*. The IUCN Red List of Threatened Species 2008:e. T15613A4908339.
Reid, F., E. Vázquez, L. Emmons, and A. D. Cuarón. 2008c. *Rheomys thomasi*. The IUCN Red List of Threatened Species 2008:e.T19486A8909070.
Reid, F., K. Helgen, and A. Kranz. 2016a. *Mustela erminea*. The IUCN Red List of Threatened Species 2016:e.T29674A45203335.
Reid, F., J. Pino, and R. Samudio. 2016b. *Scotinomys teguina*. The IUCN Red List of Threatened Species 2016:e.T20052A22390040.
Reid, F., R. Samudio, and J. Pino. 2016c. *Reithrodontomys sumichrasti*. The IUCN Red List of Threatened Species 2016:e.T19417A115151678.
Reid, F., M. Schiaffini, and J. Schipper. 2016d. *Neovison vison*. The IUCN Red List of Threatened Species 2016:e.T41661A45214988.
Reid, F., J. Schipper, and R. Timm. 2016e. *Bassariscus astutus*. The IUCN Red List of Threatened Species 2016:e.T41680A45215881.
Reid, F., E. Vázquez, and L. Emmons. 2016f. *Reithrodontomys gracilis*. The IUCN Red List of Threatened Species 2016:e.T19408A115150933.
Repenning, C. A. 1967. Subfamilies and genera of the Soricidae. Geological Survey Paper 565:1–74.
Repenning, C. A., R. S. Peterson, and C. L. Hubbs. 1971. Contributions to the systematics of the southern fur seals, with particular reference to the Juan Fernández and Guadalupe species. Pp. 1–34, *in* Antarctic Pinnipedia, Antarctic Research Series 18, (Burt, W. H., ed.). American Geophysical Union. New York, U.S.A.
Rezsutek, M., and G. N. Cameron. 1993. *Mormoops megalophylla*. Mammalian Species 448:1–5.
Rheingantz, M. L., and C. S. Trinca. 2015. *Lontra longicaudis*. The IUCN Red List of Threatened Species 2015:e.T12304A21937379.
Rheingantz, M. L., P. Rosas-Ribeiro, J. Gallo-Reynoso, V. C. Fonseca da Silva, R. Wallace, V. Utreras, and P. Hernández-Romero. 2021. *Lontra longicaudis*. The IUCN Red List of Threatened Species 2021:e.T12304A164577708.
Rhoads, S. N. 1894. Description of a New Genus and Species of Arvicoline Rodent from the United States. American Naturalist 28:182–185.
Rhymer, J. M., J. M. Barbay, and H. L. Givens. 2004. Taxonomic relationship between *Sorex dispar* and *S. gaspensis*: inferences from mitochondrial DNA sequences. Journal of Mammalogy 85:331–337.
Rice, D. W., K. Kenyon, and B. D. Lluch 1965. Pinniped populations at Islas Guadalpe, San Benito, and Cedros, Baja California, in 1965. Transactions of the San Diego Society of Natural History 14:73–84.
Rice, D. W. 1998. Marine Mammals of the World: Systematics and Distribution. Society for Marine Mammalogy, Special Publication Number 4. Lawrence, U.S.A.
Rickart, E. A. 1982. Annual cycles of activity and body composition in *Spermophilus townsendii mollis*. Canadian Journal of Zoology 60:3298–3306.
Rickart, E. A. 1986. Postnatal growth of the Piute ground squirrel (*Spermophilus mollis*). Journal of Mammalogy 67:412–416.
Rickart, E. A. 1987. *Spermophilus townsendii*. Mammalian Species 268:1–6.

Rickart, E. A., and P. B. Robertson. 1985. *Peromyscus melanocarpus*. Mammalian Species 241:1–3.
Rickart, E. A., and E. Yensen. 1991. *Spermophilus washingtoni*. Mammalian Species 371:1–5.
Rickart, E. A., L. R. Heaney, and R. S. Hoffman. 2004. First record of *Sorex tenellus* from the central Great Basin. The Southwestern Naturalist 49:132–134.
Riddle, B. R. 1999. Mearns's grasshopper mouse *Onychomys arenicola*. Pp. 588, *in* The Smithsonian book of the North American mammals (Wilson, D., and S. Ruff, eds.). Smithsonian Institution Press. Washington, U.S.A.
Riddle, B. R., D. J. Hafner, and L. F. Alexander. 2000a. Comparative phylogeography of Baileys' pocket mouse (*Chaetodipus baileyi*) and the *Peromyscus eremicus* species group: historical vicariance of the Baja California Peninsular desert. Molecular Phylogentic and Evolution 17:161—172.
Riddle, B. R., D. J. Hafner, and L. F. Alexander. 2000b. Phylogeography and systematics of *Peromyscus eremicus* species group and historical biogeography of North American warm regional deserts. Molecular Phylogentic and Evolution 17:145—160.
Riddle, B. R., T. Jezkova, M. E. Eckstut, V. Oláh-Hemmings, and L. N. Carraway. 2014. Cryptic divergence and revised species taxonomy within the Great Basin pocket mouse, *Perognathus parvus* (Peale, 1848), species group. Journal of Mammalogy 95:9–25.
Rideout, C. B., and R. S. Hoffmann. 1975. *Oreamnos americanus*. Mammalian Species 63:1–6.
Riley, G. A., and R. T. McBride. 1975. A survey of the red wolf (*Canis rufus*). Pp. 263–277, *in* The wild canids (Fox, M. W., ed.), Van Nostrand Reinhold. New York, U.S.A.
Riley S. P. D., E. E. Boydston, K. R. Crooks, and L. M. Lyren. 2010. Bobcats (*Lynx rufus*). Pp. 121–140, *in* Urban carnivores: ecology, conflict, and conservation (Gehrt, S. D., S. P. D. Riley, and B. L. Cypher, eds). The Johns Hopkins Univeristy Press. Baltimore, U.S.A.
Rios, E., and S. T. Álvarez-Castañeda. 2010. Phylogeography and systematics of the San Diego pocket mouse (*Chaetodipus fallax*). Journal of Mammalogy 91:293–301.
Rios, E., and S. T. Álvarez-Castañeda. 2011. *Peromyscus guardia*. Mammalian Species 43:172–176.
Rios, E., and S. T. Álvarez-Castañeda. 2013. Nomenclatural change of *Chaetodipus dalquesti*. Western North America Naturalist 73:399–400.
Rios, E., C. Lorenzo, and S. T. Álvarez-Castañeda. 2016. Genetic evaluation of a microendemic and threatened rodent (*Heteromys nelsoni*) from Southern Mexico. Mammalia 81:289–296.
Risch, D., C. W. Clarke, P. J. Cockeron, A. Elepfandt, K. M. Kovacs, C. Lydersen, I. Stirling, and S. M. Van Parijs. 2007. Vocalizations of male bearded seals, *Erignathus barbatus*: classification and geographical variation. Animal Behavior 73:747–762.
Rizo-Aguilar, A., C. Delfín-Alfonso, A. González-Romero, and J. A. Guerrero. 2016. Distribution and density of the zacatuche rabbit (*Romerolagus diazi*) at the Protected Natural Area "Corredor Biológico Chichinautzin". Therya 7:333–342.
Rivas-Camo, N. A., P. A. Sabido-Villanueva, C. R. Peralta-Muñoz, and R. A. Medellin. 2020. Cuba in Mexico: first record of *Phyllops falcatus* (Gray, 1839) (Chiroptera, Phyllostomidae) for Mexico and other new records of bats from Cozumel, Quintana Roo. ZooKeys 973:153–162.
Roach, N. 2016. *Microdipodops megacephalus*. The IUCN Red List of Threatened Species 2016:e.T42606A115193770.
Roach, N. 2017a. *Marmota vancouverensis*. The IUCN Red List of Threatened Species 2017:e.T12828A22259184.
Roach, N. 2017b. *Chaetodipus ammophilus*. The IUCN Red List of Threatened Species 2017:e.T96812094A22226639.
Roach, N. 2018a. *Geomys tropicalis*. The IUCN Red List of Threatened Species 2018:e.T9056A22218038.
Roach, N. 2018b. *Cynomys parvidens*. The IUCN Red List of Threatened Species 2018:e.T6090A22260975.
Roach, N. 2018c. *Dipodomys ingens*. The IUCN Red List of Threatened Species 2018:e.T6678A22227241.
Roach, N. 2018d. *Dipodomys nitratoides*. The IUCN Red List of Threatened Species 2018:e.T6683A22228395.
Roach, N. 2018e. *Dipodomys stephensi*. The IUCN Red List of Threatened Species 2018:e.T6682A22228640.
Roach, N. 2020. *Microtus breweri*. The IUCN Red List of Threatened Species 2020:e.T13417A22349291.
Roach, N., and L. Naylor. 2016a. *Cryptotis obscura*. The IUCN Red List of Threatened Species 2016:e.T136462A22286153.
Roach, N., and L. Naylor. 2016b. *Xerospermophilus mohavensis*. The IUCN Red List of Threatened Species 2016:e.T20474A22266305.
Roach, N., and L. Naylor. 2017. *Cryptotis griseoventris*. The IUCN Red List of Threatened Species 2017:e.T48269619A123794519.
Roberts, J. D., and G. Mills. 1983. Funny lookin' rats. Texas Parks and Wildlife 41:12–15.
Roberts, H. R., D. J. Schmidly, and R. D. Bradley. 1988. *Peromyscus spicilegus*. Mammalian Species 596:1–4.
Roberts, H. R., D. J. Schmidly, and R. D. Bradley. 2001. *Peromyscus simulus*. Mammalian Species 669:1–3.
Robertson, P. B., and G. G. Musser. 1976. A new species of *Peromyscus* (Rodentia: Cricetidae), and a new specimen of *P. simulatus* from southern Mexico, with comments on their ecology. Occasional Papers of the Museum of Natural History, the University of Kansas 47:1–8.
Robertson, P. B., and E. A. Rickart. 1976. *Cryptotis magna*. Mammalian Species 61:1–2.
Robinson, J. W., and R. S. Hoffmann. 1975. Geographical and interspecific cranial vanation in big-eared ground squirrels (*Spermophilus*): a multivariate study. Systematic Zoology 24:79–88.
Robinson, T. J., F. F. B. Elder, and J. A. Chapman. 1983. Evolution of chromosomal variation in cottontails, genus *Sylvilagus* (Mammalia: Lagomorpha): *S. aquaticus, S. florindanus*, and *S. transitionalis*. Cytogenetics and Cell Genetics 35:216–222.
Robinson, T. J., F. F. B. Elder, and J. A. Chapman. 1984. Evolution of chromosomal variation in cottontails, genus *Sylvilagus* (Mammalia: Lagomorpha). II. *Sylvilagus audubonii*, *S. idahoensis*, *S. nuttallii* and *S. palustris*. Cytogenetics and Cell Genetics 38:282–289.
Rodriguez, B., and J. Cajas. 2015. *Dermanura tolteca*. The IUCN Red List of Threatened Species 2015:e.T2140A21997479.
Rodriguez, B., and B. Miller. 2015. *Cynomops mexicanus*. The IUCN Red List of Threatened Species 2015:e.T136611A21987867.
Rodriguez, B., and W. Pineda. 2015. *Macrophyllum macrophyllum*. The IUCN Red List of Threatened Species 2015:e.T12615A22025883.
Rodriguez, R. M., and L. K. Ammerman, 2004. Mitochondrial DNA divergence does not reXect morphological diVerence between *Myotis californicus* and *Myotis ciliolabrum*. Journal of Mammalogy 85:842–851.
Roe, F. G. 1970. The North American buffalo. Second edition, University of Toronto Press, Toronto, Canada.
Røed, K. H., M. A. D. Ferguson, M. Çrête, and T. A. Bergerud. 1991. Genetic variation in transferrin as a predictor for differentiation and evolution of caribou from Eastern Canada. Rangifer 11:65–74.
Roehrs, Z. P., J. B. Lack, and R. A. Van Den Bussche. 2010. Tribal phylogenetic relationships within Vespertilioninae (Chiroptera: Vespertilionidae) based on mitochondrial and nuclear sequence data. Journal of Mammalogy 91:1073–1092.
Roemer, G., B. Cypher, and R. List. 2016. *Urocyon cinereoargenteus*. The IUCN Red List of Threatened Species 2016:e.T22780A46178068.
Roest, A. I. 1964. A ribbon seal from California. Journal of Mammalogy 45:416–420.

Rogers, D. S. 1989. Evolutionary implications of chromosomal variation among spiny pocket mice, genus *Heteromys* (Order Rodentia). The Southwestern Naturalist 34:85–100.
Rogers, D. S. 1990. Genetic evolution, historial biogeography, and systematic relationships amongs spiny pocket mice (subfamily Heteromynae). Journal of Mammalogy 71:668–685.
Rogers, D. S., and J. E. Rogers. 1992. *Heteromys nelsoni*. Mammalian Species 397:1–2.
Rogers, D. S., and J. A. Skoy. 2011. *Peromyscus furvus*. Mammalian Species 43:209–215.
Rogers, D. S., and M. W. González. 2010. Phylogenetic relationships among spiny pocket mice (*Heteromys*) inferred from mitochondrial and nuclear sequence data. Journal of Mammalogy 91:914–930.
Rogers, D. S., and D. J. Schmidly. 1982. Systematics of spiny pocket mice (genus *Heteromys*) of the *desmarestianus* species group from Mexico and northern Central America. Journal of Mammalogy 63:375–386.
Rogers, D. S., and V. L. Vance. 2005. Phylogenetics of Spiny Pocket Mice (Genus *Liomys*): analysis of cytochrome b based on multiple heuristic approaches. Journal of Mammalogy 86:1085–1094.
Rogers, D. S., E. J. Heske, and D. A. Coon. 1983. Karyotype and a range extension of *Reithrodontomys tenuirostris*. The Southwestern Naturalist 21:372–374.
Rogers, D. S., R. N. Leite, and R. J. Reed. 2011. Molecular phylogenetics of an endangered species: the Tamaulipan woodrat (*Neotoma angustapalata*). Conservation Genetics 12:1035–1048.
Rogers, M. A. 1991a. Evolutionary differentiation within the northern great basin pocket gopher, *Thomomys townsendii*. I. Morphological variation. The Great Basin Naturalist 51:109–126.
Rogers, M. A. 1991b. Evolutionary differentiation within the northern great basin pocket gopher, *Thomomys townsendii*. II. Genetic variation and biogeographic considerations. The Great Basin Naturalist 51:127–152.
Romo-Vázquez, E., L. León, and O. Sánchez. 2005. A new species of *Habromys* (Rodentia: Sigmodontinae) from México. Proceedings of the Biological Society of Washington 118:605–611.
Roots, E. H., and R. J. Baker. 1998. *Rhogeessa genowaysi*. Mammalian Species 589:1–3.
Roots, E. H., and R. J. Baker. 2007. *Rhogeessa parvula*. Mammalian Species 804:1–4.
Rose, R. K. 1981. *Synaptomys* not extinct in the Dismal Swamp. Journal Mammalogy 62:844–845.
Rose, R. K., and A. V. Linzey. 2021. *Synaptomys cooperi* (Rodentia: Cricetidae). Mammalian Species 53:95–111.
Rossi, R. V., R. S. Voss, and D. P. Lunde. 2010. A revision of the didelphid marsupial genus *Marmosa*. Part 1. The species in Tate's "*mexicana*" and "*mitis*" sections and other closely related forms. Bulletin of the American Museum of Natural History 334:1–83.
Rostland, E. 1960. The geographic range of the historic bison in the southeast. Annals Association American Geographers 50:395–407.
Roth, E. L. 1976. A new species of pocket mouse (*Perognathus:* Heteromyidae) from the cape region of Baja California Sur, Mexico. Journal of Mammalogy 57:562–566.
Rowe, D. L., and R. L. Honeycutt. 2002. Phylogenetic relationships, ecological correlates, and molecular evolution within the Cavioidea (Mammalia, Rodentia). Molecular Biology and Evolution 19:263–277.
Ruedas, L. A. 1986. Chromosomal variability in the New England cottontail, *Sylvilagus transitionalis* (Bangs)[sic], 1895 with evidence of recognition of a new species. Master Thesis. Fordham University. New York, U.S.A.
Ruedas, L. A. 1998. Systematics of *Sylvilagus* Gray, 1867 (Lagomorpha: Leporidae) from Southwestern North America. Journal of Mammalogy 79:1355–1378.
Ruedas, L.A. 2016. *Rattus norvegicus* (errata version published in 2020). The IUCN Red List of Threatened Species 2016: e.T19353A165118026.
Ruedas, L. A. 2018a. *Sylvilagus cognatus*. Pp. 130–131, *in* Pikas, rabbits and hares of the world (Smith, A., C. H. Johnston, P. C. Alves, and K. Hackländer, eds.). The Johns Hopkins University Press/IUCN/SSC. Baltimore, U.S.A.
Ruedas, L. A. 2018b. *Sylvilagus gabbi*. Pp. 140–142, *in* Pikas, rabbits and hares of the world (Smith, A., C. H. Johnston, P. C. Alves, and K. Hackländer, eds.). The Johns Hopkins University Press/IUCN/SSC. Baltimore, U.S.A.
Ruedas, L. A., and R. C. Dowler. 2018. *Sylvilagus robustus*. Pp. 154–155, *in* Pikas, rabbits and hares of the world (Smith, A., C. H. Johnston, P. C. Alves, and K. Hackländer, eds.). The Johns Hopkins University Press/IUCN/SSC. Baltimore, U.S.A.
Ruedas, L. A., and J. Salazar-Bravo. 2007. Morphological and chromosomal taxonomic assessment of *Sylvilagus brasiliensis gabbi* (Leporidae). Mammalia 71:63–69.
Ruedas, L., and A. T. Smith. 2019a. *Sylvilagus gabbi*. The IUCN Red List of Threatened Species 2019:e.T87491157A87491160.
Ruedas, L. A., and A. T. Smith. 2019b. *Sylvilagus robustus*. The IUCN Red List of Threatened Species 2019:e.T41310A165116781.
Ruedas, L. A., R. C. Dowler, and E. Aita. 1989. Chromosomal variation in the New England cottontail, *Sylvilagus transitionalis*. Journal of Mammalogy 70:860–864.
Ruedas, L. A., *et al.* 2017. A prolegomenon to the systematics of South American cottontail rabbits (Mammalia, Lagomorpha, Leporidae: *Sylvilagus*): designation of a neotype for *S. brasiliensis* (Linnaeus, 1758), and restoration of *S. andinus* (Thomas, 1897) and *S. tapetillus* Thomas, 1913. Miscellaneous Publications Museum of Zoology, University of Michigan 205:1–67.
Ruedas, L. A., L. López, and J. M. Mora. 2023. A propaedeutic to the taxonomy of the Eastern cottontail rabbit (Lagomorpha: Leporidae: *Sylvilagus floridanus*) from Central America. Therya 14:99–119.
Ruedi, M., and F. Mayer. 2001. Molecular systematics of bats of the genus *Myotis* (Vespertilionidae) suggests deterministic ecomorphological convergences. Molecular Phylogenetics and Evolution 21:436–448.
Ruiz-Soberanes, J. A., and G. Gómez-Álvarez. 2010. Estudio mastofaunístico del Parque Nacional Malinche, Tlaxcala, México. Therya 1:97–110.
Ruschi, A. 1951. Morcegos do Estado do Espírito Santo: Descricão de *Diphylla ecaudata* Spix e algumas observacões a seu respeito. Boletin do Museu de Biologia. "Prof. Mello Leitão" (S. Teresa, Brasil) 3:1–7.
Russell, R. J. 1968a. Evolution and classification of the pocket gophers of the subfamily Geomyinae. University of Kansas Publications, Museum of Natural History 16:473–579.
Russell, R. J. 1968b. Revision of the pocket gophers of the genus *Pappogeomys*. University of Kansas Publications, Museum of Natural History 16:581–776.

Ryan, J. M. 1989. Comparative myology and phylogenetic systematics of the Heteromydae (Mammalia, Rodentia). Miscellaneous Publications, Museum of Zoology, University of Michigan 176:1–103.

Ryan, R. M. 1968. Mammals of Deep Canyon, Colorado-Desert, California. The Desert Museum, Palm Springs, California, U.S.A.

Rydell, J., H. T. Arita, M. Santos, and J. Granados 2002. Acoustic identification of insectivorous bats (order Chiroptera) of Yucatan, Mexico. Journal of the Zoological Society of London 257:27–36.

Sakahira, F., and M. Niimi. 2007. Ancient DNA Analysis of the Japanese Sea Lion (*Zalophus japonicus japonicus* Peters, 1866): Preliminary results using mitochondrial control-region sequences. Zoological Sciences 24:81–85.

Sampaio, E., B. Lim, and S. Peters. 2016a. *Diphylla ecaudata*. The IUCN Red List of Threatened Species 2016:e.T6628A22040157.

Sampaio, E., B. Lim, and S. Peters. 2016b. *Micronycteris schmidtorum*. The IUCN Red List of Threatened Species 2016:e.T13383A22124156.

Sampaio, E., B. Lim, S. Peters, B. Miller, A. D. Cuarón, and P. C. de Grammont. 2016c. *Lophostoma brasiliense*. The IUCN Red List of Threatened Species 2016:e.T21984A115164165.

Sampaio, E., B. Lim, and S. Peters. 2017. *Molossus coibensis*. The IUCN Red List of Threatened Species 2017:e.T102208365A22106904.

Samudio, R., J. Pino, and F. Reid. 2016. *Nyctomys sumichrasti*. The IUCN Red List of Threatened Species 2016:e.T14999A115124642.

Samudio, R., Jr. 2016. Family Cuniculidae (Pacas). Pp. 398–405, *in* Handbook of the mammals of the world 6. Lagomorpha and Rodentia I (Wilson, D. E., T. E. Lacher, Jr., and R. A. Mittermeier, eds.). Lynx edicions. Barcelona, Spain.

Sanborn, C. C. 1932. The bats of the genus *Eumops*. Journal of Mammalogy 13:347–357.

Sanborn, C. C. 1936. Records and measurements of Neotropical bats. Field Museum of Natural History, Zoological series 20:93–106.

Sanborn, C. C. 1937. American bats of the subfamily Emballonurinae. Field Museum of Natural History, Zoological series 20:321–354.

Sanborn, C. C. 1941. Descriptions and records of neotropical bats. Field Museum of Natural History, Zoology Series 27:371–387.

Sanborn, C. C. 1949. Bats of the genus *Micronycteris* and its sub-genera. Fieldiana Zoology 31:215–233.

Sánchez-Casas, N., and T. Álvarez. 2000. Palinofagia de los murciélagos del género *Glossophaga* (Mammalia: Chiroptera) en México. Acta Zoológica Mexicana (n. s.) 81:23–62.

Sánchez-Cordero, V., C. Bonilla, and E. Cisneros. 1993. Thomas' mastiff bat, *Promops centralis* (Molossidae) in Oaxaca, Mexico. Bat Reserch News 34:65.

Sánchez-Cordero, V., L. Guevara, C. Lorenzo, and S. Ortega-García. 2014. Noteworthy records of an endemic shrew from Mexico (Mammalia, Soricomorpha, *Cryptotis griseoventris*), with comments on taxonomy. Mammalia 78:405–408.

Sánchez-Hernández, C., M. de la L. Romero-Almaraz, and C. García-Estrada. 2005. Mamíferos. Pp. 283–303, *in* Biodiversidad del estado de Tabasco, México (Bueno, J., F. Álvarez, and S. Santiago, eds.). Instituto de Biología, UNAM–CONABIO. Distrito Federal, México.

Sánchez-Hernández, C., and G. Gavino-de la Torre. 1988. Registros de murciélagos para la Isla de Pefia, Nayarit, México. Anales del Instituto de Biologia, Universidad Nacional Autónoma de México, Serie Zoología 58:939–940.

Sánchez-Hernández, C., and M. de la L. Romero-Almaraz. 1995. Murciélagos de Tabasco y Campeche, una propuesta para su conservación. Cuadernos del Instituto de Biología, Universidad Nacional Autónoma de México 24:1–213.

Sánchez-Hernández, C., M. de la L. Romero-Almaraz, R. D. Owen, A. Núñez-Garduño, and R. López-Wilchis. 1999. Noteworthy records of mammals from Michoacan, Mexico. The Southwestern Naturalist 44:231–235.

Sánchez-Hernández, C., G. D. Schnell, and M. de la L. Romero-Almaraz. 2009. *Peromyscus perfulvus* (Rodentia: Cricetidae). Mammalian Species 833:1–8.

Sánchez-Hernández, C., M. de la L. Romero-Almaraz, G. D. Schnell, M. L. Kennedy, T. L. Best, R. D. Owen, and C. López-González. 2002. Bats of Colima, Mexico: new records, geographic distribution, and reproductive condition. Occasional Papers Sam Noble, Oklahoma Museum of Natural History 12:1–23.

Sanchez-Herrera, O., G. Tellez-Giron, R. A. Medellín, and G. Urbano-Vidales. 1986. New records of mammals from Quintana Roo. Mammalia 50:275–278.

Sanchez-Rojas, G., and S. Gallina-Tessaro. 2016. *Odocoileus hemionus*. The IUCN Red List of Threatened Species 2016:e.T42393A22162113.

Santos-Moreno, A., and A. M. Alfaro-Espinoza. 2009. Mammalian prey of barn owl (*Tyto alba*) in southeastern Oaxaca, Mexico. Acta Zoológica Mexicana (n. s.) 25:143–149.

Santos-Moreno, A., M. Briones-Salas, G. González-Pérez, and T. de J. Ortiz. 2003. Noteworthy records of two rare mammals in Sierra Norte de Oaxaca, Mexico. The Southwestern Naturalist 48:312–313.

Santos-Moreno, A., S. García-Orozco, and E. E. Pérez-Cruz. 2010. Records of bats from Oaxaca, Mexico. The Southwestern Naturalist 55:454–456.

Schai-Braum, S., and K. Hackländer. 2018. *Lepus arcticus*. Pp. 165–168, *in* Pikas, rabbits and hares of the world (Smith, A., C. H. Johnston, P. C. Alves, and K. Hackländer, eds.). The Johns Hopkins University Press/IUCN/SSC. Baltimore, U.S.A.

Schaldach, W. J. 1960. *Xenomys nelson* Merriam, sus relaciones y sus hábitos. Revista de la Sociedad de Historia Natural 21:425–434.

Schaldach, W. J. 1964. Notas breves sobre algunos mamíferos del sur de México. Anales del Instituto de Biología, Universidad Nacional Autónoma de México 35:129–137.

Schaldach, W. J. 1966. New forms of mammals from southern Oaxaca, Mexico, with notes on some mammals of the coastal range. Säugetierkundliche Mitteilungen 14:286–297.

Schaldach, W. J., and C. A. McLaughlin. 1960. A new genus and species of glossophagine bat from Colima, Mexico. Contributions in Science, Los Angeles County Museum 37:1–8.

Scheel D. M., G. J. Slater, S-O. Kolokotronis, C. W. Potter, D. S. Rotstein, K. Tsangaras, A. D. Greenwood, and K. M. Helgen. 2014. Biogeography and taxonomy of extinct and endangered monk seals illuminated by ancient DNA and skull morphology. ZooKeys 409:1–33.

Scheffer, T. H. 1941. Ground squirrel studies in the Four-Rivers country, Washington. Journal of Mammalogy 22:270–279.

Scheffer, V. B. 1958. Seals, sea lions, and walruses, a review of the Pinnipedia. Stanford University Press. Stanford, U.S.A.

Scheick, B. K., and W. McCown. 2014. Geographic distribution of American black bears in North America. Ursus 25:24–33.

Scheuering, E. 2018. *Arborimus longicaudus*. The IUCN Red List of Threatened Species 2018:e.T42615A22389366.

Schmidly, D. J. 1972. Geographic variation in the white-ankled mouse, *Peromyscus pectoralis*. Southwestern Naturalist 17:113–138.

Schmidly, D. J. 1973. Geographic variation and taxonomy of *Peromyscus boylii* from Mexico and the southern United States. Journal of Mammalogy 54:111–130.

Schmidly, D. J. 1974. *Peromyscus pectoralis*. Mammalian Species 49:1–3.

Schmidly, D. J. 1977. The Mammals of Trans-Pecos Texas. Texas A & M University Press. College Station, U.S.A.

Schmidly, D. J. 2004. The Mammals of Texas. Revised Edition. University of Texas Press. Austin, U.S.A.

Schmidly, D. J. and F. S. Hendricks. 1976. Systematics of the southern races of Ord's kangaroo rat, *Dipodomys ordii.* Bulletin of the Southern California Academy of Sciences 75:225–237.

Schmidly, D. J., R. D. Bradley, and P. S. Cato. 1988. Morphometric differentiation and taxonomy of three chromosomally characterized groups of *Peromyscus boylii* from east-central Mexico. Journal of Mammalogy 69:460–480.

Schmidly, D. J., M. R. Lee, W. S. Modi, and E. G. Zimmerman. 1985. Systematics and notes on the biology of *Peromyscus hooperi.* Occasional Papers, the Museum, Texas Tech University 97:1–40.

Schmidt, C. A., and M. D. Engstrom. 1994. Genic variation and systematics of rice rats (*Oryzomys palustris* species group) in southern Texas and northeastern Tamaulipas, Mexico. Journal of Mammalogy 75:914–928.

Schmidt, C. A., M. D. Engstrom, and H. H. Genoways. 1989. *Heteromys gaumeri*. Mammalian Species 345:1–4.

Schmidt, U. 1978. Vampirfledermäuse. Die Neue Brehm-BÚcherei. Ziemsen Verlag, Wittenberg Lutherstadt, Germany.

Schnell, G. D., T. L. Best, and M. L. Kennedy. 1978. Interspecific morphologic variation in kangaroo rats (*Dipodomys*): degree of concordance with genic variation. Systematic Biology 27:34–48.

Schwartz, A. 1953. A systematic study of the water rat (*Neofiber alleni*). Occasional Papers Museum Zoology, University of Michigan 547:1–27.

Schwartz, A. 1956. A new subspecies of the Longtail Shrew (*Sorex dispar* Batchelder) from the Southern Appalachian Mountains. Journal of the Elisha Mitchell Scientific Society 72:24–30.

Schwartz, A., and E. P. Odum. 1957. The woodrats of the eastern United States. Journal of Mammalogy 38:197–206.

Schwartz, C. C., S. D. Miller, and M. A. Haroldson. 2003a. Grizzly bear. Pp. 556–586, *in* Wild mammals of North America: biology, management, and conservation (Feldhamer, G. A., B. C. Thompson, and J. A. Chapman, eds). Johns Hopkins University Press. Baltimore, U.S.A.

Schwartz, C. W., and E. R. Schwartz. 1981. The wild mammals of Missouri. Revised ed. University of Missouri Press and Missouri Department of Conservation. Columbia, U.S.A.

Schwartz, M. K., K. L. Pilgrim, K. S. Mckelvey, E. L. Lindquist, J. J. Claar, S. Loch, and L. F. Ruggiero. 2003b. Hybridization between Canada lynx and bobcats: Genetic results and management implications. Conservation Genetics 5:349.

Scott, F. W. 1987. First record of the long-tailed shrew, *Sorex dispar*, from Nova Scotia. Canadian Field-Naturalist 101:404–407.

Scott, F. W., and C. G. van Zyll de Jong. 1989. New Nova Scotia records of the Long-Tailed shrew, *Sorex dispar*, with comments on the taxonomic status of *Sorex dispar* and *Sorex gaspensis*. Le Naturaliste Canadien 116:145–154.

Sealander, J. A. 1979. A guide to Arkansas mammals. River Road Press. Conway, U.S.A.

Segura, V., F. Prevosti, and G. Cassini. 2013. Cranial ontogeny in the Puma lineage, *Puma concolor*, *Herpailurus yagouaroundi*, and *Acinonyx jubatus* (Carnivora: Felidae): a three-dimensional geometric morphometric approach. Zoological Journal of the Linnean Society 169:235–250.

Segura-Trujillo, C. A., and S. Navarro-Pérez. 2010. Escenario y problemática de conservación de los murciélagos (Chiroptera) cavernícolas del Complejo Volcánico de Colima, Jalisco-Colima, México. Therya 1:189–206.

Segura-Trujillo, C. A., A. L. Trujano-Álvarez and S. T. Álvarez-Castañeda. 2021. *Peromyscus ochraventer.* Mammalian Species, in press.

Segura-Trujillo, C. A., N. González-Ruiz, and S. T. Álvarez-Castañeda. In press. *Peromyscus mekisturus* (Rodentia: Cricetidae). Mammalian Species.

Sera, W. E., and C. N. Early. 2003. *Microtus montanus*. Mammalian Species 716:1–10.

Serfass, T. 2021. *Lontra canadensis*. The IUCN Red List of Threatened Species 2021:e.T12302A164577078.

Seton, E. T. 1929. Lives of game animals. Doubleday, Doran Co. Carden City, U. S. A.

Setzer, H. W. 1949. Subspeciation in the kangaroo rat, *Dipodomys ordii*. University of Kansas Publications, Museum of Natural History 1:473–573.

Severinghaus, W. D. 1977. Description of a new subspecies of prairie vole, *Microtus ochrogaster.* Proceedings of the Biological Society of Washington 90:49–54.

Severinghaus, W. D., and D. F. Hoffmeister. 1978. Qualitative cranial characters distinguishing *Sigmodon hispidus* and *Sigmodon arizonae* and the distribution of these two species in northern Mexico. Journal of Mammalogy 59:868–870.

Seymour, K. L. 1989. *Panthera onca*. Mammalian Species 340:1–9.

Shackleton, D. M. 1985. *Ovis canadensis*. Mammalian Species 230:1–9.

Shafer, A. B. A., and D. T. Stewart. 2007. Phylogenetic relationships among Nearctic shrews of the genus *Sorex* (Insectivora, Soricidae) inferred from combined cytochrome *b* and inter-SINE fingerprint data using Bayesian analysis. Molecular Phylogenetics and Evolution 44:192–203.

Shamel, H. H. 1931. Notes in the North American bats of the genus *Tadarida*. Proceedings of the United States National Museum 78:1–27.

Shaughnessy, P. D., and F. H. Fay. 1977. A review of the taxonomy and nomenclature of North Pacific harbour seals. Journal of Zoology (London) 182:385–419.

Shaw, W. T. 1934. The ability of the giant kangaroo rat as a harvester and storer of seeds. Journal of Mammalogy 15:275–286.

Sheeler-Gordon, L., and J. S. Smith. 2001. Survey of bat populations from Mexico and Paraguay for rabies. Journal of Wildlife Diseases 37:582–593.

Sheffield, S. R., and C. M. King. 1994. *Mustela nivalis*. Mammalian Species 454:1–10.

Sheffield, S. R., and H. H. Thomas. 1997. *Mustela frenata*. Mammalian Species 570:1–9.

Shellhammer, H. 1982. *Reithrodontomys raviventris*. Mammalian Species 169:1–3.

Shohfi, H. A., C. J. Conroy, A. R. Wilhelm, and J. L. Patton. 2006. New records of *Sorex preblei* and *S. tenellus* in California. The Southwestern Naturalist 51:108–111.

Shump, K. A., and R. H. Baker. 1978a. *Sigmodon alleni*. Mammalian Species 95:1–2.

Shump, K. A., and R. H. Baker. 1978b. *Sigmodon leucotis*. Mammalian Species 96:1–2.

Shump, K. A., and A. U. Shump. 1982a. *Lasiurus borealis*. Mammalian Species 183:1–6.

Shump, K. A., and A. U. Shump. 1982b. *Lasius cinereus*. Mammalian Species 185:1–5.

Silva-Caballero, A., and O. C. Rosas-Rosas. 2022. Rediscovery of the Tamaulipas white-sided jackrabbit (*Lepus altamirae*) after a century from its description. Therya Notes 3:1–3.

Simmons, N. B. 1998. A reappraisal of interfamilial relationships of bats. Pp. 1–26, *in* Bat biology and conservation (Kunz, T. H., and P. A. Racey, eds.). Smithsonian Institution Press. Washington, U.S.A.

Simmons, N. B. 2005. Order Chiroptera. Pp. 312–529, *in* Mammal species of the world. A taxonomic and geographic reference (Wilson, D. E., and D. A. Reeder, eds.). Third edition. The Johns Hopkins University Press. Baltimore, U.S.A.
Simmons, N. B., and T. M. Conway. 2001. Phylogenetic relationships of mormoopid bats (Chiroptera: Mormoopidae) based on morphological data. Bulletin of the American Museum of Natural History 258:1–97.
Simmons, N. B., and C. O. Handley, Jr. 1998. A revision of *Centronycteris* Gray (Chiroptera: Emballonuridae) with notes on natural history. American Museum Novitates 3239:1–28.
Simmons, N. B., and R. S. Voss. 1998. The mammals of Paracou, French Guiana: a neotropical lowland rainforest fauna part I. Bats. Bulletin of the American Museum of Natural History 237:1–219.
Simons, L. H., and D. F. Hoffmeister. 2003. *Sorex arizonae*. Mammalian Species 732:1–3.
Simons, L. H., and W. E. Van Pelt. 1999. Occurrence of *Sorex arizonae* and other shrews (Insectivora) in southern Arizona. The Southwestern Naturalist 44:334–342.
Simpson, G. G. 1945. The principles of classification and a classification of mammals. Bulletin American Museum Natural History 85:1–350.
Simpson, M. R. 1993. *Myotis californicus*. Mammalian Species 428:1–4.
Singh, M., A. Kumar, and H. N. Kumara. 2020. *Macaca mulatta*. The IUCN Red List of Threatened Species e.T12554A17950825.
Skinner, M. F., and O. C. Kaisen. 1947. The fossil *Bison* of Alaska and preliminary revision of the genus. Bulletin American Museum Natural History 89:123–256.
Smith, A. T., and D. E. Brown. 2019a. *Sylvilagus audubonii*. The IUCN Red List of Threatened Species 2019:e.T41297A45190821.
Smith, A. T., and D. E. Brown. 2019b. *Sylvilagus nuttallii*. The IUCN Red List of Threatened Species 2019:e.T41300A45192243.
Smith, A.T., and E. Beever. 2016. *Ochotona princeps*. The IUCN Red List of Threatened Species 2016:e.T41267A45184315.
Smith, A.T., and C. H. Johnston. 2019a. *Lepus arcticus*. The IUCN Red List of Threatened Species 2019:e.T41274A45185887.
Smith, A.T., and C. H. Johnston. 2019b. *Lepus othus*. The IUCN Red List of Threatened Species 2019:e.T11795A45178124.
Smith, A. T., and A. R. Ruedas. 2016. *Sylvilagus cognatus*. The IUCN Red List of Threatened Species 2016:e.T41309A45193738.
Smith, A. T., and M. L. Weston. 1990. *Ochotona princeps*. Mammalian Species 352:1–8.
Smith, A. T., E. A. Beever, and C. Ray. 2018. *Ochotona prínceps*. Pp. 67–72, *in* Pikas, rabbits and hares of the world (Smith, A., C. H. Johnston, P. C. Alves, and K. Hackländer, eds.). The Johns Hopkins University Press/IUCN/SSC. Baltimore, U.S.A.
Smith, F. A. 1997. *Neotoma cinerea*. Mammalian Species 564:1–8.
Smith, H. C. 1988. The wandering shrew, *Sorex vagrans,* in Alberta. Canadian Field-Naturalist 102:254–256.
Smith, H. C. 1993. Alberta mammals: an atlas and guide. Provincial Museum of Alberta, Edmonton, Canada.
Smith, J. D. 1970. The Systematic status of the black howler monkey, *Aloutta pigra* Lawrence. Journal of Mammalogy 51:358–369.
Smith, J. D. 1972. Systematics of the chiropteran Family Mormoopidae. Miscellaneous Publications, University of Kansas Publications, Museum of Natural History 56:1–132.
Smith, J. E., J. D. Long, D. I. Russell, K. L. Newcomb, and V. D. Muñoz. 2016. *Otospermophilus beecheyi* (Rodentia: Sciuridae). Mammalian Species 48:91–108.
Smith, M. E., and M. C. Belk. 1996. *Sorex monticolus*. Mammalian Species 528:1–5.
Smith, M. F. 1999. *Thomomys townsendii*. Pp. 477–478, *in* The Smithsonian book of North American Mammals (Wilson, D. E., and S. Ruff). Smithsonian Institution Press. Washington, U.S.A.
Smith, W. P. 1991. *Odocoileus virginianus*. Mammalian Species 388:1–13.
Smith, S. A., I. F. Greenbaum, D. 1. Schmidly, K. M. Davis, and T. W. Houseal. 1989. Additional notes on karyotypic variation in the *Peromyscus boylii* species group. Journal of Mammalogy 70:603–608.
Smolen, M. J. 1981. *Microtus pinetorum*. Mammalian Species 147:1–7.
Smolen, M. J., and B. L. Keller. 1987. *Microtus longicaudus*. Mammalian Species 271:1–7.
Snow, J. L., J. K. Jones, Jr., W. D. Webster. 1980. *Centurio senex*. Mammalian Species 138:1–3.
Snyder, D. P. 1982. *Tamias striatus*. Mammalian Species 168:1–8.
Sokolov, V. E., and V. N. Orlov. 1980. Opredelitel' mlekopitayushchickh Mongol'skoi Narodnoi Respubliki [Guide to the mammals of the Mongolian People's Republic]. Nauka, Moscow, Russia.
Solari, S. 2015a. *Chiroderma villosum*. The IUCN Red List of Threatened Species 2015:e.T4668A22037709.
Solari, S. 2015b. *Lonchorhina aurita*. The IUCN Red List of Threatened Species 2015:e.T12270A22039503.
Solari, S. 2015c. *Phylloderma stenops*. The IUCN Red List of Threatened Species 2015:e.T17168A22134036.
Solari, S. 2015d. *Saccopteryx bilineata*. The IUCN Red List of Threatened Species 2015:e.T19804A22004716.
Solari, S. 2015e. *Saccopteryx leptura*. The IUCN Red List of Threatened Species 2015:e.T19807A22005807.
Solari, S. 2015f. *Uroderma magnirostrum*. The IUCN Red List of Threatened Species 2015:e.T22783A22048094.
Solari, S. 2016a. *Anoura geoffroyi*. The IUCN Red List of Threatened Species 2016:e.T88109511A88109515.
Solari, S. 2016b. *Dermanura azteca*. The IUCN Red List of Threatened Species 2016:e.T2123A22000362.
Solari, S. 2016c. *Dermanura watsoni*. The IUCN Red List of Threatened Species 2016:e.T99586593A21997358.
Solari, S. 2016d. *Eumops floridanus*. The IUCN Red List of Threatened Species 2016:e.T136433A21984011.
Solari, S. 2016e. *Pteronotus mesoamericanus*. The IUCN Red List of Threatened Species 2016:e.T88018392A88018395.
Solari, S. 2016f. *Pteronotus parnellii*. The IUCN Red List of Threatened Species 2016:e.T88017638A22077695.
Solari, S. 2016g. *Vampyrodes major*. The IUCN Red List of Threatened Species 2016:e.T88151984A88151987.
Solari, S. 2017a. *Eumops nanus*. The IUCN Red List of Threatened Species 2017:e.T87994060A87994063.
Solari, S. 2017b. *Rhogeessa bickhami*. The IUCN Red List of Threatened Species 2017:e.T88151726A88151729.
Solari, S. 2017c. *Sturnira hondurensis*. The IUCN Red List of Threatened Species 2017:e.T88154577A88154581.
Solari, S. 2018a. *Bauerus dubiaquercus*. The IUCN Red List of Threatened Species 2018:e.T1789A22129523.
Solari, S. 2018b. *Choeronycteris mexicana*. The IUCN Red List of Threatened Species 2018:e.T4776A22042479.
Solari, S. 2018c. *Enchisthenes hartii*. The IUCN Red List of Threatened Species 2018:e.T2130A21996891.
Solari, S. 2018d. *Glyphonycteris sylvestris*. The IUCN Red List of Threatened Species 2018:e.T13384A22123687.
Solari, S. 2018e. *Lampronycteris brachyotis*. The IUCN Red List of Threatened Species 2018:e.T13376A22131330.

Solari, S. 2018f. *Lichonycteris obscura*. The IUCN Red List of Threatened Species 2018:e.T88120245A22057648.
Solari, S. 2018g. *Lophostoma evotis*. The IUCN Red List of Threatened Species 2018:e.T21986A22041302.
Solari, S. 2018h. *Macrotus californicus*. The IUCN Red List of Threatened Species 2018:e.T12652A22031754.
Solari, S. 2018i. *Macrotus waterhousii*. The IUCN Red List of Threatened Species 2018:e.T12653A22032004.
Solari, S. 2018j. *Myotis grisescens*. The IUCN Red List of Threatened Species 2018:e.T14132A22051652.
Solari, S. 2018k. *Myotis leibii*. The IUCN Red List of Threatened Species 2018:e.T14172A22055716.
Solari, S. 2018l. *Myotis occultus*. The IUCN Red List of Threatened Species 2018:e.T136650A21990499.
Solari, S. 2018m. *Myotis septentrionalis*. The IUCN Red List of Threatened Species 2018:e.T14201A22064312.
Solari, S. 2018n. *Perimyotis subflavus*. The IUCN Red List of Threatened Species 2018:e.T17366A22123514.
Solari, S. 2018o. *Tonatia saurophila*. The IUCN Red List of Threatened Species 2018:e.T41530A22004890.
Solari, S. 2018p. *Vampyrum spectrum*. The IUCN Red List of Threatened Species 2018:e.T22843A22059426.
Solari, S. 2019a. *Baeodon alleni*. The IUCN Red List of Threatened Species 2019:e.T19679A21989577.
Solari, S. 2019b. *Baeodon gracilis*. The IUCN Red List of Threatened Species 2019:e.T19681A22007578.
Solari, S. 2019c. *Corynorhinus mexicanus*. The IUCN Red List of Threatened Species 2019:e.T17599A21976792.
Solari, S. 2019d. *Eumops ferox*. The IUCN Red List of Threatened Species 2019:e.T87994072A87994075.
Solari, S. 2019e. *Gardnerycteris crenulatum*. The IUCN Red List of Threatened Species 2019:e.T13560A88177260.
Solari, S. 2019f. *Lasionycteris noctivagans*. The IUCN Red List of Threatened Species 2019:e.T11339A22122128.
Solari, S. 2019g. *Lasiurus seminolus*. The IUCN Red List of Threatened Species 2019:e.T11353A22119113.
Solari, S. 2019h. *Myotis nigricans*. The IUCN Red List of Threatened Species 2019:e.T14185A22066939.
Solari, S. 2019i. *Myotis volans*. The IUCN Red List of Threatened Species 2019:e.T14210A22069325.
Solari, S. 2019j. *Myotis yumanensis*. The IUCN Red List of Threatened Species 2019:e.T14213A22068335.
Solari, S. 2019k. *Natalus mexicanus*. The IUCN Red List of Threatened Species 2019:e.T123984355A22011975.
Solari, S. 2019l. *Nycticeius humeralis*. The IUCN Red List of Threatened Species 2019:e.T14944A22015223.
Solari, S. 2019m. *Pipistrellus hesperus*. The IUCN Red List of Threatened Species 2019:e.T17341A22129352.
Solari, S. 2019n. *Pteronotus gymnonotus*. The IUCN Red List of Threatened Species 2019:e.T18706A22077065.
Solari, S. 2019o. *Rhogeessa aeneus*. The IUCN Red List of Threatened Species 2019:e.T136810A22043785.
Solari, S. 2019p. *Rhogeessa parvula*. The IUCN Red List of Threatened Species 2019:e.T19684A22007495.
Solari, S. 2019q. *Sturnira parvidens*. The IUCN Red List of Threatened Species 2019:e.T88154376A88154380.
Solari, S. 2019r. *Uroderma bilobatum*. The IUCN Red List of Threatened Species 2019:e.T22782A22048748.
Solari, S. 2019s. *Molossus aztecus*. The IUCN Red List of Threatened Species 2019:e.T13645A22107522.
Solari, S. 2019t. *Nyctinomops aurispinosus*. The IUCN Red List of Threatened Species 2019:e.T14993A22010682.
Solari, S. 2019u. *Promops centralis*. The IUCN Red List of Threatened Species 2019:e.T88087651A22036112.
Solari, S. 2021. *Myotis lucifugus*. The IUCN Red List of Threatened Species 2021:e.T14176A208031565.
Solari, S., and M. Camacho. 2019. *Micronycteris microtis*. The IUCN Red List of Threatened Species 2019:e.T136424A21985267.
Solari, S., and L. Davalos. 2019. *Pteronotus davyi*. The IUCN Red List of Threatened Species 2019:e.T18705A22077399.
Solari, S., and D. Lew. 2015. *Caluromys derbianus*. The IUCN Red List of Threatened Species 2015:e.T3650A22175821.
Solari, S., S. R. Hoofer, P. A. Larsen, A. D. Brown, R. J. Bull, J. A. Guerrero, J. Ortega, J. P. Carrera, R. D. Bradley, and R. J. Baker. 2009. Operational criteria for genetically defined species: analysis of the diversification of the small fruit–eating bats, *Dermanura* (Phyllostomidae: Stenodermatinae). Acta Chiropterologica 11:279–288.
Solari, S., C. Mancina, and L. Davalos. 2019. *Phyllops falcatus*. The IUCN Red List of Threatened Species 2019:e.T17176A22133485.
Soler-Frost, A., R. A. Medellín, and G. N. Cameron. 2003. *Pappogeomys bulleri*. Mammalian Species 717:1–3.
Solmsen, E.-H., and H. Schilemann. 2007. *Choeroniscus minor*. Mammalian Species 822:1–6.
Sosa, V. J., A. Hernández-Huerta, and J. A. Vargas-Contreras 2005. Los mamíferos. Pp. 522–537, *in* Historia Natural de la Reserva de la Biosfera El Cielo, Tamaulipas, México (Sánchez-Ramos, G., P. Reyes-Castillo, and R. Dirzo, eds.). Universidad Autónoma de Tamaulipas. Tamaulipas, México.
Spencer, A. W. 1966. Identification of the dwarf shrew, *Sorex nanus*. Journal of Colorado-Wyoming Academy of Sciences 5:89.
Spencer, S. R., and G. N. Cameron. 1982. *Reithrodontomys fulvescens*. Mammalian species174:1–7.
Spitsyn, V. M., I. N. Bolotov, A. V. Kondakov, A. L. Klass, I. A. Mizin, A. A. Tomilova, N. A. Zubrii, and M. Y. Gofarov. 2021. A new Norwegian Lemming subspecies from Novaya Zemlya, Arctic Russia. Ecologica Montenegrina 40:93–117.
Spradling, T. A., J. W. Demastes, D. J. Hafner, P. L. Milbach, F. A. Cervantes, and M. S. Hafner. 2016. Systematic revision of the pocket gopher genus *Orthogeomys*. Journal of Mammalogy 97:405–423.
Stadelmann, B., L. G. Herrera, J. Arroyo-Cabrales, J. J. Flores-Martínez, B. P. May, and M. Ruedi. 2004. Molecular systematics of the Fishing bat *Myotis* (*Pizonyx*) *vivesi*. Journal of Mammalogy 85:133–139.
Stadelmann, B., L. K. Lin, T. H. Kunz, and M. Ruedi. 2007. Molecular phylogeny of New World *Myotis* (Chiroptera, Vespertilionidae) inferred from mitochondrial and nuclear DNA genes. Molecular Phylogenetics and Evolution 43:32–48.
Stangl, F. B., Jr., B. F. Koop, and C. S. Hood. 1983. Occurrence of *Baiomys taylori* (Rodentia: Cricetidae) on the Texas High Plains. Occasional Papers Museum, Texas Tech University 85:1–4.
Stalling, D. T. 1990. *Microtus ochrogaster*. Mammalian Species 355:1–9.
Statham, M., J. Murdoch, J. Janecka, K. Aubry, C. Edwards, C. Soulsbury, O. Berry, Z. Wang, D. Harrison, M. Pearch, L. Tomsett, J. Chupasko, and B. Sacks. 2014. Range-wide multilocus phylogeography of the red fox reveals ancient continental divergence, minimal genomic exchange and distinct demographic histories. Molecular Ecology 23:4813–4830.
Steele, M. A. 1998. *Tamiasciurus hudsonicus*. Mammalian Species 586:1–9.
Steele, M. A. 1999. *Tamiasciurus douglasii*. Mammalian Species 630:1–8.
Steiner, C. C., and O. A. Ryder, 2011. Molecular phylogeny and evolution of the Perissodactyla. Zoological Journal of the Linnean Society 163:1289–1303.
Stephen, C. T. 1988. *Spermophilus saturatus*. Mammalian Species 322:1–4.
Stephen, L. W. 1982. *Geomys personatus*. Mammalian Species 170:1–5.

Stephen, L. W., J. Ramírez-Pulido, and R. J. Baker. 1985. *Peromyscus alstoni*. Mammalian Species 242:1–4.
Stephen, L. W., and R. J. Baker. 1974. *Geomys arenarius*. Mammalian Species 36:1–3.
Steppan, S. J., M. R. Akhverdyan, E. A. Lyapunova, D. G. Fraser, N. N. Vorontov, R. S. Hoffmann, and M. J. Braun. 1999. Molecular Phylogeny of the Marmots (Rodentia: Sciuridae): Tests of Evolutionary and Biogeographic Hypotheses. Systematic Biology 48:715–734.
Stewart, B. S., and H. R. Huber. 1993. *Mirounga angustirostris*. Mammalian Species 449:1–10.
Stewart, D. T., and A. J. Baker. 1994a. Patterns of sequence variation in the mitochondrial D-Loop region of shrews. Molecular Biology and Evolution 11:9–21.
Stewart, D. T., and A. J. Baker. 1994b. Evolution of mtDNA D-loop sequences and their use in phylogenetic studies of shrews in the subgenus *Otisorex* (*Sorex*: Soricidae: Insectivora). Molecular Phylogenetics and Evolution 3:38–46.
Stewart, D. T., and A. J. Baker. 1997. A phylogeny of some taxa of Masked Shrews (*Sorex cinereus*) based on mitochondrial DNA D-loop sequences. Journal of Mammalogy 78:361–376.
Stewart, D. T., T. B. Herman, and T. Teferi. 1989. Littoral feeding in a high-density insular population of *Sorex cinereus*. Canadian Journal of Zoology 67: 2074–2077.
Stewart, D. T., A. J. Baker, and S. P. Hindocha. 1993. Genetic differentiation and population-structure in *Sorex haydeni* and *S. cinereus*. Journal of Mammalogy 74:21–32.
Stewart, D. T., N. D. Perry, and L. Fumagalli. 2002. The maritime shrew, *Sorex maritimensis* (Insectivora: Soricidae): A newly recognized Canadian endemic. Canadian Journal of Zoology 80:94–99.
Stewart, D. T., M. McPherson, J. Robichaud, and L. Fumagalli. 2003. Are there two species of pygmy shrews (*Sorex*)? Revisiting the question using DNA sequence data. Canadian Field-Naturalist 117: 82–88.
Stewart, R. E. A., K. M. Kovacs, and M. Acquarone. 2014. Walrus of the North Atlantic. NAMMCO Scientific Publications 9:7–12.
Stock, A. D. 1971. Chromosome evolution in the genus *Dipodomys* and its phylogenetic implications. Mammalian Chromosomes Newsletter 12:122–128.
Stock, A. D. 1974. Chromosome evolution in the genus *Dipodomys* and its taxonomic and phylogenetic implications. Journal of Mammalogy 55:505–526.
Stone, W., and J. A. G. Rehn. 1903. On the terrestrial vertebrates of portions of southern New Mexico and western Texas. Proceedings of the Academy of Natural Sciences of Philadelphia 55:16–34.
Stone, K. D., R. W. Flynn, and J. A. Cook. 2002. Post-glacial colonization of northwestern North America by the forest associated American marten (*Martes americana*). Molecular Ecology 11:2049–2064.
Stoner, K. E., M. Quesada, V. Rosas-Guerrero, and J. A. Lobo. 2002. Effects of forest fragmentation on the Colima long-nosed bat (*Musonycteris harrisoni*) foraging in tropical dry forest of Jalisco, Mexico. Biotropica 34:462–467.
Straney, D. O., and J. L. Patton. 1980. Phylogenetic and environmental determinants of geographic variation of the pocket mouse *Perognathus goldmani* Osgood. Evolution 34:888–903.
Streubel, D. P., and J. P. Fitzgerald. 1978a. *Spermophilus spilosoma*. Mammalian Species 101:1–4.
Streubel, D. P., and J. P. Fitzgerald. 1978b. *Spermophilus tridecemlineatus*. Mammalian Species 103:1–5.
Strickland, M. A., C. W. Douglas, M. Novak, and N. P. Huzinger. 1982. Fisher *Martes pennanti*. Pp. 586–598, *in* Wild Mammals of North America: biology, management, and economics (Chapman, J. A., and G.A. Feldhamer, eds.). Johns Hopkins University Press. Baltimore, U.S.A.
Sudman, P. D., J. K. Wickliffe, P. Horner, M. J. Smolen, J. W. Bickham, and R. D. Bradley. 2006. Molecular systematics of pocket gophers of the genus *Geomys*. Journal of Mammalogy 87:668–676.
Sudman, P. D., J. R. Choate, and E. G. Zimmerman. 1987. Taxonomy of chromosomal races of *Geomys bursarius lutescens* Merriam. Journal of Mammalogy 68:526–543.
Sulentich, J. M., L. R. Williams, and G. N. Cameron. 1991. *Geomys breviceps*. Mammalian Species 383:1–4.
Sullivan, E. G. 1956. Gray fox reproduction, denning, range, and weights in Alabama. Journal of Mammalogy 37:346–351.
Sullivan, J. K., and C. W. Kilpatrick. 1991. Biochemical systematics of the *Peromyscus aztecus* assemblage. Journal of Mammalogy 72:681–696.
Sullivan, J. M., C. W. Kilpatrick, and P. D. Rennert. 1991. Biochemical systematics of the *Peromyscus boylii* species group. Journal of Mammalogy 72:669–680.
Sullivan, J. K., J. A. Markert, and C. W. Kilpatrick. 1997. Phylogeography and molecular systematics of the *Peromyscus aztecus* species group (Rodentia: Muridae) inferred using parsimony and likelihood. Systematic Biology 46:426–440.
Sullivan, J. K., E. Arellano, and D. Rogers. 2000. Comparative phylogeography of Mesoamerica highland rodents: concerted versus independent response to past climatic fluctuations. The American Naturalist 155:755–768.
Sullivan, R. M., and T. L. Best. 1997a. Systematics and morphological variation in two chromosomal forms of the agile kangaroo rat (*Dipodomys agilis*). Journal of Mammalogy 78:775–797.
Sullivan, R. M., and T. L. Best. 1997b. Effects of environment on phenotypic variation and sexual dimorphism in *Dipodomys simulans* (Rodentia: Heteromyidae). Journal of Mammalogy 78:798–810.
Sullivan, R. M., D. J. Hafter, and T. L. Yates. 1986. Genetic of a contact zone between three chromosomal forms of the grasshopper mouse of the genus (*Onychomys*): a reassessment. Journal of Mammalogy 67:640–659.
Sundt, R. C., G. Dahle, and G. Naevdal. 1994. Genetic-variation in the hooded seal, *Cystophora cristata*, based on enzyme polymorphism and multilocus DNA-fingerprinting. Hereditas 121:147–155.
Sunqist, M., and F. Sunquist. 2002. Wild cats of the world. University of Chicago Press. Chicago, U.S.A.
Sutton, D. A. 1953. A systematic review of Colorado chipmunks (genus *Eutamias*). PhD. Desertation. University of Colorado, Boulder.
Sutton, D. A. 1982. The female genital bone of chipmunks, genus *Eutamias*. The Southwestern Naturalist 27:393–402.
Sutton, D. A. 1992. *Tamias amoenus*. Mammalian Species 390:1–8.
Sutton, D. A. 1993. *Tamias townsendii*. Mammalian Species 435:1–6.
Sutton, D. A., and C. F. Nadler. 1974. Systematic revision of three Townsend chipmunks (*Eutamias townsendii*). The Southwestern Naturalist 19:199–211.
Sutton, D. A., and B. D. Patterson. 2000. Geographic variation of the western chipmunks *Tamias senex* and *T. siskiyou,* with two new subspecies from California. Journal of Mammalogy 81:299–316.

Swanepoel, P., and H. H. Genoways. 1979. Morphometrics. Pp. 13–106, *in* Biology of the bats of the New World family Phyllostomidae. Part III (Baker, R. J., J. K. Jones, Jr., and D. C. Carter, eds.). Special Publications, The Museum, Texas Tech University 16:1–441.

Swank, W.G., and J. G. Teer. 1989. Status of the Jaguar. Oryx 23:14–21.

Sweeney, J. R., J. M. Sweeney, and S. W. Sweeney. 2003. Feral hog, Sus scrofa. Pp. 1164–1179, *in* Wild mammals of North America: Biology, management, and conservation (Feldhammer, G. A, B. C. Thompson, and J. A. Chapman, eds.). The Johns Hopkins University Press. Baltimore, U.S.A.

Tamarin, R. H., and T. H. Kunz. 1974. *Microtus breweri*. Mammalian Species 45:1–3.

Tamsitt, J. R. 1954. The mammals of two areas in the Big Bend region of Trans-Pecos Texas. The Texas Journal of Science 6:33–61.

Tamsitt, J. R., and D. Valdivieso. 1963. Records and observations on Colombian bats. Journal of Mammalogy 44:168–180.

Tarver, J. E., M. dos Reis, S. Mirarab, R. J. Moran, S. Parker, J. E. O'Reilly, B. L. King, M. J. O'Connell, R. J. Asher, T. Warnow, K. J. Peterson, P. C. J. Donoghue, and D. Pisani. 2016. The interrelationships of placental mammals and the limits of phylogenetic inference. Genome Biology and Evolution 8:330–44.

Tate, C. M., J. F. Pagels, and C. O. Handley, Jr. 1980. Distribution and systematic relationship of two kinds of short-tailed shrews (Soricidae: *Blarina*) in south-central Virginia. Proceedings of the Biological Society of Washington 93:50–60.

Tate, G. H. H. 1933. A systematic revision of the marsupial genus *Marmosa,* with a discussion of the adaptive radiation of the murine opossums (*Marmosa*). Bulletin of the American Museum of Natural History 66:137–144.

Tate, G. H. H. 1942. Results of the Archbold expeditions. No. 47: review of the Vespertilioninae bats, with special attention to genera and species of the Archbold collections. Bulletin of the American Museum of Natural History 80:221–296.

Tate, G. H. H. 1947. Albino prairie dog. Journal of Mammalogy 28:62.

Taub, D.M., and P. T. Mehlman. 1989. Development of the Morgan Island rhesus monkey colony. Puerto Rico Health Sciences Journal 8:159–69.

Tavares, V., and S. Burneo. 2015. *Trinycteris nicefori*. The IUCN Red List of Threatened Species 2015:e.T13381A22123365.

Tavares, V., and H. Mantilla. 2015. *Thyroptera tricolor*. The IUCN Red List of Threatened Species 2015:e.T21879A97207863.

Tavares, V., and J. Molinari. 2015. *Choeroniscus godmani*. The IUCN Red List of Threatened Species 2015:e.T4772A22041805.

Tavares, V., A. Muñoz, and J. Arroyo-Cabrales. 2015. *Vampyressa thyone*. The IUCN Red List of Threatened Species 2015:e.T136671A21989318.

Taylor, C. L., and R. F. Wilkinson Jr. 1988. First record of *Sorex longirostris* (Soricidae) in Oklahoma. The Southwestern Naturalist 33:248.

Taylor, F. H. C., M. Fujinaga, and F. Wilke. 1955. Distribution and food habits of the fur seals of the North Pacific Ocean; Report of cooperative investigations by the Governments of Canada, Japan, and the United States of America, February-July 1952. United States Fish and Wildlife Service, Washington, U.S.A.

Taylor, W. P. 1916. The status of the beavers of western North America, with a consideration of the factors in their speciation. University of California Publications in Zoology 12:413–495.

Taylor, W. P. 1918. Revision of the rodent genus Aplodontia. University of California Publications in Zoology 17:435–504.

Taylor, W. P. 1956. The deer of North America. The Stackpole Company. Harrisburg, U.S.A.

Teer, J. G., J. W. Thomas, and E. A. Walker. 1965. Ecology and management of white-tailed deer in the Llano Basin of Texas. Wildlife Monographs 15:1–62.

Tejedor, A. 2005. A new species of funnel–eared bat (Natalidae: *Natalus*) from Mexico. Journal of Mammalogy 86:1109–1120.

Tejedor, A. 2006. The type locality of *Natalus stramineus* (Chiroptera: Natalidae): implications for the taxonomy and biogeography of the genus *Natalus*. Acta Chiropterologica 8:361–380.

Tejedor, A. 2011. Systematics of funnel-eared bats (Chiroptera: Natalidae). Bulletin of the American Museum of Natural History 353:1–140.

Tellez, G., and J. Ortega. 1999. *Musonycteris harrisoni*. Mammalian Species 622:1–3.

Tener, J. S. 1965. Muskoxen in Canada: a biological and taxonomic review. Queen's Printer and Controller of Stationery. Ottawa, Canada.

Thaeler, C. S., Jr. 1968a. An analysis of three hybrid populations of pocket gophers (genus *Thomomys*). Evolution 22:543–555.

Thaeler, C. S., Jr. 1968b. An analysis of the distribution of pocket gopher species in northeastern California (genus *Thomomys*). University of California Publlications, Zoology 86:1–46.

Thaeler, C. S., Jr. 1972. Taxonomic Status of the Pocket Gophers, *Thomomys idahoensis* and *Thomomys pygmaeus* (Rodentia, Geomyidae). Journal of Mammalogy 53:417–428.

Thaeler, C. S., Jr. 1977. Taxonomic status of *Thomomys talpoides confinus*. Murrelet 58:49–50.

Thaeler, C. S., Jr. 1980. Chromosome numbers and systematics relations in the genus *Thomomys* (Rodentia: Geomyidae). Journal of Mammalogy 61:414–422.

Thaeler, C. S., Jr., and L. L. Hinesley. 1979. *Thomomys clusius*, a rediscovered species of pocket gopher. Journal of Mammalogy 60:480–488.

Thomas, D. C., and P. Everson. 1982. Geographic variation in caribou on the Canadian Arctic islands. Canadian Journal of Zoology 60:2442–2454.

Thomas, H. H., and T. L. Best. 1994. *Lepus insularis*. Mammalian Species 465:1–3.

Thomas, O. 1888. Catalogue of the Marsupialia and Monotremata in the collection of the British Museum (Natural History). British Museum. London, United Kingdom.

Thomas, O. 1903. New mammals from Chiriqui. Annals and Magazine of Natural History, including Zoology, Botany and Geology, London 7:376–382.

Thomson, C. E. 1982. *Myotis sodalis*. Mammalian Species 163:1–5.

Thorington, R. W., Jr., and R. S. Hoffmann. 2005. Family Sciuridae. Pp. 754–818, *in* Mammal species of the world: a taxonomic and geographic reference (Wilson, D. E., and D. M. Reeder, eds.). Third edition. Johns Hopkins University Press, Baltimore, U.S.A.

Thorington, Jr., R. W., J. L. Koprowski, M. A. Steele, and J. F. Whatton. 2012. Squirrels of the world. The John Hopkins University Press. Baltimore, U.S.A.

Thornton, W. A., and G. C. Creel. 1975. The taxonomic status of kit foxes. Texas Journal of Sciences 26:127–136.

Thornton, W. A., G. C. Creel, and R. E. Trimble. 1971. Hybridization in the fox genus *Vulpes* in west Texas. The Southwestern Naturalist 15:473–484.

Tiemann-Boege, I., C. W. Kilpatrick, D. J. Schmidly, and R. D. Bradley. 2000. Molecular phylogenetics of *Peromyscus boylii* species group (Rodentia: Muridae) based on mitichondrial cytochrome *b* sequences. Molecular Phylogenetics and Evolution 16:366–378.

Timm, R. M. 1985. *Artibeus phaeotis*. Mammalian Species 235:1–6.

Timm, R. 2016. *Onychomys leucogaster*. The IUCN Red List of Threatened Species 2016:e.T15338A115127288.
Timm, R. M., and H. H. Genoways. 2003. West Indian mammals from the Albert Schwartz Collection: Biological and historical information. Scientific Papers, Natural History Museum, University of Kansas 29:1–47.
Timm, R. M., and H. H. Genoways. 2004. The Florida bonneted bat, *Eumops floridanus* (Chiroptera: Molossidae): distribution, morphometrics, systematics, and ecology. Journal of Mammalogy 85:852–865.
Timm, R., and J. Matson. 2018. *Notiosorex villai*. The IUCN Red List of Threatened Species 2018:e.T136688A22293097.
Timm, R., S. T. Álvarez-Castañeda, I. Castro-Arellano, and T. Lacher. 2016a. *Ammospermophilus harrisii*. The IUCN Red List of Threatened Species 2016:e.T42399A115189204.
Timm, R., S. T. Álvarez-Castañeda, I. Castro-Arellano, and T. Lacher. 2016b. *Ammospermophilus interpres*. The IUCN Red List of Threatened Species 2016:e.T42451A115189324.
Timm, R., S. T. Álvarez-Castañeda, I. Castro-Arellano, and T. Lacher. 2016c. *Baiomys taylori*. The IUCN Red List of Threatened Species 2016:e. T2466A115062269.
Timm, R., S. T. Álvarez-Castañeda, J. Frey, and T. Lacher. 2016d. *Dipodomys spectabilis*. The IUCN Red List of Threatened Species 2019:e. T6693A22229212.
Timm, R., S. T. Álvarez-Castañeda, and T. Lacher. 2016e. *Dipodomys merriami*. The IUCN Red List of Threatened Species 2016:e. T92465716A115515430.
Timm, R., S. T. Álvarez-Castañeda, and T. Lacher. 2016f. *Neotoma leucodon*. The IUCN Red List of Threatened Species 2016:e. T136793A115212444.
Timm, R., S. T. Álvarez-Castañeda, and T. Lacher. 2016g. *Otospermophilus beecheyi*. The IUCN Red List of Threatened Species 2016:e. T20481A22263743.
Timm, R., S. T. Álvarez-Castañeda, and T. Lacher. 2016h. *Peromyscus fraterculus*. The IUCN Red List of Threatened Species 2016:e. T136412A115207364.
Timm, R., A. D. Cuarón, F. Reid, K. Helgen, and J. F. González-Maya. 2016i. *Procyon lotor*. The IUCN Red List of Threatened Species 2016:e. T41686A45216638.
Timm, R., J. Matson, D. Tirira, C. Boada, M. Weksler, R. P. Anderson, B. Rivas, C. Delgado, and M. Gómez-Laverde. 2016j. *Handleyomys alfaroi*. The IUCN Red List of Threatened Species 2016:e.T15585A115127897.
Timm, R., J. Matson, Woodman, N., and I. Castro-Arellano. 2016k. *Notiosorex crawfordi*. The IUCN Red List of Threatened Species 2016:e. T41456A115187458.
Timm, R., F. Reid, N. Woodman, T. McCarthy, and J. Matson. 2016l. *Otonyctomys hatti*. The IUCN Red List of Threatened Species 2016:e. T15664A115129163.
Timm, R., S. T. Álvarez-Castañeda, and T. Lacher. 2017a. *Oryzomys nelsoni*. The IUCN Red List of Threatened Species 2017:e.T15583A22388135.
Timm, R., J. Matson, N. Woodman, P. C. de Grammont, and I. Castro-Arellano. 2017b. *Notiosorex cockrumi*. The IUCN Red List of Threatened Species 2017:e.T136666A22293361.
Timm, R., S. T. Álvarez-Castañeda, and T. Lacher. 2018a. *Rheomys mexicanus*. The IUCN Red List of Threatened Species 2018:e.T19484A22354600.
Timm, R., T. Lacher, and S. T. Álvarez-Castañeda. 2018b. *Reithrodontomys bakeri*. The IUCN Red List of Threatened Species 2018:e. T136821A22385642.
Timm, R., S. T. Álvarez-Castañeda, J. Frey, and T. Lacher. 2019. *Dipodomys spectabilis*. The IUCN Red List of Threatened Species 2019: e.T6693A22229212.
Tirira, D. G., J. Díaz-N., M. Superina, and A. M. Abba. 2014. *Cabassous centralis*. The IUCN Red List of Threatened Species 2014:e. T3412A47437304.
Torres-Flores, J. W., and R. López-Wilchis. 2018. *Microtus quasiater* (Rodentia: Cricetidae). Mammalian Species 50:59–66.
Torres-Morales, L., D. F. García-Mendoza, C. López-González, and R. Muñiz-Martínez. 2010. Bats of northwestern Durango, Mexico: species richness at the interface of two biogeographic regions. The Southwestern Naturalist 55:347–362.
Trouessart, E. L. 1905. Catalogus mammalium tam viventium quam fossilium. Quinquennale supplementium (1899–1904). Cetacea, Edentata, Marsupialia, Allotheria, Monotremata. R. Friedlander & Sons. Berlin, Germany.
Trefethen, J. B. 1975. The wild sheep in modern North America. Winchester Press. New York, U.S.A.
Tremor, S., S. Vanderplank, and E. Mellink. 2019. The San Quintín Kangaroo rat is not extinct. Bulletin Southern California Academy of Sciences 118:71–75.
Treviño-Villareal, J.1991. The annual cycle of the Mexican prairie dog (*Cynomys mexicanus*). Occasional Papers of the Museum of Natural History, University of Kansas 139:1–27.
Trombulak, S. C. 1988. *Spermophilus saturatus*. Mammalian Species 322:1–4.
True, F. W. 1896. A revision of the American moles. Proceeding of the United States Natural Museum 19:1–111.
Trujano-Álvarez, A. L., and S. T. Álvarez-Castañeda. 2007. Taxonomic revision of *Thomomys bottae* in the Baja California Sur lowlands. Journal of Mammalogy 88:343–350.
Trujano-Álvarez, A. L., and S. T. Álvarez-Castañeda. 2010. *Peromyscus mexicanus*. Mammalian Species 42:111–118.
Trujano-Álvarez, A. L., and S. T. Álvarez-Castañeda. 2013. Phylogenetic structure among pocket gopher populations, genus *Thomomys* (Rodentia: Geomyidae), on the Baja California Peninsula. Zoological Journal of the Linnean Society 168:873–891.
Tucker, P. K., and D. J. Schmidly. 1981. Studies on a contact zone among three chromosomal races of *Geomys bursarius* in east Texas. Journal of Mammalogy 62:258–272.
Tumlison, R. 1987. *Felis lynx*. Mammalian Species 269:1–8.
Tumlison, R. 1991. Bats of the genus *Plecotus* in Mexico: discrimination and distribution. Occasional Papers, Museum of Texas Tech University 140:1–19.
Tumlison, R. 1992. *Plecotus mexicanus*. Mammalian Species 401:1–3.
Tsytsulina, K., N. Formozov, B. Sheftel, M. Stubbe, R. Samiya, J. Ariunbold, and V. Buuveibaatar. 2016. *Sorex tundrensis*. The IUCN Red List of Threatened Species 2016:e.T41422A115185726.
Udina, I. G., A. A. Danilkin, and G. G. Boeskorov. 2002. Genetic Diversity of Moose (*Alces alces* L.) in Eurasia. Russian Journal of Genetics 38:951–957.

Vallejo, R. M., and F. X. González-Cózatl. 2012. Phylogenetic affinities and species limits within the genus *Megadontomys* (Rodentia: Cricetidae) based on mitochondrial sequence data. Journal Zoological Systematics and Evolutionary Research 50:67.

Vallejo, R. M., J. A. Guerrero, and F. X. González-Cózatl. 2017. Patterns of differentiation and disparity in cranial morphology in rodent species of the genus *Megadontomys* (Rodentia: Cricetidae). Zoological Studies 7:56:e14.

Vannatta, J. M., J. A. Gore, V. L. Mathis, and B. D. Carver. 2020. *Eumops floridanus* (Chiroptera: Molossidae). Mammalian Species 53:125–133.

Van Den Bussche, R. A., and R. J. Baker. 1993. Molecular phylogenetics of the New World bat genus *Phyllostomus* based on cytochrome-*b* DNA sequence variation. Journal of Mammalogy 74:793–802.

Van Den Bussche, R. A., R. J. Baker, H. A. Wichman, and M. J. Hamilton. 1993. Molecular phylogenetics of stenodermatine bat genera: Congruence of data from nuclear and mitochondrial DNA. Molecular Biology and Evolution 10:944–959.

Van den Bussche, R. A., J. L. Hudgeons, and R. J. Baker. 1998. Phylogenetic accuracy, stability, and congruence. Relationships within and among the New World bat genera *Artibeus*, *Dermanura*, and *Koopmania*. Pp. 59—71, *in* Bat biology and conservation (Thomas H., K., and P. A. Racey, eds.). Smithsonian Institution Press. Washington, U.S.A.

Van de Graff, K. M. 1975. Reproductive ecology of some Sonoran desert rodents. Ph.D. dissertation, Northern Arizona University. Flagstaff, U.S.A.

Van Gelder, R. G. 1959a. Results of the Puritan-Amer ican Museum of Natural History expedition to western Mexico. 8, A new *Antrozous* (Mammalia, Vespertilionidae) from the Tres Marías Islands, Nayarit, Mexico. American Museum Novitates 1973:1–14.

Van Gelder, R. G. 1959b. A taxonomic revision of the spotted skunks (genus *Spilogale*). Bulletin of the American Museum of Natural History 117:229–392.

Van Gelder, R. G. 1960. Results of the Puritan-American Museum of Natral History Expedition to western Mexico. 10, Marine mammals from the coast of Baja California and the Tres Marías Islands, México. American Museum Novitates 1992:1–27.

Van Gelder, R. G. 1977. Mammalian hybrids and generic limits. American Museum Novitates 2635:1–25.

Van Gelder, R. G. 1978. A review of canid classification. American Museum Novitates 2646:1–25.

van Zyll de Jong, C. G. 1972. A systematic review of the Nearctic and Neotropical river otters (genus *Lutra,* Mustelidae, Carnivora). Life Sciences Contributions, Royal Ontario Museum 80:1–104.

van Zyll de Jong, C. G. 1982. Relationships of amphiberingian shrews of the *Sorex cinereus* group. Canadian Journal of Zoology 60:1580–1587.

van Zyll de Jong, C. G. 1983a. Handbook of Canadian Mammals. National Museums of Canada, Ottawa, Canada.

van Zyll de Jong, C. G. 1983b. Handbook of Canadian mammals: bats. National Museums of Canada, Ottawa, Ontario 2:1–212.

van Zyll de Jong, C. G. 1983c. A morphometric analysis of North American shrews of the *Sorex arcticus* group, with special consideration of the taxonomic status of *S. a. maritimensis.* Le Naturaliste Canadien 110:373–378.

van Zyll de Jong, C. G. 1984. Taxonomic relationships of Nearctic small-footed bats of the *Myotis leibii* group (Chiroptera: Vespertilionidae). Canadian Journal of Zoology 62:2519–2526.

van Zyll de Jong, C. G. 1987. A phylogenetic study of the *Lutrinae* (Carnivora; Mustelidae) using morphological data. Canadian Journal of Zoology 65:2536–2544.

van Zyll de Jong, C. G., and G. L. Kirkland, Jr. 1989. A morphometric analysis of the *Sorex cinereus* group in central and Eastern North America. Journal of Mammalogy 70:110–122.

van Zyll de Jong, C. G. 1991a. Speciation of the *Sorex cinereus* group. Pp. 65–73, *in* The Biology of the Soricidae (Findley, J. S., and T. L. Yates, eds). Museum of Southwestern Biology. Albuquerque, U.S.A.

van Zyll de Jong, C. G. 1991b. St. Lawrence Island shrew *Sorex jacksoni*. Pp. 28, *in* The Smithsonian book of North American Mammals (Wilson, D. E., and S. Ruff. 1999). Smithsonian Institution Press. Washington, U.S.A.

van Zyll de Jong, C. G. 1999a. Tundra shrew, *Sorex tundrensis*. Pp. 44–45, *in* The Smithsonian book of North American Mammals (Wilson, D. E., and S. Ruff, eds.). Smithsonian Institution Press. Washington, U.S.A.

van Zyll de Jong, C. G. 1999b. Barren ground shrew, *Sorex uyunak*. Pp. 45, *in* The Smithsonian book of North American Mammals (Wilson, D. E., and S. Ruff, eds.). Smithsonian Institution Press. Washington, U.S.A.

Van Staaden, M. J., and J. K. Jones, Jr. 1997. Comparative morphology of dorsal hair of New World bats of the Family Molossidae. Pp. 373–391, *in* Homenaje al Profesor Ticul Álvarez (Arroyo-Cabrales, J., and O. J. Polaco, coords.). Instituto Nacional de Antropología e Historia, Colección Científica. Distrito Federal, México.

Vargas, K., D. Brown, E. Wisely, and M. Culver. 2019. Reinstatement of the Tamaulipas white-sided jackrabbit, *Lepus altamirae*, based on DNA sequence data. Revista Mexicana de Biodiversidad 90:2–9.

Vashon, J. 2016. *Lynx canadensis*. The IUCN Red List of Threatened Species 2016:e.T12518A101138963.

Vázquez, A., and F. Gopar-Merino. 2018. *Romerolagus diazi.* Pp. 198–200, *in* Pikas, Rabbits and Hares of the World (Smith, A., C. H. Johnston, P. C. Alves, and K. Hackländer, eds.). The Johns Hopkins University Press/IUCN/SSC. Baltimore, U.S.A.

Vázquez, E. 2016a. *Peromyscus guatemalensis*. The IUCN Red List of Threatened Species 2016:e.T16665A22361195.

Vázquez, E. 2016b. *Peromyscus hylocetes*. The IUCN Red List of Threatened Species 2016:e.T136416A22361620.

Vázquez, E. 2017a. *Habromys delicatulus*. The IUCN Red List of Threatened Species 2017:e.T136683A22376548.

Vázquez, E. 2017b. *Heterogeomys lanius*. The IUCN Red List of Threatened Species 2017:e.T42591A22215417.

Vázquez, E. 2017c. *Chaetodipus lineatus*. The IUCN Red List of Threatened Species 2017:e.T136361A22225840.

Vázquez, E. 2018a. *Habromys simulatus*. The IUCN Red List of Threatened Species 2018:e.T9611A22376731.

Vázquez, E. 2018b. *Handleyomys chapmani*. The IUCN Red List of Threatened Species 2018:e.T15591A22328808.

Vázquez, E. 2018c. *Xenomys nelsoni*. The IUCN Red List of Threatened Species 2018:e.T23115A22359234.

Vázquez, E. 2019a. *Handleyomys rhabdops*. The IUCN Red List of Threatened Species 2019:e.T15611A22328169.

Vázquez, E. 2019b. *Reithrodontomys tenuirostris*. The IUCN Red List of Threatened Species 2019:e.T19418A22385808.

Vázquez, E., and S. T. Álvarez-Castañeda. 2016. *Peromyscus melanophrys*. The IUCN Red List of Threatened Species 2016:e.T16677A115136842.

Vázquez, E., and F. Reid. 2016. *Peromyscus gymnotis*. The IUCN Red List of Threatened Species 2016:e.T16666A22361063.

Vázquez, E., L. Emmons, F. Reid, and A. D. Cuarón. 2008. *Dasyprocta mexicana*. The IUCN Red List of Threatened Species 2008:e. T6285A12596623.

Vázquez, E., L. Emmons, and T. McCarthy. 2016a. *Orthogeomys grandis*. The IUCN Red List of Threatened Species 2016:e.T42590A115193029.

Vázquez, E., L. Emmons, F. Reid, and A. D. Cuarón. 2016b. *Heteromys salvini*. The IUCN Red List of Threatened Species 2016:e.T12076A22225346.
Vázquez, E., L. Emmons, F. Reid, and A. D. Cuarón. 2016c. *Orthogeomys hispidus*. The IUCN Red List of Threatened Species 2016:e.T15549A115127668.
Vázquez, E., L. Emmons, F. Reid, and A. D. Cuarón. 2016d. *Sciurus yucatanensis*. The IUCN Red List of Threatened Species 2016:e.T20026A115156290.
Vázquez, E., F. Reid, and A. D. Cuarón. 2016e. *Coendou mexicanus*. The IUCN Red List of Threatened Species 2016:e.T20629A22214103.
Vázquez, E., L. Emmons, and T. McCarthy. 2017. *Tylomys nudicaudus*. The IUCN Red List of Threatened Species 2017:e.T22573A22340628.
Vázquez, E., P. C. de Grammont, and A. D. Cuarón. 2018. *Reithrodontomys spectabilis*. The IUCN Red List of Threatened Species 2018:e.T19416A22386261.
Vázquez, L. B., G. N. Cameron, and R. A. Medellín. 2001. *Peromyscus aztecus*. Mammalian Species 649:1–4.
Veal, R., and W. Caire. 1979. *Peromyscus eremicus*. Mammalian Species 118:1–6.
Velazco, P. M. 2005. Morphological phylogeny of the bat genus *Platyrrhinus* Saussure, 1860 (Chiroptera: Phyllostomidae) with the description of four new species. Fieldiana Zoology (n. s.) 105:1–54.
Velazco, P. M., and B. D. Patterson. 2013. Diversification of the yellow-shouldered bats, genus *Sturnira* (Chiroptera, Phyllostomidae), in the New World tropics. Molecular Phylogenetics and Evolution 68:683–698.
Velazco, P. M., and B. D. Patterson. 2014. Two new species of yellow-shouldered bats, genus *Sturnira* Gray, 1842 (Chiroptera, Phyllostomidae) from Costa Rica, Panama and western Ecuador. ZooKeys 402:43–66.
Velazco, P. M., and N. B. Simmons. 2011. Systematics and taxonomy of great striped-faced bats of the genus *Vampyrodes* Thomas, 1900 (Chiroptera: Phyllostomidae). American Museum Novitates 3710:1–35.
Velázquez, A. 1994. Distribution and population size of *Romerolagus diazi* on El Pelado volcano, Mexico. Journal of Mammalogy 75:743–749.
Velázquez, A., and J. A. Guerrero. 2019. *Romerolagus diazi*. The IUCN Red List of Threatened Species 2019:e.T19742A45180356.
Velázquez, A., F. A. Cervantes, and C. Galindo-Leal. 1993. The volcano rabbit *Romerolagus diazi*, a peculiar lagomorph. Lutra 36:62–70.
Verts, B. J., and L. N. Carraway. 1984. Keys to the mammals of Oregon. O.S.U. Book Stores, Inc. Corvallis, U.S.A.
Verts, B. J., and L. N. Carraway. 1987a. *Thomomys bulbivorus*. Mammalian Species 273:1–4.
Verts, B. J., and L. N. Carraway. 1987b. *Microtus canicaudus*. Mammalian Species 267:1–4.
Verts, B. J., and L. N. Carraway. 1995. *Phenacomys albipes*. Mammalian Species 494:1–5.
Verts, B. J., and L. N. Carraway. 1998. Land mammals of Oregon. University of California Press. Berkeley, U.S.A.
Verts, B. J., and L. N. Carraway. 1999. *Thomomys talpoides*. Mammalian Species 618:1–11.
Verts, B. J., and L. N. Carraway. 2000. *Thomomys mazama*. Mammalian Species 641:1–7.
Verts, B. J., and L. N. Carraway. 2001a. *Scapanus latimanus*. Mammalian Species 666:1–7.
Verts, B. J., and L. N. Carraway. 2001b. *Tamias minimus*. Mammalian Species 653:1–10.
Verts, B. J., and L. N. Carraway. 2002. *Neotoma lepida*. Mammalian Species 699:1–12.
Verts, B. J., and L. N. Carraway. 2003. *Thomomys townsendii*. Mammalian Species 719:1–6.
Verts, B. J., and G. L. Kirkland, Jr. 1988. *Perognathus parvus*. Mammalian Species 318:1–8.
Verts, B. J., L. N. Carraway, and A. Kinlaw. 2001. *Spilogale gracilis*. Mammalian Species 674:1–10.
Vianna, J. A., R. K. Bonde, S. Caballero, J. P. Giraldo, R. P. Lima, A. M. Clark, M. Marmontel, B. Morales-Vela, M. J. De Souza, L. Parr, M. A. Rodríguez-Lopez, A. A. Mignucci-Giannoni, J. A. Powell, and F. R. Santos. 2006. Phylogeography, phylogeny, and hybridization in trichechid sirenians: implications on manatee conservation. Molecular Ecology 15:433–447.
Villa-R., B. 1958. El mono araña (*Ateles geoffroyi*) encontrado en la costa de Jalisco y en la region central de Tamaulipas. Anales de Instituto de Biología, Universidad Nacional Autónoma de México, Serie Zoología 28:345–347.
Villa-R., B. 1960. *Tadarida yucatanica* in Tamaulipas. Journal of Mammalogy 41:314–319.
Villa-R., B. 1967. Los murciélagos de Mexico. Instituto de Biología, Universidad Nacional Autónoma de México. Distrito Federal, México.
Villa-R., B., and E. L. Cockrum. 1962. Migration in the guano bat *Tadarida brasilensis mexicana* (Saussure). Journal of Mammalogy 43:43–64.
Villa, B., and E. R. Hall. 1949. Subspeciation of the Pocket Gophers of Kansas. University Kansas publications, Museum of Natural History 1:219–234.
Villa-R., B., G. M. Ruiz, B. O. Bonilla, and B. V. Cornejo. 1967. Rabia en dos especies de rnurciélagos insectivoros género *Pteronotus*, en condiciones naturales, colectados en Jalisco, México. Anales de Instituto de Biología, Universidad Nacional Autónoma de México, Serie Zoología 38:9–16.
Villafuerte, R, and M. Delibes-Mateos. 2019. *Oryctolagus cuniculus*. The IUCN Red List of Threatened Species 2019:e.T41291A170619657.
Villalpando, J. A., and T. Álvarez. 2000. Murciélagos (Chiroptera) de Michoacán en la Colección de Mamíferos de la Escuela Nacional de Ciencias Biológicas del IPN, México. Anales de la Escuela Nacional de Ciencias Biológicas, México 46:119–188.
Villalpando-R., J. A., and J. Arroyo-Cabrales. 1996. Una nueva localidad para *Rhogeessa mira* (Chiroptera: Vespertilionidae) en la cuenca baja del río Balsas, Michoacán, México. Vertebrata Mexicana 2:9–11.
Vitaly, M., I. N. Bolotov, A. V. Kondakov, A. L. Klass, I. A. Mizin, A. A. Tomilova, N. A. Zubrii, and M. Y. Gofarov. 2021. A new Norwegian Lemming subspecies from Novaya Zemlya, Arctic Russia, pp. 93-117 in Ecologica Montenegrina 40:116–117.
Voge, M., and H. A. Bern. 1949. Cecal villi in the red tree mouse, *Phenacomys longicaudus*. Anatomical Record 104:477–482.
Voigt, D. R. 1987. Red fox. Pp. 379–392, *in* Wild furbearer management and conservation in North America (Nowak, M., J. A. Baker, M. E. Obbard, and B. Malloch, eds.). Ontario Ministry of Natural Resources. Ontario, Canada.
Volobouev, V. T., and C. G. van Zyll de Jong. 1988. The Karyotype of *Sorex articus maritimensis* (Insectivora, Soricidae) and its systematic implications. Canadian Journal of Zoology 66:1968–1972.
Vonhof, M. J. 2000. *Rhogeessa tumida*. Mammalian Species 633:1–3.
Voss, R. S. 1988. Systematics and ecology of *Ichthyomyine* rodents (Muroidea): patterns of morphological evolution in a small adaptive radiation. American Museum of Natural History 188:259–493.
Voss, R. S. 2011. Revision notes on Neotropical porcupines (Rodentia: eretizontidae) 3. An annoted checklist of the species of Coendou Lacépède, 1799. Annals Museum Novitates 3720:1–36.

Voss, R. S., and S. A. Jansa. 2003. Phylogenetic studies on didelphid marsupials II. Nonmolecular data and new IRBP sequences: separate and combined analyses of didelphine relationships with denser taxon sampling. Bulletin American Museum of Natural History 276:1—82.

Voss, R. S., M. Gómez-Laverde, and V. Pacheco. 2002. A new genus for *Aepeomys fuscatus* Allen, 1912, and *Oryzomys intectus* Thomas, 1921:enigmatic murid rodents from Andean cloud forests. American Museum Novitates 3373:1–42.

Voss, R. S., C. Hubbard, and S. A. Jansa. 2013. Phylogenetic relationships of the new world porcupines (Rodentia:eretizontidae): Implication for taxonomy, morphological evolution and biogeography. American Museum Novitates 3769:1–36.

Voss, R. S., E. E. Gutiérrez, S. Solari, R. V. Rossi, and S. A. Jansa. 2014. Phylogenetic relationships of mouse opossums (Didelphidae, *Marmosa*) with a revised subgeneric classification and notes on sympatric diversity. American Museum novitates 3817:1–27.

Voss, R. S., J. Díaz-Nieto, and S. A. Jansa. 2018. A Revision of *Philander* (Marsupialia: Didelphidae), Part 1: *P. quica*, *P. canus*, and a New Species from Amazonia. American Museum Novitates 3891:1–70.

Voss, R. S., D. W. Fleck, and S. A. Jansa. 2019. Mammalian diversity and Matses ethnomammalogy in Amazonian Peru Part 3: Marsupials (Didelphimorphia). Bulletin of the American Museum of Natural History 432:1–90.

Wade-Smith J., and B. J. Verts. 1982. *Mephitis mephitis*. Mammalian Species 173:1–7.

Wahle, R., E. Roth, and P. Horner. 2018. *Dipodomys elator*. The IUCN Red List of Threatened Species 2018:e.T6675A22227507.

Walker, P. L. 1980. Archeological evidence for the recent extinction of three terrestrial mammals on San Miguel Island. Pp. 703–713, *in* The California Islands: proceedings of a multidisciplinary symposium (Power, D. M. ed.). Santa Barbara Museum Natural History, Santa Barbara, California, U.S.A.

Wang, H. G., R. D. Owen, C. Sánchez-Hernández, and M. de la L. Romero-Almaraz. 2003. Ecological characterization of bat species distributions in Michoacan, Mexico, using a geographic information system. Global Ecology and Biogeography 12:65–85.

Wang, K., J. A. Lenstra, L. Liu, Q. Hu, T. Ma, Q. Qiu, and J. Liu. 2018. Incomplete lineage sorting rather than hybridization explains the inconsistent phylogeny of the wisent. Communications Biology 1:1–9.

Warner, R. M. 1982. *Myotis auriculus*. Mammalian Species 191:1–3.

Warner, R. M., and N. J. Czaplewski. 1984. *Myotis volans*. Mammalian Species 224:1–4.

Warren, E. R.1936. The marmots of Colorado. Journal of Mammalogy 17:392–398.

Warren, E. R. 1942. The mammals of Colorado. 2nd ed. University Oklahoma Press, Norman, U.S.A.

Watkins, L. C. 1969. Observations on the distribution and natural history of the evening bat (*Nycticeius humeralis*) in northwestern Missouri and adjacent Iowa. Transactions of Kansas Academy of Science 72:330–336.

Watkins, L. C. 1972. *Nycticeius humeralis*. Mammalian Species 23:1–4.

Watkins, L. C. 1997. *Euderma maculatum*. Mammalian Species 77:1–4.

Watkins, L. C., J. K. Jones, Jr., and H. H. Genoways. 1972. Bats of Jalisco, Mexico. Special Publications, Museum of Texas Tech University 1:1–44.

Wayne, R. K., and S. M. Jenks. 1991. Mitonchondrial DNA analysis implying extensive hybridization of the endangered red wolf *Canis rufus*. Nature 351:565–568.

Wayne, R. K., S. B. George, D. Gilbert, and P. W. Collins. 1991a. The channel island fox (*Urocyon littoralis*) as a model of genetic change in small populations. Pp. 639–649, *in* The unity of evolutionary biology (Dudley, E. C., ed.). Proceedings of the fourth international congress of systematic and evolutionary biology. Dioscorides Press, Portland, Oregon 2:1–1048.

Wayne, R. K., S. George, D. Gilbert, P. Collins, and D. Girman. 1991b. A morphologic and genetic study of the island fox, *Urocyon littoralis*. Evolution 45:1849–1868.

Weber, M., and R. A. Medellín. 2010. Yucatán brown brocket deer *Mazama pandora* (Merriam 1901). Pp. 211–216, *in* (Duarte, J. M. B., and S. González, eds.), Neotropical Cervidology: Biology and Medicine of Latin American Deer. Funep, in collaboration with IUCN. Jaboticabal, Brazil and Gland, Switzerland.

Weber, M., P. C. de Grammont, and A. D. Cuarón. 2016. *Mazama pandora*. The IUCN Red List of Threatened Species 2016:e.T29622A22154219.

Webster, D. 1968. Comparative middle and inner ear structure of Heteromyidae. The Anatomical Record 160:447.

Webster, W. D. 1993. Systematics and evolution of the bats of the genus *Glossophaga*. Special Publications, Museum ofTexas Tech University 36:1–184.

Webster, W. D., and J. K. Jones, Jr. 1980. Taxonomic and nomenclatorial notes on the bats of the genus *Glossophaga* in North America, with description of a new species. Occasional Papers, Museum of Texas Tech University 71:1–12.

Webster, W. D., and J. K. Jones, Jr. 1982a. A new subspecies of *Glossophaga commissarisi* (Chiroptera: Phyllostomidae) from western Mexico. Occasional Papers, Museum of Texas Tech University 76: 1–6.

Webster, W. D., and J. K. Jones, Jr. 1982b. *Reithrodontomys megalotis*. Mammalian Species 167:1–5.

Webster, W. D., and J. K. Jones, Jr. 1982c. *Artibeus aztecus*. Mammalian Species 177:1–3.

Webster, W. D., and J. K. Jones, Jr. 1982d. *Artibeus toltecus*. Mammalian Species 178:1–3.

Webster, W. D., and J. K. Jones, Jr. 1983. *Artibeus hirsutus* and *Artibeus inopinatus*. Mammalian Species 199:1–3.

Webster, W. D., and J. K. Jones, Jr. 1984a. *Glossophaga leachii*. Mammalian Species 226:1–3.

Webster, W. D., and J. K. Jones, Jr. 1984b. A new subspecies of *Glossophaga mexicana* (Chiroptera: Phyllostomidae) from southern Mexico. Occasional Papers, Museum of Texas Tech University 91:1–5.

Webster, W. D., and J. K. Jones, Jr. 1985. *Glossophaga mexicana*. Mammalian Species 245:1–2.

Webster, W. D., and J. K. Jones, Jr. 1993. *Glossophaga commissarisi*. Mammalian Species 446:1–4.

Webster, W. D., J. K. Jones, Jr. and R. J. Baker. 1980. *Lasiurus intermedius*. Mammalian Species 132:1–3.

Webster W. D., L. W. Robbins, R. L. Robbins, and R. J. Baker. 1982. Comments on the status of *Mussonycteris harrisoni* (Chiroptera: Phyllostomidae). Occasional Papers, Museum of Texas Tech University 78:1–5.

Webster, W. D., N. D. Moncrief, J. R. Choate, and H. H. Genoways. 2011. Systematic revision of the northern short-tailed shrew, *Blarina brevicauda* (Say). Virginia Museum of Natural History Memoir 10:1–77.

Weksler, M. 2003. Phylogeny of Neotropical oryzomyine rodents (Muridae: Sigmodontinae) based on the nuclear IRBP exon. Molecular Phylogenetics and Evolution 29: 331–349.

Weksler, M., A. R. Percequillo, and R. S. Voss. 2006. Ten new genera of oryzomyine rodents (Cricetidae: Sigmodontinae). American Museum Novitates 3537:1–29.

Weksler, M., H. C. Lanier, and L. E. Olson. 2010. Eastern Beringian biogeography: historical and spatial genetic structure of singing voles in Alaska. Journal of biogeography 37:1414–1431.
Weksler, M., M. Aguilera, and F. Reid. 2016. *Oligoryzomys fulvescens*. The IUCN Red List of Threatened Species 2016:e.T15248A115126740.
Wells-Gosling, N., and L. R. Heaney. 1984. *Glaucomys sabrinus*. Mammalian Species 229:1–8.
Werbitsky, D., and C. W. Kilpatrick. 1987. Genetic variation and genetic differentiation among allopatric populations of *Megadontomys*. Journal of Mammalogy 68:305–312.
Weston, M. L. 1981. The *Ochonta alpina* complex: A statistical re-evaluation. Pp. 73–89, *in* Proceedings of the World lagomorph conference (Myers, K. K., and C. D. MacInnes, eds.). Guelp University Press, Guelph, Canada.
Wetzel, R. M. 1955. Speciation and dispersal of the southern bog lemming, *Synaptomys cooperi* (Baird). Journal of Mammalogy 36:1–20.
Wetzel, R. M. 1975. The species of *Tamandua* Gray (Edentata, Myrmecophagidae). Proceedings of Biological Society of Washington 88:95–112.
Wetzel, R. M. 1977. The Chacoan peccary, *Catagonus wagneri* (Rusconi). Bulletin of the Carnegie Museum of Natural History 3:1–36.
Wetzel, R. M. 1980. Revision of the naked-tailed armadillos, genus *Cabassous* McMurtrie. Annals of Carnegie Museum 49:323–357.
Wetzel, R. M. 1982. Systematics, distribution, ecology, and conservation of South American edentates. Pp. 345–375, *in* Mammalian biology in South America (Mares, M. A., and H. H. Genoways, eds.). Pymatuning Symposia in Ecology. Vol. 6. Special Publication Series. Pymatuning Laboratory of Ecology, University of Pittsburgh. Pittsburgh, U.S.A.
Wetzel, R. M. 1985a. The identification and distribution of Recent Xenarthra (= Edentata). Pp. 5–21, *in* The evolution and ecology of armadillos, sloths, and vermilinguas (Montgomery, G. G. ed.). Smithsonian Institution Press. Washington, U.S.A.
Wetzel, R. M. 1985b. Taxonomy and distribution of armadillos, Dasypodidae. Pp. 23–48, *in* The evolution and ecology of armadillos, sloths, and vermilinguas (Montgomery, G. G., ed.). Smithsonian Institution Press, Washington, U.S.A.
Wetzel, R. M., and E. Mondolfi. 1979. The subgenera and species of long-nosed armadillos, genus *Dasypus* L. Pp. 43–63, *in* Vertebrate ecology in the northern neotropics (Eisenberg, J. F., ed.). Smithsonian Institution Press, Washington, U.S.A.
Weyandt, S. E., and R. A. Van Den Bussche. 2007. Phylogeographic structuring and mammals: the case of the pallid (*Antrozous pallidus*). Journal of Biogeography 34:1233–1245.
Wharton, C. H. 1968. First records of *Microsorex hoyi* and *Sorex cinereus* from Georgia. Journal of Mammalogy 49:158–159.
Whitaker, J. O. Jr. 1972. *Zapus hudsonius*. Mammalian Species 11:1–7.
Whitaker, J. O. Jr. 1974. *Cryptotis parva*. Mammalian Species 43:1–8.
Whitaker, Jr. J. O. 1999a. Smoky Shrew (*Sorex fumeus*). Pp. 22–23, *in* The Smithsonian book of North American Mammals (Wilson, D. E., and S. Ruff, eds). Smithsonian Institution Press, Washington, U.S.A.
Whitaker, J. O., Jr. 1999b. Woodland jumping mouse *Napaeozapus insignis*. Pp. 665–666, *in* The Smithsonian book of North American Mammals (Wilson, D. E., and S. Ruff, eds.). Smithsonian Institution Press. Washington, U.S.A.
Whitaker, J. O., Jr. 1999c. Meadow jumping mouse *Zapus hudsonius*. Pp. 666–667, *in* The Smithsonian Book of North American Mammals (Wilson, D. E., and S. Ruff, eds.). Smithsonian Institution Press. Washington, U.S.A.
Whitaker, J. O. Jr, and T. W. French. 1984. Foods of six species of sympatric shrews from New Brunswick. Canadian Journal of Zoology 62:622–626.
Whitaker, J. O. Jr., and W. J. Hamilton. 1998. Mammals of the eastern United States. Cornell University Press. Ithaca, U.S.A.
Whitaker, J. O. Jr., and NatureServe 2018. *Reithrodontomys raviventris*. The IUCN Red List of Threatened Species 2018:e.T19401A22385344.
Whitaker, J. O. Jr., and R. E. Wrigley. 1972. *Napeozapus insignis*. Mammalian Species 14:1–6.
Whittaker, J. C., G. Hammerson, L. Master, and S. J. Norris. 2016. *Sorex dispar*. The IUCN Red List of Threatened Species 2016:e. T41394A115183478.
White, J. A. 1953. Genera and subgenera of chipmunks. University of Kansas Publications, Museum of Natural History 5:543–561.
Whorley, J. R., S. T. Álvarez-Castañeda, and G. J. Kenagy. 2004. Genetic structure of desert ground squirrels over a 20-degree-latitude transect from Oregon through the Baja California peninsula. Molecular Ecology 13:2709–2720.
Wiig, Ø., and R. W. Lie. 1984. An analysis of the morphological relationships between the hooded seals (*Cystophora cristata*) of Newfoundland, the Denmark Strait and Jan Mayen. Journal of Zoology (London) 203:227–240.
Wiig, Ø., S. Amstrup, T. Atwood, K. Laidre, N. Lunn, M. Obbard, E. Regehr, and G. Thiemann. 2015. *Ursus maritimus*. The IUCN Red List of Threatened Species 2015:e.T22823A14871490.
Wiley, R. W. 1980. *Neotoma floridana*. Mammalian Species 139:1–7.
Wilkins, K. T. 1986. *Reithrodontomys montanus*. Mammalian Species 257:1–5.
Wilkins, K. T. 1987. *Lasiurus seminolus*. Mammalian Species 280:1–5.
Wilkins, K. T. 1989. *Tadarida brasiliensis*. Mammalian Species 331:1–10.
Wilkins, K. T., and D. J. Schmidly. 1979. Identification and distribution of three species of pocket mice (genus *Perognathus*) in Trans-Pecos Texas. The Southwestern Naturalist 24:17–32.
Williams, D. F. 1978a. Taxonomic and karyologic comments on small brown bats, genus *Eptesicus*, from South America. Annals of Carnegie Museum 47:361–383.
Williams, D. F. 1978b. Systematics and ecogeographic variation of the Apache pocket mouse (Rodentia: Heteromyidae). Bulletin of Carnegie Museum of Natural History 10:1–57.
Williams, D. F. 1979. Checklist of California mammals. Annals Carnegie Museum Natural History 48:425–433.
Williams, D. F. 1984. Habitat associations of some rare shrews (*Sorex*) from California. Journal of Mammalogy 65:325–328.
Williams, D. F., and J. S. Findley. 1979. Sexual and size dimorphism in vespertilionid bats. The American Midland Naturlist 102:113–216.
Williams, D. F., and H. H. Genoways. 1979. A systematic review of the olive-backed pocket mouse, *Perognathus fasciatus* (Rodentia: Heteromyidae). Annals Carnegie Museum 48:73102.
Williams, D. F., and K. S. Kilburn. 1991. *Dipodomys ingens*. Mammalian Species 377:1–7.
Williams, D. F., H. H. Genoways, and S. E. Braun. 1993. Taxonomy. Pp. 38–196, *in* Biology of the Heteromyidae (Genoways, H. H., and J. H. Brown, eds.). American Society of Mammalogists.
Williams, L. R. 1999. *Geomys attwateri*. Pp. 481–483, *in* The Smithsonian book of North American Mammals (Wilson, D. E., and S. Ruff, eds.). Smithsonian Institution Press. Washington, U.S.A.
Williams, L. R., and G. N. Cameron. 1991. *Geomys attwateri*. Mammalian Species 382:1–5.

Williams, S. L., and H. H. Genoways. 1981. Systematic review of the Texas pocket gopher, *Geomys personatus* (Mammalia: Rodentia). Annals of Carnegie Museum of Natural History 50:435–473.
Williams, S. L., and J. Ramírez-Pulido. 1984. Morphometric variation in the volcano mouse, *Peromyscus alstoni* (Mammalia:Cricetidae). Annals Carnegie Museum 53:163–183.
Williams, S. L., and H. H. Genoways. 2008. Subfamily Phyllostominae Gray, 1825. Pp. 255–300, *in* Mammals of South America. Vol. 1. Marsupials, xenarthrans, shrews, and bats (Gardner, A. L., ed.). University of Chicago Press. Chicago, U.S.A.
Williams-Guillén, K., and E. I. Perfecto. 2010. Effects of agricultural intensification on the assemblage of leaf–nosed bats (Phyllostomidae) in a coffee landscape in Chiapas, Mexico. Biotropica 42:605–613.
Williamson, D. F. 2002. In the black. Status, management, and trade of the American black bear (*Ursus americanus*) in North America, World Wildlife Fund. Washington, U.S.A.
Willig, M. R. 1983. Composition, microgeographic variation, and sexual dimorphism in Caatingas and Cerrado bat communities from northeast Brazil. Bulletin of the Carnegie Museum of Natural History 23:1–131.
Willig, M. R. 1985. Reproductive activity of female bats from Northeast Brazil. Bat Research News 26:17–20.
Willner, G. R., G. A. Feldhamer, E. E. Zucker, and J. A. Chapman. 1980. *Ondatra zibethicus*. Mammalian Species 141:1–8.
Wilson, D. E. 1973. The systematic status of *Perognathus merriami* Allen. Proceedings of the Biological Society of Washington 86:175–192.
Wilson, D. E. 1976. Cranial variation in polar bears. Pp. 447–453, *in* Bears –their biology and management (Pelton, R., J. W. Lentfer, and G. E. Folk, eds.). International Union for the Conservation of Nature 40:1–467.
Wilson, D. E. 1982. Wolverine *Gulo gulo*. Pp. 644–652, *in* Wild Mammals of North America (Chapman, J. A., and G. A. Feldhamer, eds.). The Johns Hopkins University Press. Baltimore, U.S.A.
Wilson, D. E. 1991. Mammals of the Tres Marias Islands. Bulletin of the American Museum of Natural History 206:214–250.
Wilson, D. E., and R. K. LaVal. 1974. *Myotis nigricans*. Mammalian Species 39:1–3.
Wilson, D. E., and D. A. Reeder. 1993. Mammal species of the World. A taxonomic and geographic reference. Second edition. The Smithsonian Institution Press. Washington, U.S.A.
Wilson, G. M., and J. R. Choate. 1997. Taxonomic status and biogeography of the southern bog lemming, *Synaptomys cooperi*, on the central Great Plains. Journal of Mammalogy 78:444–458.
Whitaker, J. O., and W. J. Hamilton. 1998. Mammals of the Eastern United States. 3rd ed. Cornell University Press. Ithaca, U.S.A.
Wolf, J. B.W., D. Taut, and F. Trillmich. 2007. Galapagos and Californian sea lions are separate species: genetic analysis of the genus *Zalophus* and its implications for conservation management. Frontiers in Zoology 4:1–13.
Wolfe, J. L. 1982. *Oryzomys palustris*. Mammalian Species 176:1–6.
Wolfe, J. L., and A. V. Linzey. 1977. *Peromyscus gossypinus*. Mammalian Species 70:1–5.
Wood, A. E. 1935. Evolution and relationships of the Heteromyid rodents with new form from the Tertiary of western North America. Annals of Carnegie Museum 24:73–262.
Woods, C. A. 1973. *Erethizon dorsatum*. Mammalian Species 29:1–6.
Woods, C. A. 1985. Hystricognatha rodents. Pp 389–446, *in* Orders and families of recent mammals of the world (Anderson, S., and J. K. Jones, Jr. eds.). John Wild and Sons, New York.
Woods, C. A. 1992. Endangered: Florida saltmarsh vole (*Microtus pennsylvanicus dukecampbelli*). Pp. 131–139, *in* Rare and endangered biota of Florida: vol. 1. Mammals (Humphrey, S. R., ed.). University Press of Florida. Gainesville, Florida.
Woods, C. A., 1993. Suborder Hystricognathi. Pp.771–806, *in* Mammal species of the World. A taxonomic and geographic reference. Second ed. (Wilson, D. E., and D.A.M. Reeder, eds.). Smithsonian Institution Press, Washington and London in assoc. American Society of Mammalogists.
Wood, G. W., and R. H. Barrett. 1979. Status of wild pigs in the United States. Wildlife Society Bulletin 7:237–246.
Woods, C. A., and C. W. Kilpatrick. 2005. Infraorder Hystricognathi. Pp. 1538–1599, *in* Mammal Species of the World (Wilson, D. E., and D. M. Reeder, eds). Third edition. The Johns Hopkins University Press, Baltimore, U.S.A.
Woods, C. A., W. Post, and C. W. Kilpatrick. 1982. *Microtus pennsylvanicus* (Rodentia: Muridae) in Florida: a Pleistocene relict in a coastal salt-marsh. Bulletin of the Florida State Museum Biological Science 28:25–52.
Woodburne, M. O. 1968. The cranial myology and osteology of *Dicotyles tajacu*, the collared peccary, and its bearing on classification. Memoirs of the Southern California Academy of Sciences 7:1–48.
Woodhouse, S. W. 1852. Description of a new species of Pouched Rat of the genus *Geomys*, Raf. Proceedings of the Academy of Natural Sciences of Philadelphia 6:201–202.
Woodman, N. 2005. Evolution and Biogeography of the Mexican small-eared shrew of the *Cryptotis mexicana* group (Insectivora: Soricidae). Pp. 523–534, *in* Contribuciones mastozoologicas en Homenaje a Bernardo Villa (Sánchez-Cordero, V., and R. A. Medellin, eds.). Instituto de Biología e Instituto de Ecología, Universidad Nacional Autónoma de México y Comisión Nacional para el Conocimiento y Uso de la Biodiversidad. Distrito Federal, México.
Woodman, N. 2010. Two new species of shrews (Soricidae) from the western highlands of Guatemala. Journal of Mammalogy 91:566–579.
Woodman, N. 2012. Taxonomic status and relationships of *Sorex obscurus parvidens* Jackson, 1921, from California. Journal of Mammalogy 93:826–838.
Woodman, N. 2018. American Recent Eulipotyphla: Nesophontids, Solenodons, Moles, and Shrews in the New World. Smithsonian Contributions to Zoology, number 650:1–107.
Woodman, N. 2019. *Cryptotis tropicalis*. The IUCN Red List of Threatened Species 2019:e.T136757A22286227.
Woodman, N., and A. W. Ferguson. 2021. The relevance of a type locality: the case of *Mephitis interrupta* Rafinesque, 1820 (Carnivora: Mephitidae). Journal of Mammalogy 102: 1583–1591.
Woodman, N., and R. B. Stephens. 2009. At the foot of the shrew: manus morphology distinguishing closely related *Cryptotis goodwini* and *Crytotis griseoventris* (Mammaliae Soricidae) in Central America. Biological Journal of the Linnean Society 99:118–134.
Woodman, N., and R. M. Timm. 1999. Geographyc variation and evolutionary relationships among broad-clawed shrews of *Cryptotis goldmani*-group (Mammalia: Insectivore: Soricidae). Fieldiana: Zoology: New Series 91:1–35.
Woodman, N., and R. M. Timm. 2000. Taxonomy and evolutionary relationships of Phillips' small-eared shrew, *Cryptotis phillipsii* (Schaldach, 1966), from Oaxaca, Mexico (Mammalia: Insectivora: Soricidae). Proceedings of the Biological Society of Washington 113:339–355.

Woodman, N., J. O. Matson, J. J. McCarthy, R. P. Eckerlin, W. Bulmer, and N. Ordóñez-Garza. 2012. Distributional records of shrews (Mammalia, Soricomorpha, Soricidae) from northern Central America with the first records of *Sorex* from Honduras. Annals of Carnegie Museum 80:207–237.

Woodman, N., S. T. Álvarez-Castañeda, I. Castro-Arellano, and P. C. de Grammont. 2016a. *Megasorex gigas*. The IUCN Red List of Threatened Species 2016:e.T41454A22319710.

Woodman, N., J. Matson, A. D. Cuarón, and P. C. de Grammont. 2016b. *Cryptotis merriami*. The IUCN Red List of Threatened Species 2016:e. T136398A115207240.

Woodman, N., J. Matson, A. D. Cuarón, and P. C. de Grammont. 2016c. *Cryptotis parva*. The IUCN Red List of Threatened Species 2016:e. T41377A115182514.

Woodman, N., J. Matson, and I. Castro-Arellano. 2016d. *Sorex arizonae*. The IUCN Red List of Threatened Species 2016:e.T20396A115158374.

Woodman, N., J. Matson, A. D. Cuarón, and P. C. de Grammont. 2019. *Cryptotis alticola*. The IUCN Red List of Threatened Species 2019:e. T136789A22284844.

Woodward, S. M. 1994. Identification of *Sorex monticolus* and *Sorex vagrans* from British Columbia. Northwest Science 68:277–283.

Worley, K., C. Strobeck, S. Arthur, J. Carey, H. Schwantje, A. Veitch, and D. W. Coltman. 2004. Population genetic structure of North American thinhorn sheep (*Ovis dalli*). Molecular Ecology 13:2545–2556.

Wozencraft, W. C. 1989. Classification of the recent carnivora. Pp. 569–594, *in* Carnivore Behaviour, Ecology and Evolution (Gittleman, J. L., ed.). Chapman, and Hall. London, United Kingdom.

Wozencraft, W. C. 1993. Order Carnivora, Family Mustelidae. Pp. 309–325, *in* Mammal species of the world. A taxonomic and geographic reference (Wilson, D. E., and D. A. M. Reeder, eds.). Second edition. The Johns Hopkins University Press. Baltimore, U.S.A.

Wozencraft, W. C. 2005. Order Carnivora. Pp. 512–628, *in* Mammal species of the world. A taxonomic and geographic reference (Wilson, D. E., and D. A. M. Reeder, eds.). Third edition. The Johns Hopkins University Press. Baltimore, U.S.A.

Wright, D. B. 1989. Phylogenetic relationships of *Catagonus wagneri*: Sister taxa from the Tertiary of North America. Pp. 281–308, *in* Advances in Neotropical mammalogy (Redford, K. H., and J. F. Eisenberg, eds). Sandhill Crane Press, Gainesville, U.S.A.

Wright, P. L. 1953. Intergradation between *Martes americana* and *Martes caurina* in Western Montana. Journal of Mammalogy 34:70–87.

Wynen, L. P., S. D. Goldsworthy, S. J. Insley, M. Adams, J. W. Bickham, J. Francis, J. P. Gallo, A. R. Hoelzel, P. Majluf, R.W.G. White, and R. Slade. 2001. Phylogenetic relationships within the eared seals (Otariidae: Carnivora): Implications for the historical biogeography of the family. Molecular Phylogenetics and Evolution 21:270–284.

Yamaguchi, N., A. Kitchener, C. Driscoll, and B. Nussberger. 2015. *Felis silvestris*. The IUCN Red List of Threatened Species 2015:e. T60354712A50652361.

Yancey, F. D., II. 1997. The mammals of Big Bend Ranch State Park. Museum of Texas Tech University, Lubbock.

Yancey, F. D., J. R. Goetze, and C. Jones. 1998a. *Saccopteryx bilineata*. Mammalian Species 581:1–5.

Yancey, F. D., J. R. Goetze, and C. Jones. 1998b. *Saccopteryx leptura*. Mammalian Species 582:1–3.

Yates, T. L. 1978. The systematics and evolution of North American moles (Insectivora: Talpidae). Ph. D. thesis. Texas Tech University. Lubbock, U.S.A.

Yates, T. L. 1999a. American shrew mole *Neurotrichus gibbsii*. Pp. 56–57, *in* The Smithsonian book of North American Mammals (Wilson, D. E., and S. Ruff, eds.). Smithsonian Institution Press. Washington, U.S.A.

Yates, T. L. 1999b. Broad-footed mole *Scapanus latimanus*. Pp. 57–58, *in* The Smithsonian book of North American Mammals (Wilson, D. E., and S. Ruff, eds.). Smithsonian Institution Press. Washington, U.S.A.

Yates, T. L. 1999c. Townsend's mole *Scapanus townsendii*. Pp. 60–61, *in* The Smithsonian book of North American Mammals (Wilson, D. E., and S. Ruff, eds.). Smithsonian Institution Press. Washington, U.S.A.

Yates, T. L. 1999d. Eastern mole *Scalopus aquaticus*. Pp. 63–65, *in* The Smithsonian book of North American Mammals (Wilson, D. E., and S. Ruff, eds.). Smithsonian Institution Press. Washington, U.S.A.

Yates, T. L., and I. F. Greenbaum. 1982. Biochemical Systematics of North American Moles (Insectivora: Talpidae). Journal of Mammalogy 63:368–374.

Yates, T. L., and R. J. Pedersen. 1982. Moles. Pp. 37–51, *in* Wild mammals of North America: biology, management, and economics (Chapman, J. A., and G. A. Feldhamer, eds.). The Johns Hopkins University Press, Baltimore.

Yates, T. L., and J. Salazar-Bravo. 2005. A revision of *Scapanus latimanus*, with the revalidation of a species of Mexican mole. Pp. 489–506, *in* Contribuciones mastozoológicas en homenaje a Bernardo Villa (Sánchez-Cordero, V., and R. A. Medellín, eds.). Instituto de Biología e Instituto de Ecología, Universidad Nacional Autónoma de México y Comisión Nacional para el Conocimiento y Uso de la Biodiversidad. Distrito Federal, México.

Yates, T. L., and D. J. Schmidly. 1975. Karyotype of the Eastern mole (*Scalopus aquaticus*), with comments on the karyology of the family Talpidae. Journal of Mammalogy 56:902–905.

Yates, T. L., and D. J. Schmidly. 1977. Systematics of *Scalopus aquaticus* (Linnaeus) in Texas and adjacent states. Occasinal Papers, Museum of Texas Tech University 45:1–36.

Yates, T. L., and D. J. Schmidly. 1978. *Scalopus aquaticus*. Mammalian Species 105:1–4.

Yates, T. L., R. J. Baker, and R. K. Barnett. 1979a. Phylogenetic analysis of karyological variation in three species of peromyscine rodents. Systematic Zoology 28:40–48.

Yates, T. L. H. H. Genoways, and J. K. Jones, Jr. 1979b. Rabbits (genus *Sylvilagus*) of Nicaragua. Mammalia 43:113–124.

Yee, D. A. 2000. *Peropteryx macrotis*. Mammalian Species 643:1–4.

Yensen, E. 1991. Taxonomy and distribution of the Idaho ground squirrel, *Spermophilus brunneus*. Journal of Mammalogy 72: 583–600.

Yensen, E. 2000. *Urocitellus brunneus*. The IUCN Red List of Threatened Species 2000:e.T20497A9196134.

Yensen, E. 2018. *Urocitellus endemicus*. The IUCN Red List of Threatened Species 2018:e.T20498A117636227.

Yensen, E. 2019. *Urocitellus mollis*. The IUCN Red List of Threatened Species 2019:e.T116989381A116989399.

Yensen, E., and NatureServe. 2017a. *Urocitellus canus*. The IUCN Red List of Threatened Species 2017:e.T42465A22265551.

Yensen, E., and NatureServe. 2017b. *Urocitellus elegans*. The IUCN Red List of Threatened Species 2017:e.T42467A22265347.

Yensen, E., and P. W. Sherman. 1997. *Spermophilus brunneus*. Mammalian Species 560:1–5.

Yensen, E., and T. Tarifa. 2003. *Galictis vittata*. Mammalian Species 727:1–8.

Yensen, E., and M. Valdés-Alarcón. 1999. Family Sciuridae. Pp. 239–320, *in* Mamíferos del noroeste de México (Álvarez-Castañeda, S. T. and J. L. Patton, eds.). Centro de Investigaciones Biológicas del Noroeste, S. C. La Paz, México.
Young, C. J. and J. K. Jones, Jr. 1982. *Spermophilus mexicanus*. Mammalian Species 164:1–4.
Young, C. J. and J. K. Jones, Jr. 1983. *Peromyscus yucatanicus*. Mammalian Species 196:1–3.
Young, C. J., and J. K. Jones, Jr. 1984. *Reithrodontomys gracilis*. Mammalian Species 218:1–3.
Young, S. P., and E. A. Goldman. 1944. The wolves of North America. The American Wildlife Institution. Washington, U.S.A.
Young, S. P., and E. A. Goldman. 1946. The puma: Mysterious American cat. American Wildlife Institute. Washington, U.S.A.
Young, S. P., and H. H. T. Jackson. 1951. The clever coyote. American Wildlife Management Institute. Washington, U.S.A.
Youngman, P. M. 1967. A new subspecies of varying lemming, *Dicrostonyx torquatus* (Pallas), from Yukon Territory (Mammalia, Rodentia). Proceedings of the Biological Society of Washington 80:31–34.
Youngman, P. M. 1975. Mammals of the Yukon Territory. National Museum Canada, Publication Zoology 10:1–192.
Youngman, P. M. 1982. Distribution and systematics of the European mink, *Mustela lutreola* Linnaeus, 1761. Acta Zoologica Fennica 166:1–48.
Zagorodnyuk, I. V. 1990. Kariotipicheskaya izmenchuivost'I systematika serykh polevok (Rodentia, Arvicolini). Soobshchenic 1. Vidovoi sostav I khromosomnye chisla. Vestnik Zoologii 2:26–37.
Zarza, H., G. Ceballos, and M. A. Steele. 2003. *Marmosa canescens*. Mammalian Species 725:1–4.
Zamora-Gutierrez, V., and J. Ortega. 2020. *Lichonycteris obscura* (Chiroptera: Phyllostomidae). Mammalian Species 52:165–172.
Zegers, D. A. 1984. *Spermophilus elegans*. Mammalian Species 214:1–7.
Zimmerman, E. G. 1970. Karyology, systematics, and chromosomal evolution in the rodent genus, *Sigmodon*. Publications of the Museum, Michigan State University, Biological Series 4:385–454.
Zimmerman, E. G. 1999. *Geomys bursarius*. Pp. 485–486, *in* The Smithsonian book of North American Mammals (Wilson, D. E., and S. Ruff). Smithsonian Institution Press. Washington, U.S.A.
Zimmerman, E. G., and M. R. Lee. 1968. Variation in the chromosomes of the cotton rat, *Sigmodon hispidus*. Chromosoma 24:243–250.
Zimmerman, E. G., B. J. Hart, and C. W. Kilpatrick. 1975. Biochemical genetics of the *boylii* and *truei* groups of the genus *Peromyscus* (Rodentia). Comparative Biochemistry and Physiology 52B:541–545.
Zimmerman, E. G., C. W. Kilpatrick, and B. J. Hart. 1978. The genetics of speciation in the rodents genus *Peromyscus*. Evolution 32:565–579.
Zeppelin, T. K., and A. J. Orr. 2010. Stable isotope and scat analyses indicate diet and habitat partitioning in northern fur seals *Callorhinus ursinus* across the eastern Pacific. Marine Ecology Progress Series 409:241–253.

# Addenda

Page 500. *Chaetodipus siccus* (Osgood 1907). A new subspecies was described as *Chaetodipus siccus liae* (Alvarez-Castañeda 2024). Lia's Pocket Mouse, ratón de abazones de Isla Cerralvo is known from Los Planes Basing in the Baja California Peninsula, state of Baja California Sur, México. *Chaetodipus siccus siccus* is restricted to Cerralvo Island or Jaque Coustou Island, in the Gulf of California, México.

## References


Álvarez-Castañeda, S. T. 2024. Morphological variation in the Cerralvo Island pocket mouse *Chaetodipus siccus* from the Baja California Peninsula, México. Therya 15:218–229.

Osgood, W. H. 1907. Four new pocket mice. Proceedings of the Biological Society of Washington 20:19–22.




S. T. Álvarez-Castañeda, *Mammals of North America - Volume 2*, https://doi.org/10.1007/978-3-031-50825-7

# Glossary

**Abdomen** The part of the body (excepting the back) between the thorax (rib basket) and the pelvis.
**Accessory cusp** A small cusp usually situated peripheral to main biting or crushing surface of a tooth.
**Aerial** Pertaining to flying. Bats are the only truly aerial mammals.
**Alisphenoid channel** A passage in the sphenoid bone of the skull leading from just anterior to the foramen lacerum in the middle cranial fossa to the pterygopalatine fossa.
**Alisphenoid** A bony process of the sphenoid bone; there is one on each side, extending from the side of the body of the sphenoid and curving upward, laterally, and backward.
**Allopatric** Mode that occurs when biological populations become geographically isolated from each other.
**Alveolar (plural, alveoli)** Small cavity or pit, as a socket for a tooth. Alveolar length of a tooth-row therefore denotes the length of the row of the teeth, taken from the posteriormost place where the back tooth emerges from the bone to the anteriormost point where the front tooth in the row emerges from the bone.
**Alveolus** See alveolar.
**Ambulatory** Pertaining to a walking habit, as in bears and raccoons.
**Angular process** On a mandible, a third process at the back of the mandible and at the ramus level.
**Annulation** A circular or ringlike formation, as of the dermal scales on the tail of a mammal where one ring of scales that extends entirely around the tail is succeeded, posteriorly, by other rings.
**Antebrachial membrane** Wing membrane of bats between the arm and the anterior edge of the wing.
**Anterior palatine foramen** See incisive foramen.
**Anterocone** A cusp at the front of the tooth that may be divided into anterolabial and anterolingual conules in an upper teeth.
**Anteroconid** A cusp at the front of the tooth that may be divided into anterolabial and anterolingual conulids in a lower teeth.
**Anterolabial cingulum** A crest before the protoconid and protoflexid in a lower teeth.
**Anteroloph** A crest between the paracone and the anterolabial conule that may be connecting to a parastyle in an upper teeth.
**Anterolophid** A crest between the metaconid and the anterolabial conulid that may be connecting to a metastylid in a lower teeth.
**Antitragus** The antitragus is located just above the earlobe and points anteriorly. It is separated from the tragus by the intertragic notch.
**Antler** Antlers are a single structure composed of bone, cartilage, fibrous tissue, skin, nerves, and blood vessels, shed and regrow each year.
**Arboreal** Inhabiting or frequenting trees-contrasted with fossorial, aquatic, and cursorial.
**Articular process** On a mandible, the process ends in the articular condyle.
**Articulate** To join or connect two adjacent bones.
**Auditory bulla (plural, auditory bullae)** A hollow, bony prominence of rounded form (in most mammals formed by the tympanic bone) partly enclosing structures of the middle and inner ear.
**Auditory canal** Bony tubular passage between tympanic membrane and external auditory meatus.
**Auditory meatus (external)** Pathway running from the outer ear to the middle ear.

 
S. T. Álvarez-Castañeda, *Mammals of North America - Volume 2*, https://doi.org/10.1007/978-3-031-50825-7

**Auditory meatus (internal)** Pathway within the petrous part of the temporal bone of the skull between the posterior cranial fossa and the inner ear.
**Baculum** Bone found in the penis of many placental mammals.
**Basal length** Distance on skull from the anteriormost inferior border of the foramen magnum to a line connecting the anteriormost parts of the premaxillary bones.
**Basihyal** Helps them make their loud vocalizations.
**Basilar length** Distance on skull from the anteriormost inferior border of the foramen magnum to a line connecting the posteriormost margins of the alveoli of the first upper incisors.
**Basioccipital** Unpaired bone at base of occipital complex.
**Basisphenoid** Part of the sphenoid bone which formed part of the floor of the braincase and lay immediately above.
**Bead** A salient, rounded cordlike projecting ridge of bone, as in certain rodents where the superior border of the orbit is beaded.
**Bone** Hard supportive tissue cons1sting of cells distributed in a matrix of fibrous protein (collagen) and salt (chiefly calcium and phosphate).
**Brachydont** Type of dentition characterized by low-crowned teeth.
**Braincase** The part of the skull enclosing the brain.
**Bristle** Stiff hair.
**Browtine** First tine above the base on an antler.
**Bunodont** Molars with low cusps and rounded hills, which are effective crushing devices and often basically quadrate in shape.
**Calcar** In bats a process connected with the calcaneum (heel bone), helping to support the edge of the fold of skin that extends between the leg and tail.
**Canal** Perforation, or foramen, that tends to be elongated into a tube. Examples are alisphenoid canal, infraorbital canal.
**Cancellous** Having a spongy or porous structure.
**Canine** One of four basic kinds of mammalian teeth; anteriormost in maxilla (and counterpart of dentary); frequently elongate, unicuspid, and single-rooted; never more than one per quadrant.
**Caniniform** Having the general shape of a canine.
**Carnassial pair** Pair of large bladelike teeth (last upper premolar and first lower molar) that occlude with scissorlike (shearing) action, possessed by most modem carnivores.
**Carnassial** Blade-like teeth especially adapted for slicing and chopping.
**Carnivore** An animal that preys on other animals; an animal that eats the flesh of other animals; especially any mammal of the order carnivora.
**Cartilage** Relatively soft supportive tissue consisting of rounded cells in a matrix of polysaccharides and fibrous protein (collagen).
**Cheekteeth** Teeth behind the canines can be premolar and molars.
**Cingulum** A shelf at the margin of a tooth.
**Claw** Curved, pointed appendage found at the end of a toe or finger.
**Condylar (articular) process** On a mandible, the process ends in the articular condyle.
**Condylobasal length** Least distance on skull from a line connecting the posteriormost projections of the exoccipital condyles to connecting the anteriormost projections of the premaxillary bones.
**Condyloid process** See articular process.
**Conspecific** Belong to the same species.
**Coronoid process** The upward projecting process of the posterior part of the mandible, giving attachment on its outward side to the masseter muscle and on its inner side to the temporal muscle.
**Cranium** Collectively, bones that form upper part of skull (contains upper teeth and braincase); lower part of skull is the mandible.
**Crepuscular** Active at twilight, i.e., at dusk or at dawn.
**Cursorial** Pertaining to a running habit, as in artiodactyls and perissodactyls.
**Cusp** A pointed structure on a tooth.
**Cuspidate** Presence of cusps on a tooth.
**Cusplet (or secondary cusp)** A small cusp.
**Dactylopatagium** Membranous structure that assists an animal in flight (bats), is the portion found within the digits.
**Deciduous dentition (or milk teeth)** Juvenile teeth, those that appear first in lifetime of a mammal, consisting (if complete) of incisors, canines, and premolars; generally replaced by adult (permanent) dentition.
**Decurve** To curve downward.
**Deflected** Bent outward or laterally.

**Dental formula (plural, formulae)** A brief method for expressing the number and kind of teeth of mammals. The abbreviations i. (incisor), c. (canine), p. or pm. (premolar), and m. (molar) indicate the kinds in the permanent dentition, and the number in each jaw is written like a fraction, the figures above the horizontal line showing the number in the upper jaw, and those below, the number in the lower jaw. Uppercase is used for upper teeth and lowercase for lower. The dental formula of an adult coyote is i. 3/3 c. 1/1; p. 4/4; m. 2/3. [d.i., d.c. And d.p. (in place of i., c. And p.) Designate deciduous ("milk") teeth.]
**Dentary** Bone of lower jaw, forming half of mandible.
**Dentine** A calcareous material, harder and denser than bone, which composes the principal mass of a tooth.
**Dentition** The teeth, considered collectively, of any animal.
**Depth of skull at anterior margin of basioccipital** Measured from anterior end of ventral face of basioccipital, excluding median ridge, vertically to dorsal face of parietal excluding sagittal crest.
**Depth of skull at posterior borders of first upper molar** Measured from ventral face of palatine bones at posterior edge of upper molars to dorsal face of frontals in plane of postorbital processes of frontals.
**Dermal plates** Bony structure derived from intramembranous ossification forming components of the vertebrate skeleton including much of the skull, jaws, shoulder girdle and fin spines rays (lepidotrichia), and the shell of armadillos. Name to as a dermal bone; investing bone; membrane bone.
**Deuterocone** One of the cusps of a premolar tooth of a mammal corresponding in position (anteromedial) to the protocone of a true molar.
**Dewclaws (or dewhoofs)** Clawed or hoofed remnants of side toes in many carnivores and artiodactyls, located just above the main functional digits.
**Diastema** A vacant place or gap between teeth in a jaw.
**Digital web** Membranes between the fingers.
**Digitigrade** Walking on the toes with the heel and wrist permanently raised.
**Digits** Toes.
**Dilambdodont** Molars having a distinct ectoloph, shaped like two lambdas or a W shape.
**Distribution range** The geographic limit of a particular taxon's distribution is its range.
**Diurnal** Active by day-opposed to nocturnal.
**Ectoloph** A crest connecting the ectostyle to the mesocone in a upper teeth.
**Ectolophid** A crest connecting the ectostylid to the mesoconid in a lower teeth.
**Ectostyle** A stylid between the protocone and the hypocone, in the hypoflexe in an upper teeth.
**Ectostylid** A stylid between the protoconid and the hypoconid, in the hypoflexid in a lower teeth.
**Enamel** Hardest substance of the mammalian body and forming a thin layer that caps or partly covers a tooth.
**Enamel plate** A segment or portion of a tooth that is heavily invested with enamel (e.g., on pocket gopher teeth).
**Endemic** State of a species being found in a single defined geographic location.
**Enteroloph** A crest connecting the enterostyle to the mesocone in an upper teeth.
**Enterostyle** A style between the protocone and the hypocone, in the hypoflexus in an upper teeth.
**Entoconid** One of the main cusps, at the posterolingual side in a lower teeth.
**Entolophid** A crest attaching the entoconid to the hypoconid or median murid in a lower teeth.
**Ethmoid** Unpaired bone in the skull that separates the nasal cavity from the brain
**External auditory meatus** Round bony oriface or opening that is covered by eardrum.
**Faeces** (singular and plural). Intestinal excrement.
**Feces** (see faeces).
**Femur** (plural, femora). The proximal bone of the hindlimb.
**Fenestrated** Having a network of irregular perforations or holes.
**Fenestration** Opening; in current work applied to specialized openings in crania of lagomorphs and cervids.
**Flange** A laterally compressed or flattened portion of bone that increases the surface area.
**Flipper** A forelimb of an aquatic animal.
**Foramen (pl. foramina)** A hole, opening, or perforation through bone. An example is the foramen magnum.
**Foramen lacerum** A hole in the base of the skull located between the sphenoid bone, the apex of the petrous part of the temporal bone, and the basilar part of the occipital bone.
**Foramen magnum** The large opening in the back of a skull through which the spinal cord passes to become the medulla oblongata of the brain.
**Foramen** An open hole that is present in a bone, typically allow muscles, nerves, arteries, veins, or other structures to connect one part of the body with another.
**Foramens** Plural foramen.
**Forearm** The part of the forelimb between the elbow and wrist.

**Forelimb** Paired articulated appendages (limbs) attached on the cranial (anterior) end of a terrestrial tetrapod vertebrate's torso.

**Formina** Plural foramen.

**Fossa (pl. fossae)** A depression, usually forming a site of muscular attachment or bone articulation. Examples are glenoid fossa, mandibular fossa, masseteric fossa, temporal fossa.

**Fossil** Any preserved remains, impression, or trace of any once-living thing from a past geological age.

**Fossorial** Fitted for digging.

**Frontal** Pertaining to or designating the bone (paired) immediately in front of the parietal bone and behind the nasal.

**Frontal appendages** Bone growth (horns or antlers) arising from frontal bones.

**Gestation period** The period of carrying young in the uterus, as applied to placental mammals; the period of pregnancy.

**Gland** Synthesizes substances (such as hormones) for release into the bloodstream (endocrine gland) or into cavities inside the body or its outer surface (exocrine gland).

**Glandular sac** A sac which contains a gland.

**Glenoid fossa** A shallow, pyriform articular surface, which is located on the lateral angle of the scapula and articulates with the head of the humerus.

**Guard-hairs** The stiffer, longer hairs that grow up through the limber, shorter hairs (fur) of a mammal's pelage.

**Habitat** The kind of environment in which a species of organism is normally found.

**Hallux** Used to refer to the big toe of the hindfeet.

**Hamular process of pterygoid** A hook like process on the pterygoid bone.

**Heterodont** A dentition in which there are teeth of different forms (e.g., incisors and canines).

**Heterothermic** Having a variable body temperature.

**Hibernation** Torpidity especially in winter; the bodily temperature approximates that of the surroundings; the rate of respiration and the heart beat ordinarily are much slower than in an active mammal.

**Hindlimb** One of the paired articulated appendages (limbs) attached on the caudal (posterior) end of a terrestrial tetrapod.

**Homodont** Animal which possesses more than a single tooth morphology.

**Hoof (pl. hoofs or hooves)** Tip of a toe of an ungulate mammal, strengthened by a thick and horny keratin covering.

**Horizontal ramus** Lower jaw, the ramus baring the teeth, and anterior to the vertical ramus.

**Horn** A pointed, bony projection on the head of various animals, either the "true" horn, or other horn-like growths.

**Hybrid** An offspring resulting from cross-breeding.

**Hyoid bone** Helps them make their loud vocalizations.

**Hypocone** One of the main cusps, at the posterolingual side in an upper teeth.

**Hypoconid** One of the main cusps, at the posterolabial side in a lower teeth.

**Hypsodont** Dentition is characterized by high-crowned teeth and enamel that extends far past the gum line, which provides extra material for wear and tear.

**Incisiform** Having the general form of an incisor.

**Incisive foramen** (also referred to as anterior palatine foramen or palatine slit).

**Incisive foramina** The anterior palatine foramina (singular, foramen), of which there are two, in the bony roof of the anterior part of the cavity of the mouth at the juncture of the premaxillary bones and maxillary bones; transmit nasal branches of palatine arteries and nasopalatine ducts of jacobson.

**Incisor** Pertaining to or designating one of the teeth in front of the canine tooth; those in the upper jaw invariably are in the premaxillary bone.

**Inflated** Enlarged or expanded, not flattened or compressed.

**Inflected** Bent inward or medially. insertion – the site of attachment of a muscle (usually on a bone) on the more movable of the two elements, or bones, that are joined by the muscle.

**Infraorbital canal** A canal through the maxillary bone from the orbit to the face.

**Infraorbital foramen** Opening in maxilla from orbit onto the face (rostrum of cranium).

**Inguinal** Pertaining to or in the region of the groin.

**Insectivorous** Eating insects; preying or feeding on insects.

**Interfemoral membrane** In a bat the fold of skin stretching from hind legs to tail. The uropatagium.

**Interorbital breadth** Least distance across top of skull between orbits (eye sockets).

**Interorbital constriction** The least distance across the top of the skull between the orbits (eye sockets).

**Interorbital region** The region between the eye sockets; the region of the skull between the rostrum and the braincase.

**Interparietal** Pertaining to or designating the bone (rarely paired) immediately in front of the supraoccipital bone and between the two parietal bones.

**Interpterygoidea fossa** It is located in the mid-ventral part of the skull, posterior to the bony palate. Variations in it are important for the identification of some species.

**Jugal** Often called the malar or zygomatic. It is connected to the quadratojugal and maxilla, as well as other bones, which may vary by species.

**Jugo-maxillary suture** Juncture between jugal and maxillary bones at anterior margin of zygomatic arch.

**Jugular foramen** See posterior lacerate foramen.

**Karyotype** A preparation of the complete set of metaphase chromosomes in the cells of a species or in an individual organism, sorted by length, centromere location and other features.

**Labial** Pertaining to lips, for example labial side of tooth is that side nearer lips rather than tongue; lateral surface of a tooth.

**Lacerum posterior** Triangular hole in the base of skull, located between the sphenoid bone, the apex of the petrous part of the temporal bone, and the basilar part of the occipital bone.

**Lachrymal** Small and fragile bone of the facial skeleton, situated at the front part of the medial wall of the orbit.

**Lacrimal** See lachrymal.

**Lacrimal pit** Pit or opening in lacrimal bone containing tear duct.

**Lambdoidal ridge** Bony ridge formed at juncture of occiput and parietal bones.

**Lamboidal crest** See lambdoidal ridge

**Lingual** Pertaining to tongue, for example, lingual side of tooth is that side nearer tongue rather than lips, also medial surface of a tooth.

**Litter** The two or more young brought forth at one birth by a female mammal.

**Loph** A combining form used as the terminal part of certain words and denoting the ridges (or areas) composed of several cusps and styles on the occlusal face of a tooth, as protoloph.

**Lophodont** Teeth with differentiating patterns of ridges or lophs of enamel interconnecting the cusps on the crowns.

**Lyre-shaped** Shaped as the lyre, a musical instrument.

**Malar** See jugal.

**Mamma** The glandular organs for secreting milk.

**Mammae (singular, mamma)** The glandular organs for secreting milk.

**Mandible tooth-row** The row of teeth in one mandible bone; in most mammals all the premolars and molars on one side of the lower jaw.

**Mandible** Movable part of the jaw.

**Mandibular fossa (or glenoid fossa)** Concavity on ventral surface of zygomatic arm of squamosal with which the dentary articulates.

**Marsupium** The pouch is a fold of skin with a single opening that covers the teats in marsupials. Inside the pouch, the blind offspring attaches itself to one of the mother's teats and remains attached for as long as it takes to grow and develop to a juvenile stage.

**Mask** Coloration pattern in the rostrum of the species contrasting with the general color of the head.

**Masseter muscle** A muscle often consisting of several bands that originates on and adjacent to the zygomatic arch and inserts mostly in the masseteric fossa of the lower jaw.

**Mastoid breadth** Greatest distance across mastoid bones perpendiciJar to long axis of skull.

**Mastoid bulla** The part of bullar region covered by mastoid (mastoidal) bone.

**Mastoid** Designating or pertaining: to the mastoid bone (paired) or its process. This bone is bounded by the squamosal bone, the exoccipital bone, and the tympanic bone.

**Maxilla** Bone of the jaw formed from the fusion of two maxillary bones.

**Maxillary breadth** Width of skull from some designated place on the lateral face of the right maxillary bone (maxilla) to the corresponding place on the left maxillary bone; in shrews, across the ends of the zygomatic processes of the two maxillary bones.

**Maxillary tooth-row** The row of teeth in one maxillary bone; in most mammals all the premolars and molars on one side of the upper jaw.

**Mesaxonic** Having the axis of the foot formed by the middle digit (relating to the Perissodactyla).

**Mesocone** A conule in the median mure where the mesoloph is attached to it in an upper teeth.

**Mesoconid** A conulid in the median murid where the mesolophid is attached to it in a lower teeth.

**Mesolophid** A crest in front of the entoflexid, connected to the median murid.

**Mesopterygoid fossa** Shallow area posterior to internal nares and between pterygoid bones.

**Mesostyle** A style at the labial margin between the paracone and metacone in a upper teeth.

**Metabolic water** Water formed as an end product of combustion of foodstuffs in an animal's body.

**Metacarpal** Bone of the hand or forefoot between the wrist and fingers; when all the digits are present there are five more or less elongated metacarpal bones, one at the base of each digit.

**Metacone** One of the main cusps, at the posterolabial side in an upper teeth.

**Metaconid** One of the main cusps, at the anterolingual side in a lower teeth.
**Metastylid** A stylid in front of the metaconid in a lower teeth.
**Milk teeth** See deciduous dentition.
**Miocene** See geologic time.
**Globular** Globe-like, spherical.
**Molar** Teeth behind the premolar teeth, used primarily to grind food during chewing.
**Molariform** Molar-like teeth, either premolars or true molars.
**Molt (moult)** In a mammal, the act or process of shedding or casting off the hair, or outer layer of skin or horns; most mammals shed the hair once, twice, or three times annually. The cast off covering (obsolete). As a verb: to be shed (intransitive) or to shed (transitive).
**Muzzle** Projecting parts of the face (including the nose and mouth) of an animal.
**Nail** Keratinized projection at the tips of the digits; usually short, flat, and blunt.
**Nares** Openings, external and internal, of the nasal passage.
**Nasal (paired)** On the dorsal surface of the skull at its anterior end, form the bridge of the upper one third of the nose.
**Nasal leaf** Often large, lance-shaped nose, found in bats.
**Nasal** Paired bone of cranium situated on anterodorsal surface of skull.
**Nocturnal** Active by night-opposed to diurnal.
**Nostril** Either of the two orifices of the nose.
**Occipital condyle** Surface of articulation between cranium and first cervical vertebra (atlas); two such condyles in mammals, on either side of foramen magnum.
**Occipital crest** Bony ridge formed at occipital bones.
**Occipital** Cranial bone and the main bone of the back and lower part of the braincase.
**Occipitonasal length** Least distance between two vertical lines, one touching the posteriormost part of the skull above the foramen magnum (opening for the spinal cord) and the other touching the anteriormost part of the nasal bones or a nasal bone.
**Occlusal** Of or pertaining to the grinding or biting (occluding) surface of a tooth.
**Opposable** Capable of being placed opposite something else; said of the first toe of an opossum in the sense that it can be placed opposite each of the other toes on that same foot.
**Orbit** The cavity in the skull in which the eye and its appendages are situated; the eye socket.
**Orbital ring** Circular spot around the eye that generally contrasts with the rest of the rostrum coloration.
**Orbitonasal length** Distance on anterior part of skull from posterior margin of base of postorbital process of frontal bone to posteriormost part of anterior border of nasal bone on same side of skull.
**Origin** The site of attachment of a muscle (usually on a bone) on the less movable of the two elements, or bones, that are joined by the muscle.
**Os clitoris** Bone found in the clitoris of many placental mammals.
**Os penis** Bone found in the penis of many placental mammals.
**Osseous** Composed of, or resembling, bone; bony. Osseous tissue is bony tissue.
**Overhairs** The longer hairs of the pelage of a mammal that project above the fur (shorter hairs).
**Palatal bridge** Bony tooth-bearing plate (somewhat raised, part of paired maxillae) behind diastema, especially well developed in lagomorphs and rodents.
**Palatal emargination** Indentation at anterior margin of rostrum (in premaxillae), particularly in bats.
**Palatal** Bony roof of the mouth, made up of two palatine bones, two maxillary bones, and two premaxillary bones.
**Palate** The roof of the mouth, consisting of the structures that separate the mouth from the nasal cavity. The bony palate is composed of the following bones: premaxillae, maxillae, and palatines.
**Palatine slit** See incisive foramen.
**Palmated** Form like the palm of a hand; term applied to antlers in which at least some spaces between tines are filled with bony growth.
**Paracone** One of the main cusps, at the anterolabial side in an upper teeth.
**Parastyle** A style in front of the paracone in an upper teeth.
**Paraxonic** Having the axis of the foot between the third and fourth digits
**Parietal** Pertaining to or designating the parietal bone (paired) roofing the braincase. This bone is behind the frontal bone and in front of the occipital bones.
**Paroccipital process** Bony projection extending ventrally from, or located ventrally on, parocciptal bone.
**Patagium** Membranous structure that assists an animal in gliding or flight and stretches between an animal's hind limbs is called the uropatagium or the interfemoral membrane.
**Pectoral** Pertaining to, or situated or occurring in or on, the chest.

**Pelage** Hair, fur.
**Pencil** Tuft of fur or hair, as a black pencil on the end of the tail of a mammal.
**Penis bone** Bone found in the penis of many placental mammals.
**Pentadactyl** Five toed.
**Perforate** Pierced by an opening or openings.
**Petrous** Part of the temporal bone is pyramid-shaped and is wedged in at the base of the skull between the sphenoid and occipital bones.
**Phalanx (plural, phalanges)** A bone, in a finger, distal to the metacarpus or a bone, in a toe, distal to the metatarsus.
**Pinna (plural, pinnae)** The projecting part of an ear.
**Plagiopatagium** Membranous structure that assists an animal in flight (bats), is the portion found between the last digit and the hindlimbs.
**Plantar tubercles** Glandular and swollen areas found on the mammals soles. The morphological relationships of such integumentary annexes is a taxonomic characteristic.
**Plantigrade** Walking with the toes and metatarsals flat on the ground.
**Pollex** Used to refer to the big toe of the forefeet.
**Population** A group of organisms of the same species who inhabit the same particular geographical area and are capable of interbreeding.
**Postauricular** Situated behind the auricle (pinna) of the ear, as a postauricular patch (ordinarily referring to a patch of fur differing in color from surrounding fur).
**Postcanine teeth** Collectively, teeth behind canines (premolars and molars).
**Posterior lacerate foramen (or jugular foramen)** Opening in basicranial region between tympanic bulla and basioccipital.
**Posterior palatine foramen** Small opening in hard palate near juncture of maxillae and palatines.
**Posteroloph** A crest at the back of the molar, connected to the hypocone in an upper teeth.
**Posterostylid** A crest on the posterolingual corner of the molar in a lower teeth.
**Postorbital** Situated behind the eye, as the postorbital process of the frontal bone or postorbital process of the jugal bone.
**Postorbital bar** Complete dorsoventral connection (bar) of bone posterior to orbit; results from fusion of postorbital process of frontal and postorbital process of jugal.
**Postorbital process of frontal** Bony projection of frontal bone posterior to orbit.
**Postorbital process of jugal** Bony projection of jugal bone posterior to orbit.
**Pouch** Fold of skin with a single opening that covers the teats in marsupials. Inside the pouch, the blind offspring attaches itself to one of the mother's teats and remains attached for as long as it takes to grow and develop to a juvenile stage.
**Preglenoid crest** Bony ridge or shelf on anterior part of glenoid (mandibular) fossa, especially well developed in camivores.
**Prehensile** Quality of an appendage or organ that has adapted for grasping or holding.
**Premaxilla (or premaxillary)** Paired bone in anterior of cranium; point of origin of upper incisors (when present).
**Premaxillary** Bone (paired) in the mammalian skull bearing the incisor teeth of the upper jaw; the premaxilla is situated in front of the maxilla.
**Premolar** Designating or pertaining to one of the teeth (a maximum of four on each side of upper jaw and lower jaw of mammals), in front of the true molars.
**Prismatic** Cheekteeth (especially of rodents) with well-developed triangles (or prisms) of enamel surrounding basins of dentine.
**Proboscis** Elongated nose or snout.
**Process** Small bony projection.
**Procumbent** Horizontally protruding condition of the incisors in some mammals, notably many marsupials, insectivores, and primates.
**Propatagium** Membranous structure that assists an animal in flight (bats), is the patagium present from the neck to the first digit.
**Protocone** One of the main cusps, at the anterolingual side in an upper teeth.
**Protoconid** One of the main cusps, at the anterolabial side in a lower teeth.
**Protostyle** A style in front of the protocone, in the protoflexus in an upper teeth.
**Protostylid** A stylid in front of the protoconid, in the protoflexid in a lower teeth.
**Pterygoid** Paired bone on ventral surface of cranium, posterior to palatine and anterior to alisphenoid; forms border of internal nares.
**Pterygoid canal** A passage in the sphenoid bone of the skull leading from just anterior to the foramen lacerum in the middle cranial fossa to the pterygopalatine fossa.
**Quadrant** One-fourth of the total complement of teeth; one side of upper or one side of lower jaw.

**Re-entrant angle** An infold of the enamel layer on the side, front, or back of a cheektooth, as in a molar of a muskrat or wood rat.
**Rhinarium** Furless skin surface surrounding the external openings of the nostrils in many mammals.
**Rosette** French diminutive of rose. For example markings like those of a jaguar.
**Rostrum** Part of the skull projecting in front of the orbits.
**Rostrum breadth** Least distance from lateral base of hamular process of lacrimal bone to corresponding point on opposite side of skull.
**Ruminant** Ungulate with specialized four-chambered digestive system; cud-chewing mammals.
**Rut** The breeding period, as in deer.
**Sagittal crest** The ridge of bone at the juncture of the two parietal bones resulting from the coalescence of the temporal ridges; in old individuals of many species of mammals the crest extends from the middle of the lambdoidal crest anteriorly onto the frontal bones and divides there into two temporal ridges, each of which extends anterolaterally on the posterior edge of the postorbital process of the frontal bone.
**Saliva** The fluid secreted by the glands discharging into the mouth.
**Saltatorial** Pertaining to hopping, as in kangaroo rats.
**Scansorial** Pertaining to mammals that climb by use of claws, such as squirrels and cats.
**Scapula** Bone that connects the humerus (upper arm bone) with the clavicle (collar bone).
**Scutes** Flat bony plates of dermal tissue covered by epidermis forming the outer shell of armadillos.
**Secodont** Blade-like teeth especially adapted for slicing and chopping, called carnassials.
**Secondary cusp** See cusplet.
**Selenodont** Cusp pattern of molars in which individual cusps are cresent shaped; highly developed in ungulates.
**Selenodont** Teeth with the major cusp elongated into a crescent-shaped ridge.
**Semiprismatic** Cheekteeth (especially of rodents) with partially closed (slight re-entrant angles) triangles (or prisms) of enamel surrounding dentine.
**Septum** A wall, dividing a cavity or structure into smaller ones.
**Sesamoid** A bone formed in a tendon.
**Sexual dimorphism** Conditions where the sexes of the same species exhibit different characteristics, particularly characteristics not directly involved in reproduction. Differences may include secondary sex characteristics, size, weight, color, markings, or behavioral or cognitive traits.
**Skull** Cranium plus mandible.
**Snout** Projecting parts of the face (including the nose and mouth) of an animal.
**Sphenoid** Unpaired bone of the neurocranium. It is situated in the middle of the braincase toward the front, in front of the basilar part of the occipital bone.
**Squamosal** Bones form the cheek series of the skull, bordered anteroventrally by the jugal and ventrally by the quadratojugal.
**Subauricular spot** A spot, patch of hair, distinctively colored immediately below the ear.
**Subgenus** Taxonomic rank directly below genus.
**Supraorbital process of frontal** The process of the frontal bone on the top rim of the orbit, as in a rabbit.
**Supraorbital process** Bony projection above orbit on frontal bone; especially well developed in lagomorphs.
**Supraorbital shelf** Small bony ridge on dorsal margin of orbit on frontal and parietal bones.
**Suture** Fairly rigid joint between two or more hard elements of an organism, with or without significant overlap of the elements.
**Suture** Point of contact (or juncture) and fusion between adjacent bones.
**Sympatry** Two related species or populations which exist in the same geographic area and thus frequently encounter one another.
**Syndactylous** A condition in which two or more digits are bound together in a common tube of skin. The underlying bones remain distinct.
**Talon (id)** A posterior "tail" or expansion on an upper (lower) cheektooth which produces a square outline to the tooth and expands the crushing surface.
**Talonid** Crushing heel toward the rear.
**Tarsus** The ankle.
**Temporal muscle** The muscle that originates on the posterodorsal and lateral portions of the braincase and inserts chiefly on the coronoid process of the lower jaw. It is especially large in carnivores.
**Temporal ridge (paired)** A curved, raised line on the side of the braincase marking the upper limit of attachment of the fascia of the temporal muscle. The temporal ridge is prominent on the parietal bone, frequently extends forward onto the

frontal bone, and in some kinds of mammals extends backward onto the interparietal bone. When present, the sagittal crest is formed by the coalescence of the two temporal ridges.

**Terrestrial** Inhabiting the land, rather than the water, trees or air.

**Tibia (plural, tibiae)** The inner and usually the larger of the two bones of the hindlimb (leg) between the knee and the ankle.

**Tooth-row** Row of teeth in one mandible or maxilla bone; in most mammals all the premolars and molars on one side of the lower or upper jaw.

**Torpid** Having lost most of the power of exertion; dormant. A ground squirrel is torpid when it is hibernating.

**Tragus** Small pointed eminence of the outer ear.

**Tricolor** Having three colors. Said of hair on the back of mammals when hair has three bands each of a different color.

**Trigonid** Shearing ends toward the front of the jaw.

**Trochanter** Tubercle of the femur near its joint with the hip bone.

**Trochater** Tubercle of the femur near its joint with the hip bone.

**Truncated** Abruptly or sharply marked, having a square or broad end; sometimes appearing as cut off.

**Tuberculo-sectorial** A tritubercular tooth with sharp cutting edges (cusps).

**Tympanic bulla** See auditory bulla.

**Tympanic bulla breadth** From bottom of pit immediately posterior to external auditory meatus to medial face of bulla at right angle with longitudinal axis of skull.

**Tympanic bulla depth** Least distance from ventral face of basioccipital, excluding median ridge, to line touching ventral-most points of the two bullae.

**Tympanic bulla length** From posterior face to most anterior part of anterior border.

**Tympanum** Of, or pertaining to, bony ring, as in shrews, that does not form a complete bulla.

**Type locality** The place where a type specimen was obtained.

**Ulna** Long bone found in the forearm that stretches from the elbow to the smallest finger, and when in anatomical position, is found on the medial side of the forearm.

**Underfur** The short hair of a mammal; in temperate and boreal climates the underfur ordinarily is denser, made up of more hairs, than the longer and coarser overhair.

**Underparts** Underneath (ventral) side of a mammal (not the back or sides), as of a woods mouse with white underparts.

**Ungulates** Hoofed mammals, such as artiodactyls.

**Unguligrade** Walking on the nail or nails of the toes (hoof) with the heel/wrist and the digits permanently raised.

**Unicuspid** With single well-developed cusp.

**Upper parts** The top (dorsal) surface and all of the sides (not the belly, chest, or throat), as of a woods mouse with reddish-brown upper parts.

**Uropatagium** The interfemoral membrane of a bat; that is to say, the fold of skin that stretches from the hind legs to the tail.

**Uterus** Main female hormone-responsive, secondary sex organ of the reproductive system in most mammals.

**Vibrissa** Type of stiff, functional hair used by animals to sense their environment.

**Vomer** Unpaired facial bones of the skull located in the midsagittal line and articulates with the sphenoid, the ethmoid, the left and right palatine bones, and the left and right maxillary bones.

**Whiskers** Type of stiff, functional hair used by animals to sense their environment.

**Zygomatic arch** In the skull formed by the zygomatic process of the temporal bone and the temporal process of the zygomatic bone, the two being united by an oblique suture.

**Zygomatic breadth** Greatest distance across zygomatic arches of cranium at right angles to the long axis of skull.

**Zygomatic notch** A small notch in the zygomatic plate, a bony plate derived from the flattened front part of the zygomatic arch.

**Zygomatic plate** Platelike extension of maxilla in anterior part of zygoma.

# Index




S. T. Álvarez-Castañeda, *Mammals of North America - Volume 2*, https://doi.org/10.1007/978-3-031-50825-7

**I**

**J**



**K**



**L**

**P**

**R**

## S

## V



## W